Intermediate Algebra
for College Students

Intermediate Algebra for College Students

Second Edition

Robert Blitzer
Miami-Dade Community College

PRENTICE HALL
Upper Saddle River, New Jersey 07458

Library of Congress Cataloging-in-Publication Data

Blitzer, Robert.
    Intermediate algebra for college students / Robert Blitzer.—2nd ed.
    p.   cm.
    Includes index.
    ISBN 0-13-275181-X (Student Edition)
    1. Algebra.   I. Title.
QA154.2.B55 1997   97-15279
512.9—dc21   CIP

Editorial Director: Tim Bozik
Editor-in-Chief: Jerome Grant
Acquisitions Editor: Karin E. Wagner
Editorial Assistant/Supplements Editor: April Thrower/Audra Walsh
Assistant Vice President of Production and Manufacturing: David W. Riccardi
Executive Managing Editor: Kathleen Schiaparelli
Senior Managing Editor: Linda Mihatov Behrens
Manufacturing Manager: Trudy Pisciotti
Manufacturing Buyer: Alan Fischer
Director of Marketing: John Tweeddale
Marketing Manager: Jolene Howard
Marketing Assistants: Diana Penha, Jennifer Pan
Creative Director: Paula Maylahn
Art Manager: Gus Vibal
Text/Cover Design and Project Management: Elm Street Publishing Services, Inc.
Art Studio: Academy Artworks/Laurel Technical Services
Photo Researcher: Francelle Carapetyan
Cover image: John Scott "Invoking the Googul Plex" 1981 watercolour on paper
24 × 18 in./Photo courtesy The Gallery, Stratford, Ontario

Printed in the United States of America
10 9 8 7 6 5 4 3 2

ISBN 0-13-275181-X (Student Edition)

Prentice-Hall International (UK) Limited, *London*
Prentice-Hall of Australia Pty. Limited, *Sydney*
Prentice-Hall Canada Inc., *Toronto*
Prentice-Hall Hispanoamericana, S.A., *Mexico*
Prentice-Hall of India Private Limited, *New Delhi*
Prentice-Hall of Japan, Inc., *Tokyo*
Simon & Schuster Asia Pte. Ltd., *Singapore*
Editora Prentice-Hall do Brasil, Ltda., *Rio de Janeiro*

Contents

Systems of Linear Equations and Inequalities   223

Polynomials, Polynomial Functions, and Factoring   315

Rational Expressions, Functions, and Equations   411

Radicals, Radical Functions, and Rational Exponents   501

Quadratic Equations and Functions   591

Exponential and Logarithmic Functions   673

Conic Sections and Nonlinear Systems of Equations   761

Sequences, Series, and the Binomial Theorem   811

APPENDIX

Review Problems Covering the Entire Book   A1

Preface

Intermediate Algebra for College Students, Second Edition, provides comprehensive, in-depth coverage of the topics required in a one-term course in intermediate algebra. The book is written for college students who have had a course in introductory algebra. The primary goals of the Second Edition are to help students acquire a solid foundation in intermediate algebra and to show how algebra can model and solve authentic real-world problems.

New to the Second Edition

The Second Edition is a significant revision of the First Edition, with increased emphasis on problem solving, graphing, functions, mathematical modeling, technology, discovery approaches, critical thinking, geometry, collaborative learning, and contemporary applications that use real data. The book's changes are based on the recommendations of the *Curriculum and Evaluation Standards for School Mathematics* published by the National Council of Teachers of Mathematics and *Standards for Introductory College Mathematics* published by the American Mathematical Association of Two-Year Colleges. Following are the new features in the Second Edition.

Readability and Level. The chapters have been extensively rewritten to make them more accessible. The Second Edition pays close attention to ensuring that the amount of detail and depth of coverage is appropriate for this level. Every section has been rewritten to contain a better range of simple, intermediate, and challenging examples.

Problem Solving. As with the First Edition, the emphasis of the book is on learning to use the language of algebra as a tool for solving problems related to everyday life. Chapter 1 now contains much less overlap between introductory algebra topics and intermediate algebra topics, so that students get to study strategies for solving problems in intermediate algebra earlier in the course. Problem-solving steps introduced in Chapter 1 are simply and explicitly described, and used regularly. Since students have such difficulty translating word problems, increased emphasis has been placed on translating the words and phrases of verbal models into algebraic equations. The extensive collection of applications, including environmental and financial issues, promotes the problem-solving theme and demonstrates the usefulness of mathematics to students. The range of new problems, from selecting a mortgage to aiding earthquake victims to modeling the spread of AIDS, are intended to appeal to a substantial cross section of readers. Aiding earthquake victims is discussed in a new section on linear programming in Chapter 3.

Graphing. Graphing is now introduced in Chapter 1 and is integrated throughout the book. Most examples and exercises use graphs to explore relationships between data and to provide ways of visualizing a problem's solution. Line, bar, circle, and rectangular coordinate graphs that use real data appear in nearly every section and problem set.

Functions and Modeling. Increased emphasis has been placed on the use of formulas and functions that describe interesting and relevant quantitative relations in the real world. Functions are introduced in Chapter 2, with an integrated functional approach emphasized throughout the book. Old-fashioned, routine word problems have been replaced by an extensive collection of contemporary applications from a wide range of disciplines, many of which are unique. These applications are listed in the index of applications that appears inside the front cover.

Interactive Learning. Discover for yourself exercises encourage students to actively participate in the learning process as they read the book. This new feature encourages students to read with a pen in hand and interact with the text. Through the discovery exercises, they can explore problems in order to better understand them and their solutions.

Technology. The Second Edition offers the option of using graphing utilities, without requiring their use. Graphing utilities are utilized in Using technology boxes to enable students to visualize, discover, and explore procedures for manipulating algebraic expressions and solving equations. Use of graphing utilities is also reinforced in the technology problems appearing in the problem sets for those who want this option. With the book's early introduction to graphing, students can look at the calculator screens in the Using technology boxes and gain an increased understanding of an example's solution even if they are not actually using a graphing utility in the course.

Study Tips. Study tip boxes offer suggestions for problem solving, point out common student errors, and provide informal tips and suggestions. These invaluable hints appear in abundance throughout the book.

Contemporary Fine Art. Algebra and fine art enable us to view the world in new and exciting ways. An extensive collection of contemporary, thought-provoking images selected by the author provides visual commentary to the book's unique collection of contemporary applications. The art adds an aesthetic sense to the book's pages, while visually reminding students of how algebra is connected to the whole spectrum of learning.

New and Reorganized Problem Sets. Problem sets are revised, expanded, and reorganized for easy use in the Second Edition. Problem sets are organized into eight categories:

- *Practice Problems:* These problems give students an opportunity to practice the concepts that have been developed in the section. Many new problems have been added, with attention paid to making sure that the problems are appropriate for the level and graded in difficulty.

- *Application Problems:* Up to 70 percent of the application problems are new to the Second Edition. Included are many relevant, up-to-date applications that will provoke student interest. Many of these problems offer students the opportunity to construct mathematical models from data.
- *True-False Critical Thinking Problems:* Several true-false problems that take students beyond the routine application of basic algebraic concepts are included in nearly every problem set. The true-false format is less intimidating than a more open-ended format, helping students gain confidence in divergent thinking skills.
- *Technology Problems:* These problems, also new to the Second Edition, enable students to use graphing utilities to explore algebraic concepts and relevant mathematical models.
- *Writing in Mathematics:* These exercises are intended to help students communicate their mathematical knowledge by thinking and writing about algebraic topics.
- *Critical Thinking Problems:* This category contains the most challenging exercises in the problem sets. These open-ended problems were written to explore concepts while stimulating student thinking.
- *Group Activity Problems:* These collaborative activities give students the opportunity to work cooperatively as they think and talk about mathematics. There are enough of these problems in each chapter to allow instructors to use collaborative learning as an instructional format quite extensively. It is hoped that many of these problems will result in interesting group discussions.
- *Review Problems:* As with the First Edition, each problem set concludes with three review problems.

Chapter Introductions. Chapter introductions present fine art that is related either to the general idea of the chapter or to an application of algebra contained within the chapter.

Learning Objectives. Learning objectives open every section. The objectives are restated in the margin at their point of use.

New and Revised Enrichment Essays. As with the First Edition, interspersed throughout the book are enrichment essays that germinate from ideas appearing in expository sections. Most of the essays are new to the Second Edition, and stimulating fine art has been added to many.

Expanded Use of Tables. Tables that summarize the procedures discussed in the book, with supporting examples, now appear throughout.

Chapter Projects. Also new to the Second Edition are projects at the end of each chapter that use challenging and interesting applications of mathematics that not only stand alone as ways to stimulate class discussions on a variety of topics, but also cultivate an interest in independent explorations of mathematics on the Worldwide Web. Using the Worldwide Web, with links to many countries, as well as links to art, music, and history, students are encouraged to develop a multicultural, multidisciplinary approach to the study of algebra.

Chapter Tests. New to the Second Edition is a test at the end of each chapter, following the comprehensive collection of chapter review problems. The chapter tests focus on the review problems so that students can see if they are prepared for an actual class test.

Preserved and Expanded from the First Edition

The features described below that helped make the First Edition so popular continue in the Second Edition. However, they have been modified by the book's increased attention to the issue of ensuring that the amount of detail and depth of coverage is appropriate for this level. Modification of these features also reflects the book's increased emphasis on problem solving and modeling with multidisciplinary, relevant applications.

Detailed Step-by-Step Explanations. Illustrative examples are still presented one step at a time. No steps are omitted, and each step is clearly explained. Where applicable, the detailed explanations appearing to the right of each mathematical step have been improved and expanded. A second color has been added to the mathematics to show precisely where this explanation applies.

Example Titles. All examples have titles so that students immediately see the purpose of each example.

Extensive Application to Geometric Problem Solving. Problem solving using geometric formulas and concepts is emphasized throughout the Second Edition. There is also increased emphasis on geometric models that allow students to visualize algebraic formulas.

Screened Boxes. Screened boxes are used to highlight all important definitions, formulas, and procedures.

Chapter Summaries. Inclusive summaries appear at the conclusion of each chapter, helping students to bring together what they have learned after reading the chapter.

Review Problems. A comprehensive collection of review problems follows the summary at the end of each chapter. (A chapter test, new to the Second Edition, follows this collection of review problems.) In addition, Chapters 3–9 conclude with cumulative review problems. Cumulative review problems covering the entire book appear in the appendix. The appendix has been completely rewritten and reformatted to emphasize the changes throughout the book.

Supplements for the Instructor

Printed Supplements

Instructor's Edition   (0-13-860164-X)   Consists of the complete student text, with a special Instructor's answer section at the back of the text containing answers to all exercises.

Instructor's Solutions Manual   (0-13-860180-1)
- Step-by-step solutions for every even-numbered exercise.
- Step-by-step solutions (even and odd) of the Chapter Review Problems, Chapter Tests and Cumulative Reviews.

Test Item File   (0-13-860172-0)
- 6 tests per chapter, consisting of 20 questions each.
    4 free-response tests
    2 multiple-choice tests

- 4 final exams
    2 free-response tests
    2 multiple-choice tests

Media Supplements

TestPro3 Computerized Testing
IBM Single-User (0-13-860230-1)
IBM Online (0-13-897976-6)
MAC (0-13-860248-4)
- Allows instructors to generate tests or drill worksheets from algorithms keyed to the text by chapter, section, and learning objective.
- Instructors select from thousands of test questions and hundreds of algorithms which generate different but equivalent equations.
- A user-friendly expression-building toolbar, editing and graphing capabilities are included.
- Customization toolbars allow for customized headers and layout options which provide instructors with the ability to add or delete workspace or add columns to conserve paper.

Supplements for the Student

Printed Supplements

Student's Solution Manual   (0-13-860321-9)
- Contains complete step-by-step solutions for every odd-numbered exercise and
- Contains complete step-by-step solutions for all (even and odd) Chapter Review Problems, Chapter Tests and Cumulative Reviews.

How to Study Math   (ISBN 0-13-020884-1)
- Free booklet which gives developmental math students strategies for preparing for class, studying and taking exams and improving grades.

Life On the Internet: Mathematics   (ISBN 0-13-268616-3)
- Free guide which provides a brief history of the Internet, discusses the use of the Worldwide Web, and describes how to find your way within the Internet and how to find others on it. Contact your local Prentice Hall Representative for *Life on the Internet: Mathematics*.

NY Times Themes of the Times
- A free newspaper, created new each year, from Prentice Hall and *The New York Times*
- Interesting and current articles on mathematics
- Invites discussion and writing about mathematics

Media Supplements

MathPro Tutorial Software
IBM Single-User (0-13-860214-X)
IBM Network (0-13-860198-1)
MAC Single-User (0-13-860222-0)
MAC Network (0-13-899162-6)
- Fully networkable Windows-based tutorial package for campus labs or individual use

- Designed to generate practice exercises based on the exercise sets in the text
- Algorithmically driven, providing the student with unlimited practice
- Generates graded and recorded practice problems with optional step-by-step tutorial
- Includes a complete glossary including graphics and cross-references to related words

Videotapes (0-13-860313-8)
- Instructional tapes in a lecture format featuring worked-out examples and exercises taken from each section of the text.
- Presentation by Professors Michael C. Mayne and (Biff) John D. Pietro of Riverside Community College in Riverside, California.

Review Video (0-13-901067-X)
- Contains an end-of-chapter summary for every chapter in the text (10 in all).
- Each 5 minute summary highlights the most important features learned in each chapter.
- Excellent preparation for final exams.

Acknowledgments

I wish to express my appreciation to all the reviewers, of both the current and new edition, for their helpful criticisms and suggestions. In particular I would like to thank:

Howard Anderson	*Skagit Valley College*
John Anderson	*Illinois Valley Community College*
Michael H. Andreoli	*Miami-Dade Community College—North Campus*
Warren J. Burch	*Brevard Community College*
Alice Burstein	*Middlesex Community College*
Sandra Pryor Clarkson	*Hunter College*
Robert A. Davies	*Cuyahoga Community College*
Ben Divers, Jr.	*Ferrum College*
Irene Doo	*Austin Community College*
Charles C. Edgar	*Onondaga Community College*
Susan Forman	*Bronx Community College*
Donna Gerken	*Miami-Dade Community College*
Jay Graening	*University of Arkansas*
Robert B. Hafer	*Brevard Community College*
Mary Lou Hammond	*Spokane Community College*
Donald Herrick	*Northern Illinois University*
Beth Hooper	*Golden West College*
Tracy Hoy	*College of Lake County*
Judy Kasabian	*Lansing Community College*
Gary Knippenberg	*Lansing Community College*
Mary Koehler	*Cuyahoga Community College*
Hank Martel	*Broward Community College*
Diana Martelly	*Miami-Dade Community College*
John Robert Martin	*Tarrant County Junior College*
Irwin Metviner	*State University of New York at Old Westbury*

Mark Naber	*Monroe County Community College*
Kamilia Nemri	*Spokane Community College*
Allen R. Newhart	*Parkersburg Community College*
Scott W. Satake	*Eastern Washington University*
Kathy Shepard	*Monroe County Community College*
Gayle Smith	*Lane Community College*
Linda Smoke	*Central Michigan University*
Dick Spangler	*Tacoma Community College*
Janette Summers	*University of Arkansas*
Robert Thornton	*Loyola University*
Lucy C. Thrower	*Francis Marion College*
Andrew Walker	*North Seattle Community College*

Additional acknowledgments are extended to Professor John (Biff) Pietro and Professor Michael C. Mayne of Riverside Community College, for creating the videotapes for each section of the book; Donna Gerken of Miami-Dade Community College, for writing the chapter projects; Phyllis Barnidge and the team at Laurel Technical Services, for the Herculean task of solving all the book's problems, preparing the answer section and the solutions manuals, as well as serving as accuracy checker; Amy Mayfield, whose meticulous work as copy editor put me at my syntactical best; Francelle Carapetyan of Image Research, photo researcher, for playing detective and pursuing the book's photographs and contemporary art across the globe; Paula Maylahn and Gus Vibal for contributing to the book's wonderful look; the team of graphic artists at Academy Artworks, whose superb illustrations provide visual support to the verbal portions of the text; Progressive Information Technologies, the book's compositor, for inputting hundreds of pages with hardly an error; and especially, Ingrid Mount of Elm Street Publishing Services, whose talents as supervisor of production resulted in a book that looks even more wonderful than the first edition.

Most of all I wish to thank Karin Wagner and Tony Palermino. Tony, my developmental editor, contributed invaluable edits and suggestions that resulted in a finished product that is both accessible and up to date. His influence on the second edition is extraordinary, with the improved pace in the text and problem sets a result of his remarkable talents. Karin, my editor at Prentice Hall, guided and coordinated every detail of the project, overseeing both text and supplements. From the inclusion of chapter tests to the quality of the videos to the choice of the book's cover art, Karin's influence can be seen. She is the key person in making this book a reality, and I am grateful to have had an editor with her experience and professionalism.

Karin Wagner is a part of the terrific team at Prentice Hall who made this book possible, including her assistant April Thrower and supplements editor Audra Walsh. Editor-in-Chief Jerome Grant urged me onward in my quest to create the first math textbook with an extensive collection of art, always providing support and commitment. Linda Behrens, managing editor, and Alan Fischer, manufacturing buyer, kept an ever-watchful eye on the production process. Jolene Howard, marketing manager, and Jennifer Pan, marketing assistant, I thank for their outstanding sales force and very impressive marketing efforts.

Finally, as I did in the First Edition, I must conclude by extending my heartfelt thanks to the gifted artists who gave me permission to share their exciting works, and, ultimately, their humanity within the pages of this book.

Robert Blitzer

T O THE STUDENT:

Here are some suggestions for using *Intermediate Algebra for College Students* to acquire a solid algebraic foundation and solve authentic real-world problems.

Selections of fine art by contemporary artists introduce each chapter. Enjoy the art while you obtain a general idea of the chapter's concept.

C H A P T E R 2

Functions, Linear Functions, and Inequalities

ATMOSPHERE

Neil Jenney, *Atmosphere*, 1976

The amount of carbon dioxide in the atmosphere has increased by half again since 1900, as a result of the burning of oil and coal. Burning fossil fuel results in a buildup of gases and particles that trap heat and raise the planet's temperature. The resultant gradually increasing temperature is called the greenhouse effect. Carbon dioxide accounts for about half of the warming.

Figure 2.1 shows an overall trend in the increase of average global temperature. The data points can be modeled approximately by the blue line in the graph. If further measurements indicate that these points begin to fall on or near a line, the equation of this line would provide environmentalists with a way to predict average global temperature in the future.

Figure 2.1 establishes a correspondence between time and average global temperature. Each year after 1975 corresponds to exactly one temperature. Under these circumstances, we say that global temperature is a *function* of time.

The concept of a function is fundamental to all mathematics. In this chapter we use functions, linear functions, and inequalities as three powerful tools for modeling the world and solving some of its problems.

Figure 2.1

Average Global Temperature in Degrees Fahrenheit

Number of Years After 1975

SECTION 2.7 — Linear Inequalities Containing Two Variables

Solutions Manual — **Tutorial** — **Video 3**

Objectives
1 Graph linear inequalities using test points.
2 Graph linear inequalities without using test points.

Begin each section by reading the learning objectives. These objectives will tell you what you should be able to do once you have completed the section.

Supplement icons appear at the beginning of every section. Ask your instructor how you can gain access to the supplements. Use these supplements to reinforce the topics in the text.

Become a problem solver! Apply the algebra presented in *Intermediate Algebra for College Students* to model and solve a wide variety of problems.

V. Ivelva/Magnum Photos, Inc.

Chernobyl

The spread of radioactive fallout from the Chernobyl accident is expected to result in at least 100,000 deaths from cancer in the Northern Hemisphere.

users, and even the familiar bell-shaped curve. Many of these situations will be presented in the problem set that follows.

Our final example applies modeling with exponential functions to the Chernobyl disaster.

EXAMPLE 7 **Modeling the Chernobyl Disaster with an Exponential Function**

The 1986 disaster in the former Soviet Union at the Chernobyl nuclear power plant explosion sent about 1000 kilograms of radioactive cesium-137 into the atmosphere. The formula

$$A = 1000(0.5)^{t/30}$$

models the amount (A, in kilograms) of cesium-137 remaining after t years in the area surrounding Chernobyl. If even 100 kilograms of cesium-137 remain in the atmosphere, the area is considered unsafe for human habitation. Will people be able to live in the area 80 years after the accident?

Solution

We substitute 80 for t in the given model. If the resulting value of A is 100 or greater, the area will still be unsafe for human habitation.

$$A = 1000(0.5)^{t/30} \quad \text{This is the given model.}$$
$$= 1000(0.5)^{80/30} \quad \text{Substitute 80 for } t.$$
$$\approx 157 \quad \text{Use a calculator.}$$

After 80 years, 157 kilograms of cesium-137 will be in the atmosphere. Since this exceeds 100, even by the year 2066 the Chernobyl area will remain a ghost town. ■

Many examples include solving problems related to everyday life.

PROBLEM SET 1.3

Practice Problems

In Problems 1–8, plot the given points in a rectangular coordinate system and, where applicable, name the quadrant in which the point is located.

1. $(1, 4)$ **2.** $(-2, 3)$ **3.** $(-3, -5)$ **4.** $(4, -1)$

Work problems every day and check your answers. Practice Problems in each problem set give you the chance to practice the section's concepts.

Application problems provide additional practice using up-to-date sources and data.

Application Problems

It is quite likely that you have seen at least one of the ten films mentioned in the bar graph shown, since these were the top ten money earners for the period from 1990 to 1995. If x represents millions of dollars grossed by a film, write the film title or titles described in Problems 71–79.

71. $x \in [173, 206)$
72. $x \in (285, 338)$
73. $x \in [206, \infty)$
74. $x \in (-\infty, 179)$
75. $184 < x \leq 217$
76. $x \geq 338$
77. $x \leq 165$
78. $x > 338$
79. $x < 165$

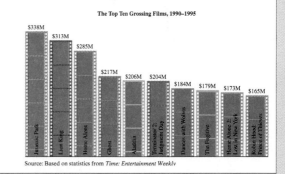

The Top Ten Grossing Films, 1990–1995

$338M, $313M, $285M, $217M, $206M, $204M, $184M, $179M, $173M, $165M

Jurassic Park, Lion King, Home Alone, Ghost, Aladdin, Terminator 2: Judgment Day, Dances with Wolves, The Fugitive, Home Alone 2: Lost in New York, Robin Hood: Prince of Thieves

Source: Based on statistics from *Time: Entertainment Weekly*

76. In 1995, the world population was approximately 5.702 billion, with a relative growth rate of 1.6% per year ($k = 0.016$), down from the high of 2% per year in 1969.
  a. What will be the population of the world in the year 2000?
  b. In what year will Earth's carrying capacity of 10 billion be reached?

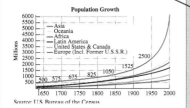

Population Growth

Asia
Oceania
Africa
Latin America
United States & Canada
Europe (Incl. Former U.S.S.R.)

Millions — 500, 575, 675, 825, 1050, 1525, 2500

1650 1700 1750 1800 1850 1900 1950 2000

Source: U.S. Bureau of the Census

Construct your own mathematical models from data and experience how the math you learn in class can help you make better decisions.

Gain confidence and develop your critical thinking skills. True/False questions help you *develop* and apply thinking skills.

Enhance your understanding by exploring algebraic and graphical concepts using a graphing utility.

Communicate your mathematical knowledge by writing and thinking about algebraic topics.

Challenge yourself! Critical Thinking Problems are designed to really get you thinking.

In the workplace, problems and projects are often solved with a collaborative effort. Group Activity Problems encourage cooperative work in and out of the classroom setting.

Take the opportunity to connect all the concepts you have learned and reinforce problem-solving strategies throughout the course as you work *review problems*.

Enrich your mathematical experience by enjoying the Enrichment Essays and the book's contemporary art.

Enrichment Essays make your reading more interesting and show you how algebra is connected to the whole spectrum of learning.

ENRICHMENT ESSAY

Linear Programming and the Cold War

The Berlin Airlift (1948–1949) was an operation put into effect by the United States and Great Britain after the former Soviet Union closed all roads and rail lines between West Germany and Berlin, cutting off supply routes to the city. The Allies used linear programming to break the blockade. The 11-month airlift, in 272,264 flights, provided basic necessities to blockaded Berlin. A simplified version describing how this was done can be found in Problem 30 of Problem Set 3.6.

Today, linear programming is one of the most widely used tools in management science, helping businesses allocate resources on hand (the constraints) to manufacture a particular array of products that will maximize profit (the objective function). Linear programming accounts for as much as 90% of all computing time used for management decisions in business.

Jasper Johns (b. 1930) "Three Flags" 1958, encaustic on canvas, $30\frac{7}{8} \times 45\frac{1}{2} \times 5$ in. (78.4 × 115.6 × 12.7 cm). 50th Anniversary Gift of the Gilman Foundation, Inc., The Lauder Foundation, A. Alfred Taubman, an anonymous donor, and purchase. 80.32. Collection of Whitney Museum of American Art, New York. Photo by Geoffrey Clements. © 1998 Jasper Johns/Licensed by VAGA, New York, NY.

Enhance your preparation with discovery exercises, study tips, and using technology boxes.

Discover for Yourself exercises encourage your active participation in the learning process as you read the book.

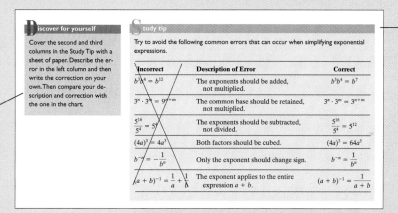

Discover for yourself

Cover the second and third columns in the Study Tip with a sheet of paper. Describe the error in the left column and then write the correction on your own. Then compare your description and correction with the one in the chart.

Study tip

Try to avoid the following common errors that can occur when simplifying exponential expressions.

Incorrect	Description of Error	Correct
$b^3 b^4 = b^{12}$	The exponents should be added, not multiplied.	$b^3 b^4 = b^7$
$3^n \cdot 3^k = 9^{n+m}$	The common base should be retained, not multiplied.	$3^n \cdot 3^m = 3^{n+m}$
$\dfrac{5^{16}}{5^4} = 5^4$	The exponents should be subtracted, not divided.	$\dfrac{5^{16}}{5^4} = 5^{12}$
$(4a)^3 = 4a^3$	Both factors should be cubed.	$(4a)^3 = 64a^3$
$b^{-n} = -\dfrac{1}{b^n}$	Only the exponent should change sign.	$b^{-n} = \dfrac{1}{b^n}$
$(a+b)^{-1} = \dfrac{1}{a} + \dfrac{1}{b}$	The exponent applies to the entire expression $a + b$.	$(a+b)^{-1} = \dfrac{1}{a+b}$

Study tip boxes offer suggestions for problem solving and point out common errors to avoid.

Increase your visual understanding of an example's solution using a *graphing utility*.

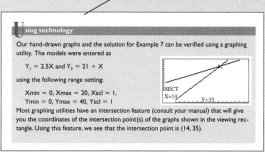

Using technology

Our hand-drawn graphs and the solution for Example 7 can be verified using a graphing utility. The models were entered as

$$Y_1 = 2.5X \text{ and } Y_2 = 21 + X$$

using the following range setting:

Xmin = 0, Xmax = 20, Xscl = 1,
Ymin = 0, Ymax = 40, Yscl = 1

Most graphing utilities have an intersection feature (consult your manual) that will give you the coordinates of the intersection point(s) of the graphs shown in the viewing rectangle. Using this feature, we see that the intersection point is (14, 35).

HAPTER PROJECT

How Far?

We developed a formula in this chapter to find the distance between two points, assuming we know the coordinates of the points on the Cartesian plane. It seems that this technique for finding distance will work wherever we can lay down a coordinate grid similar to the Cartesian system. We used this approach on page 603, when we superimposed a coordinate system over a map of Thailand to give us an estimate of how far a plane would fly between two cities. However, one of the fundamental aspects of the Cartesian coordinate system was missing … we were no longer measuring distance on a flat plane.

1. Select a city far from where you live as a destination for a trip. Using a globe with a scale for distances, locate the starting point and ending point for your trip. Use a thin wire or pipe cleaner to connect the two points on the globe, making sure you follow the contours of the globe. After you have marked the distance, carefully lift the wire off the globe and examine it. What kind of curve do you see? Does it look like a line would approximate this distance? Consult an atlas or an airline guide to see what is listed as the mileage between the selected cities. Does it match your mileage approximation?
2. Use a flat, scaled map of the world and repeat your measurement. How does this estimate compare to the one from the globe and other sources? If you can, find other styles of scaled maps and compare the measurements on each. Does the style of flat map used affect your measurements? How do you think cruiselines or airlines determine the mileage from one destination to another?
3. Using the globe, select numerous cities as destinations from your original starting point and measure the distances. Compare the global measures to a flat measure and see if you can determine at what distance the curvature of the earth seems to play a great part in affecting the measurement.
4. Obtain a scaled map of your city and the surrounding area and select two places far apart and over different terrain. Plan a route from one location to the other, on the map, and measure the distance using the scale. Follow the same route in your car, taking along another member of the class to record the mileage shown on the odometer. Does the map give the same result as the mileage from your car? Do you think the map considered differences in elevation, such as hills or valleys, when listing the scale you should use to calculate distances?

We know a straight line is the shortest distance between two points. If we mark two points on a sphere, the shortest distance between them, in space, would go through the sphere and join one point to another. However, if we restrict ourselves to moving on the surface of a sphere, the shortest distance between two points is found by traveling on a great circle. A great circle can be defined as a circle on a sphere that divides the sphere into two portions of equal area. The equator of the earth is a great circle, as are lines of longitude. Lines of latitude, although circles, are not great circles.

Link up to the Worldwide Web and surf the 'Net through the Prentice Hall website to explore multidisciplinary material that is related to the material in the Chapter Project.

• www.prenhall.com/blitzer

U se these text supplements as your resources. Ask your instructor how to obtain these items to complement your learning style.

Instructional videos feature selected worked out examples and exercises from every section of the text. A separate "Graphing Utilities" video and "Review Video" provide further video instruction.

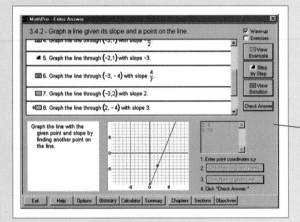

Math Pro Explorer includes explorations as well as algorithmically generated practice exercises. Available in Macintosh and Windows platforms.

A lso available:

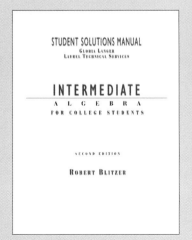

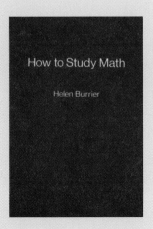

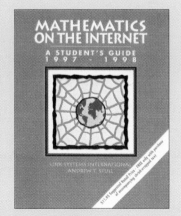

To the Student

The process of learning mathematics requires that you do at least three things—read the book, work the problems, and get your questions answered if you are stuck. This book has been written so that you can learn directly from its pages. All concepts are carefully explained, important definitions and procedures are set off in boxes, and worked-out examples that present solutions in a step-by-step manner appear throughout. Study tip boxes offer hints and suggestions, and often point out common errors to avoid. Discovery boxes encourage you to actively participate in the learning process as you read the book. A great deal of attention has been given to show you the vast and unusual applications of algebra in order to make your learning experience both interesting and relevant. As you begin your studies, I would like to offer some specific suggestions for using this book and for being successful in algebra.

- *Read the book.*
  - **a.** Begin with the chapter introduction. Enjoy the art while you obtain a general idea of what the chapter is about.
  - **b.** Move on to the objectives and the introduction to a particular section. The objectives will tell you exactly what you should be able to do once you have completed the section. Each objective is restated in the margin at the point in the section where the objective is taught.
  - **c.** At a slow and deliberate pace, read the section with pen (or pencil) in hand. Move through the illustrative examples with great care. These worked-out examples provide a model for doing the problems in the problem sets. Be sure to read all the hints and suggestions in the Study tip boxes. Your pen is in hand for the Discover for yourself exercises that are intended to encourage you to actively participate in the learning process as you read the book. The Discover for yourself exercises let you explore problems in order to understand them and their solutions better, so be sure not to jump over these valuable discovery experiences in your reading.
  - **d.** Enjoy the Enrichment essays and the contemporary art that is intended to make your reading more interesting and show you how algebra is connected to the whole spectrum of learning.

As you proceed through the reading, do not give up if you do not understand every single word. Things will become clearer as you read on and see how various procedures are applied to specific worked-out examples.

- *Work problems every day and check your answers.* The way to learn mathematics is by *doing* mathematics, which means by *solving problems.* The more problems you work, the better you will become at solving problems which, in turn, will make you a better algebra student.

  a. Work the assigned problems in each problem set. Problem sets are organized into eight categories. Minimally, you should work all odd-numbered problems in the first two categories (Practice Problems and Application Problems), and all three review problems at the end of the problem set. Answers to most odd-numbered problems and all review problems are given in the back of the book. Once you have completed a problem, be sure to check your answer. If you made an error, find out what it was. Ask questions in class about homework problems you don't understand.

  b. Problem sets also include critical thinking problems, technology problems, writing exercises, and group activity learning experiences. Don't panic! You are not expected to work every problem, or even all the odd-numbered problems, in each problem set. This vast collection of problems provides options for your learning style and your instructor's teaching methods. You may be assigned some problems from one or more of these categories. Problems in the critical thinking categories are the most difficult, intended to stimulate your ability to think and reason. Thinking about a particular question, even if you are confused and somewhat frustrated, can eventually lead to new insights.

- *Prepare for chapter exams.* After completing a chapter, study the summary, work assigned problems from the chapter review problems, and work all the problems in the chapter test.

- *Review continuously.* Working review problems lets you remember the algebra you learned for a much longer period of time. Cumulative review problems appear at the end of each chapter, beginning with Chapter 3. The book's appendix contains review problems covering the entire course. By working the appendix problems assigned by your professor, you will be able to bring together the procedures and problem-solving strategies learned throughout the course.

- *Attend all lectures.* No book is intended to be a substitute for the valuable insights and interactions that occur in the classroom. In addition to arriving for a lecture on time and prepared, you might find it helpful to read the section that will be covered in class beforehand so that you have a clear idea of the new material that will be discussed.

- *Use the supplements that come with this book.* A solutions manual that contains worked-out solutions to the book's odd-numbered problems and all review problems, as well as a series of videotapes created for every section of the book, are among the supplements created to help you learn algebra. Ask your instructor what supplements are available and where you can find them.

Algebra is often viewed as the foundation for more advanced mathematics. It is my hope that this book will make algebra accessible, relevant, and an interesting body of knowledge in and of itself.

Discover for yourself

What conclusion can you draw from these circle graphs about student success and attending lectures?

Successful Students

Sometimes absent 8%
Often absent 8%
Always or almost always in class 84%

Unsuccessful Students

Often absent 45%
Sometimes absent 8%
Always or almost always in class 47%

Source: *The Psychology of College Success: A Dynamic Approach,* by permission of H. C. Lindgren, 1969

1

Algebra and Problem Solving

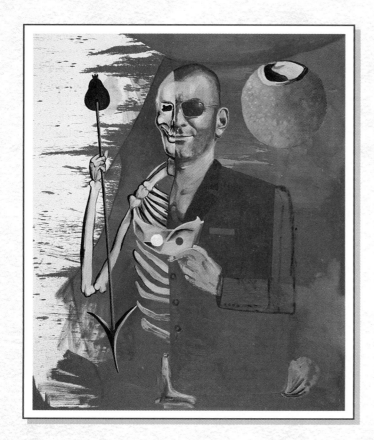

Martin Kippenberger "Ohne Titel" (Untitled) 1988, oil on canvas, 240 × 200 cm. Private Collection. Courtesy Galerie Gisela Capitain, Cologne.

The length of the humerus, the bone from the elbow to the shoulder, can be used to find a person's height. For an adult man, if the humerus length, in centimeters, is multiplied by 2.89 and 70.64 is added, the resulting value approximates his height in centimeters. Using the compact, symbolic notation of algebra, we express this relationship by

$$M = 2.89h + 70.64$$

where h represents humerus length and M represents height.

An equation such as this, which describes the variables in our physical world, is called a mathematical model.

Modeling the world's variables helps to solve some of its problems. For example, a person who is 196 centimeters tall has been missing for over a year. Police find the remains of a body in a wooded area near the missing person's residence. Although the body is badly decomposed, the humerus is intact and measures 42 centimeters. Can the remains of the body be those of the missing person?

Using algebra to model the world and solve its problems is the aim of this book. In this chapter, we begin looking at the details of how this is done.

Solutions   Tutorial   Video
Manual   1

The Real Numbers and the Number Line

Objectives

1 Use the roster method to write sets in set-builder notation.
2 Recognize subsets of the real numbers.
3 Use inequality symbols to order numbers.
4 Use interval notation.
5 Recognize properties of real numbers.

Jasper Johns "0–9" (1959–62), plastic paint on canvas, $20\frac{1}{2} \times 35\frac{1}{2}$ in./photo by Glen Steigelman, courtesy of the Leo Castelli Gallery. © 1998 Jasper Johns/Licensed by VAGA, New York, NY.

Numbers and their properties have intrigued humankind since the beginning of civilization. Pythagoras discovered that harmony in music was the result of ratios of whole numbers, developing a philosophy of the universe based on these ratios. Thousands of years later, mathematician Leopold Kronecker said, "God made the whole numbers: Everything else is the work of man."

Volumes have been written dealing with curious and interesting numbers. Although it is beyond the scope of this book to discuss these curiosities in detail, in this section we define precisely what we mean by the real numbers and examine their representation on the real number line.

Set Notation

Since the term *set* is used often in mathematics, we define this term first.

> **Definition of a set**
>
> A *set* is a collection of objects. The objects in a set are the *elements* or *members* of the set.

1 Use the roster method to write sets in set-builder notation.

The *roster method* of writing a set encloses the elements of the set in braces, { }. For example, the set of positive odd numbers less than 11 is written {1, 3, 5, 7, 9}. This set has a limited number of elements and is an example of a *finite set*.

To express the fact that 7 is an element of the set {1, 3, 5, 7, 9}, we use the symbol $\in$.

> **The symbol $\in$**
>
> The symbol $\in$ means *is an element of.*

Consequently,

$$7 \in \{1, 3, 5, 7, 9\} \quad \text{and} \quad 3 \in \{1, 3, 5, 7, 9\}.$$

In algebra, letters (called variables) are used to represent numbers. Variables are used to express sets in *set-builder notation*. The set {1, 3, 5, 7, 9} can be written using this notation as

$$\{x \mid x \text{ is an odd number between 1 and 9, inclusively}\}$$

which is read, "the set of all elements x such that x is an odd number between 1 and 9 inclusively." (The word *inclusively* includes both 1 and 9 as elements of the set.)

Table 1.1 represents three sets in set-builder and roster notations. The sets in each row are *equal* because they contain the *same* elements.

TABLE 1.1 Sets in Set-Builder and Roster Notations

Set-Builder Notation	Roster Notation
$\{x \mid x$ is an even counting number between 2, inclusively, and 14, exclusively$\}$	$\{2, 4, 6, 8, 10, 12\}$
$\{x \mid x$ is a counting number less than 8$\}$	$\{1, 2, 3, 4, 5, 6, 7\}$
$\{x \mid x$ is a counting number greater than 8$\}$	$\{9, 10, 11, 12, 13, \ldots\}$

study tip

The dots . . . within a set mean that the pattern continues in the indicated direction.

Observe that the last set in the table contains an unlimited number of elements; it is an example of an *infinite set*.

 Recognize subsets of the real numbers.

The Set of Real Numbers

The sets that make up the real numbers are summarized in Table 1.2. We refer to these sets as *subsets* of the real numbers, meaning that all elements in each subset are also elements in the set of real numbers.

study tip

The symbol $\approx$ is read "is approximately equal to." The reason that

$$\sqrt{2} \approx 1.414214$$

is because $(1.414214)^2$ is not exactly 2.

TABLE 1.2 Important Subsets of the Real Numbers

Name	Set; Description	Examples
Natural numbers	$\{1, 2, 3, 4, 5, \ldots\}$ These numbers are used for counting and are also called the *counting numbers.*	87 2369 $\sqrt{36}$ (i.e., 6)
Whole numbers	$\{0, 1, 2, 3, 4, 5, \ldots\}$ The whole numbers add 0 to the set of natural numbers.	$0, 87, \frac{18}{3}$ (i.e., 6)
Integers	$\{\ldots, -4, -3, -2, -1, 0, 1, 2, 3, 4, \ldots\}$ The integers add the opposites of the natural numbers to the set of whole numbers.	$-87, -14,$ $-\sqrt{25}$ (i.e., -5), $0, 14, 87$
Rational numbers	These numbers can be expressed as an integer divided by a non-zero integer. $\{\frac{a}{b} \mid a$ and b are integers, and $b \neq 0\}$ In decimal form, rational numbers either terminate or repeat.	$\frac{3}{4}$ or 0.75 $(a = 3, b = 4)$ $-\frac{3}{11} = -0.2727 \ldots = -0.\overline{27}$ $(a = -3, b = 11)$ $87 = \frac{87}{1}$ $(a = 87, b = 1)$
Irrational numbers	The set of numbers whose decimal representations do not repeat and do not terminate. Irrational numbers cannot be expressed as an integer divided by an integer.	$\sqrt{2} \approx 1.414214$ $-\sqrt{3} \approx -1.73205$ $\pi \approx 3.142$ $-\frac{\pi}{2} \approx -1.571$

using technology

When using a calculator, the displays for the irrational numbers are only approximations. A calculator might display 1.4142136 for $\sqrt{2}$ despite the fact that $\sqrt{2}$ cannot be expressed as a terminating decimal.

When we combine the set of all irrational numbers with the set of all rational numbers, we obtain the set of all *real numbers.*

> **Definition of the real numbers**
>
> The set of *real numbers* consists of both the set of rational numbers and the set of irrational numbers, represented in set-builder notation as
>
> $$\{x \mid x \text{ is rational or } x \text{ is irrational}\}.$$

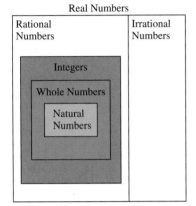

Real Numbers

Figure 1.1

The set of real numbers

Figure 1.1 is a representation of the set of real numbers. The diagram indicates that:

1. The set of irrational numbers and the set of rational numbers are subsets of the set of real numbers. Every irrational number is real and every rational number is real.
2. The set of integers is a subset of the set of rational numbers. Every integer can be expressed in the form of a rational number:

$$-5 = \frac{-5}{1} \quad -3 = \frac{-3}{1} \quad 0 = \frac{0}{1} \quad 8 = \frac{8}{1}$$

3. Every whole number is an integer, so the set of whole numbers is a subset of the set of integers.
4. Every natural number is a whole number.

The Real Number Line

The *real number line* is a graph we use to visualize the set of real numbers (Figure 1.2).

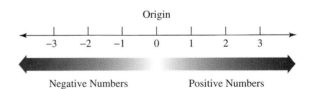

Figure 1.2

The real number line

The line extends infinitely in both directions. An arbitrary point, called the *origin,* is labeled 0, with positive units to the right of the origin and negative units to the left of the origin. Figure 1.3 on page 6 illustrates that every point on the real number line can be made to correspond to a real number, and every real number can be made to correspond to a point. If you draw a point on the real number line corresponding to a real number, you are *plotting* the real number.

ENRICHMENT ESSAY

Shocked by Irrational Numbers

Perhaps Pythagoras was a kind of magician to his followers because he taught them that nature is commanded by numbers. There is a harmony in nature, he said, a unity in her variety, and it has a language: numbers are the language of nature.

Jacob Bronowski, *The Ascent of Man*

Followers of the Greek mathematician Pythagoras in the sixth century B.C. were convinced that properties of whole numbers were the key to understanding the universe. Shown here is Renaissance artist Raphael Sanzio's (1483–1526) image of Pythagoras from *The School of Athens* mural.

Pythagoras thought that numbers that were not whole numbers could be represented as the ratio of whole numbers. It came as a shock to the Pythagoreans to discover that the square root of 2 could only be approached by the whole number ratios 14/10, 141/100, 1414/1000, and so on, but regardless how large the numbers in these fractions, the square root of 2 would never be revealed.

Since the Pythagoreans viewed numbers with reverence and awe, the consequence for disclosing this discovery was death. When a follower of Pythagoras did just this, his subsequent demise in a shipwreck was viewed as divine retribution.

Pythagoras, from *The School of Athens,* by Raphael Sanzio. (Scala/Art Resource, New York.)

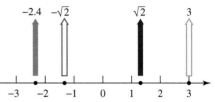

(a) Every real number corresponds to exactly one point on the real number line.

(b) Every point on the real number line corresponds to exactly one real number.

Figure 1.3

A one-to-one correspondence between real numbers and points on the real number line

study tip

A right triangle with two sides of length 1 and a third side of length $\sqrt{2}$ can be used to plot $\sqrt{2}$ on a number line.

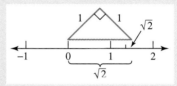

3 Use inequality symbols to order numbers.

Figure 1.4

a is less than *b*: *a* is to the left of *b*

Ordering the Real Numbers

The real number line is useful in demonstrating the *order* of real numbers. We say that real number *a is less than* real number *b*, written $a < b$, if *a* is to the left of *b* on the number line. Equivalently, *b is greater than a*, written $b > a$, if *b* is to the right of *a* on the number line (Figure 1.4).

EXAMPLE 1   **Ordering Real Numbers**

Write out the meaning of each statement, and then determine if the statement is true.

a. $-3 < -1$ **b.** $3 > -1$ **c.** $-3 > 2$

Solution

The solution is illustrated by the number line in Figure 1.5.

Figure 1.5

Ordering four real numbers

Ordering Numbers	Meaning
a. $-3 < -1$	-3 is less than -1 is true because -3 lies to the left of -1 on the number line.
b. $3 > -1$	3 is greater than -1 is true because 3 lies to the right of -1 on the number line.
c. $-3 > 2$	-3 is greater than 2 is false because -3 is to the left of 2.

S tudy tip

The symbols $<$ and $>$ are sometimes combined with an equal sign:

$-3 \leqslant 2$ -3 *is less than* or equal to 2.

$-1 \geqslant -1$ -1 is greater than or *equal to* -1.

Statements such as those in Example 1 and the Study Tip are called *inequalities*. Following are some properties of inequalities.

Properties of inequality

1. If a and b are real numbers, then one and only one of the following statements is true:

$$a < b, \qquad a > b, \qquad a = b$$

2. If $a < b$ and $b < c$, then $a < c$.
3. If $a > b$ and $b > c$, then $a > c$.
4. If a is a negative real number, then $a < 0$.
5. If a is a positive real number, then $a > 0$.

4 Use interval notation.

Intervals on the Real Number Line

Real world applications frequently require that we focus on ranges of numbers, represented as *intervals* on the real number line. For example, the range of temperature (in degrees Fahrenheit) for water to remain a liquid is greater than 32° but less than 212°. This is shown on the number line as follows:

Water remains a liquid.

— 212°F
— 32°F

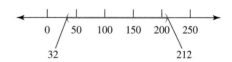

We use a parenthesis at 32 and a parenthesis at 212 to indicate that these values are *not* to be included. Symbolically, we can convey this information in set-builder notation, writing

$$\{x \mid 32 < x < 212\}$$

and reading this as "all x such that x is greater than 32 and less than 212." A second notation, called *interval notation,* can also be used to represent the interval between 32 and 212, written as

$$(32, 212).$$

The parentheses at the left and right of this notation indicate that 32 and 212 are *not* included.

The inequality $32 < x < 212$ represents two inequalities—namely, $x > 32$ (x is greater than 32) *and* $x < 212$ (x is less than 212). In general, the two inequalities $x > a$ and $x < b$ may be written as $a < x < b$. Similar ideas apply when one or both of the inequality symbols in $a < x < b$ are replaced by $\leqslant$. Thus, $a \leqslant x < b$ means that $x \geqslant a$ and $x < b$. Inequalities like these can be

S tudy tip

Sometimes an open dot ∘ is used on the number line instead of a parenthesis to exclude an endpoint and a closed dot • is used instead of a bracket to include an endpoint. In this book, we will use parentheses and brackets.

used to define *bounded intervals* on the number line, shown in Table 1.3. In each case, we have graphed all real numbers between *a* and *b*. We then use the given inequality to determine whether to include or exclude *a* and *b*. Both on the number line and in interval notation, a *parenthesis excludes* an endpoint of an interval, whereas a *bracket includes* an endpoint of an interval.

TABLE 1.3 Bounded Intervals on the Real Number Line

Let *a* and *b* represent any two real numbers, where *a* < *b*.

Inequality	Set-Builder Notation	Interval Notation	Graph
$a \leq x \leq b$	$\{x \mid a \leq x \leq b\}$: the set of real numbers x such that x is greater than or equal to a and less than or equal to b	$[a, b]$	 ![graph a to b, brackets]
$a < x < b$	$\{x \mid a < x < b\}$: the set of real numbers x such that x is greater than a and less than b	(a, b)	 ![graph a to b, parentheses]
$a \leq x < b$	$\{x \mid a \leq x < b\}$: the set of real numbers x such that x is greater than or equal to a and less than b	$[a, b)$	 ![graph a to b]
$a < x \leq b$	$\{x \mid a < x \leq b\}$: the set of real numbers x such that x is greater than a and less than or equal to b	$(a, b]$	 ![graph a to b]

EXAMPLE 2   Sketching Graphs of Bounded Intervals

Sketch the graphs of the following sets and express each in interval notation:

a. $\{x \mid -2 < x < 1\}$ **b.** $\{x \mid -1 < x \leq 2\}$ **c.** $\{x \mid -1 \leq x \leq 1\}$

Solution

a. We want to graph all real numbers greater than -2 and less than 1. Therefore, we graph all real numbers between -2 and 1 on the number line, using parentheses to exclude -2 and 1. Figure 1.6 shows the graph of $\{x \mid -2 < x < 1\}$. The interval notation for this graph is $(-2, 1)$.

b. We want to graph all real numbers greater than -1 and less than or equal to 2. Therefore, we graph all real numbers between -1 and 2 on the number line, using a parenthesis to exclude -1 and a bracket to include 2. Figure 1.7 shows the graph of $\{x \mid -1 < x \leq 2\}$. The interval notation for the graph is $(-1, 2]$. The parenthesis at -1 shows that the interval does not contain the left endpoint -1. The bracket at 2 shows that the interval does contain the right endpoint 2.

c. We want to graph all real numbers between -1 and 1, inclusively. Therefore, we graph all real numbers between -1 and 1 on the number line, using brackets to include -1 and 1. Figure 1.8 shows the graph of $\{x \mid -1 \leq x \leq 1\}$. The interval notation for the graph is $[-1, 1]$. The brackets indicate that this interval contains its left endpoint -1 and its right endpoint 1.

Figure 1.6

The graph of real numbers greater than -2 and less than 1

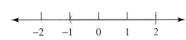

Figure 1.7

The graph of real numbers greater than -1 and less than or equal to 2

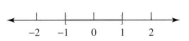

Figure 1.8

The graph of real numbers between -1 and 1, inclusively

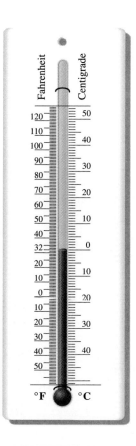

Suppose we wish to discuss the temperature range in which water is in a solid state. This occurs for temperatures *less than or equal to* 32°F, shown on the number line as follows:

The bracket used at 32 shows that the point representing 32 *is* to be included. Symbolically, we can represent this *unbounded interval* in set-builder and interval notation. Using set-builder notation, we write

$$\{x \mid x \le 32\}$$

reading this as "all x such that x is less than or equal to 32." Using interval notation, we write

$$(-\infty, 32].$$

The negative infinity symbol $-\infty$ does not represent a real number. It indicates that the interval includes all real numbers less than or equal to 32.

Intervals such as $(-\infty, 32]$ are called *unbounded intervals* and are summarized in Table 1.4. The symbols ∞ (infinity) and $-\infty$ (negative infinity) are compact, symbolic notations and do *not* represent real numbers. In interval notation, ∞ indicates that an interval extends indefinitely to the right, whereas $-\infty$ indicates that an interval extends indefinitely to the left. A parenthesis is always used next to an infinity symbol in interval notation.

TABLE 1.4 Unbounded Intervals on the Real Number Line

Let a and b represent any two real numbers, where $a < b$.

Inequality	Set-Builder Notation	Interval Notation	Graph
$x > b$	$\{x \mid x > b\}$: the set of real numbers x such that x is greater than b	(b, ∞)	
$x \ge b$	$\{x \mid x \ge b\}$: the set of real numbers x such that x is greater than or equal to b	$[b, \infty)$	
$x < a$	$\{x \mid x < a\}$: the set of real numbers x such that x is less than a	$(-\infty, a)$	
$x \le a$	$\{x \mid x \le a\}$: the set of real numbers x such that x is less than or equal to a	$(-\infty, a]$	
$-\infty < x < \infty$	$\{x \mid x \in R\}$: the set of real numbers x such that x is any real number; the set of all real numbers	$(-\infty, \infty)$	

EXAMPLE 3   Sketching Graphs of Unbounded Intervals

Sketch the graphs of the following sets and express each in interval notation:

a. $\{x \mid x \le 7\}$ **b.** $\{x \mid x > 3\}$

Figure 1.9

The graph of real numbers less than or equal to 7

Figure 1.10

The graph of real numbers greater than 3

Solution

a. We must graph all real numbers less than or equal to 7—that is, all real numbers to the left of 7 and including 7. Figure 1.9 shows the graph of $\{x \mid x \le 7\}$. The interval notation for this graph is $(-\infty, 7]$, using $-\infty$ to show the interval extends indefinitely to the left.

b. This set is represented by a graph of all real numbers greater than 3—that is, all real numbers to the right of 3 and excluding 3. Figure 1.10 shows the graph of $\{x \mid x > 3\}$. The interval notation for this graph is $(3, \infty)$, using ∞ to show the interval extends indefinitely to the right.

In our next example, we will rewrite given English expressions in both set-builder and interval notations.

| **EXAMPLE 4** | **Compact, Symbolic Representations of English Phrases** |

Rewrite each English phrase that appears in the left column in both set-builder and interval notations.

Solution

	Set-Builder Notation	Interval Notation
x is less than 5.	$\{x \mid x < 5\}$	$(-\infty, 5)$
x is greater than or equal to 3.	$\{x \mid x \ge 3\}$	$[3, \infty)$
x lies between -2 and 5, excluding -2 and 5.	$\{x \mid -2 < x < 5\}$	$(-2, 5)$
x is greater than or equal to 2 and less than 7.	$\{x \mid 2 \le x < 7\}$	$[2, 7)$
x is nonnegative. "Nonnegative" indicates a number is greater than or equal to 0.	$\{x \mid x \ge 0\}$	$[0, \infty)$
x is at most 4. "At most" means $\le$.	$\{x \mid x \le 4\}$	$(-\infty, 4]$
x is at least 2. "At least" means $\ge$.	$\{x \mid x \ge 2\}$	$[2, \infty)$
x is positive but not more than 5. "Not more than" means $\le$.	$\{x \mid 0 < x \le 5\}$	$(0, 5]$

The California grey whale's length, in meters, is $L \in [10, 15]$. Describe the whale's length in words.

Rita Fattaruso/Frank S. Balthis Photography

5 Recognize properties of real numbers.

Basic Properties of the Real Numbers

The following list summarizes the basic properties of addition and multiplication. You are probably familiar with these properties from your work in introductory algebra. For the properties listed below, a, b, and c represent any real numbers.

Properties of Real Numbers

Property	**Verbal Description and Examples**
Closure property of addition and multiplication: $a + b$ is a real number. ab is a real number.	The sum and product of two real numbers are real numbers. EXAMPLES: $1 + 8 = 9; 9$ is a unique real number. $7 \cdot 4 = 28; 28$ is a unique real number.
Commutative properties of addition and multiplication: $a + b = b + a$ $a \cdot b = b \cdot a$	Two real numbers can be added or multiplied in any order. EXAMPLES: $4 + 3 = 3 + 4$ $3 \cdot 7 = 7 \cdot 3$
Associative property of addition and multiplication: $(a + b) + c = a + (b + c)$ $(ab)c = a(bc)$	If three real numbers are added or multiplied, it makes no difference which two are added or multiplied first. EXAMPLES: $(1 + 3) + 4 = 1 + (3 + 4)$ $(2 \cdot 3) \cdot 5 = 2 \cdot (3 \cdot 5)$
Distributive property: $a(b + c) = ab + ac$ $(b + c)a = ba + ca$	Multiplication distributes over addition. EXAMPLE: $4(3 + 5) = 4 \cdot 3 + 4 \cdot 5$ $(2 + 5)6 = 2 \cdot 6 + 5 \cdot 6$
Identity property of addition: $a + 0 = 0 + a = a$	The sum of 0 and a real number equals the number itself. EXAMPLE: $7 + 0 = 0 + 7 = 7$ NOTE: Zero is called the identity of addition or the additive identity.
Identity property of multiplication: $a \cdot 1 = 1 \cdot a = a$	The product of 1 and a real number equals the number itself. EXAMPLE: $7 \cdot 1 = 1 \cdot 7 = 7$ NOTE: One is called the identity of multiplication or the multiplicative identity.
Inverse property of addition: $a + (-a) = (-a) + a = 0$	a is called the additive inverse or opposite of $-a$. Similarly, $-a$ is the additive inverse or opposite of a. The sum of a real number and its additive inverse is 0. EXAMPLE: $5 + (-5) = (-5) + 5 = 0$
Inverse property of multiplication: $a \cdot \dfrac{1}{a} = \dfrac{1}{a} \cdot a = 1, \quad a \neq 0$	$1/a$ is called the multiplicative inverse or reciprocal of a. The product of a nonzero real number and its reciprocal is 1. EXAMPLE: $8 \cdot \dfrac{1}{8} = \dfrac{1}{8} \cdot 8 = 1$

ENRICHMENT ESSAY

Art and Irrational Numbers

A $\sqrt{2}$ rectangle has adjacent sides in the ratio $1:\sqrt{2}$. The photograph shows an ancient food bowl that fits perfectly inside a $\sqrt{2}$ rectangle.

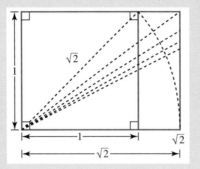

Reproduced with permission of Dover Publications, "Authentic Indian Design", edited by Maria Naylor.

Properties of Zero

The additive identity property for zero,

$$a + 0 = 0 + a = a$$

was included in our list of basic properties of real numbers. The following list summarizes other properties of zero that you studied in beginning algebra.

Properties of zero

Let a represent any real number.

Property	Examples
1. $a + 0 = a$	$\pi + 0 = \pi$
2. $a - 0 = a$	$\sqrt{2} - 0 = \sqrt{2}$
3. $a \cdot 0 = 0$	$3 \cdot 0 = 0$
$0 \cdot a = 0$	$0\left(\dfrac{\pi}{2}\right) = 0$
4. $\dfrac{0}{a} = 0, \quad a \neq 0$	$\dfrac{0}{-53} = 0$
5. $\dfrac{a}{0}$ is undefined.	$\dfrac{-53}{0}$ is undefined.

Study tip

The division property of zero does not permit division by zero, which is undefined. If $a \neq 0$, $\frac{a}{0}$ is undefined because no real number multiplied by 0 will give a. Furthermore, $\frac{0}{0}$ is *indeterminate* because any real number multiplied by 0 will give 0.

ENRICHMENT ESSAY

A Commutative Theme: Palindromes

The theme of the commutative property—a change in order makes no difference in an outcome—is found in palindromes. It makes no difference whether a palindrome is read from left to right or from right to left. The outcome is always the same. This is because a palindrome is a word, verse, phrase, sentence, or number that reads the same forward as backward. Examples are the following:

1. dad
2. repaper
3. 10,233,201
4. never odd or even
5. Doc, note, I dissent. A fast never prevents a fatness. I diet on cod.

Some sophisticated numerical palindromes involve two calculations:

$$9 + 9 = 18 \qquad 81 = 9 \times 9$$
$$24 + 3 = 27 \qquad 72 = 3 \times 24$$
$$47 + 2 = 49 \qquad 94 = 2 \times 47$$
$$497 + 2 = 499 \qquad 994 = 2 \times 497$$

An interesting number curiosity involves adding to any natural number the number formed by reversing its digits. To this sum, add the number formed by reversing the sum's digits. By continuing this process, a numerical palindrome is achieved:

$$
\begin{array}{r}
38 \\
+ 83 \\
\hline
121
\end{array}
\qquad
\begin{array}{r}
139 \\
+ 931 \\
\hline
1070 \\
+0701 \\
\hline
1771
\end{array}
\qquad
\begin{array}{r}
48,017 \\
+ 71,084 \\
\hline
119,101 \\
+ 101,911 \\
\hline
221,012 \\
+ 210,122 \\
\hline
431,134
\end{array}
$$

Some numbers are easier to work with than others. Computers have failed to end up with a palindrome for 196 after performing this process thousands of times.

This palindromic product involves every digit but 9:

$$
\begin{array}{r}
12345679 \\
\times 99999999 \\
\hline
111111111 \\
111111111 \\
111111111 \\
111111111 \\
111111111 \\
111111111 \\
111111111 \\
111111111 \\
\hline
1234567887654321
\end{array}
$$

These palindromic squares even manage to include 9:

$$11^2 = 121$$
$$111^2 = 12321$$
$$1111^2 = 1234321$$
$$11111^2 = 123454321$$
$$111111^2 = 12345654321$$
$$1111111^2 = 1234567654321$$
$$11111111^2 = 123456787654321$$
$$111111111^2 = 12345678987654321$$

Before we leave the subject, here's something to ponder: Can you think of a palindromic way of expressing the number 12?

PROBLEM SET 1.1

Practice Problems _____

In Problems 1–16, use the roster method to write each set.

1. $\{x \mid x$ is an even natural number between 14 and 20, inclusively$\}$

2. $\{x \mid x$ is an odd natural number between 7, exclusively, and 21, inclusively$\}$

3. $\{x \mid x$ is a natural number that is divisible by 5$\}$

4. $\{x \mid x$ is a natural number that is both even and divisible by 5$\}$

5. $\{x \mid x$ is a whole number but is not a natural number$\}$

6. $\left\{\dfrac{a}{b} \mid a = 1 \text{ or } 4 \text{ and } b = 2 \text{ or } 4\right\}$

7. $\{x \mid x$ is an integer and $x^2 = 1\}$

8. $\{x \mid x$ is a natural number and $x^2 = 9\}$

9. $\{x \mid x$ is a positive real number and $x^2 = 5\}$

10. $\{x \mid x$ is an integer but not a whole number$\}$

11. $\{x \mid x$ is an integer but not a natural number$\}$

12. $\{x \mid x$ is both an integer and a rational number$\}$

13. $\{x \mid \sqrt{x}$ is a natural number less than or equal to 2$\}$

14. $\{x \mid \sqrt{x}$ is a whole number less than or equal to 3$\}$

15. $\{x \mid x$ is the fractional form of $0.\overline{6}\}$

16. $\{x \mid x$ is the fractional form of $-0.25\}$

Determine which elements of the set

$$A = \left\{-10, -\sqrt{2}, -\frac{3}{4}, 0, \frac{4}{5}, \sqrt{3}, \sqrt{4}, \pi, 7, \frac{18}{2}, 100\right\}$$

are elements of the sets in Problems 17–22.

17. Natural numbers

18. Whole numbers

19. Integers

20. Rational numbers

21. Irrational numbers

22. Real numbers

In Problems 23–34, write out the meaning of each inequality statement, and then determine if the statement is true.

23. $-13 \leqslant -2$

24. $-1 \leqslant -13$

25. $-6 > 2$

26. $8 \geqslant -3$

27. $4 \geqslant -7$

28. $11 \leqslant 11$

29. $-13 < -5$

30. $8 \geqslant -9$

31. $-\pi \geqslant -\pi$

32. $\sqrt{3} < \sqrt{3}$

33. $-\sqrt{2} < -\sqrt{2}$

34. $-3 > -13$

Graph each set in Problems 35–48 and express each in interval notation.

35. $\{x \mid 0 < x < 3\}$

36. $\{x \mid 1 < x < 4\}$

37. $\{x \mid -2 \leqslant x < 1\}$

38. $\{x \mid -3 < x \leqslant 2\}$

39. $\{x \mid -2 \leqslant x \leqslant -1\}$

40. $\{x \mid -5 \leqslant x \leqslant -2\}$

41. $\{x \mid x \leqslant 2\}$

42. $\{x \mid x < 6\}$

43. $\{x \mid x > -3\}$

44. $\{x \mid x \geqslant -4\}$

45. $\{x \mid x < -1\}$

46. $\{x \mid x < -2\}$

47. $\{x \mid x \leqslant 0\}$

48. $\{x \mid x > 0\}$

Express each English phrase in Problems 49–60 in both set-builder and interval notation.

49. x lies between 5 and 12, excluding 5 and 12.

50. x lies between -4 and 7, excluding -4 and 7.

51. x lies between 2 and 13, excluding 2 and including 13.

52. x lies between -7 and 2, including -7 and excluding 2.

53. x is at most 6.

54. x is at least 3.

55. x is at least 2 and at most 5.

56. x is at least -3 and at most 2.

57. x is not more than 60.

58. x is not more than 32.

59. x is negative and at least -2.

60. x is not positive and at least -5.

State the property of algebra illustrated in Problems 61–70.

61. $3 + 7 = 7 + 3$

62. $-14 + 14 = 0$

63. $2(7 + 4) = 2 \cdot 7 + 2 \cdot 4$

64. $6 + (3 + 8) = (6 + 3) + 8$

65. $6 + (3 + 8) = 6 + (8 + 3)$

66. $\frac{1}{3}(3) = 1$

67. $2 \cdot (3 \cdot 4) = (2 \cdot 3) \cdot 4$

68. $2(3 \cdot 4) = 2(4 \cdot 3)$

69. $(15 + 6) + 0 = 15 + 6$

70. $1\sqrt{3} = \sqrt{3}$

Application Problems

It is quite likely that you have seen at least one of the ten films mentioned in the bar graph shown, since these were the top ten money earners for the period from 1990 to 1995. If x represents millions of dollars grossed by a film, write the film title or titles described in Problems 71–79.

71. $x \in [173, 206)$

72. $x \in (285, 338)$

73. $x \in [206, \infty)$

74. $x \in (-\infty, 179)$

75. $184 < x \leq 217$

76. $x \geq 338$

77. $x \leq 165$

78. $x > 338$

79. $x < 165$

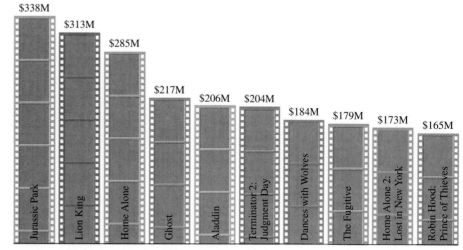

The Top Ten Grossing Films, 1990–1995

Source: Based on statistics from *Time; Entertainment Weekly*

True–False Critical Thinking Problems

80. Which one of the following is true?
   a. Zero divided by any real number is zero.
   b. Every rational number is an integer.
   c. $\sqrt{36}$ is an irrational number.
   d. Zero plays the same role in subtraction as it does in addition.

81. Which one of the following is true?
   a. Using interval notation, the set $\{x \mid x < 3\}$ can be expressed as $[-\infty, 3)$.
   b. The same set of real numbers is described by $\{x \mid 2 < x < 5\}$ and $\{x \mid x < 2 \text{ or } x > 5\}$.
   c. The symbol ∞ represents the real number infinity.
   d. Using interval notation, any unbounded interval is expressed in terms of $-\infty$ or ∞.

Writing in Mathematics

82. Describe the difference between representing a set by the roster method and set-builder notation. Provide an example of each in your description.

83. Why is every natural number a rational number, but not every rational number is a natural number?

84. How do bounded and unbounded intervals differ?

85. Describe the difference between the commutative and associative properties.

86. How can you show that the set of real numbers is not associative with respect to division?

87. The following poem was written by A. C. Orr and appeared in the *Literary Digest* in 1906. Explain how the poem allows us to write the first thirty digits of the decimal representation of π. Write the representation.

Now I, even I, would celebrate
In rhymes unapt, the great
Immortal Syracusan, rivaled nevermore,
Who in his wondrous lore
Passed on before,
Left men his guidance
How to circles mensurate.

Critical Thinking Problems

The difference between what the government receives in taxes and what it pays out for various government services is the deficit. The graph shows the deficit projection in billions of dollars for seven years.

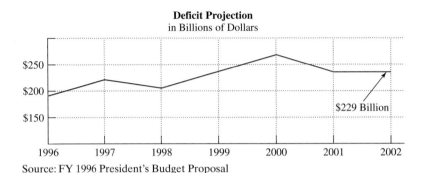

Deficit Projection
in Billions of Dollars

$229 Billion

1996   1997   1998   1999   2000   2001   2002

Source: FY 1996 President's Budget Proposal

Write a bounded interval that describes the budget deficit for the years specified in Problems 88–91.

88. The years 1996, 1997, and 1998.

89. The year 2000.

90. All of the years shown in the graph.

91. None of the years shown in the graph.

92. Instead of the set of real numbers, consider the set $\{S, L, A, R\}$ consisting of the four directions

S = Stand tall

L = Left face

A = About face

R = Right face

Let's replace addition and multiplication by a new operation, ∘, meaning "and then followed by," so that $L \circ A$ means "left face" and then follow this by "about face." The overall effect of this composite order is equivalent to "right face." Thus, $L \circ A = R$.

Similarly, $R \circ R$ means "right face" followed by "right face," which would put one in an "about face" position. Thus, $R \circ R = A$.

Instead of an addition or multiplication table, we can construct a table for the operation ∘ on the elements S, L, A, and R.

∘	S	L	A	R
S	S	L	A	R
L	L	A	R	S
A	A	R	S	L
R	R	S	L	A

a. Why is closure satisfied in this situation?

b. Use the table to find $(L \circ A) \circ R$ and $L \circ (A \circ R)$. What property does this illustrate?

c. Use the table to show that $L \circ R = R \circ L$. What property is illustrated?

d. What member of the set plays the same role as 0 does for ordinary addition or as 1 does for ordinary multiplication? What can we call this member?

e. Use the table to fill in the following:

$S \circ \underline{\hspace{1cm}} = S$

$L \circ \underline{\hspace{1cm}} = S$

$A \circ \underline{\hspace{1cm}} = S$

$R \circ \underline{\hspace{1cm}} = S$

What property is illustrated?

This example hints at the beginning of a new abstract algebra, sometimes called *group theory,* with emphasis placed on any set (rather than the set of real numbers) and an abstract operation on that set (rather than the usual operations of addition and multiplication).

93. Suppose we define $a \circ b = ab + a$. (For example, $7 \circ 4 = 7 \cdot 4 + 7 = 28 + 7 = 35$.) Is ∘ a commutative operation? (Is $a \circ b = b \circ a$?)

94. Suppose we define $a \circ b = aa + bb$. Is ∘ a commutative operation?

SECTION 1.2

Solutions   Tutorial   Video
Manual   1

Operations with the Real Numbers and Algebraic Expressions

Objectives

1 Find the absolute value of a real number.
2 Perform operations with real numbers.
3 Evaluate exponential expressions.
4 Apply the order of operations agreement.
5 Simplify algebraic expressions.
6 Evaluate mathematical models.

In this section, we review the familiar rules from introductory algebra for adding, subtracting, multiplying, and dividing real numbers. We also review the order of operations agreement for computations involving a variety of different operations.

Since a basic characteristic of algebra is the use of variables that represent numbers, we conclude by reviewing how to combine and simplify expressions containing variables and numbers. The importance of these expressions is their appearance in formulas describing all aspects of our physical world. These formulas, called *mathematical models,* are introduced in this section and form the basis of our efforts throughout the book to describe the world in a compact, meaningful way.

1 Find the absolute value of a real number.

Absolute Value

The concept of absolute value is used to describe how to operate with positive and negative numbers.

> **Absolute value**
>
> The symbol $|a|$ is read "the absolute value of a." Geometrically, the absolute value of a is the distance between the number a and 0 on the real number line.

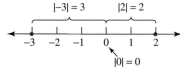

Absolute value describes distance from 0 on a number line.

EXAMPLE 1   **Finding Absolute Value**

Find the absolute value:

a. $|-3|$ **b.** $|2|$ **c.** $|0|$

Solution

a. $|-3| = 3$ The absolute value of -3 is 3 because -3 is 3 units from 0.
b. $|2| = 2$ 2 is 2 units from 0.
c. $|0| = 0$ 0 is 0 units from itself. ∎

In reviewing the properties of real numbers, we considered the inverse property of addition

$$a + (-a) = 0.$$

The real numbers a and $-a$ are *opposites* or *additive inverses*, lying the same distance from, but on opposite sides of, 0 on the number line. Using opposites, we can formally define the absolute value of a real number x.

Definition of absolute value

$$|x| = \begin{cases} x & \text{if } x \geq 0 \\ -x & \text{if } x < 0 \end{cases}$$

The first row of this definition tells us that the absolute value of a nonnegative number is the number itself. Thus, $|5| = 5$. The second row tells us that the absolute value of a negative number is the opposite of the number. For example,

$$\left| -\frac{1}{3} \right| = -\left(-\frac{1}{3} \right) = \frac{1}{3}.$$

2 Perform operations with real numbers.

Operations with Real Numbers

Operations with real numbers can be described in terms of absolute value. Table 1.5 reviews how to perform these operations.

TABLE 1.5 Operations with Real Numbers

Addition

Rule	Examples
To add two real numbers with the same sign, add their absolute values and attach the common sign to the sum.	$(-7) + (-4) = -(\lvert -7 \rvert + \lvert -4 \rvert)$ $\qquad\qquad = -(7 + 4)$ $\qquad\qquad = -11$
To add two real numbers with different signs, subtract the smaller absolute value from the greater absolute value, and attach the sign of the number with the greater absolute value. If the two numbers have the same absolute value, the sum is 0.	$7 + (-15) = -(\lvert -15 \rvert - \lvert 7 \rvert)$ $\qquad\qquad = -(15 - 7)$ $\qquad\qquad = -8$ $(-19) + 19 = 0$

Subtraction

$a - b = a + (-b)$; a minus b is the sum of a and the opposite of b.

Rule	Examples
To subtract the real number b from the real number a, add the opposite of b to a. The result is the difference of a and b.	$7 - 13 = 7 + (-13) = -6$ $32 - (-10) = 32 + 10 = 42$ $-22 - (-15) = -22 + 15 = -7$

Multiplication and Division

Rule	Examples
To multiply or divide two real numbers with the same sign, multiply or divide their absolute values, and make the sign of the product or quotient positive.	$(-5)(-11) = 55$ $\dfrac{-48}{-4} = 12$

TABLE 1.5 Operations with Real Numbers (continued)

Multiplication and Division (continued)

Rule	Examples
To multiply or divide two real numbers with different signs, multiply or divide their absolute values, and make the sign of the product or quotient negative.	$7(-5) = -35$ $$-8 \div 4 = -8 \cdot \frac{1}{4} = -2$$ $$-40 \div \frac{1}{8} = -40 \cdot 8 = -320$$
To multiply more than two nonzero real numbers, multiply their absolute values, and make the sign positive with an even number of negative factors and negative with an odd number of negative factors.	$$(-2)(-3)(-5)\left(-\frac{1}{5}\right)(4) = 24$$ four (even) negative factors $$(-4)(-3)(-7) = -84$$ three (odd) negative factors

Using technology

Try verifying some of the operations shown in Table 1.5 with a graphing calculator. Some verifications are shown here:

$-7 + (-4)$: $\boxed{(-)}$ $\boxed{7}$ $\boxed{+}$ $\boxed{(-)}$ $\boxed{4}$ $\boxed{\text{ENTER}}$

$-22 - (-15)$: $\boxed{(-)}$ $\boxed{22}$ $\boxed{-}$ $\boxed{(-)}$ $\boxed{15}$ $\boxed{\text{ENTER}}$

$-40 \div \dfrac{1}{8}$: $\boxed{(-)}$ $\boxed{40}$ $\boxed{\div}$ $\boxed{(}$ $\boxed{1}$ $\boxed{\div}$ $\boxed{8}$ $\boxed{)}$ $\boxed{\text{ENTER}}$

Don't confuse the subtraction key $\boxed{-}$ with the sign change or opposite key $\boxed{(-)}$. Describe what happens if you do. Also describe what happens if you fail to place parentheses around $\frac{1}{8}$. How can you explain the result that the calculator gives when parentheses are omitted?

The examples in Table 1.5 suggest the following basic properties of negatives.

Properties of negatives

Let a and b represent real numbers.

Property	Examples
1. $(-1)a = -a$	$(-1)3 = -3$
2. $-(-a) = a$	$-(-8) = 8$
3. $(-a)b = -(ab)$	$(-5)8 = -(5 \cdot 8)$ or -40
$a(-b) = -(ab)$	$7(-3) = -(7 \cdot 3)$ or -21
4. $(-a)(-b) = ab$	$(-6)(-9) = 6 \cdot 9$ or 54
5. $\dfrac{-a}{b} = \dfrac{a}{-b} = -\dfrac{a}{b}; \quad b \neq 0$	$\dfrac{-8}{2} = \dfrac{8}{-2} = -\dfrac{8}{2} = -4$
6. $\dfrac{-a}{-b} = \dfrac{a}{b}; \quad b \neq 0$	$\dfrac{-8}{-2} = \dfrac{8}{2} = 4$
7. $-(a + b) = -a - b$	$-(3 + x) = -3 - x$
8. $-(a - b) = -a + b$	$-(x - 7) = -x + 7$
$= b - a$	$= 7 - x$

3 Evaluate exponential expressions.

Exponential Expressions

Repeated multiplication of the same factor can be expressed using an exponent.

> **Definition of a natural number exponent**
>
> If b is a real number and n is a natural number,
>
> Exponent
> $$b^n = \underbrace{b \cdot b \cdot b \cdots b}_{}.$$
> Base b appears as a factor n times
>
> b^n is read "the nth power of b" or "b to the nth power." Thus, the nth power of b is defined as the product of n factors of b. Furthermore, $b^1 = b$.

Using technology

Verify Example 2(d) with a graphing calculator and the exponential key $\boxed{\wedge}$. Here are the keystroke sequences for parts (a)–(c).

a. $(-3)^4$:
   $\boxed{(}\ \boxed{(-)}\ \boxed{3}\ \boxed{)}\ \boxed{\wedge}\ \boxed{4}\ \boxed{\text{ENTER}}$

b. -3^4: $\boxed{(-)}\ \boxed{3}\ \boxed{\wedge}\ \boxed{4}\ \boxed{\text{ENTER}}$

c. 7^2: $\boxed{7}\ \boxed{\wedge}\ \boxed{2}\ \boxed{\text{ENTER}}$ or
   $\boxed{7}\ \boxed{x^2}\ \boxed{\text{ENTER}}$

EXAMPLE 2   **Evaluating Exponential Expressions**

Evaluate:

a. $(-3)^4$ **b.** -3^4 **c.** 7^2 **d.** $(-2)^3 \cdot 3^2$

Solution

a. $(-3)^4 = (-3)(-3)(-3)(-3) = 81$ The negative of a number is taken to a power only when the negative is inside the parentheses.

b. $-3^4 = -(3 \cdot 3 \cdot 3 \cdot 3) = -81$

c. $7^2 = 7 \cdot 7 = 49$ 7^2 is read "7 squared."

d. $(-2)^3 \cdot 3^2 = (-2)(-2)(-2)(3)(3)$ $(-2)^3$ is read "-2 cubed."
$$= -8 \cdot 9 = -72$$

4 Apply the order of operations agreement.

Order of Operations

Given a computation such as $3 + 7 \cdot 5$, we must establish an agreement as to the order of operations. Depending on whether we perform the addition or multiplication first, without such an agreement we will obtain two different answers to the same problem.

> **The order of operations agreement**
>
> 1. Perform operations above and below any fraction bar following steps 2 to 5 below.
> 2. Perform operations inside grouping symbols following steps 3 to 5. Work from innermost grouping symbols, parentheses ( ), to outermost grouping symbols, brackets [ ].
> 3. Simplify exponential expressions.
> 4. Do multiplication or division as they occur, working from left to right.
> 5. Do addition and subtraction as they occur, working from left to right.

Examples 3 through 6 illustrate the order of operations agreement.

EXAMPLE 3 Using the Order of Operations

Simplify: $3 + 7 \cdot 5$

Solution

$$3 + 7 \cdot 5 = 3 + 35 \quad \text{First multiply.}$$
$$= 38 \quad \text{Add.}$$

Using technology

Example 4 can be checked using a graphing calculator as follows:

6 ∧ 2 − 24 ÷ 2 ∧ 2 × 3
− 1 ENTER

EXAMPLE 4 Using the Order of Operations

Simplify: $6^2 - 24 \div 2^2 \cdot 3 - 1$

Solution

$$6^2 - 24 \div 2^2 \cdot 3 - 1 \quad \text{Simplify the exponential expressions.}$$
$$= 36 - 24 \div 4 \cdot 3 - 1 \quad \text{Perform the multiplication and division in the order in}$$
$$\text{which they occur from left to right. Begin with } 24 \div 4.$$
$$= 36 - 6 \cdot 3 - 1 \quad 24 \div 4 = 6. \text{ Now find } 6 \cdot 3.$$
$$= 36 - 18 - 1 \quad \text{Perform the subtraction from left to right.}$$
$$= 18 - 1$$
$$= 17$$

EXAMPLE 5 Using the Order of Operations

Discover for yourself

Use a graphing calculator to verify the answers in Examples 5 and 6.

Simplify: $\dfrac{1}{2} \cdot 10 + [4(6 \div 3) - 15]$

Solution

$$\frac{1}{2} \cdot 10 + [4(6 \div 3) - 15]$$

$$= \frac{1}{2} \cdot 10 + [4(2) - 15] \quad \text{Work inside the parentheses.}$$

$$= \frac{1}{2} \cdot 10 + [8 - 15] \quad \text{Work inside the brackets, performing the multiplication}$$
$$\text{before the subtraction.}$$

$$= \frac{1}{2} \cdot 10 + (-7) \quad \text{Complete the subtraction inside the brackets.}$$

$$= 5 + (-7) \quad \text{Multiply.}$$
$$= -2 \quad \text{Add.}$$

EXAMPLE 6 Using the Order of Operations

Simplify: $12 - \dfrac{4^2 + 3^2}{2^2 + 1} \div 2^2$

Solution

$$12 - \frac{4^2 + 3^2}{2^2 + 1} \div 2^2$$

$$= 12 - \frac{16 + 9}{4 + 1} \div 2^2 \quad \text{Perform operations above and below the fraction bar. Begin}$$
$$\text{with evaluating exponential expressions.}$$

$$= 12 - \frac{25}{5} \div 2^2 \qquad \text{Complete the operations above and below the fraction bar.}$$

$$= 12 - 5 \div 2^2 \qquad \text{Simplify exponential expressions.}$$

$$= 12 - 5 \div 4 \qquad \text{Do multiplications or divisions from left to right.}$$

$$= 12 - 1.25 \qquad \text{Divide.}$$

$$= 10.75 \qquad \text{Subtract.} \qquad \blacksquare$$

5 Simplify algebraic expressions.

Algebraic Expressions

A basic characteristic of algebra is the use of variables to stand for various numbers. An expression that combines numbers and variables using the operations of addition, subtraction, multiplication, or division as well as powers or roots, is called an *algebraic expression*. The following are examples of algebraic expressions:

$$4x^3 - 7x^2 + 5 \qquad 5(x - y) - (3x - 2y)$$

$$2L + 2W \qquad \frac{-b \pm \sqrt{b^2 - 4ac}}{2a}$$

The *terms* of an algebraic expression are those parts that are separated by addition. For example, the algebraic expression $4x^3 - 7x^2 + 5$ contains three terms: $4x^3$, $-7x^2$, and 5. Observe that $-7x^2$, rather than $7x^2$, is a term because we can express the algebraic expression as

$$4x^3 - 7x^2 + 5 = 4x^3 + (-7x^2) + 5.$$

Notice that a term indicates a product and may contain any number of factors. The terms $4x^3$ and $-7x^2$ are the *variable terms* of the expression, and 5 is the *constant term*. The numerical part of a term, such as the -7 in $-7x^2$, is called the *numerical coefficient*.

Table 1.6 gives examples of algebraic expressions and their terms.

TABLE 1.6 The Vocabulary of Algebraic Expressions

Algebraic Expression	Variable Term(s)	Numerical Coefficient(s)	Constant Term
$3x - 7 = 3x + (-7)$	$3x$	3	-7
$7x^2 - 4x + 3$	$7x^2, -4x$	$7, -4$	3
$13x^3 - x^2 + x + 5$ or $13x^3 - x^2 + 1x + 5$	$13x^3, -x^2, x$	$13, -1, 1$	5
$-9x^2y + 14xy - 3$	$-9x^2y, 14xy$	$-9, 14$	-3

The basic properties of the real numbers can be applied to algebraic expressions. Two terms that have the same variables to the same powers are called *like* or *similar* terms. Thus, $3xy$ and $5xy$ are like terms, but $5xy$ and $3xz$ are not. The distributive property enables us to *collect* or *combine* like terms.

EXAMPLE 7 Combining Like Terms

Combine like terms:

a. $4x + 15x$ **b.** $3x^2y - 7x^2y$ **c.** $-5ab^3 - ab^3$

Solution

a. $4x + 15x = (4 + 15)x = 19x$
b. $3x^2y - 7x^2y = (3 - 7)x^2y = -4x^2y$
c. $-5ab^3 - ab^3 = (-5 - 1)ab^3 = -6ab^3$

In all parts of Example 7, the algebraic expression that we obtain is *equivalent* to the given expression.

> **Equivalent algebraic expressions**
>
> Two expressions that have the same value for all possible replacements are called *equivalent expressions.*

Finding an equivalent algebraic expression for a given expression usually involves simplifying that expression. In simplified form, an algebraic expression contains no grouping symbols and all like terms are combined.

EXAMPLE 8 Simplifying Algebraic Expressions

Use the basic properties of the real numbers to obtain a simplified expression equivalent to each of the following.

a. $5x(2y + 9)$ **b.** $-\dfrac{1}{3}(3y)$ **c.** $-(-6x^4 - 9x + 17)$

Solution

a. We can remove parentheses by applying the distributive property.

$$5x\,(2y + 9) = 5x \cdot 2y + 5x \cdot 9 \qquad \text{Apply the distributive property.}$$
$$= 5 \cdot 2xy + 5 \cdot 9x \qquad \text{Use the commutative property to change the order of the multiplication.}$$
$$= 10xy + 45x \qquad \text{Multiply constant factors.}$$

b. $-\dfrac{1}{3}(3y) = \left(-\dfrac{1}{3} \cdot 3\right)y \qquad \text{Apply the associative property.}$
$$= -1y$$
$$= -y \qquad \text{Use the property of negatives: } (-1)a = -a.$$

c. The property of negatives

$$-(a - b) = -a + b$$

can be extended to more than two terms within parentheses and formulated as the following rule: If a negative appears outside the parentheses, drop the parentheses and change the sign of every term within the parentheses. Thus,

$$-(-6x^4 - 9x + 17) = 6x^4 + 9x - 17 \qquad \text{Drop the parentheses and change signs.}$$

EXAMPLE 9   Simplifying an Algebraic Expression

Simplify: $3(x + y) + 5(y - 4x)$

Solution

$$3(x + y) + 5(y - 4x)$$
$$= 3x + 3y + 5y - 20x \qquad \text{Use the distributive property to remove the parentheses.}$$
$$= -17x + 8y \qquad \text{Combine like terms.}$$

Simplification of an algebraic expression often requires that we extend the distributive property to cover more than two terms, or to distribute multiplication over subtraction.

EXAMPLE 10   Simplifying an Algebraic Expression

Simplify: $-4(5x - z) - (z - 2y) + 6(x + 2y - z)$

Solution

$$-4(5x - z) - (z - 2y) + 6(x + 2y - z)$$
$$= -20x + 4z - z + 2y + 6x + 12y - 6z \qquad \text{Use the distributive property to remove the parentheses.}$$
$$= -14x + 14y - 3z \qquad \text{Combine like terms.}$$

In Examples 9 and 10, we used the distributive property to remove parentheses. If an algebraic expression contains other kinds of grouping symbols, such as brackets [ ] or braces { }, simplification should be performed by working from the innermost to the outermost grouping symbols.

EXAMPLE 11   Simplifying an Algebraic Expression

Simplify: $7 - 5[4x - 3y - 2(1 - 3x - y)]$

Solution

$$7 - 5[4x - 3y - 2(1 - 3x - y)]$$
$$= 7 - 5[4x - 3y - 2 + 6x + 2y] \qquad \text{Use the distributive property to remove the parentheses.}$$
$$= 7 - 5[10x - y - 2] \qquad \text{Combine like terms.}$$
$$= 7 - 50x + 5y + 10 \qquad \text{Use the distributive property to remove the brackets.}$$
$$= -50x + 5y + 17 \qquad \text{Combine like terms.}$$

6 Evaluate mathematical models.

Mathematical Models

Throughout this book, we will be working with equations. An *equation* is a statement of equality between two expressions. The basic statement

$$a = b$$

ENRICHMENT ESSAY

Einstein's Famous Mathematical Model: $E = mc^2$

One of the most famous mathematical models in the world is the formula $E = mc^2$, formulated by Albert Einstein. Einstein showed that any form of energy has mass and that mass itself is a form of energy. In this formula, E represents energy in ergs, m represents mass in grams, and c represents the speed of light in centimeters/second. Because light travels at 30 billion centimeters/second, the formula indicates that 1 gram of mass will produce 900 billion ergs of energy.

Einstein's model implies that the mass of a golf ball could provide the daily energy needs of the metropolitan Boston area. Mass and energy are equivalent, and the transformation of even a tiny amount of mass releases an enormous amount of energy. If this energy is released suddenly, a destructive force is unleashed, as in an atom bomb. When the release is gradual and controlled, the energy can be used to generate power.

The theoretical results implied by Einstein's mathematical model $E = mc^2$ have not been realized because humankind has yet to develop any way of converting a mass completely to energy. The best that has been done so far, in nuclear fission reactors, is to convert about 0.1% of a mass to energy. With only 0.1% efficiency, it would take about 30 kilograms of mass (approximately 70 pounds) to power metropolitan Boston for a day. By contrast, the burning of fossil fuels—including coal, oil, and gas—requires about 10^8 kilograms, or approximately 100,000 tons.

Numerous problems stand in the way of nuclear power expansion, not the least of which is what to do with nuclear waste. By mid-1990 more than 23,000 tons of intensely radioactive nuclear waste had accumulated around the 111 operating nuclear power plants in the United States. Despite the implications of Einstein's mathematical model, if the waste-storage issue is not resolved, our spiraling demand for electricity will not be realized by the nuclear option.

Radioactive Cats © 1980 Sandy Skoglund. Cibachrome 30 × 40 inches.

Ed Paschke "The Triangle" 1991 o/l 40 × 36 in. Photo courtesy Phyllis Kind Gallery, New York & Chicago

("a is equal to b" or "a equals b") means that a and b represent the same real number.

A *formula* is an equation that uses letters to express relationships between quantities. Formulas are used in almost all academic disciplines as well as in everyday life. One of the aims of applied mathematics is to find formulas that describe the physical world. These formulas are called *mathematical models*.

In Example 12, a mathematical model is given, along with the value of one of the variables in the model. We can use the order of operations to evaluate the formula by substituting the numerical value for the given variable.

EXAMPLE 12 **An Application: Systolic Blood Pressure**

The mathematical model $P = (25t^2 + 125t) \div (t^2 + 1)$ describes systolic blood pressure P (measured in millimeters of mercury) t seconds after blood

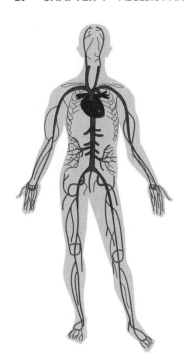

leaves the heart. The model applies for $0 \leq t \leq 10$. Find systolic pressure 3 seconds after the blood moves from the heart.

Solution

We must substitute 3 for t in our formula.

$$P = (25t^2 + 125\,t) \div (t^2 + 1)$$

becomes

$$
\begin{aligned}
P &= (25 \cdot 3^2 + 125 \cdot 3) \div (3^2 + 1) \\
  &= (25 \cdot 9 + 125 \cdot 3) \div (9 + 1) \qquad \text{Work inside the parentheses.} \\
  &= (225 + 375) \div (9 + 1) \\
  &= 600 \div 10 \\
  &= 60
\end{aligned}
$$

Consequently, 3 seconds after leaving the heart, systolic blood pressure measurement is 60 millimeters of mercury.

PROBLEM SET 1.2

Practice Problems

Use the agreed-upon order of operations to find the value of each expression in Problems 1–44. If applicable, verify your answers with a calculator.

1. $7 + 6 \cdot 3$

2. $-5 + (-3) \cdot 8$

3. $4(-5) - 6(-3)$

4. $-8(-3) - 5(-6)$

5. $6 - 4(-3) - 5$

6. $3 - 7(-1) - 6$

7. $3 - 5(-4 - 2)$

8. $3 - 9(-1 - 6)$

9. $(2 - 6)(-3 - 5)$

10. $9 - 5(6 - 4) - 10$

11. $3(-2)^2 - 4(-3)^2$

12. $5(-3)^2 - 2(-2)^3$

13. $(2 - 6)^2 - (3 - 7)^2$

14. $(4 - 6)^2 - (5 - 9)^3$

15. $6(3 - 5)^3 - 2(1 - 3)^3$

16. $3(-6 + 8)^3 - 5(-3 + 5)^3$

17. $8^2 - 16 \div 2^2 \cdot 4 - 3$

18. $10^2 - 100 \div 5^2 \cdot 2 - (-3)$

19. $\dfrac{4^2 + 3^3}{5^2 - (-18)}$

20. $\dfrac{(-3)^3 + (-2)^2}{5^2 - (-21)}$

21. $20 - 4\left(\dfrac{8 - 2}{3 - 6}\right) \div \dfrac{1}{2}$

22. $30 \div 5\left(\dfrac{8 + 16}{2^3 - 2^2}\right) - 5$

23. $\left(\dfrac{1}{2}\right)^2 + \left(\dfrac{6 - 4}{5}\right)^2 + \left(\dfrac{5 + 2}{10}\right)^2$

24. $\left(\dfrac{2^3}{2^3 + 1}\right)^2 \div \left(\dfrac{8 - (-2)}{3^2}\right)^2$

25. $-3[8 + (-6)] \div [-4 - (-5)]$

26. $-4[3 + (-4)] \div [-8 - (-10)]$

27. $(\frac{1}{2} - \frac{7}{4}) \div (1 - \frac{3}{8})$

28. $2\frac{1}{4} - 7 \div \frac{1}{4} \cdot \frac{1}{2}$

29. $\frac{1}{4} - 6(2 + 8) \div (-\frac{1}{3})(-\frac{1}{9})$

30. $\frac{2}{3} - 4(3 + 5) \div (-\frac{1}{2})(-\frac{1}{8})$

31. $6.8 - (0.3)^2 \div 0.09$

32. $0.4(1.2 - 2.3)^2 + 4.6$

33. $\frac{1}{2} - (\frac{2}{3} \cdot \frac{9}{5}) + \frac{3}{10}$

34. $\dfrac{7}{12} - \dfrac{\frac{5}{4}}{2 - \frac{7}{2}} \cdot \dfrac{1}{2}$

35. $8 - 3[-2(2 - 5) - 4(8 - 6)]$

36. $8 - 3[-2(5 - 7) - 5(4 - 2)]$

37. $\dfrac{2(-2) - 4(-3)}{5 - 8}$

38. $\dfrac{6(-4) - 5(-3)}{9 - 10}$

39. $10 - (-8)\left[\dfrac{2(-3) - 5(6)}{7 - (-1)}\right]$

40. $-12 - 4\left[\dfrac{12(-1) - 8}{4(-3) + 2(5)}\right]$

41. $6 - (-12)\left[\dfrac{2 - 4(3 - 7)}{-4 - 5(1 - 3)}\right]$

42. $8 - (-20)\left[\dfrac{6 - 1(6 - 10)}{14 - 3(6 - 8)}\right]$

43. $2[-5 - \frac{1}{3}(17 + 4)]$

44. $\dfrac{2(-6 + 4)}{-4} - \dfrac{(-10 + 3) \cdot 3}{5 - (-2)}$

In Problems 45–78, simplify each algebraic expression.

45. $-3x + 7x - 6x$

46. $3x - 10x + 4x$

47. $4x^2y - 8x^2y$

48. $3ab^3 - 5ab^3$

49. $-2x + 7y + 9x$

50. $5x - 8y - 12x$

51. $5x^2y - 3xy^2 + 2x^2y$

52. $4a^3b - 5ab^3 - 7a^3b$

53. $-4x^2 + 5y^2 - 3x^2 - 6y^2$

54. $xy - z + 8xy + 4z$

55. $7x^3y - 3 + 6x^3y - 8$

56. $4xy^2 - 2 + 3xy^2 - 3 - 5xy^2 - 8$

57. $2(x + 3) + 3(x + 2)$

58. $6(x - 3) + 4(x - 2)$

59. $-4(b - 3) - 2(b + 1)$

60. $-5(c + 2) - 6(c - 7)$

61. $5(x^2 + 3) - 7(x^2 - 4)$

62. $3(x^2 + 5) + (x^2 - 3)$

63. $-3(y^2 - 1) - (y^2 - 7)$

64. $6(x + y) - 7(x - y)$

65. $-2(x^2 - 1) - 4(3x^2 - 5)$

66. $-3(2y^2 - 6) - 4(y^2 - 8)$

67. $4(2x + 5y) - 6(3x - 2y)$

68. $-5(4x - 5y) + 3(6x - y)$

69. $2x - 5(x - 6y)$

70. $3x - 2(5x - 4)$

71. $-3(5x - z) - (z - 4y) + 5(x + 3y - z)$

72. $-5(3a^2 - c) - (c - 7b) + 6(a^2 - 3b - c)$

73. $3[x - 5(5 - 3x)]$

74. $4[-3 - 6(y - 4)]$

75. $4[x - 2(x - 3y)]$

76. $5[a - 3(a - 2b)]$

77. $10x - 6x \cdot 3 + 15y^2 \div 5 \cdot 3$

78. $8y - 2y \cdot 5 + 6y^2 \div 3 \cdot 2$

Application Problems

79. a. The model $E = 155x^2 - 65x + 150$ describes the energy (E, in watts) used by a person walking at x meters/second, where $0.5 \leq x \leq 2.5$. How much energy is used by a person walking 2 meters/second?

b. Runners require a model involving a different energy pattern. For a runner, $E = 250x + 100$, where $x \geq 2.5$. How much energy is used by a runner moving at 4 meters/second?

80. The mathematical model $T = 32.8 + 0.27(T_E - 20)$ describes the temperature of the skin (T, in degrees Celsius) in terms of the temperature of the environment (T_E), also in degrees Celsius. If the environmental temperature is 68°F, use the formula $C = \frac{5}{9}(F - 32°)$ to convert this temperature to Celsius and then find the skin temperature in degrees Celsius.

81. The model $d = 0.042s^2 + 1.1s$ describes the number of feet (d) needed to stop a car traveling at s mph on a concrete road. If a car is going 50 mph instead of 30 mph, how much farther will it travel before stopping?

82. THI, temperature–humidity index, describes the degree of discomfort we experience on hot, humid days, with discomfort experienced when THI ≥ 75. The mathematical model is THI $= 15 + 0.4(T_D + T_W)$, where T_D is the dry-bulb temperature and T_W is the wet-bulb temperature. Is there discomfort on a day when $T_D = 80$ and $T_W = 60$?

83. Rudolf Flesch developed a mathematical model for the readability (R) of a piece of writing. According to Flesch,

$$R = 206.835 - (1.015w + 0.846s)$$

where w represents the average number of words/sentence (in a 100-word sample) and s represents the average number of syllables in the 100-word sample. When $R = 0$, an article is almost unreadable; when $R = 100$, any literate person can read the article. Determine the readability level for a 100-word sample having an average syllable level of 2.5 and an average of 10 words/sentence.

84. In a study of the winter moth in Nova Scotia, the number of eggs (N) in a female moth depended on her abdominal width (W, in millimeters). The exact relationship was described by $N = 14W^3 - 17W^2 - 16W + 34$ for the interval $1.5 \leq W \leq 3.5$. How many eggs are there in a moth with an abdominal width of 2 millimeters?

85. The mathematical model

$$N = (14{,}400 + 120t + 100t^2) \div (144 + t^2)$$

describes the number of bacteria (N) t hours after an antibacterial agent is introduced to a population of 100 bacteria. How many bacteria are present after 2 hours?

86. A rock on the moon and on Earth is thrown into the air by a 6-foot person with a velocity of 48 feet/second. The height reached by the rock (in feet) after t seconds is described by

$$h_{\text{Moon}} = -2.7t^2 + 48t + 6$$
$$h_{\text{Earth}} = -16t^2 + 48t + 6$$

After 3 seconds, how much higher is the rock on the moon than on Earth? (The acceleration of gravity on the moon is approximately one-sixth of that on Earth.)

87. The model $T = 3(A - 20)^2 \div 50 + 10$ describes the time (T, in seconds) for a person who is A years old to run the 100-yard dash. Find the percent increase in time from age 30 to age 40.

88. The height (H, in feet) of a tree after t years is given by the mathematical model

$$H = 63 + 20\left(\frac{t - 120}{40}\right) - 0.4\left(\frac{t - 120}{40}\right)^2 - 1.2\left(\frac{t - 120}{40}\right)^3$$

where $0 \leq t \leq 240$. How tall is the tree after 140 years?

89. Reread the chapter introduction and use the given model to answer the question about whether the remains could be those of the missing person.

True–False Critical Thinking Problems

90. Which one of the following is true?
  a. $7 - 5 + 2 = 7 - 7 = 0$
  b. $16 \div 4 \cdot 2 = 16 \div 8 = 2$
  c. A formula that describes a practical situation is called a mathematical model.
  d. None of the above is true.

91. Which one of the following is true?
  a. $3 \cdot 5^2 = 15^2 = 225$
  b. $6 - 2(4 + 3) = 4(4 + 3) = 4(7) = 28$
  c. Calculators use subtraction keys to change the sign of a number.
  d. None of the above is true.

92. Which one of the following is true?
  a. $3 + 7x = 10x$
  b. $b \cdot b = 2b$
  c. $(3y - 4) - (8y - 1) = -5y - 3$
  d. $-4y + 4 = -4(y + 4)$

93. Which one of the following is true?
  a. $5x^2 + 4x^2 = 9x^4$
  b. $7 - 4(x - 2y) = 3x - 6y$
  c. $-y - y = 2y$
  d. Grouping symbols are not necessary in the algebraic expression $(8x^2 - 5x) - 3y$, and the expression cannot be simplified.

Writing in Mathematics

94. Explain why it is necessary to establish an order of operations agreement.

95. A bird lover visited a pet shop where there were twice 4 and 20 parrots. The bird lover purchased $\frac{1}{7}$ of the birds. English, of course, can be ambiguous and "twice 4 and 20" can mean $2(4 + 20)$ or $2 \cdot 4 + 20$. Explain how the conditions of the situation determine if "twice 4 and 20" means $2(4 + 20)$ or $2 \cdot 4 + 20$.

96. Write out a step-by-step procedure for simplifying an algebraic expression.

97. Most students simplify $(3x + 4y) + 6x$ mentally, immediately writing $9x + 4y$. Describe how the associative and distributive properties are used "between the lines" in this simplification, showing the algebraic details as part of your description.

98. An algebra student incorrectly used the distributive property and wrote $3(7 + 5x + y) = 21 + 15x + y$. If you were that student's teacher, what would you say to help the student avoid this kind of error?

Critical Thinking Problems

For Problems 99–101, insert parentheses to make each calculation correct.

99. $8 - 2 \cdot 3 - 4 = 10$

100. $8 - 2 \cdot 3 - 4 = 14$

101. $2 \cdot 5 - \frac{1}{2} \cdot 10 \cdot 9 = 45$

102. Find the value of $\dfrac{9[4 - (1 + 6)] - (3 - 9)^2}{5 + \dfrac{12}{5 - \dfrac{6}{2 + 1}}}$

103. Try the following trick with at least two different numbers. Then prove that the trick will always work by using x for the number. Think of any number and multiply it by 2. Now add 9. Add your original number. Divide by 3. Add 4. Subtract your original number. Your result is 7.

104. One person has more than five pencils, and a second person has three times as many. The first person gives five pencils to the second. Then the second person gives the first person three times as many pencils as the first person had after the initial exchange. How many pencils does the second person have after both exchanges?

Group Activity Problems

Problems 105–106 can be categorized as "contest problems." Since two (or more) heads are better than one, try solving each problem in a small group. Group members should present as many ideas as possible that may contribute to the problem's solution.

105. The formula for finding the sum of the first n integers is given by

$$1 + 2 + 3 + 4 + \cdots + n = \frac{n(n + 1)}{2}.$$

Use this formula to find the difference between the sum of the first 100 positive multiples of 3 and the sum of the first 100 positive even integers.

106. Use each of the numbers $2, 4, 9, 14,$ and 17 exactly once. Using addition, subtraction, multiplication, and grouping symbols, show how you can obtain an answer of 2.

Review Problems

From here on, each problem set will contain three review problems. It is essential to review previously covered topics to improve your understanding of the topics and to help you maintain your mastery of the material.

107. Express the following English phrase in both set-builder and interval notation: x is at most 2.

108. Use the roster method to write this set:

$\{x \mid x \text{ is a negative even integer}\}$.

109. What property of algebra is illustrated by $(x + 4) + [-(x + 4)] = 0$?

SECTION 1.3

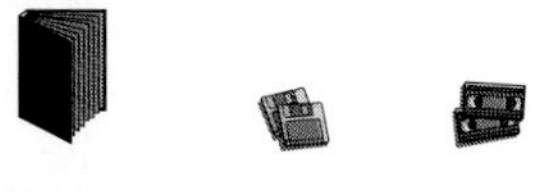

Solutions Manual **Tutorial** **Video I**

Graphing Equations

Objectives

1. Plot ordered pairs in the rectangular coordinate system.
2. Find coordinates of points in the rectangular coordinate system.
3. Interpret information given by graphs.
4. Graph equations and mathematical models.

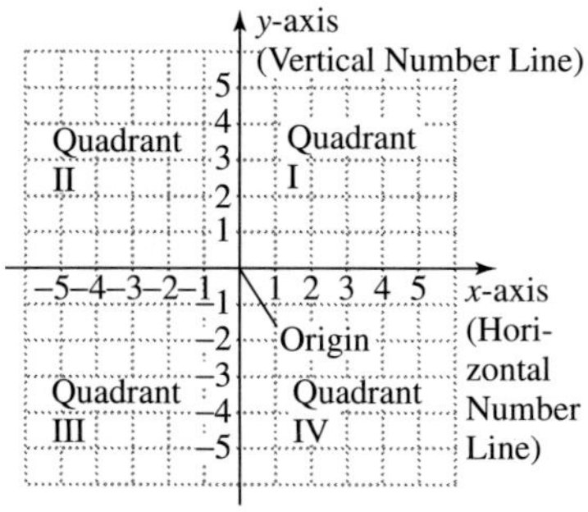

Figure 1.11

The rectangular (Cartesian) coordinate system

In Section 1.1, we saw that any real number could be represented by a point on the number line. We are now ready to graph ordered pairs of real numbers in a plane called the *rectangular coordinate system.* The ideas discussed in this section are those of the French mathematician and philosopher René Descartes (1596–1650). For this reason, the rectangular coordinate system is also called the *Cartesian coordinate plane.*

Points and Ordered Pairs

Descartes used two number lines that intersect at right angles at their zero points, as shown in Figure 1.11. Each line is called a *coordinate axis.* The horizontal axis is the *x-axis* and the vertical axis is the *y-axis.* The point of intersection of these axes is the *origin.* The horizontal and vertical number lines divide the plane into four quarters, called *quadrants,* numbered in a counterclockwise direction beginning with the upper right. The points located on the axes are not in any quadrant.

Each point in the plane corresponds to an *ordered pair* of real numbers, (x, y). Examples of such pairs are $(4, 2)$ and $(-5, -3)$. The first number in each pair, called the *x-coordinate,* denotes the distance and direction from the origin along the x-axis. The second number, called the *y-coordinate,* denotes vertical distance and direction along a line parallel to the y-axis or along the y-axis itself.

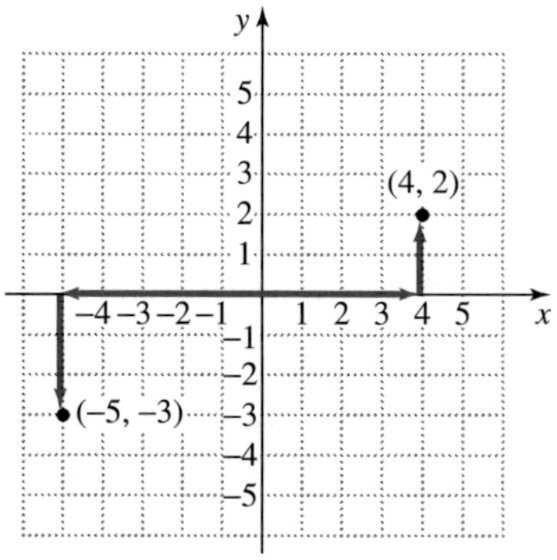

Figure 1.12

Plotting (4, 2) and (− 5, − 3)

Figure 1.12 shows how we locate or *plot* the points corresponding to the ordered pairs $(4, 2)$ and $(-5, -3)$. We plot $(4, 2)$ by going 4 units from 0 to the right along the x-axis and then 2 units up parallel to the y-axis. We plot $(-5, -3)$ by going 5 units from 0 to the left along the x-axis and 3 units down parallel to the y-axis. The phrase "the point corresponding to the ordered pair $(-5, -3)$" is often abbreviated as "the point $(-5, -3)$."

1 Plot ordered pairs in the rectangular coordinate system.

EXAMPLE 1 **Plotting Points in the Rectangular Coordinate System**

Plot the ordered pairs: $(2, 3)$, $(-2, 3)$, $(-2, -3)$, $(2, -3)$, $(2, 0)$, $(0, 1)$, $(-2, 0)$, $(0, -3)$, $(0, 0)$

ENRICHMENT ESSAY

Cartesion Coordinates: An Historical Perspective

The beginning of the 17th century was a time of innovative ideas and enormous intellectual progress in Europe. English theatergoers enjoyed a succession of exciting new plays by Shakespeare. William Harvey proposed the radical notion that the heart was a pump for blood rather than the center of emotion. Galileo, with his new-fangled invention called the telescope, supported the theory of Polish astronomer Copernicus that the sun, not the earth, was the center of the solar system. Monteverdi was writing the world's first grand operas. French mathematicians Pascal and Fermat invented a new structure of mathematics known as the theory of probability.

Into this arena of intellectual electricity stepped French aristocrat René Descartes (1596–1650). Descartes, propelled by the creativity surrounding him, felt that it was his destiny to discover a method that would bring together all thought and knowledge using the deductive system of mathematics. Beginning with a simple foundation of rules (axioms), Descartes believed that all truth regarding nature could be proved in much the same way that the ancient Greek mathematicians proved geometric theorems.

The idea that all knowledge should be presented in the framework of mathematical reasoning appeared in Descartes' book, *A Discourse on the Method of Rightly Conducting the Reason and Seeking Truth in the Sciences*. Descartes concluded his book with three specific examples of how the method could be applied. The third example, a 106-page footnote called *La Géométrie (The Geometry)*, involved the development of a new branch of mathematics that brought together arithmetic, algebra, and geometry in unified way—a way that visualized numbers as points on a graph, equations as geometric figures, and geometric figures as equations. This new branch of mathematics, called *analytic geometry*, established Descartes as one of the founders of modern thought and among the most original mathematicians and philosophers of any age.

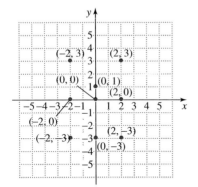

Figure 1.13

Plotting points

Solution

See Figure 1.13. We plot the points in the following way:

$(2, 3)$: 2 units right, 3 units up (in quadrant I)
$(-2, 3)$: 2 units left, 3 units up (in quadrant II)
$(-2, -3)$: 2 units left, 3 units down (in quadrant III)
$(2, -3)$: 2 units right, 3 units down (in quadrant IV)
$(2, 0)$: 2 units right, 0 units up or down (on the x-axis)
$(0, 1)$: 0 units right or left, 1 unit up (on the y-axis)
$(-2, 0)$: 2 units left, 0 units up or down (on the x-axis)
$(0, -3)$: 0 units right or left, 3 units down (on the y-axis)
$(0,0)$: 0 units right or left, 0 units up or down (at the origin)

Observe that *any point on the x-axis has a y-coordinate of* 0 and *any point on the y-axis has an x-coordinate of* 0. ■

2 Find coordinates of points in the rectangular coordinate system.

In the rectangular coordinate system, each ordered pair corresponds to exactly one point. Example 2 illustrates that each point in the plane corresponds to exactly one ordered pair.

EXAMPLE 2 **Finding Coordinates of Points**

Go into a room where there are 23 people chosen at random and there is a 50% chance that 2 people in the room share the same birthday (month and day). The graph in Figure 1.14 shows the increasing probability of a shared birthday as the size of the group increases. The graph is drawn only in quadrant I because the number of persons and probabilities are both nonnegative.

The graph shows the increasing probability of the shared birthday from the slim chance—when there are fewer than 10 people present—to near certainty (probability = 1 represents certainty)—when there are six times that number present.

Determine or estimate the coordinates for points A, B, and C and interpret the coordinates in practical terms.

Solution

Point	Position	Coordinates	Interpretation
A	15 units right, 0.25 units up	$(15, 0.25)$	With 15 people present, the probability is 25% that two people will share a birthday.
B	approximately 23 units right, 0.5 units up	$(23, 0.5)$	With 23 people present, there is 50% chance of a shared birthday.
C	30 units right, 0.7 units up	$(30, 0.7)$	When 30 people are brought together, the probability of a shared birthday is 70%.

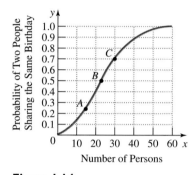

Figure 1.14

The chances of a shared birthday increases as the group's size increases.

3 Interpret information given by graphs.

Interpreting Information Given by Graphs

Magazines and newspapers often display information using graphs in the first quadrant of the rectangular coordinate system.

EXAMPLE 3 **Energy Consumption in the United States**

Figure 1.15 shows energy consumption in the United States from 1860 through 1993, along with the major primary sources of energy.

a. For the period from 1880 through 1940, when was water power as an energy source increasing and when was it decreasing?
b. For the period from 1880 through 1940, when was water power as an energy source at a maximum? What is a reasonable estimate for the maximum water power energy consumption?
c. Estimate the coordinates for point D, and interpret the coordinates in practical terms.
d. What does the graph indicate about the use of oil after 1945? What happened during this period to account for the trend shown by the graph?
e. What does the "total" portion of the graph show about energy consumption in the United States?

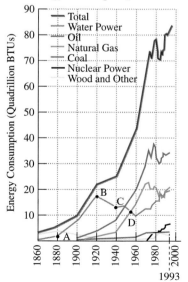

Figure 1.15

Source: *Statistical Abstracts of the United States,* U.S. Department of Commerce, 1995.

Solution

a. The graph representing water power as an energy source is rising from point A to point B, so this energy source was increasing from 1880 through 1920. The graph is falling from point B to point C, indicating a decrease in water power as an energy source from 1920 through 1940.

U.S. Energy Consumption, 1860—1993

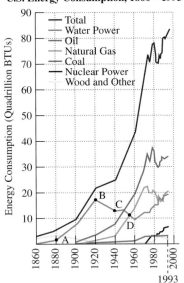

Figure 1.15 (repeated from previous page)

Source: *Statistical Abstracts of the United States,* U.S. Department of Commerce, 1995.

4 Graph equations and mathematical models.

John Fekner 1985 "Toxic", Collection Martin Wong.

b. For the period 1880 through 1940, point B indicates when water power as an energy source was at a maximum. The coordinates of B are approximately (1920, 18), meaning that water power as an energy source reached a maximum of 18 quadrillion Btus in 1920.

c. A reasonable estimate for the coordinates of point D is (1955, 12). The graphs for natural gas consumption and water power consumption intersect at this point. This means that in 1955, energy consumption was the same for natural gas and water power, namely, 12 quadrillion Btus. Before 1955 (to the left of point D), water power consumption exceeded that for natural gas. After 1955 and up to about 1984, this trend reversed.

d. After 1945 and up until about 1976, the graph representing oil consumption is rapidly rising from left to right. This means that there was a skyrocketing increase in use of oil after 1945. A possible explanation for this trend is that cars and car-dependent commuting became common after World War II.

e. The graph representing total energy consumption increases quite rapidly, especially after 1945, until the late 1970s. In spite of some fluctuations, with the graph decreasing quite dramatically over short intervals, the graph does increase when 1993 is compared to any year in the late 1970s. This means that total energy consumption has continued to grow, although the mix of primary sources has changed over the years. Decreases in consumption occurred in the late 1970s and early 1980s, but in recent years consumption has resumed its upward trend. ■

Graphing Equations

The rectangular coordinate system allows us to visualize relationships between two variables by connecting any equation in two variables with a geometric figure and vice versa. Ordered pairs present useful visual pictures of mathematical models, as shown in Example 4.

EXAMPLE 4 **Visualizing a Mathematical Model**

The cost y of removing $x\%$ of pollutants in a lake is given by the mathematical model

$$y = \frac{80{,}000x}{100 - x}.$$

Complete the accompanying table of coordinates by substituting each of the eight given values for x into the formula and finding the corresponding value for y.

x	0	20	40	50	60	80	90	99
y								

Plot the eight ordered pairs from the table, connecting the points with a smooth curve to get a visual idea of the mathematical model.

Solution

The first step is to take each given value for x, substitute that value into the model, find the corresponding value for y, and then form the ordered pairs.

x	Calculate y Using $y = \dfrac{80,000x}{100 - x}$	Ordered Pair (x, y)
$x = 0$	$y = \dfrac{80,000(0)}{100 - 0} = \dfrac{0}{100} = 0$	$(0, 0)$
$x = 20$	$y = \dfrac{80,000(20)}{100 - 20} = \dfrac{1,600,000}{80} = 20,000$	$(20, 20,000)$
$x = 40$	$y = \dfrac{80,000(40)}{100 - 40} = \dfrac{3,200,000}{60} = 53,333\frac{1}{3}$	$\left(40, 53,333\frac{1}{3}\right)$
$x = 50$	$y = \dfrac{80,000(50)}{100 - 50} = \dfrac{4,000,000}{50} = 80,000$	$(50, 80,000)$
$x = 60$	$y = \dfrac{80,000(60)}{100 - 60} = \dfrac{4,800,000}{40} = 120,000$	$(60, 120,000)$
$x = 80$	$y = \dfrac{80,000(80)}{100 - 80} = \dfrac{6,400,000}{20} = 320,000$	$(80, 320,000)$
$x = 90$	$y = \dfrac{80,000(90)}{100 - 90} = \dfrac{7,200,000}{10} = 720,000$	$(90, 720,000)$
$x = 99$	$y = \dfrac{80,000(99)}{100 - 99} = \dfrac{7,920,000}{1} = 7,920,000$	$(99, 7,920,000)$

The substitution of any of the ordered pairs in the right column into $y = 80,000x/(100 - x)$ produces a true statement. We say that these ordered pairs *satisfy* the equation.

The completed table of coordinates appears as follows:

x	0	20	40	50	60	80	90	99
y	0	20,000	$53,333\frac{1}{3}$	80,000	120,000	320,000	720,000	7,920,000

Notice how rapidly the values for y increase as x gets larger and larger. The ordered pair (80, 320,000) indicates that the cost of removing 80% of the lake's pollutants is \$320,000. The ordered pair (90, 720,000) means that a 90% cleanup will cost \$720,000. The model indicates that removing all (100%) pollutants from the lake is impossible. Substituting 100 for x in the formula results in

$$y = \frac{80,000x}{100 - x} = \frac{80,000(100)}{100 - 100} = \frac{8,000,000}{0}$$

and division by 0 is undefined.

With the completed table of values, we can get a visual idea of the model $y = \dfrac{80,000x}{100 - x}$ by plotting the ordered pairs in the rectangular coordinate system and connecting points with a smooth curve. Figure 1.16 shows the graph.

The placement of points in Figure 1.16 is approximate. For example, (20, 20,000) is plotted by going 20 units to the right and then $\frac{1}{5}$ of the distance between 0 and 100,000. Because of the scale chosen on the y-axis, the point (99, 7,920,000) cannot be shown, although the arrow at the right of the graph indicates the dramatically increasing cost. The graph approaches but never touches a vertical line drawn through 100 on the x-axis. This is a way to show

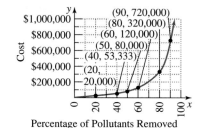

Figure 1.16

The graph of $y = \dfrac{80,000x}{100 - x}$

iscover for yourself

Find another solution of

$$y = \frac{80{,}000x}{100 - x}$$

and then locate the solution on the graph in Figure 1.16. Now select any point that does *not* lie on the graph and show that its ordered pair is not a solution of

$$y = \frac{80{,}000x}{100 - x}.$$

that x cannot equal 100; as x approaches 100, however, the values of y grow larger with no limit in sight. In other words, the cost of removing almost all pollutants is astronomical.   ■

We can think of the mathematical model $y = \frac{80{,}000x}{100 - x}$ as an equation in two variables, x and y. A *solution to an equation in two variables* is an ordered pair of real numbers with the following property: When the x-coordinate is substituted for x and the y-coordinate is substituted for y in the equation, we obtain a true statement. For example, $(90, 720{,}000)$ is an ordered-pair solution of $y = \frac{80{,}000x}{100 - x}$ because it satisfies the equation. We can generate as many ordered-pair solutions as desired by substituting numbers for one of the variables and solving the resulting equation for the other variable.

> **Equations as geometric figures**
>
> The *graph of an equation* involving two variables (usually x and y) is the set of all points whose coordinates satisfy the equation.

The method we used for sketching the graph of $y = \frac{80{,}000x}{100 - x}$ is called the *point-plotting method.* Plotting several ordered pairs that are solutions to the equation and connecting them with a smooth curve or line often gives us a picture of all ordered pairs that satisfy the equation.

Here is a summary of the steps in the point-plotting method.

> **Graphing by the point-plotting method**
>
> *Graphing an equation in two variables*
>
> 1. If necessary and if possible rewrite the equation by isolating one of the variables.
> 2. Construct a table of coordinates showing several ordered pairs that are solutions to the equation.
> 3. Plot these ordered pairs in the rectangular coordinate system and connect them with a smooth curve or line.

EXAMPLE 5   **Graphing an Equation Using the Point-Plotting Method**

Graph: $y = x^2 - 4$

Solution

We select numbers for x and find the corresponding values for y.

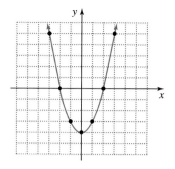

Figure 1.17

The graph of $y = x^2 - 4$

x	$y = x^2 - 4$	Ordered Pair (x, y)
-3	$y = (-3)^2 - 4 = 9 - 4 = 5$	$(-3, 5)$
-2	$y = (-2)^2 - 4 = 4 - 4 = 0$	$(-2, 0)$
-1	$y = (-1)^2 - 4 = 1 - 4 = -3$	$(-1, -3)$
0	$y = 0^2 - 4 = 0 - 4 = -4$	$(0, -4)$
1	$y = 1^2 - 4 = 1 - 4 = -3$	$(1, -3)$
2	$y = 2^2 - 4 = 4 - 4 = 0$	$(2, 0)$
3	$y = 3^2 - 4 = 9 - 4 = 5$	$(3, 5)$

Now we plot the seven points and join them with a smooth curve, as shown in Figure 1.17. The graph of $y = x^2 - 4$ is a curve where the part of the graph to the right of the y-axis is a reflection of the part to the left of it and vice versa. ■

EXAMPLE 6   **Graphing an Equation Using the Point-Plotting Method**

Graph: $y = |x|$

Solution

As we did in Example 5, we select numbers for x and find the corresponding values for y.

x	$y = |x|$	Ordered Pair (x, y)
-3	$y = |-3| = 3$	$(-3, 3)$
-2	$y = |-2| = 2$	$(-2, 2)$
-1	$y = |-1| = 1$	$(-1, 1)$
0	$y = |0| = 0$	$(0, 0)$
1	$y = |1| = 1$	$(1, 1)$
2	$y = |2| = 2$	$(2, 2)$
3	$y = |3| = 3$	$(3, 3)$

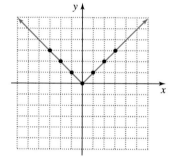

Figure 1.18

The graph of $y = |x|$

We plot the points and connect them, resulting in the graph shown in Figure 1.18. The graph is V-shaped and centered at the origin. For every point (x, y) on the graph, the point $(-x, y)$ is also on the graph, showing that the absolute value of a positive number is the same as the absolute value of its opposite. ■

Using technology

Graphing calculators or graphing software packages for computers are referred to as graphing utilities or graphers. The point-plotting method is used by all graphing utilities. A graphing utility displays only a portion of the rectangular coordinate system, called a *viewing rectangle*. The viewing rectangle is determined by six values: the minimum x-value (Xmin), the maximum x-value (Xmax), the x-scale (Xscl), the minimum y-value (Ymin), the maximum y-value (Ymax), and the y-scale (Yscl). By entering these six values into a graphing utility, you set the range of the viewing rectangle.

The standard viewing rectangle for many graphing utilities is shown in the accompanying figure.

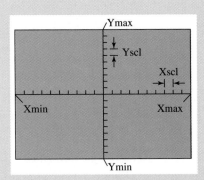

RANGE

Xmin = -10
Xmax = 10
Xscl = 1
Ymin = -10
Ymax = 10
Yscl = 1

This viewing rectangle can be described as $[-10, 10]$ by $[-10, 10]$ and in general is described as [Xmin, Xmax] by [Ymin, Ymax].

Graphing an Equation in x and y Using a Graphing Utility

1. If necessary, solve the equation for y in terms of x.
2. Enter the equation into the graphing utility.
3. Use the standard viewing rectangle or set the range to determine a viewing rectangle that will show a complete picture of the equation's graph.
4. Activate the graphing utility.

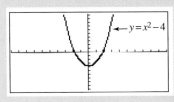

Shown on the left is the graph of $y = x^2 - 4$, the graph we drew by hand in Example 5. The equation was entered as $Y = X^2 - 4$. The viewing rectangle is $[-10, 10]$ by $[-10, 10]$. What do you observe about the spacing of the horizontal and vertical tick marks on the axes?

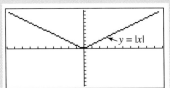

Shown also on the left is the graph of $y = |x|$. On our utility we entered the equation as $Y = |X|$. (Consult your manual for the location of the absolute value key.) Once again, a standard viewing rectangle was used.

Part of the beauty of the rectangular coordinate system is that it allows us to "see" mathematical models and visualize the solution to a problem. This idea is demonstrated in Example 7.

EXAMPLE 7 **An Application Using the Rectangular Coordinate System**

The toll to a bridge costs $2.50. Commuters who use the bridge frequently have the option of purchasing a monthly coupon book for $21.00. With the coupon book, the toll is reduced to $1.00. The monthly cost, y, of using the bridge x times can be modeled by the following formulas.

Without the coupon book:

$y = 2.50x$ The cost is $2.50 times the number of times, x, that the bridge is used.

With the coupon book:

$y = 21 + 1 \cdot x$ The cost is $21 for the book plus $1 times the number of times, x, that
$ = 21 + x$ the bridge is used.

a. Let $x = 0, 2, 4, 10, 12, 14$, and 16. Make a table of coordinates for each of the models.

b. Graph the models in the same rectangular coordinate system.

c. What are the coordinates of the intersection point for the two graphs? Interpret the coordinates in practical terms.

Solution

a. A table of coordinates for each model follows.

<table>
<tr><td colspan="3" align="center">**Without the Coupon Book**</td><td colspan="3" align="center">**With the Coupon Book**</td></tr>
<tr><td>x</td><td>$y = 2.5x$</td><td>(x, y)</td><td>x</td><td>$y = 21 + x$</td><td>(x, y)</td></tr>
<tr><td>0</td><td>$y = 2.5(0) = 0$</td><td>$(0, 0)$</td><td>0</td><td>$y = 21 + 0 = 21$</td><td>$(0, 21)$</td></tr>
<tr><td>2</td><td>$y = 2.5(2) = 5$</td><td>$(2, 5)$</td><td>2</td><td>$y = 21 + 2 = 23$</td><td>$(2, 23)$</td></tr>
<tr><td>4</td><td>$y = 2.5(4) = 10$</td><td>$(4, 10)$</td><td>4</td><td>$y = 21 + 4 = 25$</td><td>$(4, 25)$</td></tr>
<tr><td>10</td><td>$y = 2.5(10) = 25$</td><td>$(10, 25)$</td><td>10</td><td>$y = 21 + 10 = 31$</td><td>$(10, 31)$</td></tr>
<tr><td>12</td><td>$y = 2.5(12) = 30$</td><td>$(12, 30)$</td><td>12</td><td>$y = 21 + 12 = 33$</td><td>$(12, 33)$</td></tr>
<tr><td>14</td><td>$y = 2.5(14) = 35$</td><td>$(14, 35)$</td><td>14</td><td>$y = 21 + 14 = 35$</td><td>$(14, 35)$</td></tr>
<tr><td>16</td><td>$y = 2.5(16) = 40$</td><td>$(16, 40)$</td><td>16</td><td>$y = 21 + 16 = 37$</td><td>$(16, 37)$</td></tr>
</table>

b. Using the two tables of coordinates, we construct the graphs of $y = 2.5x$ and $y = 21 + x$ shown in Figure 1.19.

c. The graphs intersect at $(14, 35)$. This means that if the bridge is used 14 times in a month, the total monthly cost without the coupon book is the same as the total monthly cost with the coupon book, namely $35. If the bridge is used more than 14 times in a month, the coupon book is more economical. This is shown by the graph of $y = 21 + x$ lying below the graph of $y = 2.5x$ for $x > 14$. ■

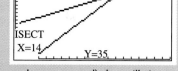

Figure 1.19

Options for a toll

Using technology

Our hand-drawn graphs and the solution for Example 7 can be verified using a graphing utility. The models were entered as

$Y_1 = 2.5X$ and $Y_2 = 21 + X$

using the following range setting:

Xmin $= 0$, Xmax $= 20$, Xscl $= 1$,
Ymin $= 0$, Ymax $= 40$, Yscl $= 1$

Most graphing utilities have an intersection feature (consult your manual) that will give you the coordinates of the intersection point(s) of the graphs shown in the viewing rectangle. Using this feature, we see that the intersection point is $(14, 35)$.

PROBLEM SET 1.3

Practice Problems

In Problems 1–8, plot the given points in a rectangular coordinate system and, where applicable, name the quadrant in which the point is located.

1. $(1, 4)$ **2.** $(-2, 3)$ **3.** $(-3, -5)$ **4.** $(4, -1)$

5. $(-\frac{3}{2}, 0)$ **6.** $(0, -\frac{5}{2})$ **7.** $(-3, -\frac{3}{2})$ **8.** $(-5, -\frac{5}{2})$

Determine the coordinates for each of the points shown in the graphs of Problems 9–12.

9.

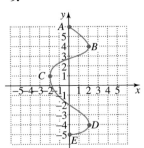

10.

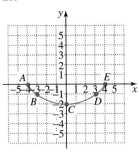

11.

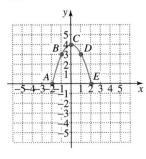

12.

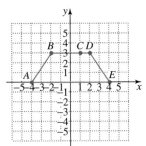

Construct a table of coordinates for each equation in Problems 13–24.

x	-3	-2	-1	0	1	2	3
y							

Then use the ordered pairs (x, y) from the table to graph the equation. If applicable, verify your hand-drawn graph with a graphing utility.

13. $y = x^2 - 2$ **14.** $y = x^2 + 2$ **15.** $y = 2x + 1$ **16.** $y = 2x - 4$

17. $y = -\frac{1}{2}x$ **18.** $y = -\frac{1}{2}x + 2$ **19.** $y = |x| + 1$ **20.** $y = |x| - 1$

21. $y = |x + 1|$ **22.** $y = |x - 1|$ **23.** $y = x^3$ **24.** $y = x^3 - 1$

Application Problems

25. The graph in the figure shows the number of people with tuberculosis in the United States.

   a. For the period shown in the graph, approximately when did the number of people infected reach a maximum? What is a reasonable estimate for the number of people with tuberculosis for that year?

   b. For the period shown in the graph, approximately when did the number of people infected reach a minimum? What is a reasonable estimate for the number of people with tuberculosis for that year?

   c. Describe the trend in the number of cases of tuberculosis in the United States from 1960 through 1993.

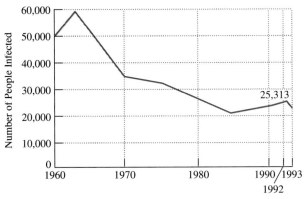

Number of People with Tuberculosis in the United States 1960–1993

Source: U.S. Centers for Disease Control

26. The graph in the figure shows the number of deaths for major U.S. airlines operating scheduled flights from 1988 through 1993.

    a. For the period shown, when did the number of airline deaths reach a maximum? What is a reasonable estimate for the number of deaths in that year?

    b. Between which two consecutive years did the number of airline deaths increase most rapidly?

    c. Between which two consecutive years did the number of airline deaths decrease by approximately 60?

    d. If x represents the year and y the number of airline deaths, for what values of x is $y \geq 200$? What does this mean in practical terms?

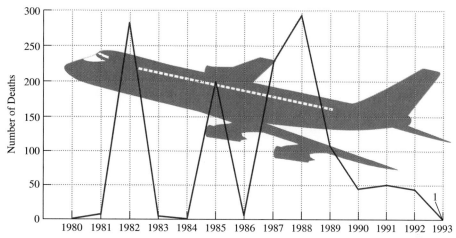

Source: National Transportation Safety Board

27. The figure indicates the distance above the ground of an object projected directly upward with an initial velocity of 16 feet/second.

    a. What is the maximum height reached by the object? After how many seconds does this occur?

    b. When does the object strike the ground?

a. How much of the drug is present after 2 hours?

b. How much of the drug is present after 8 hours?

c. After how many hours is the maximum amount of the drug present? What is this maximum?

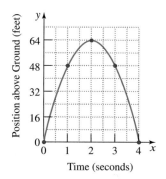

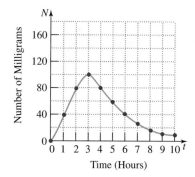

28. The figure on the right shows the number of milligrams of a drug (N) in the bloodstream t hours after taking the drug. Observe that the axes are labeled with t and N rather than the usual x and y.

29. The graph on the next page indicates the changing age structure of the world population.

    a. Estimate the coordinates for point A, and interpret the coordinates in practical terms.

b. Describe the trends indicated by the graph for the twenty-first century.

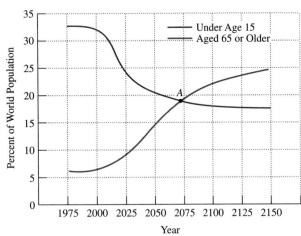

Source: United Nations Population Division, *Long-Range World Population Projections: Two Centuries of Population Growth*, 1950–2150, United Nations, New York

30. The graph indicates the percentages of rural and urban world population.

a. Estimate the coordinates for point A, and interpret the coordinates in practical terms.

b. Describe the trends shown by the graph.

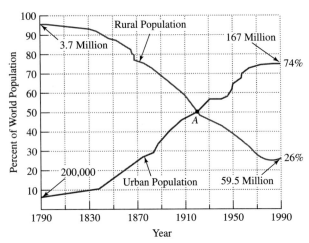

Source: U.S. Bureau of the Census

31. The graph at the top of the next column shows the gasoline mileage for compact and medium-sized cars traveling at various constant speeds.

a. Determine the coordinates for points A and B. Interpret the coordinates in practical terms.

b. Find the difference in the y-coordinates for points A and B. What exactly does this mean?

c. Find a reasonable estimate for the speed at which gasoline mileage is at a maximum for both compact and medium-sized cars.

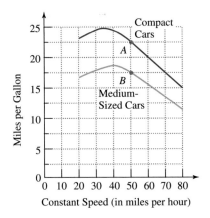

32. The graphs below show the number of heroin and cocaine emergencies in U.S. hospitals from 1988 through 1992.

a. In what year for the period shown is the number of cocaine emergencies at a minimum? What is a reasonable estimate for the number of hospital emergencies for that year?

b. In what year for the period shown is the number of heroin emergencies at a maximum? What is a reasonable estimate for the number of hospital emergencies for that year?

c. What is the percent increase in cocaine emergencies from 1991 through 1992? (*Hint:* The fraction for percent increase is the amount of increase divided by the original amount.)

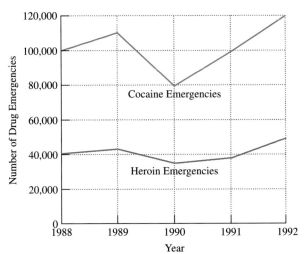

Source: Drug Abuse Warning Network

33. The formula $y = -143x^3 + 1810x^2 - 187x + 2331$ approximately models the number of new cases of AIDS (y) reported each year in the United States x

years after 1983. The model applies only for $x \in [0, 8]$. Construct a table of values and then graph the model. A calculator might be helpful in finding values of y.

x	0	1	2	3	4	5	6	7	8
y									

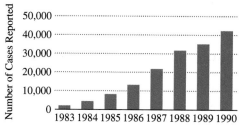

AIDS Cases 1983–1990

Data Source: U.S. Centers for Disease Control

34. The formula $y = -3.1x^2 + 51.4x + 4024.5$ approximately models the average number of cigarettes (y) consumed by Americans 18 and older x years after 1950. The model applies only for $x \in [0, 40]$. Construct a table of values and then graph the model. A calculator might be helpful in finding values for y. Round these values to the nearest integer.

x	0	5	10	15	20	30	40
y							

How does your graph compare to the one shown here, which depicts actual data points? What does this tell you about mathematical formulas that attempt to model variables in the real world?

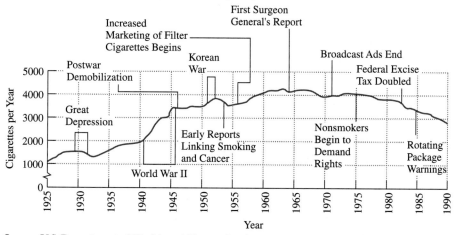

Cigarette Consumption per U.S. Adult

Source: U.S. Department of Health and Human Services

35. A rental company charges $40.00 a day plus $0.35 per mile to rent a moving truck. The total cost (y) for a day's rental if x miles are driven is modeled by $y = 40 + 0.35x$. A second company charges $36.00 a day plus $0.45 per mile, so the daily cost (y) if x miles are driven is modeled by $y = 36 + 0.45x$. The graphs of the two models are shown on page 42.

a. What is the x-coordinate of the intersection point of the graphs? Describe what this means in practical terms.

b. What is a reasonable estimate for the y-coordinate of the intersection point?

c. Substitute the x-coordinate of the intersection point into each of the mathematical models and find the corresponding value for y. Describe what this value represents in practical terms. How close is this value to your estimate from part (b)?

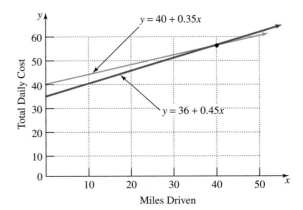

Technology Problems

36. a. Use a graphing utility to graph the AIDS model in Problem 33 using the following range setting:

   Xmin = 0, Xmax = 8, Xscl = 1, Ymin = 0,
   Ymax = 50,000, Yscl = 10,000

b. Change the range setting so that Xmax = 13. Use your graphing utility to graph the AIDS model with this new range setting. Based on the shape of the graph, does the given formula model reality for $x \in [8, 13]$? Explain. What does this tell you about formulas that model variables over time?

37. a. Graph the cigarette consumption model in Problem 34 using a graphing utility and the following range setting:

   Xmin = 0, Xmax = 40, Xscl = 1, Ymin = 0,
   Ymax = 5000, Yscl = 1000

b. Change the range setting so that Xmax = 45. Use your graphing utility to graph the model with this new range setting. Based on the shape of the graph, does the given formula model reality for $x \in (40, 45)$? Explain. What does this tell you about formulas that model variables over time?

38. Use a graphing utility to graph the models in Problem 35. Enter the equations as $Y_1 = 40 + .35X$ and $Y_2 = 36 + .45X$ using the following range setting:

   Xmin = 0, Xmax = 50, Xscl = 5, Ymin = 0,
   Ymax = 60, Yscl = 3

Now use the utility's trace feature to trace along the curves or the intersection feature to verify that the graphs intersect at $(40, 54)$.

39. a. Graph the pollution cleanup model,

$$y = \frac{80{,}000x}{100 - x}$$

discussed in Example 4, using each of the following range settings:

   $[0, 100]$ by $[0, 1{,}000{,}000]$;
      Xscl = 10, Yscl = 100,000

   $[0, 50]$ by $[0, 100{,}000]$;
      Xscl = 5, Yscl = 10,000

   $[0, 20]$ by $[0, 25{,}000]$;
      Xscl = 1, Yscl = 5000

Describe what each of these graphs illustrate in terms of the cost (y) of removing $x\%$ of the pollutants from the lake.

b. As noted in Example 4, the cost of removing almost all pollutants is astronomical. Select an appropriate range setting that illustrates this fact, focusing on the cost of removing at least 90% of the lake's pollutants.

You can determine if an algebraic expression in one variable has been simplified correctly by using a graphing utility to graph each side of the equation in the same viewing rectangle. If the graphs are identical, the simplification is correct. For example,

$$4[x + 2(1 - x)] = -4x + 8$$

is correct because the graphs of

$$y_1 = 4[x + 2(1 - x)] \quad \text{and} \quad y_2 = -4x + 8$$

are the same, as shown in the figure. Use a graphing utility to determine which of the simplifications in Problems 40–43 are correct. If the simplification is incorrect, correct the right-hand side and then use your graphing utility to verify the correction.

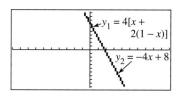

40. $-2(x + 3) - 5(x - 4) = -7x + 14$

41. $7 - (2x - 5) - 3(2x + 4) = -8x$

42. $4 - [6(3x + 2) - x] + 4 = -15x - 4$

43. $4x^2 - x^2 - 3x - 3x^2 + 4 = x^2 - 3x + 4$

Writing in Mathematics

44. A psychologist showed subjects an array of letters and then asked them to repeat as many letters from the array as possible. The graph of the data is shown in the figure, where values on the x-axis represent the number of letters in the array and values on the y-axis stand for the number of letters that the subjects remembered. Write a brief description of the results of this experiment.

45. The parentheses used to represent an ordered pair are also used to represent an open interval. If you are reading a section of a math book and (3, 4) is mentioned, how will you know if they are discussing an ordered pair or the interval for which $3 < x < 4$?

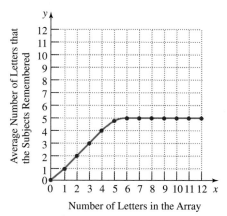

Number of Letters in the Array

Critical Thinking Problems

Construct a table of coordinates for each equation in Problems 46–49. Then use the ordered pairs (x, y) from the table to graph the equation. If applicable, verify your hand-drawn graph with a graphing utility.

46. $y = \dfrac{1}{x}$

47. $y = \dfrac{1}{x + 2}$

48. $y = \sqrt{x}$

49. $y = \sqrt{x - 2}$

50. A particle starts at $(0, 0)$ and moves along the path shown in the figure. If 1 unit is covered every second, find the coordinates of the particle after 225 seconds.

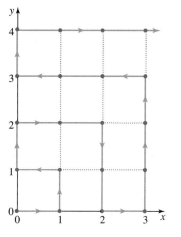

51. A particle starts at $(0, 0)$ and moves along the path shown in the figure. The first move takes the particle to $(1, -1)$, the second to $(3, 1)$, the third to $(0, 4)$, the fourth to $(-4, 0)$, and so on. What are the coordinates of the particle after 220 moves?

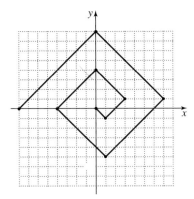

Group Activity Problem

52. Plot the points $(-3, 5), (4, 7), (5, -9)$, and $(-4, -6)$. Suppose that we want to find the area of the quadrilateral having those four points as vertices. Members of the group should offer as many suggestions as possible as to how to do this. Using the best suggestion, find the area.

Review Problems

53. Use the roster method to write the set

$\{x \,|\, x$ is a rational number with a denominator of $2, 4,$ or $6,$ equivalent to $-\frac{1}{2}\}$.

54. Perform the indicated operations:
$[12 - (13 - 17)] - [9 - (6 - 10)]$

55. Use set-builder notation and interval notation to rewrite the following: x is positive and at most 7.

SECTION 1.4

Solutions Manual   Tutorial   Video 1

Properties of Integral Exponents

Objectives

1 Multiply exponential expressions.
2 Simplify powers of exponential expressions.
3 Simplify powers of products.
4 Simplify powers of quotients.
5 Divide exponential expressions.
6 Use zero exponents.
7 Use negative exponents.
8 Simplify exponential expressions.

We have seen that exponents are used to indicate repeated multiplication. Recall that if n is any natural number, then

$$\underset{\text{Base}}{b}{}^{\overset{\text{Exponent}}{n}} = \underbrace{b \cdot b \cdot b \cdots \cdots b}_{\substack{b \text{ appears as a} \\ \text{factor } n \text{ times}}}.$$

A number of properties of exponents, sometimes called laws of exponents, follow directly from this definition.

Multiply exponential expressions.

Multiplying Exponential Expressions

The following examples suggest a rapid method for multiplying with exponential notation when the bases are the same.

$$x^3 \cdot x^4 = (x \cdot x \cdot x)(x \cdot x \cdot x \cdot x) = x^7$$
$$a^4 \cdot a^2 = (a \cdot a \cdot a \cdot a)(a \cdot a) = a^6$$

We can obtain these results if we add the exponents:

$$x^3 \cdot x^4 = x^{3+4} = x^7$$
$$a^4 \cdot a^2 = a^{4+2} = a^6$$

Multiplying exponential expressions with like bases

If n and m are natural numbers ($m, n = 1, 2, 3, \ldots$), then

$$b^n \cdot b^m = b^{n+m}.$$

When multiplying exponential expressions with the same base, retain the common base and add the exponents. Use this sum as the exponent of the common base.

Study tip

We read x^2 as "x-squared" because the area of a square with side x is

$$x \cdot x = x^2.$$

We read x^3 as "x-cubed" because the volume of a cube with side x is

$$x \cdot x \cdot x = x^3.$$

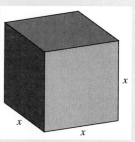

Continuing this logic, x^4 suggests a four-dimensional hypercube!

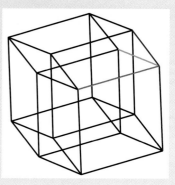

EXAMPLE 1   **Using the Property for Multiplying Exponential Expressions**

a. $2^2 \cdot 2^3 = 2^{2+3} = 2^5 = 32$
b. $(-1)^2(-1)^5 = (-1)^{2+5} = (-1)^7 = -1$
c. $y^7 \cdot y^9 = y^{7+9} = y^{16}$
d. $(3x^4)(-2x^7) = 3(-2)(x^4 \cdot x^7) = -6x^{4+7} = -6x^{11}$
e. $(-5a^4b^5)(6a^3b^{11}) = -5 \cdot 6(a^4 \cdot a^3)(b^5 \cdot b^{11}) = -30a^{4+3}b^{5+11} = -30a^7b^{16}$
f. $x^3 \cdot y^5$ cannot be simplified because the bases (x and y) are not the same. ■

2 Simplify powers of exponential expressions.

Powers of Exponential Expressions

The following examples lead to a second property for exponents.

$$(x^3)^4 = \underbrace{x^3 \cdot x^3 \cdot x^3 \cdot x^3}_{\substack{x^3 \text{ appears as a} \\ \text{factor 4 times}}} = x^{3+3+3+3} = x^{12}$$

$$(2^3)^5 = \underbrace{2^3 \cdot 2^3 \cdot 2^3 \cdot 2^3 \cdot 2^3}_{\substack{2^3 \text{ appears as a} \\ \text{factor 5 times}}} = 2^{3+3+3+3+3} = 2^{15}$$

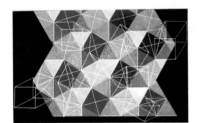

The expressions, $x^2, x^3, x^4, x^5, \dots$ suggest higher dimensions. Artist Tony Robbin combines painting and wire sculpture in this 1983 work *Simplex* to depict geometric forms and their shadows in various dimensions.

Copyright © by Tony Robbin. Reprinted by permission of Tony Robbin and his agent, Robin Straus Agency, Inc.

We can obtain these results if we multiply the exponents.

$$(x^3)^4 = x^{3 \cdot 4} = x^{12} \qquad (2^3)^5 = 2^{3 \cdot 5} = 2^{15}$$

Powers of exponential expressions

If n and m are natural numbers, then

$$(b^n)^m = b^{nm}.$$

When an exponential expression is raised to a power, multiply the exponents. This product is the exponent that appears on the base in the exponential expression once the parentheses have been removed.

Example 2 illustrates this "power to a power" situation.

EXAMPLE 2   **Using the Property for Powers of Exponential Expressions**

a. $(x^6)^3 = x^{6 \cdot 3} = x^{18}$ **b.** $(2^7)^4 = 2^{7 \cdot 4} = 2^{28}$
c. $(-1^4)^2 = (-1)^{4 \cdot 2} = (-1)^8 = 1$ ■

3 Simplify powers of products.

Powers of Products

The following examples can be generalized into a third property for exponents.

$$\begin{aligned} (2x)^4 &= (2x)(2x)(2x)(2x) \\ &= (2 \cdot 2 \cdot 2 \cdot 2)(x \cdot x \cdot x \cdot x) \\ &= 2^4x^4 \end{aligned}$$

$$(ab)^5 = (ab)(ab)(ab)(ab)(ab)$$
$$= (a \cdot a \cdot a \cdot a \cdot a)(b \cdot b \cdot b \cdot b \cdot b)$$
$$= a^5 b^5$$

We can obtain each of these results in one step by raising each factor in the product to the power.

$$(2x)^4 = 2^4 x^4 \qquad (ab)^5 = a^5 b^5$$

Powers of products

If n is a natural number, then

$$(ab)^n = a^n b^n. \qquad (a \neq 0, b \neq 0)$$

When a product is raised to a power, raise each factor in the product to the power.

EXAMPLE 3 **Using the Property for Powers of Products**

$$(5y)^3 = 5^3 y^3 = 125 y^3$$

The third property of exponents can be extended to cover more than two factors.

EXAMPLE 4 **Extending the Powers of Products Property for More Than Two Factors**

$$(2xyz)^5 = 2^5 x^5 y^5 z^5$$
$$= 32 x^5 y^5 z^5$$

The simplification of some exponential expressions requires that we use more than one of the properties of exponents.

EXAMPLE 5 **Simplifying an Exponential Expression Using More Than One Property**

Simplify: $(2x^3 y^4)^5$

Solution

$$(2x^3 y^4)^5 = 2^5 (x^3)^5 (y^4)^5 \qquad \text{Raise each factor to the fifth power.}$$
$$= 2^5 x^{3 \cdot 5} y^{4 \cdot 5} \qquad (b^n)^m = b^{nm}, \text{ so multiply the exponents.}$$
$$= 32 x^{15} y^{20}$$

EXAMPLE 6 **Simplifying an Exponential Expression**

Simplify: $(3xy)(3x)^2 + 4x(2x^2 y)$

Solution

$$(3xy)(3x)^2 + 4x(2x^2 y)$$
$$= (3xy)(9x^2) + 4x(2x^2 y) \qquad \text{Simplify } (3x)^2 \text{ by squaring both 3 and } x.$$
$$= 27x^3 y + 8x^3 y \qquad \text{In both terms, add exponents on the common base } x.$$
$$= 35x^3 y \qquad \text{Combine like terms.}$$

4 Simplify powers of quotients.

Powers of Quotients

Powers of quotients are simplified exactly like powers of products.

> **Powers of quotients**
>
> If n is a natural number, then
>
> $$\left(\frac{a}{b}\right)^n = \frac{a^n}{b^n}. \qquad (b \neq 0)$$
>
> When a quotient is raised to a power, raise both the numerator and the denominator to the power.

EXAMPLE 7   Using the Property for Powers of Quotients

a. $\left(\dfrac{a}{b}\right)^7 = \dfrac{a^7}{b^7}$

b. $\left(\dfrac{2}{x}\right)^4 = \dfrac{2^4}{x^4} = \dfrac{16}{x^4}$

c. $\left(\dfrac{x^2}{y^3}\right)^6 = \dfrac{(x^2)^6}{(y^3)^6} = \dfrac{x^{12}}{y^{18}}$

d. $\left(\dfrac{-3x^2}{2y}\right)^4 = \dfrac{(-3x^2)^4}{(2y)^4} = \dfrac{(-3)^4(x^2)^4}{2^4 y^4} = \dfrac{81x^8}{16y^4}$ ∎

5 Divide exponential expressions.

Dividing Exponential Expressions

Our final property deals with dividing exponential expressions. Once again, we can use an example from which to generalize the property:

$$\frac{b^5}{b^3} = \frac{\cancel{b} \cdot \cancel{b} \cdot \cancel{b} \cdot b \cdot b}{\cancel{b} \cdot \cancel{b} \cdot \cancel{b}} = b^2$$

We can obtain this result if we subtract the exponents.

$$\frac{b^5}{b^3} = b^{5-3} = b^2$$

> **Dividing exponential expressions**
>
> If n and m are natural numbers and $n > m$, then
>
> $$\frac{b^n}{b^m} = b^{n-m}. \qquad (b \neq 0)$$
>
> When dividing exponential expressions with the same nonzero base, retain the base and subtract the exponent of the denominator from the exponent of the numerator.

EXAMPLE 8   Dividing Exponential Expressions

Divide and simplify:

a. $\dfrac{x^{12}}{x^3}$ **b.** $\dfrac{27x^{14}y^8}{3x^3y^5}$

Solution

a. $\dfrac{x^{12}}{x^3} = x^{12-3} = x^9$ Subtract the exponents.

b. $\dfrac{27x^{14}y^8}{3x^3y^5} = \dfrac{27}{3} \cdot \dfrac{x^{14}}{x^3} \cdot \dfrac{y^8}{y^5}$ This step can be done mentally.

$\quad\qquad\qquad = 9x^{14-3}y^{8-5}$ Subtract the exponents when dividing.

$\quad\qquad\qquad = 9x^{11}y^3$ Simplify. ■

6 Use zero exponents.

The Zero Exponent

Consider applying the rule for dividing exponential expressions when the nonzero bases in the numerator and denominator are both raised to the same power:

$\dfrac{b^n}{b^n} = 1$ Any nonzero expression divided by itself is equal to 1.

$\dfrac{b^n}{b^n} = b^{n-n} = b^0$ Subtract the exponents.

To continue subtracting exponents with like bases, we must define b^0 to equal 1.

> **The definition of zero as an exponent**
>
> If b is any real number other than 0,
>
> $\quad b^0 = 1.$

EXAMPLE 9 **Zero Exponents**

a. $7^0 = 1$ **b.** $\pi^0 = 1$ **c.** $5x^0 = 5 \cdot 1 = 5$ **d.** $(5x)^0 = 1$

e. 0^0 is undefined. The definition does not apply because the base is 0. ■

7 Use negative exponents.

Negative Integers as Exponents

The rule for dividing exponential expressions can lead us to the definition of a negative exponent. Consider, for example,

$\dfrac{b^3}{b^7}$

which can be simplified in two ways. First, we can use the definition of an exponent:

$\dfrac{b^3}{b^7} = \dfrac{\cancel{b} \cdot \cancel{b} \cdot \cancel{b}}{\cancel{b} \cdot \cancel{b} \cdot \cancel{b} \cdot b \cdot b \cdot b \cdot b} = \dfrac{1}{b^4} \quad (b \neq 0)$

We can also apply the rule for dividing exponents by subtracting powers:

$\dfrac{b^3}{b^7} = b^{3-7} = b^{-4} \quad (b \neq 0)$

ENRICHMENT ESSAY

Interpreting Negative Exponents in the Real World

If you have $100 in a savings account at a simple interest rate of 9%, it will be worth $100(1.09)^t$ in t years. For example, in 2 years the investment will be worth $100(1.09)^2 = 100(1.1881) = \$118.81$. At the beginning of the investment, $t = 0$, so the worth is $100(1.09)^0 = 100(1) = \$100$. But how can we interpret *negative* exponents in this situation?

Consider this question: To have $100 in a savings account now, how much would you have needed to invest 3 years ago if the rate is 9%?

$$100(1.09)^{-3} = 100 \cdot \frac{1}{(1.09)^3} = \frac{100}{1.295029}$$

which is approximately $77.22.

From these two expressions for $\frac{b^3}{b^7}$, we can conclude that

$$b^{-4} = \frac{1}{b^4}.$$

Generalizing this last equation leads to the definition of negative exponents.

The definition of a negative exponent

If b is any real number other than 0,

$$b^{-n} = \frac{1}{b^n}.$$

This definition gives us a way of writing an expression with a negative exponent as an equivalent expression with a positive exponent.

Using technology

Verify Example 10 with a graphing calculator. Here's part (b):

((−) 5) ∧
(−) 3 ENTER

Discover for yourself

Simplify

$$\frac{1}{3^{-4}}$$

by first writing

$$\frac{1}{3^{-4}} = \frac{1}{\frac{1}{3^4}}.$$

Now repeat this process for $\frac{1}{2^{-5}}$.

In general, if $b \neq 0$, how can $\frac{1}{b^{-n}}$ be written without b in the denominator?

EXAMPLE 10   Using Negative Exponents

a. $4^{-2} = \frac{1}{4^2} = \frac{1}{16}$

b. $(-5)^{-3} = \frac{1}{(-5)^3} = \frac{1}{-125} = -\frac{1}{125}$

c. $-5^{-2} = -(5^{-2}) = -\left(\frac{1}{5^2}\right) = -\frac{1}{25}$

d. $\left(\frac{4}{5}\right)^{-2} = \frac{1}{\left(\frac{4}{5}\right)^2} = \frac{1}{\frac{16}{25}} = \frac{25}{16}$ ∎

Study tip

Be careful when simplifying $(-5)^{-3}$. Change only the sign of the exponent and not the base when moving -5 to the denominator.

Correct:

$$(-5)^{-3} = \frac{1}{(-5)^3}$$

Incorrect:

ENRICHMENT ESSAY

Exponents and Large Numbers

Names of Large Numbers

10^2	hundred	10^{18}	quintillion
10^3	thousand	10^{21}	sextillion
10^6	million	10^{24}	septillion
10^9	billion	10^{27}	octillion
10^{12}	trillion	10^{30}	nonillion
10^{15}	quadrillion	10^{100}	googol

- Assuming that you can count to 200 in 1 minute, that you decide to count for 12 hours a day at this rate, and that you don't have any pressing engagements in the foreseeable future, it would take you in the region of 19,024 years, 68 days, 10 hours, and 40 minutes to count to 1 billion (10^9).
- The number of snow crystals necessary to form the Ice Age was approximately 10^{30}.
- If the entire universe were filled with protons and electrons, the total number would be 10^{110}.

- The number of grains of sand on the beach at Coney Island is about 10^{20}.
- Although people do a great deal of talking, the total output since the beginning of gabble to the present day, including all baby talk, love songs, and congressional debates, only amounts to about 10^{16} (10 million billion) words.
- A googol is 10^{100}. The number of raindrops falling on New York in a century is much less than a googol.
- A googol times a googol is $10^{100} \cdot 10^{100}$, or 10^{200}, which is much less than a googolplex! A googolplex is 10 to the googol power, or $10^{10^{100}}$, or 1 with a googol of zeros! If you tried to write a googolplex without using exponential notation, there would not be enough room to write it if you went to the farthest star, touring all the nebulae and putting down zeros every inch of the way.

John Scott "Invoking the Googul Plex" 1981 water-colour on paper 24 × 18 in./Photo courtesy The Gallery, Stratford, Ontario

Now take a moment to work the Discover for Yourself box in the margin on page 50. Were you able to discover the second equation in the box that follows?

> **Negative exponents in numerators and denominators**
>
> If $b \neq 0$,
>
> $$b^{-n} = \frac{1}{b^n} \qquad \text{and} \qquad \frac{1}{b^{-n}} = b^n.$$

EXAMPLE 11 **Evaluating Expressions Containing Negative Exponents**

a. $\dfrac{1}{4^{-3}} = 4^3 = 64$

b. $\dfrac{1}{(-2)^{-5}} = (-2)^5 = -32$ ∎

8 Simplify exponential expressions.

Simplifying Exponential Expressions

Our definition of negative exponents came from applying the rule for dividing with like bases to exponents that were integers. Although we first presented the five properties of exponents with natural number exponents, they can be extended to all integral exponents. The properties of exponents are summarized on the next page.

Properties of exponents

For any integers n and m, and for any real numbers a and b ($b \neq 0$ whenever it appears in a denominator):

1. Multiplying with like bases: $b^n \cdot b^m = b^{n+m}$
2. Raising a power to a power: $(b^n)^m = b^{nm}$
3. Raising a product to a power: $(ab)^n = a^n b^n$
4. Raising a quotient to a power: $\left(\dfrac{a}{b}\right)^n = \dfrac{a^n}{b^n}$
5. Dividing with like bases: $\dfrac{b^n}{b^m} = b^{n-m}$
6. Zero as an exponent: $b^0 = 1$ ($b \neq 0$)
7. Negative integers as exponents: $b^{-n} = \dfrac{1}{b^n}, \quad \dfrac{1}{b^{-n}} = b^n$

Study tip

An exponential expression is in simplified form if:

1. No parentheses appear.
2. Each base appears once.
3. No zero or negative exponents appear.

EXAMPLE 12   **Simplifying an Exponential Expression**

Simplify: $\dfrac{-16x^3y^9}{4x^{10}y^3}$

Solution

$$\dfrac{-16x^3y^9}{4x^{10}y^3} = \left(\dfrac{-16}{4}\right)\left(\dfrac{x^3}{x^{10}}\right)\left(\dfrac{y^9}{y^3}\right)$$ This step can be done mentally.

$$= -4x^{3-10}y^{9-3}$$ Subtract the exponents when dividing.

$$= -4x^{-7}y^6$$ Simplify.

$$= -4 \cdot \dfrac{1}{x^7} \cdot y^6$$ $b^{-n} = \dfrac{1}{b^n}$, so put x in the denominator and change the sign of the exponent.

$$= \dfrac{-4y^6}{x^7}$$

Discover for yourself

There is often more than one correct way to simplify an exponential expression. Example 13a can be solved by starting with

$$x^3 \cdot x^{-7} = \dfrac{x^3}{x^7}.$$

In Example 13b, you can begin by writing

$$(x^5)^{-4} \quad \text{as} \quad \dfrac{1}{(x^5)^4}.$$

Try solving all parts of Example 13 by a different method from the one shown.

EXAMPLE 13   **Simplifying Exponential Expressions**

Simplify: **a.** $x^3 \cdot x^{-7}$ **b.** $(x^5)^{-4}$ **c.** $\dfrac{x^5}{x^{-12}}$ **d.** $\dfrac{y^{-7}}{y^{-3}}$

Solution

a. $x^3 \cdot x^{-7} = x^{3+(-7)} = x^{-4} = \dfrac{1}{x^4}$ Add the exponents when multiplying. Then simplify.

b. $(x^5)^{-4} = x^{5(-4)} = x^{-20} = \dfrac{1}{x^{20}}$ Multiply the exponents in the "power to a power" situation. Then simplify.

c. $\dfrac{x^5}{x^{-12}} = x^{5-(-12)} = x^{5+12} = x^{17}$ Subtract the exponents when dividing. Then simplify.

d. $\dfrac{y^{-7}}{y^{-3}} = y^{-7-(-3)} = y^{-7+3} = y^{-4} = \dfrac{1}{y^4}$ Subtract the exponents when dividing. Then simplify.

Simplification of exponential expressions can be a bit tricky in situations where a number of exponential properties must be used. As indicated in the Discover for Yourself, these expressions can often be simplified correctly in more than one way.

EXAMPLE 14   Simplifying Exponential Expressions

Simplify:

a. $(-2xy^{-14})(-3x^4y^5)^3$ **b.** $\left(\dfrac{25x^2y^4}{-5x^6y^{-8}}\right)^2$

Solution

a. $(-2xy^{-14})(-3x^4y^5)^3$

$\qquad = (-2xy^{-14})(-3)^3(x^4)^3(y^5)^3$ Cube each factor in the second parentheses.

$\qquad = (-2xy^{-14})(-27)x^{12}y^{15}$ Multiply the exponents when raising a power to a power.

$\qquad = (-2)(-27)x^{1+12}y^{-14+15}$ Mentally rearrange factors and multiply like bases by adding the exponents.

$\qquad = 54x^{13}y$ Simplify.

b. $\left(\dfrac{25x^2y^4}{-5x^6y^{-8}}\right)^2$

$\qquad = (-5x^{2-6}y^{4-(-8)})2$ Simplify inside the parentheses by subtracting the exponents when dividing.

$\qquad = (-5x^{-4}y^{12})^2$ Simplify.

$\qquad = (-5)^2(x^{-4})^2(y^{12})^2$ Square each factor in parentheses.

$\qquad = 25x^{-8}y^{24}$ Multiply the exponents when raising a power to a power.

$\qquad = \dfrac{25y^{24}}{x^8}$ Simplify x^{-8} using $b^{-n} = \dfrac{1}{b^n}$. ■

Discover for yourself

Cover the second and third columns in the Study Tip with a sheet of paper. Describe the error in the left column and then write the correction on your own. Then compare your description and correction with the one in the chart.

Study tip

Try to avoid the following common errors that can occur when simplifying exponential expressions.

Incorrect	Description of Error	Correct
$b^3b^4 = b^{12}$	The exponents should be added, not multiplied.	$b^3b^4 = b^7$
$3^n \cdot 3^m = 9^{n+m}$	The common base should be retained, not multiplied.	$3^n \cdot 3^m = 3^{n+m}$
$\dfrac{5^{16}}{5^4} = 5^4$	The exponents should be subtracted, not divided.	$\dfrac{5^{16}}{5^4} = 5^{12}$
$(4a)^3 = 4a^3$	Both factors should be cubed.	$(4a)^3 = 64a^3$
$b^{-n} = -\dfrac{1}{b^n}$	Only the exponent should change sign.	$b^{-n} = \dfrac{1}{b^n}$
$(a+b)^{-1} = \dfrac{1}{a} + \dfrac{1}{b}$	The exponent applies to the entire expression $a + b$.	$(a+b)^{-1} = \dfrac{1}{a+b}$

PROBLEM SET 1.4

Practice Problems

In Problems 1–10, multiply and simplify. Express each answer in exponential notation.

1. $2^4 \cdot 2^2$

2. $3^3 \cdot 3^4$

3. $5^3 \cdot 5$

4. $8 \cdot 8^6$

5. $x^2 \cdot x^5$

6. $y^4 \cdot y^7$

7. $2x^5 \cdot 3x^8$

8. $5x^4 \cdot 6x^{12}$

9. $(-5x^3y^2)(2xy^{17})$

10. $(-3xy^6)(15x^8y^2)$

In Problems 11–14, raise powers to powers and simplify. Express each answer in exponential notation.

11. $(2^4)^3$

12. $(3^2)^5$

13. $(x^4)^8$

14. $(y^7)^{10}$

In Problems 15–36, simplify each expression by first applying the properties for powers of products and quotients.

15. $(4x)^3$

16. $(-2x)^5$

17. $(3xy)^4$

18. $(-2xy)^5$

19. $(2xy^2)^3$

20. $(3a^2b)^4$

21. $(-3x^2y^5)^2$

22. $(-3x^4y^6)^3$

23. $(2xy)(4x)^2$

24. $(-3ab)(2b)^3$

25. $(4xy)(-2x^2y) + 17x^3y^2$

26. $(-5ab)(-3a^2b) - 19a^3b^2$

27. $(2x)^3(-3xy) + 25x^4y$

28. $(-2bc)(5b)^2 + 52b^3c$

29. $\left(\dfrac{x}{y}\right)^6$

30. $\left(\dfrac{3}{y}\right)^4$

31. $\left(\dfrac{-3x}{y}\right)^4$

32. $\left(\dfrac{-2}{xy}\right)^5$

33. $\left(\dfrac{x^4}{y^2}\right)^3$

34. $\left(\dfrac{a^7}{b^4}\right)^4$

35. $\left(\dfrac{-5x^3}{2y}\right)^3$

36. $\left(\dfrac{-3x^6}{5y^7}\right)^2$

In Problems 37–48, divide and simplify. Express each answer in exponential notation.

37. $\dfrac{5^6}{5^3}$

38. $\dfrac{2^8}{2^4}$

39. $\dfrac{x^{16}}{x^8}$

40. $\dfrac{y^{20}}{y^4}$

41. $\dfrac{8x^7}{2x^4}$

42. $\dfrac{20a^{13}}{4a^6}$

43. $\dfrac{-100x^{18}}{25x^{17}}$

44. $\dfrac{-50c^{11}}{10c^{10}}$

45. $\dfrac{-50x^2y^7}{5xy^4}$

46. $\dfrac{-36x^{12}y^4}{4x^2y^2}$

47. $\dfrac{56a^{12}b^{10}c^8}{-7ab^2c^4}$

48. $\dfrac{66a^9b^7c^6}{-6a^3bc^2}$

Use the definitions for zero as an exponent and negative integers as exponents to evaluate Problems 49–70. If applicable, use a calculator to verify your answer.

49. 6^0

50. $6x^0$

51. 17^0

52. $9y^0$

53. $(6x)^0$

54. $(9y)^0$

55. 5^{-2}

56. 4^{-3}

57. $(-4)^{-3}$

58. $(-2)^{-3}$

59. $(-4)^{-2}$

60. $(-6)^{-2}$

61. -4^{-2}

62. -6^{-2}

63. $\left(\frac{3}{4}\right)^{-2}$

64. $\left(\frac{7}{9}\right)^{-2}$

65. $\dfrac{1}{5^{-3}}$

66. $\dfrac{1}{2^{-5}}$

67. $\dfrac{1}{(-3)^{-4}}$

68. $\dfrac{1}{(-2)^{-4}}$

69. $\dfrac{1}{-3^{-4}}$

70. $\dfrac{1}{-2^{-4}}$

Simplify each exponential expression in Problems 71–112. Be sure that no zero or negative exponents appear in your final simplification.

71. $\dfrac{20x^4y^3}{5xy^3}$

72. $\dfrac{-36c^{16}d^{14}}{12c^{16}d}$

73. $\dfrac{x^3}{x^9}$

74. $\dfrac{y^6}{y^{10}}$

75. $\dfrac{20x^3}{-5x^4}$

76. $\dfrac{-9y^5}{\frac{1}{3}y^6}$

77. $\dfrac{16x^3}{8x^{10}}$

78. $\dfrac{24y^2}{48y^{11}}$

79. $\dfrac{20a^3b^8}{2ab^{13}}$

80. $\dfrac{72x^5y^{11}}{9xy^{17}}$

81. $\dfrac{1}{b^{-5}}$

82. $\dfrac{1}{c^{-13}}$

83. $x^3 \cdot x^{-12}$

84. $z^{11} \cdot z^{-20}$

85. $(2a^5)(-3a^{-7})$

86. $(-4x^2)(-2x^{-5})$

87. $(3a^7)(-2a^{-3})$

88. $(11x^{31})(-3x^{-29})$

89. $(x^3)^{-6}$

90. $(x^2)^{-11}$

91. $(x^{-3})^{-6}$

92. $(x^{-2})^{-11}$

93. $\dfrac{x^3}{x^{-7}}$

94. $\dfrac{y^4}{y^{-10}}$

95. $\dfrac{6y^2}{2y^{-8}}$

96. $\dfrac{12y^5}{3y^{-10}}$

97. $\dfrac{4x^6}{-20x^{-11}}$

98. $\dfrac{6y^3}{-18y^{-22}}$

99. $\dfrac{x^{-7}}{x^3}$

100. $\dfrac{y^{-10}}{y^4}$

101. $\dfrac{x^{-7}}{x^{-3}}$ **102.** $\dfrac{y^{-10}}{y^{-4}}$ **103.** $\dfrac{30x^2y^5}{-6x^8y^{-3}}$ **104.** $\dfrac{24a^2b^{13}}{-2a^{15}b^{-2}}$ **105.** $\dfrac{25a^{-8}b^2}{-75a^{-3}b^4}$ **106.** $\dfrac{-3x^{-5}b^6}{12x^{-2}b^{13}}$

107. $\left(\dfrac{x^3}{x^{-5}}\right)^2$ **108.** $\left(\dfrac{a^4}{a^{-11}}\right)^3$ **109.** $\left(\dfrac{15a^4b^2}{-5a^{10}b^{-3}}\right)^3$ **110.** $\left(\dfrac{10x^{14}y^8}{-30x^{17}y^{-2}}\right)^3$

111. $\left(\dfrac{10y^2}{y}\right) + \left(\dfrac{4xy^4}{xy^3}\right)$ **112.** $\left(\dfrac{12x^3y^2}{xy}\right) - \left(\dfrac{14x^5y^5}{x^3y^4}\right)$

Application Problems

113. The threshold weight for a person is defined as the crucial weight, above which the mortality risk for that particular person rises astronomically. The threshold weight (T, in pounds) for men between the ages of 40 and 49 is described by the model

$$T = \left(\dfrac{h}{12.3}\right)^3$$

where h is the height in inches. What is the threshold weight, to the nearest pound, of a man whose age is 45 and who is 5 feet 10 inches tall?

114. The figure shows the relative distances of nine planets from the sun. The distance (d) of the nth planet from the sun is described by the model

$$d = \dfrac{3(2^{n-2}) + 4}{10}$$

where d is measured in astronomical units. How much closer to the sun is Venus than Earth?

The figure shows the relative distances of nine planets from the Sun.

· Pluto

 Neptune

 Uranus

 Saturn

 Jupiter

● Mars

● Earth

◉ Venus

· Mercury

Sun

True–False Critical Thinking Problems

115. Which one of the following is true?
  a. $4 \cdot 4 \cdot 4^{-1} = \frac{1}{64}$ **b.** $5^3 \cdot 2^2 = 10^5$

  c. $4 \cdot 3^{-5} = \dfrac{1}{4 \cdot 3^5}$ **d.** $-3^{-2} = -\frac{1}{9}$

116. Which one of the following is true?
  a. $5x^{-1}y^{-3} = \dfrac{1}{5xy^3}; x \neq 0$ and $y \neq 0$

  b. $\dfrac{3x^{-6}y^{-1}}{6x^{-5}y^{-2}} = \dfrac{y}{2x}; x \neq 0$ and $y \neq 0$

  c. $\dfrac{5^{-4}}{x^{-3}} = \dfrac{x^3}{20}; x \neq 0$

  d. $3^{-1} + 4^{-1} = \frac{1}{7}$

117. Which one of the following is true?
  a. $2^2 \cdot 2^4 = 2^8$ **b.** $5^6 \cdot 5^2 = 25^8$
  c. $2^3 \cdot 3^2 = 6^5$ **d.** None of the above is true.

118. Which one of the following is true?

  a. $\dfrac{1}{(-2)^3} = 2^{-3}$ **b.** $\dfrac{2^8}{2^{-3}} = 2^5$

  c. $2^4 + 2^5 = 2^9$ **d.** None of the above is true.

Technology Problem

119. The formulas

$$y = 67.0166(1.00308)^x \quad \text{and}$$
$$y = 74.9742(1.00201)^x$$

model the life expectancy (y) for white males and females, respectively, in the United States whose present age is x.

a. Use a graphing utility to graph both models in the same viewing rectangle. Enter the formulas as

$$\text{Y}_1 \boxed{=} \; 67.0166 \boxed{\times} 1.00308 \boxed{\wedge} \boxed{\text{X}}$$
$$\text{Y}_2 \boxed{=} \; 74.9742 \boxed{\times} 1.00201 \boxed{\wedge} \boxed{\text{X}}$$

with the following range setting:

$$\text{Xmin} = 0, \text{Xmax} = 80, \text{Xscl} = 1,$$
$$\text{Ymin} = 0, \text{Ymax} = 100, \text{Yscl} = 1$$

b. Use the trace feature of your utility to trace along each curve until you reach $x = 40$. What is the corresponding value for y on Y_1 and Y_2? How long can a 40-year-old white male expect to live? How long can a 40-year-old white female expect to live?

c. Explain how each graph indicates that life expectancy increases as one gets older. Use your utility's trace feature to reinforce this observation.

Writing in Mathematics

120. If $b \neq 0$, we know that

$$b^{-n} = \frac{1}{b^n} \quad \text{and} \quad \frac{1}{b^{-n}} = b^n.$$

Express this idea strictly in words rather than using the compact, symbolic notation of algebra.

121. Describe the differences, if any, among, b^2, b^{-2}, $-b^2$, and $(-b)^2$.

Critical Thinking Problems

In Problems 122–125, simplify the expression. Assume that all variables used as exponents represent integers and that all other variables represent nonzero real numbers.

122. $x^{n-1} \cdot x^{3n+4}$

123. $(x^{-4n} \cdot x^n)^{-3}$

124. $\left(\dfrac{x^{3-n}}{x^{6-n}} \right)^{-2}$

125. $\left(\dfrac{x^n y^{3n+1}}{y^n} \right)^3$

126. If $x = (1 - 9 + 8 + 1)^{1989}$ and $y = (1 - 9 + 8 - 1)^{1989}$, what is the value of $(1 + 9 + 8 + 1)^{x+y}$? (Do *not* use a calculator!)

127. Here are the names of several large numbers:

10^{10}	dekaplex, also known as 10 billion
10^{100}	hectoplex, also known as googol
10^{1000}	kiloplex
$10^{1,000,000}$	megaplex
$10^{1,000,000,000}$	gigaplex
$10^{1,000,000,000,000}$	teraplex
10^{googol}	googolplex, also known as hectoplexplex

a. Show that megaplex^{1000} = gigaplex.

b. The brain has approximately 3 billion synapses. A thought, or state of mind, can be characterized by focusing on which synapses are firing, specifying a state of mind by determining whether the first synapse is on or off, whether the second synapse is on or off, and so on. Thus, there are around $2^{3,000,000,000}$ states of mind. What number in the above list comes closest to describing possible states of mind?

Review Problems

128. Simplify:
$$[6(xy - 1) - 2xy] - 4[(2xy + 3) - 2(xy + 2)].$$

129. What is the multiplicative inverse of $-\frac{3}{4}$?

130. Simplify: $\dfrac{3}{7^2 + (-7)(-5)} - \left[\left(\dfrac{2}{3} + \dfrac{1}{4} \right) \cdot \dfrac{6}{7} \right].$

Solutions **Tutorial** **Video**
Manual I

Francesco Clemente (Italian, b. 1952) "Sun" 1980, gouache on twelve sheets of handmade Pondicherry paper, joined by cotton strips, 1980, 91 × 95 in./Philadelphia Museum of Art: Purchased: Edward & Althea Budd Fund, Katharine Levin Farrell Fund, & funds contributed by Mrs. H. Gates Lloyd.

Scientific Notation

Objectives

1 Convert from scientific to decimal notation and vice versa.
2 Perform calculations using scientific notation.
3 Solve applied problems using scientific notation.

We frequently encounter very large and very small numbers. For example, the number of particles of tar and other pollutants in 1 cubic inch of cigarette smoke is approximately 82,000,000,000. Even worse, the mass (in grams) of one electron is approximately

0.0000000000000000000000000009

The large number of zeros in these numbers make them difficult to read, write, or say. *Scientific notation* gives us a compact way to display and say these numbers.

> **Definition of scientific notation**
>
> A scientific notation numeral has the form
>
> $$c \times 10^n,$$
>
> where $1 \leq c < 10$ and n is an integer. This means that scientific notation for a number appears as the product of two factors. The first factor is a number greater than or equal to 1 but less than 10. The second factor is base 10 raised to a power that is an integer.

EXAMPLE 1   **Examples of Scientific Notation**

a. A light-year is the number of miles that light travels in one year, approximately 5.88×10^{12} miles. (It is customary to use × rather than a dot to indicate multiplication in scientific notation.)
b. An electrical impulse travels through 9 inches of wire in a computer in 1×10^{-9} second.
c. Each day Earth is inundated with 2.6×10^7 pounds of dust from the atmosphere.
d. The diameter of a hydrogen atom is 1.016×10^{-8} centimeter.   ∎

The following example illustrates some numbers converted from scientific notation to decimal notation.

1 Convert from scientific to decimal notation and vice versa.

EXAMPLE 2   **Converting from Scientific to Decimal Notation**

a. $1.7 \times 10^3 = 1.7 \times 1000 = 1700$
b. $2.31 \times 10^0 = 2.31 \times 1 = 2.31$
c. $7.43 \times 10^{-2} = 7.43 \times \frac{1}{100} = 0.0743$
d. $6.153 \times 10^{-4} = 6.153 \times \frac{1}{10,000} = 0.000\ 615\ 3$ ∎

Observe that when multiplying by 10 to a power, we move the decimal point the same number of places as the exponent of 10. If the exponent is *positive,* we move the decimal point in the first factor to the *right.* If the exponent is *negative,* we move the decimal point in the first factor to the *left.*

The following procedure can be used to change from decimal notation to scientific notation.

Writing a number in scientific notation

Write the number as the product of two factors:

$$c \times 10^n.$$

The first factor, c, is a number greater than or equal to 1 and less than 10. n is an integer.

1. First factor: Move the decimal point in the original number to the right of the first nonzero digit to obtain a number greater than or equal to 1 and less than 10.
2. Second factor: 10^n. Count the number of places you moved the decimal point. This is the absolute value of n. If the original number is 10 or greater, then n is positive. If the original number is less than 1, then n is negative. If the original number is between 1 and 10, then the decimal point does not have to be moved, so $n = 0$.

EXAMPLE 3   **Converting from Decimal Notation to Scientific Notation**

Write each of the following real numbers in scientific notation.
a. 82,000,000,000   **b.** 0.000 000 000 000 000 000 160 2   **c.** 3.7284

Solution

a. 82,000,000,000
$\quad = 8.2 \times 10^n$ Move the decimal point in 82,000,000,000 to the right of 8, the first nonzero digit. Since the original number is greater than 10, the power of 10 is positive.

$\quad = 8.2 \times 10^{10}$ In changing 82,000,000,000 to 8.2, the decimal point was moved ten places, so $n = 10$.

$$82{,}000{,}000{,}000. = 8.2 \times 10^{10}$$

Count 10 places to
the right to get
to the decimal point.

b. 0.000 000 000 000 000 000 160 2
$\quad = 1.602 \times 10^n$ Move the decimal point in the given number to the right of 1, the first nonzero digit. Since the original number is less than 1, n is negative.

$\quad = 1.602 \times 10^{-19}$ In changing the given number to 1.602, the decimal point was moved 19 places, so $n = -19$.

$$0.000\ 000\ 000\ 000\ 000\ 000\ 160\ 2 = 1.602 \times 10^{-19}$$

Count 19 places to the left
to get to the decimal point.

c. $3.7284 = 3.7284 \times 10^n$ The first factor is 3.7284, a number between 1 and 10.

           $= 3.7284 \times 10^0$ Since the decimal point was not moved, $n = 0$.

$3.7284 = 3.7284 \times 10^0$

    ^

    ↑

No counting is necessary to get to the decimal point.   ■

U sing technology

You can change the mode setting on a graphing calculator so that numbers are displayed in scientific notation. (Consult your manual.) Once you're in the scientific notation mode, simply enter a number and then press ENTER .

Number	ENTER		Display
82,000,000,000	ENTER		8.2E10
.0000000000000000001602	ENTER		1.602E-19
3.7284	ENTER		3.7284E0

S tudy tip

In converting positive numbers from decimal notation to scientific notation, notice the following:

	Have	
Numbers ≥ 10	←——→	Positive powers of 10
0 < Numbers < 1	←——→	Negative powers of 10
Numbers between 1 and 10	←——→	Zero power of 10

2 Perform calculations using scientific notation.

Calculations Using Scientific Notation

Because scientific notation numerals involve powers of 10, we can multiply or divide numbers in scientific notation using special cases of two exponential properties.

Multiplying and dividing exponential expressions: base 10

$10^n \cdot 10^m = 10^{n+m}$ Add the exponents when multiplying.

$\dfrac{10^n}{10^m} = 10^{n-m}$ Subtract the exponents when dividing.

EXAMPLE 4 **Using Scientific Notation to Multiply Real Numbers**

Use scientific notation to find the product of 0.000 64 and 9,400,000,000 and write the answer in scientific notation.

Even if you do not set your graphing calculator to a scientific notation mode, your calculator will automatically switch to scientific notation when displaying large or small numbers that exceed the display range. Try multiplying

79,000 × 3,400,000,000.

The display shows

2.686 E 14

so that the product is

2.686×10^{14}.

If you set your calculator to the scientific notation mode, answers to all computations will be displayed in scientific notation even if they do not exceed the display range.

Solution

$0.000\,64 \times 9{,}400{,}000{,}000$
$= 6.4 \times 10^{-4} \times 9.4 \times 10^{9}$ Write each number in scientific notation.
$= (6.4 \times 9.4) \times (10^{-4} \times 10^{9})$ Rearrange factors.
$= 60.16 \times 10^{5}$ $10^n \cdot 10^m = 10^{n+m}$
$= 6.016 \times 10 \times 10^{5}$ Express 60.16 in scientific notation.
$= 6.016 \times 10^{6}$ Add the exponents when multiplying. ■

EXAMPLE 5 **Using Scientific Notation to Divide Real Numbers**

Use scientific notation to find the quotient of 0.000 001 05 and 4,200,000 and write the answer in scientific notation.

Solution

$\dfrac{0.00000105}{4{,}200{,}000}$

$= \dfrac{1.05 \times 10^{-6}}{4.2 \times 10^{6}}$ Write each number in scientific notation.

$= \dfrac{1.05}{4.2} \times \dfrac{10^{-6}}{10^{6}}$ This step can be done mentally.

$= 0.25 \times 10^{-12}$ Subtract the exponents when dividing.
$= (2.5 \times 10^{-1}) \times 10^{-12}$ Express 0.25 in scientific notation.
$= 2.5 \times 10^{-13}$ Add the exponents when multiplying. ■

Discover for yourself

Use the scientific notation feature of a graphing calculator to verify the computations in Examples 4 and 5.

3 Solve applied problems using scientific notation.

Applications of Scientific Notation

If the values of variables in mathematical models are relatively large or small, we often express these values in scientific notation before substituting them into the formula.

Example 6 can be solved using the *uniform motion model*

distance = rate · time   or $d = rt$.

The formula applies to objects moving at a constant (or uniform) rate and to objects whose average speed is involved. For example, if a car averages 45 miles per hour for 3 hours, the distance that it covers is given by

$$d = rt = 45 \frac{\text{miles}}{\text{hour}} \cdot 3 \text{ hours} = 45 \cdot 3 = 135 \text{ miles}.$$

EXAMPLE 6 **Scientific Notation and the Uniform Motion Model**

Light travels at a rate of approximately 1.86×10^{5} miles per second. It takes light 5×10^{2} seconds to travel from the sun to Earth. What is the distance between Earth and the sun?

ENRICHMENT ESSAY

Situations Giving Rise to Scientific Notation Numerals

Situation	Scientific Notation
X-ray frequencies	1×10^{16} vibrations/second
Bottles of champagne consumed by the French yearly	1.245×10^8
Number of toothbrushes purchased by Americans yearly	2.8×10^8
Number of candles to produce the same amount of light as the sun	3.01×10^{27}
Grains of sand on Miami Beach	1.7×10^{14}
Weight (in tons) of all water on Earth	1.97×10^{18}
Number of people who speak Mandarin, the language of the world spoken by the greatest number of people	6.5×10^8
Ergs of energy released by the 1906 San Francisco earthquake	6×10^{25}
Length of a hemoglobin molecule	6.8×10^{-6}
Number of wives of polygamist King Mongut of Siam (the king in *The King and I*)	9.2×10^3
Warren G. Harding's shoe size (the American president with the largest feet	1.4×10
The amount of insurance money for the legs of dancer Fred Astaire	6.5×10^5
The mass of the hydrogen atom (in grams)	1.66×10^{-24}
Number of ways of arranging the 52 cards in an ordinary deck of playing cards	8.066×10^{67}

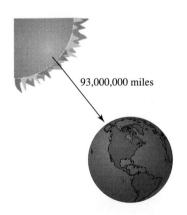

93,000,000 miles

Solution

$$d = rt$$ Use the uniform motion model.

$$d = (1.86 \times 10^5) \times (5 \times 10^2)$$ Substitute the given values.

$$d = (1.86 \times 5) \times (10^5 \times 10^2)$$ Rearrange factors.

$$d = 9.3 \times 10^7$$ Add the exponents when multiplying.

The distance between Earth and the sun is approximately 9.3×10^7 miles, or 93 million miles. ■

EXAMPLE 7   The National Debt

In 1994 the national debt was 4.644×10^{12} dollars. At that time, the U.S. population was 2.6×10^8 people. If the national debt were evenly divided among every individual in the United States, how much debt would be assigned to each person?

Solution

The problem's conditions can be modeled with an equation.

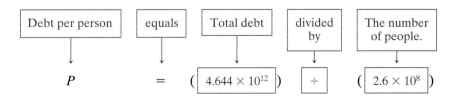

Debt per person	equals	Total debt	divided by	The number of people.
P	$=$	(4.644×10^{12})	$\div$	(2.6×10^8)

$$\frac{4.644 \times 10^{12}}{2.6 \times 10^8}$$ This is the division implied by the problem's conditions.

$$= \frac{4.644}{2.6} \times \frac{10^{12}}{10^8}$$ Rearrange factors.

$$\approx 1.786 \times 10^4$$ Subtract the exponents when dividing.

Divided among every person in the country, each individual in the United States would be assigned a debt of approximately 1.786×10^4 dollars or $17,860. ■

Study tip

Significant Digits

The number of *significant digits* in a number is determined by counting the digits from left to right, beginning with the leftmost nonzero digit. For example, 4.644 has four significant digits, and 2.6 has two. Significant digits in scientific notation are determined by counting the number of significant digits in the factor between 1 and 10, so that 4.644×10^{12} has four significant digits, whereas 2.6×10^4 has two. In Example 7, we rounded per capita debt, 1.786×10^4, to four significant digits to get a handle on how much debt each individual would be assigned. However, the following rules are usually applied:

1. Round products and quotients so that they contain the same number of significant digits as the number in the calculation with the fewest significant digits.
2. Round sums and differences so that they contain the same number of significant digits to the right of the decimal as the number in the calculation with the fewest significant digits to the right of the decimal.

Using rule 1, Example 7 would be rounded as follows:

4 digits ⟶ $$\frac{4.644 \times 10^{12}}{2.6 \times 10^8} \approx 1.8 \times 10^4$$
2 digits ⟶

Round to two significant digits.

PROBLEM SET 1.5

Practice Problems

In Problems 1–8, convert to decimal notation.

1. 6×10^{-3} **2.** 8×10^{-5} **3.** 5.93×10^6 **4.** 8.17×10^5

5. 6.284×10^{-7} **6.** 8.032×10^{-5} **7.** 4.003×10^{10} **8.** 3.008×10^9

In Problems 9–16, convert to scientific notation.

9. 98,000,000,000   **10.** 7,400,000,000,000   **11.** 746,000,000,000,000,000

12. 237,000,000,000,000,000   **13.** 0.000000023   **14.** 0.00000074

15. 0.0007924   **16.** 0.00083105

In Problems 17–32, perform the indicated operation and express the product or quotient in scientific notation. If applicable, use a graphing calculator to verify your answers.

17. $(2.8 \times 10^4)(3.2 \times 10^3)$

18. $(5.2 \times 10^6)(4.1 \times 10^3)$

19. $(4.3 \times 10^{15})(6.5 \times 10^{-11})$

20. $(3.4 \times 10^{14})(2.1 \times 10^{-9})$

21. $(6.03 \times 10^6)(2.01 \times 10^{-9})$

22. $(4.03 \times 10^7)(3.12 \times 10^{-10})$

23. $(8.04 \times 10^{-8})(3.01 \times 10^{-16})$

24. $(5.02 \times 10^{-13})(4.16 \times 10^{-12})$

25. $\dfrac{9.9 \times 10^8}{1.1 \times 10^5}$

26. $\dfrac{7.5 \times 10^9}{2.5 \times 10^3}$

27. $\dfrac{6.12 \times 10^7}{3.06 \times 10^{-4}}$

28. $\dfrac{1.05 \times 10^8}{4.2 \times 10^{-6}}$

29. $\dfrac{1.5 \times 10^{-4}}{5.5 \times 10^7}$

30. $\dfrac{3.2 \times 10^{-6}}{8 \times 10^8}$

31. $\dfrac{1.21 \times 10^{-4}}{2.42 \times 10^{-7}}$

32. $\dfrac{2.7 \times 10^{-7}}{7.5 \times 10^{-4}}$

In Problems 33–48, perform the indicated computations using scientific notation. Express the answer in scientific notation as well. If applicable, use a graphing calculator to verify your answers.

33. $(82{,}000{,}000)(3{,}000{,}000{,}000)$

34. $(94{,}000{,}000)(6{,}000{,}000{,}000)$

35. $(0.00037)(8{,}300{,}000)$

36. $(0.025)(9{,}400{,}000{,}000{,}000)$

37. $(150{,}000{,}000)(0.00005)(30{,}000)(0.002)$

38. $(32{,}000{,}000)(0.00005)(1{,}500{,}000)(0.005)$

39. $\dfrac{95{,}000{,}000{,}000}{5{,}000{,}000}$

40. $\dfrac{12{,}000{,}000{,}000}{24{,}000{,}000}$

41. $\dfrac{480{,}000{,}000{,}000}{0.000\,12}$

42. $\dfrac{77{,}000{,}000{,}000}{0.022}$

43. $\dfrac{0.000\,000\,096}{16{,}000}$

44. $\dfrac{0.000\,000\,000\,084}{0.000\,024}$

45. $\dfrac{(90{,}000)(0.004)}{(0.0003)(120)}$

46. $\dfrac{(0.000\,35)(120{,}000)}{(1500)(0.000\,002\,8)}$

47. $\dfrac{(0.000\,035)(40{,}000)}{(14{,}000)(0.000\,25)}$

48. $\dfrac{(0.000\,14)(200)(0.036)}{0.042}$

Application Problems

The third largest expense for the federal government is paying interest on our debt. This interest is the financing cost of the debt, including the interest paid to people, corporations, and institutions holding government securities. The graph shows the net interest on the federal debt in billions of dollars, including projected figures for the years 2000 and 2005. Use the graph to answer Problems 49–50.

49. Write an estimate for the net interest on the federal debt for the year 1990, expressing the answer in scientific notation.

50. Write an estimate for the projected net interest on the federal debt for the year 2000, expressing the answer in scientific notation.

51. If the length of a hydrogen atom is 0.000 000 03 millimeter and the average human foot measures 200 millimeters, how many times as large as the hydrogen atom is the human foot?

52. The tallest tree known is the Howard Libbey redwood in Redwood Grove, California. The tree measures 100,000 millimeters. The distance from Earth to the sun is approximately 150,000,000,000,000 millimeters. How many of the tallest redwood trees would it take to span this distance?

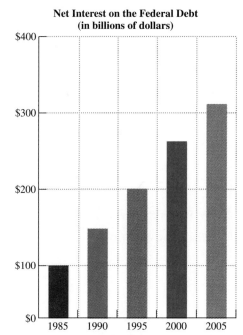

Net Interest on the Federal Debt (in billions of dollars)

Source: Fiscal Year 1996 Budget of the US Government CBO 1995

53. The chirp of a cricket releases 9000 ergs of energy. The energy release of a 100-megaton H-bomb is approximately

10,000,000,000,000,000,000,000,000,000 ergs.

How many times greater than the cricket's chirp is the energy released by the H-bomb? (*Note:* To put these numbers into perspective, when you pronounce an average syllable of a word, you release 200 ergs of energy.)

54. Poiseuille's law states that the speed of blood (*S*, in centimeters/second) located *r* centimeters from the central axis of an artery is

$$S = (1.76 \times 10^5)[(1.44 \times 10^{-2}) - r^2].$$

Find the speed of blood at the central axis of this artery.

Arteries Surrounding the Heart

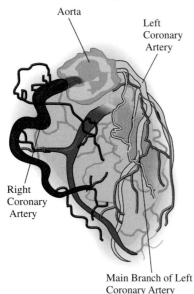

The flow of blood in a blood vessel can be modeled by Poiseuille's law.

55. Our galaxy measures approximately 1.2×10^{17} kilometers across. If a space vehicle were capable of moving at half the speed of light (approximately half of 3×10^5, or 1.5×10^5 kilometers/second), how many years would it take for the vehicle to cross the galaxy?

56. The ratio of the tide-raising forces of the moon (F_M) to the tide-raising forces of the sun (F_S) is described by the model

$$\frac{F_M}{F_S} = \frac{M_M r_S^3}{M_S r_M^3}$$

where

M_M = Mass of the moon, 7.35×10^{22} kilograms

r_s = Distance from Earth to the sun,
1.5×10^8 kilometers

M_S = Mass of the sun, 1.99×10^{30} kilograms

r_M = Distance from Earth to the moon,
3.84×10^5 kilometers

Compute $\dfrac{F_M}{F_S}$, showing that the tidal force exerted by the moon is approximately 2.2 times that exerted by the sun on the Earth.

57. The U.S. federal deficit in the year 2000 is projected to be 194.4 billion dollars. If a person spent $1000 every second, how long (in seconds) would it take to spend the projected deficit? Express the answer in scientific notation. Then convert your answer from seconds to years.

58. Our hearts beat about 70 times per minute. Express in scientific notation how many times the heart beats over a lifetime of 80 years.

True–False Critical Thinking Problems _____

59. Which one of the following is true?

a. $534.7 = 5.347 \times 10^3$

b. $\dfrac{8 \times 10^{30}}{4 \times 10^{-5}} = 2 \times 10^{25}$

c. $(7 \times 10^5) + (2 \times 10^{-3}) = 9 \times 10^2$

d. $(4 \times 10^3) + (3 \times 10^2) = 43 \times 10^2$

60. Which one of the following is true?

a. $10^{-3} = 0.0001$

b. $56.7 \times 10^5 = 5.67 \times 10^6$

c. $879.6 = 8.796 \times 10^3$

d. $(4 \times 10^5)(5 \times 10^8) = 2.0 \times 10^{13}$

Technology Problems

Use a calculator to solve Problems 61–62.

61. The model for gravitation force (*F*, measured in newtons) between two large objects is given by

$$F = \frac{(6.67 \times 10^{-11})(M_1)(M_2)}{d^2}$$

where M_1 = the mass of one object
M_2 = the mass of the second object
d = the mean distance between the objects

The mass of Earth is 5.97×10^{24} kilograms, the mass of the moon is 7.35×10^{22} kilograms, and the distance between them is 3.84×10^8 meters. What is the gravitational force between the Earth and the moon?

62. The distance (*d*) of the *n*th planet from the sun is described by the model

$$d = \frac{3(2^{n-2}) + 4}{10}$$

where *d* is measured in astronomical units.

If it is known that Pluto is approximately 4.6×10^9 miles from the sun, approximately how many miles are there in an astronomical unit?

Writing in Mathematics

63. Explain how to express 0.000 763 in scientific notation.

64. Discuss one advantage of expressing a number in scientific notation over decimal notation.

65. Give an example of a number where there is no advantage in using scientific notation over decimal notation. Explain why this is the case.

66. Describe in words the two laws of exponents needed to compute $(2.31 \times 10^6)^2$.

67. When using scientific notation, why might a person be more interested in the power on 10 than the other factor in the number?

Critical Thinking Problems

68. Consult an almanac or a relevant source in your library to find examples of numbers that can be written in scientific notation. Use two or more of these numbers to write and solve a problem like Example 7.

69. Find the product and quotient of $c \times 10^n$ and $d \times 10^m$.

70. Express in scientific notation a number that is 1000 times smaller than 6.08×10^{14}.

Review Problems

71. Graph: $y = x^2 + 3$.

72. Write $\{x \mid -3 \le x < 5\}$ in interval notation and graph the set on a number line.

73. Simplify: $-18 - (-12 + 2 \cdot 3^2)$.

SECTION 1.6

Solutions Manual · Tutorial · Video

Solving Linear Equations

Objectives

1 Solve linear equations.
2 Solve linear equations containing fractions or decimals.
3 Identify equations with no solutions or infinitely many solutions.

Linear algebra is a branch of algebra that was developed in an attempt to solve equations containing *x* to large integral powers, such as $17x^{43} + 19x^{17} - 3x^{14} = 28$. Equations involving only *x* to the first power are *first-degree* or *linear* equations. Second-degree equations, containing x^2 and no powers higher than 2, are *quadratic equations* (from quadratic, meaning "squared numbers"). Equations containing x^3 (and no higher powers) are *cubic equations,* and those containing x^4 are *quartic equations.*

In this section, we establish the basic techniques for solving linear equations. Solving an equation is the process of finding the set of numbers that will make the equation a true statement. These numbers are called the *solutions* or *roots* of the equation, and we say that they *satisfy* the equation. The set of all solutions is called the *solution set* of the equation.

The most common type of equation in algebra is a *linear equation* in one variable.

Definition of a linear equation

A *linear equation* in one variable x is any equation that can be written in the form

$$ax + b = c$$

where $a, b,$ and c are real numbers, and $a \neq 0$.

An example of a linear equation in one variable is $2x + 3 = 17$, in which $a = 2, b = 3,$ and $c = 17$. Because the greatest power on the variable is 1, a linear equation is also called a *first-degree equation*.

The equations $2x + 3 = 17$ and $x = 7$ are *equivalent equations* because they have the same solution set, namely, $\{7\}$. Replacing x with 7 in the first equation leads to a true statement.

$$2x + 3 = 17$$
$$2(7) + 3 = 17$$
$$14 + 3 = 17$$
$$17 = 17 \quad \checkmark \quad \text{A true statement}$$

We can see by inspection that $\{7\}$ is also the solution set for the equation $x = 7$.

Solve linear equations.

Solving Linear Equations

The process of solving a linear equation involves writing a series of equivalent equations. The last of the equivalent equations should be in the form

$$x = d$$

where d is a constant. Using inspection, we see that the solution set for this equation is $\{d\}$.

The addition and multiplication properties of equality are used to produce equivalent equations.

Properties used to solve linear equations

The Addition Property of Equality

For all real numbers, $a, b,$ and c,

If $a = b$, then $a + c = b + c$.

ENRICHMENT ESSAY

A 3600-Year-Old Problem

This problem is from a 3600-year-old Egyptian papyrus. Written in hieroglyphics, its translation into modern algebraic notation is

$$x + \frac{2}{3}x - \frac{1}{3}\left(x + \frac{2}{3}x\right) = 10.$$

From the time of the Pharaohs, humankind has aimed at solving mathematical problems involving an unknown number.

The same real number may be added to both sides of an equation without changing the solution set.

The Multiplication Property of Equality

For all real numbers a, b, and c (where $c \neq 0$),

If $a = b$, then $ac = bc$.

The same nonzero real number may multiply both sides of an equation without changing the solution set.

Since subtraction is defined in terms of addition, and division is defined in terms of multiplication, these properties can be extended to include those operations.

More properties used to solve linear equations

The Subtraction Property of Equality

The same real number may be subtracted from both sides of an equation without changing the solution set.

The Division Property of Equality

Both sides of an equation may be divided by the same nonzero real number without changing the solution set.

The easiest equations to solve involve situations where by using just one of these properties we immediately isolate the equation's variable. We review how to solve these equations in Table 1.7.

Example 1 illustrates that it is usually necessary to use more than one property of equality to solve an equation.

TABLE 1.7 Using Properties of Equality to Solve Linear Equations

Equation	How to Isolate x	Solving the Equation	The Equation's Solution Set
$x - 3 = 8$	Add 3 on both sides.	$x - 3 + 3 = 8 + 3$ $x = 11$	$\{11\}$
$x + 7 = -15$	Subtract 7 on both sides.	$x + 7 - 7 = -15 - 7$ $x = -22$	$\{-22\}$
$6x = 30$	Divide both sides by 6 $\left(\text{or multiply by } \dfrac{1}{6} \text{ on both sides}\right).$	$\dfrac{6x}{6} = \dfrac{30}{6}$ $x = 5$	$\{5\}$
$\dfrac{x}{5} = 9$	Multiply both sides by 5.	$5 \cdot \dfrac{x}{5} = 9 \cdot 5$ $x = 45$	$\{45\}$

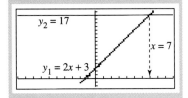
EXAMPLE 1   **Solving a Linear Equation**

Solve for x: $2x + 3 = 17$

Solution

Our goal is to obtain an equivalent equation with x isolated on the left side and a number on the other side.

$2x + 3 = 17$	This is the given equation.
$2x + 3 - 3 = 17 - 3$	Subtract 3 from both sides.
$2x = 14$	Combine like terms.
$\dfrac{2x}{2} = \dfrac{14}{2}$	Divide both sides by 2.
$x = 7$	Simplify.

Be sure to complete the solution by checking the proposed solution, 7, in the original equation.

Check

$2x + 3 = 17$	This is the original equation.
$2(7) + 3 \overset{?}{=} 17$	Substitute 7 for x.
$14 + 3 \overset{?}{=} 17$	Simplify.
$17 = 17$ ✓	This true statement indicates that 7 is the solution.

This verifies that the solution set is $\{7\}$. ∎

To solve some equations, we must simplify one or both sides before using the properties of equality.

EXAMPLE 2   **Solving a Linear Equation by First Simplifying**

Solve for x: $2(x - 4) + 5x = -22$

Solution

$$2(x - 4) + 5x = -22 \quad \text{This is the given equation.}$$

$$2x - 8 + 5x = -22 \quad \text{Apply the distributive property on the left.}$$

$$7x - 8 = -22 \quad \text{Combine like terms on the left.}$$

$$7x - 8 + 8 = -22 + 8 \quad \text{Add 8 to both sides.}$$

$$7x = -14 \quad \text{Combine like terms.}$$

$$\frac{7x}{7} = \frac{-14}{7} \quad \text{Divide both sides by 7.}$$

$$x = -2 \quad \text{Simplify.}$$

Check

Check by substituting the proposed solution, -2, into the original equation.

$$2(x - 4) + 5x = -22 \quad \text{This is the original equation.}$$

$$2(-2 - 4) + 5(-2) \overset{?}{=} -22 \quad \text{Substitute } -2 \text{ for } x.$$

$$2(-6) + 5(-2) \overset{?}{=} -22 \quad \text{Simplify inside parentheses.}$$

$$-12 + (-10) \overset{?}{=} -22 \quad \text{Multiply.}$$

$$-22 = -22 \quad \checkmark \quad \text{This true statement indicates that } -2 \text{ is the solution.}$$

The solution set is $\{-2\}$. ∎

Using technology

The graphs of

$$y_1 = 2(x - 4) + 5x$$

and $y_2 = -22$

have an intersection point whose x-coordinate is -2. This verifies that $\{-2\}$ is the solution set for

$$2(x - 4) + 5x = -22.$$

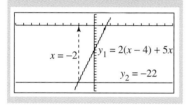

Study tip

With increased practice in solving linear equations, try working some steps mentally.

$$7x - 8 = -22 \quad \text{Simplified equation.}$$

$$7x = -14 \quad \text{Add 8 to both sides.}$$

$$x = -2 \quad \text{Divide both sides by 7.}$$

In Example 3, we solve a linear equation by subtracting the same algebraic expression from both sides.

EXAMPLE 3 **Solving a Linear Equation**

Solve for x: $\quad 5x - 12 = 8x + 24$

Solution

We will transform the equation into an equivalent equation in the form $x = d$. Since each side is simplified, we will collect all terms involving the variable on one side and all other terms on the other side. Let's collect terms with the variable on the left, accomplished by subtracting $8x$ from both sides.

$$5x - 12 = 8x + 24 \quad \text{This is the given equation.}$$

$$5x - 12 - 8x = 8x + 24 - 8x \quad \text{Subtract } 8x \text{ from both sides.}$$

$$-3x - 12 = 24 \quad \text{Simplify.}$$

$$-3x - 12 + 12 = 24 + 12 \quad \text{Collect numerical terms on the right, adding 12 to both sides.}$$

The graphs of

$$y_1 = 5x - 12$$
$$\text{and} \quad y_2 = 8x + 24$$

have an intersection point whose x-coordinate is -12. This verifies that $\{-12\}$ is the solution set for

$$5x - 12 = 8x + 24$$

We used the following range setting:

Xmin $= -14$, Xmax $= 2$,

Xscl $= 1$, Ymin $= -85$,

Ymax $= 10$, Yscl $= 1$

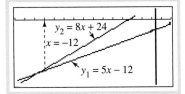

Discover for yourself

Solve the equation in Example 3 by collecting terms with the variable on the right and numerical terms on the left. Compare this solution with the steps in Example 3.

Discover for yourself

Use a graphing utility to verify that the solution set is $\{6\}$. Graph

$$y_1 = 2(x - 3) - 17$$
$$\text{and} \quad y_2 = 13 - 3(x + 2)$$

in the same viewing rectangle that shows both lines and their intersection point. The x-coordinate of the intersection point should be 6.

$$-3x = 36 \qquad \text{Simplify.}$$

$$\frac{-3x}{-3} = \frac{36}{-3} \qquad \text{Isolate } x \text{ by dividing both sides by } -3.$$

$$x = -12 \qquad \text{Simplify.}$$

Check

The solution, -12, should be checked in the original equation.

$$5x - 12 = 8x + 24 \qquad \text{This is the original equation.}$$
$$5(-12) - 12 \stackrel{?}{=} 8(-12) + 24 \qquad \text{Substitute } -12 \text{ for } x.$$
$$-60 - 12 \stackrel{?}{=} -96 + 24 \qquad \text{Multiply.}$$
$$-72 = -72 \quad \checkmark \qquad \text{This true statement indicates that } -12 \text{ is the solution.}$$

The solution set is $\{-12\}$.

Let's summarize the major steps involved in the solution of a linear equation. Not all of these steps are necessary in every equation.

Solving a linear equation

1. Simplify the algebraic expression on each side.
2. Collect all the variable terms on one side and all the constant terms on the other side.
3. Isolate the variable and solve.
4. Check the proposed solution in the original equation.

EXAMPLE 4 **Solving a Linear Equation by First Simplifying Each Side**

Solve for x: $2(x - 3) - 17 = 13 - 3(x + 2)$

Solution

We begin by simplifying each side:

$$2(x - 3) - 17 = 13 - 3(x + 2) \qquad \text{This is the given equation.}$$
$$2x - 6 - 17 = 13 - 3x - 6 \qquad \text{Use the distributive property.}$$
$$2x - 23 = -3x + 7 \qquad \text{Combine numerical terms.}$$

At this point, we apply the properties of equality.

$$2x - 23 + 3x = -3x + 7 + 3x \qquad \text{Add } 3x \text{ to both sides, isolating the variable terms on the left.}$$
$$5x - 23 = 7 \qquad \text{Simplify.}$$
$$5x - 23 + 23 = 7 + 23 \qquad \text{Add 23 to both sides.}$$
$$5x = 30 \qquad \text{Simplify.}$$
$$\frac{5x}{5} = \frac{30}{5} \qquad \text{Divide both sides by 5.}$$
$$x = 6 \qquad \text{Simplify.}$$

Check

We now check the solution, 6, by substitution into the original equation.

$$2(x - 3) - 17 = 13 - 3(x + 2)$$ This is the original equation.

$$2(6 - 3) - 17 \overset{?}{=} 13 - 3(6 + 2)$$ Substitute 6 for x.

$$2(3) - 17 \overset{?}{=} 13 - 3(8)$$ Simplify inside parentheses.

$$6 - 17 \overset{?}{=} 13 - 24$$ Multiply.

$$-11 = -11 \quad \checkmark$$ This true statement indicates that 6 is the solution.

The solution set is {6}.   ■

2 Solve linear equations containing fractions or decimals.

Linear Equations with Fractions and Decimals

Equations are easier to solve when they do not contain fractions or decimals. Equations involving fractions can be written as equivalent equations by applying the multiplication property of equality. Multiplying every term on both sides of an equation by the *least common multiple* (LCM) of all the denominators in the equation clears the equation of fractions. The least common multiple is the smallest number that is divisible by all the denominators. This idea is illustrated in Example 5.

EXAMPLE 5 **Solving a Linear Equation Containing Fractions**

Solve: $\dfrac{3y}{5} = \dfrac{2y}{3} + 1$

Solution

Eliminate fractions by multiplying both sides of the equation by the LCM of the denominators. That is, multiply both sides by the smallest number that is divisible by 5 and 3, namely 15.

$$\frac{3y}{5} = \frac{2y}{3} + 1$$ This is the given equation.

$$15\left(\frac{3y}{5}\right) = 15\left(\frac{2y}{3} + 1\right)$$ Multiply both sides by 15, the LCM of 5 and 3.

$$15\left(\frac{3y}{5}\right) = 15\left(\frac{2y}{3}\right) + 15(1)$$ Apply the distributive property on the right.

$$\overset{3}{\cancel{15}}\left(\frac{3y}{\cancel{5}}\right) = \overset{5}{\cancel{15}}\left(\frac{2y}{\cancel{3}}\right) + 15$$ Simplify by canceling identical factors in the numerators and denominators.

$$9y = 10y + 15$$ Multiply. The equivalent equation, with fractions cleared, has integers as coefficients.

$$9y - 10y = 10y + 15 - 10y$$ Collect the variable terms on the left, subtracting $10y$ from both sides.

$$-y = 15$$ Equivalently, $-1y = 15$.

$$(-1)(-y) = (-1)15$$ Multiply both sides by -1.

$$y = -15$$ Observe that if $-y = a$, then $y = -a$.

Is it always necessary to multiply both sides of an equation by the LCM of the denominators to clear fractions? Try solving

$$\frac{1}{2}(2x - 8) + 4 = \frac{2}{3}(3x + 12)$$

by first applying the distributive property rather than multiplying both sides by 6.

Check by substituting the proposed solution, -15, into the original equation. You should obtain the true statement $-9 = -9$, verifying that the solution set is $\{-15\}$.

EXAMPLE 6   Solving a Linear Equation Containing Fractions

Solve: $\dfrac{y - 8}{5} + \dfrac{y}{3} = -\dfrac{8}{5}$

Solution

With denominators of 5, 3, and 5, the LCM is 15. We multiply both sides by this LCM.

$$15\left(\frac{y - 8}{5} + \frac{y}{3}\right) = 15\left(-\frac{8}{5}\right)$$

We now apply the distributive property to the left side:

$$15\left(\frac{y - 8}{5}\right) + 15\left(\frac{y}{3}\right) = 15\left(-\frac{8}{5}\right)$$

$$\overset{3}{\cancel{15}}\left(\frac{y - 8}{\cancel{5}_{1}}\right) + \overset{5}{\cancel{15}}\left(\frac{y}{\cancel{3}_{1}}\right) = \overset{3}{\cancel{15}}\left(-\frac{8}{\cancel{5}_{1}}\right)$$ Cancel identical factors in numerators and denominators.

$$3(y - 8) + 5y = 3(-8)$$ This equivalent equation contains only integers.

$$3y - 24 + 5y = -24$$ Apply the distributive property.

$$8y - 24 = -24$$ Combine like terms.

$$8y - 24 + 24 = -24 + 24$$ Add 24 to both sides.

$$8y = 0$$ Simplify.

$$\frac{8y}{8} = \frac{0}{8}$$ Divide both sides by 8.

$$y = 0$$ Simplify.

Use a graphing utility to verify that the solution set is $\{0\}$. Enter the equations as

$Y_1 = (X - 8) \div 5 + X \div 3$

and $Y_2 = (-)8 \div 5.$

The first coordinate of the intersection point should be 0.

Check by substituting the proposed solution, 0, into the original equation. You should obtain the true statement $-\frac{8}{5} = -\frac{8}{5}$, verifying that the solution set is $\{0\}$.

To clear an equation of decimals, keep in mind that in fractional notation decimals have denominators of 10, 100, 1000, and so on. Count the greatest number of decimal places in any term of the equation. If this number is 1, multiply both sides by 10^1, or 10. If this number is 2, multiply both sides by 10^2, or 100, and so on.

EXAMPLE 7   Solving a Linear Equation by First Clearing Decimals

Solve: $0.04x + 0.09(15 - x) = 0.07(25)$

Solution

Because $0.04 = \frac{4}{100}$, $0.09 = \frac{9}{100}$, and $0.07 = \frac{7}{100}$, the LCM of the denominators is 100. We will multiply both sides by 100.

Try solving the equation in Example 7 without first multiplying both sides by 100. Describe what happens. What is the advantage to first multiplying by 100? Can you think of any disadvantages to this method? What are they?

$$0.04x + 0.09(15 - x) = 0.07(25)$$ This is the given equation.

$$100[0.04x + 0.09(15 - x)] = 100[(0.07)(25)]$$ Multiply both sides by 100.

$$100(0.04x) + 100[0.09(15 - x)] = 100[(0.07)(25)]$$ Apply the distributive property on the left.

$$[100(0.04)]x + [100(0.09)](15 - x) = [100(0.07)](25)$$ Regroup factors.

$$4x + 9(15 - x) = 7(25)$$ Perform each multiplication by 100.

$$4x + 135 - 9x = 175$$ Apply the distributive property.

$$-5x + 135 = 175$$ Combine like terms.

$$-5x + 135 - 135 = 175 - 135$$ Subtract 135 from both sides.

$$-5x = 40$$ Simplify.

$$\frac{-5x}{-5} = \frac{40}{-5}$$ Divide both sides by −5.

$$x = -8$$ Simplify.

Check to verify that the solution set is $\{-8\}$. ■

3   Identify equations with no solutions or infinitely many solutions.

Types of Linear Equations

Up to this point we have been solving *conditional equations*. These equations are true only when the variable in the equation is equal to a particular real number. Linear conditional equations contain exactly one real number in their solution sets.

Now let's consider another possibility. Some philosophers tell us that the answer is "there is no answer." This theme sometimes emerges in solving linear equations.

The graphs of

$$y_1 = -4x - 3[2(1 - x)]$$

and $y_2 = 7 + 2x$

are parallel lines. With no intersection point, the equation

$$-4x - 3[2(1 - x)] = 7 + 2x$$

has no solution.

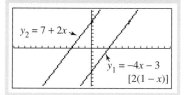

$y_2 = 7 + 2x$

$y_1 = -4x - 3$
$[2(1 - x)]$

EXAMPLE 8   **An Equation with No Solution**

Solve: $-4x - 3[2(1 - x)] = 7 + 2x$

Solution

We begin by simplifying the left side, considering innermost grouping symbols first. The distributive property will be used twice.

$$-4x - 3[2(1 - x)] = 7 + 2x$$ First, distribute 2 throughout the parentheses.

$$-4x - 3[2 - 2x] = 7 + 2x$$ Next, distribute −3 throughout the brackets.

$$-4x - 6 + 6x = 7 + 2x$$ Now −3 has been distributed.

$$2x - 6 = 7 + 2x$$ Combine like terms.

$$2x - 6 - 2x = 7 + 2x - 2x$$ Subtract 2x from both sides.

$$-6 = 7$$ Contradiction! Because this equation has no solution, no value of x satisfies the original equation.

We have a *contradiction* because we know that -6 is *not* equal to 7. This contradiction indicates that the given equation has no solution. The solution set contains no elements and is called the *null set* (written $\varnothing$) or the *empty set*. ■

Linear equations with no solution

If a contradiction (such as $-6 = 7$) is obtained in solving an equation, the equation has no solution. An equation whose solution set is the empty set is called an *inconsistent equation*.

Other philosophers assert that "any answer at all is the answer," an idea that appears in the following example.

EXAMPLE 9   **A Linear Equation That Is Satisfied by All Real Numbers**

Solve: $3x - 3(2 - x) = 6(x - 1)$

Solution

$$3x - 3(2 - x) = 6(x - 1) \qquad \text{This is the given equation.}$$
$$3x - 6 + 3x = 6x - 6 \qquad \text{Distribute } -3 \text{ on the left and 6 on the right.}$$
$$6x - 6 = 6x - 6 \qquad \text{Combine like terms. Notice that this equation is true for all real numbers.}$$
$$6x - 6 - 6x = 6x - 6 - 6x \qquad \text{Let's see what happens if we continue and subtract } 6x \text{ from both sides.}$$
$$-6 = -6$$

We have a trivial statement of equality in the final step, namely, $-6 = -6$. The left and right sides are equal regardless of what number is substituted for the variable. The solution set consists of the set of all real numbers. Based on our work in Section 1.1, we can express this as

$\{x \mid x \in R\}$ The set x such that x is any element of the set of real numbers.

or $(-\infty, \infty)$. ■

Using technology

The graphs of

$$y_1 = 3x - 3(2 - x)$$
and $y_2 = 6(x - 1)$

are the *same line*. With infinitely many intersection points, the equation

$$3x - 3(2 - x) = 6(x - 1)$$

is satisfied by all real numbers.

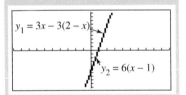

Linear equations with infinitely many solutions

If a statement in the form $a = a$ (such as $-6 = -6$) is obtained in solving an equation, this indicates that any real number will satisfy the equation. The equation is called an *identity*.

Table 1.8 summarizes the three types of linear equations.

ENRICHMENT ESSAY

The Empty Set in the Arts

In Example 8, we discovered that

$$-4x - 3[2(1 - x)] = 7 + 2x$$

has $\varnothing$, the empty set, as its solution. The empty set denotes the bizarre philosophic notion of nothingness, such as the set of people in an empty room.

Nothingness appears not only in solution sets of equations but also in the arts. John Cage's *4'33"* is a musical work written for the piano that requires 4 minutes and 33 seconds of total silence as a musician sits frozen at a piano stool. (Some critics, not impressed with Cage's philosophy that the ultimate order is chaos, rate *4'33"* as his best composition!) In the Gershwin folk opera *Porgy and Bess,* Ira Gershwin wrote, "I've got plenty of nothing." And the painting shown, a 60″ by 60″ all-black canvas by the American abstract painter Ad Reinhardt, is a visual depiction of nothingness on permanent display at New York's Museum of Modern Art.

Ad Reinhardt, a New York artist of the 1960s, found the all-black painting to be the perfect metaphor for the ultimate statement about reality. The all-black canvas said everything there was to say because it said nothing.

Ad Reinhardt, *Abstract Painting,* 1960–61. Oil on canvas, 60 in × 60 in. Courtesy of The Museum of Modern Art, New York.

TABLE 1.8 Types of Linear Equations		
Type of Equation	**Solution Set**	**Example**
Conditional equation	One real number	$3x + 1 = 7$ $3x = 6$ $x = 2$ $\{2\}$
Inconsistent equation	No real numbers	$2x + 1 = 2(x + 3)$ $2x + 1 = 2x + 6$ $1 = 6$ $\varnothing$
Identity	All real numbers	$2x + 1 = 2\left(x + \dfrac{1}{2}\right)$ $2x + 1 = 2x + 1$ $1 = 1$ $\{x \mid x \in R\}$ or $(-\infty, \infty)$

PROBLEM SET 1.6

Practice Problems _____

In Problems 1–48, solve each equation. Check proposed solutions algebraically or, if applicable, use a graphing utility.

1. $5x + 3 = 18$

2. $3x + 8 = 50$

3. $6x - 3 = 63$

4. $5x - 8 = 72$

5. $4x - 14 = -82$

6. $9x - 14 = -77$

7. $14 - 5x = -41$

8. $25 - 6x = -83$

9. $9(5x - 2) = 45$

10. $10(3x + 2) = 70$

11. $5x - (2x - 10) = 35$

12. $11x - (6x - 5) = 40$

13. $3x + 5 = 2x + 13$

14. $2x - 7 = 6 + x$

15. $8x - 2 = 7x - 5$

16. $13x + 14 = -5 + 12x$

17. $7x + 4 = x + 16$

18. $8x + 1 = x + 43$

19. $8y - 3 = 11y + 9$

20. $5y - 2 = 9y + 2$

21. $8z + 11.7 = 9z - 15$

22. $7z + 1.03 = 8z - 13.3$

23. $\frac{1}{6}y + \frac{1}{2} = \frac{1}{3}$

24. $\frac{1}{8}y + \frac{1}{4} = \frac{1}{2}$

25. $3(y + 4) + (y - 2) = 2 - (2y - 14)$

26. $2 - (7y + 5) = 13 - 3y$

27. $16 = (3y - 3) - (y - 7)$

28. $3(y - 3) - 2(y - 2) = 7$

29. $2(y + 1) - 3y = 3(3 + 2y)$

30. $4(2y + 1) - 29 = 3(2y - 5)$

31. $5(x - 2) - 2(2x + 1) = 2 + 5x$

32. $2(3x - 7) - (6 - x) = 4 - (-x + 6)$

33. $7(x + 1) = 4[x - (3 - x)]$

34. $3(2x - 1) = 5[3x - (2 - x)]$

35. $\frac{y}{4} = 2 + \frac{y - 3}{3}$

36. $5 - \frac{2 - z}{3} = \frac{z + 3}{8}$

37. $\frac{y - 3}{12} + \frac{y - 1}{6} = \frac{y + 2}{9} - 1$

38. $\frac{3z}{5} - \frac{z - 3}{2} = \frac{z + 2}{3}$

39. $\frac{x}{5} - 6 = x - 14$

40. $\frac{5x}{9} - \frac{2x - 3}{6} = 0$

41. $0.35y - 0.1 = 0.15y + 0.2$

42. $0.25y - 0.15 = 0.2y + 0.05$

43. $0.02(y - 100) = 62 + 0.06y$

44. $0.003(y + 200) = 86 - 0.05y$

45. $0.09(-y + 5000) = 513 - 0.12y$

46. $0.13(y + 300) = 61 - 0.09y$

47. $0.2z = 35 + 0.05(z - 100)$

48. $0.1z = 105 + 0.05(300 - z)$

For Problems 49–59, indicate whether each has no solution or is true for all values of the variable. If neither is the case, solve for the variable.

49. $10x - 2(4 + 5x) = -8$

50. $9x - 3(-5 + 3x) = 15$

51. $10x - 2(4 + 5x) = 8$

52. $9x - 3(-5 + 3x) - -15$

53. $7x - 2[3(1 - x)] = -(4 + 2 - 9x - 4x)$

54. $7x - 2[3(1 - x)] = -(5 + 2 - 9x - 4x)$

55. $12\left(\frac{y}{3} - \frac{1}{2}\right) = 8\left(\frac{y}{2} - 1\right)$

56. $2(3y + 4) - 4 = 9y + 4 - 3y$

57. $3(z - 4) - 5 = -2z + 5z - 9$

58. $-11p + 4(p - 3) + 6p = 4p - 12$

59. $-9 + 2(1 - 15k) = 7[2 - (3 + 4k)]$

True–False Critical Thinking Problems _____

60. Which one of the following is true?
   a. The solution set to $6x = 0$ is $\varnothing$.
   b. If $3x + 7 = 0$, then $x = \frac{7}{3}$.
   c. The equation $-2x + 4 = 6$ is equivalent to $x = -1$.
   d. The equation $2y + 1 = 2(y + 4) - 8$ is a conditional equation.

61. Which one of the following is true?
   a. Because the equation $2(y + 4) - 8 = 2y$ results in a true statement when y is replaced by 3, the solution set is $\{3\}$.
   b. The solution set for $\frac{x}{x} = 1$ is $\{x \mid \in R\}$.
   c. The equation $3x^2 + 7 = 10$ is a linear equation.
   d. The solution set for $7(y + 1) + 5(-y + 5) = 2(y + 3) - 7$ is $\varnothing$.

Technology Problems

For Problems 62–65, use your graphing utility to graph each side of the equations in the same viewing rectangle. Based on the resulting graph, label each equation as conditional, inconsistent, or an identity. If the equation is conditional, use the first coordinate of the intersection point to find the solution set. Verify this value by direct substitution into the equation.

62. $2(x - 6) + 3x = x + 6$

63. $9x + 3 - 3x = 2(3x + 1)$

64. $2(x + \frac{1}{2}) = 5x + 1 - 3x$

65. $\dfrac{2x - 1}{3} - \dfrac{x - 5}{6} = \dfrac{x - 3}{4}$

Writing in Mathematics

66. How do you know if a linear equation has one solution, no solution, or infinitely many solutions?

67. What is the difference between solving an equation such as $2(y - 4) + 5y = 34$ and simplifying an algebraic expression such as $2(y - 4) + 5y$? If there is a difference, which topic should be taught first? Why?

68. Why does the multiplication principle for solving equations exclude multiplying both sides of the equation by 0?

69. Describe what happens if you try solving $\frac{2}{3}y + \frac{1}{2} = -\frac{3}{8}$ without first multiplying both sides by 24, the LCM of the denominators.

70. Suppose that you solve $\dfrac{x}{5} - \dfrac{x}{2} = 1$ by multiplying both sides by 20, rather than the LCM of 5 and 2 (namely, 10). Describe what happens. If you get the correct solution, why do you think we clear the equation of frac-

tions by multiplying by the *least* common multiple of the denominators?

71. Suppose you were an algebra teacher grading the following solution on an examination:

$$-3(x - 6) = 2 - x$$
$$-3x - 18 = 2 - x$$
$$-2x - 18 = 2$$
$$-2x = -16$$
$$x = 8$$

You should note that 8 checks, and the solution set is {8}. The student who worked the problem therefore wants full credit. Can you find any errors in the solution? If full credit is 10 points, how many points should you give the student? Justify your position.

Critical Thinking Problems

72. Solve for x: $ax + b = c$.

73. Write three equations that are equivalent to $x = 5$.

74. If x represents a number, write an English sentence about the number that results in an inconsistent equation.

Review Problems

75. Graph: $y = 2x - 4$.

76. Simplify, expressing the answer with no negative exponents: $\dfrac{-30x^2y^8}{10x^{-4}y^{11}}$.

77. Simplify and write the quotient in scientific notation: $\dfrac{5 \times 10^{-6}}{20 \times 10^{-3}}$.

SECTION 1.7

Solutions Manual

Tutorial

Video 2

Mathematical Models

Objectives

1 Answer questions about mathematical models.

2 Solve a mathematical model for a specified variable.

In Section 1.2, we considered a variety of mathematical models that describe practical situations. Linear models can be written in the form $y = ax + b$, although the variables might not be x and y. In Section 1.3, we saw how to use the rectangular coordinate system to create graphs for mathematical models.

The ability to solve linear equations enables us to solve problems in situations that are modeled by these formulas.

■ Answer questions about mathematical models.

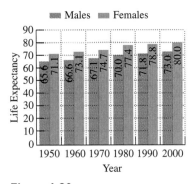

Figure 1.20

Life expectancy by year of birth
Source: U.S. Bureau of the Census
Statistical Abstract 1993

| EXAMPLE 1 | **Modeling Life Expectancy of U.S. Women** |

The graph in Figure 1.20 shows life expectancy in the United States by year of birth. The data for U.S. women can be approximated by the mathematical model

$$E = 0.215t + 71.05$$

where the variable E represents life expectancy in years and t represents the number of years after 1950. Use the given formula to determine in what year of birth life expectancy is 77.5 years.

Solution

We are given that the life expectancy for women is 77.5 years, so substitute 77.5 for E in the formula and solve for t.

$E = 0.215t + 71.05$ This is the given formula.

$77.5 = 0.215t + 71.05$ Replace E by 77.5 and solve for t.

$77{,}500 = 215t + 71{,}050$ Multiply each term by 1000 to clear decimals.

$6450 = 215t$ Subtract 71,050 from both sides.

$30 = t$ Divide both sides by 215.

The model indicates that 30 years after 1950, or in 1980, life expectancy for U.S. women is 77.5 years. Figure 1.20 indicates that the formula describes reality fairly accurately in this case, since the actual life expectancy for women born in 1980 is 77.4 years. ∎

Mathematical models are indispensable for grasping the possibilities of nature. This illustration shows models on the chalkboard used by Albert Einstein at a lecture on relativity.
Corbis-Bettmann

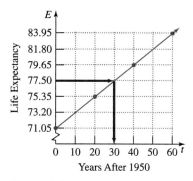

Figure 1.21

Life expectancy is 77.5 years for women born 30 years after 1950.

Mel Ramos "Man of Steel" 1962, oil on canvas, 50 × 44 in. © 1998 Mel Ramos/Licensed by VAGA, New York, NY. Photo by Steven Lopez. Courtesy Louis K. Meisel Gallery, New York.

We can visualize the solution to Example 1 by graphing the model $E = 0.215t + 71.05$ in the rectangular coordinate system. We begin with a table of coordinates.

t	$E = 0.215t + 71.05$	(t, E)
0	$E = 0.215(0) + 71.05 = 71.05$	$(0, 71.05)$
20	$E = 0.215(20) + 71.05 = 75.35$	$(20, 75.35)$
40	$E = 0.215(40) + 71.05 = 79.65$	$(40, 79.65)$
60	$E = 0.215(60) + 71.05 = 83.95$	$(60, 83.95)$

Graphing the resulting four data points and connecting them with a smooth curve indicates that they fall on a line, shown in Figure 1.21. To visualize the solution, we locate 77.5 on the vertical life expectancy axis. Then we draw a horizontal line from this point to the graph, shown with the red line in Figure 1.21. We then draw a vertical line from the point of intersection of the graph to the horizontal axis. The value on this axis is 30. We can see from the graph of the model that the life expectancy of 77.5 years is 30 birth years after 1950, or in 1980.

Techniques for solving linear equations provide algebraic methods for solving problems. The beauty of the rectangular coordinate system is that it allows us to visualize a problem's solution. This idea is demonstrated in Example 2.

EXAMPLE 2 **Competing Fitness Clubs**

Membership in Superfit, a fitness club, is $500 per year. The club charges $1.00 for each hour the facility is used. A competing club, Healthy Bodies, charges $440 yearly plus $1.75 per hour used. The yearly cost y of using each club for x hours can be modeled by the following formulas.

For Superfit:

$$y = 500 + 1x \qquad \text{The cost is \$500 plus \$1 times the number of hours of use, } x.$$
$$\text{Equivalently, } y = 500 + x.$$

For Healthy Bodies:

$$y = 440 + 1.75x \qquad \text{The cost is \$440 plus \$1.75 times the number of hours of use, } x.$$

How many hours of use each year will result in identical total yearly costs for the two fitness clubs?

Solution

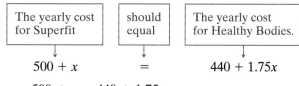

$$500 + x = 440 + 1.75x \qquad \text{This is the linear equation implied by the problem's conditions.}$$

$$500 + x - x = 440 + 1.75x - x \qquad \text{Subtract } x \text{ from both sides, isolating terms with } x \text{ on the right. If you prefer, clear decimals by first multiplying by 100.}$$

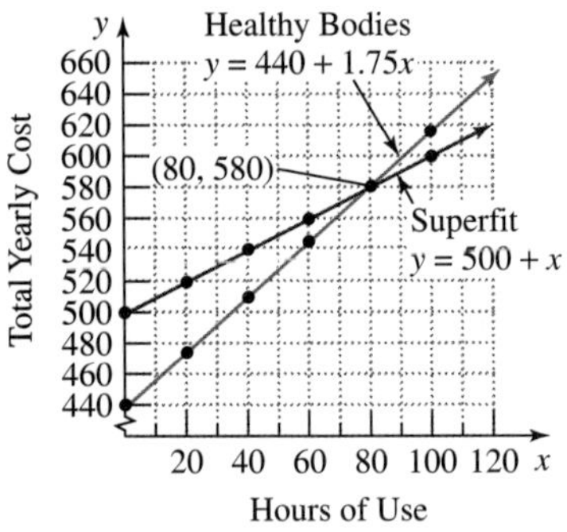

Figure 1.22

Yearly costs for competing fitness clubs.

$$500 = 440 + 0.75x \qquad \text{Simplify.}$$

$$500 - 440 = 440 + 0.75x - 440 \qquad \text{Subtract 440 from both sides, isolating numerical terms on the left.}$$

$$60 = 0.75x \qquad \text{Simplify.}$$

$$\frac{60}{0.75} = \frac{0.75x}{0.75} \qquad \text{Divide both sides by 0.75.}$$

$$80 = x \qquad \text{Simplify.}$$

The solution indicates that 80 hours of use each year will result in identical total yearly costs for the two clubs.

Let's visualize the solution to Example 2. We begin by setting up a table of coordinates for each model. Their graphs are shown in Figure 1.22.

Superfit		
x	$y = 500 + x$	(x, y)
0	$y = 500 + 0 = 500$	$(0, 500)$
20	$y = 500 + 20 = 520$	$(20, 520)$
40	$y = 500 + 40 = 540$	$(40, 540)$
60	$y = 500 + 60 = 560$	$(60, 560)$
100	$y = 500 + 100 = 600$	$(100, 600)$

Healthy Bodies		
x	$y = 440 + 1.75x$	(x, y)
0	$y = 440 + 1.75(0) = 440$	$(0, 440)$
20	$y = 440 + 1.75(20) = 475$	$(20, 475)$
40	$y = 440 + 1.75(40) = 510$	$(40, 510)$
60	$y = 440 + 1.75(60) = 545$	$(60, 545)$
100	$y = 440 + 1.75(100) = 615$	$(100, 615)$

Discover for yourself

Use a graphing utility to verify the graphs in Figure 1.22 and their intersection point.

The graphs in Figure 1.22 intersect at $(80, 580)$. This means that if each club is used 80 hours per year, the total yearly cost for each will be $580. If more than 80 hours are spent at the club, Superfit is the better deal. This is shown by the graph of $y = 500 + x$ lying below the graph of $y = 440 + 1.75x$ for $x > 80$.

2 Solve a mathematical model for a specified variable.

Solving Models for Specified Variables

One of the interesting aspects of algebra is its ability to handle infinitely many cases in one equation. For example, given the formula

$$F = \frac{9}{5}C + 32$$

we can compute the Celsius temperature, C, corresponding to a Fahrenheit temperature, F, of $77°$. We could conceivably carry out infinitely many computations. Each time we were given a value for F, we would have to solve for C. With infinitely many different values for F, we could wind up solving infinitely many equations.

Another approach would be to solve $F = \frac{9}{5}C + 32$ for C in terms of any value of F, thereby describing Celsius temperature in terms of Fahrenheit temperature. This approach is illustrated in Example 3.

EXAMPLE 3 **Solving for a Specified Variable**

Solve the model: $F = \frac{9}{5}C + 32$ for C

Solution

$$F = \frac{9}{5}C + 32$$ This is the given model.

$$5F = 5\left(\frac{9}{5}C + 32\right)$$ Multiply both sides by 5.

$$5F = 9C + 160$$ Apply the distributive property, and simplify.

$$5F - 160 = 9C + 160 - 160$$ Subtract 160 from both sides.

$$5F - 160 = 9C$$ Simplify.

$$\frac{5F - 160}{9} = \frac{9C}{9}$$ Divide both sides by 9.

$$\frac{5F - 160}{9} = C$$ Simplify.

We can write $5F - 160$ as $5(F - 32)$, using the distributive property. Thus,

$$C = \frac{5F - 160}{9} = \frac{5(F - 32)}{9}.$$

The formula usually appears as $C = \frac{5}{9}(F - 32)$. ■

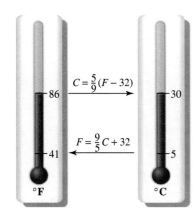

On the °F thermometer: 86, 41. On the °C thermometer: 30, 5.

$C = \frac{5}{9}(F - 32)$

$F = \frac{9}{5}C + 32$

°F   °C

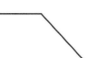

EXAMPLE 4   **Solving for a Specified Variable**

$A = \frac{1}{2}h(B + b)$ is the formula for the area of a trapezoid (Figure 1.23). Solve for B.

Solution

$$A = \frac{1}{2}h(B + b)$$ This is the given formula.

$$2A = 2\left[\frac{1}{2}h(B + b)\right]$$ Multiply both sides by 2, the LCM.

$$2A = h(B + b)$$ Simplify.

$$2A = hB + hb$$ Apply the distributive property.

$$2A - hb = hB + hb - hb$$ Isolate the term with B on the right by subtracting hb from both sides.

$$2A - hb = hB$$ Simplify.

$$\frac{2A - hb}{h} = \frac{hB}{h}$$ Solve for B by dividing both sides by h.

$$\frac{2A - hb}{h} = B$$ Simplify.

Equivalently,

$$B = \frac{2A - hb}{h}.$$ ■

In graphing an equation in two variables, x and y, it is sometimes necessary to begin by solving the equation for the variable y. Example 5 illustrates a procedure for doing this.

Figure 1.23 (caption at left):

Figure 1.23

The trapezoid's area is $A = \frac{1}{2}h(B + b)$.

Trapezoid labeled with top side b, height h, and base B.

"As far as mathematical models refer to reality, they are not certain, and as far as they are certain, they do not refer to reality."   Albert Einstein

Frances Broomfield's original paintings are available through Portal Gallery Ltd., London, England.

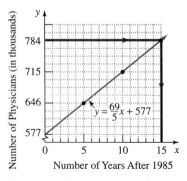

There are 784 thousand physicians predicted by the year 2000.

EXAMPLE 5   A Model Involving x and y

The model $69x - 5y = -2885$ describes the number of physicians in the United States, y (in thousands) x years after 1985.

a. Solve the model for y.
b. Use the model to predict the number of physicians in the year 2000.

Solution

a. We will isolate y on the left side.

$$69x - 5y = -2885 \qquad \text{This is the given model.}$$

$$69x - 5y - 69x = -2885 - 69x \qquad \text{Subtract } 69x \text{ from both sides of the equation.}$$

$$-5y = -2885 - 69x \qquad \text{Simplify.}$$

$$\frac{-5y}{-5} = \frac{-2885 - 69x}{-5} \qquad \text{Divide both sides of the equation by } -5.$$

$$y = \frac{-2885 - 69x}{-5} \qquad \text{Simplify.}$$

Since algebraic fractions are usually not written with negative numbers in the denominator, multiply the numerator and denominator by -1 to obtain

$$y = \frac{2885 + 69x}{5} \quad \text{or} \quad y = \frac{69}{5}x + 577.$$

b. We now use the model to predict the number of physicians in the year 2000. Since 2000 is 15 years after 1985, we substitute 15 for x and solve for y.

$$y = \frac{69}{5}x + 577 \qquad \text{This is the model from part (a).}$$

$$y = \frac{69}{5}(15) + 577 \qquad \text{Substitute 15 for } x.$$

$$y = 207 + 577 \qquad \text{Multiply: } \frac{69}{5}\overset{3}{(15)} = 207.$$

$$y = 784 \qquad \text{Add.}$$

The model predicts that there will be 784 thousand physicians in the United States by the year 2000. ■

Our next two examples use the mathematical model

$$A = P + Prt$$

which describes the amount (A) that an investment of P dollars at simple interest rate (r) will be worth in t years. Note that the variable P occurs twice in the formula.

EXAMPLE 6   Planning for a Child's Future Education

Parents of a newborn child decide that they would like to have $80,000 for their child's education in 18 years. If the interest rate is 6%, how much money should be invested?

Solution

$$A = P + Prt \qquad \text{This is the given model.}$$

$$80,000 = P + P(0.06)(18) \qquad \text{We are given that } A = 80,000, t = 18 \text{ (years)},$$
$$\text{and } r = 0.06 \quad (6\% = 0.06). \text{ We must solve}$$
$$\text{for } P.$$

$$80,000 = P + 1.08P \qquad \text{Multiply on the right.}$$

$$80,000 = 2.08P \qquad \text{Add like terms.}$$

$$\frac{80,000}{2.08} = \frac{2.08P}{2.08} \qquad \text{Divide both sides by 2.08.}$$

$$38,461.54 \approx P \qquad \text{The value for } P \text{ is rounded to the nearest cent.}$$

Substitution into the model tells us that approximately $38,461.54 should be invested now to produce $80,000 in 18 years. ■

EXAMPLE 7 **Solving for a Specified Variable that Occurs Twice**

Solve: $A = P + Prt$ for P

Solution

Because all terms with P already occur on the right side of the equation, we must use the distributive property to convert the two occurrences of P into one.

$$A = P + Prt \qquad \text{This is the given model.}$$

$$A = P(1 + rt) \qquad \text{Use the distributive property. We now have only a single occur-}$$
$$\text{rence of } P. \text{ This is called } \textit{factoring out } P.$$

$$\frac{A}{1 + rt} = \frac{P(1 + rt)}{1 + rt} \qquad \text{To isolate } P, \text{ divide each side by } 1 + rt.$$

$$\frac{A}{1 + rt} = P \qquad \text{Simplify.}$$

Equivalently,

$$P = \frac{A}{1 + rt}.$$ ■

In developing formulas that model data, mathematicians strive for both accuracy and simplicity. For example, the model that we encountered describing life expectancy for U.S. women, $E = 0.215t + 71.05$, is relatively simple to use but as we can see from Table 1.9, it is not an entirely accurate description of the data. Furthermore, we may not want to project the model past the year 2000 since unforseen progress in conquering breast cancer and other diseases that affect women could have an impact on the formula's predictions. Mathematical models may have certain restrictions, as discussed in our final example.

You cannot solve $A = P + Prt$ for P by subtracting Prt from both sides and writing

$$A - Prt = P.$$

When a formula is solved for a specified variable, that variable must be isolated on one side. The variable P occurs on both sides of

$$A - Prt = P.$$

TABLE 1.9
Life Expectancy for U.S. Women

Birth Year	Actual Value	Value Predicted by $E = 0.215t + 71.05$
1950	71.1	71.05
1960	73.1	73.2
1970	74.7	75.35
1980	77.4	77.5
1990	78.8	79.65
2000	80.0	81.8

EXAMPLE 8 **A Problem Open for Discussion: Limitations of Mathematical Models**

The model $R = -\frac{35}{2}L + 195$ describes the respiratory rate, R, for sheep whose wool length is L centimeters. Discuss possible limitations of this model.

Solution

We'll list a few of the formula's limitations. You may be able to think of others.

1. As the formula is presented, it appears that as wool length, L, increases, then respiratory rate, R, decreases. Is this true up to a certain point? What happens if wool length is too long? This information is not given with the model.
2. Does the model only apply to certain intervals of wool length?
3. Does the model apply to *every* sheep, or do other factors influence respiratory rate? Does $R = -\frac{35}{2}L + 195$, or is R only approximately equal to $-\frac{35}{2}L + 195$ $(R \approx -\frac{35}{2}L + 195)$?
4. How is R measured?
5. The model tells us nothing about at what point the respiration of the sheep becomes too high, or too low, to support life.
6. Who, if anybody, uses this formula? How is it used?

What else can you think of?

In short, when given a mathematical model, be aware of what information is conveyed by the formula but also consider what the formula does *not* convey. Does the model still apply as the variables become very large or very small?

PROBLEM SET 1.7

Practice and Application Problems

1. The model $D = 0.2F - 1$ describes age-adjusted death rate from breast cancer per 100,000 women (D) and daily fat intake (F) in grams. The death rate of American women from breast cancer is 19 women per 100,000. What is the daily fat intake for women in America?

2. The model $p = -0.5d + 100$ describes the percentage of lost hikers found in search and rescue missions (p) when members of the search team walk parallel to one another at a fixed distance (d, in yards) between searchers in the area of the search and rescue operation. If a search and rescue team finds 70% of lost hikers, what is the parallel distance of separation between members of the search party?

3. The formula $d = 5000c - 525,000$ models the relationship between the annual number of deaths (d) in the United States from heart disease and the average adult cholesterol level (c) in milligrams per deciliter of blood. In 1990, 500,000 Americans died from heart disease. What was the average cholesterol level at that time? If the United States could reduce its average cholesterol level to 180, how many lives can be saved compared to 1990?

4. The model $C = 1.44t + 280$ describes carbon dioxide concentration (C, in parts per million) t years after 1939. The preindustrial carbon dioxide concentration of 280 parts per million remained fairly constant until

World War II. When will the concentration be double the preindustrial level?

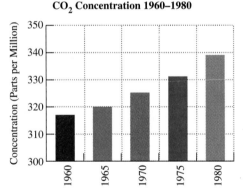

CO₂ Concentration 1960–1980

Data Source: *World Resources 1992–1993*
A Report by the World Resources Institute
Dr. Allen L. Hammond, Ed
Oxford University Press, 1992

5. Forestry officials use a mathematical model to determine the damage potential to a particular area in case of a fire. The formula is $D = 2A + V$, where $D =$ damage potential, $A =$ average age in years of brush in the area, and $V =$ value class of the area (the higher the number, the greater the value). Rank the

following regions in order of average brush age in the area:

	Damage Potential	Value Class
Region 1	8	6
Region 2	23	3
Region 3	17	7

6. The weight (W) of a car is usually nearly evenly distributed over the area (A) of contact of each of the four wheels with the ground. If P is the tire pressure in pounds/square inch, then

$$\frac{W}{4A} = P.$$

If the tire pressure in each of the four tires is 28 pounds/square inch and the area of contact for each tire is 24 square inches, what is the weight of the car?

7. The three temperature scales in common use are Celsius (C), Fahrenheit (F), and Kelvin (K). Conversions can be carried out using the models $F = \frac{9}{5}C + 32$ and $K = C + 273$. A chemistry textbook describes an experiment in which a reaction takes place at a temperature of 400 kelvin. At what Fahrenheit temperature (to the nearest degree) will the reaction take place?

8. The Internal Revenue Service approves of linear depreciation as one of several methods for depreciating business property. If the original cost of the property is C dollars and it is depreciated linearly (steadily) for N years, its value V at the end of n years is described by

$$V = C\left(1 - \frac{n}{N}\right).$$

Equipment having an original cost of $10,000 is depreciated linearly over 20 years. At the end of how many years will its value be $6500?

9. An arithmetic progression is one in which each term is obtained from the previous term by adding some positive or negative number. For example, 1, 3, 5, 7, 9, 11 is an arithmetic progression because each term is obtained from the previous number by adding 2. The first term in the progression, designated by a, is 1 ($a = 1$). The number of terms, n, is 6 ($n = 6$). The difference

between terms, d, is 2 ($d = 2$). The last term, L, which we can see is 11, can be found by the formula $L = a + (n - 1)d$. In this case,

$$L = 1 + (6 - 1) \cdot 2 = 1 + 5 \cdot 2 = 1 + 10 = 11.$$

   a. Use the formula $L = a + (n - 1)d$ to find the 23rd term in the arithmetic progression 1, 3, 5, 7, 9, 11, ...
   b. Solve the equation $L = a + (n - 1)d$ for n.

10. The value (V) of an investment (P) after 1 year at interest rate (r) is $V = P(1 + r)$. Solve the equation for r.

11. The formula for the sales price (P) of an article with cost (C) and markup (M) is $P = C + MC$. Solve the equation for M.

12. The formula $HL = KA(t_2 - t_1)$ is used in physics to describe heat flow. Solve the equation for t_1.

13. The model $7x - 3y = -715$ describes the U.S. population (y, in millions) x years after 1985.
   a. Solve the model for y.
   b. Use the model to find the U.S. population in 1992.

14. The model $4.98x - y = 41.34$ describes the percentage (y) of the American adult population with x years of education doing volunteer work.
   a. Solve the model for y.
   b. If 48.3% of the population does volunteer work, how many years of education does this group have?

15. The mathematical model

$$N = \frac{20Ld}{600 + s^2}$$

is used to estimate the number (N) of cars on the road if they are traveling at safe distances from one another, where

   L = the number of lanes of road

   d = the length of the road (in feet)

   s = the average speed of the cars
       (in miles per hour)

Use the model to find the length of a portion of a four-lane highway on which 845 cars are moving at an average speed of 20 miles per hour.

In Problems 16–21, solve each equation for y.

16. $2x - 3y = 9$

17. $3(x - 2) + 3y = 9x$

18. $\frac{1}{2}x + 2y = 8$

19. $\frac{1}{3}x - 2y = 5$

20. $x(y - 4) = 2y + 8$

21. $x(y + 3) = 4y + 2$

In Problems 22–39, solve each equation for the specified variable.

22. $ax - bx = 13$ for x

23. $7y - ay = 5$ for y

24. $ax + 13 = bx - 12$ for x

25. $xr - 11 = yr + 15$ for r

26. $B = \frac{1}{7}ac$ for a

27. $A = \frac{c}{3}(d + b)$ for b

28. $a = \frac{d}{7}(B + x)$ for B

29. $\frac{3}{4}(x + y) = 2(x - z)$ for x

30. $\frac{2}{7}(c + d) = 4(c - a)$ for d

31. $R = \frac{1}{3}a(x + c)$ for x

32. $A = \frac{1}{7}B(C + D)$ for D

33. $\frac{3ab}{c} = 7$ for a

34. $\frac{4xy}{z} = -5$ for x

35. $\frac{1}{2}(a - bx) + b = \frac{1}{2}(2b - x)$ for x

36. $\frac{1}{3}(c - dx) + 2d = \frac{1}{6}(6d - x)$ for x

37. $S = \frac{n}{2}(a + 1)$ for n

38. $\frac{acx}{3} + \frac{4c}{5} = \frac{2acx}{3}$ for x

39. $\frac{bcx}{7} + \frac{3b}{2} = \frac{4bcx}{14}$ for x

True–False Critical Thinking Problems

40. Which one of the following is true?
  a. It is not necessary to know how to solve linear equations to solve $F = \frac{9}{5}C + 32$ for C in terms of F.
  b. If a fitness club charges $80 per year plus $2.50 for each hour the facility is used, the yearly cost (y) of using this club for x hours can be described with complete accuracy by the model $y = 80 + 2.5x$.
  c. If a mathematical model gives a fairly accurate description of a situation through the year 2000, it au-
tomatically can be applied with accuracy to the entire twenty-first century.
  d. None of the above is true.

41. Which one of the following is true?
  a. Solving $S = P + Prt$ for P, the result is $P = S - Prt$.
  b. If we solve $I = Prt$ for t, we obtain $t = I - Pr$.
  c. In solving $S = 2LW + 2LH + 2WH$ for H, it is not necessary to use the distributive property.
  d. None of the above is true.

Technology Problems

Use a calculator to solve Problems 42–48.

42. The percentage share of the national radio audience for FM radio stations in any year since 1972 is described by the mathematical model $S = 3.514\,268Y - 226.2953$, where S is a percentage and Y is the last two digits in the years. (For example, for the year 1978, $y = 78$.) In what year was the percentage share of the national radio audience for FM radio 93.5%?

43. The mathematical model $y = -0.358\,709x + 256.835$ is used to project the world record for the mile run (y, in seconds), where x is the last two digits of any year. In what year was the world record for the mile run 224.91 seconds?

Archaeologists, paleontologists, and forensic scientists use the lengths of human bones to estimate the heights of individuals. A person's height h is determined from the length of the femur f (the bone from the knee to the hip socket) by two mathematical models.

For women: $h = 61.412 + 2.317f$
For men: $h = 69.089 + 2.238f$

All measurements are in centimeters.

44. What is the length of the femur for a woman who is 160 centimeters tall?

45. A partial skeleton is found of a man in which the femur is 50 centimeters long. Near the skeleton, footprints are found that indicate a man is 180 centimeters tall. Could this be a match?

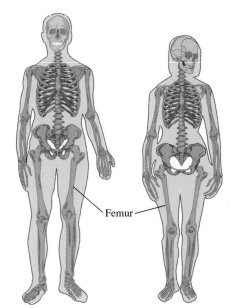
Femur

46. Chamberlain's model describes the number of years (N) that you should drive your car before purchasing a new one. The formula states

$$N = \frac{GMC}{(G - M)DP}$$

where G = new car's gas mileage, M = your present car's gas mileage, C = the cost of the new car, D = number of miles driven yearly, and P = price of gasoline/gallon. What should you pay for a new car if it gets 24 mpg, your old car has a gas mileage of 12 mpg, you drive 12,000 miles/year, gasoline costs \$1.40/gallon, and you have owned your present car for 5 years? (Round your answer to the nearest \$10.)

47. The model

$$u = \frac{fk(k + 1)}{n(n + 1)}$$

describes the money saved (u) when a loan subject to a finance charge (f) over a period of n payments is paid off k payments ahead of schedule. A loan that was scheduled to run 36 payments was paid off 12 payments ahead of schedule. If the money saved (in terms of the finance charge that need not be paid) came to \$42.16, what was the finance charge on the original loan?

48. If s is the speed of sound in a medium, v your velocity toward the source of the sound, and f the frequency of the sound emitted by the source, you will hear the frequency f', where

$$f' = f\left(1 + \frac{v}{s}\right)$$

This effect is known as the Doppler effect. The frequency of the C above middle C on the piano is about 512 cycles/second; C sharp, the next note on the piano, has a frequency of 542 cycles/second. If the speed of sound is about 760 mph, how fast must you travel toward a sound source to hear a C as a C sharp?

49. The model $y = -0.00949x + 0.479$ estimates the percentage of American smokers y (as a decimal) aged 18 to 25 who use cigarettes x years after 1974.

a. Use a graphing utility to graph the model with the following range setting:

Xmin = 0, Xmax = 40, Xscl = 5, Ymin = 0, Ymax = 0.5, Yscl = 0.05

b. What does the shape of the graph indicate about the percentage of Americans in this age group who continued to use cigarettes after 1974?

c. In what year will only 15% ($y = 0.15$) of this age group use cigarettes? Find the solution algebraically and then trace along the graph to verify your solution.

d. Extend the range setting for x to estimate the year in which no people in this age group will smoke. Do you think that the given formula accurately models reality over this extended period of time? Explain.

Andrea Badami "The Boss and His Wife" © 1973, oil on canvas, 32 × 34$\frac{1}{8}$ in. Chuck and Jan Rosenak. Photograph courtesy Museum of American Folk Art.

Writing in Mathematics

50. We discussed mathematical models in this section after we considered procedures for solving linear equations. Doesn't working with mathematical models simply mean substituting given numbers into a formula and using the order of operations? Is it really necessary to know how to solve equations to work with mathematical models? Explain.

Critical Thinking Problems

51. A local bank charges $8/month plus 5¢/check. The credit union charges $2/month plus 8¢/check.
   a. Write models for both institutions that describe total monthly cost (y) if x checks are written.
   b. How many checks written each month will result in identical total monthly costs for the two institutions?

52. The quantity of a product that consumers are willing to purchase depends on its price, with higher prices leading to less demand for the product. The quantity of a product that a supplier is willing to make available also depends on price, supplying more of a product at higher prices.

On a particular day, the demand-and-supply models for grapes in a city are given by

$$p = -0.2q + 4 \qquad \text{Consumer-demand model}$$

and

$$p = 0.07q + 0.76 \qquad \text{Supply model}$$

where q represents the quantity of grapes in thousands of pounds and p represents the price/pound.
   a. At what price will consumers be willing to purchase 5000 pounds of grapes?
   b. How many pounds of grapes will suppliers be willing to make available at this price?

53. Firefighters use the model $S = 0.5N + 26$ to calculate the maximum horizontal range S (in feet) of water coming from a $\frac{3}{4}$-inch nozzle with N pounds of pressure. They add 5 feet to S for every $\frac{1}{8}$-inch increase in nozzle diameter over $\frac{3}{4}$ of an inch. If the maximum horizontal range of a stream of water coming from a $1\frac{3}{8}$-inch diameter nozzle is 94 feet, what is the nozzle pressure?

Group Activity Problems

Problems 54–55 are appropriate for small-group discussion.

54. The world record for the mile run has decreased with surprising regularity since 1954, as shown by the listings in the table. The mathematical model $y = -0.358\,709x + 256.835$ describes the projected time for the mile run (y, in seconds) x years after 1900. Discuss possible limitations of the model.

55. The mathematical models

$$C = \frac{DA}{A + 12} \quad \text{and} \quad C = \frac{DA + D}{24}$$

relate the proper dose of medication (C) for a child whose age is A when the appropriate adult dosage is C. Discuss possible limitations of both models. Does one formula specify a larger dose of medication than the other? Is there an age when a child's dose is the same as an adult's dose? What other factors might physicians use to determine the appropriate dosage of medication for a child?

World Record for the Mile Run

Name (Country)	Year	Time
Roger Bannister (Great Britain)	1954	3:59.4
John Landy (Australia)	1954	3:58
Derek Ibbotson (Great Britain)	1957	3:57.2
Herb Elliott (Australia)	1958	3:54.5
Peter Snell (New Zealand)	1962	3:54.4
Peter Snell (New Zealand)	1964	3:54.1
Michael Jazy (France)	1965	3:53.6
Jim Ryun (United States)	1966	3:51.3
Jim Ryun (United States)	1967	3:51.1
Filbert Bayi (Tanzania)	1975	3:50
John Walker (New Zealand)	1975	3:49.4
Sebastian Coe (Great Britain)	1979	3:49.1
Steve Ovett (Great Britain)	1980	3:48.8
Sebastian Coe (Great Britain)	1981	3:48.53
Steve Ovett (Great Britain)	1981	3:48.4
Sebastian Coe (Great Britain)	1981	3:47.33
Steve Cram (Great Britain)	1985	3:46.31

Review Problems

56. Use set-builder notation and inequality symbols to rewrite the following: x is at least 7.

57. Solve for x: $\frac{2}{5}(x - 3) - 4 = \frac{1}{3}x$.

58. Use the roster method to write the following set:

$\{x \mid x$ is a natural number less than 17 that is divisible by 3$\}$.

S E C T I O N   1 . 8

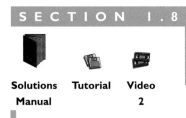

Solutions Manual	**Tutorial**	**Video 2**

▌ Solve problems using linear equations.

Strategies for Solving Problems

Objectives

1 Solve problems using linear equations.
2 Solve problems using critical thinking strategies.

Problem solving is the central theme of algebra. The problems, of course, are presented in English. We must *translate* from the ordinary language of English into the language of algebraic equations. To translate, however, we must understand the English prose and be familiar with the forms of algebraic language. Here are some general steps we will follow in solving word problems.

THE FAR SIDE   By GARY LARSON

Hell's library

> **Strategy for solving word problems**
>
> **Step 1.** Read the problem and determine the quantities that are involved. Let x (or any variable) represent one of the quantities in the problem.
> **Step 2.** If necessary, write expressions for any other unknown quantities in the problem in terms of x.
> **Step 3.** Write an equation in x that describes the verbal conditions of the problem.
> **Step 4.** Solve the equation written in step 3 and answer the problem's question.
> **Step 5.** Check the solution *in the original wording* of the problem, not in the equation obtained from the words.

Take great care with step 1. Reading mathematics is not the same as reading a newspaper. Reading the problem involves slowly working your way through its parts, making notes on what is given, and perhaps rereading the problem a few times. Only at this point should you let x represent one of the quantities.

The most difficult step in this process is step 3 because it involves translating verbal conditions into an algebraic equation. In some situations, the conditions are given explicitly. In other instances, the conditions are only implied, making it necessary to use one's knowledge about the type of word problem to generate an English sentence that must then be translated into an equation.

Translations of some commonly used English phrases are listed in Table 1.10.

Discover for yourself

Cover the right column in Table 1.10 with a sheet of paper and attempt to formulate the algebraic expression in the column on your own. Then slide the paper down and check your answer. Work through the entire table in this manner.

TABLE 1.10 Algebraic Translations of English Phrases

English Phrase	Algebraic Expression
Addition	
The sum of a number and 7	$x + 7$
Five more than a number	$x + 5$
A number increased by 6	$x + 6$
Subtraction	
A number minus 4	$x - 4$
A number decreased by 5	$x - 5$
A number subtracted from 8	$8 - x$
The difference between a number and 6	$x - 6$
The difference between 6 and a number	$6 - x$
Seven less than a number	$x - 7$
Seven minus a number	$7 - x$
Nine fewer than a number	$x - 9$
Multiplication	
Five times a number	$5x$
The product of 3 and a number	$3x$
Two-thirds of a number (used with fractions)	$\frac{2}{3}x$
Seventy-five percent of a number (used with decimals)	$0.75x$
Thirteen multiplied by a number	$13x$
A number multiplied by 13	$13x$
Twice a number	$2x$
Division	
A number divided by 3	$\dfrac{x}{3}$
The quotient of 7 and a number	$\dfrac{7}{x}$
The quotient of a number and 7	$\dfrac{x}{7}$
The reciprocal of a number	$\dfrac{1}{x}$
More than one operation	
The sum of twice a number and 7	$2x + 7$
Twice the sum of a number and 7	$2(x + 7)$
Three times the sum of 1 and twice a number	$3(1 + 2x)$
Nine subtracted from 8 times a number	$8x - 9$
Twenty-five percent of the sum of 3 times a number and 14	$0.25(3x + 14)$
Seven times a number increased by 24	$7x + 24$
Seven times the sum of a number and 24	$7(x + 24)$

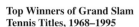

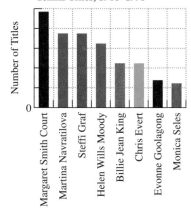

Figure 1.24

Source: The National Tennis Association

Solving Problems by Translating Given Conditions into an Equation

These types of problems contain explicit verbal conditions that can be translated into algebraic equations. We use our five-step strategy to solve these problems.

EXAMPLE 1 **Winners of Tennis Titles, 1968–1995**

The graph in Figure 1.24 shows the top winners of grand slam tennis titles among women from 1968 through 1995. Unfortunately, there are no numbers along the vertical axis, so it is impossible to tell how many titles are held by the eight athletes. However, this much is known about the number of titles won by Monica Seles: When 9 is subtracted from 8 times the number of titles she has won, the result is 3 times the sum of 1 and twice the number of titles. Find the number of tennis titles won by Monica Seles from 1968 through 1995.

Solution

Step 1. Let x represent the number. (Omit Step 2; there is only one unknown.)

Step 3. Write an equation that describes the verbal conditions.

Let $x =$ the number of tennis titles won by Monica Seles.

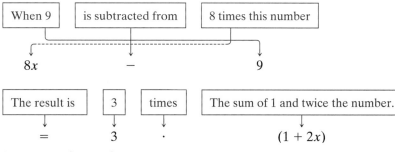

The result is	3	times	The sum of 1 and twice the number.

$$= \qquad 3 \qquad \cdot \qquad (1 + 2x)$$

Step 4. Solve the equation and answer the problem's question.

$8x - 9 = 3(1 + 2x)$ This is the algebraic equation for the given conditions.

$8x - 9 = 3 + 6x$ Apply the distributive property on the right.

$2x - 9 = 3$ Subtract $6x$ from both sides.

$2x = 12$ Add 9 to both sides.

$x = 6$ Divide both sides by 2.

The number of tennis titles won by Monica Seles from 1968 through 1995 is 6.

Step 5. Check the solution in the original wording of the problem.

Check

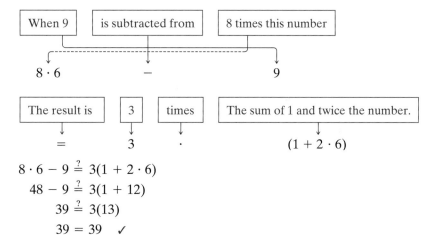

$8 \cdot 6 - 9 \stackrel{?}{=} 3(1 + 2 \cdot 6)$

$48 - 9 \stackrel{?}{=} 3(1 + 12)$

$39 \stackrel{?}{=} 3(13)$

$39 = 39$ ✓

Solving Problems by Creating Verbal Models and Then Translating into an Equation

In Example 1, the conditions necessary for writing an equation were clearly given. A more difficult situation is one in which the conditions are only implied. The first step is to write an English sentence that clearly identifies the operations involved. This sentence should contain a word such as *is* or *equals*. Such a sentence serves as a *verbal model* that is then translated into an equation.

EXAMPLE 2 **Modeling a Price Reduction**

The price of a dress is reduced by 40%. When the dress still does not sell, it is reduced by 40% of the reduced price. If the price of the dress after both reductions is $72, what was the original price?

Solution

Steps 1 and 2. Represent unknowns in terms of *x*.

Let x = the original price of the dress. Implied within this problem is the following statement.

Step 3. Write an equation that describes the problem's conditions.

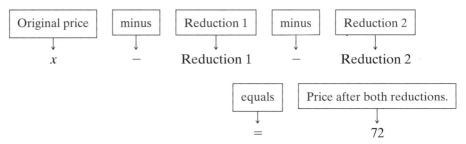

We now need algebraic expressions for reduction 1 and reduction 2.

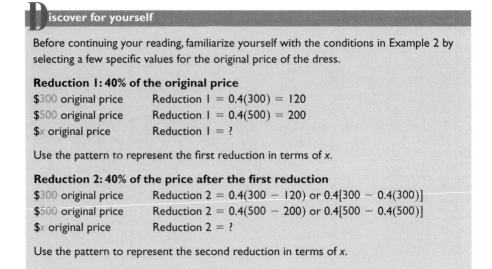

iscover for yourself

Before continuing your reading, familiarize yourself with the conditions in Example 2 by selecting a few specific values for the original price of the dress.

Reduction 1: 40% of the original price
$300 original price   Reduction 1 = 0.4(300) = 120
$500 original price   Reduction 1 = 0.4(500) = 200
$x original price   Reduction 1 = ?

Use the pattern to represent the first reduction in terms of x.

Reduction 2: 40% of the price after the first reduction
$300 original price   Reduction 2 = 0.4(300 − 120) or 0.4[300 − 0.4(300)]
$500 original price   Reduction 2 = 0.4(500 − 200) or 0.4[500 − 0.4(500)]
$x original price   Reduction 2 = ?

Use the pattern to represent the second reduction in terms of x.

In the Discover for Yourself, were you able to obtain the following algebraic expressions for the two reductions?

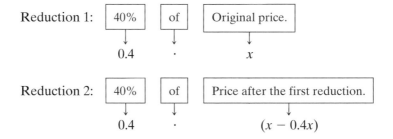

The expression for reduction 2 is a bit tricky. The price after the first reduction is the original price (x) minus 40% of the original price ($x - 0.4x$), and the second reduction is 40% of the price after the first reduction.

We now use the fact that the price after the two reductions is $72.

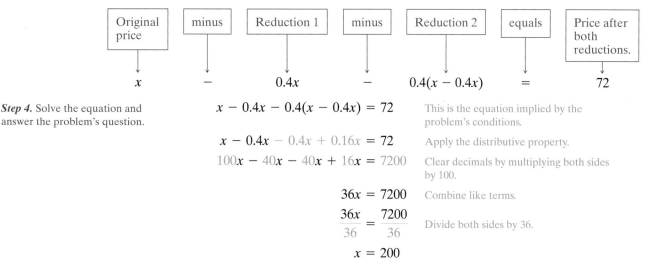

Step 4. Solve the equation and answer the problem's question.

$$x - 0.4x - 0.4(x - 0.4x) = 72 \qquad \text{This is the equation implied by the problem's conditions.}$$

$$x - 0.4x - 0.4x + 0.16x = 72 \qquad \text{Apply the distributive property.}$$

$$100x - 40x - 40x + 16x = 7200 \qquad \text{Clear decimals by multiplying both sides by 100.}$$

$$36x = 7200 \qquad \text{Combine like terms.}$$

$$\frac{36x}{36} = \frac{7200}{36} \qquad \text{Divide both sides by 36.}$$

$$x = 200$$

The original price was $200.00.

Step 5. Check the solution in the original wording of the problem.

Check

Reduction 1: 40% of 200 = (0.4)(200) = 80
Price after reduction 1: $200 − $80 = $120
Reduction 2: 40% of price after first reduction = (0.4)(120) = 48
Price after both reductions: $200 − $80 − $48 = $120 − $48 = $72

This checks with the conditions of the problem. ■

EXAMPLE 3 **Constructing Models to Solve a Problem**

The costs for two different kinds of heating systems for a three-bedroom home are given below.

System	Cost to Install	Operating Cost/Year
Solar	$29,700	$150
Electric	$5000	$1100

After how many years will total costs for solar heating and electric heating be the same? What will be the cost at that time?

Solution

Steps 1 and 2. Represent unknowns in terms of x.

Let x = the number of years where total costs for the two systems will be the same.

ENRICHMENT ESSAY

Innumeracy

The inability to understand numbers and their meanings is called *innumeracy* by mathematics professor John Allen Paulos. Paulos has written a book about mathematical illiteracy. Titled *Innumeracy,* it seeks to explain why so many people are numerically inept and to show how the problem can be corrected. Paulos writes, "The topic of percentage is continually being misapplied. A dress whose price has been 'slashed' 40 percent and then another 40 percent has been reduced by 64 percent, not 80 percent." (p. 122) We can verify this by our work in Example 2.

$$\begin{aligned}
\text{Total reduction} &= \text{Reduction 1} + \text{Reduction 2}\\
&= 0.4x + 0.4(x - 0.4x)\\
&= 0.4x + 0.4x - 0.16x\\
&= 0.8x - 0.16x\\
&= 0.64x
\end{aligned}$$

The total reduction is $0.64x$, or 64%, of the original price, not 80%.

Jasper Johns *0 Through 9,* 1961. Oil on canvas, 137 × 105 cm.
The Saatchi Collection, courtesy of the Leo Castelli Gallery.
© 1998 Jasper Johns/Licensed by VAGA, New York.

Step 3. Write an equation that describes the problem's conditions.

We begin by constructing models for the total costs for the two systems.

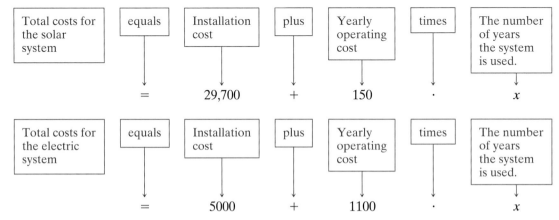

Our total cost models are

$$\text{Solar} = 29{,}700 + 150x$$
$$\text{Electric} = 5000 + 1100x$$

Now that we have modeled the conditions, we can answer the problem's question: After how many years will

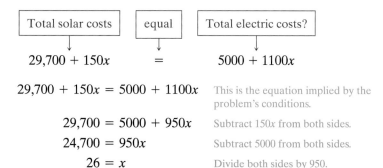

Total solar costs	equal	Total electric costs?
↓	↓	↓
$29,700 + 150x$	$=$	$5000 + 1100x$

Step 4. Solve the equation and answer the problem's question.

$$29,700 + 150x = 5000 + 1100x \qquad \text{This is the equation implied by the problem's conditions.}$$
$$29,700 = 5000 + 950x \qquad \text{Subtract } 150x \text{ from both sides.}$$
$$24,700 = 950x \qquad \text{Subtract 5000 from both sides.}$$
$$26 = x \qquad \text{Divide both sides by 950.}$$

After 26 years the total costs for the two systems will be the same.

Step 5. Check the solution in the original wording of the problem.

Check

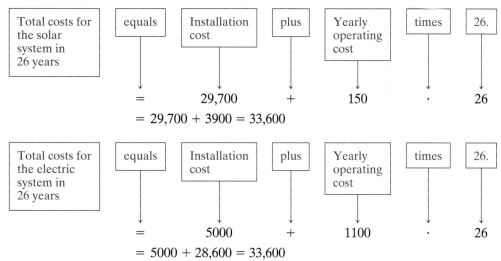

Total costs for the solar system in 26 years	equals	Installation cost	plus	Yearly operating cost	times	26.
	↓	↓	↓	↓	↓	↓
	$=$	$29,700$	$+$	150	$\cdot$	26

$$= 29,700 + 3900 = 33,600$$

Total costs for the electric system in 26 years	equals	Installation cost	plus	Yearly operating cost	times	26.
	↓	↓	↓	↓	↓	↓
	$=$	5000	$+$	1100	$\cdot$	26

$$= 5000 + 28,600 = 33,600$$

After 26 years the two systems will cost the same amount, namely $33,600. ■

We can visualize the solution to Example 3 by graphing the two models in the same rectangular coordinate system. Representing total costs by y, the models are

$$y = 29,700 + 150x \text{ and } y = 5000 + 1100x.$$

Figure 1.25 shows the graphs obtained on a graphing utility using

$$\text{Xmin} = 0, \text{Xmax} = 40, \text{Xscl} = 2, \text{Ymin} = 5000,$$
$$\text{Ymax} = 50,000, \text{Yscl} = 5000$$

The graphs intersect at $(26, 33,600)$, meaning that after 26 years total costs for both systems will be $33,600.

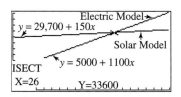

Figure 1.25

After 26 years, total costs for solar and electric systems are the same, namely $33,600.

EXAMPLE 4　Selecting a Mortgage

Table 1.11 gives the monthly mortgage payment for each $1000 of mortgage at various mortgage rates for different periods of time. For example, for a mortgage for 30 years at 8.5%, the monthly payment for principal and interest is $7.69 per thousand. A mortgage for 30 years at 7.25% results in monthly payments for principal and interest of $6.83 per thousand. Using these numbers, for a mortgage of $80,000 at 8.5%, the monthly payment is $80(7.69) = \$615.20$. For a mortgage of $80,000 at 7.25%, the monthly payment is $80(6.83) = \$546.40$.

TABLE 1.11 Any Bank, USA: Equal Monthly Payment to Amortize a Loan of $1000

Rate	Payment for a Mortgage Period (years) of:				Rate	Payment for a Mortgage Period (years) of:			
(%)	15	20	25	30	(%)	15	20	25	30
4.500	7.65	6.33	5.56	5.07	8.625	9.93	8.76	8.14	7.78
4.625	7.71	6.39	5.63	5.14	8.750	10.00	8.84	8.23	7.87
4.750	7.78	6.46	5.70	5.22	8.875	10.07	8.92	8.31	7.96
4.875	7.84	6.53	5.77	5.29	9.000	10.15	9.00	8.40	8.05
5.000	7.91	6.60	5.85	5.37	9.125	10.22	9.08	8.48	8.14
5.125	7.97	6.67	5.92	5.44	9.250	10.30	9.16	8.57	8.23
5.250	8.04	6.73	6.00	5.52	9.375	10.37	9.24	8.66	8.32
5.375	8.10	6.81	6.07	5.60	9.500	10.45	9.33	8.74	8.41
5.500	8.17	6.88	6.14	5.68	9.625	10.52	9.41	8.83	8.50
5.625	8.24	6.95	6.22	5.76	9.750	10.60	9.49	8.92	8.60
5.750	8.30	7.02	6.29	5.84	9.875	10.67	9.57	9.00	8.69
5.875	8.37	7.09	6.37	5.92	10.000	10.75	9.66	9.09	8.78
6.000	8.44	7.16	6.44	6.00	10.125	10.83	9.74	9.18	8.87
6.125	8.51	7.24	6.52	6.08	10.250	10.90	9.82	9.27	8.97
6.250	8.57	7.31	6.60	6.16	10.375	10.98	9.90	9.36	9.06
6.375	8.64	7.38	6.67	6.24	10.500	11.06	9.99	9.45	9.15
6.500	8.71	7.46	6.75	6.32	10.625	11.14	10.07	9.54	9.25
6.625	8.78	7.53	6.83	6.40	10.750	11.21	10.16	9.63	9.34
6.750	8.85	7.60	6.91	6.49	10.875	11.29	10.24	9.72	9.43
6.875	8.92	7.68	6.99	6.57	11.000	11.37	10.33	9.81	9.53
7.000	8.99	7.76	7.07	6.66	11.125	11.45	10.41	9.90	9.62
7.125	9.06	7.83	7.15	6.74	11.250	11.53	10.50	9.99	9.72
7.250	9.13	7.91	7.23	6.83	11.375	11.61	10.58	10.08	9.81
7.375	9.20	7.98	7.31	6.91	11.500	11.69	10.67	10.17	9.91
7.500	9.28	8.06	7.39	7.00	11.625	11.77	10.76	10.26	10.00
7.625	9.35	8.14	7.48	7.08	11.750	11.85	10.84	10.35	10.10
7.750	9.42	8.21	7.56	7.17	11.875	11.93	10.93	10.44	10.20
7.875	9.49	8.29	7.64	7.26	12.000	12.01	11.02	10.54	10.29
8.000	9.56	8.37	7.72	7.34	12.125	12.09	11.10	10.63	10.39
8.125	9.63	8.45	7.81	7.43	12.250	12.17	11.19	10.72	10.48
8.250	9.71	8.53	7.89	7.52	12.375	12.25	11.28	10.82	10.58
8.375	9.78	8.60	7.97	7.61	12.500	12.33	11.37	10.91	10.68
8.500	9.85	8.68	8.06	7.69					

A person purchasing a home is considering two options for an $80,000 mortgage:

Option 1:   8.5% interest for a 30-year loan, 1 point (a point is a one-time charge of 1% of the mortgage amount), and a $300 loan application fee

Option 2:   7.25% interest for a 30-year loan, 5 points, and a $1500 loan application fee.

Approximately how many months will it take for the total cost of each mortgage option to be the same?

Solution

Steps 1 and 2. Represent unknowns in terms of x.

Let x = the number of months when the total costs for the two mortgage options will be the same.

As we did in the previous example, we construct models for the total costs for each mortgage option over x months, setting the models equal to each other.

Step 3. Write an equation that describes the problem's conditions.

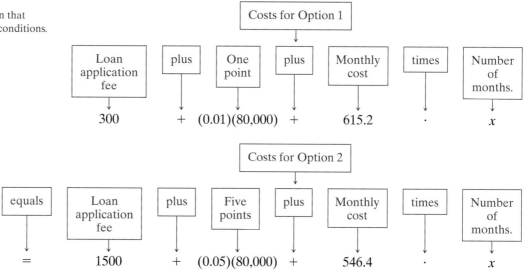

Step 4. Solve the equation and answer the problem's question.

$$300 + (0.01)(80{,}000) + 615.2x = 1500 + (0.05)(80{,}000) + 546.4x$$

     This is the equation implied by the given conditions.

$$1100 + 615.2x = 5500 + 546.4x$$

     Multiply and add as indicated.

$$1100 + 68.8x = 5500$$

     Subtract 546.4x from both sides.

$$68.8x = 4400$$

     Subtract 1100 from both sides.

$$x = \frac{4400}{68.8}$$

     Divide both sides by 68.8.

$$x \approx 64$$

Step 5. Check the solution in the original wording of the problem.

After approximately 64 months, or $\frac{64}{12} = 5\frac{1}{3}$ years, the total costs of each mortgage option will be the same. Notice that option 1 involves less money up front but greater monthly payments. Prior to 64 months, option 1 will result in lower total costs. After that time, option 2 makes more sense. If the person choosing between the options knows how long the house will be kept, this knowledge would result in selecting the loan with lower total costs for that situation. ■

Using technology

Use a graphing utility to visualize the solution to Example 4. Graph $y_1 = 1100 + 615.2x$ and $y_2 = 5500 + 546.4x$ (the simplified models) in the same viewing rectangle. Use a setting such as

Xmin = 0, Xmax = 70, Xscl = 1, Ymin = 0, Ymax = 45,000, Yscl = 300.

Then use the trace of intersection feature to find the point of intersection of the two lines. The *x*-coordinate of the intersection point should be approximately 64. What is the *y*-coordinate and what does this represent?

Solving Problems by Using Known Formulas

Some word problems often do not call for the construction of models, but rather that we bring a knowledge of certain formulas to the problem to obtain a solution. This is particularly true of geometric word problems. A knowledge of the formulas in Table 1.12 should be helpful in solving problems involving perimeter, area, and volume.

TABLE 1.12 Common Formulas for Area, Perimeter, and Volume

Square	Rectangle	Circle	Triangle	Trapezoid
$A = s^2$ $P = 4s$	$A = lw$ $P = 2l + 2w$	$A = \pi r^2$ $C = 2\pi r$	$A = \frac{1}{2}bh$	$A = \frac{1}{2}h(a + b)$

Cube	Rectangular Solid	Circular Cylinder	Sphere	Cone
$V = s^3$	$V = lwh$	$V = \pi r^2 h$	$V = \frac{4}{3}\pi r^3$	$V = \frac{1}{3}\pi r^2 h$

Vincent van Gogh (Dutch, 1853–1890). Irises, 1889, oil on canvas, 90.PA.20/The J. Paul Getty Museum, Los Angeles, California.

Steps 1 and 2. Represent unknowns in terms of x.

EXAMPLE 5 **The Price of Art**

The Van Gogh painting *Irises* has a length that is 24 inches shorter than twice the width and a perimeter of 120 inches. In 1987, the painting was sold for $53,900,000! Although it's difficult to place a price on art, in this case we can. Specifically, what did the purchaser pay for each square inch of the painting?

Solution

We need to find the painting's dimensions. Let x = the width.

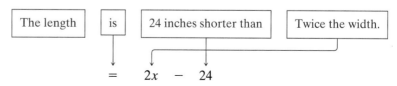

The length	is	24 inches shorter than	Twice the width.

$$= \quad 2x \quad - \quad 24$$

Thus, $2x - 24$ = the length.
Since $P = 2l + 2w$,

Step 3. Write an equation that describes the problem's conditions.

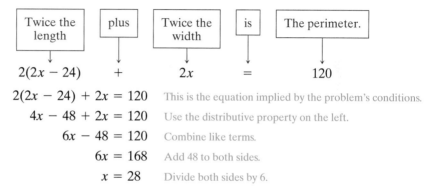

Twice the length	plus	Twice the width	is	The perimeter.

$$2(2x - 24) \quad + \quad 2x \quad = \quad 120$$

Step 4. Solve the equation and answer the problem's question.

$2(2x - 24) + 2x = 120$ This is the equation implied by the problem's conditions.
$4x - 48 + 2x = 120$ Use the distributive property on the left.
$6x - 48 = 120$ Combine like terms.
$6x = 168$ Add 48 to both sides.
$x = 28$ Divide both sides by 6.

Thus,

Width = x = 28
Length = $2x - 24 = 2(28) - 24 = 32$

The painting's dimensions are 32 inches by 28 inches. Since we are interested in its price per square inch, we now need to find the painting's area. The area of a rectangle is the product of its length and its width.

$$A = lw = (32)(28) = 896 \text{ square inches}$$

We are given that the painting was sold for $53,900,000.

$$\text{The price per square inch} = \frac{\text{the selling price}}{\text{the painting's area}}$$

$$= \frac{53,900,000}{896} = 60,156.25$$

Thus, Van Gogh's *Irises* sold for $60,156.25 per square inch.

Step 5. Check the solution in the original wording of the problem.

We can check both the painting's perimeter and its selling price.

$$\text{Perimeter} = 2l + 2w = 2(32) + 2(28) = 120 \text{ inches} \quad ✓$$

$$\text{Selling price} = (\text{price per square inch})(\text{area})$$
$$= (60,156.25)(896) = \$53,900,000 \quad ✓$$

Solving Problems by Using Tables to Organize Information

In Section 1.5, we encountered the uniform motion model $d = rt$ (distance equals rate times time). Problems involving this model can frequently be organized using a table.

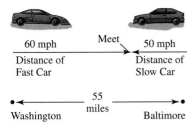

60 mph   Meet   50 mph

Distance of   Distance of
Fast Car   Slow Car

55 miles

Washington   Baltimore

EXAMPLE 6   **Solving a Uniform Motion Problem**

A car leaves Washington for Baltimore at the same time that another car leaves Baltimore for Washington, traveling the same highway. The car from Washington travels at a rate of 60 mph. The car from Baltimore travels at a rate of 50 mph. Washington and Baltimore are 55 miles apart.

a. How soon will the cars meet?
b. How far from Washington will they meet?

Solution

Steps 1 and 2. Represent unknowns in terms of x.

a. Let x = the number of hours it takes for the cars to meet. Organize the given information in a table:

	Rate, r	·	Time, t	=	Distance, d
Car from Washington	60		x		$60x$
Car from Baltimore	50		x		$50x$

Step 3. Write an equation that describes the problem's condition.

At the moment the cars meet, the sum of their distances ($60x + 50x$) is equal to the distance they were originally apart (55 miles).

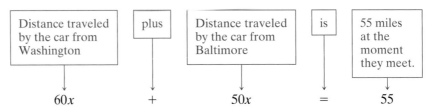

Distance traveled by the car from Washington	plus	Distance traveled by the car from Baltimore	is	55 miles at the moment they meet.
$60x$	$+$	$50x$	$=$	55

Step 4. Solve the equation and answer the problem's questions.

$60x + 50x = 55$ This is the equation implied by the problem's conditions.

$110x = 55$ Combine like terms.

$x = \dfrac{55}{110} = \dfrac{1}{2}$ Divide both sides by 110.

The cars will meet in $\frac{1}{2}$ hour.

b. We know from part (a) that each car is traveling for $\frac{1}{2}$ hour before they meet. Since the distance of the car from Washington is $60x$, we obtain $60 \cdot \frac{1}{2} = 30$ miles. Thus, the cars will meet 30 miles from Washington.

Step 5. Check the solution in the original wording of the problem.

Check

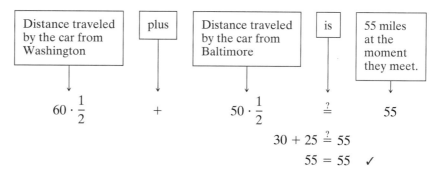

Distance traveled by the car from Washington	plus	Distance traveled by the car from Baltimore	is	55 miles at the moment they meet.

$$60 \cdot \frac{1}{2} \qquad + \qquad 50 \cdot \frac{1}{2} \qquad \overset{?}{=} \qquad 55$$

$$30 + 25 \overset{?}{=} 55$$

$$55 = 55 \quad \checkmark \qquad \blacksquare$$

Otis Kaye "A Fool and His Money". Photograph Courtesy Berry-Hill Galleries, New York.

We solved Example 6 by using a known model ($d = rt$) and by organizing information in a table. Problems involving investments and simple interest can also be solved using this strategy. Suppose $6000 is invested in an account paying 4% annual interest. After 1 year, the interest earned is $6000(0.04) = $240. If this interest is withdrawn but the original principal of $6000 is left intact, the investment will also earn $240 in interest after the second year, giving a total of $480 interest over 2 years. The withdrawing of interest after each year and earning interest on the $6000 principal is called *simple annual interest.* The mathematical model in this situation is as follows.

> **Simple annual interest**
>
> $$I = Prt$$
>
> where I = interest earned, P = principal, r = annual interest rate, and t = time of the investment in years.

Consequently, if $P = \$6000$ is deposited at $r = 4\% = 0.04$ for $t = 2$ years, then the interest earned is

$$I = Prt = \$6000(0.04)(2) = \$480.$$

EXAMPLE 7 **Solving an Investment Problem**

Suppose $8000 is invested, part in a stock paying 5% interest for 2 years and the balance in a bond paying 7% interest for 3 years. If the total interest earned is $1020, how much is invested at each rate?

Solution

Steps 1 and 2. Represent unknowns in terms of x.

Let

$$x = \text{Amount invested at } 5\%$$

$$8000 - x = \text{Amount invested at } 7\%$$

Once again we can simplify the problem by using a table.

	P	$\cdot$	r	$\cdot$	t	Interest
5% Stock	x		0.05		2	$0.05(2)x$
7% Bond	$8000 - x$		0.07		3	$0.07(3)(8000 - x)$

Step 3. Write an equation that describes the problem's conditions.

Because the total interest earned is $1020, we have

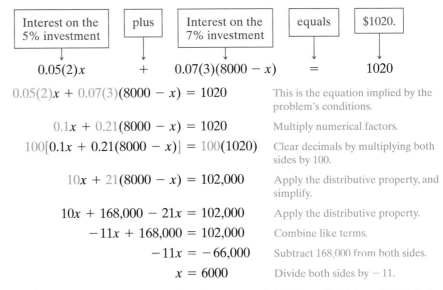

Interest on the 5% investment	plus	Interest on the 7% investment	equals	$1020.

$$0.05(2)x \quad + \quad 0.07(3)(8000 - x) \quad = \quad 1020$$

Step 4. Solve the equation and answer the problem's question.

$$0.05(2)x + 0.07(3)(8000 - x) = 1020 \qquad \text{This is the equation implied by the problem's conditions.}$$

$$0.1x + 0.21(8000 - x) = 1020 \qquad \text{Multiply numerical factors.}$$

$$100[0.1x + 0.21(8000 - x)] = 100(1020) \qquad \text{Clear decimals by multiplying both sides by 100.}$$

$$10x + 21(8000 - x) = 102{,}000 \qquad \text{Apply the distributive property, and simplify.}$$

$$10x + 168{,}000 - 21x = 102{,}000 \qquad \text{Apply the distributive property.}$$

$$-11x + 168{,}000 = 102{,}000 \qquad \text{Combine like terms.}$$

$$-11x = -66{,}000 \qquad \text{Subtract 168,000 from both sides.}$$

$$x = 6000 \qquad \text{Divide both sides by } -11.$$

Thus, $6000 is invested at 5% for 2 years, and $8000 − $6000 = $2000 is invested at 7% for 3 years.

Step 5. Check the solution in the original wording of the problem.

Check

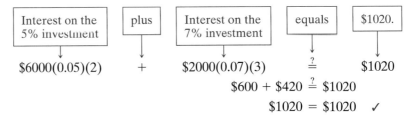

Interest on the 5% investment	plus	Interest on the 7% investment	equals	$1020.

$$\$6000(0.05)(2) \quad + \quad \$2000(0.07)(3) \quad \overset{?}{=} \quad \$1020$$

$$\$600 + \$420 \overset{?}{=} \$1020$$

$$\$1020 = \$1020 \quad \checkmark$$

2 Solve problems using critical thinking strategies.

Solving Problems Using Critical Thinking Strategies

The problems considered so far can all be solved by writing an equation. However, many mathematical problems cannot be solved by translating sentences into equations and so require different approaches. The following examples show how other strategies may be used to solve problems.

Todd Siler "Elements of Thought" (1986), mixed media maquette/Courtesy Ronald Feldman Fine Arts, New York.

EXAMPLE 8   **Solving a Problem Using Trial and Error**

In a code, each letter is assigned its numerical position in the alphabet, with $A = 1, B = 2, C = 3, \ldots , Z = 26$. An operator received a one-word message in this code but lost it, remembering only that the message had the form $x, x + 7, x + 6, x + 5$. The second letter was a vowel, and the message was an English word. What was the word?

Solution

We will use the trial-and-error (guess-and-check) strategy.

Because x represents the letter in first position (A or B or . . . or Z), $x \geq 1$. Vowels in second position can be A(1), E(5), I(9), O(15), or U(21). Using trial and error,

$$x + 7 = 1 \quad \text{or} \quad x + 7 = 5 \quad \text{or} \quad x + 7 = 9 \quad \text{or} \quad x + 7 = 15 \quad \text{or} \quad x + 7 = 21$$
$$x = -6 \qquad\qquad x = -2 \qquad\qquad x = 2 \qquad\qquad x = 8 \qquad\qquad x = 14$$
$$\text{Impossible} \qquad \text{Impossible}$$

Now consider each possibility.

1. If $x = 2$, the word x, $x + 7$, $x + 6$, $x + 5$ is 2, 9, 8, 7 = BIHG, which is not an English word.
2. If $x = 8$, the word x, $x + 7$, $x + 6$, $x + 5$ is 8, 15, 14, 13 = HONM, which is not an English word.
3. If $x = 14$, the word x, $x + 7$, $x + 6$, $x + 5$ is 14, 21, 20, 19 = NUTS, which is an English word.

The word was NUTS. ■

Example 9 is an age problem that cannot be solved with an equation. The solution of the problem is based on the concept of *prime numbers*—natural numbers greater than 1 and divisible, without a remainder, only by themselves and 1.

EXAMPLE 9 **Solving a Problem by Making a List**

A middle-age man observed that his present age was a prime number. He also noticed that the number of years in which his age would again be prime was equal to the number of years ago in which his age was prime. How old is the man?

Solution

One way to solve this problem is to make a list of possibilities for the man's present age. Since he is middle-aged and his present age is a prime number, we will list all prime numbers that are 67 or less. We list the difference between these successive prime ages on the line below:

Prime ages: 2 3 5 7 11 13 17 19 23 29 31 37 41 43 47 53 59 61 67
Difference between primes: 2 2 4 2 4 2 4 6 2 6 4 2 4 6 6 2 6

As a young boy (age 5), the man's age was prime 2 years before that and prime again 2 years after that. The situation corresponds to cases where consecutive differences are equal. This occurs again at 53. Six years ago the man's age was prime (47), and 6 years in the future it will be again prime (59). We conclude that the man is 53 years old. (The next occurrence is at 157, appropriate for an announcement during the weather segment on the "Today" show but hardly middle-aged!) ■

Example 10 is based on recognizing patterns.

EXAMPLE 10   **Solving a Problem by Looking for a Pattern**

Consider each of the following sums:

$$1 + 2 = 3$$
$$1 + 2 + 3 = 6$$
$$1 + 2 + 3 + 4 = 10$$
$$1 + 2 + 3 + 4 + 5 = 15$$
$$1 + 2 + 3 + 4 + 5 + 6 = 21$$

Write a formula that models the sum of the first n natural numbers. Then use this formula to find

$$1 + 2 + 3 + 4 + \cdots + 200.$$

study tip

Problem solving often involves combinations of strategies. In Example 10, we are both organizing information in tables and looking for patterns.

Solution

Mathematics involves the study of patterns. As you look at the sums, you probably realize that a great deal of thought may be involved to actually determine a possible emerging pattern. A bit more information, organized in a table, should be helpful.

Sum of Two Numbers (3)	Sum of Three Numbers (6)	Sum of Four Numbers (10)	Sum of Five Numbers (15)	Sum of Six Numbers (21)
$2 \cdot 3 = 6$	$3 \cdot 4 = 12$	$4 \cdot 5 = 20$	$5 \cdot 6 = 30$	$6 \cdot 7 = 42$

Discover for yourself

Use the observation about the second row to write the formula for the sum of the first n natural numbers.

The second row in the table shows that the product of the number of terms in the sum and the number of terms increased by 1 is double the actual sum.

Suppose we are adding n terms. Increasing this number by 1 gives $n + 1$. The product $n(n + 1)$ is double the actual sum. Thus, if we divide this expression by 2, we will obtain the sum. This means that the sum of the first n natural numbers is modeled by the formula

$$S = \frac{1}{2}n(n + 1) \quad \text{or} \quad S = \frac{n(n + 1)}{2}$$

in which S represents the sum. Were you able to obtain this model on your own in the Discover for Yourself?

Now we will use the model to find the sum of the first two hundred natural numbers. We substitute 200 for n in the formula:

$$S = \frac{n(n + 1)}{2} = \frac{200(201)}{2} = 100(201) = 20,100$$

This means that

$$1 + 2 + 3 + 4 + \cdots + 200 = 20,100.$$ ∎

PROBLEM SET 1.8

Practice and Application Problems

Problems 1–8 contain explicit verbal conditions that can be translated into algebraic equations. Use the five-step strategy to solve each problem.

1. When 6 is subtracted from 12 times a number, the result is 7 times the number increased by 24. Find the number.

2. When 4 is subtracted from 7 times a number, the result is 4 less than 5 times the number. Find the number.

3. AIDS spending by the federal government increased from 1990 through 1994. In 1994, AIDS spending exceeded that of the previous year by $0.52 billion.

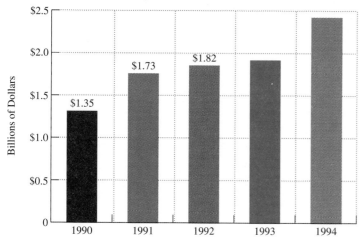

AIDS Spending by the Federal Government

Source: U.S. Department of Health and Human Services

a. Let x represent the amount (in billions) spent on AIDS by the federal government in 1993. Write an algebraic expression for the amount spent on AIDS in 1994.

b. In 1993 and 1994 combined, the federal government spent a total of $4.32 billion on AIDS. Determine the amount spent on AIDS in each individual year.

4. The number of U.S. households with dogs exceeds that with cats by 5.4 million.

a. Let x represent the number of U.S. households with cats (in millions). Write an algebraic expression for the number of households with dogs.

b. The number of households with dogs and cats combined is 63.8 million. Determine the number of U.S. households with each of these pets.

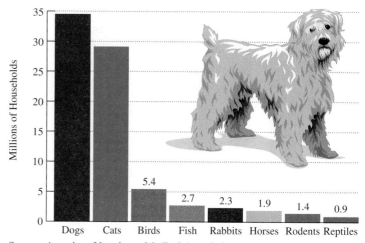

American Pet Ownership

Source: American Veterinary Medical Association

5. The graph shows the annual salaries of federal officials for 1994. The vice-president earns $23,100 more than cabinet members, and senators earn $14,800 less than cabinet members. If these salaries are combined, they exceed twice the salary of the president by $53,500. Find the annual salaries of the vice-president, cabinet members, and senators.

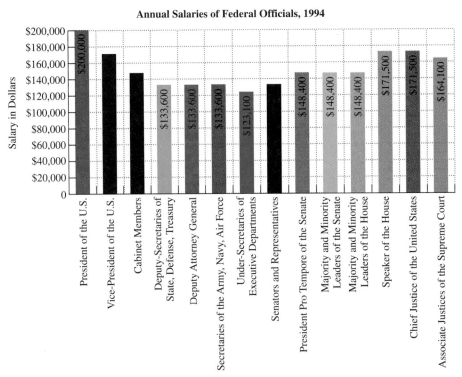

Source: Office of Personnel Management

6. The graph indicates the declining population of African elephants.

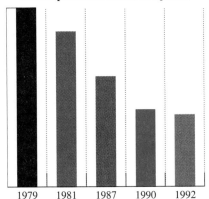

Source: World Wide Fund for Nature (WWF)

The 1987 population was 440,000 less than twice that of 1992, and the 1990 population was 10,000 more than that of 1992. If the combined population for these three years was 1,970,000, determine the population for each year. Then use the graph to determine a reasonable estimate for the African elephant population in 1979 and 1981.

7. The table at the top of page 107, indicates that annual salaries for federal officials in 1994 (Problem 5) cannot compare with those of people in the entertainment industry. Notice that three of the salaries in the table are missing: Spielberg's, Cosby's and Pink Floyd's. However, it is known that Spielberg's salary exceeded twice that of Pink Floyd's by $53 million, and Cosby earned $22 million less than Pink Floyd. The combined gross annual income for Spielberg, Pink Floyd, and Cosby was 10.2 times that of Barney the Dinosaur. Use this information to fill in the missing incomes in the table.

Entertainment Salaries

Entertainer	Gross Annual Income (in Millions)
Steven Spielberg	—
Pink Floyd	—
Oprah Winfrey	53
Eagles	52
Barbra Streisand	52
Rolling Stones	50
Bill Cosby	—
David Copperfield	29
Harrison Ford	27
Barney the Dinosaur (Creator: S. Leach; Publisher: R. Leach)	25

Source: *Forbes,* 1994

8. On the average, New York has 7 more days of rain/year than London, and Vienna has 36 more days of rain/year than half that of New York. If the number of days of rain/year in New York and London combined exceeds that of Vienna by 137 days, how many days of rain/year occur in the three cities?

Problems 9–24 can be solved by creating verbal models and translating the models into algebraic equations. Use the five-step strategy to solve each problem.

9. In 1997, Medicaid spending in the United States amounted to $219 billion. This was a 20% increase over the amount spent in 1995. How much was spent on Medicaid in 1995?

10. A real estate agent receives a commission of 6% of the selling price of a house. What should be the selling price so that the seller can get $75,200?

11. The price of a VCR is reduced by 30%. When the VCR still does not sell, it is reduced by another 40%. If the price of the VCR after both reductions is $372.40, what was the original price?

12. The price of a television is reduced by 40%. When the television still does not sell, it is reduced by another 50%. If the price of the television after both reductions is $288, what was the original price?

13. The price of an item is increased by 50%. When the item does not sell, it is reduced by 50% of the increased price. If the final price is $280, what was the original price? When the price is increased by 50% and then reduced by 50%, what percent represents the net reduction in price?

14. The price of an item is increased by 30%. When the item does not sell, it is reduced by 30% of the increased price. If the final price is $382.20, what was the original price? When the price is increased by 30% and then reduced by 30%, what percent represents the net reduction in price?

15. With a 7% sales tax, what is the maximum price of a car if the total cost, including tax, is to be $16,050?

16. The price of a meal before tax was $26. If the total price including tax was $27.43, what was the tax rate?

17. Numerous variables affect the relationship between education and income. A simplified form of two such models focuses on the effect of gender. Yearly income for men increases by $1600 for each year of education, with a zero-education yearly income of $6300. For women, the comparable model involves a $1200 increase for each year of education and a zero-education yearly income of $2100.

a. Let x = the number of years of education attained by an individual. Write models for the yearly income of men and women with x years of education.

b. Use the models to determine how many years of education a woman must achieve to earn the same yearly salary as a man with 11 years of education.

c. Let $x = 0, 5, 10, 15,$ and $20,$ and graph the two models in the same rectangular coordinate system by creating a table of coordinates for each model. Based on the graphs, is it possible for the average personal wages for women to catch up to and overtake the wages for men? Explain.

d. Describe how you can use your graphs from part (c) to illustrate the solution in part (b).

18. The bus fare in a city is $1.25. People who use the bus have the option of purchasing a monthly coupon book for $21.00. With the coupon book, the fare is reduced to $0.50.

a. Write models for the total monthly costs of using the bus x times in a month both without and with the coupon book. Be sure to include the $21.00 in the model with the coupon book.

b. Use the models to determine the number of times in a month the bus must be used so that the total monthly cost without the coupon book is the same as the total monthly cost with the coupon book.

c. Let $x = 0, 10, 20$, and 30, and graph the two models in the same rectangular coordinate system by creating a table of coordinates for each model. Then use the graphs to visualize your solution in part (b).

d. If applicable, use a graphing utility to verify your graphs in part (c) and your solution in part (b).

19. In 1980, public school bus drivers in the United States averaged $5.23/hour. This rate has increased by $0.40/hour for each year since 1980. In what year were bus drivers earning $10.03/hour?

20. In 1980, the average yearly salary for teachers in the United States was $16,116. If the salary is increasing by $1496 per year, in what year will the salary reach $44,540?

21. A person purchasing a home is considering the following two options for a $50,000 mortgage.

Option 1: 7.5% interest on a 30-year loan with no points

Option 2: 7.125% interest on a 30-year loan with 3 points

Use Table 1.11 on page 96 to determine how long it will take for the total cost of each mortgage to be the same. How much will be saved by choosing the 7.125% option over 20 years?

22. A person purchasing a home is considering the following two options for a $60,000 mortgage.

Option 1: 9.0% interest on a 30-year loan, no points, and no application fee

Option 2: 8.5% interest on a 30-year loan, 2 points, and a $200 application fee

Use Table 1.11 on page 96 to determine how long it will take for the total cost of each mortgage to be the same. How much will be saved by choosing the 8.5% option over 20 years?

23. For a long-distance person-to-person telephone call, a telephone company charges 43¢ for the first minute, 32¢ for each additional minute, and a $2.10 service charge. If the cost of a call is $5.73, how long was the call?

24. At a college, an A is worth 4 points; a B, 3 points; a C, 2 points; a D, 1 point; and an F, 0 points. Janelle has a total of 50 points. She has 2 more B's than A's, 4 C's less than 3 times the number of A's, and 4 D's less than the number of A's. Find the number of each grade that she has.

Use the five-step strategy and the appropriate geometric formula(s) to solve Problems 25–34.

25. In *Structures: The Way Things Are Built* (New York: Macmillan Co., 1990), author Nigel Hawkes tells of a 1974 discovery by Chinese workers of rectangular pits containing sculptures of warriors whose function was to guard an emperor's tomb. One of the rectangular pits had a length that was 30 yards shorter than 4 times the width. With a perimeter of 640 yards, what were the dimensions of the rectangular base of the pit?

26. The length of the rectangular tennis court at Wimbledon is 6 feet longer than twice the width. If the perimeter is 228 feet, what are the court's dimensions?

27. A bookcase is to have four shelves, including the top, as shown in the figure. The height of the bookcase is to be 3 feet more than the width, and only 30 feet of lumber is available. What should be the dimensions of the bookcase?

28. The length of a rectangular lot is 1 yard less than three times its width. If 90 yards of fencing were purchased to enclose the lot and 12 yards of fencing were not needed, find the lot's dimensions.

29. A piece of copper tubing is to be bent into the shape of a triangle such that one side measures 1 inch less than twice the length of the second side and the third side measures 1 inch more than twice the length of the second side. If the piece of tubing is 30 inches long, find the length of each side of the triangle.

30. A flag in the shape of an isosceles triangle has a base that is 2.5 inches shorter than either of the equal sides. If the perimeter of the triangle is 35 inches, what are the lengths of the three sides?

31. The lengths in centimeters of two sides of a rectangle are consecutive even integers. If 5 more than half the length of the shorter side is added to twice the length of the longer side, the result is 44 centimeters. Find the area of the rectangle.

32. The lengths in meters of two sides of a rectangle are consecutive odd integers. The perimeter is 80 meters. What is the area of the rectangle?

33. A garden is being constructed in the shape of an isosceles trapezoid, meaning that the two nonparallel sides are equal in length. Each of these sides is 1 yard shorter than 3 times the smaller base. The longer side is 2 yards more than 5 times the shorter base. If the perimeter of the garden is 36 yards, find the length of each side.

34. A room with a rectangular floor has a length that exceeds its width by 3 feet and a perimeter of 54 feet. What will it cost to carpet the floor if the carpet costs $26.50 per square yard?

Solve Problems 35–44 using either the uniform motion model (d = rt) or the simple annual interest model (I = Prt), as well as tables to organize the given information.

35. Two planes leave the same airport at the same time, flying in opposite directions. The rate of the faster plane is 300 mph. The rate of the slower plane is 200 mph. In how many hours will the planes be 2500 miles apart?

36. Two cars leave the same point, traveling in opposite directions. The rate of the faster car is 55 mph. The rate of the slower car is 50 mph. In how many hours will the cars be 420 miles apart?

37. Two trucks leave a warehouse at the same time, traveling in opposite directions. The rate of the faster truck is 5 mph faster than the rate of the slower truck. At the end of 5 hours, they are 600 miles apart. Find the rate of each truck.

38. Two buses leave a station at the same time, traveling in opposite directions. The rate of the faster bus is 15 mph faster than that of the slower bus. At the end of 3 hours, they are 345 miles apart. Find the rate of each bus.

39. A woman drove from her home to the mountains and back again along the same highway. The round-trip took 10 hours. Her average outgoing rate was 20 mph while her average rate returning was 30 mph. How long did she take in each direction? What distance did she cover each way?

40. A traveler required 6 hours on a round-trip. The average outgoing rate was 60 mph while the average rate returning was only 30 mph. How long did the traveler take in each direction? What distance was covered each way?

41. Suppose $20,000 is invested, part in a stock paying 8% interest for 2 years and the balance in a stock paying 5% interest for 3 years. If the total interest earned is $3160, how much is invested in each stock?

42. Suppose $30,000 is invested, part in a stock paying 9% interest for 2 years and the balance in a stock paying 6% interest for 3 years. If the total interest earned is $5400, how much is invested in each stock?

43. A man invested part of $35,000 in a stock paying 9% interest for 2 years and the balance in a stock paying 6% interest for 4 years. It turns out that the two investments earned equal interest. How much was invested at each rate? What is the total interest earned?

44. A woman invested part of $10,000 in a stock paying $8\frac{1}{4}$% interest and the remainder in a stock paying $5\frac{1}{2}$% interest. After 1 year, the two investments earned equal interest. How much was invested at each rate? What is the total interest earned?

Problems 45–56 call for skills that go beyond translating given or implied conditions into equations. Use other problem-solving strategies—such as making a systematic list, looking for a pattern, trial and error, working backward and eliminating possibilities, or considering special cases—to solve each problem.

45. The numbers are arranged in the following square so that all rows, columns, and diagonals contain four numbers having identical sums. The number x in the upper-right corner satisfies the equation

$$-3 - 5(2 - x) + 6x = -6 - 2[-1 + 4(x - 6)].$$

Fill in the missing numbers in the empty squares.

	4	8	x
2			14
1		6	13
12		11	0

46. The digits 1 through 9 are to be placed in the circles in the figure on the right, a different digit in each circle, so that each side of the triangle sums to 20. Fill in the appropriate number in each circle. (*Hints: x* satisfies

$5 - 2(x - 4) + 3(x - 7) = -3$; y satisfies $y/3 + 1/2 = 7/6$; z satisfies $-3(z - 6) = 2 - z$.)

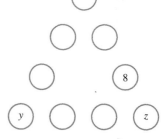

47. The following message

SEND
MORE
———
MONEY

is a correct addition sum, in which each different letter stands for a different digit, 0 possibly included. If $D = 7, E = 5$, and $R = 8$, what is the sum?

48. Some numbers in the printing of a division problem have become illegible. They are designated below by *. Fill in the blanks.

$$
\begin{array}{r}
1 \ * \ * \\
* \ * \overline{)4 \ * \ * \ *} \\
2 \ 8 \\
\hline
* \ 5 \ 6 \\
* \ * \ * \\
\hline
* \ * \ * \\
* \ * \ * \\
\hline
0
\end{array}
$$

49. You are an engineer programming the automatic gate for a 50-cent toll. The gate is programmed for exact change only and will not accept pennies. How many coin combinations must you program the gate to accept?

50. Equally priced legal pads were purchased for $3.21. If it is known that each pad cost more than 50 cents, how many pads were purchased and what did each pad cost?

51. A rectangular corral is to be fenced off with 16 yards of wood planking. What should the dimensions of the corral be to have the maximum area possible?

52. Lord Elphick was showing his guest the family portraits. Pointing to one, he remarked, "Brothers and sisters I have none, but that man's father is my father's son." What relative was represented in the portrait?

53. A company has three computers that are given to two mothers and two daughters, giving each person one computer. Explain how this can be the case.

54. In the following equation, replace *a* through *i* by the digits 1 through 9, using a different digit for each letter, so that all three factors are equal to $\frac{1}{2}$. (*Hint: a* = 3.)

$$\frac{a}{b} = \frac{c}{de} = \frac{fg}{hi}$$

55. List the four mistakes in this sentence: "This sentence contanes one misteak."

56. Consider the first four rectangular numbers (2, 6, 12, and 20):

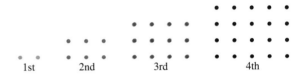

1st   2nd   3rd   4th

Use patterns to complete the following table:

Rectangular Number	1st	2nd	3rd	4th	5th	6th	12th	*n*th
Number of Dots	2	6	12	20				

Technology Problems

57. The average hourly rate (*y*) for public school cafeteria workers in the United States in 1980 was $3.82/hour. This rate has increased steadily by $0.30/hour each year since 1980.

  a. Write a mathematical model for the hourly rate *x* years after 1980.

  b. Use a graphing utility to graph the model with the following range setting:

    Xmin = 0, Xmax = 12, Xscl = 1,
    Ymin = 0, Ymax = 8, Yscl = 1

  c. Use the trace feature to trace along the curve to determine in what year the hourly wage was $6.22.

  d. Verify your observation in part (c) algebraically by setting the model equal to 6.22 and solving for *x*.

58. A tennis club offers two payment options. Members can pay a monthly fee of $30 plus $5/hour for court rental time. The second option has no monthly fee, but court time costs $7.50/hour.

  a. Write a mathematical model representing total monthly costs for each option for *x* hours of court rental time.

  b. Use a graphing utility to graph the two models in the same viewing rectangle using the following range setting:

    Xmin = 0, Xmax = 15, Xscl = 1,
    Ymin = 0, Ymax = 120, Yscl = 6

  c. Use your utility's trace or intersection feature to determine where the two graphs intersect. Describe what the coordinates of this intersection point represent in practical terms.

  d. Verify part (c) using an algebraic approach by setting the two models equal to one another and determining how many hours one has to rent the court so that the two plans result in identical monthly costs.

Writing in Mathematics

59. Many students find solving linear equations much easier than solving algebraic word problems. Discuss some of the reasons why this is the case.

60. Did you have some difficulties solving some of the problems that were assigned in this problem set? Discuss what you did if this happened to you. Did your course of action enhance your ability to solve algebraic word problems?

61. Discuss some of the reasons why students might be turned off to solving word problems. How many of these reasons apply to you?

Critical Thinking Problems

62. Solve the following puzzle by Sam Loyd (1841–1911), who is considered one of the greatest puzzle composers of all time:

"What is the age of that boy?" asked the conductor. Flattered by this interest shown in his family affairs, the suburban resident replied: "My son is five times as old as my daughter, and my wife is five times as old as the son, and I am twice as old as my wife, whereas grandmother, who is as old as all of us put together, is celebrating her eighty-first birthday today."

How old was the boy?

63. Solve the following puzzle from *The Ladies' Diary*, 1704–1841, containing mathematical puzzles and questions posed and solved by women. The problems were initially proposed, in the manner of the times, in verse, a practice that was soon abandoned.

If to my age there added be,
One half, one third, and three times three;
Six score and ten the sum you'll see,
Pray find out what my age may be.

(*Hints:* A score is 20 years. Added to my age is half my age and one-third of my age. This illustrates the difficulties that arise when mathematics is forced into versification!)

64. In a film, the actor Charles Coburn plays an elderly "uncle" character criticized for marrying a woman when he is 3 times her age. He wittily replies, "Ah, but in 20 years time I shall only be twice her age." How old is the "uncle" and the woman?

65. It was wartime when the Ricardos found out that Mrs. Ricardo was pregnant. Ricky Ricardo was drafted and made out a will, deciding that $14,000 in a savings account was to be divided between his wife and his child-to-be. Rather strangely, and certainly with gender bias, Ricky stipulated that if the child were a boy, he would get twice the amount of the mother's portion. If it were a girl, the mother would get twice the amount the girl was to receive. We'll never know what Ricky was thinking of, for (as fate would have it) he did not return from war. Mrs. Ricardo gave birth to twins—a boy and a girl. How was the money divided?

66. A mathematics tutor agrees to be paid $4.40 for every hour she works, although she must pay the math lab $6.60 for every hour that she does *not* work. At the end of 30 hours, the tutor finds that she has paid out the same amount of money as she has taken in. How many hours did she work?

67. A woman attends three bingo games. At the first, she doubles her money and spends $30 in celebration. At the second, she triples her money and spends $54 in celebration. At the third, she quadruples her money and spends $72 in celebration. Finally, feeling bingoed out, she returns home with $48. How much money did she originally have?

68. Nathan has incredible luck. He just found $2 on the sidewalk. His friend Adelaide remarked, "Now you've got 5 times as much as you'd have had if you'd lost $2." How much did Nathan have before finding $2?

69. If you spend $\frac{1}{3}$ of your money and then lose $\frac{2}{3}$ of what you still have left, leaving you with only $12, how much money did you originally have?

70. A man passed one-sixth of his life in childhood, one-twelfth in youth, and one-seventh in childless marriage. After 5 years of marriage, the man had a child. Alas! late-born wretched child; after attaining half her father's life, cruel fate overtook her, leaving the man to spend his last 4 years solving algebraic word problems. What was the man's final age?

71. A thief steals a number of rare plants from a nursery. On the way out, the thief meets three security guards, one after another. To each security guard, the thief is forced to give one-half the plants that he still has, plus 2 more. Finally, the thief leaves the nursery with 1 lone palm. How many plants were originally stolen?

72. Two avid cyclists, both traveling at 6 mph, plan to visit each other. The distance between their towns is 36 miles, connected by a straight road. As they leave their respective towns, one of the cyclists releases a pet carrier pigeon to send a message to a receiver who lives 162 miles away. The pigeon flies uniformly at a speed of 18 mph. What fractional part of the 162 miles has the pigeon covered when the two cyclists meet?

73. An athlete ran for 8 hours at a constant speed. The athlete was immediately given a bike and rode in the opposite direction for 3 hours, with twice the speed. The athlete then reversed direction again, walking for 1 hour at 40% of the initial running speed. After 12 hours, the athlete was 9.6 miles from the point where the marathon exercise excursion began. Find the athlete's three speeds.

74. Ridinghood, on the way to Grandmother's, anticipates that the 170-mile trip will take $2\frac{1}{2}$ hours. She drives along the highway at 55 mph. When she begins driving into the woods, however, a preponderance of wolves along the narrow road forces her to reduce her speed to 40 mph, and she arrives at Grandmother's 1 hour later than anticipated. Throughout the portion of the journey on the narrow road, one of the more aggressive wolves continued to grunt lasciviously, singing continuously to Ridinghood, "There's no possible way to describe what I feel when I'm chatting with my meal." How long did the wolf sing? What distance did Ridinghood cover at each speed?

75. A low-risk bond pays 5% interest, and a high-risk stock claims to yield 16%. Part of $50,000 is invested in the bond and the balance in the stock. After 1 year, the stock is actually losing 8%, so it is sold. However, the investment in the low-risk bond is continued for 2 more years. The total interest earned for both investments is $5200. How much was invested in each instrument?

Group Activity Problems

76. *A Critical Thinking Contest:* Problems 62–75 contain 14 challenging critical thinking problems. The class will be divided up into small groups. As a group, your goal is to solve as many of these problems as you can, using the input and suggestions of all group members. At the end of a designated time period, all solutions should be written up coherently and turned into your professor. Group work will be scored as follows: 10 points for each problem solved correctly, 5 points for a partial solution that contains good problem-solving work even without a final answer, and -5 points for solutions that reflect sloppy thinking or that are filled with algebraic errors. The group with the greatest number of points will be declared the winner of the critical thinking contest.

77. Some problems have no preconceived answers, asking the problem solver to think about possibilities. These problems are sometimes appropriate for group discussions, where small groups can then suggest possible solutions to the rest of the class for further input. Here's an example of what we mean. Work on this problem in small groups. The groups should then come together to share their solution with other groups to find the best solution that the class can obtain.

Suppose your parents lend you $10,000 a year, for each of 4 years, for your education. They ask that you pay them back only 75% of the loan once you have completed school. At that point, they stipulate you must pay 5% interest, but not simple interest. For each year after you complete college and the loan is outstanding, you must pay 5% on the amount your parents expect to be paid back, in addition to interest due from the preceding years the loan was outstanding.

Five years and 4 months after completing college, you are earning enough money to pay your parents back. How much do you owe them, and what will you do about the extra 4 months?

Incidentally, you are a successful engineer designing new natural gas–powered automobiles. You can give your parents a $14,000 car that you can obtain for only $5000 as a way of partial loan payments. Your parents are proud of the cars you've designed, but they must drive 50 miles from their home to find a station that has pumps dispensing natural gas. How would you factor this information into the overall picture?

Review Problems

78. Opticians use the model

$$P = \frac{100(p + q)}{pq}$$

for corrective lenses, where P is the power of the lenses (in diopters, negative for nearsighted and positive for farsighted persons), p is the object distance from the lens, and q the image distance from the lens, where both p and q are in centimeters. Find the power of corrective lenses for which $p = 300$ and $q = -50$. (Image distance is negative when the image is on the same side of the lens as the object.)

79. In physics, the model $D = A(n - 1)$ describes the law of small prisms. Solve for n.

80. Evaluate: $10^3 + 10^0 - 10^{-1}$.

CHAPTER PROJECT

Surrounded by Numbers

In the essay *Exponents and Large Numbers* on page 51, we explored the names of some very large numbers. The prefixes on the numbers read: bi, tri, quad, quint, sext, sept, oct, and non. These prefixes seem to be giving us a number sequence: 2, 3, 4, 5, 6, 7, 8, and 9. When these prefixes were first used, in the late 1400's, these numbers referred to the 2nd through 9th powers of a **million.** Thus, the first three would be:

$$\text{billion} = \text{2nd power of } 1{,}000{,}000 \text{ or } (10^6)^2 = 10^{12}$$
$$\text{trillion} = \text{3rd power of } 1{,}000{,}000 \text{ or } (10^6)^3 = 10^{18}$$
$$\text{quadrillion} = \text{4th power of } 1{,}000{,}000 \text{ or } (10^6)^4 = 10^{24}$$

Around the mid 1600's, some mathematicians began using these prefixes for the 3rd through 10th powers of a **thousand,** giving us

$$\text{billion} = \text{3rd power of } 1000 \text{ or } (10^3)^3 = 10^9$$
$$\text{trillion} = \text{4th power of } 1000 \text{ or } (10^3)^4 = 10^{12}$$
$$\text{quadrillion} = \text{5th power of } 1000 \text{ or } (10^3)^5 = 10^{15}$$

Although the later usage seems to have lost the original meaning of the prefixes, it was adopted in the United States. However, in Great Britain, the older system is still used, giving us two different systems of naming large numbers. In Great Britain, 1,000,000,000 is called a thousand-million; in the United States it is called a billion. A billionaire in Great Britain would be a trillionaire in the United States.

In addition to the prefixes given above, we also have a worldwide standard for naming powers of ten, from the very small to the very large. These prefixes may be attached to any unit. Thus, we have **kilo**meters for distance, **kilo**grams for weight, as well as **kilo**bytes for a computer. The prefixes are given in Table 1.13.

1. Choose a pair of prefixes, one large and one small, and give at least two different measurements. For example, if you use **kilo-,** you could give the weight of your Sunday paper in kilograms, and the distance from your house to school in kilometers. For extremely small or large number prefixes, you may wish to look for examples from the sciences, such as astronomical distances or atomic weights. Compare your very large or very small measurements to common things to give a better idea of scale. For example, if you write the mass of the earth in kilograms, you might want to include how many average size cars it would take to achieve the same mass.

The English language contains many words that have numerical references as a part of the word. Some of these may be fairly obvious, like a **bi**cycle with **two** wheels; others may be a bit harder to decipher. We may have to follow the history of some words before we can decide why a particular number was used as a part of a word.

2. Listed below, you will find words that make some reference to the numbers from one to ten. Working with a group, select four or five words to define and clarify how the numerical reference is part of the word. Present your findings to the class, along with any references other than the strict definition you feel may be needed to understand the word.

one

alone	union	monopoly
none	unanimous	monk
once	universal	monolithic
		monogram

two

between	dual	biscuit	par	amphitheater
twilight	duet	biceps	parallel	ambivalent
twine	doubt	binoculars	parabola	
twin	duplex		parity	
	duplicate			

three

tribe	trivium
tribute	trieme
trident	tierce

four

square	quadrilateral	trapeze
squadron	quadruped	
quadrille		

five

pentagon	quintessence
pentagram	quincunx
pentacle	

six

sextain	siesta
sextant	semesters

seven

September
hebdomadal

eight

octave
octopus

nine

enneagon	noon
Enneads	novena

ten

Decalogue	dicker
Decameron	doyen

3. Each of the following contains a reference to a number twelve from mythology, religion, or history. Select one of the following and determine why the number twelve was deemed significant.

12 ordeals of Gilgamesh

12 labors of Hercules

12 counselors of Odin

12 knights of the Round Table of King Arthur

12 gods and goddesses presided over by Zeus

12 disciples of Jesus Christ

12 tribes of Israel

12 tribes of Native America, each tribe with 12 clans

12 signs of the zodiac

TABLE 1.13 A Comprehensive Table of All SI Prefixes

prefix	symbol	power of ten	English name
yotta-	Y	+24	septillion
zetta-	Z	+21	sextillion
exa-	E	+18	quintillion
peta-	P	+15	quadrillion
tera-	τ	+12	trillion
giga-	G	+9	billion
mega-	M	+6	million
kilo-	k	+3	thousand
hecto-	h	+2	hundred
deca-	da	+1	ten
		0	one
deci-	d	−1	tenth
centi-	c	−2	hundredth
milli-	m	−3	thousandth
micro-	μ	−6	millionth
nano-	n	−9	billionth
pico-	p	−12	trillionth
femto-	f	−15	quadrillionth
atto-	a	−18	quintillionth
zepto-	z	−21	sextillionth
yocto-	y	−24	septillionth

Worldwide Web Resources

Go to the Prentice Hall website (http://www.prenhall.com/blitzer) to access other locations on the Internet that will allow you to further explore the concepts presented in this project.

Chapter Review

SUMMARY

1. The Real Numbers

The set of real numbers consists of both the set of rational numbers and the set of irrational numbers.

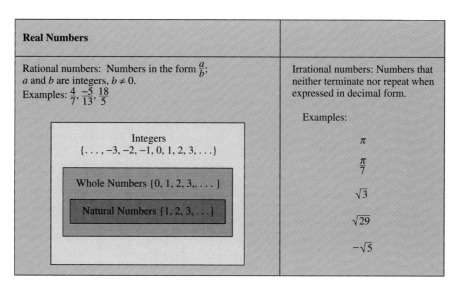

Real Numbers	
Rational numbers: Numbers in the form $\frac{a}{b}$; a and b are integers, $b \neq 0$. Examples: $\frac{4}{7}, \frac{-5}{13}, \frac{18}{5}$ Integers $\{\dots, -3, -2, -1, 0, 1, 2, 3, \dots\}$ Whole Numbers $\{0, 1, 2, 3, \dots\}$ Natural Numbers $\{1, 2, 3, \dots\}$	Irrational numbers: Numbers that neither terminate nor repeat when expressed in decimal form. Examples: π $\dfrac{\pi}{7}$ $\sqrt{3}$ $\sqrt{29}$ $-\sqrt{5}$

2. Order on the Real Number Line

a.

```
  |    |    |    |    |    |    |    |
 -3   -2   -1    0    1    2    3
```

Corresponding to every real number there is one and only one point on the number line. Corresponding to each point on the real number line, there is one and only one real number, called the coordinate of the point.

b. $a < b$ (a is less than b) means that a is to the left of b on the real number line. Equivalently, $b > a$ (b is greater than a).

c. Other symbols: $\leq$ less than or equal to
                        $\geq$ greater than or equal to

d. Set-builder notation, interval notation, and graphs:

$(a, b) = \{x \mid a < x < b\}$

$[a, b) = \{x \mid a \leq x < b\}$

$(a, b] = \{x \mid a < x \leq b\}$

$[a, b] = \{x \mid a \leq x \leq b\}$

$(-\infty, b) = \{x \mid x < b\}$

$(-\infty, b] = \{x \mid x \le b\}$

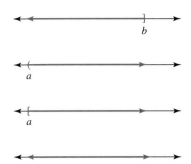

$(a, \infty) = \{x \mid x > a\}$

$[a, \infty) = \{x \mid x \ge a\}$

$(-\infty, \infty) = \{x \mid x \text{ is a real number}\} = \{x \mid x \in R\}$

3. Basic Properties of the Real Numbers
  a. *Closure property for addition:* The sum of any two real numbers is a unique real number.
  b. *Commutative property for addition:* $a + b = b + a$
  c. *Associative property for addition:*
  $a + (b + c) = (a + b) + c$
  d. *Zero is the identity element for addition:* $a + 0 = a$
  e. *Inverses for addition:* $a + (-a) = 0$
  f. *Closure property for multiplication:* The product of any two real numbers is a unique real number.
  g. *Commutative property for multiplication:* $ab - ba$
  h. *Associative property for multiplication:* $a(bc) = (ab)c$
  i. *One is the identity element for multiplication:* $a \cdot 1 = a$
  j. *Inverses for multiplication:* $a \cdot \dfrac{1}{a} = 1, a \ne 0$
  k. *Distributive property:* $a(b + c) = ab + ac$
  $(b + c)a = ba + ca$

4. Additional Properties of the Real Numbers
  a. *Properties of Zero:* $a + 0 = a; a - 0 = a; a \cdot 0 = 0;$
  $0 \cdot a = 0; \dfrac{0}{a} = 0, a \ne 0; \dfrac{a}{0}$ is undefined.
  b. *Properties of Negatives:* $(-1)a = -a; -(-a) = a;$
  $(-a)b = -(ab); a(-b) = -(ab); (-a)(-b) = ab;$
  $\dfrac{-a}{b} = \dfrac{a}{-b} = -\dfrac{a}{b} \; (b \ne 0); \quad \dfrac{-a}{-b} = \dfrac{a}{b} \; (b \ne 0);$
  $-(a + b) = -a - b;$
  $-(a - b) = -a + b = b - a.$

5. Absolute Value
  a. The absolute value of a, $|a|$, is the distance between the number a and 0 on the real number line.
  b. *Definition of Absolute Value:*
  $$|x| = \begin{cases} x & \text{if } x \ge 0 \\ -x & \text{if } x < 0 \end{cases}$$

6. Operations with Real Numbers
  a. *Addition:* The sum of two numbers with like signs has the same sign as the two numbers and is found by adding their absolute values. If the two numbers have unlike signs, the sign of the sum is the sign of the original number having the larger absolute value and is found by subtracting the smaller absolute value from the larger absolute value.
  b. *Subtraction:* By definition, $a - b = a + (-b)$. To subtract real numbers, change the sign on the second number and follow the appropriate addition rule.
  c. *Multiplication and division of two real numbers:* The product or quotient of two numbers with the same sign is positive. The product or quotient of two numbers with different signs is negative. The multiplication or division is performed by multiplying or dividing the absolute values of the two numbers and giving the answer the proper sign.
  d. A multiplication problem involving an even number of negative factors has a positive product, and one with an odd number of negative factors has a negative product.

7. The Order of Operations Agreement
  a. Perform operations above and below any fraction bar, following parts (c) to (e).
  b. Perform operations inside grouping symbols, innermost grouping symbols first, following parts (c) to (e).
  c. Simplify exponential expressions.
  d. Do multiplication or division as they occur, working from left to right.
  e. Do addition and subtraction as they occur, working from left to right.

8. Algebraic Expressions
  a. An algebraic expression combines numbers and variables. Its terms are separated by addition. Terms that have the same variables to the same powers are like terms, and they can be combined.
  b. Simplifying an algebraic expression often consists of using the distributive property to remove grouping symbols and combining like terms.

9. The Rectangular (Cartesian) Coordinate System
  a. The rectangular coordinate system is formed by placing two number lines perpendicular to each other at their zero points. Each line is called a coordinate axis. The horizontal axis is called the x-axis. The vertical axis is called the y-axis. The positive

portion of the x-axis lies to the right of the origin. The positive portion of the y-axis lies above the origin. The axes divide the coordinate plane into four regions called quadrants.

  b. Any point in the plane can be represented by a pair of numbers. The first number denotes the distance and direction from the origin along the x-axis. The second number denotes the vertical distance and direction along the y-axis or a line parallel to the y-axis.

  c. The graph of an equation involving two variables (usually x and y) is the set of all points whose coordinates satisfy the equation.

10. Integral Exponents

  a. If n is a natural number $\geqslant 1$, b^n means that b appears as a factor n times:

$$b^1 = b \qquad b^{-n} = \frac{1}{b^n} \quad (b \neq 0)$$

$$\frac{1}{b^{-n}} = b^n \quad (b \neq 0)$$

$$b^0 = 1 \quad (b \neq 0)$$

  b. Laws of exponents:

$$b^n \cdot b^m = b^{n+m} \qquad (b^n)^m = b^{nm} \qquad (ab)^n = a^n b^n$$

$$\frac{b^n}{b^m} = b^{n-m} \quad (b \neq 0)$$

$$\left(\frac{a}{b}\right)^n = \frac{a^n}{b^n} \quad (b \neq 0)$$

11. Scientific Notation

  a. A scientific notation numeral appears as the product of two factors. The first factor is a number greater than or equal to 1 but less than 10. The second factor is base 10 raised to a power.

  b. Multiplication and division in scientific notation can be performed using

$$10^n \cdot 10^m = 10^{n+m} \qquad \frac{10^n}{10^m} = 10^{n-m}.$$

12. Solving a Linear Equation $(ax + b = c)$

  a. Multiply on both sides to clear fractions or decimals.

  b. Simplify each side.

  c. Collect all variable terms on one side and all constant terms on the other side.

  d. Isolate the variable and solve.

  e. Check the proposed solution in the original equation.

  f. If a contradiction occurs, the inconsistent equation has no solution. If a statement in the form $a = a$ occurs, the identity is satisfied by all real numbers.

13. Solving Formulas and Mathematical Models for a Specified Variable

These equations contain more than one variable. Follow the procedure for solving linear equations. If the terms containing a specified variable cannot be combined as like terms, use the distributive property to isolate the variable from these terms. Then solve for the specified variable by dividing both sides of the equation by the factor that appears with the specified variable.

14. Solving Algebraic Word Problems

  a. Let x represent one of the problem's unknowns.

  b. If necessary, represent other unknowns in terms of x.

  c. Write an equation in x that describes the verbal conditions of the problem.

  d. Solve the equation and answer the problem's question.

  e. Check the solution in the original wording of the problem.

15. Strategies for Solving Problems

  a. Translate given conditions into an equation.

  b. Create a verbal model and translate the verbal model into an equation.

  c. Use known formulas.

  d. Use tables to organize information.

  e. When problems cannot be solved by translating sentences into equations, try strategies such as trial and error, making a list, looking for patterns, and working backwards to eliminate possibilities.

REVIEW PROBLEMS

In Problems 1–2, use the roster method to write each set.

1. $\{x \mid x$ is a natural number that is divisible by 4$\}$

2. $\{x \mid x$ is a whole number but not a natural number$\}$

In Problems 3–5, express each set in interval notation and graph the set on a number line.

3. $\{x \mid x \leqslant 1\}$

4. $\{x \mid x \geqslant -2\}$

5. $\{x \mid -1 < x \leqslant 2\}$

In Problems 6–7, use set-builder notation and interval notation to rewrite each sentence.

6. x lies between -3 and 6, including -3 and excluding 6.

7. x is at most 12.

8. The 1995 Internal Revenue Tax Rate Schedule X for taxpayers whose filing status is single is shown at the top of the next page. Use interval notation to write the amounts of taxable income for each of the five tax brackets.

Schedule X—Use if Your Filing Status is Single

If the Amount on Form 1040, Line 37, is: Over—	But Not Over—	Enter on Form 1040, Line 38	of the Amount Over—
$0	$23,350	. . . 15%	$0
23,350	56,550	$3,502.50 + 28%	23,350
56,550	117,950	12,798.50 + 31%	56,550
117,950	256,500	31,832.50 + 36%	117,950
256,500	. . .	81,710.50 + 39.6%	256,500

It is almost impossible to go anywhere in the United States without seeing at least one of the ten franchises listed in the bar graph, since these are the top ten franchise companies (by the number of franchises) in the country. If x *represents the number of franchises, write the name of the company or companies described by Problems 9–13.*

9. $x \in [4131, 5903)$

10. $x \in [8013, \infty)$

11. $x \in (-\infty, 3793)$

12. $8013 < x \le 10,604$

13. $x \le 4200$

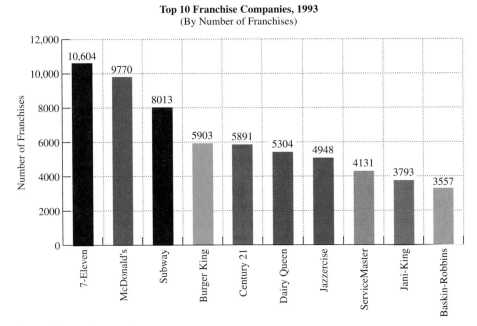

Top 10 Franchise Companies, 1993
(By Number of Franchises)

Source: *Enterprise* magazine

State the property of algebra illustrated in Problems 14–16.

14. $5(3 + 4) = 5(4 + 3)$

15. $5(3 \cdot 4) = (5 \cdot 3) \cdot 4$

16. $5(3 + 4) = 5 \cdot 3 + 5 \cdot 4$

Use the order of operations to find the value of each expression in Problems 17–38. If applicable, verify your answer with a calculator.

17. $3 + (-17) + (-25)$

18. $16 - (-14)$

19. $-11 - [-17 + (-3)]$

20. $|-17| + |3| - (|10| + |-13|)$

21. $(-0.2)(-0.5)$

22. $-\frac{1}{2}(-16)(-3)$

23. $-12 \div \frac{1}{4}$

24. $(-\frac{1}{2})^3 \cdot 2^4$

25. $-\frac{2}{7} \div (-\frac{3}{7})$

26. $-3[4 - (6 - 8)]$

27. $8^2 - 36 \div 3^2 \cdot 4 - (-7)$

28. $\dfrac{(-2)^4 + (-3)^2}{2^2 - (-21)}$

29. $25 \div \left(\dfrac{8 + 9}{2^3 - 3}\right) - 6$

30. $6 - (-20)\left[\dfrac{6 - 1(6 - 10)}{14 - 3(6 - 8)}\right]$

31. $(\frac{1}{4} - \frac{3}{8}) \div (-\frac{3}{5} - \frac{1}{4})$

32. $\frac{3}{4} - \frac{5}{8} - \left(-\frac{1}{2}\right)$

33. $\dfrac{(10 - 6)^2 + (-2)(-3)}{10 - (-4)(3)}$

34. $(2.6)(-5.4) \div (1.8) - (-5.7)$

35. $\dfrac{9(-1)^3 - 3(-6)^2}{5 - 8}$

36. $\dfrac{(7 - 9)^3 - (-4)^2}{2 + 2(8) \div 4}$

37. $2^4(-1)^{50} + |-3|^3$

38. $6 - [-3(-2)^3 \div (-8)]^2$

Simplify each algebraic expression in Problems 39–45.

39. $6(2x - 3) - 5(3x - 2)$

40. $6[b - 3(a - 6b)]$

41. $3x - [y - (2x - 3y)]$

42. $-x^2 - [3y^2 - (2x^2 - y^2)]$

43. $8x - 2x \cdot 5 + 6x^2 \div 3 \cdot 2$

44. $[6(xy - 1) - 2xy] - [4(2xy + 3) - 2(xy + 2)]$

45. $-\frac{1}{2}(x + 4) - \frac{1}{4}(-3x + 8)$

46. The model $y = 0.0005x^3 - 0.04x^2 + 2.8x + 22$ approximates California's population density y (in people per square mile) x years after 1920. Use the model to find the difference in density between 1980 and 1930. How is this result shown in the accompanying graph?

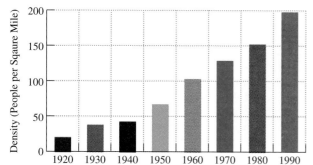
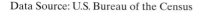
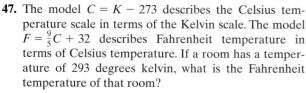
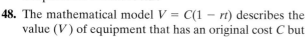

California Population Density 1920–1990

Data Source: U.S. Bureau of the Census

47. The model $C = K - 273$ describes the Celsius temperature scale in terms of the Kelvin scale. The model $F = \frac{9}{5}C + 32$ describes Fahrenheit temperature in terms of Celsius temperature. If a room has a temperature of 293 degrees kelvin, what is the Fahrenheit temperature of that room?

48. The mathematical model $V = C(1 - rt)$ describes the value (V) of equipment that has an original cost C but that depreciates steadily at rate r (in decimal form) for a period of t years. What is the value after 4 years of a ceramic kiln that originally cost $1200 if it has depreciated at a rate of 7% per year?

49. Medical researchers have found that the desirable heart rate $(R,$ in beats per minute) for beneficial exercise is approximated by the mathematical models

$$R = 0.65(220 - A) \quad \text{for women}$$
$$R = 0.75(220 - A) \quad \text{for men}$$

where A is the person's age.

a. Find the difference between the desirable heart rate for a 30-year-old man and a 30-year-old woman. How is this result shown in the graph?

b. Find the difference between the desirable heart rate for a 70-year-old man and a 20-year-old man. How is this result shown in the graph?

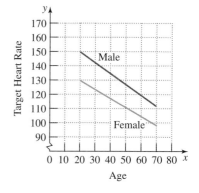

Construct a table of coordinates for each equation in Problems 50–52.

x	-3	-2	-1	0	1	2	3
y							

Then use the ordered pairs (x, y) from the table to graph the equation. If applicable, verify your hand-drawn graph with a graphing utility.

50. $y = x^2 - 1$

51. $y = |2x|$

52. $y = -2x + 3$

53. The graph shows Mexico's projected growth. Estimate the coordinates of points *A* and *C* and describe what this means in practical terms.

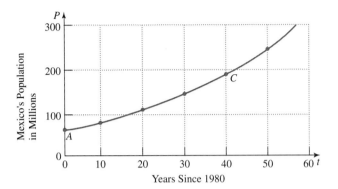

54. The graph shown here appeared in *Scientific American* and describes mortality trends in the United States from motor vehicle accidents and firearms. Estimate the coordinates of the intersection point of the two lines. What does this mean in practical terms? Describe what happens to the right of this intersection point.

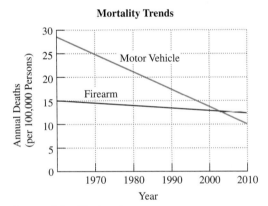

Mortality Trends

Source: *Scientific American*

55. Excessive alcohol consumption leads to more than 100,000 deaths annually in the United States. The graph shows the relation between alcohol consumption and liver cirrhosis mortality rate.
  a. In approximately what year is alcohol consumption at a maximum? What is a reasonable estimate for the number of gallons of ethanol consumed per capita for that year?
  b. For the period from 1940 onward, in approximately what year is the liver cirrhosis mortality rate at a maximum? What is a reasonable estimate for the mortality rate for that year?
  c. For the period from 1960 onward, approximately when was alcohol consumption increasing and when was it decreasing? Express your answer in interval notation.
  d. Describe one trend indicated by the graphs.

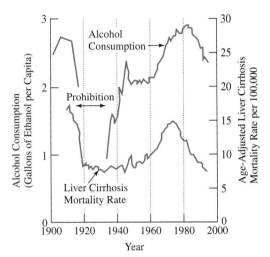

Source: National Institute on Alcohol Abuse and Alcoholism

56. The graph indicates the Fahrenheit temperature *x* hours after noon.
  a. At what time does the minimum temperature occur? What is the minimum temperature?
  b. At what time does the maximum temperature occur? What is the maximum temperature?
  c. What are the values of *x* when *y* = 0? In terms of time and temperature, what do these values indicate?
  d. What is the value of *y* when *x* = 0? What does this mean in terms of time and temperature?
  e. What is the percent increase in temperature between 7 P.M. and 8 P.M.?

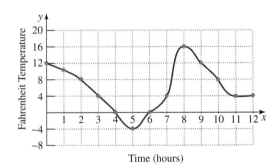

Time (hours)

57. The cost (*y*) of removing *x*% of pollutants from a lake is given by the mathematical model

$$y = \frac{10,000x}{100 - x}.$$

Complete the table of coordinates on the next page and graph the model.

x	0	20	40	50	60	80	90	99
y								

Write a brief verbal description of what the model indicates in terms of cleaning up the lake.

58. According to the National Center for Education Statistics, the model $y = 553x + 27{,}966$ approximates the average yearly salary for public school teachers in the United States. In the model, x represents the number of years after 1985 and y represents yearly salary.

a. Construct a table of coordinates and graph the model in a rectangular coordinate system.

x	0	1	2	4	6	8	10	11	12
y									

b. What does the graph indicate about salaries over time?

c. What was the average salary in 1995?

d. After what year will the average salary exceed $33,000?

In Problems 59–70, simplify each expression, expressing your answers without using negative exponents. Assume that no denominators are 0.

59. $(-3y^7)(-8y^6)$

60. $(7x^3y)^2$

61. $(-3xy)(2x^2)^3$

62. $\left(\frac{2}{3}\right)^{-2}$

63. $(-6xy)(-3x^2y) - 25x^3y^2$

64. $\dfrac{16y^3}{-2y^{10}}$

65. $(-3x^4)(-4x^{-11})$

66. $\dfrac{12x^7}{-4x^{-3}}$

67. $\dfrac{-10a^5b^6}{20a^{-3}b^{11}}$

68. $(-2)^{-3} + 2^{-2} + \frac{1}{2}x^0$

69. $\left(\dfrac{-2}{ab}\right)^5$

70. $(3x^4y^{-2})(-2x^5y^{-3})$

For Problems 71–72, convert to scientific notation.

71. 93,700,000,000,000

72. 0.000000409

In Problems 73–74, perform the indicated operation and express your answer in scientific notation.

73. $(2.8 \times 10^{13})(4.2 \times 10^{-6})$

74. $\dfrac{1.8 \times 10^{-6}}{4.8 \times 10^{-8}}$

75. Light travels at a rate of approximately 3×10^8 meters per second. It takes light from the sun 1.97×10^4 seconds to reach Pluto. What is the distance between Pluto and the sun?

76. The graph at the top of the next page shows the U.S. Defense Department Budget from 1983 through 1993. The total budget for 1989 was about $1.14 trillion and for 1993 about $1.41 trillion.

a. Use the graph to estimate the defense budget for 1989. Then write both the numbers for the defense budget and the total budget in scientific notation. Use scientific notation to find the percentage of the total that was spent on defense. Round your answer to the nearest whole percent.

b. Repeat part (a) for 1993.

c. If the graph continues to show that each year there is less money spent on the defense budget, can we assume that the percentage of the total budget spent on defense is also decreasing? Explain.

U.S. Defense Department Budget, 1983–1993
(Money in the Budget of the Department of Defense, in Billions of Dollars)

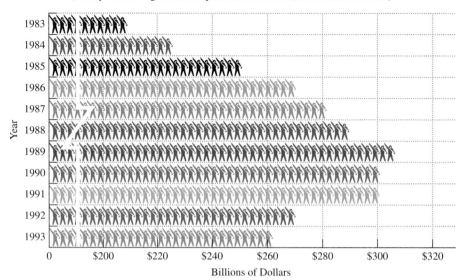

Source: U.S. Department of Defense

Solve each equation in Problems 77–85. Check proposed solutions algebraically or, if applicable, using a graphing utility.

77. $5x - 3 = x + 5$

78. $8 - (4x - 5) = x - 7$

79. $2(2x + 5) + 1 = 5 - 2(3 - x)$

80. $2(y - 2) = 2[y - 5(1 - y)]$

81. $3(x - 1) + 7(x - 3) = 5(2x - 5)$

82. $\dfrac{2y}{3} = 6 - \dfrac{y}{4}$

83. $\dfrac{3z + 1}{3} - \dfrac{13}{2} = \dfrac{1 - z}{4}$

84. $0.07y + 0.06(1400 - y) = 90$

85. $2(x + 6) + 3(x + 1) = 4x + 10 + x + 5$

In Problems 86–92, solve for the variable indicated.

86. $2x - 3y = 9$ for y

87. $y = mx + b$ for x

88. $C = \frac{5}{9}(F - 32)$ for F

89. $A = \frac{1}{2}bh$ for h

90. $A = 2HW + 2LW + 2LH$ for H

91. $x(y - 2) = 3y + 5$ for y

92. $\dfrac{ax}{3} - \dfrac{bx}{2} = \dfrac{17}{6}$ for x

93. The formula $y = 420x + 720$ is an approximate model for the amount of money (y, in millions of dollars) lost to credit card fraud worldwide x years after 1989. In what year did losses amount to 4080 million dollars?

94. Wastewater treatment operators use a mathematical model for finding the amount of chlorine to add to a basin. The formula is $A = 8.34\,FC$, where A = amount of chlorine needed in pounds, F = water flow in millions of gallons/day, and C = desired concentration of chlorine in parts per million (ppm). What is the water flow in a basin in which 2001.6 pounds of chlorine are added to obtain a concentration of 30 ppm?

95. The model $F = \frac{1}{4}C + 40$ describes the relationship between C, the number of times a cricket will chirp in 1 minute, and F, the Fahrenheit temperature of the environment in which the cricket lives.

a. Solve the formula for C.

b. Determine how temperature must be controlled to ensure that crickets will chirp 200 times/minute.

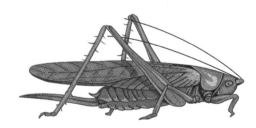

96. Membership in a fitness club is $400 per year. The club charges $2.00 for each hour the facility is used. A competing club charges $500 yearly plus $1.50/hour. The yearly cost (*y*) of using each club for *x* hours can be modeled by $y = 400 + 2x$ and $y = 500 + 1.5x$, respectively.

a. How many hours of use each year will result in identical total yearly costs for the two fitness clubs?

b. Let $x = 0, 50, 150, 200,$ and 250 and set up a partial table of coordinates for each model. Then graph the models in the same rectangular coordinate system. Where do the graphs intersect? What does this mean?

97. The model $34x - 10y = -1550$ describes the number of dentists in the United States (*y*, in thousands) *x* years after 1985.

a. Solve the model for *y*.

b. Use the model to predict the number of dentists in the year 2000.

98. When 7 times a number is decreased by 1, the result is 9 more than 5 times the number. Find the number.

Medical Dentistry—drawing a tooth with the help of pincers by Tim Bobbin, 1773.
Mary Evans Picture Library

99. Black teenagers are the most common victims of crime in the United States. The graph shows the violent crime victimization rates, per 1000, in the United States in 1995 for persons ages 12–19. The number of victims of violent crime (per 1000 persons) for black males is 9 more than twice that for white females. The number of victims among black females is 10 less than double that for white females, and the number for white males exceeds that for white females by 38. For all four groups combined, the victimization rate is 349 teenagers per 1000 persons. Find the number of crime victims per 1000 persons for each group.

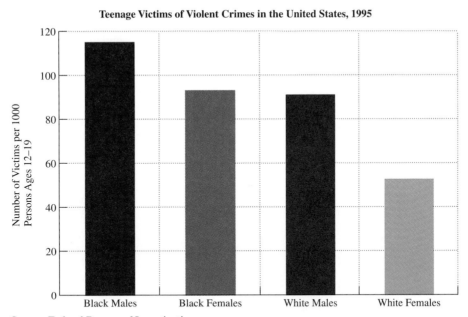

Teenage Victims of Violent Crimes in the United States, 1995

Source: Federal Bureau of Investigation

100. The price of a computer is reduced by 20%. When the computer still does not sell, it is reduced by 10% of the reduced price. If the price of the computer after both reductions is $1872, what was the original price?

101. Costs for two different kinds of heating systems for a three-bedroom house are given to the right.

System	Cost to Install	Operating Cost/Year
Electric	$5000	$1100
Gas	$12,000	$700

a. After how long will total costs for electric heating and gas heating be the same? What will be the cost at that time?

b. Let $x = 0, 5, 10, 15$, and 20, and graph the models for total electric cost and gas cost over x years. Create a table of coordinates and graph the two models in the same rectangular coordinate system. Then use the graphs to visualize your solution in part (a).

102. A person purchasing a house is considering the following two options for a $40,000 mortgage.

    Option 1:   8.5% interest on a 30-year loan and no points

    Option 2:   8% interest on a 30-year loan with 3 points

Use Table 1.11 on page 96 to determine how long it will take for the total cost of each mortgage to be the same.

103. According to the U.S. Office of Management and Budget, the U.S. gross federal debt in 1970 was $381 billion, increasing at about $43 billion each year. When will the federal debt reach $1370 billion?

104. A salesperson earns $300 per week plus 5% commission of sales. How much must be generated to earn $800 in a week?

105. The bus fare in a city is $1.50. People who use the bus have the option of purchasing a monthly coupon book for $25.00. With the coupon book, the fare is reduced to $0.25. How many times must a person use the bus in a month to make the purchase of a coupon book worthwhile?

106. A bookcase is to be constructed as shown in the figure. The length is to be 3 times the height. If 60 feet of lumber is available for the entire unit, find the length and height of the bookcase.

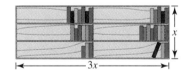

107. The most expensive painting ever sold was Vincent van Gogh's *Portrait of Dr. Gachet,* selling for $75,000,000 at Christie's on May 15, 1990. The etching's length is $1\frac{1}{4}$ inches longer than its width, and its perimeter is 26 inches.

a. What are the dimensions of the etching?

b. What was the selling price of the etching per square inch?

Vincent van Gogh (1853–1890) "Portrait of Dr. Gachet"/Musee d'Orsay, Paris, France/Scala/Art Resource, NY

108. The length of a rectangle is twice the length of a side of a square, and the width of the rectangle is 1 yard less than the side of the square. If the perimeter of the rectangle is 8 yards more than the perimeter of the square, find the length of a side of the square.

109. Two cars start from the same point and travel in opposite directions. One car travels at the rate of 50 mph and the other at the rate of 60 mph. In how many hours will the cars be 660 miles apart?

110. Two trains leave the same point, traveling in opposite directions. The rate of the faster train is 20 mph faster than the rate of the slower train. At the end of 5 hours, the trains are 500 miles apart. Find the rate of each train.

111. Suppose $5000 is invested, part at 6% for 2 years and the rest at 7% for 3 years. If the total interest is $735, how much is invested at each rate?

112. The numbers are arranged in the following square so that all rows, columns, and diagonals contain three numbers having identical sums. The number x in the upper-right corner satisfies the equation

$$\frac{x-2}{2} - \frac{x-3}{4} = \frac{7}{4}.$$

Fill in the missing numbers in the empty squares.

	3	x
5		9
6		4

113. How many ways are there of making change for a quarter using only pennies, nickels, and dimes?

114. The product of the ages of three people is 96. If only one person is a teenager and each age is an integer, what are the ages of the three people?

115. If 1 corresponds to $\frac{1}{3}$, 2 to $\frac{2}{5}$, 3 to $\frac{3}{7}$, 4 to $\frac{4}{9}$ and 5 to $\frac{5}{11}$, what does x correspond to?

CHAPTER 1 TEST

1. Use set-builder notation and interval notation to rewrite the following sentence: x is at least 6.

2. The accompanying graph shows the average size of households in the United States for five selected years. If x represents the year and y the average number of household members, write the year or years described by

$$y \in (2.62, 3.13].$$

Average Size of U.S. Household for Five Selected Years

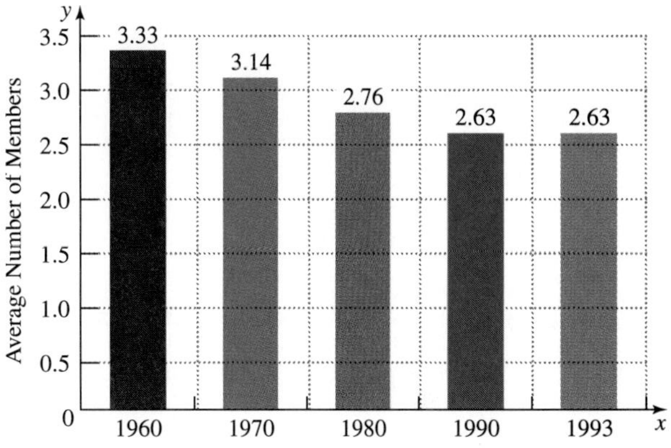

Source: U.S. Census Bureau

Use the order of operations to find the value of each expression.

3. $24 - 36 \div 4 \cdot 3$

4. $\dfrac{8 - [|-4| - (1 - 3)^2]}{5 - (-3)^2 + 6 \div 3}$

Simplify each algebraic expression.

5. $-7(x - 5) - 3(x + 6)$

6. $5y^2 - [7y - 3y(9y - 1)]$

7. The mathematical model

$$A = -24t + 624$$

describes the total number A, in thousands, of United States Air Force personnel on active duty t years after 1986. How many more people were on active duty in the Air Force in 1993 than in 1996?

8. Construct a table of coordinates for $y = 4 - x^2$ and graph the equation in a rectangular coordinate system.

x	-3	-2	-1	0	1	2	3
y							

Use the accompanying graph that shows the number of nuclear power plants in the United States to answer Problems 9 and 10.

9. Estimate the coordinates of point *A* and describe what this means in practical terms.

10. The number of nuclear power plants peaked and is now in decline. What is the reasonable estimate of the number of nuclear power plants at its peak? When did this occur?

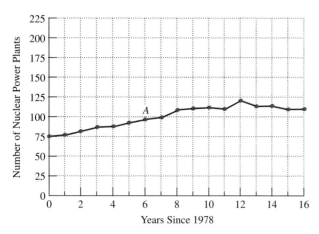

Use the accompanying graph that shows the death rate per 1000 people at various ages to answer Problems 11 and 12.

11. What age has an expected death rate of 20 people per 1000?

12. Describe one trend shown by the graph.

search the same as what NASA spent on research? Approximately how much did NASA spend for that year?

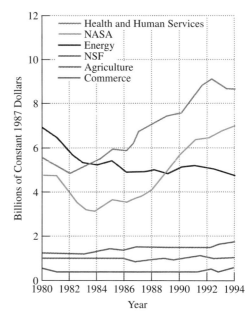

13. The graph on the right shows spending on scientific research. In what year was spending on energy re-

Simplify each expression, writing your answer without negative exponents. Assume that no denominator is 0.

14. $(-2x^5)(-7x^{-11})$

15. $\left(\dfrac{6x^5 y^{-6}}{-3xy^{-4}} \right)^2$

Perform the indicated operations and express the answer in scientific notation.

16. $(3.7 \times 10^4)(9.0 \times 10^{13})$

17. $\dfrac{6 \times 10^{-3}}{1.2 \times 10^{-8}}$

Solve.

18. $3(2x - 4) = 9 - 3(x + 1)$

19. $\dfrac{2x - 3}{4} = \dfrac{x - 4}{2} - \dfrac{x + 1}{4}$

Solve for the variable indicated.

20. $A = P + Prt$, for r

21. $2s - nf - nl = 0$, for n

22. The formula

$$y = 2350x + 22,208$$

is an approximate model for the amount of money y, in millions of dollars, spent by U.S. travelers in other countries x years after 1984. In what year does the model predict that U.S. travelers will spend $66,858 million abroad?

23. According to the U.S. Office of Management and Budget, the amount of money spent for national defense has decreased every year since 1994. In 1995, the defense budget was $14 billion less than 1994, and in 1996 it was $23 billion less than what it was in 1994. If the total defense budget for the three years combined was $623 billion, determine the output in the federal budget for national defense for 1994, 1995, and 1996.

24. With a 9% raise, a physical therapist will earn $34,880 annually. What is the physical therapist's annual salary prior to this raise?

25. Approximate population and growth figures for two states are given as follows.

	1980 Population (in Thousands)	Yearly Growth (in Thousands)
Arizona	2795	89
South Carolina	3071	43

When did Arizona have the same population as South Carolina?

26. A computer on-line service charges a flat monthly rate of $25 or a monthly rate of $10 plus 25 cents for every hour spent on-line. How many hours on-line each month will result in the same charge for the two billing options?

27. The length of a rectangular piece of property exceeds twice the width by 1 yard. If the perimeter is 302 yards, find the property's length and width.

28. A certain amount of money was invested for one year at 6% simple interest. The remainder of the money was invested at 8% under the same conditions. If $3000 more was invested at 8% than at 6% and the total yearly income from the two investments was $520, how much was invested at each rate?

29. A firefighter spraying water on a fire stood on the middle rung of a ladder. Once the smoke diminished, the firefighter moved up 4 rungs, but it got too hot, so the firefighter backed down 6 rungs. Later, the firefighter went up 7 rungs and stayed until the fire was out. Then, the firefighter climbed the remaining 4 rungs and entered the building. How many rungs does the ladder have?

Functions, Linear Functions, and Inequalities

Neil Jenney, *Atmosphere*, 1976

T he amount of carbon dioxide in the atmosphere has increased by half again since 1900, as a result of the burning of oil and coal. Burning fossil fuel results in a buildup of gases and particles that trap heat and raise the planet's temperature. The resultant gradually increasing temperature is called the greenhouse effect. Carbon dioxide accounts for about half of the warming.

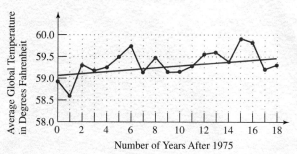

Figure 2.1

Figure 2.1 shows an overall trend in the increase of average global temperature. The data points can be modeled approximately by the blue line in the graph. If further measurements indicate that these points begin to fall on or near a line, the equation of this line would provide environmentalists with a way to predict average global temperature in the future.

Figure 2.1 establishes a correspondence between time and average global temperature. Each year after 1975 corresponds to exactly one temperature. Under these circumstances, we say that global temperature is a *function* of time.

The concept of a function is fundamental to all mathematics. In this chapter we use functions, linear functions, and inequalities as three powerful tools for modeling the world and solving some of its problems.

Solutions   Tutorial   Video
Manual   2

SECTION 2.1

Introduction to Functions

Objectives

1 Find the domain and range of a relation.
2 Determine whether a relation is a function.
3 Use $f(x)$ notation.
4 Interpret function values for mathematical models.
5 Graph functions.
6 Use the vertical line test.
7 Analyze the graph of a function.

In this section, we introduce functions, their notation, and their graphs. We will be using functions to model real world phenomena throughout the book.

Relations

The man who opened the gateway to higher mathematics, René Descartes, did so because his rectangular coordinate system made it possible to construct graphs for sets of ordered pairs. The mathematical term for a set of ordered pairs is a *relation*.

Definition of a relation

A *relation* is any set of ordered pairs. The set of all first components of the ordered pairs is called the *domain* of the relation, and the set of all second components is called the *range*.

Find the domain and range of a relation.

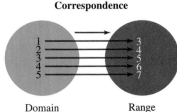

Figure 2.2

EXAMPLE 1   **Analyzing a Relation**

Find the domain and range of:

$$\{(1, 3), (2, 4), (3, 5), (4, 6), (5, 7)\}$$

Solution

The domain is the set of all first components, and the range is the set of all second components. Thus, the domain is $\{1, 2, 3, 4, 5\}$ and the range is $\{3, 4, 5, 6, 7\}$. ■

The relation in Example 1, represented by a set of ordered pairs, can also be represented by:

1. A table of coordinates.

x	1	2	3	4	5
y	3	4	5	6	7
(x, y)	$(1, 3)$	$(2, 4)$	$(3, 5)$	$(4, 6)$	$(5, 7)$

2. A graph (see Figure 2.2).
3. A rule: For each natural number from 1 to 5, inclusively, add 2 to obtain its corresponding value.
4. An equation: $y = x + 2$, for $1 \leqslant x \leqslant 5$ and $x \in \{1, 2, 3, 4, \ldots\}$.

In our first example, each member of the domain corresponded to exactly one member of the range. This is not the case in the next example.

EXAMPLE 2   Analyzing a Relation

Find the domain and range of:

$$\{(0,0), (1,1), (1,-1), (4,2), (4,-2), (9,3), (9,-3)\}$$

Solution

The domain is the set of all first components, and the range is the set of all second components.

Correspondence

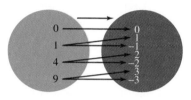

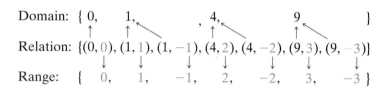

The range can equivalently be written as $\{-3, -2, -1, 0, 1, 2, 3\}$. As shown in the margin, three members of the domain correspond to more than one member of the range. ∎

The relation in Example 2, represented by a set of ordered pairs, can also be represented by:

1. A table of coordinates.

x	0	1	1	4	4	9	9
y	0	1	-1	2	-2	3	-3
(x,y)	$(0,0)$	$(1,1)$	$(1,-1)$	$(4,2)$	$(4,-2)$	$(9,3)$	$(9,-3)$

2. A graph (see Figure 2.3).
3. A rule: For the integers 0, 1, 4, and 9, assign to each its positive and negative square root:

$$0 \to +\sqrt{0} = 0 \qquad 1 \to +\sqrt{1} = +1$$
$$4 \to +\sqrt{4} = +2 \qquad 9 \to +\sqrt{9} = +3$$

4. An equation: $y = \pm\sqrt{x}$, for $x = 0, 1, 4, 9$.

Figure 2.3

The graph of the relation $\{(0,0)$ $(1,1), (1,-1), (4,2), (4,-2),$ $(9,3), (9,-3)\}$

2 Determine whether a relation is a function.

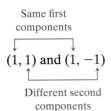

Same first components

(1, 1) and (1, −1)

Different second components

Not the ordered pairs of a function

Functions

A relation in which each member of the domain corresponds to exactly one member of the range is called a *function*. The relation in Example 1 is a function. However, the relation in Example 2 is not a function because three members of the domain correspond to more than one member of the range. Equivalently, a function is a relation in which no two ordered pairs have the same first component and different second components. The ordered pairs of Example 2 (1, 1) and (1, − 1) or (4, 2) and (4, − 2) could not be ordered pairs of a function.

> **Definition of a function**
>
> A *function* is a correspondence between a first set, called the *domain*, and a second set, called the *range*, such that each element in the domain corresponds to *exactly one* element in the range.

Notice that each element in the domain of a relation corresponds to *at least one* element in the range. Thus, a function is a special kind of relation.

EXAMPLE 3 **Determining Whether a Rule Describes a Function**

Determine whether the rule describes a function:

Domain	Rule	Range
a. The set of all U.S. states	Each state's population at a given point in time	A set of positive numbers
b. The set of real numbers	Each number's cube	The set of real numbers
c. The set of people in a town	Each person's uncle	A set of men

Solution

a. The rule does describe a function. At a given point in time, each state has exactly one population.
b. The rule does describe a function. Each real number has only one cube. We can express this function by an equation: $y = x^3$.
c. The rule does not describe a function. Some people have more than one uncle. ■

3 Use $f(x)$ notation.

Function Notation

Functions are usually given in terms of equations, rather than as sets of ordered pairs. These equations are expressed in a special notation. You can understand this notation if you think of a function as a machine into which you

Input *x*

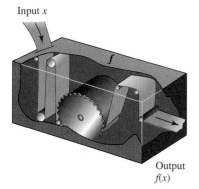

Output
f(x)

Figure 2.4

Functions as machines with inputs and outputs

put members of the domain. The machine is programmed with a rule or an equation that defines the relationship between inputs and outputs. Consequently, the machine gives you a member of the range (the output).

As in Figure 2.4, the letter *f* is frequently used to name functions. The input is represented by *x* and the output by *f(x)*. The special notation *f(x)*, read "*f* of *x*" or "*f* at *x*," represents the value of the function at *x*.

Functions that are given as equations use the *f(x)* notation. For example, *f(x)* = 3*x* − 1 describes the function *f* that takes an input *x*, multiplies it by 3, and then subtracts 1.

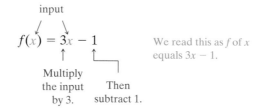

input

$$f(x) = 3x - 1$$

We read this as *f* of *x* equals 3*x* − 1.

Multiply the input by 3.   Then subtract 1.

For example, to find *f*(5) (*f* of 5), we take the input 5, multiply it by 3, and subtract 1 to get 14. Here's how it looks in function notation.

$$f(x) = 3x - 1$$ This is the given function.
$$f(5) = 3 \cdot 5 - 1$$ To find *f*(5), substitute 5 for *x*. The input is 5.
$$= 14$$

Since *f* of 5 equals 14, when the machine's input is 5, its output is 14. When the domain of the function has the value 5, the corresponding value for the range is 14.

The function *f(x)* = 3*x* − 1 can also be expressed as *y* = 3*x* − 1, replacing *f(x)* by *y*. The value of *y*, the *dependent variable,* is calculated after selecting a value for *x*, the *independent variable.*

The notation involving *f(x)* is more compact than the notation with *y*. Here's an example:

a. If *f(x)* = 3*x* − 1, then *f*(5) = 14.
b. If *y* = 3*x* − 1, then the value of *y* is 14 when *x* is 5.

EXAMPLE 4   **Using Function Notation**

Find the indicated function value.

a. *f*(4) for *f(x)* = 2*x* + 3
b. *g*(−2) for *g(x)* = 2*x*² − 1
c. *h*(−5) for *h(r)* = *r*³ − 2*r*² + 5
d. *F*(*a* + *h*) for *F(x)* = 5*x* + 7

Solution

a. $f(x) = 2x + 3$ This is the given function.
$$f(4) = 2 \cdot 4 + 3$$ To find *f* of 4, replace *x* with 4.
$$= 8 + 3$$
$$= 11$$ Thus, *f* of 4 is 11.

b. $g(x) = 2x^2 - 1$ This is the given function.
$$g(-2) = 2(-2)^2 - 1$$ To find *g* of −2, replace *x* with −2.
$$= 2(4) - 1$$
$$= 8 - 1$$
$$= 7$$ Corresponding to the value of the independent variable −2 is the dependent variable's value 7.

c. $h(r) = r^3 - 2r^2 + 5$ The function's name is h and r represents the independent variable.

$h(-5) = (-5)^3 - 2(-5)^2 + 5$ To find h at -5, replace r with -5.
$= -125 - 2(25) + 5$ Evaluate exponential expressions.
$= -125 - 50 + 5$ Multiply.
$= -170$ Thus, h of -5 is -170.

d. $F(x) = 5x + 7$ This is the given function.

$F(a + h) = 5(a + h) + 7$ Replace x with $a + h$.
$= 5a + 5h + 7$ Apply the distributive property. Corresponding to the input $a + h$ is the output $5a + 5h + 7$.   ∎

4 Interpret function values for mathematical models.

Modeling with Functions

Table 2.1 presents examples of mathematical models expressed in the special notation for functions. Throughout this course our focus will be on the process of creating functions that model real world phenomena.

TABLE 2.1 Mathematical Models in Function Notation

Mathematical Model and What It Describes	The Model in Function Notation	Finding an Indicated Function Value	What the Function Value Means
$S = 553.37t + 27{,}966$ Teachers' annual salaries t years after 1985	$S(t) = 553.37t + 27{,}966$ or equivalently $f(x) = 553.37x + 27{,}966$	$S(8) = 553.37(8) + 27{,}966$ $= 32{,}392.96$	8 years after 1985, in 1993, teachers earned about $32,392.96.
$N = 0.036x^2 - 2.8x + 58.14$ The number of deaths per year per thousand people in the U.S. (N) for people who are x years old, where $40 \le x \le 60$	$N(x) = 0.036x^2 - 2.8x + 58.14$ or equivalently $f(x) = 0.036x^2 - 2.8x + 58.14$	$N(60) = 0.036(60)^2$ $\quad - 2.8(60) + 58.14$ $= 19.74$	Approximately 20 people per thousand 60 year olds die annually.
$E = \dfrac{t + 66.94}{0.01t + 1}$ Life expectancy for a U.S. citizen born t years after 1920	$E(t) = \dfrac{t + 66.94}{0.01t + 1}$ or equivalently $f(x) = \dfrac{x + 66.94}{0.01x + 1}$	$E(70) = \dfrac{70 + 66.94}{0.01(70) + 1}$ ≈ 80.6	U.S. citizens born 70 years after 1920, in 1990, can expect to live about 80.6 years.
$f = 0.432h - 10.44$ The length of a woman's femur, f, in inches, as a function of her height, h, also in inches.	$f(h) = 0.432h - 10.44$ or equivalently $f(x) = 0.432x - 10.44$ Femur f in. h in.	$f(60) = 0.432(60) - 10.44$ $= 15.48$	A woman who is 5 feet (60 inches) tall has a femur that measures 15.48 inches.

5 Graph functions.

Graphs of Relations and Functions

The graph of a relation or function is the graph of its ordered pairs.

EXAMPLE 5 **Graphing Functions**

Graph the functions $f(x) = 2x$ and $g(x) = 2x + 4$ in the same rectangular coordinate system.

Solution

For each function we select numbers for x, the independent variable, and then *evaluate* $f(x)$ and $g(x)$. Then we can create a table of some of the coordinates for the functions.

x	$f(x) = 2x$	$(x, f(x))$
-4	$f(-4) = 2(-4) = -8$	$(-4, -8)$
-2	$f(-2) = 2(-2) = -4$	$(-2, -4)$
0	$f(0) = 2(0) = 0$	$(0, 0)$
2	$f(2) = 2(2) = 4$	$(2, 4)$

x	$g(x) = 2x + 4$	$(x, g(x))$
-4	$g(-4) = 2(-4) + 4 = -4$	$(-4, -4)$
-2	$g(-2) = 2(-2) + 4 = 0$	$(-2, 0)$
0	$g(0) = 2(0) + 4 = 4$	$(0, 4)$
2	$g(2) = 2(2) + 4 = 8$	$(2, 8)$

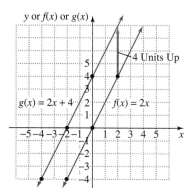

Figure 2.5

Shifting the graph of f four units up gives the graph of g.

We then plot the points and connect them, resulting in the graphs in Figure 2.5. Notice that the y-axis can also be labeled $f(x)$ or $g(x)$. Notice also that the graph of function g can be obtained from the graph of function f by moving the graph of f four units up. ■

The point-plotting method is actually not the most efficient way to graph functions f and g. If we knew in advance that the graphs were straight lines, only two ordered pairs would be necessary to graph each function. We'll concentrate on this issue in the next section as we study linear functions.

EXAMPLE 6 **Graphing a Mathematical Model**

The function $f(x) = 2x^2 + 22x + 320$ describes the number of inmates in federal and state prisons in thousands x years after 1980, where $0 \leq x \leq 20$. Graph the function and use the graph to estimate the number of prisoners for the year 1997.

Solution

We begin by creating a table of some of the coordinates for the function. The graph is the curve passing through these ordered pairs.

x	$f(x) = 2x^2 + 22x + 320$	$(x, f(x))$
0	$f(0) = 2(0)^2 + 22(0) + 320$ $= 320$	$(0, 320)$
5	$f(5) = 2(5)^2 + 22(5) + 320$ $= 480$	$(5, 480)$
10	$f(10) = 2(10)^2 + 22(10) + 320$ $= 740$	$(10, 740)$
15	$f(15) = 2(15)^2 + 22(15) + 320$ $= 1100$	$(15, 1100)$
20	$f(20) = 2(20)^2 + 22(20) + 320$ $= 1560$	$(20, 1560)$

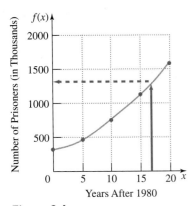

Figure 2.6

The graph of
$f(x) = 2x^2 + 22x + 320$

Plotting the five ordered pairs and connecting them with a smooth curve gives the graph shown in Figure 2.6. The graph indicates that the number of prisoners has increased at a faster rate than a steady year-by-year increase, which would be depicted by a straight line. Seeing the function's graph gives us a better understanding of the relationship between incarcerated Americans and time.

To estimate from the graph the number of prisoners in 1997, locate the point on the graph that is directly above 17. (1997 is 17 years after 1980.) Estimate the second coordinate by moving horizontally from the point to the vertical axis. As shown by the dashed blue line in Figure 2.6, the second coordinate is about three-fifths of the distance between 1000 and 1500. Thus, in the year 1997 there were approximately 1300 thousand prisoners.

Check

We can find an exact value by substituting 17 for x into the function's equation.

$$f(17) = 2(17)^2 + 22(17) + 320 = 1272$$

This indicates that our estimate was off by 28 thousand. ∎

6 Use the vertical line test.

The Vertical Line Test

Not every graph in the rectangular coordinate system is the graph of a function. The definition of a function specifies that no value of x can be paired with two or more different values of y. Consequently, if a graph contains two or more different points with the same first coordinate, the graph cannot represent a function. This is illustrated in Figure 2.7. Observe that points sharing a common first coordinate are vertically above or below each other.

This observation leads to a useful test for determining whether a relation defines y as a function of x. The test is called the *vertical line test.*

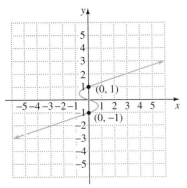

Figure 2.7

y is not a function of x because 0 is paired with two values of y, namely, 1 and -1.

The vertical line test for functions

If any vertical line intersects the graph of a relation in more than one point, the relation does not define y as a function of x.

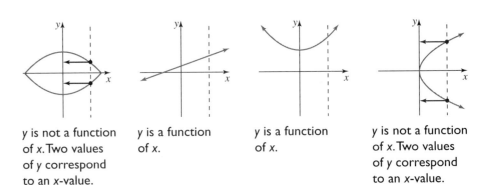

| y is not a function of x. Two values of y correspond to an x-value. | y is a function of x. | y is a function of x. | y is not a function of x. Two values of y correspond to an x-value. |

7 Analyze the graph of a function.

We have seen that functions can be described by tables of coordinates, graphs, rules, and equations. Example 7 shows that graphs of functions enable us to determine quite a bit of information even without an equation that models the data.

EXAMPLE 7 Analyzing Graphs of Functions

The graphs in Figure 2.8 indicate the number of students in U.S. high schools participating in school athletics. The graph for boys is designated by f and the graph for girls by g.

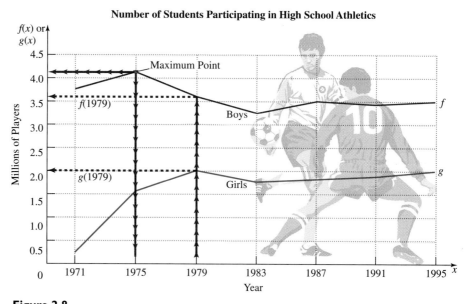

Number of Students Participating in High School Athletics

Figure 2.8

Source: National Federation of State High School statistics

a. How do we know that f and g represent graphs of functions?

b. Estimate the maximum point on the graph of function f. What exactly does this mean?

c. Find a reasonable estimate of

$$f(1979) - g(1979).$$

Describe what this means in practical terms.

d. The second coordinates of f have remained relatively constant for the period from 1987 through 1995. Write an equation that reasonably estimates the data for function f from 1987 through 1995.

e. Describe the general trend shown by the graphs of the functions.

Solution

a. By the vertical line test, f and g represent the graphs of functions. No vertical line intersects either graph more than once. Participation in high school athletics for each group is a function of time. For each year from 1971 through 1995 there corresponds one number of players.

b. The maximum point on the graph of f has an x-coordinate of 1975. We estimate the second coordinate by moving horizontally from the point to the vertical axis. This coordinate is about one-third the distance between 4.0 and 4.5, or at

$$4 + \frac{4.5 - 4.0}{3} = 4 + \frac{0.5}{3} \approx 4.17.$$

Thus, the maximum point is about (1975, 4.17). This means that participation for boys in high school athletics peaked at 4.17 million in 1975.

c. Using the same procedure as in part (b), $f(1979)$ is approximately 3.6 and $g(1979)$ is approximately 2.0. Take a moment to study the graphs in Figure 2.8 on page 137 and verify these approximations. Thus,

$$f(1979) - g(1979) \approx 3.6 - 2.0 = 1.6.$$

This means that in 1979, approximately 1.6 million more boys than girls participated in high school athletics.

d. If all of a function's inputs have the same output value, we have a *constant function*. Although the function f varies from 1971 through 1987, from 1987 through 1995 most of the output values are on 3.5 on the vertical axis. Verify this observation by looking at the graph of f on page 137. Thus, an equation that approximately models the data for the boys is

$$f(x) = 3.5 \quad \text{for} \quad 1987 \leq x \leq 1995.$$

This means that from 1987 through 1995, the number of boys participating in high school athletics remained relatively constant at 3.5 million.

e. The number of boys participating in high school athletics slipped somewhat in the period from 1971 to 1995, but girls' participation increased significantly.   ∎

PROBLEM SET 2.1

Practice Problems

Give the domain and range for each relation in Problems 1–8. Which of the relations are not functions?

1. $\{(1, 3), (1, 7), (1, 10)\}$
2. $\{(3, 1), (7, 1), (10, 1)\}$
3. $\{(2, 3), (2, 4), (3, 3), (3, 4)\}$
4. $\{(2, -1), (3, -1), (4, -1), (5, 0)\}$
5. $\{(-1, -1), (0, 0), (1, 1), (2, 2)\}$
6. $\{(-1, -2), (0, 0), (1, 2), (2, 4)\}$
7. $\{(1, -1), (2, -2), (3, -3), (4, -4)\}$
8. $\{(1, -1), (2, -2), (3, -3), (3, 3)\}$

In Problems 9–16, find the function values. (In Problems 15–16, also answer the question regarding the function.)

9. $f(x) = 3x - 2$
   a. $f(0)$ **b.** $f(-2)$ **c.** $f(7)$ **d.** $f(\frac{2}{3})$ **e.** $f(2a)$

10. $f(x) = 4x - 3$
   a. $f(0)$ **b.** $f(-2)$ **c.** $f(7)$ **d.** $f(\frac{3}{4})$ **e.** $f(2a)$

11. $g(x) = 3x^2 + 5$
   a. $g(0)$ **b.** $g(-1)$ **c.** $g(4)$ **d.** $g(-3)$ **e.** $g(4b)$

12. $g(x) = 2x^2 - 4$
   a. $g(0)$ **b.** $g(-1)$ **c.** $g(4)$ **d.** $g(-3)$ **e.** $g(4b)$

13. $h(x) = 2x^2 + 3x - 1$
   a. $h(0)$ **b.** $h(3)$ **c.** $h(-4)$ **d.** $h(\frac{1}{2})$ **e.** $h(5r)$

14. $h(x) = 3x^2 + 4x - 2$
   a. $h(0)$ **b.** $h(3)$ **c.** $h(-2)$ **d.** $h(\frac{1}{2})$ **e.** $h(5r)$

15. $f(x) = \dfrac{2x - 3}{x - 4}$

    a. $f(0)$ **b.** $f(3)$ **c.** $f(-4)$ **d.** $f(-5)$ **e.** $f(a + h)$

    f. Why must 4 be excluded from the domain of f?

16. $f(x) = \dfrac{3x - 1}{x - 5}$

    a. $f(0)$ **b.** $f(3)$ **c.** $f(-3)$ **d.** $f(10)$ **e.** $f(a + h)$

    f. Why must 5 be excluded from the domain of f?

In Problems 17–20, determine whether the rule describes a function. Explain your answer.

	Domain	**Rule**	**Range**
17.	The set of houses in a town	Each house's area	A set of positive numbers
18.	The air temperature each day at noon at an ocean beach over a one-week period	The number of people at the beach each day at noon over the one-week period	A set of positive numbers
19.	The set of numbers representing the weight of each person in a town	The number of years of education for each person in the town	A set of nonnegative numbers
20.	The set of real numbers	The square of each number	A set of nonnegative numbers

In Problems 21–28, fill in the table of coordinates for functions f, g, and h. Graph the three functions in the same rectangular coordinate system, and then answer the given question.

21. $f(x) = x^2$

x	$f(x) = x^2$
-2	
-1	
0	
1	
2	

$g(x) = x^2 - 2$

x	$g(x) = x^2 - 2$
-2	
-1	
0	
1	
2	

$h(x) = x^2 + 1$

x	$h(x) = x^2 + 1$
-2	
-1	
0	
1	
2	

Describe the relationship among the graphs of f, g, and h. Use the phrase "vertical shift" in your description. In general, if c is a positive number, what is the relationship among the graphs of $f(x)$, $f(x) - c$, and $f(x) + c$?

22. $f(x) = |x|$

x	$f(x) = |x|$
-3	
-2	
-1	
0	
1	
2	
3	

$g(x) = |x| - 2$

x	$g(x) = |x| - 2$
-3	
-2	
-1	
0	
1	
2	
3	

$h(x) = |x| + 1$

x	$h(x) = |x| + 1$
-3	
-2	
-1	
0	
1	
2	
3	

Describe the relationship among the graphs of f, g, and h. Use the phrase "vertical shift" in your description. In general, if c is a positive number, what is the relationship among the graphs of $f(x)$, $f(x) - c$, and $f(x) + c$?

23. $f(x) = \sqrt{x}$

x	$f(x) = \sqrt{x}$
0	
1	
4	
9	

$g(x) = \sqrt{x - 2}$

x	$g(x) = \sqrt{x - 2}$
2	
3	
6	
11	

$h(x) = \sqrt{x + 1}$

x	$h(x) = \sqrt{x + 1}$
-1	
0	
3	
8	

Describe the relationship among the graphs of f, g, and h. Use the phrase "horizontal shift" in your description. In general, if c is a positive number, what is the relationship among the graphs of $f(x)$, $f(x - c)$, and $f(x + c)$?

24. $f(x) = x^2$

x	$f(x) = x^2$
-2	
-1	
0	
1	
2	

$g(x) = (x - 2)^2$

x	$g(x) = (x - 2)^2$
0	
1	
2	
3	
4	

$h(x) = (x + 1)^2$

x	$h(x) = (x + 1)^2$
-3	
-2	
-1	
0	
1	

Describe the relationship among the graphs of f, g, and h. Use the phrase "horizontal shift" in your description. In general, if c is a positive number, what is the relationship among the graphs of $f(x)$, $f(x - c)$, and $f(x + c)$?

25. $f(x) = x^3$

x	$f(x) = x^3$
-2	
-1	
0	
1	
2	

$g(x) = x^3 + 1$

x	$g(x) = x^3 + 1$
-2	
-1	
0	
1	
2	

$h(x) = (x + 1)^3$

x	$h(x) = (x + 1)^3$
-3	
-2	
-1	
0	
1	

Describe the relationship among the graphs of f, g, and h. Use the phrases "vertical shift" and "horizontal shift" in your description. In general, if c is a positive number, what is the relationship among the graphs of $f(x)$, $f(x) + c$, and $f(x + c)$?

26. $f(x) = x^3$

x	$f(x) = x^3$
-2	
-1	
0	
1	
2	

$g(x) = x^3 - 1$

x	$g(x) = x^3 - 1$
-2	
-1	
0	
1	
2	

$h(x) = (x - 1)^3$

x	$h(x) = (x - 1)^3$
-1	
0	
1	
2	
3	

Describe the relationship among the graphs of f, g, and h. Use the phrases "vertical shift" and "horizontal shift" in your description. In general, if c is a positive number, what is the relationship among the graphs of $f(x)$, $f(x) - c$, and $f(x - c)$?

27. $f(x) = \sqrt{x}$

x	$f(x) = \sqrt{x}$
0	
1	
4	
9	

$g(x) = -\sqrt{x}$

x	$g(x) = -\sqrt{x}$
0	
1	
4	
9	

$h(x) = \sqrt{-x}$

x	$h(x) = \sqrt{-x}$
0	
-1	
-4	
-9	

Describe the relationship among the graphs of f, g, and h. Use the phrases "reflection about the x-axis" and "reflection about the y-axis" in your description. In general, what is the relationship among the graphs of $f(x)$, $-f(x)$, and $f(-x)$?

28. $f(x) = \sqrt{x + 1}$

x	$f(x) = \sqrt{x + 1}$
-1	
0	
3	
8	

$g(x) = -\sqrt{x + 1}$

x	$g(x) = -\sqrt{x + 1}$
-1	
0	
3	
8	

$h(x) = \sqrt{-(x + 1)}$

x	$h(x) = \sqrt{-(x + 1)}$
-1	
-2	
-5	
-10	

Describe the relationship among the graphs of f, g, and h. Use the phrases "reflection about the x-axis" and "reflection about the y-axis" in your description. In general, what is the relationship among the graphs of $f(x)$, $-f(x)$, and $f(-x)$?

For Problems 29–40, use the vertical line test to identify in which graphs y is a function of x.

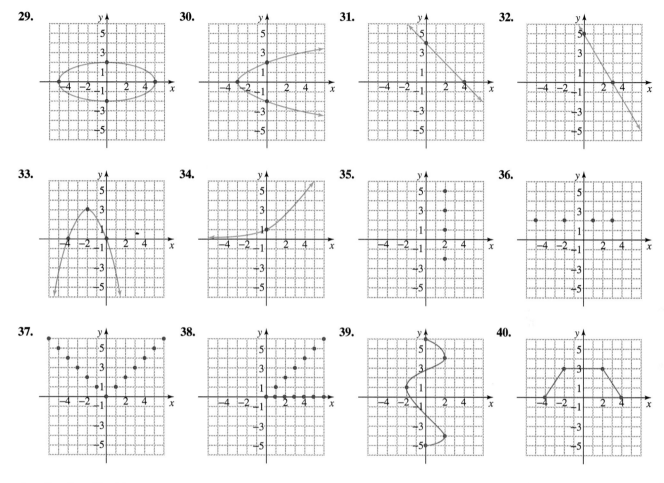

29. **30.** **31.** **32.**

33. **34.** **35.** **36.**

37. **38.** **39.** **40.**

Application Problems

41. The function $f(x) = -0.011x^2 + 1.22x - 8.5$ describes the number of union members in the United States ($f(x)$, in millions) x years after 1930. Find and interpret $f(20)$.

42. *Dimensions* is a twenty-minute piece for two dozen dancers. In this work, the dancers simulate the movement of polygons constrained within a vertical surface. A polygon is a closed figure whose sides are straight-line segments. The number of diagonals in a polygon is a function of the number of its sides n, given by $f(n) = \frac{1}{2}n^2 - \frac{3}{2}n$. How many diagonals has the pentagon simulated by the dancer on the right?

Photograph by Thomas Banchoff of the dance "Dimensions" choreographed by Julie Strandberg at Brown University. (Featured in the Scientific American Library Volume "Beyond the Third Dimension" by Thomas Banchoff.)

43. The number of eggs in a female moth is a function of her abdominal width (x, in millimeters) given by $f(x) = 14x^3 - 17x^2 - 16x + 34$ for $1.5 \leqslant x \leqslant 3.5$.
  a. Find $f(2)$ and interpret this result.
  b. Is $f(1)$ defined by this function?
  c. Write an expression in functional notation designating the difference between the number of eggs in a moth whose abdominal width is 3.2 millimeters and one whose abdominal width is 1.83 millimeters.

44. The threshold weight for a person is defined as "the crucial weight, above which the mortality risk for the patient rises astronomically." For men between the ages of 40 and 49, threshold weight in pounds is a function of height (x, in inches) given by

$$f(x) = \left(\frac{x}{12.3}\right)^3.$$

Find $f(70)$ and interpret this result.

45. The function

$$f(x) = \frac{x + 66.94}{0.01x + 1}$$

models life expectancy of an American child (at birth) born x years after 1920. Find and interpret $f(80) - f(0)$.

46. The Recording Industry Association of America modeled the number of CDs sold as a function of time. The model is given by the function $f(x) = \frac{11}{3}x^2 + \frac{94}{3}x + 23$, in which x represents the number of years after 1985, and the number of CDs sold is expressed in millions. Fill in the following table of coordinates and graph the function. What does the shape of the graph indicate about the sale of CDs over time? Could the trend be described as a steady increase? Explain.

x	$f(x) = \frac{11}{3}x^2 + \frac{94}{3}x + 23$
0	
3	
6	
9	
12	

47. The concentration of a drug (in parts per million) in a patient's bloodstream t hours after the drug is administered is given by the function $f(t) = -t^4 + 12t^3 - 58t^2 + 132t$. Fill in the following table of coordinates and graph the function. What does the shape of the graph indicate about the drug's concentration over time? After how many hours will the drug be totally eliminated from the bloodstream? How is this indicated by the graph?

t	$f(t) = -t^4 + 12t^3 - 58t^2 + 132t$
0	
1	
2	
3	
4	
5	
6	

48. Gonorrhea is the most widespread of all the reportable sexually transmitted diseases in the United States, with about 700,000 new cases annually in recent years. The graph at the top of the next page shows the number of gonorrhea cases per 100,000 people in the United States as a function of time.
  a. Why does f represent the graph of a function?
  b. Estimate the maximum point on the graph of function f. What exactly does this mean?
  c. Estimate the minimum point on the graph of function f. Describe what this means.
  d. Estimate $f(1990) - f(1942)$. What does this mean?

e. If the trend from 1975 through 1990 were to continue, in approximately what year would there be no new cases of gonorrhea?

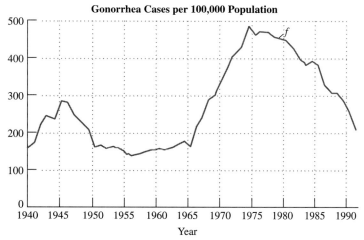

Gonorrhea Cases per 100,000 Population

Year

Source: C.D.C.

49. Cigarette smoking is a worldwide epidemic; 200 million children now under 20 years of age will eventually die from it. In the United States, cigarette smoking kills about 400,000 people every year. The graph shows cigarette consumption in thousands per capita in the United States as a function of time.

a. Why does f represent the graph of a function?

b. Estimate the maximum point on the graph of function f. What exactly does this mean?

c. Estimate the minimum point on the graph of function f. Describe what this means.

d. Estimate $f(1980) - f(1955)$. What does this mean?

e. If the trend from 1980 through 1990 were to continue, in approximately what year would there be no cigarette consumption in the United States?

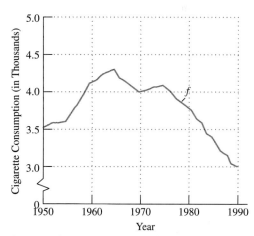

Cigarette Consumption per Capita

Cigarette Consumption (in Thousands)

Year

Source: U.S. Department of Agriculture

True–False Critical Thinking Problems

50. Which one of the following is true?

a. A function is any set of ordered pairs.

b. Some functions are not relations.

c. The set $\{(1, 7), (3, 7), (5, 7), (7, 7)\}$ is a function.

d. If $f(x) = x^2 - 3$, then $f(-2) = -7$.

51. Which one of the following is true?

a. The domain of the relation $\{(3, 1), (5, 1), (4, 1)\}$ is $\{1, 3, 4, 5\}$.

b. If $f(x) = |x + 2|$, then $f(-7) = -5$.

c. The graph of every line defines y as a function of x.

d. The vertical line test determines at a glance if a graph is the graph of a function.

Technology Problems

52. Use a graphing utility to verify the graphs that you drew by hand in Problems 21–28. Use the $\boxed{\text{TRACE}}$ feature to confirm some of the function values.

53. Use a graphing utility to verify the graphs that you drew by hand in Problems 46–47.

54. The function $f(x) = -0.00002x^3 + 0.008x^2 - 0.3x + 6.95$ models the number of annual physician visits by a person of age x.

  a. Use a graphing utility to graph the function with the following range setting:

$$\text{Xmin} = 0, \text{Xmax} = 100, \text{Xscl} = 5, \text{Ymin} = 0,$$
$$\text{Ymax} = 60, \text{Yscl} = 3$$

  b. What does the shape of the graph indicate about the relationship between one's age and the number of annual physician visits?

  c. Use the $\boxed{\text{TRACE}}$ or minimum function capability to find the coordinates of the minimum point on the graph of the function. What does this mean?

55. The rectangle shown in the figure has a perimeter of 80 meters.

  a. Show that the area of the rectangle is given by $A = x(40 - x)$, where x is its length.

  b. Graph the area function with a graphing utility.

  c. Use the graph to determine the value of x that yields the greatest area. Describe what this means in practical terms.

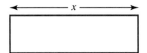

Writing in Mathematics

56. A student in introductory algebra hears that functions are studied in subsequent algebra courses. The student asks you what a function is. Provide the student with a clear, relatively concise response.

57. Describe how the vertical line test tells whether a relation is a function.

58. In a mathematical sense, a person's federal income tax is a function of that person's adjusted gross income. However, a person's caloric intake during a 24-hour period is not a function of the number of minutes that person exercises during the 24 hours. Explain the difference between these two situations, incorporating the definition of function into your explanation.

Critical Thinking Problems

59. Consider a function defined by

$$f(1) = 1$$
$$f(2) = 1 - 2 = -1$$
$$f(3) = 1 - 2 + 3 = 2$$
$$f(4) = 1 - 2 + 3 - 4 = -2$$
$$f(5) = 1 - 2 + 3 - 4 + 5 = 3$$
$$f(6) = 1 - 2 + 3 - 4 + 5 - 6 = -3$$

  a. What is the formula for $f(x)$ if x is even?

  b. What is the formula for $f(x)$ if x is odd?

  c. Find $f(20) + f(40) + f(65)$.

60. If $f(x + y) = f(x) + f(y)$ and $f(1) = 3$, find $f(2)$, $f(3)$, and $f(4)$. Is $f(x + y) = f(x) + f(y)$ for all functions?

Group Activity Problems _____

61. In your group, discuss one possible limitation for each of the four functions in Table 2.1.

62. The function $f(x) = -0.015x^3 + 1.058x$ describes approximate alcohol concentration (in tenths of a percent) in an average person's bloodstream x hours after drinking 8 ounces of 100-proof whiskey, where x is in the interval $[0, 8]$. Find $f(0), f(1), f(2), \ldots$, through $f(8)$.

Now look at the graph. This graph is not the graph of the given function f, but instead shows the risk of having an accident based on alcohol concentration in the bloodstream. As a group, use this graph to discuss what each of the nine evaluations, $f(0)$ through $f(8)$, means in terms of the risk of having an accident. *Caution:* The given function has measurements in tenths of a percent, so be sure to move the decimal point one place to the left before consulting the values on the horizontal axis of the graph.

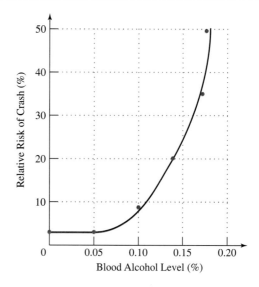

Review Problems _____

63. In 1960 the population of the United States was 179.5 million. With a steady increase of 2.35 million people per year, in what year will the population reach 297 million?

64. The model

$$P = \frac{2}{5}w\left(1 + \frac{n}{50}\right)$$

describes the monthly benefit P paid to a retired worker who had worked n years at a top monthly wage of w dollars. How many years must a person work to receive a monthly retirement benefit of $720 if the top monthly working wage had been $1000?

65. Multiply, expressing the product in scientific notation: $(8.6 \times 10^{14})(2.5 \times 10^{-2})$.

S E C T I O N   2 . 2

Solutions Manual **Tutorial** **Video 2**

Linear Functions and Slope

Objectives

1 Calculate a line's slope.
2 Use slope to show that lines are parallel or perpendicular.
3 Graph linear functions in slope-intercept form.
4 Find a line's slope and y-intercept from its equation.
5 Use intercepts to graph the standard form of a line.
6 Write equations of horizontal and vertical lines.
7 Solve applied problems that interpret slope as average rate of change.

Now that we have introduced functions, their notation, and their graphs, we turn to a special class of functions, called *linear functions*.

An Introduction to Linear Functions

The graphs of some functions are lines. These functions can be written in the form $f(x) = mx + b$ and are called linear functions.

> **Definition of a linear function**
>
> A function that can be expressed in the form
>
> $$f(x) = mx + b$$
>
> is called a *linear function*.

Table 2.2 contains examples of two linear functions. From this table we can conclude that the values of m and b have significant physical interpretations.

TABLE 2.2 Linear Mathematical Models

Linear Model and What It Describes $f(x) = mx + b$	Table of Coordinates		What m and b Signify
$f(x) = 553.37x + 27{,}966$ $m = 553.37 \quad b = 27{,}966$ Teachers' annual salaries, $f(x)$, x years after 1985	x $f(x) = 553.37x + 27{,}966$ 0 $f(0) = 27{,}966$ 1 $f(1) = 28{,}519.37$ 2 $f(2) = 29{,}072.74$ 3 $f(3) = 29{,}626.11$	$+553.37$ $+553.37$ $+553.37$	At the beginning (in 1985), teachers earned \$27,966 yearly. This has increased by \$553.37 each year.
$f(x) = -6.9x + 40.3$ $m = -6.9 \quad b = 40.3$ The percentage of men, $f(x)$, injured in the Boston Marathon by age group x: 0: under 20, 1: 20–29, 2: 30–39, 3: 40–49, 4: 50–59	x $f(x) = -6.9x + 40.3$ 0 $f(0) = 40.3$ 1 $f(1) = 33.4$ 2 $f(2) = 26.5$ 3 $f(3) = 19.6$ 4 $f(4) = 12.7$	-6.9 -6.9 -6.9 -6.9	For the first age group (men under 20), 40.3% were injured. The percentage injured decreased by 6.9% for each subsequent group.

First, we note that for the linear function $f(x) = mx + b$, the value b is the function value at 0:

$$f(0) = m(0) + b = b.$$

Values of b provided us with information about teachers' salaries or percentage of injuries when $x = 0$. Then we observe that m, the coefficient of x, seems to be related to how the function values are consistently changing. For teachers, salaries are increasing by \$553.37 each year. For Boston Marathon age groups, injuries are decreasing by 6.9% for each subsequent group.

Intercepts and Slope

As we know, the graph of the linear function $f(x) = mx + b$ is a line. The value b is related to an important point on the line and the value m is significant in terms of the line's steepness.

Study tip

If you are not using a graphing utility in this course, be sure to graph the functions in the Discover for Yourself box by hand, using tables of coordinates.

Discover for yourself

Use a graphing utility to graph the linear functions in each column in the same viewing rectangle. Use the zoom square feature to avoid possible distortions. In what way are the graphs the same? How does the value of *m* (the coefficient of *x*) affect the graphs? What effect does *b* (the constant term) have on the graphs?

$$f(x) = x + 2 \qquad\qquad f(x) = -x + 2$$
$$g(x) = 2x + 2 \qquad\qquad g(x) = -2x + 2$$
$$h(x) = 3x + 2 \qquad\qquad h(x) = -3x + 2$$

In the Discover for Yourself box, did you discover that the graph of $f(x) = mx + b$ passes through the point $(0, b)$? This is the point at which the graph intersects the *y*-axis. We refer to the number *b* as the graph's *y-intercept.*

Did you discover also that the value *m* indicates the steepness or slant of the line? For the lines in which *m* is positive, there is an upward slant from left to right; for the lines in which *m* is negative, there is a downward slant from left to right. This is illustrated in Figure 2.9.

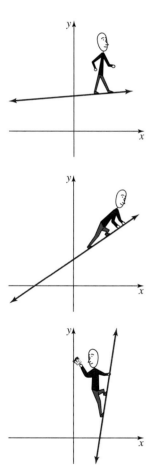

Figure 2.10

Measuring a line's steepness

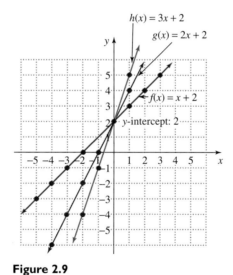

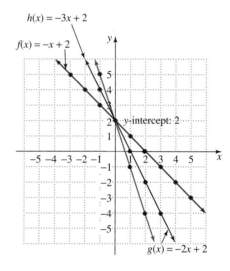

Figure 2.9

When *m* is positive, the lines slant upward.

When *m* is negative, the lines slant downward.

Mathematicians have developed a useful measure of the steepness of a line, called the *slope* of the line. Slope compares the vertical change (the *rise*) to the horizontal change (the *run*) when moving from one fixed point to another along the line (see Figure 2.10). To calculate the slope of a line, mathematicians use a ratio comparing the change in *y* (the rise) to the change in *x* (the run).

Definition of slope

The *slope* of the line through the distinct points (x_1, y_1) and (x_2, y_2) is

$$\frac{\text{Change in } y}{\text{Change in } x} = \frac{\text{Rise}}{\text{Run}}$$

$$= \frac{y_2 - y_1}{x_2 - x_1} = \frac{y_1 - y_2}{x_1 - x_2}$$

where $x_2 - x_1 \neq 0$.

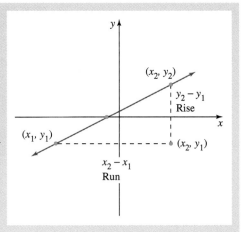

The letter m, the first letter in the French verb *monter,* meaning "to go up or to climb," is used to denote slope.

❙ Calculate a line's slope.

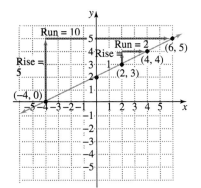

Figure 2.11

EXAMPLE 1   Using the Definition of Slope

Find the slope of the linear function whose graph is shown in Figure 2.11.

Solution

We can select any two points on the line. Using the points $(2, 3)$ and $(4, 4)$, we obtain

$$\text{Slope} = \frac{\text{Rise}}{\text{Run}} = \frac{\text{Change in } y}{\text{Change in } x} = \frac{4 - 3}{4 - 2} = \frac{1}{2}.$$

Suppose that we select another two points, such as $(-4, 0)$ and $(6, 5)$.

$$\text{Slope} = \frac{\text{Rise}}{\text{Run}} = \frac{\text{Change in } y}{\text{Change in } x} = \frac{5 - 0}{6 - (-4)} = \frac{5}{10} = \frac{1}{2}$$

Notice that between any two points on the line, the ratio of the change in y, the rise, to the corresponding change in x, the run, is always $\frac{1}{2}$. ∎

EXAMPLE 2   Finding the Slope of a Line that Falls from Left to Right

a. Graph the line connecting the points $(-2, 5)$ and $(1, -1)$.
b. Find the slope of the line.
c. Subtract the coordinates in the opposite order in the computation of slope. Describe what happens.
d. Use the graph of the linear function to find two other points along the line. Use these points to compute the line's slope.

Solution

a. Using the points $(-2, 5)$ and $(1, -1)$, the line connecting them is shown in Figure 2.12.

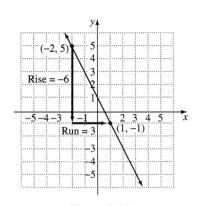

Figure 2.12

b. Using the points $(-2, 5)$ and $(1, -1)$, we compute the line's slope as follows:

$$\text{Slope} = \frac{\text{Change in } y}{\text{Change in } x} = \frac{-1 - 5}{1 - (-2)} = \frac{-6}{3} = -2 \left(\text{or } \frac{-2}{1}\right).$$

As shown in Figure 2.12, the ratio of rise (actually a fall) to run is -2.

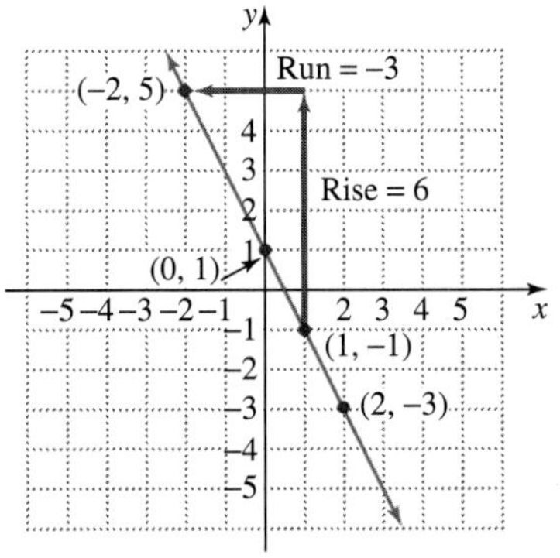

c. Let's see what happens if we subtract coordinates in the opposite order.

$$\text{Slope} = \frac{\text{Change in } y}{\text{Change in } x} = \frac{5 - (-1)}{-2 - 1} \qquad \text{We are still using} \atop (-2, 5) \text{ and } (1, -1).$$

$$= \frac{6}{-3} = -2$$

This is the same answer as our previous result. The computation is illustrated in Figure 2.13. In short, it does not matter which point is considered (x_1, y_1) and which is considered (x_2, y_2) as long as we subtract in the same order in the numerator and the denominator.

d. Let's select two other points along the line. Using Figure 2.13, we have a number of options. We'll choose $(0, 1)$ and $(2, -3)$.

$$\text{Slope} = \frac{\text{Change in } y}{\text{Change in } x} = \frac{-3 - 1}{2 - 0} = \frac{-4}{2} = -2$$

The line's slope is still -2.

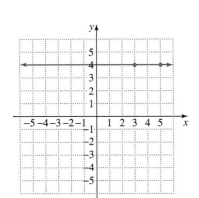

Figure 2.13

EXAMPLE 3 Finding the Slope of a Horizontal Line

Find the slope of the horizontal line connecting the points $(5, 4)$ and $(3, 4)$.

Solution

Letting $(x_1, y_1) = (5, 4)$ and $(x_2, y_2) = (3, 4)$, we obtain

$$m = \frac{y_2 - y_1}{x_2 - x_1} = \frac{4 - 4}{3 - 5} = \frac{0}{-2} = 0.$$

Notice in Figure 2.14 that a horizontal line is neither increasing nor decreasing from left to right. Thus, *the slope of any horizontal line is 0.*

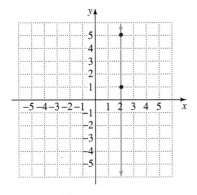

Figure 2.14

A horizontal line with zero slope

EXAMPLE 4 Finding the Slope of a Vertical Line

Find the slope of the line connecting the points $(2, 5)$ and $(2, 1)$.

Solution

Letting $(x_1, y_1) = (2, 5)$ and $(x_2, y_2) = (2, 1)$, we obtain

$$m = \frac{y_2 - y_1}{x_2 - x_1} = \frac{1 - 5}{2 - 2} = \frac{-4}{0} \qquad \text{Undefined}$$

From Figure 2.15, we see that our line is a vertical line with undefined slope. *Any vertical line has undefined slope.*

Table 2.3 summarizes the four possibilities for the slope of a line.

Figure 2.15

A vertical line with undefined slope

TABLE 2.3 Possibilities for a Line's Slope

Positive Slope	Negative Slope
$m > 0$	$m < 0$
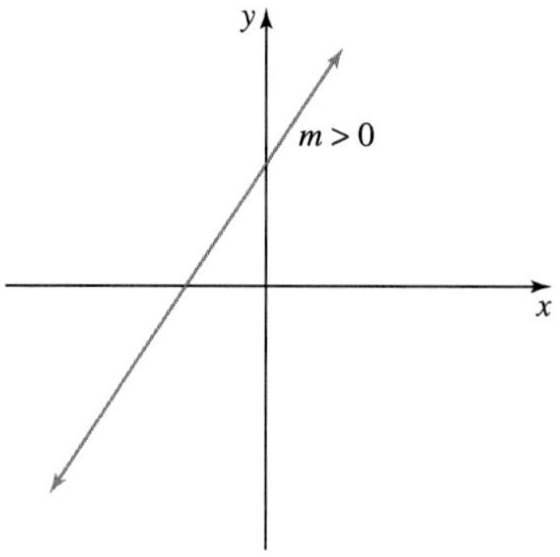 Line rises from left to right.	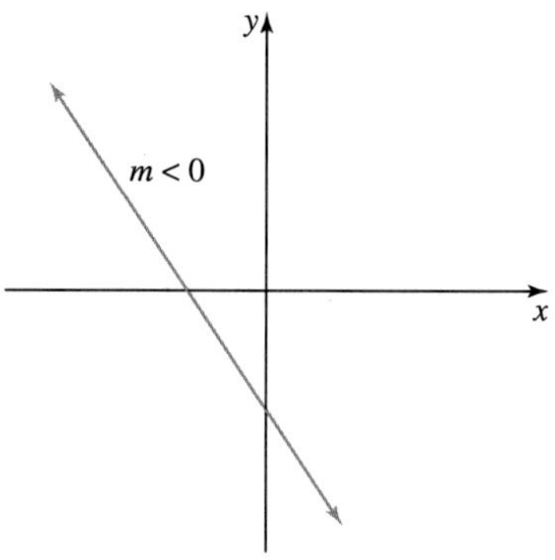 Line falls from left to right.
Zero Slope	**Undefined Slope**
$m = 0$	m is undefined.
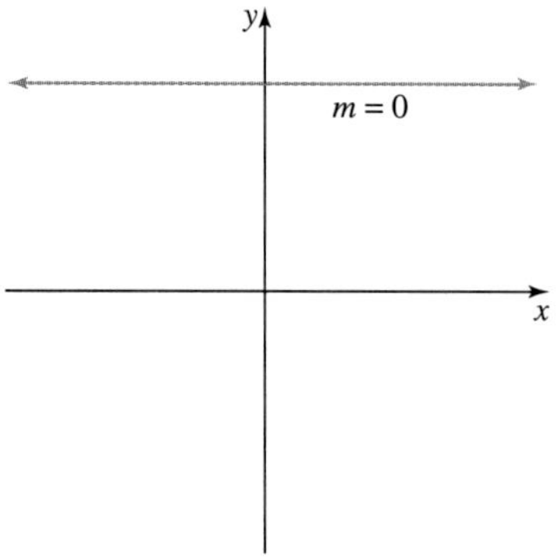 Line is horizontal.	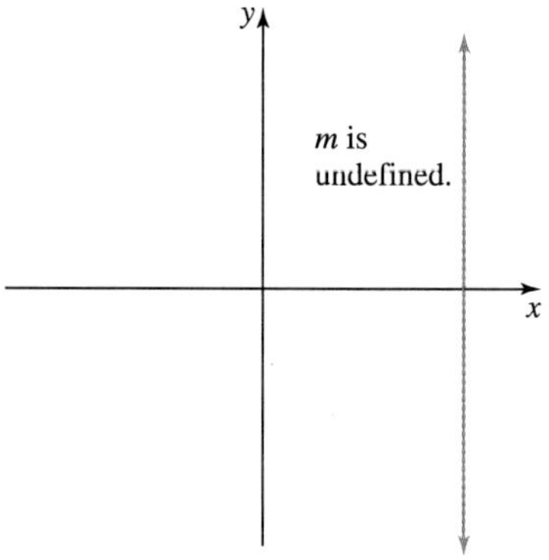 Line is vertical.

2 Use slope to show that lines are parallel or perpendicular.

Slope of Parallel and Perpendicular Lines

Discover for yourself

Use a graphing utility to graph the following linear functions in the same viewing rectangle.

$$y_1 = \frac{2}{3}x + 2, \; y_2 = \frac{2}{3}x, \; y_3 = \frac{2}{3}x - 1$$

Each function has the same slope, namely $\frac{2}{3}$. What do you observe about the graphs? In general, what appears to be true about the graphs of lines with equal slopes?

Now use a graphing utility to graph these linear functions in the same viewing rectangle.

$$y_1 = \frac{1}{2}x + \frac{1}{2}, \; y_2 = -2x - 1$$

Study tip

If you are not using a graphing utility in this course, be sure to graph the linear functions in the Discover for Yourself by hand using tables of coordinates.

Use the zoom square setting. The product of the slopes of these lines, namely $\frac{1}{2}$ and -2, is -1. What do you observe about the graphs? Repeat this process for

$$y_1 = -\frac{1}{3}x + 2, y_2 = 3x - 4$$

In general, what appears to be true about the graphs of lines with slopes whose product is -1?

Two nonintersecting lines that lie in the same plane are parallel. Two lines are parallel if the ratio of the vertical change to the horizontal change is the same for each line. Because two parallel lines have the same "steepness," they must have the same slope.

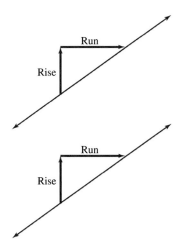

The rise-to-run ratio is the same for parallel lines, so they must have the same slope.

Slope and parallel lines

1. If two nonvertical lines are parallel, then they have the same slope.
2. If two distinct nonvertical lines have the same slope, then they are parallel.
3. Vertical lines with undefined slopes are parallel.

EXAMPLE 5 A Geometric Application of Slope

If the opposite sides of a quadrilateral are parallel, then it is a parallelogram. Determine whether $(-2, 2)$, $(0, 0)$, $(2, 6)$, and $(4, 4)$ are the vertices of a parallelogram.

Solution

Figure 2.16 shows the quadrilateral determined by these points. If the quadrilateral has opposite sides that are parallel, then it is a parallelogram. We calculate the slope of each side.

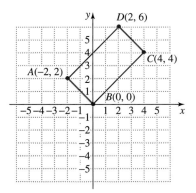

Figure 2.16
The vertices of a parallelogram

Opposite sides AB and DC:

$$m_{AB} = \frac{2 - 0}{-2 - 0} = \frac{2}{-2} = -1$$

$$m_{DC} = \frac{6 - 4}{2 - 4} = \frac{2}{-2} = -1 \qquad \text{The opposite sides have the same slope, so they are parallel.}$$

Opposite sides BC and AD:

$$m_{BC} = \frac{4 - 0}{4 - 0} = \frac{4}{4} = 1$$

$$m_{AD} = \frac{6 - 2}{2 - (-2)} = \frac{4}{4} = 1 \qquad \text{The opposite sides have the same slope, so they are parallel.}$$

With both pairs of opposite sides parallel, the vertices are those of a parallelogram.

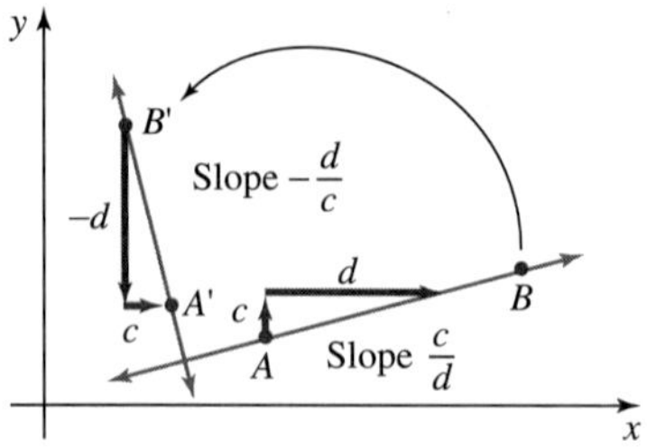

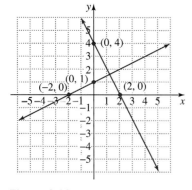

Figure 2.17

Rotate the figure 90° so that line AB coincides with line $A'B'$.

The parallelogram in Figure 2.16 appears to be a rectangle with perpendicular adjacent sides. Notice that the product of the slopes of the adjacent sides is -1. In general, if two lines (neither of which is vertical) are perpendicular, the product of their slopes is always -1. Let's take a moment to see why.

Figure 2.17 shows line AB, with slope $\frac{c}{d}$. Rotate the entire figure 90° to the left to obtain line A'B' perpendicular to line AB. The figure indicates that the rise and the run of the new line are reversed from the original line, but the rise is now negative. This means that the slope of the new line is $-\frac{d}{c}$. Notice that the product of the slopes of the two perpendicular lines is -1.

$$\left(\frac{c}{d}\right)\left(-\frac{d}{c}\right) = -1$$

This relationship holds for all perpendicular lines and is summarized below.

study tip

We can equivalently say that two lines are perpendicular if and only if their slopes are *negative reciprocals*.

Slope and perpendicular lines

1. If two nonvertical lines are perpendicular, then the product of their slopes is -1.
2. If the product of the slopes of two lines is -1, then the lines are perpendicular.
3. A horizontal line having zero slope is perpendicular to a vertical line having undefined slope.

Figure 2.18

Are these lines truly perpendicular?

EXAMPLE 6 **Verifying That Lines Are Perpendicular**

The lines shown in Figure 2.18 appear to be perpendicular. Prove that they actually are.

Solution

We must show that the product of the slopes of the lines is -1. We can determine the slope of each line by using any two points along the line. In each case, the points where the lines intersect the axes seem to be an obvious choice.

For the blue line with negative slope, we use $(0, 4)$ and $(2, 0)$:

$$m = \frac{0 - 4}{2 - 0} = \frac{-4}{2} = -2$$

For the line with positive slope, we use $(0, 1)$ and $(-2, 0)$:

$$m = \frac{0 - 1}{-2 - 0} = \frac{-1}{-2} = \frac{1}{2}$$

Since the product of the slopes is -1 (that is, $-2(\frac{1}{2}) = -1$), the lines are perpendicular. Equivalently, their slopes are negative reciprocals. ∎

The Slope-Intercept Form of a Line

We have seen that linear functions can be expressed in the form

$$f(x) = mx + b.$$

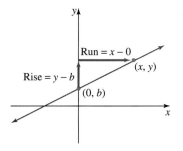

Figure 2.19

A line with slope m and y-intercept b

Now it's time to verify that m, the coefficient of x, is the slope of the line, and the constant term b is the y-intercept. We can do this by writing an equation of the line having slope m and y-intercept b.

Suppose that a line of slope m passes through the point $(0, b)$, so that b is the y-intercept (see Figure 2.19). Let (x, y) stand for all ordered pairs corresponding to points along the line other than $(0, b)$. We can begin with the definition of slope to write the line's equation:

$$\frac{y - b}{x - 0} = m \qquad \text{Slope is the rise-to-run ratio.}$$

$$\frac{y - b}{x} = m \qquad \text{Simplify.}$$

$$y - b = mx \qquad \text{Multiply by } x \text{ to clear fractions.}$$

$$y = mx + b \qquad \text{Add } b \text{ on both sides and solve for } y.$$

Slope-intercept form of the equation of a line

The equation $y = mx + b$ is called the *slope-intercept form of a line*. The coefficient of x is the slope of the line, and the constant term is its y-intercept.

The slope-intercept equation, $y = mx + b$, can be expressed in function notation, by replacing y with $f(x)$.

$$f(x) = mx + b$$

3 Graph linear functions in slope-intercept form.

EXAMPLE 7 **Using the Slope-Intercept Equation of a Line**

Graph the linear function: $f(x) = \frac{2}{3}x + 2$

Solution

The function's equation is in the form $f(x) = mx + b$, so that the slope is $\frac{2}{3}$ and the y-intercept is 2:

$$f(x) = \frac{2}{3}x + 2$$

The slope is $\frac{2}{3}$. The y-intercept is 2.

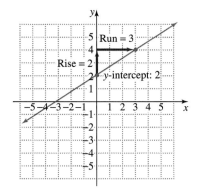

Figure 2.20

The graph of $f(x) = \frac{2}{3}x + 2$

1. We graph the function by first plotting 2, the y-intercept. This gives the point $(0, 2)$.
2. Now, using a slope of $\frac{2}{3}$,

$$m = \frac{2}{3} = \frac{\text{Rise}}{\text{Run}}$$

we locate a second point on the line by moving 2 units up and 3 units to the right, starting with the y-intercept. This puts us at $(0 + 3, 2 + 2)$ or $(3, 4)$.

3. We use a straightedge to draw a line through $(0, 2)$ and $(3, 4)$. The graph of $f(x) = \frac{2}{3}x + 2$ is shown in Figure 2.20 on the previous page. ∎

4 Find a line's slope and y-intercept from its equation.

If an equation in the form $Ax + By = C$ is solved for y, the coefficient of x is the slope and the constant term is the y-intercept. Thus, we can determine the slope and y-intercept of a line from its equation.

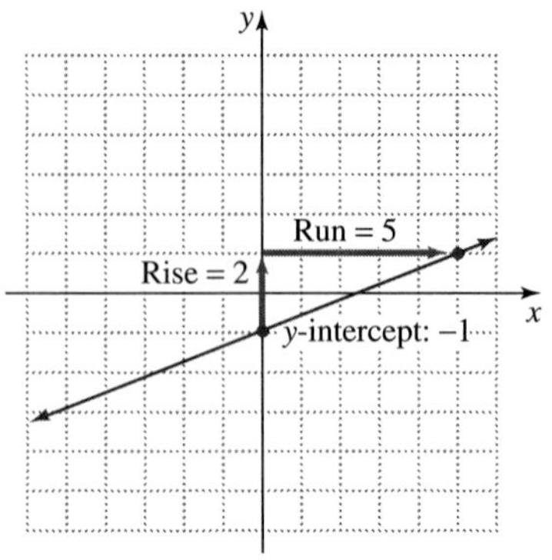

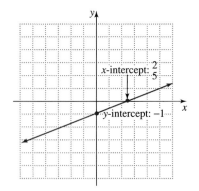

Figure 2.21
The graph of $2x - 5y = 5$ or, equivalently, $y = \frac{2}{5}x - 1$

EXAMPLE 8 **Finding the Slope and y-Intercept from a Line's Equation**

Write $2x - 5y = 5$ in slope-intercept form, and find the slope and y-intercept.

Solution

We put the equation in slope-intercept form by solving for y.

$$2x - 5y = 5 \qquad \text{This is the given equation.}$$
$$-5y = -2x + 5 \qquad \text{Subtract } 2x \text{ from both sides.}$$
$$y = \frac{2}{5}x - 1 \qquad \text{Divide both sides by } -5.$$

slope: $\frac{2}{5}$ y-intercept: -1

From the slope-intercept form, the slope is $\frac{2}{5}$ and the y-intercept is -1. The line is graphed in Figure 2.21. ∎

5 Use intercepts to graph the standard form of a line.

The Standard Form of a Line

In Example 8, we started with the equation $2x - 5y = 5$ and wrote it in slope-intercept form. The equation $2x - 5y = 5$ represents the *standard form* of a line's equation.

> **Standard form of the equation of a line**
>
> The *standard form* of the equation of a line is
>
> $$Ax + By = C$$
>
> where A and B are integers, with A and B not both zero. In this form, the coefficient of x, namely A, is usually written as a positive integer. The standard form is also called the *general form*.

Figure 2.22
The graph of $2x - 5y = 5$ with its intercepts

[figure: graph showing x-intercept $\frac{2}{5}$ and y-intercept -1]

Discover for yourself

The graph of $2x - 5y = 5$ passes through $(0, -1)$ and $(2.5, 0)$, so the y-intercept is -1 and the x-intercept is 2.5, as shown in Figure 2.22. How can you determine these intercepts using the equation $2x - 5y = 5$?

If a line's equation is in standard form, we can quickly obtain its graph by plotting the y-intercept and the x-intercept. These are the points at which the

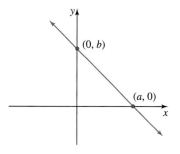

A line with x-intercept *a* and y-intercept *b*

graph intersects, respectively, the *y*- and *x*-axes. The nonzero coordinates of these points are called the graph's intercepts. In the Discover for Yourself box, did you observe the following method for locating these intercepts?

Locating intercepts

To locate an *x*-intercept *a*, let $y = 0$ in the given equation and solve for *x*. The graph passes through $(a, 0)$.

   To locate a *y*-intercept *b*, let $x = 0$ in the given equation and solve for *y*. The graph passes through $(0, b)$.

EXAMPLE 9   **Using Intercepts to Graph a Linear Equation**

Graph: $3x + 2y = 6$

Solution

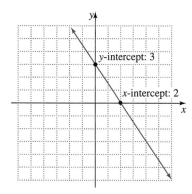

Figure 2.23
The graph of $3x + 2y = 6$

Find the x-intercept by letting $y = 0$ and solving for x.	**Find the y-intercept by letting $x = 0$ and solving for y.**
$3x + 2 \cdot 0 = 6$	$3 \cdot 0 + 2y = 6$
$3x = 6$	$2y = 6$
$x = 2$	$y = 3$

The *x*-intercept is 2, so the line passes through (2, 0). The *y*-intercept is 3, so the line passes through (0, 3). The graph of $3x + 2y = 6$ is shown in Figure 2.23. ∎

Write $3x + 2y = 6$ in slope-intercept form. Use the *y*-intercept and slope to graph the linear function. Do you obtain the same graph as the one in Figure 2.23?

6 Write equations of horizontal and vertical lines.

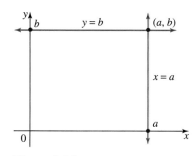

Figure 2.24
Equations of horizontal and vertical lines

Equations of Horizontal and Vertical Lines

Now let's consider equations for horizontal and vertical lines. If a line is horizontal, its slope is zero $(m = 0)$, so the equation $y = mx + b$ becomes $y = b$, where *b* is the *y*-intercept. This is illustrated in Figure 2.24.

   A vertical line has slope that is undefined. However, Figure 2.24 illustrates that the *x*-coordinate of every point on the vertical line is *a*. We can write its equation as $x = a$, where *a* is the *x*-intercept.

Vertical and horizontal lines

The equation of a horizontal line through (a, b) is $y = b$. The line's *y*-intercept is *b*.

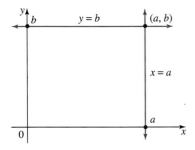

Figure 2.24 (repeated from the previous page)

Equations of horizontal and vertical lines

The equation of a vertical line through (a, b) is $x = a$. The line's x-intercept is a.

In order that you don't have to turn back a page, Figure 2.24 is shown again on the left. Look again at the vertical line in the figure. For all points along the line, the x-coordinates are a. This graph does not define y as a function of x. A vertical line drawn through a intersects the graph infinitely many times, showing that associated with the domain value a are infinitely many range values.

Now look at the horizontal line in Figure 2.24. Any vertical line intersects the graph of $y = b$ only once, so that a horizontal line represents the graph of a function. This means that we can express the equation as $f(x) = b$, called a *constant function*. Our next example illustrates how some data can be modeled with this function.

EXAMPLE 10 **Modeling Data with a Constant Function**

From 1975 to the present, per capita consumption of beer in the United States has remained relatively constant, indicated by the graph shown in Figure 2.25. Write an equation that approximately models per capita beer consumption from 1975 through 1990.

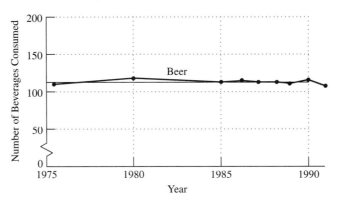

Figure 2.25

Per capita beer consumption

Solution

Shown in Figure 2.25 is a horizontal line that passes through most of the data points. The line intersects the vertical axis about one-fourth of the distance between 100 and 150, or at 112. Thus, an equation that models beer consumption is $y = 112$ or $f(x) = 112$. ■

7 Solve applied problems that interpret slope as average rate of change.

An Application of Slope: Slope as the Average Rate of Change

Slope is defined as the ratio of a change in y to a corresponding change in x. Consequently, in applied situations slope can be thought of as the *average change in y per unit of change in x,* where y is a function of x. This idea is illustrated in Example 11.

Jack Levine "Election Night" 1954, oil on canvas, $63\frac{1}{8}$ in. × 6 ft. $\frac{1}{2}$ in. (160.3 × 184.1 cm). The Museum of Modern Art, New York. Gift of Joseph H. Hirshhorn. Photograph. © 1997 The Museum of Modern Art, New York. © 1998 Jack Levine/Licensed by VAGA, New York, NY.

EXAMPLE 11 **Slope as the Average Rate of Change**

The increasing prosperity of upper-income America is illustrated in the graph in Figure 2.26, which shows a dramatic increase in the number of Americans with incomes of a million dollars or more.

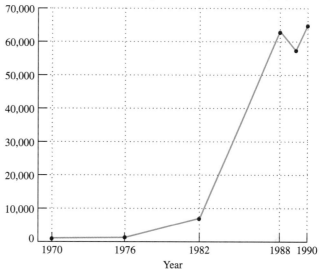

Figure 2.26
Source: U.S. Bureau of the Census

a. Estimate the slope of the line segments between 1976 and 1982 and between 1982 and 1988. Compare the two slopes and explain what this means in terms of average rate of change.

b. By looking at the graph, what can you say about the slope of the line segment between 1988 and 1989? Estimate the slope for this line segment.

Solution

a. We need to estimate the y-coordinates for the points on the graph above 1976, 1982, and 1988. We've also estimated the y-coordinate for 1989 in order to solve part (b).

Year	Approximate Location of Point	Estimate of y-coordinate	Point
1976	$\frac{1}{5}$ of the distance between 0 and 10,000	2000	(1976, 2000)
1982	$\frac{4}{5}$ of the distance between 0 and 10,000	8000	(1982, 8000)
1988	$\frac{1}{5}$ of the distance between 60,000 and 70,000	62,000	(1988, 62,000)
1989	$\frac{4}{5}$ of the distance between 50,000 and 60,000	58,000	(1989, 58,000)

ENRICHMENT ESSAY

Railroads and Highways

The steepest part of Mt. Washington Cog Railway in New Hampshire has a 37% grade, meaning a slope of $\frac{37}{100}$. For every horizontal change of 100 feet, the railroad ascends 37 feet vertically. Engineers denote slope by grade, expressing slope as a percentage.

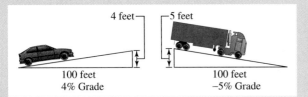

A Mount Washington Cog Railway locomotive pushing a single car up the steepest part of the road. The locomotive is about 120 years old. (Jim Zipp/Photo Researchers, Inc.)

Railroad grades are usually less than 2%, although in the mountains they may go as high as 4%. The grade of the Mt. Washington Cog Railway is phenomenal, making it necessary for locomotives to *push* single cars up its steepest part.

Study tip

We're using (1976, 2000) and (1982, 8000) to compute the slope.

Now we can estimate the slope of the line segments. Between 1976 and 1982:

$$m = \frac{\text{Change in } y}{\text{Change in } x} = \frac{\text{Change in number}}{\text{Change in time}} = \frac{8000 - 2000}{1982 - 1976} = \frac{6000}{6} = 1000$$

This means that the average number of Americans with incomes of a million dollars or more increased by 1000 people per year between 1976 and 1982. Between 1982 and 1988:

Study tip

We're using (1992, 8000) and (1988, 62,000) to compute the slope.

$$m = \frac{\text{Change in } y}{\text{Change in } x} = \frac{\text{Change in number}}{\text{Change in time}} = \frac{62,000 - 8000}{1988 - 1982}$$

$$= \frac{54,000}{6} = 9000$$

The average number of Americans with incomes of a million dollars or more increased by 9000 people per year between 1982 and 1988.

Since slope measures a rate of change, comparing the slopes for the two 6-year periods shows a much greater increase in the average rate of change in the number of million-dollar (plus) income tax returns from 1982 through 1988 than from 1976 through 1982. We can see this from the graph, where the line segment from 1982 to 1988 is much steeper than the one from 1976 through 1982.

b. The line segment between 1988 and 1989 is falling from left to right, so we can say that its slope is negative. Using our estimated points, the slope is computed as follows:

$$m = \frac{\text{Change in } y}{\text{Change in } x} = \frac{\text{Change in number}}{\text{Change in time}} = \frac{58,000 - 62,000}{1989 - 1988} = -4000$$

The number of Americans with incomes of a million dollars or more decreased by 4000 from 1988 through 1989. ■

PROBLEM SET 2.2

Practice Problems

In Problems 1–8, find the slope of the linear function whose graph is shown by first selecting any two points along the line.

1.
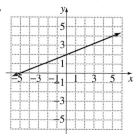

2.
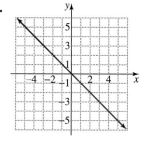

3.
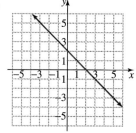

4.
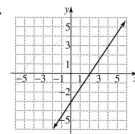

5.
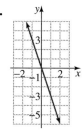

6.
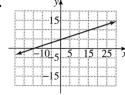

7.
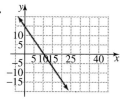

8.
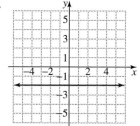

9. Determine whether the points $(-3, -3)$, $(2, -5)$, $(5, -1)$, and $(0, 1)$ are the vertices of a parallelogram.

10. Determine whether the points $(-6, 1)$, $(-2, -1)$, $(0, 3)$, and $(4, 1)$ are the vertices of a parallelogram.

11. Plot the points $A(1, 4)$, $B(3, 2)$, $C(4, 6)$, and $D(2, 8)$.
   a. Show that $ABCD$ is a parallelogram by using slope to show that both pairs of opposite sides are parallel.
   b. Show that the figure is not a rectangle by using slope to determine that AB and BC are not perpendicular.

12. Plot the points $A(-2, -1)$, $B(4, 1)$, $C(3, 4)$, and $D(-3, 2)$.
   a. Show that $ABCD$ is a parallelogram by using slope to show that both pairs of opposite sides are parallel.
   b. Show that the figure is a rectangle by using slope to show that adjacent sides are perpendicular.

13. Show that the diagonals of the quadrilateral located at $A(1, 3)$, $B(4, 3)$, $C(4, 6)$, and $D(1, 6)$ are perpendicular.

14. A triangle with one angle that measures $90°$ is a right triangle. Plot the points $(-5, -5)$, $(4, -2)$, and $(3, 1)$. Then use slope to determine if the points are the vertices of a right triangle.

15. Use the figure to list the slopes m_1, m_2, m_3, and m_4 in order of decreasing size.

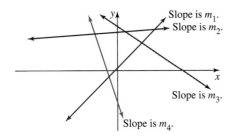

16. Use the figure to list the slopes m_1, m_2, m_3, and m_4 in order of decreasing size.

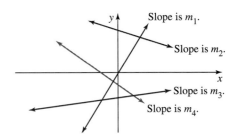

In Problems 17–34, use the slope and the y-intercept or the two intercepts to graph each linear function.

17. $y = -2x - 1$ **18.** $y = -3x - 2$ **19.** $y = \frac{1}{2}x + 1$ **20.** $y = \frac{2}{3}x + 2$

21. $y = \frac{1}{2}x - 1$ **22.** $y = \frac{2}{3}x - 2$ **23.** $y = -\frac{1}{2}x + 1$ **24.** $y = -\frac{2}{3}x + \cdot 2$

25. $y = -\frac{1}{2}x - 1$ **26.** $y = -\frac{2}{3}x - 2$ **27.** $x + y = 3$ **28.** $x + y = 2$

29. $2x + y = -1$ **30.** $3x + y = -2$ **31.** $2x + 3y = 6$ **32.** $2x - 3y = 6$

33. $-2x + 3y = 6$ **34.** $-2x - 3y = 6$

In Problems 35–42, write each equation in slope-intercept form, and find the slope and y-intercept.

35. $3x + y = 5$ **36.** $4x + y = 6$ **37.** $2x + 3y = 18$ **38.** $4x + 6y = -12$

39. $8x - 4y = 12$ **40.** $6x - 5y = 20$ **41.** $7x - 3y = 13$ **42.** $5x - 7y = 19$

In Problems 43–48, write each equation in the given pair in slope-intercept form. Find the slope of each line and use slopes to determine whether the lines are parallel, perpendicular, or neither.

43. $6x + 2y = 10$ and $3x + y = 7$ **44.** $2x - y = -3$ and $x + 2y = 14$

45. $3x - 2y = 6$ and $2x + 3y = -6$ **46.** $3x - y = 0$ and $6x - 2y = -5$

47. $2x + y = 1$ and $x - y = 2$ **48.** $2x - y = 4$ and $3x + y = 6$

Graph each equation in Problems 49–62.

49. $y = 3$ **50.** $y = 5$ **51.** $y = -2$ **52.** $y = -4$ **53.** $3y = 18$

54. $5y = -30$ **55.** $f(x) = 2$ **56.** $f(x) = 1$ **57.** $x = -5$ **58.** $x = -4$

59. $3x - 12 = 0$ **60.** $4x + 12 = 0$ **61.** $x = 0$ **62.** $y = 0$

Application Problems

If an equation in slope-intercept form models some aspect of reality, then the slope and y-intercept can be interpreted in terms of what the formula is describing. For example, the linear model $C = 1.44t + 280$ describes carbon dioxide concentration (C, in parts per million) t years after 1939. The slope is 1.44 and the y-intercept is 280. In practical terms, this means that at the onset (in 1939) carbon dioxide concentration was 280 parts per million and increased by 1.44 parts per million each year. In Problems 63–66, a linear model is given and described. For each model, find the y-intercept and the slope. Then interpret what these numbers mean in terms of what the formula is describing.

63. The model $N = 34t + 1549$ describes the number of nurses (N, in thousands) in the United States t years after 1985.

64. The model $p = -6.9A + 40.3$ describes the percentage of men injured in the Boston Marathon by age group A, where A is designated by 0 (under 20), 1 (20–29), 2 (30–39), 3 (40–49) and 4 (50–59).

65. The model $p = -0.5d + 100$ describes the percentage p of lost hikers found in search and rescue missions when members of the search team walk parallel to one another at a fixed distance of d meters between searchers in the area of the search and rescue operation.

66. The model $C = 23x + 60,000$ describes the total cost (C) for a manufacturer to produce x radios.

67. In 1996, the average total cost for a year of higher education at a state university was $9285. As shown in the graph, this is a dramatic increase over the $3200 that a year of higher education at a state university cost in 1980.

    a. Find the slope of the line segment through 1980 and 1996. Explain what this means in terms of average rate of change.

    b. What is a reasonable estimate for the total yearly cost in the year 2010?

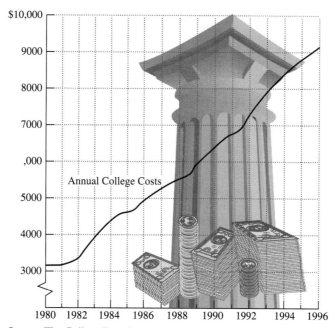

The Price of Public Universities
Average Total Yearly Expenses Estimated for a 4-Year Public Education

Source: The College Board

68. In the United States, approximately one million teenagers become pregnant annually. About 1 in 10 miscarry, 4 in 10 have abortions, and 5 in 10 give birth. Of the half-million who give birth, almost two-thirds are unwed. The number of teenagers giving birth is shown in the graph.

    a. Estimate the slope of the line segments between 1970 and 1975 and between 1985 and 1990 for the graph representing totals. Compare the two slopes and explain what this means in terms of average rate of change.

    b. Estimate the slope of the line segment between 1980 and 1985 for the graph representing totals. Then do the same thing for the same period on the graph representing unwed mothers. Compare the two slopes and explain what this means in terms of average rate of change. What conclusion can you draw with a negative slope for totals and a positive slope for unwed mothers?

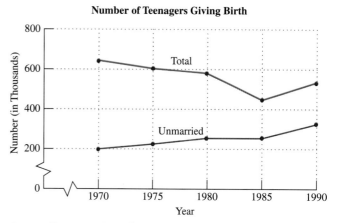

Number of Teenagers Giving Birth

Source: *Facts at a Glance* (Washington, D.C.: Child Trends, Inc., annual)

69. Men's reproductive cells seem to be in serious decline worldwide. A sperm count lower than 20 million sperm per milliliter can often mean infertility. Chemical pollution, such as pesticides sprayed by crop dusters, may be a major cause of the decline. The graph indicates the percent of men worldwide with high sperm concentration. Estimate the slopes of the four line segments and explain what this means in terms of average rate of change.

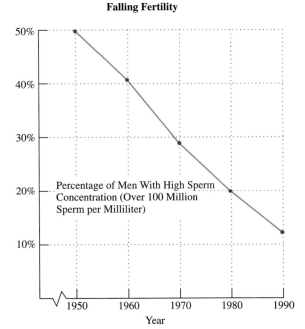

Falling Fertility

Percentage of Men With High Sperm Concentration (Over 100 Million Sperm per Milliliter)

Source: British Medical Journal

70. As shown in the graph, the percentage of people satisfied with their lives remains relatively constant for all age groups. If x represents a person's age, write a function f that reasonably models the percentage of people satisfied with their lives.

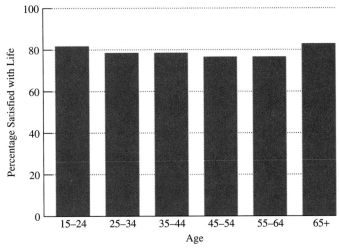

Source: Based on data reported by Ronald Inglehart in *Culture Shift in Advanced Industrial Society,* Princeton University Press, 1989

True–False Critical Thinking Problems

71. Which one of the following is true?
   a. The line through $(7, 5)$ and $(-3, 5)$ has an undefined slope.
   b. Like absolute value, slope can never be negative.
   c. The slope of the line intersecting the y-axis at -3 and the x-axis at 5 is $-\frac{3}{5}$.
   d. A line that has an undefined slope does not represent a function.

72. Which one of the following is true?
   a. Slope can be thought of as run divided by rise.
   b. The slope of a line passing through $(2, -5)$ and $(4, -3)$ is $\dfrac{-3 + 5}{2 - 4}$.
   c. A line with zero slope represents a function.
   d. If a line has a slope of $-\frac{2}{3}$, then any line perpendicular to it has a slope of $-\frac{3}{2}$.

73. Which one of the following is true?
  a. Every line has an equation that can be expressed in standard form.
  b. The line whose equation is $2y + 3x = 7$ has a slope of $-\frac{2}{3}$.
  c. The lines represented by $y = 4x + 3$ and $y = \frac{1}{4}x - \frac{1}{3}$ are perpendicular.
  d. The line whose equation is $2y = 3x + 7$ has a slope of 3 and a y-intercept of 7.

74. Which one of the following is true?
  a. There can be more than one line that passes through a given point with a given slope.
  b. Every line has an equation that can be expressed in slope-intercept form.
  c. A horizontal line has no slope.
  d. The line whose equation is $Ax + By = C$, where $B \neq 0$, has a slope of $-\frac{A}{B}$.

Technology Problems

Use a graphing utility to graph each equation in Problems 75–78. Then use the TRACE *feature to trace along the line and find the coordinates of two points. Use these points to compute the line's slope. Check your result by using the coefficient of x in the line's equation.*

75. $y = 2x + 4$ **76.** $y = -3x + 6$ **77.** $y = -\frac{1}{2}x - 5$ **78.** $y = \frac{3}{4}x - 2$

79. The lines whose equations are $y = \frac{1}{3}x + 1$ and $y = -3x - 2$ are perpendicular because the product of their slopes, $\frac{1}{3}$ and -3, respectively, is -1.
  a. Use a graphing utility to graph the equations. Do the lines appear to be perpendicular?
  b. Now use the zoom square feature of your utility. Describe what happens to the graphs. Explain why this is so.

Writing in Mathematics

80. Why is the slope of a vertical line undefined?

81. Describe why it seems visually reasonable that two parallel lines have the same slope.

82. Based on the directions given for Problems 1–8, explain why we did not include the graph of a vertical line in the problems.

83. Discuss a real world situation that can be described by the constant function $f(x) = k$. Write the function that models this situation.

Critical Thinking Problems

84. A square has vertices at points $A(a, b)$, $B(a + c, b)$, $C(a + c, b + c)$, and $D(a, b + c)$. Prove that its diagonals are perpendicular.

85. If $y = mx + b$, show that the x-intercept is $-\dfrac{b}{m}$.

86. Determine the value of b so that the line whose equation is $by = 8x - 1$ has a slope of -2.

87. Determine the value of b so that the line whose equation is $2by = 5x + 9$ has a slope of 1.

88. If two points on the nonvertical line whose equation is $Ax + By = C$ have coordinates (t, u) and (v, w), show that
$$\frac{A}{B} = \frac{w - u}{t - v}.$$

Review Problems

89. Solve for y: $ay = 3r - by$.

90. Graph: $y = |x| + 2$.

91. Solve for x: $\dfrac{3x - 2}{4} - \dfrac{x - 2}{3} = \dfrac{13}{4} - \dfrac{10x - 8}{12}$.

S E C T I O N   2 . 3

Solutions   Tutorial   Video
Manual   2

The Point-Slope Equation of a Line

Objectives

1 Write equations of lines.
2 Write equations of lines parallel or perpendicular to a given line.
3 Model data with linear functions.

In this section, we develop another form for the equation of a line. The equation is derived using the line's slope and any one point through which the line passes.

The Point-Slope Form of a Line

Another useful form of the equation of a line is the point-slope form. Suppose that a line of slope m passes through the point (x_1, y_1). Let (x, y) stand for all ordered pairs corresponding to points along the line other than (x_1, y_1), shown in Figure 2.27. By definition

$$\frac{y - y_1}{x - x_1} = m.$$

If we multiply both sides by $x - x_1$, we obtain

$$y - y_1 = m(x - x_1).$$

This last equation is called the *point-slope form* of the equation of a line.

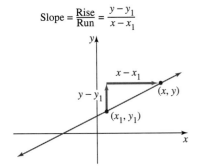

$$\text{Slope} = \frac{\text{Rise}}{\text{Run}} = \frac{y - y_1}{x - x_1}$$

Figure 2.27

A line with slope m passing through (x_1, y_1)

Point-slope form of the equation of a line

The *point-slope form* of the equation of the line that passes through the point (x_1, y_1) and has slope m is

$$y - y_1 = m(x - x_1).$$

---

1 Write equations of lines.

Using the Point-Slope Form to Write a Line's Equation

If we know the slope of a line and a point through which the line passes, the point-slope form is the equation that we should use. Once we have obtained this equation, it is customary to solve for y and write the equation in slope-intercept form. Examples 1 and 2 illustrate these ideas.

EXAMPLE 1 **Writing the Point-Slope Equation and the Slope-Intercept Equation**

a. Write the point-slope equation of the line passing through the point $(-1, 3)$ with slope $-\frac{4}{5}$.
b. Graph the equation.
c. Write the line's equation in slope-intercept form ($y = mx + b$) and in function notation.

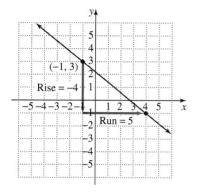

Figure 2.28

The graph of
$y - 3 = -\frac{4}{5}(x + 1)$

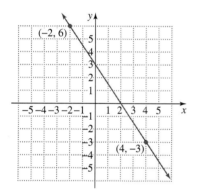

Write the point-slope equation
of this line.

Solution

a. Since $(x_1, y_1) = (-1, 3)$ and $m = -\frac{4}{5}$, we substitute -1 for x_1, 3 for y_1, and $-\frac{4}{5}$ for m in the equation.

$$y - y_1 = m(x - x_1)$$ This is the point-slope equation.

$$y - 3 = -\frac{4}{5}(x - (-1))$$ Substitute: $(x_1, y_1) = (-1, 3)$ and $m = -\frac{4}{5}$.

$$y - 3 = -\frac{4}{5}(x + 1)$$ Simplify.

b. The graph of the point-slope equation is shown in Figure 2.28. We first plot $(-1, 3)$. From this point we count off a slope of $-\frac{4}{5}$ with a rise of -4 and a run of 5.

c. By solving for y, we can express the line's equation in slope-intercept form.

$$y - 3 = -\frac{4}{5}(x + 1)$$ This is the point-slope equation found in part (a).

$$y - 3 = -\frac{4}{5}x - \frac{4}{5}$$ Use the distributive property.

$$y = -\frac{4}{5}x + \frac{11}{5}$$ Add 3 to both sides.

The slope-intercept form $(y = mx + b)$ is $y = -\frac{4}{5}x + \frac{11}{5}$. We can express this in function notation by replacing y by $f(x)$. The linear function's equation is $f(x) = -\frac{4}{5}x + \frac{11}{5}$. ■

If two points on a line are known, we can graph the line and write its equation.

EXAMPLE 2   **Using the Point-Slope Equation of a Line**

a. Find the point-slope equation of the line passing through the points $(4, -3)$ and $(-2, 6)$.

b. Express the equation in slope-intercept form and in function notation.

Solution

a. We start by finding the slope of the line:

$$m = \frac{6 - (-3)}{-2 - 4} = \frac{9}{-6} = -\frac{3}{2}$$ This is the definition of slope using $(4, -3)$ and $(-2, 6)$.

We can take either point on the line to be (x_1, y_1). Let's use $(x_1, y_1) = (4, -3)$. Now we are ready to write the point-slope equation.

$$y - y_1 = m(x - x_1)$$ This is the point-slope equation.

$$y - (-3) = -\frac{3}{2}(x - 4)$$ Substitute: $(x_1, y_1) = (4, -3)$ and $m = -\frac{3}{2}$.

$$y + 3 = -\frac{3}{2}(x - 4)$$ Simplify.

b. To express the equation in slope-intercept form, we must solve for y.

$$y + 3 = -\frac{3}{2}(x - 4)$$ This is the point-slope equation found in part (a).

$$y + 3 = -\frac{3}{2}x + 6$$ Use the distributive property.

$$y = -\frac{3}{2}x + 3$$ Subtract 3 from both sides.

The slope-intercept form $(y = mx + b)$ is $y = -\frac{3}{2}x + 3$, expressed in function notation as $f(x) = -\frac{3}{2}x + 3$. ■

Discover for yourself

Rework Example 2 using $(-2, 6)$ for (x_1, y_1). You should obtain the same equation for $f(x)$.

Summarizing the Forms for a Line's Equation

The major forms for equations of lines and methods for graphing them are summarized in Table 2.4.

TABLE 2.4 Summary of Equations of Lines and Graphing Techniques

Form	Example	How to Graph the Example
Point-Slope Form $y - y_1 = m(x - x_1)$ $m = $ slope point on the line $= (x_1, y_1)$	$y - 2 = \dfrac{3}{2}(x - 1)$ slope $= \dfrac{3}{2}$ point on the line $= (1, 2)$	Use the point on the line and the slope.
Slope-Intercept Form $y = mx + b$ $m = $ slope $b = y$-intercept	$y = -\dfrac{3}{4}x + 1$ slope $= -\dfrac{3}{4}$ y-intercept $= 1$	Use the y-intercept and slope.
Standard Form $Ax + By = C$	$2x - 4y = 8$ x-intercept ($y = 0$): $2x = 8$ $x = 4$ y-intercept ($x = 0$): $-4y = 8$ $y = -2$	Use the intercepts.

Form	Example	How to Graph the Example
Horizontal line parallel to the *x*-axis: $y = b$	$y = 5$	Draw a line parallel to the *x*-axis with *y*-intercept = 5.
Vertical line parallel to the *y*-axis: $x = b$	$x = -2$ This is the only form of a line in which *y* is not a function of *x*. Why?	Draw a line parallel to the *y*-axis with *x*-intercept = -2.

2 Write equations of lines parallel or perpendicular to a given line.

Parallel and Perpendicular Lines

In the last section, we found that parallel lines have the same slope and perpendicular lines have slopes with a product of -1. These results form the basis of Example 3.

EXAMPLE 3 **Finding Equations of Parallel and Perpendicular Lines**

Consider the line given by the equation $x - 2y = -1$.

a. Find an equation for a parallel line passing through $(1, -3)$.
b. Find an equation for a perpendicular line passing through $(1, -3)$.

Solution

The situation is illustrated in Figure 2.29. We begin by finding the slope of $x - 2y = -1$, the green line in the figure. Solve for y to obtain slope-intercept form.

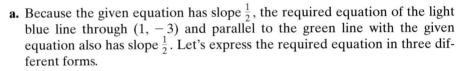

$$x - 2y = -1 \qquad \text{This is the given equation.}$$

$$-2y = -x - 1 \qquad \text{Subtract } x \text{ from both sides.}$$

$$y = \frac{1}{2}x + \frac{1}{2} \qquad \text{Divide both sides by } -2.$$

$$\uparrow \qquad \uparrow$$

Slope y-intercept

The slope of the given line is $\frac{1}{2}$.

a. Because the given equation has slope $\frac{1}{2}$, the required equation of the light blue line through $(1, -3)$ and parallel to the green line with the given equation also has slope $\frac{1}{2}$. Let's express the required equation in three different forms.

$$y - y_1 = m(x - x_1) \qquad \text{This is the point-slope form.}$$

$$y - (-3) = \frac{1}{2}(x - 1) \qquad \text{Substitute:} \quad (x_1, y_1) = (1, -3) \text{ and } m = \frac{1}{2}.$$

$$y + 3 = \frac{1}{2}(x - 1) \qquad \text{Simplify. This is the required equation in point-slope form.}$$

$$y + 3 = \frac{1}{2}x - \frac{1}{2} \qquad \text{Solve for } y \text{ to obtain slope-intercept form. Apply the distributive property.}$$

$$y = \frac{1}{2}x - \frac{7}{2} \qquad \text{Subtract 3 from both sides. This is the required equation in slope-intercept form.}$$

$$2y = 2(\tfrac{1}{2}x - \tfrac{7}{2}) \qquad \text{Finally, write the standard form. Multiply both sides by 2.}$$

$$2y = x - 7$$

$$-x + 2y = -7 \qquad \text{Subtract } x \text{ from both sides.}$$

$$x - 2y = 7 \qquad \text{Multiply both sides by } -1. \text{ This is the required equation in standard form.}$$

b. Because the given equation has slope $\frac{1}{2}$, the slope of the dark blue perpendicular line is given by the negative of the reciprocal of $\frac{1}{2}$, which is -2. Since the line passes through $(1, -3)$, the point-slope equation of the required line is

$$y - (-3) = -2(x - 1) \quad \text{or} \quad y + 3 = -2(x - 1).$$

Take a moment to show that the slope-intercept form is $y = -2x - 1$ and the standard form is $2x + y = -1$.   ■

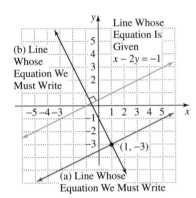

Line Whose Equation Is Given

(b) Line Whose Equation We Must Write

$x - 2y = -1$

$(1, -3)$

(a) Line Whose Equation We Must Write

Figure 2.29

3 Model data with linear functions.

Mathematical Modeling and Data Analysis

Our next example illustrates how we can use real world data to develop a mathematical model. Once the model is developed, we can *interpolate* and *extrapolate* from the data. *Interpolation* refers to finding numerical values *between* actual data measurements. *Extrapolation* refers to finding numerical values *outside* the actual data measurements.

ENRICHMENT ESSAY

Art Without Lines

The picture *Sunday Afternoon on the Island of La Grande Jatte* is painted entirely without lines. The artist, Georges Seurat (1859–1891), experimented with tiny dots of pure color, a technique that is called divisionism or pointillism. There is not a single line in the picture, although when viewed from a distance the illusion of lines becomes the controlling factor.

Georges Seurat, French, 1859–1891, A Sunday on La Grande Jatte—1884, oil on canvas, 1884–86, 207.6 × 308 cm, Helen Birch Bartlett Memorial Collection, 1926.224.

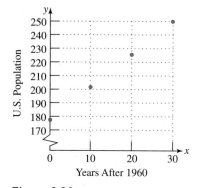

Figure 2.30

A scatter plot of data points describing U.S. population over time

EXAMPLE 4   Modeling U.S. Population

The table below gives the population of the United States (in millions) in the indicated year. The collection of data points shown in Figure 2.30 is called a *scatter plot*. The four data points appear to fall nearly on a straight line. Consequently, it seems appropriate to use a linear equation to model the data. Write a linear equation that models population (y) in terms of the year (x), letting $x = 0$ correspond to 1960. Use the model to extrapolate from the data and predict U.S. population in 2010.

Year	x (Year after 1960)	y (U.S. Population)
1960	0	179.3
1970	10	203.3
1980	20	226.5
1990	30	250.0

Solution

Let's begin by writing the equation of the line passing through the first two data points, namely, $(0, 179.3)$ and $(10, 203.3)$. We start by finding the slope:

$$m = \frac{203.3 - 179.3}{10 - 0} = \frac{24}{10} = 2.4$$

$y - y_1 = m(x - x_1)$	This is the point-slope equation.
$y - 179.3 = 2.4(x - 0)$	Either ordered pair can be (x_1, y_1). Let $(x_1, y_1) = (0, 179.3)$. From above, $m = 2.4$.
$y - 179.3 = 2.4x$	Simplify on the right.
$y = 2.4x + 179.3$	Solve for y. This is the slope-intercept form.

Now let's see how well this equation models the data.

Year	Actual U.S. Population	Population Predicted by $y = 2.4x + 179.3$	Absolute Value of the Error
1960 ($x = 0$)	179.3	$y = 2.4(0) + 179.3 = 179.3$	0
1970 ($x = 10$)	203.3	$y = 2.4(10) + 179.3 = 203.3$	0
1980 ($x = 20$)	226.5	$y = 2.4(20) + 179.3 = 227.3$	$227.3 - 226.5 = 0.8$
1990 ($x = 30$)	250.0	$y = 2.4(30) + 179.3 = 251.3$	$251.3 - 250 = 1.3$
			Total error $= 0.8 + 1.3 = 2.1$

Let's repeat this process using the data points through 1970 and 1980. The ordered pairs are $(10, 203.3)$ and $(20, 226.5)$. We start by finding the slope:

$$m = \frac{226.5 - 203.3}{20 - 10} = 2.32$$

$y - y_1 = m(x - x_1)$	This is the point-slope equation.
$y - 203.3 = 2.32(x - 10)$	Either ordered pair can be (x_1, y_1). Use $(10, 203.3)$. From above, $m = 2.32$.
$y - 203.3 = 2.32x - 23.2$	Apply the distributive property on the right.
$y = 2.32x + 180.1$	Solve for y.

Let's see how well this equation models the data.

Year	Actual U.S. Population	Population Predicted by $y = 2.32x + 180.1$	Absolute Value of the Error
1960 ($x = 0$)	179.3	$y = 2.32(0) + 180.1 = 180.1$	$180.1 - 179.3 = 0.8$
1970 ($x = 10$)	203.3	$y = 2.32(10) + 180.1 = 203.3$	0
1980 ($x = 20$)	226.5	$y = 2.32(20) + 180.1 = 226.5$	0
1990 ($x = 30$)	250.0	$y = 2.32(30) + 180.1 = 249.7$	$250 - 249.7 = 0.3$
			Total error $= 0.8 + 0.3 = 1.1$

The total error for this model is less than that of the first model. Consequently, this model appears to be a better fit than the first model.

Without the use of technology (see the Discover for Yourself box following this example), this process should be repeated using two data points at a time. In doing so, we would find that the model with the least error is the line through $(20, 226.5)$ and $(30, 250)$. Take a few minutes to verify that this model is

$$y = 2.35x + 179.5.$$

Now let's use this model to predict U.S. population in 2010.

$$f(x) = 2.35x + 179.5 \qquad \text{This is the model in function notation.}$$
$$f(50) = 2.35(50) + 179.5 \qquad \text{Since 2010 is 50 years after 1960, substitute 50 for } x.$$
$$f(50) = 297 \qquad \text{Thus, } f \text{ of 50 is 297.}$$

The model predicts that the population of the United States in the year 2010 will be 297 million. ◼

Arman "Village of the Damned" 1962—accumulation of dolls in a wood and glass store display box, $21 \times 20 \times 11$ in. $(53 \times 51 \times 28$ cm)/© 1997 Artists Rights Society (ARS), New York/ADAGP, Paris

Discover for yourself

A graphing utility will give you the equation of a line that best fits a set of data. The line of best fit is one that most closely approximates the data points and is called the *regression* line.

a. Use the statistical menu of a graphing utility to enter the four data points $((0, 179.3)$ through $(30, 250))$ of Example 4. Consult your manual for details on how to do this.

b. Use the draw menu and the scatter plot capability to draw a scatter plot of the data points.

c. Select the linear regression option. Your utility should give you the regression coefficients, a and b, for the regression equation (usually in the form $y = a + bx$). You will also be given a correlation coefficient r. Values of r close to 1 indicate that the points can be described by a linear relationship and the regression line has a positive slope. Values of r close to -1 indicate that the points can be described by a linear relationship and the regression line has a negative slope. Values of r close to 0 indicate no linear relationship between the variables.

d. Use an appropriate key sequence (consult your manual) to graph the regression equation on top of the points in the scatter plot.

How does the equation of the regression line compare to the model that we obtained in Example 4?

PROBLEM SET 2.3

Practice Problems

In Problems 1–9, find the point-slope equation of the line having the specified slope and passing through the given point. Then graph the line.

1. $m = 3; (2, 1)$ **2.** $m = 4; (3, 2)$ **3.** $m = -3; (6, 2)$ **4.** $m = -2; (3, 5)$

5. $m = 4; (-3, -5)$ **6.** $m = 1; (-4, -5)$ **7.** $m = 0; (0, -3)$ **8.** $m = 0; (0, 2)$

9. $m = \frac{4}{5}; (4, -2)$ **10.** $m = -\frac{2}{3}; (-2, 6)$

In Problems 11–22, find the point-slope equation of the line satisfying the given conditions. Then express the equation in slope-intercept form.

11. $m = 2$, passing through $(-3, 2)$

12. $m = -4$, passing through $(-1, -5)$

13. $m = \frac{1}{2}$, passing through $(1, 0)$

14. $m = -\frac{2}{3}$, passing through $(-8, 6)$

15. Passing through $(-3, 4)$ and $(1, 6)$

16. Passing through $(5, 8)$ and $(3, 16)$

17. Passing through $(-4, -2)$ and $(9, 11)$

18. Passing through $(3, 5)$ and $(2.8, 6)$

19. Passing through $(1, 3)$ and parallel to $y = 2x - 1$

20. Passing through $(2, -5)$ and parallel to $y = 3x - 2$

21. Passing through $(1, -3)$ and perpendicular to $y = 2x - 1$

22. Passing through $(2, 5)$ and perpendicular to $y = 3x - 2$

In Problems 23–28, find the equation of the linear function in slope-intercept form whose graph passes through the given pair of points.

23. $(1, -2)$ and $(0, 5)$

24. $(-2, 1)$ and $(-4, -2)$

25. $(-2, -8)$ and $(5, -8)$

26. $(-7, 6)$ and $(2, 6)$

27. $(2, 5)$ and $(-3, 6)$

28. $(1, 5)$ and $(\frac{1}{2}, 3)$

In Problems 29–38, find the point-slope equation of the line satisfying the given conditions. Then express the equation in standard form, $Ax + By = C$, with A, a positive integer.

29. Passing through $(-3, 2)$ with slope = 2

30. Passing through $(-1, -5)$ with slope = -4

31. Passing through $(1, 0)$ with slope = $\frac{1}{2}$

32. Passing through $(-5, -2)$ with slope = 0

33. Passing through $(5, 6)$ with x-intercept = 11

34. Passing through $(-2, 8)$ with y-intercept = -4

35. Passing through $(-2, -5)$ and parallel to $2x + y = 4$

36. Passing through $(-1, 0)$ and parallel to $3x + y = 7$

37. Passing through $(1, -3)$ and perpendicular to $2x - 4y = 12$

38. Passing through $(7, -2)$ and perpendicular to $3x - 2y = 6$

Application Problems

In Problems 39–40, two measurements are given for variables having a linear relationship. For each problem, write the point-slope form of the line on which these measurements fall. Then use the point-slope form of the equation to write the slope-intercept form of the equation as a function. Finally, extrapolate from the data to answer the question.

39.

x (Number of Years after 1985)	y (Total of All Health-Care Expenditures in the United States [in billions of dollars])
3	546
5	666

What will health-care expenditures in the United States be in the year 2010?

40.

x (Number of Years after 1964)	y (Time [in seconds] for the World Record for Women in Swimming for the 100-meter Freestyle Event)
0	58.9
8	58.5

In 1986, Kristin Otto of East Germany won the 100-meter freestyle with a winning time of 54.73 seconds. What time is predicted by the model for 1986?

41. A business discovers a linear relationship between the number of shirts it can sell and the price per shirt. In particular, 20,000 shirts can be sold at $19 each, and 2000 of the same shirts can be sold at $55 each. Write the slope-intercept equation of the *demand line* through the ordered pairs (20,000 shirts, $19) and (2000 shirts, $55). Then determine the number of shirts that can be sold at $50 each.

42. In 1965, radioactive wastes seeping into the Columbia River exposed citizens of eight Oregon counties and the city of Portland to radioactive contamination. In an article in the *Journal of Environmental Health* (May–June, 1965), the authors formulated an index that measured the proximity of the residents to the contamination. The ordered pair for Columbia County (6.4, 178) indicates that its index is 6.4 and there are 178 cancer deaths per 100,000 residents. The corresponding ordered pair for Clatsop County is (8.3, 210). What is the predicted number of cancer deaths for Portland, with an index of 11.6?

43. Reread the chapter introduction and refer to Figure 2.1. Approximate the coordinates on the line shown in blue corresponding to the years 1975 ($x = 0$) and 1993 ($x = 18$). Use these data measurements to write the point-slope form and then the slope-intercept form for the line's equation. What is the predicted average global temperature for the year 2010?

Technology Problem

44. The following table presents the men's and women's winning times (in seconds) in the Olympic 100-meter freestyle swimming race from 1912 through 1988.

Year	Men's	Women's	Year	Men's	Women's
1912	63.2	72.2	1960	55.2	61.2
1920	61.4	73.6	1964	53.4	59.5
1924	59.0	72.4	1968	52.2	60.0
1928	58.6	71.0	1972	51.22	58.59
1932	58.2	66.8	1976	49.99	55.65
1936	57.6	65.9	1980	50.40	54.79
1948	57.3	66.3	1984	49.80	55.92
1952	57.4	66.8	1988	48.63	54.93
1956	55.4	62.0			

Shown in the figure are graphs of the best-fitting lines (the regression lines) through the points representing the data values for the men and women.

a. Use the statistical menu of a graphing utility to enter the data values in the table for the men's winning times.

b. Select the linear regression option. Your utility should give you the regression coefficients, *a* and *b*, for the regression equation (usually in the form $y = a + bx$) and the correlation coefficient *r*. What is the value of *r*? Can you explain what this value means?

c. Repeat parts (a) and (b) for the data values in the table for the women's winning times.

d. Graph the equations of both regression lines using your graphing utility. Graph the equations in the same viewing rectangle. Use the intersection or TRACE feature to find the coordinates of the intersection point.

e. What does the intersection point mean in terms of the women's time and the men's time? What predictions could be made about the swimming race in the Olympic years to the right of the intersection point?

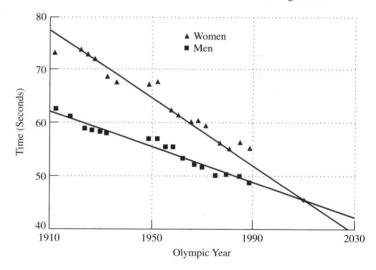

Writing in Mathematics

45. If you were asked to write an equation of a line, explain the circumstances that would lead you to begin with
a. The slope-intercept form.
b. The point-slope form.
c. Neither of these forms.

46. Describe two ways in which $x = k$ differs from all other linear relations.

47. Why is the point-slope form of a line inappropriate for writing the equation of a vertical line?

Critical Thinking Problems

For Problems 48–49, write the point-slope form, the slope-intercept form, and the standard form of the equation of each line described.

48. Having an *x*-intercept of -3 and perpendicular to the line passing through $(0, 0)$ and $(6, -2)$

49. Perpendicular to $3x - 2y = 4$ with the same *y*-intercept

50. Excited about the success of celebrity stamps, post office officials were rumored to have put forth a plan to institute two new types of thermometers. On these new scales, °*E* represents degrees Elvis and °*M* represents degrees Madonna. If it is known that $40°E = 25°M$, $280°E = 125°M$, and degrees Elvis is a linear function of degrees Madonna, write an equation expressing *E* in terms of *M*.

51. The figure represents a small piece of graph paper showing a line whose x-intercept is -3 and whose y-intercept is -6. Name the points on the line given only one of the coordinates: $(-40, \quad)$ and $(\quad, -200)$.

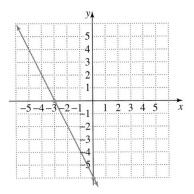

52. Determine the value of b so that the line passing through $(-2, 3)$ and $(6, b)$ has a y-intercept of 4.

53. Determine the value of a so that the line passing through $(0, 4)$ and $(a - 2, 6)$ has an x-intercept of a.

54. Show that an equation of the line with an x-intercept of a and a y-intercept of b is

$$\frac{x}{a} + \frac{y}{b} = 1 \qquad (a \neq 0 \text{ and } b \neq 0)$$

55. A sequence of numbers is represented by 17, 25, 33, 41, 49, ...

The first term of the sequence is 17: $(1, 17)$.
The second term of the sequence is 25: $(2, 25)$.
The third term of the sequence is 33: $(3, 33)$.
The fourth term of the sequence is 41: $(4, 41)$.
The fifth term of the sequence is 49: $(5, 49)$.

Write a formula for the nth term of the sequence: $(n, ?)$.

56. The rectangle in the figure represents a pool table, with six pockets as shown.
   a. A ball starts at $(8, 3)$ heading toward the x-axis with slope 2. Where will the ball strike the x-axis?
   b. A ball moving along a line with slope m will strike the table and bounce back along a line with slope $-m$. After the ball from part (a) strikes the x-axis and rebounds, will it go into one of the six pockets? If so, what are the coordinates of the pocket?

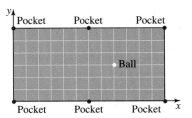

Group Activity Problem

57. The National Center for Health Statistics issued the following data for years of life expected at birth for American women and men.

Year	x (Year after 1960)	y (Years of Life Expected at Birth)
1960	0	69.7
1970	10	70.8
1980	20	73.7
1990	30	75.4

   a. Using two data points at a time, six possible linear models can be written. Each group member should write one of the possible models. In particular:
   1. Write the point-slope form of the line on which the two data points assigned to you fall.
   2. Use the point-slope form of the equation to write the slope-intercept form of the equation.
   3. Compute the error for the model using a table like the ones on page 170.

   b. Once part (a) is completed, members should return to the group and report the error for each model. Select the model with the least error.
   c. Use the model that best fits the data to predict years of life expected at birth for the year 2010. In what year will Americans be expected to live 90 years from the time of their birth?
   d. Group members should now discuss limitations of the model. What events might occur in the future to change the predictions made by the model?
   e. If your course includes the use of technology, as a group use your graphing utilities to follow the four steps in the Discover for Yourself on page 171 with the data points listed above. How does the equation of the regression line compare to the model obtained by the group working by hand? When modeling many data points that approximately fall along a line, what is the advantage of using technology rather than the approach in part (a) of this problem?

Review Problems

58. Solve: $5x - 4 = 2(6x - 1) + 12$.

59. Simplify: $\left(\dfrac{2x^5 y^{-2}}{3x^{-4}} \right)^4$.

60. Evaluate: $-2^2 - 4(1 - 3)^5$.

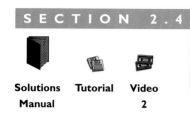

Solutions   Tutorial   Video
Manual   2

1 Solve linear inequalities.

Solving Linear Inequalities

Objectives

1 Solve linear inequalities.
2 Solve applied problems modeled by linear inequalities.

We have seen how linear equations can be used to solve problems. In this section we focus on linear inequalities and their role in modeling and solving problems.

The most common type of inequality in algebra is a *linear inequality* in one variable.

Definition of a linear inequality in one variable

A *linear inequality* in one variable x is any inequality that can be written in one of the forms

$$ax + b < c \qquad ax + b \leqslant c \qquad ax + b > c \qquad ax + b \geqslant c$$

where a, b, and c are real numbers, and $a \neq 0$.

An example of a linear inequality in one variable is $2x + 3 < 17$, in which $a = 2, b = 3$, and $c = 17$. Because the greatest power on the variable is one, a linear inequality is also called a *first-degree inequality*.

Any value of the variable for which an inequality is true is called a *solution*. The *solution set* of the inequality is the set of all such solutions.

The process of solving a linear inequality involves a series of transformations into equivalent inequalities (inequalities that have the same solution set). The last of the equivalent inequalities should be in the form

$$x > d \qquad x < d \qquad x \geqslant d \qquad x \leqslant d$$

whose solution set is obvious.

We can use properties of inequalities to isolate the variable. These properties are similar to the properties of equality that we used to solve equations. There are, however, two important exceptions: Multiplying or dividing both sides of an inequality by a negative number reverses the sense of the inequality. To illustrate, let's take three true inequality statements and multiply both sides by -3:

$$
\begin{array}{ccc}
7 < 10 & -2 < 6 & -6 < -1 \\
\downarrow & \downarrow & \downarrow \\
-3(7) > -3(10) & -3(-2) > -3(6) & -3(-6) > -3(-1) \\
-21 > -30 & 6 > -18 & 18 > 3
\end{array}
$$

In each case, the resulting inequality symbol points in the opposite direction from the original one.

Multiplying and dividing by negative numbers in inequalities

When both sides of an inequality are multiplied or divided by a negative number, the direction of the inequality symbol reverses.

To isolate the variable in a linear inequality, we use the same basic techniques used in solving linear equations. The following properties are used to create equivalent inequalities.

Study tip

These properties are valid if the symbols $<$ and $>$ are replaced by $\leq$ and $\geq$.

Properties of inequalities

Addition and Subtraction Properties

When the same quantity is added to or subtracted from both sides of an inequality, the resulting inequality is equivalent to the original.

If $a < b$, then $a + c < b + c$.
If $a < b$, then $a - c < b - c$.

Positive Multiplication and Division Properties

When multiplying or dividing both sides of an inequality by the same positive quantity, the resulting inequality is equivalent to the original one.

If $a < b$ and c is positive, then $ac < bc$.

If $a < b$ and c is positive, then $\frac{a}{c} < \frac{b}{c}$.

Negative Multiplication and Division Properties

When multiplying or dividing both sides of an inequality by the same negative quantity, the result is an equivalent inequality in which the inequality symbol is reversed.

If $a < b$ and c is negative, then $ac > bc$.

If $a < b$ and c is negative, then $\frac{a}{c} > \frac{b}{c}$.

Discover for yourself

Test various values for x and see if you can correctly guess at the set of numbers that satisfy

$2x + 3 < 7.$

Now check your guess by reading Example 1. The example illustrates the advantage of solving a linear inequality using a systematic procedure.

EXAMPLE 1 Solving a Linear Inequality

Find the solution set: $2x + 3 < 7$

Solution

$2x + 3 < 7$	This is the given inequality.
$2x + 3 - 3 < 7 - 3$	Subtract 3 from both sides.
$2x < 4$	Simplify.
$\frac{2x}{2} < \frac{4}{2}$	Divide both sides by 2.
$x < 2$	Simplify.

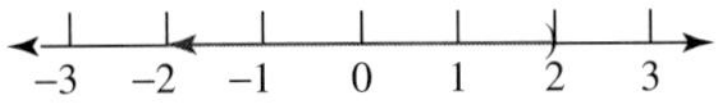

Figure 2.31
The graph of $x < 2$

The solution set consists of all real numbers that are less than 2, expressed as $\{x \mid x < 2\}$. The interval notation for the solution set is $(-\infty, 2)$. The graph is shown in Figure 2.31.

Check

You can get an idea that the solution set is correct by taking one member of the solution set—say, -1—and showing that it satisfies the inequality.

Discover for yourself

Is it possible to get an idea that the solution set is correct by using a number that is *not* in the set? Since 3 is not in the solution set of

$2x + 3 < 7,$

substitute 3 for x and describe what happens.

Using technology

You can use a graphing utility to show that the solution set for $2x + 3 < 7$ is $(-\infty, 2)$. Graph each side of the inequality, entering

$$y_1 = 2x + 3$$

$$y_2 = 7$$

As shown in the figure, notice that the graph of $y = 2x + 3$ is less than the graph of $y = 7$ when $x < 2$. Equivalently, the graph of $y = 2x + 3$ lies under the graph of $y = 7$ when $x < 2$.

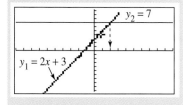

$$2x + 3 < 7 \qquad \text{This is the given inequality.}$$

$$2(-1) + 3 \overset{?}{<} 7 \qquad \text{Substitute } -1 \text{ for } x.$$

$$-2 + 3 \overset{?}{<} 7 \qquad \text{Multiply.}$$

$$1 < 7 \qquad \text{True.}$$

This true statement indicates that -1 does belong to the solution set. ■

EXAMPLE 2 Solving a Linear Inequality

Find the solution set: $3(2 - x) \le 5x - 2$

Solution

$$3(2 - x) \le 5x - 2 \qquad \text{This is the given inequality.}$$

$$6 - 3x \le 5x - 2 \qquad \text{Apply the distributive property on the left side.}$$

$$6 - 3x - 5x \le 5x - 2 - 5x \qquad \text{Collect all terms with the variable on the left. Subtract } 5x \text{ from both sides.}$$

$$6 - 8x \le -2 \qquad \text{Simplify.}$$

$$6 - 8x - 6 \le -2 - 6 \qquad \text{Subtract 6 from both sides.}$$

$$-8x \le -8 \qquad \text{Simplify.}$$

$$\frac{-8x}{-8} \ge \frac{-8}{-8} \qquad \text{Divide both sides by } -8 \text{ and reverse the sense of the inequality.}$$

$$x \ge 1 \qquad \text{Simplify.}$$

The solution set consists of all real numbers that are greater than or equal to 1, expressed as $\{x \mid x \ge 1\}$. The interval notation for the solution set is $[1, \infty)$. The graph is shown in Figure 2.32. ■

Figure 2.32

The graph of $x \ge 1$

Using technology

The graphs of $y_1 = 3(2 - x)$ and $y_2 = 5x - 2$ have an x-coordinate of intersection at 1. The graph of $y_1 = 3(2 - x)$ lies below the graph of $y_2 = 5x - 2$ when $x > 1$. This verifies that $3(2 - x) \le 5x - 2$ for $[1, \infty)$.

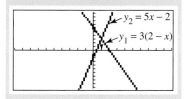

Study tip

The addition and subtraction properties of inequality provide us with the option of collecting all terms with the variable on the right and all numbers on the left. Let's apply this option to the inequality in Example 2.

$$3(2 - x) \le 5x - 2 \qquad \text{This is the given inequality.}$$

$$6 - 3x \le 5x - 2 \qquad \text{Apply the distributive property.}$$

$$6 - 3x + 3x \le 5x - 2 + 3x \qquad \text{Collect all terms with the variable on the right. Add } 3x \text{ to both sides.}$$

$$6 \le 8x - 2 \qquad \text{Simplify.}$$

$$6 + 2 \le 8x - 2 + 2 \qquad \text{Add 2 to both sides.}$$

$$8 \le 8x \qquad \text{Simplify.}$$

$$\frac{8}{8} \le \frac{8x}{8} \qquad \text{Divide both sides by 8 and preserve the sense of the inequality.}$$

$$1 \le x \qquad \text{Simplify.}$$

$1 \le x$ can be equivalently expressed as $x \ge 1$, leading to a solution set, once again, of $\{x \mid x \ge 1\}$ or $[1, \infty)$.

As is the case of solving linear equations, properties of inequalities provide us with a step-by-step procedure for solving a linear inequality. (Not all inequalities require all of these steps.)

Solving a linear inequality

1. Multiply on both sides to clear fractions or decimals.
2. Simplify the algebraic expression on both sides.
3. Collect all variable terms on one side and all constant terms on the other side.
4. Isolate the variable and solve. The direction of the inequality symbol reverses when multiplying or dividing by a negative number.
5. Express the solution in set-builder notation or interval notation, and graph the solution set on a number line.

Our next example requires all of these steps.

EXAMPLE 3 **Solving a Linear Inequality**

Find the solution set: $\dfrac{y + 3}{4} \geq \dfrac{y - 2}{3} + \dfrac{1}{4}$

Solution

$$\frac{y + 3}{4} \geq \frac{y - 2}{3} + \frac{1}{4}$$
This is the given inequality.

$$12\left(\frac{y + 3}{4}\right) \geq 12\left(\frac{y - 2}{3} + \frac{1}{4}\right)$$
Multiply both sides by 12, the LCM of 4 and 3.

$$\overset{3}{\cancel{12}}\left(\frac{y + 3}{4}\right) \geq \overset{4}{\cancel{12}}\left(\frac{y - 2}{3}\right) + \overset{3}{\cancel{12}}\left(\frac{1}{4}\right)$$
Apply the distributive property on the right. Cancel identical factors in numerators and denominators.

$$3(y + 3) \geq 4(y - 2) + 3$$
The equivalent inequality is now cleared of fractions.

$$3y + 9 \geq 4y - 8 + 3$$
Apply the distributive property.

$$3y + 9 \geq 4y - 5$$
Combine like terms.

$$3y + 9 - 4y \geq 4y - 5 - 4y$$
Collect variable terms on the left side by subtracting $4y$ from both sides.

$$-y + 9 \geq -5$$
Simplify.

$$-y + 9 - 9 \geq -5 - 9$$
Subtract 9 from both sides.

$$-y \geq -14$$
Simplify.

$$(-1)(-y) \leq (-1)(-14)$$
Multiply (or divide) both sides by -1, changing $\geq$ to $\leq$.

$$y \leq 14$$
Simplify.

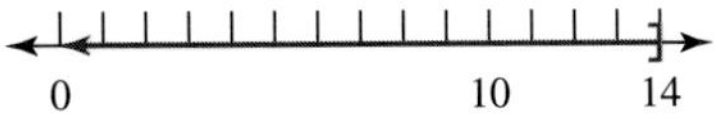

Figure 2.33

The graph of $y \leq 14$

The solution set is $\{y \mid y \leq 14\}$. The solution set in interval notation is $(-\infty, 14]$. The graph is shown in Figure 2.33. ∎

2 Solve applied problems modeled by linear inequalities.

Modeling Using Inequalities

We turn our attention to some applied problems that can be modeled using linear inequalities.

EXAMPLE 4 **Teachers' Salaries**

In 1990 the average yearly salary for teachers in the United States was $31,076. This salary has increased steadily by $1496 each year. When did average yearly salaries exceed $40,052?

Discover for yourself

Let's see what is happening to teachers' average salaries over time. Recall that the salary in 1990 was $31,076, and this has increased steadily by $1496 each year.

Number of Years after 1990	Teachers' Salaries
0	31,076
1	31,076 + 1(1496)
2	(31,076 + 1496) + 1496 = 31,076 + 2(1496)
3	[31,076 + 2(1496)] + 1496 = 31,076 + 3(1496)
4	[31,076 + 3(1496)] + 1496 = 31,076 + 4(1496)

Do you see a pattern forming? If the number of years after 1990 is represented by x, use the pattern to write an algebraic expression for annual teachers' salaries. Now write an inequality that can be used to determine when the average salaries exceeded $40,052.

Solution

We will use our five-step strategy for solving word problems.

Steps 1 and 2. Represent unknowns in terms of x.
Let x = the number of years after 1990 when salaries exceeded $40,052.
Step 3. Write an inequality that describes the verbal conditions.
Our next step is to create a model for teachers' salaries x years after 1990.

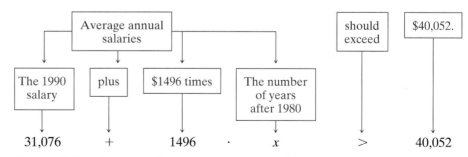

Step 4. Solve the inequality and answer the problem's question.

$$31,076 + 1496x > 40,052$$ This is the linear inequality implied by the problem's question.

$$1496x > 8976$$ Subtract 31,076 from both sides.

$$\frac{1496x}{1496} > \frac{8976}{1496} \qquad \text{Divide both sides by 1496.}$$

$$x > 6 \qquad \text{Simplify.}$$

Thus, 6 years after 1990, or after 1996, teachers' salaries exceeded $40,052.
Step 5. Check.
We can check by substituting a value for x greater than 6 in the model for annual teachers' salaries. If $x = 7$,

annual salary $= \$31,076 + (\$1496)\,(7) = \$41,548$

which exceeds $40,052.   ◼

EXAMPLE 5   Creating and Comparing Mathematical Models

Acme Car rental agency charges $4 a day plus $0.15 a mile, whereas Interstate rental agency charges $20 a day and $0.05 a mile. How many miles must be driven to make the daily cost of an Acme rental a better deal than an Interstate rental?

Solution

Steps 1 and 2. Represent unknowns in terms of x.

Let x = the number of miles driven in a day. We must now model the daily car rental cost for Acme and Interstate. Acme is a better deal than Interstate if

Step 3. Write an inequality that describes the verbal conditions.

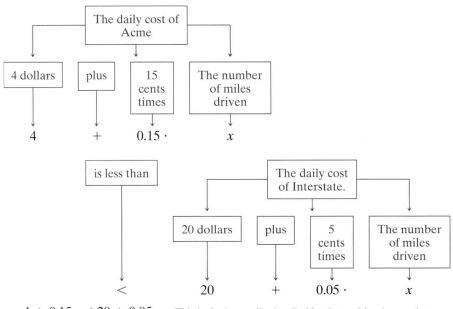

Step 4. Solve the inequality and answer the problem's question.

$4 + 0.15x < 20 + 0.05x$	This is the inequality implied by the problem's question.
$400 + 15x < 2000 + 5x$	Clear decimals, multiplying both sides by 100.
$400 + 10x < 2000$	Subtract $5x$ from both sides.
$10x < 1600$	Subtract 400 from both sides.
$x < 160$	Divide both sides by 10.

Thus, driving fewer than 160 miles per day makes Acme the better deal.

Step 5. Check

We can check by substituting a value for x less than 160 in the models for daily cost. If $x = 150$,

$$\text{Cost for Acme} = 4 + 0.15(150) = 26.50$$
$$\text{Cost for Interstate} = 20 + 0.05(150) = 27.50$$

and Acme does offer the better deal. ■

Using technology

The graphs of the daily cost models

$$y_1 = 4 + 0.15x \qquad y_2 = 20 + 0.05x$$

were obtained on a graphing utility using

$$\text{Xmin} = 0, \text{Xmax} = 300, \text{Xscl} = 10, \text{Ymin} = 0, \text{Ymax} = 40, \text{Yscl} = 4.$$

The graphs intersect at (160, 28). To the left of $x = 160$, the graph of $y = 4 + 0.15x$ lies below that of $y = 20 + 0.05x$, reinforcing that for fewer than 160 miles per day, Acme offers the better deal.

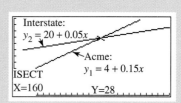

PROBLEM SET 2.4

Practice Problems

For Problems 1–32, find and graph the solution set. Express the solution set in both set-builder and interval notation.

1. $x + 7 > 10$

2. $x - 3 < 1$

3. $x - 5 \leqslant -7$

4. $6x \geqslant 24$

5. $9x < -45$

6. $0.2x < -3$

7. $0.5x > 4$

8. $-9x \geqslant 36$

9. $-5x \leqslant 30$

10. $2x + 5 < 17$

11. $5x + 11 > 26$

12. $3x - 9 > 2x - 7$

13. $2x + 3 \geqslant x + 4$

14. $8x - 11 \leqslant 3x - 13$

15. $18x + 45 \leqslant 12x - 8$

16. $3(x + 1) - 5 < 2x + 1$

17. $4(x + 1) + 2 \geqslant 3x + 6$

18. $8x + 3 > 3(2x + 1) + x + 5$

19. $7(2x - 1) < 9x + 11$

20. $8x - 2(2x + 3) \geqslant 0$

21. $-2(2x - 9) \leqslant 4 - 5(x - 2)$

22. $4.7x - 3.6 < 11.4$

23. $2.6x - 0.2 \geqslant 1.4x + 2.2$

24. $0.2x + 3 < 2.4x - 7$

25. $\dfrac{x}{3} + \dfrac{2}{5} \leqslant 4$

26. $\dfrac{3x}{2} + 4 \leqslant 6$

27. $\dfrac{x + 1}{3} - \dfrac{x - 3}{2} < \dfrac{1}{6}$

28. $\dfrac{x - 14}{3} \geqslant \dfrac{2x}{3} - \dfrac{1}{2} + \dfrac{x}{2}$

29. $3[3(y + 5) + 8y + 7] + 5[3(y - 6) - 2(3y - 5)] < 2(4y + 3)$

30. $5[3(2 - 3y) - 2(5 - y)] - 6[5(y - 2) - 2(4y - 3)] < 3y + 19$

31. $4(3y + 2) - 3y < 3(1 + 3y) - 7$

32. $3(y - 8) - 2(10 - y) > 5(y - 1)$

Application Problems

33. According to the National Safety Council, there were 49,391 motor vehicle deaths in the United States in 1988. This number declined steadily by 1546 per year from 1988 through 1990. If this trend continues, after what year will there be fewer than 35,477 motor vehicle deaths per year?

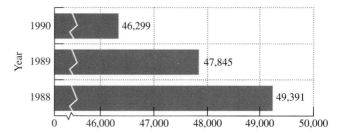

Motor Vehicle Deaths in the United States

34. According to the National Center for Education Statistics, there were 980 thousand bachelor's degrees awarded in the United States in 1985. This number has increased steadily by approximately 16 thousand degrees each year. After what year will the number of bachelor's degrees exceed one million, 188 thousand?

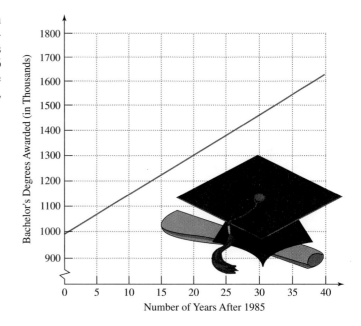

Answer the questions in Problems 35–38 based on the given graph. The first bar in each age group represents men and the second represents women.

Illicit Drug Use within the Past Month, 1992

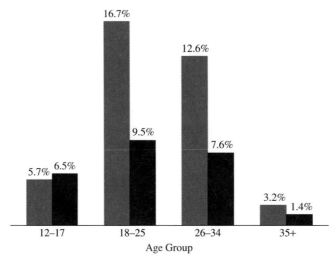

Source: *National Household Survey on Drug Abuse: 1992,* Substance Abuse and Mental Health Service Administration, U.S.H.H.S. 1993

35. For what group(s) did illicit drug use exceed 10%?

36. For what group(s) was illicit drug use at least 7% and at most 12%?

37. If x represents the percentage of people who have used illicit drugs within the past month, what group(s) is(are) described by the solution of $3x + 2 - 5 > -x + 7 + 2x$?

38. If x represents the percentage of people who have used illicit drugs within the past month, what group(s) is(are) described by the solution of $\dfrac{4 - 3x}{5} > -2$?

39. A car can be rented from Basic Rental for $260 per week with no extra charge for mileage. Continental charges $80 per week plus 25 cents for each mile driven to rent the same car. How many miles should be driven in a week to make the rental cost for Basic Rental a better deal than Continental's?

40. Membership in a fitness club costs $500 yearly plus $1 per hour spent working out. A competing club charges $440 yearly plus $1.75 per hour for use of their equipment. How many hours must a person work out yearly to make membership in the first club cheaper than membership in the second club?

41. A city commission has proposed two tax bills. The first bill requires that a homeowner pay $1800 plus 3% of the assessed home value in taxes. The second bill requires taxes of $200 plus 8% of the assessed home value. What price range of home assessment would make the first bill a better deal?

42. A local bank charges $8 per month plus 5¢ per each check written. The credit union charges only $2 per month plus 8¢ per each check written. How many checks should be written so that the bank's monthly cost is greater than that of the credit union?

43. A textbook publisher allows at least $3560 per month for advertising. Of this amount, $1600 is spent on presentations in professional journals. The remainder of the advertising money is spent on advertisements in college newspapers. If each advertisement in a college newspaper costs $25, find the minimum number of advertisements.

44. A coin collection contains 12 coins in dimes and quarters. Describe coin combinations that will make the collection less than $1.95.

True–False Critical Thinking Problems

45. Which one of the following is true?
   a. The inequality $3x > 6$ is equivalent to $2 > x$.
   b. The number 4 is a member of the solution set to the inequality $\dfrac{7 - 5x}{-2} \geq -1$.
   c. If x is no larger than 5, then $x < 5$.
   d. If x is at least 7, then $x > 7$.

46. Which one of the following is true?
   a. If x is at most 3, then $x \geq 3$.
   b. If x is no more than 5, then $x < 5$.
   c. The inequality $-3x > 6$ is equivalent to $-2 > x$.
   d. The number 0 is a member of the solution set to the inequality $\frac{1}{4}x - \frac{1}{2} < \frac{1}{2}x - \frac{2}{3}$.

Technology Problems

In Problems 47–48, solve each inequality using a graphing utility. Graph each side separately. Then determine the values of x for which the graph on the left side lies above the graph on the right side.

47. $-3(x - 6) > 2x - 2$

48. $-2(x + 4) > 6x + 16$

Use the same technique employed in Problems 47–48 to solve each inequality in Problems 49–50. In each case, what conclusion can you draw? What happens if you try solving the inequalities algebraically?

49. $12x - 10 > 2(x - 4) + 10x$

50. $2x + 3 > 3(2x - 4) - 4x$

51. A bank offers two checking account plans. Plan A has a base service charge of $4.00 per month plus 10¢ per check. Plan B charges a base service charge of $2.00 per month plus 15¢ per check.
   a. Write models for the total monthly costs for each plan if x checks are written.
   b. Use a graphing utility to graph the models in the same viewing rectangle. Use the following range setting:

   Xmin = 0, Xmax = 50, Xscl = 1, Ymin = 0,
   Ymax = 10, Yscl = 1

   c. Use the graphs (and the [TRACE] or intersection feature) to determine for what number of checks per month plan A will be better than plan B.
   d. Verify the result of part (c) algebraically by solving an inequality.

Writing in Mathematics

52. Describe ways in which solving a linear inequality is similar to solving a linear equation.

53. Describe ways in which solving a linear inequality is different than solving a linear equation.

Critical Thinking Problems

54. Eleanor's age is 3 years more than 2 times Mia's age. The sum of their ages is greater than or equal to 24 years. Which of the following is true?
   a. Mia cannot be 8 years old.
   b. Eleanor can be 19.
   c. Eleanor cannot be 17.
   d. Mia can be 6.

55. Two American tourists visiting London decide to purchase one of Mrs. Lovett's famous meat pies. Each tourist has some money, but one person needs $7 more to buy the pie and the other needs $2 more. They pool their money but still do not have enough. How much does one meat pie cost?

56. What's wrong with this argument? Suppose x and y represent two real numbers, where $x > y$:

$2 > 1$	This is a true statement.
$2(y - x) > 1(y - x)$	Multiply both sides by $y - x$.
$2y - 2x > y - x$	Use the distributive property.
$y > x$	Subtract y from both sides.
	Add $2x$ to both sides.

The final inequality, $y > x$, is impossible because we were initially given $x > y$.

Review Problems

57. Solve for the variable: $y - (3 + y) = 5 + \dfrac{2(y - 2)}{4}$.

58. The model $w = \frac{11}{2}h - 220$ describes the normal weight (w, in pounds) for a man who is h inches tall,

where $60 \leqslant h \leqslant 70$. Solve for h and then determine what the height should be for a man weighing 154 pounds.

59. Simplify: $-4x^2 - [6y^2 - (5x^2 - 2y^2)]$.

SECTION 2.5

Solutions Manual   Tutorial   Video 3

Compound Inequalities

Objectives

1 Find the intersection of two sets.
2 Solve compound inequalities with the word *and*.
3 Find the union of two sets.
4 Solve compound inequalities with the word *or*.

In this section, we consider pairs of inequalities joined by the word *and* and *or*. Such inequalities are called *compound inequalities*.

Intersection of Sets and Inequalities Linked by the Connective *And*

We begin with a hypothetical situation involving the United Cable Television Company. Let x represent the number of customers served by the company. The profit of the company is 50 times the number of customers served less $100. The company has *two* constraints placed upon it: (1) Its profits must be greater than $5000, or the corporation will be sold by the stockholders; (2) its profits must be less than $30,000, or its broadcasting license will be revoked. Since profits are $50x - 100$ (50 times the number of customers minus $100), we have

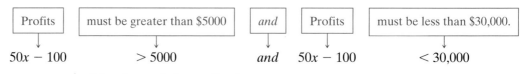

Profits	must be greater than $5000	*and*	Profits	must be less than $30,000.
↓	↓	↓	↓	↓
$50x - 100$	> 5000	*and*	$50x - 100$	$< 30,000$

We solve each inequality for x:

$$50x - 100 > 5000 \quad \text{and} \quad 50x - 100 < 30,000$$
$$50x > 5100 \quad \text{and} \quad 50x < 30,100$$
$$x > 102 \quad \text{and} \quad x < 602$$

With two conditions of inequality placed on the variable x, we now must consider $\{x \mid x > 102 \text{ and } x < 602\}$, that is, the numbers in $\{x \mid x > 102\}$ and the numbers in $\{x \mid x < 602\}$. We call the resulting set the *intersection* of the two given sets.

> ### The intersection of two sets
>
> If A and B are sets, the *intersection* of A and B consists of all elements that are in set A *and* in set B. We represent this set by $A \cap B$, read "the intersection of sets A and B."

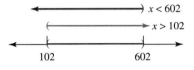

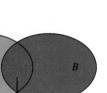

Figure 2.34

The graph of $x < 602$ and $x > 102$

The set $\{x \mid x < 602 \text{ and } x > 102\}$ can be written as

$$\{x \mid x < 602\} \cap \{x \mid x > 102\}.$$

The graph of the intersection is shown on the number line in Figure 2.34.

All real numbers between 102 and 602, excluding 102 and 602, are common to the graphs shown above the number line. The intersection of the two sets can be written as the single set

$$\{x \mid 102 < x < 602\}$$

read "all x such that x is greater than 102 *and* less than 602." Using interval notation, the intersection of the two graphs is written as $(102, 602)$. In the context of our original problem, this means that in order not to be sold or lose its license, the company must have more than 102 customers *and* fewer than 602 customers.

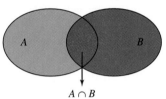

Figure 2.35

The intersection of sets

Figure 2.35 shows a useful way of picturing the intersection of sets A and B. The figure indicates that $A \cap B$ contains those elements that belong to both A and B at the same time.

1 Find the intersection of two sets.

EXAMPLE 1 Finding the Intersection of Two Sets

Let $A = \{1, 2, 3, 5\}$ and $B = \{2, 5, 7\}$. Find $A \cap B$

Solution

The numbers 2 and 5 are common to both sets. Therefore,

$$A \cap B = \{1, 2, 3, 5\} \cap \{2, 5, 7\}$$
$$= \{2, 5\}$$

2 Solve compound inequalities with the word *and*.

EXAMPLE 2 Solving a Compound Inequality Involving Intersection (and)

Solve and graph: $\{x \mid 3(x - 2) - 2x > -6\} \cap \{x \mid x - 9 \leqslant 5\}$

Solution

We must find the solutions for each of the inequalities and graph them above the same number line. First we solve

$3(x - 2) - 2x > -6$	This is the first given inequality.
$3x - 6 - 2x > -6$	Apply the distributive property on the left.
$x - 6 > -6$	Combine like terms.
$x > 0$	Add 6 to both sides.

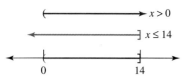

Figure 2.36

The graph of $x > 0$ and $x \leq 14$

Then we solve

$$x - 9 \leq 5 \qquad \text{This is the second given inequality.}$$
$$x \leq 14 \qquad \text{Add 9 to both sides.}$$

To find $\{x \mid x > 0\} \cap \{x \mid x \leq 14\}$, we graph each set above the number line and then locate the interval on the number line common to both graphs. The intersection of the two sets can be written as the single set $\{x \mid 0 < x \leq 14\}$, as shown on the number line in Figure 2.36. Using interval notation, the solution set is $(0, 14]$. ■

study tip

To find the intersection of two sets using a number line:

1. Graph each set above the number line, using a different color (say, red and blue) for each graph.
2. The graph on the number line represents the interval where the red and blue graphs overlap.

Let's take a moment to summarize the steps involved in the solution of a compound inequality with the word *and*.

Solving inequalities with *and*

1. Solve each inequality in the compound inequality separately.
2. Because the inequalities are connected by *and*, the solution set includes all numbers satisfying both solutions in step 1 simultaneously or the intersection of the solution sets.

EXAMPLE 3 **Solving a Compound Inequality Involving Intersection (*and*)**

Solve and graph: $-2x - 5 > -1$ and $5x + 2 < -18$

Solution

We solve each inequality in the compound inequality separately.

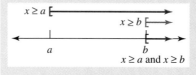

study tip

Keep writing the word *and* between inequalities.

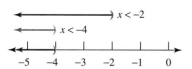

Figure 2.37

The graph of $x < -2$ and $x < -4$

The solution set is the intersection of the solution sets of the individual inequalities. In Figure 2.37, we have graphed these solution sets in red and blue above the number line. The graph on the number line, the interval where the red and blue graphs overlap, represents the numbers common to both sets, that is, numbers less than -4. Thus, the solution set is $\{x \mid x < -4\}$ or $(-\infty, -4)$. ■

Figure 2.38

A and B have no common members: $A \cap B = \varnothing$.

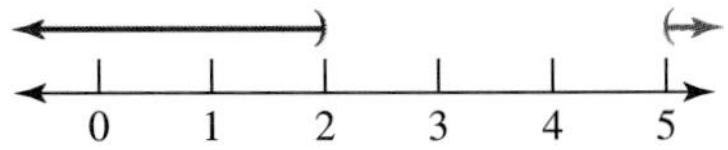

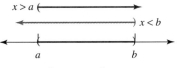

Figure 2.39

The red and blue graphs do not overlap.

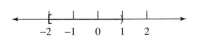

$a < x < b$ means that $x > a$ and $x < b$.

Discover for yourself

Solve Example 5 by solving the two separate inequalities

$2x + 1 \geq -3$ and $2x + 1 < 3$. Do you obtain the same solution as the one given in Example 5? Which solution method do you prefer? Why?

In our next example, the solution sets of the individual inequalities have no common members. Their intersection is the empty set, $\varnothing$, indicated by Figure 2.38.

EXAMPLE 4 **A Compound Inequality with No Solution**

Solve and graph: $2x - 7 > 3$ and $5x - 4 < 6$

Solution

We solve each inequality in the compound inequality separately.

$$2x - 7 > 3 \qquad \text{and} \qquad 5x - 4 < 6$$
$$2x > 10 \qquad \text{and} \qquad 5x < 10$$
$$x > 5 \qquad \text{and} \qquad x < 2$$

Now we graph both solution sets above the number line, as shown in Figure 2.39. We see that no real numbers are both greater than 5 and less than 2. Thus the solution set is the empty set, $\varnothing$. ∎

In Chapter 1 we studied bounded intervals on the real number line. Recall that $a < x < b$ means that x is greater than a ($x > a$) and x is less than b ($x < b$). We can now think of the notation $a < x < b$ as an abbreviation for a compound inequality linked by *and*. This idea forms the basis of our next example.

EXAMPLE 5 **Solving a Compound Inequality Where the Connective *and* Is Implied**

Solve and graph: $-3 \leq 2x + 1 < 3$

Solution

Since $-3 \leq 2x + 1 < 3$ means that

$$2x + 1 \geq -3 \qquad \text{and} \qquad 2x + 1 < 3$$

we must solve both inequalities and find their intersection. Since both inequalities require identical steps in their solutions—namely, subtracting 1 and then dividing by 2—we can deal with all three parts of the inequality at once. We would like to *isolate x by itself in the middle.*

$-3 \leq 2x + 1 < 3$	This is the given inequality.
$-3 - 1 \leq 2x + 1 - 1 < 3 - 1$	Subtract 1 from all three parts.
$-4 \leq 2x < 2$	Simplify.
$\frac{-4}{2} \leq \frac{2x}{2} < \frac{2}{2}$	Divide each part by 2.
$-2 \leq x < 1$	Simplify.

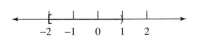

Figure 2.40

The graph of $-2 \leq x < 1$

The solution set consists of all numbers greater than or equal to -2 and less than 1, represented by $\{x \mid -2 \leq x < 1\}$ in set-builder notation and $[-2, 1)$ in interval notation. The graph is shown in Figure 2.40. ∎

sing technology

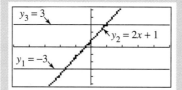

The graph of $y_2 = 2x + 1$ lies between $y_1 = -3$ and $y_3 = 3$ when x is between -2 and 1. The viewing rectangle is $[-5, 5]$ by $[-5, 5]$.

You can use a graphing utility to show that the solution set for $-3 \leq 2x + 1 < 3$ is $[-2, 1)$. Graph each part of the inequality, entering

$$Y_1 = -3$$
$$Y_2 = 2x + 1$$
$$Y_3 = 3$$

As shown in the figure, notice that the graph of $y_2 = 2x + 1$ is greater than or equal to the graph of $y_1 = -3$ when $x \geq -2$. Also, the graph of $y_2 = 2x + 1$ is less than the graph of $y_3 = 3$ when $x < 1$. Equivalently, the graph of $y_2 = 2x + 1$ lies on or above the graph of $y_1 = -3$ and under the graph of $y_3 = 3$ when $-2 \leq x < 1$.

sing technology

The graph of

$$y_1 = 3(-x - 5) - 9$$

falls within the horizontal band between $y = 0$ (the x-axis), exclusively, and $y_2 = 6$, inclusively, for $-10 \leq x < -8$.

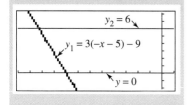

Figure 2.41

The graph of $-10 \leq x < -8$

EXAMPLE 6 **Solving a Compound Inequality**

Solve and graph: $0 < 3(-x - 5) - 9 \leq 6$

Solution

We can again use the shorter method, isolating x by itself in the middle.

$0 < 3(-x - 5) - 9 \leq 6$	This is the given inequality.
$0 < -3x - 15 - 9 \leq 6$	Apply the distributive property.
$0 < -3 - 24 \leq 6$	Combine numerical terms.
$24 < -3x \leq 30$	Add 24 to each part.
$\dfrac{24}{-3} > \dfrac{-3x}{-3} \geq \dfrac{30}{-3}$	Divide each part by -3, reversing the sense of the inequality.
$-8 > x \geq -10$	Simplify.

This inequality is correct but awkward to interpret. It can be rewritten with the lesser number, -10, on the left as

$$-10 \leq x < -8.$$

Using set notation, the solution set is $\{x \mid -10 \leq x < -8\}$. In interval notation, this can be expressed as $[-10, -8)$. The graph is shown in Figure 2.41. ∎

tudy tip

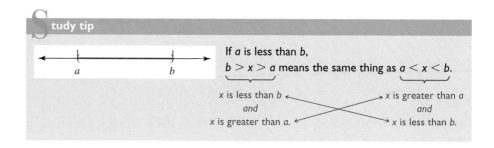

If a is less than b,
$b > x > a$ means the same thing as $a < x < b$.

3 Find the union of two sets.

Union of Sets and Inequalities Linked by the Connective *Or*

We now look at compound inequalities linked by the word *or*, which involves the *union* of sets.

> ### The union of two sets
>
> If A and B are sets, the *union* of A and B consists of all elements that are in set A or in set B, or in both A and B. We represent this set by $A \cup B$, read "the union of sets A and B."

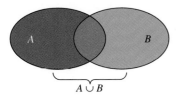

Figure 2.42

The union of sets

Figure 2.42 shows a useful way of picturing the union of sets A and B. The figure indicates that $A \cup B$ is formed by joining the sets together.

EXAMPLE 7   Finding the Union of Two Sets

Let $A = \{1, 2, 3, 5\}$ and $B = \{2, 5, 7\}$. Find $A \cup B$

First we list the elements of set A: 1, 2, 3, 5. Then we list any additional elements from set B. Since 2 and 5 are already listed, the only additional element is 7. Therefore,

$$A \cup B = \{1, 2, 3, 5\} \cup \{2, 5, 7\}$$
$$= \{1, 2, 3, 5, 7\}$$

The numbers in this set are those in either *or* both of the given sets. ∎

4 Solve compound inequalities with the word *or*.

Solving compound inequalities with the connective *or* involves the following steps.

> ### Solving inequalities with *or*
>
> 1. Solve each inequality in the compound inequality separately.
> 2. Because the inequalities are connected by *or*, the solution set includes all numbers satisfying either one *or* both of the solutions in step 1, namely, the union of the solution sets.

EXAMPLE 8   Solving a Compound Inequality Involving Union (*or*)

Solve and graph:

$$\{x \mid 2x - 6 > 10\} \cup \{x \mid -4x < -4\}$$

Solution

Since x satisfies either $2x - 6 > 10$ or $-4x < -4$, we begin by solving each inequality separately.

$$2x - 6 > 10 \quad \text{or} \quad -4x < -4 \qquad \text{Reverse the}$$
$$\qquad\qquad\qquad\qquad \frac{-4x}{-4} > \frac{-4}{-4} \qquad \text{inequality}$$
$$2x > 16 \quad \text{or} \qquad\qquad\qquad \text{sign.}$$

$$x > 8 \quad \text{or} \qquad x > 1$$

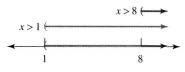

Figure 2.43

The graph of $x > 8$ or $x > 1$

Since the inequalities are joined by union (*or*), x is either greater than 8 or greater than 1. The solution set consists of all numbers that are greater than 8 *or* all numbers that are greater than 1. The solution set contains all numbers that satisfy one or both of these conditions. As shown in Figure 2.43, by graphing both these conditions above the number line and then joining the graphs together, we obtain the solution set $\{x \mid x > 1\}$ in set-builder notation and $(1, \infty)$ in interval notation. ■

Study tip

To find the union of two sets using a number line:

1. Graph each set above the number line, using a different color (say, red and blue) for each graph.
2. The graph on the number line represents *all* of the red graph and *all* of the blue graph, including the interval (if there is one) where the red and blue graphs overlap.

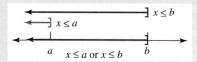

EXAMPLE 9   **Solving a Compound Inequality Involving Union (*or*)**

Solve and graph:

$$2x - 3 < 7 \qquad \text{or} \qquad 35 - 4x \leqslant 3$$

Solution

We solve each inequality in the compound inequality separately.

$$\begin{array}{rclcl} 2x - 3 &<& 7 & \text{or} & 35 - 4x \leqslant 3 \\ 2x &<& 10 & \text{or} & -4x \leqslant -32 \\ x &<& 5 & \text{or} & x \geqslant 8 \end{array}$$

Reverse the inequality sign.

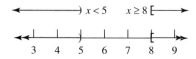

Figure 2.44

To graph $x < 5$ or $x \geqslant 8$, join the red and blue graphs together on the number line.

The solution set consists of all numbers that are less than 5 or greater than or equal to 8, shown in Figure 2.44. The solution set is expressed as $\{x \mid x < 5$ or $x \geqslant 8\}$ in set-builder notation and as $(-\infty, 5) \cup [8, \infty)$ using interval notation. There is no shortcut way to express the solution of this union when interval notation is used. ■

We have seen that some compound inequalities have no solution. Our next example involves a compound inequality that is satisfied by all real numbers.

EXAMPLE 10   **A Compound Inequality Satisfied by All Real Numbers**

Solve and graph:

$$3x - 5 \leqslant 13 \qquad \text{or} \qquad 5x + 2 > -3$$

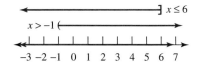

Figure 2.45

All real numbers are less than or equal to 6 or greater than − 1.

Solution

We solve each inequality in the compound inequality separately.

$$3x - 5 \le 13 \quad \text{or} \quad 5x + 2 > -3$$
$$3x \le 18 \quad \text{or} \quad 5x > -5$$
$$x \le 6 \quad \text{or} \quad x > -1$$

As shown in Figure 2.45, all real numbers are either less than or equal to 6 or greater than − 1. Thus, the solution set is $\mathbb{R}$, the set of all real numbers, expressed as $(-\infty, \infty)$ in interval notation. ■

PROBLEM SET 2.5

Practice Problems

Find the intersection of the sets in Problems 1–6.

1. $\{1, 2, 3, 4\} \cap \{2, 4, 5\}$

2. $\{1, 3, 7\} \cap \{2, 3, 8\}$

3. $\{1, 3, 5, 7\} \cap \{2, 4, 6, 8, 10\}$

4. $\{-12, -10, -8\} \cap \{-2, 0, 2\}$

5. $\{7, 9, 10\} \cap \{7, 9, 10\}$

6. $\{7, 8, 9\} \cap \varnothing$

In Problems 7–36, solve each compound inequality and graph the solution set on a number line. Express the solution set in both set-builder and interval notation.

7. $x > 3$ and $x > 6$

8. $x < -2$ and $x < -4$

9. $\{x \mid x \ge 3\} \cap \{x \mid x \le 6\}$

10. $\{x \mid x \ge -1\} \cap \{x \mid x \le 3\}$

11. $x < -3$ and $x > 1$

12. $x \le 0$ and $x \ge 2$

13. $5x < -20$ and $3x > -18$

14. $3x \le 15$ and $2x > -6$

15. $x - 4 \le 2$ and $3x + 1 > -8$

16. $3x + 2 > -4$ and $2x - 1 < 5$

17. $2x > 5x - 15$ and $7x > 2x + 10$

18. $6 - 5y > 1 - 3y$ and $4y - 3 > y - 9$

19. $\{y \mid 4(1 - y) < -6\} \cap \left\{y \mid \dfrac{y - 7}{5} \le -2\right\}$

20. $\{y \mid 5(y - 2) > 15\} \cap \left\{y \mid \dfrac{y - 6}{4} \le -2\right\}$

21. $\{x \mid x - 1 \le 7x - 1\} \cap \{x \mid 4x - 7 < 3 - x\}$

22. $\{y \mid 2y + 1 > 4y - 3\} \cap \{y \mid y - 1 \ge 3y + 5\}$

23. $\left\{x \mid \dfrac{9 + 4x}{3} > -5\right\} \cap \left\{x \mid \dfrac{x}{3} + 4 < 3\right\}$

24. $\{y \mid 6 - y > \frac{1}{2}(2 - 5y)\} \cap \{y \mid y > 7 - \frac{1}{4}(y + 8)\}$

25. $2 < x - 5 \le 7$

26. $-3 \le y + 1 < 5$

27. $-5 \le 2x - 7 < 9$

28. $-1 < 3y - 1 \le 5$

29. $-6 < \frac{1}{2}y - 4 < -3$

30. $-3 \le \frac{2}{3}x - 5 < -1$

31. $-2 < -3y \le 3$

32. $-4 \le -2x < -1$

33. $-8 < 2 - 3y < 10$

34. $-3 \le 1 - 2y \le 5$

35. $-2 \le -\frac{1}{2}y + 3 \le 6$

36. $7 < -\frac{3}{5}x + 1 \le 8$

Find the union of the sets in Problems 37–42.

37. $\{1, 2, 3, 4\} \cup \{2, 4, 5\}$

38. $\{1, 3, 7\} \cup \{2, 3, 8\}$

39. $\{1, 3, 5, 7\} \cup \{2, 4, 6, 8, 10\}$

40. $\{-12, -10, -8\} \cup \{-2, 0, 2\}$

41. $\{7, 9, 10\} \cup \{7, 9, 10\}$

42. $\{7, 8, 9\} \cup \varnothing$

In Problems 43–62, solve each compound inequality and graph the solution set on a number line. Express the solution set in both set-builder and interval notation.

43. $x > 3$ or $x > 6$

44. $x < -2$ or $x < -4$

45. $\{x \mid x \ge 3\} \cup \{x \mid x \le 6\}$

46. $\{x \mid x \ge -1\} \cup \{x \mid x \le 3\}$

47. $x < -3$ or $x > 1$

48. $x \le 0$ or $x \ge 2$

49. $5x < -20$ or $3x > -18$

50. $3x \le 15$ or $2x > -6$

51. $x - 4 \le 2$ or $3x + 1 > -8$

52. $3x + 2 > -4$ or $2x - 1 < 5$

53. $3x - 12 > 0$ or $2x - 3 \le -6$

54. $x + 1 < -6$ or $x + 1 > 2$

55. $\{x \mid 3x + 6 < 8\} \cup \{x \mid -2x + 3 > -2\}$

56. $\{y \mid 2 - y > 4\} \cup \{y \mid 6 - 3y \leqslant 2\}$

57. $\{x \mid x - 14 < 26 - 3x\} \cup \{x \mid 31 - 5x > 1 - 8x\}$

58. $\{y \mid 16 - 3y \geqslant -8\} \cup \{y \mid 13 - y > 4y + 3\}$

59. $\{y \mid 12y + 6 < 4(5y - 1)\} \cup \{y \mid 9y + 4 \geqslant 5(3y - 4)\}$

60. $\{y \mid y - 1 < 2(4 - y)\} \cup \{y \mid y + 7 \leqslant 3(2y - 1)\}$

61. $\left\{x \mid \dfrac{2x}{9} + 5 < 7\right\} \cup \left\{x \mid 3 - x > \dfrac{x - 3}{2}\right\}$

62. $\left\{y \mid \dfrac{y}{4} + 3 < 7\right\} \cup \left\{y \mid 2 - y > \dfrac{y - 8}{5}\right\}$

Application Problems

63. The formula for converting Celsius temperature (C) to Fahrenheit temperature (F) is $F = \frac{9}{5}C + 32$. If Fahrenheit temperature ranges from 77° to 104°, inclusively, what is the range for the Celsius temperature?

64. It is customary to represent the speed of a supersonic aircraft by the ratio of the speed of the aircraft A to the speed of sound s, where

$$M = \frac{A}{s}.$$

M is called the Mach number, named after Austrian physicist Ernst Mach. If the speed of sound is approximately 740 mph and the aircraft is designed to operate between Mach 1.7 and Mach 2.4, inclusively, what is the designated speed range of the aircraft in mph?

The models $R = -0.03t + 20.86$ and $R = -0.06t + 24.07$ can be used to predict the world record in the men's and women's 200-meter dash, respectively, t years after 1948. In both models, R is measured in seconds. Use these models to answer Problems 65–66.

65. a. Use the first model to find those years for which the world record for men was between 20.38 and 19.78 seconds.

b. Shown below are the actual men's records. How well does the formula and your answer in part (a) model reality? Explain.

Year	1948	1956	1964	1972	1980	1984	1988	1992
Men's Records	21.1	20.6	20.3	20	20.19	19.8	19.75	20.01

66. a. Use the second model to find those years for which the world record for women was between 22.63 and 21.67 seconds.

b. Shown below are the actual women's records. How well does the formula and your answer in part (a) model reality? Explain.

Year	1948	1956	1964	1972	1980	1984	1988	1992
Women's Records	24.4	23.4	23	22.4	22.03	21.81	21.34	21.81

67. In 1984 there were 29,750 post offices in the United States. This number has been declining by approximately 132 post offices each year. For what years were the number of post offices at least 28,694 and at most 29,222?

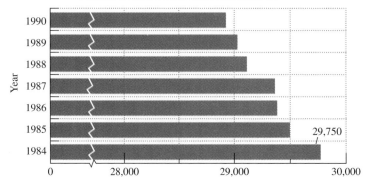

Number of Post Offices in the United States

The graph shows international child poverty rates. If x represents these rates expressed as percents, list the appropriate bar or bars by name (from the 11 bars shown in the graph) that adhere to each of the conditions in Problems 68–73.

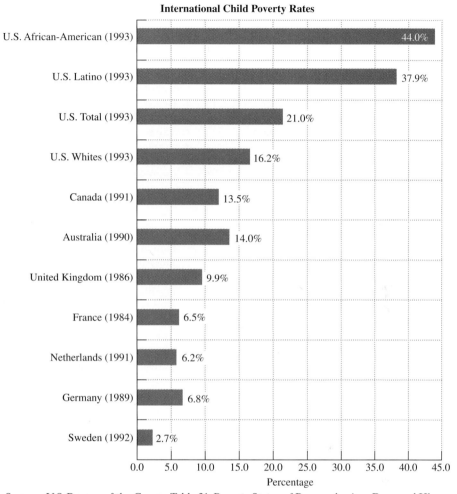

International Child Poverty Rates

U.S. African-American (1993) 44.0%
U.S. Latino (1993) 37.9%
U.S. Total (1993) 21.0%
U.S. Whites (1993) 16.2%
Canada (1991) 13.5%
Australia (1990) 14.0%
United Kingdom (1986) 9.9%
France (1984) 6.5%
Netherlands (1991) 6.2%
Germany (1989) 6.8%
Sweden (1992) 2.7%

Percentage

Sources: U.S. Bureau of the Census, Table 21: Poverty Status of Persons, by Age, Race, and Hispanic Origin: 1968–1993; Rainwater, Lee, and Timothy M. Smeeding (August 1995), *Luxembourg Income Study,* Table 3.

68. $x \geq 7$ and $x < 17$

69. $x \geq 9$ and $x \geq 21$

70. $x \geq 9$ or $x \geq 21$

71. $120 \leq 3x \leq 45$

72. $-2x \leq -28$ and $3x + 1 < 121$

73. $-\dfrac{x}{2} \geq -7$ or $2x - 1 < 41$

74. The length of a rectangle is 11 inches shorter than 3 times the width. If the perimeter of the rectangle varies between 10 and 114 inches, inclusively, find the maximum and minimum measurement for the width of the rectangle.

75. Two identical square areas are to be fenced in, as shown in the figure. If the total length of the fencing exceeds 126 meters, but is no more than 140 meters, find the dimensions of each square.

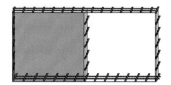

76. The two nonparallel sides of a trapezoid have equal lengths. Each of these sides is 1 inch more than twice the shorter base. The longer base is 6 inches more than

4 times the shorter base. If the perimeter of the trapezoid is between 18 and 54 inches, inclusively, what is the maximum and minimum length for the shorter base?

77. Cities A and B are between 90 and 120 miles apart, with the exact distance unknown. A car leaves A for B at the same time that a second car leaves B for A. The first car travels at 50 mph and the second at 40 mph. Find the time interval in which the cars will meet.

True–False Critical Thinking Problems

78. Which one of the following is true?
   a. $(-\infty, -1] \cap [-4, \infty) = [-4, -1]$
   b. $(-\infty, 3) \cup (-\infty, -2) = (-\infty, -2)$
   c. The union of two sets can never give the same result as the intersection of those sets.
   d. The solution set of $x < a$ and $x > a$ is the set of all real numbers excluding a.

79. Which one of the following is true?
   a. $(-1, 4) \cap (2, 7) = (2, 4)$
   b. $(-\infty, -2] \cup [-3, \infty) = [-3, -2]$
   c. The solution set of $x < a$ or $x > a$ is $\varnothing$.
   d. When solving a compound inequality, it is not necessary to reverse the sense of an inequality when multiplying by a negative number.

Technology Problem

80. The models
$$y = 0.15863x - 244.25497$$
$$\text{and } y = 0.203424x - 325.66257$$

describe the life expectancy at birth (y, in years) for American men and women, respectively, born in the year x, where $x \geq 1950$.
   a. Use a graphing utility to graph each model in the same viewing rectangle using the following range setting:
$$\text{Xmin} = 1950, \text{Xmax} = 2000, \text{Xscl} = 5,$$
$$\text{Ymin} = 60, \quad \text{Ymax} = 84, \quad \text{Yscl} = 1$$

   b. Trace along the graph representing the model for men and find x such that $69.03928 \leq y \leq 73.00503$. Describe what this result means in practical terms.
   c. What are the values for y along the graph describing life expectancy for women (as an inequality) corresponding to the values of x you found in part (c)? Answer the question by using the $\boxed{\text{TRACE}}$ feature on the women's graph. Describe what this means in practical terms.
   d. Verify part (c) algebraically by writing and solving a compound inequality. Use the model for men and a calculator for the computations.

Writing in Mathematics

81. Explain how to find the set of real numbers satisfying $x > -3$ and $x > -1$. Describe how to find the set of real numbers satisfying $x > -3$ or $x > -1$. What is the difference between the two procedures?

82. In Example 8 on page 189, we solved the compound inequality $2x - 6 > 10$ or $-4x < -4$. The solution set was $(1, \infty)$. Take a member of the solution set, such as 2. If 2 is substituted into the first inequality, here's what happens:

$2x - 6 > 10$	This is the first of the two given inequalities.
$2(2) - 6 \overset{?}{>} 10$	Since 2 is in the solution set, substitute 2 for x.
$4 - 6 \overset{?}{>} 10$	Multiply.
$-2 > 10$	Upon subtracting, we obtain a false statement.

If 2 is a member of the solution set, explain how we can obtain a false statement when 2 is substituted into the first given inequality.

Critical Thinking Problems

For Problems 83–85, solve and graph each solution set. Express the solution set in both set-builder and interval notation.

83. $-7 \leq 8 - 3y \leq 20$ and $-7 < 6y - 1 < 41$

84. $2y - 3 < 3y + 1 < 6y + 2$

85. $\{x \mid 7 < 5x + 12 < 17\} \cup \{x \mid -2 < -3x + 4 \leq 7\}$

86. Doughnuts are sold at $1 each. The doughnut shop has fixed costs of $120 per day plus 5¢ to produce each doughnut. If daily profit varies from a high of $592.50 to a low of $283.75, what is the range for the daily sales of doughnuts?

Review Problems

87. A number is decreased by 6, and this difference is then multiplied by 6. This gives the same result as decreasing the number by 9 and multiplying this difference by 9. What is the number?

88. The pressure (p, in pounds per square inch) is related to depth below the surface of the ocean (d, in feet) by the model $p = \frac{5}{11}d + 15$. At what depth is the pressure 60 pounds per square inch?

89. Graph $y = -x^3 + 1$ by filling in the following table of coordinates.

x	-2	-1	0	1	2
$y = -x^3 + 1$					

SECTION 2.6

Solutions Manual   Tutorial   Video 3

Equations and Inequalities Involving Absolute Value

Objectives

1 Solve equations involving absolute value.
2 Solve inequalities involving absolute value.

In Chapter 1, we introduced the concept of absolute value. We turn now to the study of equations and inequalities containing expressions involving absolute value.

---

1 Solve equations involving absolute value.

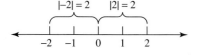

Figure 2.46
If $|x| = 2$, then $x = 2$ or $x = -2$.

Solving Equations with Absolute Value

In Section 1.2, we saw that $|a|$ (the absolute value of a) represents the distance of a from zero on the real number line. Now consider *absolute value equations,* such as

$$|x| = 2.$$

This means that we must determine real numbers whose distance from the origin on the number line is 2. Figure 2.46 shows that there are two numbers such that $|x| = 2$, namely, -2 or 2. We write $x = 2$ or $x = -2$. This observation can be generalized as follows.

> **Rewriting an absolute value equality without absolute value bars**
>
> If c is any positive number and $|X| = c$, then $X = c$ or $X = -c$.

EXAMPLE 1 **Rewriting Absolute Value Equalities Without Absolute Value Bars**

a. If $|x| = 7$, then $x = 7$ or $x = -7$.
b. If $|y| = \frac{1}{3}$, then $y = \frac{1}{3}$ or $y = -\frac{1}{3}$.
c. If $|x| = \sqrt{2}$, then $x = \sqrt{2}$ or $x = -\sqrt{2}$.
d. If $|x| = 0$, then $x = 0$. Only 0 lies 0 units from the origin.
e. There is no value of x satisfying $|x| = -4$ because the absolute value of any number, except 0, is positive. ∎

> **Solving an absolute value equation**
>
> 1. If X is an algebraic expression and c is a positive number, the solutions of $|X| = c$ are the numbers that satisfy $X = c$ or $X = -c$.
> 2. If X is an algebraic expression and c is a negative number, then $|X| = c$ has no solution.
> 3. If $|X| = 0$, then $X = 0$.

You can use a graphing utility to verify the solution set to an absolute value equation.

a. $|2x - 1| = 5$:

Graph

$y_1 = |2x - 1|$ and $y_2 = 5$. The x-coordinates of the intersection points are -2 and 3, verifying that $\{-2, 3\}$ is the solution set.

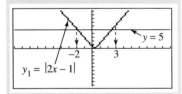

b. $|3x - 6| = -2$

Graph

$y_1 = |3x - 6|$ and $y_2 = -2$.

The graphs do not intersect, verifying that $\emptyset$ is the solution set.

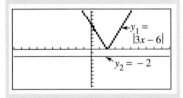

EXAMPLE 2   **Solving Absolute Value Equations**

Find the solution set: **a.** $|2x - 1| = 5$ **b.** $|3x - 6| = -2$

Solution

a.

$$\boxed{|X| = c} \quad \text{means} \quad \boxed{X = c} \quad \text{or} \quad \boxed{X = -c}$$

$$|2x - 1| = 5 \quad \text{means} \quad 2x - 1 = 5 \quad \text{or} \quad 2x - 1 = -5$$

We now solve the two equations that do not contain absolute value bars.

$$
\begin{array}{lcl}
2x - 1 = 5 & \text{or} & 2x - 1 = -5 \\
2x = 6 & \text{or} & 2x = -4 \\
x = 3 & \text{or} & x = -2
\end{array}
$$

Check 3:	Check -2:
$\lvert 2x - 1 \rvert = 5$	$\lvert 2x - 1 \rvert = 5$
$\lvert 2(3) - 1 \rvert \overset{?}{=} 5$	$\lvert 2(-2) - 1 \rvert \overset{?}{=} 5$
$\lvert 6 - 1 \rvert \overset{?}{=} 5$	$\lvert -4 - 1 \rvert \overset{?}{=} 5$
$\lvert 5 \rvert \overset{?}{=} 5$	$\lvert -5 \rvert \overset{?}{=} 5$
$5 = 5$ ✓ True	$5 = 5$ ✓ True

The check verifies that the solution set is $\{-2, 3\}$. This means that $2x - 1$ is 5 units from zero on the real number line if x is replaced by -2 or 3.

b. The solution set of $|3x - 6| = -2$ is $\emptyset$ because $|3x - 6|$ can never be negative. All values for the variable x will result in $|3x - 6|$ being positive or zero. ∎

In our next example, the given equation is not in the form $|ax + b| = c$. However, we can use the properties for solving linear equations to first isolate $|ax + b|$.

EXAMPLE 3   **Solving an Absolute Value Equation**

Solve: $2|x - 4| + 1 = 7$

Solution

$2\lvert x - 4 \rvert + 1 = 7$	This is the given equation. First isolate $\lvert x - 4 \rvert$.
$2\lvert x - 4 \rvert = 6$	Subtract 1 from both sides.
$\lvert x - 4 \rvert = 3$	Divide both sides by 2.

$$x - 4 = 3 \quad \text{or} \quad x - 4 = -3 \qquad \text{If } |X| = c \text{ (}c \text{ is positive), then } X = c \text{ or } X = -c.$$
$$x = 7 \quad \text{or} \qquad x = 1 \qquad \text{Solve each equation for } x.$$

Take a moment to check the two proposed solutions in the original equation, verifying that the solution set is $\{1, 7\}$. ■

---

sing technology

Use a graphing utility to verify Example 3. Enter

$$Y_1 = 2|x - 4| + 1 \qquad \text{and} \qquad Y_2 = 7$$

and graph each equation in the same viewing rectangle. The x-coordinates of the intersection points are 1 and 7, verifying that $\{1, 7\}$ is the solution set for $2|x - 4| + 1 = 7$.

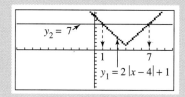

In the next example, two algebraic expressions are equal in absolute value. For this to be true, they must either be equal to each other or be the opposites of each other.

Writing an absolute value equation containing two absolute values without absolute value bars

If $|X_1| = |X_2|$, then $X_1 = X_2$ or $X_1 = -X_2$.

---

sing technology

Use a graphing utility to verify Example 4. Enter

$$Y_1 = |3x - 1|$$
$$\text{and } Y_2 = |x + 5|$$

and graph each equation in the same viewing rectangle. The x-coordinates of the intersection points are -1 and 3, verifying that $\{-1, 3\}$ is the solution set for $|3x - 1| = |x + 5|$.

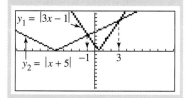

EXAMPLE 4 **Solving an Absolute Value Equation with Two Absolute Values**

Solve: $|3x - 1| = |x + 5|$

Solution

$$\boxed{|X_1| = |X_2|} \quad \text{means} \quad \boxed{X_1 = X_2} \quad \text{or} \quad \boxed{X_1 = -X_2}$$
$$|3x - 1| = |x + 5| \quad \text{means} \quad 3x - 1 = x + 5 \quad \text{or} \quad 3x - 1 = -(x + 5)$$

We now solve the two equations that do not contain absolute value bars.

$$\begin{array}{lll}
3x - 1 = x + 5 & \text{or} & 3x - 1 = -(x + 5) \\
2x - 1 = 5 & \text{or} & 3x - 1 = -x - 5 \\
2x = 6 & \text{or} & 4x - 1 = -5 \\
x = 3 & \text{or} & 4x = -4 \\
& & x = -1
\end{array}$$

Once again, take a moment to complete the solution process by checking the two proposed solutions in the original equation. The solution set is $\{-1, 3\}$. ■

2 Solve inequalities involving absolute value.

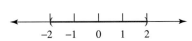

Figure 2.47

$|x| < 2$, so $-2 < x < 2$.

Figure 2.48

$|x| > 2$, so $x < -2$ or $x > 2$.

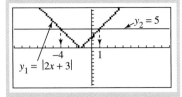
Solving Inequalities with Absolute Value

Absolute value can also arise in inequalities. Consider, for example,

$$|x| < 2.$$

This means that the distance of x from 0 is *less than* 2, as shown in Figure 2.47. The interval shows values of x that lie less than 2 units from 0. Thus, x can lie between -2 and 2. That is, x is greater than -2 and less than 2. We write $(-2, 2)$ or $\{x \mid -2 < x < 2\}$.

Some absolute value inequalities use the "greater than" symbol. For example, $|x| > 2$ means that the distance of x from 0 is *greater than* 2, as shown in Figure 2.48. Thus x can be greater than 2 *or* less than -2. We write $x > 2$ or $x < -2$.

These observations suggest the following principles for solving inequalities with absolute value.

> **Solving an absolute value inequality**
>
> If X is an algebraic expression and c is a positive number:
>
> 1. The solutions of $|X| < c$ are the numbers that satisfy $-c < X < c$.
>
> 2. The solutions of $|X| > c$ are the numbers that satisfy $X < -c$ or $X > c$.

EXAMPLE 5   **Solving an Absolute Value Inequality with $<$**

Solve and graph: $|2x + 3| < 5$

Solution

| $|X| < c$ | means | $-c < X < c$ |

$|2x + 3| < 5$ means $-5 < 2x + 3 < 5$

Now we solve the compound inequality $-5 < 2x + 3 < 5$.

$-5 < 2x + 3 < 5$ Remember that our goal is to isolate x in the middle.

$-8 < 2x < 2$ Subtract 3 from each part.

$-4 < x < 1$ Divide each part by 2.

The solution set is $\{x \mid -4 < x < 1\}$, or, in interval notation $(-4, 1)$. The graph of the solution set is shown in the figure.

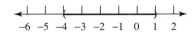

EXAMPLE 6   **Solving an Absolute Value Inequality with $\geq$**

Solve and graph: $\left|\dfrac{3x + 3}{5}\right| \geq 3$

Solution

$$\boxed{|X| \;\geqslant\; c} \quad \text{means} \quad \boxed{X \qquad \leqslant -c} \quad \text{or} \quad \boxed{X \qquad \geqslant c}$$

$$\left|\frac{3x+3}{5}\right| \geqslant 3 \quad \text{means} \quad \frac{3x+3}{5} \leqslant -3 \quad \text{or} \quad \frac{3x+3}{5} \geqslant 3$$

We will solve each of these inequalities and then find the union of their solution sets.

$$\frac{3x+3}{5} \leqslant -3 \qquad \text{or} \qquad \frac{3x+3}{5} \geqslant 3 \qquad \text{These are the inequalities without absolute value bars.}$$

$$5\left(\frac{3x+3}{5}\right) \leqslant 5(-3) \quad \text{or} \quad 5\left(\frac{3x+3}{5}\right) \geqslant 5(3) \qquad \text{Multiply both sides by 5.}$$

$$3x + 3 \leqslant -15 \quad \text{or} \quad 3x + 3 \geqslant 15 \qquad \text{Simplify.}$$

$$3x \leqslant -18 \quad \text{or} \quad 3x \geqslant 12 \qquad \text{Subtract 3 from both sides.}$$

$$x \leqslant -6 \quad \text{or} \quad x \geqslant 4 \qquad \text{Divide both sides by 3.}$$

With the connective *or*, the solution set to the original inequality is expressed as $\{x \mid x \leqslant -6 \text{ or } x \geqslant 4\}$ or $(-\infty, -6] \cup [4, \infty)$. The graph of the solution set is shown in the figure.

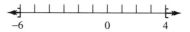

Table 2.5 summarizes what we have learned in this section. Study this table before considering the application in Example 7.

TABLE 2.5 Rewriting Absolute Value Equations and Inequalities Without Absolute Value Bars*		
Given Statement	**Rewritten Without Absolute Value**	**Graph**
$\|X\| = c$	$X = c$ or $X = -c$	![graph] $-c$ c
$\|X\| < c$	$-c < X < c$	$-c$ c Interval Notation: $(-c, c)$
$\|X\| \leqslant c$	$-c \leqslant X \leqslant c$	$-c$ c Interval Notation: $[-c, c]$
$\|X\| > c$	$X < -c$ or $X > c$	$-c$ c Interval Notation: $(-\infty, -c)$ or (c, ∞)
$\|X\| \geqslant c$	$X \leqslant -c$ or $x \geqslant c$	$-c$ c Interval Notation: $(-\infty, -c]$ or $[c, \infty)$

* In all cases, c is any positive real number and X is an algebraic expression.

Photri, Inc.

EXAMPLE 7   **Absolute Value Inequalities and Coin Tossing**

If a coin is tossed 100 times, we would expect approximately 50 of the outcomes to be heads. In statistics, it can be demonstrated that a coin is unfair if h, the number of outcomes that result in heads, satisfies

$$\left|\frac{h-50}{5}\right| \geq 1.645.$$

Find the values of h that determine an unfair coin.

Solution

$|X| \geq c$	means	$X \leq -c$	or	$X \geq c$

$$\left|\frac{h-50}{5}\right| \geq 1.645 \text{ means } \frac{h-50}{5} \leq -1.645 \text{ or } \frac{h-50}{5} \geq 1.645$$

We solve each of the resulting inequalities separately.

$$\frac{h-50}{5} \leq -1.645 \quad \text{or} \quad \frac{h-50}{5} \geq 1.645$$

$$h - 50 \leq -8.225 \quad \text{or} \quad h - 50 \geq 8.225$$

$$h \leq 41.775 \quad \text{or} \quad h \geq 58.225$$

If we toss the coin 100 times, obtaining an outcome of heads 41 or fewer times leads us to be suspicious about the coin's fairness. Similarly, we would be suspicious if an outcome of heads occurs 59 or more times.   ■

PROBLEM SET 2.6

Practice Problems_____

In Problems 1–28, find the solution set for each equation. (Some of the equations have no solution.) If applicable, use a graphing utility to verify your result.

1. $|x| = 8$
5. $|x - 2| = 7$

2. $|x| = 6$
6. $|x + 1| = 5$

3. $|x| = -8$
7. $|x + 3| = 0$

4. $|x| = -6$
8. $|x + 2| = 0$

9. $|2x - 1| = 5$
10. $|2x - 3| = 11$
11. $\left|\frac{4x - 2}{3}\right| = 2$
12. $\left|\frac{3x - 1}{5}\right| = 1$

13. $2|x + 3| - 1 = 13$
14. $3|x - 2| - 7 = -1$
15. $|2x - 4| + 5 = 2$
16. $|2x - 1| + 2 = 7$

17. $3|2x - 5| - 8 = -2$
18. $3 - 2|3x - 4| = -7$
19. $|2x + 2| = |x + 2|$

20. $|2z - 1| = |z + 3|$
21. $|3x - 3| = |x + 4|$
22. $|2y + 1| = |3y - 4|$

23. $|2x - 4| = |2x + 3|$
24. $|3y + 5| = |3y - 5|$
25. $\left|\frac{2}{3}x - 2\right| = \left|\frac{1}{3}x + 3\right|$

26. $\left|\frac{1}{2}y - 2\right| = \left|y - \frac{1}{2}\right|$
27. $|8x + 10| = |2(4x + 5)|$
28. $|7 + 2x| = |6x - 6 - 4x|$

For Problems 29–62, write the solution set in both set-builder and interval notation, and then graph the solution set on a number line. If applicable, use a graphing utility to verify your result.

29. $|x| < 4$
33. $|x - 2| < 1$
37. $|x + 3| > 1$

30. $|x| \leq 5$
34. $|x - 1| < 5$
38. $|x - 2| > 5$

31. $|x| \geq 4$
35. $|x + 2| \leq 1$
39. $|x - 4| \geq 2$

32. $|x| > 5$
36. $|x + 1| \leq 5$
40. $|x - 3| \geq 4$

41. $|2y - 6| < 8$ **42.** $|3x + 5| \leq 17$ **43.** $|2(x + 1) + 3| \leq 5$ **44.** $\left|\dfrac{7x - 2}{4}\right| < \dfrac{5}{4}$

45. $\left|\dfrac{2(3x - 1)}{3}\right| < \dfrac{1}{6}$ **46.** $|5x + 3| > 23$ **47.** $|3x - 1| > 13$ **48.** $|9x - (2x - 4)| \geq 25$

49. $|2(x - 3) + 3(x + 2) - 17| \geq 13$ **50.** $|4x - 3[2(1 + x)] - 2x| > 7$

51. $\left|\dfrac{2y + 2}{4}\right| > 2$ **52.** $\left|\dfrac{3z - 3}{9}\right| > 1$ **53.** $\left|\dfrac{7x - 2}{4}\right| \geq \dfrac{5}{4}$ **54.** $\left|\dfrac{2(3x - 1)}{3}\right| \geq \dfrac{1}{6}$

55. $|x + 2| + 9 \leq 16$ **56.** $|x - 2| + 4 \geq 5$ **57.** $2|2x - 3| + 10 > 12$ **58.** $3|2x - 1| + 2 < 8$

59. $|x - 2| < -1$ **60.** $|x - 3| < -2$ **61.** $|x - 2| > -1$ **62.** $|x - 3| > -2$

Application Problems

63. The daily production (x) at a refinery is described by $|x - 2,560,000| \leq 135,000$, where x represents the number of barrels of oil. Find the high and low production levels.

64. If the heights (h) of a population satisfy $\left|\dfrac{h - 67.5}{2.6}\right| \leq 1$, describe the interval using set-builder notation in which these heights lie.

The graph shows the percentage of American adults who smoke cigarettes divided into four groups: those with less than 12 years of education, those with exactly 12 years of education, those with 13 through 15 years of education, and those with 16 or more years of education. The percentages are given for three years: 1974, 1987, and 1990. If x represents a percent, list the appropriate bar or bars by name from the 12 bars shown in the graph (such as "adults with 16 or more years of education in 1987") that meet each condition in Problems 65–67.

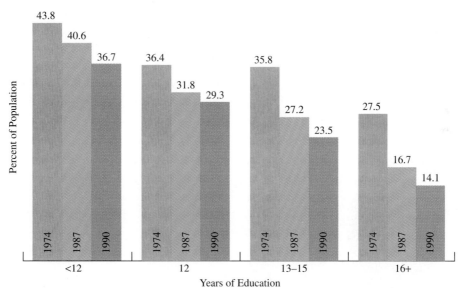

Cigarette Smoking
Adults Age 25 and Older, by Education, 1974–1990

Source: CDC, National Center for Health Statistics, Health Interview Data

65. $|2x - 30| \leq 30$ **66.** $|4x - 40| > 120$ **67.** $|x - 39.8| = 4$

True–False Critical Thinking Problems

68. Which one of the following is true?
   a. All absolute value equations have two solutions.
   b. The equation $|x| = -6$ is equivalent to $x = 6$ or $x = -6$.
   c. We can rewrite the inequality $x > 5$ or $x < -5$ more compactly as $-5 < x < 5$.
   d. Absolute value inequalities in the form $|ax + b| < c$ translate into *and* compound statements, which may be written as three-part inequalities.

69. Which one of the following is true?
   a. The equation $|3x - 1| = 6$ is equivalent to $3x - 1 = 6$ or $3x + 1 = -6$.
   b. The inequality $|x| < -4$ is equivalent to $x > -4$ or $x < 4$.
   c. Absolute value inequalities in the form $|ax + b| > c$ translate into *or* compound statements, which cannot be written as three-part inequalities.
   d. The solution set for $|6 - 3x| = 0$ is $\varnothing$.

Writing in Mathematics

70. Explain how to solve $\left|\dfrac{x}{3} - 1\right| - 9 \geq -5$, actually solving the inequality and including written explanations of what you are doing from step to step.

71. Explain why *no* real numbers satisfy $|x| = -4$ *and* $|x| < -4$.

72. Explain why *all* real numbers satisfy $|x| > -4$.

Critical Thinking Problems

In Problems 73–79, determine the value or values of x, if any, that will make the equation or inequality true.

73. $|x| = x$

74. $|x + 1| = x + 1$

75. $|x + 1| = -(x + 1)$

76. $|x| + x = 4$

77. $x - |x| = 5$

78. $|x - 2| < 4x$

79. $|x + 2| \geq 4x$

80. The percentage (p) of defective products manufactured by a company is described by $|p - 0.35\%| \leq 0.16\%$. If 100,000 products are manufactured and the company offers a \$6 refund for each defective product, use set-builder notation to describe the interval for the company's cost of refunds.

Review Problems

81. Shown here is a diagram of a bookcase that is to be constructed with four shelves and seven pieces of wood. Five of the pieces will be used for the horizontal widths and two pieces will be used for the vertical heights. If the height is to exceed the width by 3 feet, and 41 feet of wood are available to build the bookcase, what will be the dimensions of the bookcase?

82. Solve: $2(x - 3) = 4[x - (5x - 2)]$.

83. Simply: $-2[3a - (2b - 1) - 5a] + b$.

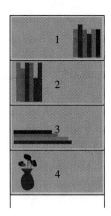

S E C T I O N   2 . 7

Solutions **Tutorial** **Video**
Manual **3**

Linear Inequalities Containing Two Variables

Objectives

1 Graph linear inequalities using test points.
2 Graph linear inequalities without using test points.

Suppose a person must get at least 1 milligram of thiamine (vitamin B$_1$) daily. One ounce of papaya juice provides 0.05 milligram of thiamine, and 1 ounce of orange juice provides 0.08 milligram. If x represents the amount of papaya juice and y the amount of orange juice consumed daily, then this person's daily intake can be modeled by the inequality

$$0.05x + 0.08y \geq 1.$$

The inequality that models this situation contains two variables. In Sections 2.4 and 2.5, we used the number line to graph linear inequalities with one variable. In this section, we use the rectangular coordinate system to graph linear inequalities in two variables.

> ### Definition of a linear inequality in two variables
>
> A *linear inequality in two variables* (x and y) is an inequality that can be written in one of the following forms:
>
> $$Ax + By > C, Ax + By \geq C, Ax + By < C, \text{ and } Ax + By \leq C$$
>
> where $A, B,$ and C are real numbers, and A and B are not both zero.

Study tip

Inequalities that appear to have only one variable can be expressed using two variables and zero coefficients. Thus, $x \geq 3$ means $x + 0y \geq 3$ and $y < -2$ means $0x + y < -2$.

Examples of linear inequalities in two variables are $x + y > 4, 2x - 3y \leq 6,$ $x \geq 3,$ and $y < -2$.

Solutions of Inequalities in Two Variables

An ordered pair (x_1, y_1) is a *solution* to an inequality in two variables if the inequality is true when x_1 is substituted for x and y_1 is substituted for y. Under these conditions, we say that (x_1, y_1) *satisfies* the inequality.

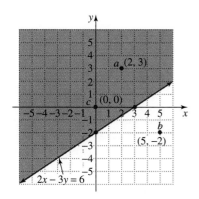

Figure 2.49

Which ordered pairs satisfy $2x - 3y \leq 6$?

EXAMPLE 1 **Deciding Whether Ordered Pairs Are Solutions of Inequalities**

Determine whether each of the following ordered pairs satisfies the inequality $2x - 3y \leq 6$.

    a. $(2, 3)$ **b.** $(5, -2)$ **c.** $(0, 0)$

Solution

The three ordered pairs are graphed and labeled in Figure 2.49. We substitute the x- and y-coordinate of each pair into the given inequality.

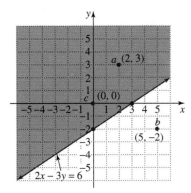

Figure 2.49 (repeated from the previous page)

Which ordered pairs satisfy $2x - 3y \leqslant 6$?

a. $(2, 3)$:

$$2x - 3y \leqslant 6$$
$$2(2) - 3(3) \overset{?}{\leqslant} 6$$
$$4 - 9 \overset{?}{\leqslant} 6$$
$$-5 \leqslant 6 \quad \text{True}$$

Thus, $(2, 3)$ is a solution.

b. $(5, -2)$:

$$2x - 3y \leqslant 6$$
$$2(5) - 3(-2) \overset{?}{\leqslant} 6$$
$$10 + 6 \overset{?}{\leqslant} 6$$
$$16 \leqslant 6 \quad \text{False}$$

Thus, $(5, -2)$ is not a solution.

c. $(0, 0)$:

$$2x - 3y \leqslant 6$$
$$2(0) - 3(0) \overset{?}{\leqslant} 6$$
$$0 - 0 \overset{?}{\leqslant} 6$$
$$0 \leqslant 6 \quad \text{True}$$

Thus, $(0, 0)$ is a solution.

The ordered pairs that satisfy the given inequality are those in parts (a) and (c). In Figure 2.49, shown again in the margin, these ordered pairs all lie in the same *half-plane* that is formed when the line whose equation is $2x - 3y = 6$ divides the plane in two. This observation will be helpful in obtaining the graph of a linear inequality. ∎

Graphing of Linear Inequalities in Two Variables

The graph of a linear inequality in two variables is the collection of all points in the rectangular coordinate system whose ordered pairs satisfy the inequality. The graph consists of an entire region rather than a line.

To sketch the graph of a linear inequality such as $2x - 3y \leqslant 6$, the inequality of Example 1, we begin by graphing the *corresponding linear equation*

$$2x - 3y = 6.$$

We then draw a *solid line* for the inequalities $\leqslant$ and $\geqslant$. Draw a dashed line for the inequalities $<$ and $>$. The line separates the plane into two half-planes. In each half-plane, one of the following statements is true:

1. All points in the half-plane satisfy the inequality.
2. No points in the half-plane satisfy the inequality.

We can determine whether the points in an entire half-plane are solutions of the inequality by testing just one point in the region.

These ideas are illustrated in the next example.

Graph linear inequalities using test points.

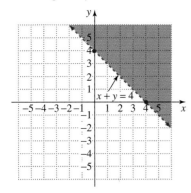

Figure 2.50

The graph of $x + y > 4$

EXAMPLE 2   Using a Line and a Test Point to Graph an Inequality

Graph: $x + y > 4$

Solution

The graph of the corresponding equation

$$x + y = 4$$

is a line with an x- and y-intercept at 4, as shown in Figure 2.50.

x-intercept: set $y = 0$.
$$x + y = 4$$
$$x + 0 = 4$$
$$x = 4$$

y-intercept: set $x = 0$.
$$x + y = 4$$
$$0 + y = 4$$
$$y = 4$$

ENRICHMENT ESSAY

Analytic Cubism

The logic and order underlying graphs in the Cartesian system appeared in an artistic style called analytic cubism. In Pablo Picasso's *Man with a Violin,* straight lines, a narrow range of color, and slicing of the plane into geometric shapes suggest a highly impersonal and rational approach to the figure. Beyond the superficial world of appearance is the same kind of mathematical order shaping regions graphed in rectangular coordinates.

Pablo Picasso, (Spanish) "Man with Violin" c. 1910, oil on canvas, $39\frac{3}{8} \times 29\frac{7}{8}$ in. Philadelphia Museum of Art: Louise and Walter Arensberg Collection. © 1997 Estate of Pablo Picasso/Artists Rights Society (ARS), New York.

The graph is indicated by a dashed line since equality is not included in $x + y > 4$. To find which half-plane is the graph, test a point from either half-plane. The origin, $(0, 0)$, is easiest.

$x + y > 4$ This is the given inequality.

$0 + 0 \overset{?}{>} 4$ Test the origin by substituting 0 for x and y.

$0 > 4$ This false statement indicates that $(0, 0)$ does not satisfy the inequality.

The graph is the half-plane not including $(0, 0)$, which is the half-plane above the line in Figure 2.50. All points in that half-plane have coordinates satisfying $x + y > 4$. (Try checking one such point.) ∎

Before considering another example, let's summarize the procedure for graphing a linear inequality in two variables.

"Senecio" is an attempt to humanize the slicing of the plane into geometric shapes. Klee's art has been called "humanized Cartesian geometry."
Paul Klee "Senecio" 1922. Kunstmuseum, Basel, Switzerland. Scala/Art Resource, NY.

Graphing a linear inequality in two variables

1. Replace the inequality symbol with an equal sign and graph the corresponding linear equation. Draw a solid boundary line if the original inequality contains a $\leq$ or $\geq$ symbol. Draw a dashed line if the original inequality contains a $<$ or $>$ symbol.
2. Choose a test point in one of the half-planes that is not on the line. Substitute the coordinates of the test point into the inequality.
3. If a true statement results, shade the half-plane containing this test point. If a false statement results, shade the half-plane not containing this test point.

Using technology

Graphing Utilities and Inequalities

Most graphing utilities can graph inequalities in two variables with the $\boxed{\text{SHADE}}$ feature. Some utilities will only shade regions between two curves, so you may need to enter a second equation such as $y = -50$ or $y = 50$ that will bound a shaded region from above or below. This is the case on a TI-85, but does not apply to the TI-92. For example, to graph $x + y > 4$ on many graphing utilities:

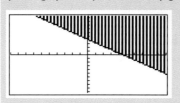

1. Select $\boxed{\text{SHADE}}$ from the $\boxed{\text{GRAPH DRAW}}$ menu. On the screen will appear: Shade (
2. Enter the following: Shade ($4 - x$, 50) The first argument $4 - x$ is obtained by solving $x + y = 4$ for y. The second argument, 50 (or $y = 50$), graphs as a horizontal line 50 units above the x-axis, which will bound the shaded region from above.
3. Press $\boxed{\text{ENTER}}$. The graph of $x + y > 4$ is shown in the figure. Another limitation is that most graphing utilities will draw only solid, and not dashed, lines.

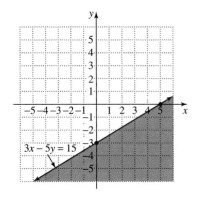

Figure 2.51

The graph of $3x - 5y \geq 15$

EXAMPLE 3   **Graphing a Linear Inequality**

Graph: $3x - 5y \geq 15$

Solution

The graph of the corresponding equation

$$3x - 5y = 15$$

is the line shown in Figure 2.51. (Verify that the x-intercept is 5 and the y-intercept is -3.) The graph is a solid line since equality is included in $3x - 5y \geq 15$. To find which half-plane belongs to the graph, test a point from either half-plane. The origin, $(0, 0)$, is easiest.

$$3x - 5y \geq 15 \quad \text{This is the given inequality.}$$
$$3(0) - 5(0) \overset{?}{\geq} 15 \quad \text{Test the origin by substituting 0 for } x \text{ and } y.$$
$$0 \geq 15 \quad \text{False; } (0, 0) \text{ is not a solution.}$$

The graph of $3x - 5y \geq 15$ is the half-plane not including $(0, 0)$ and the graph of the boundary line. All points in the half-plane below the boundary line have coordinates satisfying $3x - 5y > 15$ and all points on the boundary line have coordinates satisfying $3x - 5y = 15$. ∎

2 Graph linear inequalities without using test points.

Graphing Linear Inequalities Without Using Test Points

If a linear inequality is in slope-intercept form, it is not necessary to use a test point to obtain the graph. This is shown in Table 2.6.

TABLE 2.6 Graphing Linear Inequalities in Slope-Intercept Form

Inequality	Description of the Graph
$y < mx + b$	Half-plane *below* the line $y = mx + b$
$y \leqslant mx + b$	Half-plane *below* the line $y = mx + b$ and the boundary line $y = mx + b$
$y > mx + b$	Half-plane *above* the line $y = mx + b$
$y \geqslant mx + b$	Half-plane *above* the line $y = mx + b$ and the boundary line $y = mx + b$

EXAMPLE 4 **Graphing a Linear Inequality in Slope-Intercept Form**

Graph: $y > \frac{2}{3}x - 4$

Solution

We graph the boundary line whose equation is

$$y = \frac{2}{3}x - 4$$

with y-intercept $= -4$ and Slope $= \frac{2}{3} = \frac{\text{Rise}}{\text{Run}}$

The boundary line is shown by a dashed line. Since we are graphing $y > \frac{2}{3}x - 4$, the graph is the half-plane above the line, shown in Figure 2.52. ■

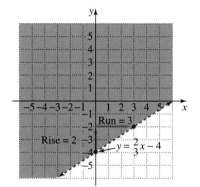

Figure 2.52
The graph of $y > \frac{2}{3}x - 4$

It is also not necessary to use test points when boundary lines are horizontal or vertical. This is shown in Table 2.7.

TABLE 2.7 Graphing Linear Inequalities with Horizontal or Vertical Boundary Lines

Inequality	Graph	Example
$x < a$	The half-plane to the *left* of the line $x = a$ 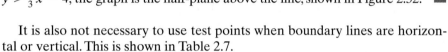	$x < 4$

TABLE 2.7 (Continued)

Inequality	Graph	Example
$x \geq a$	The half-plane to the *right* of the line $x = a$ and the boundary line	$x \geq -3$
$y \leq b$	The half-plane *below* the line $y = b$ and the boundary line	$y \leq 2$
$y > b$	The half-plane *above* the line $y = b$	$y > 4$

PROBLEM SET 2.7

Practice Problems

In Problems 1–4, determine whether the ordered pair is a solution of the given inequality.

1. $(2, -4); 2x - y > 4$ **2.** $(-4, -1); 3x - 2y < -6$ **3.** $(6, -2); 3x - 4y \leq 12$ **4.** $(7, -3); 2x \leq 5y + 10$

Use a test point to graph each linear inequality in Problems 5–18.

5. $x + y \geq 2$ **6.** $x - y \leq 1$ **7.** $3x - y \geq 6$ **8.** $3x + y \leq 3$

9. $2x + 3y > 12$ **10.** $2x - 5y < 10$ **11.** $5x + 3y \leq -15$ **12.** $3x + 4y \leq -12$

13. $2y - 3x > 6$ **14.** $2y - x > 4$ **15.** $5x + y > 3x - 2$ **16.** $x - y > 4 + y$

17. $\dfrac{x}{2} + \dfrac{y}{3} < 1$ **18.** $\dfrac{x}{4} + \dfrac{y}{2} > 1$

In Problems 19–34, graph each linear inequality without the use of a test point.

19. $y > 2x - 1$ **20.** $y \leq 3x + 2$ **21.** $y \geq \frac{2}{3}x - 1$ **22.** $y < -\frac{3}{4}x + 2$

23. $y < x$ **24.** $y > -2x$ **25.** $y > -\frac{1}{2}x$ **26.** $y < -\frac{1}{3}x$

27. $x \leq 1$ **28.** $x \leq -3$ **29.** $y > 1$ **30.** $y > -3$

31. $x \geq 0$ **32.** $y \leq 0$ **33.** $x > -5$ **34.** $x > -3$

Application Problems

35. Suppose a patient is not allowed to have more than 330 milligrams of cholesterol from a diet of eggs and meat. If each egg provides 165 milligrams of cholesterol and each ounce of meat provides 110 milligrams of cholesterol, then $165x + 110y \leq 330$, where x is the number of eggs and y the number of ounces of meat. Graph the inequality in the first quadrant. Give the coordinates of any two points in the solution set. Describe what each set of coordinates means in terms of the variables in the problem.

36. At the beginning of this section, we introduced a situation in which a person must get at least 1 milligram of thiamine daily. One ounce of papaya juice provides 0.05 milligram of thiamine and 1 ounce of orange juice provides 0.08 milligram. If x represents the amount of papaya juice and y the amount of orange juice consumed daily, then $0.05x + 0.08y \geq 1$. Graph the inequality in the first quadrant. Give the coordinates of any two points in the solution set. Describe what these mean in terms of the variables in the problem.

37. Many elevators have a capacity of 2000 pounds. If a child averages 50 pounds and an adult 150 pounds, write an inequality that describes when x children and y adults will cause the elevator to be overloaded. Graph the inequality. Select an ordered pair satisfying the inequality. Describe what this means in practical terms.

True–False Critical Thinking Problems

38. Which one of the following is true?
  a. The graph of $4x - y > 2$ is the half-plane above the line whose equation is $4x - y = 2$.
  b. The graph of $x < -2$ is the region to the left of the vertical line whose equation is $x = 2$.
  c. The graph of $-\frac{5}{12}x + \frac{1}{4}y \geq \frac{3}{4}$ includes a boundary line whose y-intercept is 3.
  d. None of the above is true.

39. Which one of the following is true?
  a. The graph of $y > -2$ is the region to the right of the vertical line whose equation is $y = -2$.

b. The graph of $2x - 5y < 10$ is the half-plane above the line whose equation is $2x - 5y = 10$.
  c. When graphing an inequality containing $>$ or $<$, it is possible that the points on the line are solutions to the inequality.
  d. The graph of a half-plane in rectangular coordinates is the graph of a function.

Technology Problems _____

Use a graphing utility with a $\boxed{\text{SHADE}}$ *feature to graph the following inequalities.*

40. $y \geq \dfrac{2}{3}x - 2$ **41.** $y \leq 6 - 1.5x$ **42.** $2x + 3y \geq 12$ **43.** $3x + 6y \leq 6$

44. Does your graphing utility have any limitations in terms of graphing linear inequalities? If so, what are they?

45. Use a graphing utility with a $\boxed{\text{SHADE}}$ feature to verify some of the graphs that you drew by hand in Problems 1–34.

Writing in Mathematics _____

46. What does a dashed boundary line mean in the graph of an inequality?

47. Under what circumstances is it more efficient to graph a linear inequality without using test points?

Critical Thinking Problems _____

Inequalities in two variables joined by the word and *can be graphed by graphing each inequality and then taking the region in rectangular coordinates where the half-planes overlap. Although we will be discussing this concept in more detail in Section 3.6, try exploring the idea on your own by graphing each pair of inequalities in Problems 48–53.*

48. $x + y \leq 3$ and $x - y \geq 3$ **49.** $x - y \geq 3$ and $x + y \leq 5$ **50.** $x - 3y \leq 6$ and $x \geq 0$

51. $y \geq 2x + 1$ and $y \leq 2x + 3$ **52.** $1 \leq x \leq 4$ **53.** $-2 < x \leq 5$

Review Problems _____

54. Solve: $3(x - 2) > 5x + 8$.

55. Solve for f: $d(f + w) = fl$.

56. Simplify: $-2[3x - (2y - 3) - 7x] + 4y$.

HAPTER PROJECT

Geometry on the Coordinate Plane

As we studied functions and their graphs in this chapter, we were relying on the work of René Descartes (1596–1650), a philosopher and mathematician who gave us the tools to visualize algebraic functions as geometric graphs on a coordinate plane. In 1637, Descartes introduced the idea of the coordinate plane and analytical geometry to the world in one of the appendices accompanying his main work in philosophy, *Discours de la method pour bien conduire sa raison et chercher la verite dans les sciences* (Discourse on the method of reasoning well and seeking truth in the sciences). The appendix was only about 100 pages long and was titled *La Geometrie*.

The key idea of analytical geometry is to describe the locations of points in the plane with two numbers (x, y). This idea supposedly occurred to Descartes as he lay in bed late one morning and watched a fly crawl across the ceiling near a corner of his room. He noticed the path of the fly could always be described if you knew its distance from the two adjacent walls. It should be noted that Descartes confined himself to the first quadrant of the coordinate plane, where all points are determined by positive numbers. He called negative numbers "false roots."

We may find it difficult to believe that someone had to discover the coordinate plane we take for granted, or that negative numbers were once considered "false," but we have had over 300 years for these ideas to settle in our culture. At the time Descartes lived, geometry was plane Euclidean geometry, and numbers were usually considered to be positive numbers. To "solve" a problem in Euclidean geometry meant you constructed a geometric figure to study. Descartes found a new meaning for "solve." He found an algebraic equation that revealed the characteristics of the geometric figure.

1. For each pair of points given below, graph each pair on a separate xy-plane and find the coordinates of the midpoint of the line joining the two points.

   $A\,(10, 3)$ and $B\,(-2, 3)$ $C\,(3, -1)$ and $D\,(3, 9)$ $E\,(-2, 5)$ and $F\,(4, 9)$

   Write down an algebraic expression for finding the coordinates of the midpoint of a line joining the two general points (x_1, y_1) and (x_2, y_2).

2. We may also find the midpoint of the line segments we graphed above using the methods of Euclidean geometry. For this exercise you will need a compass, a pencil, and a straightedge or ruler.

   a. Begin by placing the point of your compass on one end of the line segment and opening the compass until the pencil tip rests against the other end of the line segment. Pivot the compass around on its point to draw a circle with a radius the length of the line segment (Figure 2.53).

   b. Repeat the procedure above with the compass point resting on the other end of the line segment and draw a second circle (Figure 2.54).

   c. Rest the straightedge on the two intersection points of the circles and draw a line connecting those two points. The line you have drawn will pass through the midpoint of the original line segment (Figure 2.55).

   The overlapping portion of the two intersecting circles is an ancient geometric construction called the *vesica piscis* (Latin for "bladder of the fish").

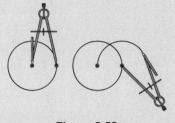

Figure 2.53   **Figure 2.54**   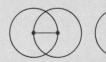 **Figure 2.55**

Next, we begin by constructing a geometric figure using the methods of Euclidean geometry and then discover the algebraic equations that would give us the same figure.

3. **a.** On a sheet of graph paper, mark a center for a circle, open your compass, place the point on the center, and swing the compass around to draw a circle. Life the compass up, without changing the size of its opening, and place the point of the compass anywhere on the edge of the circle. Turn the compass to draw a new circle (Figure 2.56).

   b. Using the straightedge, connect the two intersection points of the circles and the two centers of the circles (Figure 2.57).

   c. Place the tip of the compass where the lines you constructed intersect and open the compass until the pencil tip rests on the center of one of the circles. Turn the compass to make a circle (Figure 2.58).

   d. The smaller circle you drew in **c** intersects the lines in four points. Connect the four points to make a square (Figure 2.59).

  Figure 2.56   **Figure 2.57**   **Figure 2.58**    **Figure 2.59**

4. The square you created using the construction methods of Euclidean geometry may also be given as the intersection of four lines in the *xy*-plane. Using the square you created on graph paper, read off coordinates to allow you to obtain equations for the four lines that intersect to form the square. Write the four equations in slope-intercept form. What do you observe about the slopes of the lines?

5. Which of the methods described above gives you more information about the figures you have constructed? If you were working on a problem outside the classroom where you needed to construct a square or cut a length in half, which style of geometry would you prefer? Explain in detail why you would prefer one method over another. Can you think of any other methods, or a combination of these methods, that you would prefer?

Worldwide Web Resources

Go to the Prentice Hall website (http://www.prenhall.com/blitzer) to access other locations on the Internet that will allow you to further explore the concepts presented in this project.

Chapter Review

SUMMARY

1. Functions and Relations

a. A relation is any set of ordered pairs. A function is a relation such that for each value of the first component of the ordered pairs, there is exactly one value of the second component. The set of all first components is called the domain, and the set of all second components the range.

b. When a function is given in terms of an equation in the form $y = f(x)$, it is a relationship between two variables such that for every value of the independent variable (in this case, x), there corresponds exactly one value of the dependent variable (in this case, y). The name of the function is f, and $f(x)$ is the value of the function at x.

c. The graph of a relation or function is the graph of its ordered pairs.

d. *Vertical line test:* If any vertical line intersects a graph in the Cartesian system more than once, the graph does not define y as a function of x.

2. Linear Functions and Slope

a. Linear functions can be expressed in the form $f(x) = mx + b$.

b. *Slope*

1. The slope of a line is

$$m = \frac{y_2 - y_1}{x_2 - x_1} \quad \text{or} \quad \frac{y_1 - y_2}{x_1 - x_2}$$

where $x_2 - x_1 \neq 0$, and (x_1, y_1) and (x_2, y_2) are two distinct points on the line. Horizontal lines have zero slope, and vertical lines have undefined slope.

2. Two nonvertical lines are parallel if and only if their slopes are equal. Two nonvertical lines are perpendicular if and only if their slopes are negative reciprocals.

c. *Forms of Linear Equations*

1. *Slope-intercept equation:* $y = mx + b$ (m is the slope and b is the y-intercept.)

2. *Standard form:* $Ax + By = C$: usually graphed using the x- and y-intercepts. The x-intercept is found by setting $y = 0$, and the y-intercept is found by setting $x = 0$ in the line's equation.

3. *Horizontal and vertical line forms:* $y = b$ graphs as a line parallel to the x-axis. $x = a$ graphs as a

line parallel to the y-axis. The graph of $y = 0$ is the x-axis, and the graph of $x = 0$ is the y-axis. A vertical line is not a function.

4. *Point-slope equation:*
   $y - y_1 = m(x - x_1)$ [(x_1, y_1) is a fixed point on the line; m is the slope.]

d. *Modeling Data Points that Fall Approximately Along a Line*

1. If modeling is done without technology, one option is to use two data points at a time, writing point-slope and slope-intercept equations of the line containing these points. Then see how well the resulting linear function describes the other given data points. Select the model with the least error.

2. If modeling is done with technology, use a graphing utility to find the equation of the regression line, the line that best fits the data. Draw a scatter plot of the data and graph the regression line on the scatter plot using the utility's capabilities.

3. Once a model has been determined, interpolate and extrapolate from the data by finding values between and outside the data, respectively.

3. Solving a Linear Inequality
$(ax + b < c, ax + b \leq c, ax + b > c, ax + b \geq c)$

a. Multiply on both sides to clear fractions or decimals.

b. Simplify the algebraic expression on each side.

c. Collect all variable terms on one side and all constant terms on the other side.

d. Isolate the variable and solve. Be sure to reverse the direction of the inequality symbol when multiplying or dividing by a negative number.

e. Express the solution in set-builder notation or interval notation, and graph the solution set on a number line.

4. Solving Compound Inequalities

a. *The Intersection of Sets*
$A \cap B$ (the intersection of sets A and B) consists of all elements that are in set A and in set B.

b. *Solving Inequalities with and (intersection)*
1. Solve each inequality separately.
2. Graph the solution sets above the number line and take the interval on the number line where the graphs overlap.

c. Solve an inequality with three parts by isolating the variable in the middle.

d. *The Union of Sets*
$A \cup B$ (the union of sets A and B) consists of all elements that are in set A or in set B, or in both A and B.

e. *Solving Inequalities With or (union)*
1. Solve each inequality separately.

2. Graph the solution sets above the number line and take the interval on the number line representing each of the separate graphs, including the interval (if there is one) where the graphs overlap.

5. Equations and Inequalities Involving Absolute Value
Suppose that X represents an algebraic expression.

a. Equations
1. $|X| = c$ is equivalent to $X = c$ or $X = -c$ for a positive number c.
2. $|X| = c$ has no solution for a negative number c.
3. $|X| = 0$ is equivalent to $X = 0$.
4. $|X_1| = |X_2|$ is equivalent to $X_1 = X_2$ or $X_1 = -X_2$.

b. Inequalities
If c is a positive number:
1. $|X| < c$ is equivalent to $-c < X < c$.
2. $|X| > c$ is equivalent to $X < -c$ or $X > c$.

6. Linear Inequalities in Two Variables (x and y)

a. A linear inequality in two variables can be written in the form $Ax + By > C$ or $Ax + By \geq C$ or $Ax + By < C$, or $Ax + By \leq C$ (A and B not both zero).

b. An ordered pair (x_1, y_1) is a solution of an inequality if the inequality is true when x_1 is substituted for x and y_1 is substituted for y. Then (x_1, y_1) satisfies the inequality.

c. To graph a linear inequality, draw the graph of $Ax + By = C$, the boundary line, using a solid line for $\geq$ and $\leq$ and a dashed line for $>$ and $<$. Then choose a test point in one of the half-planes, making sure the test point is not on the line. Substitute the coordinates of the test point into the inequality. If a true statement results, shade the half-plane containing this test point. If a false statement results, shade the half-plane not containing this test point.

d. To graph a linear inequality in the form $y > mx + b$, $y < mx + b$, $y \geq mx + b$, or $y \leq mx + b$, graph $y = mx + b$, the boundary line, using the y-intercept (b) and the slope (m). Then $y > mx + b$ is the half-plane above the line and $y < mx + b$ is the half-plane below the line.

e. The graph of $x > a$ is the half-plane to the right of $x = a$ (a vertical line). The graph of $x < a$ is the half-plane to the left of $x = a$.

f. The graph of $y > b$ is the half-plane above $y = b$ (a horizontal line). The graph of $y < b$ is the half-plane below $y = b$.

REVIEW PROBLEMS

If applicable, use a graphing utility to verify as many answers as possible. For Problems 1–4, which relations are not functions? Give the domain and range for each relation.

1. $\{(2, 7), (3, 7), (5, 7)\}$ **2.** $\{(1, 10), (2, 500), (13, \pi)\}$ **3.** $\{(1, 2), (3, 4), (5, 6)\}$ **4.** $\{(12, 13), (14, 15), (12, 19)\}$

Find the function values in Problems 5–6.

5. $f(x) = -7x + 5$ **a.** $f(0)$ **b.** $f(-2)$ **c.** $f(5)$ **d.** $f(\frac{3}{7})$ **e.** $f(b + 3)$

6. $g(x) = 3x^2 - 5x + 2$ **a.** $g(0)$ **b.** $g(-2)$ **c.** $g(4)$ **d.** $g(\frac{1}{2})$ **e.** $g(3a)$

7. Fill in the table of coordinates for functions f, g, and h. Graph the three functions in the same rectangular coordinate system. Then describe the relationship among the graphs of f, g, and h.

$f(x) = x^2$

x	$f(x) = x^2$
-2	
-1	
0	
1	
2	

$g(x) = x^2 + 2$

x	$g(x) = x^2 + 2$
-2	
-1	
0	
1	
2	

$h(x) = x^2 - 1$

x	$h(x) = x^2 - 1$
-2	
-1	
0	
1	
2	

8. The graph shows the absolute value function $f(x) = |x|$. Copy the graph onto a rectangular coordinate system. Then use the pattern of Problem 7 to graph $g(x) = |x| + 2$ and $h(x) = |x| - 1$ in the same rectangular coordinate system without using tables of coordinates for functions g and h.

x	$f(x) = |x|$
-3	$f(-3) = |-3| = 3$
-2	$f(-2) = |-2| = 2$
-1	$f(-1) = |-1| = 1$
0	$f(0) = |0| = 0$
1	$f(1) = |1| = 1$
2	$f(2) = |2| = 2$
3	$f(3) = |3| = 3$

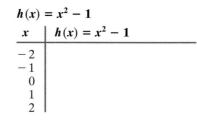

9. Fill in the table of coordinates for functions f, g, and h. Graph the three functions in the same rectangular coordinate system. Then describe the relationship among the graphs of f, g, and h.

$f(x) = \sqrt{x}$

x	$f(x) = \sqrt{x}$
0	
1	
4	
9	

$g(x) = \sqrt{x + 2}$

x	$g(x) = \sqrt{x + 2}$
-2	
-1	
2	
7	

$h(x) = \sqrt{x - 1}$

x	$h(x) = \sqrt{x - 1}$
1	
2	
5	
10	

10. Graph $f(x) = x^2 - 3x$ by first filling in the table of coordinates.

x	-2	-1	0	1	2	3	4	5
$f(x) = x^2 - 3x$								

11. Determine whether the following rule describes a function. Explain your answer.

Domain	**Rule**	**Range**
The set of all states in the United States	Each state's members of the U.S. Senate	The set of U.S. senators

For Problems 12–16, use the vertical line test to identify in which graphs y *is a function of* x.

12.

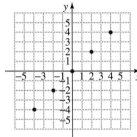

13.

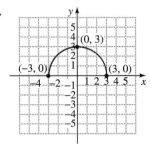

14.

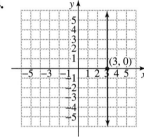

15.

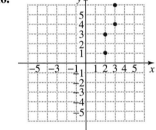

16.

17. The function $f(x) = 0.0234x^2 - 0.5029x + 12.5$ describes the number of infant deaths per 1000 live births in the United States x years after 1980. Find and interpret $f(10)$.

18. The function

$$f(x) = 0.0091x^3 + 0.1354x^2 + 2.1336x + 83.2653$$

models weekly salary for American workers x years after 1961. The graph of the function is shown in the figure.

a. Use the graph to find a reasonable estimate for $f(39)$. Describe what this means. Verify the estimate by using a calculator and substituting 39 for x.

b. Use the graph to estimate $f(29) - f(19)$. What does this mean in terms of weekly salary?

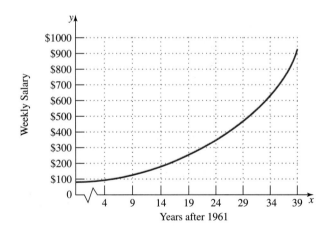

19. Violent crime, which includes murder, rape, robbery, and aggravated assault, is epidemic in America. The graph indicates violent crime rate per 1000 people in the United States from 1975 through 1992. Use the graph to answer the following questions.

   a. Why does f represent the graph of a function?

   b. Estimate the maximum point on the graph of function f. What exactly does this mean?

   c. Estimate the minimum point on the graph of function f. What exactly does this mean?

   d. If the trend from 1975 through 1992 were to continue, what is a reasonable estimate for the violent crime rate in the year 2009?

Violent Crime Rate per 1000 People

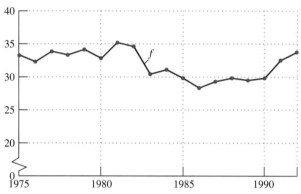

Source: F. B. I.

20. a. Graph the line connecting the points $(-3, 7)$ and $(2, -3)$.

   b. Find the slope of the line.

   c. Subtract the coordinates in the opposite order in the computation of slope. Describe what happens.

   d. Use the graph of the linear function to find two other points along the line. Use these points to compute the line's slope.

21. The percent of voting age population in the United States who reported voting from 1978 through 1992 is shown in the graph. The four years designated by * are presidential election years.

   a. If x represents the presidential election years 1980, 1984, 1988, and 1992, write an equation (as a constant function f) that approximately models the percent of voting age population who voted in election years.

   b. Consult the research department of your local library to find the percent for the 1996 presidential election. How well does your model in part (a) describe reality for 1996?

   c. Repeat part (a) for the nonelection years 1978, 1982, 1986, and 1990. Call the constant function g.

Percent of Voting Age Population Who Reported Voting, 1978–1992

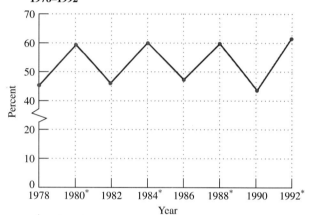

*Presidential election years

Source: U.S. Bureau of the Census, *Statistical Abstract 1993*, Table 454

22. The graph shows the percentage of U.S. households with televisions from 1950 through 1995. Estimate the slope of the line segments between 1950 and 1995, 1955 and 1960, and 1990 and 1995. Describe what each slope means in terms of the average yearly change in the percentage of households with televisions.

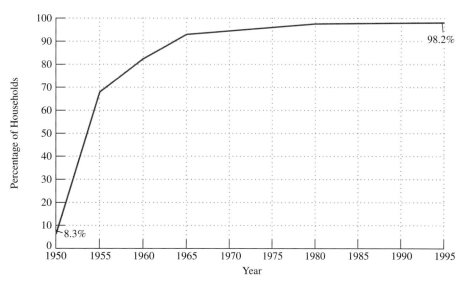

Percentage of U.S. Housholds with Televisions

Source: NMR; TBA

23. Plot the points $(12, -10)$, $(3, 4)$, $(-2, -19)$, and $(-11, -5)$.
   a. Use slope to determine if these points are the vertices of a parallelogram.
   b. If the resulting figure is a parallelogram, use slope to determine if the points are the vertices of a rectangle.

In Problems 24–27, use the most efficient method to graph each linear equation in a rectangular coordinate system.

24. $y = \frac{1}{2}x - 4$ **25.** $f(x) = -\frac{3}{4}x + 5$ **26.** $2x - 3y = 6$ **27.** $x - 3 = 2$

For Problems 28–31, write the point-slope form, the slope-intercept form, and the standard form of the equation of each line.

28. Passing through $(1, -3)$ with slope -2.

29. Passing through the points $(1, 3)$ and $(-4, 18)$

30. Passing through the point $(-1, -4)$ and parallel to the line whose equation is $2x - 3y = 6$

31. Passing through the point $(2, -3)$ and perpendicular to the line $2x - 4y = 8$

32. In 1900, the typical surfboard was 16 feet long. Since then, they have become shorter and shorter. Here are two data measurements for a typical surfboard's length.

x (Year)	y (Average Surfboard Length [in feet])
1900	16
1930	12.1

   a. Write the point-slope form of the line on which these measurements fall.
   b. Use the point-slope form of the equation to write the slope-intercept form of the equation as a function.
   c. Extrapolate from the data to find average surfboard length in 1970 and 1980.
   d. Does the model reasonably describe reality in 1990?

Based on the average length of a surfboard, write an inequality in terms of x that limits how far in the future the function can reasonably model reality.

Tom Blake with six of his surf boards
Bishop Museum.

In Problems 33–39, solve each inequality and graph the solution set. Express the solution set in both set-builder and interval notation.

33. $2x + 4 > 7$

34. $-6x + 3 \le 15$

35. $6x - 9 \ge -4x - 3$

36. $\frac{x}{4} - \frac{1}{2} < \frac{x}{2} + \frac{1}{4}$

37. $3(x + 5) \le 6(x + 1)$

38. $6x + 5 > -2(x - 3) - 25$

39. $3(2x - 1) - 2(x - 4) \ge 7 + 2(3 + 4x)$

Use an inequality to solve Problems 40–42.

40. To be listed on "The Top 50 Mutual Funds," a fund must have assets greater than $320 million. If a growth fund has current assets of $140 million and grows steadily at the rate of $9 million per year, how long will it take to make the list of the top 50 funds?

41. A person can choose between two charges on a checking account. The first method involves a fixed cost of $11 per month plus 6¢ for each check written. The sec-
ond method involves a fixed cost of $4 per month plus 20¢ for each check written. How many checks should be written to make the first method a better deal?

42. The formula for converting Fahrenheit temperature (F) to Celsius temperature (C) is $C = \frac{5}{9}(F - 32)$. If Celsius temperature ranges from $15°$ to $35°$, inclusively, what is the range for the Fahrenheit temperature?

For Problems 43–46, let $A = \{a, b, c\}$, $B = \{a, c, d, e\}$ and $C = \{a, d, f, g\}$. Find the indicated set.

43. $A \cap B$

44. $A \cap C$

45. $A \cup B$

46. $A \cup C$

In Problems 47–57, solve each compound inequality and graph the solution set on a number line. Express the solution set in both set-builder and interval notation.

47. $x \le 3$ and $x < 6$

48. $x \le 3$ or $x < 6$

49. $-2x < -12$ and $x - 3 < 5$

50. $5x + 3 \le 18$ and $2x - 7 \le -5$

51. $\{x \mid 2x - 5 > -1\} \cap \{x \mid 3x < 3\}$

52. $2x - 5 > -1$ or $3x < 3$

53. $x + 1 \le -3$ or $-4x + 3 < -5$

54. $\{x \mid 5x - 2 \le -22\} \cup \{x \mid -3x - 2 > 4\}$

55. $\{x \mid 5x + 4 \ge -11\} \cup \{x \mid 1 - 4x \ge 9\}$

56. $-1 \le 4x + 2 \le 6$

57. $-13 \le 3 - 4x < 13$

The model $w = 6t + 11$ approximates the weight (w, in pounds) for a boy t years old, where $t \le 14$. Use the model to answer Problems 58–63.

58. If weight ranges from 29 to 53 pounds, inclusively ($29 \le w \le 53$), what is the range for the boy's age?

59. What is the range for a boy's age whose weight is described by the range $w \le 65$ and $w > 41$?

60. What is the range for a boy's age whose weight is described by the range $w \le 71$ or $w \le 47$?

61. The table at the top of the next page shows actual average weights for boys ages 10 through 14. How well does the given formula model reality for these five categories? What is the relationship between your answer and the graph shown to the right of the table?

Average Weight for Boys	
Age (years)	**Weight (pounds)**
10	69
11	77
12	83
13	92
14	107

Source: *1992 World Almanac*

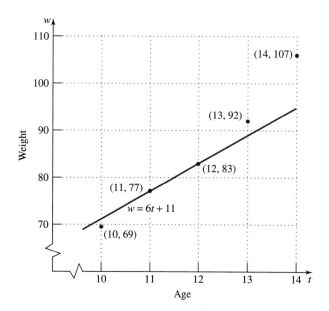

62. Substitute 30 for t, which is outside the model's data range, and find the corresponding value for w. What does this mean in practical terms? Does this seem realistic? What does this tell you about using a mathematical model outside its data interval?

63. Discuss one other possible limitation for the given model.

Solve each equation in Problems 64–67.

64. $|2x + 1| = 7$ **65.** $|3x + 2| = -5$ **66.** $2|x - 3| - 7 = 10$ **67.** $|4x - 3| = |7x + 9|$

Solve each absolute value inequality in Problems 68–71. Express the solution set in both set-builder and interval notation.

68. $|2x + 3| \leq 15$ **69.** $\left|\dfrac{2x + 6}{3}\right| > 2$ **70.** $|3 - 2x| < 7$ **71.** $|2x + 5| - 7 \geq -6$

72. Approximately 90% of the population sleeps h hours daily, where h is modeled by the inequality $|h - 6.5| \leq 1$. Write a sentence describing the range for the number of hours that most people sleep. Do *not* use the phrase "absolute value" in your description.

Graph each linear inequality in Problems 73–76.

73. $x - 3y \leq 6$ **74.** $y > -\frac{1}{2}x + 3$ **75.** $x \leq 2$ **76.** $y > -3$

CHAPTER 2 TEST

1. If $f(x) = 3x^2 - 7x + 3$, find $f(2a)$.

2. Fill in the table of coordinates for functions f and g and graph the two functions in the same rectangular coordinate system. Then describe the relationship between the two graphs.

$$f(x) = x^2 \qquad g(x) = (x - 1)^2$$

x	$f(x) = x^2$
-2	
-1	
0	
1	
2	

x	$g(x) = (x - 1)^2$
-1	
0	
1	
2	
3	

3. Graph $f(x) = x^2 + 2x + 1$ by first filling in the table of coordinates.

x	-3	-2	-1	0	1
$f(x) = x^2 + 2x + 1$					

4. List by letter all relations that are not functions.
   a. $\{(7,5), (8,5), (9,5)\}$ **b.** $\{(5,7), (5,8), (5,9)\}$

   c.

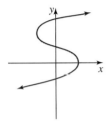

   d.

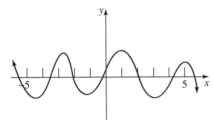

5. The function $f(x) = 0.79x^2 - 2x - 4$ models the number of board feet in a 16-foot log whose diameter averages x inches. Find and interpret $f(10)$.

6. The graph at the top of the next column shows the volume of air in human lungs over time. Which one of the following is true according to the graph?
   a. The up-and-down shape of the graph shows that the volume of air in the lungs is not a function of time.
   b. During breathing we completely fill and empty our lungs.
   c. If we take a deep breath, we can increase the air in our lungs by an extra 360 cubic inches.
   d. Normal breathing involves breathing in and out about an extra 30 cubic inches.
   e. During exercise, the volume of air involved in normal breathing increases four or five times.

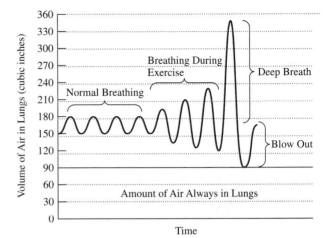

7. The graph shows world population growth over time. Find the slope of the line segment between points A and B. Describe what this means in terms of the average change in world population between 1976 and 2016.

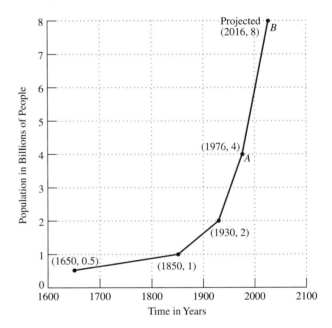

Graph these equations in a rectangular coordinate system.

8. $y = -\frac{1}{3}x + 2$

9. $4x - 3y = 12$

Write the point-slope form, the slope-intercept form, and the standard form of the equation of each line.

10. Passing through $(-1, -3)$ and $(4, 2)$

11. Passing through $(-2, 3)$ and perpendicular to the line whose equation is $y = -\frac{1}{2}x - 4$

12. Plot the points $(0, 0)$, $(5, 0)$, $(6, 2)$, and $(-1, -2)$. Then use slope to determine if these points are the vertices of a parallelogram.

13. The number of personal computers in the United States has increased steadily.

x (number of years after 1993)	y (number of personal computers in the U.S., in millions)
0	1.98
3	7.98

Write the point-slope form of the line on which these measurements fall. Then extrapolate from the data and predict the number of personal computers in the United States by the year 2010.

Solve each inequality and graph the solution set.

14. $-2x + 5 \geqslant 4 - (x - 3)$ **15.** $\frac{x}{7} - \frac{1}{5} \leqslant \frac{x}{5} - \frac{5}{7}$

16. A computer on-line service charges a flat monthly rate of \$20 or a monthly rate of \$5 plus 15 cents for every hour spent on-line. How many hours on-line each month will make the second option a better deal?

17. Find the intersection: $\{2, 4, 6, 7, 10\} \cap \{2, 4, 5, 9, 10\}$.

18. Find the union: $\{2, 4, 6, 7, 10\} \cup \{2, 4, 5, 9, 10\}$.

Solve.

19. $-2 < x - 5 \leqslant 7$

20. $-11 < 4 - 3x < 40$

21. $x + 3 \leqslant -1$ or $-4x + 3 < -5$

22. $x + 6 \geqslant 4$ and $-2x - 3 \leqslant 2$

23. $2x - 3 < 5$ or $3x - 6 \leqslant 4$

24. $-2x - 4 > -2$ and $x - 3 > -5$

25. $\left|\frac{2}{3}x - 6\right| = 2$

26. $\left|\frac{2}{3}x - 1\right| = \left|\frac{1}{3}x + 3\right|$

27. $\left|2x - 1\right| < 7$

28. $\left|2x - 3\right| \geqslant 5$

Graph each linear inequality in a rectangular coordinate system.

29. $3x - 2y < 6$

30. $y \geqslant \frac{1}{2}x - 1$

Systems of Linear Equations and Inequalities

Malcolm Morley "Los Angeles Yellow Pages" 1971 acrylic/canvas 84 × 72 in. Collection: Louisiana Museum of Modern Art, Humlebaek, Denmark. Courtesy: Mary Boone Gallery, New York.

Bottled water and medical supplies are to be shipped to victims of an earthquake by plane. How can we maximize the number of people helped given that the planes flying the supplies are limited in terms of both the weight and the volume of what they can carry?

Modeling the variables in this situation and solving the problem involves a technique called linear programming. Linear programming is one of the most widely used tools in management science, helping businesses allocate resources on hand to manufacture an array of products that will maximize profit. And as we shall see, the linear programming technique even solved the problem of the Soviet blockade on Berlin in 1948, saving one of the world's greatest cities.

SECTION 3.1

Solutions **Tutorial** **Video**
Manual **3**

Decide whether an ordered pair is a solution of a linear system.

I. Rice Pereira (1902–1971) "Oblique Progression" 1948, oil on canvas, 50×40 in. (127×101.6 cm). Purchase. 48.22. Collection of Whitney Museum of American Art, New York. Photo by Geoffrey Clements.

Linear Systems of Equations in Two Variables

Objectives

1. Decide whether an ordered pair is a solution of a linear system.
2. Solve linear systems by graphing.
3. Solve linear systems by substitution.
4. Solve linear systems by addition.
5. Identify inconsistent and dependent systems.

In this section, we consider three methods that determine when two linear equations are true at the same time.

In the previous chapter, we studied the linear equation in two variables

$$Ax + By = C$$

whose graph is a straight line. Many applied problems are modeled by two or more linear equations, such as

$$2x - 3y = -4$$
$$2x + y = 4$$

We call these equations *simultaneous linear equations* or a *system of linear equations*. A *solution* of such a system is an ordered pair of real numbers that makes *both* equations true.

Solution set of a system

The *solution set* of a system of equations in two variables is the set of all ordered pairs of values (a, b) that satisfy every equation in the system.

EXAMPLE I **Determining Whether an Ordered Pair Is a Solution of a System**

Determine whether $(1, 2)$ is a solution of the system:

$$2x - 3y = -4$$
$$2x + y = 4$$

Solution

Since 1 is the x-coordinate and 2 is the y-coordinate of $(1, 2)$, using the alphabetical order of the variables, we replace x by 1 and y by 2.

$$
\begin{array}{ll}
2x - 3y = -4 & 2x + y = 4 \\
2(1) - 3(2) \stackrel{?}{=} -4 & 2(1) + 2 \stackrel{?}{=} 4 \\
2 - 6 \stackrel{?}{=} -4 & 2 + 2 \stackrel{?}{=} 4 \\
-4 = -4 \quad \checkmark & 4 = 4 \quad \checkmark
\end{array}
$$

The pair $(1, 2)$ satisfies both equations, so it is a solution of the system. Although such a solution can be described by saying that $x = 1$ and $y = 2$, we will use set notation. The solution set to the system is $\{(1, 2)\}$—that is, the set consisting of the ordered pair $(1, 2)$. ■

2 Solve linear systems by graphing.

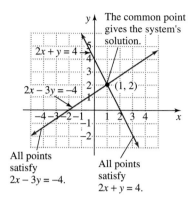

Figure 3.1

The common point gives the system's solution.

$2x + y = 4$

$2x - 3y = -4$

$(1, 2)$

All points satisfy $2x - 3y = -4$.

All points satisfy $2x + y = 4$.

Solving Linear Systems by Graphing

If we graph a system of linear equations, any point where the lines intersect is a solution of both equations. The coordinates of such a point give the solution of the system.

The system from Example 1, namely,

$$2x - 3y = -4$$
$$2x + y = 4$$

is graphed in Figure 3.1. The intersection point of the lines, $(1, 2)$, was the ordered pair that we verified as a solution in Example 1.

Solutions as intersecting graphs

The solution to a system of linear equations in two variables corresponds to the point(s) of intersection of the graphs of the equations.

Using technology

We can use a graphing utility to verify the graphs in Figure 3.1 and thereby find the system's solution set.

1. Solve each equation in the system for y.

$$2x - 3y = -4 \qquad\qquad 2x + y = 4$$
$$-3y = -2x - 4 \qquad\qquad y = -2x + 4$$
$$y = \tfrac{2}{3}x + \tfrac{4}{3}$$

2. Enter the equations.
3. Graph the equations in the same viewing rectangle.
4. Use the ☐ TRACE ☐ or intersection feature to find the intersection point.

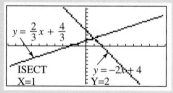

Using a standard viewing rectangle

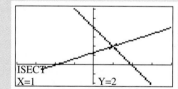

Using the viewing rectangle corresponding to Figure 3.1

EXAMPLE 2   **Solving a Linear System by Graphing**

Solve by graphing:

$$y = -x - 1$$
$$4x - 3y = 24$$

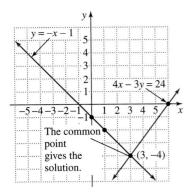

Figure 3.2

Solution

We graph each equation of the system using the methods studied in Chapter 2.

$$y = -x - 1$$

Use the y-intercept and slope to graph this equation.

Slope y-intercept:
-1 -1

$$4x - 3y = 24$$

Use intercepts to graph the equation.
x-intercept (set $y = 0$): $4x = 24$, so $x = 6$.
y-intercept (set $x = 0$): $-3y = 24$, so $y = -8$.

The system is graphed in Figure 3.2. Since graphing is not perfectly accurate, check the coordinates of the intersection point, $(3, -4)$ in both equations.

$$y = -x - 1 \qquad\qquad 4x - 3y = 24$$

Replace x with 3 and $-y$ with -4 in both equations.

$$-4 \overset{?}{=} -3 - 1 \qquad 4(3) - 3(-4) \overset{?}{=} 24$$
$$-4 = -4 \quad\checkmark \qquad\qquad 12 + 12 \overset{?}{=} 24$$
$$24 = 24 \quad\checkmark$$

Since the pair $(3, -4)$ checks in both equations, this verifies that the system's solution set is $\{(3, -4)\}$. ■

Eliminating a Variable Using the Substitution Method

Finding a system's solution using the graphing method can be awkward. For instance, a solution of $\left(-\frac{2}{3}, \frac{157}{29}\right)$ would be difficult to "see" as an intersection point.

We will now consider the substitution method, which does not depend on finding a system's solution visually. The method involves converting the system to one equation in one variable by an appropriate substitution.

3 Solve linear systems by substitution.

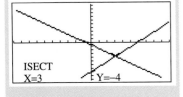
EXAMPLE 3   **Solving a System by Substitution**

Solve by the substitution method:

$$y = -x - 1$$
$$4x - 3y = 24$$

(Notice that this is the same system that we solved graphically in Example 2.)

Solution

The substitution method relies on having one variable isolated. Since y is isolated in the first equation, we can substitute the expression $-x - 1$ for y in the other equation.

$$y = \boxed{-x - 1} \qquad 4x - 3\boxed{y} = 24$$

Substitute $-x - 1$ for y.

Here are the details of the solution process.

$$4x - 3y \quad\;\; = 24 \qquad \text{This is the second equation in the given system.}$$

$$4x - 3(-x - 1) = 24 \qquad \text{Using the first equation, substitute } -x - 1 \text{ for } y. \text{ We now have one equation in one variable.}$$

$$4x + 3x + 3 = 24 \qquad \text{Apply the distributive property.}$$

$$7x + 3 = 24 \qquad \text{Combine like terms.}$$

$$7x = 21 \qquad \text{Subtract 3 from both sides.}$$

$$x = 3 \qquad \text{Divide both sides by 7.}$$

We now know that the x-coordinate of the solution is 3. To find the y-coordinate, we *back-substitute* the x-value into either one of the given equations. It is easiest to use the first equation since it is already solved for y.

$$y = -x - 1 \qquad \text{This is the first equation in the given system.}$$

$$y = -3 - 1 \qquad \text{Substitute 3 for } x.$$

$$y = -4 \qquad \text{Simplify.}$$

In Example 2 we checked $(3, -4)$ in both equations, so we know that the system's solution set is $\{(3, -4)\}$. ∎

Before considering additional examples, let's summarize the steps used in the substitution method.

> ### tudy tip
>
> *Back-substitute* means that after finding the value for a variable, we substitute that value *back* into one of the system's equations. Once we do this, we will find the value of the other variable.

> ### tudy tip
>
> Get into the habit of checking ordered-pair solutions in *both* equations of the system.

> ### tudy tip
>
> In step 1, if possible, solve for the variable whose coefficient is 1 or −1 to avoid working with fractions.

Solving linear systems by substitution

1. Solve one of the equations for one variable in terms of the other. (If one of the equations is already in this form, you can skip this step.)
2. Substitute the expression found in step 1 into the other equation. This will result in an equation in one variable.
3. Solve the equation obtained in step 2.
4. Back-substitute the value found from step 3 into the equation from step 1. Simplify and find the value of the remaining variable.
5. Check the proposed solution in both of the system's given equations.

EXAMPLE 4 **Solving a System by Substitution**

Solve by the substitution method:

$$5x - 4y = 9$$
$$x - 2y = -3$$

Solution

Step 1. Solve one of the equations for one variable in terms of the other.

We begin by isolating one of the variables in either of the equations. By solving for x in the second equation, with a coefficient of 1 we can avoid fractions.

$$x - 2y = -3 \qquad \text{This is the second equation in the given system.}$$

$$x = 2y - 3 \qquad \text{Solve for } x \text{ by adding } 2y \text{ to both sides.}$$

Step 2. Substitute the expression from step 1 into the other equation.

Since x is now isolated, we can substitute the expression for x in the first equation.

$$x = 2y - 3 \qquad 5x - 4y = 9 \quad \text{Substitute } 2y - 3 \text{ for } x.$$

$$5x \qquad - 4y = 9 \qquad \text{This is the first equation in the given system.}$$

Step 3. Solve the resulting equation containing one variable.

$$5(2y - 3) - 4y = 9 \qquad \text{Substitute } 2y - 3 \text{ for } x.$$

$$10y - 15 - 4y = 9 \qquad \text{Apply the distributive property.}$$

$$6y - 15 = 9 \qquad \text{Combine like terms.}$$

$$6y = 24 \qquad \text{Add 15 to both sides.}$$

$$y = 4 \qquad \text{Divide both sides by 6.}$$

Step 4. Back-substitute the obtained value into the equation from step 1.

Now that we have the y-coordinate of the solution, we back-substitute 4 for y in the equation $x = 2y - 3$.

$$x = 2y - 3 \qquad \text{Use the equation obtained in step 1.}$$

$$x = 2(4) - 3 \qquad \text{Substitute 4 for } y.$$

Step 5. Check.

$$x = 5 \qquad \text{Simplify.}$$

The proposed solution is $(5, 4)$. Take a moment to show that it satisfies both given equations. The solution set is $\{(5, 4)\}$. ∎

Using technology

Check Example 4 with a graphing utility. Solve each equation for y.

$$y_1 = \frac{5x - 9}{4} \quad \text{and} \quad y_2 = \frac{x + 3}{2}$$

Graph both equations in the same viewing rectangle. The solution, shown on the right, is $(5, 4)$.

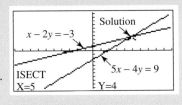

4 Solve linear systems by addition.

Eliminating a Variable Using the Addition Method

The substitution method is most useful if one of the given equations has an isolated variable. A third, and frequently the easiest, method for solving a linear system is the addition method. As with the substitution method, the addition method involves eliminating a variable and ultimately solving an equation containing only one variable. However, this time the elimination process is achieved by adding the equations. Notice how this is accomplished in our next example — when the two equations are added, the x-terms are eliminated.

EXAMPLE 5 **Solving a System by the Addition Method**

Solve by the addition method:

$$3x - 4y = 11$$
$$-3x + 2y = -7$$

Solution

Because the coefficients of x in the two equations differ only in sign, we can add the two left sides and add the two right sides, thereby eliminating the x-term.

$$
\begin{array}{rcr}
3x - 4y = & 11 \\
-3x + 2y = & -7 \\
\hline
-2y = & 4 \\
y = & -2
\end{array}
$$

Add: Solve for y, dividing both sides by -2.

Now we can back-substitute -2 for y into one of the original equations to find x. We will use both equations to show that we obtain the same value for x in either case.

Using the First Equation   **Using the Second Equation**

$3x - 4y = 11$	$-3x + 2y = -7$
$3x - 4(-2) = 11$	$-3x + 2(-2) = -7$ Replace y with -2.
$3x + 8 = 11$	$-3x - 4 = -7$ Solve for x.
$3x = 3$	$-3x = -3$
$x = 1$	$x = 1$

The solution $(1, -2)$ can be shown to satisfy both equations in the system. Consequently, the solution set is $\{(1, -2)\}$.   ∎

The object of the addition method is to obtain two equations whose sum will be an equation containing only one variable. The key step is to obtain, for one of the variables, coefficients that differ only in sign. Then when the two equations are added, this variable is eliminated.

In solving

$$3x - 4y = 11$$
$$-3x + 2y = -7$$

the x-term was eliminated when we added equations because the coefficients of the x-terms were opposites (additive inverses) of each other. When this is not the case, it is necessary to multiply one or both equations by some non-zero number so that the coefficients of one of the variables, x or y, become additive inverses. Let's see exactly how this works by considering Example 6.

EXAMPLE 6   **Solving a System by the Addition Method**

Solve by the addition method:

$$3x + 2y = 48$$
$$9x - 8y = -24$$

Solution

We must rewrite one or both equations in equivalent forms so that the coefficients of the same variable (either x or y) will be additive inverses of one another. We can accomplish this in a number of ways. Let's consider the terms in x in each equation, that is, $3x$ and $9x$. To eliminate x, we can multiply each term of the first equation by -3 and then add the equations.

$$
\begin{array}{ll}
3x + 2y = 48 & \xrightarrow{\text{Multiply by } -3.} & -9x - 6y = -144 \\
9x - 8y = -24 & \xrightarrow{\text{No change}} & \underline{9x - 8y = -24} \\
& \text{Add:} & -14y = -168 \\
& & y = 12 \quad \text{\small Solve for } y, \text{dividing} \\
& & \qquad\quad \text{\small both sides by } -14.
\end{array}
$$

Thus, $y = 12$. We back-substitute this value into either one of the given equations. We'll use the first one.

$$
\begin{array}{ll}
3x + 2y = 48 & \text{\small This is the first equation in the given system.} \\
3x + 2(12) = 48 & \text{\small Substitute 12 for } y. \\
3x + 24 = 48 & \text{\small Multiply.} \\
3x = 24 & \text{\small Subtract 24 from both sides.} \\
x = 8 & \text{\small Divide both sides by 3.}
\end{array}
$$

Finally, we complete the solution process and verify that the ordered pair $(8, 12)$ satisfies both equations in the system.

$$
\begin{array}{lll}
3x + 2y = 48 & 9x - 8y = -24 & \text{\small These are the given equations.} \\
3(8) + 2(12) \stackrel{?}{=} 48 & 9(8) - 8(12) \stackrel{?}{=} -24 & \text{\small Replace } x \text{ with 8 and } y \text{ with 12.} \\
24 + 24 \stackrel{?}{=} 48 & 72 - 96 \stackrel{?}{=} -24 & \\
48 = 48 \quad \checkmark & -24 = -24 \quad \checkmark &
\end{array}
$$

Since $(8, 12)$ checks, the system's solution set is $\{(8, 12)\}$. ∎

Before considering additional examples, let's summarize the steps involved in the solution of a system of two equations in two variables by the addition method.

> **Solving linear systems by addition**
>
> 1. If necessary, rewrite both equations in standard form
> $Ax + By = C$.
> 2. If necessary, multiply either equation or both equations by appropriate nonzero numbers so that the coefficients of x or y will have a sum of 0.
> 3. Add the equations in step 2. The sum is an equation in one variable.
> 4. Solve the equation from step 3.
> 5. Back-substitute the value obtained in step 4 into either of the given equations and solve for the other variable.
> 6. Check the solution in both of the original equations.

EXAMPLE 7 **Solving a System by the Addition Method**

Solve by the addition method:

$$
\begin{aligned}
7x &= 5 - 2y \\
3y &= 16 - 2x
\end{aligned}
$$

Solution

Step 1. Rewrite both equations in the form $Ax + By = C$.

We first arrange the system so that variable terms appear on the left and constants appear on the right. We obtain

$7x + 2y = 5$ Add $2y$ to both sides in the first equation.

$2x + 3y = 16$ Add $2x$ to both sides in the second equation.

Step 2. Multiply both equations by appropriate numbers so that y-coefficients have a zero sum.

We can eliminate x or y. Let's eliminate y by multiplying the first equation by 3 and the second equation by -2.

$$\begin{array}{c} 7x + 2y = 5 \\ 2x + 3y = 16 \end{array} \quad \begin{array}{l} \xrightarrow{\text{Multiply by 3.}} \\ \xrightarrow{\text{Multiply by } -2.} \end{array} \quad \begin{array}{rcr} 21x + 6y = & 15 \\ -4x - 6y = & -32 \\ \hline \end{array}$$

Steps 3 and 4. Add the equations and solve the resulting equation for x.

$$\text{Add:} \quad \begin{array}{rcr} 17x & = & -17 \\ x = & -1 \end{array} \quad \begin{array}{l}\text{Divide both} \\ \text{sides by 17.}\end{array}$$

Step 5. Back-substitute and find the value for y.

This is the x-coordinate of the solution to our system. Now we back-substitute -1 for x in either original equation to find the y-coordinate.

$3y = 16 - 2x$ This is the second equation in the given system.

$3y = 16 - 2(-1)$ Replace x by -1.

$3y = 16 + 2$ Multiply.

$3y = 18$ Add.

$y = 6$ Divide both sides by 3.

Step 6. Check.

The solution $(-1, 6)$ can be shown to satisfy both equations in the system. Consequently, the solution set is $\{(-1, 6)\}$. ∎

Discover for yourself

Use a graphing utility and verify the solutions to Examples 5 through 7. You'll need to begin by solving each of the six equations for y in terms of x.

EXAMPLE 8   **Solving a System by the Addition Method**

Solve by the addition method:

$$\frac{1}{2}x - 5y = 32 \qquad \frac{3}{2}x - 7y = 45$$

Solution

We can solve the system without clearing the equations of fractions. However, there is less chance for error if the coefficients for x and y are integers. Consequently, we begin by multiplying both sides of each equation by 2.

$$\frac{1}{2}x - 5y = 32 \quad \xrightarrow{\text{Multiply by 2.}} \quad x - 10y = 64$$

$$\frac{3}{2}x - 7y = 45 \quad \xrightarrow{\text{Multiply by 2.}} \quad 3x - 14y = 90$$

Now we can eliminate x by multiplying the first equation with integral coefficients by -3 and leaving the second equation unchanged.

$$\begin{array}{c} x - 10y = 64 \\ 3x - 14y = 90 \end{array} \quad \begin{array}{l} \xrightarrow{\text{Multiply by } -3.} \\ \xrightarrow{\text{No change}} \end{array} \quad \begin{array}{rcr} -3x + 30y = & -192 \\ 3x - 14y = & 90 \\ \hline \end{array}$$

$$\text{Add:} \quad \begin{array}{rcr} 16y & = & -102 \\ y & = & -\dfrac{102}{16} \end{array}$$

ENRICHMENT ESSAY

Systems of Linear Equations and Hemispheres of the Brain

The right hemisphere of our brain is mainly responsible for the holistic conceptualizations essential for art, while the left hemisphere is primarily in charge of the analytic thinking necessary for creating and understanding mathematics. Communication between the two hemispheres of the brain, between visual and logical modes, through a bundle of nerve fibers called the corpus callosum, is imperfect. Descartes' analytic geometry, by wonderfully connecting every algebraic equation with a geometric figure, and vice versa, partly enhanced this communication, synthesizing the incongruity that many people experience when mixing visual and analytic abilities.

In this section, we used both of our brain hemispheres to solve systems of equations. We began with the right hemisphere, visualizing solutions as intersecting lines. We then moved from our right hemisphere back to our left, from the visual to the analytical, as we studied algebraic methods—sub-

stitution and addition—that do not depend on visual analysis for solving linear systems. Both of these methods help us identify solutions that can only be suggested by intersecting graphs.

Alfredo Castaneda "Enajenado" 1967, oil on canvas 90 × 70.5 cm. Mary-Anne Martin/Fine Art, New York and Galeria GAM, Mexico City

---

Reducing the fraction, the y-coordinate of our solution is $-\frac{51}{8}$. Back-substitution of this value into either original equation in the system results in cumbersome arithmetic. Another option is to go back to the equations with integral coefficients and this time eliminate y instead of x.

Discover for yourself

What is the most efficient method for checking the solution of Example 8? Use this method to verify the solution.

$$
\begin{array}{ll}
x - 10y = 64 & \xrightarrow{\text{Multiply by } -7.} \quad -7x + 70y = -448 \\
3x - 14y = 90 & \xrightarrow{\text{Multiply by 5.}} \quad \underline{15x - 70y = 450} \\
& \qquad\quad \text{Add:} \quad\; 8x \qquad\quad = 2 \\
& \qquad\qquad\qquad\qquad x = \dfrac{2}{8} = \dfrac{1}{4}
\end{array}
$$

The solution to our system is $(\frac{1}{4}, \frac{-51}{8})$ and its solution set is $\{(\frac{1}{4}, \frac{-51}{8})\}$.

Comparing the Three Solution Methods

The substitution method works particularly well when one of the original equations has y expressed in terms of x (or vice versa) or when an equation has a variable with a coefficient of 1 or -1. The variable with this coefficient is the one we should solve for. If neither of those things happen, we should use the addition rather than the substitution method.

The following summary compares the graphing, addition, and substitution methods for solving linear systems of equations.

Method	Advantages	Example	Disadvantages
Graphing	You can see the solutions.	$2x + y = 6$ $x - 2y = 8$ 	If the solutions do not involve integers or are too large to be seen on the graph, it's impossible to tell exactly what the solutions are.
Addition	Gives exact solutions. Easy to use if no variable has a coefficient of 1 or -1.	$2x + 3y = -8$ $5x + 4y = -34$ Multiply by 5 and -2 respectively: $10x + 15y = -40$ $\underline{-10x - 8y = 68}$ Add: $\quad 7y = 28$ $ y = 4$ Then back-substitute. Solution: $(-10, 4)$	Solutions cannot be seen.
Substitution	Gives exact solutions. Easy to use if a variable is on one side by itself.	$y = 3x - 1$ $3x - 2y = -4$ Substitute $3x - 1$ for y: $3x - 2(3x - 1) = -4$ Solve for x: $\qquad x = 2$ Back-substitute: $\qquad y = 3(2) - 1 = 5$ Solution: $(2, 5)$	Solutions cannot be seen.

5 Identify inconsistent and dependent systems.

Linear Systems Having No Solution and Infinitely Many Solutions

Since the graph of a linear equation is a straight line, there are three possibilities for the number of solutions to a system of two linear equations.

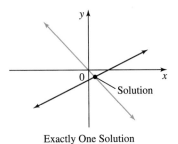

Exactly One Solution

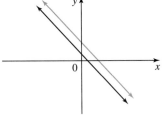

No Solution (Parallel Lines)

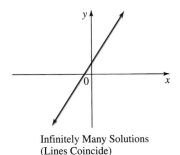

Infinitely Many Solutions
(Lines Coincide)

Figure 3.3

Possible graphs for a system
of two linear equations in
two variables

The number of solutions to a system of two linear equations

The number of solutions to a system of two linear equations in two variables is given by one of the following. (See Figure 3.3.)

Number of Solutions	What This Means Graphically
Exactly one ordered-pair solution	The two lines intersect at one point.
No solution	The two lines are parallel.
Infinitely many solutions	The two lines are identical.

Most linear systems that we will encounter have exactly one solution. This is true of all the systems that we have solved up to this point. However, let's see what occurs when we attempt to solve systems with no solution or infinitely many solutions by the addition and substitution methods.

EXAMPLE 9 Solving a System with No Solution

Solve the system:

$$4x + 6y = 12$$
$$6x + 9y = 12$$

Solution

Since no variable has a coefficient of 1 or -1, we will use the addition method. To obtain coefficients of x that differ only in sign, we multiply the first equation by 3 and multiply the second equation by -2.

$$
\begin{array}{ll}
4x + 6y = 12 & \xrightarrow{\text{Multiply by 3.}} \quad 12x + 18y = 36 \\
6x + 9y = 12 & \xrightarrow{\text{Multiply by } -2.} \quad \underline{-12x - 18y = -24} \\
& \qquad\qquad\quad\text{Add:} \qquad\quad 0 = 12
\end{array}
$$

There are no values of x and y for which $0 = 12$.

The false statement $0 = 12$ indicates that the system has no solution. The contradiction $0 = 12$ tells us that the solution set for the system is the empty set, $\varnothing$.

The lines corresponding to the two equations in the given system are shown in Figure 3.4. The lines are parallel and have no point of intersection.

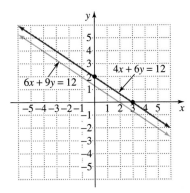

Figure 3.4

The graph of a system with no
solution

> **D**iscover for yourself
>
> Show that the graphs of $4x + 6y = 12$ and $6x + 9y = 12$ must be parallel lines by solving each equation for y. What is the slope and y-intercept for each line? What does this mean? If a linear system has infinitely many solutions, what must be true about the slopes and y-intercepts for the system's graphs?

A linear system with no solution, such as the one in Example 9, is called an *inconsistent system.* When solving such a system by either addition or substitution, both variables will be eliminated, and the resulting statement will be a *contradiction.* In Example 9, we obtained the contradiction $0 = 12$.

Now let's see what happens when we use an algebraic method (addition or substitution) to solve a linear system with infinitely many solutions. Can you guess what is going to occur?

> **EXAMPLE 10** **Solving a System with Infinitely Many Solutions**

Solve the system:

$$y = 3 - 2x$$
$$4x + 2y = 6$$

Solution

Since the variable y is isolated in the first equation, we can use the substitution method. We substitute the expression for y in the other equation.

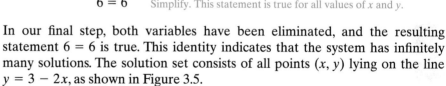

$$y = 3 - 2x \qquad 4x + 2y = 6 \qquad \text{Substitute } 3 - 2x \text{ for } y.$$

$$4x + 2y = 6 \qquad \text{This is the second equation in the given system.}$$

$$4x + 2(3 - 2x) = 6 \qquad \text{Substitute } 3 - 2x \text{ for } y.$$

$$4x + 6 - 4x = 6 \qquad \text{Apply the distributive property.}$$

$$6 = 6 \qquad \text{Simplify. This statement is true for all values of } x \text{ and } y.$$

In our final step, both variables have been eliminated, and the resulting statement $6 = 6$ is true. This identity indicates that the system has infinitely many solutions. The solution set consists of all points (x, y) lying on the line $y = 3 - 2x$, as shown in Figure 3.5.

We express the solution set for the system in one of two equivalent ways:

$$\{(x, y) \,|\, y = 3 - 2x\} \qquad \text{The set of all ordered pairs } (x, y) \text{ such that } y = 3 - 2x$$

$$\text{or} \quad \{(x, y) \,|\, 4x + 2y = 6\} \qquad \text{The set of all ordered pairs } (x, y) \text{ such that } 4x + 2y = 6 \quad \blacksquare$$

A linear system with infinitely many solutions, such as the one in Example 10, is called a *dependent system.* In a dependent system of linear equations, both equations represent the same line.

The results of Examples 9 and 10 are generalized as follows.

> **Inconsistent and dependent systems of linear equations**
>
> If both variables are eliminated when a system of linear equations is solved by substitution or addition, one of the following is true.
>
> **1.** There is no solution if the resulting statement is false. The system is inconsistent.

Figure 3.5

The graph of a system with infinitely many solutions

> **2.** There are infinitely many solutions if the resulting statement is true. The system is dependent.

Any system that has at least one solution is called a *consistent system.* Thus, a linear system with one solution and a linear system with infinitely many solutions are both said to be consistent.

PROBLEM SET 3.1

Practice Problems

If applicable, verify your answer to as many problems in this set as possible using a graphing utility.

In Problems 1–16, solve each system by graphing. Check your answer. If applicable, state that the system is inconsistent or dependent.

1. $x + y = 4$
$\quad x - y = 2$

2. $\quad x + y = 6$
$\quad -x + y = 4$

3. $x - 3y = 2$
$\quad y = 6 - x$

4. $2x + y = 4$
$\quad 4x + 3y = 10$

5. $2x = 3y - 6$
$\quad -3x + y = -5$

6. $x - y = 2$
$\quad y = 1$

7. $x + 2y = 1$
$\quad x = 3$

8. $x - y = 3$
$\quad x - y = 5$

9. $3x + y = 3$
$\quad 6x + 2y = 12$

10. $2x - 3y = 6$
$\quad 4x - 6y = 2$

11. $\dfrac{x}{3} - \dfrac{y}{4} = 1$

$\quad \dfrac{2x}{3} - \dfrac{y}{2} = 1$

12. $x - y = 2$
$\quad 3x - 3y = 6$

13. $2x + 2y = 1$

$\quad 4x + 4y = 2$

14. $\quad 3x - 2y = -1$

$\quad -6x + 4y = 2$

15. $3x - 3y = 6$

$\quad 2x - 2y = 4$

16. $\dfrac{x}{4} + \dfrac{y}{2} = 1$

$\quad 2x + 4y = 8$

Solve each system in Problems 17–24 by the substitution method.

17. $\quad x = 2y - 5$
$\quad x - 3y = 8$

18. $\quad x = 2y - 2$
$\quad 2x - 2y = 1$

19. $4x + y = 5$
$\quad 2x - 3y = 13$

20. $x - y = 4$
$\quad 2x - 5y = 8$

21. $x + y = 0$
$\quad 3x + 2y = 5$

22. $3x - 2y = 4$
$\quad 2x - y = 1$

23. $7x - 3y = 23$
$\quad x + 2y = 13$

24. $\dfrac{x}{4} = 9 + \dfrac{y}{5}$

$\quad y = 5x$

Solve each system in Problems 25–32 by the addition method.

25. $x + y = 7$
$\quad x - y = 3$

26. $2x + y = 3$
$\quad x - y = 3$

27. $12x + 3y = 15$
$\quad 2x - 3y = 13$

28. $2x + y = 3$
$\quad 2x - 3y = -41$

29. $x - 2y = 5$
$\quad 5x - y = -2$

30. $4x - 5y = 17$
$\quad 2x + 3y = 3$

31. $2x - 9y = 5$
$\quad 3x - 3y = 11$

32. $3x - 4y = 4$
$\quad 2x + 2y = 12$

Solve each system in Problems 33–56 by the method of your choice.

33. $3x - 7y = 1$
$\quad 2x - 3y = -1$

34. $2x - 3y = 2$
$\quad 5x + 4y = 51$

35. $4x + y = 2$
$\quad 2x - 3y = 8$

36. $3x + 4y = 16$
$\quad 5x + 3y = 12$

37. $2y = 5 - 5x$
$\quad 9x - 15 = -3y$

38. $5x - 40 = 6y$
$\quad 2y = 8 - 3x$

39. $9x + \frac{4}{3}y = 5$

$\quad 4x - \frac{1}{3}y = 5$

40. $-\frac{1}{6}x + \frac{1}{5}y = 4$

$\quad \frac{1}{4}x - \frac{1}{6}y = -2$

41. $y - 3x = 2$

$\quad x = \frac{1}{4}y$

42. $\quad y = 3x$

$\quad \dfrac{x}{2} + \dfrac{y}{3} = 3$

43. $x + 2y - 3 = 0$
$\quad 12 = 8y + 4x$

44. $3x + 3y = 2$
$\quad 2x + 2y = 3$

45. $2x - y - 5 = 0$
      $4x - 2y - 10 = 0$

46. $2x + y - 4 = 0$
      $2x + y - 2 = 0$

47. $2x - 3y = 7$
      $4x - 6y = 3$

48. $4x - 6y = 14$
      $2x - 3y = 7$

49. $x - 3y = 5$
      $y = \frac{2}{3}x$

50. $y = 3x + 4$
      $y = 3x + 6$

51. $y = 2x + 5$
      $y = -4x + 2$

52. $y = x$
      $x + y = 6$

53. $3(x + y) = 6$
      $3(x - y) = -36$

54. $4(x - y) = -12$
      $4(x + y) = -20$

55. $3(x - 3) - 2y = 0$
      $2(x - y) = -x - 3$

56. $5x + 2y = -5$
      $4(x + y) = 6(2 - x)$

Application Problems

57. A company has yearly fixed costs of $800,000 and a cost of $45 per item to manufacture x number of items. Thus, $y = 800,000 + 45x$. The product is to sell for $65, so that total revenue from sales is given by $y = 65x$. Solve these two equations to determine the number of units that must be sold for the company to break even, where the cost equals the revenue.

58. A nursery specializing in native plants of the Florida Keys has yearly fixed costs of $30,000 and a cost of $5 (for pots and soil) per plant. If plants sell for $15 each, determine the number of plants that must be sold for the nursery to break even.

59.

A well-known Greek paradox, dating back to 450 B.C., involves a race between Achilles and a tortoise. Achilles and the tortoise are engaged in a footrace. Achilles runs 500 yards per minute and the tortoise runs 50 yards per minute. The tortoise is given a head start of 1000 yards. The paradox is that Achilles can never overtake the tortoise. It takes Achilles 2 minutes to reach the starting point of the tortoise, but in the meantime the tortoise has crawled 100 yards ahead. Achilles will take $\frac{1}{5}$ of a minute to cover these 100 yards, but again in this $\frac{1}{5}$ of a minute the tortoise is now 10 yards ahead. This process continues ad infinitum, so that Achilles comes closer and closer to the tortoise but can never overtake it. The paradox can be resolved using linear equations and their graphs. If x represents time (in minutes) and y represents distance

(in yards), the equations become

For Achilles $y = 500x$
For the tortoise $y = 50x + 1000$
                        $\uparrow$
              This is the 1000-yard head start.

a. Graph these equations in the same rectangular coordinate system.

b. You should have a point of intersection approximately at (2.2, 1100). In terms of the paradox, what does this mean? More specifically, in how many minutes will Achilles overtake the tortoise, and approximately how far from the starting point will this take place?

c. Verify part (b) by solving the system using the substitution method.

60.

Archimedes c 287–212 BC shown holding his crown. Greek mathematician, engineer and physicist. Archive Photos

One of the greatest mathematicians of antiquity, Archimedes (287–212 B.C.), wanted a crown made from a 10-pound piece of gold. When Archimedes received his crown from the jeweler it weighed 10 pounds, but he wondered if the jeweler had cheated him by adding some silver to the crown, keeping a portion of the gold for himself.

Let x represent the amount of gold and y represent the amount of silver. Because the crown weighed 10 pounds, $x + y = 10$. Archimedes also discovered that the density of an object is related to the amount of water it displaces. He determined that x pounds of gold displaces $\frac{x}{20}$ pounds of water and y pounds of silver displaces $\frac{y}{10}$ pounds of water.

Knowing this, Archimedes dropped his crown into water and discovered that the displaced water weighed $\frac{7}{10}$ of a pound. Because water displaced by gold + water displaced by silver = $\frac{7}{10}$, he wrote

$$\frac{x}{20} + \frac{y}{10} = \frac{7}{10}.$$

Archimedes thus set up the system

$$x + y = 10$$
$$\frac{x}{20} + \frac{y}{10} = \frac{7}{10}$$

Use substitution to solve the system. (You may find it convenient to multiply both sides of the second equation by 20.) Was Archimedes cheated by the jeweler?

True–False Critical Thinking Problems

61. Which one of the following is true?
  a. The solution set to the system

$$y = 4x - 3$$
$$y = 4x + 1$$

  is the empty set.
  b. The ordered pair $(1, 2)$ is the solution to the system

$$2x + y = 4$$
$$3y - x = 1$$

  c. The graphs of dependent equations are lines that intersect in exactly one point.
  d. Both $(2, 3)$ and $(7, 8)$ are in the solution set to the system

$$x - y = -1$$
$$x + y = 5$$

62. Which one of the following is true?
  a. The solution set to the system

$$y = 4x - 3$$
$$y = 4x + 5$$

  is the empty set.
  b. The addition method cannot be used to eliminate either variable in a system of two equations in two variables.
  c. The solution set to the system

$$5x - y = 1$$
$$10x - 2y = 2$$

  is $\{(2, 9)\}$.
  d. A linear system of equations can have a solution set consisting of precisely two ordered pairs.

Technology Problem

63. Most graphing utilities can give the solution to a linear system of equations. (Consult your manual for details.) This capability is usually accessed with the SIMULT (simultaneous equations) feature. First, you will enter 2, for two equations in two variables. With each equation in standard form, you will then enter the coefficients for x and y and the constant term, one equation at a time. After entering all six numbers, press SOLVE. The solution will be displayed on the screen. (The x-coordinate may be displayed as $x_1 =$ and the y-value as $x_2 = $.) Use this capability to verify the solution to some of the problems you solved in the practice problems of this problem set. Describe what happens when you use your graphing utility on an inconsistent or dependent system.

Writing in Mathematics

64. The daily cost (y) for the Worldwide Widget Corporation to produce x widgets is given by the linear function $y = 8x + 200$. The daily revenue (y) generated by the sale of x widgets is described by another linear function, $y = 16x$. Use the graphs in the figure at the top of the next page to write a short paragraph describing how the company breaks even and under what conditions there will be either profit or loss.

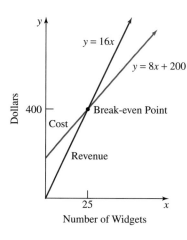

Number of Widgets

65. What is the disadvantage to solving a system of equations using the graphing method?

66. When is it easier to use the addition method rather than the substitution method when solving a system of equations?

67. When using the addition or substitution method, how can you tell if a system of linear equations has infinitely many solutions? What is the relationship between the graphs of the two equations?

68. When using the addition or substitution method, how can you tell if a system of linear equations has no solution? What is the relationship between the graphs of the two equations?

Critical Thinking Problems

69. Write a system of equations having $\{(-2, 7)\}$ as a solution set. (More than one system is possible.)

70. Write the point-slope form, the slope-intercept form, and the standard form of the line passing through the intersection of $x + 2y = 4$ and $x - y = -5$ and perpendicular to the line whose equation is $3x - 9y = 10$.

71. Consider an inconsistent system in which one equation is $3x - 2y = 4$. Which one of the following must be the other equation of the system?
a. $3x + 2y = 4$
b. $2x - 3y = 4$
c. $6x - 4y = 8$
d. $6x - 4y = 6$

72. Consider a dependent system in which one equation is $4x + 6y = 2$. Which one of the following must be the other equation of the system?
a. $4x + 6y = 4$
b. $2x + 3y = 1$
c. $6x + 4y = 2$
d. $4x - 6y = 2$

73. Find a value for k so that the following system of linear equations is inconsistent:
$$5x - 10y = 40$$
$$2x + ky = -30$$

74. Two identical twins can only be recognized by the characteristic that one always tells the truth and the other always lies. One twin tells you of a lucky num-

ber pair: "When I multiply my first lucky number by 3 and my second lucky number by 6, the addition of the resulting numbers produces a sum of 12. When I add my first lucky number and twice my second lucky number, the sum is 4." Which twin is talking?

75. If $b_1 \neq b_2$, find the solution set for the system:
$$y = m_1 x + b_1$$
$$y = m_1 x + b_2$$

76. If $\dfrac{x}{y} = \dfrac{2}{3}$, what number is represented by $\dfrac{7x - y}{3x + 4y}$?

77. Solve the system for x and y in terms of $a_1, b_1, c_1, a_2, b_2,$ and c_2:
$$a_1 x + b_1 y = c_1$$
$$a_2 x + b_2 y = c_2$$

78. Solve:
$$\frac{5}{x} + \frac{6}{y} = \frac{19}{6}$$
$$\frac{3}{x} + \frac{4}{y} = 2$$

(*Hint:* Let $a = \dfrac{1}{x}$ and $b = \dfrac{1}{y}$. Substitute these expressions into the system to find a and b. Then find x and y.)

Review Problems _____

79. The numbers are arranged in the square in the figure so that all rows, columns, and diagonals contain three numbers having identical sums. The number x in the upper right corner satisfies the equation

$$\frac{x}{3} - \frac{x-5}{5} = 3$$

and y is a two-digit number. Fill in the missing numbers in the empty squares.

y	10	x
12		16
13		11

80. The price of a detergent increased by 16% and now sells for $3.19. What was the price before the increase?

81. Solve: $|2x + 3| = |4x - 5|$.

S E C T I O N   3 . 2

Solutions Manual   **Tutorial**   **Video 3**

| Solve problems using systems of equations.

Problem Solving and Modeling Using Systems of Equations

Objectives

1 Solve problems using systems of equations.
2 Model data using systems of equations.

In this section, we focus on the central themes of algebra: problem solving and modeling.

A Problem-Solving Strategy

When we solved word problems earlier in the book, we had to translate from the language of English into an algebraic equation. In a similar way, we will now have two critical English sentences, each of which must be translated into a linear equation containing two variables. Here are some general steps we will follow in solving these verbal problems.

> **Strategy for solving word problems that result in a system of linear equations**
>
> 1. Read the problem and let x and y (or any variables) represent the quantities that are unknown.
> 2. Write a system of linear equations, in x and y, that models the verbal conditions of the problem.
> 3. Solve the system written in step 2, using the method of addition or substitution, and answer the problem's question.
> 4. Check the answers *in the original wording* of the problem, not in the system of equations obtained from the words.

Solving Problems by Translating Given Conditions into a System of Equations

These types of problems contain explicit verbal conditions that can be translated into systems of linear equations. We will use our four-step strategy to solve these problems.

Claes Oldenburg "Pie a la Mode",
1962, plaster soaked muslin,
20 × 13 × 19 in. The Museum of
Contemporary Art, Los Angeles.
The Panza Collection.

Step 1. Use variables to represent the
unknown quantities.

Step 2. Write a system of equations
modeling the problem's conditions.

Step 3. Solve the system and answer
the problem's question.

EXAMPLE 1   Cholesterol and Heart Disease

The verdict is in: After years of research, the nation's health experts agree
that high cholesterol in the blood is a major contributor to heart disease. Thus,
cholesterol intake should be limited to 300 mg or less each day. Fast foods
provide a cholesterol carnival. Two McDonald's Quarter Pounders and three
Burger King Whoppers with cheese contain 520 mg of cholesterol. Three
Quarter Pounders and one Whopper with cheese exceed the suggested daily
cholesterol intake by 53 mg. Determine the cholesterol content in each item.

Solution

Let x represent the cholesterol content of a Quarter Pounder and y the cholesterol content of a Whopper with cheese.

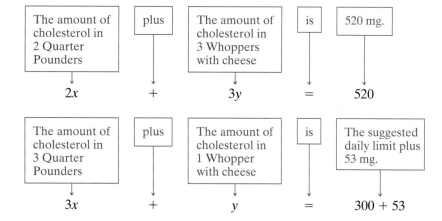

The system

$$2x + 3y = 520$$
$$3x + y = 353$$

can be solved by substitution or addition. We'll use addition, multiplying the
second equation by -3 to eliminate y.

$$
\begin{array}{ll}
2x + 3y = 520 & \xrightarrow{\text{No change}} \\
3x + y = 353 & \xrightarrow{\text{Multiply by } -3.}
\end{array}
\qquad
\begin{array}{rl}
2x + 3y = & 520 \\
-9x - 3y = & -1059 \\
\hline
\text{Add: } \quad -7x = & -539 \\
x = \dfrac{-539}{-7} = 77
\end{array}
$$

Now we can find the value of y by back-substituting 77 for x in either of the
system's equations.

$$
\begin{array}{ll}
3x + y = 353 & \text{We'll use the second equation.} \\
3(77) + y = 353 & \text{Back-substitute 77 for } x. \\
231 + y = 353 & \text{Multiply.} \\
y = 122 & \text{Subtract 231 from both sides.}
\end{array}
$$

Since $x = 77$ and $y = 122$, a Quarter Pounder contains 77 mg of cholesterol
and a Whopper with cheese contains 122 mg of cholesterol.

Step 4. Check the proposed answers in the original wording of the problem.

Two Quarter Pounders and 3 Whoppers with cheese contain $2(77) + 3(122) = 520$ mg, which checks with the given conditions. Furthermore, 3 Quarter Pounders and 1 Whopper with cheese contain $3(77) + 122 = 353$ mg, which does exceed the daily limit of 300 mg by 53 mg. ∎

Solving Problems by Creating Verbal Models and then Translating into a System of Equations

In Example 1, the conditions necessary for writing an equation were clearly given. A more difficult situation is one in which the conditions are only implied. The first step is to write two English sentences that clearly identify the operations involved. These sentences serve as verbal models that are then translated into a system of equations.

EXAMPLE 2   Energy Efficiency of Building Materials

A heat-loss survey by an electric company indicated that a wall of a house containing 40 square feet of glass and 60 square feet of plaster lost 1920 Btu (British thermal units) of heat. A second wall containing 10 square feet of glass and 100 square feet of plaster lost 1160 Btu. Determine the heat lost per square foot for the glass and for the plaster.

Solution

Step 1. Use variables to represent unknown quantities.

Let x represent the heat lost per square foot for the glass and y the heat lost per square foot for the plaster.

Step 2. Write a system of equations modeling the problem's conditions.

The most important idea in this problem is that the heat loss for each wall is the sum of the heat lost by the glass and the heat lost by the plaster.

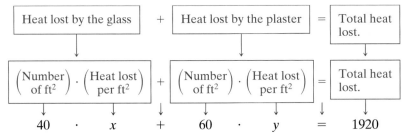

The first wall contains 40 ft^2 of glass, 60 ft^2 of plaster, and lost 1920 Btu.

Since the second wall contains 10 ft^2 of glass, 100 ft^2 of plaster, and lost 1160 Btu, the verbal model is translated by

$$10x + 100y = 1160.$$

Step 3. Solve the system and answer the problem's question.

The system

$$40x + 60y = 1920$$
$$10x + 100y = 1160$$

can be solved by addition. We'll multiply the second equation by -4 to eliminate x.

$$40x + 60y = 1920 \xrightarrow{\text{No change}} 40x + 60y = 1920$$
$$10x + 100y = 1160 \xrightarrow{\text{Multiply by } -4.} \underline{-40x - 400y = -4640}$$
$$\text{Add:} \qquad -340y = -2720$$
$$y = \frac{-2720}{-340} = 8$$

Now we can find the value of x by back-substituting 8 for y in either of the system's equations.

$$10x + 100y = 1160 \qquad \text{We'll use the second equation.}$$
$$10x + 100(8) = 1160 \qquad \text{Back-substitute 8 for } y.$$
$$10x + 800 = 1160 \qquad \text{Multiply.}$$
$$10x = 360 \qquad \text{Subtract 800 from both sides.}$$
$$x = 36 \qquad \text{Divide both sides by 10.}$$

Consequently, the heat lost through the glass is 36 Btu per square foot, and the heat lost through the plaster is 8 Btu per square foot.

Step 4. Check the proposed answers in the original wording of the problem.

The first wall, with 40 square feet of glass and 60 square feet of plaster, loses

$$40(36) + 60(8) = 1920 \text{ Btu of heat}$$

which checks with the given conditions. Furthermore, the second wall, with 10 square feet of glass and 100 square feet of plaster, loses

$$10(36) + 100(8) = 1160 \text{ Btu of heat}$$

which also checks with the given conditions. ∎

Solving Problems by Using Known Formulas

Most problems in geometry require a knowledge of formulas to obtain a solution.

Width: W

Length: L

Figure 3.6

A rectangular waterfront lot

EXAMPLE 3 **Using a Formula to Solve a Problem**

A rectangular waterfront lot has a perimeter of 1000 feet. To create a sense of privacy, the lot's owner decides to fence along three sides, excluding the side that fronts the water (see Figure 3.6). An expensive fencing along the lot's front length costs $25 per foot, and an inexpensive fencing along the two side widths costs only $5 per foot. The total cost of the fencing along all three sides comes to $9500.

a. What are the lot's dimensions?
b. The owner is considering using the expensive fencing on all three sides of the lot, but is limited by a $16,000 budget. Can this be done within the budget constraints?

Step 1. Use variables to represent unknown quantities.

Step 2. Write a system of equations modeling the problem's conditions.

Step 3. Solve the system and answer the problem's question.

Step 4. Check the proposed answers in the original wording of the problem.

Solution

a. Let L represent the lot's length and let W represent its width. The lot has a perimeter of 1000 feet.

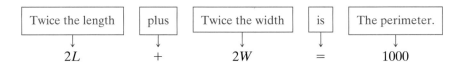

Twice the length	plus	Twice the width	is	The perimeter.
$2L$	$+$	$2W$	$=$	1000

The cost of fencing three sides of the lot is \$9500.

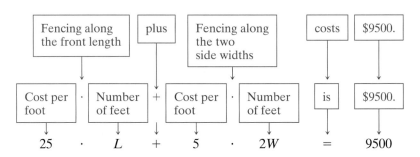

Fencing along the front length	plus	Fencing along the two side widths	costs	\$9500.
Cost per foot · Number of feet	$+$	Cost per foot · Number of feet	is	\$9500.
25 · L	$+$	5 · $2W$	$=$	9500

The system

$$2L + 2W = 1000$$
$$25L + 10W = 9500$$

can be solved by addition. We'll multiply the first equation by -5 to eliminate W.

$$
\begin{array}{ll}
2L + 2W = 1000 & \xrightarrow{\text{Multiply by } -5.} \\
25L + 10W = 9500 & \xrightarrow{\text{No change}}
\end{array}
\quad
\begin{array}{rcr}
-10L - 10W &=& -5000 \\
25L + 10W &=& 9500 \\
\hline
15L &=& 4500
\end{array}
$$

$$L = \frac{4500}{15} = 300$$

To find the value of W, we back-substitute 300 for L in either of the system's equations.

$$
\begin{array}{ll}
2L + 2W = 1000 & \text{We'll use the first equation.} \\
2(300) + 2W = 1000 & \text{Back-substitute 300 for } L. \\
600 + 2W = 1000 & \text{Multiply.} \\
2W = 400 & \text{Subtract 600 from both sides.} \\
W = 200 & \text{Divide both sides by 2.}
\end{array}
$$

The lot's dimensions are 300 feet by 200 feet.
The lot's perimeter is

$$2(300) + 2(200) = 1000 \text{ feet}$$

which checks with the given conditions.

The cost of fencing along the front (the length) and the two sides (the two widths) is

$$300(\$25) + 400(\$5) = \$9500$$

which also checks with the given conditions.

b. The owner plans to fence $300 + 2(200)$ or 700 feet. With the expensive fencing, this will cost $700(\$25) = \$17,500$. Limited by a \$16,000 budget, the owner cannot use the expensive fencing on three sides of the lot. ■

Solving Problems by Using Tables to Organize Information

In Section 1.8 we saw that problems involving the uniform motion model

$d = rt$ (distance equals rate times time)

can frequently be organized using a table. Some uniform motion problems involve airplanes that fly with or against the wind, or boats that move with or against the current. These problems contain two unknowns—the speed of the plane in still air and the speed of the wind, or the speed of the boat in still water and the speed of the current. These situations are summarized in the box.

Uniform motion involving two unknown speeds

A Plane Flying With or Against the Wind

> $x = $ Plane's speed in still air
>
> $y = $ Wind's speed
>
> $x + y = $ Plane's speed moving with the wind (the wind is called the tailwind)
>
> $x - y = $ Planes speed moving against the wind (the wind is called the headwind)

A Boat Moving With or Against the Current

> $x = $ Boat's speed in still water
>
> $y = $ Current's speed
>
> $x + y = $ Boat's speed moving with the current (the boat is moving downstream)
>
> $x - y = $ Boat's speed moving against the current (the boat is moving upstream)

Travel Downstream

Current's Direction ⟶

Travel Upstream

Current's Direction ⟶

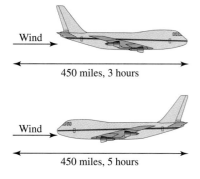

Wind ⟶

450 miles, 3 hours

Wind ⟶

450 miles, 5 hours

EXAMPLE 4   **A Uniform Motion Problem Involving Two Speeds**

When a small airplane flies with the wind, it can travel 450 miles in 3 hours. When the same airplane flies in the opposite direction against the wind, it takes 5 hours to fly the same distance. Find the speed of the plane in still air and the speed of the wind.

Solution

Let

Step 1. Use variables to represent unknown quantities.

x = Speed of plane in still air

y = Speed of wind

Step 2. Write a system of equations modeling the problem's conditions.

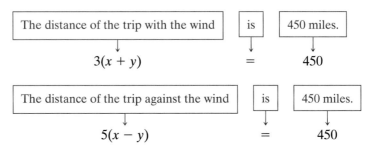

	Rate	×	**Time**	=	**Distance**
Trip with the Wind	$x + y$		3		$3(x + y)$
Trip against the Wind	$x - y$		5		$5(x - y)$

The distance of the trip with the wind	is	450 miles.
↓	↓	↓
$3(x + y)$	=	450

The distance of the trip against the wind	is	450 miles.
↓	↓	↓
$5(x - y)$	=	450

Step 3. Solve the system and answer the problem's question.

We can simplify the system that models the given conditions by dividing both sides of the equations by 3 and 5, respectively. Then we solve the resulting system by addition.

$$x + y = 150 \quad \text{This is the first equation with both sides divided by 3.}$$
$$\underline{x - y = 90} \quad \text{This is the second equation with both sides divided by 5.}$$
$$\text{Add:} \quad 2x \phantom{{}- y} = 240$$
$$x = 120$$

Back-substituting 120 for x in either of the system's equations gives $y = 30$. Since $x = 120$ and $y = 30$, the speed of the plane in still air is 120 miles per hour and the speed of the wind is 30 miles per hour.

Step 4. Check the proposed answers in the original wording of the problem.

The speed of the plane with the wind is $120 + 30 = 150$ miles per hour. In 3 hours, it travels $150(3) = 450$ miles, which checks with the given conditions. Furthermore, the speed of the plane against the wind is $120 - 30 = 90$ miles per hour. In 5 hours, it travels $90(5) = 450$ miles, the same distance as before. ■

Chemists and pharmacists often have to change the concentration of solutions and other mixtures. In these situations, the amount of a particular ingredient in the solution or mixture is expressed as a percentage of the total. For example, a 40-millimeter solution of acid in water containing 35% acid contains 35% of $40 = 0.35(40) = 14$ millimeters of acid.

Problems involving mixtures of two solutions, to strengthen or dilute a particular substance in the mixture, can be solved by using tables to organize information.

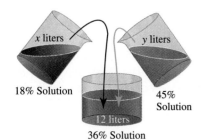

x liters

y liters

18% Solution

45% Solution

12 liters

36% Solution

EXAMPLE 5 **A Solution-Mixture Problem**

A chemist needs to mix an 18% acid solution with a 45% acid solution to obtain a 12-liter mixture consisting of 36% acid. How many liters of each of the acid solutions must be used?

Solution

Step 1. Use variables to represent unknown quantities.

Let x represent the number of liters of the 18% acid solution to be used in the mixture, and let y represent the number of liters of the 45% acid solution to be used in the mixture.

Step 2. Write a system of equations modeling the problem's conditions.

A table simplifies the problem's solution:

	Number of Liters	$\cdot$	Percentage of Acid	$=$	Amount of Acid
18% Acid Solution	x		18% (0.18)		0.18x
45% Acid Solution	y		45% (0.45)		0.45y
36% Acid Solution	12		36% (0.36)		0.36(12)

The chemist needs to obtain a 12-liter mixture.

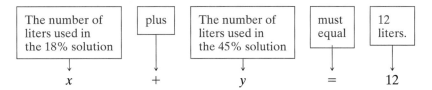

The amount of acid in the mixture must be 36% of 12, or 0.36(12) = 4.32 liters.

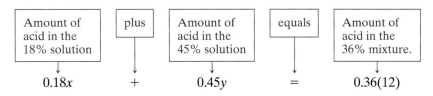

Step 3. Solve the system and answer the problem's question.

After multiplying the second equation by 100 and clearing decimals, we have the chemist's conditions modeled by the system

$$x + y = 12$$
$$18x + 45y = 432$$

The system can be solved by substitution or addition. We'll use addition, multiplying the first equation by -18 to eliminate x.

$$
\begin{array}{l}
x + y = 12 \quad \xrightarrow{\text{Multiply by } -18.} \quad -18x - 18y = -216 \\
18x + 45y = 432 \quad \xrightarrow{\text{No change}} \quad \underline{18x + 45y = 432} \\
\phantom{18x + 45y = 432 \quad \text{Add:}} \qquad\qquad 27y = 216 \\
\phantom{18x + 45y = 432 \quad \text{Add:}} \qquad\qquad\quad y = \dfrac{216}{27} = 8
\end{array}
$$

Back-substituting 8 for y in either of the system's equations gives $x = 4$.

Since $x = 4$ and $y = 8$, the chemist should mix 4 liters of the 18% acid solution with 8 liters of the 45% acid solution.

Step 4. Check the proposed answers in the original wording of the problem.

Since $4 + 8 = 12$, the chemist is indeed obtaining a 12-liter mixture, as specified by the problem's conditions. Now let's see if the 12-liter mixture is 36% acid. 36% of 12 is 4.32, so the mixture must contain 4.32 liters of acid.

$$\text{Amount of acid in the 18\% solution} = 0.18(4) = 0.72 \text{ liters}$$
$$\text{Amount of acid in the 45\% solution} = 0.45(8) = \underline{3.6 \text{ liters}}$$
$$\text{Amount of acid in mixture} = 4.32 \text{ liters} \quad \blacksquare$$

2 Model data using systems of equations.

Modeling Data Using Systems of Equations

An important economic application of systems of equations arises in connection with the *law of supply and demand.* The next two examples involve the creation of supply and demand models.

EXAMPLE 6 Modeling Demand

The quantity of a product that consumers purchase depends on its price, with higher prices leading to fewer sales. The table below shows the price of a video and the quantity of that video sold on a weekly basis in a store.

x (Price of the Video)	y (Number of Videos Sold Weekly)
$18	526
$25	435

Assuming that as price increases, the number sold weekly decreases steadily, use the linear function $y = mx + b$ to find the values of m and b, thereby modeling video demand.

Solution

Although this problem can be solved using the point-slope form of a line (as we did in the previous chapter), we will begin instead with the slope-intercept equation.

$$y = mx + b \qquad \text{This is the slope-intercept equation of a line.}$$
$$526 = m\,(18) + b \qquad \text{The table shows that when } x = 18, y = 526. \text{ Substitute these values for } x \text{ and } y.$$
$$435 = m\,(25) + b \qquad \text{The table also shows that when } x = 25, y = 435. \text{ Substitute these values for } x \text{ and } y.$$

We obtain a linear system in m and b.

$$18m + b = 526$$
$$25m + b = 435$$

We solve for m and b using the addition method.

$$
\begin{array}{l}
18m + b = 526 \\
25m + b = 435
\end{array}
\quad
\begin{array}{c}
\xrightarrow{\text{No change}} \\
\xrightarrow{\text{Multiply by } -1.}
\end{array}
\quad
\begin{array}{l}
18m + b = 526 \\
\underline{-25m - b = -435} \\
\text{Add:} \quad -7m = 91 \\
m = -13 \quad \text{Divide both sides by } -7.
\end{array}
$$

Now we find the value of b by back-substituting -13 for m in either of the system's equations.

$$18m + b = 526 \qquad \text{This is the first equation in the linear system.}$$
$$18(-13) + b = 526 \qquad \text{Substitute } -13 \text{ for } m.$$

$$-234 + b = 526 \quad \text{Multiply.}$$
$$b = 760 \quad \text{Add 234 to both sides.}$$

Substituting -13 for m and 760 for b in $y = mx + b$, the linear equation that fits the data is

$$y = -13x + 760.$$

Observe that the slope is -13. In the context of the data, this means that for every \$1 increase in the price of the video, 13 fewer videos will be sold by the store on a weekly basis. ■

Example 6 illustrates that as price goes up, demand comes down. The quantity of a product that a supplier is willing to produce also depends on price. The higher the price, the greater the supply. Mathematical models for supply and demand form the basis of Example 7.

EXAMPLE 7   Supply and Demand Models

The table below shows the price of the same video as in Example 6 and the quantity that the video manufacturer is willing to supply to the store on a weekly basis at that price.

x (Price of the Video)	y (Number of Videos Supplied by the Manufacturer Weekly)
\$18	466
\$25	480

a. Assuming that as price increases, the number of videos supplied weekly increases steadily, use the linear function $y = mx + b$ to find the values of m and b, thereby modeling video supply.

b. The point of intersection of the graphs for the supply and demand models is called the *point of market equilibrium*. The x-coordinate of this point is the *equilibrium price*. At this price, the amount that the seller will supply is the same amount that the consumer will buy. Find the point of market equilibrium for the video supply and demand models.

Solution

a. We will use the procedure of Example 6. We begin with $y = mx + b$ and substitute the values in the table.

$$466 = m(18) + b$$
$$480 = m(25) + b$$

Solving this system results in $m = 2$ and $b = 430$. The video supply model is

$$y = 2x + 430.$$

Observe that the slope is 2. This means that for every \$1 increase in the price of the video, suppliers will be willing to furnish 2 more videos on a weekly basis.

b. Here are the supply and the demand models.

$$y = -13x + 760 \qquad \text{This is the demand model from Example 6.}$$
$$y = 2x + 430 \qquad \text{This is the supply model that we derived in part (a).}$$

Supply and demand are equal at the equilibrium point. We can solve the demand-supply linear system by the substitution method. We substitute $-13x + 760$ for y in the second equation.

$$y = -13x + 760 \qquad y = 2x + 430$$

$$-13x + 760 = 2x + 430 \qquad \text{The resulting equation contains only one variable.}$$
$$-15x + 760 = 430 \qquad \text{Subtract } 2x \text{ from both sides.}$$
$$-15x = -330 \qquad \text{Subtract 760 from both sides.}$$
$$x = 22 \qquad \text{Divide both sides by } -15.$$

To find the value for y, we back-substitute 22 for x into either the demand or the supply model.

$$y = 2x + 430 \qquad \text{This is the supply model, the system's second equation.}$$
$$= 2(22) + 430 \qquad \text{Substitute 22 for } x.$$
$$= 474 \qquad \text{Simplify.}$$

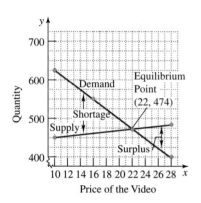

Figure 3.7

The equilibrium point, shown in Figure 3.7, is (22, 474). The equilibrium price is \$22 per video. This is the price at which supply equals demand, that is, the market price at which there is neither a surplus nor a shortage of the video. At a price of \$22, 474 units of the video will be supplied and sold weekly. ■

PROBLEM SET 3.2

Practice and Application Problems _____

Solve Problems 1–6 by translating the given conditions into a system of equations.

1. Two numbers are unknown. The second number subtracted from 2 times the first number is 9. The sum of 2 times the first and 3 times the second is -3. Find the numbers.

2. Two numbers are unknown. The difference between the first and the second is 0. The difference between 4 times the first and 3 times the second is 15. Find the numbers.

3. Cake can vary in cholesterol content. Four slices of sponge cake and 2 slices of pound cake contain 784 mg of cholesterol. One slice of sponge cake and 3 slices of pound cake contain 366 mg of cholesterol. Find the cholesterol content in each item.

4. How do the Quarter Pounder and Whopper with cheese discussed in Example 1 measure up in the calo-rie department? Actually, not too well. Two Quarter Pounders and 3 Whoppers with cheese provide 2607 calories. Even one of each provide enough calories to bring tears to Jenny Craig's eyes—9 calories in excess of what is allowed on a 1000 calorie-a-day diet. Find the caloric content in each item.

Problems 5 and 6 involve the digits of a two-digit number. If t represents the tens digit and u the units digit, the number is represented by 10t + u.

5. The sum of the digits of a two-digit number is 9. When the digits are reversed, the resulting number is 45 more than the original number. Find the number.

6. The sum of the digits of a two-digit number is 14. When the digits are reversed, the resulting number is 23 less than 2 times the original number. What is the number?

Solve Problems 7–12 by creating verbal models and then translating into a system of linear equations.

7. A department that hires tutors and graders paid a to-tal of $622.50 for 70 hours of tutoring and 50 hours of grading. The following month a total of $719 was paid for 90 hours of tutoring and 40 hours of grading. Find the hourly wage for the tutors and the graders.

8. A theater has an orchestra section and a mezzanine section. At one performance, the sale of 95 orchestra seats and 75 mezzanine seats generated a revenue of $4975. At a later performance, the sale of 60 orchestra seats and 135 mezzanine seats brought in $5070 for the theater. Find the cost for a theater ticket in the or-chestra and in the mezzanine.

9. A job in sales pays a fixed weekly salary plus a com-mission based on a percentage of sales. An applicant is told that on sales of $2000, total take-home pay is $900, and on sales of $5000, total take-home pay is $1350. What is the weekly salary and the commission rate for this sales job?

10. A car rental agency charges a daily fee plus a mileage fee on its compact cars. One customer was charged $150 for 3 days and 100 miles, and a second was charged $375 for 7 days and 400 miles. What is the daily fee and the mileage fee for compact cars?

11. A total of $1,300,000 in government funding is avail-able to divide between day-care centers, which cost $25,000 per center, and road repairs, which cost $5000 per mile. Budget allocations require that the ratio of the number of miles of road repairs to the number of day-care centers that will receive government funding must be $\frac{50}{3}$. How many day-care centers and how many road-miles will benefit from the available money?

12. One apartment is directly above a second apartment. The resident living downstairs calls his neighbor living above him and states, "If one of you is willing to come downstairs, we'll have the same number of people in both apartments." The upstairs resident responds, "We're all too tired to move. Why don't one of you come up here? Then we'll have twice as many people up here as you've got down there." How many people are in each apartment?

Most problems in geometry require a knowledge of formulas to obtain a solution. Solve Problems 13–24 using one or more of the following ideas.

- The perimeter of a rectangle is given by $P = 2L + 2W$.
- An isosceles triangle has two sides that are equal in measure. The angles opposite these sides are equal in measure.
- The sum of the measures of the angles of a triangle is 180°.
- Vertical angles have equal measure. (See figure on the right.)
- If two parallel lines are intersected by a third line, then every angle pair that is formed contains angles that are equal in measure, or angles whose measures have a sum of 180°.
- Consecutive angles of a parallelogram are supplementary (have a sum of 180°).
- Opposite angles of a parallelogram have the same measure.

13. A rectangular lot whose perimeter is 360 feet is fenced along three sides. An expensive fencing along the lot's length costs $20 per foot, and an inexpensive fencing along the two side widths costs only $8 per foot. The total cost of the fencing along the three sides comes to $3280. What are the lot's dimensions?

14. A rectangular lot whose perimeter is 320 feet is fenced along three sides. An expensive fencing along the lot's length costs $16 per foot, and an inexpensive fencing along the two side widths costs only $5 per foot. The total cost of the fencing along the three sides comes to $2140. What are the lot's dimensions?

15. In the figure, parallelogram $ABCD$ has a perimeter of 50 meters, and trapezoid $AECD$ has a perimeter of 39 meters. Triangle EBC is isosceles with EB and BC having equal measure. Furthermore, AE and EC are equal in measure. Find the lengths of AE, EB, and DC.

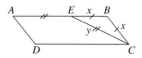

16. The perimeter of the larger rectangle in the figure is 58 meters. The combined lengths of the three sides of the smaller rectangle, excluding the side that it shares with a portion of the side of the larger rectangle, is 17.5 meters. Find x and y.

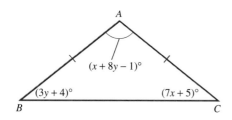

17. In the isosceles triangle shown in the figure, $AB = AC$. Find the measure of each angle in the triangle.

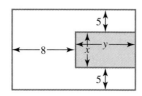

18. In the isosceles triangle shown in the figure, $AB = AC$. Find the measure of each angle in the triangle.

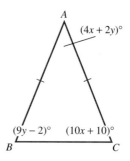

19. Find the measures of the angles marked x and y in the figure.

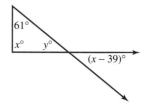

20. Find the measures of the angles marked x and y in the figure.

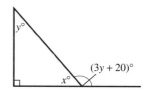

21. Find the measures of angles A, B, and C in the figure if it is known that lines L_1 and L_2 are parallel.

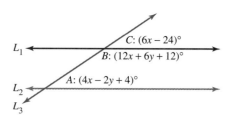

22. Find x and y in the figure, and then find the measure of each of the three angles. in which variables appear.

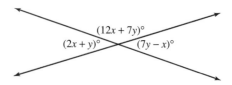

23. Find the measure of each interior angle in the parallelogram shown in the figure.

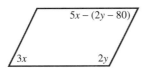

24. Find the measures of the angles marked x and y in the figure.

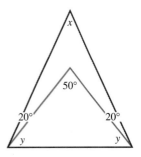

Solve Problems 25–30 by using tables to organize the information.

25. When an airplane flies with the wind, it can travel 800 miles in 4 hours. When the same airplane flies in the opposite direction against the wind, it takes 5 hours to fly the same distance. Find the speed of the plane in still air and the speed of the wind.

26. A motorboat traveling with the current can cover 120 miles in 4 hours. When the same boat travels in the opposite direction against the current, it takes 6 hours to travel the 120 miles. Find the speed of the boat in calm water and the speed of the current.

27. Solution A is 30% acid, and solution B is 60% acid. How many gallons of each solution must be used to create a 60-gallon mixture that is 50% acid?

28. Type A cream is 18% butterfat, and type B cream is 24% butterfat. How many quarts of each type of cream must be used to create a 90-quart mixture that is 22% butterfat?

29. A high school has 900 students, 40% of whom are Hispanic. A second high school has a 75% Hispanic population. The school board plans to merge the two schools into one school that will then have a 52.5% Hispanic student body. How many students are at the second school?

30. A school board plans to merge two schools into one school of 1000 students in which 42% of the students will be African-American. One of the schools has a 10% African-American student body and the other has a 90% African-American student body. What is the student population in each of the two schools?

Solve Problems 31–34 by modeling the data with a linear system of equations.

31. The table shows the average price of a house in the United States for two indicated years.

x (Years after 1980)	y (Average Price of a House)
(1988) 8	$90,597
(1995) 15	$110,428

a. Assuming a linear relationship between variables, use the linear function $y = mx + b$ to find the values of m and b, thereby modeling the data.
b. Describe what the number representing slope means in the context of the data.
c. Use your model to predict the average price of a house in the year 2000.

32. The table shows enrollment in U.S. public schools for two indicated years.

x (Years after 1985)	y (Enrollment, in Millions)
(1988) 3	40.35
(1992) 7	42.15

a. Assuming a linear relationship between variables, use the linear function $y = mx + b$ to find the values of m and b, thereby modeling the data.
b. Describe what the number representing slope means in the context of the data.
c. Use the model to determine in which year enrollment will reach 48 million students.

33. a. The table shows the price of an overhead fan and the quantity of that fan sold on a weekly basis in a chain of discount stores.

x (Price of the Fan)	y (Number of Fans Sold Weekly)
$40	6500
$60	6000

Assuming that as price increases, the number sold weekly decreases steadily, use the linear function $y = mx + b$ to find the values of m and b, thereby modeling demand for the fans.

b. The table shows the price of the same fan and the quantity that the manufacturer is willing to supply to the discount chain on a weekly basis at that price.

x (Price of the Fan)	y (Number of Fans Supplied by the Manufacturer Weekly)
$40	6200
$60	6300

Assuming that as price increases, the number of fans supplied weekly increases steadily, use the linear function $y = mx + b$ to find the values of m and b, thereby modeling fan supply.

c. Use the demand and the supply functions to find the point of market equilibrium. Describe exactly what the x- and y-coordinates of this point represent.

d. Graph the demand and the supply functions, visually illustrating your solution in part (c). If applicable, use a graphing utility to verify the graphs.

34. a. The table shows the price of an umbrella and the quantity of that umbrella sold on a weekly basis in a chain of discount stores.

x (Price of the Umbrella)	y (Number of Umbrellas Sold Weekly)
$8	1176
$12	964

Assuming that as price increases, the number sold weekly decreases steadily, use the linear function $y = mx + b$ to find the values of m and b, thereby modeling demand for the umbrellas.

b. The table shows the price of the same umbrella and the quantity that the manufacturer is willing to supply to the discount chain on a weekly basis at that price.

x (Price of the Umbrella)	y (Number of Umbrellas Supplied by the Manufacturer Weekly)
$8	920
$12	1220

Assuming that as price increases, the number of umbrellas supplied weekly increases steadily, use the linear function $y = mx + b$ to find the values of m and b, thereby modeling umbrella supply.

c. Use the demand and supply functions to find the point of market equilibrium. Describe exactly what the x- and y-coordinates of this point represent.

d. Graph the demand and supply functions, visually illustrating your solution in part (c). If applicable, use a graphing utility to verify the graphs.

Technology Problem _____

35. The demand and the supply functions for a certain commodity are given by

$$D(p) = 410 - p$$
$$S(p) = p^2 + 3p - 70$$

Use a graphing utility to graph both functions in the same viewing rectangle. Then use the $\boxed{\text{TRACE}}$ or intersection feature to complete this statement: The market for this product is at equilibrium when the unit price is $____ , and the market is in equilibrium when ____ units are supplied and demanded.

Critical Thinking Problems _____

36. The number of people in a small discussion group never changed. When they came for Wednesday evening discussions, they always sat in the same seat, on either the right or the left side of the room.

One Wednesday evening, Jane changed her mind. She had always sat on the left side, but now she wanted to sit on the right. With this change, an equal number of people were sitting on each side of the room.

The following Wednesday, Jane moved back to where she always sat. But then Ted, who always sat on the right side, moved over to the left. With this move, there were twice as many people on the left side as on the right.

How many people normally sat on the right and how many on the left?

37. A dealer paid a total of $67 for mangos and avocados. The mangos were sold at a profit of 20% on the dealer's cost, but the avocados started to spoil, resulting in a selling price of a 2% loss on the dealer's cost. The dealer made a profit of $8.56 on the total transaction. How much did the dealer pay for the mangos and the avocados?

38. A group of people agree to share in the cost of buying a boat. If 10 of them later decide not to buy in, each of those remaining would have to pay $1 more. If 25 of them decide not to buy in, each of those remaining would have to pay $3 more. How many people originally agreed to buy the boat?

39. A teacher's assistant had an income of $9800 before taxes. The assistant found that federal taxes were 20% of his income after state taxes were deducted and that state taxes were 10% of his income after federal taxes were deduced. How much federal tax and state tax did this person pay?

40. A boy has as many brothers as he has sisters. His sister has twice as many brothers as she has sisters. How many boys and girls are in this family?

41. Traveling at uniform rates of A and B mph, two people head directly toward each other across a distance of 120 miles. If both start at 10:30 A.M., they will meet at noon. If the person traveling at A mph starts at 10 A.M. and the other person starts at 11 A.M., they will also meet at noon. Find each person's rate.

42. A marching band has 52 members, and there are 24 in the pom-pom squad. They wish to form several hexagons and squares like those diagrammed below. Can it be done with no people left over?

Hexagon with pom–pom person in center

Square with band member in center

B B

B P B

B B

P P

B

P P

B = Band Member

P = Pom–pom Person

Review Problems

43. Solve $|3x + 5| \geq 2$ and graph the solution set on a number line.

44. Solve: $\dfrac{x}{2} - \dfrac{x - 4}{5} = \dfrac{23}{10}$.

45. Fill in the table of coordinates and use point-plotting to graph $f(x) = -x^2 - x + 2$.

x	-3	-2	-1	$-\frac{1}{2}$	0	1	2
$f(x)$							

Solutions Manual   **Tutorial**   **Video 4**

SECTION 3.3

Linear Systems of Equations in Three Variables

Objectives

1 Verify that an ordered triple is a solution of a linear system in three variables.
2 Solve linear systems in three variables.
3 Identify inconsistent and dependent systems.
4 Solve word problems using linear systems in three variables.

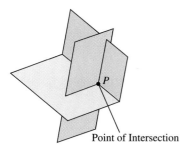

P

Point of Intersection

Figure 3.8

An equation such as $x + 2y - 3z = 9$ is called a *linear equation in three variables.* In general, any equation of the form

$$Ax + By + Cz = D$$

where $A, B, C,$ and D are real numbers such that $A, B,$ and C are not all 0, is a linear equation in the variables $x, y,$ and z. The graph of this linear equation in three variables is a plane in three-dimensional space.

The process of solving a system of three linear equations in three variables is geometrically equivalent to finding the point of intersection (assuming that there is one) of three planes in space (see Figure 3.8). In this section, we will use the addition method to find the three coordinates of this intersection point.

Verify that an ordered triple is a solution of a linear system in three variables.

Identifying Solutions

A *solution* of a system of three linear equations in three variables is an ordered triple of real numbers that satisfies all equations of the system. The *solution set* of the system is the set of all its solutions.

EXAMPLE 1 **Determining Whether an Ordered Triple Satisfies a System**

Show that the ordered triple $(-1, 2, -2)$ is a solution of the system:

$$x + 2y - 3z = 9$$
$$2x - y + 2z = -8$$
$$-x + 3y - 4z = 15$$

Solution

We must show that the numbers in $(-1, 2, -2)$ satisfy all three equations when used as replacements for $x, y,$ and z, respectively.

$$
\begin{aligned}
x + 2y - 3z &= 9 \\
-1 + 2(2) - 3(-2) &\overset{?}{=} 9 \\
-1 + 4 + 6 &\overset{?}{=} 9 \\
9 &= 9 \quad \checkmark
\end{aligned}
\qquad
\begin{aligned}
2x - y + 2z &= -8 \\
2(-1) - 2 + 2(-2) &\overset{?}{=} -8 \\
-2 - 2 - 4 &\overset{?}{=} -8 \\
-8 &= -8 \quad \checkmark
\end{aligned}
\qquad
\begin{aligned}
-x + 3y - 4z &= 15 \\
-(-1) + 3(2) - 4(-2) &\overset{?}{=} 15 \\
1 + 6 + 8 &\overset{?}{=} 15 \\
15 &= 15 \quad \checkmark
\end{aligned}
$$

The ordered triple $(-1, 2, -2)$ makes all three equations true, so it is a solution. The solution set to the system is $\{(-1, 2, -2)\}$. ∎

2 Solve linear systems in three variables.

Solving Linear Systems in Three Variables

The method for solving a system of linear equations in three variables is similar to that used on systems of linear equations in two variables. We use addition to eliminate any variable, reducing the system to two equations in two variables. Once we obtain a system of two equations in two variables, we use the addition method again to eliminate a variable. The result is a single equation in one variable that we solve to get the value of that variable. Other variable values are found by back-substitution.

This procedure is illustrated in our next example.

EXAMPLE 2 **Solving a System in Three Variables**

Solve the system:

$$
\begin{aligned}
x + 2y - 3z &= 9 && \text{Equation 1} \\
2x - y + 2z &= -8 && \text{Equation 2} \\
-x + 3y - 4z &= 15 && \text{Equation 3}
\end{aligned}
$$

ENRICHMENT ESSAY

Mathematical Blossom

This picture is the graph of

$$z = (|x| - |y|)^2 + \frac{2|xy|}{\sqrt{x^2 + y^2}}$$

in a three-dimensional Cartesian coordinate system. For many, the picture is more interesting than the compact symbolic notation of the printed equation. Students of higher mathematics often get the "feel" of the equation by seeing the picture. The transformation of pure logic into visual images hints at some of the beauty that is derivable from mathematics.

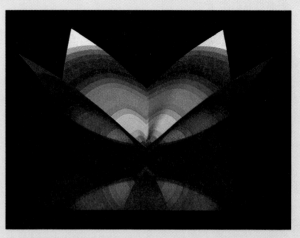

Graph of an equation in a three-dimensional Cartesian plane, Melvin L. Prueitt, Los Alamos National Laboratory

Solution

There are many ways to proceed. Because our initial goal is to reduce the system to two equations in two variables, *the central idea is to take two different pairs of equations and eliminate the same variable from each pair.*

Step 1. Reduce the system to two equations in two variables.

Choose any two equations and using the addition method eliminate a variable. Let's eliminate x by adding Equations 1 and 3.

$$
\begin{array}{rll}
x + 2y - 3z = & 9 & \text{Equation 1} \\
-x + 3y - 4z = & 15 & \text{Equation 3} \\
\hline
\text{Add:} \quad 5y - 7z = & 24 & \text{Equation 4}
\end{array}
$$

Thus, we use Equation 4 in place of Equations 1 and 3. Next we eliminate the *same* variable from two other equations. We can multiply equation 3 by 2 and add this to Equation 2.

$$
\begin{array}{rll}
-2x + 6y - 8z = & 30 & \text{Twice Equation 3} \\
2x - y + 2z = & -8 & \text{Equation 2} \\
\hline
\text{Add:} \quad 5y - 6z = & 22 & \text{Equation 5}
\end{array}
$$

Step 2. Solve the resulting system of two equations in two variables.

Thus, we use Equation 5 in place of Equations 2 and 3. Now we solve the resulting system of Equations 4 and 5 for y and z. To begin, we'll multiply Equation 5 on both sides by -1 and then add this equation to Equation 4.

$$
\begin{array}{lll}
5y - 7z = 24 & \xrightarrow{\text{No change}} & 5y - 7z = 24 \\
5y - 6z = 22 & \xrightarrow{\text{Multiply by } -1.} & -5y + 6z = -22 \\
& \text{Add:} & \overline{-z = 2} \\
& & z = -2
\end{array}
$$

Step 3. Use back-substitution in one of the equations in two variables to find the value of the second variable.

We back-substitute -2 for z in either Equation 4 or 5 to find the value of y.

$$5y - 7z = 24 \qquad \text{Equation 4}$$
$$5y - 7(-2) = 24 \qquad \text{Substitute } -2 \text{ for } z.$$
$$5y = 10 \qquad \text{Multiply and subtract 14 from both sides.}$$
$$y = 2 \qquad \text{Divide both sides by 5.}$$

Step 4. Back-substitute the values found for two variables into one of the original equations to find the value of the third variable.

We can use any one of the original equations and back-substitute the values of y and z to find the other variable. We will use Equation 1.

$$x + 2y - 3z = 9 \qquad \text{Equation 1}$$
$$x + 2(2) - 3(-2) = 9 \qquad \text{Substitute 2 for } y \text{ and } -2 \text{ for } z.$$
$$x + 10 = 9 \qquad \text{Multiply and then add.}$$
$$x = -1 \qquad \text{Subtract 10 from both sides.}$$

Step 5. Check the proposed solution in each of the original equations.

The solution is the ordered triple $(-1, 2, -2)$. The solution should be checked in all three equations. This was done in Example 1, so we know that the solution set is $\{(-1, 2, -2)\}$. ■

A number of approaches can be used to solve systems involving three variables. We are first faced with three options regarding which variable to eliminate. Then we must choose which equations to use to eliminate the desired variable. Keep in mind that the initial goal is to reduce the original system to one involving two equations in two variables.

In summary, a system of three linear equations in three variables can be solved by the addition method as follows.

> **Solving linear systems in three variables by addition**
>
> 1. Reduce the system to two equations in two variables. This is usually accomplished by taking two different pairs of equations and using the addition method to eliminate the same variable from each pair.
> 2. Solve the resulting system of two equations in two variables using addition or substitution. The result is an equation in one variable that gives the value of that variable.
> 3. Back-substitute the value of the variable found in step 2 into either of the equations in two variables to find the value of the second variable.
> 4. Use the values of the two variables from steps 2 and 3 to find the value of the third variable by back-substituting into one of the original equations.
> 5. Check the proposed solution in each of the original equations.

In some examples, one of the variables is already eliminated from an original equation. In this case, the same variable should be eliminated from the other two equations, thereby making it possible to omit one of the elimination steps. We illustrate this idea in Example 3.

| EXAMPLE 3 | **Solving a System of Equations with a Missing Term** |

Solve the system:

$$x + \quad\ z = 8 \quad \text{Equation 1}$$
$$x + \ y + 2z = 17 \quad \text{Equation 2}$$
$$x + 2y + \ z = 16 \quad \text{Equation 3}$$

Solution

Step 1. Reduce the system to two equations in two variables.

Because the first equation contains only x and z, we could eliminate y from Equations 2 and 3, resulting in two equations in two variables. To do so, we multiply Equation 2 by -2 and add Equation 3.

$$x + y + 2z = 17 \xrightarrow{\text{Multiply by } -2.} -2x - 2y - 4z = -34$$
$$x + 2y + z = 16 \xrightarrow{\text{No change}} \underline{\ x + 2y + \ z = \ \ 16}$$
$$\text{Add:} \quad -x \qquad -3z = -18$$

We have now reduced the system to

$$x + \ z = \quad 8$$
$$-x - 3z = -18 \quad \text{Equation 4}$$

Step 2. Solve the resulting system.

We now solve the system of two equations in two variables for x and z.

$$x + \ z = \quad 8 \quad \text{Equation 1}$$
$$\underline{-x - 3z = -18} \quad \text{Equation 4}$$
$$\text{Add:} \quad -2z = -10$$
$$z = \quad 5 \quad \text{Divide both sides by } -2.$$

Step 3. Use back-substitution to find the value of the second variable.

To find x, we back-substitute 5 for z in either Equation 1 or 4. Choosing Equation 1 gives

$$x + z = 8 \quad \text{Equation 1}$$
$$x + 5 = 8 \quad \text{Substitute 5 for } z.$$
$$x = 3 \quad \text{Subtract 5 from both sides.}$$

Step 4. Back-substitute into an original equation to find the value of the third variable.

To find y, we back-substitute 3 for x and 5 for z into Equation 2 or 3.

$$x + y + 2z = 17 \quad \text{Equation 2}$$
$$3 + y + 2(5) = 17 \quad \text{Substitute 3 for } x \text{ and 5 for } z.$$
$$y + 13 = 17 \quad \text{Multiply and add.}$$
$$y = 4 \quad \text{Subtract 13 from both sides.}$$

Discover for yourself

Why can't we use Equation 1?

The solution $(3, 4, 5)$ should be checked in each of the original equations of the system to verify that the solution set of the system is $\{(3, 4, 5)\}$. ■

3 Identify inconsistent and dependent systems.

Inconsistent and Dependent Systems

Because the graph of $Ax + By + Cz = D$ is a plane, three planes represented by a system need not intersect at one point. Figure 3.9 on page 360 illustrates that three planes may have no common point of intersection.

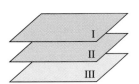

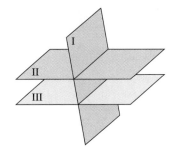

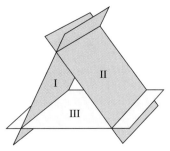

These three parallel planes have no common intersection point.

Two of the three planes are parallel with no common intersection point.

Planes intersect two at a time with no common intersection point.

Figure 3.9

Three planes may have no common point of intersection.

In each of the cases shown in Figure 3.9, the solution set for the system is the empty set, ∅, and the system is inconsistent. If at any point in the algebraic solution all variables are eliminated and a false statement (such as $0 = -10$) is obtained, this indicates that the system has no solution. This idea is illustrated in Example 4.

EXAMPLE 4   **Solving an Inconsistent System with Three Variables**

Solve the system:

$$2x + 5y + z = 12 \quad \text{Equation 1}$$
$$x - 2y + 4z = -10 \quad \text{Equation 2}$$
$$-3x + 6y - 12z = 20 \quad \text{Equation 3}$$

Solution

We begin by reducing the system to two equations in two variables. Let's agree to eliminate x. We'll do this in Equations 1 and 2 and then repeat the process for Equations 2 and 3. To eliminate x in Equations 1 and 2, we multiply both sides of Equation 2 by -2 and add the result to Equation 1.

$$
\begin{array}{ll}
2x + 5y + z = 12 & \xrightarrow{\text{No change}} \\
x - 2y + 4z = -10 & \xrightarrow{\text{Multiply by } -2.}
\end{array}
\qquad
\begin{array}{l}
2x + 5y + z = 12 \\
\underline{-2x + 4y - 8z = 20} \\
9y - 7z = 32
\end{array}
$$

Add:

Now, to eliminate x in Equations 2 and 3, we multiply both sides of Equation 2 by 3 and add the result to Equation 3.

$$
\begin{array}{ll}
x - 2y + 4z = -10 & \xrightarrow{\text{Multiply by 3.}} \\
-3x + 6y - 12z = 20 & \xrightarrow{\text{No change}}
\end{array}
\qquad
\begin{array}{l}
3x - 6y + 12z = -30 \\
\underline{-3x + 6y - 12z = 20} \\
0 = -10
\end{array}
$$

Add:

There are no values of x, y, and z for which $0 = -10$.

When solving a system of three linear equations in three variables, if a false statement occurs in any step, you can stop the solution procedure. The false statement means that two of the system's equations have graphs that are parallel planes. The solution set to the inconsistent system is $\varnothing$.

The false statement $0 = -10$ indicates that the planes represented by Equations 2 and 3 are parallel. Thus, the three planes represented by the given system can have no common intersection point. The system is inconsistent and the solution set is $\varnothing$. ∎

A system of linear equations in three variables may have infinitely many solutions, as illustrated in Figure 3.10. In each of these cases, the infinite set of solutions for the system indicates that the system includes dependent equations. Algebraic solution methods will yield a true statement such as $0 = 0$. This idea emerges in Example 5.

EXAMPLE 5 **Solving a System of Dependent Equations with Three Variables**

Solve the system

$$3x - 4y + 4z = 7 \quad \text{Equation 1}$$
$$x - y - 2z = 2 \quad \text{Equation 2}$$
$$2x - 3y + 6z = 5 \quad \text{Equation 3}$$

Solution

We begin by reducing the system to two equations in two variables. We will eliminate x, first from Equations 1 and 2, and then from Equations 2 and 3. We begin with Equations 1 and 2.

$$3x - 4y + 4z = 7 \xrightarrow{\text{No change}} 3x - 4y + 4z = 7$$
$$x - y - 2z = 2 \xrightarrow{\text{Multiply by } -3.} \underline{-3x + 3y + 6z = -6}$$
$$\text{Add:} \quad -y + 10z = 1$$

Now we repeat this process using Equations 2 and 3.

$$x - y - 2z = 2 \xrightarrow{\text{Multiply by } -2.} -2x + 2y + 4z = -4$$
$$2x - 3y + 6z = 5 \xrightarrow{\text{No change}} \underline{2x - 3y + 6z = 5}$$
$$\text{Add:} \quad -y + 10z = 1$$

Three planes coincide with infinitely many common points.

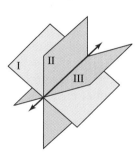

The three planes intersect along a line and have infinitely many common points.

Figure 3.10

Three planes may have infinitely many common points of intersection.

Did you notice that both equations obtained above are identical? This indicates that the three planes intersect along a line whose equation is $-y + 10z = 1$. Let's say that you didn't notice this, and you continued with the addition method on the system in two variables. Let's eliminate y.

$$-y + 10z = 1 \xrightarrow{\text{Multiply by } -1.} y - 10z = -1$$
$$-y + 10z = 1 \xrightarrow{\text{No change}} \underline{-y + 10z = 1}$$
$$\text{Add:} \quad 0 = 0$$

The true statement $0 = 0$ punctuates the fact that we have a system with infinitely many solutions. In writing the answer to this problem, we simply state that the equations are dependent. ∎

Problem Solving

4 Solve word problems using linear systems in three variables.

We now consider word problems that can be solved using a system of three equations in three variables.

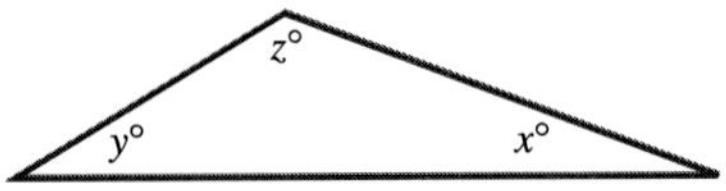

Figure 3.11

Step 1. Use variables to represent the unknown quantities.

Step 2. Write a system of equations modeling the problem's conditions.

Step 3. Solve the system and answer the problem's question.

EXAMPLE 6 **Solving a Geometry Problem**

The largest angle in a triangle measures 93° more than the smallest angle. The measure of the largest angle exceeds three times the measure of the remaining angle by 7°. Find the measure of the triangle's three interior angles.

Solution

Let x represent the measure of the smallest angle, z the largest angle, and y the remaining angle (see Figure 3.11).

The sum of the measures of the three interior angles of a triangle is 180°.

$$x + y + z = 180 \quad \text{Equation 1}$$

The largest angle measures 93° more than the smallest.

$$z = x + 93 \quad \text{Equation 2}$$

The largest angle exceeds three times the measure of the remaining angle by 7°.

$$z = 3y + 7 \quad \text{Equation 3}$$

We now have a system of three equations.

$$
\begin{array}{lll}
x + y + z = 180 & \quad & x + y + z = 180 \\
z = x + 93 \quad \text{or} & & -x \quad\quad + z = 93 \\
z = 3y + 7 & & -3y + z = 7
\end{array}
$$

Write each equation in the form $Ax + By + Cz = D$.

Because the third equation contains only y and z, let's obtain another equation in y and z.

We can do this by eliminating x from Equations 1 and 2. Since the coefficients of x are already opposites, we need only add the equations.

$$
\begin{array}{rl}
x + y + \ z = 180 & \quad \text{Equation 1} \\
-x \quad\quad + \ z = \underline{\ 93} & \quad \text{Equation 2} \\
\text{Add:} \quad\quad y + 2z = 273 & \quad \text{Equation 4}
\end{array}
$$

We have now reduced the system to

$$
\begin{array}{rl}
y + 2z = 273 & \quad \text{Equation 4} \\
-3y + \ z = \ 7 & \quad \text{Equation 3}
\end{array}
$$

We can use addition to eliminate z by first multiplying Equation 3 by -2.

$$
\begin{array}{lll}
y + 2z = 273 & \xrightarrow{\text{No change}} & y + 2z = \ 273 \\
-3y + \ z = \ 7 & \xrightarrow{\text{Multiply by } -2.} & \underline{6y - 2z = -14} \\
& \text{Add:} & 7y \quad\quad = \ 259
\end{array}
$$

$$y = \frac{259}{7} = 37$$

Using back-substitution, we find that $x = 25$ and $z = 118$.
The measures of the angles of the triangle are 25°, 37°, and 118°.

Step 4. Check the proposed answers in the original wording of the problem.

The sum of these measures is 180°.

$$25 + 37 + 118 = 180$$

The largest angle (118°) is 93° more than the smallest (25°).

    $118 = 93 + 25$

The largest angle (118°) does exceed three times the measure of the remaining angle (37°) by 7°.

    $118 = 3(37) + 7$ ■

EXAMPLE 7   Providing Nutritional Needs

A nutritionist in a hospital is arranging special diets that consist of a combination of three basic foods. It is important that patients on this diet consume exactly 310 units of calcium, 190 units of iron, and 250 units of vitamin A each day. The amounts of these nutrients in one ounce of food are given in the following table.

Units per Ounce

	Calcium	Iron	Vitamin A
Food A	30	10	10
Food B	10	10	30
Food C	20	20	20

How many ounces of each food must be used to satisfy the nutrient requirements exactly?

Solution

Step 1. Use variables to represent the unknown quantities.

Let

    $x = $ the number of ounces of Food A

    $y = $ the number of ounces of Food B

    $z = $ the number of ounces of Food C

Step 2. Write a system of equations modeling the problem's conditions.

The patients must consume 310 units of calcium daily. Using the data in the calcium column, we get

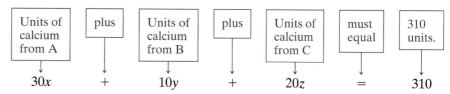

The patients must consume 190 units of iron daily. Using the data in the iron column, we get

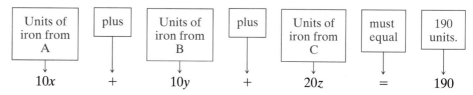

The patients must consume 250 units of vitamin A daily. Using the data in the vitamin A column, we get

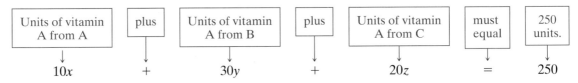

Units of vitamin A from A	plus	Units of vitamin A from B	plus	Units of vitamin A from C	must equal	250 units.
↓		↓		↓		↓
$10x$	$+$	$30y$	$+$	$20z$	$=$	250

Step 3. Solve the system and answer the problem's question.

We obtain the following linear system, which can be simplified as shown.

$$30x + 10y + 20z = 310 \quad \xrightarrow{\text{Divide by 10.}} \quad 3x + y + 2z = 31$$
$$10x + 10y + 20z = 190 \quad \xrightarrow{\text{Divide by 10.}} \quad x + y + 2z = 19$$
$$10x + 30y + 20z = 250 \quad \xrightarrow{\text{Divide by 10.}} \quad x + 3y + 2z = 25$$

Solving results in $x = 6$, $y = 3$, and $z = 5$. Patients on the special diet should be fed 6 ounces of Food A, 3 ounces of Food B, and 5 ounces of Food C every day. Take a few minutes to check this solution in terms of calcium, iron, and vitamin A requirements, respectively. ■

Principal Investigators: Donna Cox, Robert Paterson/Visualization: NCSA/UIUC

In reality, unlike the nutritionist in Example 7, dieticians are interested in dozens of foods and nutrients rather than just three. Modeling genuine problems often involves tens of thousands of equations. For instance, AT&T's domestic long distance network alone is modeled by a system of equations containing 800,000 variables.

PROBLEM SET 3.3

Practice Problems

In Problems 1–4, determine if the given ordered triple is a solution of the system.

1.
$x + y + z = 4$
$x - 2y - z = 1$
$2x - y - 2 = -1$
$(2, -1, 3)$

2.
$x + y + z = 0$
$x + 2y - 3z = 5$
$3x + 4y + 2z = -1$
$(5, -3, -2)$

3.
$x - 2y \quad = 2$
$2x + 3y \quad = 11$
$y - 4z = -7$
$(4, 1, 2)$

4. $x \quad - 2z = -5$
$y - 3z = -3$
$2x \quad - z = -4$
$(-1, 3, 2)$

Solve each system in Problems 5–30. Identify systems that are inconsistent and systems that are dependent.

5. $x + y + 2z = 11$
$x + y + 3z = 14$
$x + 2y - z = 5$

6. $2x + y - 2z = -1$
$3x - 3y - z = 5$
$x - 2y + 3z = 6$

7. $4x - y + 2z = 11$
$x + 2y - z = -1$
$2x + 2y - 3z = -1$

8. $x - y + 3z = 8$
$3x + y - 2z = -2$
$2x + 4y + z = 0$

9. $3x + 5y + 2z = 0$
$12x - 15y + 4z = 12$
$6x - 25y - 8z = 8$

10. $2x + 3y + 7z = 13$
$3x + 2y - 5z = -22$
$5x + 7y - 3z = -28$

11. $2x - 4y + 3z = 17$
$x + 2y - z = 0$
$4x - y - z = 6$

12. $x + z = 3$
$x + 2y - z = 1$
$2x - y + z = 3$

13. $2x + y = 2$
$x + y - z = 4$
$3x + 2y + z = 0$

14. $x + 3y + 5z = 20$
$y - 4z = -16$
$3x - 2y + 9z = 36$

15. $x + y = -4$
$y - z = 1$
$2x + y + 3z = -21$

16. $x + y = 11$
$y + 2z = 5$
$x - 2z = 4$

17. $x + y = 4$
$x + z = 4$
$y + z = 4$

18. $2x + y + 2z = 1$
$3x - y + z = 2$
$x - 2y - z = 0$

19. $3x + 4y + 5z = 8$
$x - 2y + 3z = -6$
$2x - 4y + 6z = 8$

20. $-\frac{1}{2}x - \frac{1}{3}y + \frac{1}{2}z = -2$
$\frac{2}{3}x - y - z = 8$
$\frac{1}{6}x + 2y + \frac{3}{2}z = 6$

21. $\begin{aligned} 6x - y + 3z &= 9 \\ \tfrac{1}{4}x - \tfrac{1}{2}y - \tfrac{1}{3}z &= -1 \\ -x + \tfrac{1}{6}y - \tfrac{2}{3}z &= 0 \end{aligned}$

22. $\begin{aligned} 3x + 9y + 6z &= 5 \\ x + 3y + 2z &= 4 \\ x - y - z &= 3 \end{aligned}$

23. $\begin{aligned} 2x + y + 4z &= 4 \\ x - y + z &= 6 \\ x + 2y + 3z &= 5 \end{aligned}$

24. $\begin{aligned} 5x - 2y - 5z &= 1 \\ 10x - 4y - 10z &= 2 \\ 15x - 6y - 15z &= 3 \end{aligned}$

25. $\begin{aligned} 3(2x + y) + 5z &= -1 \\ 2(x - 3y + 4z) &= -9 \\ 4(1 + x) &= -3(z - 3y) \end{aligned}$

26. $\begin{aligned} 7z - 3 &= 2(x - 3y) \\ 5y + 3z - 7 &= 4x \\ 4 + 5z &= 3(2x - y) \end{aligned}$

27. $\begin{aligned} 3x - 3y - z &= 0 \\ x - y + z &= 0 \\ -x + y + z &= 0 \end{aligned}$

28. $\begin{aligned} x + 2y + z &= 4 \\ 3x - 4y + z &= 4 \\ 6x - 8y + 2z &= 8 \end{aligned}$

29. $\begin{aligned} \frac{x}{2} - \frac{y}{2} + \frac{z}{4} &= 1 \\ \frac{x}{2} + \frac{y}{3} - \frac{z}{4} &= 2 \\ \frac{x}{4} - \frac{y}{2} + \frac{z}{2} &= 2 \end{aligned}$

30. $\begin{aligned} \frac{x}{2} + y + z &= \frac{5}{2} \\ \frac{x}{4} - \frac{y}{4} + \frac{z}{4} &= \frac{3}{2} \\ \frac{2x}{3} + y - \frac{z}{3} &= \frac{1}{3} \end{aligned}$

Application Problems

31. The sum of three numbers is 16. The sum of twice the first number, 3 times the second number, and 4 times the third number is 46. The difference between 5 times the first number and the second number is 31. Find the three numbers.

32. Three numbers are unknown. Three times the first number plus the second number plus twice the third number is 5. If 3 times the second number is subtracted from the sum of the first number and 3 times the third number, the result is 2. If the third number is subtracted from 2 times the first number and 3 times the second number, the result is 1. Find the numbers.

33. The bar graph indicates the ten longest rivers in the United States. The Missouri, Mississippi, and Yukon combined have a length of 6860 miles. The Missouri is 560 miles longer than the Yukon and 2140 miles shorter than twice the length of the Mississippi. Find the length of the three longest U.S. rivers.

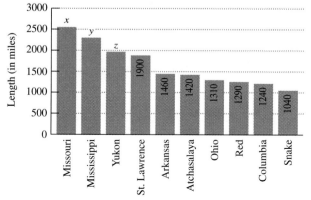

The Ten Longest Rivers in the United States

Source: U.S. Geological Survey

34. The graph shows that sexual harassment complaints are on the rise. For the period from 1991 through 1993, there were a total of 30,007 reported complaints. The number for 1993 exceeded that for 1992 by 1959 and was 1247 less than twice that for 1991. Find the number of reported sexual harassment complaints by employees for 1991, 1992, and 1993.

Reported Sexual Harassment Complaints by U.S. Employees, 1989–1993

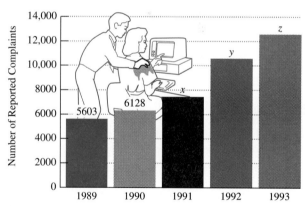

Source: Equal Employment Opportunity Commission

35. In a triangle, the largest angle is 80° greater than the smallest angle. The largest angle is also 20° less than twice as large as the remaining angle. Find the measure of each angle.

36. In a triangle, the largest angle is twice as large as the smallest angle. The largest angle is also 20° greater than the remaining angle. Find the measure of each angle.

37. A modernistic painting consists of triangles, rectangles, and pentagons, all drawn so as to not overlap or share sides. Within each rectangle are drawn 2 red roses, and each pentagon contains 5 carnations. How many triangles, rectangles, and pentagons appear in the painting if the painting contains a total of 40 geometric figures, 153 sides of geometric figures, and 72 flowers?

38. In the figure, triangle ABC is isosceles, with $AB = CB$. Point D is located midway between B and C. The perimeters of $\triangle ABC$, $\triangle ADB$, and $\triangle ACD$ are 80 feet, 70 feet, and 48 feet, respectively. Find the lengths of AB, AD, and AC.

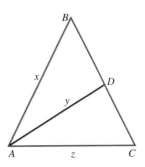

39. A rectangular solid has a base whose perimeter is 16 centimeters, a rectangular face in front whose perimeter is 18 centimeters, and a rectangular face on a side whose perimeter is 14 centimeters. What are the dimensions of the solid?

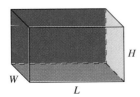

40. The sum of the measures of two angles of a triangle is equal to the measure of the third angle. The difference of the measures of these two angles is $\frac{2}{3}$ the measure of the third angle. Find the measures of the three angles.

41. Three diet foods contain grams of fat, carbohydrate, and protein as shown in the table.

	Fat (grams)	Carbohydrate (grams)	Protein (grams)
Serving of Food A	2	1	3
Serving of Food B	1	2	4
Serving of Food C	2	4	3

If a diet allows exactly 16 grams of fat, 23 grams of carbohydrate, and 29 grams of protein, how many servings of each kind of food can be eaten?

42. Three foods have the following nutritional content per ounce.

	Calories	Protein (grams)	Vitamin C (milligrams)
Food A	40	5	30
Food B	200	2	10
Food C	400	4	300

If a meal consisting of the three foods allows exactly 660 calories, 25 grams of protein, and 425 milligrams of vitamin C, how many ounces of each kind of food should be used?

43. A person invested $6700 for one year, part at 8%, part at 10%, and the remainder at 12%. The total annual income from these investments was $716. The amount of money invested at 12% was $300 more than the amount invested at 8% and 10% combined. Find the amount invested at each rate.

44. A person invested $17,000 for one year, part at 10%, part at 12%, and the remainder at 15%. The total annual income from these investments was $2110. The amount of money invested at 12% was $1000 less than the amount invested at 10% and 15% combined. Find the amount invested at each rate.

45. A certain brand of razor blades comes in packages of 6, 12, and 24 blades, costing $2, $3, and $4 per package, respectively. A store sold 12 packages containing a total of 162 razor blades and took in $35. How many packages of each type were sold?

46. A printing shop with three presses is in operation Wednesday, Thursday, and Friday. The number of hours the three presses were in operation on each of these days is as follows:

	Press I	Press II	Press III
Wednesday	8	4	7
Thursday	4	1	7
Friday	2	5	0

If Wednesday's output from the three presses was 1270 units, Thursday's was 730 units, and Friday's was 550 units, find the average output per hour for each press.

47. Three trucks were carrying cement. On the first day, truck I made 4 trips, truck II 3 trips, and truck III 5 trips. In all, the trucks carried 78 cubic yards of cement the first day. On the second day, the trucks made 5, 4, and 4 trips, respectively, carrying a total of 81 cubic yards. On the third day, the trucks made 3, 5, and 3 trips, respectively, carrying 69 cubic yards. If each truck was filled to capacity on each trip, find the number of cubic yards that each truck is capable of carrying.

48. The equation $y = \frac{1}{2}Ax^2 + Bx + C$ gives the relationship between the number of feet a car travels once the brakes are applied (y) and the number of seconds the car is in motion after the brakes are applied (x). A research firm discovered that when a car was in motion for 1 second after the brakes were applied, the car traveled 46 feet. (When $x = 1$, $y = 46$.) Similarly, it was found that when x was 2, y was 84 and when x was 3, y was 114. Use these values to find the constants, A, B, and C in the equation. What is the value for y when $x = 6$? Describe what this means in terms of the variables in the model.

True–False Critical Thinking Problems

49. Which one of the following is true?
  a. The ordered triple $(2, 15, 14)$ is the only solution to the equation $x + y - z = -1$.
  b. The substitution method cannot be used to solve three equations in three variables.
  c. If you invest x, y, and z dollars in investments paying 5%, 8%, and 9%, respectively, then the first-year interest is $0.05x + 0.08y + 0.09z$.
  d. The system

$$x - 2y + 3z = 6$$
$$2x - 4y + 6z = 12$$
$$6x - 12y + 18z = 36$$

  is inconsistent.

50. Which one of the following is true?
  a. The system

$$x + y + z = 0$$
$$-x - y + z = 0$$
$$-x + y + z = 0$$

  has $(0, 0, 0)$ as its only solution, and is therefore inconsistent.
  b. The equation $x - y - z = -6$ is satisfied by $(2, -3, 5)$.
  c. If two equations in a system are $x + y - z = 5$ and $2x + 2y - 2z = 7$, then the system must be inconsistent.
  d. An equation with four variables, such as $x + 2y - 3z + 5w = 2$, cannot be satisfied by real numbers because the algebraic limit is three equations in three variables.

Technology Problem

51. Use the simultaneous equations ($\boxed{\text{SIMULT}}$) feature of a graphing utility to verify the solution to some of the problems you solved in the practice exercises of this problem set. On most utilities, you will enter 3 (for three equations in three variables), and then enter the coefficients for x, y, z, and the constant term, one equation at a time. After entering all 12 numbers, press $\boxed{\text{SOLVE}}$. The solution will be displayed on the screen. (The x-, y-, and z-coordinates may be displayed using x_1, x_2, and x_3.) Consult your manual for details.

Writing in Mathematics

52. When using the addition method to solve three equations in three variables, a variable is eliminated from any two of the given equations. Why is it necessary to eliminate the *same* variable from any other two equations?

53. What is the geometric significance of the ordered triple that is the solution to a system of three linear equations in three variables?

54. Would it be possible to omit Section 3.1 ("Linear Systems of Equations in Two Variables") and begin by studying the solution of three linear equations in three variables? Explain. What does this tell you about learning mathematics?

Critical Thinking Problems

55. If a linear system of two equations in two variables contains dependent equations (whose graphs are the same line), the system has infinitely many solutions. However, it is possible for a linear system of three equations in three variables to contain two dependent equations (whose graphs are the same plane) and yet have no solution. Describe geometrically how this can occur.

Methods used to solve systems in three variables can be extended to solve a system of four equations in four variables. Solve the systems in Problems 56–57 by eliminating a variable from three pairs of equations, transforming the system to one involving three equations in three variables. Then proceed as you did in solving Problems 5–30. (If applicable, verify solutions using the simultaneous equations feature of a graphing utility.)

56. $x + y - z - 2w = -8$
$x - 2y + 3z + w = 18$
$2x + 2y + 2z - 2w = 10$
$2x + y - z + w = 3$

57. $x - y + 2z - 2w = -1$
$x - y - z + w = -4$
$-x + 2y - 2z - w = -7$
$2x + y + 3z - w = 6$

58. Peter's outgoing 115-kilometer bike ride involved riding uphill for 2 hours and on level ground for 3 hours. On the return trip, on a different 135-kilometer trail, he rode downhill for 2 hours and on level ground for 3 hours. Determine Peter's three speeds if his speed on level ground is the average of his uphill and downhill speeds.

59. On a particular train route, adult tickets cost $20, tickets for an adult with one child cost $30, and tickets for an adult with two children cost $35. The train's conductor collected $415 in fares from 28 passengers, 15 of whom were adults. How many tickets of each type were sold?

60. The sum of the digits of a three-digit number is 6. If the digits are reversed, the resulting number is 99 less than the original number. The tens digit is equal to the sum of the hundreds and units digits. Find the number.

61. Solve for x, y, and z in terms of the constants a, b, and c:

$$x + y = a$$
$$y + z = b$$
$$x + z = c$$

62. Two blocks of wood having the same length and width are placed on the top and bottom of a table, as shown in the left-hand figure. Length A measures 32 centimeters. The blocks are rearranged as shown in the right-hand figure. Length B measures 28 centimeters. Determine the height of the table.

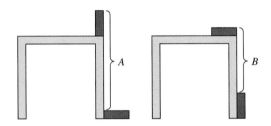

Group Activity Problems

63. In your group, write three problems that are similar to application Problems 31–48 in this problem set. Each problem should give rise to three equations in three variables. If you write problems like 33 or 34, consult the reference section of your library for appropriate graphs. Then solve each of the three problems.

64. Members of the group should research a realistic application involving foods and nutrients or any other area of interest that gives rise to a fairly large linear system. Use a graphing utility or computer to solve the system.

Review Problems

65. Write an equation in point-slope form, slope-intercept form, and standard form of the line passing through $(3, 5)$ and $(4, 2)$.

66. Graph $y = -2x + 4$ and then find the area of the triangle in the first quadrant formed by the graph of $y = -2x + 4$ and the x- and y-axes.

67. Use the vertical line test to identify in which graph(s) y is a function of x.

a.
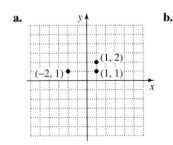

b.
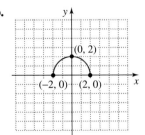

c.
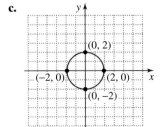

d.
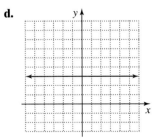

S E C T I O N   3 . 4

Solutions Tutorial Video
Manual   4

Matrix Solutions to Linear Systems

Objectives

1 Write the augmented matrix for a system.
2 Use matrices to solve systems.
3 Use matrices to identify inconsistent and dependent systems.

In this section, we look at a rather rapid technique for solving systems of linear equations. This method, using rectangular arrays of numbers called *matrices,* is essentially a streamlined way of using the addition method to eliminate variables.

Definition of a Matrix

A *matrix* (plural: *matrices*) is a rectangular array of numbers arranged in rows and columns and placed in brackets. Each number in the matrix is an *element* or *entry*.

EXAMPLE 1 **Examples of Matrices**

$$\begin{bmatrix} 3 & 4 & -1 \\ 2 & 0 & -2 \end{bmatrix}$$

This is a 2×3 (2 by 3) matrix with two rows and three columns.

$$\begin{bmatrix} 5 & -1 & -2 \\ 3 & 1 & 4 \\ 0 & 2 & -6 \\ 8 & 3 & 2 \end{bmatrix}$$

This is a 4×3 (4 by 3) matrix with four rows and three columns.

$$\begin{bmatrix} -2 & 0 \\ 1 & -4 \end{bmatrix}$$

This is a 2×2 (2 by 2) matrix with two rows and two columns. This is also a *square matrix* because it has the same number of rows as columns. ∎

1 Write the augmented matrix for a system.

Augmented Matrices

A matrix derived from a linear system of equations, each in standard form, is called the *augmented matrix* of the system. An augmented matrix has a vertical bar separating the columns of the matrix into two groups. The coefficients of each variable in a linear system are placed to the left of the vertical line, and the constants are placed to the right.

Augmented matrices

System of Linear Equations	Augmented Matrix

$$\begin{array}{c} a_1 x + b_1 y = c_1 \\ a_2 x + b_2 y = c_2 \end{array} \qquad \left[\begin{array}{cc|c} a_1 & b_1 & c_1 \\ a_2 & b_2 & c_2 \end{array}\right]$$

$$\begin{array}{c} a_1 x + b_1 y + c_1 z = d_1 \\ a_2 x + b_2 y + c_2 z = d_2 \\ a_3 x + b_3 y + c_3 z = d_3 \end{array} \qquad \left[\begin{array}{ccc|c} a_1 & b_1 & c_1 & d_1 \\ a_2 & b_2 & c_2 & d_2 \\ a_3 & b_3 & c_3 & d_3 \end{array}\right]$$

EXAMPLE 2 **Examples of Augmented Matrices**

System	Augmented Matrix

$$\begin{aligned} x + 3y &= 5 \\ 2x - y &= -4 \end{aligned} \qquad \begin{bmatrix} 1 & 3 & 5 \\ 2 & -1 & -4 \end{bmatrix}$$

$$\begin{aligned} x + 3y &= 5 \\ y &= 2 \end{aligned} \qquad \begin{bmatrix} 1 & 3 & 5 \\ 0 & 1 & 2 \end{bmatrix}$$

$$\begin{aligned} x + y - z &= 2 \\ 2x + 3y - z &= 7 \\ 3x - 2y + z &= 9 \end{aligned} \qquad \begin{bmatrix} 1 & 1 & -1 & 2 \\ 2 & 3 & -1 & 7 \\ 3 & -2 & 1 & 9 \end{bmatrix}$$

$$\begin{aligned} x + y - z &= 2 \\ y + z &= 3 \\ z &= 2 \end{aligned} \qquad \begin{bmatrix} 1 & 1 & -1 & 2 \\ 0 & 1 & 1 & 3 \\ 0 & 0 & 1 & 2 \end{bmatrix}$$

Notice how the second and fourth matrices in Example 2 contain 1s down the diagonal from upper left to lower right and 0s below the 1s. This arrangement makes it easy to find the solution of the system of equations, as Example 3 shows.

EXAMPLE 3 **Solving a System Using a Matrix**

Write the solution set for a system of equations represented by the matrix:

$$\begin{bmatrix} 1 & -2 & 2 & 9 \\ 0 & 1 & 2 & 5 \\ 0 & 0 & 1 & 3 \end{bmatrix}$$

Solution

The system represented by the given matrix is

$$\begin{bmatrix} 1 & -2 & 2 & 9 \\ 0 & 1 & 2 & 5 \\ 0 & 0 & 1 & 3 \end{bmatrix} \longrightarrow \begin{aligned} 1x - 2y + 2z &= 9 \\ 0x + 1y + 2z &= 5 \\ 0x + 0y + 1z &= 3 \end{aligned}$$

We can simplify this system to

$$\begin{aligned} x - 2y + 2z &= 9 \qquad \text{Equation 1} \\ y + 2z &= 5 \qquad \text{Equation 2} \\ z &= 3 \qquad \text{Equation 3} \end{aligned}$$

The value of z is known. We can find y by back-substitution.

$$\begin{aligned} y + 2z &= 5 && \text{Equation 2} \\ y + 2(3) &= 5 && \text{Substitute 3 for } z. \\ y + 6 &= 5 && \text{Multiply.} \\ y &= -1 && \text{Subtract 6 from both sides.} \end{aligned}$$

Now, with values for y and z, we can use back-substitution to find x.

$$x - 2y + 2z = 9 \quad \text{Equation 1}$$
$$x - 2(-1) + 2(3) = 9 \quad \text{Substitute } -1 \text{ for } y \text{ and } 3 \text{ for } z.$$
$$x + 2 + 6 = 9 \quad \text{Multiply.}$$
$$x + 8 = 9 \quad \text{Add.}$$
$$x = 1 \quad \text{Subtract 8 from both sides.}$$

We see that $x = 1$, $y = -1$, and $z = 3$. The solution set for the system is $\{(1, -1, 3)\}$. ■

If all augmented matrices could be converted to the form of Example 3, with 1s down the main diagonal and 0s below the 1s, a system's solution could easily be obtained from the matrix. As it turns out, there are ways of rewriting matrices to convert them to a more useable form.

Our goal in solving a linear system using matrices is to produce a matrix similar to the one in Example 3. In general, the matrix will be of the form

$$\begin{bmatrix} 1 & A & B \\ 0 & 1 & C \end{bmatrix} \quad \text{or} \quad \begin{bmatrix} 1 & A & B & C \\ 0 & 1 & D & E \\ 0 & 0 & 1 & F \end{bmatrix}$$

for systems with two or three equations, respectively. The last row of these matrices gives us the value of one variable. The other variables can then be found by substitution.

We wish to solve a system of linear equations by using the augmented matrix for that system. We use *row operations* on the augmented matrix. The row operations may change the numbers in the matrix, but they produce new matrices representing systems of equations with the same solution set as that of the original system.

Matrix row operations

Three row operations produce matrices that lead to systems with the same solution set as the original system.

Description of the Operation	Symbol and Meaning	Example
Interchange two rows of the matrix.	$R_i \leftrightarrow R_j$ Interchange the entries in ith and jth rows.	$\begin{bmatrix} 1 & 2 & -1 \\ 4 & -3 & -15 \end{bmatrix}$ $R_1 \leftrightarrow R_2:$ $\begin{bmatrix} 4 & -3 & -15 \\ 1 & 2 & -1 \end{bmatrix}$
Multiply the entries in any row by the same nonzero real number.	kR_i Multiply each entry in the ith row by k.	$\begin{bmatrix} 1 & 2 & 3 \\ 0 & 6 & 4 \end{bmatrix}$ $\frac{1}{6}R_2:$ $\begin{bmatrix} 1 & 2 & 3 \\ 0 & 1 & \frac{2}{3} \end{bmatrix}$

Matrix row operations (continued)

Description of the Operation	Symbol and Meaning	Example
Add the entries of any row to a multiple of the corresponding entries of another row.	$kR_i + R_j$ Add k times the entries in row i to the corresponding entries in row j.	$\begin{bmatrix} 1 & 2 & \mid -1 \\ 4 & -3 & \mid -15 \end{bmatrix}$ $-4R_1 + R_2$: $\begin{bmatrix} 1 & 2 & \mid -1 \\ (-4)(1) + 4 & (-4)(2) + (-3) & \mid (-4)(-1) + (-15) \end{bmatrix}$ which simplifies to: $\begin{bmatrix} 1 & 2 & \mid -1 \\ 0 & -11 & \mid -11 \end{bmatrix}$

Two matrices are *row-equivalent* if one can be obtained from the other by a sequence of row operations.

2 Use matrices to solve systems.

Using Matrices and Row Operations to Solve Linear Systems

Let's see how we can solve a system of two equations in two variables using matrix row operations.

EXAMPLE 4 **Using Row Operations to Solve a System**

Use matrices to solve the system:

$$4x - 3y = -15$$
$$x + 2y = -1$$

Solution

Our goal is to obtain a matrix of the form

$$\begin{bmatrix} 1 & - & \mid - \\ 0 & 1 & \mid - \end{bmatrix}$$

with 1s down the diagonal from upper left to lower right and 0 below the 1. We start with the augmented matrix of the system.

$$\begin{bmatrix} \boxed{4} & -3 & \mid -15 \\ 1 & 2 & \mid -1 \end{bmatrix}$$
Our first goal is to get a 1 where the 4 is in the upper-left cell.

To get 1 in the top position of the first column, we interchange rows 1 and 2.

$$\begin{bmatrix} 1 & 2 & \mid -1 \\ \boxed{4} & -3 & \mid -15 \end{bmatrix}$$
$R_1 \leftrightarrow R_2$
Now, this boxed entry should be 0.

We now want 0 below the 1. To get a 0 in row 2, column 1, we add to the numbers in row 2 the results of multiplying each number in row 1 by -4.

$$\begin{bmatrix} 1 & 2 & \mid -1 \\ 0 & \boxed{-11} & \mid -11 \end{bmatrix} \quad \leftarrow -4R_1 + R_2$$
Now, this boxed entry should be 1.

The last step is to obtain 1 in row 2, column 2. We multiply each number in row 2 by $-\frac{1}{11}$.

$$\begin{bmatrix} 1 & 2 & \mid -1 \\ 0 & 1 & \mid 1 \end{bmatrix} \quad \leftarrow -\frac{1}{11}R_2$$
This matrix is of the form $\begin{bmatrix} 1 & - & \mid - \\ 0 & 1 & \mid - \end{bmatrix}$.

This final augmented matrix represents the system of equations

$$1x + 2y = -1 \quad \text{or} \quad x + 2y = -1$$
$$0x + 1y = 1 \qquad \qquad \quad y = 1$$

From the second equation, $y = 1$. We back-substitute 1 for y in the first equation to obtain

$$x + 2y = -1$$
$$x + 2(1) = -1$$
$$x = -3$$

Thus, the solution set of the system is $\{(-3, 1)\}$, which can be checked by substitution in both equations. ∎

Our next example involves a system of three linear equations in three variables. Once again, we will begin with the augmented matrix and use row operations to obtain a row-equivalent matrix with 1s down the diagonal from left to right and 0s below each 1.

EXAMPLE 5 Using Row Operations to Solve a System

Use matrices to solve the system:

$$3x + y - 5z = 2$$
$$-x + 4y + 2z = 1$$
$$x + 2y - z = 3$$

Solution

We begin by writing the system in terms of an augmented matrix.

$$\left[\begin{array}{ccc|c} \boxed{3} & 1 & -5 & 2 \\ -1 & 4 & 2 & 1 \\ 1 & 2 & -1 & 3 \end{array}\right]$$ Our first goal is to get a 1 where the 3 is in the upper-left cell.

To get a 1 in the top position of the first column, we interchange rows 1 and 3.

$$\left[\begin{array}{ccc|c} 1 & 2 & -1 & 3 \\ \boxed{-1} & 4 & 2 & 1 \\ \boxed{3} & 1 & -5 & 2 \end{array}\right]$$ $R_1 \leftrightarrow R_3$
Now, these boxed entries should be 0s.

Now we want 0s in the second two positions of column 1. The first 0 (where there is now -1) can be obtained by adding to row 2 the results of multiplying each number in row 1 by 1. The second 0 (where there is now 3) can be obtained by adding to row 3 the results of multiplying each number in row 1 by -3.

$$\left[\begin{array}{ccc|c} 1 & 2 & -1 & 3 \\ 1 + (-1) & 2 + 4 & -1 + 2 & 3 + 1 \\ -3 \cdot 1 + 3 & -3 \cdot 2 + 1 & -3(-1) + (-5) & -3(3) + 2 \end{array}\right]$$ $\begin{array}{l} \leftarrow 1R_1 + R_2 \\ \leftarrow -3R_1 + R_3 \end{array}$

We can simplify this matrix.

$$\left[\begin{array}{ccc|c} 1 & 2 & -1 & 3 \\ 0 & \boxed{6} & 1 & 4 \\ 0 & -5 & -2 & -7 \end{array}\right]$$ Because we want 1s down the main diagonal, this boxed entry should be 1.

We can obtain 1 in row 2, column 2, by multiplying each number in row 2 by $\frac{1}{6}$.

$$\begin{bmatrix} 1 & 2 & -1 & \bigm| & 3 \\ 0 & 1 & \frac{1}{6} & \bigm| & \frac{2}{3} \\ 0 & \boxed{-5} & -2 & \bigm| & -7 \end{bmatrix} \quad \leftarrow \frac{1}{6}R_2 \quad \text{Now, this boxed entry should be 0.}$$

We now want 0 below the 1. Get 0 in row 3, column 2, by adding to row 3 the results of multiplying each number in row 2 by 5.

$$\begin{bmatrix} 1 & 2 & -1 & \bigm| & 3 \\ 0 & 1 & \frac{1}{6} & \bigm| & \frac{2}{3} \\ 5 \cdot 0 + 0 & 5 \cdot 1 + (-5) & 5 \cdot \frac{1}{6} + (-2) & \bigm| & 5 \cdot \frac{2}{3} + (-7) \end{bmatrix} \quad \leftarrow 5R_2 + R_3$$

Now we simplify the entries.

$$\begin{bmatrix} 1 & 2 & -1 & \bigm| & 3 \\ 0 & 1 & \frac{1}{6} & \bigm| & \frac{2}{3} \\ 0 & 0 & \boxed{-\frac{7}{6}} & \bigm| & -\frac{11}{3} \end{bmatrix} \quad \begin{array}{l} \text{Remember that we want 1s down the main diagonal. This} \\ \text{boxed entry should be 1.} \end{array}$$

The last step involves obtaining 1 in row 3, column 3. We multiply each number in row 3 by $-\frac{6}{7}$.

$$\begin{bmatrix} 1 & 2 & -1 & \bigm| & 3 \\ 0 & 1 & \frac{1}{6} & \bigm| & \frac{2}{3} \\ 0 & 0 & 1 & \bigm| & \frac{22}{7} \end{bmatrix} \quad \leftarrow -\frac{6}{7}R_3$$

Converting back to a system of equations, we obtain

$$x + 2y - z = 3$$
$$y + \frac{1}{6}z = \frac{2}{3}$$
$$z = \frac{22}{7}$$

To find y, we back-substitute $\frac{22}{7}$ for z in the second equation.

$$y + \frac{1}{6}z = \frac{2}{3}$$

$$y + \frac{1}{6}\left(\frac{22}{7}\right) = \frac{2}{3} \qquad \text{Substitute } \frac{22}{7} \text{ for } z.$$

$$y + \frac{11}{21} = \frac{2}{3} \qquad \text{Multiply.}$$

$$21y + 11 = 14 \qquad \text{Multiply both sides by 21.}$$

$$21y = 3 \qquad \text{Subtract 11 from both sides.}$$

$$y = \frac{3}{21} = \frac{1}{7} \qquad \text{Divide both sides by 21.}$$

Finally, we back-substitute $\frac{1}{7}$ for y and $\frac{22}{7}$ for z in the first equation.

$$x + 2y - z = 3$$

$$x + 2\left(\frac{1}{7}\right) - \frac{22}{7} = 3 \qquad \text{Substitute the obtained values.}$$

$$x + \frac{2}{7} - \frac{22}{7} = 3 \qquad \text{Multiply.}$$

$$x - \frac{20}{7} = 3 \qquad \text{Subtract.}$$

$$7x - 20 = 21 \qquad \text{Multiply both sides by 7.}$$

$$7x = 41 \qquad \text{Add 20 to both sides.}$$

$$x = \frac{41}{7} \qquad \text{Divide both sides by 7.}$$

The solution set of the original system is $\{(\frac{41}{7}, \frac{1}{7}, \frac{22}{7})\}$. ∎

Here's a summary of the steps that we used in Examples 4 and 5 to solve linear systems with matrices.

> ### Solving linear systems using matrices
>
> 1. Write the augmented matrix for the system.
> 2. Use matrix row operations to simplify the matrix to one with 1s down the diagonal from upper left to lower right, and 0s below the 1s.
> 3. Write the system of linear equations corresponding to the matrix in step 2, and use back-substitution to find the system's solution.

The process summarized in the box is called *Gaussian elimination* with back-substitution, after the German mathematician Carl Frederich Gauss (1777–1855).

3 Use matrices to identify inconsistent and dependent systems.

Gaussian Elimination with Inconsistent and Dependent Systems

Each linear system that we have solved so far has had precisely one solution. However, we know that a system of linear equations can have no solution (an inconsistent system) or an infinite number of solutions (a system with dependent equations). We turn now to the use of matrices for solving such systems.

EXAMPLE 6 Matrices and Inconsistent Systems

Use matrices to solve the system:

$$2x - 4y = 6$$
$$3x - 6y = 5$$

Solution

Write the augmented matrix.

$$\begin{bmatrix} \boxed{2} & -4 & | & 6 \\ 3 & -6 & | & 5 \end{bmatrix} \qquad \text{Our first goal is to get a 1 where the 2 now appears.}$$

To get 1 in row 1, column 1, we multiply the first row by $\frac{1}{2}$.

$$\begin{bmatrix} 1 & -2 & | & 3 \\ \boxed{3} & -6 & | & 5 \end{bmatrix} \qquad \begin{array}{l} \leftarrow \frac{1}{2}R_1 \\ \text{Now, this boxed entry should be 0.} \end{array}$$

To get 0 in row 2, column 1, we multiply row 1 by -3 and add the results to row 2.

$$\begin{bmatrix} 1 & -2 & 3 \\ 0 & 0 & -4 \end{bmatrix} \quad \leftarrow -3R_1 + R_2$$

The corresponding system of equations is

$$x - 2y = 3$$
$$0 = -4$$

which has no solution. The system is inconsistent, and the solution set is $\varnothing$. ∎

EXAMPLE 7 **Matrices and Systems with Dependent Equations**

Use matrices to solve the system:

$$3x + y - 4z = 2$$
$$6x + 2y - 8z = 4$$
$$-9x - 3y + 12z = -6$$

Solution

The augmented matrix is

$$\begin{bmatrix} \boxed{3} & 1 & -4 & 2 \\ 6 & 2 & -8 & 4 \\ -9 & -3 & 12 & -6 \end{bmatrix} \qquad \text{We want to get a 1 where the 3 now appears.}$$

To get 1 in row 1, column 1, we multiply the first row by $\frac{1}{3}$.

$$\begin{bmatrix} 1 & \frac{1}{3} & -\frac{4}{3} & \frac{2}{3} \\ \boxed{6} & 2 & -8 & 4 \\ \boxed{-9} & -3 & 12 & -6 \end{bmatrix} \quad \begin{array}{l} \leftarrow \frac{1}{3}R_1 \\[4pt] \text{Now, these boxed entries should be 0s.} \end{array}$$

To get 0 in row 2, column 1, we add to row 2 the results of multiplying each number in row 1 by -6. To get 0 in row 3, column 1, we add to row 3 the results of multiplying each number in row 1 by 9.

$$\begin{bmatrix} 1 & \frac{1}{3} & -\frac{4}{3} & \frac{2}{3} \\ 0 & 0 & 0 & 0 \\ 0 & 0 & 0 & 0 \end{bmatrix} \quad \begin{array}{l} \leftarrow -6R_1 + R_2 \\ \leftarrow 9R_1 + R_3 \end{array}$$

The corresponding system is

$$x + \frac{1}{3}y - \frac{4}{3}z = \frac{2}{3}$$
$$0 = 0$$
$$0 = 0$$

which has dependent equations. ∎

> **D**iscover for yourself
>
> The system
>
> $$3x + y - 4z = 2$$
> $$6x + 2y - 8z = 4$$
> $$-9x - 3y + 12z = -6$$
>
> contains dependent equations. Multiply the first equation by 2. What do you observe? Now multiply the first equation by -3. What do you notice? Based on these observations, explain why the solution set for this system can be expressed as
>
> $$\{(x, y, z) \mid 3x + y - 4z = 2\}.$$

ENRICHMENT ESSAY

Presenting Information with Matrices

Jerome Paul Witkin (American, b. 1939) "Jeff Davies" 1980 oil on canvas 72 × 48 in. Gift of the American Academy and Institute of Arts and Letters (Hassam and Speicher Purchase Fund), Palmer Museum of Art, The Pennsylvania State University

We have seen how to solve linear systems with matrices. Matrices are also used to present rectangular arrays of numbers, as shown in the following example.

How Fast You Burn Off Calories

Moderate Activity	Weight (in pounds)					
	110	132	154	176	187	209
	Calories Burned per Hour					
Housework	175	210	245	285	300	320
Cycling	190	215	245	270	280	295
Making bed	165	185	210	230	245	255
Gardening	155	190	215	250	265	280
Mowing lawn	195	225	250	280	295	305
Tennis	335	380	425	470	495	520
Watching TV	60	70	80	85	90	95
Golf (foursome)	210	240	270	295	310	325
Bowling	210	240	270	300	310	325
Jogging	515	585	655	725	760	795
Walking briskly	355	400	450	500	525	550
Dancing	350	395	445	490	575	540

Although we attempt to watch our caloric intake, an estimated 25 percent or more of adults in the United States are dangerously overweight. Perhaps successful intervention will depend on a better understanding of the genetics of obesity.

PROBLEM SET 3.4

Practice Problems

In Problems 1–8, write the system of equations corresponding to each given matrix. Once the system is written, use back-substitution to find its solution.

1. $\begin{bmatrix} 1 & -3 & | & 11 \\ 0 & 1 & | & -3 \end{bmatrix}$

2. $\begin{bmatrix} 1 & -\frac{3}{2} & | & 5 \\ 0 & 1 & | & -1 \end{bmatrix}$

3. $\begin{bmatrix} 1 & -3 & | & 1 \\ 0 & 1 & | & -1 \end{bmatrix}$

4. $\begin{bmatrix} 1 & 2 & | & 13 \\ 0 & 1 & | & 4 \end{bmatrix}$

5. $\begin{bmatrix} 1 & 2 & 1 & | & 0 \\ 0 & 1 & 0 & | & -2 \\ 0 & 0 & 1 & | & 3 \end{bmatrix}$

6. $\begin{bmatrix} 1 & 0 & -4 & | & 5 \\ 0 & -1 & 12 & | & -13 \\ 0 & 0 & 1 & | & -\frac{1}{2} \end{bmatrix}$

7. $\begin{bmatrix} 1 & 1 & 0 & | & 3 \\ 0 & 1 & \frac{3}{2} & | & -2 \\ 0 & 0 & 1 & | & 0 \end{bmatrix}$

8. $\begin{bmatrix} 1 & \frac{1}{2} & 1 & | & \frac{11}{2} \\ 0 & 1 & \frac{3}{2} & | & 7 \\ 0 & 0 & 1 & | & 4 \end{bmatrix}$

In Problems 9–20, perform each matrix row operation(s) and write the new matrix.

9. $\begin{bmatrix} 2 & 2 & | & 5 \\ 1 & -\frac{3}{2} & | & 5 \end{bmatrix} R_1 \leftrightarrow R_2$

10. $\begin{bmatrix} -6 & 9 & | & 4 \\ 1 & -\frac{3}{2} & | & 4 \end{bmatrix} R_1 \leftrightarrow R_2$

11. $\begin{bmatrix} -6 & 8 & | & -12 \\ 3 & 5 & | & -2 \end{bmatrix} -\frac{1}{6} R_1$

12. $\begin{bmatrix} -2 & 3 & | & -10 \\ 4 & 2 & | & 5 \end{bmatrix} -\frac{1}{2} R_1$

13. $\begin{bmatrix} 1 & -3 & | & 5 \\ 2 & 6 & | & 4 \end{bmatrix} -2R_1 + R_2$

14. $\begin{bmatrix} 1 & -3 & | & 1 \\ 2 & 1 & | & -5 \end{bmatrix} -2R_1 + R_2$

15. $\begin{bmatrix} 1 & -\frac{3}{2} & | & \frac{7}{2} \\ 3 & 4 & | & 2 \end{bmatrix} -3R_1 + R_2$

16. $\begin{bmatrix} 1 & -\frac{2}{5} & | & \frac{3}{4} \\ 4 & 2 & | & -1 \end{bmatrix} -4R_1 + R_2$

17. $\begin{bmatrix} 1 & -1 & 5 & | & -6 \\ 3 & 3 & -1 & | & 10 \\ 1 & 3 & 2 & | & 5 \end{bmatrix} \begin{matrix} -3R_1 + R_2 \\ \text{and} \\ -1R_1 + R_3 \end{matrix}$

18. $\begin{bmatrix} 1 & -3 & 2 & | & 0 \\ 3 & 1 & -1 & | & 7 \\ 2 & -2 & 1 & | & 3 \end{bmatrix} \begin{matrix} -3R_1 + R_2 \\ -2R_1 + R_3 \end{matrix}$

19. $\begin{bmatrix} 1 & 1 & -1 & | & 6 \\ 2 & -1 & 1 & | & -3 \\ 3 & -1 & -1 & | & 4 \end{bmatrix} \begin{matrix} -2R_1 + R_2 \\ \text{and} \\ -3R_1 + R_3 \end{matrix}$

20. $\begin{bmatrix} 1 & 3 & 4 & | & 10 \\ 0 & -5 & -15 & | & -38 \\ 4 & 8 & 4 & | & 9 \end{bmatrix} \begin{matrix} -5R_1 + R_2 \\ -4R_1 + R_3 \end{matrix}$

Solve each system in Problems 21–44 using row-equivalent matrices. Identify systems that are inconsistent and systems that are dependent.

21. $x + y = 6$
$x - y = 2$

22. $x + 2y = 11$
$x - y = -1$

23. $2x + y = 3$
$x - 3y = 12$

24. $3x - 5y = 7$
$x - y = 1$

25. $5x + 7y = -25$
$11x + 6y = -8$

26. $2x - 5y = 10$
$5x + 7y = -3$

27. $4x - 2y = 5$
$-2x + y = 6$

28. $-3x + 4y = 12$
$6x - 8y = 16$

29. $x - 2y = 1$
$-2x + 4y = -2$

30. $3x - 6y = 1$
$2x - 4y = \frac{2}{3}$

31. $x + y - z = -2$
$2x - y + z = 5$
$-x + 2y + 2z = 1$

32. $x - 2y - z = 2$
$2x - y + z = 4$
$-x + y - 2z = -4$

33. $x + 3y = 0$
$x + y + z = 1$
$3x - y - z = 11$

34. $3y - z = -1$
$x + 5y - z = -4$
$-3x + 6y + 2z = 11$

35. $2x + 2y + 7z = -1$
$2x + y + 2z = 2$
$4x + 6y + z = 15$

36. $3x + 2y + 3z = 3$
$4x - 5y + 7z = 1$
$2x + 3y - 2z = 6$

37. $x + y + z = 6$
$x - z = -2$
$y + 3z = 11$

38. $x + y + z = 3$
$-y + 2z = 1$
$-x + z = 0$

39. $x - y + 3z = 4$
$x + 2y - 2z = 10$
$3x - y + 5z = 14$

40. $3x - y + 2z = 4$
$-6x + 2y - 4z = 1$
$5x - 3y + 8z = 0$

41. $x - 2y + z = 4$
$5x - 10y + 5z = 20$
$-2x + 4y - 2z = -8$

42. $x - 3y + z = 2$
$4x - 12y + 4z = 8$
$-2x + 6y - 2z = -4$

43. $x + y = 1$
$y + 2z = -2$
$2x - z = 0$

44. $x + 3y = 3$
$y + 2z = -8$
$x - z = 7$

Application Problems

45. Two angles are complementary (the sum of their measures is 90°). The larger angle measures 15° more than twice the smaller.
   a. Represent the measures of the angles by x and y, and write a system of two equations in two variables that models the problem's conditions.
   b. Use matrices to solve the system and find the measure of each angle.

46. The perimeter of a rectangle is 140 feet and the length exceeds the width by 30 feet.
   a. Represent the length by L, the width by W, and write a system of two equations in two variables that models the problem's conditions.

   b. Use matrices to solve the system and find the dimensions of the rectangle.

47. In triangle ABC, angle A is 10° less than the sum of B and C, and angle B is 50° less than the sum of A and C.
   a. Let x represent the measure of angle A, y the measure of angle B, and z the measure of angle C. Write a system of three equations in three variables that models the problem's conditions.
   b. Use matrices to solve the system and find the measure of each of the triangle's angles.

48. A furniture company produces three types of desks: a children's model, an office model, and a deluxe model. Each desk is manufactured in three stages: cutting, construction, and finishing. The time requirements for each model and manufacturing stage are given in the following table.

	Children's Model	Office Model	Deluxe Model
Cutting	2 hr	3 hr	2 hr
Construction	2 hr	1 hr	3 hr
Finishing	1 hr	1 hr	2 hr

Each week the company has available a maximum of 100 hours for cutting, 100 hours for construction, and 65 hours for finishing. If all available time must be used, how many of each type of desk should be produced each week? Use matrices to solve the system.

True–False Critical Thinking Problems

49. Which one of the following is true?
  a. A row-equivalent matrix will not be obtained if the numbers in any row are multiplied by a negative improper fraction.
  b. The augmented matrix for the system

$$\begin{aligned} x - 3y &= 5 \\ y - 2z &= 7 \\ 2x + z &= 4 \end{aligned} \quad \text{is} \quad \begin{bmatrix} 1 & -3 & 5 \\ 1 & -2 & 7 \\ 2 & 1 & 4 \end{bmatrix}.$$

  c. In solving a linear system of three equations in three variables, we begin with the augmented matrix and use row operations to obtain a row-equivalent matrix with 0s down the diagonal from left to right and 1s below each 0.
  d. The solution set for the system represented by the matrix

$$\begin{bmatrix} 1 & -\frac{3}{2} & 5 \\ 0 & 0 & 6 \end{bmatrix}$$

  is $\varnothing$.

50. Which one of the following is true?
  a. The solution set for the system represented by the matrix

$$\begin{bmatrix} 3 & 1 & 1 \\ 0 & 0 & 0 \end{bmatrix}$$

  is $\varnothing$.
  b. The notation $R_1 \leftrightarrow R_3$ means to replace R_1 by $3 R_1$.
  c. The augmented matrix for a system of two linear equations in two variables is always a matrix with two rows and three columns, described as a 2×3 matrix.
  d. A matrix row operation such as $-\frac{4}{5} R_1 + R_2$ is not permitted because of the negative fraction.

Technology Problems

51. Most graphing utilities can perform row operations on matrices. Consult the owner's manual for your graphing utility to learn proper keystroke sequences for performing these operations. Then duplicate the elementary row operations of Example 4 on page 272 and Example 5 on page 273.

52. The final augmented matrix that we obtain when using Gaussian elimination is said to be in *row-echelon*

form. For systems of linear equations with unique solutions, this form results when each entry in the main diagonal is 1 and all entries below the main diagonal are 0s. Most graphing utilities can transform a matrix to row-echelon form. Consult your owner's manual. If your utility has this capability, use it to verify some of the solutions in Problems 21–44.

Writing in Mathematics

53. Describe what is meant by the *augmented matrix* of a system of linear equations.

54. One of the matrix row operations for producing a row-equivalent matrix states that any row may be changed by adding to the numbers of the row the product of a constant and the numbers of another row. Compare this operation with the addition method for solving a system of linear equations.

Critical Thinking Problems

Row-equivalent matrices can be used to solve systems involving more than three equations. Solve each of the systems in Problems 55–56 with matrices using row operations to obtain a matrix with 1s down the diagonal from upper left to lower right and 0s below each 1. If applicable, use the discussion in Problem 52 and a graphing utility to verify your final matrix.

55.
$$\begin{aligned}
x + y + z + w &= 4 \\
2x + y - 2z - w &= 0 \\
x - 2y - z - 2w &= -2 \\
3x + 2y + z + 3w &= 4
\end{aligned}$$

56.
$$\begin{aligned}
x + y + z + w &= 5 \\
x + 2y - z - 2w &= -1 \\
x - 3y - 3z - w &= -1 \\
2x - y + 2z - w &= -2
\end{aligned}$$

57. The function $y = ax^3 + bx^2 + cx + d$ can be used to estimate the number of bachelor's degrees (y) conferred in mathematics x years after 1970, where $0 \le x \le 20$.

  a. Use the four data values shown in the bar graph—(5, 14,685), (10, 13,140), (15, 15,095), and (20, 15,150)—to create a polynomial model that fits the data. You'll need to substitute the given data points into the function, one at a time, and solve the resulting linear system of four equations for $a, b, c,$ and d. Solve using the method of Problems 55 and 56.

Mathematics Degrees, 1970–1990

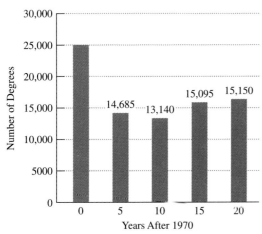

Data Source: U.S. Department of Education

b. Use a graphing utility to graph the function with the following range setting:

  Xmin = 0, Xmax = 30, Xscl = 3, Ymin = 0,
  Ymax = 25,000, Yscl = 5000

Trace along the curve and verify the four given data points. Explain, based on the graph's shape, why it is not a good model for $x > 20$.

58. The vitamin content per ounce for three foods is given in the table.

Milligrams per Ounce

	Thiamin	Riboflavin	Niacin
Food A	3	7	1
Food B	1	5	3
Food C	3	8	2

  a. Use matrices to show that no combination of these foods can provide exactly 14 mg of thiamin, 32 mg of riboflavin, and 9 mg of niacin.
  b. Use matrices to describe in practical terms what happens if the riboflavin requirement is increased by 5 milligrams and the other requirements stay the same.

Review Problems

59. Let $A = \{1, 2, 8, 9\}$ and $B = \{7, 8, 9, 11\}$. Find:
  a. $A \cup B$; **b.** $A \cap B$.

60. Solve the compound inequality, graphing the solution set on a number line. Express the solution in set-builder and interval notations:
  $-3x \ge 6$ or $3x - 1 < -10$.

61. Automobile Rental Agency A charges $30 a day plus 14 cents a mile. Agency B charges $16 a day plus 24 cents a mile. What distance would one have to drive in a day to make the daily costs the same? Illustrate your solution by graphing each of the daily cost models in a rectangular coordinate system.

Solutions Manual   **Tutorial**   **Video 4**

Solving Linear Systems of Equations Using Determinants and Cramer's Rule

SECTION 3.5

Objectives

1. Evaluate a second-order determinant.
2. Solve a linear system of equations in two variables using Cramer's rule.
3. Evaluate a third-order determinant.
4. Solve a linear system of equations in three variables using Cramer's rule.
5. Use determinants to identify inconsistent and dependent systems.

In this chapter, we have so far studied graphing, substitution, addition, and matrix methods for solving linear systems of equations. We turn now to another method for solving such systems. As with matrix methods, solutions are obtained by writing down the coefficients and constants of a linear system and operating with them. The result is known as Cramer's rule, in honor of the Swiss geometer Gabriel Cramer (1704–1752).

Evaluate a second-order determinant.

The Determinant of a 2 × 2 Matrix

If a matrix has the same number of rows and columns, it is called a square matrix. Associated with every square matrix is a real number called its *determinant.* The determinant for a 2 × 2 (2 by 2) square matrix is defined as follows.

> **S**tudy tip
>
> To evaluate a determinant, find the difference of the product of the two diagonals.
>
> $$\begin{vmatrix} a_1 & b_1 \\ a_2 & b_2 \end{vmatrix} = a_1b_2 - a_2b_1$$

Definition of the determinant of a 2 × 2 matrix

The determinant of the matrix $\begin{bmatrix} a_1 & b_1 \\ a_2 & b_2 \end{bmatrix}$ is denoted by $\begin{vmatrix} a_1 & b_1 \\ a_2 & b_2 \end{vmatrix}$ and is defined by

$$\begin{vmatrix} a_1 & b_1 \\ a_2 & b_2 \end{vmatrix} = a_1b_2 - a_2b_1.$$

We also say that the *value* of the *second-order determinant* $\begin{vmatrix} a_1 & b_1 \\ a_2 & b_2 \end{vmatrix}$ is $a_1b_2 - a_2b_1.$

EXAMPLE 1   **Evaluating 2 × 2 Determinants**

Evaluate each determinant.

a. $\begin{vmatrix} 5 & 6 \\ 7 & 3 \end{vmatrix}$ b. $\begin{vmatrix} 2 & 4 \\ -3 & -5 \end{vmatrix}$

ENRICHMENT ESSAY

Nonlinear Chaos

You have some x-and-y equation. Any value for x gives you a value for y. So you put a dot where it's right for both x and y. Then you take the next value for x, which gives you another value for y, and when you've done that a few times you join up the dots and that's your graph of whatever the equation is. . . . What she's doing is, every time she works out a value for y, she's using that as her next value for x. And so on. Like a feedback. She's feeding the solution back into the equation, and then solving it again. Iteration, you see.

Tom Stoppard, *Arcadia*

This quote is from Tom Stoppard's remarkable new play, *Arcadia*, where the characters dream and talk about mathematics from Newton to chaos theory. If the points represented in the method described above are plotted on the Cartesian plane, they seem to be scattered randomly. But by having a computer plot thousands of points, the points strangely begin to form a pattern, as shown in the graph below. One of the new frontiers of mathematics suggests that there is an underlying order in apparent chaos.

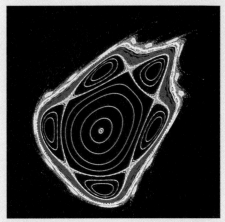

Robert S. Hotchkiss, Los Alamos National Laboratory

iscover for yourself

Example I illustrates that the determinant of a matrix may be positive or negative. The determinant can also have 0 as its value. Write and then evaluate three determinants, one whose value is positive, one whose value is negative, and one whose value is 0.

Solution

We multiply and subtract as indicated.

a. $\begin{vmatrix} 5 & 6 \\ 7 & 3 \end{vmatrix} = 5 \cdot 3 - 7 \cdot 6 = 15 - 42 = -27$ The value of the second-order determinant is -27.

b. $\begin{vmatrix} 2 & 4 \\ -3 & -5 \end{vmatrix} = 2(-5) - (-3)(4) = -10 + 12 = 2$ The value of the second-order determinant is 2.

2 Solve a linear system of equations in two variables using Cramer's rule.

Solving Linear Systems of Equations in Two Variables Using Determinants

Determinants can be used to solve a linear system in two variables. In general, such a system appears as

$$a_1 x + b_1 y = c_1$$
$$a_2 x + b_2 y = c_2.$$

Let's first solve this system for x using the addition method. We can solve for x by eliminating y from the equations. Multiply the first equation by b_2 and the second equation by $-b_1$. Then add the two equations:

$$
\begin{array}{ll}
a_1 x + b_1 y = c_1 & \xrightarrow{\text{Multiply by } b_2.} \\
a_2 x + b_2 y = c_2 & \xrightarrow{\text{Multiply by } -b_1.}
\end{array}
\qquad
\begin{array}{l}
\quad a_1 b_2 x + b_1 b_2 y = \quad b_2 c_1 \\
\underline{-a_2 b_1 x - b_1 b_2 y = -b_1 c_2} \\
\text{Add: } (a_1 b_2 - a_2 b_1)x = b_2 c_1 - b_1 c_2 \\
\qquad\qquad\quad x = \dfrac{b_2 c_1 - b_1 c_2}{a_1 b_2 - a_2 b_1}
\end{array}
$$

Because

$$
\begin{vmatrix} c_1 & b_1 \\ c_2 & b_2 \end{vmatrix} = c_1 b_2 - c_2 b_1 \qquad \text{and} \qquad \begin{vmatrix} a_1 & b_1 \\ a_2 & b_2 \end{vmatrix} = a_1 b_2 - a_2 b_1
$$

we can express our answer for x as the quotient of two determinants:

$$
x = \frac{\begin{vmatrix} c_1 & b_1 \\ c_2 & b_2 \end{vmatrix}}{\begin{vmatrix} a_1 & b_1 \\ a_2 & b_2 \end{vmatrix}}.
$$

In a similar way, we could use the addition method to solve our system for y, again expressing y as the quotient of two determinants. This method of using determinants to solve the linear system, called *Cramer's rule,* is summarized in the box.

No, not that
Kramer!

Solving a system using determinants

Cramer's Rule

If

$$
\begin{aligned}
a_1 x + b_1 y &= c_1 \\
a_2 x + b_2 y &= c_2
\end{aligned}
$$

then

$$
x = \frac{\begin{vmatrix} c_1 & b_1 \\ c_2 & b_2 \end{vmatrix}}{\begin{vmatrix} a_1 & b_1 \\ a_2 & b_2 \end{vmatrix}} \qquad \text{and} \qquad y = \frac{\begin{vmatrix} a_1 & c_1 \\ a_2 & c_2 \end{vmatrix}}{\begin{vmatrix} a_1 & b_1 \\ a_2 & b_2 \end{vmatrix}}
$$

where

$$
\begin{vmatrix} a_1 & b_1 \\ a_2 & b_2 \end{vmatrix} \neq 0.
$$

Example 2 illustrates the use of Cramer's rule.

EXAMPLE 2 **Using Cramer's Rule to Solve a Linear System**

Use Cramer's rule to solve the system:

$$5x - 4y = 2$$
$$6x - 5y = 1$$

Solution

Because

$$x = \frac{D_x}{D} \quad \text{and} \quad y = \frac{D_y}{D}$$

we will set up and evaluate the three determinants $D, D_x,$ and D_y.

1. D, the determinant in both denominators, consists of the x- and y-coefficients.

$$D = \begin{vmatrix} 5 & -4 \\ 6 & -5 \end{vmatrix} = (5)(-5) - (6)(-4) = -25 + 24 = -1$$

2. D_x, the determinant in the numerator for x, is obtained by replacing the x-coefficients in D, 5 and 6, by the constants on the right side of the equation, 2 and 1.

$$D_x = \begin{vmatrix} 2 & -4 \\ 1 & -5 \end{vmatrix} = (2)(-5) - (1)(-4) = -10 + 4 = -6$$

3. D_y, the determinant in the numerator for y, is obtained by replacing the y-coefficients in D, -4 and -5, by the constants on the right side of the equation, 2 and 1.

$$D_y = \begin{vmatrix} 5 & 2 \\ 6 & 1 \end{vmatrix} = (5)(1) - (6)(2) = 5 - 12 = -7$$

4. Thus,

$$x = \frac{D_x}{D} = \frac{-6}{-1} = 6 \qquad \text{and} \qquad y = \frac{D_y}{D} = \frac{-7}{-1} = 7.$$

As always, the solution $(6, 7)$ can be checked by substituting these values into the original equations. The solution set is $\{(6, 7)\}$. ∎

3 Evaluate a third-order determinant.

The Determinant of a 3 × 3 Matrix

The determinant for a 3×3 square matrix, called a *third-order determinant*, is defined in terms of second-order determinants.

Definition of the determinant of a 3 × 3 matrix

A third-order determinant is defined by

Subtract. Add.

$$\begin{vmatrix} a_1 & b_1 & c_1 \\ a_2 & b_2 & c_2 \\ a_3 & b_3 & c_3 \end{vmatrix} = a_1 \begin{vmatrix} b_2 & c_2 \\ b_3 & c_3 \end{vmatrix} - a_2 \begin{vmatrix} b_1 & c_1 \\ b_3 & c_3 \end{vmatrix} + a_3 \begin{vmatrix} b_1 & c_1 \\ b_2 & c_2 \end{vmatrix}.$$

Note that the a's come from the first column.

Study tip

Here are some helpful tips when evaluating a 3×3 determinant using the definition given above.

1. Each of the three terms in the definition contains two factors — a number and a second-order determinant.

2. The numerical factor in each term is an element from the first column of the third-order determinant.

3. The minus sign precedes the second term.

4. The second-order determinant that appears in each term is obtained by crossing out the row and the column containing the numerical factor.

$$a_1 \begin{vmatrix} b_2 & c_2 \\ b_3 & c_3 \end{vmatrix} - a_2 \begin{vmatrix} b_1 & c_1 \\ b_3 & c_3 \end{vmatrix} + a_3 \begin{vmatrix} b_1 & c_1 \\ b_2 & c_2 \end{vmatrix}$$

$$\begin{vmatrix} a_1 & b_1 & c_1 \\ a_2 & b_2 & c_2 \\ a_3 & b_3 & c_3 \end{vmatrix} \qquad \begin{vmatrix} a_1 & b_1 & c_1 \\ a_2 & b_2 & c_2 \\ a_3 & b_3 & c_3 \end{vmatrix} \qquad \begin{vmatrix} a_1 & b_1 & c_1 \\ a_2 & b_2 & c_2 \\ a_3 & b_3 & c_3 \end{vmatrix}$$

EXAMPLE 3   **Evaluating a Third-Order Determinant**

Evaluate: $\begin{vmatrix} 1 & -1 & 2 \\ -2 & 1 & 1 \\ 1 & -1 & 3 \end{vmatrix}$

Solution

Don't forget to supply the minus sign.

$$\begin{vmatrix} 1 & -1 & 2 \\ -2 & 1 & 1 \\ 1 & -1 & 3 \end{vmatrix} = 1\begin{vmatrix} 1 & 1 \\ -1 & 3 \end{vmatrix} - (-2)\begin{vmatrix} -1 & 2 \\ -1 & 3 \end{vmatrix} + 1\begin{vmatrix} -1 & 2 \\ 1 & 1 \end{vmatrix}$$

$$= 1(3 + 1) + 2(-3 + 2) + 1(-1 - 2)$$ Evaluate the three
$$= 1(4) + 2(-1) + 1(-3)$$ second-order
$$= 4 - 2 - 3$$ determinants.
$$= -1$$

4 Solve a linear system of equations in three variables using Cramer's rule.

Solving Linear Systems of Equations in Three Variables Using Determinants

Cramer's rule can be applied to solving linear systems of equations in three variables. The determinants in the numerator and denominator of all variables are third-order determinants.

Solving a linear system in three variables using determinants

Cramer's Rule

If

$$a_1 x + b_1 y + c_1 z = d_1$$
$$a_2 x + b_2 y + c_2 z = d_2$$
$$a_3 x + b_3 y + c_3 z = d_3$$

then

$$x = \frac{D_x}{D}, y = \frac{D_y}{D}, \text{and } z = \frac{D_z}{D}.$$

These four third-order determinants are given by

$$D = \begin{vmatrix} a_1 & b_1 & c_1 \\ a_2 & b_2 & c_2 \\ a_3 & b_3 & c_3 \end{vmatrix}$$ These are the coefficients of the variables $x, y,$ and z.
$(D \neq 0)$

$$D_x = \begin{vmatrix} d_1 & b_1 & c_1 \\ d_2 & b_2 & c_2 \\ d_3 & b_3 & c_3 \end{vmatrix}$$ Replace x-coefficients in D with the constants at the right of the three equations.

$$D_y = \begin{vmatrix} a_1 & d_1 & c_1 \\ a_2 & d_2 & c_2 \\ a_3 & d_3 & c_3 \end{vmatrix}$$ Replace y-coefficients in D with the constants at the right of the three equations.

$$D_z = \begin{vmatrix} a_1 & b_1 & d_1 \\ a_2 & b_2 & d_2 \\ a_3 & b_3 & d_3 \end{vmatrix}$$ Replace z-coefficients in D with the constants at the right of the three equations.

EXAMPLE 4   **Using Cramer's Rule to Solve a Linear System in Three Variables**

Use Cramer's rule to solve the system:

$$3x - 2y + z = 16$$
$$2x + 3y - z = -9$$
$$x + 4y + 3z = 2$$

Solution

Because

$$x = \frac{D_x}{D} \qquad y = \frac{D_y}{D} \qquad z = \frac{D_z}{D}$$

we need to set up and evaluate four determinants.

Step 1. Set up the determinants.
1. D, the determinant in all three denominators, consists of the x-, y-, and z-coefficients.

$$D = \begin{vmatrix} 3 & -2 & 1 \\ 2 & 3 & -1 \\ 1 & 4 & 3 \end{vmatrix}$$

2. D_x, the determinant in the numerator for x, is obtained by replacing the x-coefficients in D, 3, 2, and 1, by the constants on the right side of the equation, 16, -9, and 2.

$$D_x = \begin{vmatrix} 16 & -2 & 1 \\ -9 & 3 & -1 \\ 2 & 4 & 3 \end{vmatrix}$$

3. D_y, the determinant in the numerator for y, is obtained by replacing the y-coefficients in D, -2, 3, and 4, by the constants on the right side of the equation, 16, -9, and 2.

$$D_y = \begin{vmatrix} 3 & 16 & 1 \\ 2 & -9 & -1 \\ 1 & 2 & 3 \end{vmatrix}$$

4. D_z, the determinant in the numerator for z, is obtained by replacing the z-coefficients in D, 1, -1, and 3, by the constants on the right side of the equation, 16, -9, and 2.

$$D_z = \begin{vmatrix} 3 & -2 & 16 \\ 2 & 3 & -9 \\ 1 & 4 & 2 \end{vmatrix}$$

Enrichment Essay

The First Computer

The drudgery of repeated computation being relegated to a machine was the initial impetus in the invention of computers. Charles Babbage (1792–1871), an English inventor, was quite familiar with Cramer's ideas, designing a machine to calculate and print mathematical tables. His difference engine, an elaborate machine for multiplying, consisted of a series of toothed wheels on shafts. Although Babbage was ahead of his time, his machine failed because the necessary precision tools needed in its manufacture were not available at the time.

A portion of Charles Babbage's unrealized Difference Engine. David Parker/Science Museum/Science Photo Library/ Photo Researchers, Inc.

Step 2. Evaluate the four determinants.

1.
$$D = \begin{vmatrix} 3 & -2 & 1 \\ 2 & 3 & -1 \\ 1 & 4 & 3 \end{vmatrix} = 3\begin{vmatrix} 3 & -1 \\ 4 & 3 \end{vmatrix} - 2\begin{vmatrix} -2 & 1 \\ 4 & 3 \end{vmatrix} + 1\begin{vmatrix} -2 & 1 \\ 3 & -1 \end{vmatrix}$$

$$= 3(9 + 4) - 2(-6 - 4) + 1(2 - 3)$$
$$= 3(13) - 2(-10) + 1(-1)$$
$$= 39 + 20 - 1$$
$$= 58$$

Using the same technique and evaluating each determinant about the elements in the first column, we obtain

2. $D_x = 116$
3. $D_y = -174$
4. $D_z = 232$

Step 3. Substitute these four values and solve the system.

$$x = \frac{D_x}{D} = \frac{116}{58} = 2$$

$$y = \frac{D_y}{D} = \frac{-174}{58} = -3$$

$$z = \frac{D_z}{D} = \frac{232}{58} = 4$$

The solution $(2, -3, 4)$ can be checked by substitution into the original three equations. The solution set is $\{(2, -3, 4)\}$. ∎

5 Use determinants to identify inconsistent and dependent systems.

Cramer's Rule with Inconsistent and Dependent Systems

If D, the determinant in the denominator, is 0, the variables described by the quotient of determinants are not real numbers. However, when $D = 0$, this

Discover for yourself

Write a system of two equations that is inconsistent. Now use determinants and the result in the box to verify that this is truly an inconsistent system. Repeat the same process for a system with two dependent equations.

indicates that the system is inconsistent or contains dependent equations. This gives rise to the following two situations.

Determinants: inconsistent and dependent systems

1. If $D = 0$ and at least one of the determinants in the numerator is not 0, then the system is inconsistent. The solution set is $\varnothing$.
2. If $D = 0$ and all the determinants in the numerators are 0, then the equations in the system are dependent.

PROBLEM SET 3.5

Practice Problems

Evaluate each determinant in Problems 1–10.

1. $\begin{vmatrix} 5 & 7 \\ 2 & 3 \end{vmatrix}$
   2. $\begin{vmatrix} 4 & 8 \\ 5 & 6 \end{vmatrix}$
   3. $\begin{vmatrix} -4 & 1 \\ 5 & 6 \end{vmatrix}$
   4. $\begin{vmatrix} 7 & 9 \\ -2 & -5 \end{vmatrix}$
   5. $\begin{vmatrix} -7 & 14 \\ 2 & -4 \end{vmatrix}$

6. $\begin{vmatrix} 1 & -1 & 2 \\ 2 & 1 & 3 \\ 0 & -2 & 1 \end{vmatrix}$
   7. $\begin{vmatrix} 0 & -1 & 2 \\ -3 & 1 & 3 \\ 1 & -2 & -1 \end{vmatrix}$
   8. $\begin{vmatrix} 3 & 4 & -2 \\ -1 & 2 & 1 \\ -1 & -1 & 5 \end{vmatrix}$
   9. $\begin{vmatrix} 2 & 4 & 0 \\ 1 & -2 & -3 \\ 3 & 1 & 0 \end{vmatrix}$
   10. $\begin{vmatrix} 4 & 1 & 0 \\ 1 & -1 & -1 \\ -2 & -1 & 0 \end{vmatrix}$

For Problems 11–26, use Cramer's rule to solve each system or to determine that the system is inconsistent or dependent.

11. $x + y = 7$
$x - y = 3$

12. $2x + y = 3$
$x - y = 3$

13. $12x + 3y = 15$
$2x - 3y = 13$

14. $x - 2y = 5$
$5x - y = -2$

15. $4x - 5y = 17$
$2x + 3y = 3$

16. $3x + 2y = 2$
$2x + 2y = 3$

17. $x + 2y = 3$
$5x + 10y = 15$

18. $2x - 9y = 5$
$3x - 3y = 11$

19. $3x + 3y - z = 10$
$x + 9y + 2z = 16$
$x - y + 6z = 14$

20. $x - y + z = 4$
$3x - y + z = 6$
$2x + 2y - 3z = -6$

21. $x - y + 2z = 4$
$3x + 2y - 4z = 2$
$x + y + z = 3$

22. $x - y - 2z = -3$
$2x - y + z = 7$
$x + y + z = 2$

23. $x - y = 8$
$x + y - z = 1$
$x - 2z = 0$

24. $3x + z = 7$
$y + 4z = 6$
$5y - z = 9$

25. $3x + y = -5$
$2x + z = -1$
$x - 2y = -4$

26. $x + z = 0$
$x - 3y = 1$
$4y - 3z = 3$

Application Problems

Determinants are used to find the area of a triangle whose vertices are given by three points in a rectangular coordinate system. The area of a triangle with vertices (x_1, y_1), (x_2, y_2), and (x_3, y_3) is

$$\text{Area} = \pm \frac{1}{2} \begin{vmatrix} x_1 & y_1 & 1 \\ x_2 & y_2 & 1 \\ x_3 & y_3 & 1 \end{vmatrix}$$

where the symbol $(\pm)$ indicates that the appropriate sign should be chosen to yield a positive area. Use this information to work Problems 27–28.

27. a. Use determinants to find the area of the triangle whose vertices are $(3, -5)$, $(2, 6)$, and $(-3, 5)$.
   b. Graph the triangle in part (a) and then confirm your answer by using the formula for a triangle's area, $A = \frac{1}{2} bh$.

28. Find the area of the triangle whose vertices are $(1, 1)$, $(-2, -3)$, and $(11, -3)$.

Determinants are used to show that three points lie on the same line (are collinear). If

$$\begin{vmatrix} x_1 & y_1 & 1 \\ x_2 & y_2 & 1 \\ x_3 & y_3 & 1 \end{vmatrix} = 0$$

then the points (x_1, y_1), (x_2, y_2), and (x_3, y_3) are collinear. If the determinant does not equal 0, then the points are not collinear. Use this information to work Problems 29–30.

29. Do the points $(3, -1)$, $(0, -3)$, and $(12, 5)$ lie on the same line?

30. Do the points $(-4, -6)$, $(1, 0)$, and $(11, 12)$ lie on the same line?

Determinants are used to write an equation of a line passing through two points. An equation of the line passing through the distinct points (x_1, y_1) and (x_2, y_2) is given by

$$\begin{vmatrix} x & y & 1 \\ x_1 & y_1 & 1 \\ x_2 & y_2 & 1 \end{vmatrix} = 0$$

Use this information to work Problems 31–32.

31. Use the determinant to write an equation for the line passing through $(3, -5)$ and $(-2, 6)$. Then expand the determinant, expressing the line's equation in standard form.

32. Use the determinant to write an equation for the line passing through $(-1, 3)$ and $(2, 4)$. Then expand the determinant, expressing the line's equation in standard form.

True–False Critical Thinking Problems

33. Which one of the following is true?
   a. Using Cramer's rule on a system of two linear equations, both x and y consist of the quotient of two determinants. Therefore, a total of four determinants must be evaluated to solve two equations in two variables.
   b. $\begin{vmatrix} 8 & -4 \\ 4 & 2 \end{vmatrix} = 0$
   c. If the determinant $D = 0$, then a system has no solution.
   d. $\begin{vmatrix} \frac{1}{2} & \frac{1}{4} \\ \frac{1}{3} & -\frac{1}{2} \end{vmatrix} = -\frac{1}{3}$

34. Which one of the following is true?
   a. If $D = 0$, then every variable has a value of 0.
   b. Using Cramer's rule, we use $\dfrac{D}{D_y}$ to get the value of y.
   c. Since there are different determinants in the numerators of x and y, if a system is solved using Cramer's rule, x and y cannot have the same value.
   d. Only one 2×2 determinant is needed to evaluate
   $$\begin{vmatrix} 2 & 3 & -2 \\ 0 & 1 & 3 \\ 0 & 4 & -1 \end{vmatrix}.$$

Technology Problems

35. You can use your graphing utility to evaluate the determinant of a square matrix. First use the matrix feature to enter the matrix. Then evaluate the determinant using a sequence similar to det [A] $\boxed{\text{ENTER}}$. Use this feature to verify your results in Problems 1–10.

36. What is the fastest method for solving a linear system with your graphing utility?

Writing in Mathematics

37. Students usually find it easier to solve a system using Cramer's rule than to solve word problems that give rise to two equations in two variables. Why do you think that this is the case?

38. Describe how Cramer's rule is derived.

39. In applying Cramer's rule, what does it mean if $D = 0$?

40. Describe D_x and D_y in terms of the coefficients and constants in a system of two equations in two variables.

41. Some students find the process of solving a linear system in three variables using Cramer's rule rather tedious. Is there a way of speeding up this process, perhaps using Cramer's rule to find the value for only one of the variables? Describe how this process might work, presenting a specific example with your description. Remember that your goal is still to find the value for each variable in the system.

42. Describe a procedure that might be used to evaluate a fourth-order determinant such as

$$\begin{vmatrix} 1 & 2 & -1 & 1 \\ 2 & -1 & 0 & 2 \\ 0 & 1 & 3 & 0 \\ 0 & 0 & -1 & 0 \end{vmatrix}$$

In what sort of system would one use such a determinant in the solution process?

Critical Thinking Problems

43. Solve for x: $\begin{vmatrix} 2x+1 & 5 \\ x-1 & -3 \end{vmatrix} = 3x - 1$.

44. What happens to the value of a second-order determinant if the two columns are interchanged?

For Problems 45–46, consider the system

$$a_1 x + b_1 y = c_1$$
$$a_2 x + b_2 y = c_2$$

45. Use Cramer's rule to prove that if one of the equations of the system is multiplied by the nonzero constant K, the resulting system has the same solution as the original system.

46. Use Cramer's rule to prove that if the first equation of the system is replaced by the sum of the two equations, the resulting system has the same solution as the original system.

47. Show that the equation of a line through (x_1, y_1) and (x_2, y_2) is given by the determinant

$$\begin{vmatrix} x & y & 1 \\ x_1 & y_1 & 1 \\ x_2 & y_2 & 1 \end{vmatrix} = 0$$

48. Evaluate: $\begin{vmatrix} \begin{vmatrix} 2 & -1 \\ -3 & -1 \end{vmatrix} & \begin{vmatrix} -2 & 1 \\ -3 & -1 \end{vmatrix} \\ \begin{vmatrix} -1 & -2 \\ -1 & 3 \end{vmatrix} & \begin{vmatrix} 2 & -1 \\ 3 & -1 \end{vmatrix} \end{vmatrix}$

49. a. Evaluate: $\begin{vmatrix} a & a \\ 0 & a \end{vmatrix}$

b. Evaluate: $\begin{vmatrix} a & a & a \\ 0 & a & a \\ 0 & 0 & a \end{vmatrix}$

c. Evaluate: $\begin{vmatrix} a & a & a & a \\ 0 & a & a & a \\ 0 & 0 & a & a \\ 0 & 0 & 0 & a \end{vmatrix}$

d. Describe the pattern in the given determinants.
e. Describe the pattern in the evaluations.

50. Suppose that the determinant form for a triangle's area, given just before Problem 27, resulted in zero once you substituted the coordinates of the three given points. Describe what this means about the points.

51. Verify that $(-1, 0, 1)$ is a solution of the system

$$x + 2y + 6z = 5$$
$$-3x - 6y + 5z = 8$$
$$2x + 6y + 9z = 7$$

Explain why this solution cannot be found using Cramer's rule.

Review Problems

52. A taxi fare is $1.40 plus $1.60 per mile. How far can a person travel for $35?

53. Graph: $y < 3x + 2$.

54. Solve: $2[5 - 4(x + 2)] = 2(x - 3) + 5$.

S E C T I O N   3 . 6

Solutions Manual   Tutorial   Video 4

Systems of Linear Inequalities and Linear Programming

Objectives

1 Graph a system of linear inequalities.
2 Model objective functions and constraints.
3 Use linear programming to solve problems.

In this section, we consider systems of linear inequalities and their applications.

Graph a system of linear inequalities.

Systems of Linear Inequalities

We have seen that a linear system can be solved by graphing the individual equations in the system and then finding the intersection of the lines. We follow the same procedure for a system of inequalities. Graph each inequality and then find the intersection of the graphs.

Juan Gris "Breakfast" 1914, cut-and-pasted paper, crayon and oil over canvas, $37\frac{7}{8} \times 23\frac{1}{2}$ in. (80.9 × 59.7 cm). The Museum of Modern Art, New York. Acquired through the Lillie P. Bliss Bequest. Photograph © 1997 The Museum of Modern Art, New York.

EXAMPLE 1 **Graphing a System of Inequalities**

Graph the system:

$$x + 2y > 4$$
$$2x - 3y < -6$$

Solution

We'll solve this problem using five steps:

1. Graph $x + 2y = 4$.
2. Graph $x + 2y > 4$.
3. Graph $2x - 3y = -6$ in the same rectangular coordinate system.
4. Graph $2x - 3y < -6$ in the same rectangular coordinate system.
5. Find the intersection of the graphs in steps 2 and 4.

1. To graph the inequality $x + 2y > 4$, we graph $x + 2y = 4$ using a dashed line. The x-intercept is 4 and the y-intercept is 2. The dashed line is shown in Figure 3.13.
2. Consider $(0, 0)$ as a test point.

$$x + 2y > 4$$
$$0 + 2(0) > 4$$
$$0 > 4 \quad \text{False}$$

This means that $(0, 0)$ does not belong in the graph. The graph of $x + 2y > 4$ is shown in yellow in Figure 3.12.
3. To graph $2x - 3y < -6$, we graph $2x - 3y = -6$ using a dashed line. The x-intercept is -3 and the y-intercept is 2. This line should be graphed on the same rectangular coordinate system as the graph from step 2, resulting in the graphs shown in Figure 3.13.
4. Consider $(0, 0)$ as a test point.

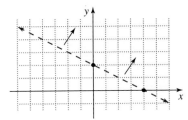

Figure 3.12

The graph of $x + 2y > 4$

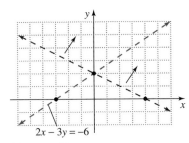

$2x - 3y = -6$

Figure 3.13

Adding the graph of $2x - 3y = -6$ to the graph of $x + 2y > 4$

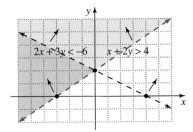

$2x + 3y < -6$ $x + 2y > 4$

Figure 3.14

Graphing both inequalities, but not specifying their intersection

$$2x - 3y < -6$$
$$2(0) - 3(0) < -6$$
$$0 < -6 \quad \text{False}$$

Once again, $(0, 0)$ does not belong in the graph. The graph of $2x - 3y < -6$ includes the half-plane above the line, shown in blue in Figure 3.14. The arrows near the ends of each line indicate the half-plane that contains solutions for each inequality.

5. The graph of the solution set for the given system is the intersection of the individual graphs. Since it is shaded in both yellow and blue, it appears to be green in Figure 3.15.

Discover for yourself

Select three points in the shaded green region in Figure 3.15. For each point, show that its coordinates satisfy both $x + 2y > 4$ and $2x - 3y < -6$.

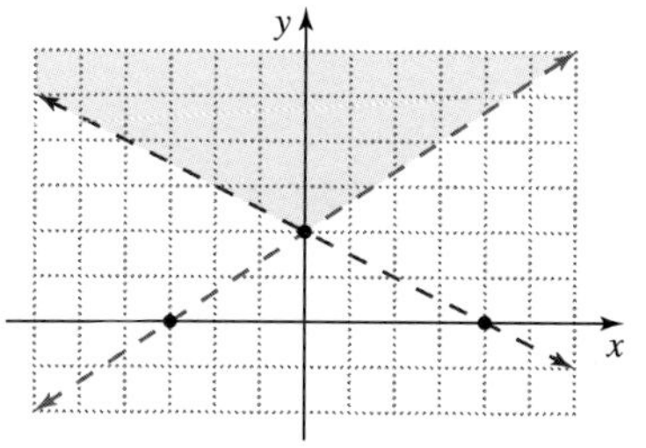

All points in the region shaded in green have coordinates satisfying both $x + 2y > 4$ and $2x - 3y < -6$.

Figure 3.15

The graph of a system of inequalities

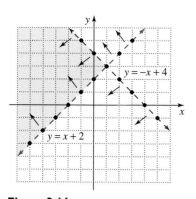

Figure 3.16

The graph of $y > x + 2$ and $y < -x + 4$

EXAMPLE 2 **Graphing a System of Inequalities**

Graph the system:

$$y > x + 2$$
$$y < -x + 4$$

Solution

We begin by graphing $y = x + 2$ (y-intercept $= 2$; slope $= 1$) and $y = -x + 4$ (y-intercept $= 4$; slope $= -1$) in the same rectangular coordinate system (see Figure 3.16). Both lines are shown as dashed lines. The arrows near each line indicate the half-plane that contains the solutions for each inequality. The intersection of these half-planes, shaded in green in Figure 3.16, is the graph of the given system.

Discover for yourself

Where do the lines in Figure 3.16 appear to intersect? Verify this observation algebraically. How can this intersection point be used to solve Example 2?

EXAMPLE 3 **Graphing a Compound Inequality**

Graph $-3 \leq x < 5$ in the rectangular coordinate system.

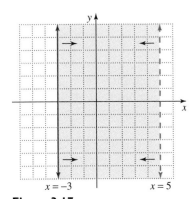

Figure 3.17

The graph of $-3 \leq x < 5$

Solution

We must graph the system

$$x \geq -3$$
$$x < 5$$

The graph of $x \geq -3$ includes the vertical line $x = -3$ and the half-plane to the right (Figure 3.17). The graph of $x < 5$ does not include the vertical line $x = 5$, but does include the half-plane to its left. The arrows near the ends of each line indicate the half-plane that contains the solutions for each inequality. The intersection of these half-planes, shaded in green in Figure 3.17, is the graph of $-3 \leq x < 5$ in the rectangular coordinate system. ∎

tudy tip

Do not confuse the graph of $-3 \leq x < 5$ in the rectangular coordinate system, shown in Figure 3.17, with the graph of $-3 \leq x < 5$ on a number line. In what ways are these graphs similar? How do they differ?

$$\xleftarrow{\quad} \underset{-5\ -4\ -3\ -2\ -1\ \ 0\ \ 1\ \ 2\ \ 3\ \ 4\ \ 5}{|\ \ |\ \ |\ \ |\ \ |\ \ |\ \ |\ \ |\ \ |\ \ |\ \ |} \xrightarrow{\quad}$$

2 Model objective functions and constraints.

Roger Brown "Midnight Tremor" 1972 o/c 72 × 48 in. Photo courtesy of Phyllis Kind Gallery, New York & Chicago.

Objective Functions and Constraints in Linear Programming

Linear programming is a method for solving problems in which a particular quantity that must be maximized or minimized is limited in some ways. For example, for the situation described in the chapter opening, we want to maximize the amount of bottled water and the number of medical kits that can be shipped to earthquake victims. However, the planes that ship these supplies are limited in terms of both the weight and the volume of what they can carry. Before we consider this problem in detail, we need to define two terms.

The vocabulary of linear programming

Objective Function

An algebraic expression in two or more variables modeling a quantity that must be maximized or minimized

Constraints

A system of inequalities that models the limitations in the situation

---

Now let's see how we can help those earthquake victims.

EXAMPLE 4 **Writing an Objective Function**

Bottled water and medical supplies are to be shipped to victims of an earthquake by plane. Each container of bottled water will serve 10 people and each medical kit will aid 6 people. If x represents the number of bottles of water to be shipped and y represents the number of medical kits, write the objective function that models the number of people that can be helped.

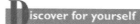

Solution

Since each bottle of water serves 10 people and each medical kit aids 6 people, we have

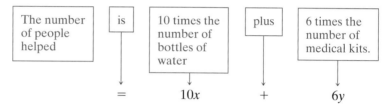

Using z to represent the objective function, we have

$$z = 10x + 6y.$$

Unlike the functions that we have seen so far, the objective function has two independent variables. ■

    From the Discover for Yourself box, it appears that x and y should be increased without limit until all victims of the earthquake are helped. However, the planes are bound by certain limitations. These limitations are considered in Examples 5 and 6.

EXAMPLE 5 Writing a Constraint

Each plane can carry no more than 80,000 pounds. The bottled water weighs 20 pounds per container and each medical kit weighs 10 pounds. If x represents the number of bottles of water to be shipped and y represents the number of medical kits, write an inequality that models this constraint.

Solution

Since each plane can carry no more than 80,000 pounds, we have:

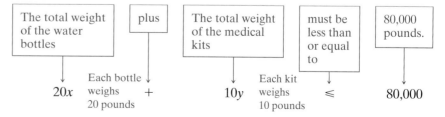

The plane's weight constraint is modeled by the inequality $20x + 10y \leq$ 80,000. ■

In addition to a weight constraint on its cargo, each plane has a limited amount of space in which to carry supplies. Example 6 demonstrates how to express this constraint.

EXAMPLE 6   **Writing a Constraint**

Planes can carry a total volume for supplies that does not exceed 6000 cubic feet. Each water bottle is 1 cubic foot and each medical kit also has a volume of 1 cubic foot. With x still representing the number of water bottles and y the number of medical kits, write an inequality that models this second constraint.

Solution

Since each plane can carry a volume for supplies that does not exceed 6000 cubic feet, we have:

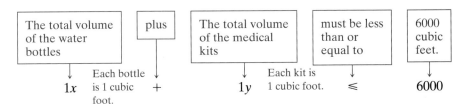

| The total volume of the water bottles | plus | The total volume of the medical kits | must be less than or equal to | 6000 cubic feet. |

$$1x \quad + \quad 1y \quad\quad \leqslant \quad\quad 6000$$

The plane's volume constraint is modeled by the inequality $x + y \leqslant 6000$. ∎

In summary, here's what we have modeled in this aid-to-earthquake-victims situation:

$z = 10x + 6y$ This is the objective function describing the number of people helped with x bottles of water and y medical kits.

$20x + 10y \leqslant 80{,}000$ These are the constraints based
$\quad\quad x + \quad y \leqslant 6000$ on each plane's weight and volume limitations.

3 Use linear programming to solve problems.

Solving Problems with Linear Programming

The problem in the situation described above is to maximize the number of earthquake victims who can be helped subject to the planes' weight and volume constraints. The process of solving this problem is called linear programming, based on a theorem that was proven during World War II.

> **Solving a linear programming problem**
>
> Let $z = ax + by$ be an objective function that depends on x and y. Furthermore, z is subject to a number of constraints on x and y. If a maximum or minimum value of z exists, it can be determined as follows:
>
> 1. Graph the system of inequalities representing the constraints.
> 2. Find the value of the objective function at each corner, or vertex, of the graphed region. The maximum and minimum of the objective function occurs at the corner points.

EXAMPLE 7 **Solving the Linear Programming Problem**

Solve the problem of determining how many bottles of water and how many medical kits should be sent on each plane to maximize the number of earthquake victims who can be helped.

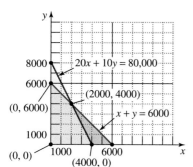

Figure 3.18

The region in quadrant I representing the constraints

$20x + 10y \leq 80,000$

$x + y \leq 6000$

Solution

We must maximize $z = 10x + 6y$ subject to the constraints

$$20x + 10y \leq 80,000$$
$$x + y \leq 6000$$

Since x (the number of bottles of water per plane) and y (the number of medical kits per plane) must be positive, we need only graph the system of inequalities in quadrant I. To graph the inequality $20x + 10y \leq 80,000$, we graph the equation $20x + 10y = 80,000$ as a solid line. Setting $y = 0$, the x-intercept is 4000 and setting $x = 0$, the y-intercept is 8000. Using $(0, 0)$ as a test point, the inequality is satisfied, so we graph below the line, as shown in yellow in Figure 3.18.

Now we graph $x + y \leq 6000$ by first graphing $x + y = 6000$ as a solid line. Setting $y = 0$, the x-intercept is 6000 and setting $x = 0$, the y-intercept is 6000. Using $(0, 0)$ as a test point, the inequality is satisfied, so we graph below the line, as shown in blue in Figure 3.18.

We use the addition method to find where the lines $20x + 10y = 80,000$ and $x + y = 6000$ intersect.

$$
\begin{array}{ll}
20x + 10y = 80,000 & \xrightarrow{\text{No change}} \\
x + y = 6000 & \xrightarrow{\text{Multiply by } -10.}
\end{array}
\qquad
\begin{array}{rl}
20x + 10y = & 80,000 \\
-10x - 10y = & -60,000 \\
\hline
10x = & 20,000 \\
x = & 2000
\end{array}
$$

Back-substituting 2000 for x in $x + y = 6000$, we find $y = 4000$, so the intersection point is $(2000, 4000)$.

With the region formed by the constraints shown in green in Figure 3.18, we must evaluate the objective function at the four corners of this region.

Corner (Vertex) (x, y)	Objective Function $z = 10x + 6y$
$(0, 0)$	$z = 10(0) + 6(0) = 0$
$(4000, 0)$	$z = 10(4000) + 6(0) = 40,000$
$(2000, 4000)$	$z = 10(2000) + 6(4000) = 44,000$ ← maximum
$(0, 6000)$	$z = 10(0) + 6(6000) = 36,000$

Thus, the maximum value of z is 44,000 and this occurs when $x = 2000$ and $y = 4000$. In practical terms, this means that the maximum number of earthquake victims who can be helped with each plane shipment is 44,000. This can be accomplished by sending 2000 water bottles and 4000 medical kits per plane.

Before considering another example, let's summarize the steps in solving a linear programming problem.

The steps in solving a linear programming problem

1. Write the objective function and all necessary constraints.
2. Graph the region for the system of inequalities representing the constraints.
3. Find the value of the objective function at each vertex or corner of the graphed region.
4. The maximum or minimum of the objective function occurs at the corner points.

EXAMPLE 8 **Solving a Linear Programming Problem**

A student earns $10 per hour for tutoring and $7 per hour as a teacher's aid. Although the student wants to maximize weekly earnings, there are two constraints in the situation:

1. To have enough time for studies, the student can work no more than 20 hours a week.
2. The tutoring center for which the student works requires that each tutor spend at least 3 hours, but no more than 8 hours, a week tutoring.

How many hours each week should the student devote to tutoring and how many hours to working as a teacher's aid to maximize weekly earnings?

Solution

Step 1. Write the objective function and all necessary constraints.

Let

x = the number of hours each week spent tutoring, and

y = the number of hours each week spent as a teacher's aid

The objective function is the function that models weekly earnings. This is the function that is to be maximized.

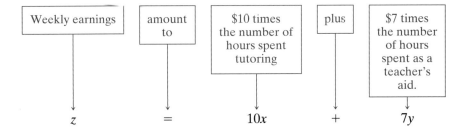

Now we must model each of the constraints. Since the student can work no more than 20 hours a week, we have

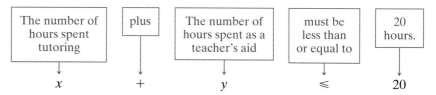

The student can spend at least 3 hours, but no more than 8 hours, a week tutoring.

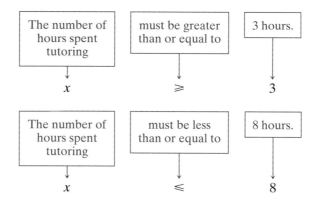

We can write this last constraint as

$$3 \leq x \leq 8$$

In summary, here's what we have:

Objective Function: $z = 10x + 7y$

Constraints: $x + y \leq 20$

$$3 \leq x \leq 8$$

We are now ready to graph the constraints. Since x and y must both be positive, we need only graph the system of inequalities in quadrant I. The graph is shown in Figure 3.19.

Using the region in Figure 3.19, we must evaluate the objective function at the four vertices.

Step 2. Graph the constraints.

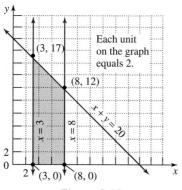

Figure 3.19

Step 3. Evaluate the objective function at each vertex.

Step 4. Solve the problem by selecting the vertex that gives the greatest value of the objective function.

Vertex (x, y)	Objective Function $z = 10x + 7y$
$(3, 0)$	$z = 10(3) + 7(0) = 30$
$(8, 0)$	$z = 10(8) + 7(0) = 80$
$(8, 12)$	$z = 10(8) + 7(12) = 164$ ← maximum
$(3, 17)$	$z = 10(3) + 7(17) = 149$

Thus, the maximum value of z is 164, and this occurs when $x = 8$ and $y = 12$. This means that the maximum amount that the student can earn is $164 per week. This can be accomplished by tutoring for 8 hours and working as a teacher's aid for 12 hours. ∎

ENRICHMENT ESSAY

Linear Programming and the Cold War

The Berlin Airlift (1948–1949) was an operation put into effect by the United States and Great Britain after the former Soviet Union closed all roads and rail lines between West Germany and Berlin, cutting off supply routes to the city. The Allies used linear programming to break the blockade. The 11-month airlift, in 272,264 flights, provided basic necessities to blockaded Berlin. A simplified version describing how this was done can be found in Problem 30 of Problem Set 3.6.

Today, linear programming is one of the most widely used tools in management science, helping businesses allocate resources on hand (the constraints) to manufacture a particular array of products that will maximize profit (the objective function). Linear programming accounts for as much as 90% of all computing time used for management decisions in business.

Jasper Johns (b. 1930) "Three Flags" 1958, encaustic on canvas, $30\frac{7}{8} \times 45\frac{1}{2} \times 5$ in. ($78.4 \times 115.6 \times 12.7$ cm). 50th Anniversary Gift of the Gilman Foundation, Inc., The Lauder Foundation, A. Alfred Taubman, an anonymous donor, and purchase. 80.32. Collection of Whitney Museum of American Art, New York. Photo by Geoffrey Clements. © 1998 Jasper Johns/Licensed by VAGA, New York, NY.

PROBLEM SET 3.6

Practice Problems

In Problems 1–18, graph the system of inequalities. Find the coordinates of any vertices that are formed.

1. $3x + 6y \leq 6$
$2x + y \leq 8$

2. $x - y \geq 4$
$x + y \leq 6$

3. $y < -2x + 3$
$x - y > 2$

4. $y < -x + 4$
$4x - 2y < 6$

5. $y < -2x + 4$
$y < x - 4$

6. $y > 2x - 3$
$y < -x + 6$

7. $-2 < x \leq 4$

8. $-3 < x \leq 6$

9. $-4 \leq y < 2$

10. $-5 \leq y < 1$

11. $4x - 5y > -20$
$x > -3$

12. $y < 3x + 4$
$y > 2$

13. $x + y \geq 0$
$y \leq 3$
$3x - y \leq 6$

14. $x + 2y \geq 8$
$x - y \leq 2$
$y \leq 4$

15. $x + 2y \leq 20$
$x + y \leq 16$
$x \geq 0$
$y \geq 0$

16. $2x + y \leq 8$
$2x + 3y \leq 12$
$x \geq 0$
$y \geq 0$

17. $x + 2y \leq 8$
$y \geq 2x$
$y \leq 3x$

18. $x + y \leq 6$
$x + 2y \leq 8$
$x \leq 4$
$x \geq 0$
$y \geq 0$

In Problems 19–22, find the maximum and minimum values of the given objective function in the indicated region.

19. $z = 5x + 6y$ **20.** $z = 3x + 2y$ **21.** $z = 40x + 50y$ **22.** $z = 30x + 45y$

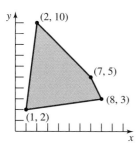
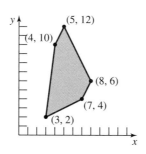
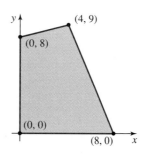
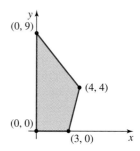

In Problems 23–26, find the maximum and minimum values of the objective function and the values of x *and* y *for which they occur.*

23. $z = x + y$
Subject to
$x \leq 6$
$y \geq 1$
$y \leq 2x + 1$

24. $z = 3x - 2y$
Subject to
$1 \leq x \leq 5$
$y \geq 2$
$y \leq x + 3$

25. $z = -x + 3y$
Subject to
$2 \leq x \leq 6$
$y \leq x + 6$
$y \geq -2x + 6$

26. $z = 6x + 10y$
Subject to
$x + y \leq 12$
$x + 2y \leq 20$
$\left.\begin{array}{l} x \geq 0 \\ y \geq 0 \end{array}\right\}$ Quadrant I

Application Problems

27. A television manufacturer makes console and wide-screen televisions. The profit per unit is $125 for the console televisions and $200 for the wide-screen televisions.

a. Let

$x =$ the number of consoles manufactured in a month and

$y =$ the number of wide screens manufactured in a month

Write the objective function that models the total monthly profit.

b. The manufacturer is bound by the following constraints:

 1. Equipment in the factory allows for making at most 450 console televisions in one month.

 2. Equipment in the factory allows for making at most 200 wide-screen televisions in one month.

 3. The cost to the manufacturer per unit is $600 for the console televisions and $900 for the wide-screen televisions. Total monthly costs cannot exceed $360,000.

Write a system of inequalities that models these constraints.

c. Graph the system of inequalities in part (b). Use only the first quadrant, since x and y must both be positive.

d. Evaluate the objective profit function at each of the five vertices of the graphed region. (The vertices should occur at (0, 0), (0, 200), (300, 200), (450, 100), and (450, 0).)

e. Complete the missing portions of this statement: The television manufacturer will make the greatest profit by manufacturing _____ console televisions each month and _____ wide-screen televisions each month. The maximum monthly profit is $_____.

Use the four steps given in the box on page 298 to solve Problems 28–32.

28. A large institution is preparing lunch menus containing foods A and B. The specifications for the two foods are given in the following table.

Food	Units of Fat per Ounce	Units of Carbohydrates per Ounce	Units of Protein per Ounce
A	1	2	1
B	1	1	1

Each lunch must provide at least 6 units of fat per serving, no more than 7 units of protein, and at least 10 units of carbohydrates. The institution can purchase food A for $0.12 per ounce and food B for $0.08 per ounce. How many ounces of each food should a serving contain to meet the dietary requirements at the least cost?

29. Food and clothing are shipped to victims of a natural disaster. Each carton of food will feed 5 people, while each carton of clothing will help 6 people. Each 30-cubic-foot box of food weighs 50 pounds and each 20-cubic-foot box of clothing weighs 5 pounds. The commercial carriers transporting food and clothing are bound by the following constraints:

    1. The total weight per carrier cannot exceed 18,000 pounds.
    2. The total volume must be less than 12,000 cubic feet.

    How many cartons of food and clothing should be sent with each plane shipment to maximize the number of people who can be helped?

30. On June 24, 1948, the former Soviet Union blocked all land and water routes through East Germany to Berlin. A gigantic airlift was organized using American and British planes to supply food, clothing, and other supplies to the more than 2 million people in West Berlin. The cargo capacity was 30,000 cubic feet for an American plane and 20,000 cubic feet for a British plane. To break the Soviet blockade, the Western Allies had to maximize cargo capacity, but were subject to the following restrictions:

1. No more than 44 planes could be used.
2. The larger American planes required 16 personnel per flight, double that of the requirement for the British planes. The total number of personnel available could not exceed 512.
3. The cost of an American flight was $9000 and the cost of a British flight was $5000. Total weekly costs could not exceed $300,000.

Find the number of American and British planes that were used to maximize cargo capacity.

31. A theater is presenting a program on drinking and driving for students and their parents. The proceeds will be donated to a local alcohol information center. Admission is $2.00 for parents and $1.00 for students. However, the situation has two constraints: The theater can hold no more than 150 people and every two parents must bring at least one student. How many parents and students should attend to raise the maximum amount of money?

32. You are about to take a test that contains computation problems worth 6 points each and word problems worth 10 points each. You can do a computation problem in 2 minutes and a word problem in 4 minutes. You have 40 minutes to take the test and may answer no more than 12 problems. Assuming you answer all the problems attempted correctly, how many of each type of problem must you do to maximize your score? What is the maximum score?

True–False Critical Thinking Problems

33. Which one of the following is true?
    a. The point $(-2, 4)$ is in the graph of the system

    $$2x + 3y \geqslant 4$$
    $$2x - y > -6$$

    b. Any rectangular coordinate system in which nothing at all is graphed could be the graph of the system

    $$y \geqslant 2x + 3$$
    $$y \leqslant 2x - 3$$

    c. In rectangular coordinates, there is no difference between the graphs of the system $x > -2$ *and* $y < 4$ and the system $x > -2$ *or* $y < 4$.
    d. The solution to a system of linear inequalities is represented by the set of points in rectangular coordinates that satisfy any one of the inequalities in the system.

34. Which one of the following is true?
    a. In linear programming, constraints are limitations placed on problem solvers without access to computer programs.
    b. A linear programming problem is solved by evaluating the objective function at the vertices of the region represented by the constraints.
    c. The graph of the system

    $$x + y \leqslant 4$$
    $$2x + y \leqslant 6$$
    $$x \geqslant 0, y \geqslant 0$$

    does not have a corner or vertex at $(2, 2)$.
    d. A student who works for x hours at $15 per hour by tutoring and y hours at $12 per hour as a teacher's aid has earnings modeled by the objective function $z = 15xy + 12xy$.

Writing in Mathematics

35. Explain how to graph a system of linear inequalities.

36. Choose a particular field that is of interest to you. Research how linear programming is used to solve problems in that field. If possible, investigate the solution of a specific practical problem. Write a report on your findings, including the contributions of George Danzig, Narendra Karmarkar, and L. G. Khachion to linear programming.

Critical Thinking Problems

37. Graph $|x| \leq 3$ in rectangular coordinates.

38. Graph $|y + 1| \geq 3$ in rectangular coordinates.

39. Graph the system of inequalities:

$$|x| \leq 3$$
$$|y + 1| \geq 3$$

40. Graph the system of inequalities:

$$y < |x|$$
$$y < 3$$

41. Consider the objective function $z = ax + by$ subject to the following constraints: $2x + 3y \leq 9$, $x - y \leq 2$, $x \geq 0$, and $y \geq 0$. Prove that the objective function will have the same minimum value at the vertices $(3, 1)$ and $(0, 3)$ if $a = \frac{2}{3} b$.

Group Activity Problem

42. Members of the group should interview a business executive who is in charge of deciding the product mix for a business. How are production policy decisions made? Are other methods used in conjunction with linear programming? What are these methods? What sort of academic background, particularly in mathematics, does this executive have? Write a group report addressing these questions, emphasizing the role of linear programming for the business.

Review Problems

43. Simplify: $-14\left(\dfrac{2}{7}\right) - 6 \div (-3)$.

44. Solve: $\dfrac{x + 2}{2} + \dfrac{x - 2}{4} = 8$.

45. Solve: $3x + 5 \geq 26$ and $-5x + 1 \leq 11$.

HAPTER PROJECT

Graphing in Three Dimensions

In Section 3.3 we introduced equations in three variables and stated that the equation

$$Ax + By + Cz = D$$

represents the graph of a plane in three-dimensional space. To graph a particular plane, we need to establish a coordinate system similar to the Cartesian system we used to graph curves in two dimensions. Once we have our system, we can plot points to determine where in space our plane lies. Two points will determine where a line should be graphed; for a plane, we will need three points.

To create a three-dimensional coordinate system, begin by imagining our xy-coordinate plane as the floor of a room. It may help to look at a corner of a room and picture floor tiles as the coordinate grid. Another axis, the z-axis, will extend up from the floor to the ceiling. We now have the x-axis running front to back, the y-axis running left to right, and the z-axis running up and down. The origin is down at the corner of the room where the walls and floor meet. We will have positive values of x coming out towards us, positive values of y going to the right, and positive values of z going upward.

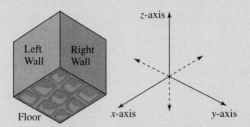

Points in three-dimensional space are given by an ordered triple (x, y, z). To plot the point $(2, 3, 4)$ we would start at the origin and count two units out on the x-axis, then three units to the right, in the direction of the positive y-axis, and four units up, in the positive z direction.

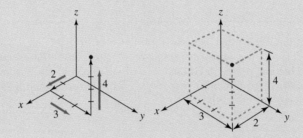

It sometimes helps to visualize a rectangular box with your point on the corner. To graph a plane, the easiest points to pick are the intercepts, $(x, 0, 0)$, $(0, y, 0)$, and $(0, 0, z)$. For example, the plane given by

$$2x - 2y + z = 6$$

would cross the x-axis when $y = 0$, $z = 0$, and $x = 3$. Similarly, it would cross the y-axis at $y = -3$ and the z-axis at $z = 6$. If we connect the points to form a triangle, it will show us how the plane lies with respect to the axes. If we were to also graph the plane given by

$$-4x + 4y - 2z = 12$$

on the same set of axes, it would cross the x-axis at $x = -3$, the y-axis at $y = 3$ and the z-axis at $z = -6$ and we would see two parallel planes.

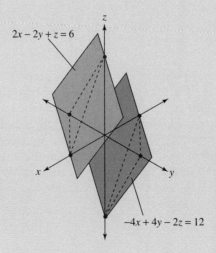

When you begin to plot points in three-dimensional space, it may help to use graph paper with a grid designed for three dimensions, rather than two. This style of graph paper is called *isometric*. Art supply stores or office supply stores that carry tools for architects will usually stock such paper. Once you have seen what isometric graph paper looks like, you may want to try and create your own, starting from ordinary graph paper.

1. Using isometric graph paper, plot the following points:

   A (4, 0, 0)   B (4, 4, 0)   C (0, 4, 0)   D (0, 0, 0)
   E (4, 0, 4)   F (4, 4, 4)   G (0, 4, 4)   H (0, 0, 4)

   Connect the following pairs of points: AB, AD, DC, BC, AE, BD, CG, HE, HG, ED, DG, HD. Identify the figure you have sketched.

2. The distance between two points in three-dimensional space is found using the following formula:

   Given points A (a, b, c) and B (d, e, f)
   the distance AB is given by $\sqrt{(a - d)^2 + (b - e)^2 + (c - f)^2}$.

   Use this formula to calculate the distances DB, DE, and DG. Are those distances what you expected? Use the formula to calculate the distance DF. Could you have predicted this distance?

3. It is sometimes very difficult to see a three-dimensional graph as a sketch on a two-dimensional piece of paper. It is much easier if we make a model of the coordinate axes and study the model in three dimensions. For this assignment, you should work in groups to create models of some of the problems you have solved in your text with three equations in three variables. Construct the three coordinate axes so that they will stand alone on a table top or hang from the ceiling. Using a problem selected by your instructor, make a model that clearly shows the intersection point, if any, or why there is no intersection point if your system has no solution.

Worldwide Web Resources

Go to the Prentice Hall website (http://www.prenhall.com/blitzer) to access other locations on the Internet that will allow you to further explore the concepts presented in this project.

Chapter Review

SUMMARY

1. Linear Systems of Equations in Two Variables

a. The solution set is the set of all ordered pairs of values (a, b) that satisfy every equation in the system.

b. *Solving by Graphing*
   1. Graph each equation in the system.
   2. The solution corresponds to the point of intersection of the graphs.
   3. Inconsistent systems have no solution. Graphs are parallel lines. Systems with dependent equations have infinitely many solutions. Graphs are the same line.

c. *Solving by Substitution*
   1. If necessary, solve one of the equations for one variable in terms of the other.
   2. Substitute the expression for that variable into the other equation.
   3. Solve the resulting equation in one variable.
   4. Back-substitute the value from step 3 into the

equation in step 1 and find the value of the remaining variable.
5. Check the solution in both of the system's given equations.

d. *Solving by Addition*
1. If necessary, rewrite both equations in the form $Ax + By = C$.
2. If necessary, multiply either or both equations by appropriate numbers so that coefficients of x or y will have a sum of 0.
3. Add equations. The sum is an equation in one variable.
4. Solve the equation in step 3.
5. Back-substitute the value from step 4 into either of the given equations and solve for the other variable.
6. Check the solution in both of the system's given equations.

e. *Inconsistent and Dependent Systems*
1. If both variables are eliminated and the resulting statement is false, the system is inconsistent.
2. If both variables are eliminated and the resulting statement is true, the system contains dependent equations.

2. Problem Solving and Modeling Using Systems in Two Variables

a. *Problem-Solving Strategy*
1. Use two variables to represent unknown quantities.
2. Write a linear system modeling the problem's conditions.
3. Solve the system and answer the problem's question.
4. Check the answers in the original wording of the problem.

b. *Problem-Solving Techniques*
These include translating given conditions into a linear system, creating verbal models and then translating into a linear system, using known formulas, and using tables to organize information.

c. *Modeling Using Systems of Equations*
1. *Linear Modeling:* Use $y = mx + b$ and substitute each of the two given data points for x and y. Solve the system for m and b. Substitute the values of m and b into $y = mx + b$ to obtain the desired model.
2. *Supply and Demand Models:* The point of intersection of the graphs is the point of market equilibrium. The x-coordinate, the equilibrium price, is the price at which supply equals demand.

3. Solving Systems in Three Variables (x, y, and z)
Use the addition method to reduce the system to two equations in two variables. Use substitution or the addition method to solve the resulting system in two variables.

4. Solving Linear Systems Using Matrices
a. Write the augmented matrix for the system:

$$\begin{aligned} a_1x + b_1y &= c_1 \\ a_2x + b_2y &= c_2 \end{aligned} \qquad \left[\begin{array}{cc|c} a_1 & b_1 & c_1 \\ a_2 & b_2 & c_2 \end{array}\right]$$

$$\begin{aligned} a_1x + b_1y + c_1z &= d_1 \\ a_2x + b_2y + c_2z &= d_2 \\ a_3x + b_3y + c_3z &= d_3 \end{aligned} \qquad \left[\begin{array}{ccc|c} a_1 & b_1 & c_1 & d_1 \\ a_2 & b_2 & c_2 & d_2 \\ a_3 & b_3 & c_3 & d_3 \end{array}\right]$$

b. Use matrix row operations to simplify the matrix to one with 1s down the diagonal from upper left to lower right, and 0s below the 1s. Matrix row operations:
1. Any two rows of the matrix may be interchanged.
2. The numbers in any row may be multiplied by any nonzero constant.
3. Any row may be changed by adding to the numbers of the row the product of a constant and the numbers of another row.

c. Write the system of linear equations corresponding to the matrix in part (b).

$$\left[\begin{array}{cc|c} 1 & A & B \\ 0 & 1 & C \end{array}\right] \quad \text{or} \quad \left[\begin{array}{ccc|c} 1 & A & B & C \\ 0 & 1 & D & E \\ 0 & 0 & 1 & F \end{array}\right]$$
$$\text{so } y = C \qquad\qquad \text{so } z = F$$

The last row of these matrices gives the value of one variable. Other variables are found by back-substitution.

5. Determinants and Cramer's Rule

a. $\begin{vmatrix} a_1 & b_1 \\ a_2 & b_2 \end{vmatrix} = a_1b_2 - a_2b_1$

b. If
$$\begin{aligned} a_1x + b_1y &= c_1 \\ a_2x + b_2y &= c_2 \end{aligned}$$

then $\qquad x = \dfrac{D_x}{D} \qquad$ and $\qquad y = \dfrac{D_y}{D}$

where

$$D = \begin{vmatrix} a_1 & b_1 \\ a_2 & b_2 \end{vmatrix} \neq 0 \qquad D_x = \begin{vmatrix} c_1 & b_1 \\ c_2 & b_2 \end{vmatrix}$$

$$D_y = \begin{vmatrix} a_1 & c_1 \\ a_2 & c_2 \end{vmatrix}$$

c. $\begin{vmatrix} a_1 & b_1 & c_1 \\ a_2 & b_2 & c_2 \\ a_3 & b_3 & c_3 \end{vmatrix} = a_1\begin{vmatrix} b_2 & c_2 \\ b_3 & c_3 \end{vmatrix} - a_2\begin{vmatrix} b_1 & c_1 \\ b_3 & c_3 \end{vmatrix}$

$$+ a_3\begin{vmatrix} b_1 & c_1 \\ b_2 & c_2 \end{vmatrix}$$

d. If
$$\begin{aligned} a_1x + b_1y + c_1z &= d_1 \\ a_2x + b_2y + c_2z &= d_2 \\ a_3x + b_3y + c_3z &= d_3 \end{aligned}$$

then $\quad x = \dfrac{D_x}{D} \qquad y = \dfrac{D_y}{D} \qquad z = \dfrac{D_z}{D}$

where

$$D = \begin{vmatrix} a_1 & b_1 & c_1 \\ a_2 & b_2 & c_2 \\ a_3 & b_3 & c_3 \end{vmatrix} \neq 0 \qquad D_x = \begin{vmatrix} d_1 & b_1 & c_1 \\ d_2 & b_2 & c_2 \\ d_3 & b_3 & c_3 \end{vmatrix}$$

$$D_y = \begin{vmatrix} a_1 & d_1 & c_1 \\ a_2 & d_2 & c_2 \\ a_3 & d_3 & c_3 \end{vmatrix} \qquad D_z = \begin{vmatrix} a_1 & b_1 & d_1 \\ a_2 & b_2 & d_2 \\ a_3 & b_3 & d_3 \end{vmatrix}$$

e. 1. If $D = 0$ and at least one of the other determinants is not 0, then the system is inconsistent.
 2. If $D = 0$ and all other determinants are also 0, then the system contains dependent equations.

6. Graphing Systems of Inequalities
 a. Graph each inequality in the system in the same rectangular coordinate system.
 b. Find the intersection of the individual graphs.

7. Linear Programming
 a. Write the objective function and all necessary constraints.
 b. Graph the region for the system of inequalities representing the constraints.
 c. Find the value of the objective function at each vertex of the graphed region.
 d. To maximize the objective function, the solution is given by the vertex producing the greatest value of the objective function. To minimize the objective function, the solution is given by the vertex producing the least value of the objective function.

REVIEW PROBLEMS

Solve each system in Problems 1–4 by graphing. If applicable, state that the system is inconsistent or contains dependent equations.

1. $\begin{aligned} x + y &= 5 \\ 3x - y &= 3 \end{aligned}$

2. $\begin{aligned} 3x - 2y &= 6 \\ 6x - 4y &= 12 \end{aligned}$

3. $\begin{aligned} y &= \tfrac{3}{5}x - 3 \\ 2x - y &= -4 \end{aligned}$

4. $\begin{aligned} y &= -x + 4 \\ 3x + 3y &= -6 \end{aligned}$

For Problems 5–9, use the substitution or addition method to solve each system. If applicable, state that the system is inconsistent or contains dependent equations.

5. $\begin{aligned} 2x - y &= 2 \\ x + 2y &= 11 \end{aligned}$

6. $\begin{aligned} y &= 4 - x \\ 3x + 3y &= 12 \end{aligned}$

7. $\begin{aligned} 5x + 3y &= -3 \\ 2x + 7y &= -7 \end{aligned}$

8. $\begin{aligned} x - 2y + 3 &= 0 \\ 2x - 4y + 7 &= 0 \end{aligned}$

9. $\begin{aligned} \tfrac{1}{8}x + \tfrac{3}{4}y &= \tfrac{19}{8} \\ -\tfrac{1}{2}x + \tfrac{3}{4}y &= \tfrac{1}{2} \end{aligned}$

10. Rollerblading is the fastest-growing sport in the United States. In 1992 and 1993 combined, there were 21.8 million rollerbladers. The difference between twice the number of skaters in 1993 and the number in 1992 exceeds the number in 1991 by 9.2 million. How many U.S. rollerbladers were there (in millions) in 1992 and 1993?

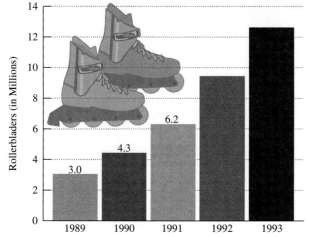

Number of Rollerbladers in the United States

Rollerbladers (in Millions)

1989: 3.0
1990: 4.3
1991: 6.2

Source: Based on statistics from *Rollerblade*

11. Health experts agree that cholesterol intake should be limited to 300 mg or less each day. Three ounces of shrimp and 2 ounces of scallops contain 156 mg of cholesterol. Five ounces of shrimp and 3 ounces of scallops contain 45 mg of cholesterol less than the suggested maximum daily intake. Determine the cholesterol content in an ounce of each item.

12. If 8 pens and 6 pads of paper cost $16.10 and 3 of the same pens and 2 of the same pads cost $5.85, find the cost of 1 pen and 1 pad.

13. The building lot shown in the figure at the top of page 308 is composed of three rectangles. The lengths of two of the sides, as shown, are designated by A and B. All other sides have lengths that can be expressed in terms of A and B. The perimeter of the lot is 310 feet. For privacy, fencing is to be placed along the three lengths designated by A, B, and A, respectively. Fencing along A costs $30 per foot, and fencing along B costs $8 per foot, with a total cost for the three sides of

$2600. Find A and B. Then write a description of the lot's shape for a perspective buyer using these values.

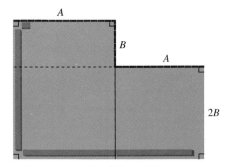

14. In the isosceles triangle, AB and AC have equal measures. Find the measure of each angle in the triangle.

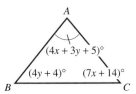

15. When a canoeist rows with the current, it takes 2 hours to row 12 miles. When the same canoeist rows in the opposite direction against the current, it takes 4 hours to travel the same distance. Find the speed of the canoeist in calm water and the rate of the current.

16. How many liters of a 90% sulfuric acid solution should be mixed with a 75% sulfuric acid solution to produce 20 liters of a 78% solution?

17. This problem is from *The Nine Chapters of Mathematical Art,* written in China in the third century B.C. This work contains the first known examples of the solution of linear systems of equations, well ahead of the West. "Suppose there are a number of rabbits and pheasants confined in a cage, in all 35 heads and 94 feet. Find the number of each."

18. a. The table below shows the price of a video and the quantity of that video sold on a weekly basis in a store.

x (Price of the Video)	y (Number of Videos Sold Weekly
$10	400
$14	160

Assuming that as price increases, the number sold weekly decreases steadily, use the linear function $y = mx + b$ to find the values of m and b, thereby modeling demand for the videos.

b. The table below shows the price of the same video and the quantity that the manufacturer is willing to supply to the store on a weekly basis at that price.

x (Price of the Video)	y (Number of Videos Supplied by the Manufacturer Weekly)
$10	240
$14	256

Assuming that as price increases, the number of videos supplied weekly increases steadily, use the linear function $y = mx + b$ to find the values of m and b, thereby modeling video supply.

c. Use the demand and the supply functions to find the point of market equilibrium. Describe exactly what the x- and y-coordinates of this point represent.

d. Graph the demand and the supply functions, visually illustrating your solution in part (c). If applicable, use a graphing utility to verify the graphs.

19. This problem mentions three functions that appear in all business situations:
1. The Cost Function, $C(x)$: This is the cost to a company of producing x units of a product.
2. The Revenue Function, $R(x)$: This is the money taken in by the company for selling x units of the product.
3. The Profit Function, $P(x)$: This is the company's total revenue minus total cost. $P(x) = R(x) - C(x)$

A gasoline station has weekly costs and revenue that are functions of the number of gallons (x) of gasoline purchased and sold. The figure shows the graphs of the cost function, $C(x) = 1.10x + 1080$, and the revenue function, $R(x) = 1.3x$. Use the graphs to answer the following questions.

a. How many gallons of gasoline must be sold for the station to break even?

b. Under what conditions will the station lose money?

c. Why are the lines not extended to the left of the y-axis?

d. What is the weekly profit if 8000 gallons of gasoline are sold? How is this shown in the figure?

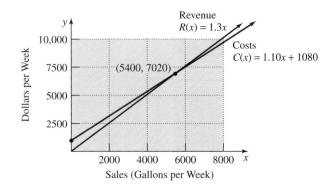

Solve each system in Problems 20–22 by the addition method.

20. $3x - y + 4z = 4$
$4x + 4y - 3z = 3$
$2x + 3y + 2z = -4$

21. $2x - 5y + 2z = -4$
$5x - 3y - 4z = -18$
$3x + 2y + 3z = 13$

22. $x - z + 2 = 0$
$y + 3z = 11$
$x + y + z = 6$

23. New York City, Chicago, and Los Angeles have the largest police forces in the United States, with a total of 50,760 officers. New York City has 22,900 more officers than Los Angeles. The number of officers in New York City exceeds twice that of Chicago by 5335. What is the size of the three largest police forces in the United States?

24. The largest angle in a triangle measures 66° more than the smallest angle. The measure of the largest angle is 17° less than three times the measure of the remaining angle. Find the measure of each angle of the triangle.

25. A dietician wants to prepare a meal that consists of foods A, B, and C. The nutritional component per ounce in these foods is given in the following table.

Nutritional Component per Ounce

	Calories	Protein (in grams)	Vitamin C (in milligrams)
Food A	200	0.2	0
Food B	50	3	10
Food C	10	1	30

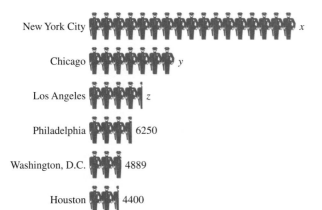

Largest Police Forces in the United States

New York City — x
Chicago — y
Los Angeles — z
Philadelphia — 6250
Washington, D.C. — 4889
Houston — 4400

Source: International Association of Chiefs of Police

A meal with foods A, B, and C must contain exactly 740 calories, 10.6 grams of protein, and 140 milligrams of vitamin C. Model the situation with a system of three linear equations and determine how many ounces of each type of food must be used. Solve the linear system in three variables using the addition method.

Solve each system in Problems 26–31 using row-equivalent matrices.

26. $x + 4y = 7$
$3x + 5y = 0$

27. $6x - 3y = 1$
$5x + 6y = 15$

28. $x + y + 2z = 0$
$2x - y - z = 1$
$x + 2y + 3z = 1$

29. $3x - y + 2z = 2$
$x + 4z = -1$
$3x - 2y = -1$

30. $2x - y - 3z = 1$
$6x - 3y - 9z = 3$
$4x - 2y - 6z = 2$

31. $3x + 2y + z = 7$
$x + y - z = 2$
$6x + 4y + 2z = 10$

For Problems 32–35, evaluate each determinant.

32. $\begin{vmatrix} 3 & 2 \\ -1 & 5 \end{vmatrix}$

33. $\begin{vmatrix} -2 & -3 \\ -4 & -8 \end{vmatrix}$

34. $\begin{vmatrix} 2 & 4 & -3 \\ -1 & 7 & -4 \\ 1 & -6 & -2 \end{vmatrix}$

35. $\begin{vmatrix} 4 & 7 & 0 \\ -5 & 6 & 0 \\ 3 & 2 & -4 \end{vmatrix}$

Use Cramer's rule to solve each system in Problems 36–41.

36. $2x - y = 2$
$x + 2y = 11$

37. $4x = 12 - 3y$
$2x - 6 = 5y$

38. $4x + 2y + 3z = 9$
$3x + 5y + 4z = 19$
$9x + 3y + 2z = 3$

39. $2x + z = -4$
$-2y + z = -4$
$2x - 4y + z = -20$

40. $3x - 2y = 16$
$12x - 8y = -5$

41. $x - y - 2z = 1$
$2x - 2y - 4z = 2$
$-x + y + 2z = -1$

Graph each system of inequalities in Problems 42–49. Find the coordinates of any vertices that are formed.

42. $3x - y \leq 6$
$x + y \geq 2$

43. $y < -x + 4$
$y > x - 4$

44. $-3 \leq x < 5$

45. $-2 < y \leq 6$

46. $x \geq 3$
$y \leq 0$

47. $2x - y > -4$
$x \geq 0$

48. $x + y \leq 6$
$y \geq 2x - 3$

49. $3x + y \leq 6$
$4x + 6y \leq 24$
$x \geq 0$
$y \geq 0$

50. Find the maximum and minimum values of the objective function $z = 2x + 3y$ on the region graphed in the figure.

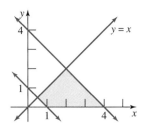

51. Find the maximum and minimum values of the objective function $z = 1260 - 10x - 15y$ subject to the constraints described by the following inequalities: $x + y \geq 12$ and $x + y \leq 15$ and $0 \leq x \leq 10$ and $0 \leq y \leq 12$.

52. A paper manufacturing company converts wood pulp to writing paper and newsprint. The profit on a unit of writing paper is $500 and the profit on a unit of newsprint is $350.

  a. Let x represent the number of units of writing paper produced daily. Let y represent the number of units of newsprint produced daily. Write the objective function that models total daily profit.

  b. The manufacturer is bound by the following constraints:

   1. Equipment in the factory allows for making at most 200 units of paper (writing paper and newsprint) in a day.

   2. Regular customers require at least 10 units of writing paper and at least 80 units of newsprint daily.

   Write a system of inequalities that models these constraints.

  c. Graph the inequalities in part (b). Use only the first quadrant, since x and y must both be positive. (*Suggestion:* Let each unit along the x- and y-axes represent 20.)

  d. Evaluate the objective profit function at each of the three vertices of the graphed region.

  e. Complete the missing portions of this statement: The company will make the greatest profit by producing ____ units of writing paper and ____ units of newsprint each day. The maximum daily profit is $____ .

Fanny Brennan "Tagged Tree" 1982, oil, $2\frac{1}{2} \times 3\frac{5}{8}$ in.

CHAPTER 3 TEST

1. Solve by graphing.

$$x + y = 6$$
$$4x - y = 4$$

Use the substitution or addition method to solve each system. If applicable, state that the system is inconsistent or contains dependent equations.

2. $5x + 4y = 10$
$3x + 5y = -7$

3. $x = y + 4$
$-3x - 7y = 18$

4. $4x = 2y + 6$
$y = 2x - 3$

5. The average annual salary for six professions is shown in the graph. Doubling the salary for teachers results in $1783 more than what attorneys earn annually. The combined salary for attorneys and teachers exceeds the combined salaries of engineers and accountants by $9322. Find the average annual salary for attorneys and teachers.

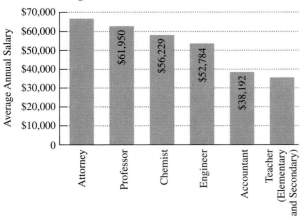

Average Annual Salaries for Six Professions

Source: American Federation of Teachers

6. The figure shows a right triangle and two vertical angles, one inside the triangle and one outside the triangle, that have the same measures. Find the measure of each angle in the triangle.

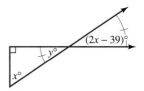

7. A theater sells all orchestra seats at one price and all mezzanine seats at another price. One person purchased 4 orchestra tickets and 3 mezzanine tickets for a total of $134.00. A second person purchased 5 orchestra tickets and 2 mezzanine tickets for $143.00. What is the price of one orchestra ticket and one mezzanine ticket?

8. The demand and supply functions for a product are given by $D(p) = 1000 - 20p$ and $S(p) = 250 + 5p$. At what price will supply equal demand? At that price, how many units of the product will be supplied and sold?

Solve by the addition method.

9.
$$x + 3y - z = 1$$
$$x - y + 2z = -4$$
$$2x + y + 3z = 2$$

10.
$$x + 2y + z = 5$$
$$y + 5z = 6$$
$$4x + 2z = 10$$

11. A department store with three branches is having a sale on three of its graphing utilities, models A, B, and C. The accompanying table shows the number of each model sold and the total income from these sales on the first day of the sale. Find the price of each model of the graphing utilities.

	Model A	Model B	Model C	Total Sales
Store 1	1	2	1	$300
Store 2	2	1	3	$465
Store 3	2	1	4	$555

Solve each system using row-equivalent matrices.

12.
$$2x + y = 6$$
$$3x - 2y = 16$$

13.
$$x - 4y + 4z = -1$$
$$2x - y + 5z = 6$$
$$-x + 3y - z = 5$$

Evaluate each determinant.

14. $\begin{vmatrix} -1 & -3 \\ 7 & 4 \end{vmatrix}$

15. $\begin{vmatrix} 3 & 4 & 5 \\ -1 & -6 & -3 \\ 4 & 9 & 5 \end{vmatrix}$

16. Solve by Cramer's rule.
$$4x - 3y = 14$$
$$3x - y = 3$$

17. Solve for x only using Cramer's rule.
$$2x + 3y + z = 2$$
$$3x + 3y - z = 0$$
$$x - 2y - 3z = 1$$

Graph each system of inequalities. Find the coordinates of any vertices that are formed.

18. $x + y \geq 2$
$x - y \geq 5$

19. $x + 2y \leq 6$
$5x + 4y \leq 20$
$x \geq 0$
$y \geq 0$

20. Find the maximum and minimum value of the objective function $z = 3x + y$ subject to the following constraints:

$$y \leq 2x + 1, \qquad 1 \leq y \leq 3, \qquad y \leq -\tfrac{1}{2}x + 6.$$

CUMULATIVE REVIEW PROBLEMS (CHAPTERS 1–3)

1. Solve: $\dfrac{3x}{5} + 4 = \dfrac{x}{3}$.

2. Solve: $-2(x - 3) - 9 = 3x - 4$.

3. Simplify: $\dfrac{-10x^2y^4}{15x^7y^{-3}}$.

4. Multiply and express the product in scientific notation: $(9.3 \times 10^{-3})(5.1 \times 10^8)$.

5. Solve, expressing the solution in both set-builder and interval notation: $2x + 5 < 1 \quad$ or $\quad 7 - 2x \leq 1$.

6. In 1989, $720 million was lost worldwide due to credit card fraud. This amount has increased by approximately $420 million per year. In what year will losses reach $7860 million ($7,860,000,000)?

7. The average number of people per family unit in the United States x years after 1940 can be modeled by the function

$$f(x) = 0.000002x^4 - 0.0002x^3 + 0.006x^2 - 0.06x + 3.76.$$

Find and interpret $f(10)$.

People per Family Unit, 1940–1980

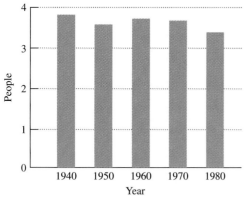

Data Source: U.S. Bureau of the Census

8. Use the number of dots in the first four terms to determine a pattern. Use this pattern to describe how many dots would occur in the tenth term and how many dots would occur in the nth term.

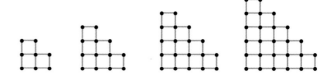

9. Graph $f(x) = x^2 - 1$ by completing the following table of coordinates:

x	-2	-1	0	1	2
$f(x)$					

Explain why the graph defines y as a function of x.

10. Solve: $|4x - 3| + 2 = 6$.

11. Point A in the figure at the top of page 313 indicates average glucose (blood sugar) level. The graph of function f shows glucose level as a function of time after eating a candy bar. The graph of function g shows glucose level as a function of time after eating an apple.
   a. Why do f and g represent the graphs of functions?
   b. Estimate the maximum point on the graph of function f. What does this mean in terms of maximum energy?
   c. Estimate the maximum point on the graph of function g. What exactly does this mean?
   d. Estimate the intersection point of the graphs. Describe what this means.

e. Which item will give you more energy over a longer period of time? How is this shown by the graphs?

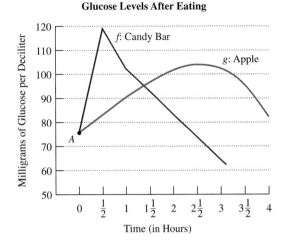

Glucose Levels After Eating

In Problems 12–13, write the point-slope form, the slope-intercept form, and the standard form of the equation of each line.

12. Passing through the points $(2, 4)$ and $(4, -2)$.

13. Passing through the point $(2, 3)$ and parallel to the line whose equation is $2x - 3y = 8$.

In Problems 14–16, use the most efficient method to graph each linear equation in a rectangular coordinate system.

14. $2x - 4y = -8$

15. $f(x) = \frac{3}{4}x - 6$

16. $y - 3 = 1$

17. Graph the compound inequality: $6x - 3y < 12$ and $x < 2$.

18. Evaluate: $24 \div 4[3 - (2 - 4)]^2 - 9$.

19. Solve, expressing the solution in both set-builder and interval notation: $|2x - 1| > 11$.

20. Solve for a: $R = 2(a - b)$.

21. The graph shows the population of Manhattan Island, a borough of New York City, every 10 years from 1790 to 1980.

  a. Compute the slope of the line passing through points A and B and interpret this number in terms of the average rate of change.

  b. Compute the slope of the line passing through points B and C and interpret this number in terms of the average rate of change.

22. Find the maximum value of the objective function $z = 3x + 2y$ on the region graphed in the figure.

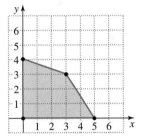

23. Solve by graphing:

$$x - y = 7$$
$$2x + y = 2$$

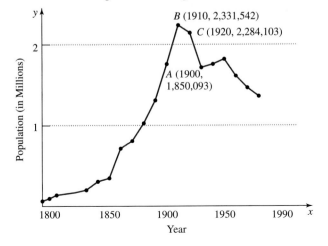

For Problems 24–25, use the addition method to solve each system.

24. $5x + 4y = 10$
$3x + 5y = -7$

25. $2x - 4y + 3z = 0$
$5x + 3y - 2z = 19$
$x - 2y - 5z = 13$

For Problems 27–28, use Cramer's rule to solve each system.

27. $6x + 7y = -9$
$4x + 6y = 0$

28. $2x - 7y - z = 35$
$x + y = 1$
$2y + z = -5$

29. The length of the basketball court in the figure is 6 feet less than twice the width. If the perimeter is 288 feet, find the court's dimensions.

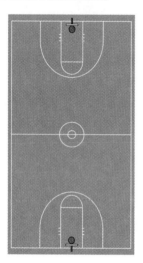

26. Solve using row-equivalent matrices:

$x - 2y - z = 6$
$y + 4z = 5$
$4x + 2y + 3z = 8$

30. A runner who is monitoring pulse rate notes a pulse of 124 after running for 30 minutes and a pulse of 136 after 40 minutes.

x (Time Spent Running)	y (Pulse)
30	124
40	136

Assuming a linear relationship between the variables, write the point-slope form and the slope-intercept form of the line on which these measurements fall. Then extrapolate from the data to find the pulse after running for 60 minutes.

Polynomials, Polynomial Functions, and Factoring

Henry Small, ceramic caricatures of classic Chevys

Many environmentalists are lobbying for automobile fuel efficiency of 45 miles per gallon by the year 2000. This is a far cry from the gas-guzzlers of the 1950s, whose spectacular but overstated designs symbolically de- fied problems of atmospheric carbon dioxide. But is a goal of 45 miles per gallon realistic? An understanding of poly- nomial functions and their graphs will enable us to model fuel efficiency of passenger cars over time and decide.

S E C T I O N   4 . 1

Solutions **Tutorial** **Video**
Manual **4**

Introduction to Polynomials and Polynomial Functions

Objectives

1 Recognize vocabulary associated with polynomials.
2 Evaluate polynomial functions.
3 Graph quadratic functions.
4 Recognize the end behavior of a cubic function's graph.
5 Add and subtract polynomials.
6 Find the sum and difference of polynomial functions.

Catherine Murphy, "Bedside
Still Life" 1982, oil on canvas,
29 × 19 in. Lennon,
Weinberg, Inc.

The common cold is caused by a rhinovirus. Although there is no cure for the cold, the virus enters our bodies, multiplies, and begins to die at a certain point. After t days of invasion by the viral particles, there are N billion particles in our bodies, where $N = -\frac{3}{4}t^4 + 3t^3 + 5$. The model enables mathematicians to determine the day on which there is a maximum number of viral particles (and, consequently, the day we feel sickest).

The model for the cold virus contains three terms. All variables have whole number exponents. Furthermore, no variable appears in a denominator. The expression

$$-\frac{3}{4}t^4 + 3t^3 + 5$$

is an example of a *polynomial*.

Definition of a polynomial

A *polynomial* is a term or a finite sum of terms in which all variables have whole number exponents and no variable appears in a denominator.

Polynomial models abound in such diverse areas as science, business, medicine, psychology, and sociology.

The Vocabulary of Polynomial Expressions

1 Recognize vocabulary associated with polynomials.

Table 4.1 contains the basic vocabulary associated with polynomials. You are likely familiar with this terminology from your work in introductory algebra.

TABLE 4.1 Polynomial Terminology

Vocabulary	Examples
A *monomial* is a constant or the product of a constant and one or more variables raised to whole number exponents, with no variables appearing in denominators.	-13; $4x^5$; $7x^3y$; $-3x^2y^5z^4$; $-abc^2$
A *polynomial* is a monomial or a finite sum of terms in which all variables have whole number exponents and no variables appear in denominators.	$7x - 6$; $3x^2 - 7x + 5$; $-3x^2y + 4xy^2$; $-5x^3y^2z^2 + 4x^2y^2z + 13xyz$
A *binomial* is a polynomial of two terms.	$3x^2 + 7x$; $-9x^2y + 4xy$

Vocabulary	Examples
A *trinomial* is a polynomial of three terms. (Polynomials of more than three terms do not have special names.)	$3x^2 + 7x - 19; 8x^2 + 7xy - 4y$
A *polynomial in x* contains only x as a variable.	$x^4 - 3x^2 + 7x + 2$
A polynomial in one variable is written in *descending powers* of the variable when the exponents on the terms decrease from left to right.	$x^4 - 3x^2 + 7x + 2$ (Recall that the term 2 can be thought of as $2x^0$.)
The *degree of a monomial* is the sum of the exponents on all the variables.	x: degree 1 $4x^3$: degree 3 $5x^2y^7$: degree 9 (2 + 7) y^n: degree n (assuming that n is a whole number) 8: degree 0 (because $8 = 8x^0$)
The *degree of a polynomial* is the greatest of the degrees of any of its terms.	$7x + 4$: degree 1 $5x^3 + 7x^2 - 9$: degree 3 $7x^3y^2 + 5x^4 - y$: degree 5 $7x^{2n} - 5x^n + 3$: degree $2n$
The *leading coefficient* of a polynomial is the numerical coefficient of the term of highest degree.	The leading coefficient of $7x^3 + 6x^2 - 5x + 3$ is 7.

EXAMPLE 1 **The Vocabulary of Polynomials**

Write each of the following polynomials in descending powers of x. Then find the degree of each term, the degree of the polynomial, and the leading coefficient.

a. $15 + 19x^3 - 8x + 4x^2$ **b.** $8x^3 - 15xy + 9x^2y^4z + 3y - 5$

Solution

	a. Polynomial in Descending Powers of x $19x^3 + 4x^2 - 8x + 15$				b. Polynomial in Descending Powers of x $8x^3 + 9x^2y^4z - 15xy + 3y - 5$				
Term	$19x^3$	$4x^2$	$-8x$	15	$8x^3$	$9x^2y^4z$	$-15xy$	$3y$	-5
Degree	3	2	1	0	3	$2 + 4 + 1 = 7$	$1 + 1 = 2$	1	0
Degree of the Polynomial			3				7		
Leading Coefficient			19				9		

Polynomial Functions

A function whose equation is defined by a polynomial is a polynomial function. Table 4.2 shows examples of polynomial functions.

TABLE 4.2 Examples of Polynomial Functions

Name	Example
First-Degree Polynomial Function (also called a linear function) $f(x) = ax + b, a \neq 0$	$f(x) = 3x + 4$
Second-Degree Polynomial Function (also called a quadratic function) $f(x) = ax^2 + bx + c, a \neq 0$	$f(x) = x^2 + 4x + 3$
Third-Degree Polynomial Function (also called a cubic function) $f(x) = ax^3 + bx^2 + cx + d, a \neq 0$	$f(x) = x^3 + x^2 - 3x - 1$
Fourth-Degree Polynomial Function $f(x) = ax^4 + bx^3 + cx^2 + dx + e, a \neq 0$	$f(x) = 3x^4 - 2x^3 + 5x^2 - 3x + 2$

D iscover for yourself

Find examples of some of the functions in Table 4.2 either from previous chapters or perhaps from subsequent chapters.

The generalization of the functions in Table 4.2 is called the *nth-degree polynomial function*. The numerical coefficients are usually written using subscripts (a_i) rather than using a, b, c, d, e, and so on.

Polynomial function

The *n*th-degree polynomial function f is

$$f(x) = a_n x^n + a_{n-1} x^{n-1} + a_{n-2} x^{n-2} + \cdots + a_1 x + a_0 \qquad (a_n \neq 0)$$

where n is a whole number and $a_n, a_{n-1}, a_{n-2}, \cdots, a_1$, and a_0 are real numbers. The number a_n is the *leading coefficient* and the number a_0 is the *constant term*.

2 Evaluate polynomial functions.

Any real number can be substituted for x in a polynomial function, and so the domain of the function is the set of all real numbers. The range of the polynomial function, designated by $f(x)$, depends on the degree of the polynomial and the numerical coefficients.

EXAMPLE 2   An Application of a Quadratic Function

The function

$$f(x) = -\frac{1}{60}x^2 + \frac{7}{6}x + 10$$

describes the weight of a hamster (in grams) as a function of time after birth (x, in days), where $0 \leq x \leq 35$. (In the model $a = -\frac{1}{60}$ because a is the coefficient of x^2; $b = \frac{7}{6}$ because b is the coefficient of x; and $c = 10$.) Find and interpret $f(10)$.

Solution

$$f(x) = -\frac{1}{60}x^2 + \frac{7}{6}x + 10 \qquad \text{This is the given quadratic function.}$$

$$f(10) = -\frac{1}{60}(10)^2 + \frac{7}{6}(10) + 10 \qquad \text{To find } f(10), \text{replace } x \text{ with } 10.$$

$$f(10) = -\frac{1}{60}(100) + \frac{7}{6}(10) + 10 \qquad 10^2 = 100$$

$$f(10) = -\frac{10}{6} + \frac{70}{6} + 10 \qquad \text{Multiply.}$$

$$f(10) = \frac{60}{6} + 10 \qquad \text{Add the first two numbers.}$$

$$f(10) = 10 + 10$$

$$f(10) = 20$$

Because $f(10) = 20$ (f of 10 is 20), after 10 days the hamster weighs 20 grams. ■

Graphing Polynomial Functions

Although we will still be using point-plotting to graph polynomial functions, it's helpful to have an idea of what these graphs will look like in advance. Let's focus on second- and third-degree polynomial functions.

3 Graph quadratic functions.

Graphs of Quadratic Functions. The graph of $f(x) = ax^2 + bx + c$ $(a \neq 0)$ is called a *parabola*. Figure 4.1 indicates that the shape of parabolas is cuplike. They open upward if a is positive and downward if a is negative. The *vertex* is the lowest point on a parabola that opens upward and the highest point on a parabola that opens downward.

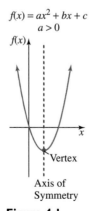

Figure 4.1

Parabolas have cuplike shapes.

Notice that the parabola is *symmetric* about a vertical line through the vertex. This means that if we folded the page along this line, called the *axis of symmetry,* the left and right sides would coincide. The location of the axis of symmetry is helpful in selecting the values of x in a table of coordinates.

> ### A parabola's axis of symmetry
>
> The graph of $f(x) = ax^2 + bx + c$ is a parabola whose axis of symmetry is the vertical line
>
> $$x = -\frac{b}{2a}.$$

From: INVERSIONS © 1989 by Scott Kim. Used with permission of W.H. Freeman and Company.

In Chapter 7, we will actually prove that $x = -\dfrac{b}{2a}$ is the formula for the parabola's axis of symmetry. For now, let's use the formula with the awareness that $-\dfrac{b}{2a}$ is also the x-coordinate of the parabola's vertex. These ideas are illustrated in the next example.

Keep in mind that all polynomial functions have graphs that are continuous curves with smooth, rounded turns. These graphs eventually rise or fall without bound. Let's see how this applies to the parabola.

EXAMPLE 3   Graphing a Parabola

Graph: $f(x) = x^2 - 3x - 4$

Solution

Before setting up a table of coordinates, we find the axis of symmetry and the coordinates of the vertex.

Jack Goldstein, "Untitled" 1981. Courtesy of John Weber Gallery, New York.

$$x = -\frac{b}{2a}$$

Use the given function: $f(x) = x^2 - 3x - 4$

$$a = 1 \quad b = -3$$

$$= \frac{-(-3)}{2(1)} = \frac{3}{2}$$

This is the x-coordinate of the vertex.

Now we find the y-coordinate of the vertex using $f(x) = x^2 - 3x - 4$.

$$y = f\left(\frac{3}{2}\right) = \left(\frac{3}{2}\right)^2 - 3\left(\frac{3}{2}\right) - 4 = \frac{9}{4} - \frac{9}{2} - \frac{4}{1}$$

$$= \frac{9}{4} - \frac{18}{4} - \frac{16}{4} = -\frac{25}{4}$$

This is the y-coordinate of the vertex.

The vertex is at $\left(\frac{3}{2}, -\frac{25}{4}\right)$. Because $a = 1$ ($a > 0$), the graph opens upward and the vertex is a minimum point.

We will now select some values of x for our table of coordinates. Since the axis of symmetry is $x = \frac{3}{2}$, we will select some integral values of x less than $\frac{3}{2}$ and some greater than $\frac{3}{2}$.

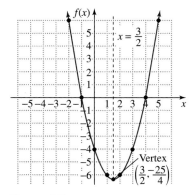

Figure 4.2

The graph of
$f(x) = x^2 - 3x - 4$

x	$f(x) = x^2 - 3x - 4$	(x,y) or $(x, f(x))$
-2	$f(-2) = (-2)^2 - 3(-2) - 4 = 6$	$(-2, 6)$
-1	$f(-1) = (-1)^2 - 3(-1) - 4 = 0$	$(-1, 0)$
0	$f(0) = 0^2 - 3(0) - 4 = -4$	$(0, -4)$
1	$f(1) = 1^2 - 3(1) - 4 = -6$	$(1, -6)$
2	$f(2) = 2^2 - 3(2) - 4 = -6$	$(2, -6)$
3	$f(3) = 3^2 - 3(3) - 4 = -4$	$(3, -4)$
4	$f(4) = 4^2 - 3(4) - 4 = 0$	$(4, 0)$
5	$f(5) = 5^2 - 3(5) - 4 = 6$	$(5, 6)$

values less than $\frac{3}{2}$ — values greater than $\frac{3}{2}$

The graph of the given polynomial function is shown in Figure 4.2. Since the vertex is a minimum point, the range of this function is $\{ f(x) \mid f(x) \geqslant \frac{-25}{4} \}$. ∎

Minimum and maximum: quadratic functions

Consider $f(x) = ax^2 + bx + c$.

1. If $a > 0$, then the function has a minimum point that occurs at
$$x = -\frac{b}{2a}.$$

2. If $a < 0$, then the function has a maximum point that occurs at
$$x = -\frac{b}{2a}.$$

EXAMPLE 4 **Alcohol Consumption**

Based on data provided by the U.S. Department of Agriculture, the model

$$f(x) = -0.053x^2 + 1.17x + 35.6$$

describes the average number of gallons of alcohol ($f(x)$) consumed by each adult in the United States as a function of time, where x represents the num-

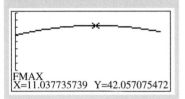

Using technology

The graph of $y = -0.053x^2 + 1.17x + 35.6$ was obtained with a graphing utility using

Xmin $= 0$, Xmax $= 20$, Xscl $= 1$,
Ymin $= 0$, Ymax $= 50$, Yscl $= 5$

The maximum function feature of the utility verifies our algebraic analysis.

FMAX
X=11.037735739 Y=42.057075472

ber of years after 1970. In what year was alcohol consumption at a maximum? What was the average per capita consumption for that year?

Solution

$$f(x) = -0.053x^2 + 1.17x + 35.6 \quad \text{This is the given function.}$$

$$a = -0.053 \qquad b = 1.17$$

Since a is negative ($a < 0$), the function has a maximum value at $x = -\dfrac{b}{2a}$.

$$x = -\frac{b}{2a} = -\frac{1.17}{2(-0.053)} \approx 11.04$$

This means that alcohol consumption was at a maximum approximately 11 years after 1970, or in 1981. The average consumption for that year was

$$f(11.04) = -0.053(11.04)^2 + 1.17(11.04) + 35.6 \approx 42.06 \quad ■$$

or approximately 42 gallons of alcohol consumed by each adult.

4 Recognize the end behavior of a cubic function's graph.

Graphs of Cubic Functions. The graph of $f(x) = ax^3 + bx^2 + cx + d$ ($a \neq 0$) falls to the left and rises to the right if a is positive. If a is negative, the graph rises to the left and falls to the right. Figure 4.3 illustrates the graphs of some cubic functions.

If a is positive ($a > 0$), the graph falls to the left and rises to the right.

If a is negative ($a < 0$), the graph rises to the left and falls to the right.

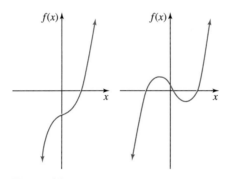
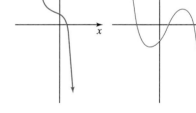

Figure 4.3
Possible graphs of $f(x) = ax^3 + bx^2 + cx + d$

EXAMPLE 5 **AIDS Cases**

The number of AIDS cases in the United States for the years 1983 through 1990 is approximated by the cubic polynomial function

$$f(x) = -143x^3 + 1810x^2 - 187x + 2331$$

where x represents years after 1983. Why is the model only valid for a limited time period?

Solution

Because the leading coefficient, a, is negative ($a = -143$), the graph of the function falls to the right. This indicates at some point the number of AIDS cases will be negative, an impossibility. No function that has a graph that falls to the right (or decreases without bound) over time can model real world phenomena over a long period. ■

Using technology

The graph of the cubic polynomial function given in Example 5 is shown in the figure on the left. The graph was obtained on a graphing utility using

Xmin = 0, Xmax = 8, Xscl = 1, Ymin = 0, Ymax = 50,000, Yscl = 5000.

Now look at the graph to the right, where we changed the range setting to

Xmin = 0, Xmax = 15, Xscl = 1, Ymin = −5000, Ymax = 50,000, Yscl = 5000.

By year 13 (1996), values of y are negative and the function no longer models reality.

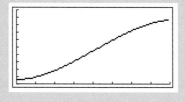 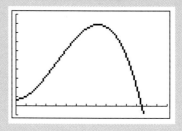

5 Add and subtract polynomials.

Adding and Subtracting Polynomials

As we know from our work in Chapter 1, we cannot combine terms in the polynomial $3x^2 + 7x - 5$. Only like (or similar) terms containing exactly the same variables to the same powers can be combined. If we want to add polynomials, we apply the commutative and associative properties to group like terms together and then use the distributive property to combine like terms.

EXAMPLE 6 **Adding Polynomials**

Add: $-9x^3 + 7x^2 - 5x + 3$ and $13x^3 + 2x^2 - 8x - 6$

Solution

$$(-9x^3 + 7x^2 - 5x + 3) + (13x^3 + 2x^2 - 8x - 6)$$
$$= (-9x^3 + 13x^3) + (7x^2 + 2x^2)$$
$$+ (-5x - 8x) + (3 - 6)$$

Use the commutative and associative properties

$$= (-9 + 13)x^3 + (7 + 2)x^2$$
$$+ (-5 - 8)x + (3 - 6)$$
$$= 4x^3 + 9x^2 - 13x - 3$$

Use the distributive property; this step is usually done mentally. ■

In Example 6, because

$$(-9x^3 + 7x^2 - 5x + 3) + (13x^3 + 2x^2 - 8x - 6) = 4x^3 + 9x^2 - 13x - 3$$

is a true statement for every replacement of the variable, the two sides of the equation are called *equivalent expressions*. Replacing the original polynomials with an equivalent polynomial in which no two of its terms are like terms is called *simplifying the polynomial*.

> ### Adding polynomials
>
> To add polynomials, combine the like terms of the polynomials.

Study tip

Here's the solution to Example 7 using a vertical format:

$$\begin{array}{r} 5x^3y - 4x^2y - 7y \\ 2x^3y + 6x^2y - 4y - 5 \\ \hline 7x^3y + 2x^2y - 11y - 5 \end{array}$$

EXAMPLE 7   Adding Polynomials

Simplify: $(5x^3y - 4x^2y - 7y) + (2x^3y + 6x^2y - 4y - 5)$

Solution

$(5x^3y - 4x^2y - 7y) + (2x^3y + 6x^2y - 4y - 5)$ The given problem involves adding polynomials in two variables.

$= 5x^3y - 4x^2y - 7y + 2x^3y + 6x^2y - 4y - 5$ Remove the parentheses.

$= \underbrace{5x^3y + 2x^3y}\ \underbrace{- 4x^2y + 6x^2y}\ \underbrace{- 7y - 4y}\ \underbrace{- 5}$ Rearrange the terms so that like terms are adjacent.

$= \quad 7x^3y \qquad + 2x^2y \qquad -11y \quad -5$ Combine like terms.   ■

In Chapter 1, we defined subtraction of real numbers by

$$a - b = a + (-b).$$

By this definition, we add the additive inverse of the second number to the first. We follow a similar definition for the subtraction of polynomials.

> ### Subtracting polynomials
>
> To *subtract* two polynomials, add the additive inverse of the second polynomial to the first polynomial. The additive inverse of the second polynomial is that polynomial with the sign of every term changed.

The following examples illustrate the subtraction of polynomials.

Study tip

Here's the solution to Example 8 using a vertical format:

$$\begin{array}{r} 7x^3 - 8x^2 + 9x - 6 \\ -(2x^3 - 6x^2 - 3x + 9) \\ \\ 7x^3 - 8x^2 + 9x - 6 \\ + -2x^3 + 6x^2 + 3x - 9 \\ \hline 5x^3 - 2x^2 + 12x - 15 \end{array}$$

EXAMPLE 8   Subtracting Polynomials

Subtract: $2x^3 - 6x^2 - 3x + 9$ from $7x^3 - 8x^2 + 9x - 6$

Solution

$(7x^3 - 8x^2 + 9x - 6) - (2x^3 - 6x^2 - 3x + 9)$

$= (7x^3 - 8x^2 + 9x - 6)$

$\quad + (-2x^3 + 6x^2 + 3x - 9)$ Rewrite subtraction as the addition of the additive inverse.

$= \underbrace{7x^3 - 2x^3}\ \underbrace{- 8x^2 + 6x^2}\ \underbrace{+ 9x + 3x}\ \underbrace{- 6 - 9}$ Rearrange terms.

$= \quad 5x^3 \qquad -2x^2 \qquad +12x \quad -15$ Combine like terms.   ■

Here's the procedure for subtracting polynomials.

> **Subtracting polynomials**
>
> To subtract polynomials:
>
> **1.** Remove parentheses from the polynomial being subtracted and change the sign of every term of this polynomial.
> **2.** Combine like terms.

Here's the subtraction using a vertical format:

$$3x^5y^2 - 4x^3y - 3$$
$$-(-2x^5y^2 - 3x^3y + 7)$$
$$\overline{}$$
$$3x^5y^2 - 4x^3y - 3$$
$$2x^5y^2 + 3x^3y - 7$$
$$\overline{5x^5y^2 - x^3y - 10}$$

6 Find the sum and difference of polynomial functions.

EXAMPLE 9 **Subtracting Polynomials**

Simplify: $(3x^5y^2 - 4x^3y - 3) - (-2x^5y^2 - 3x^3y + 7)$

Solution

$(3x^5y^2 - 4x^3y - 3) - (-2x^5y^2 - 3x^3y + 7)$
$= 3x^5y^2 - 4x^3y - 3 + 2x^5y^2 + 3x^3y - 7$ Remove the parentheses, changing the sign of every term after the subtraction sign.

$= \underbrace{3x^5y^2 + 2x^5y^2}_{} \underbrace{- 4x^3y + 3x^3y}_{} \underbrace{- 3 - 7}_{}$ Rearrange terms.

$= \underbrace{5x^5y^2}_{} \underbrace{-x^3y}_{} \underbrace{-10}_{}$ Combine like terms. ∎

Adding and Subtracting Polynomial Functions

Just as we add and subtract polynomials, we can add and subtract polynomial functions to obtain new functions. We can develop an *algebra of functions* to define two other functions as follows.

> **Adding and subtracting polynomial functions**
>
> If f and g are polynomial functions, then
>
> **1.** $(f + g)(x) = f(x) + g(x)$ The function $f + g$ is the sum of the functions f and g.
>
> **2.** $(f - g)(x) = f(x) - g(x)$ The function $f - g$ is the difference of the functions f and g.

Use the functions on the right to find $f(2)$ and $g(2)$. Now add these function values and find $f(2) + g(2)$. Finally, use

$$(f + g)(x) = 3x^2 + 2x$$

and find $(f + g)(2)$. What do you observe? Verify this observation with other function values.

EXAMPLE 10 **Applying the Algebra of Functions**

Let $f(x) = 3x + 2$ and $g(x) = 3x^2 - x - 2$.
Find: a. $(f + g)(x)$ b. $(f - g)(x)$.

Solution

a. $(f + g)(x) = f(x) + g(x)$
$ = (3x + 2) + (3x^2 - x - 2)$ Substitute the given functions.
$ = 3x^2 + 2x$ Combine like terms.

b. $(f - g)(x) = f(x) - g(x)$
$ = (3x + 2) - (3x^2 - x - 2)$ Substitute the given functions.
$ = 3x + 2 - 3x^2 + x + 2$ Remove parentheses.
$ = -3x^2 + 4x + 4$ Combine like terms. ∎

ENRICHMENT ESSAY

Visualizing Graphs of Polynomial Functions

All polynomial functions have graphs that are continuous curves with no breaks or sharp corners. Like the curves in this painting, polynomial functions appear with smooth, rounded turns. However, unlike these curves, the graphs of polynomial functions eventually rise or fall without bound.

Philip Taaffe "Yellow Painting" 1984, linoprint collage, enamel, acrylic on paper on canvas. 76 × 76 inches. Courtesy of the Artist and Gagosian Gallery, NY.

PROBLEM SET 4.1

Practice Problems

In Problems 1–6, write each polynomial in descending powers of x. Then find the degree of each term and the degree of the polynomial.

1. $x^2 - 4x^3 + 9x - 12x^4 - 6$

2. $x^2 - 8x^3 + 11x - 15x^4 - 5$

3. $7xy + 5x^2 + 6x^3y^2 + x^4 - 3$

4. $8xy - 11x^2 + 9x^3y^2 + x^4 - 12$

5. $4x^3yz^2 + 8x^5 - 12y - 3x^6y + 5x^2y^4z^3$

6. $9x^3yz^2 - 11x^5 + 13y - 2x^6y + 8x^2y^4z^3$

In Problems 7–12, a polynomial appears in each description. Where appropriate, identify the polynomial as a monomial, binomial, or trinomial. Then give the degree of each polynomial.

7. The formula $y = 0.071x^2 + 0.73x$ describes the relationship between the speed of an automobile (x, in mph) and the distance (y, in feet) that it takes to stop the car when moving at speed x.

8. The marketing research department of a record company has determined that when the company spends x thousand dollars on advertising, the number of units it sells is given by $-x^2 + 50x + 5000$.

9. A sociological study reveals that the crime rate in a certain county depends on the amount of money spent on welfare, as measured in w (hundreds of millions of dollars) and on the amount of money spent on prisons as measured by p (in hundreds of millions of dollars). The polynomial $w^3 + 2p^2 - 6w + 6p - 6wp$ expresses this relationship.

10. Ecologists discover that a nuclear power plant is leaking waste into a nearby lake. The amount (in gallons) of waste that flowed into the lake during the first t days after the leak occurred may be expressed by the polynomial $3t^2 + 2t$.

11. An efficiency expert determines that the average worker who arrives on the job at 8:30 A.M. will have assembled $15x + 6x^2 - x^3$ components x hours later.

12. Trophane is a beneficial enzyme produced by receiving medications called Curine I and Curine II. The model $A = -x^2 + 2xy - 2y^2 + 8x + 4y$ describes the number of milligrams of Trophane (A) produced by a patient receiving x grams of Curine I and y grams of Curine II.

For Problems 13–20, determine the vertex of the parabola and graph the quadratic function. Use the graph to state the function's range.

13. $f(x) = -x^2 + 2x + 3$

14. $f(x) = -x^2 + 10x - 21$

15. $f(x) = x^2 - x - 2$

16. $f(x) = x^2 + 2x - 3$

17. $f(x) = -x^2 - x + 2$

18. $f(x) = 3x^2 - 12x + 13$

19. $f(x) = 4x^2 + 8x + 4$

20. $f(x) = -2x^2 - 8x + 4$

In Problems 21–24, complete the table of coordinates and graph each cubic function. If the leading coefficient is positive, the graph should fall to the left and rise to the right. If the leading coefficient is negative, the graph should rise to the left and fall to the right.

21. $f(x) = -2x^3 + 6x^2 + 2x - 6$

x	-2	-1	0	1	2	3	4
$f(x)$							

22. $f(x) = -x^3 - x^2 + 2x$

x	-3	-2	-1	0	$\frac{1}{2}$	1	2
$f(x)$							

23. $f(x) = x^3 + 2x - 1$

x	-2	-1	0	1	2
$f(x)$					

24. $f(x) = 2x^3 - 5x^2 + x - 3$

x	-1	0	1	1.5	2	3
$f(x)$						

Add the polynomials in Problems 25–30.

25. $7x^2 - 3x$ and $4x^2 - 7x - 8$

26. $5x^4 + 6x - 3x^2 + 9$ and $-2x^2 + 5x + 3 - x^4$

27. $-7r^3 + 3r - 2 + 8r^2$ and $-3r^2 + 7r + 4$

28. $4x^5y^2 - 7x^3y + 2xy$ and $-4x^4y^2 + 3x^3y - 9xy$

29. $-3x^2y + 2xy$, $-5x^3y + 4x^2y - 7$, and $7xy + 7$

30. $-x^3y^2 + 4x^2y + 8x + 5$, $x^4y^3 - 2x^2y + 1$, and $6x^4y^3 + 5x^2y - 9$

In Problems 31–36, perform the indicated operation.

31. Subtract $5x^3 - 9x^2 - 8x + 11$ from $17x^3 - 5x^2 + 4x - 3$.

32. Subtract $9y^4 - 6y^3 - 5y + 7$ from $18y^4 - 2y^3 - 7y + 8$.

33. Subtract $-9r^5 - 7r^3 + 8r^2 + 11$ from $13r^5 + 9r^4 - 5r^2 + 3r + 6$.

34. Subtract $-4s^7 - 3s^5 + 9s - 12$ from $8s^7 + 5s^6 - 7s^2 - 5s + 12$.

35. Subtract $7x^3y^2 - 5x^2y + 9xy - 3$ from $-6x^3y^2 - 8x^2y + 11xy - 3$.

36. Subtract $-13r^5s^3 + 3r^4s^2 - 9r^3s - 1$ from $15r^5s^3 - 2r^4s^2 + 10r^3s + 1$.

Simplify each expression in Problems 37–40.

37. $(5x^2 - 7x - 8) + (2x^2 - 3x + 7) - (x^2 - 4x - 3)$

38. $(8y^2 + 7y - 5) - (3y^2 - 4y) - (-6y^3 - 5y^2 + 3)$

39. $(6y^4 - 5y^3 + 2y) - (4y^3 + 3y^2 - 1) + (y^4 - 2y^2 + 7y - 3)$

40. $(4x^2 - 5xy + 6y^2) + (7x^2 + 2xy - 4y^2) - (8x^2 - xy - 3y^2)$

For each pair of functions in Problems 41–48, find $(f + g)(x)$, $(f - g)(x)$, $f(1)$, $g(1)$, and $(f + g)(1)$.

41. $f(x) = x^2 + 2x$, $g(x) = 2x + 3$

42. $f(x) = x^2 - 3x$, $g(x) = 4x - 1$

43. $f(x) = -3x^2 + 5x + 7$, $g(x) = x^2 - 5x - 4$

44. $f(x) = -6x^2 + 7x + 3$, $g(x) = 5x^2 - 7x - 9$

45. $f(x) = 8x^2 - 5x - 4x^3$, $g(x) = 3x^2 - 9x - 7x^3$

46. $f(x) = 5x^2 - 7x - 6x^3 - 2$, $g(x) = -4x^2 - 8x - 3x^3 - 5$

47. $f(x) = 3x^3 - 7x^2 + 5x - 2$, $g(x) = -7x^3 + 4x^2 - 8x - 3$

48. $f(x) = 2x^4 - 6x^3 + 5x^2 - 9x + 3$, $g(x) = -5x^4 - 2x^3 + 4x^2 - 6x - 2$

49. a. Graph $f(x) = x^2 + 2x$, $g(x) = 2x + 3$, and $h(x) = (f + g)(x)$ in the same rectangular coordinate system.

b. Describe how the graph of h visually shows the sum of functions f and g. If you're not sure where to begin, take a value of x and observe the values of y on the graphs of f, g, and h.

Application Problems

50. The number of people who catch a cold t weeks after January 1 is $C = 5t - 3t^2 + t^3$ and the number of people who recover t weeks after January 1 is $R = t - t^2 + \frac{1}{3}t^3$. Write the model (in simplified form) for the number of people who are still ill with a cold t weeks after January 1.

51. A rock on Earth and a rock on the moon are thrown into the air by a 6-foot person, with a velocity of 48 feet per second. The height (h, in feet) reached by the rock after t seconds is given by

$$h_{\text{Earth}} = -16t^2 + 48t + 6$$
$$h_{\text{moon}} = -2.7t^2 + 48t + 6$$

Write a model expressing the difference between moon height and Earth height after t seconds.

Use the coordinates for the vertex of a parabola to answer Problems 52–53.

52. A person standing close to the edge of the top of an 80-foot building throws a ball directly upward with an initial speed of 64 feet per second. After t seconds, the height of the ball above the ground is modeled by the position function $s(t) = -16t^2 + 64t + 80$. After how many seconds does the ball reach its maximum height? What is the maximum height?

53. The polynomial function $f(x) = 0.0075x^2 - 0.2676x + 14.8$ models fuel efficiency of passenger cars ($f(x)$, measured in miles per gallon) x years after 1940. In what year (to the nearest whole year) was fuel efficiency poorest? What was the average number of miles per gallon for passenger cars in that year?

54. A company that sells radios has yearly fixed costs of $600,000. It costs the company $45 to produce each radio. Each radio will sell for $65. The company's costs and revenue are modeled by the functions

$$C(x) = 600,000 + 45x \qquad \text{This function models the company's costs.}$$

$$R(x) = 65x \qquad \text{This function models the company's revenue.}$$

Find and interpret $(R - C)(20,000)$, $(R - C)(30,000)$ and $(R - C)(40,000)$.

True–False Critical Thinking Problems

55. Which one of the following is true?

a. $4x^3 + 7x^2 - 5x + \dfrac{2}{x}$ is a polynomial containing four terms.

b. If two polynomials of degree 2 are added, their sum must be a polynomial of degree 2.

c. $(x^2 - 7x) - (x^2 - 4x) = -11x$ for any values of x.

d. All terms of a polynomial are monomials.

56. Which one of the following is true?

a. The polynomial $4x^6 - 7x^4 + 3x^2 - 9$ is in descending powers, has degree 6, and is a trinomial.

b. If two polynomials of degree 3 are subtracted, their difference might not be a polynomial of degree 3.

c. There is no such thing as a polynomial in x having four terms, written in descending powers, and lacking a third-degree term.

d. The expression $5x^{-2} - 7x + 3$ is a trinomial.

Technology Problems

57. Use a graphing utility to verify Problems 13–24 and Problem 49.

58. a. Use a graphing utility to graph $y = 2x^2 - 82x + 720$ in a standard viewing rectangle. What do you observe?

b. Find the coordinates of the vertex for the given quadratic function.

c. The answer to part (b) is $(20.5, -120.5)$. Since the leading coefficient of the given function (2) is positive, the vertex is a minimum point on the graph. Use this fact to help find a range setting that will give a relatively complete picture of the parabola.

With an axis of symmetry at $x = 20.5$, the setting for x should extend past this, so try Xmin = 0 and Xmax = 30. The setting for y should include (and probably go below) the y-coordinate of the graph's minimum point, so try Ymin = -130. Experiment with Ymax until your utility shows the parabola's major features.

d. In general, explain how knowing the coordinates of a parabola's vertex can help determine a reasonable range setting on a graphing utility for obtaining a complete picture of the parabola.

In Problems 59–62, find the vertex for each parabola. Then determine a range setting on your graphing utility and use the utility to graph the parabola.

59. $y = -0.25x^2 + 40x$

60. $y = -4x^2 + 20x + 160$

61. $y = 5x^2 + 40x + 600$

62. $y = 0.01x^2 + 0.6x + 100$

63. Use a graphing utility to graph the cubic function $y = x^3 + 60x^2 + 800$ for each range setting given on the right. How do you know which setting gives a relatively complete picture of the polynomial's graph?

a. Xmin = -10, Xmax = 10, Xscl = 1,
Ymin = -10, Ymax = 1000, Yscl = 1

b. Xmin = -10, Xmax = 10, Xscl = 1,
Ymin = -10, Ymax = 5000, Yscl = 500

c. Xmin = -80, Xmax = 30, Xscl = 10,
Ymin = $-10{,}000$, Ymax = $50{,}000$, Yscl = $10{,}000$

Use a graphing utility to graph each cubic function in Problems 64–65. Experiment with the range setting based on your knowledge of the graph's behavior at the left and right end until you have obtained a relatively complete graph.

64. $y = 2x^3 - 9x^2 - 10x - 12$

65. $y = -x^3 + 8x^2 + 10x - 15$

You can use your graphing utility to check the results of polynomial addition and subtraction for polynomials in one variable. For example, to verify that

$$(-9x^3 + 7x^2 - 5x + 3) + (13x^3 + 2x^2 - 8x - 6) = 4x^3 + 9x^2 - 13x - 3$$

graph the equations

$$y = (-9x^3 + 7x^2 - 5x + 3) + (13x^3 + 2x^2 - 8x - 6)$$
$$\text{and} \quad y = 4x^3 + 9x^2 - 13x - 3$$

on the same screen. (Do this now.) The graphs should coincide, verifying that the polynomials on each side of the equation are equivalent. (Try to obtain reasonably complete graphs.) Use this idea to determine if the polynomial operations in Problems 66–69 are correct. If there is an error, correct the right side of the equation and then use your graphing utility to verify the correction.

66. $(6x^2 - 6x + 4) - (-2x^2 - x + 7) = 8x^2 - 5x - 3$

67. $(x^2 - 6x + 3) - (2x + 5) = x^2 - 4x - 2$

68. $(5x^2 - x - 1) - (-3x^2 - 2x - 5) = 8x^2 + x + 4$

69. $(9x^3 + 5x - 4x^2) - (2x^3 - 5x^2) = 7x^3 - 9x^2 + 5x$

70. The polynomial function $f(x) = 0.013x^2 - 0.96x + 25.4$ describes the average yearly consumption of whole milk per person in the United States x years after 1970. The polynomial function $g(x) = 0.41x + 6.03$ describes the average yearly consumption of low-fat milk per person in the United States x years after 1970.

a. Use a graphing utility to graph each function in the same viewing rectangle for the years 1970 through 1987.

b. Use the graphs to describe the trend in consumption for both types of milk. What possible explanations are there for these consumption patterns?

Writing in Mathematics

71. Describe in your own words what is meant by a polynomial in x. Use your description to explain why $7x^{-3} + 4x^{-2} + 5x^{-1} + 6$ is *not* a polynomial.

72. Can a quadratic function with a negative leading coefficient be used to model data that are increasing over time, x, where x is greater than the x-coordinate of the parabola's vertex? Explain your answer.

Critical Thinking Problems

In Problems 73–74, simplify, assuming that all exponents represent whole numbers.

73. $(x^{2n} - 3x^n + 5) + (4x^{2n} - 3x^n - 4) - (2x^{2n} - 5x^n - 3)$

74. $(y^{3n} - 7y^{2n} + 3) - (-3y^{3n} - 2y^{2n} - 1) + (6y^{3n} - y^{2n} + 1)$

75. From what polynomial must $4x^2 + 2x - 3$ be subtracted to obtain $5x^2 - 5x + 8$?

In Problems 76–78, refer to the graphs of the quadratic functions $f(x) = x^2 - 3x - 4$ *and* $g(x) = x^2 + 3x - 4$ *shown in the figure below. Use only the graphs to answer each question.*

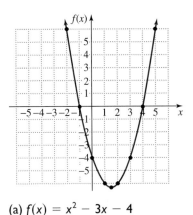

(a) $f(x) = x^2 - 3x - 4$

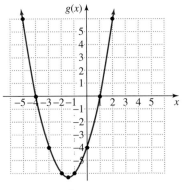

(b) $g(x) = x^2 + 3x - 4$

76. Which is greater, $f(4)$ or $g(-1)$?

77. Which is greater, $f(0)$ or $g(-6)$?

78. Which one of the following is true?
  a. The two functions do not have identical domains and ranges.
  b. The coordinates of the turning point of

$$f(x) = x^2 - 3x - 4$$

  are $(-1, -6)$.
  c. $f(0) + g(0) = -7$
  d. $|f(3)| = 4$

79. a. Graph $f(x) = x^4$, $g(x) = (x - 1)^4$, and $h(x) = x^4 - 1$ in the same rectangular coordinate system.
  b. Describe the relationship among the graphs of f, g, and h.
  c. Generalize from your graphs and complete the following statement using the words *rises* or *falls:* The graph of a fourth-degree polynomial function whose leading coefficient is positive _____ to the left and _____ to the right.

80. a. Graph $f(x) = -x^4$, $g(x) = -(x + 1)^4$, and $h(x) = -x^4 + 1$ in the same rectangular coordinate system.

  b. Describe the relationship among the graphs of f, g, and h.
  c. Generalize from your graphs and complete the following statement using the words *rises* or *falls:* The graph of a fourth-degree polynomial function whose leading coefficient is negative _____ to the left and _____ to the right.

81. The figure shows the graphs of six functions for the interval $[0, 1]$. The functions are $f(x) = x$, $g(x) = x^2$, $h(x) = x^3$, $F(x) = x^4$, $G(x) = x^5$, and $H(x) = x^6$. The graphs are not labeled. Which is which?

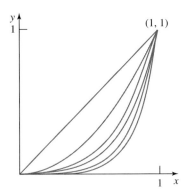

Fill in the numbers in the fifth, sixth, and seventh positions in each number sequence in Problems 82–87 by using the emerging patterns. Then write a polynomial in x for the number that appears in the xth position.

82. $1, 4, 9, 16, \_\_\_, \_\_\_, \_\_\_$

83. $1, 8, 27, 64, \_\_\_, \_\_\_, \_\_\_$

84. $0, 3, 8, 15, \_\_\_, \_\_\_, \_\_\_$

85. $-1, 6, 25, 62, \_\_\_, \_\_\_, \_\_\_$

86. $2, 6, 12, 20, \_\_\_, \_\_\_, \_\_\_$

87. $2, 10, 30, 68, \_\_\_, \_\_\_, \_\_\_$

Review Problems

88. Solve: $9(x - 1) = 1 + 3(x - 2)$.

89. Graph: $2x - 3y < -6$.

90. Write the point-slope, slope-intercept, and standard equations for a line passing through the point $(-2, 5)$ and perpendicular to the line whose equation is $3x - y = 9$.

SECTION 4.2

Solutions Tutorial Video
Manual 5

Multiplication of Polynomials

Objectives

1 Multiply monomials.
2 Multiply a monomial and a polynomial.
3 Multiply polynomials.
4 Multiply binomials using the FOIL method.
5 Square binomials.
6 Multiply the sum and difference of two terms.
7 Multiply polynomial functions.
8 Use polynomial multiplication in geometric modeling.
9 Use polynomial multiplication to evaluate a function.

The construction industry uses a polynomial to determine the number of board feet (N) that can be manufactured from a tree with a diameter of x inches and a length of y feet. This polynomial,

$$N = \left(\frac{x - 4}{4}\right)^2 y$$

is called the Doyle log model. The right side of the model might not look like a polynomial in x and y because it is not expressed as a sum of terms. However, by multiplying out the right side, we can express the model in a format similar to the polynomials discussed in the previous section.

$$N = \left(\frac{x - 4}{4}\right)\left(\frac{x - 4}{4}\right)y$$

In this section, we turn our attention to polynomial multiplication, such as that shown in the Doyle log model.

1 Multiply monomials.

Multiplying Monomials

Here's a step-by-step procedure for multiplying monomials.

> **Multiplication of monomials**
>
> **1.** Multiply the coefficients.
> **2.** Multiply variable factors with the same base by adding exponents.

EXAMPLE 1 **Multiplying Monomials**

Find the products:

a. $(5x^3y^4)(-6x^7y^8)$ **b.** $(4x^3y^2z^5)(2x^5y^2z^4)$

Solution

a. $(5x^3y^4)(-6x^7y^8) = 5(-6)x^3 \cdot x^7 \cdot y^4 \cdot y^8$ Rearrange factors. This step is usually done mentally.

$$= -30x^{3+7}y^{4+8}$$ Multiply coefficients and add exponents.

$$= -30x^{10}y^{12}$$ Simplify.

b. $(4x^3y^2z^5)(2x^5y^2z^4) = 4 \cdot 2 \cdot x^3 \cdot x^5 \cdot y^2 \cdot y^2 \cdot z^5 \cdot z^4$ Rearrange factors.

$$= 8x^{3+5}y^{2+2}z^{5+4}$$ Multiply coefficients and add exponents.

$$= 8x^8y^4z^9$$ Simplify. ∎

2 Multiply a monomial and a polynomial.

Multiplying a Monomial and a Polynomial Other Than a Monomial

We use the distributive property to multiply a polynomial by a monomial. Once the monomial factor is distributed, we multiply the resulting monomials using the procedure shown in Example 1.

> **EXAMPLE 2**   **Multiplying a Monomial and a Trinomial**

Find the products:

a. $4x^3(\frac{1}{2}x^5 - 2x^2 + 3)$ **b.** $5x^3y^4(2x^7y - 6x^4y^3 - 3)$

Solution

a. $4x^3\left(\dfrac{1}{2}x^5 - 2x^2 + 3\right) = 4x^3 \cdot \dfrac{1}{2}x^5 - 4x^3 \cdot 2x^2 + 4x^3 \cdot 3$ Use the distributive property.

$$= 2x^8 - 8x^5 + 12x^3$$ Multiply coefficients and add exponents.

b. $5x^3y^4 (2x^7y - 6x^4y^3 - 3)$

$$= 5x^3y^4(2x^7y) - 5x^3y^4(6x^4y^3) - 5x^3y^4(3)$$ Use the distributive property.

$$= 10x^{10}y^5 - 30x^7y^7 - 15x^3y^4$$ Multiply coefficients and add exponents. ∎

3 Multiply polynomials.

Multiplying Polynomials That Are Not Monomials

The product of two polynomials is the polynomial obtained by multiplying each term of one polynomial by each term of the other polynomial and then combining like terms.

> **EXAMPLE 3**   **Multiplying a Binomial and a Trinomial**

Multiply: $(2x + 3)(x^2 + 4x + 5)$

Solution

$$(2x + 3)(x^2 + 4x + 5)$$

$$= 2x(x^2 + 4x + 5) + 3(x^2 + 4x + 5)$$ Use the distributive property to multiply the trinomial by each term of the binomial.

$$= 2x^3 + 8x^2 + 10x + 3x^2 + 12x + 15$$ Use the distributive property.

$$= 2x^3 + 11x^2 + 22x + 15$$ Combine like terms. ∎

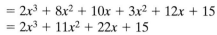 **tudy tip**

Another method is to use a vertical format similar to that used for multiplying whole numbers. For Example 3,

$$
\begin{array}{r}
x^2 + 4x + 5 \\
2x + 3 \\
\hline
3x^2 + 12x + 15 \\
2x^3 + 8x^2 + 10x \\
\hline
2x^3 + 11x^2 + 22x + 15
\end{array}
$$

Like terms are written in the same column.

← $3(x^2 + 4x + 5)$

← $2x(x^2 + 4x + 5)$

Combine like terms.

EXAMPLE 4 **Multiplying a Binomial and a Polynomial**

Multiply: $(5x^2y + 3xy)(7x^3y - 2x^2y + 5x - 1)$

Solution

$$(5x^2y + 3xy)(7x^3y - 2x^2y + 5x - 1)$$
$$= 5x^2y(7x^3y - 2x^2y + 5x - 1)$$
$$+ 3xy(7x^3y - 2x^2y + 5x - 1)$$

Multiply the second polynomial by each term of the binomial.

$$= 35x^5y^2 - 10x^4y^2 \qquad\qquad + 25x^3y - 5x^2y$$
$$+ 21x^4y^2 - 6x^3y^2 \qquad\qquad + 15x^2y - 3xy$$

Use the distributive property and align like terms.

$$= 35x^5y^2 + 11x^4y^2 - 6x^3y^2 + 25x^3y + 10x^2y - 3xy$$

Combine like terms. ∎

---

iscover for yourself

Look at the way the product in the Study Tip for Example 4 is expressed. Is the polynomial in descending powers of x? Why is $-6x^3y^2$ written before $25x^3y$? Is this the only way to write the polynomial in descending powers of x?

 tudy tip

Here's the solution to Example 4 using a vertical format.

$$
\begin{array}{r}
7x^3y - 2x^2y + 5x - 1 \\
5x^2y + 3xy \\
\hline
21x^4y^2 - 6x^3y^2 + 15x^2y - 3xy \\
35x^5y^2 - 10x^4y^2 + 25x^3y - 5x^2y \\
\hline
35x^5y^2 + 11x^4y^2 - 6x^3y^2 + 25x^3y + 10x^2y - 3xy
\end{array}
$$

$3xy(7x^3y - 2x^2y + 5x - 1) \longrightarrow$

$5x^2y(7x^3y - 2x^2y + 5x - 1) \longrightarrow$

4 Multiply binomials using the FOIL method.

Multiplying Two Binomials Using the FOIL Method

The product of two binomials occurs quite frequently in algebra. The product can be found using a method called FOIL, which is based on the distributive property. For example, we can find the product of the binomials $(3x + 2)$ and $(4x + 5)$ as follows.

$$\begin{aligned}(3x + 2)(4x + 5) &= 3x(4x + 5) + 2(4x + 5) \\ &= 3x(4x) + 3x(5) + 2(4x) + 2(5) \\ &= 12x^2 + 15x + 8x + 10\end{aligned}$$

Before combining like terms, let's consider the origin of each of the four terms in the sum.

Origin of	Terms of $(3x + 2)(4x + 5)$	Result of Multiplying Terms	
$12x^2$	$(3x + 2)(4x + 5)$	$(3x)(4x) = 12x^2$	**F**irst terms
$15x$	$(3x + 2)(4x + 5)$	$(3x)(5) = 15x$	**O**uter terms
$8x$	$(3x + 2)(4x + 5)$	$(2)(4x) = 8x$	**I**nner terms
10	$(3x + 2)(4x + 5)$	$(2)(5) = 10$	**L**ast terms

The product is obtained by combining these four results.

$$\begin{aligned}(3x + 2)(4x + 5) &= 12x^2 + 15x + 8x + 10 \\ &= 12x^2 + 23x + 10\end{aligned}$$

We see, then, that two binomials can be quickly multiplied by using the FOIL method, in which *F* represents the product of the *first* two terms in each binomial, *O* represents the product of the two *outside* or *outermost* terms, *I* represents the product of the two *inside* or *innermost* terms, and *L* represents the product of the *last* two terms in each binomial.

$$\begin{array}{cc} \text{F} & \text{L} \\ & \\ (3x + 2) \quad (4x + 5) \end{array} \begin{array}{cccc} \text{F} & \text{O} & \text{I} & \text{L} \\ = 12x^2 + 15x + 8x + 10 \\ = 12x^2 + 23x + 10 \end{array}$$

The FOIL method

Consider $(3x + 2)(4x + 5)$.

F | **1.** Multiply *first* terms of each binomial.

$$\text{F}$$
$$(3x + 2)(4x + 5) \quad \text{Product: } (3x)(4x) = 12x^2$$

O | **2.** Multiply *outer* terms of each binomial.

$$\text{O}$$
$$(3x + 2)(4x + 5) \quad \text{Product: } (3x)(5) = 15x$$

I | **3.** Multiply *inner* terms of each binomial.

$$\text{I}$$
$$(3x + 2)(4x + 5) \quad \text{Product: } (2)(4x) = 8x$$

L | **4.** Multiply *last* terms of each binomial.

$$\text{L}$$
$$(3x + 2)(4x + 5) \quad \text{Product: } (2)(5) = 10$$

The product of two binomials is the sum of these four products.

$$
\begin{array}{c}
\text{F} \quad \text{L} \\
\text{F} \qquad \text{O} \qquad \text{I} \qquad \text{L} \\
(3x + 2)(4x + 5) = (3x)(4x) + (3x)(5) + (2)(4x) + (2)(5) \\
\text{I} \\
= 12x^2 + 15x + 8x + 10 \\
\text{O} \\
= 12x^2 + 23x + 10 \quad \text{Collect like terms.}
\end{array}
$$

EXAMPLE 5 **Using the FOIL Method**

Multiply using the FOIL method: $(5x^2 - 6)(4x^2 + 2)$

Solution

$$
\begin{array}{c}
\text{F} \qquad \text{L} \qquad\qquad \text{F} \qquad\quad \text{O} \qquad\quad \text{I} \qquad\quad \text{L} \\
(5x^2 - 6)(4x^2 + 2) = (5x^2)(4x^2) + (5x^2)(2) + (-6)(4x^2) + (-6)(2) \\
\text{I} \qquad\qquad = 20x^4 + 10x^2 - 24x^2 - 12 \\
\text{O} \qquad\qquad = 20x^4 - 14x^2 - 12
\end{array}
$$

EXAMPLE 6 **Using the FOIL Method**

Multiply using the FOIL method: $(7x^3 - 2x)(3x^3 - 5x)$

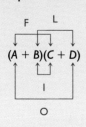

Solution

$$F \qquad L \qquad\qquad F \qquad O \qquad I \qquad L$$

$$(7x^3 - 2x)(3x^3 - 5x) = (7x^3)(3x^3) + (7x^3)(-5x) + (-2x)(3x^3) + (-2x)(-5x)$$

$$= 21x^6 - 35x^4 - 6x^4 + 10x^2$$

$$= 21x^6 - 41x^4 + 10x^2$$

$$I$$

$$O$$

5 Square binomials.

The Square of a Binomial

Using FOIL, we can determine a pattern for the square of a binomial.

Squaring a Binomial Sum:	Squaring a Binomial Difference:
$(A + B)^2 = (A + B)(A + B)$	$(A - B)^2 = (A - B)(A - B)$
$= A^2 + AB + AB + B^2$	$= A^2 - AB - AB + B^2$
$= A^2 + 2AB + B^2$	$= A^2 - 2AB + B^2$

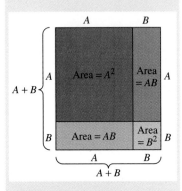

Squaring a binomial

$$(A + B)^2 = A^2 + 2AB + B^2$$

To square a binomial sum, square the first term (A^2), add twice the product of the two terms $(2AB)$, then add the square of the last term (B^2).

$$(A - B)^2 = A^2 - 2AB + B^2$$

To square a binomial difference, square the first term (A^2), subtract twice the product of the two terms $(2AB)$, and add the square of the last term (B^2).

EXAMPLE 7   Squaring Binomials

Multiply:

a. $(y + 5)^2$ **b.** $(x - 8)^2$ **c.** $(4x^2y + xy)^2$ **d.** $(3m - 4n)^2$

Solution

$$(A + B)^2 = A^2 + 2 \cdot A \cdot B + B^2$$

a. $(y + 5)^2 = y^2 + 2 \cdot y \cdot 5 + 5^2$

$= y^2 + 10y + 25$

$$(A - B)^2 = A^2 - 2 \cdot A \cdot B + B^2$$

b. $(x - 8)^2 = x^2 - 2 \cdot x \cdot 8 + 8^2$

$= x^2 - 16x + 64$

$$(A + B)^2 = A^2 + 2 \cdot A \cdot B + B^2$$

c. $(4x^2y + xy)^2 = (4x^2y)^2 + 2 \cdot 4x^2y \cdot xy + (xy)^2$

$= 16x^4y^2 + 8x^3y^2 + x^2y^2$

Caution! The square of a sum is *not* the sum of the squares.

$$(A + B)^2 \neq A^2 + B^2$$

The middle term $2AB$ is missing.

For example,

$$(x + 1)^2 \neq x^2 + 1.$$

Show that $(x + 1)^2$ and $x^2 + 1$ are not equal by substituting 5 for x in each expression and simplifying.
In general:

$$(A + B)^n \neq A^n + B^n$$

$$(A - B)^n \neq A^n - B^n$$

$$(A - B)^2 = A^2 - 2 \cdot A \cdot B + B^2$$
$$\quad\downarrow\qquad\;\downarrow\qquad\quad\downarrow\quad\;\downarrow\;\;\downarrow\;\;\downarrow\qquad\downarrow$$

d. $(3m - 4n)^2 = (3m)^2 - 2 \cdot 3m \cdot 4n + (4n)^2$
$$= 9m^2 - 24mn + 16n^2 \qquad\blacksquare$$

Using technology

We can use a graphing utility to show that the square of a sum is not the sum of the squares.

$$(x + 1)^2 \neq x^2 + 1$$

The graphs of $y_1 = (x + 1)^2$ and $y_2 = x^2 + 1$ are shown below. Clearly, they are not the same.

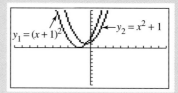

EXAMPLE 8 **Squaring a Binomial**

Multiply: $[(2x - 3) + 5y]^2$

Solution

Although this expansion appears to be more complicated than the previous example, we can multiply using both formulas for squaring a binomial.

$$[\quad A\; +\; B\;]^2 = \quad A^2\quad +\; 2 \cdot\quad A\quad \cdot B\; +\quad B^2$$

$$[(2x-3) + 5y]^2 = (2x-3)^2 + 2 \cdot (2x-3) \cdot 5y + (5y)^2$$

Now use the
formula for
$(A - B)^2$, with
$A = 2x$ and $B = 3$

$$= (2x)^2 - 2 \cdot 2x \cdot 3 + 3^2 + 2(2x - 3)5y + (5y)^2$$
$$= 4x^2 - 12x + 9 + (4x - 6)5y + 25y^2$$
$$= 4x^2 - 12x + 9 + 20xy - 30y + 25y^2 \qquad\blacksquare$$

6 Multiply the sum and difference of two terms.

Multiplying the Sum and Difference of Two Terms

Using FOIL, we can also determine a pattern for the product of the sum and difference of the same two terms.

$$\qquad\qquad\;\; F\quad\; O\quad\; I\quad\; L$$
$$\qquad\qquad\;\;\downarrow\quad\;\downarrow\quad\;\downarrow\quad\;\downarrow$$
$$(A + B)(A - B) = A^2 - AB + AB - B^2$$
$$= A^2 - B^2$$

The product of the sum and difference of two terms

$$(A + B)(A - B) = A^2 - B^2$$

The product of the sum and difference of the same two terms is the difference of the squares of the terms.

| EXAMPLE 9 | **Multiplying the Sum and Difference of Two Terms** |

Multiply:

a. $(y + 8)(y - 8)$ **b.** $(9x + 5y)(9x - 5y)$

c. $(6a^2b + 5b)(6a^2b - 5b)$ **d.** $(-5x^5y^4 + 3z)(5x^5y^4 + 3z)$

Solution

$$(A \quad + \quad B) \quad (A \quad - \quad B) \quad = \quad A^2 \quad - \quad B^2$$

a. $(y + 8)(y - 8) = y^2 - 8^2 = y^2 - 64$

b. $(9x + 5y)(9x - 5y) = (9x)^2 - (5y)^2 = 81x^2 - 25y^2$

c. $(6a^2b + 5b)(6a^2b - 5b) = (6a^2b)^2 - (5b)^2 = 36a^4b^2 - 25b^2$

d. $(3z + 5x^5y^4)(3z - 5x^5y^4) = (3z)^2 - (5x^5y^4)^2 = 9z^2 - 25x^{10}y^8$

Let's stop to summarize the FOIL method and the two special products.

FOIL and special products

Let A and B be real numbers, variables, or algebraic expressions.

| **Method** | **Example** |

Foil

$$\overset{F \quad\quad O \quad\quad I \quad\quad L}{(A + B)(C + D) = AC + AD + BC + BD}$$

$$\overset{F \quad\quad\quad O \quad\quad\quad I \quad\quad\quad L}{(2x + 3)(4x + 5) = (2x)(4x) + (2x)(5) + (3)(4x) + (3)(5)}$$
$$= 8x^2 + 10x + 12x + 15$$
$$= 8x^2 + 22x + 15$$

Square of a Binomial

$$(A + B)^2 = A^2 + 2AB + B^2$$

$$(2x + 3)^2 = (2x)^2 + 2(2x)(3) + 3^2$$
$$= 4x^2 + 12x + 9$$

$$(A - B)^2 = A^2 - 2AB + B^2$$

$$(2x - 3)^2 = (2x)^2 - 2(2x)(3) + 3^2$$
$$= 4x^2 - 12x + 9$$

Sum and Difference of Two Terms

$$(A + B)(A - B) = A^2 - B^2$$

$$(2x + 3)(2x - 3) = (2x)^2 - 3^2$$
$$= 4x^2 - 9$$

These patterns can sometimes be used with more complicated products, as illustrated in Example 10.

EXAMPLE 10 **Using the Special Products**

Multiply: **a.** $(7x + 5 + 4y)(7x + 5 - 4y)$ **b.** $(3x + y + 1)^2$

Solution

a. $(7x + 5 + 4y)(7x + 5 - 4y)$

$= (7x + 5 + 4y)(7x + 5 - 4y)$ ← Product of the sum and difference of the same two expressions

$= (7x + 5)^2 - (4y)^2$ ← Difference of their squares

$= (7x)^2 + 2 \cdot 7x \cdot 5 + 5^2 - (4y)^2$ Use the formula for the square of a binomial.

$= 49x^2 + 70x + 25 - 16y^2$

b. $(3x + y + 1)^2$

$= [(3x + y) + 1]^2$ Group the terms so that the formula for the square of a binomial can be applied.

$= (3x + y)^2 + 2 \cdot (3x + y) \cdot 1 + 1^2$

$= 9x^2 + 6xy + y^2 + 6x + 2y + 1$ ∎

7 Multiply polynomial functions.

Multiplying Polynomial Functions

Just as we multiply polynomials, we can multiply polynomial functions to obtain new functions. We can add to our previous discussion of the algebra of functions by defining the multiplication of polynomial functions.

> **Multiplying polynomial functions**
>
> If f and g are polynomial functions, then
>
> $$(fg)(x) = f(x)g(x).$$
>
> The function fg is the product of the functions f and g.

EXAMPLE 11 **Applying the Algebra of Functions**

Let $f(x) = x - 5$ and $g(x) = x - 2$. Find: $(fg)(x)$

Solution

$(fg)(x) = f(x)g(x)$

$= (x - 5)(x - 2)$ Substitute the given functions.

$= x^2 - 2x - 5x + 10$ Multiply using FOIL.

$= x^2 - 7x + 10$ Simplify.

The graph of $y = x^2 - 7x + 10$, the quadratic function obtained by multiplying the two given linear functions, is shown in Figure 4.4. ∎

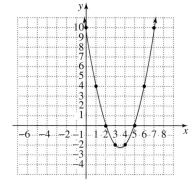

Figure 4.4

The graph of $y = x^2 - 7x + 10$

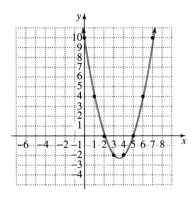

Use the graph of $y = x^2 - 7x + 10$ in Figure 4.4 shown again on the left.

1. Verify the coordinates of the parabola's vertex.
2. What is the y-intercept? How can this be determined from the function's equation?
3. What are the x-intercepts? What are the x-intercepts for the linear functions in Example 11? What do you observe?
4. Describe how a knowledge of a parabola's intercepts and the coordinates of its vertex can be used to determine its graph fairly quickly.

Polynomial Multiplication and Modeling

8 Use polynomial multiplication in geometric modeling.

Our work with multiplication can be used to find a function that models a geometric situation.

EXAMPLE 12   **A Geometric Application**

A rectangular piece of cardboard measuring 10 inches by 8 inches has a square cut out at each corner (see Figure 4.5). The sides are then turned up to form an open box. If x represents the length of the side of the square cut from each corner of the rectangle, express the volume of the box as a polynomial function written in descending powers of x.

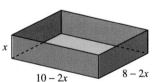

Figure 4.5

Constructing an open box

Solution

Let

$$x = \text{the height of the box}$$
$$10 - 2x = \text{the length of the box}$$
$$8 - 2x = \text{the width of the box}$$

We can find the volume of the box using $V = LWH$, where L, W, and H represent its length, width, and height, respectively. We will express the volume as $V(x)$ to show that the volume of the box is a function of x, the length of the square cut from each corner of the cardboard.

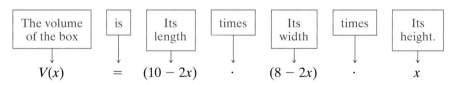

$$V(x) = (10 - 2x)(8 - 2x)x \qquad \text{This is the function that models the problem's conditions.}$$
$$= (80 - 36x + 4x^2)x \qquad \text{Multiply binomials using FOIL.}$$
$$= 80x - 36x^2 + 4x^3 \qquad \text{Use the distributive property.}$$
$$= 4x^3 - 36x^2 + 80x \qquad \text{Write the trinomial in descending powers of } x.$$

Thus, the volume of the box is modeled by the polynomial function $V(x) = 4x^3 - 36x^2 + 80x$, where $V(x)$ is expressed in cubic inches. ∎

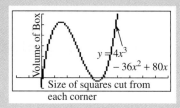

Discover for yourself

The graph of the volume function in Example 12 was obtained with a graphing utility and the following range setting:

Xmin = −2, Xmax = 10, Xscl = 1, Ymin = −10, Ymax = 60, Yscl = 10.

1. How does this graph illustrate the behavior of a cubic function?

2. In terms of the conditions given in Example 12, how much of the graph models the problem? (Recall that the cardboard is 10 inches by 8 inches and x represents the length of the side of the squares cut from each corner.) (See Figure 4.5.) What is the largest value that x can be? What sort of volume would the resulting box have under these conditions? How is this illustrated in the graph?

3. Use your graphing utility to reproduce this graph. Then use the $\boxed{\text{TRACE}}$ or maximum function feature to find the highest point on the graph between $x = 0$ and $x = 4$. Describe the significance of the coordinates of this point in terms of the situation modeled by the function.

9 Use polynomial multiplication to evaluate a function.

Evaluating Quadratic Functions

In our final example, we use our work with multiplying polynomials to evaluate a quadratic function.

$\boxed{\text{EXAMPLE 13}}$ **Evaluating a Quadratic Function**

Given $f(x) = x^2 - 7x + 3$, find and simplify:

a. $f(a + 4)$ **b.** $f(a + h) - f(a)$

Solution

a. We can find $f(a + 4)$ (read "f at a plus 4") by replacing each occurrence of x with $a + 4$. Then we multiply and simplify.

$$f(x) = x^2 - 7x + 3 \qquad \text{This is the given quadratic function.}$$

$$f(a + 4) = (a + 4)^2 - 7(a + 4) + 3 \qquad \text{Replace each occurrence of } x \text{ with } a + 4.$$
$$= a^2 + 8a + 16 - 7a - 28 + 3 \qquad \text{Multiply as indicated.}$$
$$= a^2 + a - 9 \qquad \text{Combine like terms.}$$

b. To find $f(a + h) - f(a)$, we first replace each occurrence of x by $a + h$ and then replace each occurrence of x by a. Then we multiply and simplify.

$$\overbrace{f(a+h)}^{} \quad - \quad \overbrace{f(a)}^{}$$
$$\downarrow$$
$$= (a+h)^2 - 7(a+h) + 3 - (a^2 - 7a + 3) \qquad \text{Use } f(x) = x^2 - 7x + 3, \text{ replacing}$$
$$x \text{ with } a + h \text{ and } a, \text{ respectively.}$$
$$= a^2 + 2ah + h^2 - 7a - 7h + 3 - a^2 + 7a - 3 \qquad \text{Multiply as indicated.}$$
$$= 2ah + h^2 - 7h \qquad \text{Combine like terms. Observe}$$
$$\text{that } a^2 - a^2, \ -7a + 7a, \text{ and}$$
$$3 - 3 \text{ all have zero sums.} \quad \blacksquare$$

PROBLEM SET 4.2

Practice Problems

Find the product of the monomials in Problems 1–8.

1. $y^7 \cdot y^5$
2. $x^{13} \cdot x^8$
3. $(3x^8)(5x^6)$
4. $(-2y^7)(-4y)$
5. $(3x^2y^4)(5xy^7)$
6. $(7x^4y)(6xy^5)$
7. $(-3xy^2z^5)(2xy^7z^4)$
8. $(-11x^3y^2z^4)(-3xy^5z^6)$

Find the product of the monomial and the polynomial in Problems 9–20.

9. $4x^2(3x - 2)$
10. $5x^3(7x^2 + 4)$
11. $-2x^2(5x^3 - 8x^2 + 7x - 3)$
12. $-6x^2(4x^3 - 11x^2 + 3x - 5)$
13. $2xy(4x^2y + 7x - 2y - 3)$
14. $5xy(3x^3y^2 - 3x + 4y - 2)$
15. $\frac{1}{3}x^3y^7(\frac{1}{2}xy^6 + \frac{2}{5}x^4y^2 + 6)$
16. $8a^2b^3c(2ab^4c - 3ab + 6a^4b^3c^2)$
17. $16x^4x^5z^3(-\frac{1}{8}xz + \frac{1}{16}x^4yz^2 - \frac{1}{32}x^6y^2z)$
18. $(4x^3yz^2 - 8xy + x^5)(-3x^2y^4z)$
19. $(6uv^3w - 8uv + w^4)(-5u^5v^3w)$
20. $(a^2c^3 + b^2c^3 - a^2b^2)(-3a^2c^3)$

In Problems 21–32, find each product using either a horizontal or a vertical format.

21. $(x - 3)(x^2 + 2x + 5)$
22. $(x + 4)(x^2 - 5x + 8)$
23. $(x + 5)(x^2 - 5x + 25)$
24. $(x - 1)(x^2 + x + 1)$
25. $(a - b)(a^2 + ab + b^2)$
26. $(a + b)(a^2 - ab + b^2)$
27. $(x^2 + 2x - 1)(x^2 + 3x - 4)$
28. $(x^2 - 2x + 3)(x^2 + x + 1)$
29. $(xy + 2)(x^2y^2 - 2xy + 4)$
30. $(xy - 3)(4x^2y^2 - 3xy + 2)$
31. $(5a^2b - 3ab + 2b^2)(ab - 3b + a)$
32. $(4a^2 + b^2 - ab)(a^2 - 3b^2 - 2ab)$

Use the FOIL method to multiply the binomials in Problems 33–56.

33. $(x + 4)(x + 7)$
34. $(x + 7)(x + 11)$
35. $(y + 5)(y - 6)$
36. $(y + 8)(y - 12)$
37. $(5x + 3)(7x + 1)$
38. $(2x + 5)(7x + 2)$
39. $(3y - 11)(2y - 3)$
40. $(5y - 8)(7y - 6)$
41. $(7x - 12)(3x + 8)$
42. $(11x - 5)(4x - 3)$
43. $(9x^2 - 4)(3x^2 + 5)$
44. $(8y^2 - 11)(2y^2 + 3)$
45. $(8x^3 - 3x)(5x^3 - 2x)$
46. $(5y^3 - 7y)(2y^3 + 5y)$
47. $(5z^2 + 1)(3z - 4)$
48. $(3z^2 + 7)(2z - 13)$
49. $(9x - 5)(7 + 4x)$
50. $(11x - 3)(8 + 7x)$
51. $(9x^2 + 4)(9x^3 - 4)$
52. $(8y^2 + 3)(8y^3 - 3)$
53. $(3x + 7y)(2x - 5y)$
54. $(9x - 11y)(5x + 4y)$
55. $(3x^2yz - 2y)(5x^2yz + 7y)$
56. $(4x^2yz - 3y)(9x^2yz - 5y)$

In Problems 57–98, multiply using the special products for the square of a binomial or the product of the sum and difference of two terms.

57. $(x + 3)^2$
58. $(y + 11)^2$
59. $(y - 5)^2$
60. $(x - 12)^2$
61. $(2x + y)^2$
62. $(4x + y)^2$
63. $(5x - 3y)^2$
64. $(3x - 4y)^2$
65. $(2x^2 - 3y)^2$
66. $(3x^2 + 4y)^2$
67. $(4x^3 + 2y^5)^2$
68. $(2x^3 - 3y^4)^2$
69. $(4ab^2 + ab)^2$
70. $(5m^2n^3 - 3mn^2)^2$
71. $(y + 7)(y - 7)$
72. $(y + 11)(y - 11)$
73. $(5x + 3)(5x - 3)$
74. $(7x + 5)(7x - 5)$
75. $(3x + 7y)(3x - 7y)$
76. $(8x + 3y)(8x - 3y)$
77. $(x^2 + yz)(x^2 - yz)$
78. $(a^3 + bc)(a^3 - bc)$
79. $(7x^2y - 3z)(7x^2y^2y + 3z)$
80. $(9x^2y - 2z)(9x^2y + 2z)$

81. $(3x^3y^2z - 1)(3x^3y^2z + 1)$

82. $(4a^2b^3c - 7ab)(4a^2b^3c + 7ab)$

83. $(-xy + y^2)(xy + y^2)$

84. $(-2xy + y^2)(y^2 + 2xy)$

85. $(3x + 7 + 5y)(3x + 7 - 5y)$

86. $(5x + 7y + 2)(5x - 7y - 2)$

87. $[5y - (2x + 3)][5y + (2x + 3)]$

88. $[8y + (7 - 3x)][8y - (7 - 3x)]$

89. $(2x + y + 1)^2$

90. $(5x + 1 + 6y)^2$

91. $[(3x - 1) + y]^2$

92. $[(2x - y) + 8]^2$

93. $[(3x - 1) + y][(3x - 1) - y]$

94. $(y + 1)(y - 1)(y^2 + 1)$

95. $(x + 2)(x - 2)(x^2 + 4)$

96. $(x - y)(x + y)(x^2 + y^2)$

97. $(3x - y)(3x + y)(9x^2 + y^2)$

98. $(5x - y)(5x + y)(25x^2 + y^2)$

For each pair of functions in Problems 99–102, find $(f + g)(x)$, $(f - g)(x)$, and $(fg)(x)$.

99. $f(x) = 2x^2 + 3, g(x) = x^2 - 4$

101. $f(x) = x^2 - 2x + 1, g(x) = x^3 + x^2 + x + 3$

100. $f(x) = 3x^2 - 7, g(x) = 4x^2 - 5x + 3$

102. $f(x) = x^2 + x + 5, g(x) = x^2 - 3x + 1$

For each function in problems 103–106, find $f(a + 3)$, $f(2a - 1)$, and $f(a + h) - f(a)$. Simplify.

103. $f(x) = x^2 - 2x + 5$

105. $f(x) = 3x^2 + 5x - 2$

104. $f(x) = x^2 - 5x + 4$

106. $f(x) = -2x^2 + 4x - 3$

Application Problems

In Problems 107–110, write a polynomial function, $A(x)$, that models the area of each shaded region. Express the function's formula in descending powers of x.

107.

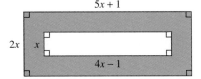

108.

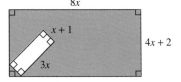

109.

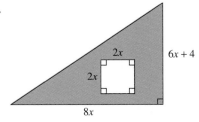

110.

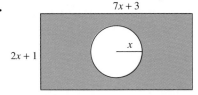

In Problems 111–112, write a polynomial function, $V(x)$, that models the volume of each solid. Express the function's formula in descending powers of x.

111. This solid shows a metal casting $x + 10$ centimeters square and 2 centimeters thick with a hole in the center $x + 5$ centimeters square.

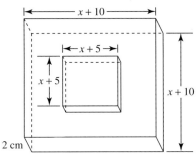

112. This solid is formed by attaching two cubes.

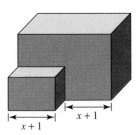

113. Multiply the right side of the Doyle log model

$$N = \left(\frac{x - 4}{4}\right)^2 y$$

discussed in the introduction to this section.

114. French number theorist Pierre de Fermat (1601–1665) proved that every odd prime number can be expressed as the difference of two squares in one and only one way. If p is an odd prime,

$$p = \left(\frac{p+1}{2}\right)^2 - \left(\frac{p-1}{2}\right)^2.$$

a. Let $p = 5$ and substitute into the formula, writing 5 as the difference of squares.
b. Repeat part (a) for $p = 13$.
c. Verify the formula by working with the right side and simplifying to obtain p.

True–False Critical Thinking Problems

115. Which one of the following is true?
a. $(x - 5)^2 = x^2 - 5x + 25$
b. Suppose a square garden has an area represented by $9x^2$ square feet. If one side is made 7 feet longer and the other side is made 2 feet shorter, then the trinomial that represents the area of the larger garden is $9x^2 + 15x - 14$ square feet.
c. $(x^m + 2)(x^m + 4) = x^{m^2} + 6x^m + 8$
d. The expression $7x^2 \cdot 4x^3$ is the product of two binomials.

116. Which one of the following is true?
a. $(4x^3 + 5)(4x^3 - 5) = 16x^9 - 25$

b. The FOIL method can also be used to add binomials.
c. $(xy + 2)^2 - x^2y^2 = 4$
d. This box, with an open top and dimensions as shown, has a surface area, $A(x)$ given by $A(x) = 5x^2 - 8x$.

Technology Problems

In Problems 117–120, use a graphing utility to graph the functions y_1 and y_2. What can you conclude? Verify your conclusion using polynomial multiplication.

117. $y_1 = (x - 2)^2$
$y_2 = x^2 - 4x + 4$

118. $y_1 = (x - 4)^3$
$y_2 = x^3 - 12x^2 + 48x - 64$

119. $y_1 = (x^2 - 3x + 2)(x - 4)$
$y_2 = x^3 - 7x^2 + 14x - 8$

120. $y_1 = (x + 1.5)(x - 1.5)$
$y_2 = x^2 - 2.25$

Writing in Mathematics

121. Explain how to multiply a binomial and a trinomial.

122. Explain why $(7x + 5)^2$ is *not* equal to $49x^2 + 25$.

Critical Thinking Problems

123. Simplify: $(y^n + 2)(y^n - 2) - (y^n - 3)^2$.

124. Simplify: $2y^{2n}(3y^{3n} + 4y^n - 1) - 5y^{2n}(y^{2n} - 3)$.

125. What polynomial when divided by $4x - 5$ has a quotient of $3x + 1$?

126. Evaluate and simplify: $\begin{vmatrix} x+5 & x-4 \\ x-3 & x-2 \end{vmatrix}$.

Represent the area of each plane figure in Problems 127–130 as a polynomial in descending powers of x.

127.

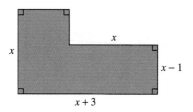

128.

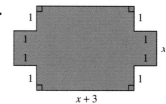

129.

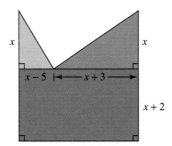

130.

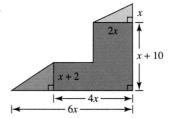

For Problems 131–132, represent the volume of each figure as a polynomial in descending powers of x.

131.

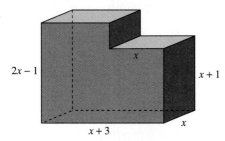

132.

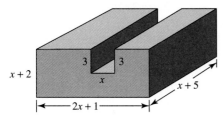

133. Perform each multiplication.
   a. $(y - 1)(y + 1)$
   b. $(y - 1)(y^2 + y + 1)$

   c. $(y - 1)(y^3 + y^2 + y + 1)$
   d. Use the emerging pattern to predict the product of $(y - 1)(y^4 + y^3 + y^2 + y + 1)$ without actually multiplying.

134. a. Find the sum of the coefficients of $(x + y)^2 + (x - y)^2$.
   b. Find the sum of the coefficients of $(x + y)^3 + (x - y)^3$.
   c. Find the sum of the coefficients of $(x + y)^4 + (x - y)^4$.
   d. Use the emerging pattern to predict the sum of the coefficients of $(x + y)^5 + (x - y)^5$.

135. If

$$(5 - 2)^2 = 25 - 20 + 4$$

and

$$(7 - 3)^2 = 49 - 42 + 9$$

write a similar expression for $(11 - 5)^2$.

Group Activity Problems

136. In this activity, you will be developing a geometric model for cubing a binomial. You will need to know that the volume of a rectangular solid is the product of its length, width, and height. Refer to the figure shown on the right.

   The first figure shows a cube that has a volume of $(A + B)(A + B)(A + B)$ or $(A + B)^3$. Also shown are eight rectangular solids that make up the large cube. Find an expression using A, B, or both for the volume of each of these eight solids. Find the sum of these volumes, add like terms, and write the expression in descending powers of A. Then set this polynomial equal to $(A + B)^3$, the volume of the large cube. You should now have a formula for cubing a binomial.

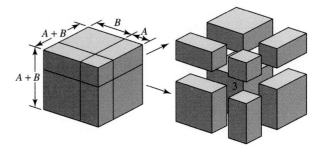

137. Try working on this problem in small groups. Groups should then compare the solution and the solution method. Is there a "best" way to arrive at the answer? If $xy \neq 0$, find the number represented by

$$\left(x + \frac{1}{x}\right)^2 + \left(y + \frac{1}{y}\right)^2 + \left(xy + \frac{1}{xy}\right)^2$$

$$- \left(x + \frac{1}{x}\right)\left(y + \frac{1}{y}\right)\left(xy + \frac{1}{xy}\right).$$

Review Problems_____

138. Solve and graph the solution set on the number line:
$|4 - 3x| \geq 10$.

139. Subtract $y^2 + 3y - 1$ from the sum of $2y^2 - 3y - 1$
and $3y^2 + y - 1$.

140. Solve the system using matrices:

$$x - y + z = 4$$
$$3x - y + z = 6$$
$$2x + 2y - 3z = -6$$

SECTION 4.3

Solutions Manual **Tutorial**   **Video 5**

Greatest Common Factors and Factoring by Grouping

Objectives

1 Factor out the greatest common factor of a polynomial.
2 Factor by grouping.

In general, factoring involves the reverse of the multiplication process. For example, because we know that $(2x + 1)(3x - 2) = 6x^2 - x - 2$, we can also say that

$$6x^2 - x - 2 = (2x + 1)(3x - 2).$$

When we write the polynomial $6x^2 - x - 2$ as a product, we are *factoring* the polynomial. The *linear factors* of $6x^2 - x - 2$ are $2x + 1$ and $3x - 2$.

Factoring is the process of writing a polynomial as the product of two or more polynomials. Factoring is one of the most important tools in algebra, necessary for working with algebraic fractions and useful for solving many equations and inequalities.

In this section, we limit our discussion to factoring polynomials with integral coefficients, called *factoring over the integers*.

1 Factor out the greatest common factor of a polynomial.

Terms with Common Factors

The *greatest common factor* (GCF) for a polynomial is the largest expression that divides each term of the polynomial. For example, the GCF for the polynomial $15x^5 + 20x^3 - 25x^2$ is $5x^2$ because it is the largest monomial that divides each of the three terms. We can use the distributive property and our knowledge of exponents to write

$$15x^5 + 20x^3 - 25x^2 = 5x^2(3x^3) + 5x^2(4x) + 5x^2(-5)$$
$$= 5x^2(3x^3 + 4x - 5)$$

The last line is written in factored form. When we express the polynomial as the product of the GCF $5x^2$ and $3x^3 + 4x - 5$ we are *factoring out the greatest common factor* or *removing the greatest common factor*.

The following steps are useful in finding the GCF of a polynomial.

Finding the GCF of a polynomial

1. Find the coefficient of the GCF by determining the largest integer that will divide into the numerical coefficients of each term.
2. List, to the lowest power to which it occurs, each variable factor that occurs in every term of the polynomial.
3. The GCF of the polynomial is the product of the GCF of the coefficients and the variable factors from step 2.

EXAMPLE 1   **Factoring Out the GCF**

Find the GCF of the polynomial and then factor out the GCF:

a. $5y^2 + 15$ **b.** $9x^5 - 15x^3$

c. $16x^2y^3 - 24x^3y^4$ **d.** $12x^5y^4 - 4x^4y^3 + 2x^3y^2$

Solution

a. $5y^2 + 15$. The GCF of $5y^2$ and 15 is 5.

$$5y^2 + 15 = 5(y^2) + 5(3) \quad \text{Express each term with the GCF as a factor.}$$
$$= 5(y^2 + 3) \quad \text{Use the distributive property to factor out the GCF:}$$
$$ab + ac = a(b + c).$$

b. $9x^5 - 15x^3$. The numerical part of the GCF is 3, the greatest integer that divides into both 9 and 15. For the variable parts, x^5 and x^3, we use the least exponent that appears on x, namely, 3. The GCF is $3x^3$.

$$9x^5 - 15x^3 = 3x^3(3x^2) - 3x^3(5) \quad \text{Express each term with the GCF as a factor.}$$
$$= 3x^3(3x^2 - 5) \quad \text{Factor out the GCF.}$$

c. $16x^2y^3 - 24x^3y^4$. The numerical part of the GCF is 8, the greatest integer that divides into both 16 and 24. For the variable parts, x^2 and x^3, use the lesser exponent, so x^2 is a factor of the GCF. Similarly, for y^3 and y^4, use the lesser exponent, so y^3 is a factor of the GCF. Thus, the GCF is $8x^2y^3$.

$$16x^2y^3 - 24x^3y^4 = 8x^2y^3(2) - 8x^2y^3(3xy) \quad \text{Express each term with the GCF as a factor.}$$
$$= 8x^2y^3(2 - 3xy) \quad \text{Factor out the GCF.}$$

d. $12x^5y^4 - 4x^4y^3 + 2x^3y^2$. The numerical part of the GCF is 2, the greatest integer that divides into 12, 4, and 2. The lowest power of x that occurs is 3, and the lowest power of y that occurs is 2. Thus, the GCF is $2x^3y^2$.

$$12x^5y^4 - 4x^4y^3 + 2x^3y^2$$
$$= 2x^3y^2(6x^2y^2) - 2x^3y^2(2xy) + 2x^3y^2(1) \quad \text{Express each term with the GCF as a factor.}$$
$$= 2x^3y^2(6x^2y^2 - 2xy + 1) \quad \text{Factor out the GCF.} \quad ■$$

The polynomials in Example 1 were factored using a common monomial factor that had a positive coefficient. However, in various situations there may be a reason to factor a negative number out of a polynomial. This is illustrated in our next example.

EXAMPLE 2   **Factoring Out a Negative GCF**

Factor each polynomial in two ways, first with a GCF having a positive coefficient and then with a GCF having a negative coefficient.

a. $-18x^2 + 9$ **b.** $-3x^3 + 12x^2 - 15x$

S tudy tip

Check your factorization by using the distributive property to multiply the factors. This should give you the original polynomial.

D iscover for yourself

Change the last term of the polynomial in Example 1d to $5xy^3$ and consider

$$12x^5y^4 - 4x^4y^3 + 5xy^3.$$

What happens to the GCF by making this change to the third term? Factor out the GCF and compare the factorization to the one given in Example 1d. Now write a polynomial whose GCF does not change when you change the last term to a monomial of your choice.

Solution

a. $-18x^2 + 9 = 9(-2x^2) + 9(1)$ The GCF is positive.
$= 9(-2x^2 + 1)$ Factor out the positive GCF.

or

$-18x^2 + 9 = -9(2x^2) + (-9)(-1)$ Now the GCF is negative.
$= -9(2x^2 - 1)$ Factor out the negative GCF.

Thus, $-18x^2 + 9 = 9(-2x^2 + 1) = -9(2x^2 - 1)$.

b. $-3x^3 + 12x^2 - 15x = 3x(-x^2) + 3x(4x) + 3x(-5)$ The GCF has a positive coefficient.
$= 3x(-x^2 + 4x - 5)$ Factor out the GCF.

or

$-3x^3 + 12x^2 - 15x = -3x(x^2) - 3x(-4x) - 3x(5)$ The GCF now has a negative coefficient.
$= -3x(x^2 - 4x + 5)$ Factor out the GCF.

Thus, $-3x^3 + 12x^2 - 15x = 3x(-x^2 + 4x - 5) = -3x(x^2 - 4x + 5)$. ∎

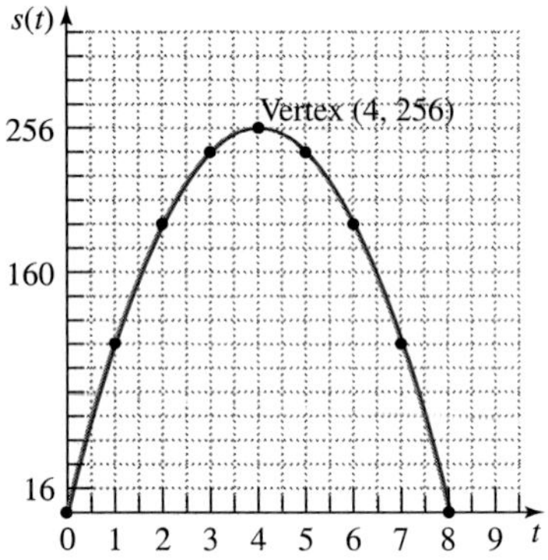

EXAMPLE 3 **An Application Involving a Negative GCF**

If an object is projected directly upward from the ground with an initial velocity of 128 feet per second, its height ($s(t)$) above the ground after t seconds is modeled by the quadratic function $s(t) = -16t^2 + 128t$. Find an equivalent expression for the function's formula by factoring out the GCF.

Solution

We will factor out $-16t$, the GCF with a negative coefficient.

$s(t) = -16t^2 + 128t$ This is the given position function.
$= -16t(t) - 16t(-8)$ Express each term with the GCF as a factor.
$= -16t(t - 8)$ Factor out the GCF.

Thus, $s(t) = -16t(t - 8)$ is an equivalent expression for the function. ∎

The graph of $s(t) = -16t^2 + 128t$ or, equivalently, $s(t) = -16t(t - 8)$ is shown in Figure 4.6. The coordinates of the parabola's vertex indicate that a maximum height of 256 feet is reached after 4 seconds.

Some polynomials have common factors with two or more terms.

EXAMPLE 4 **Factoring Out a Binomial GCF**

Factor: $x^2(x - 3) + 2(x - 3)$

Solution

In this situation, the GCF is the common binomial factor $(x - 3)$. We factor out this common factor as follows.

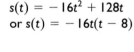

$x^2(x - 3) + 2(x - 3) = (x - 3)(x^2 + 2)$ Factor out the common binomial factor. ∎

---

s(t) graph

Vertex (4, 256)

256

160

16

0 1 2 3 4 5 6 7 8 9 *t*

Figure 4.6

The graph of
$s(t) = -16t^2 + 128t$
or $s(t) = -16t(t - 8)$

Discover for yourself

1. Use algebraic methods to verify the coordinates of the vertex for the position function in Example 3.

2. Look at the factored form of the function. What does it mean about the position of the object when each factor is equal to 0?

3. Use a graphing utility to verify the hand-drawn graph in Figure 4.6.

2  Factor by grouping.

Factoring by Grouping

Some polynomials only have a GCF of 1. However, by a suitable rearrangement of the terms, it still may be possible to factor. This process, called *factoring by grouping,* is illustrated in Examples 5 through 9.

EXAMPLE 5 **Factoring by Grouping**

Factor: $x^3 + 4x^2 + 3x + 12$

Solution

First we group terms that have a common factor.

$$\boxed{x^3 + 4x^2} + \boxed{3x + 12}$$

Common factor is x^2.

Common factor is 3.

We can now factor the given polynomial.

$$
\begin{aligned}
& x^3 + 4x^2 + 3x + 12 && \text{This is the given polynomial.}\\
& = (x^3 + 4x^2) + (3x + 12) && \text{Group terms with common factors.}\\
& = x^2(x + 4) + 3(x + 4) && \text{Factor out the GCF from the grouped terms. The remaining two terms have } (x + 4) \text{ as a common binomial factor.}\\
& = (x + 4)(x^2 + 3) && \text{Factor out } (x + 4) \text{ from both terms.}
\end{aligned}
$$

Thus, $x^3 + 4x^2 + 3x + 12 = (x + 4)(x^2 + 3)$. Check the factorization by multiplying the right side using the FOIL method. You should obtain the polynomial on the left. ■

Discover for yourself

In Example 5, try grouping the terms as follows:

$$(x^3 + 3x) + (4x^2 + 12)$$

Factor out the GCF from each group and complete the factoring process. Describe what happens. What can you conclude?

Using technology

If a polynomial contains one variable, you can check your factorization by graphing the given polynomial and the factored form of the polynomial in the same viewing rectangle. The graphs should be identical.

We did this for Example 5, graphing

$$y_1 = x^3 + 4x^2 + 3x + 12$$
$$\text{and}\quad y_2 = (x + 4)(x^2 + 3)$$

as shown in the figure. Knowing that a cubic function with a positive leading coefficient has a graph that falls to the left and rises to the right, we had to experiment with the range setting to obtain a relatively complete graph. We used

Xmin = −5, Xmax = 5, Xscl = 1, Ymin = −50, Ymax = 50, Yscl = 5

although, of course, other range settings are possible.

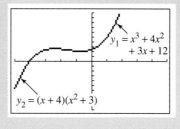

Later in this chapter, we will see that terms of a polynomial can be grouped in several ways. One grouping might lead to a factorization, whereas other groupings might not. In our present focus, the polynomial to be factored by grouping contains four terms and can usually be factored using the following method.

> **Factoring by grouping**
>
> 1. Group terms together (generally two) that contain a common factor. There will usually be two groups. Sometimes the terms must be rearranged.
> 2. Determine and factor out the common factor of each group.
> 3. Factor out the remaining factor that is common to the expression in step 2.

EXAMPLE 6   Factoring by Grouping

Factor: $3xz - yz + 6x - 2y$

Solution

$3xz - yz + 6x - 2y$	This is the given polynomial.
$= (3xz - yz) + (6x - 2y)$	Group the terms that have a GCF. The first two terms, with a GCF of z, are grouped. The last two terms, with a GCF of 2, are also grouped.
$= z(3x - y) + 2(3x - y)$	Factor out the GCF from the grouped terms. The remaining two terms have a GCF of $(3x - y)$.
$= (3x - y)(z + 2)$	Factor out $(3x - y)$ from both terms.

Thus, $3xz - yz + 6x - 2y = (3x - y)(z + 2)$. This can be checked by multiplying the right side using the FOIL method. You should obtain the polynomial on the left.

In our next example, factoring by grouping requires that the terms be rearranged before the groupings are made.

EXAMPLE 7   Factoring by Grouping

Factor: $x^2y^2 - 21 - 3y^2 + 7x^2$

Solution

The first two terms and the last two terms have no common factor other than 1. Let's rearrange the terms and try a different grouping.

$x^2y^2 - 21 - 3y^2 + 7x^2$	This is the given polynomial.
$= (x^2y^2 - 3y^2) + (7x^2 - 21)$	Rearrange and group the terms that have a GCF. The first two terms have a GCF of y^2 and the last two terms have a GCF of 7.
$= y^2(x^2 - 3) + 7(x^2 - 3)$	Factor out the GCF from the grouped terms. The remaining two terms have a GCF of $(x^2 - 3)$.
$= (x^2 - 3)(y^2 + 7)$	Factor out $(x^2 - 3)$ from both terms.

Thus, $x^2y^2 - 21 - 3y^2 + 7x^2 = (x^2 - 3)(y^2 + 7)$. Check by multiplying the right side using the FOIL method. You should obtain the polynomial on the left.

iscover for yourself

In Example 7, try grouping the terms as follows:

$(x^2y^2 + 7x^2) + (-3y^2 - 21)$.

Complete the factoring process by factoring out x^2 from the first two terms and -3 from the last two terms. Do you get the same answer as in Example 7? Describe what happens if you factor 3, rather than -3, from the second grouping.

EXAMPLE 8 **Factoring by Grouping**

Factor: $3a^2 + 12a - 2ab - 8b$

Solution

$3a^2 + 12a - 2ab - 8b$	This is the given polynomial.
$= (3a^2 + 12a) + (-2ab - 8b)$	Group the terms that have a GCF.
$= 3a(a + 4) - 2b(a + 4)$	By factoring $-2b$ from the second grouping,
$= (a + 4)(3a - 2b)$	we immediately produce a GCF of $(a + 4)$ in the two remaining terms.

In Example 8, we factored $-2b$ rather than $2b$ from the second grouping to produce a GCF. In some situations, it is not obvious if we should factor with a positive or negative coefficient. Example 9 illustrates what to do if a factoring in one of the groupings results in factors that are opposites in the two remaining terms.

EXAMPLE 9 **Factoring by Grouping**

Factor: $y + ax - x - ay$

Solution

$y + ax - x - ay$	This is the given polynomial.
$= y - ay + ax - x$	Rearrange terms so that adjacent terms have a common factor.
$= y(1 - a) + x(a - 1)$	Factor y from the first grouping and x from the second grouping.

The two remaining terms do not have a common factor, although $(1 - a)$ and $(a - 1)$ are opposites. We can create a GCF in this situation by factoring out -1 from either of the terms.

> **S**tudy tip
>
> In factoring Example 9, it is helpful to remember that
> $$a - b$$
> and $-1(-a + b)$
> and $-(b - a)$
> are all equivalent.

$y + ax - x - ay$	This is the given polynomial.
$= y(1 - a) + x(a - 1)$	This is our work from above.
$= y(1 - a) + x(-1)(1 - a)$	Factor out -1 to obtain a GCF of $(1 - a)$.
$= y(1 - a) - x(1 - a)$	Simplify.
$= (1 - a)(y - x)$	Factor out $(1 - a)$ from both terms.

Thus, $y + ax - x - ay = (1 - a)(y - x)$.

PROBLEM SET 4.3

Practice Problems _____

In Problems 1–46, factor out the GCF from each polynomial.

1. $4x - 20$ **2.** $14y - 28$ **3.** $18a + 27$ **4.** $16y + 24$

5. $12x^2 + 4x$ **6.** $6y^3 + 9y^2$ **7.** $9x - 18y$ **8.** $16a - 24b$

9. $12x^2y - 8xy^2$ **10.** $33x^2 - 22xy^4$ **11.** $4xy - 7xy^2$ **12.** $2ab - 5a^3b$

13. $18x^4 + 9x^3 - 27x^2$ **14.** $5a^5 + 15a^7 - 20a^3$ **15.** $10y^7 - 16y^4 + 8y^3$ **16.** $18p^4 - 9p^3 + 27p^5$

17. $15x^5y^3 - 25x^3y^4$ **18.** $18a^4b^3 - 27a^3b^2$ **19.** $2a^2b - 5ab^2 + 7a^2b^2$ **20.** $14x^3y^2 + 21x^2y - 7x^5y^3$

21. $24xy^3 - 36x^3y^2 + 12x^2y^4$ **22.** $9a^3b^3 - 15a^3b^4 + 6ab^3$ **23.** $26x^5y^3 + 52x^7y^2 - 39x^8y^5$

24. $32a^{11}b^4 + 144a^3b^5 - 16a^4b^5$ **25.** $30x^4y + 50x^2y^2 + 20x^3z^3$ **26.** $15a^3b - 3a^2c + 12a^2c^2$

27. $55x^2y^2z^4 - 77x^3y^2z$

28. $24a^5b^5c^3 + 16a^6b^4c^2$

29. $7x^3y^2 + 14x^2y - 42x^5y^3 + 21xy^4$

30. $36pq^3 - 12p^3q^2 + 60p^2q^4 - 24p^4q^3$

31. $70x^3y + 42x^2y - 28x^2y^2 - 84xy^3$

32. $105a^2b^2 - 84a^4b^3 + 126a^3b^4 - 63a^2b^5$

33. $7(x - 3) + y(x - 3)$

34. $9(y + 5) + a(y + 5)$

35. $-2a(x + 7y) + 4c(x + 7y)$

36. $(a + b)(7x) - (a + b)(11y)$

37. $4x(a + b - c) - 2y(a + b - c)$

38. $(5x - 3y + z)a - (5x - 3y + z)2b$

39. $(x - 4y)z^2 + (x - 4y)z + (x - 4y)$

40. $(a - 3b)x^2 + (a - 3b)x + (a - 3b)$

41. $(x - 4y)z^2 + (x - 4y)z + (4y - x)$

42. $(a - 3b)x^2 + (a - 3b)x + (3b - a)$

43. $5x^3(2a - 7b) + 15x^2(2a - 7b)$

44. $14y^2(3a - 4b) - 7y^5(3a - 4b)$

45. $77x^3y^2(7a - 5b) + 11x^2y(7a - 5b)$

46. $18x^4y^3z^2(8a - 3b) - 27x^3y^2z^2(8a - 3b)$

In Problems 47–55, factor each polynomial in two ways, first with a GCF having a positive coefficient and then with a GCF having a negative coefficient.

47. $-4x + 12$

48. $-5x + 20$

49. $-3x^2 + 27x$

50. $-5x^2 + 25x$

51. $-3x^3 + 15x^2 - 21x$

52. $-4y^4 - 40y^3 + 44y$

53. $-x^2 + 7x - 5$

54. $-y^3 - 7y^2 + 13$

55. $-b^4 + 3b^3 - 15b$

56. Look at the two factorizations for each of Problems 47–55. Describe a method for taking the first factorization and obtaining the second without having to factor the given polynomial a second time.

Factor Problems 57–76 by grouping.

57. $x^3 - 3x^2 + 4x - 12$

58. $x^3 - 2x^2 + 5x - 10$

59. $y^3 - y^2 + 2y - 2$

60. $y^3 + 6y^2 - 2y - 12$

61. $y^3 - y^2 - 5y + 5$

62. $3x^3 - 2x^2 - 6x + 4$

63. $a^2c + 5ac + 2a + 10$

64. $a + b + 4a^2 + 4ab$

65. $3Y^2 + 4YZ + 24Y + 32Z$

66. $y^2 - 4y + 3yz - 12z$

67. $2a - 6b + ac - 3bc$

68. $c + d + 2c^2 + 2cd$

69. $3ab + 3ac - b - c$

70. $4x^2y + 16xy - x - 4$

71. $5x^2 + 15 - x^2y - 3y$

72. $3ax + 3ay - 2x - 2y$

73. $2ab + ac - 6b - 3c$

74. $4x^2 + 4xy - 3xy^2 - 3y^3$

75. $a^2 - ab - a + b$

76. $4x - 3y - 12ax + 9ay$

Application Problems

77. The model $A = P + Pr + (P + Pr)r$ describes the amount of money (A) in a savings account at the end of 2 years when P dollars are placed in the account subject to an interest rate (r) compounded yearly. Use factoring by grouping to express the right side of the formula as $P(1 + r)^2$.

78. At the end of 3 years, the amount in the account in Problem 77 is $A = P(1 + r)^2 + P(1 + r)^2r$. Use factoring to express the model as $A = P(1 + r)^3$.

79. The surface area of a right circular cylinder is given by $2\pi rh + 2\pi r^2$. Factor this expression.

80. A right circular cone with radius r and slant height s has a surface area described by $\pi r^2 + \pi rs$. Factor this expression.

81. If an object is projected directly upward from the ground with an initial velocity of 64 feet per second, its height $s(t)$ above the ground after t seconds is modeled by the quadratic function $s(t) = -16t^2 + 64t$.
   a. Find an equivalent expression for the function's formula by factoring out the GCF.
   b. Find the coordinates for the parabola's vertex and graph the position function from $t = 0$ to $t = 4$.

True–False Critical Thinking Problems

82. Which one of the following is true?
   a. $2a + 4ab - 10ac = 2a(2b - 5c)$
   b. For the polynomial $4a^2b - 8ab^2$, we can factor out either $4ab$ or $-4ab$.
   c. The GCF of x^{3n+1} and x^{3n} is the monomial with the highest power of x.
   d. None of the above is true.

83. Which one of the following is true?
   a. If all terms of a polynomial contain the same letter raised to different powers, the exponent on the variable that you will factor out is the highest power that appears in all the terms.
   b. Since the GCF of $9x^3 + 6x^2 + 3x$ is $3x$, it is not necessary to write the 1 when $3x$ is factored from the last term.

c. Some polynomials with four terms, such as $x^3 + x^2 + 4x - 4$, cannot be factored by grouping.

d. Since a monomial contains one term, it follows that a monomial can be factored in precisely one way.

Technology Problems

In Problems 84–87, verify each factorization by using a graphing utility to graph the polynomial on the left side and the factored form on the right side in the same viewing rectangle. Try to show a complete picture of each quadratic or cubic function. The factorization is correct if the resulting graphs are identical. If the graphs are not the same, correct the right side (the factorization) of the given equation. Then use your graphing utility to verify that the new factorization is correct.

84. $x^2 - 2x = x(x - 2)$

85. $x^2 + 3x - 5x - 15 = (x + 5)(x - 3)$

86. $x^3 - 3x^2 + 4x - 12 = (x^2 + 4)(x - 3)$

87. $x^3 + x^2 - x - 1 = (x^2 + 1)(x - 1)$

Writing in Mathematics

88. Explain how to find the GCF of the terms of a polynomial.

89. Use two different groupings to factor $ac - ad + bd - bc$ in two ways. Then explain why the two factorizations are the same.

Critical Thinking Problems

Assuming that all variable exponents represent whole numbers, factor out the GCF from each polynomial in Problems 90–92.

90. $25y^{7n} + 20y^{5n} - 15y^{4n}$

91. $3x^{3m}y^m + 7x^{2m}y^{2m}$

92. $8y^{2n+4} + 16y^{2n+3} - 12y^{2n}$

93. The price of a computer is reduced by 5%. When the computer still does not sell, the sale price is reduced by another 5%.

a. If x represents the price before both reductions, write an expression for the final price of the computer.

b. Factor the expression in part (a). Use the factored form to write a statement describing how the final price of the computer compares to its price before the two reductions.

Review Problems

94. The length of a rectangle is 2 feet greater than twice its width. If the rectangle's perimeter is 22 feet, find the length and width.

95. Solve for L: $3A = \dfrac{2L - B}{4}$.

96. Solve the system:

$$\begin{aligned} x - y + z &= 4 \\ 2x - y + z &= 9 \\ x + 2y - z &= 5 \end{aligned}$$

 S E C T I O N   4 . 4

Solutions   Tutorial   Video
Manual   5

Factoring Trinomials

Objectives

1 Factor trinomials whose leading coefficient is 1.

2 Factor trinomials whose leading coefficient is not 1.

We now discuss methods for factoring trinomials of the form $ax^2 + bx + c$ or of the form $ax^2 + bxy + cy^2$. The techniques that we shall use will enable us to either factor a trinomial or determine that it cannot be factored.

Factor trinomials whose leading coefficient is 1.

Factoring Trinomials Whose Leading Coefficient is 1

To see how to factor trinomials of the form $x^2 + bx + c$, recall how FOIL multiplication works.

$$(x + m)(x + n) = x^2 + nx + mx + mn$$

$$= x^2 + \boxed{(n + m)}\, x + \boxed{mn}$$

$$= x^2 + b\,x + c$$

These observations provide us with a procedure for factoring $x^2 + bx + c$.

A strategy for factoring $x^2 + bx + c$

1. Enter x as the first term of each factor.

$$(x \quad)(x \quad)$$

2. Find two integers m and n whose product is c and whose sum is b. If $mn = c$ and $m + n = b$, then

$$x^2 + bx + c = (x + m)(x + n) \quad \text{or} \quad (x + n)(x + m).$$

If there are no such integers, the polynomial cannot be factored over the integers and is called *prime*.

EXAMPLE 1   Factoring a Trinomial in $x^2 + bx + c$ Form

Factor: $x^2 + 6x + 8$

Solution

We begin by writing $(x \quad)(x \quad)$. To find the second term of each factor, we must find two numbers whose product is 8 and whose sum is 6.

Possible Factors of 8	Sums of These Factors
(8)(1)	$8 + 1 = 9$
(4)(2)	$4 + 2 = 6$ ←
(−8)(−1)	$-8 + (-1) = -9$
(−4)(−2)	$-4 + (-2) = -6$

The factors whose sum is 6 are 4 and 2.

From the list, we see that 4 and 2 are the required integers. In terms of our factoring strategy this means $m = 4$ and $n = 2$. Thus,

$$x^2 + 6x + 8 = (x + 4)(x + 2).$$

Verify this result by multiplying the right side using the FOIL method. You should obtain the original trinomial. Because of the commutative property, we can also say that

$$x^2 + 6x + 8 = (x + 2)(x + 4). \qquad \blacksquare$$

EXAMPLE 2 Factoring a Trinomial Whose Leading Coefficient is 1

Factor: $x^2 - 14x + 24$

Solution

We begin by writing $(x \qquad)(x \qquad)$. To find the second term of each factor, we must find two integers (m and n) whose product is 24 and whose sum is -14. We'll look for negative integers only. (Why?)

Possible Factors of 24 That Are Negative	Sums of These Factors
$(-8)(-3)$	$-8 + (-3) = -11$
$(-12)(-2)$	$-12 + (-2) = -14$ ←
$(-6)(-4)$	$-6 + (-4) = -10$
$(-24)(-1)$	$-24 + (-1) = -25$

The factors whose sum is -14 are -12 and -2.

Thus, $m = -12$ and $n = -2$, and so

$$x^2 - 14x + 24 = (x - 12)(x - 2).$$

■

EXAMPLE 3 Factoring a Trinomial Whose Constant Term is Negative

Factor: $y^3 + 7y^2 - 60y$

Solution

As always, we begin the factoring process by looking for a common factor. In this example, the GCF is y, so we factor it out.

$$y^3 + 7y^2 - 60y = y(y^2 + 7y - 60) \qquad \text{Factor out the GCF.}$$
$$= y(y + m)(y + n) \qquad m = ? \quad n = ?$$

Now we look for two integers (m and n) whose product is -60 and whose sum is 7. (Can you immediately think of the required numbers?) We look for a factorization of -60 in which one factor is positive and the other factor is negative. Since the sum is to be positive, the positive factor must be further from 0 than the negative factor is. Consequently we will only consider pairs of factors of -60 in which the positive factor has the larger absolute value, that is, factors such as 10, 12, 15, and so on.

Some Possible Factors of −60	Sums of These Factors
$(10)(-6)$	$10 + (-6) = 4$
$(30)(-2)$	$30 + (-2) = 28$
$(12)(-5)$	$12 + (-5) = 7$ ←

The factors whose sum is 7 are 12 and -5.

Thus,

$$y^3 + 7y^2 - 60y = y(y + 12)(y - 5) \qquad \text{Be sure to write the common factor, } y, \text{ as part of the answer.}$$

■

As you become more experienced at factoring, you should be able to mentally narrow down the list of pairs of factors whose sum is the coefficient of the middle term in $x^2 + bx + c$. Here are some study tips for narrowing down the list.

Discover for yourself

Write three polynomials in the form $x^2 + bx + c$ that meet the criteria in the Study Tip. Factor the polynomials and illustrate each statement in the box.

Study tip

Factoring $x^2 + bx + c$ as

$$x^2 + bx + c = (x + m)(x + n)$$

Remembering that $mn = c$ and $m + n = b$,

1. If b and c are both positive, then m and n must be positive.
2. If c is positive and b is negative, then m and n must both be negative.
3. If c is negative, then m and n must have opposite signs.

EXAMPLE 4   **Factoring a Trinomial with a Common Factor**

Factor: $8x^3 - 40x^2 - 48x$

Solution

$$\begin{aligned}
&8x^3 - 40x^2 - 48x && \text{This is the given trinomial.}\\
&= 8x(x^2 - 5x - 6) && \text{Factor out } 8x, \text{ the GCF.}\\
&= 8x(x + m)(x + n) && m = ?\ n = ? \text{ Find two integers, } m \text{ and } n, \text{whose product is } -6\\
& && \text{and whose sum is } -5.\\
&= 8x(x - 6)(x + 1) && \text{The integers are } -6 \text{ and } 1.
\end{aligned}$$

Discover for yourself

Fill in the following table:

Possible Factors of -5	Sums of These Factors

Confirm for yourself that none of the factors has a sum of 1. Thus,

$$x^2 + x - 5$$

is prime.

EXAMPLE 5   **A Trinomial That Cannot Be Factored**

Factor: $x^2 + x - 5$

Solution

$$x^2 + x - 5 = (x + m)(x + n) \qquad m = ?\ \ n = ?$$

We look for two integers, m and n, whose product is -5 and whose sum is 1. Because no such integers exist, $x^2 + x - 5$ cannot be factored and is prime.

We can apply the factoring procedure for $x^2 + bx + c$ to a trinomial with more than one variable.

EXAMPLE 6   **Factoring a Trinomial in Two Variables**

Factor: $x^2 - 4xy - 21y^2$

Solution

$$x^2 - 4xy - 21y^2 = (x + my)(x + ny) \qquad m = ?\ \ n = ?$$

We must find two integers, m and n, whose product is -21 and whose sum is -4. The desired integers are 3 and -7. Thus,

$$x^2 - 4xy - 21y^2 = (x + 3y)(x - 7y) \quad \text{or} \quad (x - 7y)(x + 3y).$$

Verify that the factorization is correct by multiplying the right side using the FOIL method. You should obtain the original trinomial. ■

If the highest power of a trinomial is greater than 2, and the next-highest power is half this power, we can sometimes factor by substituting one variable for another. The next example illustrates factoring by substitution.

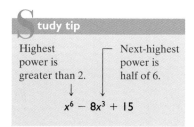
EXAMPLE 7 **Factoring by Substitution**

Factor: $x^6 - 8x^3 + 15$

Solution

Method 1. Introducing a Substitution
The trinomial $x^6 - 8x^3 + 15$ is quadratic in form. If we let $t = x^3$, the quadratic equation can be written as $t^2 - 8t + 15$. Let's see how we might use this approach to factor the given polynomial.

$(x^3)^2 - 8x^3 + 15$ This is the given polynomial, with x^6 written as $(x^3)^2$.

$= t^2 - 8t + 15$ Let $t = x^3$. Rewrite the trinomial in terms of t.

$= (t - 5)(t - 3)$ Factor.

$= (x^3 - 5)(x^3 - 3)$ Now substitute x^3 for t.

Thus, the given trinomial can be factored as

$$x^6 - 8x^3 + 15 = (x^3 - 5)(x^3 - 3).$$

Once again, check this result using FOIL multiplication on the right.

Using technology

Each graph below was obtained with a graphing utility in a standard viewing rectangle. What do you observe about the factored form of the trinomial and the x-intercepts of its graph?

$x^2 + 6x + 8 = (x + 4)(x + 2)$

Graph of $y = x^2 + 6x + 8$

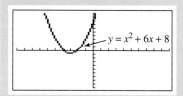

$x^2 - 6x + 5 = (x - 1)(x - 5)$

Graph of $y = x^2 - 6x + 5$

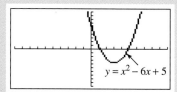

$x^2 - 3x - 4 = (x + 1)(x - 4)$

Graph of $y = x^2 - 3x - 4$

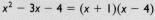
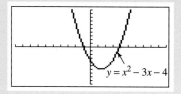

1. Describe how you can obtain the factorization for a trinomial (that is not prime) in the form $x^2 + bx + c$ from the graph of $y = x^2 + bx + c$.

2. Use your graphing utility to graph each of the following parabolas. Then use the graph to factor the trinomial in the parabola's equation.
    a. $y = x^2 - 4x - 5$ **b.** $y = x^2 - 7x + 6$ **c.** $y = x^2 + 7x + 10$

3. If $x^2 + bx + c$ is not factorable over the integers, describe how this can be seen from the graph of $y = x^2 + bx + c$. Verify your description with the graph of the prime trinomial from Example 5.

Method 2. Factoring Directly

Since $x^6 = (x^3)(x^3)$, we can factor the given trinomial without introducing a substitution.

$$x^6 - 8x^3 + 15$$

$= (x^3 + m)(x^3 + n)$ Set up the form for the factorization. Find two integers, m and n, whose product is 15 and whose sum is -8.

$= (x^3 - 5)(x^3 - 3)$ The desired integers are $m = -5$ and $n = -3$.

Which of the two solution methods do you prefer? ∎

2 Factor trinomials whose leading coefficient is not 1.

Factoring Trinomials Whose Leading Coefficient is Not 1

A generalization of the method used for factoring $x^2 + bx + c$ can be applied to factor trinomials of the form $ax^2 + bx + c$, where $a \neq 1$. Once again, our procedure is based on the FOIL method.

Consider the following FOIL multiplication:

$$
\begin{array}{cccc}
\text{F} & \text{O} & \text{I} & \text{L} \\
\downarrow & \downarrow & \downarrow & \downarrow
\end{array}
$$

$$(2x + 3)(x + 2) = 2x^2 + \underbrace{4x + 3x}_{} + 6$$

$$= 2x^2 + 7x + 6$$

Factoring $2x^2 + 7x + 6$ reverses this process. We must find two binomials whose product is $2x^2 + 7x + 6$. The product of the *First* terms must be $2x^2$. The product of the *Last* terms must be 6. The sum of the *Outer* and *Inner* products must be $7x$. Although we know from our multiplication problem that $2x^2 + 7x + 6$ factors as $(2x + 3)(x + 2)$, finding this factorization can involve a trial-and-error process.

Here's a strategy for factoring $ax^2 + bx + c$ based on reversing the process of FOIL multiplication.

A strategy for factoring $ax^2 + bx + c$ ($a \neq 1$) (Trial-and-Error FOIL)

(Assume, for the moment, that there is no GCF.)

1. Find two *First* terms whose product is ax^2.

$$(\blacksquare x + \quad)(\blacksquare x + \quad) = ax^2 + bx + c$$

2. Find two *Last* terms whose product is c.

$$(x + \blacksquare)(x + \blacksquare) = ax^2 + bx + c$$

3. By trial and error, perform steps 1 and 2 until the sum of the *Outer* product and *Inner* product is bx.

$$(\square x + \square)(\square x + \square) = ax^2 + bx + c$$

(sum of O + I)

If no such combinations exist, the polynomial is prime.

EXAMPLE 8 **Factoring a Trinomial in $ax^2 + bx + c$ Form**

Factor: $2x^2 + 7x + 3$

Solution

Step 1. First, we factor the first term, $2x^2$. Let's agree that when the first term in the trinomial is positive, we will use only positive factors of that first term. Thus, the only possibility for factors is $(2x)(x)$ (we agree to exclude $(-2x)(-x)$).

$$2x^2 + 7x + 3 = (2x + \Box)(x + \Box)$$

To complete the factorization, we must fill in the blanks with the correct numbers.

Step 2. Next, we find two last terms whose product is 3. The possibilities are $(3)(1)$ or $(-3)(-1)$. Because the signs in the given trinomial are all positive, we will use 3 and 1. Thus the possible factors are

$(2x + 3)(x + 1)$
$(2x + 1)(x + 3)$

Step 3. Since we are factoring $2x^2 + 7x + 3$, the sum of the outer and inner products must be $7x$.

$(2x + 3)(x + 1)$ $2x + 3x = 5x$, which is not what we want.

$(2x + 1)(x + 3)$ $6x + x = 7x$, which is the trinomial's middle term.

Thus, the desired factorization is

$$2x^2 + 7x + 3 = (2x + 1)(x + 3) \quad \text{or} \quad (x + 3)(2x + 1).$$

Using technology

The graphs of

$$y_1 = 2x^2 + 7x + 3$$
and $y_2 = (2x + 1)(x + 3)$

are shown in the same viewing rectangle. With identical graphs, we can conclude that

$2x^2 + 7x + 3$
$\quad = (2x + 1)(x + 3).$

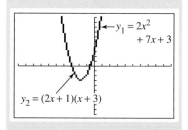

EXAMPLE 9 **Factoring a Trinomial Whose Leading Coefficient Is Not 1**

Factor: $9y^2 + 15y + 4$

Solution

Step 1. We factor the first term, $9y^2$. We have two possibilities:

$$9y^2 + 15y + 4 = (9y + \Box)(y + \Box)$$
$$9y^2 + 15y + 4 = (3y + \Box)(3y + \Box)$$

Step 2. Next, we find two last terms whose product is 4. Because the signs in the given trinomial are all positive, we will use only positive numbers for the last terms. The possibilities are

$$(4)(1) \quad \text{or} \quad (1)(4) \quad \text{or} \quad (2)(2).$$

Step 3. Since we are factoring $9y^2 + 15y + 4$, the sum of the outer and inner products must be $15y$.

	Possible Factors of $9y^2 + 15y + 4$	Sums of Outer and Inner Products (should equal $15y$)	
$(9y + \Box)(y + \Box)$ $(3y + \Box)(3y + \Box)$	$(9y + 4)(y + 1)$ $(9y + 1)(y + 4)$ $(9y + 2)(y + 2)$ $(3y + 2)(3y + 2)$ $(3y + 4)(3y + 1)$	$9y + 4y = 13y$ $36y + y = 37y$ $18y + 2y = 20y$ $6y + 6y = 12y$ $3y + 12y = 15y$	This is the required middle term.

Thus, the desired factorization is

$$9y^2 + 15y + 4 = (3y + 4)(3y + 1) \quad \text{or} \quad (3y + 1)(3y + 4).$$

EXAMPLE 10 **Factoring a Trinomial in Two Variables**

Factor: $6x^2 + 19xy - 7y^2$

Solution

Step 1. We factor the first term, $6x^2$. We have two possibilities:

$$6x^2 + 19xy - 7y^2 = (6x + \Box y)(x + \Box y)$$
$$6x^2 + 19xy - 7y^2 = (3x + \Box y)(2x + \Box y)$$

Step 2. Next, find two last terms whose product is $-7y^2$. Since we already placed the ys in the possible factorizations, we need only concentrate on -7. The possibilities include

$$(7)(-1) \quad \text{or} \quad (-1)(7) \quad \text{or} \quad (-7)(1) \quad \text{or} \quad (1)(-7).$$

Step 3. Since we are factoring $6x^2 + 19xy - 7y^2$, the sum of the outer and inner products must be $19xy$.

	Possible Factors of $6x^2 + 19xy - 7y^2$	Sums of Outer and Inner Products (should equal $19xy$)	
	Using $(7)(-1)$: $(6x + 7y)(x - y)$ $(3x + 7y)(2x - y)$	$-6xy + 7xy = xy$ $-3xy + 14xy = 11xy$	
$(6x + \Box y)(x + \Box y)$ $(3x + \Box y)(2x + \Box y)$	Using $(-1)(7)$: $(6x - y)(x + 7y)$ $(3x - y)(2x + 7y)$	$42xy - xy = 41xy$ $21xy - 2xy = 19xy$	← This is the required middle term.
	Using $(-7)(1)$: $(6x - 7y)(x + y)$ $(3x - 7y)(2x + y)$	$6xy - 7xy = -xy$ $3xy - 14xy = -11xy$	
	Using $(1)(-7)$: $(6x + y)(x - 7y)$ $(3x + y)(2x - 7y)$	$-42xy + xy = -41xy$ $-21xy + 2xy = -19xy$	

Since we found the required middle term, $19xy$, in our fourth factorization, it is not necessary to continue listing other possible factorizations. We included them here simply to show a complete list of all possibilities. The desired factorization is

$$6x^2 + 19xy - 7y^2 = (3x - y)(2x + 7y) \quad \text{or} \quad (2x + 7y)(3x - y).$$ ■

tudy tip

If factoring a trinomial with a negative leading coefficient, first factor out -1. For example,

$$-6x^2 - 19xy + 7y^2$$
$$= -1(6x^2 + 19xy - 7y^2)$$
$$= -(3x - y)(2x + 7y)$$

See Example 10.

This factored form can be written in other ways. Two of them are

$(-3x + y)(2x + 7y)$ Distribute -1 over the first factor.
$(3x - y)(-2x - 7y)$ Distribute -1 over the second factor.

EXAMPLE 11 **A Trinomial That Cannot Be Factored**

Factor: $6a^2 + 14a + 7$

Solution

Step 1. We factor the first term, $6a^2$.

$$6a^2 + 14a + 7 = (6a + \Box)(a + \Box)$$
$$6a^2 + 14a + 7 = (3a + \Box)(2a + \Box)$$

Step 2. Next, we find two last terms whose product is 7. Because the signs in the given trinomial are all positive, we will use only positive numbers for the last terms. The possibilities include

$$(7)(1) \quad \text{or} \quad (1)(7).$$

Step 3. Since we are factoring $6a^2 + 14a + 7$, the sum of the outer and inner products must be $14a$.

	Possible Factors of $6a^2 + 14a + 7$	Sums of Outer and Inner Products (should equal $14a$)
$(6a + \Box)(a + \Box)$	$(6a + 7)(a + 1)$	$6a + 7a = 13a$
$(3a + \Box)(2a + \Box)$	$(6a + 1)(a + 7)$	$42a + a = 43a$
	$(3a + 7)(2a + 1)$	$3a + 14a = 17a$
	$(3a + 1)(2a + 7)$	$21a + 2a = 23a$

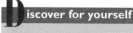iscover for yourself

Write two trinomials that are prime. Explain how you obtained these trinomials and also explain why they cannot be factored.

None of the factorizations has an outer-inner sum of $14a$. Thus, we can conclude that $6a^2 + 14a + 7$ cannot be factored using integers. The quadratic expression is prime. ■

EXAMPLE 12 **Factoring a Trinomial with a Common Factor**

Factor: $-6x^3 + 29x^2 - 13x$

Solution

We begin by factoring out the GCF. The two possibilities are

$$-6x^3 + 29x^2 - 13x = x(-6x^2 + 29x - 13)$$

and

$$-6x^3 + 29x^2 - 13x = -x(6x^2 - 29x + 13).$$

We'll work with the second possibility because it is easier to factor a trinomial with a positive leading coefficient.

Study tip

With increased practice, you will be able to mentally narrow down the list of possible factorizations to a few that can be tested by FOIL multiplication.

Possible Factors of $6x^2 - 29x + 13$	**Sums of Outer and Inner Products** (should equal $-29x$)
$(6x - 1)(x - 13)$	$-78x - x = -79x$
$(6x - 13)(x - 1)$	$-6x - 13x = -19x$
$(2x - 13)(3x - 1)$	$-2x - 39x = -41x$
$(2x - 1)(3x - 13)$	$-26x - 3x = -29x$ ← This is the required middle term.

Thus,

$$-6x^3 + 29x^2 - 13x = -x(6x^2 - 29x + 13)$$
$$= -x(2x - 1)(3x - 13)$$

Check the factorization by multiplying the factors. ∎

In certain instances, a substitution can be introduced into a complicated polynomial so that the resulting expression can be factored as a trinomial. This procedure is illustrated in Examples 13 and 14.

EXAMPLE 13   Factoring by Substitution

Factor: $6y^4 + 13y^2 + 6$

Discover for yourself

Try Example 13 without using a substitution. Start with

$6y^4 + 13y^2 + 16$
$= (\square y^2 + \square)(\square y^2 + \square)$.

Which method do you prefer?

Solution

$6(y^2)^2 + 13y^2 + 6$	This is the given polynomial, with y^4 written as $(y^2)^2$.
$= 6t^2 + 13t + 6$	Let $t = y^2$. Rewrite the trinomial in terms of t.
$= (3t + 2)(2t + 3)$	Factor the trinomial.
$= (3y^2 + 2)(2y^2 + 3)$	Now substitute y^2 for t.

Therefore, $6y^4 + 13y^2 + 6 = (3y^2 + 2)(2y^2 + 3)$. Check using FOIL multiplication. ∎

EXAMPLE 14   Factoring by Substitution

Factor: $12(x + 3)^2 + 11(x + 3) - 5$

Solution

$$12(x + 3)^2 + 11(x + 3) - 5 \qquad \text{This is the given polynomial.}$$

$$= 12t^2 \qquad + 11t - 5 \qquad \begin{array}{l}\text{Let } t = x + 3.\\ \text{Rewrite in terms of } t.\end{array}$$

$$= (4t + 5)(3t - 1) \qquad \begin{array}{l}\text{Factor the trinomial using}\\ \text{trial-and-error FOIL.}\end{array}$$

$$= [4(x + 3) + 5][3(x + 3) - 1] \qquad \text{Now substitute } x + 3 \text{ for } t.$$
$$= (4x + 12 + 5)(3x + 9 - 1) \qquad \begin{array}{l}\text{Simplify each factor, using}\\ \text{the distributive property.}\end{array}$$

$$= (4x + 17)(3x + 8) \qquad \text{Combine numerical terms.}$$

Thus,

$$12(x + 3)^2 + 11(x + 3) - 5 = (4x + 17)(3x + 8). \qquad ■$$

Factoring Trinomials by Grouping

Factoring a trinomial can involve quite a bit of trial and error. It is possible to use factoring by grouping. Factoring $ax^2 + bx + c$ by grouping depends on finding two numbers p and q for which $p + q = b$ and then factoring $ax^2 + px + qx + c$ using grouping.

Let's see how this works by looking at a particular factorization.

$$8y^2 - 10y - 3 = (2y - 3)(4y + 1)$$

If we multiply using FOIL on the right, we obtain

$$(2y - 3)(4y + 1) = 8y^2 + 2y - 12y - 3.$$

In this case, the desired numbers p and q are $p = 2$ and $q = -12$. These numbers are factors of ac, or -24, and have a sum of b, namely -10. By expressing the middle term $-10y$ in terms of these numbers, we can factor by grouping as follows:

$$8y^2 - 10y - 3$$
$$= 8y^2 + (2y - 12y) - 3 \qquad \text{Rewrite } -10y \text{ as } 2y - 12y.$$
$$= (8y^2 + 2y) + (-12y - 3) \qquad \text{Group terms.}$$
$$= 2y(4y + 1) - 3(4y + 1) \qquad \text{Factor from each group.}$$
$$= (4y + 1)(2y - 3) \qquad \text{Factor out the common binomial factor.}$$

Generalizing from this example, here's the procedure for factoring a trinomial by grouping.

Factoring $ax^2 + bx + c$ by grouping ($a \neq 1$)

1. Multiply the leading coefficient a and the constant c.
2. Find the factors of ac whose sum is b.
3. Rewrite the middle term (bx) as a sum or difference using the factors from step 2.
4. Factor by grouping.

ENRICHMENT ESSAY

Weird Numbers

Certain numbers are regarded as weird. Seventy is such a number.

The factors of 70 are 1, 2, 5, 7, 10, 14, and 35. The sum of these factors is 74, yet no group of these factors sums to 70.

The number 70 is the smallest *weird number.* Mathematicians call a number weird if the sum of its factors is greater than the number, but no partial collection of the factors adds up to the number.

Weird numbers are rare. Below 10,000, the weird numbers are 70, 836, 4030, 5830, 7192, 7912, and 9272.

It is not known whether an odd weird number exists.

Robert Blitzer, *London in the '70s*, 1972

EXAMPLE 15   **Factoring a Trinomial by Grouping**

Factor by grouping: $12x^2 - 5x - 2$

Solution

Step 1. We multiply the leading coefficient and the constant:

$$12(-2) = -24$$

Step 2. Now we find factors of -24 whose sum is the coefficient of the middle term, -5.

Factors of -24	Sums of These Factors (should equal -5)	
1, −24	−23	
−1, 24	23	
2, −12	−10	
−2, 12	10	
3, −8	−5	← This is the desired sum, so we can stop listing pairs of factors.

Step 3. Now we express the middle term of $12x^2 - 5x - 2$ as a sum or difference using the factors from step 2.

$$-5x = 3x - 8x \quad \text{Use the factors 3 and } -8.$$

Step 4. Last, we factor by grouping.

$$12x^2 - 5x - 2$$

$= 12x^2 + (3x - 8x) - 2$	Rewrite the middle term.
$= (12x^2 + 3x) + (-8x - 2)$	Group terms.
$= 3x(4x + 1) - 2(4x + 1)$	Factor from each group.
$= (4x + 1)(3x - 2)$	Factor out the common binomial factor.

The first line $12x^2 - 5x - 2$ is annotated: This is the given trinomial.

Therefore, the trinomial factors as

$$12x^2 - 5x - 2 = (4x + 1)(3x - 2).$$

As always, check the factorization by multiplying.

PROBLEM SET 4.4

Practice Problems

In Problems 1–80, factor each polynomial, if possible. Verify all factorizations by multiplying. If applicable, use a graphing utility to check factorizations for trinomials in one variable.

1. $x^2 + 5x + 6$ **2.** $x^2 + 10x + 9$ **3.** $a^2 + 8a + 15$ **4.** $S^2 + 22S + 21$

5. $x^2 + 9x + 20$ **6.** $x^2 + 11x + 24$ **7.** $d^2 + 10d + 16$ **8.** $W^2 + 9W + 18$

9. $t^2 + 6t + 9$ **10.** $a^2 + 12a + 36$ **11.** $Y^2 - 14Y + 49$ **12.** $b^2 - 2b + 1$

13. $x^2 - 8x + 15$ **14.** $x^2 - 5x + 6$ **15.** $y^2 - 12y + 20$ **16.** $P^2 - 25P + 24$

17. $y^2 + 5y - 14$ **18.** $B^2 + B - 12$ **19.** $x^2 + x - 30$ **20.** $y^2 + 14y - 32$

21. $d^2 - 3d - 28$ **22.** $y^2 - 4y - 21$ **23.** $R^2 - 5R - 36$ **24.** $x^2 - 3x - 40$

25. $-12x + 35 + x^2$ **26.** $-24 + x^2 - 5x$ **27.** $x^2 - x + 7$ **28.** $x^2 + 3x + 8$

29. $X^2 + 12XY + 20Y^2$ **30.** $R^2 - 6RS + 8S^2$ **31.** $W^2 - 10WZ - 11Z^2$ **32.** $a^2 + ab + b^2$

33. $a^2 - ab + b^2$ **34.** $5x^3 - 15x^2 - 20x$ **35.** $2x^3 + 6x^2 + 4x$ **36.** $3x^2 + 8x + 5$

37. $2M^2 + 9M + 7$ **38.** $5y^2 + 56y + 11$ **39.** $4x^2 + 9x + 2$ **40.** $8b^2 + 10b + 3$

41. $5T^2 + 17T + 6$ **42.** $10s^2 + 19s + 6$ **43.** $6b^2 + 19b + 15$ **44.** $8a^2 + 26a + 15$

45. $6x^2 + 17x + 12$ **46.** $4y^2 + 20y + 25$ **47.** $9y^2 - 30y + 25$ **48.** $8x^2 - 18x + 9$

49. $4a^2 - 27a + 18$ **50.** $6b^2 - 23b + 15$ **51.** $16S^2 - 6S - 27$ **52.** $12b^2 - 19b - 21$

53. $4y^2 - y - 18$ **54.** $2A^2 + 13AB + 15B^2$ **55.** $9M^2 - 3MN - 2N^2$ **56.** $25R^2 + 10RS - 8S^2$

57. $10x^2 + 29xy - 21y^2$ **58.** $8A^2 + 6AB - 9B^2$ **59.** $6a^2 + 14a + 3$ **60.** $4x^2 + 22xy - 5y^2$

61. $15w^3 - 25w^2 + 10w$ **62.** $10v^3 + 24v^2 + 14v$ **63.** $-x^2 + 6x + 27$ **64.** $-x^2 + 2x + 15$

65. $-12x^2 + 35x + 3$ **66.** $-7y^2 + 6y + 13$ **67.** $4a^4b - 24a^3b - 64a^2b$ **68.** $3m^4n - 3m^3n - 90m^2n$

69. $36x^3y - 6x^2y - 20xy$ **70.** $12y^5x^2 + 56y^4x^2 + 32y^3x^2$ **71.** $4x^2 - 12xy + 9y^2$ **72.** $18x^2 - 3xy - 28y^2$

73. $35a^2 - 41ab - 24b^2$ **74.** $10m^2 + mn - 3n^2$ **75.** $8x^2 - 14xy - 39y^2$ **76.** $6m^2 - 5mn - 39n^2$

77. $6x^2y^2 + 13xy + 6$ **78.** $15a^2b^2 - 22ab - 5$ **79.** $13x^3y^3 + 39x^3y^2 - 52x^3y$ **80.** $4p^3q^5 + 24p^2q^5 - 64pq^5$

In Problems 81–98, factor each polynomial. You may find it convenient first to introduce an appropriate substitution.

81. $y^4 + 5y^2 + 6$ **82.** $x^4 + 6x^2 + 8$ **83.** $5m^4 + m^2 - 6$ **84.** $2r^4 - r^2 - 3$

85. $2n^4 + mn^2 - 6m^2$ **86.** $12x^4 + 28x^2y - 5y^2$ **87.** $y^6 - 9y^3 - 36$ **88.** $2r^{10} + 5r^5 + 3$

89. $y^8 + 10y^4 - 39$ **90.** $x^{10} + 2x^5 - 48$

91. $(a - 3b)^2 - 5(a - 3b) - 36$ **92.** $(a + 4b)^2 + 29(a + 4b) + 100$

93. $5(a + b)^2 + 12(a + b) + 7$ **94.** $3(a + b)^2 - 5(a + b) + 2$

95. $18(x + y)^2 - 3(x + y)b - 28b^2$ **96.** $35(a + b)^2 - 41(a + b)y - 24y^2$

97. $6a^2(a - b)^2 - 13ab(a - b) + 6b^2$ **98.** $5a^2(a - b)^2 - ab(a - b) - 6b^2$

Application Problems

99. A diver jumps directly upward from a board that is 32 feet high. After t seconds the height of the diver above the water is modeled by the quadratic function $s(t) = -16t^2 + 16t + 32$.
  a. Find an equivalent expression for the function's formula by factoring it completely.
  b. Find the coordinates for the parabola's vertex and graph the position function from $t = 0$ to $t = 4$.

  c. Write a brief description of the height of the diver above the water during the time period that the diver is in the air.

100. If x represents a positive integer, factor $x^3 + 3x^2 + 2x$ to show that the trinomial represents the product of three consecutive integers.

True–False Critical Thinking Problems

101. Which one of the following is true?
  a. Once a GCF is factored from $6y^6 - 19y^5 + 10y^4$, the remaining trinomial factor is prime.
  b. A factor of $8y^2 - 51y + 18$ is $8y - 3$.
  c. We can immediately tell that $6x^2 - 11xy - 10y^2$ is prime because 11 is a prime number and the polynomial contains two variables.
  d. A factor of $12x^2 - 19xy + 5y$ is $4x - y$.

102. Which one of the following is true?
  a. The GCF of $x^5y - 3x^4y - 18x^3y$ is x^3y. Once this GCF is factored from the polynomial, the remaining trinomial factor is prime.
  b. If the terms of a polynomial have no common factor (except 1), then none of the terms of its factors can have a common factor.
  c. A factor of $-3x^2 + 16x + 12$ is $3x - 2$.
  d. The factorization $-1(3x + 2)(x - 6)$ cannot be expressed as $(3x + 2)(6 - x)$.

Technology Problems

In Problems 103–107, verify each factorization by using a graphing utility to graph the polynomial on the left side and the factored form on the right side in the same viewing rectangle. Try to show a complete picture of each function's main characteristics. The factorization is correct if the resulting graphs are identical. If the graphs are not the same, correct the right side (the factorization) of the given equation. Then use your graphing utility to verify that the new factorization is correct.

103. $x^2 + 7x + 12 = (x + 4)(x + 3)$

104. $-x^2 - 4x + 5 = (x - 1)(x - 5)$

105. $-3x^2 + 16x + 12 = -(3x - 2)(x + 6)$

106. $2(x + 3)^2 + 5(x + 3) - 12 = (2x + 3)(x + 7)$

107. $6x^3 + 5x^2 - 4x = x(3x + 4)(2x - 1)$

Writing in Mathematics

108. Suppose $ax^2 + bx + c$ is factored as

  $(px + r)(qx + s)$.

  Describe the conditions that must be satisfied by $p, q, r,$ and s.

109. Write an expression for the sum of the areas of the six smaller rectangles and squares in the figure. Then explain how the figure serves as a geometric model for the factorization of that expression.

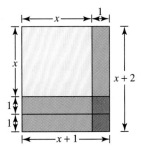

Critical Thinking Problems

Factor each polynomial in Problems 110–113. Assume that all exponents represent whole numbers.

110. $x^{2n} + 6x^n + 8$ **111.** $x^{2n} - 8x^n + 15$ **112.** $9x^{2n} + x^n - 8$ **113.** $4y^{2m} - 9y^m + 5$

For Problems 114–117, find all integers K such that the trinomial can be factored.

114. $x^2 + Kx - 6$ **115.** $y^2 + Kx + 8$ **116.** $4x^2 + Kx - 1$ **117.** $3x^2 + Kx + 5$

118. If $ax + b$ is a factor of $4x^2 + 4x + 1$ and $2x^2 - 5x + c$, find values for $a, b,$ and c.

Review Problems

119. Solve for P: $I = Prt + P$.

120. Solve $-2x \leqslant 6$ and $-2x + 3 < -7$. Express the solution set in both set-builder and interval notations.

121. Write the point-slope and slope-intercept equations of the line through $(3, -2)$ and perpendicular to the line whose equation is $x + 2y = 3$.

SECTION 4.5

Solutions Manual   **Tutorial**   **Video 5**

Factoring Special Forms

Objectives

1 Factor the difference of two squares.
2 Factor the sum and difference of two cubes.
3 Factor perfect square trinomials.

In this section, we will discuss certain forms of polynomials that occur frequently in algebra. In fact, these special forms occur so frequently that the factors of these forms should be memorized.

1 Factor the difference of two squares.

The Difference of Two Squares

Earlier in this chapter we discussed the product of the sum and difference of two terms:

$$(A + B)(A - B) = A^2 - B^2.$$

By reversing the two sides of this formula, we obtain a formula for factoring the difference of two squares.

> **Factoring the difference of two squares**
>
> Let A and B be real numbers, variables, or algebraic expressions.
>
> $$A^2 - B^2 = (A + B)(A - B)$$
>
> The difference of two squares factors as the product of the sum and the difference of the two quantities being squared.

Discover for yourself

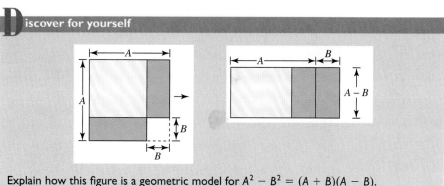

Explain how this figure is a geometric model for $A^2 - B^2 = (A + B)(A - B)$.

EXAMPLE 1   **Factoring the Difference of Squares**

Factor:

a. $x^2 - 64$ **b.** $9x^2 - 100$ **c.** $9a^2 - 16b^2$
d. $36y^6 - 49x^4$ **e.** $100y^2 - (3x + 1)^2$

Solution

We begin by rewriting each polynomial as the difference of two squares. Then we use the formula for factoring the difference of two squares.

$$A^2 \quad - \quad B^2 \quad = \quad (A + B)(A - B)$$

a. $x^2 - 64 = \quad x^2 \quad - \quad 8^2 \quad = \quad (x + 8)(x - 8)$

b. $9x^2 - 100 = \quad (3x)^2 \quad - \quad 10^2 \quad = \quad (3x + 10)(3x - 10)$

c. $9a^2 - 16b^2 = \quad (3a)^2 \quad - \quad (4b)^2 \quad = \quad (3a + 4b)(3a - 4b)$

d. $36y^6 - 49x^4 = \quad (6y^3)^2 \quad - \quad (7x^2)^2 \quad = \quad (6y^3 + 7x^2)(6y^3 - 7x^2)$

e. $100y^2 - (3x + 1)^2 = (10y)^2 - (3x + 1)^2 = [10y + (3x + 1)][10y - (3x + 1)]$
$$= (10y + 3x + 1)(10y - 3x - 1)$$

In certain situations, we can apply our technique for factoring the difference of two squares more than once.

EXAMPLE 2   **Factoring the Difference of Squares More Than Once**

Factor completely: **a.** $x^4 - 1$ **b.** $81a^4 - 16b^4$

Solution

As in Example 1, we begin by rewriting each polynomial as the difference of two squares.

$$A^2 \quad - \quad B^2 = (A \ + \ B)(A \ - \ B)$$

a. $x^4 - 1 = (x^2)^2 - 1^2 = (x^2 + 1)(x^2 - 1)$ We can still factor $x^2 - 1$.
$$= (x^2 + 1)(x^2 - 1^2)$$ Rewrite $x^2 - 1$ as the difference of two squares.

$$= (x^2 + 1)(x + 1)(x - 1)$$ Factor $x^2 - 1$.

$$A^2 \quad - \quad B^2$$

b. $81a^4 - 16b^4 = (9a^2)^2 - (4b^2)^2$
$$= (9a^2 + 4b^2)(9a^2 - 4b^2)$$ Factor as $(A + B)(A - B)$.
$$= (9a^2 + 4b^2)[(3a)^2 - (2b)^2]$$ Rewrite the second factor as the difference of two squares.
$$= (9a^2 + 4b^2)(3a + 2b)(3a - 2b)$$ Factor again.

Some polynomials require us to factor out the GCF as well as to factor the difference between squares. It is best to *begin by factoring out the GCF.*

| **EXAMPLE 3** | **Factoring by First Factoring Out the GCF** |

Factor: **a.** $3y - 3x^6y^5$ **b.** $3xy^5 - 48xy$

Solution

a. $3y - 3x^6y^5 = 3y(1 - x^6y^4)$ Factor out the GCF.
$\qquad\qquad\quad = 3y[1^2 - (x^3y^2)^2]$ Rewrite $1 - x^6y^4$ as the difference of two squares.

$\qquad\qquad\quad = 3y(1 + x^3y^2)(1 - x^3y^2)$ Factor the difference of two squares.

b. $3xy^5 - 48xy = 3xy(y^4 - 16)$ Factor out the GCF.
$\qquad\qquad\qquad = 3xy[(y^2)^2 - 4^2]$ Write as the difference of squares.
$\qquad\qquad\qquad = 3xy(y^2 + 4)(y^2 - 4)$ Factor the difference of squares.
$\qquad\qquad\qquad = 3xy(y^2 + 4)(y + 2)(y - 2)$ Factor $y^2 - 2^2$, which is also a difference of squares. ∎

In our next example, we begin by applying factoring by grouping. We can then factor further using the difference of squares.

| **EXAMPLE 4** | **Factoring Completely** |

Factor completely: $x^3 + 5x^2 - 9x - 45$

Solution

$(x^3 + 5x^2) + (-9x - 45)$ This is the given polynomial, with the terms having a GCF grouped.

$= x^2(x + 5) - 9(x + 5)$ Factor out the GCF from each group.
$= (x + 5)(x^2 - 9)$ Factor out $x + 5$ from both terms.
$= (x + 5)(x + 3)(x - 3)$ Factor $x^2 - 3^2$, the difference of two squares. ∎

2 Factor the sum and difference of two cubes.

The Sum and Difference of Two Cubes

In the Study Tip for Example 2, we noted that $x^2 + 1^2$, the sum of two squares, cannot be factored using real numbers. However, a sum of two cubes, such as $x^3 + 1^3$, can be factored over the integers. Let's look at two multiplications that lead to factoring formulas for both the sum of two cubes and the difference of two cubes.

$$(A + B)(A^2 - AB + B^2) = A(A^2 - AB + B^2) + B(A^2 - AB + B^2)$$
$$= A^3 - A^2B + AB^2 + A^2B - AB^2 + B^3$$
$$= A^3 + B^3 \qquad \text{Combine like terms,}$$

and

$$(A - B)(A^2 + AB + B^2) = A(A^2 + AB + B^2) - B(A^2 + AB + B^2)$$
$$= A^3 + A^2B + AB^2 - A^2B - AB^2 - B^3$$
$$= A^3 - B^3 \qquad \text{Combine like terms.}$$

By reversing the two sides of these formulas, we obtain formulas for factoring the sum and difference of two cubes. These formulas should be memorized.

Factoring the sum and difference of two cubes

Let A and B be real numbers, variables, or algebraic expressions.

Factoring the Sum of Cubes

Like signs

$$A^3 + B^3 = (A + B)(A^2 - AB + B^2)$$

Unlike signs

Factoring the Difference of Cubes

Like signs

$$A^3 - B^3 = (A - B)(A^2 + AB + B^2)$$

Unlike signs

Our next example uses the formula for factoring the sum of two cubes.

using technology

Graphing

$$y_1 = x^3 + 8$$

and

$$y_2 = (x + 2)(x^2 - 2x + 4)$$

results in the same graph, so

$$x^3 + 8$$
$$= (x + 2)(x^2 - 2x + 4).$$

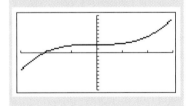

EXAMPLE 5   **Factoring the Sum of Two Cubes**

Factor:

a. $x^3 + 8$ **b.** $64z^3 + 27$ **c.** $8r^3s^6 + 125t^6$ **d.** $x^6 + y^9$

Solution

We begin by rewriting each polynomial as the sum of two cubes. Then we apply the factoring formula for $A^3 + B^3$.

$$A^3 + B^3 = (A + B)(A^2 - A B + B^2)$$

a. $x^3 + 8 = x^3 + 2^3 = (x + 2)(x^2 - x \cdot 2 + 2^2)$ Use the formula for $A^3 + B^3$.
$$= (x + 2)(x^2 - 2x + 4)$$ Simplify.

$$A^3 + B^3 = (A + B)(A^2 - A B + B^2)$$

b. $64z^3 + 27 = (4z)^3 + 3^3 = (4z + 3)[(4z)^2 - (4z)(3) + 3^2]$ Use the formula for $A^3 + B^3$.

$$= (4z + 3)(16z^2 - 12z + 9)$$ Simplify.

$$A^3 + B^3 = (A + B)(A^2 - A B + B^2)$$

c. $8r^3s^6 + 125t^6 = (2rs^2)^3 + (5t^2)^3 = (2rs^2 + 5t^2)[(2rs^2)^2 - (2rs^2)(5t^2) + (5t^2)^2]$ Use the formula for $A^3 + B^3$.
$$= (2rs^2 + 5t^2)(4r^2s^4 - 10rs^2t^2 + 25t^4)$$ Simplify.

$$A^3 + B^3 = (A + B)(A^2 - A B + B^2)$$

d. $x^6 + y^9 = (x^2)^3 + (y^3)^3 = (x^2 + y^3)[(x^2)^2 - (x^2)(y^3) + (y^3)^2]$ Use the formula for $A^3 + B^3$.

$$= (x^2 + y^3)(x^4 - x^2y^3 + y^6)$$ Simplify.

The application of the factoring formula for the difference of two cubes involves an approach almost identical to that used in Example 5. The first step is to express the binomial as the difference of cubes.

EXAMPLE 6 Factoring the Difference of Cubes

Factor:

a. $K^3 - 216$ **b.** $8 - 125x^3y^3$ **c.** $(x + y)^3 - (x - y)^3$

Solution

$$A^3 - B^3 = (A - B)(A^2 + A \cdot B + B^2)$$

a. $K^3 - 216 = K^3 - 6^3 = (K - 6)(K^2 + K \cdot 6 + 6^2)$ Use the formula for $A^3 - B^3$.

$$= (K - 6)(K^2 + 6K + 36) \quad \text{Simplify.}$$

$$A^3 - B^3 = (A - B)(A^2 + A \quad B + B^2)$$

b. $8 - 125x^3y^3 = 2^3 - (5xy)^3 = (2 - 5xy)[2^2 + 2(5xy) + (5xy)^2]$ Use the formula for $A^3 - B^3$.

$$= (2 - 5xy)(4 + 10xy + 25x^2y^2) \quad \text{Simplify.}$$

$$A^3 - B^3 = (A - B)(A^2 + A \cdot B + B^2)$$

c. $(x + y)^3 - (x - y)^3 = [(x + y) - (x - y)][(x + y)^2 + (x + y)(x - y) + (x - y)^2]$ Use the formula for $A^3 - B^3$.

$$= [x + y - x + y][x^2 + 2xy + y^2 + x^2 - y^2 + x^2 - 2xy + y^2]$$

$$= 2y(3x^2 + y^2) \qquad\qquad \text{Perform the operations inside the brackets and simplify.} ■$$

3 Factor perfect square trinomials.

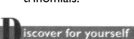

Explain how these figures serve as a geometric model for $A^2 + 2AB + B^2 = (A + B)^2$.

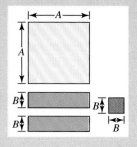

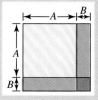

Perfect Square Trinomials

Earlier in this chapter we determined a pattern for the square of a binomial:

$$(A + B)^2 = A^2 + 2AB + B^2$$
$$(A - B)^2 = A^2 - 2AB + B^2$$

The two trinomials on the right are called *perfect square trinomials* because they are the squares of the binomials $A + B$ and $A - B$. Notice that the perfect square trinomials on the right come in two forms, one in which the middle term is positive and the other in which it is negative.

By reversing the two sides of these formulas, we obtain formulas for factoring perfect square trinomials. These formulas should be memorized.

Factoring perfect square trinomials

Let A and B be real numbers, variables, or algebraic expressions.

Factoring a Perfect Square Trinomial: Positive Middle Term

$$A^2 + 2AB + B^2 = (A + B)^2$$

Same sign

> *Factoring a Perfect Square Trinomial: Negative Middle Term*
>
> $$A^2 - 2AB + B^2 = (A - B)^2$$
>
> Same sign

Perfect square trinomials can always be factored using the trial-and-error FOIL method for factoring trinomials. However, by recognizing a perfect square trinomial, we can immediately factor by inspection.

EXAMPLE 7 **Factoring Perfect Square Trinomials**

Factor:

a. $x^2 + 6x + 9$ **b.** $16x^2 + 40xy + 25y^2$

c. $9y^4 - 12y^2 + 4$ **d.** $(m + 3)^2 - 10(m + 3) + 25$

Solution

In a perfect square trinomial, the first and last terms must be perfect squares (A^2 and B^2). The middle terms must be twice the product of A and B. We begin by rewriting each polynomial in this form. Then we apply the factoring formula for perfect square trinomials.

a. $x^2 + 6x + 9 = x^2 + 2 \cdot x \cdot 3 + 3^2 = (x + 3)^2$

b. $16x^2 + 40xy + 25y^2 = (4x)^2 + 2(4x)(5y) + (5y)^2 = (4x + 5y)^2$

c. $9y^4 - 12y^2 + 4 = (3y^2)^2 - 2(3y^2)(2) + 2^2 = (3y^2 - 2)^2$

d. $(m + 3)^2 - 10(m + 3) + 25 = (m + 3)^2 - 2(m + 3) \cdot 5 + 5^2 = [(m + 3) - 5]^2$
$$= (m - 2)^2 \quad \blacksquare$$

The special factoring formulas discussed in this section are summarized in Table 4.3. Again, you should take time to memorize these formulas.

Special forms and their factorizations are summarized below.

Elizabeth Murray "Tempest" 1979, oil on canvas, 10 ft × 14 ft 2 in. Photo courtesy of Pace Wildenstein.

TABLE 4.3 Special Factorizations

Name	Formula	Example
Difference of two squares	$A^2 - B^2 = (A + B)(A - B)$	$64x^2 - 9 = (8x)^2 - 3^2 = (8x + 3)(8x - 3)$
Sum of two cubes	$A^3 + B^3 = (A + B)(A^2 - AB + B^2)$	$x^3 + 1 = x^3 + 1^3 = (x + 1)(x^2 - x \cdot 1 + 1^2)$
		$= (x + 1)(x^2 - x + 1)$
Difference of two cubes	$A^3 - B^3 = (A - B)(A^2 + AB + B^2)$	$y^3 - 216 = y^3 - 6^3 = (y - 6)(y^2 + y \cdot 6 + 6^2)$
		$= (y - 6)(y^2 + 6y + 36)$
Perfect square trinomials	$A^2 + 2AB + B^2 = (A + B)^2$	$x^2 - 14x + 49 = x^2 - 2 \cdot x \cdot 7 + 7^2 = (x - 7)^2$
	$A^2 - 2AB + B^2 = (A - B)^2$	

The terms of a polynomial can be grouped in several ways. A successful grouping that will lead to a factorization is frequently based on one or more of the five special formulas listed in Table 4.3.

EXAMPLE 8 **Using Grouping to Obtain the Difference of Two Squares**

Factor: $x^2 - 8x + 16 - y^2$

Solution

$x^2 - 8x + 16 - y^2$	This is the given polynomial.
$= (x^2 - 8x + 16) - y^2$	Group the first three terms to obtain a perfect square trinomial.
$= (x - 4)^2 - y^2$	Factor the perfect square trinomial.
$= [(x - 4) + y][(x - 4) - y]$	Use the factoring formula for the difference of two squares.
$= (x - 4 + y)(x - 4 - y)$	Simplify. ■

EXAMPLE 9 **Using Grouping to Obtain the Difference of Two Squares**

Factor: $a^2 - b^2 + 10b - 25$

Solution

It may be tempting to group into two terms of two terms each. Try doing this. Although you can factor $a^2 - b^2$ and you can factor 5 from $10b - 25$, the resulting expression does not have a common factor.

Another option is to look for a perfect square trinomial. As shown below, the perfect square trinomial is being subtracted from a^2.

$a^2 - b^2 + 10b - 25$	This is the given polynomial.
$= a^2 - (b^2 - 10b + 25)$	Factor out -1 and rewrite as subtraction.
$= a^2 - (b - 5)^2$	Factor the perfect square trinomial.
$= [a + (b - 5)][a - (b - 5)]$	Use the factoring formula for the difference of two squares.
$= (a + b - 5)(a - b + 5)$	Simplify. ■

PROBLEM SET 4.5

Practice Problems

Use the difference of two squares formula to factor Problems 1–46 completely, or if the polynomial cannot be factored, so state.

1. $B^2 - 1$ **2.** $y^2 - 100$ **3.** $25 - a^2$ **4.** $36 - y^2$

5. $36x^2 - 49$ **6.** $16y^2 - 4$ **7.** $36x^2 - 49y^2$ **8.** $64c^2 - 25d^2$

9. $x^2y^2 - a^2b^2$ **10.** $x^2y^2 - a^4b^2$ **11.** $x^2y^6 - a^4b^2$ **12.** $x^4y^4 - a^6b^2$

13. $4x^2y^6 - 25a^4b^2$ **14.** $81x^2y^6 - 49a^4b^2$ **15.** $81a^2b^4c^6 - 49x^8y^2$ **16.** $100x^2y^2z^4 - 9a^8b^2$

17. $(x + 3)^2 - y^2$ **18.** $(x - 7)^2 - y^2$ **19.** $(x + y)^2 - 36$ **20.** $(x - y)^2 - 100$

21. $x^2 + 4$ **22.** $9y^2 - (2x + 1)^2$ **23.** $16y^2 - (3x - 1)^2$ **24.** $x^2 + 1$

25. $(x + 1)^2 - (x + 3)^2$ **26.** $(x + 5)^2 - (x + 2)^2$ **27.** $(2x - 1)^2 - (3x + 2)^2$ **28.** $(4x + 3)^2 - (2x - 1)^2$

29. $a^{14} - 9$ **30.** $x^{22} - 16$ **31.** $25x^2 + 36y^2$ **32.** $2a^2 - 25b^2$ **33.** $-16 + x^2$

34. $-100 + y^2$ **35.** $-25A^2 + x^2y^2$ **36.** $-9y^2 + x^4y^4$ **37.** $y^4 - 1$ **38.** $a^4 - 81$

39. $1 - 81b^2$ **40.** $9 - 64y^2$ **41.** $x^2y^3 - 16y$ **42.** $a^4y - y$ **43.** $3x^3 - 3x$

44. $4y^3 - 100y$ **45.** $3x^3y - 12xy$ **46.** $2x^4y - 50x^2y$

In Problems 47–70, factor completely using the sum or difference of two cubes formula, or if the polynomial cannot be factored, so state.

47. $p^3 + q^3$ **48.** $x^3 - y^3$ **49.** $27x^3 - 64y^3$ **50.** $125C^3 + 8D^3$ **51.** $27R^3 - 1$

52. $1 - 8B^3$ **53.** $125b^3 + 64$ **54.** $8 - 27y^3$ **55.** $125 + 64d^3$ **56.** $27x^3y^3 + 1$

57. $8a^3b^3 - 27$ **58.** $125 - 8p^3q^3$ **59.** $64 - 27Y^3Z^3$ **60.** $y^9 - z^6$ **61.** $8x^3 + 27y^{12}$

62. $125B^6 + 64C^6$ **63.** $a^3b^6 - c^6d^{12}$ **64.** $8p^9q^3 + 27r^6s^3$ **65.** $125x^2 + 27y^3$ **66.** $a^3 - b^8$

67. $(x + y)^3 + (x - y)^3$ **68.** $(x - y)^3 + (x + y)^3$ **69.** $(2x - y)^3 - (2x + y)^3$ **70.** $(2x - y)^3 - (x + 2y)^3$

In Problems 71–76, factor completely using the perfect square trinomial formulas.

71. $y^2 + 4y + 4$ **72.** $x^2 + 6x + 9$ **73.** $z^2 - 10z + 25$

74. $y^2 - 8y + 16$ **75.** $9x^2 - 12xy + 4y^2$ **76.** $25x^2 - 20xy + 4y^2$

In Problems 77–90, factor completely using the appropriate formula(s).

77. $(x + y)^2 + 2(x + y) + 1$ **78.** $(r + s)^2 + 6(r + s) + 9$ **79.** $(v - w)^2 + 4(v - w) + 4$

80. $(r - s)^2 + 8(r - s) + 16$ **81.** $x^2 - 6x + 9 - y^2$ **82.** $y^2 - 10y + 25 - x^2$

83. $9x^2 - 30x + 25 - 36y^2$ **84.** $25y^2 - 20y + 4 - 81x^2$ **85.** $r^2 - (16s^2 - 24s + 9)$

86. $r^2 - (16s^2 - 8s + 1)$ **87.** $y^2 - x^2 - 4x - 4$ **88.** $y^2 - x^2 - 6x - 9$

89. $z^2 - x^2 + 4xy - 4y^2$ **90.** $100x^2 - 9r^2 + 6rs - s^2$

Application Problems

In Problems 91–92, find the formula for the area of the shaded region and express it in factored form.

91.
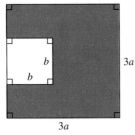

92.
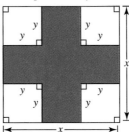

Problems 93–94 involve the creation of geometric models for special factoring formulas. Refer to the figure.

93. a. Find the area of the figure on the left by taking the difference of the area of the larger square and the smaller square that is removed.

   b. The figure on the right contains rectangles I and II. What is the area of this rectangle?

   c. Equate expressions from parts (a) and (b). What is the resulting special factoring formula?

94. Equate the areas in the figures shown. What is the resulting special factoring formula?

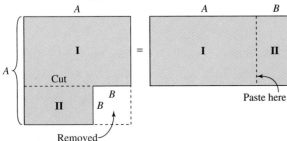

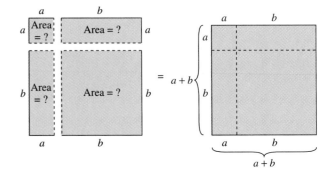

Use the special factoring formulas of this section to verify the statements in Problems 95–96, without raising any number to a power.

95. $10^2 - 9^2 = 19$

96. $1000^2 - 900^2 = 190,000$

97. Evaluate 39^2 using $39^2 = (40 - 1)^2$ and one of the special factoring formulas.

True–False Critical Thinking Problems

98. Which one of the following is true?
  a. $a^3 - b^3 = (a - b)^3$
  b. $x^2 + 8x + 64$ is a perfect square trinomial.
  c. The binomial $x + 2$ is a factor of $x^3 - 8$.
  d. The difference of two squares can be factored as a sum and a difference.

99. Which one of the following is true?
  a. $9x^2 + 15x + 25 = (3x + 5)^2$
  b. $x^3 - 27 = (x - 3)(x^2 + 6x + 9)$
  c. $x^3 - 64 = (x - 4)^3$
  d. None of the above is true.

Technology Problems

In Problems 100–107, use a graphing utility to graph the function on the left side and the function on the right side in the same viewing rectangle. Are the graphs identical? If so, this means that the polynomial on the left side has been correctly factored. If not, factor the polynomial correctly and then use your graphing utility to verify the factorization.

100. $9x^2 - 4 = (3x + 2)(3x - 2)$

101. $x^2 + 4x + 4 = (x + 4)^2$

102. $9x^2 + 12x + 4 = (3x + 2)^2$

103. $25 - (x^2 + 4x + 4) = (x + 7)(x - 3)$

104. $(2x + 3)^2 - 9 = 4x(x + 3)$

105. $(x - 3)^2 + 8(x - 3) + 16 = (x - 1)^2$

106. $x^3 - 1 = (x - 1)(x^2 - x + 1)$

107. $(x + 1)^3 + 1 = (x + 1)(x^2 + x + 1)$

Writing in Mathematics

108. Describe how to tell when a polynomial is a perfect square trinomial.

Critical Thinking Problems

109. Here is an old algebraic fallacy that relates to factoring the difference of two squares. Can you find the error?

Let $a = b$
Multiply both sides by a: $a^2 = ab$
Subtract b^2 from both sides: $a^2 - b^2 = ab - b^2$

Factor: $(a + b)(a - b) = b(a - b)$
Divide both sides by $(a - b)$: $a + b = b$
But $a = b$; therefore: $2b = b$
Divide both sides by b: $2 = 1$

Factor each polynomial in Problems 110–111. Assume that all variable exponents represent whole numbers.

110. $125x^{3n} + 27y^{12m}$

111. $4x^{2n} + 20x^n y^m + 25y^{2m}$

Find a value of k so that each polynomial in Problems 112–115 will be a perfect square trinomial.

112. $x^2 - 12x + k$

113. $x^2 - 6xy + ky^2$

114. $kx^2 + 8xy + y^2$

115. $kx^2 - 112xy + 64y^2$

116. Use the formula

$$(c + d)^5 = c^5 + 5c^4d + 10c^3d^2 \\ + 10c^2d^3 + 5cd^4 + d^5$$

to factor $a^5 - 10a^4 + 40a^3 - 80a^2 + 80a - 32$.

117. Factor completely: $y^3 + x + x^3 + y$.

118. Factor $x^6 - y^6$ first as the difference of squares and then as the difference of cubes. From these two factorizations, determine that a product of two special trinomials equals another special trinomial.

Group Activity Problems

119. The ancient Greeks used geometric models such as the figure on page 376 to derive factoring formulas. The figure on the left shows two cubes: a large cube whose volume is A^3 and a smaller cube whose volume is B^3. The smaller cube is removed from the larger and the remaining solid has a volume of $A^3 - B^3$.

    The volume of the remaining solid on the left is equal to the sum of the volumes of Boxes I, II, and

III, shown on the right. Find the volume of each box, find the sum of these volumes, factor $A - B$ from this expression, and describe how the results provide a visual model for the factoring formula $A^3 - B^3 = (A - B)(A^2 + AB + B^2)$.

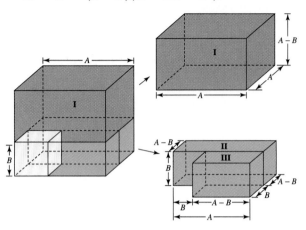

120. Consider the numbers, 1, 2, 3, 4, . . . , 50. Each member of the group should be assigned some of these numbers with the task of finding the numbers that have an odd number of factors, excluding factors of 1. Be sure that all 50 numbers are assigned. Group members should come together to formulate a complete list of all the numbers that have an odd number of factors. Then, as a group, answer this question: What kinds of numbers have an odd number of factors?

Review Problems

121. Graph in a rectangular coordinate system:

$$y > x - 2 \quad \text{and} \quad x + y > -2.$$

122. Two cars leave the same point traveling in opposite directions. The speed of the faster car is 60 mph, and the speed of the slower car is 40 mph. In how many hours will the cars be 400 miles apart?

123. Solve: $3 - 2(x + 7) \geq 3 - x$.

S E C T I O N   4 . 6

Solutions Manual   **Tutorial**   **Video 5**

| Recognize the appropriate method for factoring a polynomial.

A General Factoring Strategy

Objectives

1 Recognize the appropriate method for factoring a polynomial.
2 Factor polynomials using two or more factoring techniques.

It is important to practice factoring a wide variety of polynomials so that you can quickly select the appropriate technique. The polynomial is factored completely when all its polynomial factors, except possibly for monomial factors, are prime. Because of the commutative property, the order of the factors does not matter.

The following strategy outlines the methods of factoring covered in this chapter.

A factoring strategy

Factoring a Polynomial Over the Integers

1. Is there a common factor? If so, factor out the GCF.
2. Is the polynomial a binomial? If so, can it be factored by one of the following special formulas?

$$A^2 - B^2 = (A + B)(A - B) \qquad \text{Difference of two squares}$$

$$A^3 + B^3 = (A + B)(A^2 - AB + B^2) \qquad \text{Sum of two cubes}$$

$$A^3 - B^3 = (A - B)(A^2 + AB + B^2) \qquad \text{Difference of two cubes}$$

3. Is the polynomial a trinomial? If it is not a perfect square trinomial, use the trial-and-error FOIL method or the optional grouping method developed in Section 4.4. If it is a perfect square trinomial, use one of the following special formulas:

$$A^2 + 2AB + B^2 = (A + B)^2$$

$$A^2 - 2AB + B^2 = (A - B)^2$$

4. Does the polynomial contain four or more terms? If so, try factoring by grouping.

5. Is the polynomial factored completely? If a factor with more than one term can be factored further, do so.

---

2 Factor polynomials using two or more factoring techniques.

The following examples and those in the problem set are similar to the previous factoring problems. One difference is that although these polynomials may be factored using the techniques we have studied in this chapter, each must be factored using at least two techniques. Also different is that these factorizations are not all of the same type; they are intentionally mixed to promote the development of a general factoring strategy.

Using technology

The cubic functions

$$y_1 = 2x^3 + 8x^2 + 8x$$

and

$$y_2 = 2x(x + 2)^2$$

have identical graphs, verifying that

$$2x^3 + 8x^2 + 8x = 2x(x + 2)^2.$$

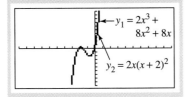

EXAMPLE 1 Factoring a Polynomial

Factor: $2a^3 + 8a^2 + 8a$

Solution

We first look for a common factor. Since $2a$ is common to all terms, we factor it out.

$$2a^3 + 8a^2 + 8a = 2a(a^2 + 4a + 4) \qquad \text{Factor out the GCF.}$$

$$= 2a(a + 2)^2 \qquad \text{Factor the perfect square trinomial.} \quad \blacksquare$$

EXAMPLE 2 Factoring a Polynomial

Factor: $4x^2 - 16x - 20$

Solution

Again, we first look for a common factor. Since 4 is common to all terms, we factor it out.

$$4x^2 - 16x - 20 = 4(x^2 - 4x - 5) \qquad \text{Factor out the GCF.}$$

$$= 4(x + 1)(x - 5) \qquad \text{Factor the remaining trinomial.} \quad \blacksquare$$

EXAMPLE 3 **Factoring a Polynomial**

Factor: $3x^4 + 24x$

Solution

We begin by factoring out $3x$, the GCF. Once we do this, the binomial factor can be factored as the sum of two cubes.

$$
\begin{aligned}
3x^4 + 24x &= 3x(x^3 + 8) && \text{Factor out the GCF.}\\
&= 3x(x^3 + 2^3) && \text{Express as the sum of two cubes.}\\
&= 3x(x + 2)(x^2 - 2x + 4) && \text{Factor the sum of two cubes.}
\end{aligned}
$$

EXAMPLE 4 **Factoring a Polynomial**

Factor: $x^2(x - 5) - 4(x - 5)$

Solution

We begin by factoring out the common binomial, $x - 5$. Then we can factor the difference of two squares.

$$
\begin{aligned}
&x^2(x - 5) - 4(x - 5)\\
&= (x - 5)(x^2 - 4) && \text{Factor out the GCF.}\\
&= (x - 5)(x + 2)(x - 2) && \text{Factor the difference of two squares.}
\end{aligned}
$$

EXAMPLE 5 **Factoring a Polynomial**

Factor: $9b^2x - 16y - 16x + 9b^2y$

Solution

There is no factor common to all four terms. Since there are four terms, we try factoring by grouping. Notice that the first and last terms have a common factor of $9b^2$ and the two middle terms have a common factor of -16.

$$
\begin{aligned}
&9b^2x - 16y - 16x + 9b^2y\\
&= (9b^2x + 9b^2y) + (-16x - 16y) && \text{Rearrange the terms for factoring by grouping.}\\
&= 9b^2(x + y) - 16(x + y) && \text{Factor within each group.}\\
&= (x + y)(9b^2 - 16) && \text{Factor out the common binomial.}\\
&= (x + y)(3b + 4)(3b - 4) && \text{Factor the difference of two squares.}
\end{aligned}
$$

EXAMPLE 6 **Factoring a Polynomial**

Factor: $x^2 + 8x + 16 - 25a^2$

Solution

There is no factor common to all four terms. Since there are four terms, we try factoring by grouping. Notice that the first three terms are a perfect square trinomial and the last term, $25a^2$, is a perfect square. This suggests that we should try grouping the four terms as the difference of two squares.

$$
\begin{aligned}
&x^2 + 8x + 16 - 25a^2\\
&= (x^2 + 8x + 16) - 25a^2 && \text{Group the perfect square trinomial.}\\
&= (x + 4)^2 - 25a^2 && \text{Factor the perfect square trinomial.}\\
&= (x + 4)^2 - (5a)^2 && \text{Express the polynomial as the difference of two squares.}\\
&= (x + 4 + 5a)(x + 4 - 5a) && \text{Factor the difference of two squares.}
\end{aligned}
$$

ENRICHMENT ESSAY

Prime-Rich Polynomials

Divisible only by themselves and the number 1, *prime numbers* are the building blocks in the mathematics of whole numbers. All other whole numbers, known as *composites,* can be written as the product of unique prime factors.

This spiral square shows the primes between 41 and 439, beginning with 41 in the center. Each square represents an increment of one. The prime-rich polynomial $x^2 + x + 41$ generates the values along the diagonal, starting with $x = 0$. For values of x up to 10,000, the polynomial $x^2 + x + 41$ produces primes almost half the time. However, beyond 10,000, the proportion of primes dramatically increases. Another prime-rich trinomial is $x^2 + x + 17$, generating primes for all values of x from 0 through 15. Mathematicians have proved that no polynomial has only prime values, making the search for a specific polynomial formula that will generate only primes fruitless.

421		419							409				
	347							337			331	401	
	281		277				271	269					
349		223							211				
	283		173			167			163				
			131			127				263	397		
				97									
	353	227			71		67	89					
				53									
	229			73	43			157					
431		179	101		41								
		137			47			257					
433		181	103		59	61							
359	233	139		79		83			389				
	293			107	109			113	199	317			
					149	151							
			191	193			197						
		239	241					251					
439							307		311	313			
	367				373			379		383			

EXAMPLE 7   Factoring a Polynomial

Factor: $-x^2 - 4x + 5$

Solution

Since we are accustomed to factoring trinomials with positive leading coefficients, we first factor out -1.

$$-x^2 - 4x + 5 = -1(x^2 + 4x - 5) \qquad \text{Factor out } -1.$$
$$= -(x - 1)(x + 5) \qquad \text{Factor the remaining trinomial.}$$

EXAMPLE 8   Factoring a Polynomial

Factor: $x^9 + 1$

Solution

There is no common factor (other than 1 or -1). However, the two terms in the binomial can be expressed as the sum of two cubes.

$$x^9 + 1 = (x^3)^3 + 1^3 \qquad \text{Express the polynomial as the sum of two cubes.}$$
$$= (x^3 + 1)[(x^3)^2 - x^3 \cdot 1 + 1^2] \qquad \text{Factor the sum of two cubes.}$$
$$= (x^3 + 1)(x^6 - x^3 + 1) \qquad \text{Simplify the second factor.}$$
$$= (x + 1)(x^2 - x + 1)(x^6 - x^3 + 1) \qquad \text{Factor completely by factoring the sum of two cubes.}$$

| EXAMPLE 9 | **Factoring a Polynomial** |

Factor: $(x^2 - 3)^2 - 7(x^2 - 3) + 6$

Solution

There is no factor common to all three terms. In a situation like this, where $x^2 - 3$ appears in two of the terms, we can simplify the factoring process by introducing a substitution. We will replace $x^2 - 3$ by t.

$$(x^2 - 3)^2 - 7(x^2 - 3) + 6$$
$$= t^2 - 7t + 6 \qquad \text{Let } t = x^2 - 3.$$
$$= (t - 6)(t - 1) \qquad \text{Factor the trinomial.}$$
$$= (x^2 - 3 - 6)(x^2 - 3 - 1) \qquad \text{Replace } t \text{ with } x^2 - 3.$$
$$= (x^2 - 9)(x^2 - 4) \qquad \text{Simplify.}$$
$$= (x + 3)(x - 3)(x + 2)(x - 2) \qquad \text{Factor completely by using the difference of two squares formula on each factor.}$$

PROBLEM SET 4.6

Practice Problems

For Problems 1–93, factor each expression completely, or state that the polynomial is prime. For polynomials in one variable, use a graphing utility to verify your factorization.

1. $c^3 - 16c$

2. $y^3 - y$

3. $3x^2 + 18x + 27$

4. $8x^2 + 40x + 50$

5. $81x^3 - 3$

6. $9b^3 - 9b^6$

7. $B^2C - 16C + 32 - 2B^2$

8. $12x^2y - 27y - 4x^2 + 9$

9. $-x^2 + 12x - 27$

10. $-x^2 + 10x - 25$

11. $4a^2b - 2ab - 30b$

12. $32y^2 - 48y + 18$

13. $a(y^2 - 4) - 4(y^2 - 4)$

14. $x^2(x^3 + y^3) - z^2(x^3 + y^3)$

15. $11x^5 - 11xy^2$

16. $4y^9 - 400y$

17. $3x^2 + 3x + 3y - 3y^2$

18. $6ab - 6a + 6cb - 6c$

19. $25x^2 - xy + 36y^2$

20. $25y^2 - 10y + 1$

21. $ax^3 + 8a$

22. $bx^3 - b$

23. $s^2 - 12s + 36 - 49t^2$

24. $b^2 - 10b + 25 - 36c^2$

25. $4m^{10} + 12m^5n^3 + 9n^6$

26. $24b^6 + 3c^6$

27. $9s^2t^2 - 36t^2$

28. $12a^3b - 12ab^3$

29. $ax + ay + bx + by$

30. $25x^2 - 30xy + 9y^2$

31. $5x^2yz - 5y^3z$

32. $35y^2 + 37y + 6$

33. $20a^3b - 245ab^3$

34. $cx^2 + cx + dxy + dy$

35. $63y^2 + 30y - 72$

36. $8x^3 - 125y^3$

37. $r^6 + 4r^3s + 4s^2$

38. $6bx^2 + 6by^2$

39. $4ax^3 - 32a$

40. $6x^2 - 66$

41. $100x^4 + 120x^3y + 36x^2y^2$

42. $63x^3y - 175xy$

43. $49x^2 + 126xy + 81y^2$

44. $5xy^3 - 5x$

45. $71bx^4 - 71b$

46. $12r^2 - 12rs + 3s^2$

47. $x^2 + 25$

48. $8r^6 - s^3$

49. $r^3 - s^3 + r - s$

50. $7by^4 - 7b$

51. $(x^2 - 3)^2 - 4(x^2 - 3) - 12$

52. $9r^2 - 6r + 1 - 25s^2$

53. $a^2 + 4a + 4 - 16b^2$

54. $2x^2 + 2x + 2xy + 2y$

55. $27r^3s + 72r^2s^2 + 48rs^3$

56. $7x^5y - 7xy^5$

57. $-6by^3 + 24by$

58. $200y^2 + 2$

59. $16x^2 + 49y^2$

60. $3bx^2 - 3by^2 + 3by - 3bx$

61. $(3x - y)^2 - 100a^2$

62. $(x + 3y)^2 - 10(x + 3y) + 25$

63. $(5x + 3y)^2 - 6(5x + 3y) + 9$

64. $21x^4 - 32x^2 - 5$

65. $6y^4 - 11y^2 - 10$

66. $24x^3y + 52x^2y^2 + 20xy^3$

67. $48y^4 - 243$

68. $7x^2y - 28w^2z^2y$

69. $20bx^4 + 220bx^2y + 605by^2$

70. $r^3 + 4r^2s + 4rs^2$

71. $18x^3 + 63x^2 - 36x$

72. $7rsy + 35ry + 7sy + 35y$

73. $4x^7 + 32xy^3$

74. $(x^2 - 8)^2 - 4(x^2 - 8) - 32$

75. $x^4 - (x - 6)^2$

76. $x^8 - y^{12}$

77. $x^{16} + 1$

78. $4x^2y + 5 - 20x^2 - y$

79. $r^4s + 3 - 3r^4 - s$

80. $x^4 - 25x^2 - 144$

81. $y^3 - 2y^2 - 4y + 8$

82. $4r^2s^2 - 4r^2 - 9s^2 + 9$

83. $ay^3 + b - by^3 - a$

84. $rs^2 - 2a - r + 2as^2$

85. $4x^2 + 4x - 1$

86. $7y^2 - 28$

87. $16x - 2x^4$

88. $r^3s^3 - r^3$

89. $y^3 - 2y^2 - y + 2$

90. $x^4 - y^4$

91. $a^6b^6 - a^3b^3$

92. $x^3y^6 - x^3$

93. $10x^3 - 6x^2 - 21x$

Application Problems

94. What are the dimensions of the solid shown in the figure if the factors of the volume are the dimensions?

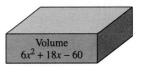

Volume
$6x^2 + 18x - 60$

In Problems 95–96, write a polynomial that models the area of the shaded region and factor the polynomial.

95.

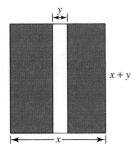

y

$x + y$

x

96.

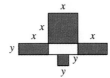

x

x x x

y y

y

In Problems 97–98, write a polynomial that models the volume outside the smaller solid and inside the larger solid. Then factor the polynomial. Express answers in terms of π.

97. Cylinder

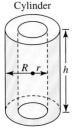

R r h

98. Sphere

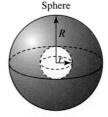

R

r

True–False Critical Thinking Problems

99. Which one of the following is true?
   a. $x^2 - 25$ factors as $(x - 5)^2$.
   b. The terms of a polynomial must always be rearranged to factor by grouping.
   c. The polynomial $4x^2 + 100$ is the sum of two squares and therefore cannot be factored.
   d. A partial check to see if a polynomial has been factored correctly can be performed by evaluating both the polynomial and its factorization for a few values of the variable. If the factorization is correct, the polynomial and its factored form will have the same value for any replacement(s) of the variable.

100. Which one of the following is true?
   a. $x^4 - 16$ is factored completely as $(x^2 + 4)(x^2 - 4)$.
   b. The trinomial $x^2 - 4x - 4$ is a prime polynomial.
   c. $x^2 + 36 = (x + 6)^2$
   d. $x^3 - 64 = (x + 4)(x^2 + 4x - 16)$

Technology Problems

101. Suppose that you are using a graphing utility, and you have factored $x^2 - 9a^2 + 12x + 36$ as $(x + 6 + 3a)(x + 6 - 3a)$. You would like to verify the factorization by graphing each side of the equation

$$x^2 - 9a^2 + 12x + 36 = (x + 6 + 3a)(x + 6 - 3a)$$

obtaining identical graphs. However, your graphing utility can only graph in one variable. What procedure might you use to be able to use your graphing utility to verify your factorization?

The answer to Problem 101 is that you can take various values of a, substitute each value into your factorization, graph both sides of the resulting equation, and see if identical graphs result. If they do, your factorization is probably correct. Use this idea to verify the factorizations in Problems 102–106. If for any value of a that you substitute into the given equation the graphs on the right and left side are not the same, the polynomial on the left side has not been correctly factored. In this case, factor correctly and then use your graphing utility to verify the revised factorization.

102. $3x + 12 + ax^2 + 4ax = (x + 4)(3 + ax)$

103. $2x^2 + 18a^2 - 12ax = 2(x - 6a)^2$

104. $x^2 - 4a^2 = (x - 2a)^2$

105. $x^3a - 16xa^3 = xa(x + 4a)(x - 4a)$

106. $a^2 - x^2 - 6x - 9 = (a - x + 3)(a + x + 3)$

Writing in Mathematics

107. Suppose you are factoring $x^2 - 2x - 8$. Why must the second terms of the two binomial factors have opposite signs?

108. If the product $(4x + 10)(x - 2)$ is $4x^2 + 2x - 20$, is $4x^2 + 2x - 20$ factored completely? In your answer, describe what is meant by *factored completely*.

109. Describe the factoring technique called factoring by grouping. In your description, explain whether the expression $x^2 + 3x - bx + 3b$ can be factored by this technique.

Critical Thinking Problems

In Problems 110–115, factor each expression completely. If applicable, use a graphing utility to verify your result.

110. $(1 - x)^3 - (x - 1)^6$

111. $5r^3s^2 - 5rs^4 - 5r^2s^2 + 5rs^3$

112. $x^{4n+1} - xy^{4n}$

113. $(25x^2 - 10xy + y^2) + (10xz - 2yz) - 24z^2$

114. $3x^{n+2} + 4x^{n+1} - 4x^n$

115. $x^4 - y^4 - 2x^3y + 2xy^3$

116. In certain circumstances, the sum of two perfect squares can be factored by adding and subtracting the same perfect square. For example,

$$x^4 + 4 = x^4 + 4x^2 + 4 - 4x^2 \quad \text{Add and subtract } 4x^2.$$

Use this first step to factor $x^4 + 4$.

117. Express $x^3 + x + 2x^4 + 4x^2 + 2$ as the product of two polynomials of degree 2.

Group Activity Problem

118. Study the factoring strategy in the box on pages 376–377. Without looking at any factoring problems in the book, create five factoring problems that cover each of the first four strategies in the box. Create problems that each require at least two factoring strategies. Once you have written your five factoring problems, describe to the group how you created factorable polynomials. Group members should then exchange factoring problems and work on factoring the problems. Share with the person whose problems you are working on your response to the required factorizations: Are they too easy? too difficult? Can the polynomials really be factored or is there a "kink" in the problem that the person who wrote the problem is not aware of? What can you learn by creating problems on material that you have been studying? Finally, grade each other's work (20 points per factoring problem), awarding partial credit. If you take off points, explain why points are deducted and why you decided to take off a particular number of points in terms of the error(s) that you found.

Review Problems

119. Evaluate: $\left[\dfrac{7 + (-16)}{|7 - 10|}\right]\left[\dfrac{12 + (-2)}{3 + (-2)^3}\right]$.

120. Solve: $\dfrac{3x - 1}{5} + \dfrac{x + 2}{2} = -\dfrac{3}{10}$.

121. Simplify: $(4x^3y^{-1})^2(2x^{-3}y)^{-1}$.

SECTION 4.7

Solutions Manual **Tutorial** **Video 5**

Bachmann/Photri, Inc.

Polynomial Equations, Modeling, and Problem Solving

Objectives

1 Solve polynomial equations by factoring.
2 Graph parabolas.
3 Solve problems using quadratic models.

Throughout this chapter, we have seen *quadratic functions* in the form $f(x) = ax^2 + bx + c$ $(a \neq 0)$ and the parabolas that are their graphs. We now turn our attention to *quadratic* or *second-degree equations*. These equations can be written in the form $ax^2 + bx + c = 0$, where a, b, and c are real numbers and $a \neq 0$. (If we allowed a to equal 0, the equation would not be quadratic. The resulting equation, $bx + c = 0$, would be a linear equation.)

> **Definition of a quadratic equation**
>
> A *quadratic equation* in x is an equation that can be written in the standard form
>
> $$ax^2 + bx + c = 0$$
>
> where a, b, and c are real numbers with $a \neq 0$. A quadratic equation in x is also called a *second-degree polynomial equation* in x.

Both quadratic equations and functions derive their name from the Latin word *quadrus* meaning "square." Both are in standard form when the polynomial is in descending order of powers and, for a quadratic equation, is equal to 0.

1 Solve polynomial equations by factoring.

The Zero Product Principle

Some quadratic equations can be solved by factoring, using the *zero product principle*. This principle states that if a product is 0, then at least one of the factors must be zero.

> **The zero product principle**
>
> Let A and B be real numbers, variables, or algebraic expressions.
>
> **1.** If $AB = 0$, then $A = 0$ or $B = 0$ (or both).
> **2.** If $A = 0$ or $B = 0$, then $AB = 0$.

EXAMPLE 1   **Solving a Quadratic Equation Using the Zero Product Principle**

Solve: $x^2 - 7x = -10$

Solution

To use the zero product principle to solve a quadratic equation, we first write the equation in standard form. Since we want zero on one side, we begin by adding 10 to both sides.

$$x^2 - 7x = -10$$ The zero product principle requires a product of two factors equal to 0. To obtain 0 in the right member, add 10 to both sides, writing the quadratic equation in standard form.

$$x^2 - 7x + 10 = 0$$

$$(x - 5)(x - 2) = 0$$ Factor on the left.

$$x - 5 = 0 \quad \text{or} \quad x - 2 = 0$$ Set each factor equal to 0, using the zero product principle.

$$x = 5 \quad \text{or} \quad x = 2$$ Solve the two resulting equations.

Check each proposed solution by substituting that value into the original equation.

Check 5:
$$x^2 - 7x = -10$$
$$5^2 - 7(5) \stackrel{?}{=} -10$$
$$25 - 35 \stackrel{?}{=} -10$$
$$-10 = -10 \quad \checkmark$$

Check 2:
$$x^2 - 7x = -10$$
$$2^2 - 7(2) \stackrel{?}{=} -1 - 0$$
$$4 - 14 \stackrel{?}{=} -10$$
$$-10 = -10 \quad \checkmark$$

The resulting true statements indicate that the solutions are 5 and 2. The solution set is $\{2, 5\}$.   ∎

There is a relationship between a quadratic *equation,* such as $x^2 - 7x + 10 = 0$, and a quadratic *function,* such as $f(x) = x^2 - 7x + 10$. The solutions of the equation $x^2 - 7x + 10 = 0$ are 2 and 5. Figure 4.7 shows that the graph of $f(x) = x^2 - 7x + 10$ crosses the *x*-axis at 2 and 5. The real solutions to the quadratic equation $ax^2 + bx + c = 0$ are the *x*-intercepts for the quadratic function $f(x) = ax^2 + bx + c$. We say that 2 and 5 are *roots* of the quadratic equation $x^2 - 7x + 10 = 0$. We also call 2 and 5 the *zeros* of the quadratic function $f(x) = x^2 - 7x + 10$ because these are the values that result in the function being equal to 0. The zeros of the quadratic function are the *x*-intercepts of its graph.

Let's summarize the steps involved in solving a quadratic equation by factoring.

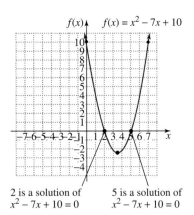

2 is a solution of
$x^2 - 7x + 10 = 0$

5 is a solution of
$x^2 - 7x + 10 = 0$

Figure 4.7

The real solutions of the quadratic equation $ax^2 + bx + c = 0$ are the *x*-intercepts of the parabola described by $f(x) = ax^2 + bx + c.$

Solving a quadratic equation by factoring

1. If necessary, rewrite the equation in the form $ax^2 + bx + c = 0$, setting one side equal to 0.
2. Factor.
3. Apply the zero product principle, setting each factor equal to 0.

4. Solve the equations in step 3.
5. Check the solutions in the original equation.

EXAMPLE 2 **Solving Quadratic Equations by Factoring**

Solve:

a. $5x^2 = 20x$ **b.** $x^2 + 4 = 8x - 12$ **c.** $(x - 7)(x + 5) = -20$

Solution

a.

$5x^2 = 20x$	This is the given equation.
$5x^2 - 20x = 0$	Subtract $20x$ from both sides, getting 0 on the right.
$5x(x - 4) = 0$	Factor.
$5x = 0$ or $x - 4 = 0$	Set each factor equal to 0.
$x = 0$ or $x = 4$	Solve the resulting equations.

Check by substituting 0 and 4 into the given equation. The graph of $y = 5x^2 - 20x$, obtained with a graphing utility, is shown in Figure 4.8. The x-intercepts are 0 and 4, also verifying that the solution set is $\{0, 4\}$.

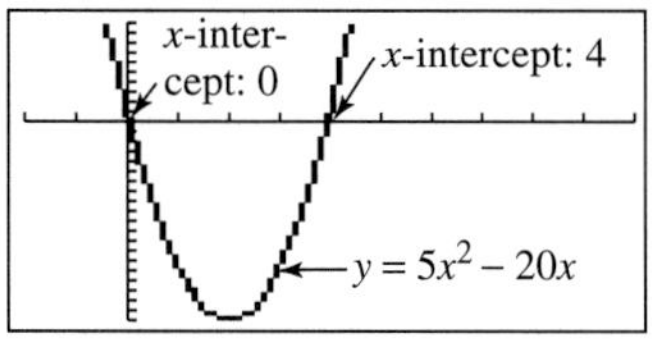

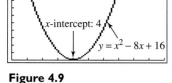

Figure 4.8

The solution set of $5x^2 - 20x = 0$ is $\{0, 4\}$.

b.

$x^2 + 4 = 8x - 12$	This is the given equation.
$x^2 - 8x + 16 = 0$	Write the equation in standard form by subtracting $8x$ and adding 12 on both sides.
$(x - 4)(x - 4) = 0$	Factor.
$x - 4 = 0$ or $x - 4 = 0$	Set each factor equal to 0.
$x = 4$ or $x = 4$	Solve the resulting equations.

Notice that there is only one solution (or, if you prefer, a repeated solution.) The trinomial $x^2 - 8x + 16$ is a perfect square trinomial that could have been factored as $(x - 4)^2$. The graph of $y = x^2 - 8x + 16$, obtained with a graphing utility and shown in Figure 4.9, is a parabola with only one x intercept at 4. This verifies that the equation's solution set is $\{4\}$.

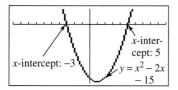

Figure 4.9

The solution set of $x^2 - 8x + 16 = 0$ is $\{4\}$.

c. Be careful! Although the left side is factored in $(x - 7)(x + 5) = -20$, we cannot use the zero product principle because the right side of the equation is not 0. So we begin by multiplying the factors on the left side of the equation. Then we add 20 to both sides to obtain 0 on the right side.

$(x - 7)(x + 5) = -20$	This is the given equation.
$x^2 - 2x - 35 = -20$	Use the FOIL method to multiply on the left side.
$x^2 - 2x - 15 = 0$	Add 20 to both sides.
$(x + 3)(x - 5) = 0$	Factor.
$x + 3 = 0$ or $x - 5 = 0$	Set each factor equal to 0.
$x = -3$ or $x = 5$	Solve the resulting equations.

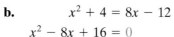
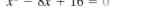

Check by substituting -3 and 5 into the given equation. The graph of $y = x^2 - 2x - 15$, obtained with a graphing utility, is shown in Figure 4.10. The x-intercepts are -3 and 5, verifying that the solution set is $\{-3, 5\}$. ■

Figure 4.10

The solution set of $x^2 - 2x - 15 = 0$ is $\{-3, 5\}$.

Study tip

Avoid the following errors:

$$5x^2 = 20x$$

$$\frac{5x^2}{x} = \frac{20x}{x}$$

$$5x = 20$$

$$x = 4$$

Never divide both sides of an equation by x. Division by zero is undefined and x may be zero. Indeed, the solution set for this equation (Example 2a) is $\{0, 4\}$. Dividing both sides by x does not permit us to find both solutions.

$$(x - 7)(x + 5) = -20$$

$$x - 7 = -20 \quad \text{or} \quad x + 5 = -20$$

$$x = -13 \quad \text{or} \quad x = -25$$

The zero product principle cannot be used because the right side of the equation is not equal to 0.

EXAMPLE 3 Solving a Quadratic Equation Requiring Rewriting

Solve: $4(x - 4) + 4(x - 1) = 5(x - 1)(x - 4)$

Solution

$4(x - 4) + 4(x - 1) = 5(x - 1)(x - 4)$ — This is the given equation. To put the equation in standard form, multiply on each side.

$4(x - 4) + 4(x - 1) = 5(x^2 - 5x + 4)$ — Use the FOIL method on the right.

$4x - 16 + 4x - 4 = 5x^2 - 25x + 20$ — Use the distributive property.

$8x - 20 = 5x^2 - 25x + 20$ — Combine like terms.

$8x - 8x - 20 + 20 = 5x^2 - 25x - 8x + 20 + 20$ — Because it is easier to factor $ax^2 + bx + c$ when $a > 0$, collect $8x - 20$ on the right by subtracting $8x$ and adding 20 on both sides.

$0 = 5x^2 - 33x + 40$ — Combine like terms.

$5x^2 - 33x + 40 = 0$ — The equation is now in standard form. If $A = B$, then $B = A$. This step is optional.

$(5x - 8)(x - 5) = 0$ — Factor.

$5x - 8 = 0 \quad \text{or} \quad x - 5 = 0$ — Set each factor equal to zero.

$5x = 8 \quad \text{or} \quad x = 5$ — Solve the resulting equations.

$x = \frac{8}{5}$

Checking these values in the original equation confirms that the solution set is $\{\frac{8}{5}, 5\}$. The roots of the equation are $\frac{8}{5}$ and 5. ■

Using technology

The solution set for $5x^2 - 33x + 40 = 0$, namely $\{\frac{8}{5}, 5\}$ can be verified by graphing the parabola whose equation is $y = 5x^2 - 33x + 40$. As shown below, the x-intercepts appear to be $\frac{8}{5}$ (or 1.6) and 5. You can use the TRACE feature to verify the intercept on the left.

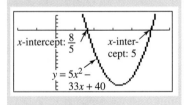

x-intercept: $\frac{8}{5}$ x-intercept: 5

$y = 5x^2 - 33x + 40$

Solving Higher-Order Polynomial Equations

The zero product principle can be extended to three or more factors.

Extending the zero product principle

Let $A, B, C,$ and D be real numbers, variables, or algebraic expressions.

    1. If $ABC = 0$, then $A = 0$ or $B = 0$ or $C = 0$.
    2. If $ABCD = 0$, then $A = 0$ or $B = 0$ or $C = 0$ or $D = 0$.

EXAMPLE 4   **Solving a Polynomial Equation with Three Factors**

Solve: $x^3 + x^2 = 4x + 4$

Solution

$x^3 + x^2 = 4x + 4$	This is the given equation.
$x^3 + x^2 - 4x - 4 = 0$	Set the equation equal to zero.
$x^2(x + 1) - 4(x + 1) = 0$	Factor on the left using grouping.
$(x + 1)(x^2 - 4) = 0$	Factor out the common binomial from each term.
$(x + 1)(x + 2)(x - 2) = 0$	Factor completely by factoring $x^2 - 4$ as the difference of two squares.
$x + 1 = 0$ or $x + 2 = 0$ or $x - 2 = 0$	Set each factor equal to 0.
$x = -1$ or $x = -2$ or $x = 2$	Solve.

Check by substituting -2, -1, and 2 into the given equation. The graph of the cubic function $y = x^3 + x^2 - 4x - 4$, obtained with a graphing utility, is shown in Figure 4.11. The x-intercepts are -2, -1, and 2, verifying that the solution set is $\{-2, -1, 2\}$. ∎

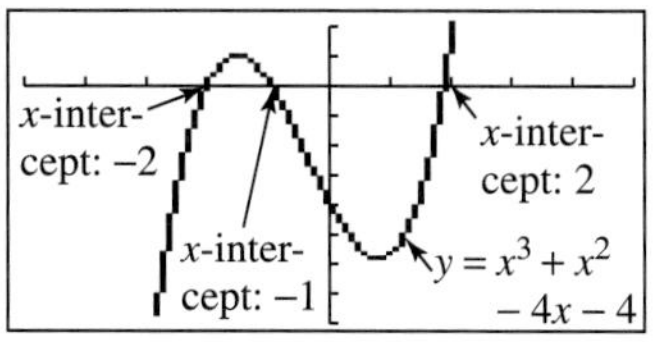

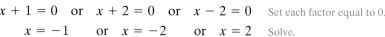

The equation in Example 4 is a third-degree polynomial equation and has three solutions. A polynomial equation can have *at most* as many real number solutions as its degree. This means that a quadratic equation can have no real solutions, one real solution, or two real solutions. We have seen examples of quadratic equations with one and two real solutions. In Chapter 7 we will be studying quadratic equations with no real solutions.

Graphs of Quadratic Functions

We have seen that the graph of $y = ax^2 + bx + c$ or, equivalently, $f(x) = ax^2 + bx + c$, is called a parabola and is cuplike in shape. If $ax^2 + bx + c$ is a factorable trinomial, we know that the real solutions of the quadratic equation $ax^2 + bx + c = 0$ are the x-intercepts of the parabola described by $y = ax^2 + bx + c$. We also know (but have not yet proved) that a

---

Figure 4.11

The solution set of
$x^3 + x^2 - 4x - 4 = 0$ is
$\{-2, -1, 2\}$.

Discover for yourself

Sketch a graph of a quadratic function $y = ax^2 + bx + c$ such that the equation $ax^2 + bx + c = 0$ has no real numbers as solutions.

 Graph parabolas.

parabola has a vertex when the x-coordinate is $-\dfrac{b}{2a}$. Putting all of this together gives us a procedure for graphing certain parabolas by hand.

> **Graphing the quadratic function $y = ax^2 + bx + c$ or $f(x) = ax^2 + bx + c$, whose graph is called a parabola**
>
> 1. Find any x-intercepts by replacing y or $f(x)$ with 0.
> 2. Find the y-intercept by replacing x with 0.
> 3. Find the vertex. The x-coordinate of the vertex is $-\dfrac{b}{2a}$. The y-coordinate is found by substituting $-\dfrac{b}{2a}$ for x in the quadratic function.
> 4. Plot the intercepts and the vertex.
> 5. If needed, find and plot additional ordered pairs located near the vertex and intercepts, connecting points with a smooth curve.
> 6. The graph opens upward if $a > 0$ and downward if $a < 0$.

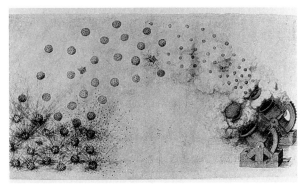

A cannon in action. The flight of exploding mortar shells, drawn by Leonardo da Vinci. Scala/Art Resource, New York

EXAMPLE 5   **Using Intercepts and the Vertex to Graph a Quadratic Function**

Graph: $f(x) = x^2 - 2x - 3$

Solution

Step 1. Find the x-intercepts. Replace $f(x)$ with 0.

$f(x) = x^2 - 2x - 3$ This is the given quadratic function.

$0 = x^2 - 2x - 3$ Replace $f(x)$ with 0.

$0 = (x - 3)(x + 1)$ Factor.

$x - 3 = 0$ or $x + 1 = 0$ Set each factor equal to 0.

$x = 3$ $x = -1$ The x-intercepts are 3 and -1. The parabola passes through $(3, 0)$ and $(-1, 0)$.

Step 2. Find the *y*-intercept. Replace *x* with 0.

$$f(x) = x^2 - 2x - 3 \qquad \text{This is the given quadratic function.}$$
$$f(0) = 0^2 - 2 \cdot 0 - 3 \qquad \text{Replace } x \text{ with 0.}$$
$$= -3 \qquad \text{The } y\text{-intercept is } -3. \text{ The parabola passes through } (0, -3).$$

Step 3. Find the vertex.

$$f(x) = x^2 - 2x - 3 \qquad \text{This is the given function, with } a = 1, b = -2, \text{ and } c = -3.$$

$$a = 1 \quad b = -2 \quad c = -3$$

$$x = -\frac{b}{2a} \qquad \text{Find the } x\text{-coordinate of the vertex.}$$

$$= \frac{-(-2)}{2(1)} = \frac{2}{2} = 1 \qquad \text{Replace } a \text{ by 1 and } b \text{ by } -2. \text{ The } x\text{-coordinate of the vertex is 1.}$$

$$f(1) = 1^2 - 2 \cdot 1 - 3 \qquad \text{Substitute 1 for } x \text{ in the given function to find the } y\text{-coordinate of the vertex.}$$

$$= 1 - 2 - 3$$
$$= -4 \qquad \text{The } y\text{-coordinate is } -4. \text{ The coordinates of the vertex are } (1, -4).$$

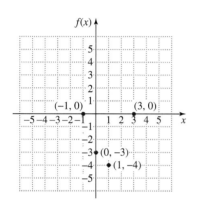

Figure 4.12

Useful points for graphing
$f(x) = x^2 - 2x - 3$

Step 4. Plot the intercepts and vertex. The intercepts and vertex are shown in Figure 4.12.

Step 5. Find additional ordered pairs. Let's find two additional points located near the *x*-intercepts by letting $x = -2$ and $x = 4$.

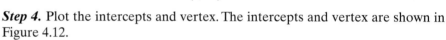

x	$f(x) = x^2 - 2x - 3$	$(x, f(x))$ or (x, y)
-2	$f(-2) = (-2)^2 - 2(-2) - 3 = 5$	$(-2, 5)$
4	$f(4) = 4^2 - 2 \cdot 4 - 3 = 5$	$(4, 5)$

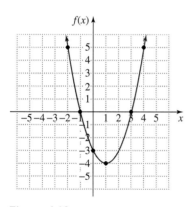

Figure 4.13

The graph of
$f(x) = x^2 - 2x - 3$

The points $(-2, 5)$ and $(4, 5)$ are added to those in Figure 4.12 and all points are connected with a smooth curve. The parabola, the graph of $f(x) = x^2 - 2x - 3$, is shown in Figure 4.13. Because $a = 1$, which is greater than 0, the parabola opens upward and the vertex is a minimum point. ∎

Using technology

Use a graphing utility to verify the hand-drawn graph in Figure 4.13. Use the TRACE or minimum function feature to check the coordinates of the vertex.

3 Solve problems using quadratic models.

Using Quadratic Models

Many situations in physics, biology, business, economics, and psychology can be modeled by quadratic functions and equations. Factoring techniques can be used to answer questions about variables contained in quadratic models.

Gene Bodio, *New City*, 1992.

EXAMPLE 6 Modeling with the Position Function

An arrow is shot directly upward from the top of a 112-foot tall building with an initial velocity of 96 feet per second. The height of the arrow above the ground after t seconds is given by the position function

$$s(t) = -16t^2 + 96t + 112.$$

a. After how many seconds will the arrow strike the ground?

b. When does the arrow reach its maximum height? What is the maximum height?

Solution

a. At the moment the arrow strikes the ground, its height above the ground is 0 feet. Thus, its position, denoted by $s(t)$, is 0.

$$s(t) = -16t^2 + 96t + 112 \quad \text{This is the given position function.}$$

$$-16t^2 + 96t + 112 = 0 \quad \text{We want the value of } t \text{ for which } s(t) = 0.$$

$$-16(t^2 - 6t - 7) = 0 \quad \text{Factor out } -16, \text{ the GCF.}$$

$$-16(t + 1)(t - 7) = 0 \quad \text{Factor the trinomial.}$$

$$t + 1 = 0 \quad \text{or} \quad t - 7 = 0 \quad \text{Set each variable factor equal to zero.}$$

$$t = -1 \quad \text{or} \quad t = 7 \quad \text{Solve the resulting equations.}$$

Since t cannot be negative, we disregard -1 and check 7.

$$s(t) = -16t^2 + 96t + 112 \quad \text{Use the original function.}$$

$$s(7) = -16(7)^2 + 96(7) + 112 \quad \text{Replace } t \text{ with 7.}$$

$$= -784 + 672 + 112 \quad \text{Simplify.}$$

$$= 0$$

This verifies that the arrow will hit the ground after 7 seconds.

b. Since the parabola whose equation is $s(t) = -16t^2 + 96t + 112$ opens downward, the vertex is the maximum point on the graph. To find when the arrow reaches its maximum height and what the maximum height is, we must determine the coordinates of the vertex.

$$s(t) = -16t^2 + 96t + 112 \quad \text{This is the given function.}$$

$$t = -\frac{b}{2a} \quad \text{With } a = -16 \text{(negative) and } b = 96, -\frac{b}{2a} \text{ tells when the arrow reaches its maximum height.}$$

$$t = -\frac{96}{2(-16)}$$

$$t = 3$$

The arrow reaches its maximum height after 3 seconds. The maximum height is $s(3)$:

$$s(3) = -16(3)^2 + 96(3) + 112$$

$$= -144 + 288 + 112$$

$$= 256$$

The arrow's maximum height is 256 feet.

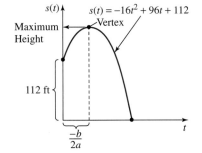

$s(t)$

Maximum Height

$s(t) = -16t^2 + 96t + 112$

Vertex

112 ft

$\dfrac{-b}{2a}$

t

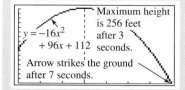

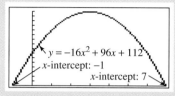

The solution set for
$-16x^2 + 96x + 112 = 0$ is
$\{-1, 7\}$.

We can use the results of Example 6 to obtain the graph of the given position function

$$s(t) = -16t^2 + 96t + 112$$

with a graphing utility. Enter the function as

$$y = -16x^2 + 96x + 112.$$

Since the arrow hits the ground after 7 seconds, we let x range from 0 to 7 (Xmin = 0 *and* Xmax = 7). Since the arrow's maximum height is 256 feet, we let y range from 0 to 256 (Ymin = 0, Ymax = 256), with Yscl = 16. The graph is shown in the top figure.

Both solutions obtained in Example 6a are shown in the bottom figure. We extended the range setting to include Xmin = -1. However, since the arrow's position is only being described for positive values of t, the portion of the parabola to the left of the vertical axis is not part of the situation we are interested in modeling.

Creating Mathematical Models

In Example 6 we were given the function that modeled the arrow's position over time. A more difficult situation is to use a problem's conditions to create a mathematical model. In Examples 7 and 8 we use our five-step problem-solving strategy and methods for solving quadratic equations to solve the problems.

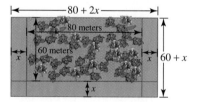

Figure 4.14

The garden's area is to be doubled by adding the path.

Steps 1 and 2: Represent unknown quantities in terms of x

EXAMPLE 7 Solving a Geometric Word Problem

A rectangular garden measures 80 by 60 meters. A large path of uniform width is to be added along both shorter sides and one longer side of the garden. The landscape artist doing the work wants to double the garden's area with the addition of this path. How wide should the path be?

Solution

We begin by making a sketch similar to the one in Figure 4.14, labeling the drawing with both known and unknown information. In particular, we let

$x =$ the width of the path
$80 + 2x =$ the length of the new (larger) rectangle
$60 + x =$ the width of the new (larger) rectangle

Step 3. Write an equation modeling the problem's conditions.

The area of the rectangle is doubled by the addition of the path.

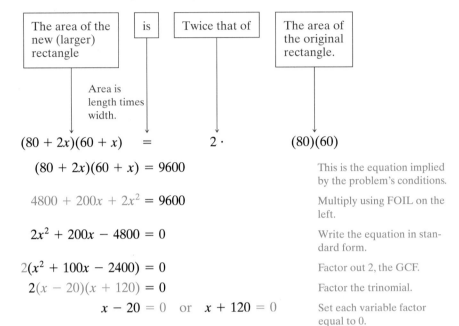

The area of the new (larger) rectangle	is	Twice that of	The area of the original rectangle.

Area is length times width.

$$(80 + 2x)(60 + x) \quad = \quad 2 \cdot \quad (80)(60)$$

Step 4. Solve the equation and answer the problem's question.

$(80 + 2x)(60 + x) = 9600$ This is the equation implied by the problem's conditions.

$4800 + 200x + 2x^2 = 9600$ Multiply using FOIL on the left.

$2x^2 + 200x - 4800 = 0$ Write the equation in standard form.

$2(x^2 + 100x - 2400) = 0$ Factor out 2, the GCF.

$2(x - 20)(x + 120) = 0$ Factor the trinomial.

$x - 20 = 0$ or $x + 120 = 0$ Set each variable factor equal to 0.

$x = 20$ $x = -120$ Solve the resulting equations.

The solution -120 is geometrically impossible, so the width of the path is 20 meters.

Step 5. Check the proposed solution in the wording of the problem.

$80 + 2x = 80 + 2(20) = 120$ meters is the larger rectangle's length

$60 + x = 60 + 20 = 80$ meters is the larger rectangle's width

The area of the original rectangle is $(80)(60)$ or 4800 square meters. The area of the new (larger) rectangle is $(120)(80)$ or 9600 square meters. This checks with the condition that the area of the garden be doubled by adding the path. ■

We could not have solved Example 7 without knowing the formula for the area of a rectangle. Geometric problem solving often requires a knowledge of certain formulas or concepts to obtain a solution. The solution to our next problem relies on knowing the *Pythagorean Theorem*, which relates the lengths of the sides of a right triangle. The side opposite the 90° angle is called the *hypotenuse*. The other sides are called *legs*.

Pythagorean theorem

In any right triangle, the sum of the squares of the legs is equal to the square of the hypotenuse.

$$a^2 + b^2 = c^2$$

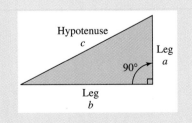

The Pythagorean Theorem is mentioned on this stamp issued by Nicaragua as one of the "ten mathematical formulas that changed the face of Earth."

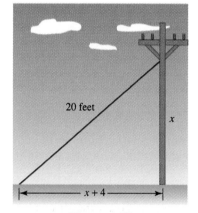

Figure 4.15

EXAMPLE 8 Solving a Problem Using the Pythagorean Theorem

A piece of wire measuring 20 feet is attached to a telephone pole as a guy wire. The distance along the ground from the bottom of the pole to the end of the wire is 4 feet greater than the height where the wire is attached to the pole. How far up the pole does the guy wire reach?

Solution

Steps 1 and 2. Represent unknown quantities in terms of x.

We begin by making a sketch similar to the one in Figure 4.15. We label the drawing with both known and unknown information. In particular, we let

x = the distance up the pole that the wire reaches.

We are given that the distance along the ground from the wire to the pole is 4 feet greater than the distance that the wire reaches up the pole, so we let

$x + 4$ = this distance.

Step 3. Write an equation describing the problem's conditions.

With the legs represented by x and $x + 4$, and a hypotenuse measuring 20 feet, we are now in a position to apply the Pythagorean Theorem:

$$\boxed{(\text{Leg})^2} \quad \boxed{+} \quad \boxed{(\text{Leg})^2} \quad \boxed{=} \quad \boxed{(\text{Hypotenuse})^2}$$

$$x^2 \quad + \quad (x + 4)^2 \quad = \quad 20^2$$

Step 4. Solve the equation and answer the problem's question.

$$x^2 + (x + 4)^2 = 20^2 \qquad \text{This is the equation arising from the Pythagorean Theorem.}$$

$$x^2 + x^2 + 8x + 16 = 400 \qquad \text{Square } x + 4 \text{ and } 20.$$

$$2x^2 + 8x + 16 = 400 \qquad \text{Combine like terms.}$$

$$2x^2 + 8x - 384 = 0 \qquad \text{Subtract 400 from both sides.}$$

$$2(x^2 + 4x - 192) = 0 \qquad \text{Factor out 2, the GCF.}$$

$$2(x + 16)(x - 12) = 0 \qquad \text{Factor the trinomial.}$$

$$x + 16 = 0 \quad \text{or} \quad x - 12 = 0 \qquad \text{Set variable factors equal to 0.}$$
$$x = -16 \quad \text{or} \quad x = 12 \qquad \text{Solve the resulting equations.}$$

The solution -16 is geometrically impossible. Thus, $x = 12$, meaning that the guy wire reaches 12 feet up the pole.

Step 5. Check the proposed solution in the implied wording (the Pythagorean Theorem) of the problem.

The distance from the bottom of the pole to the end of the wire is represented by $x + 4$. Since $x = 12$, this distance is 16 feet. However, in Pythagorean Theorem problems, we should also check that the Theorem is satisfied for the lengths of the right triangle's sides, in this case 12, 16, and 20.

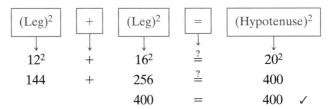

(Leg)²	+	(Leg)²	=	(Hypotenuse)²
12^2	$+$	16^2	$\stackrel{?}{=}$	20^2
144	$+$	256	$\stackrel{?}{=}$	400
		400	$=$	$400 \quad \checkmark$

In Examples 7 and 8, we used verbal models to create equations. However, we can also use real world data to develop quadratic models. Consider, for example, the data below showing fuel efficiency in miles per gallon for all U.S. automobiles in the indicated years. Figure 4.16 shows the data plotted on a rectangular coordinate system.

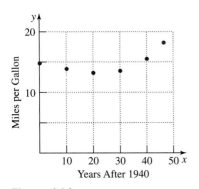

Figure 4.16

Data points for automobile fuel efficiency

x (Years after 1940)	y (Average Number of Miles/Gallon for U.S. Automobiles)
1940: 0	14.8
1950: 10	13.9
1960: 20	13.4
1970: 30	13.5
1980: 40	15.5
1986: 46	18.3

Source: Statistical Abstracts of the United States

A straight line will not fit the data in Figure 4.16. With the way values of y are decreasing and then increasing, the cuplike shape of a parabola provides a better fit. In Example 9, we use a quadratic function to model three of the data points.

EXAMPLE 9 **Modeling Automobile Fuel Efficiency**

Use the quadratic function $y = ax^2 + bx + c$ to model the data shown below. In what year was automobile fuel efficiency at its worst? What was the average number of miles per gallon for that year?

x (Years after 1940)	y (Average Number of Miles/Gallon for U.S. Automobiles)
0	14.8
30	13.5
46	18.3

ENRICHMENT ESSAY

Polynomials, the Presidency, and the Pythagorean Theorem

The Pythagorean Theorem for the right triangle is one of the best known ideas in mathematics. Several hundred different proofs of this theorem have been recorded. James Garfield, the 20th president of the United States, used his knowledge of polynomials and the trapezoid shown in the diagram to prove the theorem. See if you can reproduce Garfield's proof by finding the area of the trapezoid in two ways, establishing that $a^2 + b^2 = c^2$. (*Hint:* Find the area of the trapezoid using

$$A = \tfrac{1}{2} \cdot \text{Height} \cdot (\text{Sum of the lengths of the two parallel sides})$$

Set this equal to the sum of the areas of the three triangles shown in the figure.)

President James Garfield. Photri, Inc.

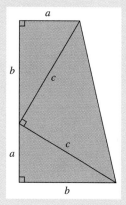

Solution

$$y = ax^2 + bx + c \qquad \text{Use the quadratic function to model the data.}$$
$$14.8 = a \cdot 0^2 + b \cdot 0 + c \qquad \text{Substitute the first data value: } x = 0 \text{ and } y = 14.8.$$
$$14.8 = c \qquad \text{Simplify.}$$

We see that $c = 14.8$, so we can write the developing model as

$$y = ax^2 + bx + 14.8.$$

We now substitute the second data value into this equation.

$$y = ax^2 + bx + 14.8 \qquad \text{This is the model that we've obtained up to this point.}$$
$$13.5 = a(30)^2 + b(30) + 14.8 \qquad \text{Substitute the second data value: } x = 30 \text{ and } y = 13.5.$$
$$13.5 = 900a + 30b + 14.8 \qquad \text{Simplify.}$$
$$-1.3 = 900a + 30b \qquad \text{Subtract 14.8 from both sides.}$$

Finally, we substitute the third data value into $y = ax^2 + bx + 14.8$.

$$18.3 = a(46)^2 + b(46) + 14.8 \qquad \text{Substitute the third data value: } x = 46 \text{ and } y = 18.3.$$
$$18.3 = 2116a + 46b + 14.8 \qquad \text{Simplify.}$$
$$3.5 = 2116a + 46b \qquad \text{Subtract 14.8 from both sides.}$$

The last two substitutions result in a linear system in a and b. We now use addition to solve for a.

iscover for yourself

Real world data can lead to systems with "messy" solutions. Try solving this system using the simultaneous equation feature of a graphing utility.

$$2116a + 46b = 3.5 \xrightarrow{\text{Multiply by 30.}}$$
$$900a + 30b = -1.3 \xrightarrow{\text{Multiply by } -46.}$$

$$63{,}480a + 1380b = 105$$
$$\underline{-41{,}400a - 1380b = 59.8}$$
Add: $\quad 22{,}080a = 164.8$

$$a = \frac{164.8}{22{,}080}$$
$$\approx 0.0075$$

We substitute 0.0075 for a in either of the system's equations. Using a calculator, we find that

$$b \approx -0.2672.$$

We now substitute the values for a and b into $y = ax^2 + bx + 14.8$. The function that models the given data is

$$y = 0.0075x^2 - 0.2672x + 14.8 \quad \text{or} \quad f(x) = 0.0075x^2 - 0.2672x + 14.8.$$

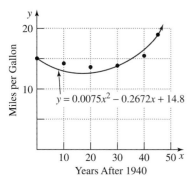

Figure 4.17

Fitting a quadratic model to data points

Figure 4.17 shows the graph of the model—a parabola—in the same rectangular coordinate system as the original six data points. Notice that the parabola passes through the three points representing the data values that we used to derive the model.

Now that we have our model, we can determine the worst year for automobile fuel efficiency.

$$f(x) = 0.0075x^2 - 0.2672x + 14.8$$

This function models miles per gallon x years after 1940.

$$a = 0.0075 \qquad b = -0.2672$$

iscover for yourself

Environmentalists are lobbying for fuel efficiency of 45 miles per gallon by the year 2000. Substitute 60 for x (2000 is 60 years after 1940) into the model

$$f(x) = 0.0075x^2 - 0.2672x + 14.8$$

Does this goal seem realistic by the year 2000? Explain.

Since a is positive, the function has a minimum value at $x = -\dfrac{b}{2a}$.

$$x = -\frac{b}{2a} = -\frac{(-0.2672)}{2(0.0075)} \approx 18$$

This means that automobile fuel efficiency was at a minimum approximately 18 years after 1940, or in 1958. The average number of miles per gallon for U.S. automobiles in that year was

$$f(18) = 0.0075(18)^2 - 0.2672(18) + 14.8$$
$$\approx 12.4$$

or a dismal 12.4 miles per gallon. ∎

sing technology

The graph at the right shows the automobile fuel efficiency model

$$y = 0.0075x^2 - 0.2672x + 14.8$$

obtained using a graphing utility. Using

Xmin = 0, Xmax = 40, Xscl = 2,
Ymin = 0, Ymax = 20, Yscl = 1

FMIN
X=17.813342603 Y=12.420138667

and the minimum point feature of the utility, we see that the minimum gas mileage of 12.4 miles per gallon occurred 17.8 years after 1940, or between 1957 and 1958.

P R O B L E M   S E T   4 . 7

Practice Problems

Solve the equations in Problems 1–56 and check the solutions.

1. $(x - 7)(x + 3) = 0$
2. $(x + 5)(x - 18) = 0$
3. $(2x + 3)(5x - 1) = 0$
4. $(7x - 4)(8x + 16) = 0$

5. $x^2 + x - 12 = 0$
6. $x^2 - 2x - 15 = 0$
7. $x^2 + 6x - 7 = 0$
8. $x^2 - 4x - 45 = 0$

9. $3x^2 + 10x - 8 = 0$
10. $2x^2 - 5x - 3 = 0$
11. $5x^2 - 8x + 3 = 0$
12. $7x^2 - 30x + 8 = 0$

13. $6x^2 - x - 35 = 0$
14. $7x^2 - 13x - 2 = 0$
15. $5x^2 + 26x + 5 = 0$
16. $3x^2 + x - 4 = 0$

17. $5x^2 - 3x - 2 = 0$
18. $6x^2 - 5x - 4 = 0$
19. $3x^2 + 5x - 2 = 0$
20. $15x^2 + 14x - 8 = 0$

21. $5x^2 - 8x - 21 = 0$
22. $7x^2 - 31x + 12 = 0$
23. $x^2 - x = 2$
24. $x^2 + 8x = -15$

25. $3x^2 - 17x = -10$
26. $4x^2 - 11x = -6$
27. $x(x - 3) = 54$
28. $x(2x - 5) = -3$

29. $x(2x + 1) = 3$
30. $x(x - 6) = 16$
31. $x^2 = \frac{5}{6}x + \frac{2}{3}$
32. $x^2 = -\frac{5}{2}x + 6$

33. $(x + 1)^2 - 5(x + 2) = 3x + 7$
34. $x + 4(x + 2) = (x + 1)^2 - 147$
35. $\frac{1}{6}x^2 + x - \frac{1}{2} = -2$

36. $\frac{1}{2}y^2 - y = 4$
37. $x + (x + 2)^2 = 130$
38. $(x + 1)^2 = 2(x + 5)$

39. $3(x^2 - 4x - 1) = 2(x + 1)$
40. $-3x - 2 = 2(x^2 - 4x - 7)$
41. $9x^2 + 6x = -1$

42. $4x^2 + 4x = -1$
43. $25 = 30x - 9x^2$
44. $4 = 28x - 49x^2$

45. $(x + 2)(x - 5) = 8$
46. $(x + 1)(2x - 3) = 3$
47. $2x - [(x + 2)(x - 3) + 8] = 0$

48. $4x - [(x + 1)(x - 2) + 6] = 0$
49. $3[(x + 2)^2 - 4x] = 15$
50. $(x + 1)^2 - 2x = 10$

51. $x^3 + 4x^2 - 25x - 100 = 0$
52. $x^3 - 2x^2 - x + 2 = 0$
53. $x^3 - x^2 = 25x - 25$

54. $x^3 + 2x^2 = 16x + 32$
55. $x^4 + x^3 = 4x^2 + 4x$
56. $x^4 - 5x^2 + 4 = 0$

Use x-intercepts, the y-intercept, the vertex, and if necessary, one or two additional points to graph the quadratic functions in Problems 57–64. If applicable, verify your hand-drawn graph using a graphing utility.

57. $f(x) = x^2 + 6x + 5$
58. $f(x) = x^2 + x - 6$
59. $f(x) = x^2 + 4x + 3$

60. $f(x) = x^2 - 2x - 8$
61. $y = -x^2 - 4x - 5$
62. $y = -x^2 - x + 6$

63. $f(x) = -x^2 - 4x - 3$
64. $f(x) = -x^2 + 2x + 8$

Application Problems

65. A person standing close to the edge on the top of an 80-foot building throws a ball upward with an initial speed of 64 feet per second. After t seconds, the height of the ball above the ground is $s(t) = -16t^2 + 64t + 80$.
  a. How long does it take for the ball to reach the ground?
  b. What is the maximum height reached by the ball? After how many seconds does this occur?
  c. Graph the position function, with meaningful values of t along the horizontal axis and values of $s(t)$ along the vertical axis.
  d. The ball passes the edge of the top of the building from which it was thrown as it falls to the ground. After how many seconds does this occur? How is this situation illustrated on your graph from part (c)?
  e. If applicable, use a graphing utility to verify all parts of this problem.

66. The function $P(I) = -5I^2 + 80I$ describes the power ($P(I)$) of an 80-volt generator subject to a current (I) of electricity given in amperes.

  a. What current is necessary to produce a power of 75 volts?
  b. What current produces maximum power? What is the maximum power?
  c. Graph the function, using values of I along the x-axis and values of $P(I)$ along the y-axis.

67. Up to a point, the more fired up we get, the better we perform. Of course, after this point is reached, one can become so aroused that a high level of anxiety impedes performance. Coaches have established a 1 to 100 scale for both arousal and performance, using a formula to relate the two variables. The performance level (P) of an athlete is related to arousal level (A) by the formula $P = -\frac{1}{50}A^2 + 2A + 22$.
  a. How should the arousal level be controlled to guarantee a performance level of 72?
  b. What arousal level will result in maximum performance?

68. The formula $D = \frac{7}{10}x^2 + \frac{3}{4}x$ describes the distance in feet (D) that it takes to stop a vehicle traveling x

miles per hour. How should the car's speed be controlled so that it can stop in exactly 295 feet?

69. The formula $W = 10 + \frac{7}{6}t - \frac{1}{60}t^2$ describes the weight in grams (W) of a hamster t days after birth, with $0 \le t \le 35$. After how many days does the hamster weigh 20 grams?

70. The crocodile, an endangered species, is subject to a protection program. The formula $P = 3500 + 475t - 10t^2$ describes the crocodile population (P) after t years of the program, where $0 \le t \le 20$. How long will the program have to be continued to bring the crocodile population up to 7250?

71. The supply and demand functions for a commodity are given by $S(p) = p^2 + 3p - 70$ and $D(p) = 410 - p$, respectively. In both models, p represents the price of the commodity in dollars. Find the equilibrium price, the price at which supply and demand are equal.

72. The revenue R from the sale of x units of a product is given by $R = 3x^2 + x$. The cost of producing x units of the product is given by $C = 2x^2 - 2x + 18$. How many units must the company produce and sell to break even?

73. A pool measuring 10 meters by 20 meters is surrounded by a path of uniform width, as shown in the figure. If the area of the pool and the path combined is 600 square meters, what is the width of the path?

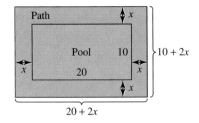

74. A garden measuring 6 yards by 4 yards is to be increased in both length and width, as shown in the figure. The length and width are to be increased by the same amount, resulting in a garden whose area is 48 square yards. By how much should the length and width be increased?

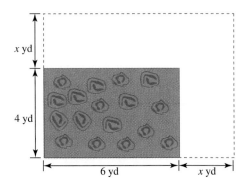

75. From each of the four corners of a square piece of cardboard, a square piece 2 inches on a side is cut out. The flaps are then turned up to form an open box, as shown. If the volume of the box is 128 cubic inches, find the dimensions of the piece of square cardboard.

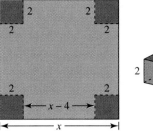
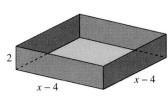

76. The rectangular floor of a closet is divided into two right triangles by drawing a diagonal, as shown in the figure. One leg of the right triangle is 2 feet more than twice the other leg. The hypotenuse is 13 feet. Determine the closet's length and width.

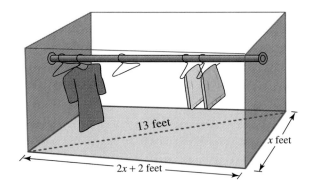

77. A ladder leaning against a wall is 10 meters from a wall. The ladder is 2 meters longer than the distance it reaches up the wall. What is this distance?

78. One leg of a right triangle is 3 centimeters shorter than the other leg. If the hypotenuse is 15 centimeters, find the lengths of the two legs.

79. The longer leg of a right triangle is 1 inch shorter than twice the shorter leg, and the hypotenuse is 1 inch longer than twice the shorter leg. Find all three sides of the triangle.

80. The table shows the number of inmates in federal and state prisons for three selected years.

x (Years after 1980)	y (Number of Inmates, in Thousands)
0	320
4	440
10	740

Use the quadratic function $y = ax^2 + bx + c$ to find the values of a, b, and c, thereby modeling the data. The actual number of inmates in 1985 was 480,000. How well does the function model reality for that year?

81. The table shows per capita consumption of cigarettes by Americans 18 and older for three selected years.

x (Years after 1960)	y (Per Capita Cigarette Consumption)
(1960) 0	4025
(1970) 10	4235
(1980) 20	3845

a. Use the quadratic function $y = ax^2 + bx + c$ to find the values of a, b, and c, thereby modeling the data.

b. In what year (to the nearest whole year) was cigarette consumption at a maximum? What was the per capita cigarette consumption for that year?

c. Graph the model that you obtained in part (a) in a rectangular coordinate system, letting $x = 0, 3, 6, 9, \ldots, 30$. Find the corresponding values of y, plot the 11 points, and connect them with a smooth curve.

d. Describe the trends in cigarette consumption based on the parabola that you graphed in part (c).

True–False Critical Thinking Problems

82. Which one of the following is true?

a. Quadratic equations solved by factoring always have two different solutions.

b. If $4x(x^2 + 49) = 0$, then

$$4x = 0 \quad \text{or} \quad x^2 + 49 = 0$$
$$x = 0 \qquad\qquad x = 7 \quad \text{or} \quad x = -7$$

c. If -4 is a solution to $7y^2 + (2k - 5)y - 20 = 0$, then k must equal 14.

d. Some quadratic equations have more than two solutions.

83. Which one of the following is true?

a. If $(x - 2)(x + 5) = 12$, then $x - 2 = 6$ or $x + 5 = 2$.

b. The solution set to $6(x - 2)(x - 3)$ is $\{6, 2, 3\}$.

c. Suppose 3 inches of matting is placed around a square picture with sides each x inches long, and the area of the matting is 60 square inches. The sides of the square picture can be found using the equation $(x + 6)^2 - x^2 = 60$.

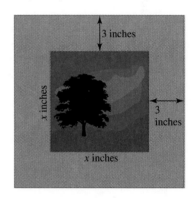

d. If $ab = 0$, then a and b are both equal to 0.

Technology Problems

84. Use your graphing utility to visualize the solutions in Problems 67–70. Begin by graphing the parabola(s) whose equation is given. Then $\boxed{\text{TRACE}}$ along the curve and find the coordinates of the point corresponding to the solution.

In Problems 85–89, use a graphing utility to solve each equation graphically.

85. $x^2 - 7x + 6 = 0$

86. $(x - 3)^2 - 9 = 0$

87. $x^3 - 9x = 0$

88. $x^3 + 2x^2 - 5x - 6 = 0$

89. $x^4 - 8x^3 + 7x^2 + 72x - 144 = 0$

90. Most graphing calculators will give the solutions to quadratic equations. Generally, this can be done using the $\boxed{\text{POLY}}$ (polynomial equation) feature. The order of a quadratic equation is 2, so you will probably need to enter order = 2. With the equation in standard form, enter the coefficients of x^2, x, and the constant term, often denoted by $a_2 =$, $a_1 =$, and $a_0 =$. After entering these three numbers, press $\boxed{\text{SOLVE}}$. The solutions should be displayed on the screen. In the case of a cubic equation, enter order = 3 and then enter the four numbers representing coefficients of the variables and the constant term. Consult your manual for details. On some calculators you will use the $\boxed{\text{SOLVE}}$ feature, input the entire equation, and then simply press $\boxed{\text{ENTER}}$. Use this feature to check the solutions for some of the equations that you solved in Problems 1–56.

Writing in Mathematics

91. Describe two ways in which a quadratic equation differs from a linear equation.

92. Describe how the zero product principle is used to solve quadratic equations.

Critical Thinking Problems

93. Solve for x: $\left| x^2 + 2x - 36 \right| = 12$.

94. Solve for x in terms of b and d:

$$x^2 - 3dx + 5bx - 15bd = 0.$$

95. Solve for x: $\begin{vmatrix} 2x & x \\ -1 & x \end{vmatrix} = 1$.

96. A path of uniform width is constructed around a 20-by 25-meter rectangular garden. If 196 square meters of brick are used to construct the path, what is its width?

97. A rectangular field that is adjacent to a river is to be fenced on three sides with the side bordered by the river left open. If the fence is to enclose an area of 1200 square hectometers and 100 hectometers of fencing are to be used, what are the dimensions of the enclosure?

98. The cross section of the house shown in the figure has the shape of a rectangle on the bottom and a triangle on the top. The length of the rectangle is 4 times its height, and the altitude of the triangular cross section is 1 yard more than the height of the rectangle. If the area of the entire cross section is 60 square yards, determine the height of the rectangle.

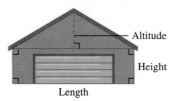

Altitude

Height

Length

99. As it swings from its highest point to its lowest point, the end of a pendulum drops 2 meters vertically while moving 6 meters horizontally (see the figure). How long is the pendulum?

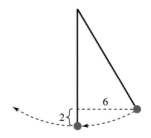

6

2

100. A large rectangular lake is 4 kilometers wide and 12 kilometers long. The lake is surrounded by a grass border whose width is uniform and whose area is 4 times the area of the lake. What is the width of the border?

101. Find all values of x that satisfy $\dfrac{x^3 - x^2 - x + 1}{x^3 - x^2 + x - 1} = 0$.

Review Problems

102. Factor completely: $4x^2 - 25y^2 - 4x + 10y$.

103. Evaluate: $\begin{vmatrix} 5 & 2 & 34 \\ -1 & 3 & 22 \\ 0 & 0 & 4 \end{vmatrix}$.

104. Solve the system:

$$
\begin{aligned}
2x - 3y + z &= 0 \\
3x + y + 2z &= -2 \\
x - 2y + z &= -2
\end{aligned}
$$

C HAPTER PROJECT

Simple Numbers and Complex Problems

The mathematician and philosopher Pythagoras was once asked "What is a friend?" He answered, "It is another such as I, such as 220 and 284." To this day, mathematicians refer to number pairs such as 220 and 284 as amicable numbers. In this chapter, we introduced weird numbers (page 364) and added to a list of common words such as *rational*, *irrational*, *real*, *imaginary*, and *complex* that are examples of words given a precise mathematical definition quite different from the meaning assigned in everyday conversation.

1. Look up the words rational, irrational, real, imaginary, and complex in a standard dictionary and give the definition of each. Come up with your own theory on why those particular words were selected to be used as mathematical terms. Using a mathematics dictionary, or a reference on mathematical history, find the reasons why those words were selected. Does your theory agree with history?

Amicable and weird numbers do share one common trait. Both are defined in terms of properties involving their proper divisors. In this case, proper divisors include one, but exclude the number itself. Two numbers are amicable if each is the sum of the proper divisors of the other. The proper divisors of 220 are 1, 2, 4, 5, 10, 11, 20, 22, 44, 55, and 110. The sum of these divisors is 284. Likewise, the sum of the divisors of 284, $1 + 2 + 4 + 71 + 142$, gives 220. However, knowing the definition of an amicable pair does not give you a way of finding a pair. The next pair of amicable numbers were not discovered until hundreds of years later by the Arabic mathematician Thabit ibn Qurra. They were 17,296 and 18,416.

2. 17,296 and 18,416 were the next pair discovered, but they are not the next pair in numerical order. The next smallest pair is 1184 and 1210. If we know the prime factorization of a number, we are able to write down all of its factors. Use the information given below to verify the following pairs of numbers are amicable.

$$1184 = 2^5 \times 37 \quad \text{and} \quad 1210 = 2 \times 5 \times 11^2$$

$$10,744 = 2^3 \times 17 \times 79 \quad \text{and} \quad 10,856 = 2^3 \times 23 \times 59$$

3. Numbers may also be classified as *perfect*, *abundant*, or *deficient*. The first few perfect numbers are 6, 28, and 496. Use a mathematics dictionary or resources on the Worldwide Web to discover the definition of a perfect number and show that the numbers listed fit the definition. Additionally, find the definitions for abundant and deficient numbers and give at least two examples of each.

We may also see curious connections between numbers by looking at a "geometry" of numbers. Numbers may be classified as triangular, square, pentagonal, hexagonal, and so on. Triangular numbers follow the pattern $1, 1 + 2, 1 + 2 + 3, 1 + 2 + 3 + 4, \ldots$ giving us the sequence of numbers $1, 3, 6, 10, \ldots$. Using coins, marbles, or any other small objects, we can create geometric views of these numbers.

4. Build triangular numbers in the following way: place one object on the first row, two on the second, three on the third, etc. You may wish to begin each row on the left edge, rather than stacking like a set of billiard balls. (See Figure 4.18.) Display square numbers in the obvious way, by using parallel rows of matching length. Show that the sum of two consecutive triangular numbers will always yield a square number, using your geometric forms.

```
    X                       X
    X X                   X   X
    X X X              X   X   X
```

Figure 4.18

5. Another way to describe our "geometric" numbers is by noticing how we go from one number to another in the list of numbers we will be summing. Begin with the natural numbers 1, 2, 3, 4, 5, 6, 7, 8, 9, 10, 11, 12, . . .

Triangular numbers: 1, 3, 6, 10, 15, . . . Add each number to the next.
1, 1 + 2, 1 + 2 + 3, 1 + 2 + 3 + 4, . . .
Square numbers: 1, 4, 9, 16, 25, . . . Add every second number.
1, 1 + 3, 1 + 3 + 5, 1 + 3 + 5 + 7, . . .
Pentagonal numbers: 1, 5, 12, 22, 22, 35, . . . Add every third number.
1, 1 + 4, 1 + 4 + 7, 1 + 4 + 7 + 10, . . .

Following this pattern, give the first six hexagonal, heptagonal, and octagonal numbers.

6. Using your markers, discover a way to display pentagonal and hexagonal numbers in a geometric form. Display at least the first four numbers in each case. Compare your results with those of others in your class. Is there more than one way to display these numbers? Compare your hexagonal numbers to the triangular numbers. What do you conclude? Show your conclusion geometrically. Combine three identical pentagonal numbers and compare this new number to the triangular numbers. What do you observe?

7. When taking the sum of odd numbers, there are two ways to perform the addition. The one listed above, 1, 1 + 3, 1 + 3 + 5, 1 + 3 + 5 + 7, . . . gives us square numbers. What number pattern do you obtain if the odd numbers are added like this:

1, 3 + 5, 7 + 9 + 11, 13 + 15 + 17 + 19, . . . ?

The study of numbers is called Number Theory. Unlike many other branches of mathematics, problems in Number Theory may be very simple to state. Anyone can read the question, but finding the answer is an entirely different matter. Mathematicians suspect the following statements may be true, and have searched for many years for a proof.

- Every even number, other than 2, may be written as the sum of two primes.
- Every perfect number is even.

From the examples seen above, we might conclude that Number Theory is more of an amusement than anything else and does not seem to have any practical application. Nothing could be further from the truth. We should not be deceived by the simplicity of the numbers. Studying these patterns, and others like them, may lead to very complex mathematics. As an example, studying prime numbers helps in creating security codes used by credit card companies, banks and the federal government. The military and intelligence communities rely on extremely large prime numbers to create virtually unbreakable codes for secure communications.

Worldwide Web Resources

Go to the Prentice Hall website (http://www.prenhall.com/blitzer) to access other locations on the Internet that will allow you to further explore the concepts presented in this project.

Chapter Review

SUMMARY

1. Basic Vocabulary and Notation
  a. A polynomial is a monomial or a finite sum of terms in which all variables contain exponents that are whole numbers. A polynomial cannot contain variables in a denominator.
  b. Monomials contain one term, binomials contain two terms, and trinomials contain three terms.
  c. The degree of a monomial is the sum of the exponents on all the variables. The degree of a polynomial is the greatest of the degrees of any of its terms.

2. Sums and Differences of Polynomials
  a. Polynomials are added by combining like terms.
  b. Polynomials are subtracted by removing parentheses from the polynomial being subtracted, changing the sign of every term of this polynomial, and combining like terms.

3. Polynomial Functions
  a. *Vocabulary*
    1. First-Degree Polynomial Function (Linear Function)

$$f(x) = ax + b$$

    2. Second-Degree Polynomial Function (Quadratic Function)

$$f(x) = ax^2 + bx + c \quad (a \neq 0)$$

    3. Third-Degree Polynomial Function (Cubic Function)

$$f(x) = ax^3 + bx^2 + cx + d \quad (a \neq 0)$$

    4. Fourth-Degree Polynomial Function

$$f(x) = ax^4 + bx^3 + cx^2 + dx + e \quad (a \neq 0)$$

    5. nth-Degree Polynomial Function

$$f(x) = a_n x^n + a_{n-1} x^{n-1} + a_{a-2} x^{n-2}$$
$$+ \cdots + a_1 x + a_0 \quad (a_n \neq 0)$$

The number a_n is the leading coefficient and the number a_0 is the constant term.

  b. *Graphs of Polynomial Functions*
    1. All polynomial functions have graphs that are smooth, continuous curves with no sharp corners and no jumps or breaks.
    2. Graphing quadratic functions:
$f(x) = ax^2 + bx + c$ or $y = ax^2 + bx + c$
      a. Find x-intercept(s) by setting y or $f(x)$ equal to 0, solving $ax^2 + bx + c = 0$.

  b. Find the y-intercept setting x equal to 0 ($y = c$).

  c. Find the vertex. The x-coordinate of the vertex is $-\dfrac{b}{2a}$.

  d. The vertex is a minimum point if $a > 0$ and a maximum point if $a < 0$.

  e. The graph of a quadratic function is called a parabola.

3. Graphing cubic functions: $f(x) = ax^3 + bx^2 + cx + d$ or $y = ax^3 + bx^2 + cx + d$
  a. If a is positive, the graph falls to the left and rises to the right.
  b. If a is negative, the graph rises to the left and falls to the right.

4. Multiplying Polynomials
  a. Use $b^n \cdot b^m = b^{n+m}$ to multiply monomials.
  b. Use the distributive property and $b^n \cdot b^m = b^{n+m}$ to find the product of a monomial and a polynomial other than a monomial.
  c. Use the FOIL method to multiply two binomials. (*F*irst terms multiplied, *O*uter terms multiplied, *In*ner terms multiplied, *L*ast terms multiplied.)

$$(A + B)(C + D) = AC + AD + BC + BD$$

  d. *Squaring a binomial:*

$$(A + B)^2 = A^2 + 2AB + B^2$$

  and

$$(A - B)^2 = A^2 - 2AB + B^2$$

  The trinomials that are produced in each instance are called perfect square trinomials.

  e. *Product of the sum and difference of two terms:*

$$(A + B)(A - B) = A^2 - B^2$$

  The product of the sum and difference of two expressions is the difference of their squares.

  f. To multiply two polynomials involving factors of at least a binomial and at least a trinomial, multiply each term of one factor by each term in the other factor. (Then combine like terms.) In this case, you may want to use a vertical format.

5. **The Algebra of Polynomial Functions** If f and g are polynomial functions, then
   a. $(f + g)(x) = f(x) + g(x)$
   b. $(f - g)(x) = f(x) - g(x)$
   c. $(fg)(x) = f(x) \cdot g(x)$

6. **Factoring over the Integers**
   a. If possible, factor out the GCF.
   b. If the polynomial is a binomial, try factoring by one of the special formulas:

   $$A^2 - B^2 = (A + B)(A - B)$$
   $$A^3 + B^3 = (A + B)(A^2 - AB + B^2)$$
   $$A^3 - B^3 = (A - B)(A^2 + AB + B^2)$$

   c. If the polynomial is a trinomial, factor (if possible) using the trial-and-error FOIL or grouping method. If it is a perfect square trinomial, use one of the following special forms:

$$A^2 + 2AB + B^2 = (A + B)^2$$
$$A^2 - 2AB + B^2 = (A - B)^2$$

d. If the polynomial contains four or more terms, try factoring by grouping.

7. **Solving Quadratic Equations by Factoring**
   a. The standard form of a quadratic equation is $ax^2 + bx + c = 0$ $(a \neq 0)$.
   b. Some quadratic equations can be solved by factoring.
   1. Write the equation in standard form.
   2. Factor $ax^2 + bx + c$.
   3. Apply the zero product principle, setting each factor equal to 0.
   4. Solve the equations in step 3.
   5. Check solutions in the original equation.

REVIEW PROBLEMS

1. Where appropriate, identify the given polynomial as a monomial, binomial, or trinomial. Give the degree of each polynomial.

   a. $4y^2 - 8y^3 + 9y$ **b.** $12x^4y^3z$
   c. $8x^4y^2 - 7xy^6$ **d.** $7x^5 + 3x^3 - 2x^2 + x - 3$

In Problems 2–3, add the polynomials.

2. $(-8x^3 + 5x^2 - 7x + 4) + (9x^3 - 11x^2 + 6x - 13)$
3. $(7x^3y - 13x^2y - 6y) + (5x^3y + 11x^2y - 8y - 17)$

In Problems 4–6, subtract the polynomials.

4. $(5x^2 - 7x + 3) - (-6x^2 + 4x - 5)$
5. $(7y^3 - 6y^2 + 5y - 11) - (-8y^3 + 4y^2 - 6y - 17)$
6. $(4x^5y^2 - 7x^3y - 4) - (-8x^5y^2 - 3x^3y + 4)$

7. Subtract $10 - 6a^4b^3 - 2ab^3$ from the sum of $6 + 3a^4b^3 + 4ab^3$ and $-5 - 2a^4b^3 - 3ab^3$.

8. Find the coordinates of the vertex for the quadratic function $f(x) = x^2 - 8x + 15$.

9. According to the Bicycle Institute of America, the number of mountain bike owners $(f(x)$, in millions) in the United States is modeled by the quadratic function $f(x) = 0.337x^2 - 2.265x + 3.962$. In the model, $x = 3$ corresponds to the year 1983, and the model is valid for $3 \leqslant x \leqslant 10$. Find and interpret $f(10)$. If applicable, use a graphing utility to graph the model and show $f(10)$ on the graph.

10. The U.S. Center for Disease Control modeled the average annual per capita consumption of cigarettes $f(x)$ by Americans 18 and older as a function of time. The function is $f(x) = -3.1x^2 + 51.4x + 4024.5$, where x represents the number of years after 1960. According to this model, in what year (to the nearest whole year) did per capita consumption reach a maximum? What was the average per capita consumption for that year?

11. Graph each cubic function in the same rectangular coordinate system. What relationship do you observe among the graphs? How do the graphs exhibit the known end behavior of cubic functions?
    a. $f(x) = x^3$ **b.** $g(x) = (x - 1)^3$
    c. $h(x) = -x^3$ **d.** $k(x) = -x^3 + 1$

12. Explain why a cubic function with a negative leading coefficient cannot model real world phenomena over long periods of time.

In Problems 13–28, multiply the polynomials.

13. $(4x^2yz^5)(-3x^4yz^2)$
14. $6x^3(\frac{1}{3}x^5 - 4x^2 - 2)$
15. $7x^3y^4(3x^7y - 5x^4y^3 - 6)$
16. $(2x + 5)(3x^2 + 7x - 4)$

17. $(3x^2 - 4x + 2)(5x^2 + 7x - 8)$

18. $(2xy + 2)(x^2y - 3y + 4)$

19. $(4x - 2)(3x - 5)$

20. $(4xy + 5z)(3xy - z)$

21. $(8x^3 - 3x)(7x^3 + 9x)$

22. $(3x + 7y)^2$

23. $(2x^2y - 3z)^2$

24. $[(3x - 5) + 8y]^2$

25. $(3x + y - 2)^2$

26. $(2x + 7y)(2x - 7y)$

27. $(7a^2b + 3b)(7a^2b - 3b)$

28. $[5y - (2x + 7)][5y + (2x + 7)]$

29. If $f(x) = 5x - 3$ and $g(x) = 8x^2 - 7x + 9$, find $(f + g)(x)$, $(f - g)(x)$, and $(fg)(x)$.

30. As shown in the figure, a rectangular piece of cardboard measuring 5 feet by 4 feet has a square cut out of each corner. The sides are then turned up to form an open box. If x represents the length of the side of the square cut from each corner of the rectangle, express the volume of the box as a polynomial function, $V(x)$, written in descending powers of x.

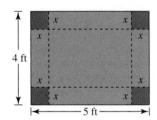

31. The larger rectangle shown in the figure at the top of the next column has dimensions of 26 meters by 18

meters. Write a polynomial in descending powers of x that represents the area of the smaller shaded rectangle.

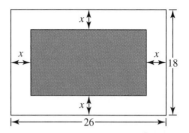

32. In the figure, the area of the shaded region is 17 square meters. Find the dimensions of the larger rectangle.

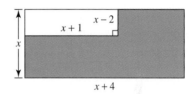

33. Given $f(x) = 4x^2 - 5x + 2$, find and simplify:
  a. $f(a + 6)$
  b. $f(a + h) - f(a)$

34. The height in feet of a particular tree is a function of the tree's age (x, in years), defined by

$$f(x) = 63 + 20\left(\frac{x - 120}{40}\right) - 0.4\left(\frac{x - 120}{40}\right)^2$$
$$- 1.2\left(\frac{x - 120}{40}\right)^3$$

where $0 \le x \le 240$. Find and interpret $f(140)$.

For Problems 35–73, factor each polynomial completely, or indicate that the polynomial is prime.

35. $15x^2 + 3x$

36. $5x^4y^2 - 20x^3y^3 + 15x^6y^2$

37. $x^3 + 5x^2 - 2x - 10$

38. $x^2y^2 - 36 - 4y^2 + 9x^2$

39. $bc - d - bd + c$

40. $x^2 + 37x + 36$

41. $x^3 - 15x^2 + 26x$

42. $-2x^3 + 36x^2 - 64x$

43. $8y^4 - 14y^2 - 15$

44. $6x^2 - 11x - 35$

45. $-2a^4 + 24a^3 - 54a^2$

46. $6y^6 + 13y^3 - 5$

47. $3(x + 2)^2 + 14(x + 2) + 8$

48. $4x^2 - 16$

49. $81x^4 - 100y^4$

50. $(x + 4)^2 - (3x - 1)^2$

51. $x^2(b^2 - 9) - 25(b^2 - 9)$

52. $x^4 - 16$

53. $4a^3 + 32$

54. $1 - 64y^3$

55. $4x^2 + 12x + 9$

56. $x^4 + 49$

57. $y^5 - y$

58. $9x^2 - 30x + 25$

59. $x^2 + 6x + 9 - 4a^2$

60. $9x^2 - 21xy + 10y^2$

61. $a^2 - b^2 + 4b - 4$

62. $9x^2 + 30xy + 25y^2$

63. $2a^3 + 12a^2 + 18a$

64. $-x^2 + 4x + 21$

65. $5x^4 - 40x$

66. $2x^3 - x^2 - 18x + 9$

67. $(x^2 - 1)^2 - 11(x^2 - 1) + 24$

68. $x^4 - 6x^2 + 9$

69. $x^3 + y + y^3 + x$

70. $27b^3 - 125c^3$

71. $10x^3y + 22x^2y - 24xy$

72. $x^4 - x^3 - x + 1$

73. $(x + y)^3 - (2x + y)^3$

Solve the equations in Problems 74–79.

74. $x^2 + 6x + 5 = 0$

75. $x(x - 12) = -20$

76. $(y - 2)(2y + 1) = -3$

77. $3x^2 = 12x$

78. $x^2 + 2 = 6x - 7$

79. $x^3 + 5x^2 = 9x + 45$

Use x-intercepts, the y-intercept, the vertex, and if necessary one or two additional points to graph the functions in Problems 80–81.

80. $y = x^2 + 5x + 4$

81. $f(x) = -2x^2 + 12x - 10$

82. A ball is thrown directly upward from the top of a 640-foot building with an initial velocity of 48 feet per second. The height of the ball above the ground after t seconds is given by the position function $s(t) = -16t^2 - 48t + 640$.

   a. How long does it take for the ball to reach the ground?

   b. What is the maximum height reached by the ball? After how many seconds does this occur?

   c. Graph the position function, with meaningful values of t along the horizontal axis and values of $s(t)$ along the vertical axis.

83. The formula $P = 100 + 25x - 5x^2$ describes systolic blood pressure (P, measured in millimeters of mercury) 1 hour after a patient is administered x milligrams of a drug. At the very most, how many milligrams of the drug were given if a patient has a blood pressure of 120 millimeters of mercury 1 hour later?

84. The larger leg of a right triangle is 7 meters longer than the shorter leg. If the hypotenuse measures 13 meters, find the length of the larger leg.

85. The length of a rectangular room exceeds the width by 10 feet. If the room's area is 1200 square feet, what are its dimensions?

86. A painting measuring 10 centimeters by 16 centimeters is surrounded by a frame of uniform width. If the combined area of the painting and frame is 280 square centimeters, determine the width of the frame.

87. A painting measuring 4 centimeters by 7 centimeters is surrounded by a frame of uniform width. If the area of the frame alone is 26 square centimeters, determine the width of the frame.

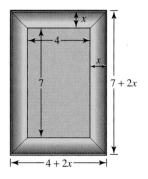

88. A rectangular piece of tin is twice as long as it is wide. Squares 2 inches on a side are cut out of each corner, and the ends are turned up to make a box whose volume is 480 cubic inches. Find the dimensions of the piece of tin.

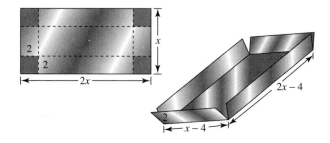

89. In a study relating sleep and death rate, the following data were obtained.

x (Average Number of Hours of Sleep)	y (Death Rate per Year Per 100,000 Males)
4	1682
7	626
9	967

Use the function $y = ax^2 + bx + c$ to model the data. Then find the death rate of males who sleep 5 hours, 6 hours, and 11 hours.

CHAPTER 4 TEST

Perform the indicated operations.

1. $(4x^3y - 19x^2y - 7y) + (3x^3y + x^2y + 6y - 9)$
2. $(6x^2 - 7x - 9) - (-5x^2 + 6x - 3)$
3. $(-7x^3y)(-12x^4y^2)$
4. $(7x - 9y)(3x + y)$
5. $(5ab + 4c)(3ab - c)$
6. $(x - y)(x^2 - 3xy - y^2)$
7. $(5x + 9y)^2$
8. $[(2x - 7) - 3y]^2$
9. $(7x - 3y)(7x + 3y)$
10. Write a polynomial in descending powers of x that represents the volume of the box in the figure shown.

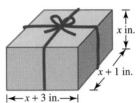
x in.
$x + 1$ in.
$x + 3$ in.

11. Write a polynomial in descending powers of x that represents the area of the shaded right triangle in the figure shown.

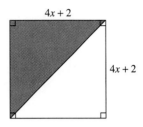
$4x + 2$
$4x + 2$

Use the model below to answer Problems 12 and 13.

The function

$$f(x) = -1.14x^3 - 4.82x^2 + 16.28x - 87.49$$

describes the U.S. trade balance, $f(x)$, in billions of dollars, x years after 1990. Negative values of $f(x)$ indicate trade deficits.

12. What does this cubic model indicate is happening to the U.S. trade balance over time?

13. The actual U.S. trade balance in 1995 was $-\$159.6$ billion. Find $f(5)$. How well does the model estimate the actual value for 1995?

Factor completely or indicate that the polynomial is prime.

14. $14x^3 - 15x^2$
15. $81y^2 - 25$
16. $x^3 + 3x^2 - 25x - 75$
17. $25x^2 + 9 - 30x$
18. $x^2 + 10x + 25 - 9y^2$
19. $x^4 + 1$
20. $y^2 - 16y - 36$
21. $14x^2 + 41x + 15$
22. $5p^3 - 5$
23. $12x^2 - 3y^2$
24. $12x^2 - 34x + 10$
25. $3x^4 - 3$
26. $27a^3b^6 - 8b^6$

Solve.

27. $3x^2 - 5x - 2 = 0$
28. $x(x + 6) = -5$
29. $x^3 - x = 0$
30. Graph: $f(x) = x^2 - 6x + 8$.
31. A diver jumps from a diving board that is 32 feet above the water. The model

$$s(t) = -16t^2 + 16t + 32$$

gives the diver's height above the water $s(t)$, measured in feet, after t seconds. When does the diver hit the water?

32. A dynamite blast blows a rock vertically upward, as shown in the figure at the top of page 408. The figure, however, does not indicate how high up the rock goes. The model

$$s(t) = -16t^2 + 160t$$

gives the rock's height above the ground $s(t)$, measured in feet, after t seconds. When does the rock hit its maximum height? How high does it go?

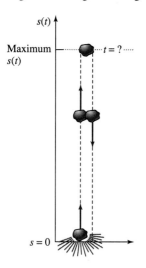

33. The length of a rectangle is 2 meters greater than 3 times the width. If the area of the rectangle is 56 square meters, find its length and width.

CUMULATIVE REVIEW PROBLEMS (CHAPTERS 1–4)

1. What property is shown by
$$(x + 1) + y = x + (1 + y)?$$

2. Use scientific notation to compute the following:
$$\frac{242{,}000}{0.0605}.$$

3. Simplify: $\dfrac{6 - 2^2 - (-2)^3}{8 - 3(2) - (-3)}$.

4. Solve: $|2x - 8| = |6x - 7|$.

5. Solve: $8(y + 2) - 3(2 - y) = 4(2y + 6) - 2$.

6. Solve, expressing the solution set in both set-builder and interval notation: $2x + 4 < 10$ and $3x - 1 > 5$.

7. Solve for x: $\quad x = \dfrac{ax + b}{c}$.

8. The figure shows a square with four identical equilateral triangles attached. If the distance around the figure is 24 inches, what is the value of k?

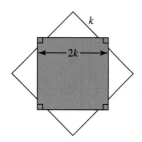

9. In 1996, there were approximately 700,000 Americans infected with HIV. About 41,000 new infections occur annually. If the growth continues in this linear fashion, when will 1,479,000 Americans be infected? (These figures are from the federal Centers for Disease Control and Prevention. In the late 1980s, models used by the CDC put the total number of infected individuals as high as 1.2 million in 1996. List one factor that might contribute to a greater rate of increase than that predicted by the model in this problem. List one factor that might contribute to a much lower rate of increase.)

10. Write the slope-intercept equation of the line passing through $(-2, -3)$, and $(2, 5)$.

11. Use the graph of function f in the figure to find
$$f(5) - f(2) - |f(-3) - f(-2)|.$$

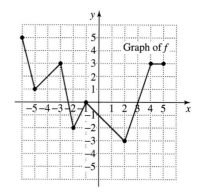

12. Solve the system:

$$x = 2(y - 5)$$
$$4x + 40 = y - 7$$

13. Solve the system:

$$6x + 4y + 4z = 2$$
$$7x + 5y + z = 14$$
$$5x + 4y + 3z = 4$$

14. Solve using Cramer's rule:

$$4x - 5y = 2$$
$$6x + 2y + 1 = 0$$

15. Solve using row-equivalent matrices:

$$6x - y - 3z = 2$$
$$-3x + y - 3z = 1$$
$$-2x + 3y + z = -6$$

16. To publish a new textbook, a company has fixed costs of $400,000 and $10 per book. Each book will be sold for $30. How many books must be produced and sold for the company to break even? Illustrate your solution by graphing the company's cost and revenue functions in the same rectangular coordinate system.

17. Suppose 2 cans of paint and 3 paintbrushes can be purchased for $53. Furthermore, 3 cans of the same paint and 1 paintbrush cost $55. Find the cost of each can of paint and each paintbrush.

18. In this figure, all vertical lines are parallel, all angles are right angles, and all horizontal lines are equally distant. What fractional part of the figure is shaded?

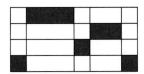

19. In a campuswide election for student government president, 2800 votes were cast for the two candidates. If the winner had 160 more votes than the loser, how many votes were cast for each candidate?

20. The preindustrial carbon dioxide (CO_2) level of concentration in the atmosphere was 280 parts per million. It is estimated that double this amount will cause an average global temperature increase of 5.4°F. (An increase as small as 1.8° in global temperature can cause a 1-foot rise in ocean levels.) Use the model

$C = 1.44t + 318.1$, which describes CO_2 concentration (C, in parts per million) as a function of time t years after 1965, to predict in what year CO_2 levels will be double the preindustrial level of concentration. List one social, political, or physical change that might affect the accuracy of the given model.

21. Multiply: $(6x + 1)(2x^2 + 2x - 7)$.

22. Find a polynomial in descending powers of x that models the area of the shaded region in the figure.

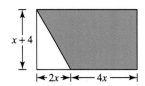

23. Solve, expressing the solution set in both set-builder and interval notation: $|2x - 5| \geq 9$.

24. Simplify: $\dfrac{-8x^3y^6}{16x^9y^{-4}}$.

25. Graph the compound inequality: $2x - y < -4$ and $x < -2$.

26. Factor completely: $x^4 - x^2 - 12$.

27. Factor completely: $x^3 - 3x^2 - 9x + 27$.

28. The area of the square shown in the figure is $(a + b + c)^2$. Write another polynomial that models the area by finding the sum of the areas of the nine smaller rectangles. Simplify by combining like terms. By setting this expression equal to $(a + b + c)^2$, you have developed a formula for squaring a trinomial.

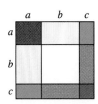

29. The price of a computer is reduced by 35% to $2080. What was the original price?

30. Consider the set $\{\frac{1}{4}, -5, 6, 0, \sqrt{9}, \sqrt{10}, -3.72, 1\}$. List the elements of the set that are:

   a. natural numbers   **d.** rational numbers
   b. whole numbers   **e.** irrational numbers
   c. integers   **f.** real numbers

Rational Expressions, Functions, and Equations

5

Jeff Hester and Paul Scowen (Arizona State University), and NASA

At first glance the image on this page looks like columns of smoke rising from a fire into a starry sky. Those are, indeed, stars in the background, but you are not looking at ordinary smoke columns. These stand at almost 6 trillion miles high, and are 7000 light-years from Earth— more than 400 million times as far away as the Sun.

The photograph, released by NASA in 1995, is one of a series of stunning images captured from the ends of the universe by the Hubble Space Telescope. The image shows infant star systems the size of our solar system emerging from the gas and dust that shrouded their creation. Astronomers call these formations "evaporating gaseous globules," or EGGs. From them, it appears, much of the universe may have hatched.

Without the Hubble Space Telescope, this discovery would not have been possible. With still at least a decade of useful life, the Hubble is expected to answer many of the most profound mysteries of the cosmos: How big and how old is the universe? What is it made of? How did the galaxies come to exist? Do other Earth-like planets orbit other sunlike stars?

In this chapter, we will use algebraic fractions, or *rational expressions,* to gain further insight into the Hubble's capabilities. Rational expressions model phenomena as diverse as the distance the Hubble can see, the cost of removing environmental pollutants, the average cost for a business to manufacture each unit of a commodity, and human memory. Because one aim of algebra is a compact, symbolic description of reality, the time has come to move beyond the rational numbers of ordinary arithmetic into the realm of algebraic fractions.

Solutions   Tutorial   Video
Manual   6

Rational Expressions and Functions: Multiplying and Dividing

Objectives

1 Define a rational expression.
2 Define a rational function.
3 Find the domain of a rational function.
4 Simplify rational expressions.
5 Multiply rational expressions.
6 Divide rational expressions.
7 Divide polynomial functions.
8 Model with rational functions.

The ratio of two integers is called a rational number; the ratio of two polynomials is called a rational expression. Rational expressions model phenomena as diverse as the cost of manufacturing products, the cost of removing pollutants from our environment, time in uniform motion situations, and human memory. In our quest to model reality in a compact, symbolic manner, we turn now to rational expressions and their operations. In this section we will focus on multiplication and division.

1 Define a rational expression.

Rational Expressions

Rational numbers are an important part of the study of arithmetic. As you know, a rational number is defined as any number that can be expressed as the quotient of two integers. In a similar way, a *rational expression* is any expression that can be written as the quotient of two polynomials. The following are examples of rational expressions.

$$\frac{1}{x} \qquad \frac{x}{x^2 - x - 2} \qquad \frac{x^3 - 9x}{x^2 + 2x - 15} \qquad \frac{3x^2 + 12xy - 15y^2}{6x^3 - 6xy^2}$$

2 Define a rational function.

Rational Functions

A function given by a formula that is a rational expression is called a *rational function*.

> **Definition of a rational function**
>
> A *rational function* is one that can be written in the form
>
> $$f(x) = \frac{p(x)}{q(x)}$$
>
> where $p(x)$ and $q(x)$ are polynomial functions, and $q(x) \neq 0$.

The definition of a rational function excludes $q(x) = 0$ to avoid dividing by zero. Let's see what this means for a specific rational function and its graph.

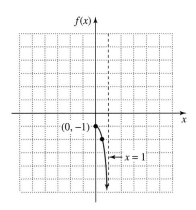

Figure 5.1

As x approaches 1 from the left, values of $f(x)$ are large negative numbers.

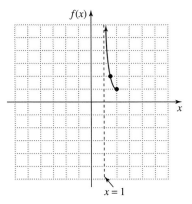

Figure 5.2

As x approaches 1 from the right, values of $f(x)$ are large positive numbers.

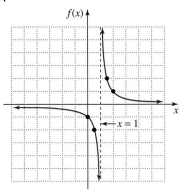

Figure 5.3

The graph of $f(x) = \dfrac{1}{x-1}$ and the vertical asymptote $x = 1$.

EXAMPLE 1 **A Rational Function and Its Graph**

A specific example of a rational function is

$$f(x) = \frac{1}{x-1}.$$

Rational functions are not defined for those values of x for which the denominator is 0. By inspection we see that this occurs when $x = 1$, so we must exclude the value $x = 1$. Construct a table of coordinates for values of x that get close to 1 and graph f near this excluded x-value.

Solution

Let's begin with values of x that approach 1 from the left, such as 0, 0.5, 0.9, 0.99, and 0.999.

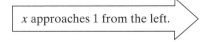

x approaches 1 from the left.

x	0	0.5	0.9	0.99	0.999
$f(x) = \dfrac{1}{x-1}$	-1	-2	-10	-100	-1000
$(x, f(x))$	$(0, -1)$	$(0.5, -2)$	$(0.9, -10)$	$(0.99, -100)$	$(0.999, -1000)$

Take a moment to verify the values of $f(x)$ (or y) in the second row of the table. It appears that as x gets closer and closer to 1 from the left, the values of $f(x)$ (or y) are getting to be very large negative numbers, decreasing without bound. Notice how this is shown in the partial graph of $f(x) = \dfrac{1}{x-1}$ in Figure 5.1.

Now let's see what happens to $f(x)$ as x approaches 1 from the right.

x approaches 1 from the right.

x	1.001	1.01	1.1	1.5	2
$f(x) = \dfrac{1}{x-1}$	1000	100	10	2	1
$(x, f(x))$	$(1.001, 1000)$	$(1.01, 100)$	$(1.1, 10)$	$(1.5, 2)$	$(2, 1)$

This time it appears that as x gets closer and closer to 1 from the right, the values of $f(x)$ (or y) are getting to be very large positive numbers, increasing without bound. Notice how this is shown in the partial graph of $f(x) = \dfrac{1}{x-1}$ in Figure 5.2.

Figure 5.3 shows the graph of f. We say that $x = 1$ is a *vertical asymptote* for the graph of $f(x)$ because the graph of f gets closer and closer to the vertical line whose equation is $x = 1$ even though it does not touch it. ∎

The vertical asymptote, the line $x = 1$ shown in Figure 5.3, is not part of the graph, but is certainly helpful in obtaining the graph of the rational function.

Definition of asymptote

A line that a graph approaches but never touches is called an *asymptote*.

Using technology

The graph of $y = \dfrac{1}{x - 1}$, shown on the left, was obtained using a graphing utility and the same range setting as in Figure 5.3. It appears that the utility is also drawing the asymptote whose equation is $x = 1$, but this is not the case. The utility is in a mode called the *connected mode,* connecting all points it plots. Just to the left of 1, the value of y is a very large negative number, and just to the right of 1 the value of y is a very large positive number. The vertical line is the utility's attempt to connect the point with this very large negative y-value with the point with this very large positive y-value.

Shown on the right is the same graph using the utility's *DOT mode.* In this mode the utility displays unconnected points that have been calculated. Note that the vertical line no longer appears in the DOT mode. By zooming in repeatedly or adjusting the range setting, you can see that the x-value 1 is not in the domain of $y = \dfrac{1}{x - 1}$.

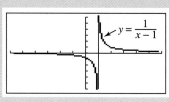

Connected mode   DOT mode

Our work in Example 1 demonstrates that unlike polynomial functions, many rational functions have graphs with breaks in their curves. These breaks occur at those values of x for which the denominator in the function's equation is 0.

Recall that the set of all first components of the ordered pairs of a function is called its *domain.* For rational functions, these first components should not include values for which the denominator in the function's equation is 0.

3 Find the domain of a rational function.

Definition of the domain of a rational function

The domain of a rational function is the set of all real numbers except those for which the denominator is zero.

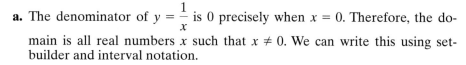

EXAMPLE 2 **Finding Domains for Rational Functions**

Find the domain:

a. $y = \dfrac{1}{x}$ **b.** $f(x) = \dfrac{x}{x^2 - x - 2}$ **c.** $g(x) = \dfrac{2}{x^2 + 1}$

Solution

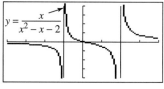

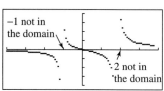

Figure 5.4

a. The denominator of $y = \dfrac{1}{x}$ is 0 precisely when $x = 0$. Therefore, the domain is all real numbers x such that $x \neq 0$. We can write this using set-builder and interval notation.

Domain of $y = \dfrac{1}{x}$ is $\{x \mid x \neq 0\}$ or $(-\infty, 0) \cup (0, \infty)$.

The graph of $y = \dfrac{1}{x}$, obtained with a graphing utility in both the connected and DOT modes, is shown in Figure 5.4. The figure confirms that 0 is excluded from the function's domain and that there is a vertical asymptote at $x = 0$.

b. To avoid division by 0 in $f(x) = \dfrac{x}{x^2 - x - 2}$, we must determine the values of x that cause $x^2 - x - 2$ to be 0. Since this is difficult to do by inspection, we set the denominator equal to 0 and solve.

$$x^2 - x - 2 = 0 \quad \text{Set the denominator equal to 0.}$$
$$(x + 1)(x - 2) = 0 \quad \text{Factor.}$$
$$x + 1 = 0 \quad \text{or} \quad x - 2 = 0 \quad \text{Set each factor equal to 0.}$$
$$x = -1 \qquad \quad x = 2 \quad \text{Solve for } x.$$

Both -1 and 2 are excluded from the domain of f. Thus, in set-builder notation,

Domain of $f = \{x \mid x \neq -1 \quad \text{and} \quad x \neq 2\}$.

Using interval notation,

Domain of $f = (-\infty, -1) \cup (-1, 2) \cup (2, \infty)$.

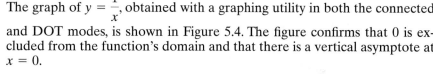

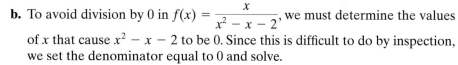

Figure 5.5

The graph of $y = \dfrac{x}{x^2 - x - 2}$, obtained with a graphing utility in both the connected and DOT modes, is shown in Figure 5.5. The DOT mode graph shows that -1 and 2 are excluded from the domain of f, and the connected mode leads us to suspect that there are vertical asymptotes at $x = -1$ and $x = 2$.

c. No real number substituted in $g(x) = \dfrac{2}{x^2 + 1}$ will cause $x^2 + 1$ to be 0, so no values of x need to be excluded from the domain of g. Thus,

Domain of $g = \{x \mid x \in R\}$ or $(-\infty, \infty)$.

The graph of $y = \dfrac{2}{x^2 + 1}$, obtained with a graphing utility in the connected

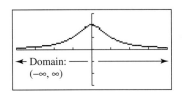

Figure 5.6
The graph of $y = \dfrac{2}{x^2 + 1}$

mode, is shown in Figure 5.6. There are no breaks in the curve and consequently no excluded values of x. The domain of this function is the set of all real numbers. ■

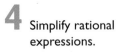

4 Simplify rational expressions.

M. C. Escher, (1898–1972) "Sky and Water I" © 1997 Cordon Art-Baarn-Holland. All rights reserved.

D **iscover for yourself**

Example 2c shows that some rational functions have graphs that are continuous curves with no breaks or sharp corners. Since this is true of all polynomial functions, how does the graph in Figure 5.6 differ from a parabola that opens down?

Simplifying Rational Expressions and Functions

A rational expression is *simplified* or *reduced to its lowest terms* if its numerator and denominator have no common factors other than 1 or -1. The following procedure can be used to simplify rational expressions.

Simplifying rational expressions

Reducing Rational Expressions to Lowest Terms

1. Factor the numerator and denominator completely.
2. Divide both the numerator and denominator by the common factors.

EXAMPLE 3 **Simplifying a Rational Function**

Simplify and graph: $f(x) = \dfrac{x^2 + 4x + 3}{x + 1}$

Solution

$$f(x) = \frac{x^2 + 4x + 3}{x + 1}$$ 　　This is the given function.

$$= \frac{(x + 1)(x + 3)}{x + 1}$$ 　　Factor. Observe that $x \neq -1$.

$$= \frac{x + 1}{x + 1} \cdot \frac{x + 3}{1}$$ 　　This step is usually done mentally, but notice that $\dfrac{x + 1}{x + 1} = 1$ for $x \neq -1$.

$$= x + 3$$ 　　We usually write $\dfrac{(x + 1)(x + 3)}{x + 1} = x + 3$.

Dividing both the numerator and denominator of a rational expression by the common factors can change the domain of the function defined by the expression. For example, the domain of the given function f is all real numbers x such that $x \neq -1$. To make the domains of the given function and its simplified form agree, we should write

$$f(x) = x + 3, \quad x \neq -1.$$

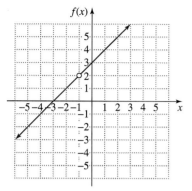

Figure 5.7

The graph of

$$f(x) = \frac{x^2 + 4x + 3}{x + 1}, \text{ with } -1$$

excluded from the domain

Thus, the graph of f is a line with y-intercept $= 3$ and slope $= 1$, shown in Figure 5.7. The open dot, or hole, at $x = -1$ shows that -1 is excluded from the function's domain. ▪

Using technology

Use a graphing utility to graph $y = \dfrac{x^2 + 4x + 3}{x + 1}$ in a $[-3, 3]$ by $[-1, 4]$ viewing

rectangle. Can you tell that -1 is excluded from the function's domain? Now TRACE along the curve until you get to $x = -1$. What do you observe?

EXAMPLE 4 **Simplifying Rational Expressions**

Simplify:

a. $\dfrac{3x^3 - 15x}{6x^2}$ **b.** $\dfrac{x^2 - 7x - 18}{x^2 + 3x + 2}$ **c.** $\dfrac{x^3 - 9x}{x^2 + 2x - 15}$

Solution

a. $\dfrac{3x^3 - 15x}{6x^2} = \dfrac{3x(x^2 - 5)}{3x(2x)}$ Factor the numerator and denominator. Observe that $x \neq 0$.

$\qquad = \dfrac{3\cancel{x}(x^2 - 5)}{3\cancel{x}(2x)}$ Divide by the common factors.

$\qquad = \dfrac{x^2 - 5}{2x}$ Simplified form

Notice that 0 must be excluded from both the given rational expression and its simplified form.

b. $\dfrac{x^2 - 7x - 18}{x^2 + 3x + 2} = \dfrac{(x - 9)(x + 2)}{(x + 1)(x + 2)}$ Factor the numerator and denominator. Observe that $x \neq -1$ and $x \neq -2$.

$\qquad = \dfrac{(x - 9)\cancel{(x + 2)}}{(x + 1)\cancel{(x + 2)}}$ Divide by the common factors.

$\qquad = \dfrac{x - 9}{x + 1}, \quad x \neq -2$ Simplified form

Notice that -1 and -2 must be excluded from the given rational expression. It's obvious that -1 must also be excluded from its simplified form since -1 would result in a denominator of 0. Since it is not obvious by looking at the simplified form that x cannot equal -2, we've written this exclusion next to the simplified form. In this way we have not changed the permissible values of x for the given rational expression and its simplification.

c. $\dfrac{x^3 - 9x}{x^2 + 2x - 15} = \dfrac{x(x^2 - 9)}{(x + 5)(x - 3)}$ Partially factor.

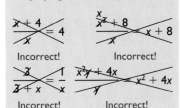
$$= \frac{x(x + 3)(x - 3)}{(x + 5)(x - 3)} \qquad \text{Factor the numerator completely. Observe that } x \neq -5 \text{ and } x \neq 3.$$

$$= \frac{x(x + 3)(x - 3)}{(x + 5)(x - 3)} \qquad \text{Divide by the common factor.}$$

$$= \frac{x(x + 3)}{x + 5}, \quad x \neq 3 \qquad \text{Simplified form.}$$

We will soon see that it is best to leave a simplified rational expression in factored form.

The process of simplifying a rational expression comes down to canceling identical factors in the numerator and denominator. In some situations, factors in the numerator and denominator have exactly opposite signs and are additive inverses, such as $x - 4$ and $4 - x$. One approach to reducing the expression $\dfrac{x - 4}{4 - x}$ involves factoring -1 from the numerator.

$$\frac{x - 4}{4 - x} = \frac{-(-x + 4)}{4 - x} \qquad \text{Factor } -1 \text{ from the numerator.}$$

$$= -\frac{(4 - x)}{(4 - x)} \qquad \text{Divide by the common factor.}$$

$$= -1 \qquad \text{Simplified form}$$

Generalizing from this situation suggests the following property.

Rational expressions simplifying to -1

The quotient of two polynomials that have exactly opposite signs and are additive inverses is -1.

Let's use this property in Example 5.

EXAMPLE 5 **Simplifying Rational Expressions That Have Polynomials with Opposite Signs**

Simplify: $\dfrac{x^3 - 125}{5x - x^2}$

Solution

$$\frac{x^3 - 125}{5x - x^2} = \frac{(x - 5)(x^2 + 5x + 25)}{x(5 - x)} \qquad \text{Factor the numerator and denominator. The numerator is the difference of two cubes, } x^3 - 5^3. \ x \neq 0 \text{ and } x \neq 5.$$

$$= \frac{\overset{(-1)}{(x - 5)}(x^2 + 5x + 25)}{x(5 - x)} \qquad \text{Two polynomials that have opposite signs and are additive inverses have a quotient of } -1.$$

$$= \frac{-(x^2 + 5x + 25)}{x}$$

$$= \frac{-x^2 - 5x - 25}{x}, \quad x \neq 5 \qquad \text{Simplified form}$$

In the next example, the procedure for simplifying a rational expression is applied to rational expressions that involve more than one variable.

EXAMPLE 6 Simplifying Rational Expressions That Have Two Variables

Simplify:

a. $\dfrac{4xy - y^2}{5y}$ **b.** $\dfrac{3x^2 + 12xy - 15y^2}{6x^3 - 6xy^2}$

Solution

a. $\dfrac{4xy - y^2}{5y} = \dfrac{y(4x - y)}{5y}$ Factor the numerator.

$\phantom{\dfrac{4xy - y^2}{5y}} = \dfrac{\cancel{y}(4x - y)}{5\cancel{y}}$ Divide by (or cancel) the common factor y. $y \neq 0$.

$\phantom{\dfrac{4xy - y^2}{5y}} = \dfrac{4x - y}{5}, \quad y \neq 0$ Simplified form

b. $\dfrac{3x^2 + 12xy - 15y^2}{6x^3 - 6xy^2} = \dfrac{3(x^2 + 4xy - 5y^2)}{6x(x^2 - y^2)}$ Partially factor.

$\phantom{\dfrac{3x^2 + 12xy - 15y^2}{6x^3 - 6xy^2}} = \dfrac{3(x + 5y)(x - y)}{3 \cdot 2x(x + y)(x - y)}$ Factor completely.

$\phantom{\dfrac{3x^2 + 12xy - 15y^2}{6x^3 - 6xy^2}} = \dfrac{\cancel{3}(x + 5y)\cancel{(x - y)}}{\cancel{3} \cdot 2x(x + y)\cancel{(x - y)}}$ Divide by the common factors.

$\phantom{\dfrac{3x^2 + 12xy - 15y^2}{6x^3 - 6xy^2}} = \dfrac{x + 5y}{2x(x + y)}, \quad x \neq y$ Simplified form   ■

5 Multiply rational expressions.

Multiplying Rational Expressions

In arithmetic we know that the product of two rational numbers equals the product of their numerators divided by the product of their denominators. This is shown symbolically in the following equation.

$$\frac{a}{b} \cdot \frac{c}{d} = \frac{ac}{bd}, \quad b \neq 0 \quad \text{and} \quad d \neq 0$$

In a similar manner, the product of two rational expressions is the product of their numerators over the product of their denominators.

> **Procedure for multiplying rational expressions**
>
> *Multiplying Rational Expressions*
>
> 1. Factor all numerators and denominators completely.
> 2. Write the product of the numerators over the product of the denominators.
> 3. Divide both numerator and denominator by common factors.
> 4. Multiply remaining factors in the numerator and multiply remaining factors in the denominator.

M. C. Escher, (1898–1972) "Smaller and Smaller" © 1997 Cordon Art-Baarn-Holland. All rights reserved.

EXAMPLE 7 **Multiplying Rational Expressions**

Multiply: $(4x^2 - 25) \cdot \dfrac{14}{2x + 5}$

Solution

$$(4x^2 - 25) \cdot \frac{14}{2x + 5} = \frac{4x^2 - 25}{1} \cdot \frac{14}{2x + 5}$$

Write $4x^2 - 25$ with a denominator of 1.

$$= \frac{(2x + 5)(2x - 5)}{1} \cdot \frac{14}{2x + 5}$$

Factor.

$$= \frac{(2x + 5)(2x - 5)14}{1(2x + 5)}$$

Multiply numerators and denominators.

$$= \frac{(2x + 5)(2x - 5)14}{1(2x + 5)}$$

Simplify. $x \neq -\dfrac{5}{2}$

$$= 14(2x - 5)$$

or $\quad 28x - 70, \quad x \neq -\dfrac{5}{2}$

Multiply the remaining factors in the numerator and denominator. ∎

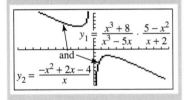

EXAMPLE 8 **Multiplying Rational Expressions**

Multiply: $\dfrac{a^3 + 8}{a^3 - 5a} \cdot \dfrac{5 - a^2}{a + 2}$

Solution

$$\frac{a^3 + 8}{a^3 - 5a} \cdot \frac{5 - a^2}{a + 2} = \frac{(a + 2)(a^2 - 2a + 4)}{a(a^2 - 5)} \cdot \frac{5 - a^2}{a + 2}$$

Factor.

$$= \frac{(a + 2)(a^2 - 2a + 4)(5 - a^2)}{a(a^2 - 5)(a + 2)}$$

Multiply numerators and denominators.

$$= \frac{(a + 2)(a^2 - 2a + 4)(5 - a^2)}{a(a^2 - 5)(a + 2)}$$

Simplify. Recall that polynomials with opposite signs throughout have a quotient of -1. $a \neq -\sqrt{5}, a \neq \sqrt{5}, a \neq -2$

$$= \frac{-1(a^2 - 2a + 4)}{a}$$

Multiply the remaining factors.

$$= \frac{-a^2 + 2a - 4}{a}, a \neq -\sqrt{5}, a \neq \sqrt{5}, a \neq -2$$ ∎

---

6 Divide rational expressions.

Dividing Rational Expressions

In arithmetic we know that the quotient of two rational numbers is found by multiplying the first number by the reciprocal of the second. This is shown symbolically in the following equation.

$$\frac{a}{b} \div \frac{c}{d} = \frac{a}{b} \cdot \frac{d}{c}, \quad b \neq 0, d \neq 0, c \neq 0$$

In a similar manner, the quotient of two rational expressions is the product of the first and the reciprocal of the second. As suggested by the arithmetic procedure, the *reciprocal* of a rational expression is found by interchanging numerator and denominator. For example,

The reciprocal of $\dfrac{x}{x^2 + 5}$ is $\dfrac{x^2 + 5}{x}$.

The reciprocal of $x + 7$ is $\dfrac{1}{x + 7}$.

In each case, the product of the rational expression and its reciprocal is 1.

EXAMPLE 9 **Dividing Rational Expressions**

Divide:

a. $\dfrac{x^2 + 3x - 10}{2x} \div \dfrac{x^2 - 5x + 6}{x^2 - 3x}$ **b.** $\dfrac{x^2 - y^2}{8a^3} \div \dfrac{(x - y)^2}{16a^4}$

Solution

a. $\dfrac{x^2 + 3x - 10}{2x} \div \dfrac{x^2 - 5x + 6}{x^2 - 3x}$

$= \dfrac{x^2 + 3x - 10}{2x} \cdot \dfrac{x^2 - 3x}{x^2 - 5x + 6}$ Multiply by the reciprocal of the divisor.

$= \dfrac{(x + 5)(x - 2)x(x - 3)}{2x(x - 3)(x - 2)}$ Factor and multiply.

$= \dfrac{(x + 5)(x - 2)x(x - 3)}{2x(x - 3)(x - 2)}$ Simplify.

$= \dfrac{x + 5}{2}, \quad x \neq 0, x \neq 3, x \neq 2$ Multiply the remaining factors.

b. $\dfrac{x^2 - y^2}{8a^3} \div \dfrac{(x - y)^2}{16a^4} = \dfrac{x^2 - y^2}{8a^3} \cdot \dfrac{16a^4}{(x - y)^2}$ Multiply by the reciprocal of the divisor.

$= \dfrac{(x + y)(x - y)\,16a^4}{8a^3(x - y)(x - y)}$ Factor and multiply.

$= \dfrac{(x + y)(x - y)\overset{2\ a}{16a^4}}{\underset{1 \cdot 1}{8a^3}\,(x - y)(x - y)}$ Simplify.

$= \dfrac{2a(x + y)}{x - y}, \quad a \neq 0$ Multiply the remaining factors. ∎

7 Divide polynomial functions.

Extending the Algebra of Functions

In Chapter 4, we discussed adding, subtracting, and multiplying polynomial functions. Now that we can work with rational expressions, we can include the division of polynomial functions in this algebra of functions.

Dividing polynomial functions

If f and g are polynomial functions, then

$$\left(\frac{f}{g}\right)(x) = \frac{f(x)}{g(x)}, \quad \text{provided} \quad g(x) \neq 0.$$

EXAMPLE 10 **Applying the Algebra of Functions**

Let $f(x) = x^2 + 3x - 10$ and $g(x) = x - 2$. Find:

a. $\left(\dfrac{f}{g}\right)(x)$ **b.** $\left(\dfrac{f}{g}\right)(4)$ **c.** $\dfrac{f(4)}{g(4)}$

Solution

a. $\left(\dfrac{f}{g}\right)(x) = \dfrac{f(x)}{g(x)}$

$\qquad\qquad = \dfrac{x^2 + 3x - 10}{x - 2}$ Substitute each function's equation.

$\qquad\qquad = \dfrac{(x + 5)(x - 2)}{x - 2}$ Simplify the resulting rational expression.

$\qquad\qquad = x + 5, \quad x \neq 2$

Thus, $\left(\dfrac{f}{g}\right)(x) = x + 5, x \neq 2$.

b. To find $\left(\dfrac{f}{g}\right)(4)$, we substitute 4 for x in $\left(\dfrac{f}{g}\right)(x)$.

$\left(\dfrac{f}{g}\right)(x) = x + 5$ This is the function from part (a).

$\left(\dfrac{f}{g}\right)(4) = 4 + 5$ Substitute 4 for x.

$\qquad\qquad = 9$ The function value is 9.

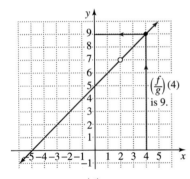

The graph of $\left(\dfrac{f}{g}\right)(x) = x + 5$,

$x \neq 2$

c. To find $\dfrac{f(4)}{g(4)}$, we substitute 4 for x in both $f(x)$ and $g(x)$, and then divide the function values.

$\dfrac{f(4)}{g(4)} = \dfrac{4^2 + 3 \cdot 4 - 10}{4 - 2}$ Use $f(x) = x^2 + 3x - 10$ and $g(x) = x - 2$, replacing x by 4.

$\qquad\quad = \dfrac{18}{2} = 9$

We see by comparing parts (b) and (c) that

$$\left(\frac{f}{g}\right)(4) = \frac{f(4)}{g(4)}.$$ ∎

We now have a complete algebra of polynomial functions, summarized as follows:

Operations with polynomial functions: the algebra of functions

If f and g are functions, then

$$(f + g)(x) = f(x) + g(x)$$ The function $f + g$ is the sum of the functions f and g.

$$(f - g)(x) = f(x) - g(x)$$ The function $f - g$ is the difference of the functions f and g.

$$(fg)(x) = f(x)g(x)$$ The function fg is the product of the functions f and g.

$$\left(\frac{f}{g}\right)(x) = \frac{f(x)}{g(x)}$$ The function $\frac{f}{g}$ is the quotient of the functions f and g, where $g(x) \neq 0$.

8 Model with rational functions.

Modeling With Rational Functions

Our next example uses a rational function to model a business situation. If the function $C(x)$ describes the cost for a company of producing x units of a product, then the average cost per unit is the total cost function $(C(x))$ divided by the number of units produced (x). This rational function is often denoted by $\overline{C}(x)$.

Costs: $100,000 plus $100 per bicycle

EXAMPLE 11   **Modeling Average Cost**

A bicycle manufacturing business required an initial investment of $100,000. This initial investment represents the fixed costs for the business. However, there are also variable costs of $100 for each bicycle manufactured.

a. Find the cost function $C(x)$ that models the cost of producing x bicycles.
b. Find a function $\overline{C}$ that models the average cost per bicycle for x bicycles.
c. Find and interpret $\overline{C}(500)$, $\overline{C}(1000)$, $\overline{C}(2000)$, and $\overline{C}(4000)$.

Solution

a. We begin with $C(x)$, the total cost function.

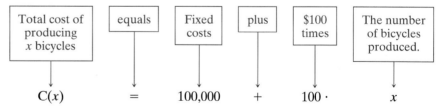

Total cost of producing x bicycles	equals	Fixed costs	plus	$100 times	The number of bicycles produced.
$C(x)$	=	100,000	+	100 ·	x

The cost function that models the cost to the company of producing x bicycles is

$$C(x) = 100,000 + 100x.$$

b. The average cost per bicycle is the total cost function $(C(x))$ divided by the number of bicycles produced (x). The function that models the average cost per bicycle for x bicycles is the rational function

$$\overline{C}(x) = \frac{100x + 100,000}{x}$$ The average cost function ($\overline{C}$) is found by dividing the cost function by x.

c. We now use the average cost function, replacing x with the four specified values.

$$\overline{C}(x) = \frac{100x + 100,000}{x}$$

This is the rational function that models the average cost per bicycle of producing x bicycles.

$$\overline{C}(500) = \frac{100(500) + 100,000}{500}$$

Replace x with 500.

$$= 300$$

The average cost per bicycle of producing 500 bicycles is $300.

$$\overline{C}(1000) = \frac{100(1000) + 100,000}{1000}$$

Replace x with 1000.

$$= 200$$

The average cost per bicycle of producing 1000 bicycles is $200.

$$\overline{C}(2000) = \frac{100(2000) + 100,000}{2000}$$

Replace x with 2000.

$$= 150$$

The average cost per bicycle of producing 2000 bicycles is $150.

$$\overline{C}(4000) = \frac{100(4000) + 100,000}{4000}$$

Finally, replace x with 4000.

$$= 125$$

The average cost per bicycle of producing 4000 bicycles is $125.

If we continued substituting values for x into the average cost function and organized our results in a table, we would obtain the following table of coordinates.

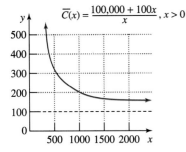

Figure 5.8

High production levels lower the cost of manufacturing each bicycle.

x	500	1000	2000	4000	10,000	20,000	50,000	100,000	1,000,000
$\overline{C}(x)$	300	200	150	125	110	105	102	101	100.10

It appears from the table that the more bicycles produced, the closer the average cost per bicycle comes to $100. Notice how this is shown by the graph of $\overline{C}$ in Figure 5.8. Since x represents the number of bicycles produced, $x > 0$. Thus the graph of the rational function appears only in the first quadrant. The graph illustrates that competitively low prices take place with high production levels, posing a major problem for small businesses. ∎

Using technology

Use a graphing utility to reproduce the hand-drawn graph in Figure 5.8. Then adjust the range setting and use the TRACE feature to include increasingly larger values of x. What is the practical significance of the horizontal asymptote whose equation is $y = 100$?

Our next example models a uniform motion situation. Recall that the basic equation used to model uniform motion is

$$d = rt \qquad (\text{Distance} = \text{Rate} \times \text{Time})$$

An equivalent equation can be written by solving for t:

$$t = \frac{d}{r} \qquad \left(\text{Time} = \frac{\text{Distance}}{\text{Rate}} \right)$$

In Example 12, the latter form of the equation will be useful because something will be known about the time of travel.

EXAMPLE 12 **Modeling Uniform Motion**

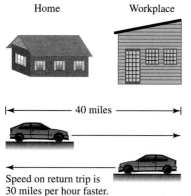

Home   Workplace

|← 40 miles →|

Speed on return trip is
30 miles per hour faster.

A commuter drove to work a distance of 40 miles and then returned on the same highway. The average speed on the return trip was 30 miles per hour faster than the average speed on the outgoing trip. If

x = the average speed on the outgoing trip

and

$x + 30$ = the average speed on the return trip

write a rational function that models the total time t required to complete the round trip.

Solution

The total time t required to complete the round trip is a function of the average speed on the outgoing trip. The critical sentence that we must model is

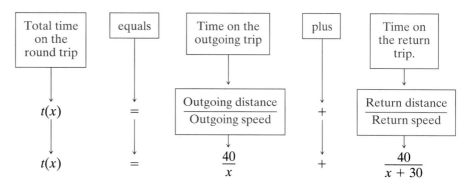

Thus, the function that models total time to and from work is

$$t(x) = \frac{40}{x} + \frac{40}{x + 30}.$$ ∎

Let's take some time to look at the graph of the function modeling total time to and from work to see what we can discover geometrically. Figure 5.9 shows the graph of

$$y = \frac{40}{x} + \frac{40}{x + 30}$$

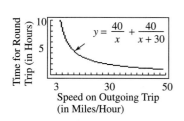

Figure 5.9

The length of time decreases as speed increases.

obtained on a graphing utility. Remember that x represents the average speed on the outgoing trip and y represents the total time for the round trip. Since it seems unlikely that an average outgoing speed exceeds 60 miles per hour with an average return speed that is 30 miles per hour faster, we used

Xmin = 0, Xmax = 60, Xscl = 3.

It also seems unlikely that a round trip of 40 miles in each direction would take more than 10 hours at a reasonable speed, so we used

Ymin = 0, Ymax = 10, Yscl = 1.

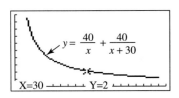

Figure 5.10

At an outgoing speed of 30 miles per hour, the round trip will take 2 hours.

The graph that we obtained, shown again in Figure 5.10, is falling from left to right, showing that the length of time decreases as speed increases. As you move along the graph to your left, the values of y are getting to be very large positive numbers, increasing without bound. This indicates that close to an outgoing rate of zero miles per hour, the round trip will take nearly forever.

EXAMPLE 13 **Interpreting the Graph of a Rational Function**

The graph of the rational function modeling time to and from work is shown in Figure 5.10. The [TRACE] feature was used to trace along the curve, as shown. Describe what these numbers mean in terms of the variables modeled by the function.

Solution

Remember that

$$x = \text{average speed on the outgoing trip}$$
$$\text{and} \quad y = \text{total time for the round trip}$$

Since we are given that $x = 30$ and $y = 2$, this means that at an outgoing speed of 30 miles per hour, the round trip will take 2 hours.

Let's see if these numbers make sense by finding total time for an outgoing speed of 30 miles per hour. Since the average speed on the return trip is 30 miles per hour faster, the average speed on the return trip is 60 miles per hour. Thus, for the 40-mile trip in each direction, we have

$$\text{Time on outgoing trip} = \tfrac{40}{30} \text{ or } 1\tfrac{1}{3} \text{ hours}$$
$$\text{Time on return trip} = \tfrac{40}{60} = \tfrac{2}{3} \text{ hour}$$
$$\text{Total time} = 1\tfrac{1}{3} + \tfrac{2}{3} = 2 \text{ hours}$$

This checks with $y = 2$ given by the graphing utility in Figure 5.10. ∎

Using technology

Use a graphing utility to reproduce the graph in Figure 5.9. [TRACE] along the curve as in Figure 5.10 and determine another value for x and for y. What do these numbers mean? Verify these numbers as in Example 13.

PROBLEM SET 5.1

Practice Problems

Find the domain of each rational function in Problems 1–8. Use both set-builder and interval notation.

1. $f(x) = \dfrac{3}{x - 4}$

2. $g(x) = \dfrac{5}{x + 2}$

3. $h(x) = \dfrac{x + 7}{x^2 - 3x}$

4. $r(x) = \dfrac{4 - x}{x^2 - 5x}$

5. $f(x) = \dfrac{3}{x^2 - 1}$

6. $g(x) = \dfrac{7}{x^2 - 16}$

7. $f(x) = \dfrac{x + 8}{x^2 - x - 12}$

8. $f(x) = \dfrac{x + 7}{x^2 + 8x + 15}$

9. Match each function with its appropriate graph.

$$f(x) = \frac{2}{(x-1)^3}$$

$$g(x) = \frac{3}{x^2+1}$$

$$h(x) = \frac{3}{x^2-1}$$

a.

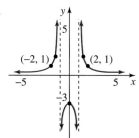

b.

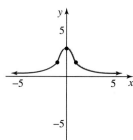

c.

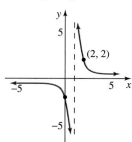

10. Match each function with its appropriate graph.

$$f(x) = \frac{2x}{4-x^2}$$

$$g(x) = \frac{2}{x^3+1}$$

$$h(x) = \frac{x^2-x-2}{9-x^2}$$

a.

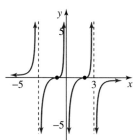

b.

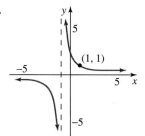

c.

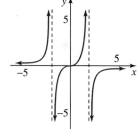

11. Consider the function defined by $f(x) = \dfrac{2}{x-2}$.

  a. What is the domain of f?

  b. Fill in the following tables.

x approaches 2 from the left.

x	1	1.5	1.9	1.99	1.999
$f(x) = \dfrac{2}{x-2}$					

x approaches 2 from the right.

x	2.001	2.01	2.1	2.5	3
$f(x) = \dfrac{2}{x-2}$					

  c. Graph the line whose equation is $x = 2$ and then graph f [using the tables from part (b)] near the excluded x-value.

12. Consider the function defined by $f(x) = \dfrac{1}{x+1}$.

  a. What is the domain of f?

  b. Fill in the following tables.

x approaches -1 from the left.

x	-2	-1.5	-1.1	-1.01	-1.001
$f(x) = \dfrac{1}{x+1}$					

x approaches -1 from the right.

x	-0.999	-0.99	-0.9	-0.5	0
$f(x) = \dfrac{1}{x+1}$					

  c. Graph the line whose equation is $x = -1$ and then graph f [using the tables from part (b)] near the excluded x-value.

For Problems 13–44, simplify each rational expression, if possible.

13. $\dfrac{y - 1}{y^2 - 1}$

14. $\dfrac{x^2 - 4}{2x - 4}$

15. $\dfrac{3a + 9}{a^2 + 6a + 9}$

16. $\dfrac{4b - 8}{b^2 - 4b + 4}$

17. $\dfrac{c^2 - 12c + 36}{4c - 24}$

18. $\dfrac{y^2 - 8y + 16}{3y - 12}$

19. $\dfrac{a^2 - 1}{a^2 + 2a + 1}$

20. $\dfrac{x^2 - 2xy + y^2}{x^2 - y^2}$

21. $\dfrac{x^4 - y^4}{x^4 + 2x^2y^2 + y^4}$

22. $\dfrac{a^4 - b^4}{a^2 - 2ab + b^2}$

23. $\dfrac{a^2 + 7a - 18}{a^2 - 3a + 2}$

24. $\dfrac{y^2 - 4y - 5}{y^2 + 5y + 4}$

25. $\dfrac{2x^2 - 5x + 2}{2x^2 - 7x + 3}$

26. $\dfrac{6b^2 - b - 2}{3b^2 + 4b - 4}$

27. $\dfrac{c^2 - 14c + 49}{c^2 - 49}$

28. $\dfrac{x^2 - 9}{x^2 + x - 6}$

29. $\dfrac{x^2 - 9}{x^3 - 27}$

30. $\dfrac{a^3 + 64}{a^2 - 16}$

31. $\dfrac{y^2 - 9y + 18}{y^3 - 27}$

32. $\dfrac{x^3 - 8}{x^2 + 2x - 8}$

33. $\dfrac{x^3 + 12x^2 + 35x}{x^3 + 4x^2 - 21x}$

34. $\dfrac{x^3 + x^2 - 20x}{x^3 + 2x^2 - 15x}$

35. $\dfrac{x^3 - 9x}{x^3 + 5x^2 + 6x}$

36. $\dfrac{2x + 3}{2x - 5}$

37. $\dfrac{x - 4}{4x - 1}$

38. $\dfrac{x}{x + y}$

39. $\dfrac{x^2 - 4}{x^2 - 9}$

40. $\dfrac{x^2 - 5x + 6}{x^2 - 7x - 18}$

41. $\dfrac{x^2 + 5x + 2xy + 10y}{x^2 - 25}$

42. $\dfrac{a^2 - 16}{a^2 - 4a + 3ab - 12b}$

43. $\dfrac{a^2 - 4}{2 - a}$

44. $\dfrac{b^2 - b - 12}{4 - b}$

45. Simplify and graph: $f(x) = \dfrac{x^2 - x - 2}{x + 1}$.

47. Simplify and graph: $f(x) = \dfrac{x^2 - 5x - 14}{x + 2}$.

46. Simplify and graph: $f(x) = \dfrac{x^2 - 3x + 2}{x - 2}$.

48. Simplify and graph: $f(x) = \dfrac{x^2 - x - 12}{x - 4}$.

In Problems 49–64, multiply, and simplify if possible.

49. $\dfrac{y^2 - 4}{y - 2} \cdot \dfrac{y - 2}{y^2 + y - 6}$

50. $\dfrac{x + 1}{x^2 + x - 6} \cdot \dfrac{x^2 - 9}{x + 1}$

51. $\dfrac{5x + 5y}{x - y} \cdot \dfrac{3x - 3y}{10}$

52. $\dfrac{4x^2 + 10}{x - 3} \cdot \dfrac{x^2 - 9}{6x^2 + 15}$

53. $\dfrac{b^2 + b}{b^2 - 4} \cdot \dfrac{b^2 + 5b + 6}{b^2 - 1}$

54. $\dfrac{(a + y)^2}{a^2 - y^2} \cdot \dfrac{a^2 - 2ay + y^2}{ab - by}$

55. $\dfrac{m^2 - 4}{m^2 + 4m + 4} \cdot \dfrac{2m + 4}{m^2 + m - 6}$

56. $(a - 3) \cdot \dfrac{a^2 + a + 1}{a^2 - 5a + 6}$

57. $\dfrac{b + 2}{b^2 + 7b + 6} \cdot (b + 1)$

58. $\dfrac{y + 4}{y - 2} \cdot (y^2 + 2y - 8)$

59. $\dfrac{x^3 - 8}{x^2 - 4} \cdot \dfrac{x + 2}{3x}$

60. $\dfrac{x^2 + 6x + 9}{x^3 + 27} \cdot \dfrac{1}{x + 3}$

61. $\dfrac{2a^2 - 13a - 7}{a^2 - 6a - 7} \cdot \dfrac{a^2 - a - 2}{2a^2 - 5a - 3}$

62. $\dfrac{2a^2 - 3ab - 2b^2}{3a^2 - 4ab + b^2} \cdot \dfrac{3a^2 - 2ab - b^2}{a^2 + ab - 6b^2}$

63. $\dfrac{2y^2 + 9y - 35}{6y^2 - 13y - 5} \cdot \dfrac{3y^2 + 10y + 3}{y^2 + 10y + 21}$

64. $\dfrac{bx - by + 2x - 2y}{(b + 2)^2} \cdot \dfrac{bx + by + 2x + 2y}{x^2 - y^2}$

In Problems 65–80, divide, and simplify if possible.

65. $\dfrac{2a + 4b}{3ab} \div \dfrac{6a + 12b}{6a^2b}$

66. $\dfrac{4x - 4}{3y - 3} \div \dfrac{x - 1}{y - 1}$

67. $\dfrac{x + y}{x^2 - xy} \div \dfrac{3x + 3y}{x - y}$

68. $\dfrac{a^2 - b^2}{3y^2} \div \dfrac{(a - b)^2}{9y}$

69. $\dfrac{x^2 - y^2}{(x + y)^2} \div \dfrac{x - y}{4x + 4y}$

70. $\dfrac{(a + 2)^2}{cx - cy} \div \dfrac{a^2 + 2a}{a^2x - a^2y}$

71. $\dfrac{x^3 - 27}{a^3 + 8} \div \dfrac{x - 3}{a + 2}$

72. $\dfrac{x^2 + 2x - 8}{3a} \div \dfrac{x^2 - 4}{9a^3}$

73. $\dfrac{4b^2 + 20b + 25}{5b} \div \dfrac{4b^2 - 25}{4b^2}$

74. $\dfrac{p^2 - q^2}{4p^2} \div \dfrac{p^2 - 2pq + q^2}{2q^2}$

75. $\dfrac{a^2 - 4a - 21}{a^2 - 10a + 25} \div \dfrac{a^2 + 2a - 3}{a^2 - 6a + 5}$

76. $\dfrac{b^2 + b - 2}{b^2 + 6b - 7} \div \dfrac{b^2 - 3b - 10}{b^2 + 5b - 14}$

77. $\dfrac{9x^2 - 12x + 4}{2x^2 + 3x - 5} \div \dfrac{3x^2 - 8x + 4}{2x^2 + 7x + 5}$

78. $\dfrac{4a^2 + 12a + 9}{3a^2 + 2a - 1} \div \dfrac{4a^2 - 9}{3a^2 + 14a - 5}$

79. $(bx + by + 3x + 3y) \div \dfrac{x^2 - y^2}{b + 3}$

80. $\dfrac{a^3 - 8}{a - 3} \div \dfrac{2 - a}{ab - 3b + ac - 3c}$

81. Given $f(x) = x^2 - 9x + 14$ and $g(x) = x - 2$, find the following.

   a. $\left(\dfrac{f}{g}\right)(x)$ b. $\left(\dfrac{f}{g}\right)(3)$ c. $\dfrac{f(3)}{g(3)}$

82. Given $f(x) = x^2 - 8x + 15$ and $g(x) = x - 3$, find the following.

   a. $\left(\dfrac{f}{g}\right)(x)$ b. $\left(\dfrac{f}{g}\right)(7)$ c. $\dfrac{f(7)}{g(7)}$

83. Given $f(x) = x^3 - 16x$ and $g(x) = x^2 - 2x - 8$, find the following.

   a. $\left(\dfrac{f}{g}\right)(x)$ b. $\left(\dfrac{f}{g}\right)(8)$ c. $\dfrac{f(8)}{g(8)}$

84. Given $f(x) = x^2 - 4$ and $g(x) = 2x^2 - 3x - 2$, find the following.

   a. $\left(\dfrac{f}{g}\right)(x)$ b. $\left(\dfrac{f}{g}\right)(4)$ c. $\dfrac{f(4)}{g(4)}$

Application Problems

85. The cost ($C(x)$, in thousands of dollars) of eliminating $x\%$ of pollutants from a stream is given by the rational function

$$C(x) = \frac{4x}{100 - x}.$$

   a. What is the cost to remove 80% of pollutants from the stream?
   b. Find and interpret $C(95)$.
   c. What is the domain of the function? What does this mean in terms of the variables modeled by the function?
   d. What happens to the cost as x approaches 100 (percent)?

86. A company that manufactures small canoes has fixed costs of $20,000 and variable cost of $20 per canoe.
   a. Find the cost function $C(x)$ that models the total cost to the company (including fixed costs) of manufacturing x canoes.
   b. Find a function $\overline{C}$ that models the average cost per canoe for x canoes.
   c. Find and interpret $\overline{C}(50)$, $\overline{C}(100)$, $\overline{C}(1000)$, and $\overline{C}$ (10,000). Then fill in the missing portion of this statement: The more canoes that are produced, the closer the average cost per canoe comes to $\$\underline{\quad}$.

87. The polynomial model $P = -0.14t^2 + 0.51t + 31.6$ describes the U.S. population (P, in millions) age 65 and older t years after 1990. Furthermore, the polynomial model $C = 0.54t^2 + 12.64t + 107.1$ describes the total yearly cost of Medicare (C, in billions of dollars) t years after 1990. Write a rational model expressing the average cost of Medicare per person age 65 or older t years after 1990. Then find this average cost for the years 1994 through 1997.

88. It is estimated in a certain community that the number of hours required to distribute new telephone books to $x\%$ of the households can be represented by $\dfrac{600x}{300x - 300}$. Simplify this rational expression. Use this

simplified form to find the number of hours necessary to distribute telephone books to $\frac{4}{5}$ of the population.

89. A person drove from home to a vacation resort 600 miles away and returned on the same highway. The average speed on the return trip was 10 miles per hour slower than the average speed on the outgoing trip.
   a. If x = the average speed on the outgoing trip, write a rational function that models the total time t required to complete the round trip.
   b. Find and interpret $t(50)$.

90. The student government at a college is producing a handbook on student rights and responsibilities. Using their own graphic artists, photographers, and computers, they will need to spend $850 plus $3.25 for printing each handbook. The graph of the average cost per handbook for printing x handbooks is shown in the figure.
   a. Use the graph to find a reasonable estimate for the average cost of producing 1000 and 2000 handbooks.
   b. Find a function $\overline{C}$ that models the average cost per handbook for x handbooks. Compute $\overline{C}(1000)$ and $\overline{C}(2000)$, comparing these values with your estimates in part (a).
   c. What does the graph indicate about the average cost as the number of handbooks increases? How is this shown by the graph?

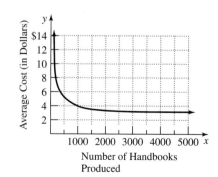

Number of Handbooks
Produced

91. Rational functions are often used to model how much we remember over time. In an experiment on memory, students in a language class are asked to memorize 40 vocabulary words in Latin, a language with which the students are not familiar. After studying the words for one day, the class is tested each day thereafter to see how many words they remember. The class average is taken and the results are graphed as shown below.
  a. Use the graph to find a reasonable estimate of the number of Latin words remembered after 1 day, 5 days, and 15 days.
  b. The function that models the number of Latin words remembered by the students after t days is given by

$$N(t) = \frac{5t + 30}{t}, \quad \text{where } t \geq 1.$$

  Find $N(1)$, $N(5)$, and $N(15)$, comparing these values with your estimates from part (a).
  c. What does the graph indicate about the number of Latin words remembered by the group over time?
  d. The graph approaches but never touches the horizontal line whose equation is $y = 5$. Describe what this horizontal asymptote means in terms of the variables modeled in this situation.

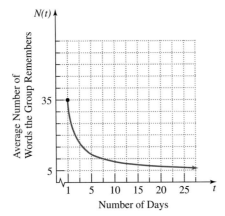

92. Manufacturers of cylindrical cans for fruit juice are to construct cans that can hold 21.6 cubic inches of juice. The graph below shows the surface area of these cans as a function of their radius. The coordinates for one point along this surface area function are shown.
  a. Describe what these numbers mean in terms of the variables modeled by the function.
  b. The function that models the surface area of the cylindrical can in terms of its radius is

$$A(r) = \frac{2\pi r^3 + 43.2}{r}.$$

  Use this function to verify the coordinates shown on the graph.
  c. Describe how the point shown on the graph can be useful to the manufacturers of the cans.

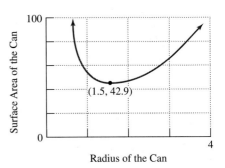

In Problems 93–94, write a function $R(x)$ that models the ratio of the unshaded area to total area of each figure. Simplify each function's formula.

93.

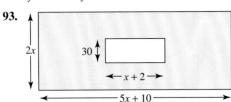

94.

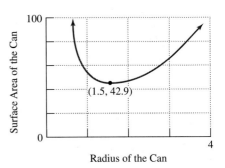

True–False Critical Thinking Problems _____

95. Which one of the following is true?

  a. $\dfrac{x^2 - 25}{x - 5} = x - 5$

  b. $\dfrac{x^2 + 7}{7} = x^2 + 1$

c. All rational functions have vertical asymptotes defined by $x = a$ when a causes the denominator of a rational function to equal 0.

d. None of the above is true.

96. Which one of the following is true?

a. $f(x) = \dfrac{x^2}{x^2 + x - 6}$ has a graph with a hole at $x = 2$.

b. $\dfrac{x^2 + 3x}{x^2 - 3x} = -1$

c. The domain of $f(x) = \dfrac{x - 3}{x + 5}$ is

$$\{x \mid x \neq -5 \text{ and } x \neq 3\}.$$

d. None of the above is true.

97. Which one of the following is true?

a. Only rational expressions with identical denominators can be multiplied.

b. $x \div y = \dfrac{1}{x} \cdot y$, if $x \neq 0$.

c. $\dfrac{x}{y} \div \dfrac{y}{x} = 1$, if $x \neq 0$ and $y \neq 0$.

d. None of the above is true.

Technology Problems

98. Graph the cost function in Problem 85 for each of the following range settings: $[0, 50]$ by $[0, 5]$; $[0, 80]$ by $[0, 16]$; $[0, 95]$ by $[0, 80]$. Explain how the three graphs visualize the answer to Problem 85d. Describe what happens if you attempt to graph the cost function and include $Xmax = 100$ in the range setting.

99. Use your graphing utility to graph the average cost function C from Problem 86b. Use expanding range settings to illustrate the solution to Problem 86c.

100. Use your graphing utility to graph each rational function in Problems 1–8. Use both a connected and a dot format. Explain how the graph verifies the domain of each rational function.

101. a. Graph $f(x) = \dfrac{x^2 - x - 2}{x - 2}$ and $g(x) = x + 1$ in the same viewing rectangle. What do you observe?

b. Simplify the formula in the definition of function f. Do f and g represent exactly the same function? Explain.

c. Display the graphs of f and g separately in the utility's viewing rectangle. TRACE along each of the curves until you get to $x = 2$. What difference do you observe? What does this mean?

102. a. Use a graphing utility to graph the function in Problem 89 for meaningful values of x.

b. TRACE along the curve until you reach $x = 40$. What is the corresponding value of y? Describe what these numbers mean in terms of the variables modeled by the function.

Writing in Mathematics

103. In a get-tough drug policy, a politician promises to spend "whatever it takes" to seize all illegal drugs as they enter the country. If the cost of this venture is modeled by the rational function $C(x) = \dfrac{Ap}{100 - p}$, where A is a positive constant, $C(x)$ is expressed in millions of dollars, and p is the percent of illegal drugs seized, use this model to write a statement that critiques the politician's promise.

104. Explain how the graph of $f(x) = \dfrac{x^2 - x - 2}{x - 2}$ differs from the graph of $g(x) = x + 1$.

105. Why is it necessary to restrict the domain for certain rational functions?

106. Explain what is meant by reducing a rational expression to lowest terms.

107. What is the major difference in the procedures for multiplying and dividing rational expressions?

Critical Thinking Problems

In Problems 108–109, simplify.

108. $\dfrac{x^{2n} - y^{2n}}{x^{2n} + 2x^n y^n + y^{2n}}$

109. $\dfrac{6x^{3n} + 6x^{2n}y^n}{x^{2n} - y^{2n}}$

Perform the indicated operations in Problems 110–111.

110. $\dfrac{y^{2n} - 1}{y^{2n} + 3y^n + 2} \cdot \dfrac{y^{2n} - y^n - 6}{y^{2n} + y^n - 12}$

111. $\dfrac{4y^3 - 12y^2}{4y^n + 8} \div \dfrac{y^{n+1} - 2y}{y^{2n} - 4}$

Perform the indicated operations in Problems 112–113.

112. $\dfrac{a - b}{4c} \div \left(\dfrac{b - a}{c} \div \dfrac{a - b}{c^2} \right)$

113. $\left(\dfrac{a - b}{4c} \div \dfrac{b - a}{c} \right) \div \dfrac{a - b}{c^2}$

114. a. Simplify: $\dfrac{9x^2 - (x - 4)^2}{(x - 1)(x + 2)}$.

   b. Find $\dfrac{9(3351)^2 - (3347)^2}{3350 \cdot 3353}$ without using a calcula-

   tor.

115. Simplify $\dfrac{c^4 + 64d^4}{c^2 + 4cd + 8d^2}$ by first adding and subtract-

   ing $16c^2d^2$ in the numerator, as shown:

   $$\dfrac{c^4 + 16c^2d^2 + 64d^4 - 16c^2d^2}{c^2 + 4cd + 8d^2}$$

116. If the solution set to the equation $f(x) = 0$ is
$\{2, 3, 5, 7, 9\}$ and the solution set for the equation
$g(x) = 0$ is $\{-1, 3, 5, 7, 8\}$, what is the solution set of
the equation $\dfrac{f(x)}{g(x)} = 0$?

117. Write a simplified expression in terms of x that models the quotient of the area of the trapezoid and the
area of the rectangle shown in the figure.

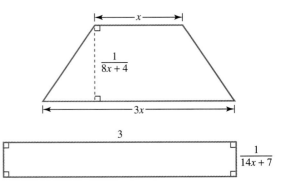

118. Find the product of the following 199 factors:

$$\left(1 - \dfrac{1}{2}\right)\left(1 - \dfrac{1}{3}\right)\left(1 - \dfrac{1}{4}\right)\cdots\left(1 - \dfrac{1}{x + 1}\right)\cdots\left(1 - \dfrac{1}{200}\right)$$

Group Activity Problem

119. Group members are sales analysts for a home video
game company. It has been determined that the ra-
tional function $f(x) = \dfrac{200x}{x^2 + 100}$ defines the monthly
sales ($f(x)$, in thousands of games) of a new video
game as a function of the number of months (x) after
the game is introduced. The figure shows the graph of
the function. What are your recommendations to the
company in terms of how long the video game should
be on the market before another new video game is
introduced? What other factors might you want to
take into account in terms of your recommendation?
What will eventually happen to sales, and how is this
indicated by the graph? What does this have to do
with a horizontal asymptote? What could the company do to change the behavior of this function and
continue generating sales? Would this be cost effective?

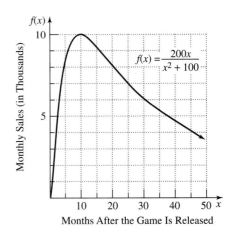

Review Problems

120. Multiply: $(2x - 5)(3x^2 - 5x + 4)$.

121. Use scientific notation to find the value:
$\dfrac{(0.000012)(400,000)}{0.000006}$.

122. Graph: $y \geq 3x - 2$.

Solutions Manual   **Tutorial**   **Video 6**

Adding and Subtracting Rational Expressions

Objectives

1   Add and subtract rational expressions with the same denominator.
2   Add and subtract rational expressions with different denominators.
3   Simplify complex rational expressions.

In the previous section we used the rational function

$$t(x) = \frac{40}{x} + \frac{40}{x + 30}$$

to model the total time to and from work. In this situation, you may recall that the commuter drove a distance of 40 miles, returned home on the same highway, and averaged 30 miles per hour faster on the return trip than on the outgoing trip.

We now turn our attention to adding $\dfrac{40}{x}$ and $\dfrac{40}{x + 30}$ with the goal of obtaining a single rational expression in which the numerator and denominator are polynomials. We first look at expressions having the same denominator.

▮ Add and subtract rational expressions with the same denominator.

Adding and Subtracting Rational Expressions with the Same Denominator

Rational numbers containing the same denominator are added and subtracted using the following properties.

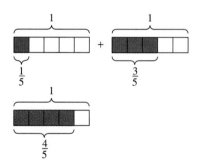

Adding and subtracting rational numbers
$\dfrac{a}{b} + \dfrac{c}{b} = \dfrac{a + c}{b} \qquad (b \neq 0)$
$\dfrac{a}{b} - \dfrac{c}{b} = \dfrac{a - c}{b} \qquad (b \neq 0)$

To add or subtract fractions with the same denominator, we add or subtract the numerators and put this result over the common denominator. Rational expressions are handled in an identical manner.

Adding or subtracting rational expressions with the same denominator
1. Add or subtract the numerators.
2. Place the result over the common denominator.
3. If possible, simplify the answer by reducing to lowest terms.

The following examples illustrate this procedure.

As always, use your graphing utility to check results. Graph the equations

$$y_1 = \frac{5}{2x} + \frac{1}{2x}$$

and

$$y_2 = \frac{3}{x}$$

on the same screen. As shown below, the graphs coincide, so the two functions are equal. Neither graph crosses the vertical line $x = 0$ because 0 is not in the domain of either function.

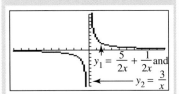

Use a graphing utility to check Example 1b if $x \neq -3$.

EXAMPLE 1 **Adding Rational Expressions with the Same Denominator**

Add:

a. $\dfrac{5}{2x} + \dfrac{1}{2x}$ **b.** $\dfrac{x}{x^2 - 9} + \dfrac{3}{x^2 - 9}$

Solution

a. $\dfrac{5}{2x} + \dfrac{1}{2x} = \dfrac{5 + 1}{2x}$ Add numerators. Place this sum over the common denominator.

$= \dfrac{6}{2x}$ Combine numerical terms.

$= \dfrac{3}{x}$ Simplify: $\dfrac{6}{2x} = \dfrac{2 \cdot 3}{2x} = \dfrac{3}{x}$.

b. $\dfrac{x}{x^2 - 9} + \dfrac{3}{x^2 - 9} = \dfrac{x + 3}{x^2 - 9}$ Add numerators. Place this sum over the common denominator.

$= \dfrac{(x + 3)}{(x + 3)(x - 3)}$ Factor. $x \neq -3$ and $x \neq 3$.

$= \dfrac{1}{x - 3}, \quad x \neq -3$ Simplified form ∎

EXAMPLE 2 **Subtracting Rational Expressions with the Same Denominator**

Subtract: $\dfrac{3y^3 - 5x^3}{x^2 - y^2} - \dfrac{4y^3 - 6x^3}{x^2 - y^2}$

Solution

$\dfrac{3y^3 - 5x^3}{x^2 - y^2} - \dfrac{4y^3 - 6x^3}{x^2 - y^2} = \dfrac{3y^3 - 5x^3 - (4y^3 - 6x^3)}{x^2 - y^2}$ The parentheses around $4y^3 - 6x^3$ are essential to indicate that the entire quantity is subtracted.

$= \dfrac{3y^3 - 5x^3 - 4y^3 + 6x^3}{x^2 - y^2}$ Remove parentheses and change the sign of each term in the parentheses.

$= \dfrac{x^3 - y^3}{x^2 - y^2}$ Combine like terms in the numerator.

$= \dfrac{(x - y)(x^2 + xy + y^2)}{(x + y)(x - y)}$ Factor and simplify.

$= \dfrac{x^2 + xy + y^2}{x + y}$ Simplified form ∎

2 Add and subtract rational expressions with different denominators.

Adding and Subtracting Rational Expressions with Different Denominators

Rational numbers containing different prime denominators are added and subtracted using the following properties.

$$\frac{a}{b} + \frac{c}{d} = \frac{ad + bc}{bd} \qquad (b \neq 0, d \neq 0)$$

$$\frac{a}{b} - \frac{c}{d} = \frac{ad - bc}{bd} \qquad (b \neq 0, d \neq 0)$$

The expression bd is the *least common denominator* (abbreviated LCD) for $\dfrac{a}{b}$ and $\dfrac{c}{d}$ because it is the smallest expression that is divisible by each of the denominators, namely, b and d.

The first step in combining two rational expressions is to find the LCD. Because the LCD can be described as the least common multiple (LCM) of all polynomials in the denominators, we can find the LCM by factoring these denominators completely. *The least common multiple is the product of each factor the greatest number of times it occurs in any one factorization.* Let's see exactly what this means.

EXAMPLE 3 **Finding the LCM**

Find each LCM:

a. $4x^3y$ and $6xy^2$
b. x and $x + 2$
c. $x^2 - x - 20$ and $x^2 + 9x + 20$
d. $3x^2 - 6xy + 3y^2$ and $5x^2 - 5y^2$

Solution

a. The LCM for $4x^3y$ and $6xy^2$ is the smallest polynomial divisible by both $4x^3y$ and $6xy^2$. It must contain a factor of 12 to be divisible by both 4 and 6. It must also contain a factor of x^3 (divisible by both x^3 and x) and of y^2 (divisible by both y and y^2). The LCM is $12x^3y^2$.
b. The LCM, the polynomial divisible by x and $x + 2$, is $x(x + 2)$.
c. We begin by factoring each polynomial.

$$x^2 - x - 20 = (x - 5)(x + 4)$$
$$x^2 + 9x + 20 = (x + 5)(x + 4)$$

The LCM, divisible by the given polynomials, is the product of the factors

$$(x - 5)(x + 4)(x + 5).$$

d. $3x^2 - 6xy + 3y^2 = 3(x^2 - 2xy + y^2) = 3(x - y)^2$ Factor.

$$5x^2 - 5y^2 = 5(x^2 - y^2) = 5(x + y)(x - y)$$

The LCM, the product of all different factors from each polynomial, with each factor raised to the highest power that occurs in any polynomial, is

$$3 \cdot 5(x - y)^2(x + y) \qquad \text{or} \qquad 15(x - y)^2(x + y) \qquad \blacksquare$$

Example 4 illustrates the role of the least common denominator in the addition process.

EXAMPLE 4 **Adding Rational Expressions with Different Denominators**

Add: $\dfrac{x}{x - 3} + \dfrac{x - 1}{x + 3}$

Solution

The LCM of $x - 3$ and $x + 3$, is $(x - 3)(x + 3)$. Thus, $(x - 3)(x + 3)$ is the least common denominator.

$$\frac{x}{x - 3} + \frac{x - 1}{x + 3}$$

$$= \frac{x(x + 3)}{(x - 3)(x + 3)} + \frac{(x - 1)(x - 3)}{(x - 3)(x + 3)}$$

Rewrite each rational expression with the LCD. Multiply the numerator and denominator by whatever extra factors are required to form the LCD.

$$= \frac{x(x + 3) + (x - 1)(x - 3)}{(x - 3)(x + 3)}$$

Add numerators, putting this sum over the LCD.

$$= \frac{x^2 + 3x + x^2 - 4x + 3}{(x - 3)(x + 3)}$$

Multiply in the numerator.

$$= \frac{2x^2 - x + 3}{(x - 3)(x + 3)}$$

Combine like terms in the numerator. ■

Before considering additional examples, let's summarize the steps involved in Example 4.

Discover for yourself

Use this procedure to write the total-time-to-and-from-work model,

$$t(x) = \frac{40}{x} + \frac{40}{x + 30}$$

as a function with a single rational expression. Check your work by using a graphing utility to graph both functions.

Procedure for adding and subtracting rational expressions with different denominators

1. Find the LCD, the product of all different factors from each denominator, with each factor raised to the highest power occurring in any denominator.
2. Write all rational expressions as equivalent rational expressions having the LCD as the specified denominator. To do this, insert into the numerator and denominator of each rational expression whatever extra factors are required to form the LCD.
3. Add or subtract the numerators, placing the resulting expression over the LCD.
4. If necessary, simplify the resulting rational expression.

The following examples illustrate this procedure.

EXAMPLE 5 **Subtracting Rational Expressions with Different Denominators**

Subtract: $\dfrac{x - 4}{8x} - \dfrac{x - 2}{4x^2}$

Solution

The LCD is $8x^2$. We must rewrite each rational expression with this denominator.

$$\frac{x - 4}{8x} - \frac{x - 2}{4x^2} = \frac{(x - 4) \cdot x}{8x \cdot x} - \frac{(x - 2) \cdot 2}{4x^2 \cdot 2}$$

Rewrite each rational expression in terms of the LCD.

$$= \frac{x(x - 4) - 2(x - 2)}{8x^2}$$

Subtract.

$$= \frac{x^2 - 4x - 2x + 4}{8x^2} \qquad \text{Apply the distributive property.}$$

$$= \frac{x^2 - 6x + 4}{8x^2} \qquad \text{Combine like terms.} \qquad ∎$$

EXAMPLE 6 **Subtracting Rational Expressions with Different Denominators**

Subtract: $\dfrac{3y}{y - 5} - \dfrac{y - 2}{y + 5}$

Solution

The LCD is $(y - 5)(y + 5)$. We rewrite each rational expression with this denominator.

$$\frac{3y}{y - 5} - \frac{y - 2}{y + 5} = \frac{3y(y + 5)}{(y - 5)(y + 5)} - \frac{(y - 2)(y - 5)}{(y - 5)(y + 5)} \qquad \begin{array}{l}\text{Rewrite each rational}\\\text{expression in terms of}\\\text{the LCD.}\end{array}$$

$$= \frac{3y(y + 5) - (y - 2)(y - 5)}{(y - 5)(y + 5)} \qquad \text{Subtract.}$$

$$= \frac{3y^2 + 15y - (y^2 - 7y + 10)}{(y - 5)(y + 5)} \qquad \begin{array}{l}\text{Multiply in the}\\\text{numerator.}\end{array}$$

$$= \frac{3y^2 + 15y - y^2 + 7y - 10}{(y - 5)(y + 5)} \qquad \begin{array}{l}\text{Distribute } -1 \text{ through-}\\\text{out the grouping}\\\text{symbols.}\end{array}$$

$$= \frac{2y^2 + 22y - 10}{(y - 5)(y + 5)} \qquad \text{Combine like terms.}$$

$$= \frac{2(y^2 + 11y - 5)}{(y - 5)(y + 5)} \qquad \begin{array}{l}\text{Factor the}\\\text{numerator.}\end{array} \qquad ∎$$

EXAMPLE 7 **Subtracting Rational Expressions with Different Denominators**

Subtract: $\dfrac{2y^2 - y + 2}{y^3 - 1} - \dfrac{1}{y - 1}$

Solution

Because $y^3 - 1 = (y - 1)(y^2 + y + 1)$ and $y - 1 = 1(y - 1)$, the LCD is $(y - 1)(y^2 + y + 1)$.

$$\frac{2y^2 - y + 2}{y^3 - 1} - \frac{1}{y - 1} = \frac{2y^2 - y + 2}{(y - 1)(y^2 + y + 1)} - \frac{1(y^2 + y + 1)}{(y - 1)(y^2 + y + 1)} \qquad \begin{array}{l}\text{Rewrite each rational expression in terms}\\\text{of the LCD.}\end{array}$$

$$= \frac{2y^2 - y + 2 - (y^2 + y + 1)}{(y - 1)(y^2 + y + 1)} \qquad \text{Subtract.}$$

$$= \frac{2y^2 - y + 2 - y^2 - y - 1}{(y - 1)(y^2 + y + 1)}$$

$$= \frac{y^2 - 2y + 1}{(y - 1)(y^2 + y + 1)} \qquad \text{Combine like terms in the numerator.}$$

$$= \frac{(y-1)(y-1)}{(y-1)(y^2+y+1)}$$

Factor and simplify.

$$= \frac{y-1}{y^2+y+1}$$

$\blacksquare$

EXAMPLE 8 **Adding and Subtracting Rational Expressions with Different Denominators**

Combine: $\dfrac{3y+2}{y-5} + \dfrac{4}{3y+4} - \dfrac{7y^2+24y+28}{3y^2-11y-20}$

Solution

Because $3y^2 - 11y - 20 = (3y+4)(y-5)$ and these are the factors in the denominators of the other two rational expressions in the indicated sum, the LCD is $(3y+4)(y-5)$.

$$\frac{3y+2}{y-5} + \frac{4}{3y+4} - \frac{7y^2+24y+28}{3y^2-11y-20}$$

$$= \frac{(3y+2)(3y+4)}{(3y+4)(y-5)} + \frac{4(y-5)}{(3y+4)(y-5)} - \frac{7y^2+24y+28}{(3y+4)(y-5)}$$

Rewrite each rational expression in terms of the LCD.

$$= \frac{(3y+2)(3y+4) + 4(y-5) - (7y^2+24y+28)}{(3y+4)(y-5)}$$

Add and subtract as indicated.

$$= \frac{9y^2 + 18y + 8 + 4y - 20 - 7y^2 - 24y - 28}{(3y+4)(y-5)}$$

$$= \frac{2y^2 - 2y - 40}{(3y+4)(y-5)}$$

Combine like terms in the numerator.

$$= \frac{2(y-5)(y+4)}{(3y+4)(y-5)}$$

Factor and simplify.

$$= \frac{2(y+4)}{3y+4}$$

$\blacksquare$

Example 9 differs from the ones we have presented up to this point. The two factors in the denominators are additive inverses.

EXAMPLE 9 **Adding Rational Expressions Whose Denominators Have Opposite Signs**

Add: $\dfrac{y^2}{y-5} + \dfrac{4y+5}{5-y}$

Solution

With denominators that are additive inverses, we can multiply the numerator and denominator of either rational expression by -1 and immediately obtain a common denominator.

$$\frac{y^2}{y-5} + \frac{4y+5}{5-y} \cdot \frac{-1}{-1} = \frac{y^2}{y-5} + \frac{-4y-5}{y-5}$$

$$= \frac{y^2 - 4y - 5}{y-5}$$

Add numerators.

$$= \frac{(y-5)(y+1)}{y-5}$$

Factor and simplify.

$$= y + 1$$

$\blacksquare$

3
Simplify complex rational expressions.

Simplifying Complex Rational Expressions

Complex rational expressions have numerators or denominators containing one or more rational expressions, such as

$$\frac{\dfrac{1}{x-2} - \dfrac{1}{x-3}}{1 + \dfrac{1}{x^2 - 5x + 6}}.$$

One way to simplify this expression is to first combine both the numerator and the denominator into single rational expressions. Then we can find the quotient by multiplying the numerator by the reciprocal of the denominator.

EXAMPLE 10   **Simplifying a Complex Rational Expression: Method 1**

Simplify: $\dfrac{\dfrac{1}{x-2} - \dfrac{1}{x-3}}{1 + \dfrac{1}{x^2 - 5x + 6}}$

Solution

We begin by subtracting in the numerator and adding in the denominator.

$$\frac{\dfrac{1}{x-2} - \dfrac{1}{x-3}}{1 + \dfrac{1}{x^2 - 5x + 6}} = \frac{\dfrac{1(x-3)}{(x-2)(x-3)} - \dfrac{1(x-2)}{(x-2)(x-3)}}{\dfrac{x^2 - 5x + 6}{x^2 - 5x + 6} + \dfrac{1}{x^2 - 5x + 6}}$$

The LCD is $(x-2)(x-3)$ in the numerator. The LCD in the denominator is $x^2 - 5x + 6$.

$$= \frac{\dfrac{(x-3) - (x-2)}{(x-2)(x-3)}}{\dfrac{x^2 - 5x + 6 + 1}{x^2 - 5x + 6}}$$

Subtract in the numerator and add in the denominator.

$$= \frac{\dfrac{-1}{(x-2)(x-3)}}{\dfrac{x^2 - 5x + 7}{x^2 - 5x + 6}}$$

Simplify by removing parentheses around $x - 2$, and combine like terms.

$$= \frac{-1}{(x-2)(x-3)} \cdot \frac{x^2 - 5x + 6}{x^2 - 5x + 7}$$

Multiply the numerator by the reciprocal of the denominator.

$$= \frac{-1}{(x-2)(x-3)} \cdot \frac{(x-2)(x-3)}{x^2 - 5x + 7}$$

Factor and multiply. Cancel identical factors in the numerator and the denominator.

$$= \frac{-1}{x^2 - 5x + 7}, \quad x \neq 2, x \neq 3$$

Multiply the remaining factors.   ∎

A second method for simplifying a complex rational expression is to multiply the numerator and denominator by the least common denominator of every rational expression in the numerator and denominator. We see how this is done in Example 11.

EXAMPLE 11 **Simplifying a Complex Rational Expression: Method 2**

Simplify:
$$\dfrac{\dfrac{1}{x-2}-\dfrac{1}{x-3}}{1+\dfrac{1}{x^2-5x+6}}$$

Solution

We begin by factoring $x^2 - 5x + 6$ as a way to determine the LCD of every fraction in the numerator and denominator.

$$\dfrac{\dfrac{1}{x-2}-\dfrac{1}{x-3}}{1+\dfrac{1}{x^2-5x+6}} = \dfrac{\dfrac{1}{x-2}-\dfrac{1}{x-3}}{1+\dfrac{1}{(x-2)(x-3)}}$$

Factor $x^2 - 5x + 6$ as $(x - 2)(x - 3)$. The LCD of every fraction is $(x - 2)(x - 3)$.

$$= \dfrac{\left(\dfrac{1}{x-2}-\dfrac{1}{x-3}\right)}{\left(1+\dfrac{1}{(x-2)(x-3)}\right)} \cdot \dfrac{(x-2)(x-3)}{(x-2)(x-3)}$$

Multiply the numerator and denominator by the LCD of the denominators.

$$= \dfrac{\dfrac{1}{x-2}(x-2)(x-3)-\dfrac{1}{x-3}(x-2)(x-3)}{1(x-2)(x-3)+\dfrac{1}{(x-2)(x-3)}(x-2)(x-3)}$$

Apply the distributive property.

$$= \dfrac{(x-3)-(x-2)}{(x-2)(x-3)+1}$$

Divide the numerators and denominators of the terms by the common factors.

$$= \dfrac{x-3-x+2}{x^2-3x-2x+6+1}$$

Perform the indicated operations.

$$= \dfrac{-1}{x^2-5x+7}, \quad x \neq 2, x \neq 3$$

Combine like terms. ∎

The two methods for simplifying a complex rational expression are summarized as follows.

> **Methods for simplifying a complex rational expression**
>
> *Method 1.* Add or subtract, as indicated, to obtain a single rational expression in the numerator and denominator. Then find the quotient by inverting and multiplying. Simplify, if possible.
>
> *Method 2.* Find the LCD of all expressions within the given complex rational expression. Multiply the numerator and denominator (all fractions) by this LCD by distributing and simplifying. If possible, simplify the resulting expression.

Which of the two methods do you prefer? Let's try them both in Example 12.

EXAMPLE 12 **Simplifying a Complex Rational Expression: Comparing Methods**

Simplify: $\dfrac{1 - \dfrac{1}{x^2}}{1 - \dfrac{4}{x} + \dfrac{3}{x^2}}$

Solution

Method 1. First add and subtract

$$\frac{1 - \dfrac{1}{x^2}}{1 - \dfrac{4}{x} + \dfrac{3}{x^2}} = \frac{\dfrac{x^2}{x^2} - \dfrac{1}{x^2}}{\dfrac{x^2}{x^2} - \dfrac{4}{x} \cdot \dfrac{x}{x} + \dfrac{3}{x^2}}$$ ← Find a common ← denominator.

$$= \frac{\dfrac{x^2 - 1}{x^2}}{\dfrac{x^2 - 4x + 3}{x^2}}$$ ← Subtract

← Subtract and add.

$$= \frac{x^2 - 1}{x^2} \cdot \frac{x^2}{x^2 - 4x + 3}$$ Invert and multiply.

$$= \frac{(x + 1)(x - 1)}{x^2} \cdot \frac{x^2}{(x - 1)(x - 3)}$$ Factor and simplify.

$$= \frac{x + 1}{x - 3}$$ Multiply remaining factors in numerator and denominator.

Method 2. Multiply by the LCD

$$\frac{1 - \dfrac{1}{x^2}}{1 - \dfrac{4}{x} + \dfrac{3}{x^2}} = \frac{\left(1 - \dfrac{1}{x^2}\right) x^2}{\left(1 - \dfrac{4}{x} + \dfrac{3}{x^2}\right) x^2}$$ Multiply the numerator and denominator by x^2, the LCD of all fractions.

$$= \frac{1 \cdot x^2 - \dfrac{1}{x^2} \cdot x^2}{1 \cdot x^2 - \dfrac{4}{x} \cdot x^2 + \dfrac{3}{x^2} \cdot x^2}$$ Apply the distributive property.

$$= \frac{x^2 - 1}{x^2 - 4x + 3}$$ Simplify.

$$= \frac{(x + 1)(x - 1)}{(x - 3)(x - 1)}$$ Factor and simplify.

$$= \frac{x + 1}{x - 3}$$

Our work in Example 12 shows that the graphs of

$$f(x) = \frac{1 - \dfrac{1}{x^2}}{1 - \dfrac{4}{x} + \dfrac{3}{x^2}} \quad \text{and} \quad g(x) = \frac{x + 1}{x - 3}$$

← Denominator is zero if $x = 0, 1,$ or 3.   ← Denominator is zero if $x = 3$.

are identical except that the domain of f excludes 0 and 1. Both functions exclude 3 from their domains. This is shown by the graphs in Figure 5.11 at the top of the next page. Therefore, if we want to set the given complex rational expression equal to its simplified form, we should write

$$\frac{1 - \dfrac{1}{x^2}}{1 - \dfrac{4}{x} + \dfrac{3}{x^2}} = \frac{x + 1}{x - 3}, \quad x \neq 0, x \neq 1.$$

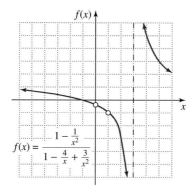

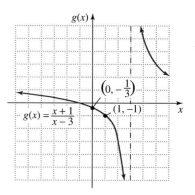

f is undefined at 0, 1, and 3.

Figure 5.11

g is undefined at 3. However, $g(0) = -\frac{1}{3}$ and $g(1) = -1$.

Using technology

Use a graphing utility to verify the hand-drawn graphs in Figure 5.11. TRACE along each curve and verify that *f* is undefined at 0 and 1, whereas on *g*, if $x = 0$, $y = -\frac{1}{3}$ and if $x = 1$, $y = -1$.

PROBLEM SET 5.2

Practice Problems

In Problems 1–18, perform the indicated operations. These problems involve addition and subtraction with the same denominators. If applicable, use a graphing utility to verify your work on all problems involving one variable.

1. $\dfrac{3}{2x} + \dfrac{7}{2x}$

2. $\dfrac{4}{3x} + \dfrac{11}{3x}$

3. $\dfrac{8}{5xy^2} - \dfrac{3}{5xy^2}$

4. $\dfrac{23}{7x^2y} - \dfrac{2}{7x^2y}$

5. $\dfrac{x + 5y}{x + y} + \dfrac{x - 3y}{x + y}$

6. $\dfrac{a + 7b}{a + b} + \dfrac{a - 5b}{a + b}$

7. $\dfrac{3x + 2}{x - 4} - \dfrac{x - 6}{x - 4}$

8. $\dfrac{4x - 10}{x - 2} - \dfrac{x - 4}{x - 2}$

9. $\dfrac{a^2 + 7a + 3}{a^2 + 9a + 9} + \dfrac{2a + 6}{a^2 + 9a + 9}$

10. $\dfrac{y^2 + y - 7}{y^2 + 3y - 10} + \dfrac{y^2 + 5y - 13}{y^2 + 3y - 10}$

11. $\dfrac{a^2 - 4a}{a^2 - a - 6} - \dfrac{a - 6}{a^2 - a - 6}$

12. $\dfrac{x^2 + 6x + 2}{(x + 3)(x - 2)} - \dfrac{2x - 1}{(x + 3)(x - 2)}$

13. $\dfrac{9x}{10} - \dfrac{7x}{10} + \dfrac{3x}{10}$

14. $\dfrac{6}{15x} + \dfrac{11}{15x} - \dfrac{2}{15x}$

15. $\dfrac{3a + b}{a + b} + \dfrac{2a + 3b}{a + b} - \dfrac{4a + 3b}{a + b}$

16. $\dfrac{x^2 - 7}{x - 2} - \dfrac{x - 1}{x - 2} + \dfrac{x + 2}{x - 2}$

17. $\dfrac{3a^3 + 4b^3}{a^2 - b^2} - \dfrac{5b^3 + 2a^3}{a^2 - b^2}$

18. $\dfrac{2x^3 - 3y^3}{x^2 - y^2} - \dfrac{x^3 - 2y^3}{x^2 - y^2}$

In Problems 19–74, perform the indicated operations. These problems involve addition and subtraction with different denominators.

19. $\dfrac{3}{2x} + \dfrac{4}{3x}$

20. $\dfrac{10}{7y} - \dfrac{5}{2y}$

21. $\dfrac{5}{2x^2} - \dfrac{2}{7x}$

22. $\dfrac{7}{2x^2} - \dfrac{3}{5x}$

23. $\dfrac{3}{6x^3} - \dfrac{2}{9x^2}$

24. $\dfrac{3c}{8a^3b} + \dfrac{c}{2ab^2}$

25. $\dfrac{2a}{3c^2} - \dfrac{3b}{4cd}$

26. $\dfrac{5x}{2c^3} - \dfrac{2x^2}{3c^2} - \dfrac{5x^3}{4c}$

27. $\dfrac{3a}{2c^2} - \dfrac{2a}{3cd} + \dfrac{a}{6d^2}$

28. $\dfrac{bx + 1}{a^2} - \dfrac{b + a}{a}$

29. $\dfrac{2b - 2c}{b^2c} + \dfrac{b - c}{bc^2}$

30. $\dfrac{a - b}{ax} - \dfrac{c - b}{cx}$

31. $\dfrac{b - 2y}{4b^2y} + \dfrac{2b + y}{6by^2}$

32. $\dfrac{x + y}{2xy^2} + \dfrac{x^2 - 2y^2}{3x^2y^2} - \dfrac{2x - y}{4x^2y}$

33. $\dfrac{4x - 3y}{6xy} - \dfrac{x - 4z}{8xz} - \dfrac{3y - z}{4yz}$

34. $\dfrac{6}{y + 5} + \dfrac{3}{y + 4}$

35. $\dfrac{10}{x + 4} - \dfrac{2}{x - 6}$

36. $\dfrac{8}{c - 2} + \dfrac{2}{c - 3}$

37. $\dfrac{b}{b - c} - \dfrac{c}{b + c}$

38. $\dfrac{4}{x} - \dfrac{3}{x + 3}$

39. $\dfrac{3}{a + 1} - \dfrac{3}{a}$

40. $\dfrac{6}{y} + \dfrac{2}{y + 2}$

41. $\dfrac{5x}{x - 2} - \dfrac{x - 1}{x + 2}$

42. $\dfrac{2a}{a + 2} + \dfrac{a + 2}{a - 2}$

43. $\dfrac{3x}{x - 3} - \dfrac{x + 4}{x + 2}$

44. $\dfrac{x + 3}{x - 3} + \dfrac{x - 3}{x + 3}$

45. $\dfrac{a - b}{a + b} - \dfrac{a + b}{a - b}$

46. $\dfrac{4}{x} + \dfrac{2}{x - y} - \dfrac{1}{x + y}$

47. $\dfrac{4}{x + 2} - \dfrac{3}{x + 1} + \dfrac{2}{x}$

48. $\dfrac{3}{5a + 2} + \dfrac{5a}{25a^2 - 4}$

49. $\dfrac{5}{2b - 8} + \dfrac{3}{4b - 2}$

50. $\dfrac{5}{y^2 - 4} - \dfrac{3}{y + 2}$

51. $\dfrac{4}{x^2 + 6x + 9} + \dfrac{4}{x + 3}$

52. $2 + \dfrac{x - 4}{x + 1}$

53. $4 + \dfrac{x + 3}{x - 5}$

54. $2 + \dfrac{x}{x - 3} - \dfrac{3}{x^2 - 9}$

55. $3 + \dfrac{1}{x + 2} - \dfrac{2}{x^2 - 4}$

56. $\dfrac{2}{x - 3} - \dfrac{x + 3}{x^2 - x - 6}$

57. $\dfrac{c}{c^2 - 10c + 25} - \dfrac{c - 4}{2c - 10}$

58. $\dfrac{4}{b^2 - 4b + 4} - \dfrac{2}{b^2 - 4}$

59. $\dfrac{a - b}{3a + 3b} + \dfrac{a + b}{2a - 2b}$

60. $\dfrac{y - 4}{y + 6} - \dfrac{y - 6}{y + 4}$

61. $\dfrac{b + 2}{b^2 + b - 2} + \dfrac{2}{b^2 - 1}$

62. $\dfrac{x - 1}{x^2 - 9} - \dfrac{3x + 2}{x^2 - x - 6}$

63. $\dfrac{y + 3}{y^2 - y - 2} - \dfrac{y - 1}{y^2 + 2y + 1}$

64. $\dfrac{c - 1}{c^2 + 3c + 2} - \dfrac{c + 7}{c^2 + 5c + 6}$

65. $\dfrac{x^2 + x + 2}{x^3 - 1} - \dfrac{1}{x - 1}$

66. $\dfrac{-a^2 - a - 6}{a^3 - 8} + \dfrac{1}{a - 2}$

67. $\dfrac{1}{y - x} + \dfrac{1}{x - y}$

68. $\dfrac{1}{a - b} - \dfrac{1}{b - a}$

69. $\dfrac{y^2}{y - 7} + \dfrac{6y + 7}{7 - y}$

70. $\dfrac{3b}{2a - b} + \dfrac{2a - 1}{b - 2a}$

71. $x - \dfrac{3}{x - 2}$

72. $y - \dfrac{25y - 5y^2}{y^2 - 10y + 25}$

73. $\dfrac{x + 3y}{x^2 - 7xy + 12y^2} - \dfrac{x - 3y}{x^2 - xy - 12y^2}$

74. $\dfrac{x + 5}{x^2 - x - 6} + \dfrac{x + 3}{x^2 + 7x + 10}$

Simplify each complex rational expression in Problems 75–108 by either method discussed in this section. If applicable, use a graphing utility to verify your work on problems involving one variable.

75. $\dfrac{\dfrac{3}{y}}{y - \dfrac{1}{y}}$

76. $\dfrac{\dfrac{3}{4} - x}{\dfrac{3}{4} + x}$

77. $\dfrac{3 - \dfrac{2}{b}}{\dfrac{2}{b} + \dfrac{3}{b}}$

78. $\dfrac{\dfrac{1}{a} + \dfrac{1}{b}}{\dfrac{1}{a} - \dfrac{1}{b}}$

79. $\dfrac{\dfrac{1}{y} + \dfrac{1}{y^2}}{1 + \dfrac{1}{y}}$

80. $\dfrac{\dfrac{1}{3} - \dfrac{4}{a}}{\dfrac{3}{4} - \dfrac{1}{a}}$

81. $\dfrac{\dfrac{x}{y} + \dfrac{y}{x}}{\dfrac{1}{y} + \dfrac{1}{x}}$

82. $\dfrac{a^2 - \dfrac{b^2}{4}}{a + \dfrac{b}{2}}$

83. $\dfrac{\dfrac{b^2 - c^2}{b}}{\dfrac{b - c}{b^2}}$

84. $\dfrac{x - \dfrac{y^2}{x}}{\dfrac{x}{y} - \dfrac{y}{x}}$

85. $\dfrac{\dfrac{x}{y} - \dfrac{y}{x}}{\dfrac{x^2}{y} - y}$

86. $\dfrac{\dfrac{m^2}{4} - 4m^2}{\dfrac{m}{2} + 2m}$

87. $\dfrac{y + 5 + \dfrac{6}{y}}{y - \dfrac{9}{y}}$

88. $\dfrac{c + \dfrac{9}{c} - 6}{\dfrac{c}{2} - \dfrac{9}{2c}}$

89. $\dfrac{1 - \dfrac{1}{a}}{\dfrac{a + 1}{a}}$

90. $\dfrac{a + b}{\dfrac{1}{a^2} - \dfrac{1}{b^2}}$

91. $\dfrac{\dfrac{1}{x} + \dfrac{1}{y}}{x + y}$

92. $\dfrac{1 - \dfrac{1}{x}}{xy}$

93. $\dfrac{b^2 - c^2}{\dfrac{1}{b} + \dfrac{1}{c}}$

94. $\dfrac{\dfrac{m + 2}{2m}}{m^2 - 4}$

95. $\dfrac{3 - \dfrac{1}{c}}{3c - 1}$

96. $\dfrac{c - \dfrac{c}{c + 3}}{c + 2}$

97. $\dfrac{b - 3}{b - \dfrac{3}{b - 2}}$

98. $\dfrac{x + \dfrac{x}{x + 2}}{x - \dfrac{x}{x + 2}}$

99. $\dfrac{y + \dfrac{12}{y - 7}}{y - 3}$

100. $\dfrac{\dfrac{2}{a} + \dfrac{1}{a + 1}}{3a + 2}$

101. $\dfrac{\dfrac{3}{y - 2} - \dfrac{4}{y + 2}}{\dfrac{7}{y^2 - 4}}$

102. $\dfrac{\dfrac{x^2 - y^2}{a + b}}{\dfrac{x + y}{a^2 - b^2}}$

103. $\dfrac{\dfrac{4}{b - 2} + 1}{\dfrac{3}{b^2 - 4} + 1}$

104. $\dfrac{\dfrac{1}{m + 1}}{\dfrac{1}{m^2 - 2m - 3} + \dfrac{1}{m - 3}}$

105. $\dfrac{\dfrac{6}{x^2 + 2x - 15} - \dfrac{1}{x - 3}}{\dfrac{1}{x + 5} + 1}$

106. $\dfrac{x + \dfrac{2x}{x - 2}}{1 + \dfrac{4}{x^2 - 4}}$

107. $\dfrac{\dfrac{3}{x + 2y} - \dfrac{2y}{x^2 + 2xy}}{\dfrac{3y}{x^2 + 2xy} + \dfrac{5}{x}}$

108. $\dfrac{\dfrac{1}{x^3 - y^3}}{\dfrac{1}{x - y} - \dfrac{1}{x^2 + xy + y^2}}$

Application Problems

109. A small boat averages 10 miles per hour in still water. The boat travels 24 miles upstream (against the current) and then reverses this direction, traveling 24 miles downstream.
  a. If x represents the speed of the current, write a rational function that models the total time t upstream and back.
  b. Add the two rational expressions in the model, writing the total time function with a single rational expression.
  c. Find and interpret $t(2)$.

110. A rat that has been given x previous trials of running through a maze can run through the maze in t minutes, where

$$t = \dfrac{n^2 - 4}{10n^3}\left(6 + \dfrac{20}{n + 2}\right).$$

Write this formula as a single rational expression in simplified form.

111. The average speed on a round trip having a one-way distance d is given by the complex rational expression

$$\dfrac{2d}{\dfrac{d}{r_1} + \dfrac{d}{r_2}}$$

in which r_1 and r_2 are the speeds on the outgoing and return trips respectively. Simplify the expression. Then find the average speed for a person who drives from home to work at 30 miles per hour and returns on the same route averaging 20 miles per hour. Explain why the answer is not 25 miles per hour.

112. If three resistors with resistances R_1, R_2, and R_3 are connected in parallel, their combined resistance R is given by the mathematical model

$$R = \dfrac{1}{\dfrac{1}{R_1} + \dfrac{1}{R_2} + \dfrac{1}{R_3}}.$$

Simplify the complex rational expression on the right. Then find R when R_1 is 4 ohms. R_2 is 8 ohms, and R_3 is 12 ohms.

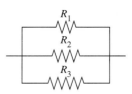

True–False Critical Thinking Problems

113. Which one of the following is true?
  a. Because $\dfrac{a}{c} + \dfrac{b}{c} = \dfrac{a + b}{c}$, then $\dfrac{a}{b} + \dfrac{a}{c} = \dfrac{a}{b + c}$.
  b. After adding or subtracting rational expressions, we should simplify the resulting expression by eliminating common variables in the numerator and denominator.
  c. $6 + \dfrac{1}{x} = \dfrac{7}{x}$
  d. None of the above is true.

114. Which one of the following is true?
  a. $5^{-1} + 6^{-1} = (5 + 6)^{-1}$
  b. To simplify

$$\frac{\dfrac{1}{x} + \dfrac{1}{y}}{\dfrac{2}{x} + \dfrac{3}{y}}$$

we multiply the numerator and denominator by $\dfrac{xy}{xy}$.

  c. $\dfrac{1 - x^{-1}}{1 - x^{-2}} = \dfrac{x}{x + 1}$
  d. None of the above is true.

Technology Problems

115. An architect is constructing a house whose cross section up to the roof is in the shape of a rectangle with an area of 2500 square feet. The figure on the right shows the width of the rectangle represented by x and the length by $\dfrac{2500}{x}$.

  a. Write a function that models the perimeter of the rectangle in terms of x. Simplify the function's formula by adding the rational expressions in the perimeter model.
  b. Use your graphing utility to graph the function using the following range setting:

  Xmin = 0, Xmax = 100, Xscl = 10, Ymin = 0,
  Ymax = 400, Yscl = 20

  c. Use the |TRACE| or |fMIN| feature to find the coordinates of the function's minimum point.
  d. A minimum perimeter will reduce construction costs for the house. The x-coordinate that you identified in part (c) represents the width of the house. What length will result in the least possible

perimeter? Complete this statement: The minimum perimeter occurs when the rectangle is a _____.

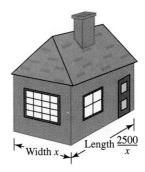

Width x ⟶ Length $\dfrac{2500}{x}$

116. a. Use a graphing utility to graph the function in Problem 109 for meaningful values of x.
  b. |TRACE| along the curve until you reach $x = 4$. What is the corresponding value of y? Describe what these numbers mean in terms of the variables modeled by the function.

Writing in Mathematics

Explain the error in Problems 117–119. Then rewrite the right side of the equation to correct the error that now exists.

117. $\dfrac{1}{a} + \dfrac{1}{b} = \dfrac{1}{a + b}$

118. $\dfrac{1}{x} + 7 = \dfrac{1}{x + 7}$

119. $\dfrac{a}{x} + \dfrac{a}{b} = \dfrac{a}{x + b}$

120. It is possible to add $\dfrac{2}{x + 1}$ and $\dfrac{5}{x}$ using a common denominator other than the LCD. Perform the addition $\dfrac{2}{x + 1} + \dfrac{5}{x}$ *twice*, first using the LCD and then again using $x(x + 1)^2$ as the common denominator. Describe in words what happens.

121. In one short sentence, five words or less, explain what

$$\frac{\dfrac{1}{x} + \dfrac{1}{x^2} + \dfrac{1}{x^3}}{\dfrac{1}{x^4} + \dfrac{1}{x^5} + \dfrac{1}{x^6}}$$

does to each number x.

122. Suppose you are a teacher grading an examination. Each question is worth 10 points, with partial credit to be given at your discretion. One question is to simplify $\dfrac{a^{-1} + b}{b^{-1} + a}$. Take a moment to simplify this complex fraction correctly. Now suppose that you are grading a student's paper and the work is done as follows:

$$\frac{a^{-1} + b}{b^{-1} + a} = \frac{b + b}{a + a} = \frac{2b}{2a} = \frac{b}{a}$$

Notice that the answer is correct. How many points out of 10, if any, would you give this student? Write a justification of your position.

123. Without actually graphing, describe how the graphs of

$$f(x) = \frac{1}{x} \quad \text{and} \quad g(x) = \frac{\dfrac{1}{x-1}}{\dfrac{x}{x-1}}$$

differ.

Critical Thinking Problems

124. "This is a curious thing," a friend said to me. "There are two numbers whose sum equals their product. They are 2 and 2, for if you add them or multiply them, the result is 4." Then my friend tripped badly, adding, "These are the only two numbers that have this property."

   a. Let $f(x) = x$ and $g(x) = \dfrac{x}{x-1}$. Show that

$$f + g = fg$$

   b. Use the functions f and g to show how my friend "tripped badly" by finding three additional number pairs whose sum equals their product.

Perform the indicated operations in Problems 126–127.

126. $\dfrac{1}{x^n - 1} - \dfrac{1}{x^n + 1} - \dfrac{1}{x^{2n} - 1}$

127. $(x - y)^{-1} + (x + y)^{-1}$

128. Find the sum of the areas of the shaded regions shown in the figure.

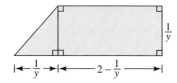
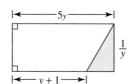

129. Find the sum of the areas of the shaded regions shown in the figure.

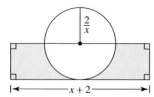
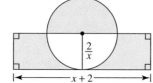

130. a. Simplify:

$$\left(1 - \frac{1}{x}\right)\left(1 - \frac{1}{x+1}\right)\left(1 - \frac{1}{x+2}\right)\left(1 - \frac{1}{x+3}\right).$$

125. Consider the rational expressions

$$\frac{y - z}{1 + yz} \qquad \frac{z - x}{1 + xz} \qquad \frac{x - y}{1 + xy}$$

   a. Let $x = 1$, $y = 2$, and $z = 3$. Find the sum of the three resulting rational numbers and their product. What pattern do you notice?

   b. Show that the sum and product of the three given rational expressions are equal.

(Problem 130 continued)

   b. Simplify: $\left(1 - \dfrac{1}{x}\right)\left(1 - \dfrac{1}{x+1}\right)\left(1 - \dfrac{1}{x+2}\right)$

$$\left(1 - \frac{1}{x+3}\right)\left(1 - \frac{1}{x+4}\right).$$

   c. Use the pattern in parts (a) and (b) to find

$$\left(1 - \frac{1}{x}\right)\left(1 - \frac{1}{x+1}\right)\left(1 - \frac{1}{x+2}\right) \cdots$$

$$\left(1 - \frac{1}{x+99}\right)\left(1 - \frac{1}{x+100}\right).$$

131. The harmonic mean of the two numbers, x and y, is found by taking their reciprocals, finding their average, and taking the reciprocal of the result.

   a. Show that the harmonic mean of x and y can be simplified to $\dfrac{2xy}{x + y}$.

   b. Fill in the missing numbers so that the cells in the middle of each side and the center cell are each the harmonic means of the numbers sandwiching them.

1260		
		360
315	280	

Group Activity Problem

132. Group members should study the solutions in Examples 10 and 11 on pages 439 and 440. As a group, list one major advantage and one major disadvantage of each method. Make up an example that would be easier to simplify by method 1, and do the same for method 2.

Review Problems

133. Solve: $|3x - 1| \leq 14$.

134. Factor completely: $2x^4 - 9x^2$.

135. Solve: $y^2 + 27 = 12y$.

SECTION 5.3

Solutions Manual   **Tutorial**   **Video 6**

Dividing Polynomials

Objectives

1 Divide monomials.
2 Divide a polynomial by a monomial.
3 Divide polynomials using long division.
4 Divide polynomials using synthetic division.

In the previous chapter, we studied the addition, subtraction, and multiplication of polynomials. Since rational expressions indicate polynomial division, we now turn our attention to division of polynomials.

1 Divide monomials.

Dividing a Monomial by a Monomial

Division of monomials can be performed using

$$\frac{b^n}{b^m} = b^{n-m}, \quad b \neq 0.$$

Here's a step-by-step procedure.

> **Dividing monomials**
>
> 1. Divide the coefficients.
> 2. Divide variable factors with the same base by subtracting exponents.

EXAMPLE 1   Dividing Monomials

Find each quotient:

a. $\dfrac{x^7}{x^3}$ **b.** $\dfrac{25x^8}{5x^6}$ **c.** $\dfrac{10y^6}{-2y^{13}}$ **d.** $\dfrac{-30x^4y^3}{60x^2y}$

Solution

a. $\dfrac{x^7}{x^3} = x^{7-3} = x^4$ **b.** $\dfrac{25x^8}{5x^6} = \dfrac{25}{5} \cdot \dfrac{x^8}{x^6} = 5x^{8-6} = 5x^2$

c. $\dfrac{10y^6}{-2y^{13}} = \dfrac{10}{-2} \cdot \dfrac{y^6}{y^{13}} = -5y^{6-13} = -5y^{-7} = -\dfrac{5}{y^7}$

This result is not a monomial because the variable appears in the denominator. *The quotient of two monomials need not be a monomial.*

d. $\dfrac{-30x^4y^3}{60x^2y} = \dfrac{-30}{60} \cdot \dfrac{x^4}{x^2} \cdot \dfrac{y^3}{y} = -\dfrac{1}{2}x^{4-2}y^{3-1} = -\dfrac{1}{2}x^2y^2$

The quotient can be written equivalently as $\dfrac{-x^2y^2}{2}$. ∎

2 Divide a polynomial by a monomial.

Dividing a Polynomial by a Monomial

In this type of division, we can use the rule for adding with the same denominator:

$$\frac{a}{b} + \frac{c}{b} = \frac{a + c}{b}$$

Reversing this rule gives

$$\frac{a + c}{b} = \frac{a}{b} + \frac{c}{b}.$$

This gives us a step-by-step procedure for dividing a polynomial by a monomial.

> **Dividing a polynomial by a monomial**
>
> **1.** Divide each term in the polynomial by the monomial.
> **2.** Simplify terms using the rule for division of monomials.

EXAMPLE 2   **Dividing a Trinomial by a Monomial**

Find the quotient of: $12x^6 + 8x^5 + 4x^3$ and $2x^2$

Solution

$$\frac{12x^6 + 8x^5 + 4x^3}{2x^2}$$

$$= \frac{12x^6}{2x^2} + \frac{8x^5}{2x^2} + \frac{4x^3}{2x^2} \qquad \text{Divide each term of the trinomial by the monomial.}$$

$$= 6x^{6-2} + 4x^{5-2} + 2x^{3-2} \qquad \text{Divide coefficients and subtract exponents.}$$

$$= 6x^4 + 4x^3 + 2x$$ ∎

EXAMPLE 3   **Dividing a Polynomial by a Monomial**

Divide, expressing each quotient with positive exponents:

a. $\dfrac{12x^6y^4 - 8x^4y^2 + 4x^3y^3}{-4x^2y}$ **b.** $\dfrac{8m^4n^2 + 6mn^3 - 10m^3n - 12mn}{4m^2n^2}$

Solution

a. $\dfrac{12x^6y^4 - 8x^4y^2 + 4x^3y^3}{-4x^2y}$

$$= \frac{12x^6y^4}{-4x^2y} + \frac{-8x^4y^2}{-4x^2y} + \frac{4x^3y^3}{-4x^2y} \qquad \text{Divide each term of the trinomial by the monomial.}$$

$$= -3x^{6-2}y^{4-1} + 2x^{4-2}y^{2-1} - x^{3-2}y^{3-1} \qquad \text{Divide coefficients and subtract exponents.}$$
$$= -3x^4y^3 + 2x^2y - xy^2$$

This quotient is a polynomial in x and y because all exponents on the variables consist of whole numbers.

b. $\dfrac{8m^4n^2 + 6mn^3 - 10m^3n - 12mn}{4m^2n^2}$

$$= \frac{8m^4n^2}{4m^2n^2} + \frac{6mn^3}{4m^2n^2} - \frac{10m^3n}{4m^2n^2} - \frac{12mn}{4m^2n^2} \qquad \begin{array}{l}\text{Divide each term of the}\\ \text{polynomial by the}\\ \text{monomial.}\end{array}$$

$$= 2m^{4-2}n^0 + \frac{3}{2}m^{1-2}n^{3-2} - \frac{5}{2}m^{3-2}n^{1-2} - 3m^{1-2}n^{1-2} \qquad \begin{array}{l}\text{Divide coefficients and}\\ \text{subtract exponents.}\end{array}$$

$$= 2m^2 + \frac{3}{2}m^{-1}n - \frac{5}{2}mn^{-1} - 3m^{-1}n^{-1}$$

$$= 2m^2 + \frac{3}{2}\cdot\frac{1}{m}\cdot n - \frac{5}{2}\cdot m\cdot\frac{1}{n} - 3\cdot\frac{1}{m}\cdot\frac{1}{n}$$

$$= 2m^2 + \frac{3n}{2m} - \frac{5m}{2n} - \frac{3}{mn}$$

This quotient is not a polynomial in m and n because the last three terms contain variables in the denominators. ∎

3 Divide polynomials using long division.

Dividing by a Polynomial Containing More Than One Term

When a divisor has more than one term, the four steps used to divide whole numbers—*divide, multiply, subtract, bring down* the next term—form the repetitive procedure for polynomial long division.

Discover for yourself

Before reading the solution to Example 4, find the quotient

$$\frac{x^2 + 10x + 21}{x + 3}$$

by factoring and simplifying. Under what conditions would this method not work, making the long division process shown in Example 4 necessary? If you're not sure of the answer to this question, first read the solution and then reconsider the question.

EXAMPLE 4   **Dividing a Trinomial by a Binomial**

Divide: $x^2 + 10x + 21$ by $x + 3$

Solution

$$x + 3\overline{)x^2 + 10x + 21}$$

Arrange the terms of the dividend ($x^2 + 10x + 21$) and the divisor ($x + 3$) in descending powers of x.

$$x + 3\overline{)\stackrel{\textstyle x}{x^2 + 10x + 21}}$$

Divide x^2 (the first term in the dividend) by x (the first term in the divisor): $\dfrac{x^2}{x} = x$. Align similar terms.

Multiply.

$$x + 3\overline{)\stackrel{\textstyle x}{x^2 + 10x + 21}}$$
$$\underline{x^2 + \ 3x}$$

Multiply each term in the divisor by $(x + 3)$ by x, aligning under similar terms in the dividend.

$$
\begin{array}{r}
x \\
x + 3 \overline{)\, x^2 + 10x + 21} \\
\ominus x^2 + \ominus 3x \\
\hline
7x
\end{array}
$$

Subtract $x^2 + 3x$ from $x^2 + 10x$ by changing the sign of each term in the lower expression and adding.

$$
\begin{array}{r}
x \\
x + 3 \overline{)\, x^2 + 10x + 21} \\
x^2 + 3x \downarrow \\
\hline
7x + 21
\end{array}
$$

Bring down 21 from the original dividend and add algebraically to form a new dividend.

$$
\begin{array}{r}
x + 7 \\
x + 3 \overline{)\, x^2 + 10x + 21} \\
x^2 + 3x \\
\hline
7x + 21
\end{array}
$$

Find the second term of the quotient. *Divide* the first term of $7x + 21$ by x, the first term of the divisor: $\dfrac{7x}{x} = 7$.

Multiply.

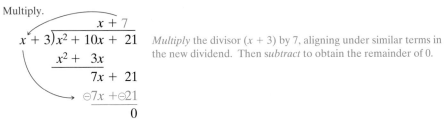

$$
\begin{array}{r}
x + 7 \\
x + 3 \overline{)\, x^2 + 10x + 21} \\
x^2 + 3x \\
\hline
7x + 21 \\
\ominus 7x + \ominus 21 \\
\hline
0
\end{array}
$$

Multiply the divisor $(x + 3)$ by 7, aligning under similar terms in the new dividend. Then *subtract* to obtain the remainder of 0.

Because the remainder is 0, we say that $x + 3$ is a *factor* of $x^2 + 10x + 21$. The answer is

$$(x^2 + 10x + 21) \div (x + 3) = x + 7.$$

| Dividend | Divisor | Quotient |

Check

We can check our division by observing that

$$\underbrace{(x + 3)(x + 7)}_{} = \underbrace{x^2 + 10x + 21}_{}$$

| (Divisor)(Quotient) | = | Dividend |

Before considering additional examples, let's summarize the general procedure for dividing one polynomial by another.

Long division of polynomials

1. *Arrange the terms* of both the dividend and the divisor in descending powers of any variable.
2. *Divide* the first term in the dividend by the first term in the divisor. The result is the first term of the quotient.
3. *Multiply* every term in the divisor by the first term in the quotient. Write the resulting product beneath the dividend with similar terms under each other.
4. *Subtract* the product from the dividend.

5. *Bring down* the next term in the original dividend and write it next to the remainder to form a new dividend.
6. Use this new expression as the dividend and repeat this process until the remainder can no longer be divided. This will occur when the degree of the remainder (the highest exponent on a variable of the remainder) is smaller than the degree of the divisor.

EXAMPLE 5   Long Division of Polynomials

Divide using long division: $4 - 5x + 6x^3 - x^2$ by $3x - 2$

Solution

We begin by writing the divisor and dividend in descending powers of x.

Multiply.

$$\begin{array}{r} 2x^2 \\ 3x - 2 \overline{)6x^3 - x^2 - 5x + 4} \\ \ominus 6x^3 - \oplus 4x^2 \\ \hline 3x^2 - 5x \end{array}$$

Divide: $\frac{6x^3}{3x} = 2x^2$.
Multiply: $2x^2(3x - 2) = 6x^3 - 4x^2$.
Subtract $6x^3 - 4x^2$ from $6x^3 - x^2$ and bring down $-5x$.

Now we divide $3x^2$ by $3x$ to obtain x, multiply x and the divisor, and subtract.

Multiply.

$$\begin{array}{r} 2x^2 + x \\ 3x - 2 \overline{)6x^3 - x^2 - 5x + 4} \\ 6x^3 - 4x^2 \\ \hline 3x^2 \ominus 5x \\ 3x^2 \oplus 2x \\ \hline -3x + 4 \end{array}$$

Divide: $\frac{3x^2}{3x} = x$.
Multiply: $x(3x - 2) = 3x^2 - 2x$.
Subtract $3x^2 - 2x$ from $3x^2 - 5x$ and bring down 4.

Now we divide $-3x$ by $3x$ to obtain -1, multiply -1 and the divisor, and subtract.

Multiply.

$$\begin{array}{r} 2x^2 + x - 1 \\ 3x - 2 \overline{)6x^3 - x^2 - 5x + 4} \\ 6x^3 - 4x^2 \\ \hline 3x^2 - 5x \\ 3x^2 - 2x \\ \hline -3x \oplus 4 \\ -3x \ominus 2 \\ \hline 2 \end{array}$$

Divide: $\frac{-3x}{3x} = -1$.
Multiply: $-1(3x - 2) = -3x + 2$.
Subtract $-3x + 2$ from $-3x + 4$, leaving a remainder of 2.

Because there is a remainder of 2, $3x - 2$ is *not* a factor of $6x^3 - x^2 - 5x + 4$.

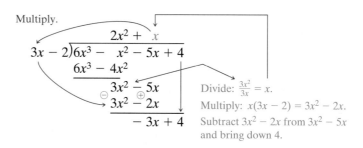

The answer is:

$$\frac{6x^3 - x^2 - 5x + 4}{3x - 2} = 2x^2 + x - 1 + \frac{2}{3x - 2}$$

The remainder of 2 is the numerator, and the divisor is the denominator.

Check

$$(3x - 2)\left[2x^2 + x - 1 + \frac{2}{3x - 2}\right]$$

$$= (3x - 2)(2x^2 + x - 1) + (3x - 2)\left(\frac{2}{3x - 2}\right)$$

Distribute the factor $3x - 2$, and then perform the multiplications.

$$= (6x^3 - x^2 - 5x + 2) + 2$$
$$= 6x^3 - x^2 - 5x + 4 \quad \checkmark$$

Because the result is the dividend, the answer checks. ■

Using technology

In addition to the algebraic check shown in the solution to Example 5, you can verify the result using a graphing utility. The figure below shows the graphs of

$$y_1 = \frac{6x^3 - x^2 - 5x + 4}{3x - 2} \quad \text{and} \quad y_2 = 2x^2 + x - 1 + \frac{2}{3x - 2}.$$

The graphs coincide, so the expressions are equivalent.

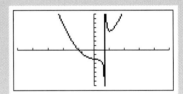

EXAMPLE 6 Long Division of Polynomials

Divide using long division: $6x^4 + 5x^3 + 3x - 5$ by $3x^2 - 2x$

Solution

We write the dividend, $6x^4 + 5x^3 + 3x - 5$ as $6x^4 + 5x^3 + 0x^2 + 3x - 5$ so as to keep all similar terms aligned.

Multiply.

$$
\begin{array}{r}
2x^2 + 3x + 2 \\
3x^2 - 2x \overline{)6x^4 + 5x^3 + 0x^2 + 3x - 5} \\
\ominus\ 6x^4 \overset{\oplus}{-} 4x^3 \\
\hline
9x^3 + 0x^2 \\
\ominus\ 9x^3 \overset{\oplus}{-} 6x^2 \\
\hline
6x^2 + 3x \\
\ominus\ 6x^2 \overset{\oplus}{-} 4x \\
\hline
7x - 5
\end{array}
$$

$2x^2(3x^2 - 2x)$

$3x(3x^2 - 2x)$

$2(3x^2 - 2x)$

Study tip

If a power of x is missing in either a dividend or a divisor, add that power of x with a coefficient of 0 and then divide. In this way, similar terms will be aligned as you carry out the division process.

The division process is finished because the degree of $7x - 5$, which is 1, is less than the degree of the divisor $3x^2 - 2x$, which is 2.
The answer is

$$\frac{6x^4 + 5x^3 + 3x - 5}{3x^2 - 2x} = 2x^2 + 3x + 2 + \frac{7x - 5}{3x^2 - 2x}.$$

■

4 Divide polynomials using synthetic division.

Dividing Polynomials Using Synthetic Division

There is a faster method, called *synthetic division,* for division by polynomials of the form $x - c$. Let's compare the two methods showing $x^3 + 4x^2 - 5x + 5$ divided by $x - 3$.

Long Division

$$
\begin{array}{r}
x^2 + 7x + 16 \quad \leftarrow \text{Partial quotient} \\
x - 3{\overline{\smash{\big)}\,x^3 + 4x^2 - 5x + 5}} \quad \leftarrow \text{Dividend} \\
\ominus\, \underline{x^3 \overset{\oplus}{-} 3x^2} \\
7x^2 - 5x \\
\ominus\, \underline{7x^2 \overset{\oplus}{-} 21x} \\
16x + 5 \\
\ominus\, \underline{16x \overset{\oplus}{-} 48} \\
53 \quad \leftarrow \text{Remainder}
\end{array}
$$

Divisor

Synthetic Division

$$
\begin{array}{r|rrrr}
+3 & 1 & 4 & -5 & 5 \\
& \downarrow & 3 & 21 & 48 \\
\hline
& 1 & 7 & 16 & 53
\end{array}
$$

Discover for yourself

Study the two methods and answer these questions.

1. How is the dividend represented in the synthetic division process?
2. How is the divisor represented in the synthetic division process? If the divisor is $x - c$, what number would appear to the left of the coefficients of the dividend?
3. How is the partial quotient represented in the synthetic division process?
4. Where is the remainder in the synthetic division process?
5. The magenta arrow indicates that the 1 is brought down to the last row. Describe how you might obtain the other three numbers in this row (7, 16, and 53) using a series of multiplications and additions.

You can check your answers to the questions in the Discover for Yourself box as we now outline the steps in the synthetic division process. As you follow the process that appears in the left column in the box, pay close attention to the example on the right.

David Hockney, "The Crossword Puzzle, Minneapolis, Jan. 1983," 1983, photographic collage, 33 × 46 in. © David Hockney

Synthetic Division

To divide a polynomial by $x - c$:

1. Arrange polynomials in descending powers, with a 0 coefficient for any missing term.
2. Write $+c$ for the divisor, $x - c$. To the right, write the coefficients of the dividend.
3. Write the leading coefficient of the dividend on the bottom row.

4. Multiply c (in this case, 3) times the value just written on the bottom row. Write the product in the next column in the second row.
5. Add the values in this new column, writing the sum in the bottom row.

6. Repeat this series of multiplications and additions until all columns are filled in.

Example

$$x - 3 \overline{)x^3 + 4x^2 - 5x + 5}$$

$$\underline{+3}\ \ \begin{array}{cccc} 1 & 4 & -5 & 5 \end{array}$$

$$\underline{+3}\ \ \begin{array}{cccc} 1 & 4 & -5 & 5 \end{array}$$
$$\downarrow$$
$$\ \ 1$$

$$\underline{+3}\ \ \begin{array}{cccc} 1 & 4 & -5 & 5 \end{array}$$
Multiply. $\downarrow\quad 3$
$$\ \ 1$$

$$\underline{+3}\ \ \begin{array}{cccc} 1 & 4 & -5 & 5 \end{array}$$
$$\downarrow\quad 3\ |\text{Add.}$$
$$\ \ 1\quad 7$$

$$\underline{+3}\ \ \begin{array}{cccc} 1 & 4 & -5 & 5 \end{array}$$
Multiply. 3 $\quad$ 21
$$\ \ 1\quad 7\quad 16$$

$$\underline{+3}\ \ \begin{array}{cccc} 1 & 4 & -5 & 5 \end{array}$$
$$\quad\quad 3\quad 21\quad 48$$
$$\ \ 1\quad 7\quad 16\quad 53$$

Multiply.

| Remainder |

7. Use the last row to write the quotient. The final value in this row is the remainder.

$$\frac{x^3 + 4x^2 - 5x + 5}{x - 3}$$

$$= x^2 + 7x + 16 + \frac{53}{x - 3}$$

EXAMPLE 7 **Using Synthetic Division**

Divide using synthetic division: $3x^3 - x^2 + 2x + 5$ by $x - 3$

Solution

All polynomials are in descending powers.

$$\underline{+3}\ \ \begin{array}{cccc} 3 & -1 & 2 & 5 \end{array}$$

Write $+3$ for the divisor. Then write the coefficients of $3x^3 - x^2 + 2x + 5$, placing $+3$ to the left.

$$\underline{+3}\ \ \begin{array}{cccc} 3 & -1 & 2 & 5 \end{array}$$
$$\downarrow\quad 9$$
$$\ \ 3$$

Bring down the leading coefficient to the bottom row. Multiply: $3 \cdot 3 = 9$. Write this product in the next column in the second row.

$$\underline{+3}\ \ \begin{array}{cccc} 3 & -1 & 2 & 5 \end{array}$$
$$\downarrow\quad 9\ \downarrow$$
$$\ \ 3\quad 8$$

Add the values in the second column: $-1 + 9 = 8$. Write this result in the bottom row.

$$\begin{array}{r|rrrr} +3 & 3 & -1 & 2 & 5 \\ & \downarrow & 9 & 24 & \\ \hline & 3 & 8 & & \end{array}$$

Multiply 3 by the value in the bottom row: $3 \cdot 8 = 24$. Write this product in the next column in the second row.

$$\begin{array}{r|rrrr} +3 & 3 & -1 & 2 & 5 \\ & \downarrow & 9 & 24 \downarrow & \\ \hline & 3 & 8 & 26 & \end{array}$$

Add the values in the third column: $2 + 24 = 26$. Write this result in the bottom row.

$$\begin{array}{r|rrrr} +3 & 3 & -1 & 2 & 5 \\ & \downarrow & 9 & 24 & 78 \\ \hline & 3 & 8 & 26 & \end{array}$$

Multiply 3 by the value in the bottom row: $3 \cdot 26 = 78$. Write this product in the last column in the second row.

$$\begin{array}{r|rrrr} +3 & 3 & -1 & 2 & 5 \\ & & 9 & 24 & 78 \downarrow \\ \hline & 3 & 8 & 26 & 83 \end{array}$$

Add the values in the final column: $5 + 78 = 83$.

$$\frac{3x^3 - x^2 + 2x + 5}{x - 3}$$

Write the quotient from the last row, using the final value as the remainder. The degree of the first term of the quotient is always one degree less than the degree of the first term of the dividend.

$$= 3x^2 + 8x + 26 + \frac{83}{x - 3}$$

Notice that the process is a series of multiplications and additions, indicated in the following diagram by the arrows.

$$\begin{array}{r|rrrr} +3 & 3 & -1 & 2 & 5 \\ & \downarrow & 9 \downarrow & 24 \downarrow & 78 \downarrow \\ \hline & 3 & 8 & 26 & 83 \end{array}$$

EXAMPLE 8   Using Synthetic Division

Find the quotient using synthetic division:

$$\frac{x^4 + 2x^2 + 1}{x + 4}$$

Solution

Because we are dividing by $x - c$, we can rewrite this problem as

$$\frac{x^4 + 0x^3 + 2x^2 + 0x + 1}{x - (-4)}.$$

We see that $c = -4$ and proceed.

$$\begin{array}{r|rrrrr} -4 & 1 & 0 & 2 & 0 & 1 \\ & \downarrow & -4 & 16 & -72 & 288 \\ \hline & 1 & -4 & 18 & -72 & 289 \end{array}$$

Thus,

$$\frac{x^4 + 2x^2 + 1}{x + 4} = x^3 - 4x^2 + 18x - 72 + \frac{289}{x + 4}.$$

PROBLEM SET 5.3

Practice Problems

Find the quotient of the monomials in Problems 1–8.

1. $y^7 \div y^5$ **2.** $x^{13} \div x^8$ **3.** $\dfrac{15x^8}{5x^6}$ **4.** $\dfrac{-16y^9}{8y^3}$ **5.** $\dfrac{25x^3y^7}{-5xy^5}$ **6.** $\dfrac{17x^8y^4}{-34x^4y^2}$

7. $(-54x^7y^4z^5) \div (3x^3yz^2)$ **8.** $(-100a^{12}b^6c^8) \div (-300a^6b^2c^4)$

In Problems 9–24, divide and express each quotient with positive exponents.

9. $\dfrac{24x^7 - 15x^4 + 18x^3}{3x}$

10. $\dfrac{81a^8 - 36a^5 + 64a^3}{9a}$

11. $(28x^3 - 14x^2 - 35x) \div (7x)$

12. $(24x^4 - 18x^2 - 12x) \div (6x)$

13. $(18x^6 - 12x^4 - 36x^2) \div (-6x)$

14. $(27x^7 - 15x^4 - 3x^2) \div (-3x^2)$

15. $\dfrac{6x^7 - 3x^4 + x^2 - 5x + 2}{3x^3}$

16. $\dfrac{16y^8 - 14y^5 - 3y^4 - 2y^2 + 5}{-2y^4}$

17. $\dfrac{x^2y - x^3y^3 - x^5y^5}{x^2y}$

18. $\dfrac{a^3b^2 - a^3b^3 - a^4b^2}{a^2b^2}$

19. $\dfrac{16x^3y^2 - 28x^2y^3 - 20x^2y^5}{4x^2y^2}$

20. $\dfrac{16x^4y^2 - 8x^6y^4 + 12x^8y^3}{-4x^4y^2}$

21. $\dfrac{36x^4y^3 - 18x^3y^2 - 12x^2y}{6x^3y^3}$

22. $\dfrac{40x^4y^3 - 20x^3y^2 - 50x^2y}{-10x^3y^3}$

23. $\dfrac{8x^4y^2 - 6xy^3 - 14x^3y - 12xy}{-4x^2y^2}$

24. $\dfrac{-75x^5y^4 - 90y^4z + 105x^3z^2}{-15xy^2z}$

For Problems 25–50, divide using long division.

25. $(a^2 + 8a + 15) \div (a + 5)$

26. $(y^2 + 3y - 10) \div (y - 2)$

27. $(b^2 - 4b - 12) \div (b - 6)$

28. $(8 + 2x - x^2) \div (2 + x)$

29. $(24 - 10c - c^2) \div (2 - c)$

30. $(a^3 + 5a^2 + 7a + 2) \div (a + 2)$

31. $(b^3 - 2b^2 - 5b + 6) \div (b - 3)$

32. $(6x^3 + 7x^2 + 12x - 5) \div (3x - 1)$

33. $(6b^3 + 17b^2 + 27b + 20) \div (3b + 4)$

34. $(15x^3 + 41x^2 + 2x - 28) \div (3x + 7)$

35. $(a^3 + 3a^2b + 2ab^2) \div (a + b)$

36. $(9x^3 - 21x^2y + 16xy^2 - 4y^3) \div (3x - 2y)$

37. $(12x^2 + x - 4) \div (3x - 2)$

38. $(4a^2 - 8a + 6) \div (2a - 1)$

39. $\dfrac{2y^3 + 7y^2 + 9y - 20}{y + 3}$

40. $\dfrac{3a^2 - 2a + 5}{a - 3}$

41. $\dfrac{4x^4 - 4x^2 + 6x}{x - 4}$

42. $\dfrac{x^2 - 25}{x + 5}$

43. $\dfrac{x^3 - 1}{x - 1}$

44. $\dfrac{b^4 - 81}{b - 3}$

45. $\dfrac{6a^3 + 13a^2 - 11a - 15}{3a^2 - a - 3}$

46. $\dfrac{x^4 + 2x^3 - 4x^2 - 5x - 6}{x^2 + x - 2}$

47. $\dfrac{y^4 + y^3 - 3y^2 - y + 2}{y^2 + 3y + 2}$

48. $\dfrac{4y - 3 + y^3 + 2y^4}{3 + 2y^2 - y}$

49. $\dfrac{18y^4 + 9y^3 + 3y^2}{3y^2 + 1}$

50. $\dfrac{2x^5 - 8x^4 + 2x^3 + x^2}{2x^3 + 1}$

Use synthetic division to determine each quotient in Problems 51–66.

51. $(2x^2 + x - 10) \div (x - 2)$

52. $(x^2 + x - 2) \div (x - 1)$

53. $(3x^2 + 7x - 20) \div (x + 5)$

54. $(5x^2 - 12x - 8) \div (x + 3)$

55. $(4x^3 - 3x^2 + 3x - 1) \div (x - 1)$

56. $(5x^3 - 6x^2 + 3x + 11) \div (x - 2)$

57. $(6y^5 - 2y^3 + 4y^2 - 3y + 1) \div (y - 2)$

58. $(y^5 + 4y^4 - 3y^2 + 2y + 3) \div (y - 3)$

59. $(x^2 - 5x - 5x^3 + x^4) \div (5 + x)$

60. $(x^2 - 6x - 6x^3 + x^4) \div (6 + x)$

61. $\dfrac{z^5 + z^3 - 2}{z - 1}$

62. $\dfrac{z^7 + z^5 - 10z^3 + 12}{z + 2}$

63. $\dfrac{y^4 - 256}{y - 4}$

64. $\dfrac{z^7 - 128}{z - 2}$

65. $\dfrac{2y^5 - 3y^4 + y^3 - y^2 + 2y - 1}{y + 2}$

66. $\dfrac{y^5 - 2y^4 - y^3 + 3y^2 - y + 1}{y - 2}$

Application Problems

67. The amount of nuclear waste (in gallons) flowing into a stream t days after a leak in the cooling system of a nuclear power plant is described by

$$\frac{16t^4 + 8t^3}{2t^2}.$$

Find the quotient of these polynomials and then determine the amount of waste in the stream 4 days after the leak occurred.

68. The Fahrenheit temperature at a ski resort t hours after noon is described by

$$\frac{10t^5 - 240t^4 + 1440t^3}{72t^3}$$

where $0 < t \le 10$. Find the quotient of these polynomials and then determine the difference in temperature between 3 P.M. and 6 P.M.

69. The area of a rectangle is represented by $6x^2 + 11x - 35$ and the width by $2x + 7$. Represent the length in terms of x.

70. The volume of a rectangular solid is the product of its length, width, and height ($V = LWH$). If the length is represented by $2y - 1$, the width by $y + 3$, and the volume by $11y^2 + 6y^3 + 6 - 19y$ express the height in terms of y.

71. The volume of a right pyramid is $\frac{1}{3}$ the product of the area of its base and its height. If the pyramid has a volume given by $8y^3 - 12y^2 - 18y - 5$ and a base area of $12y^2 + 12y + 3$, express the height in terms of y.

72. If a car travels $6x^3 + 7x^2 - 11x - 12$ miles in $2x + 3$ hours, find a polynomial representing its speed in miles per hour.

True–False Critical Thinking Problems

73. Which one of the following is true?
a. If a trinomial in x of degree 6 is divided by a trinomial in x of degree 3, the degree of the quotient is 2.
b. Synthetic division could not be used to find the quotient of $10x^3 - 6x^2 + 4x - 1$ and $x - \frac{1}{2}$.
c. Any problem that can be done by synthetic division can also be done by the method for long division of polynomials.
d. If a polynomial long-division problem results in a remainder that is a whole number, then the divisor is a factor of the dividend.

74. Which one of the following is true?
a. All long-division problems can be done instead by factoring the dividend and canceling identical factors in the dividend and the divisor.
b. Polynomial long division always shows that the quotient of two polynomials is a polynomial.
c. The long division process should be continued until the degree of the remainder is the same as the degree of the divisor.
d. We can use our knowledge of factoring to determine that a remainder of 0 must result when $125x^3 - 8$ is divided by $5x - 2$.

Technology Problems

In Problems 75–78, use a graphing utility to graph the function on each side of the given equation. If the graphs coincide, this verifies that the expressions are equivalent and the division has been performed correctly. If the graphs do not coincide, correct the expression on the right by performing the division. Then use your graphing utility to verify your result.

75. $\dfrac{x^4 + 6x^3 + 6x^2 - 10x - 3}{x^2 + 2x - 3} = x^2 + 4x + 1,$

$x \ne -3, x \ne 1$

76. $\dfrac{2x^3 - 3x^2 - 3x + 4}{x - 1} = 2x^2 - x + 4, \quad x \ne 1$

77. $\dfrac{3x^4 + 4x^3 - 32x^2 - 5x - 20}{x + 4} = 3x^3 + 8x^2 - 5,$

$x \ne -4$

78. $\dfrac{10x^3 - 26x^2 + 17x - 13}{5x - 3} = 2x^2 - 4x + 1 - \dfrac{10}{5x - 3}$

Writing in Mathematics

79. Describe how to check polynomial long division.

80. Describe when it is necessary to place zeros in the dividend when using synthetic division.

81. Challenge problem: Synthetic division is a process for dividing a polynomial by $x - a$. The coefficient of x must be one. How might synthtic division be used if dividing by $2x - 4$?

Critical Thinking Problems

82. Consider the polynomial $x^3 - 4x^2 + 5x + 3$.
   a. Evaluate the polynomial for each value of x in the following table.

x	Value of $x^3 - 4x^2 + 5x + 3$
-2	
-1	
1	
2	

   b. Use synthetic division to find the remainder when $x^3 - 4x^2 + 5x + 3$ is divided by $x + 2$, $x + 1$, $x - 1$, and $x - 2$. Enter the values in the following table.

x	Divisor	Remainder When $x^3 - 4x^2 + 5x + 3$ Is Divided by the Indicated Divisor
-2	$x + 2$	
-1	$x + 1$	
1	$x - 1$	
2	$x - 2$	

   c. Compare the values of the polynomial from the table in part (a) with the remainders in part (b). Draw a conclusion, expressing the conclusion in a clearly written statement.

   d. Use the conclusion in part (c) and synthetic division to find the value of $x^3 - 4x^2 + 5x + 3$ when $x = 4$.

83. When $2x^2 - 7x + 9$ is divided by a polynomial, the quotient is $2x - 3$ and the remainder is 3. Find the polynomial.

84. Find the quotient of $x^{3n} + 1$ and $x^n + 1$.

85. Find the quotient of $27y^{3n} - 1$ and $3y^n - 1$.

86. Find the quotient of $15 - x^n - 10x^{2n} - 5x^{3n} + x^{4n}$ and $x^n - 1$.

In Problems 87–88, determine K so that the remainder for each division is 0.

87. $(18x^2 + 27x + K) \div (6x + 5)$

88. $(20y^3 + 23y^2 - 10y + K) \div (4y + 3)$

In Problems 89–90, use synthetic division to answer each question.

89. When $-2x^3 + 3x^2 + x + K$ is divided by $x + 1$, what value of K will give a remainder of 3?

90. For what value of K will the quotient of $x^4 + 2x + K$ and $x + 3$ have a remainder of 76?

Review Problems

91. Solve using Cramer's rule:

$$3x + 5y = 0$$
$$x + 4y = 7$$

92. Factor completely: $16x^3 + 250$.

93. Solve the system:

$$3x - y + 3z = 4$$
$$x + 2y + z = -1$$
$$2x - 3y + z = 1$$

S E C T I O N   5 . 4

Solutions   **Tutorial**   **Video**
Manual   **6**

Solve rational equations.

Rational Equations, Modeling, and Problem Solving

Objectives

1 Solve rational equations.
2 Solve problems using rational models.

Solving Rational Equations

A rational equation is an equation that contains one or more rational expressions. Here are some examples.

$$\frac{3}{4} + \frac{2}{3x} = \frac{7}{3x} - \frac{1}{12}, \qquad \frac{6x - 12}{x + 3} + \frac{5}{x - 2} = 6, \qquad \frac{x}{3} + \frac{9}{x} = 4$$

One method for solving a rational equation is to find the least common multiple (LCM) of all denominators in the equation. Multiply on both sides by this expression to clear the equation of all fractions. Then solve for the variable.

EXAMPLE 1   Solving a Rational Equation

Solve: $\dfrac{3}{4} + \dfrac{2}{3x} = \dfrac{7}{3x} - \dfrac{1}{12}$

Solution

The LCM of the denominators is $12x$. Because the multiplication property of equality does not allow multiplying both sides of an equation by 0, we will multiply both sides by $12x$ with the restriction that $x \neq 0$.

$$\frac{3}{4} + \frac{2}{3x} = \frac{7}{3x} - \frac{1}{12}$$ This is the given equation.

$$12x\left(\frac{3}{4} + \frac{2}{3x}\right) = 12x\left(\frac{7}{3x} - \frac{1}{12}\right)$$ Multiply both sides by the LCM of the denominators.

$$12x \cdot \frac{3}{4} + 12x \cdot \frac{2}{3x} = 12x \cdot \frac{7}{3x} - 12x \cdot \frac{1}{12}$$ Apply the distributive property.

$$\overset{3}{12x} \cdot \frac{3}{4} + \overset{4}{12x} \cdot \frac{2}{3\overset{}{x}} = \overset{4}{12x} \cdot \frac{7}{3\overset{}{x}} - 12x \cdot \frac{1}{12}$$ Simplify, dividing numerators and denominators by common factors.

$$9x + 8 = 28 - x$$ This is the simplified equation, cleared of fractions.

$$10x + 8 = 28$$ Solve the resulting equation. Add x to both sides.

$$10x = 20 \qquad \text{Subtract 8 from both sides.}$$
$$x = 2 \qquad \text{Divide both sides by 10.}$$

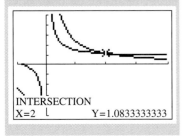
As always, check the proposed solution by substituting it into the given equation.

$$\frac{3}{4} + \frac{2}{3x} = \frac{7}{3x} - \frac{1}{12} \qquad \text{This is the original equation.}$$

$$\frac{3}{4} + \frac{2}{3(2)} \overset{?}{=} \frac{7}{3(2)} - \frac{1}{12} \qquad \text{Substitute the proposed solution, 2, for } x.$$

$$\frac{3}{4} + \frac{1}{3} \overset{?}{=} \frac{7}{6} - \frac{1}{12} \qquad \text{Simplify.}$$

$$\frac{9}{12} + \frac{4}{12} \overset{?}{=} \frac{14}{12} - \frac{1}{12} \qquad \text{Express each fraction with a denominator of 12.}$$

$$\frac{13}{12} = \frac{13}{12} \quad \checkmark$$

This true statement verifies that 2 is the solution. As shown in the Using Technology, you can also check the proposed solution graphically. The solution set is {2}. ■

EXAMPLE 2   **Solving a Rational Equation**

Solve: $\dfrac{6x - 12}{x + 3} + \dfrac{5}{x - 2} = 6$

Solution

The LCM of the denominators is $(x + 3)(x - 2)$. We will multiply both sides of the equation by $(x + 3)(x - 2)$ under the restriction that this expression is not 0. This means that $x \neq -3$ and $x \neq 2$.

$$\frac{6x - 12}{x + 3} + \frac{5}{x - 2} = 6 \qquad \text{This is the given equation.}$$

$$(x + 3)(x - 2)\left[\frac{6x - 12}{x + 3} + \frac{5}{x - 2}\right] = (x + 3)(x - 2) \cdot 6 \qquad \text{Multiply both sides by the LCM of the denominators.}$$

$$(x + 3)(x - 2) \cdot \frac{6x - 12}{x + 3} + (x + 3)(x - 2) \cdot \frac{5}{x - 2} = (x + 3)(x - 2) \cdot 6 \qquad \text{Apply the distributive property.}$$

$$(x + 3)(x - 2) \cdot \frac{6x - 12}{x + 3} + (x + 3)(x - 2) \cdot \frac{5}{x - 2} = (x + 3)(x - 2) \cdot 6 \qquad \text{Simplify, dividing numerators and denominators by common factors.}$$

$$(x - 2)(6x - 12) + 5(x + 3) = 6(x + 3)(x - 2) \qquad \text{This is the simplified equation cleared of fractions.}$$

$$6x^2 - 24x + 24 + 5x + 15 = 6x^2 + 6x - 36 \qquad \text{Multiply.}$$

$$6x^2 - 19x + 39 = 6x^2 + 6x - 36 \qquad \text{Combine like terms.}$$

$$-19x + 39 = 6x - 36 \qquad \text{Subtract } 6x^2 \text{ from both sides.}$$

$$-25x + 39 = -36 \qquad \text{Solve the resulting equation. Subtract } 6x \text{ from both sides.}$$

$$-25x = -75 \qquad \text{Subtract 39 from both sides.}$$

$$x = 3 \qquad \text{Divide both sides by } -25.$$

Because we had the restriction that $x \neq -3$ and $x \neq 2$, we can substitute 3 for x in the original equation, showing that it does check. (Do this.) The solution set is {3}. ■

Using technology

The graphs of

$$y_1 = \frac{6x - 12}{x + 3} + \frac{5}{x - 2}$$

and $y_2 = 6$

were obtained with a graphing utility in the DOT mode. The solution of

$$\frac{6x - 12}{x + 3} + \frac{5}{x - 2} = 6$$

is the x-coordinate of the intersection point, which is 3. This verifies the solution to Example 2.

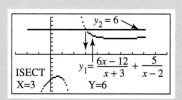

ISECT
X=3   Y=6

$y_2 = 6$

$y_1 = \dfrac{6x - 12}{x + 3} + \dfrac{5}{x - 2}$

EXAMPLE 3   **A Rational Equation with No Solution**

Solve: $\dfrac{x}{x - 3} = \dfrac{3}{x - 3} + 9$

Solution

The LCM of the denominators is $x - 3$. We will multiply both sides of the equation by $x - 3$ under the restriction that $x \neq 3$.

$$\frac{x}{x - 3} = \frac{3}{x - 3} + 9 \qquad \text{This is the given equation.}$$

$$(x - 3)\frac{x}{x - 3} = (x - 3)\left(\frac{3}{x - 3} + 9\right) \qquad \begin{array}{l}\text{Multiply both sides by the}\\ \text{LCM of the denominators.}\end{array}$$

$$(x-3)\frac{x}{x-3} = (x-3) \cdot \frac{3}{(x-3)} + (x - 3) \cdot 9 \qquad \begin{array}{l}\text{Apply the distributive prop-}\\ \text{erty and simplify.}\end{array}$$

$$x = 3 + 9(x - 3) \qquad \begin{array}{l}\text{This is the simplified equation,}\\ \text{cleared of fractions.}\end{array}$$

$$x = 3 + 9x - 27 \qquad \text{Distribute again.}$$

$$x = 9x - 24 \qquad \text{Combine numerical terms.}$$

$$-8x = -24 \qquad \text{Subtract } 9x \text{ from both sides.}$$

$$x = 3 \qquad \begin{array}{l}\text{Solve for } x, \text{ dividing both sides}\\ \text{by } -8.\end{array}$$

The proposed solution, 3, is *not* a solution because of the restriction $x \neq 3$. If we substitute 3 for x in the original equation, we obtain

$$\frac{3}{3 - 3} \stackrel{?}{=} \frac{3}{3 - 3} + 9$$

$$\frac{3}{0} \stackrel{?}{=} \frac{3}{0} + 9$$

Using technology

The graphs of

$$y_1 = \frac{x}{x - 3}$$

and $y_2 = \frac{3}{x - 3} + 9$

were obtained using a graphing utility. The graphs do not intersect. Thus,

$$\frac{x}{x - 3} = \frac{3}{x - 3} + 9$$

has no solution. The graphs indicate that

$$\frac{x}{x - 3} \quad \text{and} \quad \frac{3}{x - 3} + 9$$

are *approximately* equal as x gets closer and closer to 3.

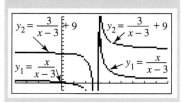

$y_2 = \dfrac{3}{x-3} + 9$

$y_1 = \dfrac{x}{x-3}$

$y_2 = \dfrac{3}{x-3} + 9$

$y_1 = \dfrac{x}{x-3}$

The term $\frac{3}{0}$ is undefined. The proposed solution does not check in the original equation. There is *no solution* to this equation. The solution set is $\varnothing$, the empty set.   ∎

Examples 4 and 5 involve rational equations that become quadratic after clearing fractions.

study tip

If a rational equation contains a variable in any denominator, check all proposed solutions in the original equation. If a proposed solution makes any denominator equal to zero, that value is not a solution to the equation.

EXAMPLE 4   **A Rational Equation That Leads to a Quadratic Equation**

Solve: $\dfrac{x}{3} + \dfrac{9}{x} = 4$

Solution

$$\dfrac{x}{3} + \dfrac{9}{x} = 4 \qquad \text{This is the given equation.}$$

$$3x\left(\dfrac{x}{3} + \dfrac{9}{x}\right) = 4(3x) \qquad \text{Multiply both sides by the LCM of the denominators. } x \ne 0.$$

$$3x \cdot \dfrac{x}{3} + 3x \cdot \dfrac{9}{x} = 12x \qquad \text{Apply the distributive property.}$$

$$3x \cdot \dfrac{x}{3} + 3\cancel{x} \cdot \dfrac{9}{\cancel{x}} = 12x \qquad \text{Simplify, dividing numerators and denominators by common factors.}$$

$$x^2 + 27 = 12x \qquad \text{The simplified equation, cleared of fractions, is quadratic.}$$

$$x^2 - 12x + 27 = 0 \qquad \text{Write the quadratic equation in standard form.}$$

$$(x - 9)(x - 3) = 0 \qquad \text{Factor.}$$

$$x - 9 = 0 \quad \text{or} \quad x - 3 = 0 \qquad \text{Apply the zero product principle.}$$

$$x = 9 \qquad\qquad x = 3 \qquad \text{Solve the resulting equations.}$$

Both 9 and 3 can be shown to check as solutions. This does not interfere with the original restriction that $x \ne 0$.

sing technology

The graphs of

$$y_1 = \dfrac{x}{3} + \dfrac{9}{x}$$

and $y_2 = 4$

have two intersection points. The x-coordinates of the points are 3 and 9. This verifies that

$$\dfrac{x}{3} + \dfrac{9}{x} = 4$$

has both 3 and 9 as solutions.

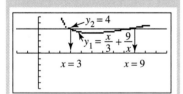

Check 9:

$$\dfrac{x}{3} + \dfrac{9}{x} = 4$$

$$\dfrac{9}{3} + \dfrac{9}{9} \overset{?}{=} 4$$

$$3 + 1 \overset{?}{=} 4$$

$$4 = 4 \quad \checkmark$$

Check 3:

$$\dfrac{x}{3} + \dfrac{9}{x} = 4$$

$$\dfrac{3}{3} + \dfrac{9}{3} \overset{?}{=} 4$$

$$1 + 3 \overset{?}{=} 4$$

$$4 = 4 \quad \checkmark$$

The solution set is $\{3, 9\}$.   ∎

EXAMPLE 5   **A Rational Equation That Leads to a Quadratic Equation**

Solve: $\dfrac{1}{y^2 + 3y + 2} + \dfrac{1}{y - 1} = \dfrac{2}{y^2 - 1}$

Solution

$$\frac{1}{(y + 2)(y + 1)} + \frac{1}{y - 1} = \frac{2}{(y + 1)(y - 1)}$$

Factor the denominators. The LCM is $(y + 2)(y + 1)(y - 1)$. Multiply both sides by the LCM. $y \neq -2$, $y \neq -1$, and $y \neq 1$.

$$(y + 2)(y + 1)(y - 1)\left[\frac{1}{(y + 2)(y + 1)} + \frac{1}{y - 1}\right] = \left[\frac{2}{(y + 1)(y - 1)}\right](y + 2)(y + 1)(y - 1)$$

$$\cancel{(y + 2)}\cancel{(y + 1)}(y - 1) \cdot \frac{1}{\cancel{(y + 2)}\cancel{(y + 1)}} + (y + 2)(y + 1)\cancel{(y - 1)} \cdot \frac{1}{\cancel{y - 1}}$$

Apply the distributive property and simplify.

$$= \frac{2}{\cancel{(y + 1)}\cancel{(y - 1)}} \cdot (y + 2)\cancel{(y + 1)}\cancel{(y - 1)}$$

$$y - 1 + (y + 2)(y + 1) = 2(y + 2)$$

This is the simplified equation, cleared of fractions.

$$y - 1 + y^2 + 3y + 2 = 2y + 4$$

Multiply.

$$y^2 + 4y + 1 = 2y + 4$$

Combine like terms on the left side.

$$y^2 + 2y - 3 = 0$$

Write the quadratic equation in standard form by subtracting $2y + 4$ from both sides.

$$(y + 3)(y - 1) = 0$$

Factor.

$$y + 3 = 0 \quad \text{or} \quad y - 1 = 0$$

Apply the zero product principle.

$$y = -3 \qquad y = 1$$

Solve the resulting equations.

One of the proposed solutions, 1, is *not* a solution because of the restriction that $y \neq 1$. Take time to verify that the solution -3 satisfies the original equation. The solution set is $\{-3\}$. ∎

Using technology

Change the variable from y to x in Example 5. Use a graphing utility to graph both sides of the equation in the same viewing rectangle. Based on the equation's solution, what would you expect to happen with the graphs? Try using the best range setting possible to verify your observation.

2 Solve problems using rational models.

Mathematicians have models for learning and for forgetting what we've learned.

Alfredo Castañeda, "Figura No. 14" 1982 © Christie's Images

Using Rational Models

Many situations in physics, biology, business, economics, and psychology can be modeled by rational functions and equations. The procedure that we have been using to solve rational equations can be applied to answering questions about variables contained in these models.

EXAMPLE 6 **Modeling Human Learning**

Psychologists have developed mathematical models to predict the percent of correct responses as a function of the number of trials of a particular task. One such model, called a learning curve, is

$$P = \frac{0.9n - 0.4}{0.9n + 0.1}$$

where P is the percent of correct responses after n trials. The model is developed so that P is expressed in decimal form. Using this model, how many learning trials are necessary for 95% of the responses to be correct?

Solution

$$P = \frac{0.9n - 0.4}{0.9n + 0.1}$$ This is the given model.

$$0.95 = \frac{0.9n - 0.4}{0.9n + 0.1}$$ Since we want 95% of the responses to be correct, we substitute 0.95 for P. We must solve for n.

$$0.95(0.9n + 0.1) = \frac{0.9n - 0.4}{0.9n + 0.1}(0.9n + 0.1)$$ Multiply both sides by $0.9n + 0.1$.

$$0.95(0.9n + 0.1) = \frac{0.9n - 0.4}{0.9n + 0.1}(0.9n + 0.1)$$ Simplify.

$$0.855n + 0.095 = 0.9n - 0.4$$ Apply the distributive property.

$$0.495 = 0.045n$$ Collect terms with n on the right and constant terms on the left. Subtract $0.855n$ and add 0.4 on both sides.

$$n = \frac{0.495}{0.045} = 11$$ Divide both sides by 0.45.

Our solution indicates that 11 learning trials are necessary for 95% of the responses to be correct. ■

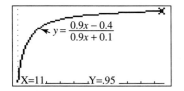

Figure 5.12

The graph of a learning model using Xmin = 0, Xmax = 11, Xscl = 1, Ymin = 0, Ymax = 1, Yscl = 0.05

The graph of the learning model from Example 6 is shown in Figure 5.12. Also shown is the solution point $(11, 0.95)$, verifying that 11 learning trials are necessary for 95% of the responses to be correct.

Notice how the graph increases quite rapidly at first, and then tends to taper off. Psychologists have discovered that when we learn new tasks, learning is initially quite rapid but tends to taper off as time increases. Once we have mastered a task, our performance level approaches peak efficiency, with additional practice having minimal effect on performance.

The disadvantage of the method used in Example 6 is that each time we are given a value of P, we must solve for n. A more efficient approach would be to solve the rational model for n in terms of P. This forms the basis of Example 7.

EXAMPLE 7   **Solving a Rational Model for a Specified Variable**

Solve for n: $P = \dfrac{0.9n - 0.4}{0.9n + 0.1}$

Solution

We first clear fractions by multiplying both sides by $0.9n + 0.1$.

$$P = \frac{0.9n - 0.4}{0.9n + 0.1}$$ This is the given model.

$$P(0.9n + 0.1) = \frac{0.9n - 0.4}{0.9n + 0.1}(0.9n + 0.1)$$ Multiply both sides by $0.9n + 0.1$.

$$P(0.9n + 0.1) = \frac{0.9n - 0.4}{0.9n + 0.1}(0.9n + 0.1)$$ Simplify.

$$0.9nP + 0.1P = 0.9n - 0.4$$ Use the distributive property.

We must now get all terms containing n alone on one side.

$$0.1P + 0.4 = 0.9n - 0.9nP \qquad \text{Collect terms with } n \text{ on the right and all other terms on the left. Subtract } 0.9nP \text{ and add } 0.4 \text{ on both sides.}$$

$$0.1P + 0.4 = (0.9 - 0.9P)n \qquad \text{Use the distributive property to isolate } n \text{ by factoring it out on the right.}$$

$$\frac{0.1P + 0.4}{0.9 - 0.9P} = \frac{(0.9 - 0.9P)n}{0.9 - 0.9P} \qquad \text{Divide both sides by } 0.9 - 0.9P.$$

$$\frac{0.1P + 0.4}{0.9 - 0.9P} = n \qquad \text{Simplify.}$$

We now have n alone on one side, and n does not appear on the other side, so we have solved the model for n. Thus,

$$n = \frac{0.1P + 0.4}{0.9 - 0.9P}$$ ■

We can use the procedure of Example 7 to solve any model containing a rational expression for a specified variable.

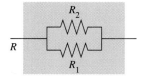

iscover for yourself

In the form of the model obtained in Example 7, namely,

$$n = \frac{0.1P + 0.4}{0.9 - 0.9P}$$

we can determine n for any value of P. Let $P = 0.95$ and find the number of learning trials necessary to produce a 95% correct response rate. Compare this approach with the one used in Example 6.

| EXAMPLE 8 | **Solving a Mathematical Model for a Specified Variable** |

An electric circuit is an unbroken path of wire and other material that carries electricity. Some wires are harder to push electrons through than others because they have a greater resistance, measured in a unit called the ohm. The higher the resistance, the more of the electron energy is converted into heat. Resistors are specifically placed in some circuits to add resistance. For example, toasters and space heaters need high-resistance wires so they will glow and give off large amounts of heat when current flows through them.

The total resistance (R) of two resistors connected in a parallel circuit is determined by the model

$$\frac{1}{R} = \frac{1}{R_1} + \frac{1}{R_2}$$

where R_1 and R_2 are the resistances of the individual resistors (measured in ohms) in the circuit.

a. Solve for R_1.
b. Find R_1 when $R = 120$ ohms and $R_2 = 300$ ohms.

Solution

a. To solve for R_1, we begin by multiplying both sides of the model by the LCM of the denominators, RR_1R_2. This will clear the equation of fractions.

$$\frac{1}{R} = \frac{1}{R_1} + \frac{1}{R_2} \qquad \text{This is the given model.}$$

$$RR_1R_2 \cdot \frac{1}{R} = RR_1R_2\left(\frac{1}{R_1} + \frac{1}{R_2}\right) \qquad \text{Multiply both sides by the LCM of the denominators.}$$

$$RR_1R_2 \cdot \frac{1}{R} = RR_1R_2 \cdot \frac{1}{R_1} + RR_1R_2 \cdot \frac{1}{R_2} \qquad \text{Apply the distributive property and simplify.}$$

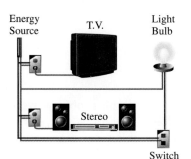

In a series circuit, electrical current flows through one element, then through the next, and so on. In the parallel circuit shown here, each element has its own loop of wire to carry current. If one element is defective, the others will receive current and continue working.

$$R_1 R_2 = RR_2 + RR_1$$

This is the simplified equation, cleared of fractions.

We must now get all terms containing R_1 alone on one side.

$$R_1 R_2 - RR_1 = RR_2$$

Since we want to solve for R_1, collect terms with R_1 on the left. Subtract RR_1 from both sides.

$$R_1(R_2 - R) = RR_2$$

Isolate R_1 by factoring it out on the left.

$$\frac{R_1(R_2 - R)}{R_2 - R} = \frac{RR_2}{R_2 - R}$$

Divide both sides by $R_2 - R$.

$$R_1 = \frac{RR_2}{R_2 - R}$$

Simplify.

b. Now that we have solved for R_1, we use this equation, substituting 120 for R and 300 for R_2.

$$R_1 = \frac{RR_2}{R_2 - R} = \frac{120(300)}{300 - 120} = \frac{36,000}{180} = 200$$

The first resistor has a resistance of 200 ohms. ◼

Here's a step-by-step procedure that you can use to solve any model containing a rational expression for a specified variable.

Solving a formula containing rational expressions for a specified variable

1. Multiply both sides by the LCM of all denominators to clear fractions.
2. Collect all terms with the specified variable on one side of the equation and all other terms on the other side.
3. If the specified variable appears in more than one term, factor it out.
4. Isolate the specified variable by dividing both sides of the equation by the factor that appears with the variable.

Creating Mathematical Models

Cost and Revenue of Sales

Units of Production

In Examples 6–8, the models that we needed to solve the problems were given. A more difficult situation is to use a problem's conditions to create a rational model. In Example 9 we use our five-step problem-solving strategy and methods for solving rational equations to solve the problem. The verbal model leading to an equation is based on the total cost of producing x units of a product. Total cost is the sum of a company's fixed costs plus the cost of manufacturing the product. The average cost per unit is the total cost divided by the number of units produced.

EXAMPLE 9 **Modeling Average Cost**

A manufacturing plant can produce x units of a product for $25 per unit in addition to an initial investment of $60,000. How many units must be produced to have an average cost of $29 per unit?

Solution

Steps 1 and 2. Represent unknown quantities in terms of x.

Let x = the number of units that must be produced to have an average cost of $29 per unit.

Step 3. Write an equation modeling the problem's conditions.

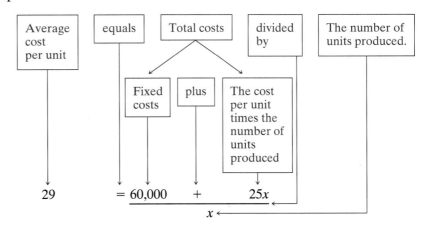

Step 4. Solve the equation and answer the problem's question.

$$29 = \frac{60{,}000 + 25x}{x}$$ This is the equation based on average cost as the quotient of total cost and the number of units produced.

$$29x = \frac{60{,}000 + 25x}{x} \cdot x$$ Multiply by the LCM of the denominators, x.

$$29x = 60{,}000 + 25x$$ The equation is now cleared of fractions.

$$4x = 60{,}000$$ Subtract $25x$ from both sides.

$$x = 15{,}000$$ Divide both sides by 4.

Step 5. Check the proposed solution in the original wording of the problem.

The plant should produce 15,000 units. Check this number in the original wording of the problem. ∎

We can visually display the solution to Example 9 using a graphing utility. The graph of the average cost function

$$y = \frac{60{,}000 + 25x}{x}$$

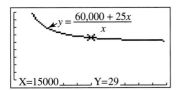

Figure 5.13

The graph of $y = \dfrac{60{,}000 + 25x}{x}$ using Xmin = 0, Xmax = 30,000, Xscl = 3000, Ymin = 0, Ymax = 50, Yscl = 5. Producing 15,000 units, the average cost per unit is $29.

is shown in Figure 5.13. Tracing along the curve verifies that if the plant produces 15,000 units, the average cost per unit is $29. The average cost function shown in the graph illustrates that high production levels lead to lower costs for the company to manufacture each unit.

In our next problem, we will be modeling a statement about time in a uniform motion situation. Recall that

$$\text{Time} = \frac{\text{Distance}}{\text{Rate}} \left(\text{or } \frac{\text{Distance}}{\text{Traveling speed}}\right).$$

EXAMPLE 10 **Modeling a Uniform Motion Problem**

After riding at a steady speed for 40 miles, a bicyclist had a flat tire and walked 5 miles to a repair shop. The cycling rate was 4 times faster than the walking rate. If the time spent cycling and walking was 5 hours, at what rate was the cyclist riding?

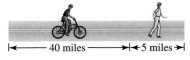

Time cycling + walking
= 5 hours

Solution

Steps 1 and 2. Represent unknown quantities in terms of x.

Let

$$x = \text{Walking rate}$$
$$4x = \text{Cycling rate}$$

The time spent cycling is

$$t = \frac{d}{r} = \frac{40}{4x}$$

and the time spent walking is

$$t = \frac{d}{r} = \frac{5}{x}$$

We summarize this information in the following table.

	d	r	$t = \dfrac{d}{r}$
Cycling	40	$4x$	$\dfrac{40}{4x}$ ←
Walking	5	x	$\dfrac{5}{x}$ ←

Total time spent is 5 hours.

Step 3. Write an equation modeling the problem's conditions.

The total time is 5 hours, so

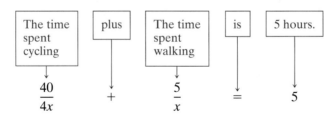

The time spent cycling	plus	The time spent walking	is	5 hours.
$\dfrac{40}{4x}$	$+$	$\dfrac{5}{x}$	$=$	5

Step 4. Solve the equation and answer the problem's question.

We simplify the first rational expression by dividing the numerator and denominator by 4, and solve for x.

$$\frac{10}{x} + \frac{5}{x} = 5 \qquad \text{This is the equation implied by the problem's conditions.}$$

$$x\left(\frac{10}{x} + \frac{5}{x}\right) = x \cdot 5 \qquad \text{Multiply both sides by } x, \text{ the LCM.}$$

$$x \cdot \frac{10}{x} + x \cdot \frac{5}{x} = 5x \qquad \text{Apply the distributive property and simplify.}$$

$$15 = 5x \qquad \text{Combine numerical terms on the left.}$$

$$3 = x \qquad \text{Divide both sides by 5.}$$

We see that

$$x = \text{Walking rate} = 3 \text{ miles per hour}$$
$$4x = \text{Cycling rate} = 4 \cdot 3 = 12 \text{ miles per hour}$$

The cyclist was riding at 12 miles per hour.

Check

Time spent walking: $\dfrac{\text{Distance}}{\text{Rate}} = \dfrac{5}{3}$ hours

Time spent cycling: $\dfrac{\text{Distance}}{\text{Rate}} = \dfrac{40}{12} = \dfrac{10}{3}$ hours

Total time: $\dfrac{5}{3} + \dfrac{10}{3} = \dfrac{15}{3} = 5$ hours   ✓

This checks with the problem's conditions.   ■

Once again we can use a graphing utility to visualize the solution to Example 10. The graph of the total time function

$$y = \frac{10}{x} + \frac{5}{x} \quad \text{or} \quad y = \frac{15}{x}$$

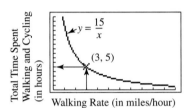

Figure 5.14

Walking at 3 miles per hour, the trip takes 5 hours.

is shown in Figure 5.14. Since x represents the walking rate, we can see, based on the point $(3, 5)$, that at a walking rate of 3 miles per hour, the trip takes 5 hours. The graph is falling from left to right, showing decreasing time with increasing walking speed. As you move along the graph to your left, the values of y are getting to be very large. This indicates that close to a walking speed of zero miles per hour (a stupified crawl), the trip to the repair shop will take nearly forever.

Rational equations are useful in modeling situations that involve the amount of work completed by people or machines. In these examples, it will be assumed that workers and machines operate at a constant rate.

We begin with one observation about work. If a person can do a job in 5 hours, then the fractional part of the job completed in 1 hour is $\frac{1}{5}$; in 2 hours, $\frac{2}{5}$; and in 3 hours, $\frac{3}{5}$. In 4 hours, that person can do $\frac{4}{5}$ of the job. Consequently, in x hours, that person can accomplish $\dfrac{x}{5}$ of the job.

EXAMPLE 11   **Solving a Work Problem**

Working together, Louis and Lestat can complete a job in 4 hours. Working alone, Louis requires 6 hours more than Lestat to do the job. How many hours does it take Lestat to do the job if he works alone?

Solution

Let

$x =$ Number of hours it takes Lestat to do the job alone

$x + 6 =$ Number of hours it takes Louis to do the job alone

We see that if it takes Lestat x hours to do the job, in 1 hour $\dfrac{1}{x}$ of the job is done. (If it took him 5 hours to do the job, in 1 hour $\frac{1}{5}$ of the job would be done.) Similarly, if it takes Louis $x + 6$ hours to do the job, in 1 hour $\dfrac{1}{x + 6}$ of the job is done. Consequently, in 4 hours, $\dfrac{4}{x}$ of the job is done by Lestat and $\dfrac{4}{x + 6}$ of the job is done by Louis.

Let's summarize these observations in a table.

	Fractional Part of the Job Completed in 1 Hour	**Time Spent Working Together**	**Fractional Part of the Job Completed in 4 Hours**
Lestat (takes x hours working alone)	$\dfrac{1}{x}$	4	$\dfrac{4}{x}$
Louis (takes $x + 6$ hours working alone)	$\dfrac{1}{x + 6}$	4	$\dfrac{4}{x + 6}$

Because Louis and Lestat can complete the job in 4 hours,

Step 3. Write an equation modeling the problem's conditions.

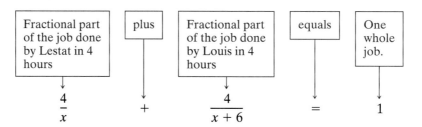

Fractional part of the job done by Lestat in 4 hours	plus	Fractional part of the job done by Louis in 4 hours	equals	One whole job.
$\dfrac{4}{x}$	$+$	$\dfrac{4}{x + 6}$	$=$	1

Step 4. Solve the equation and answer the problem's question.

$$\frac{4}{x} + \frac{4}{x + 6} = 1$$ This is the equation implied by the problem's conditions.

$$x(x + 6)\left(\frac{4}{x} + \frac{4}{x + 6}\right) = x(x + 6) \cdot 1$$ Multiply both sides by the LCM.

$$\not{x}(x + 6) \cdot \frac{4}{\not{x}} + x(\not{x + 6}) \cdot \frac{4}{\not{x + 6}} = x(x + 6)$$ Apply the distributive property and simplify.

$$4(x + 6) + 4x = x(x + 6)$$ This is the simplified equation, cleared of fractions.

$$4x + 24 + 4x = x^2 + 6x$$ Distribute again.

$$8x + 24 = x^2 + 6x$$ Combine like terms.

$$0 = x^2 - 2x - 24$$ Write the quadratic equation in standard form.

$$0 = (x - 6)(x + 4)$$ Factor.

$$x - 6 = 0 \quad \text{or} \quad x + 4 = 0$$ Set each factor equal to 0.

$$x = 6 \qquad\qquad x = -4 \quad \text{Reject}$$ Solve the resulting equations.

Working alone, Lestat takes 6 hours to do the job.

Step 5. Check the proposed solution in the original wording of the problem.

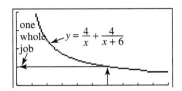

Since it takes Lestat 6 hours to do the job alone, he and Louis complete the whole job in 4 hours.

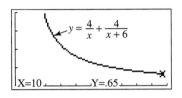

Figure 5.15

Check

Working alone, Lestat takes 6 hours, and Louis takes $x + 6 = 6 + 6 = 12$ hours.

In 4 hours, Lestat completes $\frac{4}{6} = \frac{2}{3}$ of the job.

In 4 hours, Louis completes $\frac{4}{12} = \frac{1}{3}$ of the job.

Together, they complete $\frac{2}{3} + \frac{1}{3} = 1$ whole job. This checks with the given condition that, working together, Louis and Lestat can complete a job in 4 hours. ∎

The solution to Example 11 is visualized in the upper graph in Figure 5.15. A graphing utility was used to obtain the graph for

$$y = \frac{4}{x} + \frac{4}{x + 6}$$

in which

x = the time for Lestat to do the job working alone

and

y = the fractional part of the job completed by Lestat and Louis in 4 hours

The graph is falling from left to right, showing that as it takes more and more time for Lestat to do the job working alone, the fractional part of the job done by Lestat and Louis in 4 hours decreases. For example, the bottom graph illustrates that if it took Lestat 10 hours to do the job alone, he and Louis would complete only 65% of the job in 4 hours.

> **EXAMPLE 12** **Solving a Work Problem**

Pipe A fills a pool in 4 hours, pipe B fills it in 6 hours, and pipe C drains it in 12 hours. Pipes A and B were turned on to fill the pool, but pipe C was accidentally left open. How long did it take both pipes to fill the pool with pipe C left open?

Solution

Steps 1 and 2. Represent unknown quantities in terms of x.

Let x = the time to fill the pool.

	Fractional Part of the Job Completed in 1 Hour	Time Spent Working Together	Fractional Part of the Job Completed in x Hours
Pipe A	$\dfrac{1}{4}$	x	$\dfrac{x}{4}$
Pipe B	$\dfrac{1}{6}$	x	$\dfrac{x}{6}$
Pipe C	$\dfrac{1}{12}$	x	$\dfrac{x}{12}$

Step 3. Write an equation modeling the problem's conditions.

Because pipe C is draining the pool, we must subtract the work it does from the sum of the work done by pipes A and B.

$$\frac{x}{4} + \frac{x}{6} - \frac{x}{12} = 1$$

We have 1 on the right side because the entire job was done (the pool was filled) despite the fact that pipe C was draining it.

Step 4. Solve the equation and answer the problem's question.

$$12\left(\frac{x}{4} + \frac{x}{6} - \frac{x}{12}\right) = 12 \cdot 1 \quad \text{Multiply both sides by 12, the LCM of 4, 6, and 12.}$$

$$3x + 2x - x = 12 \quad \text{Apply the distributive property and simplify.}$$

$$4x = 12 \quad \text{Combine like terms.}$$

$$x = 3 \quad \text{Divide both sides by 4.}$$

It took 3 hours for both pipes to fill the pool despite the fact that pipe C was draining it.

Step 5. Check the proposed solution in the original wording of the problem.

Check

In 3 hours,

Pipe A fills $\dfrac{x}{4}$ or $\dfrac{3}{4}$ of the pool.

Pipe B fills $\dfrac{x}{6}$ or $\dfrac{3}{6} = \dfrac{1}{2}$ of the pool.

Pipe C drains $\dfrac{x}{12}$ or $\dfrac{3}{12} = \dfrac{1}{4}$ of the pool.

Since

$$\frac{3}{4} + \frac{1}{2} - \frac{1}{4} = \frac{3}{4} + \frac{2}{4} - \frac{1}{4} = \frac{4}{4}, \quad \text{or} \quad 1$$

the complete job is done, and the pool is filled. The solution is visualized in Figure 5.16. ∎

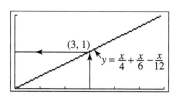

Figure 5.16

In 3 hours, one complete job is done and the pool is filled.

PROBLEM SET 5.4

Practice Problems

Find the solution set for the equations in Problems 1–46. Verify your answers algebraically or graphically using a graphing utility.

1. $\dfrac{2}{3x} + \dfrac{1}{4} = \dfrac{11}{6x} - \dfrac{1}{3}$

2. $\dfrac{1}{2} - \dfrac{3}{2b} = \dfrac{4}{b} - \dfrac{5}{12}$

3. $\dfrac{4}{5} + \dfrac{7}{2a} = \dfrac{13}{2a} - \dfrac{4}{20}$

4. $1 + \dfrac{2}{3m} + \dfrac{1}{2m} = \dfrac{13}{6m}$

5. $\dfrac{2}{y - 3} + \dfrac{3y + 1}{y + 3} = 3$

6. $\dfrac{15}{b} + \dfrac{9b - 7}{b + 2} = 9$

7. $\dfrac{2x - 4}{x - 4} - 2 = \dfrac{20}{x + 4}$

8. $\dfrac{3m - 2}{m + 1} = 4 - \dfrac{m + 2}{m - 1}$

9. $\dfrac{3c}{c + 1} = \dfrac{5c}{c - 1} - 2$

10. $\dfrac{2y}{y^2 - 1} + \dfrac{y + 3}{y + 1} = \dfrac{y + 1}{y - 1}$

11. $\dfrac{x + 5}{x^2 - 4} - \dfrac{3}{2x - 4} = \dfrac{1}{2x + 4}$

12. $\dfrac{3}{2 - b} - \dfrac{2}{2 + b} = \dfrac{7}{4 - b^2}$

13. $\dfrac{4}{a + 2} + \dfrac{2}{a - 4} = \dfrac{30}{a^2 - 2a - 8}$

14. $\dfrac{2}{m - 5} - \dfrac{1}{m - 2} = \dfrac{9}{m^2 - 7m + 10}$

15. $\dfrac{y + 5}{y + 1} - \dfrac{y}{y + 2} = \dfrac{4y + 1}{y^2 + 3y + 2}$

16. $\dfrac{12x + 19}{x^2 + 7x + 12} - \dfrac{5}{x + 4} = \dfrac{3}{x + 3}$

17. $\dfrac{c}{c + 4} = \dfrac{2}{5} - \dfrac{4}{c + 4}$

18. $\dfrac{3}{2m} = \dfrac{1}{m}$

19. $\dfrac{3y}{y - 3} - \dfrac{5y}{y - 3} = -2$

20. $\dfrac{2}{y^2 - 1} - \dfrac{1}{y - 1} = \dfrac{1}{y + 1}$

21. $\dfrac{6}{x} - \dfrac{x}{3} = 1$

22. $\dfrac{x}{2} - \dfrac{12}{x} = 1$

23. $\dfrac{2}{x} + \dfrac{9}{x+4} = 1$

24. $\dfrac{60}{x} = \dfrac{60}{x+6} + \dfrac{1}{2}$

25. $\dfrac{1}{x-1} + \dfrac{1}{x-4} = \dfrac{5}{4}$

26. $x - 9 = \dfrac{72}{x-8}$

27. $\dfrac{x^2+10}{x-5} = \dfrac{7x}{x-5}$

28. $x - \dfrac{9}{x} = 8$

29. $\dfrac{7}{y+5} - \dfrac{3}{y-1} = \dfrac{8}{y-6}$

30. $\dfrac{2x^2}{x+1} + 1 = \dfrac{2}{x+1}$

31. $\dfrac{24}{10+y} + \dfrac{24}{10-y} = 5$

32. $\dfrac{y-1}{y^2-9} - \dfrac{3y+5}{y+3} = \dfrac{y+3}{y-3}$

33. $\dfrac{x}{x-5} + \dfrac{17}{25-x^2} = \dfrac{1}{x+5}$

34. $\dfrac{4}{x-1} = \dfrac{5}{2x-2} + \dfrac{3x}{4}$

35. $\dfrac{5}{y-3} = \dfrac{30}{y^2-9} + 1$

36. $\dfrac{y+7}{y-1} + \dfrac{y-1}{y+1} = \dfrac{4}{y^2-1}$

37. $\dfrac{8}{x+1} + \dfrac{2}{1-x^2} = \dfrac{x}{x-1}$

38. $\dfrac{y}{y-3} + \dfrac{4}{y+1} = \dfrac{y^2-2y+2}{y^2-2y-3}$

39. $\dfrac{x}{x+5} + \dfrac{x}{5-x} = \dfrac{15+5x}{x^2-25}$

40. $\dfrac{y}{y+3} + \dfrac{3}{y-3} = \dfrac{y^2+9}{y^2-9}$

41. $\dfrac{x+2}{x^2-x} - \dfrac{6}{x^2-1} = 0$

42. $\dfrac{x+3}{x^2-x} - \dfrac{8}{x^2-1} = 0$

43. $\dfrac{1}{x^3-8} - \dfrac{2}{x^2+2x+4} = \dfrac{3}{(2-x)(x^2+2x+4)}$

44. $\dfrac{2}{a^3-1} = \dfrac{4}{(1-a)(a^2+a+1)} - \dfrac{1}{a^2+a+1}$

45. $5y^{-2} + 1 = 6y^{-1}$

46. $3y^{-2} + 1 = 4y^{-1}$

Application Problems

47. The model

$$C = \dfrac{4p}{100 - p}$$

describes the cost (C, in millions of dollars) to remove p percent of pollution from a river due to pesticide runoff from area farms. What percent of pollution can be removed at a cost of $196 million dollars? Obtain your answer algebraically and then explain how it is visualized in the bar graph.

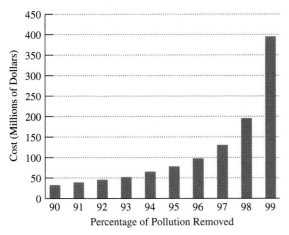

48. The function

$$\overline{C}(x) = \dfrac{100x + 100{,}000}{x}$$

models the average cost per bicycle, $\overline{C}(x)$, for a manufacturing business that produces x bicycles. How many

bicycles must be manufactured to bring the average cost per bicycle down to $125?

49. The function

$$t(x) = \dfrac{40}{x} + \dfrac{40}{x+30}$$

models the total time t required to complete a round trip that is 40 miles in each direction, where

$x =$ the average speed on the outgoing trip

and $x + 30 =$ the average speed on the return trip

What outgoing speed will result in the round trip taking 5 hours?

50. The model

$$N = \dfrac{10(4 + 3t)}{1 + 0.02t}$$

describes the number of deer (N) in a park t years after 40 deer are introduced into the region. Approximately how long (to the nearest year) will it take the deer population to reach 196?

51. After a drug is injected into a patient, the drug's concentration ($C(t)$, in milligrams per liter) after t minutes is modeled by

$$C(t) = \dfrac{30t}{t^2 + 2}.$$

How long after the injection will the concentration be 10 milligrams per liter?

52. The function

$$f(x) = \frac{200x}{x^2 + 100}$$

models the monthly sales ($f(x)$, in thousands of games) of a new video game in terms of the number of months (x) after the game is introduced. After how many months will 8 thousand games be sold?

53. In t years from 1990, the population (P, in thousands) of a community will be

$$P = 20 - \frac{4}{t + 1}.$$

When will the population of the community be 19,000?

54. If x prey are available per unit of area, a predator will consume $\dfrac{0.8x}{1 + 0.03x}$ prey daily. How should the prey per unit-area be controlled if wildlife managers want a predator to consume 20 prey daily?

55. The formula $y = \dfrac{100}{x}$ describes the relationship in shallow water between the number of species per thousand individuals (x) and the total organic matter (y, in milligrams per liter) in the water. In a controlled shallow-water environment, how many species per thousand individuals should be present to ensure 4 milligrams of organic matter per liter?

56. In economics, a demand formula presents the number of units that will sell at x dollars per unit. An example of a demand formula for barbecue grills is

$$D = \frac{600}{x} + 40.$$

How must the price of each grill be controlled to ensure a demand for 70 barbecue grills?

57. Solve the model in Problem 47,

$$C = \frac{4p}{100 - p}$$

for p. Then use this form of the model to verify the result you obtained in Problem 47.

58. The formula

$$W = \frac{10x}{150 - x}$$

describes the number of weeks (W) it takes to raise $x\%$ of a campaign's financial goal. Solve for x and then find the percentage of the campaign's financial goal that can be raised in 5 weeks.

59. A rat given n trials in a maze can run through the maze in t minutes, where

$$t = 6 + \frac{20}{n + 2}.$$

a. Solve the model for n.
b. Use the formula in part (a) to determine how many previous trials are necessary so that the rat can run through the maze in 7 minutes.

60. Solve the model in Problem 50,

$$N = \frac{10(4 + 3t)}{1 + 0.02t}$$

for t. Then use this form of the model to verify the result you obtained in Problem 50.

In Problems 61–72, solve each model for the variable indicated.

61. $\dfrac{1}{p} + \dfrac{1}{q} = \dfrac{1}{f}$ for q (optics)

62. $\dfrac{1}{R} = \dfrac{1}{R_1} + \dfrac{1}{R_2}$ for R (electronics)

63. $F = \dfrac{Gm_1m_2}{d^2}$ for G (physics)

64. $F = \dfrac{Gm_1m_2}{d^2}$ for m_1 (physics)

65. $d = \dfrac{fl}{f + w}$ for f (physics)

66. $A = \dfrac{Rr}{R + r}$ for r (engineering)

67. $v = \dfrac{d_2 - d_1}{t_2 - t_1}$ for t_2 (physics)

68. $T = \dfrac{D - d}{L}$ for d (used by machinists)

69. $x = \dfrac{F_G(r + p)}{F_G - F_S}$ for F_G (used in business)

70. $P = \dfrac{96.3F}{SL}$ for S (used in designing sprinkler systems)

71. $P = \dfrac{t^2 dN}{3.78}$ for d (called Pomeroy's formula, used by machinists)

72. $z = \dfrac{\bar{x} - \mu}{\dfrac{\sigma}{\sqrt{n}}}$ for $\bar{x}$ (statistics)

73. The opposition to an electric current offered by some components is called resistance, measured in units called ohms. Resistors are specifically placed in a circuit to add resistance. The formula

$$\frac{1}{R} = \frac{1}{R_1} + \frac{1}{R_2}$$

describes the total resistance (R) of a circuit that consists of two resistors of resistance R_1 and R_2 connected in parallel. If the total resistance in a circuit is 60 ohms and R_1 has twice the resistance as R_2, what is the resistance of the two resistors?

74. Use the model described in Problem 73 to answer this question: If R_1 has three times the resistance of R_2, and if the total resistance is 15 ohms, find the number of ohms in R_1 and R_2.

75. A company that manufactures graphing calculators has fixed costs of $80,000 and variable costs of $20 per calculator. How many calculators must be manufactured to have an average cost of $22 per calculator?

76. A company that manufactures pens has fixed costs of $150,000 and variable costs of $0.25 per pen. How many pens must be manufactured to have an average cost of $1.75 per pen?

77. A baseball player's batting average is the quotient of the total number of hits and the total times at bat. A player has been at bat 185 times and has hit the ball 45 times. How many consecutive hits must this player have to obtain a batting average of .300?

78. A baseball player's batting average is the quotient of the total number of hits and the total times at bat. A player has eight hits after 50 times at bat. How many additional consecutive times must this player hit the ball to achieve a batting average of .250?

79. Russ drives 72 miles, has a flat tire and no spare, and walks 4 miles to a gas station. His driving rate was 12 times faster than his walking rate. If Russ spent $2\frac{1}{2}$ hours driving and walking, at what rate did he walk?

80. An executive traveled 900 miles by commercial jet and an additional 600 miles by helicopter. The rate of the jet was 3 times the rate of the helicopter. If the entire trip took 6 hours, what was the rate of the jet?

81. A cyclist and a jogger start at the same point and time for a destination 40 miles away. The cyclist averages twice the speed as the jogger, both following an identical route. If the cyclist arrives at the destination 4 hours earlier than the jogger, what is the rate of the cyclist?

82. An engine pulls a train 140 miles. Then a second engine whose average speed is 5 miles per hour faster than the first engine takes over and pulls the train 200 miles. The total time required for the 340 miles is 9 hours. Find the average speed of each engine.

83. Working together, Lou and Bud can complete a job in 6 days. Working alone, it takes Lou 5 days longer than Bud to do the job. How long does it take Bud to do the job if he works alone?

84. It takes Juan 10 more days than it takes Evita to do the work necessary to assume dictatorial control. Working together, the duo can assume absolute power in 12 days. Working alone, how long would it take Evita to assume dictatorial control?

85. A faucet can fill a sink in 5 minutes. It takes twice that long for the drain to empty the sink. How long will it take to fill the sink if the drain is open and the faucet is on?

86. A pool can be filled by a pipe in 3 hours. It takes 3 times as long for another pipe to empty the pool. How long will it take to fill the pool if both pipes are open?

87. Find the value of x in this geometric situation. The figures shown are not drawn to scale. Three-fourths of the perimeter of the first square equals the sum of the reciprocal of the second square's area and half the perimeter of the third square.

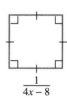
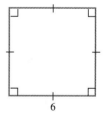
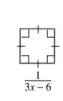

88. Two women were standing at the seashore when a large flock of sea gulls flew overhead. "There must be a hundred sea gulls there," said one woman. The other woman just happened to be a mathematics professor and said, "A hundred? Ridiculous! Before there could be a hundred, there must be again as many as there are now and then half as many more than there are now and then a quarter as many as there are now and still one more sea gull." If x equals the number of sea gulls in the flock, set up the rational equation and determine the actual number of birds in the flock.

True–False Critical Thinking Problems

89. Which one of the following is true?
  a. The best way to solve

$$\frac{2x}{x^2 - 4} + \frac{1}{x - 2} = \frac{2}{x + 2}$$

is to multiply both sides of the equation by $(x^2 - 4)(x - 2)(x + 2)$.
  b. Once a rational equation is cleared of fractions, all solutions of the resulting equation are also solutions of the rational equation.

c. The best way to subtract

$$\frac{4}{y} - \frac{2}{y+1}$$

is to multiply by $y(y+1)$.

d. If $\dfrac{a}{b} = \dfrac{c}{d}$, $b \neq 0$ and $d \neq 0$, then the cross multiplication $ad = bc$ is the result of multiplying both sides by the LCM of the denominators.

90. Which one of the following is true?

a. If 60 miles are covered in x hours, then the rate of travel is $\dfrac{60}{x}$ miles per hour.

b. If $\dfrac{300}{x}$ hours is 2 hours less than $\dfrac{300}{x-3}$ hours, then these conditions are represented by

$$\frac{300}{x} - 2 = \frac{300}{x-3}.$$

c. If a pipe can fill a pool in $x + 7$ hours, then it fills $\dfrac{1}{x+7}$ of the pool in 1 hour.

d. None of the above is true.

Technology Problems

91. One of the exciting aspects of using a graphing utility is the ability to visualize solutions to problems on graphs. Try doing this in Problems 47–56 by:

a. Representing each of the given models with x and y, using your graphing utility to graph the model. Use the algebraic solution that you obtained in the problem to determine a reasonable range setting.

b. TRACE along the graph and identify the point whose coordinates correspond to the problem's solution.

As you look at each graph, have you gained insight into the behavior modeled by the rational function that you did not get by looking at the equation? If so, what helpful characteristics were shown in the graph?

92. The rational function

$$y = \frac{72,900}{100x^2 + 729}$$

models the 1992 unemployment rate in the United States as a function of years of education.

a. Graph the function using your graphing utility and

$$\text{Xmin} = 0, \text{Xmax} = 16, \text{Xscl} = 1,$$
$$\text{Ymin} = 0, \text{Ymax} = 100, \text{Yscl} = 10$$

b. What does the shape of the graph indicate about increasing levels of education?

c. Is there an education level that leads to guaranteed employment? If not, how is this indicated on the graph?

d. Use the TRACE feature to solve the equation

$$\frac{72,900}{100x^2 + 729} = 10.2.$$

What does your solution mean in terms of the variables in the model?

93. For Problems 75–76, use a graphing utility to graph the average cost function described by the problem's conditions. Then TRACE along the curve and find the point that visually shows the problem's solution.

94. For Problems 79–80, use a graphing utility to graph the total time function described by the problem's conditions. Then TRACE along the curve and find the point that visually shows the problem's solution.

95. For Problems 83–84, use a graphing utility to graph the function representing the sum of the fractional parts of the job done by the two people for the period of time in which they can complete the job together. Then TRACE along the curve and find the point that visually shows the problem's solution.

96. A boat can travel 10 miles per hour in still water. The boat travels 24 miles upstream (against the current) and then 24 miles downstream (with the current).

a. If x = the current's speed, write a function in terms of x that models the total time for the boat to travel upstream and downstream.

b. Use a graphing utility to graph the total time model in part (a).

c. TRACE along the curve and determine the speed of the current if the time for the trip is 5 hours.

d. Verify part (c) algebraically.

Writing in Mathematics

97. Why is it necessary to check proposed solutions to an equation containing rational expressions?

98. Describe the major difference in the solution process between solving $\dfrac{x}{2} + \dfrac{x}{5} = 3$ and adding $\dfrac{x}{2} + \dfrac{x}{5}$.

Critical Thinking Problems

99. The formula

$$V = C\left(1 - \frac{T}{N}\right)$$

describes the value at the end of T years of an article that was originally worth C dollars but depreciated steadily over a period of N years. Solve the formula for N.

100. a. Solve for y: $\dfrac{3y + 1}{y + 5} = \dfrac{y - 1}{y + 1} + 2$.

   b. What is wrong with the following argument?

   $\dfrac{3y + 1}{y + 5} = \dfrac{y - 1}{y + 1} + 2$

   $\dfrac{3y + 1}{y + 5} = \dfrac{y - 1}{y + 1} + \dfrac{2(y + 1)}{y + 1}$ Write the right side with a denominator of $y + 1$.

   $\dfrac{3y + 1}{y + 5} = \dfrac{3y + 1}{y + 1}$ Add the rational expressions on the right.

   $y + 5 = y + 1$ If two equal rational expressions have equal numerators, their denominators must also be equal.

   $5 = 1$ Subtract y from both sides of the equation.

101. If $\dfrac{13w - 6z}{w + 3z} = 3$, what is the value of $\dfrac{w + z}{w - z}$?

102. Solve:

$$\left(\frac{1}{y + 1} + \frac{y}{1 - y}\right) \div \left(\frac{y}{y + 1} - \frac{1}{y - 1}\right) = -1.$$

103. Solve: $\left|\dfrac{y + 1}{y + 8}\right| = \dfrac{2}{3}$.

104. Solve $\left(\dfrac{3x - 2}{2x - 3}\right)^2 + \left(\dfrac{3x - 2}{2x - 3}\right) - 12 = 0$ by letting $y = \dfrac{3x - 2}{2x - 3}$.

105. If $x = \dfrac{1 - a}{1 + a}$ and $a = \dfrac{1 + y}{1 - y}$, show that $x + y = 0$.

106. Find b such that $\dfrac{7x + 4}{b} + 13 = x$ will have a solution set given by $\{-6\}$.

107. Find b such that $\dfrac{4x - b}{x - 5} = 3$ will have a solution set given by $\varnothing$.

108. It takes Mr. Todd 4 days longer to prepare an order of pies than it takes Mrs. Lovett. They bake together for 2 days when Mrs. Lovett leaves. Mr. Todd takes 7 additional days to complete the work. Working alone, how long does it take Mrs. Lovett to prepare the pies?

109. Two investments have interest rates that differ by 1%. An investment for 1 year at the lower rate earns $96. The same principal amount invested for a year at the higher rate earns $108. What are the two interest rates?

110. If $\dfrac{a}{b} = r$, express $\dfrac{a + b}{a - b}$ in terms of a rational expression containing only r as the variable.

111. A new schedule for a train requires it to travel 351 miles in $\frac{1}{4}$ hour less time than before. To accomplish this, the speed of the train must be increased by 2 miles per hour. What should the average speed of the train be so that it can keep on the new schedule?

112. This problem was part of a contest for the Interscholastic Mathematics League in New York City. Only 17% of the contestants correctly solved the problem, so this may qualify as one of the more challenging problems in the book. If

$$R = \frac{x^a}{1 - x^a} \qquad S = \frac{x^b}{1 - x^b} \qquad T = \frac{x^{a+b}}{1 - x^{a+b}}$$

write a simplified expression for T in terms of R and S. (*Hint:* Solve the first equation for x^a and the second equation for x^b.)

Review Problems

113. Graph: $f(x) = x^2 - 6x + 8$.

114. Subtract: $7x^2 - 8x - 9 - (9 - 8x - 7x^2)$.

115. Multiply: $(4x^2 - y)^2$.

SECTION 5.5

Solutions Manual　Tutorial　Video 6

Modeling Using Variation

Objectives

1　Model with direct variation.
2　Model with inverse variation.
3　Model with combined variation.
4　Model with joint variation.

Certain mathematical models occur so frequently in the natural and social sciences that they are given special names. *Variation* models show how one quantity varies in relation to another quantity. Quantities can vary *directly, inversely,* or *jointly.*

Model with direct variation.

Walter De Maria "The Lightning Field" 1977, Quemado, New Mexico. Photo credit: John Cliett. All reproduction rights reserved: © Dia Center For The Arts

Modeling with Direct Variation

Because light travels faster than sound, during a thunderstorm we see lightning before hearing thunder. The model

$$d = 1080t$$

describes the time t, in seconds, that it takes to hear thunder in a storm whose center is d feet away. Thus,

If $t = 1, d = 1080$:　It takes one second to hear thunder when the storm's center is 1080 feet away.

If $t = 2, d = 2160$:　It takes two seconds to hear thunder when the storm's center is 2160 feet away.

If $t = 3, d = 3240$:　It takes three seconds to hear thunder when the storm's center is 3240 feet away.

As the linear function $d = 1080t$ illustrates, the distance to the storm's center is a constant multiple of how long it takes to hear the thunder. When the time is doubled, the storm's distance is doubled; when the time is tripled, the storm's distance is tripled; and so on. Because of this, the distance is said to *vary directly* as the time. The *equation of variation* is

$$d = 1080t.$$

Generalizing, we obtain the following statement.

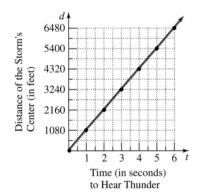

The graph of $d = 1080t$. Distance to a storm's center varies directly as the time it takes to hear thunder.

Direct variation

If a situation is modeled by an equation in the form

$$y = kx$$

where k is a constant, we say that y *varies directly as* x. We also say that y *is proportional to* x. The number k is called the *constant of variation* or the *constant or proportionality.*

iscover for yourself

If the time interval between the lightning and thunder is 9 seconds, how far away is the storm's center? What is the speed of sound?

EXAMPLE 1 **Direct Variation Relationships**

a. A person earns \$25 per hour. The person's total wage (w) is proportional to the number of hours worked (h). The equation of variation is $w = 25h$. The constant of variation or the constant of proportionality is 25.

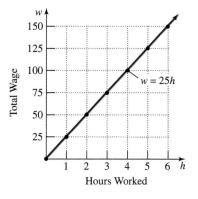

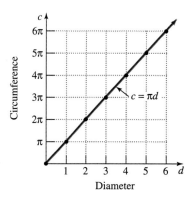

b. The circumference (c) of a circle varies directly as the diameter (d). The direct variation equation is $c = \pi d$, where π is the constant of variation. ■

If we know one pair of values in a direct variation, then we can find k, the constant of variation. Once k is known, we can write the equation of variation and other values can be determined.

> **Solving a variation problem**
>
> 1. Write an equation that models the given English statement.
> 2. Substitute the given pair of values into the equation in step 1 and find the value of k.
> 3. Substitute the value of k into the equation in step 1.
> 4. Use the equation from step 3 to answer the problem's question.

EXAMPLE 2 **Solving a Direct Variation Problem**

The volume of a gas (V) at constant pressure varies directly as the temperature (T). The gas has a volume of 250 cubic meters at a temperature of 30 Kelvin. Predict the volume of the gas at a temperature of 90 Kelvin.

Solution

Step 1. Write an equation.

$V = kT$ Translate "volume varies directly as temperature" into an equation.

Step 2. Use the given values to find k.

$250 = k \cdot 30$ We are given that when $V = 250$, $T = 30$.

$\dfrac{25}{3} = k$ Divide both sides by 30 and simplify.

Ice Water 270 K

Gas Expands and
the Balloon Inflates

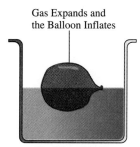

Hot Water 324 K

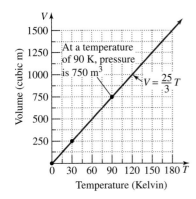

At a temperature of 90 K, pressure is 750 m³

$V = \frac{25}{3} T$

Temperature (Kelvin)

Step 3. Substitute k into the equation.

$V = kT$ Use the equation from step 1.

$V = \dfrac{25}{3}T$ Replace k, the constant of variation, with $\dfrac{25}{3}$.

Step 4. Answer the problem's question.

$V = \dfrac{25}{3}T$ Use the equation from step 3.

$V = \dfrac{25}{3} \cdot 90$ Predict volume ($V = ?$) at a temperature of 90 Kelvin ($T = 90$). Substitute 90 for T.

$V = 750$

At a temperature of 90 Kelvin, the volume of the gas is 750 cubic meters.   ■

A direct variation situation can involve variables to higher powers. For example, y can vary directly as x^2 ($y = kx^2$) or as x^3 ($y = kx^3$).

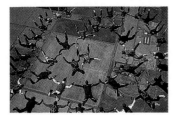

Tom Sanders/Photri, Inc.

Study tip

The direct variation equation $y = kx$ is a linear function, namely,

$f(x) = kx.$

If $k > 0$, then the slope of the line is positive. Consequently, as x increases, y also increases. On the other hand, if $k < 0$, then the slope of $y = kx$ is negative. For $k < 0$, the line represented by $y = kx$ goes down from left to right, so that as x increases, y decreases.

Direct variation with powers

y varies directly as the nth power of x (or *y is proportional to the nth power of x*) if there exists some nonzero constant k such that

$y = kx^n$

Direct variation with powers is modeled by polynomial functions. In our next example, the graph of the variation model is the familiar parabola.

EXAMPLE 3   **Modeling Direct Variation with a Quadratic Function**

The distance (s) that a body falls from rest varies directly as the square of the time (t) of the fall. If a body falls 64 feet in 2 seconds, how far will it fall in 4.5 seconds?

Solution

Step 1. Write an equation.

$s = kt^2$ Translate "distance (s) varies directly as the square of time (t)."

Step 2. Use the given values to find k.

$64 = k \cdot 2^2$ Find k. Because a body falls 64 feet in 2 seconds, when $t = 2$, $s = 64$.

$64 = 4k$ Square 2.

$16 = k$ Divide both sides by 4.

Step 3. Substitute k into the equation.

$s = kt^2$ becomes $s = 16t^2$.

Step 4. Answer the problem's question.

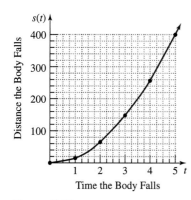

Figure 5.17

The graph of $s(t) = 16t^2$

2 Model with inverse variation.

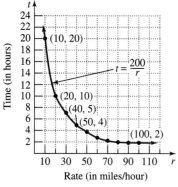

Figure 5.18

A trip's time is a function of speed, modeled by the inverse variation equation $t = \dfrac{200}{r}$.

$s = 16t^2$ Use the quadratic function from step 3.

$s = 16(4.5)^2$ How far ($s = ?$) will the body fall in 4.5 seconds? Substitute 4.5 for t.

$s = 16(20.25)$

$s = 324$

Thus, in 4.5 seconds, a body will fall 324 feet. ■

We can express the variation model from Example 3 in functional notation, writing

$s(t) = 16t^2$.

The distance that a body falls from rest is a function of the time (t) of the fall. The parabola that is the graph of the model is shown in Figure 5.17. The graph increases rapidly from left to right, showing the effects of the acceleration of gravity.

Modeling with Inverse Variation

Suppose you plan to make a 200-mile trip by car. The time it takes is a function of your speed, modeled by the formula

$$t = \frac{200}{r}$$ Since $rt = d$, then $t = \dfrac{d}{r}$ and $d = 200$.

Possible values of r and t are given in the following table of values.

r (miles/hour)	10	20	40	50	100
$t = \dfrac{200}{r}$ (hours)	20	10	5	4	2

The graph of $t = \dfrac{200}{r}$ is shown in Figure 5.18. Notice that as your speed increases, the trip's time decreases. Time is said to *vary inversely* as speed, and the equation of variation is $t = \dfrac{200}{r}$.

Generalizing, we obtain the following statement.

Inverse variation

If a situation is modeled by an equation in the form

$$y = \frac{k}{x}$$

where k is a constant, we say that y *varies inversely as* x. We also say that y *is inversely proportional to* x. The number k is called the *constant of variation* or *the constant of proportionality*.

Notice that the inverse variation model

$$y = \frac{k}{x} \quad \text{or} \quad f(x) = \frac{k}{x}$$

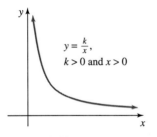

Figure 5.19

The graph of the inverse variation model

is a rational function. The graph of the function takes on the familiar shape shown in Figure 5.19. We use graphs shaped like this to model time in uniform motion situations and the average cost of a commodity for a manufacturer.

EXAMPLE 4 Solving an Inverse Variation Problem

Figure 5.20 illustrates that when the pressure exerted by the piston is doubled, the gas volume is halved. This is an example of Boyle's law, which states that the pressure P of a sample of gas at a constant temperature varies inversely as the volume V. This relationship is illustrated by the graph in Figure 5.21. The ordered pair $(8, 12)$ indicates that the pressure of a gas sample in a container whose volume is 8 cubic inches is 12 pounds per square inch. Suppose that the sample expands to a volume of 22 cubic inches. Find the new pressure of the gas.

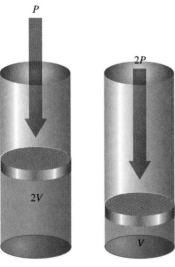

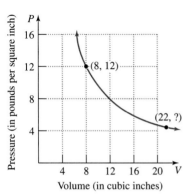

Figure 5.21

Pressure varies inversely as volume.

Figure 5.20

Doubling the pressure halves the volume.

Solution

Step 1. Write an equation.

$$P = \frac{k}{V} \quad \text{Translate "pressure } (P) \text{ varies inversely as volume } (V)."$$

Step 2. Use the given values to find k.

$$12 = \frac{k}{8} \quad \text{The ordered pair in Figure 5.21 shows that if } V = 8, \text{ then } P = 12.$$

$$96 = k \quad \text{Multiply both sides by 8.}$$

Step 3. Substitute k into the equation.

$$P = \frac{k}{V} \text{ becomes } P = \frac{96}{V}.$$

A compressed gas is a source of energy, doing useful work when it expands to atmospheric pressure, as in pneumatic drills, jackhammers, and spray cans.

Photri, Inc.

Step 4. Answer the problem's question.

$$P = \frac{96}{V}$$ Use the rational function from step 3.

$$P = \frac{96}{22}$$ Substitute 22 for V.

$$P = 4\frac{4}{11}$$ Simplify.

When the volume is 22 cubic inches, the pressure of the gas is $4\frac{4}{11}$ pounds per square inch. ■

Predictable numbers emerge from the model in Example 4 for certain values of V.

Volume is tripled.

Volume is doubled.

V	8	16	24
$P = \dfrac{96}{V}$	12	6	4

Pressure is halved.

Pressure is $\frac{1}{3}$ of what it had been.

3 Model with combined variation.

Modeling with Combined Variation

If we combine our work from Examples 2 and 4 on the effect of temperature and pressure on the volume of a gas, we can say that the pressure P of a sample of gas is directly proportional to the temperature T and inversely proportional to the volume V, modeled by

$$P = \frac{kT}{V}.$$

In this *combined variation,* two types of variation occur at the same time. Combined variation models numerous real life situations, illustrated in our next example.

EXAMPLE 5 **Solving a Combined Variation Problem**

A company determines that the monthly sales (S) of one of its products varies directly with their advertising budget (A) and inversely with the price of the product (P). When $60,000 is spent on advertising and the price of the product is $40, the monthly sales is 12,000 units.

a. Write an equation of variation that models this situation.
b. Determine monthly sales if the amount of the advertising budget is increased to $70,000.
c. Determine monthly sales if the amount of the advertising budget is increased to $70,000 and the unit price is increased to $50.

Solution

a. Write an equation.

$$S = \frac{kA}{P} \quad \begin{array}{l} \leftarrow \text{Translate "sales vary directly with the advertising budget} \\ \leftarrow \text{and inversely with the product's price."} \end{array}$$

Use the given values to find k.

$$12{,}000 = \frac{k(60{,}000)}{40} \qquad \begin{array}{l} \text{When \$60,000 is spent on advertising } (A = 60{,}000) \text{ and the} \\ \text{price is \$40 } (P = 40), \text{monthly sales are 12,000 units} \\ (S = 12{,}000). \end{array}$$

$$12{,}000 = 1500k \qquad \text{Divide numerator and denominator by 40 on the right.}$$

$$8 = k \qquad\qquad \text{Divide both sides of the equation by 1500.}$$

Therefore, the equation of variation that models monthly sales is

$$S = \frac{8A}{P}.$$

Discover for yourself

Use the model

$$S = \frac{8A}{P}$$

to describe what happens to sales if the unit price is fixed but the advertising budget increases. Describe what happens to sales if the unit price increases but the advertising budget is fixed.

b. The advertising budget is increased to \$70,000, so $A = 70{,}000$. The product's unit price is still \$40, so $P = 40$.

$$S = \frac{8A}{P} \qquad \text{This is the model from part (a).}$$

$$S = \frac{8(70{,}000)}{40} \qquad \text{Substitute 70,000 for } A \text{ and 40 for } P.$$

$$S = 14{,}000$$

With a \$70,000 advertising budget, the company can expect to sell 14,000 units of its product in a month (up from 12,000).

c. This time the advertising budget is increased to \$70,000 ($A = 70{,}000$), but the unit price is also increased to \$50 ($P = 50$).

$$S = \frac{8A}{P} \qquad \text{This is the model from part (a).}$$

$$S = \frac{8(70{,}000)}{50} \qquad \text{Substitute 70,000 for } A \text{ and 50 for } P.$$

$$S = 11{,}200$$

With an increase in both advertising budget and the product's price, the company can expect to sell 11,200 units monthly (down from the 12,000 originally given). ∎

4 Model with joint variation.   **Modeling with Joint Variation**

Joint variation is a variation in which a variable varies directly as the product of two or more other variables. Thus, the equation $y = kxz$ is read "y varies jointly as x and z." Example 6 illustrates a joint variation.

EXAMPLE 6   **Solving a Joint Variation Problem**

The volume of a cone (V) varies jointly as its height (h) and the square of its radius (r). A cone with a radius measuring 6 feet and a height measuring 10 feet has a volume of 120π cubic feet. Find the volume for a cone having a radius of 12 feet and a height of 2 feet.

Solution

Step 1. Write an equation.

$\qquad V = khr^2$ Translate "V varies jointly as h and the square of r."

Step 2. Use the given values to find k.

$\qquad 120\pi = k(10)(6)^2$ Find k. When $r = 6$ and $h = 10$, $V = 120\pi$.

$\qquad 120\pi = 360k$

$\qquad \dfrac{120\pi}{360} = k$

$\qquad \dfrac{\pi}{3} = k$

Step 3. Substitute k into the equation.

$\qquad V = khr^2$ becomes $V = \dfrac{\pi}{3}hr^2$.

Step 4. Answer the problem's question.

$\qquad V = \dfrac{\pi}{3}hr^2$ Use the equation from step 3.

$\qquad V = \dfrac{\pi}{3}(2)(12)^2$ Substitute the given values.

$\qquad V = \dfrac{\pi}{3}(2)(144)$

$\qquad V = 96\pi$

Thus, a cone having a radius of 12 feet and a height of 2 feet has a volume of 96π cubic feet. ∎

Summary of variation models

Direct

$\qquad y = kx$ y varies directly as x.

Inverse

$\qquad y = \dfrac{k}{x}$ y varies inversely as x.

Joint

$\qquad y = kxz$ y varies jointly as x and z.

Galileo's telescope brought about revolutionary changes in astronomy. A comparable leap in our ability to observe the universe is now taking place as a result of the Hubble Space Telescope. The telescope has thrown Big Bang theorists a curve by suggesting that some stars in the universe are older than

ENRICHMENT ESSAY

Challenging Variation

Isaac Newton's (1642–1727) Law of Gravitation says that two objects with masses m_1 and m_2 attract each other with a force F that varies jointly as their masses and inversely as the square of the distance d between them. The formula

$$F = G\frac{m_1 m_2}{d^2}$$

models the force of gravitation. (G is the gravitational constant.) The rational expression on the right indicates that gravitational force exists between any two objects in the universe, increasing as the distance between the bodies decreases. Consequently, the pull of the moon on the oceans is greater on one side of Earth than on the other. This gravitational imbalance is what produces tides.

Before Einstein's revelations about gravity, Newton's model was believed to be an ideal beyond dispute. As Einstein presented the possibility that alternate explanations exist regarding the interaction of space and mass, artist René Magritte created a visual image brazenly violating Newton's basic law of gravitation. In *Le Château des Pyrénées*, a levitating mountain crowned with a fortress provides a visual disregard for the model

$$F = G\frac{m_1 m_2}{d^2},$$

complementing the obscure formulas of the new physics.

René Magritte, "Le Château des Pyrénées" (1959), oil on canvas. (Israel Museum, Jerusalem. Giraudon/Art Resource, New York/© 1995 C. Herscovici, Brussels/ARS, New York.)

the universe itself. At the core of one galaxy, the telescope has found a black hole as massive as 3 billion suns. It has made scientists front-row spectators at the collision of comet Shoemaker-Levy 9 and the planet Jupiter. And it has begun to unravel the riddle of the brilliant beacons of cosmic light known as quasars. "The Hubble," declares University of Arizona astronomer Rodger Thompson, "is fundamentally altering our view of the universe."

EXAMPLE 7 **The Hubble Space Telescope**

Because it is above Earth's atmosphere, the Hubble Space Telescope can see stars and galaxies that are $\frac{1}{50}$ as bright as the faintest objects now observable using ground-based telescopes. How many times farther can the space telescope see than a ground-based telescope? Answer the question using the following inverse variation statement.

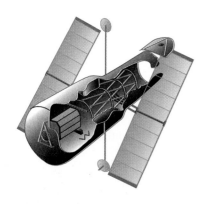

The brightness of a point source (B), such as a star, varies inversely as the square of its distance (D), from the observer. Thus,

$$B = \frac{K}{D^2}.$$

Solution

We apply the given inverse variation model to both ground-based telescopes and the Hubble Space Telescope.

For the ground-based telescope: $\quad B_G = \dfrac{K}{D_G^2}$

For the Hubble Space Telescope: $\quad B_S = \dfrac{K}{D_S^2}$

We are given that the Hubble Space Telescope can see objects $\frac{1}{50}$ the brightness of ground-based telescopes. This means that $B_S = \frac{1}{50}B_G$.

Gravitational Lensing of a Quasar, *Hubble Space Telescope,* 1990.

European Space Agency/NASA Headquarters

$B_S = \dfrac{K}{D_S^2}$	This is the inverse variation model for the Hubble Space Telescope.
$\dfrac{1}{50}B_G = \dfrac{K}{D_S^2}$	We let B_S equal $\frac{1}{50}B_G$ because the space telescope can see objects $\frac{1}{50}$ the brightness of ground-based telescopes.
$\dfrac{1}{50}\boxed{\dfrac{K}{D_G^2}} = \dfrac{K}{D_S^2}$	Substitute $\frac{K}{D_G^2}$ for B_G, obtained from the model for ground-based telescopes.
$\dfrac{1}{50D_G^2} = \dfrac{1}{D_S^2}$	Divide both sides by K.
$50D_G^2 D_S^2 \cdot \dfrac{1}{50D_G^2} = 50D_G^2 D_S^2 \cdot \dfrac{1}{D_S^2}$	Multiply by the LCM.
$D_S^2 = 50D_G^2$	Simplify.
$D_S \approx 7.1D_G$	

Thus, the distance the Hubble Space Telescope can see is approximately 7.1 times more than ground-based telescopes can see. ∎

PROBLEM SET 5.5

Practice Problems

For Problems 1–14, a. find the constant of variation, b. write the equation of variation, and c. find the quantity indicated.

1. y varies directly as x. $y = 35$ when $x = 5$. Find y when $x = 12$.

2. y varies directly as x. $y = 55$ when $x = 5$. Find y when $x = 13$.

3. y varies directly as x^2. $y = 48$ when $x = 3$. Find y when $x = 7$.

4. y varies directly as x^2. $y = 96$ when $x = 4$. Find y when $x = 7$.

5. y varies inversely as x. $y = 10$ when $x = 5$. Find y when $x = 2$.

6. y varies inversely as x. $y = 5$ when $x = 3$. Find y when $x = 9$.

7. y varies inversely as x^2. $y = 3$ when $x = 5$. Find y when $x = 10$.

8. y varies inversely as x^3. $y = 6$ when $x = 2$. Find y when $x = \frac{1}{2}$.

9. z varies directly as x and inversely as y. $z = \frac{1}{3}$ when $x = 10$ and $y = 2$. Find z when $x = 30$ and $y = 5$.

10. y varies directly as x and inversely as the square of z. $y = 20$ when $x = 50$ and $z = 5$. Find y when $x = 3$ and $z = 6$.

11. y varies jointly as x and z. $y = 25$ when $x = 2$ and $z = 5$. Find y when $x = 8$ and $z = 12$.

12. y varies jointly as x and z. $y = 10$ when $x = 5$ and $z = \frac{1}{4}$. Find y when $x = \frac{1}{16}$ and $z = 24$.

13. y varies jointly as x^2 and z and inversely as R^3. $y = \frac{1}{3}$ when $x = 2$, $z = 6$, and $R = 2$. Find y when $x = 3$, $z = 2$, and $R = 1$.

14. F varies jointly as m_1 and m_2 and inversely as the square of d. $F = 8$ when $m_1 = 2$, $m_2 = 8$, and $d = 4$. Find F when $m_1 = 28$, $m_2 = 12$, and $d = 2$.

Application Problems

15. The circumference (C) of a circle varies directly as the diameter (d). When the diameter is 2 feet, the circumference is 2π feet. Find the circumference when the radius is 8 feet.

16. The cost (C) of an airplane ticket is directly proportional to the number of miles in the trip (M). A 3000-mile trip costs $400. How many miles can be covered at a cost of $20?

17. The distance required to stop a car (d) varies directly as the square of its speed (r). If 200 feet are required to stop a car traveling 60 miles per hour, how many feet are required to stop a car traveling 100 miles per hour?

18. The distance (d) that a body falls varies directly as the square of the time (t) in which it falls. A body falls 64 feet in 2 seconds. Predict the distance that the same body will fall in 10 seconds.

19. The volume (V) of a gas in a container at a constant temperature varies inversely as the pressure (p). If the volume is 32 cubic centimeters at a pressure of 8 pounds, find the pressure when the volume is 40 cubic centimeters.

20. The current in a circuit is inversely proportional to the resistance. The current is 20 amperes when the resistance is 5 ohms. Predict the current for a resistance of 16 ohms.

21. The intensity of illumination on a surface varies inversely as the square of the distance of the light source from the surface. The illumination from a source is 25 foot-candles at a distance of 4 feet. What is the illumination when the distance is 6 feet?

22. The gravitational force with which Earth attracts an object varies inversely with the square of the distance from the center of Earth. If a gravitational force of 160 pounds acts on an object 4000 miles from Earth's center, predict the force of attraction on an object 6000 miles from the center of Earth.

23. The volume of a gas is directly proportional to its temperature and inversely proportional to the pressure. At a temperature of 100 Kelvin and a pressure of 15 kilograms per square meter, the gas occupies a volume of 20 cubic meters. Find the volume at a temperature of 150 Kelvin and a pressure of 30 kilograms per square meter.

24. The cephalic index is used by anthropologists to study differences among races of human beings. The index varies directly as the width of the head and inversely as the length of the head. If the cephalic index is 75 for a width of 6 inches and a length of 8 inches, find the index for a head width of 7 inches and a length of 10 inches.

25. Simple interest varies jointly as the interest rate and the time that money is invested. An investment at 12% for 2 years earns $280. How much will the investment yield at 16% for 4 years?

26. The volume of a cylinder varies jointly as its height and the square of its radius. A cylinder with a height of 6 feet and a radius of 2 feet has a volume of 24π cubic feet. Find the volume of a cylindrical refinery having a height of 10 feet and a diameter of 6 feet.

27. The force of attraction between two bodies varies jointly as the product of their masses and inversely as the square of the distance between them. Masses of 4 units and 2 units, separated by 3 feet, have a force of 16 units. What is the force for masses of 5 and 3 units separated by 2 feet?

28. If all men had identical body types, their weight would vary directly as the cube of their height. Shown at the top of page 489 is Robert Wadlow, who reached a record height of 8 feet 11 inches before his death at age 22. If a man who is 5 feet 10 inches tall with the same body type as Mr. Wadlow weighs 170 pounds, what was Robert Wadlow's weight shortly before his death?

UPI/Corbis-Bettmann

29. A person's body-mass index is used to assess levels of fatness, with an index from 20 to 26 considered in the desirable range. The index varies directly as one's weight in pounds and inversely as one's height in inches. A person who weighs 150 pounds and is 70 inches tall has an index of 21. What is the body-mass index of a person who weighs 240 pounds and is 74 inches tall? Since the index is rounded to the nearest whole number, do so and then determine if this person's level of fatness is in the desirable range.

30. Kinetic energy varies jointly as the mass and the square of the velocity. A mass of 8 grams and velocity of 3 centimeters per second has a kinetic energy of 36 ergs. Predict the kinetic energy for a mass of 4 grams and velocity of 6 centimeters per second.

31. The electrical resistance of a wire varies directly as its length and inversely as the square of its diameter. A wire of 720 feet with $\frac{1}{4}$-inch diameter has a resistance of $1\frac{1}{2}$ ohms. Predict the resistance for 960 feet of the same kind of wire if its diameter is doubled.

32. The centrifugal force of a body moving in a circle varies jointly with the radius of the circular path and the body's mass and inversely with the square of the time it takes to move about one full circle. A 6-gram body moving in a circle with radius 100 centimeters that completes a revolution in 2 seconds has a centrifugal force of 6000 dynes. Find the centrifugal force of an 18-gram body moving in a circle with radius 100 centimeters and completing each revolution in 3 seconds.

True–False Critical Thinking Problems _____

33. Which one of the following is true?
 a. It seems reasonable that the demand for a product varies directly as the price of the product.
 b. If a varies directly as b and inversely as c, then $a = \dfrac{kb}{c}$.
 c. If a varies inversely as the square of b, then $a = kb^2$.
 d. The domain of $y = \dfrac{1}{x}$ is the set of all real numbers.

34. Which one of the following is true?
 a. If a varies inversely as the square of b, then $a = \dfrac{k}{\sqrt{b}}$.
 b. Some inverse relations $\left(y = \dfrac{k}{x}\right)$ are not functions.

 c. The direct variation equation, $y = kx$, is an example of a linear function.
 d. It seems reasonable that the weight of an iguana varies inversely as its length.

35. Which one of the following is not true?
 a. The force needed to lift an object varies inversely as its weight.
 b. The area of a rectangle whose length remains constant varies directly as its width.
 c. Distance traveled varies directly as the time spent traveling.
 d. The area of a circle varies directly as the square of its radius.

Technology Problem _____

36. Use a graphing utility to graph the models in Problems 17–22. Then TRACE along each curve and identify the point that corresponds to the problem's solution.

Writing in Mathematics

In Problems 37–38, describe in words the variation modeled by each equation.

37. $y = \dfrac{k\sqrt{x}}{y^2}$

38. $y = kx^2\sqrt{y}$

39. Describe what is wrong with this argument: A 2-inch-long grasshopper can jump 40 inches. Since jumping ability varies directly as length, a 6-foot-tall human should be able to jump 1440 inches.

40. In a hurricane, the wind pressure varies directly as the square of the wind velocity. If wind pressure is a measure of a hurricane's destructive capacity, what happens to this destructive power when the wind speed doubles?

41. The illumination from a light source varies inversely as the square of the distance from the light source. If you raise a lamp from 15 inches to 30 inches over your desk, what happens to the illumination?

Critical Thinking Problems

42. If y varies directly as x and y increases by 18 when x is increased by 3, find the constant of variation.

43. If y is inversely proportional to x $\left(y = \dfrac{K_1}{x}\right)$ and x is inversely proportional to $z\left(x = \dfrac{K_2}{z}\right)$, describe the relationship of y to z. What is the constant of variation?

44. The heat generated by a stove element varies directly as the square of the voltage and inversely as the resistance. If the voltage remains constant, what needs to be done to triple the amount of heat generated?

45. Reread Problem 27. Describe what happens to the force of attraction if both masses are doubled and the distance between the objects is halved.

Group Activity Problems

46. Group members are in charge of advertising for the Worldwide Widget Corporation. Members were told that the demand for widgets varies directly as the amount spent on advertising. The demand for widgets varies inversely as the price of the product. However, as more money is spent on advertising, the price of a widget rises. Under what conditions would you recommend an increased expense in advertising? (Write mathematical models for the given conditions and experiment with hypothetical numbers.) What other factors might you take into consideration as part of your recommendation? How do these other factors affect the demand for widgets?

47. Use the model from Problem 29 to determine the body-mass index of members of the group. An index of 27–29 indicates a slight risk of health problems related to weight and an index of 30 or more indicates a high risk of weight-related health problems.

48. The daily number of phone calls between two cities varies jointly as their populations and inversely as the square of the distance between them.

a. Consult the research department of your library, select two U.S. cities of your choice, and find their populations, the distance separating them, and the average number of daily phone calls between them. Use this information to write the variation equation that models this situation.

b. Use the model from part (a) to predict the average number of phone calls in a day between another two cities. You'll need to know the population of each city and the distance between them. (Continue using the research department of your library to determine the actual number of phone calls.) How accurate is the model in this situation?

c. The model under consideration is sometimes called the *gravity model*. Compare the model to the one discussed in the enrichment essay on page 486 and describe why this name applies.

Review Problems

49. Factor completely: $y^4 - 3y^3 + 2y^2 - 6y$.

50. Graph: $3x - 6y < 12$.

51. Solve the system:

$$x + 3y + 2z = 5$$
$$x + 5y + 5z = 6$$
$$3x + 3y - z = 10$$

CHAPTER PROJECT

Tiling the plane

Tiling the plane, in a mathematical sense, is very similar to tiling a floor. Tiling, or *tessellating,* the plane means covering the entire plane with endless repetitions of a basic design, without leaving any space empty. If we restrict ourselves to geometric figures composed of equal sides and equal angles, these are called regular polygons. Only three shapes will tile the plane: the square, the equilateral triangle, and the hexagon. However, it is not necessary to have such rigid restrictions on the shapes we will use. In fact, *any* triangle and *any* quadrilateral will cover the plane. We have seen examples in this chapter of patterns used by the artist M. C. Escher (pages 416 and 419) which seem to cover the plane with interesting shapes and designs.

1. Use a piece of graph paper to make a series of different four-sided figures. Cut out the originals and use each of them to make a cardboard stencil. Convince yourself that each stencil, no matter the exact shape of the four-sided figure, will allow you to create a repetitive pattern to cover the plane. Share your results with the class.

We can extend the idea of a quadrilateral covering the plane to a slightly irregular shape, one which will allow us to create more fanciful images. We begin with a simple four-sided figure and draw a design on one edge of the figure. Cut out the design and reapply it to the opposite side. We may repeat the same sequence with the other two straight sides if we wish.

2. Working in a group, create a figure and use it to cover the plane. You may choose to make random shapes or think of a shape you would like to see repeated and try to create it. In either case, notice how the same shape may be rotated to fit in different ways. Vary the style of your original geometric figure before you settle on one design. For example, experiment with long, thin shapes as well as more compact figures. You may also wish to use colors or interesting lines inside the shape to enhance your finished pattern.

3. Many cultures have used tessellations of the plane not only on tiles, but also on fabrics, wall coverings, window screen designs, baskets and pottery, to list but a few. Working in a group, select a culture to study and reproduce some of the patterns you discover. A partial list would include Celtic, Greek, Roman, or Egyptian border designs; Mexican, Native American, South or Central American pottery or basket designs; Japanese, Russian, or Hungarian fabric designs; Arabic or African tile, rug, and lattice designs as well as traditional early American quilt designs. Share your research with the class.

Escher's work was influenced by the geometric designs he observed while visiting the Alhambra, a 14th century palace in Granada, Spain. The intricately designed tilework was left by the Moorish conquerors of Spain when they occupied the castle. Escher was inspired to experiment with a simple regular figure, such as a square, covered with a basic design which would match other tiles at specific points. We can use a simple procedure to create our own sets of tiles.

4. Begin by constructing a square on a sheet of graph paper with five units to each side. Mark off the four corners and the middle grid on each side (Figure 5.22). Connect the four corners and the centers of the edges of the square in any way you wish (Figure 5.23). Also, create a mirror image of the design when you are finished (Figure 5.24). Use a machine to photocopy multiple copies of each square to use in creating your tiling pattern. Using your "tiles," create at least two different patterns to cover the plane.

| **Figure 5.22** | **Figure 5.23** | **Figure 5.24** |

5. Design a 3-4-5 right triangle in a similar way as the square above and use it to create a new, triangular, tiling of the plane.

Worldwide Web Resources

Go to the Prentice Hall website (http://www.prenhall.com/blitzer) to access other locations on the Internet that will allow you to further explore the concepts presented in this project.

Chapter Review

SUMMARY

1. **Rational Expressions and Rational Functions**
   A rational expression is the quotient of two polynomials. A rational function is of the form $f(x) = \dfrac{p(x)}{q(x)}$, where $p(x)$ and $q(x)$ are polynomials and $q(x) \neq 0$.

2. **Domain of a Rational Function**
   The domain of a rational function is the set of all real numbers excluding numbers that make the polynomial in the denominator 0. The graph of a rational function has either a vertical asymptote or a hole at any value of x excluded from its domain.

3. **Reducing Rational Expressions to Lowest Terms (Simplifying Rational Expressions)**
   a. Factor the numerator and denominator completely.
   b. Divide both the numerator and denominator by the common factors.
   c. The quotient of two polynomials that have exactly opposite signs and are additive inverses is -1.

4. **Multiplying Rational Expressions**
   a. Factor all numerators and denominators completely.
   b. Write the product of the numerators over the product of the denominators.
   c. Divide both numerator and denominator by common factors.
   d. Multiply remaining factors in the numerator and multiply remaining factors in the denominator.

5. **Dividing Rational Expressions**
   The quotient of two rational expressions is the product of the first rational expression and the reciprocal of the second rational expression, the divisor.

6. **Adding and Subtracting Rational Expressions**
   a. To add or subtract rational expressions with the same denominator, add or subtract the numerators and place the result over the common denominator. If possible, simplify the resulting rational expression.

b. To add or subtract rational expressions that have different denominators:
   1. Find the LCD, the product of all different factors from each denominator, with each factor raised to the highest power occurring in any denominator.
   2. Write all rational expressions as equivalent rational expressions having the LCD as the specified denominator.
   3. Add or subtract numerators, placing the resulting expression over the LCD.
   4. If necessary, simplify the resulting rational expression.

7. Simplifying Complex Rational Expressions
   a. A complex rational expression is a rational expression whose numerator or denominator contains one or more fractions.
   b. To simplify a complex rational expression, use one of the following methods.
      1. Simplify the numerator and denominator of the complex expression. Then find the quotient by multiplying the numerator by the reciprocal of the denominator.
      2. Multiply the numerator and denominator of the complex expression by the LCM of all the denominators. Then simplify the resulting expression.

8. Dividing Polynomials
   a. Use $\dfrac{b^n}{b^m} = b^{n-m}$ to divide monomials. Divide coefficients and divide variable factors with the same base by subtracting exponents.
   b. To divide a polynomial containing more than one term by a monomial, divide each term of the polynomial by the monomial. Then use the procedure in part (a).
   c. To divide polynomials using long division:
      1. Arrange terms in the dividend and the divisor in descending powers of any variable.
      2. Divide the first term in the dividend by the first term in the divisor. The result is the first term in the quotient.
      3. Multiply the divisor by the first term in the quotient. Write the resulting product beneath the dividend.
      4. Subtract the product from the dividend.
      5. Bring down the next term in the original dividend and write it next to the remainder to form a new dividend.
      6. Use this new expression as the dividend and repeat this process until the degree of the remainder is less than the degree of the divisor.
   d. Synthetic division can be used to divide a polynomial by the binomial $x - c$. Using $+c$ and the coefficients of the dividend, a series of multiplications and additions produces the coefficients of the quotient.

9. Solving Equations Containing Rational Expressions
   Multiply both sides of the equation by the LCM of the denominators. Solve the resulting equation. Any value of a variable that makes any denominator of the original equation equal 0 is not a solution.

10. Solving a Formula Containing Rational Expressions for a Specified Variable
   a. Multiply both sides by the LCM of all denominators to clear fractions.
   b. Collect all terms with the specified variable on one side of the equation and all other terms on the other side.
   c. If the specified variable appears in more than one term, factor it out.
   d. Isolate the specified variable by dividing both sides of the equation by the factor that appears with the variable.

11. Creating Rational Models
   a. Problems involving the average cost per unit for a company to produce x units of a product can be modeled using

   $$\text{Average cost} = \frac{\text{Total cost}}{\text{Number of units}}$$

   $$= \frac{\text{Fixed costs} + (\text{Cost per unit})x}{x}$$

   Problems involving uniform motion and work often result in rational equations.
   b. In the case of motion problems, because $rt = d$, then $t = \dfrac{d}{r}$. Find two rational expressions for time and then use the verbal conditions of the problem to write an equation.
   c. In the case of work problems, if a person can do a job in x hours and works for b hours, then the fractional part of the job completed is $\dfrac{b}{x}$. If two people work together to complete a job, the sum of the fractional parts of the job done by each person is 1.

12. Variation
   a.

English Statement	Equation
y varies directly as x. y is proportional to x.	$y = kx$
y varies directly as x^n y is proportional to x^n.	$y = kx^n$
y varies inversely as x. y is inversely proportional to x.	$y = \dfrac{k}{x}$
y varies inversely as x^n. y is inversely proportional to x^n.	$y = \dfrac{k}{x^n}$
y varies jointly as x and z.	$y = kxz$

b. To solve a variation problem,
  1. Translate from an English statement into an equation.
  2. Find the value of k, the variation constant.

  3. Substitute the value for k into the equation in step 1.
  4. Use the equation from step 3 to answer the given question.

REVIEW PROBLEMS

In this problem set, verify your answers using a graphing utility if applicable. Describe the domain of the rational functions in Problems 1–4, using both set-builder and interval notation.

1. $f(x) = \dfrac{7x}{9x - 18}$

2. $f(x) = \dfrac{x + 3}{(x - 1)(x + 5)}$

3. $f(x) = \dfrac{7x + 14}{2x^2 + 5x - 3}$

4. $f(x) = \dfrac{x^2 - 25}{x^2 + 4}$

5. Consider the function defined by $f(x) = \dfrac{1}{x + 2}$.

  a. What is the domain of f?
  b. Fill in the following tables.

x approaches −2 from the left.

x	−3	−2.5	−2.1	−2.01	−2.001
$f(x) = \dfrac{1}{x + 2}$					

x approaches −2 from the right.

x	−1.999	−1.99	−1.9	−1.5	−1
$f(x) = \dfrac{1}{x + 2}$					

  c. Graph the line whose equation is $x = -2$ and then graph f.

6. Which one of the following functions is graphed in the figure?

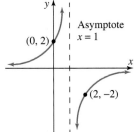

  a. $y = \dfrac{2}{x - 1}$

  b. $y = \dfrac{2}{x + 1}$

  c. $y = \dfrac{-2}{x - 1}$

  d. $y = \dfrac{-2}{x + 1}$

7. A business that sells small alarm clocks has fixed costs of $250,000 and variable costs of $3 per clock.
  a. Find the cost function $C(x)$ that models the cost (fixed costs plus variable costs) of producing x alarm clocks.
  b. Find a function $\overline{C}$ that models the average cost per clock for x clocks.
  c. Find and interpret $\overline{C}(1000)$, $\overline{C}(10,000)$, and $\overline{C}(100,000)$. What conclusion can you draw?

Simplify each rational expression in Problems 8–12.

8. $\dfrac{5x^3 - 35x}{15x^2}$

9. $\dfrac{x^2 + 6x - 7}{x^2 - 49}$

10. $\dfrac{6m^2 + 7m + 2}{2m^2 - 9m - 5}$

11. $\dfrac{y^3 - 8}{y^2 - 4}$

12. $\dfrac{3x^2 + 15xy + 12y^2}{3x^3 + 9x^2y - 12xy^2}$

13. Simplify, and graph the resulting linear function:

$$f(x) = \dfrac{x^2 - 7x + 12}{x - 4}.$$

Be sure to show an open dot, or hole, on the graph corresponding to the value of x excluded from the function's domain.

In Problems 14–17, multiply the rational expressions.

14. $\dfrac{5x^2 - 5}{3x + 12} \cdot \dfrac{x + 4}{x - 1}$

15. $\dfrac{x^2 - 9x + 14}{x^3 + 2x^2} \cdot \dfrac{x^2 - 4}{(x - 2)^2}$

16. $\dfrac{y^4 - 81}{y^2 + 9} \cdot \dfrac{4y - 20}{y^2 - 8y + 15}$

17. $\dfrac{5xy - 10y^2}{x^2 - 3xy + 2y^2} \cdot \dfrac{x + 2y}{x^2} \cdot \dfrac{3x^2 - 3xy}{xy + 2y^2}$

In Problems 18–21, divide the rational expressions.

18. $\dfrac{25x^2 - 1}{x^2 - 25} \div \dfrac{5x + 1}{x + 5}$

19. $\dfrac{x^2 + 16x + 64}{2x^2 - 128} \div \dfrac{3x^2 + 30x + 48}{x^2 - 6x - 16}$

20. $\dfrac{a^3 - 27}{a^2 + 3a + 9} \div (ab + ac - 3b - 3c)$

21. $\dfrac{y^3 - 8}{y^4 - 16} \div \dfrac{y^2 + 2y + 4}{y^2 + 4}$

22. Given $f(x) = 2x^2 + x - 15$ and $g(x) = x + 3$, find the following.

  a. $\left(\dfrac{f}{g}\right)(x)$ **b.** $\left(\dfrac{f}{g}\right)(6)$ **c.** $\dfrac{f(6)}{g(6)}$

23. A commuter drove to work a distance of 30 miles and then returned on the same highway. The average speed on the return trip was 10 miles per hour faster than the average speed on the outgoing trip.

  a. Let x = the average speed on the outgoing trip. Write a function using the addition of rational expressions that models the total time t required to complete the round trip.

  b. Add the two terms in the round-trip model of part (a).

In Problems 24–31, add or subtract, then simplify where possible.

24. $\dfrac{x^3}{x^3 + 125} + \dfrac{5}{x^3 + 125}$

25. $\dfrac{2x - 7}{x^2 - 9} - \dfrac{x - 4}{x^2 - 9}$

26. $\dfrac{1}{x} + \dfrac{2}{x - 5}$

27. $\dfrac{3x^2}{9x^2 - 16} - \dfrac{x}{3x + 4}$

28. $\dfrac{7}{x^2 y^3} - \dfrac{5}{xy^3} + \dfrac{4}{x^2 y}$

29. $\dfrac{y}{y^2 + 5y + 6} - \dfrac{2}{y^2 + 3y + 2}$

30. $\dfrac{x}{x + 3} + \dfrac{x}{x - 3} - \dfrac{9}{x^2 - 9}$

31. $\dfrac{4}{a^2 + a - 2} - \dfrac{2}{a^2 - 4} + \dfrac{3}{a^2 - 4a + 4}$

Simplify each complex rational expression in Problems 32–35.

32. $\dfrac{4 + \dfrac{1}{x}}{x^2}$

33. $\dfrac{4 - \dfrac{1}{y^2}}{4 + \dfrac{4}{y} + \dfrac{1}{y^2}}$

34. $\dfrac{3 - \dfrac{1}{x + 3}}{3 + \dfrac{1}{x + 3}}$

35. $\dfrac{\dfrac{1}{y + 5} + 1}{\dfrac{6}{y^2 + 2y - 15} - \dfrac{1}{y - 3}}$

In Problems 36–39, divide.

36. $(35x^2 y^3 - 25x^2 y^2 - 15x^4 y^3) \div 5xy^2$

37. $(6x^2 - 5x + 5) \div (2x + 3)$

38. $(10x^3 - 26x^2 + 17x - 13) \div (5x - 3)$

39. $(4x^4 + 6x^3 + 3x - 1) \div (2x^2 + 1)$

In Problems 40–41, divide using synthetic division.

40. $\dfrac{4x^3 - 3x^2 - 2x + 1}{x + 1}$

41. $(3y^4 - 2y^2 - 10y) \div (y - 2)$

Solve the equations in Problems 42–47.

42. $\dfrac{2y}{y - 2} - 3 = \dfrac{4}{y - 2}$

43. $\dfrac{1}{y - 5} - \dfrac{3}{y + 5} = \dfrac{6}{y^2 - 25}$

44. $\dfrac{x + 5}{x + 1} - \dfrac{x}{x + 2} = \dfrac{4x + 1}{x^2 + 3x + 2}$

45. $\dfrac{2}{3} - \dfrac{5}{3y} = \dfrac{1}{y^2}$

46. $\dfrac{2}{y - 1} = \dfrac{1}{4} + \dfrac{7}{y + 2}$

47. $\dfrac{2y + 7}{y + 5} - \dfrac{y - 8}{y - 4} = \dfrac{y + 18}{y^2 + y - 20}$

48. The model

$$C = \frac{4p}{100 - p}$$

describes the cost (C, in millions of dollars) to remove p percent of pollution from a river due to pesticide runoff from area farms.

  a. What percent of pollution can be removed at a cost of $16 million?

  b. Solve the model for p. Then use this form of the model to verify the result you obtained in part (a).

  c. What value of p must be excluded from the domain of the given rational function? What does this mean in terms of the variables described by the model?

  d. Suppose that you were to graph the given model, with values of p along the horizontal axis (from 0 to 100) and values of C along the vertical axis. Without actually graphing the rational function, describe the graph's shape, particularly as p gets close to 100. What does this mean about removing all the pollution? If applicable, verify your observations by using a graphing utility to graph the model.

49. In t years from 1985, the population of a community will be $P = 30 - \dfrac{9}{t + 1}$, where P is expressed in thousands. When will the community have a population of 29,000?

50. A drug is injected into a patient and the concentration of the drug in the bloodstream is monitored. The drug's concentration ($C(t)$, in milligrams per liter) after t hours is modeled by

$$C(t) = \frac{5t}{t^2 + 1}.$$

The graph of this rational function, obtained with a graphing utility, is shown in the figure.

  a. How long after the injection will the concentration be 2 milligrams per liter? Identify your solution with a point on the function's graph.

  b. Use the graph to obtain a reasonable estimate of the drug's concentration after 3 hours. Then verify this estimate algebraically.

  c. As you look at the graph, describe one characteristic of the drug's concentration in the patient's bloodstream over time that is not apparent by looking at the given equation.

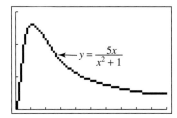

In Problems 51–54, solve each model for the indicated variable.

51. $P = \dfrac{R - C}{n}$ for C **52.** $T = \dfrac{A - p}{pr}$ for p **53.** $\dfrac{1}{p} + \dfrac{1}{q} = \dfrac{1}{f}$ for p **54.** $I = \dfrac{En}{Rn + r}$ for n

55. Clear, sharp pictures require precision camera focus. A camera lens has a characteristic measurement (f) called its focal length. When an object is in focus, its distance from the lens (p) and the distance from the lens to the film (q) (see the figure) satisfy the model

$$\frac{1}{p} + \frac{1}{q} = \frac{1}{f}.$$

Find the focal length of a lens such that an object located from the lens at a distance that is 4 times the focal length of the lens is in focus when the distance from the lens to the film is 3 centimeters greater than the focal length.

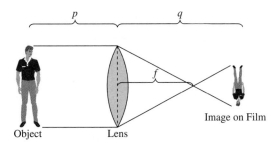

Object   Lens   Image on Film

56. A furniture company can produce x units of a desk for $60 per desk in addition to an initial investment of $100,000. How many units must be produced to have an average cost of $80 per desk?

57. After riding at a steady speed for 60 miles, a bicyclist had a flat tire and walked 8 miles to a repair shop. The cycling rate was 3 times faster than the walking rate. If the time spent cycling and walking was 7 hours, at what rate was the cyclist riding?

58. The current of a stream moves at 3 miles per hour. It takes a boat a total of 3 hours to travel 12 miles upstream (against the current) and return the same distance traveling downstream (with the current). What is the boat's speed in still water?

59. Working together, Norman and his mother can paint the Bates Motel in 20 days. Norman can paint the motel by himself in 9 days less time than it takes his mother. How many days would it take each of them to paint the motel alone?

60. Pipe A can fill a pool in 8 hours, and pipe B can drain the pool in 12 hours. Pipe A was turned on to fill the pool, but pipe B was accidently left open to drain the pool. After a period of time, the mistake was found, and pipe B was closed. At this point, the pool was already one-half full. How much time had elapsed before the mistake was found?

61. The weight of an object on the moon varies directly as its weight on Earth. An object that weighs 17.6 kilograms on the moon has a weight of 110 kilograms on Earth. Find the Earth weight of an object that weighs 28.8 kilograms on the moon.

62. The weight that can be supported by a 2 inch by 4 inch piece (called a 2-by-4) of pine varies inversely as its length. A 10-foot pine 2-by-4 can support 500 pounds. What weight can be supported by a 125-foot pine 2-by-4?

63. The distance that an object falls varies directly as the square of the time it has been falling. An object falls 144 feet in 3 seconds. Predict how far it will fall after 7 seconds.

64. The force of wind blowing on a window positioned at a right angle to its direction varies jointly with the area of the window and the square of the wind's speed. It is known that a wind of 30 miles per hour blowing on a window measuring 4 feet by 5 feet has a force of 150 pounds. During a storm with winds of 60 miles per hour, should hurricane shutters be placed on a window that measures 3 feet by 4 feet and is capable of withstanding 300 pounds of force?

CHAPTER 5 TEST

1. Find the domain of $f(x) = \dfrac{x^2 - 2x}{x^2 - 7x + 10}$.
Then simplify the function's equation.

2. Explain why the graph in the figure cannot be the graph of $f(x) = \dfrac{1}{x + 2}$.

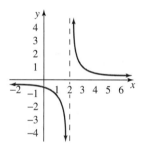

In Problems 3–10, perform the indicated operations. Simplify where possible.

3. $\dfrac{x^2}{x^2 - 6} \cdot \dfrac{x^2 + 7x + 12}{x^2 + 3x}$

4. $\dfrac{x + 2y}{x + 3y} \div \dfrac{3x + 6y}{x^2 - 9y^2}$

5. $\dfrac{x^3 - 1}{(x - 1)^3} \div \dfrac{x^2 + x + 1}{x^2 - 1}$

6. $\dfrac{x}{x + 3} + \dfrac{5}{x - 3}$

7. $\dfrac{2}{x^2 - 4x + 3} + \dfrac{3x}{x^2 + x - 2}$

8. $\dfrac{5x}{x^2 - 4} - \dfrac{2}{x^2 + x - 2}$

9. $\dfrac{xy}{x^2 - y^2} - \dfrac{y}{x + y}$

10. $\dfrac{1}{a} + \dfrac{2a - 5}{6a + 9} - \dfrac{4}{2a^2 + 3a}$

11. Simplify, and graph the resulting linear function:

$$f(x) = \dfrac{x^2 - 2x - 15}{x - 5}.$$

Use an open dot, or hole, on the graph corresponding to the value of x excluded from the function's domain.

In Problems 12–13, simplify.

12. $\dfrac{\dfrac{x}{4} - \dfrac{1}{x}}{1 + \dfrac{x + 4}{x}}$

13. $\dfrac{\dfrac{x + 2}{x^2 - 9}}{\dfrac{4}{x - 3} + \dfrac{x}{x^2 - 6x + 9}}$

14. Divide: $(16x^2y^3 - 10x^2y^2 + 12x^4y^3) \div 4x^2y$.

15. Divide: $(9x^3 - 3x^2 - 3x + 4) \div (3x + 2)$.

16. Divide using synthetic division:
$(3x^4 + 11x^3 - 20x^2 + 7x + 35) \div (x + 5)$.

Solve each equation in Problems 17–18.

17. $x + \dfrac{6}{x} = -5$

18. $\dfrac{2}{x-3} - \dfrac{4}{x+3} = \dfrac{8}{x^2-9}$

19. The average cost per manual for a company to produce x graphing utility manuals is given by

$$\overline{C} = \frac{6x + 3000}{x}.$$

How many manuals must be produced so that the average cost per manual for the company is $10?

20. The model

$$E = \frac{t + 67}{0.01t + 1}$$

describes the life expectancy (E, in years) for a child born in the United States t years after 1950. If life expectancy at birth is 78 years, in what year will that child be born?

21. Solve for a:

$$R = \frac{as}{a + s}.$$

22. Given $f(x) = 3x^2 + 11x + 10$ and $g(x) = x + 2$, find $\left(\dfrac{f}{g}\right)(-3)$.

23. It takes one pipe 3 hours to fill a pool and a second pipe 4 hours to drain the pool. The pool is empty and the first pipe begins to fill it. The second pipe is accidently left open, so the water is also draining out of the pool. Under these conditions, how long will it take to fill the pool?

24. A motorboat averages 20 miles per hour in still water. It takes the boat the same amount of time to travel 3 miles with the current as it does to travel 2 miles against the current. What is the current's speed?

25. The intensity of light received at a source varies inversely as the square of the distance from the source. A particular light has an intensity of 20 foot-candles at 15 feet. What is the light's intensity at 10 feet?

CUMULATIVE REVIEW PROBLEMS (CHAPTERS 1–5)

1. Solve: $2(x + 1) - 7(3 + 2x) = -4(2x + 1)$.

2. Evaluate: $\dfrac{8(-6) - (-7)(4) - (-2)}{6 - (3)(4)}$.

3. Solve, expressing the solution set in both set-builder and interval notation:

$$\{x \mid 2x + 5 \le 11\} \cap \{x \mid -3x > 18\}.$$

4. A ball is thrown directly upward from a height of 64 feet with an initial velocity of 48 feet per second. The ball's height above the ground ($s(t)$) after t seconds is modeled by $s(t) = -16t^2 + 48t + 64$.
a. When will the ball hit the ground?
b. After how many seconds does the ball reach its maximum height above the ground? What is the maximum height?
c. Use your results from parts (a) and (b), as well as any other useful points, to graph the parabola whose equation is given in the problem. Select values of t that begin at 0 and end at the time that the ball strikes the ground.

5. Use the graph to find the slope of the line connecting the years 1976 and 1995. Interpret the slope in terms of the variables appearing in the graph.

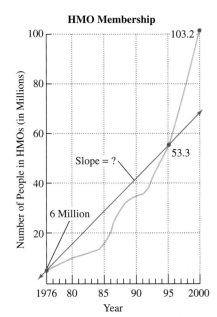

HMO Membership

Source: Newsweek, June 24, 1996.
1996–2000 are projections. GHAA National Directory of HMOs, Interstudy

6. Suppose Ana weighs less than Juan. Ana's weight plus Juan's weight is 230 pounds. Ana's weight plus Mike's weight is 274 pounds. Is it possible that Mike weighs the most? Is this necessarily the case or only possibly true?

7. The graph shows the number of executions in the United States from 1930 through 1992.

a. The graph levels off to what almost appears to be zero executions from 1972 through 1977. Does this mean that f does not represent the graph of a function? Explain.

b. Estimate the maximum point on the graph. What does this mean?

c. If the trend from 1977 through 1992 were to continue, what is a reasonable estimate for the year in which there will be 100 executions?

Executions in the United States

Source: "Death Row, U.S.A." (New York: NAACP Legal Defense Fund, 1992)

8. The figure shows the graph of f. According to the graph, which one of the following is true?

a. There are no values of x for which $f(x) = 0$.

b. As x increases from c to d, the corresponding values of $f(x)$ first increase and then decrease.

c. $f(a) = f(d)$

d. $f(a) + f(b) + f(c) + f(d) = 5$

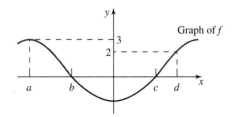

9. Solve by graphing:
$$2x - y = 4$$
$$x + y = 5$$

10. Solve the system:
$$4x + 3y + 3z = 4$$
$$3x \quad\quad + 2z = 2$$
$$2x - 5y \quad\quad = -4$$

11. Traveling against the wind, it takes an airplane 3 hours to travel 1950 kilometers. The return trip with the wind takes 2 hours. Find the speed of the plane in still air and the speed of the wind.

12. Evaluate:
$$\begin{vmatrix} 5 & 2 & 1 \\ 3 & 0 & -2 \\ -4 & -1 & 2 \end{vmatrix}.$$

13. Solve using row-equivalent matrices:
$$x - y = 1$$
$$2y + 3z = 8$$
$$2x + z = 6$$

14. The illumination produced by a light source varies inversely as the square of the distance from the source. The illumination produced 4 feet from a light source is 75 foot-candles. Determine the illumination produced 9 feet from the same source.

15. Multiply: $(3x - 4)(2x^2 - 5x + 3)$.

16. Solve: $2x^2 = 7x + 4$.

17. Write a polynomial in descending powers of x that models the area of the shaded region in the figure.

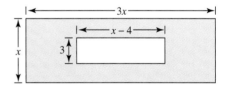

18. A copy machine can be rented for a five-year period for $4500 plus 5 cents per copy. The same machine can be purchased for $7000. How many copies would have to be made over the five-year period to result in the total rental cost being the same as the purchase price?

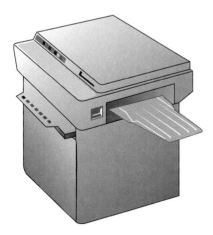

19. Factor completely: $6x^2 + 11x - 10$.

20. Factor completely: $3(x + 5)^2 - 11(x + 5) - 4$.

21. The length of a rectangle is 3 yards more than twice the width. If the rectangle has an area of 65 square yards, find the rectangle's dimensions.

22. The function $f(x) = 0.011x^2 - 0.097x + 4.1$ models the number of Americans ($f(x)$, in millions) holding more than one job x years after 1970. Find and interpret $f(10) - f(0)$.

23. Write the special product that is modeled geometrically in the figure.

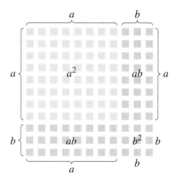

24. Find the indicated sum: $\dfrac{2x - 10}{x + 2} + \dfrac{x + 4}{x - 2}$.

25. Simplify: $\dfrac{\dfrac{1}{x} - \dfrac{1}{3}}{x + 3}$.

26. Perform the indicated operations: $(-7x^3y^2 - 5xy^3) - (-5x^2y^2 + x^3y^2) - (2xy^3 - 6x^2y^2)$.

27. Solve: $x + \dfrac{12}{x} = -7$.

28. Simplify and graph: $f(x) = \dfrac{2x^2 + x - 3}{x - 1}$.

29. Divide: $(3x^2 + 10x + 10) \div (x + 2)$.

30. Is the following set of order pairs a function? Explain your answer. $\{(3, 4), (5, 6), (7, 8), (9, 11), (3, 11)\}$

Radicals, Radical Functions, and Rational Exponents

Marisol Escobar "The Family" 1963, mixed media assemblage, 79 in. height. Collection of the Robert B. Mayer Family, Chicago on loan to the Milwaukee Art Museum © 1998 Marisol Escobar/Licensed by VAGA, New York, NY

I love Euclidean geometry, but it is quite clear that it does not give a reasonable presentation of the world. Mountains are not cones, clouds are not spheres, trees are not cylinders, neither does lightning travel in a straight line. Almost everything around us is non-Euclidean.

- Benoit Mandelbrot (1924–)

How can we use equations to model nature's complexity in objects such as ferns, coastlines, or the human circulatory system? The answer is provided by fractal geometry, developed over a period of years by the IBM mathematician Benoit Mandelbrot. The newly discovered mathematics of fractals, discussed on page 577 can match nature's fabric with equations that generate graphs showing both infinite repetition and unpredictable complexity.

Another new development in mathematical modeling involves virtual reality systems where users enter a computer-created reality and interact with it. The bridge to these new plateaus in mathematics is an understanding of radicals and rational exponents.

S E C T I O N   6 . 1

Solutions Manual   **Tutorial**   **Video 7**

Radicals and Radical Functions

Objectives

1 Find square roots.
2 Graph square root functions.
3 Find the domain of a square root function.
4 Evaluate and interpret radical functions.
5 Find *n*th roots.
6 Graph radical functions.

Radical expressions frequently appear in models of the physical world. Table 6.1 contains a few examples of such models. In this section, we turn our attention to radical expressions and mathematical models that contain these expressions. We also expand our work with functions to include radical functions and their graphs.

TABLE 6.1 Applied Mathematical Models Containing Radical Expressions

Users	Description	The Formula
Racing cyclists	The maximum velocity for a cyclist to turn a corner without tipping over	$v = 4\sqrt{r}$ r = radius of the corner in feet v = maximum velocity in miles per hour
Police	The speed of a car based on the length of its skid marks on dry pavement when brakes are applied	$v = \sqrt{24L}$ L = length of skid marks in feet v = speed of the car in miles per hour
Medical specialists	The pulse rate of an adult in terms of that person's height	$P = \dfrac{600}{\sqrt{h}}$ h = person's height in inches P = pulse rate in beats per minute
Environmentalists	The evaporation of a large body of water on a particular day	$E = \dfrac{w}{20\sqrt{a}}$ a = surface area of the water in square miles w = average wind speed of the air over the water in miles per hour E = evaporation, in inches per day
Oceanographers	The velocity of a tsunami, a great sea wave produced by underwater earthquakes	$v = 3\sqrt{d}$ d = depth of the water in feet v = velocity, in feet per second, of a tsunami as it approaches land

TABLE 6.1 (Continued)		
Users	**Description**	**The Formula**
Hikers, mountain climbers	The distance that can be seen to the horizon from a given height	$d = \sqrt{1.5h}$ h = height of the observer, in feet d = distance the observer can see, in miles
Scientists working with man-made satellites	How fast man-made satellites have to travel to stay in orbit	$s = \sqrt{\dfrac{1.24 \times 10^{12}}{r + a}}$ r = radius of the earth (in miles) ≈ 3960 miles a = altitude in miles above Earth of an orbiting satellite s = speed of the satellite in miles per hour
Botanists	The number of plant species on the various islands of the Galápagos Islands	$S = 28.6\sqrt[3]{A}$ A = area of the island in square miles S = number of plant species on the island

Giant Galápagos tortoise

Brandon D. Cole/Ellis Nature Photography

Find square roots.

Finding Square Roots

From our earlier work with exponents, we know that

$$5^2 = 25 \qquad \text{and} \qquad (-5)^2 = 25.$$

Consequently, 5 is a *square root* of 25 and -5 is a *square root* of 25. In our work with radicals, we will use the following notation and conventions.

The symbol $\sqrt{}$ is called the *radical sign* because the word *radical* comes from the Latin word *radix,* which means "root." Late medieval Latin writers wrote ℞ 25 for $\sqrt{25}$ (that is, Radix 25), a bit confusing today because ℞ also stands for the familiar *recipe* in a physician's prescription. Credit for inventing the radical sign is given to Christoff Rudolff, a German mathematician of the 16th century, although Hindu–Arabic mathematicians used $\sqrt{}$ to indicate square roots beginning around A.D. 800. The expression under the radical sign is referred to as the *radicand.* (In $\sqrt{25}$, 25 is the radicand.) Finally, the entire expression (such as $\sqrt{25}$) is called the *radical.*

EXAMPLE 1 **Finding Square Roots**

Find each square root, or state that the square root is not a real number.

a. $\sqrt{49}$ **b.** $\pm\sqrt{100}$ **c.** $-\sqrt{100}$ **d.** $\sqrt{0}$ **e.** $\sqrt{-25}$

Solution

a. $\sqrt{49} = 7$ because $7^2 = 49$.
b. $\pm\sqrt{100} = \pm 10$ because $10^2 = 100$ and $(-10)^2 = 100$.
c. $-\sqrt{100} = -(\sqrt{100}) = -10$.

d. $\sqrt{0} = 0$. We do not write $\pm\sqrt{0}$ because 0 is neither positive nor negative.

e. $\sqrt{-25}$ is not a real number because there is no real number whose square is -25. In general, $\sqrt{N}$ is not a real number when N is negative. ■

Before turning to square root functions, let's take a moment to summarize our discussion of square roots.

Square roots

1. The number a is a square root of b if $a^2 = b$.
2. Every positive real number has two real number square roots.
3. Zero has only one square root: $\sqrt{0} = 0$
4. Negative numbers do not have real number square roots.
5. The principal square root of a positive real number b, written $\sqrt{b}$, is that positive number whose square equals b.

---

2 Graph square root functions.

The Square Root Function

We now define a special function called the *square root function*. The function is defined as the principal square root of a polynomial.

Square root function

Any function in the form

$$f(x) = \sqrt{P}$$

where P is a polynomial in x, $P \geq 0$, is called a *square root function*.

EXAMPLE 2 **Graphing Square Root Functions**

Use the point-plotting method to graph each function. State the function's domain and range.

a. $f(x) = \sqrt{x}$ **b.** $g(x) = \sqrt{x-2}$ **c.** $h(x) = \sqrt{x} - 2$

Solution

a. Integers such as 1, 4, 9, 16, 49, and 81 are called *perfect squares* because they have integer square roots. To graph $f(x) = \sqrt{x}$, we construct a table of coordinates by choosing values of x that are perfect squares.

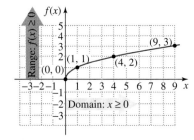

Figure 6.1

The graph of $f(x) = \sqrt{x}$

x	$f(x) = \sqrt{x}$	$(x, f(x))$
0	$f(0) = \sqrt{0} = 0$	$(0, 0)$
1	$f(1) = \sqrt{1} = 1$	$(1, 1)$
4	$f(4) = \sqrt{4} = 2$	$(4, 2)$
9	$f(9) = \sqrt{9} = 3$	$(9, 3)$

The graph of $f(x) = \sqrt{x}$ is shown in Figure 6.1. The arrows in the figure show that the domain of the function is the set of all nonnegative real numbers and the range is also the set of all nonnegative real numbers.

$$\text{Domain of } f = \{x \mid x \geq 0\} = [0, \infty)$$
$$\text{Range of } f = \{f(x) \mid f(x) \geq 0\} = [0, \infty)$$

b. To construct a table of coordinates for $g(x) = \sqrt{x - 2}$, we must avoid values of x that result in square roots of negative numbers. For example, $g(1) = \sqrt{1 - 2} = \sqrt{-1}$, which is not a real number. Thus, 1 is not in the domain of g. To graph this function, then, it would be best to first find the domain of g.

To determine the domain of g, we must find all x-values for which the radicand $x - 2$ is nonnegative. To do this, we solve the inequality $x - 2 \geq 0$.

$$x - 2 \geq 0 \quad \text{Set the radicand} \geq 0.$$
$$x \geq 2 \quad \text{Solve for } x.$$

The domain of g is $\{x \mid x \geq 2\}$.

We can now construct a table of coordinates by choosing values of $x \geq 2$ that make $x - 2$ a perfect square.

x	$g(x) = \sqrt{x - 2}$	$(x, g(x))$
2	$g(2) = \sqrt{2 - 2} = \sqrt{0} = 0$	$(2, 0)$
3	$g(3) = \sqrt{3 - 2} = \sqrt{1} = 1$	$(3, 1)$
6	$g(6) = \sqrt{6 - 2} = \sqrt{4} = 2$	$(6, 2)$
11	$g(11) = \sqrt{11 - 2} = \sqrt{9} = 3$	$(11, 3)$

The graph of $g(x) = \sqrt{x - 2}$ is shown in Figure 6.2. The arrows in the figure show that the domain of g consists of all real numbers greater than or equal to 2 and the range of g consists of all real numbers greater than or equal to 0.

$$\text{Domain of } g = \{x \mid x \geq 2\} = [2, \infty)$$
$$\text{Range of } g = \{g(x) \mid g(x) \geq 0\} = [0, \infty)$$

c. We graph $h(x) = \sqrt{x} - 2$ by first constructing a table of coordinates.

x	$h(x) = \sqrt{x} - 2$	$(x, h(x))$
0	$h(0) = \sqrt{0} - 2 = -2$	$(0, -2)$
1	$h(1) = \sqrt{1} - 2 = 1 - 2 = -1$	$(1, -1)$
4	$h(4) = \sqrt{4} - 2 = 2 - 2 = 0$	$(4, 0)$
9	$h(9) = \sqrt{9} - 2 = 3 - 2 = 1$	$(9, 1)$

The graph of $h(x) = \sqrt{x} - 2$ is shown in Figure 6.3. The arrows in the figure show that the domain of h consists of all real numbers greater than or equal to 0 and the range of h consists of all real numbers greater than or equal to -2.

$$\text{Domain of } h = \{x \mid x \geq 0\} = [0, \infty)$$
$$\text{Range of } h = \{h(x) \mid h(x) \geq -2\} = [-2, \infty)$$

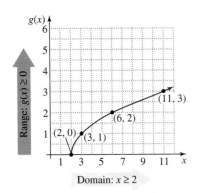

Domain: $x \geq 2$

Figure 6.2

The graph of $g(x) = \sqrt{x - 2}$

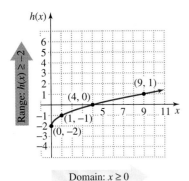

Domain: $x \geq 0$

Figure 6.3

The graph of $h(x) = \sqrt{x} - 2$

Study tip

Be careful with grouping symbols when you use a graphing utility to graph square root functions. For example, to graph $g(x) = \sqrt{x - 2}$, enter

$$y_1 = \boxed{\sqrt{}}\,\boxed{(}\,x\,\boxed{-}\,2\,\boxed{)}$$

and to graph $h(x) = \sqrt{x} - 2$, enter

$$y_1 = \boxed{\sqrt{}}\,x\,\boxed{-}\,2$$

Discover for yourself

1. Study the graphs in Example 2 on pages 504 and 505. What relationship do you observe between the graphs of $y = \sqrt{x}$ and $y = \sqrt{x} - 2$? Based on your observation, what do you think the graph of $y = \sqrt{x} - k$ does to the graph of $y = \sqrt{x}$ if k is positive? Use your graphing utility to verify your observation by graphing $y = \sqrt{x} - k$ and $y = \sqrt{x}$ in the same viewing rectangle for various positive values of k.

2. Study the graphs in Example 2. What relationship do you observe between the graphs of $y = \sqrt{x}$ and $y = \sqrt{x} - 2$ Based on your observation, what do you think the graph of $y = \sqrt{x} - k$ does to the graph of $y = \sqrt{x}$ if k is positive? Use your graphing utility to verify your observation by graphing $y = \sqrt{x} - k$ and $y = \sqrt{x}$ in the same viewing rectangle for various positive values of k.

(If your course does not utilize graphing utilities, draw these graphs by hand.)

In part (1) of the Discover for Yourself box, did you discover that the graph of $y = \sqrt{x} - 2$ is the graph of $y = \sqrt{x}$ shifted 2 units to the right? In general, the following is true.

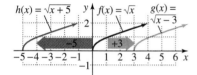

Shifting to the right and the left

Let k be a positive real number.

1. The graph of $y = \sqrt{x - k}$ is the graph of $y = \sqrt{x}$ shifted k units to the right.
2. The graph of $y = \sqrt{x + k}$ is the graph of $y = \sqrt{x}$ shifted k units to the left.

In part (2) of the Discover for Yourself box, did you observe that the graph of $y = \sqrt{x} - 2$ is the graph of $y = \sqrt{x}$ shifted downward 2 units? Generalizing from this situation gives the following results.

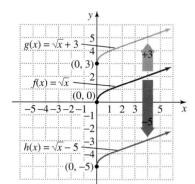

Upward and downward shifts

Let k be a positive real number.

1. The graph of $y = \sqrt{x} + k$ is the graph of $y = \sqrt{x}$ shifted k units upward.
2. The graph of $y = \sqrt{x} - k$ is the graph of $y = \sqrt{x}$ shifted k units downward.

3 Find the domain of a square root function.

In Example 2b we determined the domain of a square root function by finding all x-values for which the radicand was nonnegative. We see this situation again in Example 3.

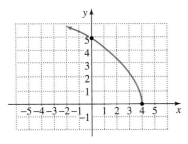

Figure 6.4

The graph of $f(x) = \sqrt{24 - 6x}$.
The domain is $\{x \mid x \leq 4\}$.

4 Evaluate and interpret radical functions.

| EXAMPLE 3 | **Finding the Domain of a Square Root Function** |

If $f(x) = \sqrt{24 - 6x}$, find the domain.

Solution

The radicand $24 - 6x$ must be greater than or equal to 0.

$$24 - 6x \geq 0 \qquad \text{Set the radicand} \geq 0.$$
$$-6x \geq -24 \qquad \text{Subtract 24 from both sides.}$$
$$x \leq 4 \qquad \text{Divide both sides by } -6, \text{ reversing the sense of the inequality.}$$

Thus, the domain of f is $\{x \mid x \leq 4\}$ or, in interval notation $(-\infty, 4]$. This is illustrated by the graph in Figure 6.4. ∎

An Application: Einstein's Relativity

Understanding that the speed of light (approximately 186,000 miles per second) is the speed limit of the universe is basic to understanding Albert Einstein's concept of the universe. Einstein's functions explain what happens to vehicles that travel at velocities approaching the speed of light. His equations model the universe as a closed four-dimensional sphere (time is the fourth dimension) with a diameter of about 100 million light-years. The next example illustrates one of Einstein's functions.

| EXAMPLE 4 | **Square Roots and Relativistic Squashing** |

The function $L(v) = 600\sqrt{1 - \dfrac{v^2}{c^2}}$ describes the length of a 600-meter-tall starship moving at velocity v from the perspective of an observer at rest. (c, the speed of light, is approximately 186,000 miles per second). Find and interpret $L(184{,}140)$.

Solution

$$L(v) = 600\sqrt{1 - \frac{v^2}{c^2}} \qquad \text{This is the given model.}$$

$$L(184{,}140) = 600\sqrt{1 - \frac{(184{,}140)^2}{(186{,}000)^2}} \qquad \text{Substitute 184,140 for } v \text{ and 186,000 for } c, \text{ the speed of light.}$$

$$= 600\sqrt{1 - \left(\frac{184{,}140}{186{,}000}\right)^2}$$

$$= 600\sqrt{1 - (0.99)^2} \qquad \text{Since } \frac{v}{c} = 0.99, \text{ at a speed of 184,140 miles per second, the starship is cruising at 99\% the speed of light.}$$

$$= 600\sqrt{0.0199} \qquad \text{Simplify the radicand.}$$

$$\approx 600(0.14) \qquad \text{Use a calculator:}$$
$$\boxed{\sqrt{}}\ .0199\ \boxed{\text{ENTER}}$$

$$= 84$$

600 meters

Starship at Rest

ENRICHMENT ESSAY

Using Art to Express Ideas Contained in Mathematical Models

M. C. Escher (1898–1972) "Up and Down." © 1997 Cordon Art-Baarn-Holland. All rights reserved.

Einstein's function in Example 4 suggests that what an observer sees is influenced by context and vantage point. This idea is illustrated in M.C. Escher's lithograph, *High and Low*. In the upper half the viewer is looking down; in the lower half the viewer is on the patio, looking up. Do your eyes see the tiled diamond in the center as a floor or as a ceiling? Describe what happens as you draw back from the print. *High and Low* gives visual expression to the abstract concepts of Einstein's mathematical models.

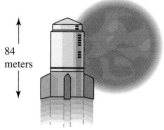

84 meters

Change in starship's dimensions when moving at 99% of light's speed when viewed by an observer

5 Find *n*th roots.

Since $L(184, 140) = 84$, when the 600-meter-tall starship is moving at 184,140 miles per second, it measures only 84 meters tall from the perspective of an observer at rest. As shown on the left, the length along the direction of the starship's motion is compressed. However, lengths perpendicular to the direction of motion are unchanged by the effect of speed, so from the point of view of a stationary observer who looks up and sees the ship cruising by, its width remains unchanged. ∎

Finding Even and Odd *n*th Roots

Roots in general are related to repeated factors. For example

$$2^3 = 2 \cdot 2 \cdot 2 = 8$$

and so 2 is the *cube root* of 8, written

$$\sqrt[3]{8} = 2.$$

As another example,

$$3^4 = 3 \cdot 3 \cdot 3 \cdot 3 = 81$$

and so 3 is the *fourth root* of 81, written

$$\sqrt[4]{81} = 3.$$

Let's generalize from these examples, breaking down our discussion into even and odd nth roots. We begin with the definition of even roots.

> **Definition of even roots**
>
> If n is an even natural number and $b > 0$, $\sqrt[n]{b}$ (the nth root of b) is the positive real number whose nth power is b. Thus, $\sqrt[n]{b} = a$ if $a^n = b$.
>
> $(\sqrt[n]{b})^n = b$, when $b > 0$

EXAMPLE 5 **Finding nth Roots When n Is an Even Number**

Find each root, or state that the root is not a real number.

a. $\sqrt[4]{81}$ **b.** $-\sqrt[4]{81}$ **c.** $\sqrt[4]{-81}$ **d.** $\sqrt[6]{64}$

e. $(\sqrt[4]{6})^4$ **f.** $\sqrt[4]{\dfrac{81}{16}}$

Solution

a. $\sqrt[4]{81} = 3$ because $3^4 = 81$.

b. $-\sqrt[4]{81} = -(\sqrt[4]{81}) = -3$.

c. $\sqrt[4]{-81}$ is not a real number. There is no real number we can raise to the fourth power and obtain -81.

d. $\sqrt[6]{64} = 2$ because $2^6 = 64$.

e. $(\sqrt[4]{6})^4 = 6$.

f. $\sqrt[4]{\dfrac{81}{16}} = \dfrac{3}{2}$ because $\left(\dfrac{3}{2}\right)^4 = \dfrac{81}{16}$.   ∎

If n is an odd natural number, it is not necessary to restrict b to positive real numbers.

Using technology

Verify the results of Examples 5 and 6 using a graphing calculator, if possible. First consult your manual to locate the root $\sqrt[x]{\ \ }$ key.

Verifying 5a: $\sqrt[4]{81}$

   4 $\boxed{\sqrt[x]{\ }}$ 81 $\boxed{\text{ENTER}}$

Verifying 6c: $\sqrt[5]{-32}$

   5 $\boxed{\sqrt[x]{\ }}$ $\boxed{(-)}$ 32 $\boxed{\text{ENTER}}$

> **Definition of odd roots**
>
> If n is an odd natural number, $\sqrt[n]{b}$ is the positive or negative real number whose nth power is b. Thus, $\sqrt[n]{b} = a$ if $a^n = b$.
>
> $(\sqrt[n]{b})^n = b$

EXAMPLE 6 **Finding nth Roots when n Is an Odd Number**

Find each root:

a. $\sqrt[3]{27}$ **b.** $\sqrt[3]{-64}$ **c.** $\sqrt[5]{-32}$ **d.** $(\sqrt[7]{-6})^7$ **e.** $\sqrt[7]{-\dfrac{1}{128}}$

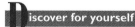

If n is even or odd, the nth root of 1 is 1.

$$\sqrt{1} = 1 \quad \sqrt[3]{1} = 1$$
$$\sqrt[4]{1} = 1 \quad \sqrt[5]{1} = 1$$
$$\sqrt[6]{1} = 1 \dots$$

In general

$$\sqrt[n]{1} = 1.$$

Can you think of another number all of whose nth roots equal the number itself?

Solution

a. $\sqrt[3]{27} = 3$ because $3^3 = 27$.

b. $\sqrt[3]{-64} = -4$ because $(-4)^3 = -64$.

c. $\sqrt[5]{-32} = -2$ because $(-2)^5 = -32$.

d. $(\sqrt[7]{-6})^7 = -6$ because $(\sqrt[n]{b})^n = b$.

e. $\sqrt[7]{-\dfrac{1}{128}} = -\dfrac{1}{2}$ because $\left(-\dfrac{1}{2}\right)^7 = -\dfrac{1}{128}$. ■

Now take a moment to work the Discover for Yourself in the margin. The number you are asked to find is one with many special properties.

The nth root of zero

If n is an even or odd root,

$$\sqrt[n]{0} = 0.$$

(Were you able to discover that all nth roots of zero are zero before this relationship was stated in the box?)

Mathematicians have a special vocabulary related to even and odd roots. We mentioned some of these definitions in our discussion of square roots on page 503.

The vocabulary of radicals

Given $\sqrt{b}$, $\sqrt[3]{b}$, $\sqrt[4]{b}$, $\sqrt[5]{b}$, and so on, the numbers $2, 3, 4, 5, \dots$ are called the *index* of the root. Observe that an index of 2, indicating the square root, is omitted. $\sqrt{}$ is called the *radical sign*. The expression under the radical, b, is the *radicand*. The expressions $\sqrt{b}$, $\sqrt[3]{b}$, $\sqrt[4]{b}$, and $\sqrt[5]{b}$ are *radicals*.

Before turning to radical functions, let's take a moment to summarize our discussion of nth roots.

---

nth roots

1. Let n be even: $\sqrt[n]{b} = a$ ($b > 0$) if $a^n = b$.
   a. Every positive number has two nth roots that are opposites of each other. For example, the fourth roots of 16 are 2 and -2.
   b. If $b < 0$, then $\sqrt[n]{b}$ is not a real number.
2. Let n be odd: $\sqrt[n]{b} = a$ if $a^n = b$.
   a. Every real number has one nth root. For example, $\sqrt[3]{8} = 2$ and $\sqrt[3]{-27} = -3$.
   b. The nth root of a positive number is positive, and the nth root of a negative number is negative.
3. Let n be even or odd: $\sqrt[n]{0} = 0$.

Radical Functions

Just as we began with square roots and generalized to nth roots, we can move from the square root function to a general radical function.

> **Radical function**
>
> A radical (nth root) function
>
> $$f(x) = \sqrt[n]{P}$$
>
> where P is a polynomial in x. If n is even, $P \geqslant 0$.

6 Graph radical functions.

EXAMPLE 7 **Graphing Cube Root Functions**

Use the point-plotting method to graph:

a. $f(x) = \sqrt[3]{x}$ **b.** $g(x) = \sqrt[3]{x + 2}$ **c.** $h(x) = \sqrt[3]{x} + 2$

Solution

a. Integers such as 1, 8, 27, 64, and 125 are called *perfect cubes* because they have integer cube roots. To graph $f(x) = \sqrt[3]{x}$, we construct a table of coordinates by choosing values of x that are perfect cubes.

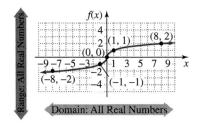

Figure 6.5

The graph of $f(x) = \sqrt[3]{x}$

x	$f(x) = \sqrt[3]{x}$	$(x, f(x))$
-8	$f(-8) = \sqrt[3]{-8} = -2$	$(-8, -2)$
-1	$f(-1) = \sqrt[3]{-1} = -1$	$(-1, -1)$
0	$f(0) = \sqrt[3]{0} = 0$	$(0, 0)$
1	$f(1) = \sqrt[3]{1} = 1$	$(1, 1)$
8	$f(8) = \sqrt[3]{8} = 2$	$(8, 2)$

The graph of $f(x) = \sqrt[3]{x}$ is shown in Figure 6.5. Notice that the domain and range of $f(x) = \sqrt[3]{x}$ is the set of all real numbers. This is true for all cube root functions.

b. We graph $g(x) = \sqrt[3]{x + 2}$ by selecting values of x that make $x + 2$ a perfect cube.

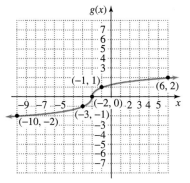

Figure 6.6

The graph of $g(x) = \sqrt[3]{x + 2}$ shifts the graph of $f(x) = \sqrt[3]{x}$ 2 units to the left.

x	$g(x) = \sqrt[3]{x + 2}$	$(x, f(x))$
-10	$g(-10) = \sqrt[3]{-10 + 2} = \sqrt[3]{-8} = -2$	$(-10, -2)$
-3	$g(-3) = \sqrt[3]{-3 + 2} = \sqrt[3]{-1} = -1$	$(-3, -1)$
-2	$g(-2) = \sqrt[3]{-2 + 2} = \sqrt[3]{0} = 0$	$(-2, 0)$
-1	$g(-1) = \sqrt[3]{-1 + 2} = \sqrt[3]{1} = 1$	$(-1, 1)$
6	$g(6) = \sqrt[3]{6 + 2} = \sqrt[3]{8} = 2$	$(6, 2)$

The graph of $g(x)$ is shown in Figure 6.6. The graph of $g(x) = \sqrt[3]{x + 2}$ is the graph of $f(x) = \sqrt[3]{x}$ shifted 2 units to the left.

ENRICHMENT ESSAY

Radicals and Windchill

The way that we perceive the temperature on a cold day depends on both air temperature and wind speed. The windchill temperature is what the air temperature would have to be with no wind to achieve the same chilling effect on the skin. The model that describes windchill temperature (W) in terms of the velocity of the wind (v, in miles per hour) and the actual air temperature (t, in degrees Fahrenheit) is

$$W = 9.14 - \frac{(10.5 + 6.7\sqrt{v} - 0.45v)(457 - 5t)}{110}.$$

Use your calculator to describe how cold the air temperature feels (that is, the windchill temperature) when the temperature is 15° Fahrenheit and the wind is 5 miles per hour. Contrast this with a temperature of 40° Fahrenheit and a wind blowing at 50 miles per hour.

Paul Katz/The Image Bank

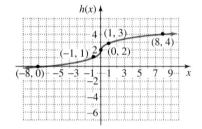

Figure 6.7

The graph of $h(x) = \sqrt[3]{x} + 2$ shifts the graph of $f(x) = \sqrt[3]{x}$ upward 2 units.

c. We graph $h(x) = \sqrt[3]{x} + 2$ by first constructing a table of coordinates.

x	$h(x) = \sqrt[3]{x} + 2$	$(x, h(x))$
-8	$h(-8) = \sqrt[3]{-8} + 2 = -2 + 2 = 0$	$(-8, 0)$
-1	$h(-1) = \sqrt[3]{-1} + 2 = -1 + 2 = 1$	$(-1, 1)$
0	$h(0) = \sqrt[3]{0} + 2 = 0 + 2 = 2$	$(0, 2)$
1	$h(1) = \sqrt[3]{1} + 2 = 1 + 2 = 3$	$(1, 3)$
8	$h(8) = \sqrt[3]{8} + 2 = 2 + 2 = 4$	$(8, 4)$

The graph of $h(x)$ is shown in Figure 6.7. The graph of $h(x) = \sqrt[3]{x} + 2$ is the graph of $f(x) = \sqrt[3]{x}$ shifted upward 2 units. ∎

Using technology

Use a graphing utility to verify the three graphs in Example 7. Consult your manual for the location of the root $\boxed{\sqrt[x]{}}$ key.

EXAMPLE 8　An Application of Radical Functions

The number of plant species on the various islands of the Galápagos Islands is a function of the area (A, in square miles) of each island, given by $f(A) = 28.6\sqrt[3]{A}$. Find and interpret $f(8)$.

Solution

$$f(A) = 28.6\sqrt[3]{A} \quad \text{This is the given function.}$$
$$f(8) = 28.6\sqrt[3]{8} \quad \text{Replace } A \text{ by 8.}$$
$$= 28.6(2) \quad \sqrt[3]{8} = 2$$
$$= 57.2 \quad \text{Multiply.}$$

An island of the Galápagos chain whose area is 8 square miles has (approximately) 57 plant species.

PROBLEM SET 6.1

Practice Problems

In Problems 1–22, find each root or state that the root is not a real number.

1. $\sqrt{49}$ **2.** $\sqrt{100}$ **3.** $-\sqrt{49}$ **4.** $-\sqrt{100}$ **5.** $\sqrt{-49}$ **6.** $\sqrt{-100}$ **7.** $\sqrt[4]{16}$ **8.** $\sqrt[4]{81}$
9. $\sqrt[5]{-1}$ **10.** $\sqrt[7]{-1}$ **11.** $\sqrt{\frac{1}{9}}$ **12.** $\sqrt{\frac{4}{9}}$ **13.** $\sqrt[3]{-\frac{1}{64}}$ **14.** $\sqrt[3]{\frac{-27}{64}}$ **15.** $\sqrt[5]{\frac{1}{32}}$ **16.** $\sqrt[5]{-\frac{1}{32}}$
17. $\sqrt[4]{-16}$ **18.** $\sqrt[4]{-81}$ **19.** $(\sqrt[3]{2})^3$ **20.** $(\sqrt[3]{4})^5$ **21.** $(\sqrt[5]{-3})^5$ **22.** $(\sqrt[5]{-\frac{1}{2}})^5$

In Problems 23–38, graph each pair of functions in the same rectangular coordinate system. State each function's domain and range. Then describe the relationship of the graph of function g to the graph of function f. Use one of the following phrases: shift to the right; shift to the left; shift upward; shift downward.

23. $f(x) = \sqrt{x}, g(x) = \sqrt{x} + 3$ **24.** $f(x) = \sqrt{x}, g(x) = \sqrt{x} + 1$ **25.** $f(x) = \sqrt{x}, g(x) = \sqrt{x + 3}$
26. $f(x) = \sqrt{x}, g(x) = \sqrt{x + 1}$ **27.** $f(x) = \sqrt{x}, g(x) = \sqrt{x} - 4$ **28.** $f(x) = \sqrt{x}, g(x) = \sqrt{x} - 5$
29. $f(x) = \sqrt{x}, g(x) = \sqrt{x - 4}$ **30.** $f(x) = \sqrt{x}, g(x) = \sqrt{x - 5}$ **31.** $f(x) = \sqrt[3]{x}, g(x) = \sqrt[3]{x} - 1$
32. $f(x) = \sqrt[3]{x}, g(x) = \sqrt[3]{x} - 2$ **33.** $f(x) = \sqrt[3]{x}, g(x) = \sqrt[3]{x - 1}$ **34.** $f(x) = \sqrt[3]{x}, g(x) = \sqrt[3]{x - 2}$
35. $f(x) = \sqrt[3]{x}, g(x) = \sqrt[3]{x + 3}$ **36.** $f(x) = \sqrt[3]{x}, g(x) = \sqrt[3]{x + 1}$ **37.** $f(x) = \sqrt[4]{x}, g(x) = \sqrt[4]{x - 1}$
38. $f(x) = \sqrt[4]{x}, g(x) = \sqrt[4]{x + 1}$

Determine the domain of each function in Problems 39–50.

39. $f(x) = \sqrt{x + 5}$ **40.** $g(x) = \sqrt{x - 7}$ **41.** $h(x) = \sqrt{4 - x}$ **42.** $f(x) = \sqrt{7 - x}$
43. $g(x) = \sqrt{4x - 12}$ **44.** $h(x) = \sqrt{3x - 9}$ **45.** $r(x) = \sqrt{12 - 6x}$ **46.** $f(x) = \sqrt{18 - 3x}$
47. $f(x) = \sqrt[3]{x + 5}$ **48.** $g(x) = \sqrt[3]{x - 7}$ **49.** $f(x) = \sqrt[4]{x + 5}$ **50.** $g(x) = \sqrt[4]{x - 7}$

51. Evaluate $2\sqrt{x} - \sqrt[3]{y} + 4\sqrt[5]{z}$ when $x = 36, y = -8$, and $z = 1$.

52. Evaluate $4\sqrt{x} + 3\sqrt[3]{y} - 2\sqrt[4]{z}$ when $x = 16, y = -1$, and $z = 81$.

53. Use the formula

$$A = \sqrt[3]{-\frac{b}{2} + \sqrt{\frac{b^2}{4} + \frac{a^3}{27}}}$$

to find A when $b = 2$ and $a = -3$.

54. Use the formula $S = \pi(r_1 + r_2)\sqrt{h^2 + (r_1 - r_2)^2}$ to find S when $h = 6, r_1 = 10$, and $r_2 = 2$. Express S in terms of π.

55. Evaluate $\dfrac{-b \pm \sqrt{b^2 - 4ac}}{2a}$ when $a = 5, b = 3$, and $c = -8$.

56. Evaluate $\dfrac{-b \pm \sqrt{b^2 - 4ac}}{2a}$ when $a = 6, b = -13$, and $c = 6$.

Application Problems

57. The function

$$L(v) = 400\sqrt{1 - \frac{v^2}{c^2}}$$

describes the length of a 400-meter-tall starship mov-

ing at velocity v (in miles per second) from the perspective of an observer at rest. (c, the speed of light, is approximately 186,000 miles per second.) Find and interpret $L(148,800)$.

58. The function

$$L(v) = 800\sqrt{1 - \frac{v^2}{c^2}}$$

describes the length of an 800-meter-tall starship moving at velocity v, (in miles per second) from the perspective of an observer at rest. (c, the speed of light, is approximately 186,000 miles per second.) Find and interpret $L(93,000)$.

59. Psychologists use the formula $N = 2\sqrt{Q} - 9$ to determine the number of nonsense syllables (N) a subject with an IQ of Q can repeat. How many syllables can be repeated by a subject whose IQ is 121?

60. A bamboo plant has its height (h, in inches) given by $h = 3\sqrt{t} - 0.23t$, where t is measured in weeks after the plant comes through the soil. What is the height of the bamboo 9 weeks after breaking through the soil?

61. The function $f(x) = 4(\sqrt{x})^5 + 17,300$ describes yearly income (in dollars) as a function of x years of education. Find and interpret $f(16)$.

62. For a group of 50,000 births, the number of people surviving (N) to age x is given by the formula $N = 5000\sqrt{100 - x}$. How many people in the group will not survive for 36 years?

63. The formula $H = (10.45 + \sqrt{100w} - w)(33 - t)$ describes the rate of heat loss (H, measured in kilocalories per square meter per hour) when the air temperature is t degrees Celsius and the wind speed is w meters per second. When H is 2000 kilocalories per square meter per hour, exposed flesh will freeze in 1 minute. Will this occur when the wind is blowing at 4 meters per second and the temperature is 0° Celsius?

64. The formula $I = \sqrt{x} + \sqrt{y} + 2xy$ describes the pollution index (I) when x milligrams of nitrous oxide and y milligrams of carbon monoxide are in a sample of air. What is the pollution index on a day in which 9 milligrams of nitrous oxide and 25 milligrams of carbon monoxide are contained in an air sample?

True–False Critical Thinking Problems

65. Which one of the following is true?
   a. The functions $f(x) = \sqrt[3]{x} + 1$ and $g(x) = \sqrt[3]{x + 1}$ have identical graphs.
   b. If n is odd and b is negative, then $\sqrt[n]{b}$ is not a real number.
   c. The expression $\sqrt[n]{4}$ represents increasingly larger numbers for $n = 2, 3, 4, 5, 6,$ and so on.
   d. None of the above is true.

66. Which one of the following is true?
   a. The functions $f(x) = \sqrt{x - 1}$ and $g(x) = \sqrt{x} - 1$ have identical graphs.
   b. If $f(x) = \sqrt[3]{x + 5}$, then this function's domain is $\{x \mid x \geq -5\}$.
   c. The index of $\sqrt{x}$ is understood to be 1.
   d. None of the above is true.

Technology Problems

67. Use a graphing utility to verify some of the graphs that you drew by hand in Problems 23–38.

68. Use a graphing utility to graph the functions in Problems 39–50. Then use the graph to verify the function's domain.

Writing in Mathematics

69. Why is it that the cube root of a negative number is a real number, but the square root of a negative number is not real?

Critical Thinking Problems

70. Evaluate: $\sqrt[3]{\sqrt[4]{16} + \sqrt{625}}$.

71. Evaluate: $\sqrt[3]{\sqrt{\sqrt{169} + \sqrt{9}} + \sqrt{\sqrt[3]{1000} + \sqrt[3]{216}}}$.

72. The table on the right shows the number of square roots, cube roots, and fourth roots lying between 1 and 2, where the radicand is a natural number.

There are 2 square roots, 6 cube roots, and 14 fourth roots lying between 1 and 2. Use the pattern to determine the number of fifth roots lying between 1 and 2. How many irrational nth roots lie between 1 and 2 if the radicand is a natural number?

73. Use your calculator to find $\sqrt{2}, \sqrt{\sqrt{2}}, \sqrt{\sqrt{\sqrt{2}}}$, and so on. Describe the two emerging patterns. One pattern is far less obvious than the other.

$\sqrt{1} = 1.00 \ldots$	$\sqrt[3]{1} = 1.00 \ldots$	$\sqrt[4]{1} = 1.00 \ldots$
		$\sqrt[4]{2} = 1.18 \ldots$
	$\sqrt[3]{2} = 1.25 \ldots$	$\sqrt[4]{3} = 1.31 \ldots$
		$\sqrt[4]{4} = 1.41 \ldots$
	$\sqrt[3]{3} = 1.44 \ldots$	$\sqrt[4]{5} = 1.49 \ldots$
$\sqrt{2} = 1.41 \ldots$		$\sqrt[4]{6} = 1.56 \ldots$
	$\sqrt[3]{4} = 1.58 \ldots$	$\sqrt[4]{7} = 1.62 \ldots$
		$\sqrt[4]{8} = 1.68 \ldots$
		$\sqrt[4]{9} = 1.73 \ldots$
	$\sqrt[3]{5} = 1.70 \ldots$	$\sqrt[4]{10} = 1.77 \ldots$
$\sqrt{3} = 1.73 \ldots$		$\sqrt[4]{11} = 1.82 \ldots$
	$\sqrt[3]{6} = 1.81 \ldots$	$\sqrt[4]{12} = 1.86 \ldots$
		$\sqrt[4]{13} = 1.89 \ldots$
	$\sqrt[3]{7} = 1.91 \ldots$	$\sqrt[4]{14} = 1.93 \ldots$
		$\sqrt[4]{15} = 1.96 \ldots$
$\sqrt{4} = 2.00 \ldots$	$\sqrt[3]{8} = 2.00 \ldots$	$\sqrt[4]{16} = 2.00 \ldots$

Graph each pair of functions in Problems 74–75 in the same rectangular coordinate system. State each function's domain and range. Then describe the relationship of the graph of function g to the graph of function f using one of the following phrases: reflection about the x-axis; reflection about the y-axis.

74. $f(x) = \sqrt{x}, g(x) = -\sqrt{x}$

75. $f(x) = \sqrt{x}, g(x) = \sqrt{-x}$

For Problems 76–77, use the graph of $f(x) = \sqrt{x}$ to write the equation of the function that represents the graph.

76.

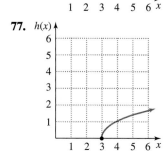

77. h(x)

78. Einstein's model in Example 4 indicates that the length of a moving object decreases from the perspective of an observer at rest. If this theory is true, why does a moving automobile not decrease in length as you watch it pass?

Review Problems _____

79. Simplify: $\dfrac{x^3 + 64}{x^2 - 16}$.

80. Factor: $125 - 8x^3$.

81. Solve the system:

$$\begin{aligned} x + 3y - z &= 5 \\ 2x - 5y - z &= -8 \\ -x + 2y + 3z &= 13 \end{aligned}$$

SECTION 6.2

Solutions Manual **Tutorial** **Video 7**

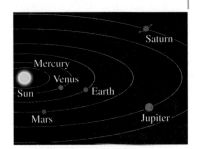

Inverse Properties of *n*th Powers and *n*th Roots; Rational Exponents

Objectives

1 Evaluate $\sqrt[n]{x^n}$ for even and odd *n*.
2 Use rational exponents.
3 Simplify expressions with rational exponents.
4 Use rational exponents to simplify radical expressions.
5 Factor expressions with rational exponents.

The average radius *R* of the orbit of a planet around the sun is modeled by

$$R = T^{2/3}$$

where *T* is the number of years for one orbit and *R* is measured in astronomical units. To use this model, we must understand the meaning of $T^{2/3}$, an expression with a rational exponent. In this section, our focus is on rational exponents and their relationship to roots of real numbers.

Evaluate $\sqrt[n]{x^n}$ for even and odd *n*.

Inverse Properties of *n*th Powers and *n*th Roots

Raising a number to the *n*th power and taking the principal *n*th root can be thought of as *inverse* operations. For example,

$$4^3 = 64 \quad \text{and} \quad \sqrt[3]{64} = 4. \text{ Thus, } \sqrt[3]{4^3} = 4.$$
$$2^5 = 32 \quad \text{and} \quad \sqrt[5]{32} = 2. \text{ Thus, } \sqrt[5]{2^5} = 2.$$
$$7^2 = 49 \quad \text{and} \quad \sqrt{49} = 7. \text{ Thus, } \sqrt{7^2} = 7.$$

Discover for yourself

Is $\sqrt[n]{x^n}$ always equal to *x*? What happens in each of the following cases?

$$(-3)^2 = 9 \qquad \text{and} \qquad \sqrt{9} = 3. \text{ Thus, } \sqrt{(-3)^2} = \underline{\quad}.$$
$$(-2)^4 = 16 \qquad \text{and} \qquad \sqrt[4]{16} = 2. \text{ Thus, } \sqrt[4]{(-2)^4} = \underline{\quad}.$$

Can you use absolute value to express $\sqrt[n]{x^n}$ if *n* is even and *x* is negative?

In the Discover for Yourself box, did you discover that $\sqrt[n]{x^n}$ is not equal to *x* if *n* is even and $x < 0$? For example,

$$\sqrt{(-3)^2} = \sqrt{9} = 3 \qquad \text{The answer is not } -3, \text{ but } |-3| = 3.$$
$$\sqrt[4]{(-2)^4} = \sqrt[4]{16} = 2 \qquad \text{The answer is not } -2, \text{ but } |-2| = 2.$$

Based on these observations, we can make the following generalizations. Did you discover the second generalization on your own?

Inverse properties of *n*th powers and *n*th roots

1. If *n* is odd, then $\sqrt[n]{x^n} = x$.
2. If *n* is even, then $\sqrt[n]{x^n} = |x|$.

| **EXAMPLE 1** | **Evaluating Radical Expressions** |

Evaluate:

a. $\sqrt[3]{2^3}$ **b.** $\sqrt[3]{(-4)^3}$ **c.** $\sqrt[4]{3^4}$ **d.** $\sqrt{(-6)^2}$ **e.** $\sqrt[6]{(-5)^6}$

Solution

a. $\sqrt[3]{2^3} = 2$ $\sqrt[n]{x^n} = x$ when n is odd.
b. $\sqrt[3]{(-4)^3} = -4$ $\sqrt[n]{x^n} = x$ when n is odd.
c. $\sqrt[4]{3^4} = |3| = 3$ $\sqrt[n]{x^n} = |x|$ when n is even.
d. $\sqrt{(-6)^2} = |-6| = 6$ $\sqrt[n]{x^n} = |x|$ when n is even.
e. $\sqrt[6]{(-5)^6} = |-5| = 5$ $\sqrt[n]{x^n} = |x|$ when n is even.   ◼

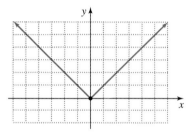

Figure 6.8

The graphs of $y = \sqrt{x^2}$ and $y = |x|$ are the same, so $\sqrt{x^2} = |x|$.

Using technology

You can use a graphing utility to show that

$$\sqrt{x^2} = |x|$$

by graphing

$$y_1 = \sqrt{x^2}$$
and $$y_2 = |x|$$

in the same viewing rectangle. The graphs should be the same and look like the one in Figure 6.8. Now use this procedure to show that

$$\sqrt[3]{x^3} = x.$$

2 Use rational exponents.

Rational Numbers as Exponents

We are now ready to give meaning to expressions that contain rational numbers as exponents such as $27^{1/3}$. Let's begin with the equation $y = 27^{1/3}$. Although we cannot interpret the fractional exponent, we can apply the laws of exponents and cube both sides of the equation.

$$y = 27^{1/3}$$
$$y^3 = (27^{1/3})^3 \quad \text{Cube both sides of the equation.}$$
$$y^3 = 27^{(1/3)(3)} \quad (b^m)^n = b^{mn}. \text{ To raise a power to a power, multiply the exponents.}$$
$$y^3 = 27$$

We see that y is a number whose cube is 27, so y is the cube root of 27.

$$y = \sqrt[3]{27}$$

Because we began with $y = 27^{1/3}$, we now have

$$27^{1/3} = \sqrt[3]{27}.$$

Extending this result, we can set up a definition in which rational exponents indicate roots.

Definition of $b^{1/n}$

$b^{1/n} = \sqrt[n]{b}$, where b is a real number and n is a positive integer greater than 1. (If n is even, $b \geq 0$.)

EXAMPLE 2 **Using the Rational Exponent Definition**

Rewrite without rational exponents:

a. $x^{1/2}$ **b.** $(-125)^{1/3}$ **c.** $(xyz)^{1/4}$

Solution

a. $x^{1/2} = \sqrt{x}$ **b.** $(-125)^{1/3} = \sqrt[3]{-125} = -5$

c. $(xyz)^{1/4} = \sqrt[4]{xyz}$

In Example 2, we started with expressions with rational exponents and rewrote them as equivalent expressions in radical form. In Example 3, we reverse this process.

EXAMPLE 3 **Using the Rational Exponent Definition**

Rewrite with rational exponents:

a. $\sqrt[5]{13ab}$ **b.** $\sqrt[7]{\dfrac{xy^2}{17}}$

Solution

Note that the parentheses indicate the base and the radical's index indicates the exponent's denominator.

a. $\sqrt[5]{13ab} = (13ab)^{1/5}$ **b.** $\sqrt[7]{\dfrac{xy^2}{17}} = \left(\dfrac{xy^2}{17}\right)^{1/7}$

By extending the laws of exponents to cover rational exponents, we can discuss the meaning of rational exponents whose numerators are not 1. In particular,

$$b^{m/n} = (b^m)^{1/n} = \sqrt[n]{b^m}.$$

Furthermore,

$$b^{m/n} = (b^{1/n})^m = (\sqrt[n]{b})^m.$$

We have obtained the following result.

Definition of $b^{m/n}$

$b^{m/n}$ means $\sqrt[n]{b^m}$ or $(\sqrt[n]{b})^m$.

(Again, b is a real number. If n is even, $b \geq 0$. n is a positive integer greater than 1.)

We refer to $b^{m/n}$ as the *exponential form* and $\sqrt[n]{b^m}$ or $(\sqrt[n]{b})^m$ as the *radical form.* Using our result, we can say that

$$125^{2/3} = \sqrt[3]{125^2} = \sqrt[3]{15{,}625} = 25 \qquad \text{Second power, then cube root}$$

or

$$125^{2/3} = (\sqrt[3]{125})^2 = 5^2 = 25 \qquad \text{Cube root, then second power}$$

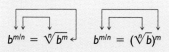
Study tip

If b is a relatively large number, the form

$$(\sqrt[n]{b})^m$$

is easier to work with. In some situations, either form will suffice. For example,

$$8^{2/3} = \sqrt[3]{8^2} = \sqrt[3]{64} = 4 \qquad \text{Second power, then cube root}$$

or

$$8^{2/3} = (\sqrt[3]{8})^2 = 2^2 = 4 \qquad \text{Cube root, then second power}$$

EXAMPLE 4 **Using the Rational Exponent Definition**

Rewrite in radical form and evaluate:

a. $64^{2/3}$ **b.** $16^{5/6}$ **c.** $(-125)^{4/3}$ **d.** $-125^{4/3}$

Solution

a. $64^{2/3} = (\sqrt[3]{64})^2 = 4^2 = 16$
b. $16^{5/2} = (\sqrt{16})^5 = 4^5 = 1024$
c. $(-125)^{4/3} = (\sqrt[3]{-125})^4 = (-5)^4 = 625$
d. $-125^{4/3} = -(\sqrt[3]{125})^4 = -(5)^4 = -625$

Using technology

There are two ways to evaluate radicals on graphing calculators. The first method is to use the appropriate root key. For square roots, you can use the square root key $\boxed{\sqrt{}}$ and for roots having an index greater than 2, you can use the root key $\boxed{\sqrt[x]{}}$. The second method is to use the exponential key $\boxed{\wedge}$. Here are the keystroke sequences for Example 4 with the exponential key.

a. $64^{2/3}$: 64 $\boxed{\wedge}$ $\boxed{(}$ 2 $\boxed{\div}$ 3 $\boxed{)}$ $\boxed{\text{ENTER}}$
b. $16^{5/2}$: 16 $\boxed{\wedge}$ $\boxed{(}$ 5 $\boxed{\div}$ 2 $\boxed{)}$ $\boxed{\text{ENTER}}$
c. $(-125)^{4/3} = [(-125)^{1/3}]^4$: $\boxed{(}$ $\boxed{(-)}$ 125 $\boxed{)}$ $\boxed{\wedge}$ $\boxed{(}$ 1 $\boxed{\div}$ 3 $\boxed{)}$ $\boxed{\wedge}$ 4 $\boxed{\text{ENTER}}$
d. $-125^{4/3}$: $\boxed{(-)}$ 125 $\boxed{\wedge}$ $\boxed{(}$ 4 $\boxed{\div}$ 3 $\boxed{)}$ $\boxed{\text{ENTER}}$

Discover for yourself

Observe the keystroke sequence for $(-125)^{4/3}$ shown in the Using Technology. In general, consider the function $y = x^{4/3}$.

a. The domain of this function consists of the set of all real numbers. Use a graphing utility to graph the function by entering

$$y_1 = x \boxed{\wedge} \boxed{(}\; 4 \boxed{\div} 3 \boxed{)}.$$

Do you obtain a complete graph? Explain.

b. Now graph the function by entering

$$y_1 = \boxed{(}\; x \boxed{\wedge} \boxed{(}\; 1 \boxed{\div} 3 \boxed{)} \boxed{)} \boxed{\wedge} 4.$$

Describe what happens.

c. Explain why the keystroke sequence in part (b) produces a complete graph, whereas the sequence in part (a) results in only half the graph.

We can use the rational exponent definition to change expressions from radical form to exponential form.

EXAMPLE 5   **Using the Rational Exponent Definition**

Rewrite in exponential form:

a. $\sqrt[3]{7^5}$ **b.** $(\sqrt[4]{13xy})^9$

Solution

The index is the denominator of the rational exponent.

a. $\sqrt[3]{7^5} = 7^{5/3}$
b. $(\sqrt[4]{13xy})^9 = (13xy)^{9/4}$ ■

Negative rational exponents are defined just like negative integer exponents.

Definition of $b^{-m/n}$

$$b^{-m/n} = \frac{1}{b^{m/n}} \quad (b \neq 0) \text{ provided that } b^{m/n} \text{ is a real number.}$$

EXAMPLE 6   **Using Negative Rational Exponents**

Rewrite with positive exponents, and evaluate if possible:

a. $(81)^{-3/4}$ **b.** $(7xy)^{-4/5}$

Solution

a. $81^{-3/4} = \dfrac{1}{(81)^{3/4}} = \dfrac{1}{(\sqrt[4]{81})^3} = \dfrac{1}{3^3} = \dfrac{1}{27}$

b. $(7xy)^{-4/5} = \dfrac{1}{(7xy)^{4/5}}$ ■

Rachel Whiteread "House" Oct 93–Jan 94. Courtesy Artangel. Photo: Sue Ormerod.

EXAMPLE 7 **An Investment Application**

The model

$$r = \left(\frac{A}{P}\right)^{1/t} - 1$$

describes the annual rate of return (r) of an investment of P dollars that is worth A dollars after t years. What is the annual rate of return on a condominium that is purchased for $60,000 and sold 5 years later for $80,000?

Solution

We use $A = 80{,}000$, $P = 60{,}000$, and $t = 5$ in the formula.

$$r = \left(\frac{80{,}000}{60{,}000}\right)^{1/5} - 1$$

$$= \left(\frac{4}{3}\right)^{1/5} - 1$$

$$\approx 1.059 - 1 \qquad \text{Use a calculator:} \quad \boxed{(}\;4\;\boxed{\div}\;3\;\boxed{)}\;\boxed{\wedge}\;\boxed{(}\;1\;\boxed{\div}\;5\;\boxed{)}\;\boxed{\text{ENTER}}$$

$$= 0.059, \text{ or } 5.9\%$$

The annual rate of return on the condominium is 5.9%. ■

3 Simplify expression with rational exponents.

Properties of Rational Exponents

The same properties apply to rational exponents as to integer exponents. The following is a summary of these properties, with supporting examples.

Properties of rational exponents

Let r and s be rational numbers, and let a and b represent real numbers, variables, or algebraic expressions. Assume that $b \neq 0$ whenever it appears in a denominator, and all roots are real numbers (which excludes even roots of negative numbers).

Property	What the Property Says	Example
1. $b^r \cdot b^s = b^{r+s}$	In multiplying, add exponents if the bases are the same.	$3^{1/2} \cdot 3^{1/4} = 3^{(1/2)+(1/4)} = 3^{3/4}$ (or $\sqrt[4]{3^3}$)
2. $(b^r)^s = b^{rs}$	To raise a power to a power, multiply the exponents.	$(2^{2/3})^{3/4} = 2^{(2/3)\cdot(3/4)} = 2^{6/12} = 2^{1/2}$ (or $\sqrt{2}$)
3. $(ab)^r = a^r b^r$	To raise a product to a power, raise each factor to the power and multiply.	$(8x)^{1/3} = 8^{1/3} \cdot x^{1/3} = 2x^{1/3}$ (or $2\sqrt[3]{x}$)
4. $\left(\dfrac{a}{b}\right)^r = \dfrac{a^r}{b^r}$	To raise a quotient to a power, raise numerator and denominator to the power and divide.	$\left(\dfrac{x}{81}\right)^{1/4} = \dfrac{x^{1/4}}{81^{1/4}} = \dfrac{x^{1/4}}{3}$ $\left(\text{or } \dfrac{\sqrt[4]{x}}{3}\right)$
5. $\dfrac{b^r}{b^s} = b^{r-s}$	In dividing, subtract exponents if the bases are the same.	$\dfrac{5^{1/2}}{5^{1/4}} = 5^{1/2-1/4} = 5^{1/4}$ (or $\sqrt[4]{5}$)

We can use these properties to simplify exponential expressions containing rational exponents. Once again, only positive exponents will appear in the final result. Let's see how this works.

EXAMPLE 8 **Simplifying Expressions with Rational Exponents**

Simplify using properties of exponents:

a. $(5x^{1/2})(7x^{3/4})$ **b.** $\dfrac{32x^{1/2}}{16x^{3/4}}$ **c.** $(10x^{1/3}y^2z^{2/3})^3$ **d.** $\left(\dfrac{7x^{3/5}}{4y^{1/3}}\right)^3$

Solution

a. $(5x^{1/2})(7x^{3/4}) = 5 \cdot 7 \cdot x^{1/2} \cdot x^{3/4}$

  Rearrange factors. This step is usually done mentally.

$= 35x^{(1/2)+(3/4)}$ Multiply coefficients and add exponents.

$= 35x^{(2/4)+(3/4)}$ Write exponents in terms of a LCD.

$= 35x^{5/4}$ Perform the addition.

b. $\dfrac{32x^{1/2}}{16x^{3/4}} = \dfrac{32}{16}x^{(1/2)-(3/4)}$

  Divide coefficients and subtract exponents.

$= 2x^{(2/4)-(3/4)}$ Write exponents in terms of a LCD.

$= 2x^{-1/4}$ Perform the subtraction.

$= \dfrac{2}{x^{1/4}}$ Use $b^{-r} = \dfrac{1}{b^r}$ to write the result with a positive exponent.

c. $(10x^{1/3}y^2z^{2/3})^3 = 10^3(x^{1/3})^3(y^2)^3(z^{2/3})^3$

  Raise each factor in parentheses to the third power.

$= 10^3x^{1/3\cdot3}y^{2\cdot3}z^{2/3\cdot3}$ Raise exponential expressions to powers by multiplying exponents.

$= 1000xy^6z^2$ Simplify.

d. $\left(\dfrac{7x^{3/5}}{4y^{1/3}}\right)^3 = \dfrac{(7x^{3/5})^3}{(4y^{1/3})^3}$

  Raise the numerator and denominator to the third power.

$= \dfrac{7^3(x^{3/5})^3}{4^3(y^{1/3})^3}$ Raise each factor in both parentheses to the third power.

$= \dfrac{7^3 \, x^{3/5\cdot3}}{4^3 \, y^{1/3\cdot3}}$ Raise powers to powers by multiplying exponents.

$= \dfrac{343x^{9/5}}{64y}$ Simplify. ∎

4 Use rational exponents to simplify radical expressions.

Using Rational Exponents to Simplify Radical Expressions

Some radical expressions can be simplified using rational exponents. We will use the following procedure.

Simplifying radical expressions using rational exponents

1. Rewrite each radical in its equivalent fractional exponential form.
2. Simplify using properties of exponents.
3. Rewrite in radical notation when rational exponents still appear.

EXAMPLE 9 **Simplifying Radical Expressions Using Rational Exponents**

Use rational exponents to simplify:

a. $\sqrt[10]{x^5}$ **b.** $\sqrt[3]{27a^{15}}$ **c.** $\sqrt[8]{x^4y^2}$

d. $\sqrt[6]{16x^2y^4}$ **e.** $\sqrt{5}\cdot\sqrt[3]{5}$ **f.** $\dfrac{\sqrt[5]{2}}{\sqrt[7]{2}}$

Solution

a. $\sqrt[10]{x^5} = x^{5/10}$ — Rewrite in exponential form.
$= x^{1/2}$ — Simplify the exponent.
$= \sqrt{x}$ — Rewrite in radical form.

b. $\sqrt[3]{27a^{15}} = (27a^{15})^{1/3}$ — Rewrite in exponential form.
$= 27^{1/3}(a^{15})^{1/3}$ — Raise each factor in parentheses to the $\frac{1}{3}$ power.
$= 3a^5$ — $27^{1/3} = \sqrt[3]{27} = 3; (a^{15})^{1/3} = a^{15\cdot1/3} = a^5$

c. $\sqrt[8]{x^4y^2} = (x^4y^2)^{1/8}$ — Rewrite in exponential form.
$= (x^4)^{1/8}(y^2)^{1/8}$ — Raise each factor in parentheses to the $\frac{1}{8}$ power.
$= x^{4\cdot1/8}\cdot y^{2\cdot1/8}$ — Raise powers to powers by multiplying exponents.
$= x^{1/2}\cdot y^{1/4}$ — Simplify.
$= x^{2/4}\cdot y^{1/4}$ — To rewrite in radical form, express rational exponents in terms of the LCD of the fractions.
$= (x^2y^1)^{1/4}$ — Since $(ab)^r = a^rb^r$, then $a^rb^r = (ab)^r$. We've used the property for raising a product to a power in reverse.
$= \sqrt[4]{x^2y}$ — Rewrite in radical form.

d. $\sqrt[6]{16x^2y^4} = \sqrt[6]{2^4x^2y^4}$ — Since 16 does not have a recognizable 6th root, rewrite 16 in exponential form using the least base that is greater than 1.
$= (2^4x^2y^4)^{1/6}$ — Rewrite in exponential form.
$= (2^4)^{1/6}(x^2)^{1/6}(y^4)^{1/6}$ — Raise each factor in parentheses to the $\frac{1}{6}$ power.
$= 2^{2/3}x^{1/3}y^{2/3}$ — Multiply exponents.
$= (2^2xy^2)^{1/3}$ — Use the property for raising a product to a power in reverse.
$= \sqrt[3]{4xy^2}$ — Rewrite in radical form.

e. $\sqrt{5}\sqrt[3]{5} = 5^{1/2}\cdot5^{1/3}$ — Rewrite in exponential form.
$= 5^{(1/2)+(1/3)}$ — Add exponents.
$= 5^{(3/6)+(2/6)}$ — Write exponents in terms of the LCD.
$= 5^{5/6}$ — Perform the addition.
$= \sqrt[6]{5^5}$ or $\sqrt[6]{3125}$ — Rewrite in radical form.

f. $\dfrac{\sqrt[5]{2}}{\sqrt[7]{2}} = \dfrac{2^{1/5}}{2^{1/7}}$ — Rewrite in exponential form.
$= 2^{(1/5)-(1/7)}$ — Subtract exponents.
$= 2^{(7/35)-(5/35)}$ — Write exponents in terms of the LCD.
$= 2^{2/35}$ — Perform the subtraction.
$= \sqrt[35]{2^2}$ or $\sqrt[35]{4}$ — Rewrite in radical form. ∎

Discover for yourself

Use some specific values of x to convince yourself that

$$\sqrt[10]{x^5} = \sqrt{x}$$

Start with $x = 4$. What is $\sqrt{4}$? Is this the same number as $\sqrt[10]{4^5}$ or $\sqrt[10]{1024}$?

5 Factor expressions with rational exponents.

Factoring Expressions Containing Rational Exponents

Expressions with rational exponents that are not integers are not polynomials. However, factoring these expressions (assuming, of course, that they are factorable) should remind you of the procedures for factoring polynomials.

For example, consider the polynomial

$$x^2 + x^9.$$

We can factor out the variable x to the second power because 2 is the lesser exponent on both terms. Doing so, we obtain

$$x^2 + x^9 = x^2(1 + x^7).$$
$$\overset{\displaystyle\lceil 9 - 2 = 7}{}$$

The exponent of the x in the parentheses can be determined by subtracting the exponent on the x being factored out, the 2, from the exponent on the x it is being factored from, the 9 ($9 - 2 = 7$).

We can use this same procedure to factor an expression such as

$$x^{1/3} + x^{4/3}.$$

Since the lesser exponent is $\frac{1}{3}$, we factor $x^{1/3}$ from both terms. This forms the basis of Example 10.

EXAMPLE 10   Factoring with Rational Exponents

Factor: $\quad x^{1/3} + x^{4/3}$

Solution

We factor $x^{1/3}$ from both terms. When we factor $x^{1/3}$ from $x^{4/3}$, what is left? We can determine the exponent on this term by subtracting the smaller exponent (the exponent on the variable being factored out, the $\frac{1}{3}$) from the larger exponent (the exponent on the variable it is being factored from, the $\frac{4}{3}$).

$$\frac{4}{3} - \frac{1}{3} = \frac{3}{3} = 1$$

$$x^{1/3} + x^{4/3} = x^{1/3}(1 + x^1)$$ Factor out $x^{1/3}$. The exponent on the second term in parentheses can be determined by inspection or subtraction.

$$= x^{1/3}(1 + x)$$ Use the distributive property to check this factorization. ∎

EXAMPLE 11   Factoring with Rational Exponents

Factor: $\quad 5x^{1/2} - 10x^{-1/2}$

Solution

Since $-\frac{1}{2}$ is less than $\frac{1}{2}$, we factor $x^{-1/2}$ from each term. We can also factor 5 from each term.

$$-\frac{1}{2} - \left(-\frac{1}{2}\right) = 1$$

$$5x^{1/2} - 10x^{-1/2} = 5x^{-1/2}(x^1 - 2)$$ Factor out $5x^{-1/2}$. The exponent on the first term in parentheses can be determined by inspection or subtraction.

$$= \frac{5(x - 2)}{x^{1/2}}$$ Write the factored form with positive exponents. ∎

PROBLEM SET 6.2

Practice Problems_____

Evaluate each radical expression in Problems 1–18.

1. $\sqrt[3]{4^3}$

2. $\sqrt[3]{2^3}$

3. $\sqrt[3]{(-2)^3}$

4. $\sqrt[3]{(-4)^3}$

5. $\sqrt[4]{2^4}$

6. $\sqrt[4]{3^4}$

7. $\sqrt[4]{(-2)^4}$

8. $\sqrt[4]{(-3)^4}$

9. $\sqrt[5]{(-2)^5}$

10. $\sqrt[5]{(-3)^5}$

11. $\sqrt[17]{(-9)^{17}}$

12. $\sqrt[23]{(-11)^{23}}$

13. $\sqrt[18]{(-9)^{18}}$

14. $\sqrt[24]{(-11)^{24}}$

15. $\sqrt[n]{(-8)^n}$ (*n* is odd.)

16. $\sqrt[n]{(-6)^n}$ (*n* is odd.)

17. $\sqrt[n]{(-8)^n}$ (*n* is even.)

18. $\sqrt[n]{(-6)^n}$ (*n* is even.)

In Problems 19–42, rewrite each expression in radical form and evaluate where possible.

19. $x^{1/4}$

20. $x^{1/5}$

21. $36^{1/2}$

22. $121^{1/2}$

23. $8^{1/3}$

24. $(-27)^{1/3}$

25. $(xy)^{1/6}$

26. $(xy)^{1/7}$

27. $16^{5/2}$

28. $125^{2/3}$

29. $(8x)^{2/3}$

30. $(-27x)^{4/3}$

31. $-27^{4/3}$

32. $-25^{3/2}$

33. $64^{-1/2}$

34. $8^{-1/3}$

35. $\left(\frac{1}{64}\right)^{-1/3}$

36. $\left(\frac{1}{32}\right)^{-1/5}$

37. $(-64)^{-1/3}$

38. $\left(\frac{1}{64}\right)^{-2/3}$

39. $\left(\frac{1}{32}\right)^{-3/5}$

40. $32^{-4/5}$

41. $16^{-5/2}$

42. $\left(-\frac{27}{64}\right)^{-1/3}$

In Problems 43–56, rewrite each expression with rational exponents.

43. $\sqrt[3]{7}$

44. $\sqrt[3]{5}$

45. $\sqrt{x^3}$

46. $\sqrt{x^5}$

47. $\sqrt[5]{x^3}$

48. $\sqrt[7]{x^4}$

49. $\sqrt[4]{ab}$

50. $\sqrt[4]{cd}$

51. $\sqrt[5]{x^2yz^4}$

52. $\sqrt[7]{xy^3z^4}$

53. $(\sqrt{19xy})^3$

54. $(\sqrt[3]{11xy})^5$

55. $(\sqrt[6]{11xy^2})^5$

56. $(\sqrt[7]{3x^5y})^9$

Use the properties of exponents to simplify Problems 57–86. Final results should be written in terms of positive exponents.

57. $3^{3/4} \cdot 3^{1/4}$

58. $5^{2/3} \cdot 5^{1/3}$

59. $\dfrac{16^{3/4}}{16^{1/4}}$

60. $\dfrac{100^{3/4}}{100^{1/4}}$

61. $x^{3/4} \cdot x^{1/3}$

62. $x^{2/3} \cdot x^{1/4}$

63. $\dfrac{x^{4/5}}{x^{1/5}}$

64. $\dfrac{x^{3/7}}{x^{1/7}}$

65. $(x^{2/3})^3$

66. $(x^{4/5})^5$

67. $\dfrac{x^{1/3}}{x^{3/4}}$

68. $\dfrac{x^{3/5}}{x^{1/4}}$

69. $(7y^{1/3})(2y^{1/4})$

70. $(3x^{2/3})(4x^{3/4})$

71. $(3x^{3/4})(-5x^{-1/2})$

72. $(-6x^{2/3})(-8x^{-1/4})$

73. $\dfrac{20x^{1/2}}{5x^{1/4}}$

74. $\dfrac{72y^{3/4}}{9y^{1/3}}$

75. $\dfrac{80y^{1/6}}{10y^{1/4}}$

76. $\dfrac{-60y^{1/3}}{20y^{3/4}}$

77. $(2x^{1/5}y^2z^{2/5})^5$

78. $(3x^{1/4}y^3z^{3/4})^4$

79. $(25x^4y^6)^{1/2}$

80. $(125x^9y^6)^{1/3}$

81. $(16xy^{1/4}z^{2/3})^{1/4}$

82. $(27x^{1/4}y^{2/3}z^3)^{1/3}$

83. $\left(\dfrac{2x^{1/4}}{5y^{1/3}}\right)^3$

84. $\left(\dfrac{7x^{2/3}}{3y^{1/2}}\right)^2$

85. $\left(\dfrac{x^3}{y^5}\right)^{-1/2}$

86. $\left(\dfrac{a^5}{b^{-4}}\right)^{-1/5}$

Use rational exponents to simplify Problems 87–102, expressing the final answer in radical form.

87. $\sqrt[9]{a^3}$

88. $\sqrt[9]{y^6}$

89. $\sqrt[3]{8x^6}$

90. $\sqrt[4]{81x^{12}}$

91. $\sqrt[5]{x^{10}y^{15}}$

92. $\sqrt[4]{x^{12}y^{16}}$

93. $\sqrt[9]{2^3x^3y^6}$

94. $\sqrt[12]{2^4x^4y^8}$

95. $\sqrt[9]{27x^3y^6}$

96. $\sqrt[4]{81x^2y^6}$

97. $\sqrt[3]{3}\sqrt{3}$

98. $\sqrt{2}\sqrt[3]{2}$

99. $\sqrt[4]{2}\sqrt{2}$

100. $\sqrt[3]{6}\sqrt{6}$

101. $\dfrac{\sqrt{3}}{\sqrt[3]{3}}$

102. $\dfrac{\sqrt[3]{2}}{\sqrt[4]{2}}$

Factor Problems 103–110, writing the answer without negative exponents.

103. $x^{3/2} + x^{1/2}$

104. $x^{1/4} + x^{5/4}$

105. $6x^{1/3} + 3x^{4/3}$

106. $9x^{4/3} + 18x^{1/3}$

107. $x^{2/5} + x^{-3/5}$

108. $x^{1/2} + x^{-1/2}$

109. $15x^{-1/2} - 20x^{-5/2}$

110. $8x^{-9/5} - 12x^{-4/5}$

Application Problems

111. After t years in operation, a factory dumps P tons of pollutants into a river, where $P = 4t^{3/2} + 60$. During its fourth year in operation, the factory is fined $10,000 for each ton of pollutants deposited into the river beyond 62 tons. What will the fines come to during the fourth year?

112. After t weeks of studying a foreign language, the number of vocabulary words (N) memorized by the average student is $N = 30t^{3/4} - 10$. Pierre decides to visit Paris after he has mastered at least 250 words of French. Will Pierre be ready for his trip after studying the language for 16 weeks?

113. The rate of increase of pollution in a river after t years is described by the formula $R = \frac{1}{4}(t^{1/4} + 3)t^{-3/4}$. Find the rate of increase (in units of pollution per year) after 16 years.

114. The maximum velocity (v, in miles per hour) that an automobile can travel around a curve with radius r (in feet) without skidding is described by the model $v = \left(\dfrac{5r}{2}\right)^{1/2}$. If the curve has a radius of 250 feet, find the maximum speed a car can travel around it without skidding.

Use the investment model $r = \left(\dfrac{A}{P}\right)^{1/t} - 1$ (described in Example 7) to answer Problems 115–116.

115. What is the annual rate of return on a house that is purchased for $80,000 and sold 4 years later for $120,000?

116. What is the annual rate of return on a collection of musical theater cast albums that cost an investor $500 and that were sold 5 years later for $800?

117. The function $f(d) = 0.07d^{3/2}$ is used by meteorologists to determine the duration of a storm ($f(d)$ in hours) whose diameter is d miles. Find and interpret $f(9)$.

True–False Critical Thinking Problems

118. Which one of the following is true?
a. If n is odd, then $(-b)^{1/n} = -b^{1/n}$.
b. $(a + b)^{1/n} = a^{1/n} + b^{1/n}$
c. $8^{-2/3} = -4$
d. None of the above is true.

119. Which one of the following is true?
a. $b^{1/2}(b^{1/4} - b^{1/2}) = b^{1/8} - b^{1/4}$

b. $(-32x^5)^{-3/5} = -\dfrac{1}{8x^3} \quad (x \neq 0)$

c. $(a^{1/2} + b^{-1/2})^2 = a + \dfrac{1}{b} \quad (a > 0 \text{ and } b > 0)$

d. $\dfrac{x^{1/3}}{x^{1/4}} = x^{4/3} \quad (x > 0)$

Technology Problems

120. Use a graphing calculator to verify your results in Problems 21–24, 27–28, and 31–42.

121. Use a graphing utility to graph $y_1 = \sqrt{x^2}$ and $y_2 = x$ in the same viewing rectangle.
a. For what values of x is $\sqrt{x^2} = x$?
b. For what values of x is $\sqrt{x^2} \neq x$?

122. Use a graphing utility to graph $y_1 = \sqrt{x^2}$ and $y_2 = -x$ in the same viewing rectangle.
a. For what values of x is $\sqrt{x^2} = -x$?
b. For what values of x is $\sqrt{x^2} \neq -x$?

123. The graph of $y = x^{2/3}$ is shown in the figure on the right.
a. Use a graphing utility to obtain this graph by entering

$$Y_1 = X \boxed{\wedge} (2 \boxed{\div} 3).$$

Describe what happens.
b. Repeat part (a), but this time enter

$$Y_1 = (X \boxed{\wedge} (1 \boxed{\div} 3)) \boxed{\wedge} 2.$$

Describe what happens.

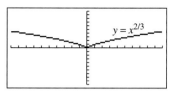
$y = x^{2/3}$

124. a. Graph $y = x^{2/6}$ by entering

$$Y_1 = (X \boxed{\wedge} (1 \boxed{\div} 6)) \boxed{\wedge} 2.$$

b. Graph $y = x^{1/3}$ by entering

$$Y_1 = X \boxed{\wedge} (1 \boxed{\div} 3).$$

c. Does your graphing utility produce the same graph in parts (a) and (b)? Can you explain what is happening?

125. The space shuttle *Challenger* exploded 73 seconds into flight on January 28, 1986. The tragedy involved O-rings used to seal the connections between different sections of the shuttle engines. The number of O-rings damaged increases dramatically with falling Fahrenheit temperature t, modeled by $N = 13.49(0.967)^t - 1$.
   a. On the morning the *Challenger* was launched, the temperature was 31°, colder than any previous experience with a space shuttle. Find the number of O-rings expected to fail at this temperature.
   b. Use a graphing utility to graph the function. If NASA engineers had used this function and its graph, is it likely they would have allowed the *Challenger* to be launched? Explain.

In January 1986 the Orbiter *Challenger* blew up 73 seconds after take-off, killing all seven crew on board and suspending the space shuttle program for two and a half years.

Photri, Inc.

Writing in Mathematics

126. What is the meaning of a rational exponent?

127. Under what conditions is $\sqrt[n]{x^n}$ not equal to x? Give an example.

128. How does the definition of $b^{1/n}$ bring together the concepts of exponents and roots?

129. Graph $y = \sqrt{x}$ and $y = x^2$ on the same axes. Describe as many relationships as possible that you notice between the two curves. (The relationships can include differences between the behavior of the graphs.)

Critical Thinking Problems

130. A mathematics professor recently purchased a birthday cake for her son with the inscription:

Happy $(2^{5/2} \cdot 2^{3/4} \div 2^{1/4})$th Birthday.

How old is the son?

131. The birthday boy in Problem 130, excited by the inscription on the cake, tried immediately to consume the whole thing. Professor Mom, concerned about the possible metamorphosis of her son into a blimp, exclaimed, "Hold on! It is your birthday, so why not take $\dfrac{8^{-4/3} + 2^{-2}}{16^{-3/4} + 2^{-1}}$ of the cake? I'll eat half of what's left over." How much of the cake did the professor eat?

132. Factor: $x^{1/2} + x^{1/4} - 6$.

133. Factor: $6x^{2/3} - 17x^{1/3} + 12$.

134. Factor: $4x^{2/9} - 9$.

135. Simplify: $[3 + (27^{2/3} + 32^{2/5})]^{3/2} - 9^{1/2}$.

136. The population (P, in thousands) of a community t years after 1970 is given by $P = 2x + 6x^{3/2} + 700$. What is the percent increase in population from 1974 to 1986?

137. The following "proof" was written on the wall of the math building at Cornell University. Underneath it, someone wrote "Extremely improbable." See if you can discover the error in the proof.

$$(x + 1)^2 = x^2 + 2x + 1$$
$$(x + 1)^2 - (2x + 1) = x^2$$
$$(x + 1)^2 - (2x + 1) - x(2x + 1) = x^2 - x(2x + 1)$$
$$(x + 1)^2 - (x + 1)(2x + 1) + \tfrac{1}{4}(2x + 1)^2 = x^2 - x(2x + 1) + \tfrac{1}{4}(2x + 1)^2$$
$$[(x + 1) - \tfrac{1}{2}(2x + 1)]^2 = [x - \tfrac{1}{2}(2x + 1)]^2$$
$$(x + 1) - \tfrac{1}{2}(2x + 1) = x - \tfrac{1}{2}(2x + 1)$$
$$x + 1 = x$$
$$1 = 0$$

138. If $x \uparrow y = x^y$ and $x \downarrow y = y^x$, evaluate
$$[(64 \uparrow \tfrac{1}{2}) \uparrow \tfrac{1}{3}] \downarrow 5.$$

139. Which of the following is greater: $81^{1/4}$ or $81^{1/2}$? In general, if $n > m$ and $b > 1$, which is greater: $b^{1/n}$ or $b^{1/m}$?

140. Which of the following numbers is greater: $(\tfrac{1}{81})^{1/4}$ or $(\tfrac{1}{81})^{1/2}$? In general, if $n > m$ and $0 < b < 1$, which is greater: $b^{1/n}$ or $b^{1/m}$?

141. If the cube root of a certain number is divided into that number raised to the negative $\tfrac{2}{3}$ power, the quotient is $\tfrac{1}{4}$. What is the number?

142. Describe what is wrong with the following:
$$-1 = (-1)^1 = (-1)^{2/2} = [(-1)^2]^{1/2} = 1^{1/2}$$
$$= \sqrt{1} = 1$$

Review Problems_____

143. Solve using matrix methods:
$$x + 4y - 2z = -3$$
$$2x + y + z = 3$$
$$-5x - 2y + 3z = -14$$

144. Factor: $x^6 - x^2$.

145. Solve for y: $\dfrac{5}{y+2} - \dfrac{3}{y+5} = \dfrac{9}{y^2 + 7y + 10}$.

S E C T I O N   6 . 3

Solutions Manual   Tutorial   Video 7

Multiplying and Simplifying Radical Expressions

Objectives

1 Use the product rule to multiply radicals.
2 Use the product rule to simplify radicals.
3 Use the product rule to multiply and simplify radicals.

In this section, we concentrate on multiplying and simplifying radical expressions. Writing radicals in simplified form will provide the basis for adding and subtracting radical expressions in Section 6.5.

1 Use the product rule to multiply radicals.

The Product Rule for Radicals

A rule for multiplying radicals can be generalized by comparing $\sqrt{25} \cdot \sqrt{4}$ and $\sqrt{25 \cdot 4}$. Notice that

$$\sqrt{25} \cdot \sqrt{4} = 5 \cdot 2 = 10 \qquad \text{and} \qquad \sqrt{25 \cdot 4} = \sqrt{100} = 10.$$

Since we obtain 10 in both situations, the original radical expressions must be equal. That is,

$$\sqrt{25} \cdot \sqrt{4} = \sqrt{25 \cdot 4}.$$

This result is a special case of the *product rule for radicals* that can be generalized as follows.

Multiplying radicals: the product rule

Let M and N be real numbers, variables, or algebraic expressions. If the nth roots of M and N are real numbers, then

$$\sqrt[n]{M} \cdot \sqrt[n]{N} = \sqrt[n]{M \cdot N}.$$

The product of two nth roots is the nth root of the product.

We can verify the product rule using rational exponents. Since $\sqrt[n]{M} = M^{1/n}$ and $\sqrt[n]{N} = N^{1/n}$,

$$\sqrt[n]{M} \cdot \sqrt[n]{N} = M^{1/n} \cdot N^{1/n} = (MN)^{1/n} = \sqrt[n]{MN}$$

Study tip

The product rule can be used only when the radicals have the same index. If indices differ, rational exponents can be used, as in $\sqrt{5} \cdot \sqrt[3]{5}$, which was Example 9e in the previous section.

EXAMPLE 1 **Using the Product Rule for Radicals**

Multiply:

a. $\sqrt{3} \cdot \sqrt{7}$ **b.** $\sqrt{x + 7} \cdot \sqrt{x - 7}$ **c.** $\sqrt{11} \cdot \sqrt{3xy}$
d. $\sqrt[3]{7} \cdot \sqrt[3]{9}$ **e.** $\sqrt[8]{10x} \cdot \sqrt[8]{8x^4}$

Solution

a. $\sqrt{3} \cdot \sqrt{7} = \sqrt{3 \cdot 7} = \sqrt{21}$
b. $\sqrt{x + 7} \cdot \sqrt{x - 7} = \sqrt{(x + 7)(x - 7)}$ $\quad$ Study Tip
$\qquad\qquad\qquad\qquad\quad = \sqrt{x^2 - 49}$ $\quad \leftarrow \sqrt{x^2 - 49} \neq \sqrt{x^2} - \sqrt{49}$
c. $\sqrt{11} \cdot \sqrt{3xy} = \sqrt{11 \cdot 3xy} = \sqrt{33xy}$
d. $\sqrt[3]{7} \cdot \sqrt[3]{9} = \sqrt[3]{7 \cdot 9} = \sqrt[3]{63}$
e. $\sqrt[8]{10x} \cdot \sqrt[8]{8x^4} = \sqrt[8]{10x \cdot 8x^4} = \sqrt[8]{80x^5}$ ■

2 Use the product rule to simplify radicals.

Using the Product Rule to Simplify Radicals

The product rule, read from right to left, enables us to simplify radicals without changing their value.

> **Simplifying radicals: the product rule**
>
> $$\sqrt[n]{M \cdot N} = \sqrt[n]{M} \cdot \sqrt[n]{N}$$

Simplified radical expressions contain no factors that are perfect nth powers, where n is the index. For example, consider $\sqrt{300}$. The radicand 300 has the factor 100, which is a perfect square. Therefore,

$\sqrt{300} = \sqrt{100 \cdot 3}$ $\quad$ Factor the radicand. Notice that 100 is the largest perfect square factor of 300.

$\qquad = \sqrt{100} \cdot \sqrt{3}$ $\quad$ Factor into two radicals using the product rule.

$\qquad = 10\sqrt{3}$ $\quad$ Take the square root of 100.

Using technology

You can use a calculator to provide numerical support that $\sqrt{300} = 10\sqrt{3}$. For $\sqrt{300}$:

$\boxed{\sqrt{}}$ 300 $\boxed{\text{ENTER}}$ ≈ 17.32

For $10\sqrt{3}$:

10 $\boxed{\sqrt{}}$ 3 $\boxed{\text{ENTER}}$ ≈ 17.32

Use this technique to support the numerical results in Example 2 on page 530.

> **Simplifying radicals with constant radicands**
>
> A radical expression whose index is n is simplified when its radicand has no factors that are perfect nth powers. To simplify radicals with constant radicands:
>
> **1.** Write the radicand as the product of two numbers, one of which is the largest perfect nth power.
> **2.** Use the product rule to write the radical expression as the product of roots.
> **3.** Find the nth root of any perfect nth powers.

EXAMPLE 2 **Simplifying Radicals with Constant Radicands**

Write each radical in simplified form by factoring the radicand and removing as many perfect nth powers as possible.

a. $\sqrt{75}$ **b.** $\sqrt[3]{54}$ **c.** $\sqrt[5]{64}$ **d.** $6\sqrt{80}$

Solution

a. $\sqrt{75} = \sqrt{(25)(3)} = \sqrt{25}\sqrt{3} = 5\sqrt{3}$ 25 is the largest perfect square factor of 75.

b. $\sqrt[3]{54} = \sqrt[3]{(27)(2)} = \sqrt[3]{27}\sqrt[3]{2} = 3\sqrt[3]{2}$ 27 is the largest perfect cube factor of 54.

c. $\sqrt[5]{64} = \sqrt[5]{(32)(2)} = \sqrt[5]{32}\sqrt[5]{2} = 2\sqrt[5]{2}$ 32 is the largest perfect fifth-power factor of 64.

d. $6\sqrt{80} = 6\sqrt{(16)(5)} = 6\sqrt{16}\sqrt{5} = 6(4)\sqrt{5} = 24\sqrt{5}$ ∎

$\mathcal{S}$**tudy tip**

Here are some perfect nth powers that you should learn to recognize.

Squares of natural numbers:	$1^2, 2^2, 3^2, 4^2, 5^2, 6^2, 7^2, 8^2, 9^2, \ldots$
Perfect squares:	$1, 4, 9, 16, 25, 36, 49, 64, 81, \ldots$
Cubes of natural numbers:	$1^3, 2^3, 3^3, 4^3, 5^3, 6^3, \ldots$
Perfect cubes:	$1, 8, 27, 64, 125, 216, \ldots$
Fourth powers of natural numbers:	$1^4, 2^4, 3^4, 4^4, \ldots$
Perfect fourth powers:	$1, 16, 81, 256, \ldots$
Fifth powers of natural numbers:	$1^5, 2^5, 3^5, \ldots$
Perfect fifth powers:	$1, 32, 243, \ldots$

Things get a bit more complicated if a radical contains variables in the radicand. For example, we know that $\sqrt{x^2} = |x|$ because the index (2) is even. However, suppose we are simplifying $\sqrt{x^3}$. Because $\sqrt{x^3}$ is not a real number if $x < 0$, we assume that the variable x represents a positive number. Thus,

$$\sqrt{x^3} = \sqrt{x^2 \cdot x} = \sqrt{x^2}\sqrt{x} = x\sqrt{x}.$$

Because x is positive, it is not necessary to include absolute value bars.

For the remainder of this chapter, let's assume that *all variables in any radicand represent positive real numbers*. Based on this assumption, we can state the following.

If x is a positive real number

$$\sqrt[n]{x^n} = x.$$

It appears that we must make the exponent in the radicand the same number as the index of the radical to use this idea. For example,

$$\sqrt{x^2} = x, \qquad \sqrt[3]{y^3} = y, \qquad \sqrt[4]{z^4} = z, \qquad \text{and} \qquad \sqrt[5]{y^{10}} = \sqrt[5]{(y^2)^5} = y^2.$$

Consider again $\sqrt[5]{y^{10}}$, which simplifies to y^2. The radicand, y^{10}, is a perfect fifth power because it is a multiple of 5. The radicand can be simplified using exponential form.

$$\sqrt[5]{y^{10}} = (y^{10})^{1/5} = y^{10 \cdot 1/5} = y^2$$

EXAMPLE 3 **Simplifying Radicals with Variables in the Radicand**

Simplify: **a.** $\sqrt{x^6}$ **b.** $\sqrt[3]{y^{21}}$ **c.** $\sqrt[6]{z^{24}}$

Solution

a. $\sqrt{x^6} = (x^6)^{1/2} = x^{6 \cdot 1/2} = x^3$
b. $\sqrt[3]{y^{21}} = (y^{21})^{1/3} = y^{21 \cdot 1/3} = y^7$
c. $\sqrt[6]{z^{24}} = (z^{24})^{1/6} = z^{24 \cdot 1/6} = z^4$ ∎

Notice that a factor in a radicand is a perfect nth power when its exponent is a multiple of n.

Perfect squares:	$x^2, x^4, x^6, x^8, x^{10}, \ldots$
Perfect cubes:	$x^3, x^6, x^9, x^{12}, x^{15}, \ldots$
Perfect fourth powers:	$x^4, x^8, x^{12}, x^{16}, x^{20}, \ldots$
Perfect fifth powers:	$x^5, x^{10}, x^{15}, x^{20}, x^{25}, \ldots$
Perfect nth powers:	$x^n, x^{2n}, x^{3n}, x^{4n}, x^{5n}, \ldots$

As when simplifying radicals with constant radicands, our goal is to write the radicand as a product, where one of the factors is the largest perfect nth power.

EXAMPLE 4 **Simplifying Radicals with Variables in the Radicand**

Simplify: **a.** $\sqrt{x^{11}}$ **b.** $\sqrt[3]{y^{29}}$ **c.** $\sqrt[5]{z^{28}}$

Solution

a. The largest perfect square factor of x^{11} is x^{10}.

$$\sqrt{x^{11}} = \sqrt{x^{10} \cdot x} = \sqrt{x^{10}} \cdot \sqrt{x} = (x^{10})^{1/2}\sqrt{x} = x^5\sqrt{x}$$

b. The largest perfect cube factor of y^{29} is y^{27}.

$$\sqrt[3]{y^{29}} = \sqrt[3]{y^{27} \cdot y^2} = \sqrt[3]{y^{27}} \cdot \sqrt[3]{y^2} = (y^{27})^{1/3}\sqrt[3]{y^2} = y^9\sqrt[3]{y^2}$$

c. The largest perfect fifth power factor of z^{28} is z^{25}.

$$\sqrt[5]{z^{28}} = \sqrt[5]{z^{25} \cdot z^3} = \sqrt[5]{z^{25}} \cdot \sqrt[5]{z^3} = (z^{25})^{1/5} \cdot \sqrt[5]{z^3} = z^5\sqrt[5]{z^3}$$ ∎

Example 4 illustrates that a radical expression whose index is n can be simplified when its radicand has factors with exponents greater than or equal to n.

> **Simplifying radicals with variable radicands**
>
> 1. Where possible, write each variable as the product of two factors, one of which is the largest perfect nth power.
> 2. Use the product rule to write the radical expression as the product of roots. Put all perfect nth powers under one radical sign.
> 3. Find the roots of any perfect powers.

EXAMPLE 5 **Simplifying Radicals with Variables in the Radicand**

Simplify: **a.** $\sqrt{x^{14}y^{15}}$ **b.** $\sqrt[3]{x^3y^{23}z^2}$

Solution

a. The largest perfect square factor of $x^{14}y^{15}$ is $x^{14}y^{14}$.

$$\sqrt{x^{14}y^{15}} = \sqrt{x^{14}\cdot y^{14}\cdot y} = \sqrt{x^{14}y^{14}}\cdot\sqrt{y}$$

Put all perfect square factors under one radical sign.

$$= (x^{14}y^{14})^{1/2}\sqrt{y}$$

You may eventually omit this step.

$$= x^7y^7\sqrt{y}$$

$(x^{14}y^{14})^{1/2} = (x^{14})^{1/2}(y^{14})^{1/2}$
$= x^7y^7$

b. The largest perfect cube factor of $x^3y^{23}z^2$ is x^3y^{21}.

$$\sqrt[3]{x^3y^{23}z^2} = \sqrt[3]{x^3\cdot y^{21}\cdot y^2\cdot z^2} = \sqrt[3]{x^3y^{21}}\cdot\sqrt[3]{y^2z^2}$$

Put all perfect cube factors under one radical sign.

$$= (x^3y^{21})^{1/3}\sqrt[3]{y^2z^2}$$

Write the perfect cube factor in exponential notation.

$$= xy^7\sqrt[3]{y^2z^2}$$

$(x^3y^{21})^{1/3} = (x^3)^{1/3}(y^{21})^{1/3}$
$= xy^7$

In our next example, the radicand contains both a numerical factor and variable factors. We will place all perfect nth powers (numbers and variables) under the same radical sign.

EXAMPLE 6 **Simplifying Radicals with Constant and Variable Factors**

Simplify: **a.** $\sqrt{48x^5y^6}$ **b.** $\sqrt[5]{64x^3y^7z^{26}}$

Solution

a. 16 is the largest perfect square factor of 48. The largest perfect square factor of x^5y^6 is x^4y^6.

$$\sqrt{48x^5y^6} = \sqrt{16\cdot 3\cdot x^4\cdot x\cdot y^6} = \sqrt{16x^4y^6}\cdot\sqrt{3x}$$

Put all perfect square factors under one radical sign.

$$= 4(x^4y^6)^{1/2}\sqrt{3x}$$

Use exponential notation on the variable perfect square factor.

$$= 4x^2y^3\sqrt{3x}$$

Simplify.

b. 32 is the largest perfect fifth-power factor of 64. ($2^5 = 32$). The largest perfect fifth-power factor of $x^3y^7z^{26}$ is y^5z^{25}.

$$\sqrt[5]{64x^3y^7z^{26}} = \sqrt[5]{32 \cdot 2 \cdot x^3 \cdot y^5 \cdot y^2 \cdot z^{25} \cdot z}$$

$$= \sqrt[5]{32y^5z^{25}} \cdot \sqrt[5]{2x^3y^2z} \qquad \text{Put all perfect fifth-power factors under one radical sign.}$$

$$= 2(y^5z^{25})^{1/5}\sqrt[5]{2x^3y^2z} \qquad \text{Use exponential notation on the variable perfect fifth-power factor.}$$

$$= 2yz^5\sqrt[5]{2x^3y^2z} \qquad \text{Simplify.} \qquad ■$$

3 Use the product rule to multiply and simplify radicals.

Multiplying and Simplifying Radicals

We have seen how to use the product rule when multiplying radicals with the same index. Sometimes after multiplying, we can simplify the resulting radical.

EXAMPLE 7 **Multiplying Radicals and then Simplifying**

Multiply and simplify:

a. $\sqrt{15}\sqrt{3}$ **b.** $7\sqrt[3]{4} \cdot 5\sqrt[3]{6}$ **c.** $\sqrt[3]{5x^4} \cdot \sqrt[3]{50x^{22}}$

d. $\sqrt[4]{8x^3y} \cdot \sqrt[4]{8x^{10}y^2}$

Solution

a. $\sqrt{15}\sqrt{3} = \sqrt{15 \cdot 3}$ Use the product rule.

$\qquad\qquad = \sqrt{45}$ Multiply.

$\qquad\qquad = \sqrt{9 \cdot 5}$ 9 is the largest perfect square factor of 45.

$\qquad\qquad = 3\sqrt{5}$ Simplify: $\sqrt{9 \cdot 5} = \sqrt{9}\sqrt{5} = 3\sqrt{5}$.

b. $7\sqrt[3]{4} \cdot 5\sqrt[3]{6} = 7 \cdot 5\sqrt[3]{4 \cdot 6}$ Use the product rule.

$\qquad\qquad = 35\sqrt[3]{24}$ Multiply.

$\qquad\qquad = 35\sqrt[3]{8 \cdot 3}$ 8 is the largest perfect cube factor of 24.

$\qquad\qquad = 35 \cdot 2\sqrt[3]{3}$ Simplify: $\sqrt[3]{8 \cdot 3} = \sqrt[3]{8} \cdot \sqrt[3]{3} = 2\sqrt[3]{3}$.

$\qquad\qquad = 70\sqrt[3]{3}$ Multiply.

c. $\sqrt[3]{5x^4} \cdot \sqrt[3]{50x^{22}} = \sqrt[3]{5x^4 \cdot 50x^{22}}$ Use the product rule.

$\qquad\qquad = \sqrt[3]{250x^{26}}$ Multiply.

$\qquad\qquad = \sqrt[3]{125 \cdot 2x^{24} \cdot x^2}$ 125 is the largest perfect cube factor of 250 $(5^3 = 125)$. The largest perfect cube factor of x^{26} is x^{24}.

$\qquad\qquad = \sqrt[3]{125x^{24}} \cdot \sqrt[3]{2x^2}$ Put all perfect cube factors under one radical sign.

$\qquad\qquad = 5(x^{24})^{1/3}\sqrt[3]{2x^2}$ Use exponential notation on the variable perfect cube factor.

$\qquad\qquad = 5x^8\sqrt[3]{2x^2}$ Simplify: $(x^{24})^{1/3} = x^{24 \cdot 1/3} = x^8$.

d. $\sqrt[4]{8x^3y} \cdot \sqrt[4]{8x^{10}y^2} = \sqrt[4]{8x^3y \cdot 8x^{10}y^2}$ Use the product rule.

$\qquad\qquad = \sqrt[4]{64x^{13}y^3}$ Multiply.

$\qquad\qquad = \sqrt[4]{16 \cdot 4x^{12} \cdot xy^3}$ 16 is the largest perfect fourth-power factor of 64 $(2^4 = 16)$. The largest perfect fourth-power factor of $x^{13}y^3$ is x^{12}.

$\qquad\qquad = \sqrt[4]{16x^{12}} \cdot \sqrt[4]{4xy^3}$ Put all perfect fourth-power factors under one radical sign.

$\qquad\qquad = 2(x^{12})^{1/4}\sqrt[4]{4xy^3}$ Use exponential notation on the variable perfect fourth-power factor.

$\qquad\qquad = 2x^3\sqrt[4]{4xy^3}$ Simplify: $(x^{12})^{1/4} = x^{12 \cdot 1/4} = x^3$.   ■

Yves Tanguy "Multiplication of the Arcs" 1954, oil on canvas, 40 × 60 in. (101.6 × 152.4 cm.) The Museum of Modern Art, New York. Mrs. Simon Guggenheim Fund. Photograph © 1997 The Museum of Modern Art, New York. © 1998 Estate of Yves Tanguy/Artists Rights Society (ARS), New York.

PROBLEM SET 6.3

Practice Problems

In this problem set, all variables represent positive real numbers. Use the product rule for radicals to multiply Problems 1–12.

1. $\sqrt{3}\sqrt{5}$ **2.** $\sqrt{7}\sqrt{5}$ **3.** $\sqrt[3]{2}\,\sqrt[3]{9}$ **4.** $\sqrt[3]{5}\sqrt[3]{4}$ **5.** $\sqrt[4]{11}\sqrt[4]{3}$

6. $\sqrt[5]{9}\sqrt[5]{3}$ **7.** $\sqrt{x+3}\sqrt{x-3}$ **8.** $\sqrt{x+6}\sqrt{x-6}$ **9.** $\sqrt{7}\sqrt{2xy}$ **10.** $\sqrt[3]{2x}\sqrt[3]{5xy}$

11. $\sqrt[7]{7x^2y}\ \sqrt[7]{11x^3y^2}$ **12.** $\sqrt[9]{12x^2y^3}\ \sqrt[9]{3x^4y^4}$

In Problems 13–56, write each radical in simplified form.

13. $\sqrt{20}$ **14.** $\sqrt{50}$ **15.** $\sqrt{80}$ **16.** $\sqrt{12}$ **17.** $\sqrt{250}$ **18.** $\sqrt{192}$

19. $7\sqrt{28}$ **20.** $3\sqrt{52}$ **21.** $2\sqrt{98}$ **22.** $\sqrt[3]{16}$ **23.** $\sqrt[3]{54}$ **24.** $\sqrt[3]{250}$

25. $\sqrt[5]{64}$ **26.** $\sqrt[4]{32}$ **27.** $6\sqrt[3]{16}$ **28.** $4\sqrt[3]{81}$ **29.** $\sqrt{x^7}$ **30.** $\sqrt{x^5}$

31. $\sqrt[3]{y^8}$ **32.** $\sqrt[3]{y^{14}}$ **33.** $\sqrt[3]{z^{16}}$ **34.** $\sqrt[7]{z^{22}}$ **35.** $\sqrt{x^8y^9}$ **36.** $\sqrt{x^6y^7}$

37. $\sqrt[3]{x^{14}y^3z}$ **38.** $\sqrt[3]{x^3y^{17}z^2}$ **39.** $\sqrt{12y}$ **40.** $\sqrt{18x}$ **41.** $\sqrt{48x^3}$ **42.** $\sqrt{40y^3}$

43. $\sqrt[3]{32x^{13}}$ **44.** $\sqrt[3]{16y^{17}}$ **45.** $\sqrt[3]{81x^8y^6}$ **46.** $\sqrt[3]{32x^9y^{17}}$ **47.** $\sqrt[4]{80x^{10}}$ **48.** $\sqrt[4]{96y^{11}}$

49. $\sqrt[5]{64x^6y^{17}}$ **50.** $\sqrt[5]{64x^{13}y^{22}}$ **51.** $\sqrt[3]{18x^{15}y^7z^2}$ **52.** $\sqrt[3]{81x^{50}y^{21}z^2}$ **53.** $6\sqrt[4]{32x^{19}y^8z^9}$

54. $3\sqrt[4]{32x^{21}y^9z^3}$ **55.** $\sqrt[3]{(x+y)^4}$ **56.** $\sqrt[4]{(x-y)^6}$

In Problems 57–76, multiply and simplify.

57. $\sqrt{3}\sqrt{6}$ **58.** $\sqrt{12}\sqrt{2}$ **59.** $(2\sqrt{5})(3\sqrt{20})$ **60.** $(3\sqrt{15})(5\sqrt{6})$

61. $\sqrt[3]{9}\sqrt[3]{6}$ **62.** $\sqrt[3]{5}\sqrt[3]{50}$ **63.** $\sqrt{5x^3}\ \sqrt{8x^2}$ **64.** $\sqrt{2x^7}\sqrt{12x^4}$

65. $\sqrt{6xy^4}\ \sqrt{2x^3y^7}$ **66.** $\sqrt{6x^3y}\sqrt{8x^5y^2}$ **67.** $\sqrt[3]{25x^4y^2}\ \sqrt[3]{5xy^{12}}$ **68.** $\sqrt[3]{6x^7y}\sqrt[3]{9x^4y^{12}}$

69. $\sqrt[4]{8x^2y^3z^6}\ \sqrt[4]{2x^4yz}$ **70.** $\sqrt[4]{4x^2y^3z^3}\ \sqrt[4]{8x^4yz^6}$ **71.** $\sqrt[5]{8x^4y^6}\ \sqrt[5]{8xy^7}$ **72.** $\sqrt[5]{8x^4y^3}\ \sqrt[5]{8xy^9}$

73. $\sqrt[3]{(x+2)^2}\ \sqrt[3]{(x+2)^5}$ **74.** $\sqrt[3]{(x-y)}\ \sqrt[3]{(x-y)^6}$ **75.** $2\sqrt[3]{x^4y^5}\ \sqrt[3]{8x^{15}y}$ **76.** $5\sqrt[3]{16x^5y^4}\ \sqrt[3]{2x^{17}y^5}$

Application Problems

77. The formula $r = 2\sqrt{5L}$ is used to estimate the speed of a car prior to an accident (r in miles per hour) based on the length of its skid marks (L in feet). Find the speed of a car that left skid marks 40 feet long, and write the answer in simplified radical form.

78. The time (t, in seconds) that it takes an object to fall a distance (d, in feet) is given by the formula $t = \sqrt{\dfrac{d}{16}}$.

Find how long it will take a ball dropped from the top of a building 320 feet tall to hit the ground and write the answer in simplified radical form.

True–False Critical Thinking Problems

79. Which one of the following is true?
  a. $2\sqrt{5} \cdot 6\sqrt{5} = 12\sqrt{5}$
  b. $\sqrt[3]{4} \cdot \sqrt[3]{4} = 4$
  c. $\sqrt{12} = 2\sqrt{6}$
  d. $\sqrt[5]{3^{25}} = 243$

80. Which one of the following is true?
  a. $\sqrt{15} \cdot \sqrt{6} = 9\sqrt{10}$
  b. $3\sqrt[3]{25} \cdot 8\sqrt[3]{5} = 120\sqrt[3]{5}$
  c. $\sqrt{x^2 + 6x + 9} = \sqrt{(x+3)^2} = x + 3$
  d. None of the above is true.

Technology Problems

In Problems 81–84, determine if each equation is correct for all $x \geq 0$ by graphing the function on each side of the given equality with your graphing utility. The graphs should be the same. If they are not, correct the right side of the equation and then use your graphing utility to verify the result.

81. $\sqrt{x^4} = x^2 \, (x \geq 0)$ **82.** $\sqrt{x^9} = x^3 \, (x \geq 0)$ **83.** $\sqrt{18x^2} = 9\sqrt{2}x \, (x \geq 0)$ **84.** $\sqrt[3]{2x}\sqrt[3]{4x^2} = 4x$

Writing in Mathematics

85. Consider each of the following simplifications:

$$\sqrt{x^{14}y^{15}} = x^7 y^7 \sqrt{y} \qquad \sqrt[3]{x^3 y^{23} z^2} = xy^7 \sqrt[3]{y^2 z^2} \qquad \sqrt[4]{x^6 y^{23}} = xy^5 \sqrt[4]{x^2 y^3}$$

All of the steps in the simplification process have been omitted. Write a rule for obtaining the simplified form for each radical. (Your rule should involve the division of the exponents on the variables in the radicand by the index.) Then explain why your rule works.

Critical Thinking Problems

86. Simplify: $\sqrt{x^2 - 18x + 81}$.

87. Find three pairs of values that satisfy the equation $\sqrt{a + b} = \sqrt{a} + \sqrt{b}$.

88. How does doubling a number affect its square root?

89. If a number is tripled, what happens to its square root?

90. What must be done to a number so that its cube root is doubled?

91. What must be done to a number so that its cube root is tripled?

92. Simplify: $\sqrt{\sqrt{\left(\sqrt{\left(\sqrt{\left(\sqrt{2^2}\right)^2}\right)^4}\right)^2}}$.

Review Problems

93. Simplify, expressing the final quotient with positive exponents: $\dfrac{8x^3 y^2}{-24x^7 y^{-5}}$.

94. Divide: $4y^3 - 22y^2 + 44y - 35$ by $2y - 5$.

95. The model

$$H = \frac{62.4Ns}{33,000}$$

describes the horsepower (H) generated by N cubic feet of water per minute flowing over a dam that is s feet high. Solve the formula for s and then determine the height of a dam that generates a horsepower of 468 when 1500 cubic feet of water are flowing per minute.

Solutions Manual   **Tutorial**   **Video 7**

SECTION 6.4

Dividing and Simplifying Radical Expressions

Objectives

1 Use the quotient rule to divide radicals.
2 Use the quotient rule to simplify radicals.
3 Rationalize denominators of radical expressions.

In the previous section, we used the product rule to multiply and simplify radical expressions. In this section, we concentrate on dividing and simplifying radicals.

▌ Use the quotient rule to divide radicals.

The Quotient Rule for Radicals

A rule for dividing radicals can be generalized by comparing

$$\sqrt{\frac{64}{4}} \qquad \text{and} \qquad \frac{\sqrt{64}}{\sqrt{4}}.$$

Note that

$$\sqrt{\frac{64}{4}} = \sqrt{16} = 4 \qquad \text{and} \qquad \frac{\sqrt{64}}{\sqrt{4}} = \frac{8}{2} = 4.$$

Since we obtain 4 in both situations, the original radical expressions must be equal. That is,

$$\sqrt{\frac{64}{4}} = \frac{\sqrt{64}}{\sqrt{4}}.$$

This result is a special case of the *quotient rule for radicals* that can be generalized as follows.

Dividing radicals: the quotient rule

Let M and N be real numbers, variables, or algebraic expressions. If the nth roots of M and N are real numbers, then

$$\frac{\sqrt[n]{M}}{\sqrt[n]{N}} = \sqrt[n]{\frac{M}{N}} \qquad \text{and} \qquad \sqrt[n]{\frac{M}{N}} = \frac{\sqrt[n]{M}}{\sqrt[n]{N}} \quad (N \neq 0).$$

The quotient of two nth roots is the nth root of their quotient.

As we did with the product rule, we can verify the quotient rule using rational exponents. Since $\sqrt[n]{M} = M^{1/n}$ and $\sqrt[n]{N} = N^{1/n}$,

$$\frac{\sqrt[n]{M}}{\sqrt[n]{N}} = \frac{M^{1/n}}{N^{1/n}} = \left(\frac{M}{N}\right)^{1/n} = \sqrt[n]{\frac{M}{N}}.$$

EXAMPLE 1   **Using the Quotient Rule for Radicals**

Divide, and simplify if possible:

a. $\dfrac{\sqrt{16}}{\sqrt{2}}$ **b.** $\dfrac{-8\sqrt[3]{32}}{2\sqrt[3]{2}}$ **c.** $\dfrac{\sqrt[3]{108y^5}}{\sqrt[3]{4}}$ **d.** $\dfrac{\sqrt{98xy^3}}{\sqrt{2x^{-3}y^{-2}}}$

Solution

a. $\dfrac{\sqrt{16}}{\sqrt{2}} = \sqrt{\dfrac{16}{2}}$ The quotient of two square roots is the square root of their quotient.

$\qquad\qquad = \sqrt{8}$ Divide.

$\qquad\qquad = \sqrt{4 \cdot 2}$ Factor the radicand. 4 is the largest perfect square factor of 8.

$\qquad\qquad = 2\sqrt{2}$ Simplify.

b. $\dfrac{-8\sqrt[3]{32}}{2\sqrt[3]{2}} = -4\sqrt[3]{\dfrac{32}{2}}$ The quotient of two cube roots is the cube root of their quotient.

$\qquad\qquad = -4\sqrt[3]{16}$ Divide.

$\qquad\qquad = -4\sqrt[3]{8 \cdot 2}$ Factor the radicand. 8 is the largest perfect cube factor of 16.

$\qquad\qquad = -4 \cdot 2\sqrt[3]{2}$ Simplify: $\sqrt[3]{8} = 2$.

$\qquad\qquad = -8\sqrt[3]{2}$ Multiply.

c. $\dfrac{\sqrt[3]{108y^5}}{\sqrt[3]{4}} = \sqrt[3]{\dfrac{108y^5}{4}}$ The quotient of two cube roots is the cube root of their quotient.

$= \sqrt[3]{27y^5}$ $\dfrac{108}{4} = 27$

$= \sqrt[3]{27y^3 \, y^2}$ Arrange factors that are perfect cubes.

$= 3y\sqrt[3]{y^2}$ Remove perfect cube factors from the radicand. $\sqrt[3]{y^3} = y$.

d. $\dfrac{\sqrt{98xy^3}}{\sqrt{2x^{-3}y^{-2}}} = \sqrt{\dfrac{98xy^3}{2x^{-3}y^{-2}}}$ The quotient of two square roots is the square root of their quotient.

$= \sqrt{49x^{1-(-3)}y^{3-(-2)}}$ Divide factors in the radicand, subtracting exponents on common bases.

$= \sqrt{49x^4y^5}$ Simplify.

$ \sqrt{49x^4y^4}\,\sqrt{y}$ Put all perfect square factors under one radical sign.

$= 7x^2y^2\sqrt{y}$ Simplify: $\sqrt{49} = 7$ and $\sqrt{x^4y^4} = (x^4y^4)^{1/2} = (x^4)^{1/2}(y^4)^{1/2} = x^2y^2.$ ∎

2 Use the quotient rule to simplify radicals.

Using the Quotient Rule to Simplify Radicals

A radical expression is simplified when three conditions are met.

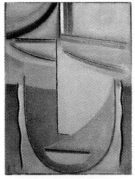

Alexej V. Jawlensky "Abstract Head" September 1927, oil on linen finish paper laid down on cardboard, 43.2 × 32.5 cm. Private Collection. Courtesy Alexej Von Jawlensky Archives, Locarno, Switzerland.

Simplifying radicals

Simplified radical form

1. When the radicand is factored to prime factors, the index of the radical is greater than any exponent that appears on a factor of the radicand.
2. There are no fractions in the radicand.
3. There are no radicals in the denominator.

In the previous section, we focused on the first condition for simplified radical form by removing as many factors from the radicand as possible. The second condition for simplified radical form states that there are no fractions in the radicand. The quotient rule for radicals can be applied in this situation if the denominator of the radicand is a perfect nth power.

EXAMPLE 2   **Simplifying Radicals by Removing Fractions from the Radicand**

Use the quotient rule to simplify:

a. $\sqrt[3]{\dfrac{16}{27}}$ **b.** $\sqrt{\dfrac{x^2}{25y^6}}$ **c.** $\sqrt[4]{\dfrac{7y^5}{x^{12}}}$

Solution

The nth root of a quotient is the quotient of the two nth roots: $\sqrt[n]{\dfrac{M}{N}} = \dfrac{\sqrt[n]{M}}{\sqrt[n]{N}}.$

a. $\sqrt[3]{\dfrac{16}{27}} = \dfrac{\sqrt[3]{16}}{\sqrt[3]{27}} = \dfrac{\sqrt[3]{8 \cdot 2}}{3} = \dfrac{\sqrt[3]{8}\sqrt[3]{2}}{3} = \dfrac{2\sqrt[3]{2}}{3}$

b. $\sqrt{\dfrac{x^2}{25y^6}} = \dfrac{\sqrt{x^2}}{\sqrt{25y^6}} = \dfrac{x}{5(y^6)^{1/2}} = \dfrac{x}{5y^3}$

c. $\sqrt[4]{\dfrac{7y^5}{x^{12}}} = \dfrac{\sqrt[4]{7y^5}}{\sqrt[4]{x^{12}}} = \dfrac{\sqrt[4]{7y^4y}}{\sqrt[4]{x^{12}}} = \dfrac{y\sqrt[4]{7y}}{(x^{12})^{1/4}} = \dfrac{y\sqrt[4]{7y}}{x^3}$ ∎

Each radical expression in Example 2 is simplified because each answer to the three parts of the problem contains no fractions in the radicand. This procedure would not be effective in simplifying

$$\sqrt{\dfrac{3}{7}} = \dfrac{\sqrt{3}}{\sqrt{7}}$$

because the radical expression on the right still has a radical in the denominator. Because simplified radical form requires that there be no radicals in a denominator, we turn now to a process for simplifying this expression, called rationalizing the denominator.

3 Rationalize denominators of radical expressions.

Rationalizing Denominators of Radical Expressions

When the denominator of a fraction contains a radicand that is not a perfect nth power, we simplify the expression by *rationalizing the denominator*. The process involves obtaining an equivalent expression without a radical in the denominator. We carry this out by multiplying by 1, the multiplicative identity, and therefore not changing the value of the given radical expression. This process is illustrated in Example 3.

EXAMPLE 3   **Rationalizing the Denominator**

Simplify: $\sqrt{\dfrac{3}{7}}$

Solution

$$\sqrt{\dfrac{3}{7}} = \dfrac{\sqrt{3}}{\sqrt{7}}$$

This expression is not simplified because a radical still appears in the denominator. We can remedy this situation by multiplying the denominator *and numerator* by $\sqrt{7}$. Because $\dfrac{\sqrt{7}}{\sqrt{7}} = 1$, the value of the irrational number $\dfrac{\sqrt{3}}{\sqrt{7}}$ will not be changed by multiplying by 1, the identity element of multiplication.

$$\dfrac{\sqrt{3}}{\sqrt{7}} = \dfrac{\sqrt{3}}{\sqrt{7}} \cdot \dfrac{\sqrt{7}}{\sqrt{7}} = \dfrac{\sqrt{3 \cdot 7}}{\sqrt{7^2}} = \dfrac{\sqrt{21}}{7}$$

$$\underset{\dfrac{\sqrt{7}}{\sqrt{7}} = 1 \quad \text{One is the multiplicative identity.}}{\uparrow}$$ ∎

The process of rationalizing the denominator can frequently be accomplished in a number of ways, although each procedure uses the identity of multiplication.

> **EXAMPLE 4** **Rationalizing a Denominator That Is a Square Root**

Simplify: $\sqrt{\dfrac{3}{8}}$

Solution

Method 1 $\quad \sqrt{\dfrac{3}{8}} = \dfrac{\sqrt{3}}{\sqrt{8}} = \dfrac{\sqrt{3}}{\sqrt{8}} \cdot \dfrac{\sqrt{8}}{\sqrt{8}} = \dfrac{\sqrt{24}}{\sqrt{64}} = \dfrac{\sqrt{4 \cdot 6}}{8} = \dfrac{2\sqrt{6}}{8} = \dfrac{\sqrt{6}}{4}$

$\qquad\qquad\qquad\qquad\qquad\qquad$ ↑
$\qquad\qquad\qquad\qquad\qquad$ Multiplicative identity $\qquad\qquad$ Simplify by dividing
$\qquad\qquad\qquad\qquad\qquad\qquad\qquad\qquad\qquad\qquad\qquad$ the numerator and
$\qquad\qquad\qquad\qquad\qquad\qquad\qquad\qquad\qquad\qquad\qquad$ denominator by 2.

Method 2 $\quad \sqrt{\dfrac{3}{8}} = \dfrac{\sqrt{3}}{\sqrt{8}} = \dfrac{\sqrt{3}}{\sqrt{8}} \cdot \dfrac{\sqrt{2}}{\sqrt{2}} = \dfrac{\sqrt{6}}{\sqrt{16}} = \dfrac{\sqrt{6}}{4}$

$\qquad\qquad\qquad\qquad\qquad\qquad$ ↑
$\qquad\qquad\qquad\qquad\qquad$ Multiplicative identity

Method 3 $\quad \sqrt{\dfrac{3}{8}} = \dfrac{\sqrt{3}}{\sqrt{8}} = \dfrac{\sqrt{3}}{\sqrt{(4)(2)}} = \dfrac{\sqrt{3}}{2\sqrt{2}} = \dfrac{\sqrt{3}}{2\sqrt{2}} \cdot \dfrac{\sqrt{2}}{\sqrt{2}} = \dfrac{\sqrt{6}}{2\sqrt{4}} = \dfrac{\sqrt{6}}{2 \cdot 2} = \dfrac{\sqrt{6}}{4}$

$\qquad\qquad\qquad\qquad\qquad\qquad\qquad\qquad\qquad\qquad$ ↑
$\qquad\qquad\qquad\qquad\qquad\qquad\qquad\qquad\qquad$ Multiplicative identity $\qquad\qquad$ ■

You may select the method with which you feel most comfortable. Many students find it easier to look for the *smallest* number that will produce a perfect nth power in the denominator. For them, method 2 proves to be the most convenient method even though method 3 ultimately involves multiplying by $\dfrac{\sqrt{2}}{\sqrt{2}}$.

When the denominator involves an nth root, we must multiply the numerator and denominator by a radical of index n that will produce a perfect nth power in the radicand of the denominator. For example, if a denominator contained $\sqrt[5]{2^3}$, we would multiply the numerator and denominator by $\sqrt[5]{2^2}$ because

$$\sqrt[5]{2^3} \cdot \sqrt[5]{2^2} = \sqrt[5]{2^5} = 2.$$

Examples 5 and 6 illustrate this idea.

> **iscover for yourself**
>
> Express the number from Example 5 as $\dfrac{\sqrt[3]{2}}{\sqrt[3]{9}}$. Describe what happens if you multiply the numerator and denominator by $\sqrt[3]{9}$. Does this produce a perfect cube in the denominator? Since $\sqrt[3]{27} = 3$, what should the numerator and denominator be multiplied by to rationalize the denominator?

> **EXAMPLE 5** **Rationalizing a Denominator That Is a Cube Root**

Simplify: $\sqrt[3]{\dfrac{2}{9}}$

Solution

$$\sqrt[3]{\dfrac{2}{9}} = \dfrac{\sqrt[3]{2}}{\sqrt[3]{9}} = \dfrac{\sqrt[3]{2}}{\sqrt[3]{3^2}} \cdot \dfrac{\sqrt[3]{3}}{\sqrt[3]{3}} = \dfrac{\sqrt[3]{6}}{\sqrt[3]{3^3}} = \dfrac{\sqrt[3]{6}}{3}$$

$\qquad\qquad\qquad\qquad\qquad$ ↑
$\qquad\qquad\qquad$ Multiplicative identity
$\qquad\qquad\qquad$ (We choose $\sqrt[3]{3}$ because the product
$\qquad\qquad\qquad$ of $\sqrt[3]{3^2}$ and $\sqrt[3]{3}$ is $\sqrt[3]{3^3}$, which is 3.) $\qquad$ ■

EXAMPLE 6 **Rationalizing a Denominator That Is a Fifth Root**

Simplify: $\dfrac{8}{3}\sqrt[5]{\dfrac{7}{16}}$

Solution

$$\frac{8}{3}\sqrt[5]{\frac{7}{16}} = \frac{8\sqrt[5]{7}}{3\sqrt[5]{16}} = \frac{8\sqrt[5]{7}}{3\sqrt[5]{2^4}} \cdot \frac{\sqrt[5]{2}}{\sqrt[5]{2}} = \frac{8\sqrt[5]{14}}{3\sqrt[5]{2^5}} = \frac{8\sqrt[5]{14}}{3(2)} = \frac{8\sqrt[5]{14}}{6} = \frac{4\sqrt[5]{14}}{3}$$

Multiplicative identity
(We choose $\sqrt[5]{2}$ because the product of $\sqrt[5]{2^4}$ and $\sqrt[5]{2}$ is $\sqrt[5]{2^5}$, which is 2.)

We summarize as follows.

Rationalizing the denominator

To rationalize a denominator, multiply both the numerator and denominator by a radical factor that creates a perfect nth power in the radicand of the denominator.

Simplifying Radicals with Variables in the Denominator

We can apply the process of rationalizing the denominator to situations where variables appear in the radicand. Let's consider a variety of examples to see how this is accomplished.

EXAMPLE 7 **Rationalizing a Variable Denominator**

Simplify: $\sqrt{\dfrac{3a}{5b}}$

Solution

$$\sqrt{\frac{3a}{5b}} = \frac{\sqrt{3a}}{\sqrt{5b}} = \frac{\sqrt{3a}}{\sqrt{5b}} \cdot \frac{\sqrt{5b}}{\sqrt{5b}} = \frac{\sqrt{15ab}}{\sqrt{25b^2}} = \frac{\sqrt{15ab}}{5b}$$

Multiplicative identity   Perfect square radicand

EXAMPLE 8 **Rationalizing a Variable Denominator**

Simplify: $\sqrt{\dfrac{7}{8x^3}}$

Solution

Method 1 First simplify and then rationalize the denominator.

$$\sqrt{\frac{7}{8x^3}} = \frac{\sqrt{7}}{\sqrt{8x^3}}$$

The square root of a quotient is the quotient of the two square roots.

$$= \frac{\sqrt{7}}{\sqrt{4 \cdot 2x^2 x}}$$

The perfect square factors are 4 and x^2.

$$= \frac{\sqrt{7}}{\sqrt{4x^2}\sqrt{2x}}$$

Put perfect square factors under one radical sign.

$$= \frac{\sqrt{7}}{2x\sqrt{2x}}$$

$\sqrt{4x^2} = 2x$; simplification is not complete because there is a radical in the denominator.

$$= \frac{\sqrt{7}}{2x\sqrt{2x}} \cdot \frac{\sqrt{2x}}{\sqrt{2x}}$$

Multiply the numerator and denominator by $\sqrt{2x}$. We are now multiplying by 1.

$$= \frac{\sqrt{14x}}{2x\sqrt{4x^2}}$$

Apply the product rule in the numerator and denominator.

$$= \frac{\sqrt{14x}}{2x(2x)}$$

$\sqrt{4x^2} = 2x$.

$$= \frac{\sqrt{14x}}{4x^2}$$

Complete the multiplication in the denominator.

Method 2 First rationalize the denominator and then simplify.

$$\sqrt{\frac{7}{8x^3}} = \frac{\sqrt{7}}{\sqrt{8x^3}}$$

The square root of a quotient is the quotient of the two square roots.

$$= \frac{\sqrt{7}}{\sqrt{8x^3}} \cdot \frac{\sqrt{2x}}{\sqrt{2x}}$$

Multiply the numerator and denominator by $\sqrt{2x}$ to produce a perfect square in the denominator. We are multiplying by 1.

$$= \frac{\sqrt{14x}}{\sqrt{16x^4}}$$

Apply the product rule in the numerator and denominator.

$$= \frac{\sqrt{14x}}{4x^2}$$

Simplify: $\sqrt{16x^4} = 4(x^4)^{1/2} = 4x^2$. ∎

EXAMPLE 9 **Rationalizing a Variable Denominator**

Simplify: $\sqrt{\dfrac{32x^3 y}{27z^5}}$

Solution

We will begin by simplifying, then we will rationalize the denominator.

$$\sqrt{\frac{32x^3y}{27z^5}} = \frac{\sqrt{32x^3y}}{\sqrt{27z^5}}$$ The square root of a quotient is the quotient of the two square roots.

$$= \frac{\sqrt{16 \cdot 2x^2xy}}{\sqrt{9 \cdot 3z^4z}}$$ The perfect square factors are $16, x^2, 9,$ and z^4.

$$= \frac{\sqrt{16x^2}\sqrt{2xy}}{\sqrt{9z^4}\sqrt{3z}}$$ Put perfect square factors under separate radical signs.

$$= \frac{4x\sqrt{2xy}}{3z^2\sqrt{3z}}$$ Simplify the perfect square radicals. Simplification is not complete with a radical in the denominator.

$$= \frac{4x\sqrt{2xy}}{3z^2\sqrt{3z}} \cdot \frac{\sqrt{3z}}{\sqrt{3z}}$$ Rationalize the denominator by multiplying the numerator and denominator by $\sqrt{3z}$.

$$= \frac{4x\sqrt{6xyz}}{3z^2\sqrt{9z^2}}$$ Apply the product rule in the numerator and denominator.

$$= \frac{4x\sqrt{6xyz}}{3z^2(3z)}$$ $\sqrt{9z^2} = 3z$

$$= \frac{4x\sqrt{6xyz}}{9z^3}$$ Complete the multiplication in the denominator. ■

EXAMPLE 10 **Simplifying a Cube Root with a Variable Denominator**

Simplify: $\sqrt[3]{\dfrac{7}{4xy^2}}$

Solution

$$\sqrt[3]{\frac{7}{4xy^2}} = \frac{\sqrt[3]{7}}{\sqrt[3]{4xy^2}} \cdot \frac{\sqrt[3]{2x^2y}}{\sqrt[3]{2x^2y}}$$ This multiplication will produce a perfect cube in the radicand in the denominator. The idea is to create factors in the denominator's radicand that are perfect cubes.

$$= \frac{\sqrt[3]{14x^2y}}{\sqrt[3]{8x^3y^3}}$$ Apply the product rule in the numerator and denominator.

$$= \frac{\sqrt[3]{14x^2y}}{2xy}$$ Simplify: $\sqrt[3]{8x^3y^3} = 2(x^3y^3)^{1/3} = 2xy.$ ■

PROBLEM SET 6.4

Practice Problems

In this problem set, all variables represent positive real numbers. In Problems 1–18, use the quotient rule to divide, and simplify if possible.

1. $\dfrac{\sqrt{40}}{\sqrt{5}}$

2. $\dfrac{\sqrt{200}}{\sqrt{10}}$

3. $\dfrac{\sqrt[3]{48}}{\sqrt[3]{3}}$

4. $\dfrac{\sqrt[3]{108}}{\sqrt[3]{2}}$

5. $\dfrac{\sqrt{54x^3}}{\sqrt{6x}}$

6. $\dfrac{\sqrt{72y^8}}{\sqrt{2y^4}}$

7. $\dfrac{-10\sqrt[3]{15}}{-2\sqrt[3]{5}}$

8. $\dfrac{-20\sqrt[3]{27}}{\frac{1}{2}\sqrt[3]{3}}$

9. $\dfrac{-12\sqrt[4]{8}}{6\sqrt[4]{2}}$

10. $\dfrac{-\frac{2}{3}\sqrt[4]{15}}{-\frac{1}{6}\sqrt[4]{5}}$

11. $\dfrac{\sqrt{200x^3}}{\sqrt{10x^{-1}}}$

12. $\dfrac{\sqrt{54x}}{\sqrt{6x^{-5}}}$

13. $\dfrac{\sqrt[3]{108x^4y^5}}{\sqrt[3]{2xy^3}}$

14. $\dfrac{\sqrt[4]{64x^7y^3}}{\sqrt[4]{2x^2y}}$

15. $\dfrac{\sqrt{98x^2y}}{\sqrt{2x^{-3}y}}$

16. $\dfrac{\sqrt[3]{32x^2y}}{\sqrt[3]{2x^{-4}y^{-4}}}$

17. $\dfrac{\sqrt{x^2-y^2}}{\sqrt{x-y}}$

18. $\dfrac{\sqrt{x^3-y^3}}{\sqrt{x-y}}$

Use the quotient rule to simplify Problems 19–34.

19. $\sqrt{\dfrac{11}{4}}$

20. $\sqrt{\dfrac{19}{25}}$

21. $\sqrt[3]{\dfrac{19}{27}}$

22. $\sqrt[3]{\dfrac{11}{64}}$

23. $\sqrt{\dfrac{x^2}{36y^8}}$

24. $\sqrt{\dfrac{x^2}{144y^{12}}}$

25. $\sqrt{\dfrac{8x^3}{25y^6}}$

26. $\sqrt{\dfrac{50x^3}{81y^8}}$

27. $\sqrt[3]{\dfrac{x^4}{8y^3}}$

28. $\sqrt[3]{\dfrac{x^5}{125y^3}}$

29. $\sqrt[3]{\dfrac{75x^8}{27y^{12}}}$

30. $\sqrt[3]{\dfrac{54x^8}{8y^{15}}}$

31. $\sqrt[4]{\dfrac{9y^6}{x^8}}$

32. $\sqrt[4]{\dfrac{13y^7}{x^{12}}}$

33. $\sqrt[5]{\dfrac{64x^{13}}{y^{20}}}$

34. $\sqrt[5]{\dfrac{64x^{14}}{y^{15}}}$

Simplify Problems 35–54 by rationalizing the denominator.

35. $\dfrac{1}{\sqrt{2}}$

36. $\dfrac{1}{\sqrt{3}}$

37. $\sqrt{\dfrac{7}{10}}$

38. $\sqrt{\dfrac{2}{9}}$

39. $\sqrt{\dfrac{5}{8}}$

40. $\sqrt{\dfrac{7}{8}}$

41. $\dfrac{1}{\sqrt[3]{2}}$

42. $\dfrac{1}{\sqrt[3]{3}}$

43. $\dfrac{6}{\sqrt[3]{4}}$

44. $\dfrac{10}{\sqrt[3]{5}}$

45. $\sqrt[3]{\dfrac{2}{3}}$

46. $\sqrt[3]{\dfrac{3}{4}}$

47. $\sqrt[3]{\dfrac{2}{9}}$

48. $\sqrt[3]{\dfrac{16}{9}}$

49. $\dfrac{1}{\sqrt[4]{3}}$

50. $\dfrac{1}{\sqrt[4]{2}}$

51. $\sqrt[4]{\dfrac{5}{2}}$

52. $\sqrt[4]{\dfrac{4}{3}}$

53. $\sqrt{1+\dfrac{1}{2}}$

54. $\sqrt{2-\dfrac{1}{3}}$

In Problems 55–84, the radical expressions contain variables in the denominator's radicand. Simplify by rationalizing the denominator.

55. $\dfrac{3}{\sqrt{x}}$

56. $\dfrac{7}{\sqrt{x}}$

57. $\dfrac{5}{\sqrt{3x}}$

58. $\dfrac{7}{\sqrt{2x}}$

59. $\sqrt{\dfrac{3}{5x}}$

60. $\sqrt{\dfrac{7}{11x}}$

61. $\sqrt{\dfrac{x}{7y}}$

62. $\sqrt{\dfrac{x}{13y}}$

63. $\sqrt{\dfrac{3x}{7y}}$

64. $\sqrt{\dfrac{4x}{5y}}$

65. $\dfrac{4}{\sqrt[3]{x}}$

66. $\dfrac{7}{\sqrt[3]{y}}$

67. $\sqrt[3]{\dfrac{2}{x^2}}$

68. $\sqrt[3]{\dfrac{5}{y^2}}$

69. $\dfrac{5}{\sqrt{12y}}$

70. $\dfrac{13}{\sqrt{18x}}$

71. $\dfrac{7}{\sqrt[3]{2x^2}}$

72. $\dfrac{10}{\sqrt[3]{4y^2}}$

73. $\dfrac{3}{\sqrt[4]{x}}$

74. $\dfrac{5}{\sqrt[5]{y}}$

75. $\dfrac{7}{\sqrt[3]{y^2}}$

76. $\dfrac{11}{\sqrt[4]{x^2}}$

77. $\sqrt[3]{\dfrac{7}{2x}}$

78. $\sqrt[3]{\dfrac{5}{2y}}$

79. $\dfrac{2}{\sqrt[3]{16x^2y}}$

80. $\dfrac{4}{\sqrt[3]{16ab^2}}$

81. $\sqrt[5]{\dfrac{3}{8x^3}}$

82. $\sqrt[5]{\dfrac{7}{16y^2}}$

83. $\sqrt[4]{\dfrac{5}{3x}}$

84. $\sqrt[4]{\dfrac{2a^3}{4b^2}}$

In Problems 85–100, the radical expressions contain variables in the denominator's radicand. Simplify either by first simplifying the radicals and then rationalizing the denominator or by first rationalizing the denominator and then simplifying the resulting radical expression.

85. $\sqrt{\dfrac{3}{32y^2}}$

86. $\sqrt{\dfrac{5}{24x^2}}$

87. $\sqrt{\dfrac{11}{20y^3}}$

88. $\sqrt{\dfrac{13}{12x^3}}$

89. $\sqrt{\dfrac{3x}{32y^3}}$

90. $\sqrt{\dfrac{5y}{24x^3}}$

91. $\dfrac{3x}{\sqrt{8x^3}}$

92. $\dfrac{5y}{\sqrt{200y^3}}$

93. $\dfrac{7x^2}{\sqrt[3]{2x^5}}$

94. $\dfrac{10y^7}{\sqrt[3]{4y^{11}}}$

95. $\dfrac{7}{\sqrt[5]{32x^7y^4}}$

96. $\dfrac{-6}{\sqrt[5]{64x^{11}y^3}}$

97. $\dfrac{\sqrt{20x^2y}}{\sqrt{5x^3y^3}}$

98. $\dfrac{\sqrt{24x^2y^3}}{\sqrt{6xy^6}}$

99. $\sqrt[3]{\dfrac{5}{36x^4y}}$

100. $\sqrt[3]{\dfrac{7}{100x^2y^5}}$

True–False Critical Thinking Problems

101. Which one of the following is true?

a. $\dfrac{6}{\sqrt{2}} = \sqrt{3}$

b. To rationalize the denominator of $\dfrac{1}{\sqrt[12]{x^7}}$, multiply the numerator and denominator by $\sqrt[12]{x^5}$.

c. $\sqrt{\dfrac{x}{y}} = \dfrac{\sqrt{xy}}{y^2}$

d. $\sqrt[3]{\dfrac{x}{y}} = \dfrac{\sqrt[3]{xy}}{y}$

102. Which one of the following is true?

a. $\dfrac{\sqrt{10}}{\sqrt{5}} = 2$

b. $\sqrt[4]{\dfrac{1}{3}} = \dfrac{\sqrt[4]{27}}{3}$

c. $\sqrt[3]{\dfrac{5}{9}} = \dfrac{\sqrt[3]{45}}{9}$

d. $\sqrt[5]{\dfrac{x}{y}} = \dfrac{\sqrt[5]{xy}}{y}$

Technology Problems

In Problems 103–105, determine if each simplification is correct by graphing the function on each side of the equality with your graphing utility. Shown in the same viewing rectangle, the graphs should be the same. If they are not, correct the right side of the equation. Then use your graphing utility to verify the result.

103. $\sqrt{\dfrac{2}{x}} = \dfrac{\sqrt{2x}}{x}, x > 0$

104. $\sqrt[3]{\dfrac{1}{x}} = \dfrac{\sqrt[3]{x}}{x}$

105. $\dfrac{x}{\sqrt[4]{2}} = \dfrac{\sqrt[4]{2x}}{2}$

106. Select any four problems from Problems 35–54. Use a calculator to provide numerical support to your work. Begin by finding a decimal approximation for the given radical expression. Then find a decimal approximation for your simplified, rationalized-denominator answer. The two decimal approximations should be the same.

Writing in Mathematics

107. We can eliminate the radical in the denominator of $\dfrac{7}{\sqrt{2}}$ by squaring the real number, obtaining $\left(\dfrac{7}{\sqrt{2}}\right)^2 = \dfrac{49}{2}$. Is this the same as rationalizing the denominator? Explain.

Critical Thinking Problems

108. Rationalize the denominator: $\sqrt{y^2 + \dfrac{1}{y^3}}$.

109. Rationalize the denominator: $\dfrac{5}{\sqrt{2x - 3y}}$.

110. Rationalize the *numerator:* $\sqrt[3]{\dfrac{5y}{7x^2}}$.

Review Problems

111. Simplify: $\dfrac{\dfrac{1}{y^2} - \dfrac{1}{100}}{\dfrac{1}{y} - \dfrac{1}{10}}$.

112. Multiply: $(2x - 3)(4x^2 - 5x - 1)$.

113. Solve: $\dfrac{4x}{x + 1} + \dfrac{5}{x} = 4$.

S E C T I O N   6 . 5

Solutions   Tutorial   Video
Manual   7

Further Operations with Radicals

Objectives

1 Add and subtract radical expressions.
2 Use the distributive property to multiply radicals.
3 Use the FOIL method to multiply radicals.
4 Use polynomial special products to multiply radicals.
5 Rationalize denominators that contain two terms.

In this section, we complete our work with operations involving radicals.

1 Add and subtract radical expressions.

Adding and Subtracting Radical Expressions

In our earlier work with polynomials, the distributive property enabled us to combine like terms. For example,

$$7x^2 + 5x^2 = (7 + 5)x^2 = 12x^2.$$

We can use the same idea to add certain radical expressions. For example,

$$7\sqrt{2} + 5\sqrt{2} = (7 + 5)\sqrt{2} = 12\sqrt{2}$$ When 5 square roots of 2 are added to 7 square roots of 2, the result is 12 square roots of 2.

Just as $7x^2$ and $5x^2$ are like terms, we call $7\sqrt{2}$ and $5\sqrt{2}$ *like* (or similar) *radicals*.

> **Definition of like (similar) radicals**
>
> Radicals are said to be *like radicals* (or *similar radicals*) if they have the same index and the same radicand.

EXAMPLE 1   **Adding and Subtracting Like Radicals**

Add and subtract as indicated:

a. $6\sqrt{7} + 11\sqrt{7}$ **b.** $5\sqrt{x} - 8\sqrt{x}$ **c.** $\sqrt[3]{7x} + 12\sqrt[3]{7x}$
d. $9\sqrt[3]{7} - 6x\sqrt[3]{7} + 12\sqrt[3]{7}$ **e.** $11\sqrt[4]{3x} - 10\sqrt[4]{3x} + 2\sqrt[3]{3x}$

Solution

a. $6\sqrt{7} + 11\sqrt{7} = (6 + 11)\sqrt{7}$ Use the distributive property. (Try to do this step mentally.)

$$= 17\sqrt{7}$$

b. $5\sqrt{x} - 8\sqrt{x} = (5 - 8)\sqrt{x}$ Use the distributive property.

$$= -3\sqrt{x}$$

c. $\sqrt[3]{7x} + 12\sqrt[3]{7x} = (1 + 12)\sqrt[3]{7x}$ Use the distributive property.

$$= 13\sqrt[3]{7x}$$

d. $9\sqrt[3]{7} - 6x\sqrt[3]{7} + 12\sqrt[3]{7} = (9 - 6x + 12)\sqrt[3]{7}$ Use the distributive property.

$$= (21 - 6x)\sqrt[3]{7}$$ Be sure to include parentheses in this result.

tudy tip

Only radicals with the same index and the same radicand can
be combined.

e. $11\sqrt[4]{3x} - 10\sqrt[4]{3x} + 2\sqrt[3]{3x}$

$= (11 - 10)\sqrt[4]{3x} + 2\sqrt[3]{3x}$ Use the distributive property on the like radicals.

$= \sqrt[4]{3x} + 2\sqrt[3]{3x}$ These are not like radicals because they have different indices. They

(or $2\sqrt[3]{3x} + \sqrt[4]{3x}$)   cannot be combined. ■

Although at first glance many radicals with the same index might not appear to have identical radicands, simplification often reveals like radicals and
enables us to add or subtract. Let's see how this works.

EXAMPLE 2   **Simplifying and Combining Like Radicals**

Simplify, and combine like radicals, if possible.

a. $\sqrt{2} + \sqrt{8}$ **b.** $7\sqrt{18} + 5\sqrt{8} - 6\sqrt{2}$

c. $3\sqrt{2x^3} - \sqrt{25x^2} + 4\sqrt{18x^3}$ **d.** $11\sqrt[3]{16x^4} - 5\sqrt[3]{2x}$

Solution

a. $\sqrt{2} + \sqrt{8}$

$= \sqrt{2} + \sqrt{4 \cdot 2}$ Simplify $\sqrt{8}$.

$= \sqrt{2} + 2\sqrt{2}$

$= (1 + 2)\sqrt{2}$ Use the distributive property.

$= 3\sqrt{2}$

b. $7\sqrt{18} + 5\sqrt{8} - 6\sqrt{2}$

$= 7\sqrt{9 \cdot 2} + 5\sqrt{4 \cdot 2} - 6\sqrt{2}$ Begin simplifying $\sqrt{18}$ and $\sqrt{8}$.

$= 7 \cdot 3\sqrt{2} + 5 \cdot 2\sqrt{2} - 6\sqrt{2}$ $\sqrt{18} = \sqrt{9 \cdot 2} = 3\sqrt{2}$ and $\sqrt{8} = \sqrt{4 \cdot 2} = 2\sqrt{2}$.

$= 21\sqrt{2} + 10\sqrt{2} - 6\sqrt{2}$ Multiply.

$= (21 + 10 - 6)\sqrt{2}$ Use the distributive property.

$= 25\sqrt{2}$

c. $3\sqrt{2x^3} - \sqrt{25x^2} + 4\sqrt{18x^3}$

$= 3\sqrt{2x^2 \cdot x} - 5x + 4\sqrt{9 \cdot 2x^2 x}$ Simplify each radical.

$= 3x\sqrt{2x} - 5x + 4 \cdot 3x\sqrt{2x}$ Complete the simplifications.

$= 3x\sqrt{2x} - 5x + 12x\sqrt{2x}$ Multiply.

$= 3x\sqrt{2x} + 12x\sqrt{2x} - 5x$ Rearrange terms so that like radicals are next to one another.

$= (3x + 12x)\sqrt{2x} - 5x$ Use the distributive property.

$= 15x\sqrt{2x} - 5x$ Add like radicals.

d. $11\sqrt[3]{16x^4} - 5\sqrt[3]{2x}$

$= 11\sqrt[3]{8x^3 \cdot 2x} - 5\sqrt[3]{2x}$ Begin by simplifying the first radical.

$= 11\sqrt[3]{8x^3}\,\sqrt[3]{2x} - 5\sqrt[3]{2x}$ Put the perfect cube factor under one radical sign.

$= 11 \cdot 2x \cdot \sqrt[3]{2x} - 5\sqrt[3]{2x}$ Take the cube root.

$= 22x\sqrt[3]{2x} - 5\sqrt[3]{2x}$ Multiply.

$= (22x - 5)\sqrt[3]{2x}$ Use the distributive property. ■

Let's now summarize our work in Example 2.

A procedure for adding and subtracting radicals

1. Simplify all terms with radicals.
2. Where possible, combine like terms.

In our next example, we first rationalize the denominators and then combine like radicals.

EXAMPLE 3   **Subtracting Radicals by First Rationalizing Denominators**

Simplify: $12\sqrt{\dfrac{1}{3}} - 4\sqrt{\dfrac{1}{12}}$

Solution

First we rationalize the denominators.

$$\sqrt{\frac{1}{3}} = \frac{1}{\sqrt{3}} \cdot \frac{\sqrt{3}}{\sqrt{3}} = \frac{\sqrt{3}}{\sqrt{9}} = \frac{\sqrt{3}}{3}$$

$$\sqrt{\frac{1}{12}} = \frac{1}{\sqrt{12}} \cdot \frac{\sqrt{3}}{\sqrt{3}} = \frac{\sqrt{3}}{\sqrt{36}} = \frac{\sqrt{3}}{6}$$

Now we rewrite the problem in terms of rationalized denominators.

$$12\sqrt{\frac{1}{3}} - 4\sqrt{\frac{1}{12}} = 12 \cdot \frac{\sqrt{3}}{3} - 4 \cdot \frac{\sqrt{3}}{6}$$ Replace radicals by the expressions with rationalized denominators.

$$= 4\sqrt{3} - \frac{2}{3}\sqrt{3}$$ $\dfrac{12}{3} = 4$ and $\dfrac{4}{6} = \dfrac{2}{3}$

$$= \left(4 - \frac{2}{3}\right)\sqrt{3}$$ Use the distributive property.

$$= \left(\frac{12}{3} - \frac{2}{3}\right)\sqrt{3}$$ $4 = \dfrac{12}{3}$

$$= \frac{10\sqrt{3}}{3}$$ Subtract.

■

More Multiplication with Radicals

To multiply a radical expression containing one term and a radical expression containing two or more terms, we use the distributive property.

2 Use the distributive property to multiply radicals.

EXAMPLE 4   **Multiplying Radicals Using the Distributive Property**

Multiply:

a. $5\sqrt{3}(4\sqrt{6} + 2\sqrt{15})$ **b.** $\sqrt{2x}(3y\sqrt{12xy} - 4\sqrt{48xy^3})$
c. $4\sqrt[3]{5}(2\sqrt[3]{50} + 3\sqrt[3]{2})$

Solution

a. $5\sqrt{3}(4\sqrt{6} + 2\sqrt{15})$

$= (5\sqrt{3})(4\sqrt{6}) + (5\sqrt{3})(2\sqrt{15})$ Use the distributive property.

$= 20\sqrt{18} + 10\sqrt{45}$ Apply the product rule of radicals.

$= 20\sqrt{(9)(2)} + 10\sqrt{(9)(5)}$ Find the perfect square factors.

$= 20 \cdot 3\sqrt{2} + 10 \cdot 3\sqrt{5}$ Take the square root of 9.

$= 60\sqrt{2} + 30\sqrt{5}$ Multiply. There are not like radicals and cannot be combined.

b. $\sqrt{2x}(3y\sqrt{12xy} - 4\sqrt{48xy^3})$

$= (\sqrt{2x})(3y\sqrt{12xy}) - (\sqrt{2x})(4\sqrt{48xy^3})$ Use the distributive property.

$= 3y\sqrt{24x^2y} \quad 4\sqrt{96x^2y^3}$ Apply the product rule of radicals.

$= 3y\sqrt{(4)(6)x^2y} - 4\sqrt{(16)(6)x^2y^2y}$ Find the perfect square factors.

$= 3y(2x)\sqrt{6y} - 4(4xy)\sqrt{6y}$ Simplify: $\sqrt{4x^2}\sqrt{6y} = 2x\sqrt{6y}$ and $\sqrt{16x^2y^2}\sqrt{6y} = 4xy\sqrt{6y}$.

$= 6xy\sqrt{6y} - 16xy\sqrt{6y}$ Multiply.

$= (6xy - 16xy)\sqrt{6y}$ Use the distributive property.

$= -10xy\sqrt{6y}$ Combine like terms.

c. $4\sqrt[3]{5}(2\sqrt[3]{50} + 3\sqrt[3]{2})$

$= (4\sqrt[3]{5})(2\sqrt[3]{50}) + (4\sqrt[3]{5})(3\sqrt[3]{2})$ Use the distributive property.

$= 8\sqrt[3]{250} + 12\sqrt[3]{10}$ Apply the product rule of radicals.

$= 8\sqrt[3]{(125)(2)} + 12\sqrt[3]{10}$ Find the perfect cube factors.

$= 8(5)\sqrt[3]{2} + 12\sqrt[3]{10}$ Simplify.

$= 40\sqrt[3]{2} + 12\sqrt[3]{10}$ Multiply. ∎

3 Use the FOIL method to multiply radicals.

     Expressions in which both factors contain two radical terms can be multiplied using the FOIL method. Let's consider an example to see how this works.

EXAMPLE 5   **Multiplying Radicals Using FOIL**

Multiply:

a. $(5\sqrt{2} + 2\sqrt{3})(4\sqrt{2} - 3\sqrt{3})$

b. $(\sqrt{x} + \sqrt{7})(\sqrt{y} + \sqrt{7})$

Solution

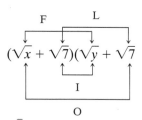

a. $(5\sqrt{2} + 2\sqrt{3})(4\sqrt{2} - 3\sqrt{3})$

$$
\begin{array}{cccc}
\text{F} & \text{O} & \text{I} & \text{L}
\end{array}
$$
$$= 20\sqrt{4} - 15\sqrt{6} + 8\sqrt{6} - 6\sqrt{9} \qquad \text{Use the FOIL method.}$$
$$= 20(2) - 15\sqrt{6} + 8\sqrt{6} - 6(3) \qquad \sqrt{4} = 2 \text{ and } \sqrt{9} = 3$$
$$= 40 - 15\sqrt{6} + 8\sqrt{6} - 18 \qquad \text{Multiply.}$$
$$= 40 - 18 + 8\sqrt{6} - 15\sqrt{6} \qquad \text{Rearrange terms with like terms next to one another. You will probably do this step mentally.}$$
$$= 22 + (8 - 15)\sqrt{6} \qquad \text{Combine numerical terms and use the distributive property.}$$
$$= 22 - 7\sqrt{6}$$

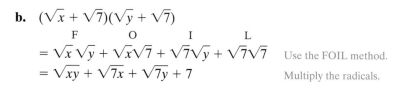

b. $(\sqrt{x} + \sqrt{7})(\sqrt{y} + \sqrt{7})$

$$
\begin{array}{cccc}
\text{F} & \text{O} & \text{I} & \text{L}
\end{array}
$$
$$= \sqrt{x}\,\sqrt{y} + \sqrt{x}\sqrt{7} + \sqrt{7}\sqrt{y} + \sqrt{7}\sqrt{7} \qquad \text{Use the FOIL method.}$$
$$= \sqrt{xy} + \sqrt{7x} + \sqrt{7y} + 7 \qquad \text{Multiply the radicals.} \qquad \blacksquare$$

4 Use polynomial special products to multiply radicals.

Some radicals can be multiplied using the special products for multiplying polynomials.

EXAMPLE 6   **Using Special Products to Multiply Radicals**

Multiply:

a. $(\sqrt{3} + \sqrt{7})^2$ **b.** $(\sqrt{7} + \sqrt{3})(\sqrt{7} - \sqrt{3})$

c. $(\sqrt{x} + \sqrt{y})(\sqrt{x} - \sqrt{y})$

Solution

a.
$$
\begin{array}{ccccccccc}
(A & + & B)^2 & = & A^2 & + & 2 \cdot A \cdot B & + & B^2 \\
\downarrow & & \downarrow & & \downarrow & & \downarrow\;\downarrow\;\downarrow & & \downarrow
\end{array}
$$
$$(\sqrt{3} + \sqrt{7})^2 = (\sqrt{3})^2 + 2 \cdot \sqrt{3} \cdot \sqrt{7} + (\sqrt{7})^2 \qquad \text{Use the special product for } (A + B)^2.$$
$$= 3 + 2\sqrt{21} + 7 \qquad \text{Multiply the radicals.}$$
$$= 10 + 2\sqrt{21}$$

b.
$$
\begin{array}{ccccccc}
(A & + & B) & (A & - & B) & = & A^2 & - & B^2 \\
\downarrow & & \downarrow & \downarrow & & \downarrow & & \downarrow & & \downarrow
\end{array}
$$
$$(\sqrt{7} + \sqrt{3})(\sqrt{7} - \sqrt{3}) = (\sqrt{7})^2 - (\sqrt{3})^2 \qquad \text{Use the special product for } (A + B)(A - B).$$
$$= 7 - 3 \qquad \text{Multiply the radicals.}$$
$$= 4$$

---

c.

$$\underset{\downarrow}{(A} + \underset{\downarrow}{B)} \; \underset{\downarrow}{(A} - \underset{\downarrow}{B)} = \underset{\downarrow}{A^2} - \underset{\downarrow}{B^2}$$

$$(\sqrt{x} + \sqrt{y})(\sqrt{x} - \sqrt{y}) = (\sqrt{x})^2 - (\sqrt{y})^2 \quad \text{Use the special product for } (A + B)(A - B).$$

$$= x - y \quad \text{Simplify.} \qquad \blacksquare$$

5 Rationalize denominators that contain two terms.

Rationalizing Binomial Denominators

Pairs of expressions in the form $\sqrt{x} + \sqrt{y}$ and $\sqrt{x} - \sqrt{y}$ are called *conjugates.* When a pair of conjugates is multiplied, the product has no radical in it.

$$(\sqrt{x} + \sqrt{y})(\sqrt{x} - \sqrt{y}) = x - y$$

Let's see how we can use this idea to rationalize denominators that contain two terms.

EXAMPLE 7 **Rationalizing a Denominator Containing Two Terms**

Rationalize the denominator: $\dfrac{20}{\sqrt{7} + \sqrt{3}}$

Solution

$\sqrt{7} + \sqrt{3}$ is a binomial. Let's multiply the numerator and denominator of $\dfrac{20}{\sqrt{7} + \sqrt{3}}$ by $\sqrt{7} - \sqrt{3}$ because $(\sqrt{7} + \sqrt{3})(\sqrt{7} - \sqrt{3}) = 7 - 3 = 4$, a term with no radical.

$$\frac{20}{\sqrt{7} + \sqrt{3}} \cdot \left(\frac{\sqrt{7} - \sqrt{3}}{\sqrt{7} - \sqrt{3}} \right) = \frac{20(\sqrt{7} - \sqrt{3})}{7 - 3}$$

$$\underset{\substack{\uparrow \\ \text{Identity of} \\ \text{multiplication}}}{} \qquad = \frac{20(\sqrt{7} - \sqrt{3})}{4} = 5(\sqrt{7} - \sqrt{3}) \qquad \blacksquare$$

Based on Example 3, we can generalize the following rule.

Rationalizing binomial denominators

To rationalize the denominator of an expression containing a binomial denominator with one or more square roots, multiply the numerator and denominator by the conjugate of the denominator.

Consequently, to rationalize the denominator of $\dfrac{\sqrt{6}}{3 - \sqrt{5}}$, we multiply the numerator and denominator by $3 + \sqrt{5}$, the conjugate of $3 - \sqrt{5}$. To rationalize the denominator of $\dfrac{7}{2\sqrt{3} + 5\sqrt{2}}$, we multiply the numerator and denominator by $2\sqrt{3} - 5\sqrt{2}$, the conjugate of $2\sqrt{3} + 5\sqrt{2}$. Let's actually carry out this process in Example 8.

EXAMPLE 8 **Rationalizing Binomial Denominators**

Rationalize the denominator:

a. $\dfrac{\sqrt{6}}{3 - \sqrt{5}}$ **b.** $\dfrac{7}{2\sqrt{3} + 5\sqrt{2}}$

Solution

a. $\dfrac{\sqrt{6}}{3 - \sqrt{5}} = \dfrac{\sqrt{6}}{3 - \sqrt{5}} \cdot \dfrac{3 + \sqrt{5}}{3 + \sqrt{5}}$ Multiply the numerator and denominator by the conjugate of the denominator.

$= \dfrac{\sqrt{6}(3 + \sqrt{5})}{(3 - \sqrt{5})(3 + \sqrt{5})}$ Multiply numerators and denominators.

$= \dfrac{3\sqrt{6} + \sqrt{30}}{9 - 5}$ Distribute in the numerator. In the denominator, we obtain $(3 - \sqrt{5})(3 + \sqrt{5}) = 3^2 - (\sqrt{5})^2 = 9 - 5$.

$= \dfrac{3\sqrt{6} + \sqrt{30}}{4}$

b. $\dfrac{7}{2\sqrt{3} + 5\sqrt{2}} = \dfrac{7}{2\sqrt{3} + 5\sqrt{2}} \cdot \dfrac{2\sqrt{3} - 5\sqrt{2}}{2\sqrt{3} - 5\sqrt{2}}$ Multiply the numerator and denominator by the denominator's conjugate.

$= \dfrac{7(2\sqrt{3} - 5\sqrt{2})}{(2\sqrt{3} + 5\sqrt{2})(2\sqrt{3} - 5\sqrt{2})}$ Multiply numerators and denominators.

$= \dfrac{14\sqrt{3} - 35\sqrt{2}}{(4)(3) - (25)(2)}$ Distribute in the numerator. In the denominator, we obtain $(2\sqrt{3})^2 - (5\sqrt{2})^2 = 4 \cdot 3 - 25 \cdot 2$.

$= \dfrac{14\sqrt{3} - 35\sqrt{2}}{-38}$

Here are three equivalent ways to express this answer:

$-\dfrac{14\sqrt{3} - 35\sqrt{2}}{38}$ $\dfrac{-14\sqrt{3} + 35\sqrt{2}}{38}$ $-\dfrac{1}{38}(14\sqrt{3} - 35\sqrt{2})$ ∎

We now consider an example with a variable in the radicand.

EXAMPLE 9 **Rationalizing a Binomial Denominator Containing a Variable**

Rationalize the denominator: $\dfrac{\sqrt{x} + 3}{\sqrt{x} - 5}$

ENRICHMENT ESSAY

Continued Fractions

Here's what is called a *continued fraction* for $\sqrt{2}$:

$$\sqrt{2} = 1 + \cfrac{1}{2 + \cfrac{1}{2 + \cfrac{1}{2 + \cfrac{1}{2 + \dots}}}}$$

We can obtain this continued (or chain) fraction in the following manner:

$$\sqrt{2} = 1 + (\sqrt{2} - 1) \qquad \text{Add and subtract 1.}$$

$$= 1 + \frac{1}{\sqrt{2} + 1} \qquad \begin{array}{l}\text{Rationalize the numer-}\\\text{ator of } \sqrt{2} - 1:\end{array}$$

$$\sqrt{2} - 1 \cdot \frac{\sqrt{2} + 1}{\sqrt{2} + 1}$$

$$= \frac{2 - 1}{\sqrt{2} + 1} = \frac{1}{\sqrt{2} + 1}$$

$$= 1 + \cfrac{1}{2 + (\sqrt{2} - 1)} \qquad \begin{array}{l}\text{Write } \sqrt{2} + 1 \text{ as}\\2 + (\sqrt{2} - 1).\end{array}$$

$$= 1 + \cfrac{1}{2 + \cfrac{1}{\sqrt{2} + 1}} \qquad \begin{array}{l}\text{Rationalize the numer-}\\\text{ator of } \sqrt{2} - 1, \text{obtain-}\\\text{ing } \frac{1}{\sqrt{2} + 1}.\end{array}$$

$$= 1 + \cfrac{1}{2 + \cfrac{1}{2 + (\sqrt{2} - 1)}} \qquad \begin{array}{l}\text{Write } \sqrt{2} + 1 \text{ as}\\2 + (\sqrt{2} - 1).\end{array}$$

$$= 1 + \cfrac{1}{2 + \cfrac{1}{2 + \cfrac{1}{\sqrt{2} + 1}}} \qquad \begin{array}{l}\text{Replace } \sqrt{2} - 1 \text{ by}\\\frac{1}{\sqrt{2} + 1}.\end{array}$$

By continually setting $\sqrt{2} + 1$ to $2 + \sqrt{2} - 1$ and replacing $\sqrt{2} - 1$ by $\dfrac{1}{\sqrt{2} + 1}$ (rationalizing the numerator of $\sqrt{2} - 1$), we obtain the continued fraction for $\sqrt{2}$. See if you can obtain the continued fraction for $\sqrt{5}$:

$$\sqrt{5} = 2 + \cfrac{1}{4 + \cfrac{1}{4 + \cfrac{1}{4 + \dots}}}$$

Solution

$$\frac{\sqrt{x} + 3}{\sqrt{x} - 5} = \frac{\sqrt{x} + 3}{\sqrt{x} - 5} \cdot \frac{\sqrt{x} + 5}{\sqrt{x} + 5} \qquad \begin{array}{l}\text{Multiply the numerator and denominator by}\\\text{the denominator's conjugate.}\end{array}$$

$$= \frac{(\sqrt{x} + 3)(\sqrt{x} + 5)}{(\sqrt{x} - 5)(\sqrt{x} + 5)} \qquad \text{Use the FOIL method in the numerator.}$$

$$= \frac{x + 5\sqrt{x} + 3\sqrt{x} + 15}{x - 25} \qquad \begin{array}{l}\text{In the denominator, we obtain}\\(\sqrt{x})^2 - 5^2 = x - 25.\end{array}$$

$$= \frac{x + 8\sqrt{x} + 15}{x - 25} \qquad \text{Combine like radicals in the numerator.} \qquad \blacksquare$$

EXAMPLE 10 **Contrasting Two Rationalization Procedures**

Rationalize the denominator:

a. $\dfrac{2}{\sqrt{x} + \sqrt{y}}$ **b.** $\dfrac{2}{\sqrt{x} + y}$

Study tip

Since

$$\sqrt{x + y} \neq \sqrt{x} + \sqrt{y}$$

pay close attention to the difference in simplifying parts (a) and (b). Only in part (a) is it necessary to use conjugates.

Solution

a. $\dfrac{2}{\sqrt{x} + \sqrt{y}} = \dfrac{2}{\sqrt{x} + \sqrt{y}} \cdot \dfrac{\sqrt{x} - \sqrt{y}}{\sqrt{x} - \sqrt{y}}$

$= \dfrac{2\sqrt{x} - 2\sqrt{y}}{x - y} \quad \left[\text{or } \dfrac{2(\sqrt{x} - \sqrt{y})}{x - y} \right]$

b. $\dfrac{2}{\sqrt{x + y}} = \dfrac{2}{\sqrt{x + y}} \cdot \dfrac{\sqrt{x + y}}{\sqrt{x + y}} = \dfrac{2\sqrt{x + y}}{\sqrt{(x + y)^2}} = \dfrac{2\sqrt{x + y}}{x + y}$

PROBLEM SET 6.5

Practice Problems

In this problem set, all variables represent positive real numbers. Add or subtract as indicated in Problems 1–42.

1. $7\sqrt{3} + 8\sqrt{3}$

2. $5\sqrt{7} + 23\sqrt{7}$

3. $8\sqrt{7x} - 7\sqrt{7x} - \sqrt{7x}$

4. $7\sqrt{11x} - 5\sqrt{11x} - \sqrt{11x}$

5. $3\sqrt{13} - 2\sqrt{5} - 2\sqrt{13} + 4\sqrt{5}$

6. $8\sqrt{17} - 5\sqrt{19} - 6\sqrt{17} + 4\sqrt{19}$

7. $\sqrt{2} - \sqrt{11} + 6\sqrt{2} + 4\sqrt{11}$

8. $\sqrt{19} - \sqrt{6} + 2\sqrt{6} + \sqrt{19}$

9. $3\sqrt{15} - 2\sqrt{7} - 2\sqrt{15} + 2\sqrt{7}$

10. $8\sqrt{21} - 5\sqrt{2} + 5\sqrt{2} - 8\sqrt{21}$

11. $\sqrt{50} + \sqrt{18}$

12. $\sqrt{28} + \sqrt{63}$

13. $3\sqrt{18} - 5\sqrt{50}$

14. $4\sqrt{12} - 2\sqrt{75}$

15. $3\sqrt{8} - \sqrt{32} + 3\sqrt{72} - \sqrt{75}$

16. $3\sqrt{54} - 2\sqrt{24} - \sqrt{96} + 4\sqrt{63}$

17. $8\sqrt{\frac{1}{2}} - \frac{1}{2}\sqrt{8}$

18. $\sqrt{\frac{5}{9}} + \sqrt{\frac{9}{5}}$

19. $\frac{\sqrt{63}}{3} + 7\sqrt{3}$

20. $\sqrt{9a} + \sqrt{4a}$

21. $\sqrt{25x} + \sqrt{16x}$

22. $\sqrt[3]{6} + \sqrt[3]{48}$

23. $\sqrt[4]{32} + 3\sqrt[4]{1250}$

24. $2\sqrt[3]{81} - 5\sqrt[3]{24}$

25. $4\sqrt[3]{40} - 3\sqrt[3]{320} + 2\sqrt[3]{625}$

26. $\frac{1}{5}\sqrt{45} - \frac{1}{3}\sqrt{20}$

27. $\frac{1}{4}\sqrt{2} + \frac{2}{3}\sqrt{8}$

28. $\frac{1}{2}\sqrt{7} - \frac{1}{3}\sqrt{28}$

29. $\frac{\sqrt{45}}{4} - \sqrt{80} + \frac{\sqrt{20}}{3}$

30. $\frac{2}{3}\sqrt{18} - \frac{\sqrt{50}}{2} + \frac{3}{5}\sqrt{8}$

31. $7\sqrt[3]{2} + 8\sqrt[3]{16} - 2\sqrt[3]{54}$

32. $5\sqrt[3]{3} + 2\sqrt[3]{81} - \sqrt[3]{54}$

33. $2\sqrt[3]{2} - \sqrt[3]{16} - \sqrt[3]{54}$

34. $6\sqrt[3]{81} + 7\sqrt[3]{3} - \sqrt[3]{24}$

35. $5\sqrt{12x} - 2\sqrt{3x}$

36. $2\sqrt{20y} - 4\sqrt{5y} + 7\sqrt{45y}$

37. $\sqrt[3]{54x^5} + 2x\sqrt[3]{16x^2} - 7\sqrt[3]{2x^5}$

38. $3x^2\sqrt{8x} + 4\sqrt{18x^5} - \sqrt{72x}$

39. $16\sqrt{\frac{5}{8}} + 6\sqrt{\frac{5}{2}}$

40. $4\sqrt{\frac{1}{12}} + 18\sqrt{\frac{1}{27}}$

41. $12\sqrt{\frac{2}{3}} + 24\sqrt{\frac{1}{6}}$

42. $10\sqrt{\frac{4}{5}} - 2\sqrt{45} + 15\sqrt{\frac{16}{5}}$

In Problems 43–78, find each product, writing all answers in simplified form.

43. $\sqrt{2}(\sqrt{3} + \sqrt{7})$

44. $\sqrt{5}(\sqrt{6} + \sqrt{3})$

45. $4\sqrt{3}(2\sqrt{5} + 3\sqrt{7})$

46. $2\sqrt{3}(7\sqrt{5} - 2\sqrt{7})$

47. $5\sqrt{6}(7\sqrt{8} - 2\sqrt{12})$

48. $3\sqrt{2}(8\sqrt{12} - 2\sqrt{6})$

49. $4\sqrt{x}(7\sqrt{2} - 3\sqrt{y})$

50. $\sqrt{3x}(3\sqrt{y} - 7\sqrt{2})$

51. $\sqrt{2x}(\sqrt{6x} - 3\sqrt{x})$

52. $\sqrt{5x}(\sqrt{10x} - 2\sqrt{x})$

53. $\sqrt[3]{2}(\sqrt[3]{6} + 4\sqrt[3]{5})$

54. $\sqrt[3]{3}(\sqrt[3]{6} - 7\sqrt[3]{4})$

55. $(\sqrt{2} + \sqrt{7})(\sqrt{3} + \sqrt{5})$

56. $(\sqrt{3} + \sqrt{2})(\sqrt{10} + \sqrt{11})$

57. $(\sqrt{2} - \sqrt{7})(\sqrt{3} - \sqrt{5})$

58. $(\sqrt{3} - \sqrt{2})(\sqrt{10} - \sqrt{11})$

59. $(4\sqrt{2} + 5\sqrt{7})(2\sqrt{3} + 3\sqrt{5})$

60. $(4\sqrt{3} + 5\sqrt{2})(2\sqrt{10} + 3\sqrt{11})$

61. $(3\sqrt{2} - 2\sqrt{8})(2\sqrt{3} - 4\sqrt{5})$

62. $(2\sqrt{2} - 6\sqrt{6})(3\sqrt{10} - \sqrt{7})$

63. $(\sqrt{5} + 7)(\sqrt{5} - 7)$

64. $(\sqrt{11} + 3)(\sqrt{11} - 3)$

65. $(2 + 5\sqrt{3})(2 - 5\sqrt{3})$

66. $(11 + 4\sqrt{2})(11 - 4\sqrt{2})$

67. $(\sqrt{3} + \sqrt{5})^2$

68. $(\sqrt{7} + \sqrt{2})^2$

69. $(2\sqrt{3} - 4\sqrt{7})^2$

70. $(5\sqrt{2} - 3\sqrt{5})^2$

71. $(\sqrt{x} + \sqrt{3})(\sqrt{x} + \sqrt{2})$

72. $(3 - \sqrt{x})(2 - \sqrt{x})$

73. $(x + \sqrt{y})^2$

74. $(x - \sqrt{y})^2$

75. $(\sqrt{x} + \sqrt{y})^2$

76. $(\sqrt{a} - \sqrt{b})^2$

77. $(x + \sqrt[3]{y^2})(2x - \sqrt[3]{y^2})$

78. $(x - \sqrt[4]{x^3})(2x + \sqrt[4]{x^3})$

Simplify Problems 79–113 by rationalizing the denominator.

79. $\dfrac{8}{\sqrt{5}-2}$

80. $\dfrac{15}{\sqrt{6}-1}$

81. $\dfrac{13}{\sqrt{11}+3}$

82. $\dfrac{3}{\sqrt{7}+3}$

83. $\dfrac{6}{\sqrt{5}+\sqrt{3}}$

84. $\dfrac{12}{\sqrt{7}+\sqrt{3}}$

85. $\dfrac{11}{\sqrt{7}-\sqrt{3}}$

86. $\dfrac{13}{\sqrt{5}-\sqrt{3}}$

87. $\dfrac{\sqrt{5}}{\sqrt{7}+\sqrt{3}}$

88. $\dfrac{\sqrt{8}}{\sqrt{5}+\sqrt{3}}$

89. $\dfrac{\sqrt{2}}{\sqrt{7}-\sqrt{2}}$

90. $\dfrac{\sqrt{5}}{\sqrt{11}-\sqrt{5}}$

91. $\dfrac{8}{3+2\sqrt{2}}$

92. $\dfrac{12}{3-2\sqrt{5}}$

93. $\dfrac{25}{5\sqrt{2}-3\sqrt{5}}$

94. $\dfrac{16}{2\sqrt{5}-4\sqrt{3}}$

95. $\dfrac{\sqrt{3}}{2\sqrt{3}+3\sqrt{5}}$

96. $\dfrac{\sqrt{5}}{7\sqrt{3}+4\sqrt{5}}$

97. $\dfrac{7}{\sqrt{x}-5}$

98. $\dfrac{11}{\sqrt{x}+7}$

99. $\dfrac{\sqrt{y}}{\sqrt{y}+3}$

100. $\dfrac{\sqrt{y}}{\sqrt{y}-2}$

101. $\dfrac{2\sqrt{3}-1}{2\sqrt{3}+1}$

102. $\dfrac{2\sqrt{5}+3}{2\sqrt{5}-5}$

103. $\dfrac{\sqrt{5}+\sqrt{3}}{\sqrt{5}-\sqrt{3}}$

104. $\dfrac{\sqrt{7}-\sqrt{2}}{\sqrt{7}+\sqrt{2}}$

105. $\dfrac{\sqrt{x}+1}{\sqrt{x}+3}$

106. $\dfrac{\sqrt{y}-2}{\sqrt{y}-5}$

107. $\dfrac{\sqrt{y}}{\sqrt{x}+\sqrt{y}}$

108. $\dfrac{3\sqrt{x}}{\sqrt{x}-\sqrt{y}}$

109. $\dfrac{3\sqrt{5}+2\sqrt{2}}{4\sqrt{5}+\sqrt{2}}$

110. $\dfrac{3\sqrt{7}+2\sqrt{3}}{2\sqrt{7}-\sqrt{3}}$

111. $\dfrac{2\sqrt{x}+\sqrt{y}}{4\sqrt{x}+3\sqrt{y}}$

112. $\dfrac{5\sqrt{x}-2\sqrt{y}}{\sqrt{x}-3\sqrt{y}}$

113. $\dfrac{x^3-y^3}{\sqrt{x}+\sqrt{y}}$

Application Problems

For Problems 114–115, express the perimeter and area of each figure in simplified radical form.

114.

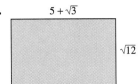

$5+\sqrt{3}$ · $\sqrt{12}$

115.

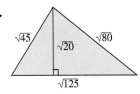

$\sqrt{45}$ · $\sqrt{20}$ · $\sqrt{80}$ · $\sqrt{125}$

116. Express the area of the trapezoid in the figure in simplified radical form. Is the figure correctly drawn to scale?

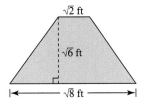

$\sqrt{2}$ ft · $\sqrt{6}$ ft · $\sqrt{8}$ ft

117. The area of a triangle can be found using the formula $A = \sqrt{s(s-a)(s-b)(s-c)}$, where a, b, and c are the lengths of the sides and s is half the triangle's perimeter. Use this formula to find the volume of the prism shown in the figure if the volume of a prism is the product of the area of its triangular base and its height. Express the answer in simplified radical form.

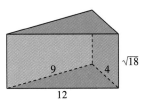

9 · 4 · $\sqrt{18}$ · 12

True–False Critical Thinking Problems

118. Which one of the following is true?
  a. The same general procedure is used to both add radicals and multiply monomials.
  b. $\sqrt{5} + \sqrt{5} = \sqrt{10}$
  c. At least two radicals can be combined in simplifying $\sqrt{300} - \sqrt{12} + \sqrt{18}$.
  d. If any two radical expressions are completely simplified, they can then be combined through addition or subtraction.

119. Which one of the following is true?
  a. $(\sqrt{x} - 7)^2 = x - 49$
  b. $(3\sqrt{5})^2 = 15$
  c. $\dfrac{\sqrt{x}}{\sqrt{y}} = \dfrac{\sqrt{xy}}{y}; x \geq 0, y > 0$
  d. The conjugate of $\sqrt{a} + \sqrt{b}$ is found by replacing b with its opposite.

Technology Problems

120. Use a calculator to provide numerical support to any four problems that you worked from Problems 1–104 that do not contain variables. Begin by finding a decimal approximation for the given radical expression. Then find a decimal approximation for your answer. The two decimal approximations should be the same.

In Problems 121–126, determine if each equation is correct by graphing the function on each side of the given equality with your graphing utility. Shown in the same viewing rectangle, the graphs should be the same. If they are not, correct the right side of the equation and then use your graphing utility to verify the result.

121. $\sqrt{8x} + \sqrt{2x} = \sqrt{10x}$

122. $(\sqrt{x} + 3)^2 = x + 9, x \geq 0$

123. $8\sqrt{x} + 2\sqrt{x} = 10\sqrt{x}, x \geq 0$

124. $(\sqrt{x} + \sqrt{3})(\sqrt{x} - \sqrt{3}) = x - 3, x \geq 0$

125. $\dfrac{\sqrt{x^7} + \sqrt{x^3}}{\sqrt{x}} = x^3 + x, x > 0$

126. $\dfrac{3}{\sqrt{x+3} - \sqrt{x}} = \sqrt{x+3} + \sqrt{x}$

Writing in Mathematics

Radical Disasters! In Problems 127–133, explain in a sentence or two the radical error that is being committed in each radical situation. Then write out the mathematics that corrects each disaster.

127. $\dfrac{\sqrt{3} + 7}{\sqrt{3} - 2} = -\dfrac{7}{2}$

128. $\dfrac{4}{\sqrt{x+y}} = \dfrac{4}{\sqrt{x+y}} \cdot \dfrac{\sqrt{x-y}}{\sqrt{x-y}}$

129. $4 - 3(\sqrt{7} + \sqrt{2}) = 1(\sqrt{7} + \sqrt{2}) = \sqrt{7} + \sqrt{2}$

130. $(\sqrt{x} + \sqrt{y})^2 = x + y$

131. $\sqrt[3]{75} = \sqrt[3]{(25)(3)} = 5\sqrt[3]{3}$

132. $\sqrt[3]{4}\sqrt[5]{2} = \sqrt[15]{8}$

133. $\sqrt{x^2} = \sqrt{y^2}$, so $x = y$

Critical Thinking Problems

134. If an irrational number is decreased by $2\sqrt{18} - \sqrt{50}$, the result is $\sqrt{2}$. What is the number?

135. Solve for x: $\left|\dfrac{3x + \sqrt{32}}{2}\right| = \sqrt{50}$.

136. Solve for x: $2(3x + \sqrt{28}) \geq 4(2x + \sqrt{175})$.

137. Italian mathematician Luca Pacioli (1445–1509) used a notation in his book *Suma* that was considered quite sophisticated for its time. Using $\bar{p}$. for addition, $\bar{m}$. for subtraction, and R. for a square root, translate the following short excerpt from Pacioli's work into modern symbolic notation:

  via. 4. $\bar{p}$. R. 6.
  4. $\bar{m}$. R. 6.
  ——————
  16. $\bar{m}$. 6.
  Productum 10

138. Show that $\dfrac{3 + \sqrt{3}}{3}$ is a solution of $3y^2 + 2 = 6y$.

139. Show that $\sqrt{2\sqrt{15} + 8} = \sqrt{3} + \sqrt{5}$ by squaring both sides of the statement.

140. Simplify: $(\sqrt{2 + \sqrt{3}} + \sqrt{2 - \sqrt{3}})^2$.

141. Rationalize the denominator: $\dfrac{1}{\sqrt{2} + \sqrt{3} + \sqrt{4}}$.

142. a. Show that $(\sqrt[3]{x} + \sqrt[3]{y})(\sqrt[3]{x^2} - \sqrt[3]{xy} + \sqrt[3]{y^2}) = x + y$.

  b. Use the result of part (a) to rationalize the denominator of $\dfrac{a^2 - 4}{\sqrt[3]{a} + 2}$.

143. The area of square $BEGC$ in the figure is 2 square meters, and the area of square $ADFB$ is 3 square meters. What is the area of the shaded region?

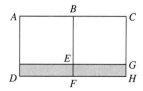

144. Find the exact value of

$$\sqrt{13 + \sqrt{2} + \frac{7}{3 + \sqrt{2}}}$$

without the use of a calculator.

145. Rationalize the *numerator* and simplify:

$$\frac{\sqrt{a + h} - \sqrt{a}}{h}$$

Group Activity Problem

146. As a group, use the errors in Problems 127–133 to list some of the common errors that can occur when simplifying radical expressions or when performing operations with radical expressions. Try to go beyond the errors that occur in these problems by having group members share the kinds of errors they made in working the practice problems. Group members should then suggest a strategy for avoiding each of the errors in the list.

Review Problems

147. Show that 2 satisfies the equation $\sqrt{5y + 6} + \sqrt{3y - 2} = 6$.

148. Solve: $4x^2 = 3x + 10$.

149. One pipe can fill a pool in 30 minutes and a second pipe can empty it in 20 minutes. When the pool is full, a worker accidentally opens both pipes. How long will it take before the pool has no water?

SECTION 6.6

**Solutions   Tutorial   Video
Manual   8**

Radical Equations

Objectives

1   Solve radical equations.
2   Answer questions about radical models.

In this section, we consider equations with variables in the radicand, called *radical equations.* Two examples of such equations are

$$\sqrt{3x - 6} = 3 \quad \text{and} \quad \sqrt{3x + 4} + \sqrt{x + 2} = 2.$$

| Solve radical equations.

Solving Radical Equations

Radical equations can be solved by isolating the radical term and raising both sides of the equation to a power such as 2, 3, 4, and so on. We can square, cube, raise to the fourth power, or raise to any positive integral power both sides of the equation as a way to eliminate the radical. This idea is called the *power rule.*

> **Taking powers of both sides of an equation**
>
> **Power Rule**
>
> If $a = b$, then $a^n = b^n$.
>
> If both sides of an equation are raised to the same power, all solutions of the original equation are also solutions of the resulting equation.

Read the power rule carefully. The statement that results by interchanging the two parts of the "if-then" rule is not necessarily true.

Not always true →   If $a^n = b^n$, then $a = b$ ← Not always true

For example, $5^2 = (-5)^2$ is true, but $5 = -5$ is not true. The power rule states that the new equation $a^n = b^n$ has all the solutions of $a = b$, but $a^n = b^n$ may have some extra solutions that do not satisfy $a = b$. Consequently, the new equation may not have the same solution set as the original equation. Because the new equation may not be equivalent to the original equation, the solution process for a radical equation must be completed by substituting all proposed solutions into the original equation to determine whether they are really solutions. Let's see exactly how this works.

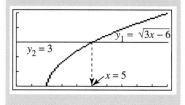

EXAMPLE 1 **Solving a Radical Equation**

Solve: $\sqrt{3x - 6} = 3$

Solution

$\sqrt{3x - 6} = 3$	This is the given equation.
$(\sqrt{3x - 6})^2 = 3^2$	Use the power rule to eliminate the radical by squaring both sides.
$3x - 6 = 9$	Simplify.
$3x = 15$	Add 6 to both sides.
$x = 5$	Divide both sides by 5.

We now must complete the solution process by substituting the proposed solution, 5, into the original equation.

$\sqrt{3x - 6} = 3$	This is the original equation.
$\sqrt{3 \cdot 5 - 6} \overset{?}{=} 3$	Substitute 5, the proposed solution, for x.
$\sqrt{15 - 6} \overset{?}{=} 3$	Multiply.
$\sqrt{9} \overset{?}{=} 3$	Subtract.
$3 = 3$ ✓	This true statement indicates that 5 is the solution.

The solution set is {5}. ■

In the Discover for Yourself box, did you find that the equation has no solution? Let's see what happens if we use the method of the previous example to solve this equation.

EXAMPLE 2 **A Radical Equation with No Solution**

Solve: $\sqrt{x - 3} + 6 = 5$

Solution

$\sqrt{x - 3} + 6 = 5$	This is the given equation.
$\sqrt{x - 3} = -1$	Isolate the radical term by subtracting 6 from both sides.

The graphs of

$$y_1 = \sqrt{x - 3} + 6$$

and $y_2 = 5$

do not intersect, as shown below. Thus,

$$\sqrt{x - 3} + 6 = 5$$

has no real number solution.

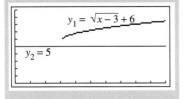

This equation cannot have a solution because a principal square root is always nonnegative. Nonetheless, we'll continue the solution process to see what occurs.

$$(\sqrt{x - 3})^2 = (-1)^2 \qquad \text{Use the power rule to eliminate the radical by squaring both sides.}$$

$$x - 3 = 1 \qquad \text{Simplify.}$$

$$x = 4 \qquad \text{Add 3 to both sides.}$$

We must now complete the solution process by substituting the proposed solution, 4, in the original equation.

$$\sqrt{x - 3} + 6 = 5 \qquad \text{This is the original equation.}$$

$$\sqrt{4 - 3} + 6 \stackrel{?}{=} 5 \qquad \text{Substitute 4, the proposed solution, for } x.$$

$$\sqrt{1} + 6 \stackrel{?}{=} 5 \qquad \text{Simplify the radicand.}$$

$$1 + 6 \stackrel{?}{=} 5 \qquad \text{Take the square root.}$$

$$7 = 5 \quad \text{False} \quad \text{This false statement indicates that 4 is not a solution.}$$

We see that 4 is not a solution. Observe that 4 *does* satisfy the equation squared $(x - 3 = 1)$, but *does not* satisfy the original equation: $\sqrt{x - 3} + 6 = 5$.

This equation has no solution. The extra value that satisfies $a^n = b^n$ but not $a = b$ is sometimes called an *extraneous solution*. In this example, 4 is an extraneous solution. The solution set is $\varnothing$. ∎

The two examples we have considered so far illustrate the general method for solving equations with radicals. Before turning to additional problems, let's summarize the method.

Procedure for solving radical equations

1. If necessary, rewrite the equation so that a radical containing a variable radicand is isolated on one side of the equation.
2. Raise both sides of the equation to a power that is the same as the index of the radical.
3. Combine like terms on both sides of the equation.
4. If the equation still contains a term with a variable in a radicand, repeat steps 1 through 3.
5. Solve the resulting equation, which no longer contains a radical, for the variable.
6. Substitute all proposed solutions in the original radical equation, eliminating extraneous solutions.

EXAMPLE 3 **Solving a Radical Equation That Becomes Quadratic**

Solve: $\sqrt{3x + 4} + x = 8$

Solution

$$\sqrt{3x + 4} + x = 8 \qquad \text{This is the given equation.}$$

$$\sqrt{3x + 4} = 8 - x \qquad \text{Isolate the radical by subtracting } x \text{ from both sides.}$$

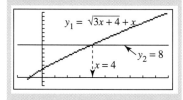
$$(\sqrt{3x + 4})^2 = (8 - x)^2$$ Use the power rule to eliminate the radical by squaring both sides.

$$3x + 4 = 64 - 16x + x^2$$ Simplify.

$$0 = x^2 - 19x + 60$$ Subtract $3x + 4$ from both sides to write the quadratic equation in standard form.

$$0 = (x - 4)(x - 15)$$ Factor.

$$x - 4 = 0 \quad \text{or} \quad x - 15 = 0$$ Set each factor equal to 0.

$$x = 4 \quad \text{or} \quad x = 15$$ Solve the resulting equations.

The possible solutions are 4 and 15. Complete the solution process by checking each in the original equation.

Check 4:

$$\sqrt{3x + 4} + x = 8$$
$$\sqrt{3 \cdot 4 + 4} + 4 \stackrel{?}{=} 8$$
$$\sqrt{16} + 4 \stackrel{?}{=} 8$$
$$4 + 4 \stackrel{?}{=} 8$$
$$8 = 8 \quad \checkmark$$

Check 15:

$$\sqrt{3x + 4} + x = 8$$
$$\sqrt{3 \cdot 15 + 4} + 15 \stackrel{?}{=} 8$$
$$\sqrt{49} + 15 \stackrel{?}{=} 8$$
$$7 + 15 \stackrel{?}{=} 8$$
$$22 = 8 \quad \text{False}$$

We see that 4 checks, but 15 does not. Thus, 15 is an extraneous solution. The solution set is {4}.

EXAMPLE 4   **Solving a Radical Equation Containing a Cube Root**

Solve: $\sqrt[3]{x^2 + 2x} - 2 = 0$

Solution

$$\sqrt[3]{x^2 + 2x} - 2 = 0$$ This is the given equation.

$$\sqrt[3]{x^2 + 2x} = 2$$ Isolate the radical term by adding 2 to both sides.

$$(\sqrt[3]{x^2 + 2x})^3 = 2^3$$ Cube both sides to eliminate the radical.

$$x^2 + 2x = 8$$ Simplify.

$$x^2 + 2x - 8 = 0$$ Solve the resulting quadratic equation.

$$(x + 4)(x - 2) = 0$$ Factor.

$$x + 4 = 0 \quad \text{or} \quad x - 2 = 0$$ Set each factor equal to 0.

$$x = -4 \quad \text{or} \quad x = 2$$ Solve the resulting equations.

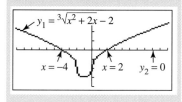
Raising both sides of an equation to an even power might produce extraneous roots. However, raising both sides to an odd power, as we've done here, will always produce an equivalent equation with no extraneous roots. Thus, we will now check just to be sure we did not make any errors in the solution process.

Check -4:

$$\sqrt[3]{x^2 + 2x} - 2 = 0$$
$$\sqrt[3]{(-4)^2 + 2(-4)} - 2 \stackrel{?}{=} 0$$
$$\sqrt[3]{16 + (-8)} - 2 \stackrel{?}{=} 0$$
$$\sqrt[3]{8} - 2 \stackrel{?}{=} 0$$
$$2 - 2 \stackrel{?}{=} 0$$
$$0 = 0 \quad \checkmark$$

Check 2:

$$\sqrt[3]{x^2 + 2x} - 2 = 0$$
$$\sqrt[3]{2^2 + 2(2)} - 2 \stackrel{?}{=} 0$$
$$\sqrt[3]{4 + 4} - 2 \stackrel{?}{=} 0$$
$$\sqrt[3]{8} - 2 \stackrel{?}{=} 0$$
$$2 - 2 \stackrel{?}{=} 0$$
$$0 = 0 \quad \checkmark$$

The solution set is $\{-4, 2\}$.

Solving Radical Equations with Two Radical Terms

In Example 5, we are asked to solve an equation in which the radicals cannot all be eliminated simply by squaring both sides of the equation. After squaring both sides of the equation, the resulting equation still contains a radical. Thus, it is necessary to square both sides again.

EXAMPLE 5 **Solving a Radical Equation Containing Two Radicals**

Solve: $\sqrt{x + 8} - \sqrt{x + 35} = -3$

Solution

$$\sqrt{x + 8} - \sqrt{x + 35} = -3$$ This is the given equation.

$$\sqrt{x + 8} = \sqrt{x + 35} - 3$$ Isolate one of the radical terms by adding $\sqrt{x + 35}$ to both sides.

$$(\sqrt{x + 8})^2 = (\sqrt{x + 35} - 3)^2$$ Square both sides.

$$x + 8 = (x + 35) - 2\sqrt{x + 35} \cdot 3 + 9$$ Simplify on the left. Square on the right using $(A - B)^2 = A^2 - 2AB + B^2$.

$$x + 8 = x + 44 - 6\sqrt{x + 35}$$ Combine numerical terms on the right.

$$-36 = -6\sqrt{x + 35}$$ Isolate the remaining radical term by subtracting $x + 44$ from both sides.

$$6 = \sqrt{x + 35}$$ Divide both sides by -6.

$$6^2 = (\sqrt{x + 35})^2$$ Since the equation still contains a radical, square both sides again.

$$36 = x + 35$$ Simplify.

$$1 = x$$ Subtract 35 from both sides.

> **S**tudy tip
>
> Here are the details for squaring $\sqrt{x + 35} - 3$.
>
> $$(A - B)^2 = A^2 - 2 \quad A \cdot B + B^2$$
>
> $$(\sqrt{x + 35} - 3)^2 = (\sqrt{x + 35})^2 - 2 \cdot \sqrt{x + 35} \cdot 3 + 3^2$$
> $$= x + 35 - 6\sqrt{x + 35} + 9$$

Complete the solution process by checking.

$$\sqrt{x + 8} - \sqrt{x + 35} = -3$$ This is the original equation.

$$\sqrt{1 + 8} - \sqrt{1 + 35} \overset{?}{=} -3$$ Substitute 1, the proposed solution, for x.

$$\sqrt{9} - \sqrt{36} \overset{?}{=} -3$$ Simplify the radicands.

$$3 - 6 \overset{?}{=} -3 \qquad \text{Take the square roots.}$$
$$-3 = -3 \quad \checkmark \qquad \text{This true statement indicates that 1 is the solution.}$$

The solution set is {1}.   ∎

The next example is, once again, an equation with two radical terms. However, once we eliminate all radicals, the resulting equation is quadratic.

EXAMPLE 6   **Solving a Radical Equation Containing Two Radicals**

Solve: $\sqrt{3x + 4} + \sqrt{x + 2} = 2$

Solution

$\sqrt{3x + 4} + \sqrt{x + 2} = 2$	This is the given equation.
$\sqrt{3x + 4} = 2 - \sqrt{x + 2}$	Isolate one of the radical terms by subtracting $\sqrt{x + 2}$ from both sides.
$(\sqrt{3x + 4})^2 = (2 - \sqrt{x + 2})^2$	Square both sides.
$3x + 4 = 4 - 2 \cdot 2\sqrt{x + 2} + (x + 2)$	Simplify on the left. Square on the right using $(A - B)^2 = A^2 - 2AB + B^2$.
$3x + 4 = x + 6 - 4\sqrt{x + 2}$	Combine numerical terms on the right.
$2x - 2 = -4\sqrt{x + 2}$	Isolate the remaining radical by subtracting $x + 6$ from both sides.
$(2x - 2)^2 = (-4\sqrt{x + 2})^2$	Square both sides.
$4x^2 - 8x + 4 = 16(x + 2)$	Perform the resulting operations.
$4x^2 - 8x + 4 = 16x + 32$	Use the distributive property on the right.
$4x^2 - 24x - 28 = 0$	Subtract $16x + 32$ from both sides to write the quadratic equation in standard form.
$4(x^2 - 6x - 7) = 0$	Factor out the GCF.
$4(x + 1)(x - 7) = 0$	Factor completely.
$x + 1 = 0$ or $x - 7 = 0$	Set each variable factor equal to 0.
$x = -1$ or $x = 7$	Solve the resulting equations.

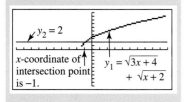
The possible solutions are −1 and 7. Take a few minutes to complete the solution process by substituting each of these values into the original equation. You should find that −1 checks and 7 does not. (This is reinforced in the Using Technology.) The solution set is {−1}.   ∎

2 Answer questions about radical models.

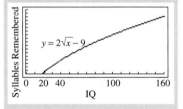
Modeling with Radical Equations

The ability to solve radical equations enables us to gain increased understanding of models that contain radicals. Let's first look at an example from psychology.

EXAMPLE 7   Using Radical Equations in Psychology

Psychologists use the formula $N = 2\sqrt{Q} - 9$ to determine the number of nonsense syllables (N) a subject with an IQ of Q can repeat. Solve the model for Q in terms of N.

Solution

$$N = 2\sqrt{Q} - 9 \qquad \text{This is the given formula.}$$
$$N + 9 = 2\sqrt{Q} \qquad \text{Isolate the radical term by adding 9 to both sides.}$$
$$(N + 9)^2 = (2\sqrt{Q})^2 \qquad \text{Square both sides.}$$
$$N^2 + 18N + 81 = 4Q \qquad \text{Perform the resulting operations.}$$
$$\frac{N^2 + 18N + 81}{4} = Q \qquad \text{Divide both sides by 4.}$$

In this form, we can compute IQ using the number of syllables a subject can repeat. Repeating 14 nonsense syllables implies an IQ of

$$Q = \frac{14^2 + (18)(14) + 81}{4} = 132\frac{1}{4}$$

or an approximate IQ of 132.   ■

EXAMPLE 8   The Length of a Pendulum

The time (T, in seconds) that it takes for a pendulum to make one complete swing back and forth, called its period, is modeled by

$$T = 2\pi\sqrt{\frac{L}{32}}$$

where L is the length of the pendulum in feet. How long is the pendulum of a grandfather clock with a period of 2.22 seconds?

Solution

We substitute 2.22 for T in the formula and solve for L.

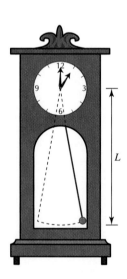

$$T = 2\pi\sqrt{\frac{L}{32}} \qquad \text{This is the given model.}$$

$$2.22 = 2\pi\sqrt{\frac{L}{32}} \qquad \text{Substitute 2.22 for } T.$$

$$\frac{2.22}{2\pi} = \sqrt{\frac{L}{32}} \qquad \text{Isolate the radical by dividing both sides by } 2\pi.$$

$$\frac{1.11}{\pi} = \sqrt{\frac{L}{32}} \qquad \text{Simplify.}$$

$$\left(\frac{1.11}{\pi}\right)^2 = \left(\sqrt{\frac{L}{32}}\right)^2 \qquad \text{Square both sides.}$$

$$\left(\frac{1.11}{\pi}\right)^2 = \frac{L}{32} \qquad \text{Simplify.}$$

$$32 \cdot \left(\frac{1.11}{\pi}\right)^2 = \frac{L}{32} \cdot 32 \qquad \text{Multiply both sides by 32 to clear fractions.}$$

$$32\left(\frac{1.11}{\pi}\right)^2 = L \qquad \text{Simplify.}$$

We now use a calculator to find an approximation for L, the length of the pendulum.

$$L = 32 \; \boxed{(} \; 1.11 \; \boxed{\div} \; \pi \; \boxed{)} \; \boxed{\wedge} \; 2 \; \boxed{\text{ENTER}}$$

$$L \approx 3.99$$

The pendulum of the grandfather clock is approximately 4 feet long. ■

PROBLEM SET 6.6

Practice Problems

Find the solution set for each equation in Problems 1–48. Check your proposed solutions by direct substitution and, if applicable, with a graphing utility.

1. $\sqrt{3x - 1} = 4$
2. $\sqrt{5y + 2} = 7$
3. $\sqrt{2x + 4} - 6 = 0$
4. $\sqrt{3y - 3} - 6 = 0$

5. $\sqrt{3x - 1} = -4$
6. $\sqrt{5y + 2} = -7$
7. $\sqrt{2x + 4} + 6 = 0$
8. $\sqrt{3y - 3} + 6 = 0$

9. $\sqrt{6x + 7} = x + 2$
10. $\sqrt{x - 3} = x - 5$
11. $\sqrt{3y + 1} - 3y + 11 = 0$
12. $\sqrt{2y - 5} + y - 2 = 0$

13. $\sqrt{5x + 9} - x + 1 = 0$
14. $\sqrt{2x + 6} - 3x = -11$
15. $\sqrt{z - 1} - 7 = -z$
16. $2(z - 4) = \sqrt{3z - 2}$

17. $z + \sqrt{5z - 1} - 5 = 0$
18. $\sqrt[3]{z^2 - 1} = 2$
19. $\sqrt[3]{4x^2 - 3x} = 1$
20. $\sqrt[3]{x^2 + 6x} = -2$

21. $\sqrt[3]{2x^2 + 3x} = -1$
22. $\sqrt{y - 7} = 7 - \sqrt{y}$
23. $\sqrt{y - 8} = \sqrt{y} - 2$
24. $\sqrt{r + 3} = \sqrt{r} - 3$

25. $\sqrt{y + 1} = \sqrt{y} + 1$
26. $\sqrt{y + 8} = \sqrt{y - 4} + 2$
27. $\sqrt{y + 5} - \sqrt{y - 3} = 2$
28. $\sqrt{p - 5} - \sqrt{p - 8} = 3$

29. $\sqrt{y - 3} - 4 = \sqrt{y - 3}$
30. $\sqrt{x + 4} + 2 = \sqrt{x + 20}$
31. $\sqrt{x + 2} = 1 - \sqrt{x - 3}$
32. $\sqrt{x + 6} = 2 - \sqrt{x - 2}$

33. $\sqrt{y + 2} + \sqrt{y - 1} = 3$
34. $\sqrt{x - 4} + \sqrt{x + 4} = 4$
35. $\sqrt{y} + \sqrt{2} = \sqrt{y + 2}$
36. $\sqrt{10y} = y$

37. $\sqrt{y + 1} = \sqrt{5y - 1}$
38. $\sqrt{2y} + 2 = \sqrt{8y + 5}$
39. $\sqrt{4z - 3} - \sqrt{8z + 1} + 2 = 0$

40. $\sqrt{3z + 3} - \sqrt{6z + 7} + 1 = 0$
41. $2\sqrt{3x - 2} + \sqrt{2x - 3} = 5$
42. $\sqrt{2x - 5} - \sqrt{x - 2} = 2$

43. $\sqrt{2y + 3} + \sqrt{y + 2} = 2$
44. $\sqrt{2y + 3} + \sqrt{3y - 5} + 1 = 0$
45. $\sqrt{3 - y} + \sqrt{2 + y} = 1$

46. $\sqrt{2z + 5} + \sqrt{z + 2} - 1 = 0$
47. $\sqrt{y + 1} + \sqrt{7y + 4} - 3 = 0$
48. $\sqrt{3y} - \sqrt{y^2 + 3} = 1$

Application Problems

49. For a group of 50,000 births, the number of people (N) surviving to age x is given by the formula $N = 5000\sqrt{100 - x}$. To what age will 40,000 people in the group survive?

50. The velocity of a falling object is given by $v = \sqrt{2gh}$, where v is the velocity in feet per second, g, the acceleration of gravity, is 32 feet per second per second, and h is the distance in feet that the object has fallen. Find the height from which a rock has been dropped if it strikes the ground with a velocity of 80 feet per second.

51. Shallow-water wave motion can be modeled by $c = \sqrt{gH}$, where c is the wave velocity in feet per second, H is the water depth in feet, and g, the acceleration of gravity, is 32 feet per second per second. If the wave velocity is 24 feet per second, find the water's depth.

52. The amount of power used by an electrical appliance is given by

$$I = \sqrt{\frac{P}{R}}$$

where I is the current (in amperes), R is the resistance (in ohms), and P is the power (in watts). What is the power consumed by an electric toaster for which $I = 10$ amps and $R = 15$ ohms?

53. The formula

$$b = \sqrt{\frac{3V}{H}}$$

describes the length b of one side of a square-based pyramid in terms of its volume V and height H. Find the volume of a square-based pyramid if each side of the base is 8 feet long and the pyramid is 6 feet high.

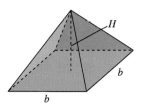

54. The model $N = 0.2\sqrt{R^3}$ describes the number of Earth days (N) in a planet's year, where R is the distance of the planet from the sun (in millions of kilometers). Use the model to find the distance of the planet Mercury from the sun if one year on Mercury is equivalent to 88 Earth days.

55. The model $N = 1220\sqrt[3]{t + 42} + 4900$ describes the number of congressional aides in the House of Representatives t years after 1930. In what year were approximately 9780 aides assigned to the House of Representatives?

56. Solve Problem 52 by first solving the given equation for P.

Solve the radical models in Problems 57–59 for the indicated variable.

57. The approximate speed of a car (S, in miles per hour) prior to an accident can be estimated by the model $S = 2\sqrt{5L}$, where L is the length of the skid marks left by the car (in feet). Solve for L.

58. The model $v = \sqrt{2gR}$ describes the velocity v needed for a spacecraft to escape a planet's gravitational field, where g is the planet's force of gravity and R is its radius. Solve for R.

59. The model

$$r = \sqrt[3]{\frac{GMt^2}{4\pi^2}}$$

is used in aerospace engineering to describe the radius of the orbit of a satellite, where t represents the time it takes the satellite to complete one orbit, G represents the constant of universal gravitation, and M is the mass of the satellite. Solve for M.

Photri, Inc.

60. Kepler's third law of planetary motion states that the ratio $\dfrac{T^3}{R^3}$ is the same for all planets in our solar system, where R is the average radius of the planet's orbit (measured in astronomical units) and T is the number of years it takes for one complete orbit around the sun. Saturn orbits the sun in 29.46 years with an average radius of 9.5 astronomical units. If Jupiter orbits the sun in 11.86 years, find the average radius of its orbit.

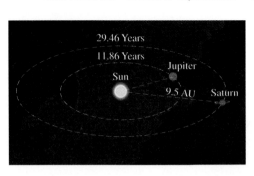

True–False Critical Thinking Problems

61. Which one of the following is true?

a. The best way to solve $\sqrt{x} + \sqrt{7} = 10$ is to square both sides, obtaining $x + 7 = 100$.

b. If $t = \sqrt{\dfrac{d}{16}}$, then $d = 4t^2$.

c. If $T = 2\pi\sqrt{\dfrac{L}{32}}$, then $L = \dfrac{8T^2}{\pi^2}$.

d. The equation $\sqrt{x^2 + 9x + 3} = -x$ has no solution because a principal square root is always non-negative.

62. Which one of the following is true?

a. The first step in solving $\sqrt{x + 6} = x + 2$ is to square both sides, obtaining $x + 6 = x^2 + 4$.

b. The equations $\sqrt{x + 4} = -5$ and $x + 4 = 25$ have the same solution set and are therefore equivalent.

c. Extraneous solutions are solutions that cannot be found.

d. The equation $-\sqrt{x} = 9$ has no solution.

Technology Problems

63. If you have not already done so, use a graphing utility to verify some of your solutions in Problems 1–48.

Use a graphing utility to solve the equations in Problems 64–67. Check by direct substitution.

64. $\sqrt{2x + 13} = x + 5$

65. $\sqrt{4 - x} - \sqrt{x + 6} = 2$

66. $\sqrt{5x + 1} = \sqrt{3x + 4} + \sqrt{x - 6}$

67. $\sqrt[3]{6x + 9} = -3$

68. Use a graphing utility to graph the models in Problems 49–51 and 54–55. Then use the $\boxed{\text{TRACE}}$ feature to trace along each curve until you reach the point that visually shows the problem's solution.

Writing in Mathematics

69. In solving $\sqrt{3x + 4} - \sqrt{2x + 4} = 2$, why is it a good idea to isolate the radical term? What if we don't do this and simply square each side? Describe what happens.

70. What is an extraneous solution to a radical equation?

71. a. Find the two values in the domain of $f(x) = \sqrt{5x - 1} - \sqrt{x + 2}$ that result in a range value of 1.

b. Because there are two values in the domain of f that result in one value in the range, doesn't this violate the definition of a function? Explain.

Critical Thinking Problems

72. Find the length of the three sides of the right triangle shown in the figure.

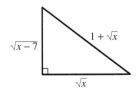

73. The model

$$T = \dfrac{T_0}{\sqrt{1 - \dfrac{v^2}{c^2}}}$$

relates the passage of time T on a futuristic starship moving at v miles per second to the passage of time for a stationary observer on Earth, T_0. (c, the speed of

light, is approximately 186,000 miles per second.) Suppose that 4 hours pass on the starship, but for you, on
Earth, 2 hours have passed. How fast is the starship
traveling?

74. Solve *without* squaring both sides: $5 - \dfrac{2}{x} = \sqrt{5 - \dfrac{2}{x}}$.

75. Solve for y: $\sqrt[3]{y\sqrt{y}} = 9$.

76. Find the smallest prime numbers x and y satisfying $y = \sqrt{x - 2} + 2$ if $x \neq y$.

77. Find the exact value of y such that

$$y = \sqrt{12 + \sqrt{12 + \sqrt{12 + \ldots}}}$$

78. Increase a number by 1 and take the square root. Increase this expression by 1 and again take the square
root. Add 1 for the third time and again take the
square root of the entire expression. The result is 2.
What was the original number?

Group Activity Problem

79. In many ways, the solution process in Example 6 on
page 561 is an algebraic minefield, filled with potential
disasters along the way. In your group, discuss errors
that can occur in the solution process. Then suggest
remedies for avoiding these errors.

Review Problems

80. Simplify: $\dfrac{\sqrt[3]{5}}{\sqrt[4]{5}}$.   Express the quotient in simplified
radical form.

81. The current I in an electrical conductor varies inversely as the resistance R of the conductor. The current is 0.5 ampere when the resistance is 240 ohms.
What is the current when the resistance is 460 ohms?

82. Factor: $x^2 - 6x + 9 - y^2$.

S E C T I O N   6 . 7

Solutions   Tutorial   Video
Manual   8

Imaginary and Complex Numbers

Objectives

1 Write the square root of a negative number in terms of *i*.
2 Add and subtract complex numbers.
3 Multiply complex numbers.
4 Verify an imaginary solution to an equation.
5 Simplify powers of *i*.
6 Divide complex numbers.

In Chapter 1, we studied the various subsets of the real numbers, including
the natural numbers, the whole numbers, the integers, the rational numbers,
and the irrational numbers. In this section, we introduce a new set of numbers
that contains the real numbers as a subset.

Imaginary and Complex Numbers

In Section 6.1, we emphasized the fact that negative numbers do not have
real number square roots. This means that as long as we confine ourselves to
the set of real numbers, the equation

$$x^2 = -1$$

has no solution, since there is no real number that when squared is -1.

To solve quadratic equations such as $x^2 = -1$, we need to define a new set
of numbers. In defining this set, we state that the number i has the following
properties.

The number *i*

The number *i* is defined to be the square root of -1. Thus,

$$i = \sqrt{-1} \quad \text{and} \quad i^2 = -1.$$

The number *i* is called the *imaginary unit*.

Write the square root of a negative number in terms of *i*.

The imaginary unit *i* enables us to express the square root of any negative number in terms of *i*.

The square root of a negative number

If *c* is a positive real number, then

$$\sqrt{-c} = \sqrt{-1 \cdot c} = \sqrt{-1} \cdot \sqrt{c} = i\sqrt{c}.$$

We can also write $\sqrt{c}i$, but note that *i* is not under the radical.

EXAMPLE 1 **Expressing Square Roots of Negative Numbers in Terms of *i***

Express in terms of *i*:

a. $\sqrt{-25}$ **b.** $\sqrt{-13}$ **c.** $\sqrt{-125}$ **d.** $-\sqrt{-17}$ **e.** $-\sqrt{-81}$

Solution

a. $\sqrt{-25} = \sqrt{-1 \cdot 25} = \sqrt{-1} \cdot \sqrt{25} = i \cdot 5 = 5i$

study tip

It is easy to confuse $\sqrt{13}i$ with $\sqrt{13i}$. Consequently, when i is multiplied by a radical factor, such as $i\sqrt{13}$, we will write the i factor first so that it is clear that the i is not under the radical. However, both $i\sqrt{13}$ and $\sqrt{13}i$ are correct.

b. $\sqrt{-13} = \sqrt{-1 \cdot 13} = \sqrt{-1} \cdot \sqrt{13} = i\sqrt{13}, \text{ or } \sqrt{13}i$

c. $\sqrt{-125} = \sqrt{-1 \cdot 125} = \sqrt{-1} \cdot \sqrt{125} = i\sqrt{125} = i\sqrt{25 \cdot 5} = i \cdot 5\sqrt{5}$
$$= 5i\sqrt{5}, \text{ or } 5\sqrt{5}i$$

d. $-\sqrt{-17} = -\sqrt{-1 \cdot 17} = -\sqrt{-1} \cdot \sqrt{17} = -i\sqrt{17}, \text{ or } -\sqrt{17}i$

e. $-\sqrt{-81} = -\sqrt{-1 \cdot 81} = -\sqrt{-1} \cdot \sqrt{81} = -i\sqrt{81} = -i \cdot 9 = -9i$ ■

Seventeenth-century French mathematician René Descartes, skeptical that these square roots of negative numbers would have any practical significance, called them *imaginary numbers*. In the 18th century, Swiss mathematician Leonhard Euler introduced the symbolism *i* to represent $\sqrt{-1}$.

Definition of an imaginary number

An imaginary number is a number that can be written in the form $a + bi$, where *a* and *b* are real numbers and $b \neq 0$.

Here are some examples of imaginary numbers:

$$a + bi$$

$$5i = 0 + 5i \qquad \text{Here, } a = 0 \text{ and } b = 5.$$
$$-9i = 0 + (-9i) \qquad \text{Here, } a = 0 \text{ and } b = -9.$$
$$i\sqrt{13} = 0 + \sqrt{13}i \qquad \text{Here, } a = 0 \text{ and } b = \sqrt{13}.$$
$$7 - 2i = 7 + (-2i) \qquad \text{Here, } a = 7 \text{ and } b = -2.$$

In the first three examples, $a = 0$. Such numbers are called *pure imaginary numbers*.

The definition of an imaginary number, $a + bi$, does not permit b to be zero. If b is allowed to be 0, the set of numbers $a + bi$ is called the *complex numbers*.

Rene Magritte (1898–1967) "La lunette d'approche" (The Field Glass) 1963, oil on canvas, 69 5/16 × 45 1/4 in. (176 × 115 cm). Photographer: Hickey–Robertson, Houston. The Menil Collection, Houston. © C. Herscovici, Brussels/Artists Rights Society (ARS), New York.

Definition of a complex number

A complex number is any number that can be put in the form $a + bi$, where a and b are real numbers and $i = \sqrt{-1}$. We call a the *real part* and b the *imaginary part*. This definition has some special cases:

1. If $b \neq 0$, $a + bi$ is called an *imaginary number*.
2. If $a = 0$ and $b \neq 0$, $a + bi = bi$ is called a *pure imaginary number*.
3. If $b = 0$, $a + bi = a$ is a *real number* (so every real number is complex).

A complex number expressed in the form $a + bi$ is in *standard form*.

Table 6.2 gives examples of complex numbers.

TABLE 6.2 Examples of Complex Numbers		
Complex Number $a + bi$	**Name**	**Values of** a **and** b
$7 + 4i$	Imaginary	$a = 7, b = 4$
$6 - i\sqrt{7}$ (or $6 + (-\sqrt{7})i$)	Imaginary	$a = 6, b = -\sqrt{7}$
17 (or $17 + 0i$)	Real	$a = 17, b = 0$
$6i$ (or $0 + 6i$)	Pure imaginary	$a = 0, b = 6$
$-i\sqrt{5}$ (or $0 + (-\sqrt{5})i$)	Pure imaginary	$a = 0, b = -\sqrt{5}$

Relationships among sets of numbers are illustrated in Figure 6.9.

ENRICHMENT ESSAY

Complex Numbers on a Postage Stamp

This stamp honors the work done by the German mathematician Carl Friedrich Gauss (1777–1855) with complex numbers. Gauss represented numbers of the form $a + bi$ as points in the plane.

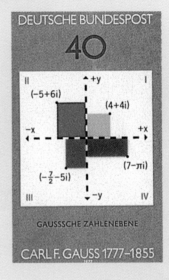

Complex Numbers

Real Numbers		Imaginary Numbers
Rational numbers $$\frac{2}{17}, \frac{-3}{25}, \frac{17}{2}$$	Irrational numbers $$\sqrt{3}, -\sqrt{7}, \pi$$	$7 + 4i$ $6 - i\sqrt{7}$

Integers
−13, −9

Whole
numbers
0, 17, 23

Pure
imaginary
numbers
$6i, -i\sqrt{5}$

Figure 6.9

Examples of Complex Numbers

2 Add and subtract complex numbers.

Adding and Subtracting Complex Numbers

Operations with complex numbers are very similar to operations with polynomials. Let's begin by considering the following rules for adding and subtracting complex numbers.

> **Adding and subtracting complex numbers**
>
> **1.** $(a + bi) + (c + di) = (a + c) + (b + d)i$
> In words, this says that complex numbers can be added by adding their real parts, adding their imaginary parts, and expressing the sum as a complex number.
> **2.** $(a + bi) - (c + di) = (a - c) + (b - d)i$
> In words, this says that complex numbers can be subtracted by subtracting their real parts, subtracting their imaginary parts, and expressing the difference as a complex number.

The following example illustrates these ideas.

EXAMPLE 2 **Adding and Subtracting Complex Numbers**

Add or subtract as indicated:

a. $(5 + 11i) + (7 + 16i)$ **b.** $8i + (-3 - i)$
c. $(18 + 13i) - (15 + 3i)$ **d.** $(-5 + 7i) - (-11 - 6i)$

Solution

a. $(5 + 11i) + (7 + 16i)$
$= (5 + 7) + (11 + 16)i$ Add real parts. Add imaginary parts.
$= 12 + 27i$ Simplify.

b. $8i + (-3 - i)$
$= (0 + 8i) + (-3 - i)$ $8i = 0 + 8i$. This step is optional.
$= [0 + (-3)] + [8 + (-1)]i$ Add real parts. Add imaginary parts.
$= -3 + 7i$ Simplify.

c. $(18 + 13i) - (15 + 3i)$
$= (18 - 15) + (13 - 3)i$ Subtract real parts. Subtract imaginary parts.
$= 3 + 10i$ Simplify.

d. $(-5 + 7i) - (-11 - 6i)$
$= [-5 - (-11)] + [7 - (-6)]i$ Subtract real parts. Subtract imaginary parts.
$= 6 + 13i$ Simplify. ∎

Imaginary numbers should be expressed in terms of i before they are added or subtracted.

EXAMPLE 3 **Adding Imaginary Numbers**

Add: $(12 - \sqrt{-32}) + (5 + \sqrt{-8})$

Solution

We begin by expressing each imaginary number in terms of i.

$$(12 - \sqrt{-32}) + (5 + \sqrt{-8})$$

$$= (12 - \sqrt{-1} \cdot \sqrt{16}\sqrt{2}) + (5 + \sqrt{-1} \cdot \sqrt{4}\sqrt{2}) \qquad \sqrt{-c} = \sqrt{-1} \cdot \sqrt{c}.$$
Simplify $\sqrt{32}$ and $\sqrt{8}$.

$$= (12 - 4i\sqrt{2}) + (5 + 2i\sqrt{2}) \qquad\qquad \text{Each imaginary number is now in terms of } i.$$

$$= (12 + 5) + (-4\sqrt{2} + 2\sqrt{2})i \qquad\qquad \text{Add real parts. Add imaginary parts.}$$

$$= 17 - 2\sqrt{2}i, \text{ or } 17 - 2i\sqrt{2} \qquad\qquad \text{Simplify.} \quad\blacksquare$$

3 Multiply complex numbers.

Multiplying Complex Numbers

Multiplication of complex numbers is performed the same way as multiplication of polynomials, using the distributive property and the FOIL method. After completing the multiplication, we replace i^2 with -1. This idea is illustrated in the next example.

> **EXAMPLE 4**   **Multiplying Complex Numbers**

Find:

a. $4i(3 - 5i)$ **b.** $(3 - 2i)(4 + 3i)$ **c.** $(4 - 7i)(4 + 7i)$

Solution

a. $4i(3 - 5i) = 4i(3) - 4i(5i)$ Distribute $4i$ throughout the parentheses.

$\qquad\qquad = 12i - 20i^2$ Multiply.

$\qquad\qquad = 12i - 20(-1)$ Replace i^2 with -1.

$\qquad\qquad = 20 + 12i$ Simplify.

b. $(3 - 2i)(4 + 3i)$

$\qquad\qquad$ F   O   I   L

$\qquad = 3(4) + 3(3i) + (-2i)(4) + (-2i)(3i)$ Use FOIL.

$\qquad = 12 + 9i - 8i - 6i^2$ Multiply.

$\qquad = 12 + i - 6(-1)$ Combine imaginary parts and replace i^2 with -1.

$\qquad = 12 + i + 6$ Simplify.

$\qquad = 18 + i$ Combine real parts.

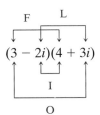

$$(3 - 2i)(4 + 3i)$$

c. $(4 - 7i)(4 + 7i)$

$\qquad\qquad$ F   O   I   L

$\qquad = 4(4) + 4(7i) + (-7i)(4) + (-7i)(7i)$ Use FOIL.

$\qquad = 16 + 28i - 28i - 49i^2$ Multiply.

$\qquad = 16 - 49(-1)$ $28i - 28i = 0i = 0$ and $i^2 = -1$

$\qquad = 16 + 49 = 65$ $\blacksquare$

Discover for yourself

Work Example 4c using

$$(A - B)(A + B) = A^2 - B^2$$

with

$$A = 4$$
and $\quad B = 7i.$

Which method do you find easier?

Study tip

Do not apply the properties

$$\sqrt{M}\,\sqrt{N} = \sqrt{MN} \quad \text{and} \quad \frac{\sqrt{M}}{\sqrt{N}} = \sqrt{\frac{M}{N}}$$

to the pure imaginary numbers because these properties can only be used when M and N are positive.

Correct:
$$\sqrt{-25}\,\sqrt{-4} = \sqrt{25}i\,\sqrt{4}i$$
$$= (5i)(2i)$$
$$= 10i^2$$
$$= 10(-1)$$
$$= -10$$

Incorrect:
$$\sqrt{-25}\,\sqrt{-4} = \sqrt{(-25)(-4)}$$
$$= \sqrt{100}$$
$$\neq 10$$

One way to avoid confusion is to *represent all imaginary numbers in terms of i before adding, subtracting, multiplying, or dividing.*

Example 5 illustrates the Study Tip.

EXAMPLE 5 **Multiplying Imaginary Numbers**

Find:

a. $\sqrt{-3}\,\sqrt{-5}$ **b.** $\sqrt{-6}\,\sqrt{-8}$ **c.** $\sqrt{-9}\,\sqrt{-100}$

d. $\sqrt{-4}(\sqrt{-2} + 9)$ **e.** $(4 - \sqrt{-2})(3 + \sqrt{-18})$

Solution

a. $\sqrt{-3}\sqrt{-5} = \sqrt{-1} \cdot \sqrt{3} \cdot \sqrt{-1} \cdot \sqrt{5}$

$\qquad\qquad = i\sqrt{3} \cdot i\sqrt{5}$ Express imaginary numbers in terms of i.

$\qquad\qquad = i^2\sqrt{15}$ Multiply.

$\qquad\qquad = -1 \cdot \sqrt{15}$ $i^2 = -1$

$\qquad\qquad = -\sqrt{15}$

b. $\sqrt{-6}\sqrt{-8} = \sqrt{-1} \cdot \sqrt{6} \cdot \sqrt{-1} \cdot \sqrt{8}$

$\qquad\qquad = i\sqrt{6} \cdot i\sqrt{8}$ Express imaginary numbers in terms of i.

$\qquad\qquad = i^2\sqrt{48}$ Multiply.

$\qquad\qquad = -1 \cdot \sqrt{16 \cdot 3}$ $i^2 = -1$; Simplify $\sqrt{48}$.

$\qquad\qquad = -4\sqrt{3}$

c. $\sqrt{-9}\,\sqrt{-100} = \sqrt{-1} \cdot \sqrt{9} \cdot \sqrt{-1} \cdot \sqrt{100}$

$\qquad\qquad\qquad = i \cdot 3 \cdot i \cdot 10$ Express imaginary numbers in terms of i.

$\qquad\qquad\qquad = i^2 \cdot 30$ Multiply.

$\qquad\qquad\qquad = -1 \cdot 30$ $i^2 = -1$

$\qquad\qquad\qquad = -30$

d. $\sqrt{-4}(\sqrt{-2} + 9)$

$= \sqrt{-1}\sqrt{4}(\sqrt{-1}\sqrt{2} + 9)$

$= 2i(i\sqrt{2} + 9)$ Express imaginary numbers in terms of i.

$= 2i(i\sqrt{2}) + 2i(9)$ Use the distributive property.

$= 2i^2\sqrt{2} + 18i$ Multiply.

$= -2\sqrt{2} + 18i$ $i^2 = -1$

e. $(4 - \sqrt{-2})(3 + \sqrt{-18})$

$= (4 - \sqrt{-1}\sqrt{2})(3 + \sqrt{-1}\sqrt{18})$

$= (4 - i\sqrt{2})(3 + i\sqrt{18})$ Express imaginary numbers in terms of i.

$= 12 + 4i\sqrt{18} - 3i\sqrt{2} - i^2\sqrt{36}$ Use FOIL.

$= 12 + 4i\sqrt{9}\sqrt{2} - 3i\sqrt{2} - (-1) \cdot 6$ Simplify $\sqrt{18}$. Replace i^2 by -1.

$= 12 + 12i\sqrt{2} - 3i\sqrt{2} + 6$ $4i\sqrt{9}\sqrt{2} = 4i \cdot 3\sqrt{2} = 12i\sqrt{2}$

$= 18 + 9i\sqrt{2}$ Combine real and imaginary parts. ∎

F   L

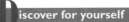

$(4 - i\sqrt{2})(3 + i\sqrt{18})$

I

O

4   Verify an imaginary solution to an equation.

In the next chapter, we will learn how to solve quadratic equations with imaginary solutions. Based on our work with adding and multiplying imaginary numbers, at this point we can verify that a given imaginary number is a solution to a quadratic equation.

EXAMPLE 6   **Verifying an Imaginary Solution to an Equation**

Show that $1 + 2i$ is a solution of the quadratic equation $x^2 - 2x + 5 = 0$.

Solution

$$x^2 - 2x + 5 = 0$$ This is the given quadratic equation.

$$(1 + 2i)^2 - 2(1 + 2i) + 5 \overset{?}{=} 0$$ Substitute $1 + 2i$ for x.

$$1 + 4i + 4i^2 - 2 - 4i + 5 \overset{?}{=} 0$$ Expand.

$$(1 - 2 + 5) + (4i - 4i) + 4i^2 \overset{?}{=} 0$$ Group terms that can be combined.

$$4 + 4i^2 \overset{?}{=} 0$$ Combine like terms.

$$4 + 4(-1) \overset{?}{=} 0$$ $i^2 = -1$

$$0 = 0 \quad \checkmark$$

This true statement verifies that $1 + 2i$ is a solution to the given equation. ∎

5   Simplify powers of i.

Powers of i

The fact that $i^2 = -1$ can be used to find higher powers of i.

Discover for yourself

Replace i^2 with -1 and express each of the following as $1, -1, i,$ or $-i$.

$i = i, i^2 = -1, i^3 = i^2 \cdot i = $ _____ $, i^4 = (i^2)^2 = $ _____ ,

$i^5 = (i^2)^2i = $ _____ $, i^6 = (i^2)^3 = $ _____ $, i^7 = (i^2)^3i = $ _____ $, i^8 = (i^2)^4 = $ _____

In the Discover for Yourself box, did you observe that the powers of i rotate through the four numbers $i, -1, -i,$ and 1?

> **Powers of i**
>
> | | | | |
> |---|---|---|---|
> | $i^1 = i$ | $i^5 = i$ | $i^9 = i$ | $i^{13} = i$ |
> | $i^2 = -1$ | $i^6 = -1$ | $i^{10} = -1$ | $i^{14} = -1$ |
> | $i^3 = -i$ | $i^7 = -i$ | $i^{11} = -i$ | $i^{15} = -i$ |
> | $i^4 = 1$ | $i^8 = 1$ | $i^{12} = 1$ | $i^{16} = 1$ |

One way to simplify higher powers of i is to write them in terms of i^2.

iscover for yourself

Since $i^4 = 1$, another method for simplifying higher powers of i is to write them in terms of i^4. Use this method to solve Example 7.

EXAMPLE 7 **Simplifying Powers of i**

Simplify: **a.** i^{12} **b.** i^{39} **c.** i^{50}

Solution

a. $i^{12} = (i^2)^6 = (-1)^6 = 1$
b. $i^{39} = i^{38} i = (i^2)^{19} i = (-1)^{19} i = (-1)i = -i$
c. $i^{50} = (i^2)^{25} = (-1)^{25} = -1$

    ■

6 Divide complex numbers.

Conjugates and Dividing Complex Numbers

An important idea in the division of complex numbers is found by multiplying $a + bi$ and $a - bi$. Use the special product $(A + B)(A - B) = A^2 - B^2$.

$$
\begin{array}{cccccc}
(A & + & B) & (A & - & B) & = & A^2 & - & B^2 \\
\downarrow & & \downarrow & \downarrow & & \downarrow & & \downarrow & & \downarrow
\end{array}
$$

$$
\begin{aligned}
(a + bi)(a - bi) &= a^2 - (bi)^2 && \text{Use the special product.} \\
&= a^2 - b^2 i^2 && \text{Square each factor in parentheses.} \\
&= a^2 - b^2(-1) && i^2 = -1 \\
&= a^2 + b^2
\end{aligned}
$$

This product eliminates i. We say that $a + bi$ and $a - bi$ are *conjugates*. The multiplication of conjugates results in a real number.

> **Conjugate of a complex number**
>
> The conjugate of the complex number $a + bi$ is $a - bi$, and the conjugate of $a - bi$ is $a + bi$. The multiplication of conjugates gives a real number.
>
> $$(a + bi)(a - bi) = a^2 + b^2$$
> $$(a - bi)(a + bi) = a^2 + b^2$$

EXAMPLE 8 **Multiplying Conjugates**

Multiply: **a.** $(3 + 4i)(3 - 4i)$ **b.** $(7 - 6i)(7 + 6i)$

Solution

$$\begin{array}{c} (a + bi)(a - bi) = a^2 + b^2 \\ \downarrow \quad \downarrow \quad \downarrow \quad \downarrow \quad\quad \downarrow \quad \downarrow \quad \downarrow \end{array}$$

a. $(3 + 4i)(3 - 4i) = 3^2 + 4^2$
$$= 9 + 16$$
$$= 25$$

$$\begin{array}{c} (a - bi)(a + bi) = a^2 + b^2 \\ \downarrow \quad \downarrow \quad \downarrow \quad \downarrow \quad\quad \downarrow \quad \downarrow \end{array}$$

b. $(7 - 6i)(7 + 6i) = 7^2 + 6^2$
$$= 49 + 36$$
$$= 85$$

Discover for yourself

Consider the division problem

$$\frac{7 + 4i}{2 - 5i}.$$

What must be done to both the numerator and denominator for the denominator to contain only a real number?

Take a moment to work the Discover for Yourself box in the margin.

Did you observe that by multiplying the numerator and denominator by $2 + 5i$, the conjugate of the denominator, you obtain a real number in the denominator?

Now let's see how conjugates are used when dividing complex numbers.

EXAMPLE 9 **Using Conjugates to Divide Complex Numbers**

Divide: $7 + 4i$ by $2 - 5i$

Solution

First we write the problem as $\dfrac{7 + 4i}{2 - 5i}$. The conjugate of the denominator, $2 - 5i$, is $2 + 5i$, so we multiply the numerator and denominator by $2 + 5i$.

$$\frac{7 + 4i}{2 - 5i} = \frac{(7 + 4i)}{(2 - 5i)} \cdot \frac{(2 + 5i)}{(2 + 5i)}$$
Multiply the numerator and denominator by the conjugate of the denominator.

$$= \frac{\overset{F}{1}4 + \overset{O}{3}5i + \overset{I}{8}i + \overset{L}{2}0i^2}{2^2 + 5^2}$$
Use the FOIL method in the numerator and $(a - bi)(a + bi) = a^2 + b^2$ in the denominator.

$$= \frac{14 + 43i + 20(-1)}{29}$$
Replace i^2 by -1.

$$= \frac{-6 + 43i}{29}$$
Combine real parts in the numerator.

$$= -\frac{6}{29} + \frac{43}{29}i$$

Observe that the quotient is expressed in the form $a + bi$, with $a = -\frac{6}{29}$ and $b = \frac{43}{29}$.

> **Using conjugates to divide complex numbers**
>
> When dividing two complex numbers, first express the indicated division as a fraction and then multiply the numerator and the denominator by the conjugate of the denominator.

EXAMPLE 10 **Dividing Complex Numbers**

Divide, and simplify to the form $a + bi$: $\dfrac{10 + 6i}{5i}$

Solution

The conjugate of $0 + 5i$ is $0 - 5i$, or $-5i$, so we will multiply the numerator and denominator by $-5i$.

$$\frac{10 + 6i}{5i} = \frac{10 + 6i}{5i} \cdot \frac{-5i}{-5i}$$ Multiply by 1 using the conjugate of the denominator.

$$= \frac{-50i - 30i^2}{-25i^2}$$ Multiply using the distributive property in the numerator.

$$= \frac{-50i - 30(-1)}{-25(-1)}$$ $i^2 = -1$

$$= \frac{-50i + 30}{25}$$ Multiply.

$$= \frac{-50}{25}i + \frac{30}{25}$$ To write the quotient in the form $a + bi$, divide each term in the numerator by 25.

$$= \frac{6}{5} - 2i$$ Reverse the order of the terms and simplify.

The quotient is in the form $a + bi$, with $a = \dfrac{6}{5}$ and $b = -2$. ■

As we have seen throughout this section, imaginary numbers should be expressed in terms of i before performing any operations. This idea is reinforced in our final example.

EXAMPLE 11 **Dividing by Expressing Imaginary Numbers in Terms of i**

Divide: **a.** $\dfrac{\sqrt{-300}}{\sqrt{3}}$ **b.** $\dfrac{\sqrt{-180}}{\sqrt{-5}}$

Solution

a. $\dfrac{\sqrt{-300}}{\sqrt{3}} = \dfrac{i\sqrt{300}}{\sqrt{3}} = i\sqrt{\dfrac{300}{3}} = i\sqrt{100} = 10i$

b. $\dfrac{\sqrt{-180}}{\sqrt{-5}} = \dfrac{i\sqrt{180}}{i\sqrt{5}} = \dfrac{\cancel{i}\sqrt{180}}{\cancel{i}\sqrt{5}} = \dfrac{\sqrt{180}}{\sqrt{5}} = \sqrt{\dfrac{180}{5}} = \sqrt{36} = 6$ ■

ENRICHMENT ESSAY

The Most Complicated Mathematical Object Ever Known

A complex number $a + bi$ can be graphed in a system that looks like rectangular coordinates. In the complex plane, the horizontal axis represents the real part of the complex number and the vertical axis the imaginary part. We've graphed the two points corresponding to $3 + 2i$ and $-2 - 4i$.

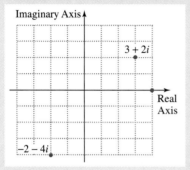

The complex plane

By graphing in the complex plane, we can construct dramatic visual images. The Mandelbrot set, named after mathematician Benoit Mandelbrot (1924–   ) has a graph that assumes an image that is "buglike" in shape. The complex numbers that appear as points inside the "bug" all have a special property. If a certain repeated process is applied to these numbers, the successive numbers that appear in this process remain small, and the sequence of numbers is bounded. Points that are not in the Mandelbrot set represent numbers for which this process results in an unbounded sequence of numbers. These points are shown in white.

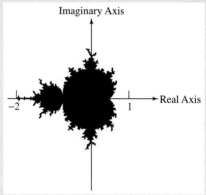

Graph of the Mandelbrot set

The process, or sequence, used to determine whether a complex number belongs to the Mandelbrot set is represented by

$$c, \quad c^2 + c, \quad (c^2 + c)^2 + c, \quad [(c^2 + c)^2 + c]^2 + c$$

Each successive number is obtained by adding c to the square of the previous result. Let's apply this process to four complex numbers to determine whether each is a member of the Mandelbrot set.

Complex Number c	c	$(c^2 + c)$	$(c^2 + c)^2 + c$	$[(c^2 + c)^2 + c]^2 + c$	Keep Squaring the Previous Result and Add c.
$c = 1$	1	$1^2 + 1 = 2$	$2^2 + 1 = 5$	$5^2 + 1 = 26$	$26^2 + 1 = 677$
		The numbers $1, 2, 5, 26, 677, \ldots$ are getting larger and larger. Thus, 1 is not a member of the Mandelbrot set.			
$c = -1$	-1	$(-1)^2 + (-1) = 0$	$0^2 + (-1) = -1$	$(-1)^2 + (-1) = 0$	$0^2 + (-1) = -1$
		The sequence of numbers $-1, 0, -1, 0, -1, \ldots$ remains small. Thus, -1 is a member of the Mandelbrot set.			
$c = i$	i	$i^2 + i = -1 + i$	$(-1 + i)^2 + i$ $= 1 - 2i + i^2 + i$ $= -i$	$(-i)^2 + i$ $= -1 + i$	$(-1 + i)^2 + i = -i$
		The sequence of numbers $i, -1 + i, -i, -1 + i, -i, \ldots$ remains small. Thus, i is a member of the Mandelbrot set.			
$c = 1 + i$	$1 + i$	$(1 + i)^2 + 1 + i = 1 + 3i$	$(1 + 3i)^2 + 1 + i$ $= -7 + 7i$	$(-7 + 7i)^2 + 1 + i$ $= 1 - 97i$	$(1 - 97i)^2 + 1 + i$ $= -9407 - 193i$
		The sequence of numbers $1 + i, 1 + 3i, -7 + 7i, 1 - 97i, -9407 - 193i$ does not remain small and is not bounded. Thus, $1 + i$ is not a member of the Mandelbrot set.			

Mathematicians and computer artists have added colors to the graph of the boundary of the Mandelbrot set. Color choices depend on how quickly the numbers in the boundary approach infinity. The magnified boundary shown in the photo yields repetition of the original overall buglike structure as well as new and interesting patterns. With each new level of magnification, repetition and unpredictable formations interact to create what has been called the most complicated mathematical object ever known.

The Mandelbrot set is a mathematical *fractal*, a set of points that exhibits increasing detail with increasing magnification. Fractals are equally complex in their details as in their overall form. The blending of infinite repetition and infinite variety is one of the hallmarks of the Mandelbrot set.

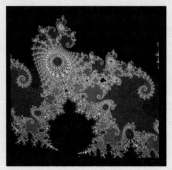

Homer Smith/Art Matrix

PROBLEM SET 6.7

Practice Problems

In Problems 1–12, express each number in terms of i, and simplify if possible.

1. $\sqrt{-4}$ **2.** $\sqrt{-9}$ **3.** $\sqrt{-17}$ **4.** $\sqrt{-23}$ **5.** $\sqrt{-28}$ **6.** $\sqrt{-32}$

7. $\sqrt{-45}$ **8.** $\sqrt{-18}$ **9.** $\sqrt{-\dfrac{4}{9}}$ **10.** $\sqrt{\dfrac{-36}{49}}$ **11.** $5\sqrt{-12}$ **12.** $4\sqrt{-27}$

In Problems 13–38, add or subtract as indicated.

13. $(3 + 2i) + (5 + i)$ **14.** $(6 - 5i) + (4 + 3i)$ **15.** $(7 + 2i) + (1 - 4i)$ **16.** $(-2 + 6i) + (4 - i)$
17. $(3 + 2i) - (5 + i)$ **18.** $(6 - 5i) - (4 + 3i)$ **19.** $(7 + 2i) - (1 - 4i)$ **20.** $(-2 + 6i) - (4 - i)$
21. $(2 + i\sqrt{3}) + (7 + 4i\sqrt{3})$ **22.** $(8 - i\sqrt{5}) + (13 + 7i\sqrt{5})$ **23.** $(5 + 2i\sqrt{32}) + (11 - 5i\sqrt{8})$
24. $(3 + 7i\sqrt{8}) + (-10 + 2i\sqrt{50})$ **25.** $(5 + 2i\sqrt{32}) - (11 - 5i\sqrt{8})$ **26.** $(3 + 7i\sqrt{8}) - (-10 - 2i\sqrt{50})$
27. $\sqrt{-49} + \sqrt{-100}$ **28.** $\sqrt{-16} + \sqrt{-81}$ **29.** $\sqrt{-64} - \sqrt{-25}$ **30.** $\sqrt{-81} - \sqrt{-144}$
31. $3\sqrt{-49} + 5\sqrt{-100}$ **32.** $5\sqrt{-16} - 3\sqrt{-81}$ **33.** $\sqrt{-72} + \sqrt{-50}$ **34.** $\sqrt{-32} - \sqrt{-50}$
35. $5\sqrt{-8} - 3\sqrt{-18}$ **36.** $3\sqrt{-12} + 9\sqrt{-27}$ **37.** $\frac{3}{5}\sqrt{-50} + \frac{1}{2}\sqrt{-32}$ **38.** $\frac{5}{6}\sqrt{-108} - \frac{1}{4}\sqrt{-48}$

In Problems 39–74, find the product, expressing the answer in the form a + bi.

39. $(7 + 3i)(5 + 2i)$ **40.** $(8 + 2i)(9 + 5i)$ **41.** $(3 + 4i)(4 - 7i)$ **42.** $(3 + i)(5 - i)$
43. $(-5 - 4i)(3 + 7i)$ **44.** $(-4 - 8i)(3 + 9i)$ **45.** $(7 - 5i)(-2 - 3i)$ **46.** $(8 - 4i)(-3 - 9i)$
47. $(3 + 5i)(3 - 5i)$ **48.** $(2 + 7i)(2 - 7i)$ **49.** $(-5 + 3i)(-5 - 3i)$ **50.** $(-7 - 4i)(-7 + 4i)$
51. $(2 + 3i)^2$ **52.** $(5 - 2i)^2$ **53.** $4i(7 - 6i)$ **54.** $-5i(-3 - 2i)$
55. $\sqrt{-7}\sqrt{-2}$ **56.** $\sqrt{-5}\sqrt{-3}$ **57.** $\sqrt{-9}\sqrt{-4}$ **58.** $\sqrt{-25}\sqrt{-36}$
59. $\sqrt{-7}\sqrt{-25}$ **60.** $\sqrt{-3}\sqrt{-36}$ **61.** $\sqrt{-8}\sqrt{-3}$ **62.** $\sqrt{-9}\sqrt{-5}$
63. $(2\sqrt{-8})(3\sqrt{-6})$ **64.** $(2\sqrt{-3})(-4\sqrt{-54})$ **65.** $(3\sqrt{-5})(-4\sqrt{-12})$ **66.** $(3\sqrt{-7})(2\sqrt{-8})$
67. $\sqrt{-2}(3 - \sqrt{-8})$ **68.** $\sqrt{-4}(7 - \sqrt{-2})$ **69.** $\sqrt{-6}(2\sqrt{3} + \sqrt{-6})$ **70.** $\sqrt{-5}(2\sqrt{8} + \sqrt{-5})$
71. $(3 - \sqrt{-8})(4 + \sqrt{-2})$ **72.** $(5 - \sqrt{-8})(6 + \sqrt{-2})$ **73.** $(8 - \sqrt{-6})(2 - \sqrt{-3})$ **74.** $(5 - \sqrt{-10})(4 - \sqrt{-2})$

Use substitution in Problems 75–80 to determine if the given value of x is a solution to the equation.

75. $x^2 - 2x + 2 = 0; x = 1 + i$

76. $x^2 - 2x + 2 = 0; x = 1 - i$

77. $x^2 - 6x + 13 = 0; x = 3 - 2i$

78. $x^2 - 6x + 13 = 0; x = 3 + 2i$

79. $x^2 - 12x + 40 = 0; x = 5 + 2i$

80. $x^2 - 12x + 4 = 0; x = 5 - 2i$

81. Find each of the following powers of i.

    a. i^{14} **b.** i^{31} **c.** i^{22} **d.** i^{37}

82. Find each of the following powers of i.

    a. i^{17} **b.** i^{98} **c.** i^{75} **d.** i^{1000}

In Problems 83–106, divide, and simplify to the form a + bi.

83. $\dfrac{2i}{3-i}$

84. $\dfrac{3i}{4+i}$

85. $\dfrac{3-i}{2i}$

86. $\dfrac{4+i}{3i}$

87. $\dfrac{1+i}{1-i}$

88. $\dfrac{1-i}{1+i}$

89. $\dfrac{2+3i}{3-i}$

90. $\dfrac{2-3i}{3+i}$

91. $\dfrac{3+2i}{2+i}$

92. $\dfrac{3-4i}{4+3i}$

93. $\dfrac{-4+7i}{-2-5i}$

94. $\dfrac{-3-i}{-4-3i}$

95. $\dfrac{8-5i}{2i}$

96. $\dfrac{3+4i}{5i}$

97. $\dfrac{4+7i}{-3i}$

98. $\dfrac{5+i}{-4i}$

99. $\dfrac{7}{3i}$

100. $\dfrac{5}{7i}$

101. $\dfrac{\sqrt{-125}}{\sqrt{5}}$

102. $\dfrac{\sqrt{-24}}{\sqrt{6}}$

103. $\dfrac{\sqrt{-24}}{\sqrt{-6}}$

104. $\dfrac{\sqrt{-125}}{\sqrt{-5}}$

105. $\dfrac{\sqrt{-200}}{\sqrt{5}}$

106. $\dfrac{\sqrt{-135}}{\sqrt{3}}$

Application Problems

Complex numbers are used in electronics to describe the current in an electric circuit. Ohm's law relates the current in a circuit (I, in amperes), the voltage of the circuit (E, in volts), and the resistance of the circuit (R, in ohms) by the model E = IR. Use this formula to solve Problems 107–108.

107. Find E if $I = 4 - 5i$ amperes and $R = 3 + 7i$ ohms.

108. Find E if $I = 2 - 3i$ amperes and $R = 3 + 5i$ ohms.

109. The mathematician Girolamo Cardano is credited with the first use (in 1545) of negative square roots in solving the now-famous problem, "Find two numbers whose sum is 10 and whose product is 40." Show that the complex numbers $5 + i\sqrt{15}$ and $5 - i\sqrt{15}$ satisfy the conditions of the problem. (Cardano did not use the symbolism $i\sqrt{15}$ or even $\sqrt{-15}$. He wrote R̶.m 15 for $\sqrt{-15}$, meaning "radix minus 15." He regarded the numbers $5 + $ R̶.m 15 and $5 - $ R̶.m 15 as "fictitious" or "ghost numbers," and considered the problem "manifestly impossible." But in a mathematically adventurous spirit, he exclaimed, "Nevertheless, we will operate.")

The reactance of an electrical circuit is represented by a pure imaginary number and is found by the formula $X = X_L - X_C$. X_L represents the inductive reactance and X_C represents the reactance from all capacitors. Use this formula to solve Problems 110–111.

110. If the reactance of a circuit is $16i$ ohms and the capacitive reactance is $7i$ ohms, find the inductive reactance.

111. If the reactance of a circuit is $22i$ ohms and the capacitive reactance is $9i$ ohms, find the inductive reactance.

True–False Critical Thinking Problems

112. Which one of the following is true?

    a. The conjugate of $7 + 3i$ is $-7 - 3i$.

    b. $(3 + 7i)(3 - 7i)$ is an imaginary number.

    c. $\dfrac{7 + 3i}{5 + 3i} = \dfrac{7}{5}$

    d. In the complex number system $x^2 + y^2$, the sum of two squares, can be factored as $(x + yi)(x - yi)$.

113. Which one of the following is true?

    a. Some irrational numbers are not complex numbers.

    b. The difference between a complex number $a + bi$ and its conjugate $(b \neq 0)$ is always an imaginary number.

    c. $(2 - i)^2 + 3(2 - i) - 5 = 4 + 5i$

    d. $i^{-1} = i$

Technology Problem

114. Use a computer with appropriate software to generate images of fractals. (This software is available free of charge from many places on the Internet.)

Writing in Mathematics

115. A stand-up comedian uses algebra in some jokes, including one about a telephone recording that announces "You have just reached an imaginary number. Please multiply by i and dial again." Explain the joke.

Problems 116–120 contain radical imaginary disasters! Explain the error in each problem.

116. $-\sqrt{-25} = -(-\sqrt{25}) = \sqrt{25} = 5$

117. $(\sqrt{-4})^2 = \sqrt{-4}\sqrt{-4} = \sqrt{16} = 4$

118. $\sqrt{-9} + \sqrt{-36} = \sqrt{-45} = i\sqrt{45} = i\sqrt{(9)(5)} = 3i\sqrt{5}$

119. $\dfrac{\sqrt{-20}}{\sqrt{-5}} = \sqrt{\dfrac{-20}{-5}} = \sqrt{4} = 2$

120. The only square root of negative 1 is i.

121. Write a research paper on fractals, consulting appropriate references. In your paper, be sure to cover the following concepts: algorithm, iterations, iteration number, chaos, and fractals in nature.

Critical Thinking Problems

122. Use the procedure shown in the table in the enrichment essay on page 577 to determine if each of the following numbers is a member of the Mandelbrot set.

    a. 0   **b.** 2   **c.** $-i$ **d.** $\dfrac{i}{2}$ **e.** $1 - i$

    f. $2 - i$

123. Write in the form $a + bi$:

$$(8 + 9i)(2 - i) - (1 - i)(1 + i).$$

124. Show that $\sqrt[3]{1} = -\dfrac{1}{2} - \dfrac{\sqrt{3}}{2}i$ by showing that

$$\left(-\dfrac{1}{2} - \dfrac{\sqrt{3}}{2}i\right)^3 = 1.$$

125. Perform the operations indicated: $\dfrac{1 + i}{1 + 2i} + \dfrac{1 - i}{1 - 2i}$.

126. Perform the operations indicated:

$$\dfrac{(1 + i)(-1 + 2i) + (2 - i)}{2 - 3i}$$

127. Three consecutive integers have a sum of $\sqrt{-36}\sqrt{-4}$. Find the integers. (*Hint:* Simplify $\sqrt{-36}\sqrt{-4}$ before writing the equation.)

128. Simplify: i^{2n+1}.

129. Show that the opposite and reciprocal of i are the same number.

130. Simplify: $\left(\dfrac{i}{1 + i} + \dfrac{2 + i}{2 - i}\right)i$.

131. Find a complex number that is its own conjugate.

132. Find the flaw in this "proof" that $1 = -1$.

$$\sqrt{(-1)(-1)} = \sqrt{-1}\sqrt{-1}$$
$$\sqrt{1} = i \cdot i$$
$$1 = i^2$$

But $i^2 = -1$. Thus, $1 = -1$.

133. Find the flaw in this "proof" that $i = -i$.

$$i = i$$
$$\sqrt{-1} = \sqrt{-1}$$
$$\sqrt{\dfrac{-1}{1}} = \sqrt{\dfrac{1}{-1}}$$
$$\dfrac{\sqrt{-1}}{\sqrt{1}} = \dfrac{\sqrt{1}}{\sqrt{-1}}$$
$$\dfrac{i}{1} = \dfrac{1}{i} = \dfrac{-(-1)}{i} = \dfrac{-i^2}{i} = -i$$
$$i = -i$$

Review Problems

134. Divide: $\dfrac{4x^2 + 8x + 3}{2x^2 - x - 1} \div \dfrac{4x^2 + 12x + 9}{x^2 - 1}$.

135. The volume of a sphere varies directly as the cube of its radius. If the volume is 36π cubic meters when the radius is 3 meters, find the volume when the radius is 5 meters.

136. Solve the system:

$$x - 2y - z = 1$$
$$2x + y - z = -1$$
$$x + y + z = 2$$

C HAPTER PROJECT

The Chaotic World

Working through a mathematics problem, we perform a series of steps, achieve a result, and are usually satisfied when we check and see the answer is correct. For numerical answers, such as the ones we discover by evaluating functions, there is usually a clear pattern in the graphical representation for our problem. The graphs are simple lines and curves and each answer we record as a point on the plane falls nicely into place. If we made an error in our calculations, the point we record catches our attention because it is plotted outside the ordered flow of our curve. We look for order in our problems. Eventually, after working enough problems, we not only look for order, we expect order in the form of smooth curves and straight lines.

Thus far in our explorations of algebraic functions, a small change in our input value generally gives us a small change in our output value. For example, we assume that using input values for x such as 1.002 and 1.0019, in a given function produces output values for y which are close together. Imagine our surprise if $x = 1.002$ yielded $y = 1.1$ and $x = 1.0019$ yielded $y = 30,000$. If this happened for each minor change we made in our x-values, we would soon lose any expectation for orderly, smooth curves. We might even expect the opposite of order … chaos.

We have a glimpse of mathematical chaos in the colorful representation of a portion of the Mandlebrot set, discussed on pages 577 and 578. The complex, lacy detail of the Mandlebrot set was created from a very simple mathematical process. In this case, a simple function did not yield a simple curve. Very small changes in input values at the beginning of the process sometimes created wildly different output values by the end of the process. For this graph, different colors were assigned to show just how different the output values would become. In mathematical terms, we say there is a sensitive dependence on initial conditions. A more poetic way to describe how one very, very small change may have consequences we can barely imagine is given by:

> *For want of a nail, the shoe was lost,*
> *For want of a shoe, the horse was lost,*
> *For want of a horse, the rider was lost,*
> *For want of a rider, the battle was lost,*
> *For want of a battle, the kingdom was lost.*

Chaos, in the mathematical sense, does not mean a complete lack of form or arrangement. In mathematics, chaos is used to describe something that only appears random, but is not random. The Mandlebrot set looks like an artistic, random splash of colors, but it is not. The Mandlebrot set is formed from a very precise, ordered process. Each time an input is used, an output will occur. What is not certain, however, is what will happen when we change an input by a very small amount. We can follow any input to a result, but we are lost when we try to predict what a new input will give us.

For this project, work in a small group on one of the real world applications of chaos described below. Each of the topics has many different avenues for exploration. As the main part of your research, conduct interviews with people actually using the applications described. You may find people to interview in your own town, or you may wish to use letters, e-mail or telephone conversations. Before you begin to talk to people, sit down with the members of your group and decide on what questions you wish to ask. You may also decide on how the interviews will be conducted. For example, will you use a tape recorder, video camera, or hand-written notes? How many people will you interview? Will all of the people in your group conduct interviews, or will some be assigned other tasks? Prepare a presentation for your class when you have completed the assignment.

1. Weather prediction is a major area of concern for many activities. The launch of a space shuttle, the evacuation of people in the path of a dangerous storm, or the routes of travel for planes and ships are

only a few of the ways in which predicting the weather might have serious consequences. Discover what role chaos plays in determining how far in advance we are able to forecast accurately for any of the examples just given, or use an example of your own.

2. As you tune a radio, searching for a weak signal, you may hear the hiss and crackle of background noises intruding on your music. Noise is a serious problem when we are trying to communicate in many different mediums. *Noise* is the interference with our communications, *signal* is what we wish to transmit in as pure a form as possible. Television pictures and phone lines used for voice communications or computer data are other examples where the signal needs to be separated from the noise. Discover what role chaos plays in the noise of a particular mode of communication.

3. Chaos may even be found inside ourselves. The human brain suffers from its own sort of "noise." A seizure, or epilepsy, may be described as an uncontrolled storm of noise in the nerve cells of the brain. Irregularities in the heartbeat, some of them severe enough to cause a heart attack, or irregularities in our sleeping patterns, such as insomnia, are other forms of unpredictable behavior. Discover what role chaos plays in finding solutions or new methods of approach to medical problems.

Worldwide Web Resources

Go to the Prentice Hall website (http://www.prenhall.com/blitzer) to access other locations on the Internet that will allow you to further explore the concepts presented in this project.

Chapter Review

SUMMARY

1. **Roots and Radicals**
   a. If n is even: $\sqrt[n]{b} = a(b \geq 0)$ means $a^n = b$. Every positive real number has two nth roots that are opposites of each other. If $b < 0$, $\sqrt[n]{b}$ is not a real number.
   b. If n is odd: $\sqrt[n]{b} = a$ means $a^n = b$. Every real number has one nth root, with positive numbers having positive roots and negative numbers having negative roots.
   c. If n is even or odd: $\sqrt[n]{0} = 0$.

2. **Radical Functions**
   a. Any function in the form $f(x) = \sqrt[n]{P}$, where P is a polynomial in x, is a radical function. If n is even, $P \geq 0$.
   b. $f(x) = \sqrt{P}$ is a square root function ($P \geq 0$).

3. **Inverse Properties of nth Powers and nth Roots**
   a. If n is odd, then $\sqrt[n]{x^n} = x$.
   b. If n is even, then $\sqrt[n]{x^n} = |x|$.
   c. If it is known that x is nonnegative, then $\sqrt[n]{x^n} = x$.

4. **Rational Exponents**
   a. $b^{1/n} = \sqrt[n]{b}$
   b. $b^{m/n} = \sqrt[n]{b^m} = (\sqrt[n]{b})^m$. In radical form $\sqrt[n]{b^m}$, n is the index, $\sqrt{}$ the radical sign, b^m the radicand, and $\sqrt[n]{b^m}$ the radical or radical expression.
   c. $b^{-m/n} = \dfrac{1}{b^{m/n}}$
   d. *Properties of Rational Exponents:* If r and s are rational numbers,

1. $b^r \cdot b^s = b^{r+s}$ 2. $(b^r)^s = b^{rs}$

3. $(ab)^r = a^r b^r$ 4. $\left(\dfrac{a}{b}\right)^r = \dfrac{a^r}{b^r}$

5. $\dfrac{b^r}{b^s} = b^{r-s}$

e. *Radicals with Different Indices*
To simplify products and quotients with different indices, rewrite each radical using rational exponents, use the properties of rational exponents (under part (d)), and write the final answer as a radical.

f. *Factoring Expressions Containing Rational Exponents*
Factor out the lowest power of the variable that occurs in all terms.

5. Multiplying and Simplifying Radical Expressions

a. *The Product Rule:* $\sqrt[n]{M} \cdot \sqrt[n]{N} = \sqrt[n]{M \cdot N}$ If the nth roots of M and N are real numbers, then the product of two nth roots is the nth root of the product.

b. *Simplifying Using the Product Rule*
A radical expression with index n can be simplified when its radicand has factors that are perfect nth powers.

1. If the radicand has a numerical factor, express it as a product of two numbers, one of which is the largest perfect nth power.
2. Write each variable factor whose exponent is greater than n as the product of two factors, one of which is the largest perfect nth power.
3. Use the product rule to write the radical expression as the product of roots. Put all perfect nth powers (numbers and variables) under one radical sign.
4. Simplify the radical containing the perfect nth powers.

6. Simplified Radical Form
For a radical to be simplified, the following things must be true.

a. All perfect nth powers of the coefficient of the radicand have been simplified using $\sqrt[n]{MN} = \sqrt[n]{M}\sqrt[n]{N}$. ($\sqrt[n]{M}$ and $\sqrt[n]{N}$ are real numbers.)

b. The index of the radical is greater than any exponent that appears as a factor of the radicand. The critical idea is that $\sqrt[n]{x^n} = x$, assuming $x \geq 0$.

c. No radical appears in the denominator of a fraction. (The denominator is *rationalized* by multiplying the numerator and denominator by an expression that will create a perfect nth power in the denominator.) The property

$$\sqrt[n]{\dfrac{M}{N}} = \dfrac{\sqrt[n]{M}}{\sqrt[n]{N}}$$

($\sqrt[n]{M}$ and $\sqrt[n]{N}$ are real numbers) is frequently used ($N \neq 0$).

7. Operations with Radicals (Same Indices)

a. To add or subtract, simplify (if necessary) and then combine terms, where possible, using the distributive property.

b. To multiply or divide, use

$$\sqrt[n]{M}\sqrt[n]{N} = \sqrt[n]{MN} \quad \text{and} \quad \dfrac{\sqrt[n]{M}}{\sqrt[n]{N}} = \sqrt[n]{\dfrac{M}{N}}$$

where $\sqrt[n]{M}$ and $\sqrt[n]{N}$ are real numbers.

c. To rationalize the denominator of a fraction when the denominator is a binomial with one or more square roots, multiply the numerator and denominator by a binomial containing the same terms but whose second term has the opposite sign (the conjugate).

8. Radical Equations

a. Radical equations contain variables in one or more radicands.

b. *Procedure for Solving Radical Equations*

1. If necessary, rewrite the equation so that a radical containing a variable radicand is isolated on one side of the equation.
2. Raise both sides of the equation to a power that is the same as the index of the radical.
3. Combine like terms on both sides of the equation.
4. If the equation still contains a term with a variable in a radicand, repeat steps 1 through 3.
5. Solve the resulting equation, which no longer contains a radical, for the variable.
6. Substitute all proposed solutions in the original radical equation, eliminating extraneous solutions.

9. Complex Numbers

a. A complex number is of the form $a + bi$, where a and b are real, and $i = \sqrt{-1}$. (Because $i = \sqrt{-1}$, $i^2 = -1$.)

$a + 0i = a$	Real number
$a + bi \, (b \neq 0)$	Imaginary number
$0 + bi = bi$	Pure imaginary number
$0 + 0i = 0$	Zero

b. Simplify pure imaginary numbers using $\sqrt{-c} = \sqrt{-1}\sqrt{c} = i\sqrt{c}$. Represent all imaginary numbers in terms of i before adding, subtracting, multiplying, or dividing.

c. Addition and subtraction of complex numbers can be accomplished by using

$$(a + bi) + (c + di) = (a + c) + (b + d)i$$
$$(a + bi) - (c + di) = (a - c) + (b - d)i$$

d. Multiplication of complex numbers can be obtained in the same way that we multiply $a + b$ $c + d$, replacing i^2 by -1 after multiplying.

e. Powers of i can be simplified by writing them in terms of i^2, replacing i^2 by -1.

f. $a + bi$ and $a - bi$ are conjugates. Division of complex numbers can be accomplished by first expressing the division as a fraction and then multiplying the numerator and denominator by the conjugate of the denominator.

REVIEW PROBLEMS

If applicable, verify your work in this problem set using a graphing utility.
In Problems 1–5, find each root.

1. $\sqrt{81}$
2. $\sqrt{\dfrac{4}{49}}$
3. $\sqrt[3]{-27}$
4. $\sqrt[3]{\dfrac{64}{125}}$
5. $\sqrt[5]{-32}$

6. The height of a bamboo plant ($h(t)$ in inches) t weeks after the plant comes through the soil is modeled by the function $h(t) = 3\sqrt{t} - 0.23t$. Find and interpret $h(25)$.

Determine the domain of each function in Problems 7–9.

7. $f(x) = \sqrt{x - 2}$
8. $g(x) = \sqrt{100 - 4x}$
9. $h(x) = \sqrt[3]{x - 2}$

In Problems 10–12, graph each pair of functions in the same rectangular coordinate system. State each function's domain and range. Then describe the relationship of the graph of function g to the graph of function f. Use one of the following phrases: shift to the right; shift to the left; shift upward; shift downward.

10. $f(x) = \sqrt{x}, g(x) = \sqrt{x} + 2$
11. $f(x) = \sqrt{x}, g(x) = \sqrt{x + 2}$
12. $f(x) = \sqrt[3]{x}, g(x) = \sqrt[3]{x - 1}$

13. The function $L(v) = 500\sqrt{1 - \dfrac{v^2}{c^2}}$ describes the length of a 500-meter-tall starship moving at velocity v (in miles per second) from the perspective of an observer at rest. (c, the speed of light, is approximately 186,000 miles per second.) Find and interpret $L\,(167{,}400)$.

Evaluate each radical expression in Problems 14–19.

14. $\sqrt[3]{5^3}$
15. $\sqrt[3]{(-5)^3}$
16. $\sqrt{5^2}$
17. $\sqrt{(-5)^2}$
18. $\sqrt[6]{(-2)^6}$
19. $\sqrt[7]{(-2)^7}$

In Problems 20–22, evaluate.

20. $16^{3/2}$
21. $8^{-2/3}$
22. $\left(\dfrac{1}{32}\right)^{4/5}$

In Problems 23–26, simplify each expression using properties of exponents.

23. $(7x^{1/3})(4x^{1/4})$
24. $\dfrac{80y^{3/4}}{-20y^{1/5}}$
25. $(9x^{-2}y^{1/2})^{3/2}$
26. $\left(\dfrac{32x^5}{y^{2/3}}\right)^{1/5}$

Use rational exponents to simplify each radical expression in Problems 27–31.

27. $\sqrt[9]{a^6}$
28. $\sqrt[8]{x^2 y^4}$
29. $\sqrt[6]{2^4 x^4 y^2}$
30. $\sqrt{3}\sqrt[3]{3}$
31. $\dfrac{\sqrt[3]{2}}{\sqrt[4]{2}}$

In Problems 32–33, factor, and write the answer without negative exponents.

32. $x^{1/2} + x^{3/4}$
33. $4x^{-1/2} - 8x^{1/2}$

34. The distance (D, in miles) that a person can see is a function of the observer's height (H, in feet), modeled by $D = (2H)^{1/2}$. A person whose eyes are 6 feet above ground level is standing on a mountain that is 794 feet over water. Approximately how far to the horizon can that person see?

35. The function $f(d) = 0.07d^{3/2}$ is used by meteorologists to determine the duration of a storm, $f(d)$ in hours, whose diameter is d miles. Find and interpret $f(16)$.

For the remainder of these review problems, assume that all variables represent positive real numbers. Write each radical in simplified form in Problems 36–39.

36. $\sqrt{20x^3}$ **37.** $\sqrt[3]{54x^8y^6}$ **38.** $\sqrt[4]{32x^3y^{10}}$ **39.** $\sqrt[5]{64x^3y^{12}z^6}$

In Problems 40–43, multiply and simplify if possible.

40. $\sqrt{6x^3} \cdot \sqrt{4x^2}$ **41.** $\sqrt[3]{4x^2y} \cdot \sqrt[3]{4xy^5}$ **42.** $\sqrt[5]{8x^4y^3} \cdot \sqrt[5]{8xy^6}$ **43.** $\sqrt{x+1} \cdot \sqrt{x-1}$

44. Oceanographers use the model $v = 3\sqrt{d}$ to describe the velocity of a tsunami, a great sea wave produced by underwater earthquakes. In the model, v represents the velocity, in feet per second, of a tsunami as it approaches land, and d is the depth of the water in feet. Express the velocity at a depth of 75 feet in simplified radical form.

In Problems 45–48, divide and simplify if possible.

45. $\dfrac{\sqrt{48}}{\sqrt{2}}$ **46.** $\dfrac{-10\sqrt[3]{32}}{2\sqrt[3]{2}}$ **47.** $\dfrac{\sqrt[4]{64x^7}}{\sqrt[4]{2x^2}}$ **48.** $\dfrac{\sqrt{200x^3y^2}}{\sqrt{2x^{-2}y}}$

Use the quotient rule to simplify each of the following in Problems 49–51.

49. $\sqrt[3]{\dfrac{16}{125}}$ **50.** $\sqrt{\dfrac{x^3}{100y^4}}$ **51.** $\sqrt[4]{\dfrac{3y^5}{16x^{20}}}$

Simplify by rationalizing the denominator in Problems 52–58.

52. $\dfrac{4}{\sqrt{6}}$ **53.** $\sqrt{\dfrac{2}{7}}$ **54.** $\dfrac{12}{\sqrt[3]{9}}$ **55.** $\sqrt{\dfrac{2x}{5y}}$ **56.** $\dfrac{14}{\sqrt[3]{2x^2}}$ **57.** $\sqrt[4]{\dfrac{7}{3x}}$ **58.** $\dfrac{5}{\sqrt[5]{32x^4y}}$

The following radical expressions contain variables in the denominator's radicand. Simplify by first simplifying the radicals and then rationalizing the denominator or by first rationalizing the denominator and then simplifying the resulting radical expression.

59. $\sqrt{\dfrac{5}{8x^3}}$ **60.** $\sqrt{\dfrac{6x^5}{y^3}}$ **61.** $\dfrac{4}{\sqrt[3]{2x^5}}$ **62.** $\dfrac{18}{\sqrt[5]{32x^7y^4}}$

Add and subtract as indicated in Problems 63–68.

63. $5\sqrt{18} + 3\sqrt{8} - \sqrt{2}$ **64.** $6\sqrt{2x^3} - \sqrt{49x^2} + 7\sqrt{18x^3}$ **65.** $2\sqrt[3]{6} + 5\sqrt[3]{48}$

66. $7\sqrt[3]{16x^4} - 3\sqrt[3]{2x}$ **67.** $2x\sqrt[4]{32y^5} - 3y\sqrt[4]{162x^4y} + \sqrt[4]{2x^4y^4}$ **68.** $6\sqrt{\dfrac{1}{3}} + 4\sqrt{\dfrac{1}{2}}$

In Problems 69–77, find each of the following products and write all answers in simplified form.

69. $5\sqrt{3}(2\sqrt{6} + 4\sqrt{15})$ **70.** $\sqrt{2x}(\sqrt{6x} + 3\sqrt{x})$ **71.** $2\sqrt[3]{5}(3\sqrt[3]{50} - 4\sqrt[3]{2})$

72. $(5\sqrt{2} - 4\sqrt{3})(7\sqrt{2} + 3\sqrt{3})$ **73.** $(\sqrt{x} + \sqrt{11})(\sqrt{y} + \sqrt{11})$ **74.** $(\sqrt{7} + \sqrt{5})^2$

75. $(2\sqrt{3} - \sqrt{10})^2$ **76.** $(\sqrt{7} + \sqrt{13})(\sqrt{7} - \sqrt{13})$ **77.** $(7 + 3\sqrt{5})(7 - 3\sqrt{5})$

For Problems 78–85, rationalize the denominator, simplifying if possible.

78. $\dfrac{6}{\sqrt{3} - 1}$ **79.** $\dfrac{\sqrt{7}}{\sqrt{5} + \sqrt{3}}$ **80.** $\dfrac{7}{2\sqrt{5} - 3\sqrt{7}}$ **81.** $\dfrac{\sqrt{y} + 5}{\sqrt{y} - 3}$

82. $\dfrac{\sqrt{7} + \sqrt{3}}{\sqrt{7} - \sqrt{3}}$ **83.** $\dfrac{2\sqrt{x}}{\sqrt{x} + \sqrt{y}}$ **84.** $\dfrac{\sqrt{3a} + \sqrt{b}}{\sqrt{5a} + \sqrt{2b}}$ **85.** $\dfrac{3\sqrt{7} + 1}{\sqrt{3} - 4\sqrt{5}}$

86. The evaporation (E, in inches per day) on a large body of water is approximated by the model

$$E = \dfrac{w}{20\sqrt{a}}$$

where a represents the area of the water in square miles and w represents the average wind speed of the air over the water (in miles per hour). What is the daily evaporation for a lake that is 8 square miles on a day where the average wind speed is 12 miles per hour? Write the answer in simplified radical form.

In Problems 87–88, find the perimeter and area of the following figures. Write your answers in simplified radical form.

87.

88.

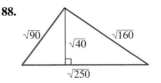

Solve each equation in Problems 89–94.

89. $\sqrt{2x + 4} = 6$

90. $\sqrt{x - 5} + 9 = 4$

91. $\sqrt{2x - 3} + x = 3$

92. $\sqrt[3]{x^2 + 6x} + 2 = 0$

93. $\sqrt{x - 4} + \sqrt{x + 1} = 5$

94. $2 - \sqrt{3y + 1} + \sqrt{y - 1} = 0$

95. The formula

$$r = \sqrt{\frac{V}{\pi h}}$$

describes the radius of a cylinder in terms its volume V and height h. A cylindrical hot water tank has a 2-foot radius and a 6-foot height.

a. Find the water tank's volume in terms of π.

b. A small business needs at least 650 gallons of hot water in storage at all times. If each cubic foot holds 8 gallons of water, will the tank described in this problem do the job, or will the business have to purchase a larger and more expensive tank?

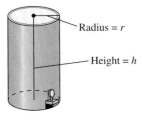

In Problems 96–97, solve for the indicated variable.

96. $t = \sqrt{\dfrac{2s}{g}}$ for s

97. $r = \sqrt[3]{\dfrac{2mM}{c}}$ for c

98. Police use the model $v = \sqrt{24L}$ to estimate the speed of a car (v in miles per hour) prior to an accident based on the length of its skid marks (L in feet) on dry pavement. If a car is traveling at 60 miles per hour, what is the length of its skid marks?

In Problems 99–101, express each number in terms of i and simplify.

99. $\sqrt{-81}$

100. $\sqrt{-63}$

101. $-\sqrt{-8}$

In Problems 102–113, perform all operations and write the answer in the form a + bi.

102. $(7 + 12i) + (5 - 10i)$

103. $(7 - 12i) - (-3 - 7i)$

104. $(7 - 5i)(2 + 3i)$

105. $(2 + 5i)^2$

106. $\dfrac{3i}{5 + i}$

107. $\dfrac{3 - 4i}{4 + 2i}$

108. $\dfrac{5 + i}{3i}$

109. i^{23}

110. $2\sqrt{-100} + 3\sqrt{-36}$

111. $\sqrt{-5}\sqrt{-9}$

112. $\dfrac{\sqrt{-24}}{\sqrt{-6}}$

113. $(2 + \sqrt{-8})(3 + \sqrt{-2})$

114. Show by substitution that $1 - 2i$ is a solution of the equation $x^2 - 2x + 5 = 0$.

115. An electrical engineer is designing the electrical circuits for a new office building. There are three things to consider in an electrical circuit: the flow of the electrical current (I), the resistance to that flow (R, called impedance), and the electromotive force (E, called voltage). These variables are related by the model $E = IR$. The current of the circuit that the engineer is designing is to be $35 - 40i$ amperes. Find the impedance of the circuit (in ohms) if the voltage is to be $430 - 330i$ volts. Express the answer in the form $a + bi$. (*Note:* If you consult an engineering textbook, you will find that electrical engineers use the letter j, rather than i, to represent the imaginary unit.)

CHAPTER 6 TEST

Assume that all variables represent positive real numbers.

1. Simplify: $\sqrt[3]{\dfrac{-8}{125}}$.

2. Determine the domain of f if $f(x) = \sqrt{8 - 2x}$.

3. Graph f and g in the same rectangular coordinate system and describe the relationship of the graph of g to the graph of f.

Use rational exponents to simplify Problems 6–7.

6. $\sqrt[8]{x^4}$

7. $\sqrt{2} \cdot \sqrt[3]{2}$

8. Factor, and write the answer without negative exponents: $6x^{1/2} + 12x^{-1/2}$.

9. The function $f(d) = 0.07d^{3/2}$ is used by meteorologists to determine the duration of a storm ($f(d)$, in hours) whose diameter is d miles. Find and interpret $f(9)$.

10. Simplify: $\sqrt[3]{16x^{10}y^{11}}$.

11. Multiply and simplify: $\sqrt[3]{4x^4} \cdot \sqrt[3]{2x^7}$.

$$f(x) = \sqrt{x} \qquad g(x) = \sqrt{x - 2}$$

4. Evaluate: $64^{-2/3}$.

5. Simplify: $(25x^{-4}y^{1/2})^{3/2}$.

12. The formula $s = \sqrt{30fd}$ is used by police to estimate the speed (s, in miles per hour) at which a car is traveling if it skids d feet after the brakes are applied. The variable f measures how slippery the road is and depends on whether the road is wet or dry and what it is made of. For example, on a tar road that is wet, $f = 0.5$. If a car skids 50 feet on such a surface, express in simplified radical form the speed at which it was moving when the brakes were applied.

13. Divide and simplify: $\dfrac{\sqrt[3]{32x^8}}{\sqrt[3]{2x^4}}$.

Rationalize each denominator in Problems 14–15.

14. $\sqrt{\dfrac{3}{5}}$

15. $\dfrac{5}{\sqrt[3]{5x^2}}$

16. Simplify and then rationalize the denominator: $\dfrac{3}{\sqrt{18x^5}}$.

In Problems 17–19, perform the indicated operations, writing all answers in simplified radical form.

17. $3\sqrt{18} + 4\sqrt{32} - 5\sqrt{8}$

18. $(4\sqrt{3} + \sqrt{2})(6\sqrt{3} - 5\sqrt{2})$

19. $(7 - \sqrt{3})^2$

20. Rationalize the denominator: $\dfrac{6}{\sqrt{5} - \sqrt{2}}$.

Solve Problems 21–22.

21. $\sqrt{x - 3} + 5 = x$

22. $\sqrt{x + 4} + \sqrt{x - 1} = 5$

23. Express the perimeter of the rectangle in the figure in simplified radical form.

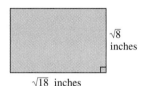

$\sqrt{8}$ inches

$\sqrt{18}$ inches

24. The period of a pendulum is given by

$$T = 2\pi\sqrt{\dfrac{L}{32}}$$

where T is the time (in seconds) and L is the length of the pendulum (in feet). If the period is $\dfrac{\pi}{4}$ seconds, find the pendulum's length.

25. Express in terms of i and simplify: $\sqrt{-75}$.

In Problems 26–30, perform the indicated operations and write in the form $a + bi$.

26. $(5 + 3i) - (-6 - 9i)$

27. $(3 - 4i)(2 + 5i)$

28. $(7 - 2i)^2$

29. $\dfrac{3 + i}{1 - 2i}$

30. $\sqrt{-25}\sqrt{-4}$

CUMULATIVE REVIEW PROBLEMS (CHAPTERS 1–6)

1. Use scientific notation to perform the following calculation. Express the answer in scientific notation:

$$\frac{(0.00072)(0.003)}{0.00024}.$$

2. Perform the operations indicated, simplifying the answer:

$$\frac{-60 \div (-2)(-3) + 5 - (-7) + 3}{81 \div (-9) + 3(-5) + 2 - (-2)}.$$

3. Solve, expressing the solution set in both set-builder and interval notation: $|16x - 8| > 4$.

4. Solve: $3x - 5 = 4(2x + 3) - 7$.

5. The graph shows U.S. cancer death rates from 1950 through 1980.
   a. Estimate the slope of the line representing lung cancer from 1960 through 1980. Interpret the slope in terms of the variables appearing in the graph.
   b. Write a constant function that serves as a reasonable model for prostate cancer death rates from 1950 through 1980.

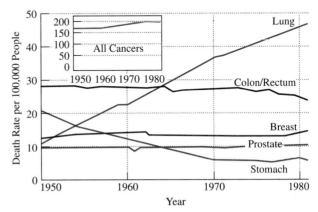

Source: *The New England Journal of Medicine,* May 1986

6. As shown in the figure, a square piece of cardboard has a 4-inch square cut out of each corner. The sides are then turned up to form an open box.
   a. If x represents the length and width of the original square piece of cardboard, express the volume of the box as a polynomial function, $V(x)$, written in descending powers of x.
   b. If the volume of the box is 256 cubic inches, find the dimensions of the piece of cardboard used to construct the box.

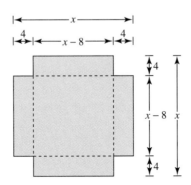

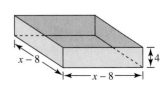

7. Write the point-slope form and then the slope-intercept form of the line passing through the points $(-4, -1)$ and $(3, 4)$.

8. Solve the system:

$$2x - y + z = -5$$
$$x - 2y - 3z = 6$$
$$x + y - 2z = 1$$

9. The graph shows mortality curves by age for men and women in the United States. Estimate the coordinates of the intersection point of the two curves. What does this mean in practical terms? Describe what happens to the right of this intersection point.

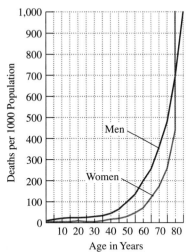

Source: U.S. Bureau of the Census

10. Graph: $x + 2y < 2$ and $2y - x > 4$.

11. Solve for W: $S = 2LW + 2LH + 2WH$.

12. The toll to a bridge is 50¢. One of the lanes has an automatic gate for exact change only. The automatic gate will not accept pennies. How many combinations of coins is the gate programmed to accept?

13. Factor completely: $3x^2y^4 - 48y^6$.

14. The model $f(t) = 0.002t^2 + 0.41t + 7.34$ describes the percent of the American population ($f(t)$) that graduated from college t years after 1960. Find and interpret $f(10)$.

15. According to the U.S. Bureau of Labor Statistics, in 1980 the average weekly earnings for workers in the United States was $270. This amount has increased by $14.60 each year. When will the average earnings reach $620.40?

16. Solve for R: $I = \dfrac{2V}{R + 2r}$.

17. An animal feed prepared by a veterinarian contains soybeans and corn. One unit of each ingredient provides units of protein and fat as shown in the table below. How many units of each ingredient should be used to make a feed that contains 14 units of protein and 18 units of fat?

	Protein	Fat
Corn	0.25	0.4
Soybeans	0.4	0.2

18. Divide and simplify: $\dfrac{8x^2}{3x^2 - 12} \div \dfrac{40}{2 - x}$.

19. Simplify: $\dfrac{x + \dfrac{1}{y}}{y + \dfrac{1}{x}}$.

20. Solve: $\dfrac{x}{x - 3} - \dfrac{3x}{x^2 - x - 6} = \dfrac{4x^2 - 4x - 18}{x^2 - x - 6}$.

21. The length of a rectangle is 1 inch more than twice the width. The perimeter is 110 inches. Find the rectangle's dimensions.

22. Multiply: $(2x - 3)(4x^2 - 5x - 2)$.

23. Use x-intercepts, the y-intercept, the vertex, and if necessary, one or two additional points to graph the parabola whose equation is $y = x^2 + 2x - 3$.

24. A business that sells small tables has fixed costs of $50,000 and variable costs of $20 per table.
  a. Find the cost function $C(x)$ that models the cost (fixed costs plus variable costs) of producing x tables.
  b. Find a function $\overline{C}$ that models the average cost per table for x tables.
  c. Find and interpret $\overline{C}(1000)$, $\overline{C}(10,000)$ and $\overline{C}(100,000)$. What conclusion can you draw?

25. The loudness of a stereo speaker varies inversely as the square of the distance of the listener from the speaker. If the loudness is 28 decibels at a distance of 8 feet, determine the loudness at a distance of 4 feet from the speaker.

26. The graph shows the suicide rate per 100,000 people in the United States from 1900 through 1991. Estimate the coordinates of the graph's maximum point. Describe what this means in terms of the variables in the graph. What was happening in the United States at that time that might account for this high suicide rate?

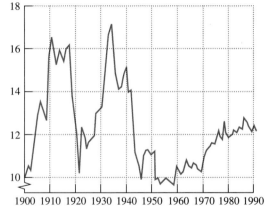

Suicides per 100,000 People

Source: National Institute of Mental Health

27. Write a polynomial in factored form (in terms of π and r) that models the area of the shaded region in the figure.

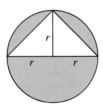

28. Find the quotient using either long division or synthetic division:

$$\frac{4x^3 + 12x^2 + x - 12}{x + 3}.$$

29. Simplify: $3\sqrt{2y^3} - y\sqrt{200y} + \sqrt{32y^3}$.

30. Perform the indicated operation and express the answer in the form $a + bi$: $\dfrac{16 - 15i}{6 - i}$.

Quadratic Equations and Functions

George Tooker (1920–) "Mirror II" 1963, egg tempera on gesso panel, 1968.4. Gift of R.H. Donnelley Erdman (PA 1956). Addison Gallery of American Art, Phillips Academy, Andover, Massachusetts. (Photo by Greg Heins).

The number of deaths per year for every thousand people and the number of automobile accidents for every 50 million miles driven are both functions of age. These functions can be written in the form $f(x) = ax^2 + bx + c$ and are quadratic.

A thorough analysis of quadratic models and their graphs is the focus of this chapter. The diversity of these models reiterates the theme that the world is astonishingly mathematical and, indeed, π is in the sky.

S E C T I O N   7 . 1

Solutions   Tutorial   Video
Manual   8

Richard Megna/Fundamental
Photographs

Solving Quadratic Equations by the Square Root Method

Objectives

1 Review quadratic functions and equations.
2 Solve quadratic equations by the square root method.
3 Solve applied problems using the square root method.
4 Determine the distance between two points.
5 Find a line segment's midpoint.

In this section, we begin by reviewing procedures for graphing quadratic functions and solving quadratic equations by factoring. We then introduce a second method for solving certain types of quadratic equations—the square root method.

Reviewing Quadratic Functions and Equations

1 Review quadratic functions and equations.

In Section 4.7 we saw that the graph of the *quadratic function* $y = ax^2 + bx + c$ or, equivalently, $f(x) = ax^2 + bx + c$, is called a *parabola* and is cuplike in shape. We also solved quadratic equations—equations that can be expressed in the form $ax^2 + bx + c = 0$—by factoring. Recall that the real solutions to the quadratic equation are the parabola's x-intercepts.

Information for graphing parabolas by hand is summarized as follows.

Pietro Longhi "Banquet
in the Casa Nani"
Ca'Rezzonico, Venice,
Italy. Scala/Art Resource,
NY.

Procedure for graphing quadratic functions

Graphing Parabolas

The graph of $y = ax^2 + bx + c, a \neq 0$, has

1. x-intercepts (if they exist) at values of x where $ax^2 + bx + c = 0$.
2. A y-intercept at $y = c$. This is found by setting x equal to 0.
3. A vertex when $x = -\dfrac{b}{2a}$. The vertex is a minimum point if $a > 0$ and a maximum point if $a < 0$.

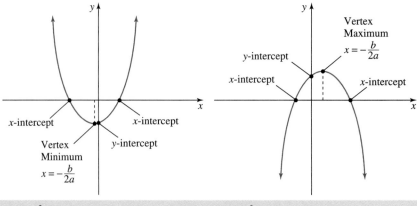

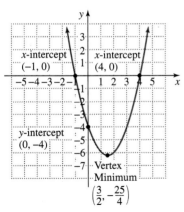

Figure 7.1

The graph of $y = x^2 - 3x - 4$

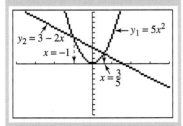

sing technology

We can verify that $\{-1, \frac{3}{5}\}$ is the solution set for $5x^2 = 3 - 2x$ in either of two ways.

Method 1. Graph $y_1 = 5x^2$ and $y_2 = 3 - 2x$ in the same viewing rectangle. The two intersection points have x-coordinates of -1 and $\frac{3}{5}$.

Method 2. Use the standard form of the equation $(5x^2 + 2x - 3 = 0)$. The x-intercepts of the resulting parabola are -1 and $\frac{3}{5}$.

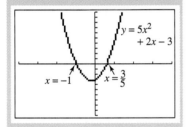

EXAMPLE 1 Using Intercepts and the Vertex to Graph a Quadratic Function

Graph by finding the intercepts and the vertex: $y = x^2 - 3x - 4$

Solution

Step 1. To find the x-intercepts, we set $y = 0$.

$$0 = x^2 - 3x - 4 \qquad \text{At each } x\text{-intercept, } y = 0.$$
$$0 = (x - 4)(x + 1) \qquad \text{Factor.}$$
$$x - 4 = 0 \quad \text{or} \quad x + 1 = 0 \qquad \text{Set each factor equal to 0.}$$
$$x = 4 \qquad\qquad x = -1 \qquad \text{Solve for } x.$$

There are two x-intercepts, one at 4 and one at -1.

Step 2. To find the y-intercept, we set $x = 0$.

$$y = x^2 - 3x - 4 = 0^2 - 3 \cdot 0 - 4 = -4$$

The y-intercept is -4.

Step 3. To find the vertex, we begin with the x-coordinate. We need to identify a and b of the form $ax^2 + bx + c$.

$$y = x^2 - 3x - 4 \qquad \text{Use the function's equation to identify } a \text{ and } b.$$
$$ \uparrow \quad\; \uparrow$$
$$ a = 1 \; b = -3$$

The x-coordinate of the vertex is

$$x = -\frac{b}{2a} = -\frac{(-3)}{2(1)} = \frac{3}{2}.$$

To find the y-coordinate of the vertex, we substitute $\frac{3}{2}$ for x into $y = x^2 - 3x - 4$, the function's equation.

$$y = x^2 - 3x - 4 = \left(\frac{3}{2}\right)^2 - 3\left(\frac{3}{2}\right) - 4 = \frac{9}{4} - \frac{9}{2} - \frac{4}{1} = \frac{9 - 18 - 16}{4} = -\frac{25}{4}$$

The vertex is $(\frac{3}{2}, -\frac{25}{4})$. Because $a = 1$ ($a > 0$), the vertex is a minimum point. The parabola, the graph of $y = x^2 - 3x - 4$, is shown in Figure 7.1. ■

In the next section we will prove that a parabola has a vertex when $x = -\dfrac{b}{2a}$. For now, let's look again at the factoring procedure for solving quadratic equations.

EXAMPLE 2 Solving a Quadratic Equation by Factoring

Solve: $5x^2 = 3 - 2x$

Solution

We begin by writing the quadratic equation in the standard form $ax^2 + bx + c = 0$. As always, we want 0 on the right side.

$$5x^2 = 3 - 2x \qquad \text{This is the given equation.}$$

$$5x^2 + 2x - 3 = 0 \qquad \text{Add } 2x \text{ and subtract 3 on both sides, getting 0 on the right.}$$

$$(5x - 3)(x + 1) = 0 \qquad \text{Factor.}$$

$$5x - 3 = 0 \quad \text{or} \quad x + 1 = 0 \qquad \text{Set each factor equal to 0.}$$

$$5x = 3 \qquad\qquad x = -1 \qquad \text{Solve the resulting equations.}$$

$$x = \frac{3}{5}$$

Check by substituting -1 and $\frac{3}{5}$ in the given equation. Two methods for verifying these solutions with a graphing utility are shown in the Using Technology box. The solution set is $\{-1, \frac{3}{5}\}$. ∎

2 Solve quadratic equations by the square root method.

The Square Root Method

Discover for yourself

Solve $x^2 = 36$ using two methods.

Method 1. Set the equation equal to 0, factor $x^2 - 36$ as the difference of squares, then set each factor equal to 0 and solve for x.

Method 2. The equation $x^2 = 36$ says that we must find the two square roots of 36. Based on this observation, what are the equation's solutions?

Which solution method is faster? What kinds of quadratic equations can be solved by the second method?

In the Discover for Yourself box, did you find out that to solve $x^2 = 36$, we can solve by factoring, but that method 2 is faster? We know that 36 has two square roots, namely -6 and 6, and these are the equation's solutions. Square roots provide a rapid method for solving equations in the form $x^2 = d$.

The square root method

If $x^2 = d$, then $x = \sqrt{d}$ or $x = -\sqrt{d}$.

EXAMPLE 3 **Solving a Quadratic Equation by the Square Root Method**

Solve: $4x^2 = 12$

Solution

To use the square root method, we must first isolate the variable on one side of the equation and then take the square root of both sides.

$$4x^2 = 12 \qquad \text{This is the given equation.}$$

$$x^2 = 3 \qquad \text{Isolate } x^2 \text{ by dividing both sides by 4.}$$

$$x = \sqrt{3} \quad \text{or} \quad x = -\sqrt{3} \qquad \text{Apply the square root method.}$$

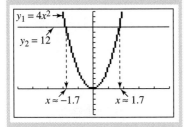

We can represent these values of x using the abbreviated notation $x = \pm\sqrt{3}$. Let's check each of these values.

Check $\sqrt{3}$:
$$4x^2 = 12$$
$$4(\sqrt{3})^2 \overset{?}{=} 12$$
$$4(3) \overset{?}{=} 12$$
$$12 = 12 \quad \checkmark$$

Check $-\sqrt{3}$:
$$4x^2 = 12$$
$$4(-\sqrt{3})^2 \overset{?}{=} 12$$
$$4(3) \overset{?}{=} 12$$
$$12 = 12 \quad \checkmark$$

This verifies that the equation's solution set is $\{-\sqrt{3}, \sqrt{3}\}$.

EXAMPLE 4 **Solving a Quadratic Equation by the Square Root Method**

Solve: $3x^2 - 16 = 0$

Solution

We begin by isolating x^2. We then apply the square root method.

$3x^2 - 16 = 0$	This is the given equation.
$3x^2 = 16$	Isolate the term with x^2 by adding 16 to both sides.
$x^2 = \dfrac{16}{3}$	Isolate x^2 by dividing both sides by 3.
$x = \sqrt{\dfrac{16}{3}} \quad \text{or} \quad x = -\sqrt{\dfrac{16}{3}}$	Apply the square root method.
$= \dfrac{4}{\sqrt{3}} \qquad\qquad = -\dfrac{4}{\sqrt{3}}$	The square root of a quotient is the quotient of the square roots.
$= \dfrac{4}{\sqrt{3}} \cdot \dfrac{\sqrt{3}}{\sqrt{3}} \quad = -\dfrac{4}{\sqrt{3}} \cdot \dfrac{\sqrt{3}}{\sqrt{3}}$	Simplify further by rationalizing denominators.
$= \dfrac{4\sqrt{3}}{3} \qquad\quad = -\dfrac{4\sqrt{3}}{3}$	We can abbreviate by writing $x = \pm\dfrac{4\sqrt{3}}{3}$.

Since the given equation has an x^2-term, but no x-term, we can check both solutions at once.

Check $\pm\dfrac{4\sqrt{3}}{3}$:

$$3x^2 - 16 = 0$$
$$3\left(\pm\frac{4\sqrt{3}}{3}\right)^2 - 16 \overset{?}{=} 0$$
$$3\left(\frac{16 \cdot 3}{9}\right) - 16 \overset{?}{=} 0$$
$$3\left(\frac{\overset{}{48}}{\underset{3}{9}}\right) - 16 \overset{?}{=} 0$$
$$16 - 16 \overset{?}{=} 0$$
$$0 = 0 \quad \checkmark$$

This verifies that the equation's solution set is $\left\{\dfrac{-4\sqrt{3}}{3}, \dfrac{4\sqrt{3}}{3}\right\}$.

Some quadratic equations have solutions that are imaginary numbers, as illustrated in Example 5.

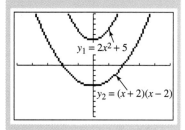

EXAMPLE 5 A Quadratic Equation with Pure Imaginary Solutions

Solve: $2x^2 + 5 = (x + 2)(x - 2)$

Solution

$2x^2 + 5 = (x + 2)(x - 2)$	This is the given equation.
$2x^2 + 5 = x^2 - 4$	Multiply the factors on the right.
$x^2 = -9$	Isolate x^2, by subtracting 5 and x^2 from both sides.
$x = \sqrt{-9} \quad$ or $\quad x = -\sqrt{-9}$	Apply the square root method.
$= 3i \qquad\qquad\quad = -3i$	$\sqrt{-9} = \sqrt{9(-1)} = \sqrt{9}\sqrt{-1} = 3i$

Let's check one of these numbers, $3i$, in the original equation.

$2x^2 + 5 = (x + 2)(x - 2)$	This is the original equation.
$2(3i)^2 + 5 \stackrel{?}{=} (3i + 2)(3i - 2)$	Replace x with $3i$.
$2(9i^2) + 5 \stackrel{?}{=} 9i^2 - 6i + 6i - 4$	Multiply.
$18i^2 + 5 \stackrel{?}{=} 9i^2 - 4$	$-6i + 6i = 0$
$18(-1) + 5 \stackrel{?}{=} 9(-1) - 4$	Since $i^2 = -1$, substitute -1 for i^2.
$-18 + 5 \stackrel{?}{=} -9 - 4$	Multiply.
$-13 = -13 \quad \checkmark$	This true statement indicates that $3i$ is a solution.

You should verify that $-3i$ is also a solution. The solution set is $\{-3i, 3i\}$. ∎

The square root method tells us that if $x^2 = d$, then $x = \sqrt{d}$ or $x = -\sqrt{d}$. We can use the method to solve equations such as $(y - 4)^2 = 49$ by substituting $(y - 4)^2$ for x^2 and 49 for d. We obtain

$y - 4 = \sqrt{49}$	or $\quad y - 4 = -\sqrt{49}$	Equivalently: $y - 4 = \pm\sqrt{49}$.
$y - 4 = 7$	$y - 4 = -7$	$\pm\sqrt{49} = \pm 7$
$y = 11$	$y = -3$	Add 4 to both sides.

Check that both 11 and -3 satisfy $(y - 4)^2 = 49$, verifying that $\{11, -3\}$ is the solution set.

EXAMPLE 6 Using the Square Root Method

Solve: $(2y - 7)^2 = 50$

Solution

$(2y - 7)^2 = 50$		This is the given equation.
$2y - 7 = \sqrt{50} \quad$ or $\quad 2y - 7 = -\sqrt{50}$		Apply the square root method.
$2y - 7 = 5\sqrt{2} \qquad\qquad 2y - 7 = -5\sqrt{2}$		$\sqrt{50} = \sqrt{25(2)} = 5\sqrt{2}$
$2y = 7 + 5\sqrt{2} \qquad\qquad 2y = 7 - 5\sqrt{2}$		Add 7 to both sides of each equation.
$y = \dfrac{7 + 5\sqrt{2}}{2} \qquad\qquad y = \dfrac{7 - 5\sqrt{2}}{2}$		Solve for y. Equivalently, $y = \dfrac{7 \pm 5\sqrt{2}}{2}$.

Check

Let's check one of these proposed solutions.

$$(2y - 7)^2 = 50 \qquad \text{This is the original equation.}$$

$$\left[2\left(\frac{7 + 5\sqrt{2}}{2}\right) - 7\right]^2 \overset{?}{=} 50 \qquad \text{Check } \frac{7 + 5\sqrt{2}}{2}, \text{substituting this value for } y.$$

$$(7 + 5\sqrt{2} - 7)^2 \overset{?}{=} 50 \qquad \text{Cancel 2 in the numerator and denominator.}$$

$$(5\sqrt{2})^2 \overset{?}{=} 50$$

$$25(2) \overset{?}{=} 50 \qquad \text{Square each factor.}$$

$$50 = 50 \quad \checkmark \quad \text{This true statement verifies that } \frac{7 + 5\sqrt{2}}{2} \text{ is a}$$

solution.

Now check $\dfrac{7 - 5\sqrt{2}}{2}$ to verify that the solution set is

$$\left\{\frac{7 + 5\sqrt{2}}{2}, \frac{7 - 5\sqrt{2}}{2}\right\}. \qquad \blacksquare$$

EXAMPLE 7 **Using the Square Root Method**

Solve: $(x + 4)^2 = -25$

Solution

$$(x + 4)^2 = -25 \qquad\qquad\quad \text{This is the given equation.}$$

$$x + 4 = \sqrt{-25} \quad \text{or} \quad x + 4 = -\sqrt{-25} \quad \text{Apply the square root method.}$$

$$x + 4 = 5i \qquad\qquad\quad x + 4 = -5i \qquad \sqrt{-25} = i\sqrt{25} = 5i$$

$$x = -4 + 5i \qquad\qquad x = -4 - 5i \qquad \text{Equivalently: } x = -4 \pm 5i$$

Check

Let's check one of these proposed solutions.

$$(x + 4)^2 = -25 \qquad \text{This is the original equation.}$$

$$(-4 + 5i + 4)^2 \overset{?}{=} -25 \qquad \text{Check } -4 + 5i, \text{substituting this value for } x.$$

$$(5i)^2 \overset{?}{=} -25 \qquad \text{Simplify in parentheses.}$$

$$25i^2 \overset{?}{=} -25 \qquad \text{Square each factor.}$$

$$25(-1) \overset{?}{=} -25 \qquad \text{Substitute } -1 \text{ for } i^2.$$

$$-25 = -25 \quad \checkmark \quad \text{This true statement indicates } -4 + 5i \text{ is a solution.}$$

Now check $-4 - 5i$ to verify that the solution set is $\{-4 + 5i, -4 - 5i\}$. $\blacksquare$

3 Solve applied problems using the square root method.

Applications of the Square Root Method

The ability to solve quadratic equations by the square root method enables us to gain increased understanding of models that contain variables that are squared. Let's first look at an example from biology.

EXAMPLE 8 **An Application: Growth of Bacteria**

At the beginning of an experiment, a culture contains 10^5 bacteria. After t hours, the number of bacteria present (B) is given by the formula

$B = 10^5(1 + 2t^2)$. After how many hours will the population contain 9×10^5 bacteria?

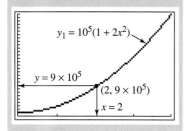
Solution

We must substitute 9×10^5 for B, solving the equation for t.

$B = 10^5(1 + 2t^2)$	This is the formula that models bacteria growth.
$9 \times 10^5 = 10^5(1 + 2t^2)$	Substitute 9×10^5 for B.
$9 = 1 + 2t^2$	Divide both sides by 10^5.
$8 = 2t^2$	Subtract 1 from both sides.
$4 = t^2$	Divide both sides by 2.

When you use the square root method to solve for the square of a variable that appears in a mathematical model, you will often take only the principal, or positive, root. In this example, a negative value for t, time, is meaningless. Thus,

$t = \sqrt{4}$ It is not necessary to write $t = \sqrt{4}$ or $t = -\sqrt{4}$ because t cannot be negative.

$t = 2$

Consequently, after 2 hours the number of bacteria present will be 9×10^5. ∎

EXAMPLE 9 Using the Square Root Method to Solve a Model for a Specified Variable

If \$400 is deposited in an account with the annual interest rate (r) compounded twice a year, the amount of money in the account after 1 year is described by the model

$$A = 400\left(1 + \frac{r}{2}\right)^2.$$

Solve the formula for r.

Solution

$400\left(1 + \dfrac{r}{2}\right)^2 = A$	Reverse the sides of the formula.
$\left(1 + \dfrac{r}{2}\right)^2 = \dfrac{A}{400}$	Divide both sides by 400.
$1 + \dfrac{r}{2} = \sqrt{\dfrac{A}{400}}$ or $1 + \dfrac{r}{2} = -\sqrt{\dfrac{A}{400}}$	Use the square root method.

Because r, the interest rate, must be positive, we will reject the equation with the negative square root.

$1 + \dfrac{r}{2} = \sqrt{\dfrac{A}{400}}$	Use the equation with the positive square root.
$1 + \dfrac{r}{2} = \dfrac{\sqrt{A}}{20}$	$\sqrt{\dfrac{A}{400}} = \dfrac{\sqrt{A}}{\sqrt{400}} = \dfrac{\sqrt{A}}{20}$
$20 + 10r = \sqrt{A}$	Clear fractions by multiplying both sides by 20.

$$10r = \sqrt{A} - 20 \qquad \text{Subtract 20 from both sides.}$$

$$r = \frac{\sqrt{A} - 20}{10} \qquad \text{Divide both sides by 10.}$$

Knowing the amount of money in the account at the end of the year, we can determine the interest rate using this result. If the amount was \$450.36, then

$$r = \frac{\sqrt{450.36} - 20}{10} \qquad \text{Use a graphing calculator:} \boxed{(} \boxed{\sqrt{}} 450.36 \boxed{-} 20 \boxed{)} \boxed{\div} 10$$
$$\boxed{\text{ENTER}}$$

$$r \approx 0.12$$

This indicates that with \$450.36 at year's end (we started with \$400), the annual interest rate is approximately 12%. ◼

The Pythagorean Theorem

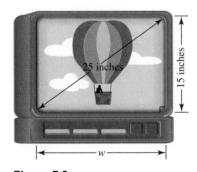

Figure 7.2

The Pythagorean Theorem:
$a^2 + b^2 = c^2$

We first studied the Pythagorean Theorem in Section 4.7. Recall that the Pythagorean Theorem expresses the relationship between the legs of a right triangle and its hypotenuse. As shown in Figure 7.2, the square of the hypotenuse is equal to the sum of the squares of the two legs.

We use the square root method to solve the Pythagorean Theorem for a, b, or c. As we did in our previous applications, we take only the positive square root since a length cannot be a negative number.

> **EXAMPLE 10**   **Using the Pythagorean Theorem**

In a 25-inch television set, the length of the screen's diagonal is 25 inches. If the screen's height is 15 inches, what is its width?

Solution

Figure 7.3 shows a right triangle that is formed by the height, width, and diagonal. We can find w, the screen's width, using the Pythagorean Theorem.

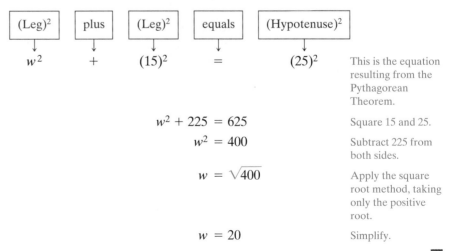

(Leg)²	plus	(Leg)²	equals	(Hypotenuse)²
w^2	$+$	$(15)^2$	$=$	$(25)^2$

This is the equation resulting from the Pythagorean Theorem.

$$w^2 + 225 = 625 \qquad \text{Square 15 and 25.}$$

$$w^2 = 400 \qquad \text{Subtract 225 from both sides.}$$

$$w = \sqrt{400} \qquad \text{Apply the square root method, taking only the positive root.}$$

$$w = 20 \qquad \text{Simplify.}$$

Figure 7.3

A right triangle is formed by the television's height, width, and diagonal.

The screen's width is 20 inches. ◼

Figure 7.4

An isosceles right triangle has legs that are the same length and acute angles each measuring 45°.

If the lengths of both legs of a right triangle are the same, the triangle is called an *isosceles right triangle,* shown in Figure 7.4. Our next example involves such a triangle.

EXAMPLE 11 **Using the Pythagorean Theorem**

A wire is attached to the ground 80 feet from the base of a building, as shown in Figure 7.5. If the wire reaches 80 feet up the building, how long is it?

Solution

Let w = the wire's length. We can find w in the isosceles right triangle using the Pythagorean Theorem.

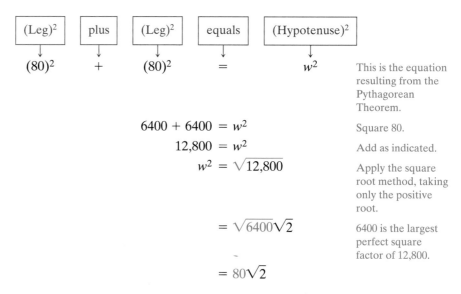

(Leg)²	plus	(Leg)²	equals	(Hypotenuse)²	
$(80)^2$	$+$	$(80)^2$	$=$	w^2	This is the equation resulting from the Pythagorean Theorem.

$$6400 + 6400 = w^2 \qquad \text{Square 80.}$$
$$12{,}800 = w^2 \qquad \text{Add as indicated.}$$
$$w^2 = \sqrt{12{,}800} \qquad \text{Apply the square root method, taking only the positive root.}$$
$$= \sqrt{6400}\sqrt{2} \qquad \text{6400 is the largest perfect square factor of 12,800.}$$
$$= 80\sqrt{2}$$

The length of the wire is $80\sqrt{2}$ feet, or (using a calculator) approximately 113 feet. ∎

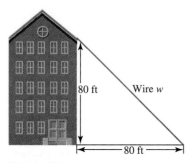

Figure 7.5

Finding a wire's length

Now take a few minutes to work the exercise in the Discover for Yourself box that appears in the margin.

In the Discover for Yourself box, did you discover the length of each hypotenuse to be $1\sqrt{2}$, $2\sqrt{2}$, and $3\sqrt{2}$? In general, the length of the hypotenuse in an isosceles right triangle is the length of a leg times $\sqrt{2}$. This relationship can be verified by the square root method. Consider the isosceles right triangle shown in Figure 7.6 in the margin at the top of the next page. We can find the length of the hypotenuse using the Pythagorean Theorem:

$$c^2 = a^2 + a^2 \qquad \text{The square of the hypotenuse equals the sum of the squares of the legs. Both legs are the same size.}$$
$$c^2 = 2a^2 \qquad \text{Add like terms.}$$
$$c = \sqrt{2a^2} \qquad \text{Apply the square root method, taking only the positive root.}$$
$$= \sqrt{a^2 \cdot 2} = a\sqrt{2} \qquad \text{Simplify.}$$

Discover for yourself

Use the Pythagorean Theorem to find the length of the hypotenuse in each isosceles right triangle.

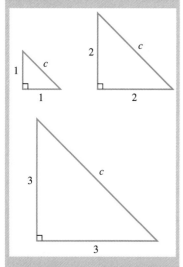

Express c in simplified radical form. In general, describe the length of the hypotenuse in terms of the length of a leg.

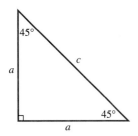

Figure 7.6

Lengths within an isosceles right triangle

The length of the hypotenuse in an isosceles right triangle is the length of a leg times $\sqrt{2}$.

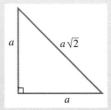

4 Determine the distance between two points.

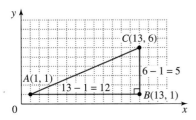

Figure 7.7

Using the Pythagorean Theorem to find the distance between $(1, 1)$ and $(13, 6)$

The Distance Formula

We now consider the problem of finding the distance between any two points in the rectangular coordinate system. When these points are located in a horizontal or a vertical manner, the problem is relatively simple.

Consider, for example, points A, B, and C shown in Figure 7.7. We can see that the (horizontal) distance from A to B is 12 units. We can find this mathematically by subtracting the x-coordinate of A from the x-coordinate of B.

$$\text{Distance} = x_{\text{right}} - x_{\text{left}} = 13 - 1 = 12$$

We can also see that the (vertical) distance from B to C is 5 units and can find this by subtracting the y-coordinate of B from the y-coordinate of C.

$$\text{Distance} = y_{\text{top}} - y_{\text{bottom}} = 6 - 1 = 5$$

We can now find the distance between A and C using the Pythagorean Theorem for the right triangle.

$$(AC)^2 = (AB)^2 + (BC)^2 \quad \text{Apply the Pythagorean Theorem.}$$
$$= 12^2 + 5^2$$
$$= 144 + 25$$
$$(AC)^2 = 169$$
$$AC = \sqrt{169} \quad \text{or} \quad AC = -\sqrt{169} \quad \text{Apply the square root method.}$$
$$= 13 \qquad\qquad\qquad\qquad\qquad \text{Take only the principal square root because distance is positive.}$$

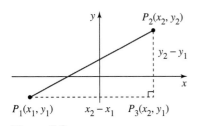

Figure 7.8

Using the Pythagorean Theorem to find the distance between two points

Let's now generalize this approach by considering any two points P_1 and P_2 in the rectangular coordinate system. P_1 is represented as (x_1, y_1) and P_2 as (x_2, y_2) (see Figure 7.8). P_3 has an x-coordinate identical to that of P_2 because P_3 and P_2 are the same distance horizontally from the y-axis. P_3 has a y-coordinate identical to that of P_1 because P_3 and P_1 are the same distance vertically from the x-axis. We see that

$$P_1P_3 = x_{\text{right}} - x_{\text{left}} = x_2 - x_1$$
$$P_3P_2 = y_{\text{top}} - y_{\text{bottom}} = y_2 - y_1$$

By the Pythagorean Theorem,

$$(P_1P_2)^2 = (P_1P_3)^2 + (P_3P_2)^2 \quad \text{The square of the hypotenuse equals the sum of the squares of the legs.}$$

$$(P_1P_2)^2 = (x_2 - x_1)^2 + (y_2 - y_1)^2 \quad \text{Substitute the expressions from above.}$$

Applying the square root method, we obtain

$$P_1P_2 = \sqrt{(x_2 - x_1)^2 + (y_2 - y_1)^2} \quad \text{or} \quad P_1P_2 = -\sqrt{(x_2 - x_1)^2 + (y_2 - y_1)^2}$$

$$P_1P_2 = \sqrt{(x_2 - x_1)^2 + (y_2 - y_1)^2}$$ Take only the principal square root because distance is positive.

We have proved the following result, which is often called the distance formula.

Finding the distance between two points

The distance formula

The distance d between the points (x_1, y_1) and (x_2, y_2) is

$$d = \sqrt{(x_2 - x_1)^2 + (y_2 - y_1)^2}.$$

EXAMPLE 12 **Using the Distance Formula**

Find the distance between $(-1, -3)$ and $(2, 3)$.

Solution

Letting $(x_1, y_1) = (-1, -3)$ and $(x_2, y_2) = (2, 3)$, we obtain

$$
\begin{aligned}
d &= \sqrt{(x_2 - x_1)^2 + (y_2 - y_1)^2} && \text{Use the distance formula.} \\
&= \sqrt{[2 - (-1)]^2 + [3 - (-3)]^2} && \text{Substitute the given values.} \\
&= \sqrt{(2 + 1)^2 + (3 + 3)^2} && \text{Perform subtractions within the grouping symbols.} \\
&= \sqrt{3^2 + 6^2} && \\
&= \sqrt{9 + 36} && \text{Square 3 and 6.} \\
&= \sqrt{45} && \text{Add.} \\
&= \sqrt{9 \cdot 5} && \text{9 is the largest perfect square factor of 45.} \\
&= 3\sqrt{5} && \text{Simplify.}
\end{aligned}
$$

The distance between the given points is $3\sqrt{5}$ units, or approximately 6.7 units.

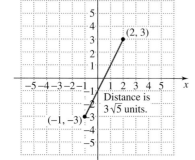

(2, 3)

Distance is $3\sqrt{5}$ units.

$(-1, -3)$

Once again, work the exercises in the Discover for Yourself box that appears in the margin.

In the Discover for Yourself, did you observe that it does not matter which point is selected to be (x_1, y_1)? It can also be verified that an equivalent form of the distance formula is

$$d = \sqrt{(x_1 - x_2)^2 + (y_1 - y_2)^2}.$$

iscover for yourself

Repeat Example 12 using $(x_1, y_1) = (2, 3)$ and $(x_2, y_2) = (-1, -3)$. Does it matter which point is called (x_1, y_1) and which is called (x_2, y_2)?

EXAMPLE 13 **An Application of the Distance Formula**

A rectangular coordinate system with coordinates in miles is superimposed over the map in Figure 7.9 on page 603. Bangkok has coordinates $(-115, 170)$ and Phnom Penh has coordinates $(65, 70)$. How long will it take an airplane averaging 400 miles per hour to fly directly from one city to the other?

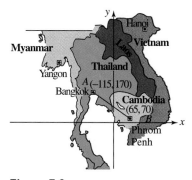

Figure 7.9

Superimposing rectangular co-ordinates over a map

5 Find a line segment's midpoint.

Solution

$$d = \sqrt{(x_2 - x_1)^2 + (y_2 - y_1)^2}$$ Use the distance formula.

$$= \sqrt{[65 - (-115)]^2 + (70 - 170)^2}$$ Let $(x_1, y_1) = (-115, 170)$ and $(x_2, y_2) = (65, 70)$.

$$= \sqrt{(180)^2 + (-100)^2}$$ Simplify.

$$= \sqrt{32,400 + 10,000}$$

$$= \sqrt{42,400}$$

$$= \sqrt{400}\sqrt{106}$$ 400 is the largest perfect square factor of 42,400.

$$= 20\sqrt{106} \approx 206$$

Because the cities are approximately 206 miles apart, the flight will take $\frac{206}{400}$, or 0.52 of an hour (approximately 31 minutes). ■

The Midpoint Formula

The distance formula can be used to prove a formula for finding the midpoint of a line segment between two given points. The formula is stated below and its proof is left for the problem set.

> **The midpoint formula**
>
> Consider a line segment whose endpoints are (x_1, y_1) and (x_2, y_2). The coordinates of the segment's midpoint are
>
> $$\left(\frac{x_1 + x_2}{2}, \frac{y_1 + y_2}{2} \right).$$
>
> To find the midpoint, take the average of the two x-values and of the two y-values.

EXAMPLE 14 **Using the Midpoint Formula**

Find the midpoint of the line segment with endpoints $(1, -6)$ and $(-8, -4)$.

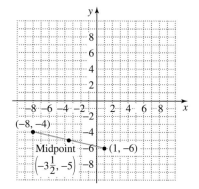

Figure 7.10

Finding a line segment's midpoint

Solution

To find the coordinates of the midpoint, we average the coordinates of the endpoints.

$$\text{Midpoint} = \left(\frac{1 + (-8)}{2}, \frac{-6 + (-4)}{2} \right) = \left(\frac{-7}{2}, \frac{-10}{2} \right) = \left(\frac{-7}{2}, -5 \right)$$

                    ↑   ↑

           Average   Average

            the   the

     x-coordinates. y-coordinates.

Figure 7.10 illustrates that the point $(-\frac{7}{2}, -5)$ is midway between the points $(1, -6)$ and $(-8, -4)$. ■

PROBLEM SET 7.1

Practice Problems

Graph Problems 1–8 using x-intercepts, the y-intercept, the vertex, and if necessary one or two additional points. Use the formula $x = -\dfrac{b}{2a}$ for the x-coordinate of the vertex.

1. $y = x^2 + 2x - 3$ **2.** $y = x^2 - 6x + 8$ **3.** $y = x^2 - 9x + 14$ **4.** $y = x^2 + 5x + 4$

5. $f(x) = 3x^2 - 8x + 4$ **6.** $f(x) = 2x^2 + 5x - 3$ **7.** $g(x) = 2x^2 + 5x - 7$ **8.** $g(x) = 2x^2 + x - 3$

Use factoring to solve each quadratic equation in Problems 9–18.

9. $8x^2 = 15 - 14x$ **10.** $12x^2 = 16x + 3$ **11.** $7x + 10 = 6x^2$ **12.** $15x - 2 = 28x^2$

13. $x(3x + 1) = 2$ **14.** $x(3x - 7) = 20$ **15.** $2(x^2 - 3) = 2(x + 3) + 3x$ **16.** $(3x + 2)(x - 1) = 4x$

17. $(x + 3)(x - 2) = 2(x - 2)(x + 2) + 2$ **18.** $(x + 1)^2 - 7 = 3x$

Use the square root method to solve each quadratic equation in Problems 19–60. Check any two of the solutions that are irrational and any two of the solutions that are not real numbers.

19. $y^2 = 100$ **20.** $x^2 = 144$ **21.** $y^2 = 7$ **22.** $x^2 = 17$

23. $x^2 = 75$ **24.** $x^2 = 24$ **25.** $z^2 = -4$ **26.** $z^2 = -25$

27. $4y^2 = 100$ **28.** $6y^2 = -24$ **29.** $3x^2 = 25$ **30.** $5x^2 = 36$

31. $7x^2 + 2 = 13$ **32.** $3x^2 - 5 = 19$ **33.** $4x^2 + 7 = 3(x^2 + 1)$ **34.** $5x^2 + 92 = 4(x^2 - 2)$

35. $4(x^2 + 2x) + 7 = 3x^2 + 8x + 2$ **36.** $6x^2 + 2x + 3 = (5x + 7)(x - 1)$

37. $(x + 4)(x + 1) = 5x - 71$ **38.** $-3x + (x - 10)(x + 10) + 17x = 14x - 300$

39. $(x + 7)^2 = 9$ **40.** $(x - 3)^2 = 25$ **41.** $(3y - 1)^2 = 16$ **42.** $(2y + 5)^2 = 36$

43. $(2x + 7)^2 = 5$ **44.** $(4y - 3)^2 = 11$ **45.** $(5y - 4)^2 = 24$ **46.** $(7z + 2)^2 = 8$

47. $(x - 3)^2 = -4$ **48.** $(x - 5)^2 = -49$ **49.** $(2y + 5)^2 = -5$ **50.** $(3y + 2)^2 = -3$

51. $(3z - 2)^2 = -50$ **52.** $(4z - 1)^2 = -18$ **53.** $3(5x - 4)^2 - 81 = 0$ **54.** $2(5x - 1)^2 - 300 = 0$

55. $4(2x + 5)^2 + 100 = 0$ **56.** $2(2x + 7)^2 + 400 = 0$ **57.** $4(5x - 3)^2 - 1 = 0$ **58.** $9(5y - 6)^2 - 1 = 0$

59. $(x - \frac{1}{3})^2 = \frac{1}{3}$ **60.** $(x - \frac{1}{2})^2 = \frac{1}{2}$

For Problems 61–70, plot both ordered pairs and find the distance between each of the points. Express answers involving radicals in simplified radical form.

61. $(4, -3)$ and $(-6, 2)$ **62.** $(6, -3)$ and $(-4, -5)$ **63.** $(3, 2)$ and $(6, 7)$ **64.** $(-3, 7)$ and $(5, 7)$

65. $(0, 0)$ and $(5, -12)$ **66.** $(\frac{7}{4}, -\frac{3}{2})$ and $(\frac{5}{2}, -\frac{3}{4})$ **67.** $(0, -3)$ and $(-3, 3)$ **68.** $(-3, 6)$ and $(3, 4)$

69. $(1, -2)$ and $(-3, 6)$ **70.** $(5, 7)$ and $(2, 3)$

71. Plot the following ordered pairs: $(5, 7)$, $(1, 10)$, and $(-3, -8)$. Then find the perimeter of the triangle whose vertices are the three points.

72. The vertices of a quadrilateral are $(-1, -5)$, $(-8, 2)$, $(7, 10)$, and $(4, -3)$. Find the length of the diagonals.

73. Plot the points $(2, 3)$, $(-1, -1)$, and $(3, -4)$. Then prove that the triangle connecting these points is isosceles.

74. Find the length of the radius of a circle with center at $(2, 3)$ and passing through $(8, -5)$.

75. Show that the points $A(-4, -6)$, $B(1, 0)$, and $C(11, 12)$ are collinear (lie along a straight line) by showing that the distance from A to B plus the distance from B to C equals the distance from A to C.

In Problems 76–89, determine the midpoint of the line segment with the given endpoints.

76. $(-2, 7)$ and $(3, -9)$ **77.** $(-8, -2)$ and $(-2, -6)$ **78.** $(-3, 1)$ and $(7, 4)$ **79.** $(4, -3)$ and $(-2, 8)$

80. $(-12.2, 8.2)$ and $(-6.4, 15.3)$ **81.** $(2.8, -8.6)$ and $(-3.2, 12.4)$ **82.** $(5, 0)$ and $(0, 8)$ **83.** $(3, 5)$ and $(-3, -5)$

84. $(\frac{3}{4}, \frac{1}{8})$ and $(-\frac{2}{3}, -\frac{4}{5})$ **85.** $(\frac{5}{6}, -\frac{1}{3})$ and $(-\frac{3}{4}, \frac{1}{2})$ **86.** $(8, 3\sqrt{5})$ and $(-6, 7\sqrt{5})$

87. $(7\sqrt{3}, -6)$ and $(3\sqrt{3}, -2)$ **88.** $(3\sqrt{18}, -4)$ and $(-5\sqrt{2}, 4)$ **89.** $(-8, 2\sqrt{27})$ and $(8, -4\sqrt{3})$

Application Problems _____

90. The weight of a human fetus is given by the formula $W = 3t^2$, where W is the weight in grams and t is the time in weeks, $0 \le t \le 39$. After how many weeks does the fetus weigh 300 grams?

91. At the beginning of an experiment, a culture contains 10^5 bacteria. After t hours, the number of bacteria present (B) is given by the formula $B = 10^5(1 + 2t^2)$. After how many hours will the population contain 33×10^5 bacteria?

92. When traveling at s miles per hour, the cost per hour (C) of operating a truck is given by the formula

$$C = 12 + \frac{s^2}{60}.$$

During a particular hour, the cost for truck operation was $72. What was the speed of the truck during that hour?

93. The area of a circle is 100π square feet. What is the radius of the circle? Express your answer in terms of π.

94. The volume of a right circular cylinder is the product of π, the square of the radius of its circular base and its height $(V = \pi r^2 h)$. Find the measure of the radius of a right circular cylinder having a volume of 75π cubic yards and a height of 3 yards.

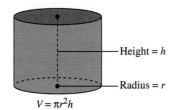

Height = h

Radius = r

$V = \pi r^2 h$

95. The length of a rectangle is 3 times its width. If the width is decreased by 1 yard and the length is increased by 3 yards, the area will be 72 square yards. Find the dimensions of the original rectangle.

96. A free-falling body will travel $16t^2$ feet in t seconds. When will an object dropped from a 300-meter cliff be 200 meters above the ground?

97. In Einstein's famous formula $E = mc^2$, one can obtain energy (E) equal to the quantity of the mass (m) multiplied by the square of the speed of light (c) (c is 186,000 miles per second). This equation indicates that 1 pound of any kind of matter contains enough energy to send a large vessel on at least 100 ocean voyages. Solve the formula for c.

98. The surface area of a sphere with radius r is given by the formula $A = 4\pi r^2$. Express the radius in terms of the area. (Solve for r.)

In Problems 99–102, solve each formula for the indicated variable. Use only the positive root.

99. $A = \pi r^2$ for r

100. $V = \pi r^2 h$ for r

101. $C = 12 + \dfrac{s^2}{60}$ for s

102. $a^2 + b^2 = c^2$ for b

Use the Pythagorean Theorem to solve Problems 103–110. Express irrational answers in simplified radical form. Then use a calculator to approximate the answer to the nearest tenth.

103. The length of a diagonal of a square is $3\sqrt{2}$ feet. Find the square's dimensions.

104. The hypotenuse of a right triangle measures 3 feet, and the legs are equal in length. Find the length of a leg of the triangle.

105. A baseball diamond is actually a square with 90-foot sides. What is the distance from home plate to second base?

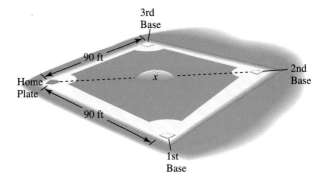

106. A 20-foot ladder is 15 feet from the house. How far up the house does the ladder reach?

107. The base of a 18-foot-long guy wire is 12 feet from a telephone pole that it is helping to support. How far up the pole does the wire reach?

108. A rectangular park is 4 miles long and 2 miles wide. How long is a pedestrian route that runs diagonally across the park?

109. Two flag poles are 42 feet and 49 feet high, respectively, and are 24 feet apart. How long is a wire from the top of one pole to the top of the second pole?

110. During the summer heat, concrete slabs tend to expand and will crack and buckle without proper expansion joints. The figure shows a concrete slab 6 feet long that expands $\frac{1}{4}$ inch due to the heat. Find the height, x, (to the nearest inch) to which the slab will rise.

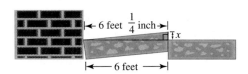

111. Plot the points $A(-3, 6)$, $B(2, -3)$, $C(11, 2)$, and $D(6, 11)$.
  a. Show that $ABCD$ is a parallelogram by using slope to show that both pairs of opposite sides are parallel.
  b. Show that $ABCD$ is a rhombus (a parallelogram with four equal sides) by using the distance formula.
  c. Show that $ABCD$ is a square by using slope to determine that AB and BC are perpendicular.

112. Consider a physician who is looking for the effect of a certain medical treatment for two consecutive days. The variables in this instance are as follows.

> Systolic blood pressure, P: *the pressure exerted by the heart and arteries to keep the blood circulating in the blood vessels throughout the body. The approximate range is 0 to 280, with an average value of P being 127 millimeters of mercury. (We write $\overline{P} = 127$.)*

> Serum creatinine, C: *a measure of kidney functions. The approximate range is 0 to 25, with an average value of C being 1 milligram percent. (We write $\overline{C} = 1$.) A high level of creatinine indicates kidney dysfunction.*

Two formulas can be used to denote difference from normal for these two variables. They are

$$P_D = |P - \overline{P}| = |P - 127|$$
$$C_D = |C - \overline{C}| = |C - 1|.$$

For example, a patient with $P = 139$ and $C = 5$ has values of P_D and C_D given by

$$P_D = |130 - 127| = 3$$
$$C_D = |5 - 1| = 4.$$

The ordered pair $(3, 4)$ can be interpreted in a $P_D C_D$ coordinate plane, with an x-axis labeled by P_D and a y-axis by C_D. In such a plane, the origin $(0, 0)$ corresponds to a healthy person. The greater the distance between the point (P_D, C_D) and the origin, the farther a patient is from normal.
  a. Consider the following data:

  Day 1: $P = 149, C = 3$
  Day 2: $P = 138, C = 2$

For each day, compute P_D and C_D. Then find the distance between the ordered pair (P_D, C_D) and the origin. What does this indicate about the effects of the medical treatment for this patient for 2 days?
  b. Repeat part (a) for the following data:

  Day 1: $P = 120, C = 2$
  Day 2: $P = 143, C = 3$

True–False Critical Thinking Problems_____

113. Which one of the following is true?
  a. All equations can be solved using the square root method, even though other methods might lead more rapidly to the solution set.
  b. The equation $x^2 = 18$ is equivalent to $x = 3\sqrt{2}$.
  c. A person's age can be computed using the following method. Ask that person not to tell you his or her age, but simply to tell you the product of the correct age 1 year from now and 1 year ago. If you add 1 to this product and take the square root of this result, you'll determine the correct age of the person.
  d. None of the above is true.

114. Which one of the following is true?
 a. The equation $(x - 5)^2 = 12$ is equivalent to $x - 5 = 2\sqrt{3}$.
 b. Every equation in the form $(x + 5)^2 = a$, where a is any real number, has two solutions because the square root method involves finding both $\sqrt{a}$ and $-\sqrt{a}$.

 c. If an equation contains an expression that is squared, the square root method should always be applied as the very first step.
 d. None of the above is true.

Technology Problems

115. The model $y = 10.675x^2 + 1007.775$ describes the value (y, in millions of dollars) of private-property loss to fire damage in the United States x years after 1970. What year had a private-property loss of approximately $1392 million due to fire? (Use a calculator.)

116. If $900 is deposited in an account with an annual interest rate of r, the amount of money accumulated in the account after 2 years is described by the mathematical model $A = 900(1 + r)^2$. Solve for r. Then, using a calculator with a square root key ($\sqrt{}$), find the interest rate if the accumulated amount in the account after 2 years is $980.36.

117. Use a graphing utility to verify your hand-drawn parabolas in Problems 1–8.

118. Use a graphing utility to verify your solutions for any two problems in Problems 9–18.

119. Use a graphing utility to verify the real solutions that you obtained for any three problems in Problems 19–60.

120. Use a graphing utility to graph the given models in Problems 90, 91, 92, and 115. Then $\boxed{\text{TRACE}}$ along each curve until you can see the point whose coordinates correspond to the problem's solution.

Writing in Mathematics

121. Describe two applied situations in which we can solve $x^2 = d$ using $x = \sqrt{d}$ or $x = -\sqrt{d}$ but will then only retain the principal root ($x = \sqrt{d}$).

122. Do all equations in the form $x^n = d$ result in $x = \sqrt[n]{d}$ or $x = -\sqrt[n]{d}$? Explain.

123. Why is it that when we derived the distance formula using the square root method we did not take both positive and negative roots?

124. What's wrong with this statement: If we solve $(x + 5)^2 = 100$ using the square root method, we must find both the square root of 100 and the square root of -100?

125. In the 1939 movie *The Wizard of Oz,* upon being presented with a Th.D (Doctor of Thinkology), the Scarecrow proudly exclaims, "The sum of the square roots of any two sides of an isosceles triangle is equal to the square root of the remaining side." Did the

Scarecrow get the Pythagorean Theorem right? In particular, describe four errors in the Scarecrow's statement.

"That's Dancing!" from MGM/UA Entertainment Co. Courtesy The Kobal Collection.

Critical Thinking Problems

126. Solve for y: $\dfrac{x^2}{a^2} + \dfrac{y^2}{b^2} = 1$.

127. Point $P(x_1, y_1)$ is located such that its distance to $(-1, 0)$ is equal to its distance to $(-1, 5)$. Find y_1.

128. Consider the parallelogram shown in the figure at the top of the next page.
 a. Use slope to show that AB and CD are parallel. Then show that AC and BD are parallel.

 b. Prove that if the diagonals of a parallelogram are perpendicular, the parallelogram is a rhombus. (*Hints:* (1) Use the distance formula to find the lengths of AB and AC; (2) use slope, and write an equation resulting from the fact that $m_{AD} \cdot m_{BC} = -1$; (3) substitute the equation from part (2) into the formula representing the length of AC and show that $AC = AB$. Thus, $ABCD$ is a parallelogram with two equal adjacent sides, which is a rhombus.)

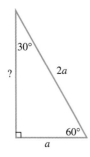

C(b, d)   D(a + b, d)

A(0, 0)   B(a, 0)   x

129. In this section we saw that if both legs of a right triangle are the same size, the triangle is an isosceles right triangle. A second special triangle is the 30-60-90 right triangle, named because of its angle measures. If the hypotenuse of this triangle is twice as long as the shorter leg, prove that the length of the longer leg is the length of the shorter leg times $\sqrt{3}$.

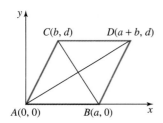

30°

?   2a

60°

a

130. Show that the points $A(1, 1 + d)$, $B(3, 3 + d)$, and $C(6, 6 + d)$ are collinear. (*Hint:* Follow the procedure outlined in Problem 75.)

131. Prove the midpoint formula by using the following procedure.

a. Show that the distance between (x_1, y_1) and $\left(\dfrac{x_1 + x_2}{2}, \dfrac{y_1 + y_2}{2}\right)$ is equal to the distance between (x_2, y_2) and $\left(\dfrac{x_1 + x_2}{2}, \dfrac{y_1 + y_2}{2}\right)$.

b. Use the procedure outlined in Problem 75 to show that the points (x_1, y_1), $\left(\dfrac{x_1 + x_2}{2}, \dfrac{y_1 + y_2}{2}\right)$, and (x_2, y_2) are collinear.

Review Problems

132. Simplify:

$$\frac{y^2 - 2y + 1}{3y^2 + 7y - 20} \cdot \frac{3y^2 - 2y - 5}{y^2 + 3y - 4} \div \frac{y^2 - 4y + 3}{y + 4}.$$

133. Simplify: $\dfrac{2}{\sqrt{3} + 1}$.

134. The length of a rectangle is 4 yards longer than the width. If the area of the rectangle is 96 square yards, what are its dimensions?

SECTION   7.2   **Completing the Square and Graphs of Quadratic Functions**

Solutions   Tutorial   Video
Manual   8

Objectives

| Complete the square for a given binomial.
2 Solve quadratic equations by completing the square.
3 Graph quadratic functions in the form $y = a(x - h)^2 + k$.

In the previous section we solved equations such as

$$(x - 3)^2 = 5$$

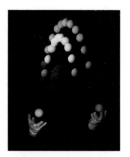

Ken Regan/Camera 5

| Complete the square for a given binomial.

by the square root method, obtaining the solutions $3 \pm \sqrt{5}$. Can we solve such an equation in its standard form $x^2 - 6x + 4 = 0$? The left side of the equation is a prime polynomial, so factoring does not work. But there is a way.

In this section, we will study a technique for rewriting $x^2 - 6x + 4 = 0$ as $(x - 3)^2 = 5$. This technique, called *completing the square,* can be used to solve all quadratic equations and is helpful in graphing quadratic functions.

Solving Quadratic Equations by Completing the Square

The goal of completing the square is to express a quadratic equation in the form $(x + d)^2 = e$ and then apply the square root method. To change a quadratic equation in standard form

$$ax^2 + bx + c = 0$$

to an equivalent equation in the form

$$(x + d)^2 = e$$

we need a method for constructing trinomials in the form $(x + d)^2$, or perfect square trinomials. For example, the perfect square trinomial whose first two terms are $x^2 + 8x$ is $x^2 + 8x + 16$. The trinomial $x^2 + 8x + 16$ is a perfect square trinomial because it is the square of the binomial $x + 4$.

$$x^2 + 8x + 16 = (x + 4)^2$$

Let's consider this problem geometrically to gain some insights about the process. Adding 16 to $x^2 + 8x$ is modeled geometrically in Figure 7.11. Without the small purple dark square on the bottom right, the area of the figure is $x^2 + 4x + 4x$, or $x^2 + 8x$. By adding the square whose area is 16, we create a square whose area is $(x + 4)^2$. This small purple square fills in the missing portion of the large square and completes the square.

Without the geometric model in Figure 7.11, you may wonder how we know what to add to $x^2 + 8x$ to create a perfect square trinomial and thereby complete the square. The answer is that we *take half the coefficient of x and square it.*

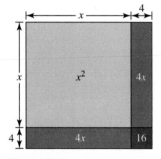

Figure 7.11

Adding 16 to $x^2 + 8x$ fills in the missing portion of the large square and completes the square.

Given trinomial   Completing the square
$$x^2 + 8x \qquad\qquad x^2 + 8x + 16$$

$$\tfrac{1}{2}(8) = 4 \quad \text{and} \quad 4^2 = 16$$

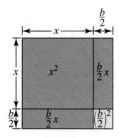

Adding $(\frac{b}{2})^2$ to $x^2 + bx$ completes the square.

Constructing a perfect square trinomial to complete the square

Completing the Square

If $x^2 + bx$ is a binomial, then adding $\left(\dfrac{b}{2}\right)^2$, which is the square of half the coefficient of x, produces a perfect square trinomial. That is,

$$x^2 + bx + \left(\frac{b}{2}\right)^2 = \left(x + \frac{b}{2}\right)^2.$$

EXAMPLE 1 **Completing the Square**

Find the perfect square trinomial whose first two terms are given. Then write the perfect square trinomial in factored form.

a. $x^2 + 6x$　　**b.** $x^2 - 8x$　　**c.** $x^2 - 7x$　　**d.** $x^2 + \dfrac{3}{5}x$

Solution

In each case, we add the square of half the coefficient of x.

a. $x^2 + 6x$
$\quad\quad\quad \longrightarrow$ Half of 6 is 3 and $3^2 = 9$.
$\quad\quad\quad\quad\quad$ We add 9.

Thus, the perfect square trinomial and its factored form are

$$x^2 + 6x + 9 = (x + 3)^2.$$

b. $x^2 - 8x$
$\quad\quad\quad \longrightarrow \frac{1}{2}(-8) = -4$ and $(-4)^2 = 16$.
$\quad\quad\quad\quad\quad$ We add 16.

The perfect square trinomial and its factored form are

$$x^2 - 8x + 16 = (x - 4)^2.$$

c. $x^2 - 7x$
$\quad\quad\quad \longrightarrow \frac{1}{2}(-7) = -\frac{7}{2}$ and $(-\frac{7}{2})^2 = \frac{49}{4}$.
$\quad\quad\quad\quad\quad$ We add $\frac{49}{4}$.

The perfect square trinomial and its factored form are

$$x^2 - 7x + \frac{49}{4} = \left(x - \frac{7}{2}\right)^2.$$

d. $x^2 + \frac{3}{5}x$
$\quad\quad\quad \longrightarrow \frac{1}{2}(\frac{3}{5}) = \frac{3}{10}$ and $(\frac{3}{10})^2 = \frac{9}{100}$.
$\quad\quad\quad\quad\quad$ We add $\frac{9}{100}$.

The perfect square trinomial and its factored form are

$$x^2 + \frac{3}{5}x + \frac{9}{100} = \left(x + \frac{3}{10}\right)^2.$$

Study tip

Note the pattern in each factorization in Example 1. The first term in the parentheses is x. The second term is half the coefficient of x.

$x^2 + 6x + 9 = (x + 3)^2$
$\quad\quad\quad \frac{1}{2} \cdot 6 = 3$

$x^2 - 8x + 16 = (x - 4)^2$
$\quad\quad\quad \frac{1}{2}(-8) = -4$

$x^2 - 7x + \dfrac{49}{4} = \left(x - \dfrac{7}{2}\right)^2$
$\quad\quad\quad \frac{1}{2}(-7) = -\frac{7}{2}$

$x^2 + \dfrac{3}{5}x + \dfrac{9}{100} = \left(x + \dfrac{3}{10}\right)^2$
$\quad\quad\quad \frac{1}{2} \cdot \frac{3}{5} = \frac{3}{10}$

2 Solve quadratic equations by completing the square.

Let's see how we can apply this process to solving a quadratic equation.

EXAMPLE 2 **Solving a Quadratic Equation by Completing the Square**

Solve by completing the square:　$x^2 - 6x + 2 = 0$

Use a graphing utility to verify the solution set of $x^2 - 6x + 2 = 0$. First find decimal approximations of the solutions:

$$\{3 + \sqrt{7}, 3 - \sqrt{7}\} \approx \{5.6, 0.4\}$$

The x-intercepts of the graph of $y = x^2 - 6x + 2$ are approximately 5.6 and 0.4.

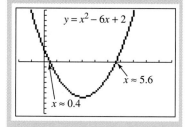

Solution

$x^2 - 6x + 2 = 0$	This is the given equation.
$x^2 - 6x = -2$	To isolate the binomial $x^2 - 6x$ so that we can complete the square, we subtract 2 from both sides.
$x^2 - 6x + 9 = -2 + 9$	Add 9 on both sides to complete the square: $\frac{1}{2}(-6) = -3$ and $(-3)^2 = 9$.
$(x - 3)^2 = 7$	Factor the perfect square trinomial.
$x - 3 = \pm\sqrt{7}$	Apply the square root method.
$x = 3 \pm \sqrt{7}$	Solve by adding 3 on both sides.

The solutions are $3 + \sqrt{7}$ and $3 - \sqrt{7}$. Check these in the original equation, verifying that the solution set is $\{3 + \sqrt{7}, 3 - \sqrt{7}\}$. ∎

To rewrite a quadratic equation in the form $(x + d)^2 = e$, the coefficient of x^2 must be 1. If it is not 1, begin by dividing both sides of the equation by the x^2-coefficient.

EXAMPLE 3   **Completing the Square: Leading Coefficient is Not 1**

Solve by completing the square: $2x^2 + 5x - 4 = 0$

Use a graphing utility to verify the solution set of $2x^2 + 5x - 4 = 0$. First find decimal approximations of the solutions. For

$$\frac{-5 + \sqrt{57}}{4}$$

((−) 5 + √ 57)

÷ 4 ENTER

$$\frac{-5 + \sqrt{57}}{4} \approx 0.6$$

and (similarly)

$$\frac{-5 - \sqrt{57}}{4} \approx -3.1$$

The x-intercepts of the graph of $y = 2x^2 + 5x - 4$ are approximately 0.6 and -3.1.

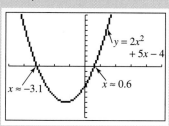

Solution

$2x^2 + 5x - 4 = 0$	This is the given equation.
$\dfrac{2x^2}{2} + \dfrac{5x}{2} - \dfrac{4}{2} = \dfrac{0}{2}$	Divide both sides by 2 so that the coefficient of x^2 is 1.
$x^2 + \dfrac{5}{2}x - 2 = 0$	Simplify.
$x^2 + \dfrac{5}{2}x = 2$	Isolate the binomial by adding 2 to both sides.
$x^2 + \dfrac{5}{2}x + \dfrac{25}{16} = 2 + \dfrac{25}{16}$	Add $\frac{25}{16}$ on both sides to complete the square: $\frac{1}{2}\left(\frac{5}{2}\right) = \frac{5}{4}$ and $\left(\frac{5}{4}\right)^2 = \frac{25}{16}$.
$\left(x + \dfrac{5}{4}\right)^2 = \dfrac{57}{16}$	Factor the perfect square trinomial. On the right, $2 + \frac{25}{16} = \frac{32}{16} + \frac{25}{16} = \frac{57}{16}$.
$x + \dfrac{5}{4} = \pm\dfrac{\sqrt{57}}{4}$	Apply the square root method. On the right, $\sqrt{\frac{57}{16}} = \frac{\sqrt{57}}{\sqrt{16}} = \frac{\sqrt{57}}{4}$.
$x = -\dfrac{5}{4} \pm \dfrac{\sqrt{57}}{4}$	Solve for x by subtracting $\frac{5}{4}$ from both sides.
$x = \dfrac{-5 \pm \sqrt{57}}{4}$	Simplify.

Since the proposed solutions are difficult to check by direct substitution, use a graphing utility or carefully recheck your solution. Unless there is an error in your work, completing the square will never result in numbers that are not solutions of the given quadratic equation. The equation's solution set is

$$\left\{\frac{-5 + \sqrt{57}}{4}, \frac{-5 - \sqrt{57}}{4}\right\}.$$

■

Before considering another example, let's summarize the process used in the preceding two examples.

Study tip

In step 3, be sure to add the square of half the coefficient of x to *both sides*.

Incorrect:

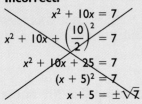

Solving a quadratic equation by completing the square

1. If the leading coefficient is not 1, divide both sides by the coefficient of the squared term.
2. Rewrite the equation with the constant term by itself on the right side of the equation.
3. Complete the square by adding the square of half the coefficient of the first-degree term to both sides of the equation.
4. Factor the left side, writing it as the square of a binomial, and simplify the right side if possible.
5. Apply the square root method and solve.

Using technology

The graph of $y = x^2 + 4x + 8$ has no x-intercepts, verifying that $x^2 + 4x + 8 = 0$ has imaginary solutions.

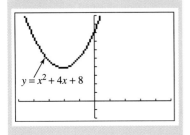

EXAMPLE 4   **Completing the Square: Imaginary Solutions**

Solve by completing the square: $x^2 + 4x + 8 = 0$

Solution

We can omit the first step in the preceding box because the leading coefficient is 1.

$x^2 + 4x + 8 = 0$ This is the given equation.

$x^2 + 4x = -8$ Isolate the binomial by subtracting 8 from both sides.

$x^2 + 4x + 4 = -8 + 4$ Add 4 on both sides to complete the square: $\frac{1}{2}(4) = 2$ and $2^2 = 4$.

$(x + 2)^2 = -4$ Factor the perfect square trinomial.

$x + 2 = \pm\sqrt{-4}$ Apply the square root method.

$x + 2 = \pm 2i$ Express the imaginary number in terms of i: $\sqrt{-4} = \sqrt{4}\sqrt{-1} = 2i$.

$x = -2 \pm 2i$ Solve for x by subtracting 2 from both sides.

Try checking one of the solutions by direct substitution into the given equation. The equation's solution set is $\{-2 - 2i, -2 + 2i\}$.   ■

Proving That a Parabola Has a Vertex When $x = -\dfrac{b}{2a}$

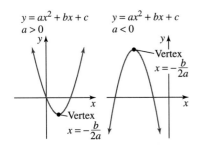

Completing the square can help us in our work with quadratic functions and their cuplike graphs. In particular, we can now prove that the graph of

$y = ax^2 + bx + c$ has a vertex whose x-coordinate is $x = -\dfrac{b}{2a}$. The idea is to complete the square on the right side of $y = ax^2 + bx + c$.

$y = ax^2 + bx + c$	This is the equation of the quadratic function.
$y = a\left(x^2 + \dfrac{b}{a}x\right) + c$	Factor out a from the first two terms so that the coefficient of x^2 is 1.
$y = a\left(x^2 + \dfrac{b}{a}x + \dfrac{b^2}{4a^2} - \dfrac{b^2}{4a^2}\right) + c$	Add and subtract $\dfrac{b^2}{4a^2}$ to complete the square: $\dfrac{1}{2} \cdot \dfrac{b}{a} = \dfrac{b}{2a}$ and $\left(\dfrac{b}{2a}\right)^2 = \dfrac{b^2}{4a^2}$.
$y = a\left(x^2 + \dfrac{b}{a}x + \dfrac{b^2}{4a^2}\right) - \dfrac{b^2}{4a} + c$	Bring $a\left(-\dfrac{b^2}{4a^2}\right) = -\dfrac{b^2}{4a}$ outside the parentheses.
$y = a\left(x + \dfrac{b}{2a}\right)^2 - \dfrac{b^2}{4a} + c \cdot \dfrac{4a}{4a}$	Factor the perfect square trinomial. Express the last two terms with an LCD.
$y = a\left(x + \dfrac{b}{2a}\right)^2 + \dfrac{4ac - b^2}{4a}$	Simplify the last two terms.

The value of y depends on the value of x in the expression

$$\left(x + \frac{b}{2a}\right)^2.$$

This expression is positive or 0; it has a value of 0 when $x = -\dfrac{b}{2a}$. Because $x = -\dfrac{b}{2a}$ results in the smallest value of this term *and* the term is multiplied by a, we see that

1. If a is positive, then $x = -\dfrac{b}{2a}$ will yield a minimum value of y.

2. If a is negative, then $x = -\dfrac{b}{2a}$ will yield a maximum value for y.

By definition, the vertex is the lowest point on a parabola that opens upward or the highest point on a parabola that opens downward. By items 1 and 2 listed above, this means that the x-coordinate of the vertex is $x = -\dfrac{b}{2a}$. This is precisely what we were trying to prove.

It is not necessary to memorize the equation of a parabola in the form

$$y = a\left(x + \frac{b}{2a}\right)^2 + \frac{4ac - b^2}{4a}.$$

However, this form of the equation will help us to gain insight into another useful form of a parabola's equation.

3 Graph quadratic functions in the form $y = a(x - h)^2 + k$.

Graphing Quadratic Functions in the Form $y = a(x - h)^2 + k$

The graphs of $y = x^2$, $y = (x - 2)^2$, $y = x^2 + 3$, and $y = (x - 2)^2 + 3$ are shown in Figure 7.12. These graphs all have the same shape. The difference is their position.

S tudy tip

When completing the square with a quadratic equation, add the square of half the coefficient of x to both sides:

$$x^2 + 6x = 7$$
$$x^2 + 6x + 9 = 7 + 9$$
$$(x + 3)^2 = 16$$

When completing the square with a quadratic function, add and subtract the square of half the coefficient of x on the right side:

$$y = x^2 + 6x + 7$$
$$y = x^2 + 6x + 9 - 9 + 7$$
$$y = \underbrace{}_{(x+3)^2} \underbrace{}_{-2}$$
$$y = \quad (x + 3)^2 \quad - 2$$

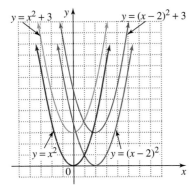

Figure 7.12

Graphs with the same shape, but in different positions

Equation	Vertex	Description of Graph
$y = x^2$	$(0,0)$	Basic parabola that opens up, with vertex at the origin
$y = (x - 2)^2$	$(2,0)$	Looks like $y = x^2$, but shifted 2 units to the right
$y = x^2 + 3$	$(0,3)$	Looks like $y = x^2$, but shifted 3 units up
$y = (x - 2)^2 + 3$	$(2,3)$	Looks like $y = x^2$, but shifted 2 units to the right and 3 units up

In Table 7.1, we generalize from these results, starting with the equation $y = ax^2$.

TABLE 7.1 The Graph of $y = a(x - h)^2 + k$

Equation	Vertex	Description of Graph
$y = ax^2$	$(0,0)$	Parabola opens upward if a is positive and opens downward if a is negative
$y = a(x - h)^2$	$(h,0)$	Looks like $y = ax^2$, but shifted h units to the right if h is positive, and $|h|$ units to the left if h is negative
$y = ax^2 + k$	$(0,k)$	Looks like $y = ax^2$, but shifted k units up if k is positive, and $|k|$ units down if k is negative
$y = a(x - h)^2 + k$	(h,k)	Looks like $y = ax^2$, but shifted horizontally and vertically as described above

Notice that in the form $y = a(x - h)^2 + k$, a minus sign is needed in the parentheses and a plus sign is needed before the constant k. Thus, $y = (x + 3)^2 - 4$ must be rewritten, as shown below, to be in this form.

Equation	Equation in the Form $y = a(x - h)^2 + k$	Vertex (h, k)	Description of Graph
$y = (x + 3)^2 - 4$	$y = (x - (-3))^2 + (-4)$ $\quad\uparrow\qquad\quad\uparrow$ $\quad h = -3 \quad k = -4$	$(-3, -4)$	Looks like $y = x^2$, but shifted $|-3|$, or 3, units to the left and $|-4|$, or 4, units down.

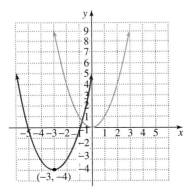

Graphs of $y = x^2$ and $y = (x - (-3))^2 - 4$

Let's take a moment to prove that the parabola whose equation is $f(x) = a(x - h)^2 + k$ has a vertex at (h, k). As shown in the next example, the proof depends on our knowledge that the x-coordinate of the vertex for the graph of $y = ax^2 + bx + c$ is $x = -\dfrac{b}{2a}$.

EXAMPLE 5 **Finding the Vertex of a Parabola**

Find the coordinates of the vertex for the parabola whose equation is:

$$f(x) = a(x - h)^2 + k$$

Arthur Gurmankin/Mary Morina/Visuals Unlimited

Solution

$$f(x) = a(x - h)^2 + k \qquad \text{This is the given quadratic function. We will write the function in the form } f(x) = ax^2 + bx + c.$$

$$= a(x^2 - 2hx + h^2) + k \qquad \text{Square } x - h.$$

$$= ax^2 - 2ahx + ah^2 + k \qquad \text{Apply the distributive property.}$$

In this form, a (the coefficient of x^2) is a, and b (the coefficient of x) is $-2ah$. The x-coordinate of the vertex is

$$x = -\frac{b}{2a}$$

$$= -\frac{(-2ah)}{2a} \qquad \text{Substitute } -2ah \text{ for } b, \text{ the coefficient of } x.$$

$$= \frac{2ah}{2a}$$

$$= h \qquad \text{Simplify.}$$

Because the x-coordinate of the vertex is h, we can find the y-coordinate by substituting h for x in the quadratic function.

$$y = f(h) \qquad h \text{ is the } x\text{-coordinate of the vertex.}$$

$$= a(h - h)^2 + k \qquad \text{We were given that } f(x) = a(x - h)^2 + k.$$

$$= a \cdot 0^2 + k$$

$$= k$$

Thus, the coordinates of the vertex for the parabola whose equation is in the form $f(x) = a(x - h)^2 + k$ are (h, k). ■

We now have two forms of the equation of a parabola. In both forms, if $a > 0$ the parabola opens upward, and if $a < 0$ the parabola opens downward.

Forms of the equation of a parabola

$$f(x) = ax^2 + bx + c$$
$$y = ax^2 + bx + c$$

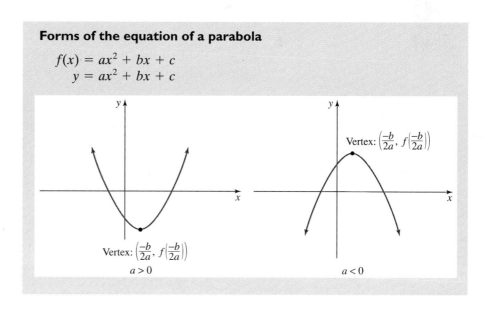

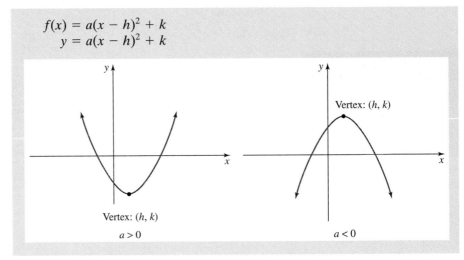

$$f(x) = a(x - h)^2 + k$$
$$y = a(x - h)^2 + k$$

Vertex: (h, k)

Vertex: (h, k)

$a > 0$

$a < 0$

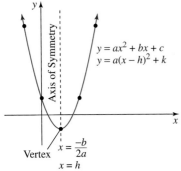

$$y = ax^2 + bx + c$$
$$y = a(x - h)^2 + k$$

Axis of Symmetry

Vertex $\quad x = \dfrac{-b}{2a}$

$x = h$

Figure 7.13

The equation of the axis of symmetry is $x = -\dfrac{b}{2a}$ or $x = h$.

Observe that for all four parabolas, a vertical line drawn through the vertex divides the parabola into two identical (mirror-image) parts. The line is called the *axis of symmetry* and is shown in Figure 7.13. For each point on the parabola to the left of the axis, there is a corresponding point to the right of the axis. These points are equidistant from the axis of symmetry and have the same second component.

EXAMPLE 6 **Graphing a Parabola**

Graph: $f(x) = (x + 1)^2 + 3$

Solution

We begin by expressing the function's equation in the form
$$f(x) = a(x - h)^2 + k.$$

$f(x) = (x + 1)^2 + 3$ This is the given equation.

$f(x) = (x - (-1))^2 + 3$ A minus sign is needed in the parentheses.

$\qquad\qquad h = -1 \quad k = 3$

The vertex is (h, k) or $(-1, 3)$. The axis of symmetry is $x = -1$. The graph looks like $y = x^2$, but is shifted 1 unit to the left and 3 units up.

Now we compute a few points located near the vertex. The parabola is shown in Figure 7.14.

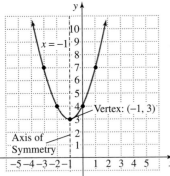

$x = -1$

Vertex: $(-1, 3)$

Axis of Symmetry

Figure 7.14

The graph of
$f(x) = (x + 1)^2 + 3$

x	$f(x) = (x + 1)^2 + 3$	$(x, f(x))$	
-3	$f(-3) = (-3 + 1)^2 + 3 = 7$	$(-3, 7)$	
-2	$f(-2) = (-2 + 1)^2 + 3 = 4$	$(-2, 4)$	
-1	$f(-1) = (-1 + 1)^2 + 3 = 3$	$(-1, 3)$	← Vertex
0	$f(0) = (0 + 1)^2 + 3 = 4$	$(0, 4)$	← y-intercept is 4.
1	$f(1) = (1 + 1)^2 + 3 = 7$	$(1, 7)$	

EXAMPLE 7 **Graphing a Parabola by Completing the Square**

a. Write the equation $y = -2x^2 + 4x + 5$ in the form $y = a(x - h)^2 + k$.
b. Use the form from part (a) to graph the function.

ENRICHMENT ESSAY

Symmetry

A major characteristic of a parabola's shape is its symmetry about a vertical line drawn through the vertex. The symmetry is bilateral, meaning that the two halves of the whole are each other's mirror images. The movements of gymnasts, divers, and swimmers approximate bilateral symmetry. Find an example of this form of symmetry in nature, architecture, art, or other graphs of mathematical formulas.

Simon Bruty/Allsport
Photography (USA), Inc.

A parabola rotated about its axis of symmetry forms a curved surface called a paraboloid, used in constructing searchlights and radar dishes.

Ernesto
Burciaga/Photri, Inc.

Solution

a. Since we want the coefficient of x^2 to be 1, we begin by factoring -2 from the first two terms. Then we complete the square on the expression within parentheses.

$$y = -2x^2 + 4x + 5 \qquad \text{This is the given equation.}$$
$$= -2(x^2 - 2x) + 5 \qquad \text{Factor out } -2 \text{ from the first two terms so that the coefficient of } x^2 \text{ is 1.}$$
$$= -2(x^2 - 2x + 1 - 1) + 5 \qquad \text{Complete the square: } \tfrac{1}{2}(-2) = -1 \quad \text{and } (-1)^2 = 1 \quad \text{Add and subtract 1.}$$
$$= -2(x^2 - 2x + 1) + 2 + 5 \qquad \text{Bring } (-2)(-1) = 2 \text{ outside the parentheses.}$$
$$= -2(x - 1)^2 + 7 \qquad \text{Factor.}$$

b. The last equation is in the form $y = a(x - h)^2 + k$, where $a = -2, h = 1$, and $k = 7$. Since $a < 0$, the parabola opens downward. The vertex is (h, k), or $(1, 7)$. The axis of symmetry is $x = 1$.

Now we compute a few points located near the vertex. The parabola is shown in Figure 7.15.

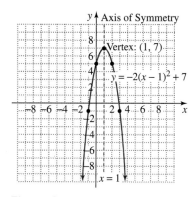

Figure 7.15

The graph of $y = -2x^2 + 4x + 5 = -2(x - 1)^2 + 7$

x	$y = -2(x - 1)^2 + 7$	(x, y)	
-2	$y = -2(-2 - 1)^2 + 7 = -11$	$(-2, -11)$	
-1	$y = -2(-1 - 1)^2 + 7 = -1$	$(-1, -1)$	
0	$y = -2(0 - 1)^2 + 7 = 5$	$(0, 5)$	← y-intercept is 5.
1	$y = 7$	$(1, 7)$	← Vertex
2	$y = -2(2 - 1)^2 + 7 = 5$	$(2, 5)$	
3	$y = -2(3 - 1)^2 + 7 = -1$	$(3, -1)$	

Using technology

Use a graphing utility to verify the hand-drawn graphs in Examples 6 and 7.

PROBLEM SET 7.2

Practice Problems _____

In Problems 1–12, complete the square. Then write the perfect square trinomial in factored form.

1. $x^2 + 12x$ **2.** $x^2 + 14x$ **3.** $x^2 - 16x$ **4.** $x^2 - 10x$ **5.** $x^2 + 7x$ **6.** $x^2 + 9x$

7. $x^2 - 3x$ **8.** $x^2 - 5x$ **9.** $x^2 + \dfrac{1}{3}x$ **10.** $x^2 + \dfrac{1}{4}x$ **11.** $x^2 - \dfrac{2}{3}x$ **12.** $x^2 - \dfrac{4}{5}x$

Solve the quadratic equations in Problems 13–36 by completing the square. Check any two irrational solutions and any two imaginary solutions by direct substitution in the given equation.

13. $x^2 - 4x = 21$ **14.** $x^2 + 8x = -15$ **15.** $x(x - 6) = 16$ **16.** $y(y - 2) = 8$

17. $y^2 - 6y + 2 = 0$ **18.** $y^2 + 4y - 16 = 0$ **19.** $x^2 + x - 1 = 0$ **20.** $y^2 + 3y - 1 = 0$

21. $2y^2 - 5y = 3$ **22.** $2y^2 + y = 3$ **23.** $9z^2 - 30z + 25 = 0$ **24.** $4z^2 + 4z + 1 = 0$

25. $2z^2 + z = 5$ **26.** $3z^2 + 2z = 2$ **27.** $2x^2 - 2x = 3$ **28.** $3x^2 - 2x = 6$

29. $y^2 + 2y + 2 = 0$ **30.** $y^2 - 4y + 8 = 0$ **31.** $x^2 - x + 1 = 0$ **32.** $x^2 + x + 1 = 0$

33. $8z^2 - 4z = -1$ **34.** $9z^2 - 6z = -5$ **35.** $3y^2 + 2y + 4 = 0$ **36.** $2y^2 + y + 1 = 0$

For each function in Problems 37–40,
     a. *Identify the coordinates of the vertex for the function's graph.*
     b. *Describe the graph in relationship to the graph of $y = x^2$. Use two of the following phrases in your description: shifted* _____ *units to the right; shifted* _____ *units to the left; shifted* _____ *units up; shifted* _____ *units down.*
     c. *Compute a few points as needed and graph the parabola.*

37. $y = (x - 2)^2 + 1$ **38.** $y = (x - 3)^2 + 2$ **39.** $y = (x + 1)^2 - 2$ **40.** $y = (x + 3)^2 - 1$

In Problems 41–48, graph the parabola whose equation is given.

41. $f(x) = -(x - 3)^2 - 1$ **42.** $f(x) = -(x - 2)^2 - 4$ **43.** $y = 2(x + 1)^2 - 3$ **44.** $y = 2(x + 2)^2 - 4$

45. $g(x) = -2(x - 4)^2 + 3$ **46.** $g(x) = -2(x - 2)^2 + 5$ **47.** $y = -3(x + 1)^2 + 4$ **48.** $y = -3(x + 2)^2 + 3$

For Problems 49–56,
     a. *Complete the square and write the function's equation in the form $y = a(x - h)^2 + k$.*
     b. *Use the form from part (b) to graph the function.*

49. $y = x^2 + 6x + 5$ **50.** $y = x^2 + 4x + 3$ **51.** $y = -x^2 - 4x - 3$ **52.** $y = -x^2 + 2x + 8$

53. $y = 2x^2 + 6x + 8$ **54.** $y = 4x^2 + 8x - 3$ **55.** $y = 3x^2 - 12x + 13$ **56.** $y = -3x^2 + 6x + 1$

Each graph in Problems 57–60 has the same shape as the graph of $y = x^2$, but the graph is shifted. For each graph,
     a. *Write the equation of the quadratic function in the form $y = a(x - h)^2 + k$.*
     b. *Multiply and combine like terms on the right side to express the equation in the form $y = ax^2 + bx + c$.*

57. 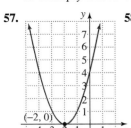   **58.** 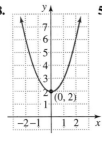   **59.** 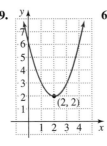   **60.**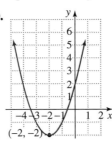

Application Problems _____

61. According to the U.S. Department of Labor, union membership in the United States can be modeled by the quadratic function $f(x) = -0.011x^2 + 1.22x - 8.5$, in which x is the number of years since 1930 and $f(x)$ is the number of union members (in millions). The model approximates reality for $20 \leqslant x \leqslant 60$.

a. Find the x-coordinate of the vertex. Round the answer to the nearest whole number and describe what this means in terms of the variables modeled by the function.

b. Find the y-coordinate of the vertex. Describe what this means in practical terms.

62. Suppose that 100 yards of fencing is used to fence the rectangular region shown in the figure on the right. The area of the region whose length is represented by x can be modeled by the quadratic function

$$A(x) = x(50 - x)$$

or

$$A(x) = -x^2 + 50x.$$

What is the vertex for the graph of this function? Use the vertex to complete the following statement: To fence in the largest area possible with the 100 yards of fencing, the rectangle's length (x) should be ____ yards and its width ($50 - x$) should be ____ yards. The rectangle that gives the maximum area is actually a ____ with an area of ____ square yards.

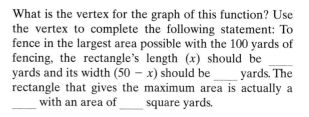
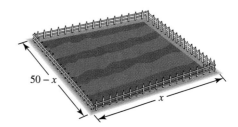

True–False Critical Thinking Problems

63. Which one of the following is true?
  a. To complete the square for $3x^2 - 4x = 15$, add 4 to both sides.
  b. The graph of $y = (x - 5)^2 + 4$ looks like the graph of $y = x^2$, but shifted 5 units to the left and 4 units up.
  c. A parabola whose equation has a positive leading coefficient and whose vertex is at $(2, 3)$ has no x-intercepts.
  d. Completing the square means drawing in a line segment when three line segments have been drawn, resulting in a perfect four-sided square.

64. Which one of the following is true?
  a. If a quadratic equation cannot be solved by factoring, it cannot be solved by completing the square.
  b. To complete the square for $3x^2 + 2x = 1$, add 4 to both sides.
  c. The graph of $y = (x - 2)^2 + 3$ cannot have x-intercepts.
  d. The graphs of $y = x^2 - 6x + 8$ and $y = (x - 3)^2 - 1$ are different parabolas because their equations are different.

Technology Problems

65. Use a graphing utility to verify the real solutions that you obtained for any four problems in Problems 13–36.

66. Use a graphing utility to verify any four of the graphs that you drew by hand in Problems 37–56.

67. Use a graphing utility to verify the equations that you wrote in Problems 57–60. Graph both equations in the same viewing rectangle. The graphs should be identical, and should look just like the hand-drawn graph given in the problem.

68. Use a graphing utility to graph the given models in Problems 61–62. Then $\boxed{\text{TRACE}}$ along each curve until

you can see the point whose coordinates correspond to the problem's solution.

69. Graph $y = x^2$, $y = 4x^2$, and $y = \frac{1}{4}x^2$ in the same viewing rectangle. Then repeat for $y = x^2$, $y = 6x^2$, and $y = \frac{1}{6}x^2$. Write a statement describing the relationship between the graphs of $y = ax^2$ and $y = x^2$ for $a > 1$ and for $0 < a < 1$.

70. Write a statement describing the relationship between the graphs of $y = ax^2$ and $y = x^2$ for $a < -1$ and for $-1 < a < 0$. Use a procedure similar to the one for Problem 69, graphing each equation that you select using your graphing utility.

Writing in Mathematics

71. Describe the steps that are needed to solve $x^2 + 6x + 5 = 0$ by completing the square.

72. Consider the equation $x^2 - 2x + 1 = d$ or $(x - 1)^2 = d$. Describe how the solutions depend on d, discussing values of d that will result in rational, irrational, or nonreal solutions.

73. What must be done to complete the square for $x^2 + 6x$? Describe how this process is modeled geometrically in the figures.

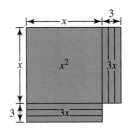
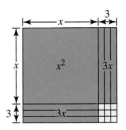

Critical Thinking Problems

74. The towers of a suspension bridge are 800 feet apart and rise 160 feet above the road. The cable between the towers has the shape of a parabola, and the cable just touches the sides of the road midway between the towers. What is the height of the cable 100 feet from a tower?

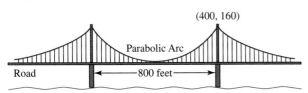

(400, 160)

Parabolic Arc

Road ← 800 feet →

75. Write the equation of a parabola whose vertex and y-intercept are the same point.

Review Problems

76. Solve the system:

$$3x - 2y = 1$$
$$5y + 3z = -7$$
$$2x + 5y = 45$$

77. Solve: $\sqrt{2x + 3} - \sqrt{x + 1} = 1$.

78. Simplify: $\dfrac{\dfrac{x}{y} - \dfrac{y}{x}}{\dfrac{x}{y} + 1}$.

SECTION 7.3

Solutions Manual   Tutorial   Video 8

The Quadratic Formula

Objectives

1 Solve quadratic equations using the quadratic formula.
2 Determine the nature of a quadratic equation's solutions.
3 Use the quadratic formula to answer questions involving modeling.
4 Determine the most efficient technique to use when solving a quadratic equation.

The method of completing the square is really a means to an end. The end is a compact formula that can be used to solve every quadratic equation. In this section, we will derive and use this formula, called the *quadratic formula*.

Deriving the Quadratic Formula

Mathematicians often like to generalize a procedure to arrive at a formula. Since completing the square is a method that can be used to solve all quadratic equations, let's apply this method to the general quadratic equation in standard form, $ax^2 + bx + c = 0$. Assume that $a > 0$. In our derivation, we also show a particular quadratic equation $3x^2 - 2x - 4 = 0$ to specifically illustrate what we are doing.

ENRICHMENT ESSAY

Supercomputing and Modeling

The length of an ocean wave ($f(x)$, in feet) is a function of its speed (x, in knots) modeled by the quadratic function

$$f(x) = 0.6x^2.$$

The wave shown in this picture is a function of a supercomputer's ability to model complex phenomena. This computer replica of Katsushika Hokusai's *The Great Wave Off Kanagawa* illustrates that the finer the grid, the closer the simulation is to reality.

The Wave of the Future—designed by Grafik, illustrated in part by B. Pomoroy, published by Nokes Berry Graphics

TABLE 7.2 Deriving the Quadratic Formula

Standard Form of a Quadratic Equation	Comment	A Specific Example
$ax^2 + bx + c = 0, a > 0$	This is the given equation.	$3x^2 - 2x - 4 = 0$
$x^2 + \dfrac{b}{a}x + \dfrac{c}{a} = 0$	Divide both sides by the coefficient of x^2.	$x^2 - \dfrac{2}{3}x - \dfrac{4}{3} = 0$
$x^2 + \dfrac{b}{a}x = -\dfrac{c}{a}$	Isolate the binomial by adding $-\dfrac{c}{a}$ on both sides.	$x^2 - \dfrac{2}{3}x = \dfrac{4}{3}$
$x^2 + \dfrac{b}{a}x + \dfrac{b^2}{4a^2} = -\dfrac{c}{a} + \dfrac{b^2}{4a^2}$	Complete the square: $\dfrac{1}{2} \cdot \dfrac{b}{a} = \dfrac{b}{2a}$ and $\left(\dfrac{b}{2a}\right)^2 = \dfrac{b^2}{4a^2}$. Add the square of half the coefficient of x to both sides.	$x^2 - \dfrac{2}{3}x + \dfrac{1}{9} = \dfrac{4}{3} + \dfrac{1}{9}$
$\left(x + \dfrac{b}{2a}\right)^2 = -\dfrac{c}{a} \cdot \dfrac{4a}{4a} + \dfrac{b^2}{4a^2}$	Factor on the left and obtain a common denominator on the right.	$\left(x - \dfrac{1}{3}\right)^2 = \dfrac{4}{3} \cdot \dfrac{3}{3} + \dfrac{1}{9}$
$\left(x + \dfrac{b}{2a}\right)^2 = \dfrac{-4ac + b^2}{4a^2}$ $\left(x + \dfrac{b}{2a}\right)^2 = \dfrac{b^2 - 4ac}{4a^2}$	Add fractions on the right.	$\left(x - \dfrac{1}{3}\right)^2 = \dfrac{12 + 1}{9}$ $\left(x - \dfrac{1}{3}\right)^2 = \dfrac{13}{9}$
$x + \dfrac{b}{2a} = \pm\sqrt{\dfrac{b^2 - 4ac}{4a^2}}$	Apply the square root method.	$x - \dfrac{1}{3} = \pm\sqrt{\dfrac{13}{9}}$
$x + \dfrac{b}{2a} = \pm\dfrac{\sqrt{b^2 - 4ac}}{2a}$	Take the square root of the quotient, simplifying the denominator.	$x - \dfrac{1}{3} = \pm\dfrac{\sqrt{13}}{3}$
$x = \dfrac{-b}{2a} \pm \dfrac{\sqrt{b^2 - 4ac}}{2a}$	Solve for x by subtracting $\dfrac{b}{2a}$ from both sides.	$x = \dfrac{1}{3} \pm \dfrac{\sqrt{13}}{3}$
$x = \dfrac{-b \pm \sqrt{b^2 - 4ac}}{2a}$	Combine fractions on the right.	$x = \dfrac{1 \pm \sqrt{13}}{3}$

A similar derivation gives us the same formula in the last step when a is negative. The derived formula is called the *quadratic formula* and indicates that the two solutions of

$$ax^2 + bx + c = 0$$

are

$$x = \frac{-b + \sqrt{b^2 - 4ac}}{2a} \quad \text{and} \quad x = \frac{-b - \sqrt{b^2 - 4ac}}{2a}.$$

▌ Solve quadratic equations using the quadratic formula.

Solving Equations Using the Quadratic Formula

Here's a step-by-step method for using the quadratic formula.

Solving a quadratic equation using the quadratic formula

1. If necessary, write the quadratic equation in standard form, $ax^2 + bx + c = 0$. Determine the numerical values for a, b, and c.
2. Substitute the values for a, b, and c in the quadratic formula

$$x = \frac{-b \pm \sqrt{b^2 - 4ac}}{2a}.$$

3. Evaluate the formula and obtain the quadratic equation's solutions.

EXAMPLE 1 **Using the Quadratic Formula**

Solve using the quadratic formula: $2x^2 + 9x = 5$

Solution

We must first write the quadratic equation in standard form to determine the values of a, b, and c.

$$2x^2 + 9x = 5 \qquad \text{This is the given equation.}$$
$$2x^2 + 9x - 5 = 0 \qquad \text{Subtract 5 on both sides.}$$
$$\updownarrow \quad \updownarrow \quad \updownarrow$$
$$ax^2 + bx + c = 0$$

We see that $a = 2, b = 9$, and $c = -5$. Substituting these values into the quadratic formula and simplifying gives the equation's solutions.

$$x = \frac{-b \pm \sqrt{b^2 - 4ac}}{2a} \qquad \text{Use the quadratic formula.}$$

$$= \frac{-9 \pm \sqrt{9^2 - 4(2)(-5)}}{2(2)} \qquad \text{Substitute the values for } a, b, \text{ and } c.$$

$$= \frac{-9 \pm \sqrt{81 + 40}}{4} \qquad 9^2 - 4(2)(-5) = 81 - (-40) = 81 + 40$$

$$= \frac{-9 \pm \sqrt{121}}{4} \qquad \text{Add under the radical sign.}$$

$$= \frac{-9 \pm 11}{4} \qquad \sqrt{121} = 11$$

ENRICHMENT ESSAY

Modeling Quadratic Functions

The quadratic function $S = 4\pi r^2$ describes the surface area of a sphere of radius r. The same function, $A = 4\pi r^2$ gives the area of a circle of radius $2r$. How are these identical functions modeled in the diagram shown to the right?

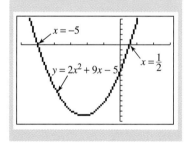
$$x = \frac{-9 + 11}{4} \quad \text{or} \quad x = \frac{-9 - 11}{4}$$

$$= \frac{2}{4} = \frac{1}{2} \qquad\qquad = \frac{-20}{4} = -5$$

The solution set is $\{-5, \frac{1}{2}\}$.

Example 1 can also be solved by factoring.

$2x^2 + 9x = 5$	This is the given equation.
$2x^2 + 9x - 5 = 0$	Write the equation in standard form.
$(2x - 1)(x + 5) = 0$	Factor.
$2x - 1 = 0 \quad \text{or} \quad x + 5 = 0$	Set each factor equal to 0.
$2x = 1 \quad \text{or} \qquad x = -5$	Solve for x.
$x = \dfrac{1}{2} \quad \text{or} \qquad x = -5$	

Factoring appears to provide a somewhat shorter solution than using the quadratic formula. However, the solutions of a quadratic equation cannot always be found by factoring, whereas they can always be determined using the quadratic formula.

For example, quadratic equations with irrational solutions cannot be solved by factoring. However, as shown in our next example, they can be readily solved using the quadratic formula.

EXAMPLE 2 **Using the Quadratic Formula**

a. Solve using the quadratic formula: $2x^2 - 4x - 1 = 0$
b. Use the result of part (a) to graph the quadratic function

$$y = 2x^2 - 4x - 1.$$

Solution

a. The quadratic equation is given in standard form, so we can determine values for $a, b,$ and c by inspection.

$$2x^2 - 4x - 1 = 0$$
$$\updownarrow \qquad \updownarrow \quad \updownarrow$$
$$ax^2 + bx + c = 0$$

We substitute $a = 2, b = -4,$ and $c = -1$ into the quadratic formula.

$$x = \frac{-b \pm \sqrt{b^2 - 4ac}}{2a} \qquad \text{Use the quadratic formula.}$$

$$= \frac{-(-4) \pm \sqrt{(-4)^2 - 4(2)(-1)}}{2(2)} \qquad a = 2, b = -4, \text{ and } c = -1$$

$$= \frac{4 \pm \sqrt{16 - (-8)}}{4} \qquad \text{Multiply under the radical sign.}$$

$$= \frac{4 \pm \sqrt{24}}{4} \qquad \text{Subtract under the radical sign.}$$

$$= \frac{4 \pm 2\sqrt{6}}{4} \qquad \text{Simplify: } \sqrt{24} = \sqrt{4 \cdot 6} = \sqrt{4}\sqrt{6} = 2\sqrt{6}.$$

The solutions $\dfrac{4 + 2\sqrt{6}}{4}$ and $\dfrac{4 - 2\sqrt{6}}{4}$ can be simplified. First we factor out 2 from both terms in the numerator. Then we divide the numerator and the denominator by the common factor.

$$x = \frac{4 \pm 2\sqrt{6}}{4} = \frac{\cancel{2}(2 \pm \sqrt{6})}{\cancel{2} \cdot 2} = \frac{2 \pm \sqrt{6}}{2} \qquad \begin{array}{l}\text{Divide the numerator and denominator}\\ \text{by the common factor, 2.}\end{array}$$

In simplified radical form, the equation's solution set is

$$\left\{ \frac{2 - \sqrt{6}}{2}, \frac{2 + \sqrt{6}}{2} \right\}.$$

b. Now we will graph the parabola whose equation is $y = 2x^2 - 4x - 1$.

Step 1. To find the x-intercepts, we set $y = 0$. We obtain $2x^2 - 4x - 1 = 0$, the equation that we just solved. The parabola has x-intercepts at $\dfrac{2 - \sqrt{6}}{2}$ and $\dfrac{2 + \sqrt{6}}{2}$, that is, approximately -0.2 and 2.2.

Step 2. To find the y intercept, we set $x = 0$.

$$y = 2(0)^2 - 4 \cdot 0 - 1 = -1$$

The parabola has a y-intercept at -1.

Step 3. To find the vertex, we begin with the x-coordinate.

$$x = -\frac{b}{2a} = -\frac{(-4)}{2(2)} = 1$$

Now we find the y-coordinate.

$$y = 2x^2 - 4x - 1 = 2(1)^2 - 4(1) - 1 = -3$$

The vertex is $(1, -3)$.

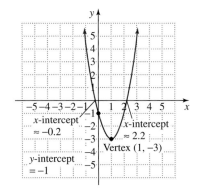

Figure 7.16

The graph of $y = 2x^2 - 4x - 1$

Using the x-intercepts, the y-intercept, and the vertex, we graph the quadratic function $y = 2x^2 - 4x - 1$ as shown in Figure 7.16 on page 624. ■

Our next example shows how the quadratic formula can be used to find solutions that are imaginary numbers.

EXAMPLE 3   Using the Quadratic Formula

Solve using the quadratic formula: $\dfrac{4}{z} + 3 = -\dfrac{2}{z^2}$

Solution

We begin by writing the equation in standard form.

$$z^2\left(\frac{4}{z} + 3\right) = \left(\frac{-2}{z^2}\right)z^2 \qquad \text{Multiply both sides by } z^2, \text{ the LCM of the denominators.}$$

$$4z + 3z^2 = -2 \qquad \text{Apply the distributive property and simplify.}$$

$$3z^2 + 4z + 2 = 0 \qquad \text{Add 2 to both sides.}$$

From the standard form $az^2 + bz + c = 0$, we see that $a = 3$, $b = 4$, and $c = 2$. Now we use the quadratic formula.

$$z = \frac{-b \pm \sqrt{b^2 - 4ac}}{2a} \qquad \text{Use the quadratic formula.}$$

$$= \frac{-4 \pm \sqrt{4^2 - 4(3)(2)}}{2(3)} \qquad \text{Substitute the values for } a, b, \text{ and } c.$$

$$= \frac{-4 \pm \sqrt{16 - 24}}{6} \qquad \text{Multiply under the radical.}$$

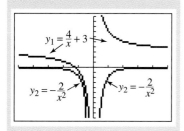

$$= \frac{-4 \pm \sqrt{-8}}{6}$$

Subtract under the radical sign.

$$= \frac{-4 \pm 2i\sqrt{2}}{6}$$

$\sqrt{-8} = \sqrt{-1}\sqrt{4}\sqrt{2} = i(2\sqrt{2}) = 2i\sqrt{2}$

$$= \frac{2(-2 \pm i\sqrt{2})}{6}$$

Factor out 2 from the numerator.

$$= \frac{\cancel{2}(-2 \pm i\sqrt{2})}{\cancel{2} \cdot 3}$$

Cancel identical factors of 2 in the numerator and denominator.

$$= \frac{-2 \pm i\sqrt{2}}{3}$$

You can also write this as $\frac{-2 \pm \sqrt{2}i}{3}$.

The solution set is $\left\{ \dfrac{-2 - i\sqrt{2}}{3}, \dfrac{-2 + i\sqrt{2}}{3} \right\}$. ∎

2 Determine the nature of a quadratic equation's solutions.

The Nature of the Solutions to the Quadratic Equation

The Greek mathematician Diophantus (A.D. 200) is believed to have known the quadratic formula, but used only

$$x = \frac{-b + \sqrt{b^2 - 4ac}}{2a}$$

and rejected any negative rational solutions to the quadratic equation. Furthermore, if an irrational or imaginary number appeared as a solution, Diophantus rejected these numbers as "impossible." In short, he only accepted solutions to $ax^2 + bx + c = 0$ that were positive rational numbers.

As you observed in the previous examples, the nature of the solutions (the kinds of solutions) one obtains to the quadratic equation depends on $b^2 - 4ac$, the expression under the radical sign. This expression is called the *discriminant*.

The discriminant and the nature of a quadratic equation's solutions

The expression $b^2 - 4ac$ is called the *discriminant* of the quadratic equation $ax^2 + bx + c = 0$. Because the solutions are given by

$$x = \frac{-b \pm \sqrt{b^2 - 4ac}}{2a}$$

$b^2 - 4ac$ determines the nature of the solutions to the quadratic equation.

1. If $b^2 - 4ac$ is negative, the solutions are not real numbers.
2. If $b^2 - 4ac$ is a positive perfect square, the solutions are rational numbers.
3. If $b^2 - 4ac$ is a positive number that is not a perfect square, the solutions are irrational numbers.

Table 7.3 revisits the three quadratic equations in Examples 1–3 that we solved using the quadratic formula.

TABLE 7.3 Examples of the Discriminant

Equation in Standard Form	Value of the Discriminant $b^2 - 4ac$	Description of the Discriminant	What the Discriminant Tells Us
$2x^2 + 9x - 5 = 0$ $a = 2, b = 9, c = -5$	$b^2 - 4ac$ $= 9^2 - 4(2)(-5)$ $= 81 + 40$ $= 121$	Perfect square	The solutions to $2x^2 + 9x - 5 = 0$ are rational numbers.
$2x^2 - 4x - 1 = 0$ $a = 2, b = -4, c = -1$	$b^2 - 4ac$ $= (-4)^2 - 4(2)(-1)$ $= 16 + 8$ $= 24$	Positive, but not a perfect square	The solutions to $2x^2 - 4x - 1 = 0$ are irrational numbers.
$3z^2 + 4z + 2 = 0$ $a = 3, b = 4, c = 2$	$b^2 - 4ac$ $= 4^2 - 4(3)(2)$ $= 16 - 24$ $= -8$	Negative	The solutions to $3z^2 + 4z + 2 = 0$ are not real numbers. They are imaginary numbers.

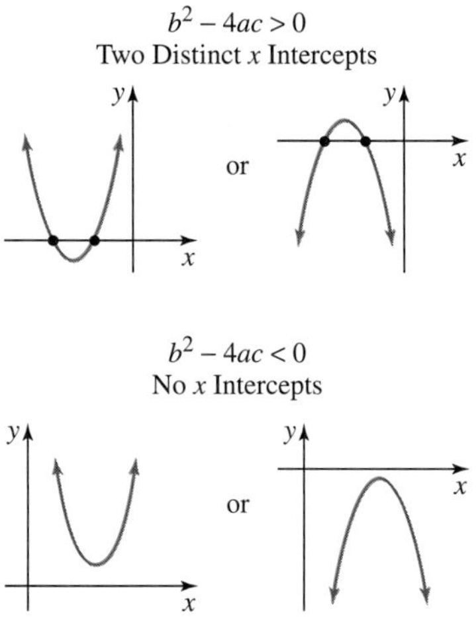

$b^2 - 4ac > 0$
Two Distinct x Intercepts

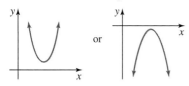

or

$b^2 - 4ac < 0$
No x Intercepts

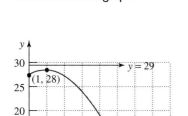

or

Figure 7.17

The discriminant and a quadratic function's graph

There is also a relationship between the discriminant of the equation $ax^2 + bx + c = 0$ and the graph of the function $y = ax^2 + bx + c$. The real solutions of the equation are the x-intercepts of the parabola. If the discriminant is negative, the equation has no real solutions, which means the parabola has no x-intercepts. This relationship is illustrated in Figure 7.17.

D iscover for yourself

What happens to the graph of $y = ax^2 + bx + c$ if $b^2 - 4ac$, the discriminant, is equal to zero? Answer this question by solving $4x^2 - 4x + 1 = 0$ using the quadratic formula. What is the value of the discriminant? Now graph $y = 4x^2 - 4x + 1$ by hand or using a graphing utility. What do you observe?

EXAMPLE 4 **Applying the Discriminant**

The model $y = -x^2 + 2x + 27$ describes the height (y, in meters) of a diver (after x seconds) who jumps from a cliff that is 27 meters above the water. Will the diver ever reach a height of 29 meters?

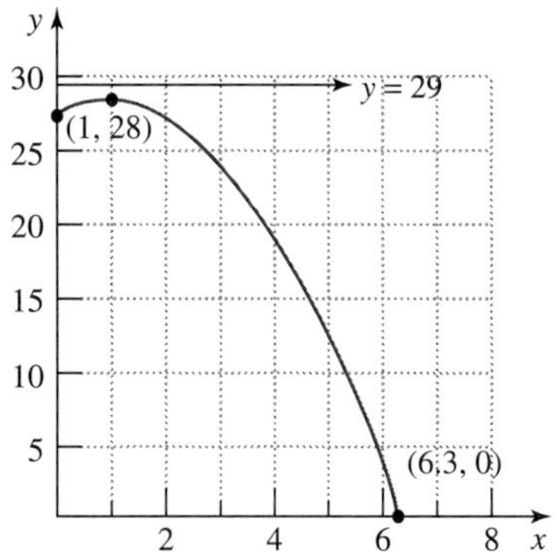

Solution

$$y = -x^2 + 2x + 27 \quad \text{This is the given model.}$$
$$29 = -x^2 + 2x + 27 \quad \text{Replace } y \text{ with 29. We must find a value of } x \text{ that makes the expression on the right equal 29.}$$
$$x^2 - 2x + 2 = 0 \quad \text{Write the equation in standard form. } a = 1, b = -2, \text{ and } c = 2.$$

Rather than solving the equation, we will compute the discriminant.

$$b^2 - 4ac = (-2)^2 - 4(1)(2) = 4 - 8 = -4$$

The discriminant is negative. This indicates that $29 = -x^2 + 2x + 27$ has no

Figure 7.18

The diver's height, modeled by $y = -x^2 + 2x + 27$, will never reach 29 meters.

real solution. In terms of the problem's question, this means that the diver will never reach a height of 29 meters.

Figure 7.18 at the bottom of page 627, shows the graph of $y = -x^2 + 2x + 27$. The graph indicates that the diver reaches a maximum height of 28 meters after 1 second, but that a height of 29 meters is never achieved. ■

3 Use the quadratic formula to answer questions involving modeling.

Some formulas do not model reality as precisely as we might like.

Arman "Successive-ment" 1983, cast bronze cello and bow, welded, 135 × 80 × 35 cms, 53 × $31\frac{1}{2}$ × $13\frac{3}{4}$ in. Cast of 8. © 1997 Artists Rights Society (ARS), New York/ADAGP, Paris.

John Chamberlain "Nanoweap" 1969, painted and chromium-plated steel, 54 × 63 in. (137.1 × 160 cm) 74–129 DJ. Photographer: Hickey-Robertson, Houston. The Mentil Collection, Houston. © 1997 John Chamberlain/Artists Rights Society (ARS), New York

Modeling and the Quadratic Formula

The quadratic formula can give us insight into situations modeled by quadratic functions.

EXAMPLE 5 **A Modeling Application**

The function $N = 0.4x^2 - 36x + 1000$ approximates the number of accidents (N) per 50 million miles for a driver who is x years old. The formula models reality for $16 \leq x \leq 74$ (that is, for drivers between ages 16 and 74, inclusively).

a. What is the age of a driver predicted to have 312 accidents per 50 million miles driven?

b. What is the age of a driver predicted to have the least number of accidents per 50 million miles driven? What is the least number of accidents?

Solution

a. 　$N = 0.4x^2 - 36x + 1000$　　This is the given model.

　　$312 = 0.4x^2 - 36x + 1000$　　Substitute 312 for N. We must find x, a driver's age, with 312 accidents per 50 million miles.

　　　$0 = 0.4x^2 - 36x + 688$　　Subtract 312 on both sides.
　　　　　↕　　　　↕　　　↕
　　　$0 = ax^2 + bx + c$

We see that $a = 0.4$, $b = -36$, and $c = 688$, so we substitute these values into the quadratic formula.

$$x = \frac{-b \pm \sqrt{b^2 - 4ac}}{2a} = \frac{-(-36) \pm \sqrt{(-36)^2 - 4(0.4)(688)}}{2(0.4)}$$

$$= \frac{36 \pm \sqrt{195.2}}{0.8}$$

Thus,

$$x = \frac{36 + \sqrt{195.2}}{0.8} \quad \text{or} \quad x = \frac{36 - \sqrt{195.2}}{0.8}$$

$$\approx 62 \qquad\qquad\qquad \approx 28$$

Use a calculator to obtain an approximation to the nearest whole number.

Drivers who are about 28 and 62 years old are predicted to have 312 accidents per 50 million miles driven.

b. The leading coefficient of the given model, 0.4, is positive, so the vertex is a minimum point on the model's graph. To find the vertex, we begin with the x-coordinate.

$$x = -\frac{b}{2a} = -\frac{(-36)}{2(0.4)} = 45$$

Use $N = 0.4x^2 - 36x + 1000$, with $a = 0.4$ and $b = -36$.

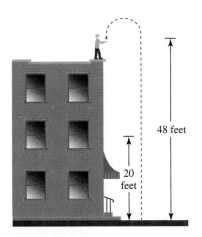

The graph of

$$y = 0.4x^2 - 36x + 1000$$

shown here was obtained with a graphing utility using

Xmin = 0, Xmax = 75,
Xscl = 5, Ymin = 0,
Ymax = 400, Yscl = 20

Number of Accidents per 50 Million Miles

$y = 0.4x^2$
$- 36x + 1000$

$y = 190$

$x = 45$

5   45

Drivers who are 45 years old are predicted by the model to have the least number of accidents. To find this number, we substitute 45 for x and solve for N.

$$N = 0.4x^2 - 36x + 1000 \qquad \text{This is the given model.}$$
$$= 0.4(45)^2 - 36(45) + 1000 \qquad \text{Substitute 45 for } x.\text{ Compute by hand or with a calculator.}$$
$$= 190$$

Thus, drivers who are 45 years old are predicted to have the least number of accidents, namely 190 per 50 million miles driven. ■

The graph in the Using Technology visually confirms what we discovered algebraically in Example 5. The parabola's shape indicates that the number of accidents decreases until age 45, at which point it reaches a minimum. After this, the number of accidents increases.

Do you think that this model accurately describes reality? Why do 45-year-olds have the least number of accidents?

If we know that a quadratic function models a situation, we can use three data points to find the equation of the function that models the data. Let's see how this is done.

EXAMPLE 6 **Modeling Data with a Quadratic Function**

The height (s, in feet) of a falling object above the ground after t seconds can be modeled by a quadratic function. A ball is thrown directly upward from a height of 48 feet, as shown in Figure 7.19. The ball's position above the ground after 1 and 2 seconds is shown in the following table.

t (Time)	s (Position above Ground)
0	48
1	64
2	48

a. Find a quadratic function that fits the data.
b. How long after the ball is thrown, to the nearest tenth of a second, will it be 20 feet above the ground?
c. How long after the ball is thrown will it hit the ground?

Figure 7.19

Throwing a ball directly upward from a height of 48 feet

48 feet

20 feet

Solution

a. Since the height s is a quadratic function of time t, we can express this situation as

$$s = at^2 + bt + c. \qquad \text{We can also write } s(t) = at^2 + bt + c.$$

The given data values for t and s will enable us to determine values of $a, b,$ and c.

Data Value (t, s)	Substitute into $s = at^2 + bt + c$	The Resulting Equation
$(0, 48)$	$48 = a \cdot 0^2 + b \cdot 0 + c$ $48 = c$	$c = 48$
$(1, 64)$	$64 = a \cdot 1^2 + b \cdot 1 + c$ $64 = a + b + c$	$a + b + c = 64$
$(2, 48)$	$48 = a \cdot 2^2 + b \cdot 2 + c$ $48 = 4a + 2b + c$	$4a + 2b + c = 48$

We can now find values of $a, b,$ and c using the equations in the third column. Since $c = 48$, the second and third equations become

$$\begin{array}{c} a + b + 48 = 64 \\ 4a + 2b + 48 = 48 \end{array} \quad \text{or} \quad \begin{array}{c} a + b = 16 \\ 4a + 2b = 0 \end{array} \quad \begin{array}{l} \leftarrow \text{Solve this system} \\ \text{and show that} \\ a = -16 \text{ and } b = 32. \end{array}$$

We now return to the form for our quadratic model.

$$\begin{aligned} s &= at^2 + bt + c &&\text{Position (height) is a quadratic function of time.} \\ &= -16t^2 + 32t + 48 &&\text{Substitute the values for } a, b, \text{ and } c: \\ &&& a = -16, b = 32, c = 48. \end{aligned}$$

Thus, the position function that fits the given data is

$$s = -16t^2 + 32t + 48$$

or, equivalently,

$$s(t) = -16t^2 + 32t + 48$$

b. We now want to know when the ball will be 20 feet above the ground.

$$\begin{aligned} s &= -16t^2 + 32t + 48 &&\text{This is the position function from part (a).} \\ 20 &= -16t^2 + 32t + 48 &&\text{Substitute 20 for } s \text{ and solve the resulting quadratic} \\ &&&\text{equation for } t. \\ 0 &= -16t^2 + 32t + 28 &&\text{Subtract 20 from both sides.} \end{aligned}$$

We can simplify the numbers if we divide both sides of the equation by -4.

$$\frac{-16t^2}{-4} + \frac{32t}{-4} + \frac{28}{-4} = \frac{0}{-4}$$

$$4t^2 - 8t - 7 = 0$$

$$\updownarrow \quad \updownarrow \quad \updownarrow$$

$$at^2 + bt + c = 0$$

We see that $a = 4, b = -8,$ and $c = -7$, so we substitute these values into the quadratic formula.

$$t = \frac{-b \pm \sqrt{b^2 - 4ac}}{2a} = \frac{-(-8) \pm \sqrt{(-8)^2 - 4(4)(-7)}}{2(4)} = \frac{8 \pm \sqrt{176}}{8}$$

$$= \frac{8 \pm \sqrt{16}\sqrt{11}}{8} = \frac{8 \pm 4\sqrt{11}}{8} = \frac{4(2 \pm \sqrt{11})}{4 \cdot 2} = \frac{2 \pm \sqrt{11}}{2}$$

Thus,

$$t = \frac{2 + \sqrt{11}}{2} \quad \text{or} \quad t = \frac{2 - \sqrt{11}}{2}$$

$$\approx 2.7 \qquad\qquad \approx -0.7$$

Since negative time is meaningless in this situation, the only realistic solution is 2.7. In terms of the problem's question, this means that about 2.7 seconds after the ball is thrown, it will be 20 feet above the ground.

c. At the moment the ball strikes the ground, its height above the ground is 0 feet. We substitute 0 for s in the model and solve for t.

$$s = -16t^2 + 32t + 48 \qquad \text{Use the position function from part (a).}$$
$$0 = -16t^2 + 32t + 48 \qquad \text{Substitute 0 for } s.$$

We can simplify by dividing both sides by -16.

$$\frac{-16t^2}{-16} + \frac{32t}{-16} + \frac{48}{-16} = \frac{0}{-16}$$
$$t^2 - 2t - 3 = 0$$

Although we can solve for t by the quadratic formula, factoring is much faster in this situation.

$$(t - 3)(t + 1) = 0 \qquad\qquad \text{Factor.}$$
$$t - 3 = 0 \quad \text{or} \quad t + 1 = 0 \qquad \text{Set each factor equal to 0.}$$
$$t = 3 \qquad\qquad t = -1 \qquad \text{Solve.}$$

The only meaningful solution is 3. The ball will hit the ground 3 seconds after it was thrown. ■

The graph of the position function from Example 6, obtained with a graphing utility, is shown in Figure 7.20. The vertex $(1, 64)$ indicates that the ball reaches its maximum height of 64 feet above the ground 1 second after it is thrown. (Use the formula $t = -\dfrac{b}{2a}$ to verify this.)

Study tip

Once a quadratic equation is in standard form, try factoring. Turn to the quadratic formula if factoring is impossible or too difficult.

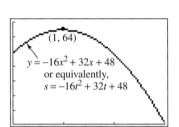

(1, 64)

$y = -16x^2 + 32x + 48$
or equivalently,
$s = -16t^2 + 32t + 48$

Figure 7.20

The graph of $y = -16x^2 + 32x + 48$ using Xmin = 0, Xmax = 3, Xscl = 1, Ymin = 0, Ymax = 70, Yscl = 10

Determine the most efficient technique to use when solving a quadratic equation.

Discover for yourself

What happens to the model $s = -16t^2 + 32t + 48$ if $t > 3$? Explore the model both numerically and graphically for $t > 3$. Why does the function no longer describe the physical situation?

Determining Which Method to Use

Although all quadratic equations can be solved by the quadratic formula, if an equation is in the form $u^2 = e$, such as $x^2 = 5$ or $(2x + 3)^2 = 8$, it is faster to use the square root property, taking the square root of both sides. If the equation is not in the form $u^2 = e$, write the quadratic equation in standard form $(ax^2 + bx + c = 0)$. Try to solve the equation by the factoring method. If $ax^2 + bx + c$ cannot be factored, then solve the quadratic equation by the quadratic formula.

Since we used the method of completing the square to derive the quadratic formula, we no longer need it for solving quadratic equations.

These observations are summarized in Table 7.4 on page 633.

ENRICHMENT ESSAY

Quadratic Equations and Art

The golden rectangle *ABCD,* shown below, is defined in such a way that if square *APQD* is removed, the ratio of the sides of the remaining rectangle $\dfrac{BC}{QC}$ is the same as the ratio of the sides of the original rectangle $\dfrac{AB}{AD}$.

$$\frac{BC}{QC} = \frac{AB}{AD}$$ This defines the golden rectangle.

$$\frac{1}{x-1} = \frac{x}{1}$$ See the figure. Assume that the width is 1.

$$(x-1)\cdot\frac{1}{x-1} = (x-1)\cdot\frac{x}{1}$$ Multiply by the LCD or cross multiply.

$$1 = x^2 - x$$

$$0 = x^2 - x - 1$$

Using the quadratic formula, the positive solution (representing side *AB*) is $\dfrac{1+\sqrt{5}}{2}$ which is approx-

imately 1.618. Adjacent sides of the golden rectangle are in the ratio 1.618 to 1.

Place de la Concorde, an abstract painting by Piet Mondrian, uses rectangles in a 1.618 to 1 ratio. Asked if he always painted rectangles in the golden ratio, Mondrian was deliberately vague. "Rectangles? I see no rectangles in my painting."

Find another example of the golden rectangle in art, architecture, or nature.

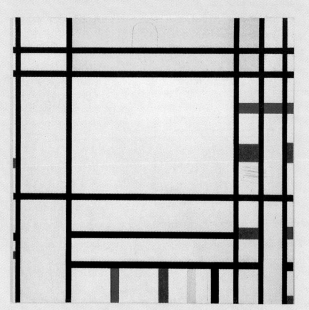

Piet Mondrian (Dutch, 1872–1944). "Place de la Concorde," oil on canvas. Courtesy Dallas Museum of Art, Foundation for the Arts Collection, gift of the James H. and Lillian Clark Foundation.

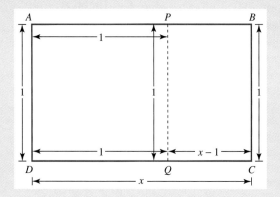

TABLE 7.4 Determining the Most Efficient Technique to Use When Solving a Quadratic Equation

Description and Form of the Quadratic Equation	Most Efficient Solution Method	Example
$ax^2 + c = 0$ The quadratic equation has no linear (x) term.	Solving for x^2 and using the square root method	$4x^2 - 7 = 0$ $4x^2 = 7$ $x^2 = \dfrac{7}{4}$ $x = \pm\dfrac{\sqrt{7}}{2}$
$u^2 = e$; u is a linear expression.	The square root method	$(x + 4)^2 = 5$ $x + 4 = \pm\sqrt{5}$ $x = -4 \pm \sqrt{5}$
$ax^2 + bx + c = 0$ and $ax^2 + bx + c$ can be obviously factored.	Factoring and the zero-product principle	$3x^2 + 5x - 2 = 0$ $(3x - 1)(x + 2) = 0$ $3x - 1 = 0$ or $x + 2 = 0$ $x = \dfrac{1}{3}$ $x = -2$
$ax^2 + bx + c = 0$ and $ax^2 + bx + c$ cannot be factored or the factoring is too difficult.	The quadratic formula: $x = \dfrac{-b \pm \sqrt{b^2 - 4ac}}{2a}$	$x^2 - 2x - 6 = 0$ $x = \dfrac{2 \pm \sqrt{4 - 4(1)(-6)}}{2(1)}$ $= \dfrac{2 \pm \sqrt{28}}{2} = \dfrac{2 \pm \sqrt{4}\sqrt{7}}{2}$ $= \dfrac{2 \pm 2\sqrt{7}}{2} = \dfrac{2(1 \pm \sqrt{7})}{2}$ $= 1 \pm \sqrt{7}$

PROBLEM SET 7.3

Practice Problems

In Problems 1–20, solve each equation using the quadratic formula.

1. $x^2 - 9x + 20 = 0$ **2.** $x^2 - 6x + 5 = 0$ **3.** $x^2 - 4x - 60 = 0$

4. $x^2 + 3x - 40 = 0$ **5.** $2x^2 - 7x = -5$ **6.** $6x^2 + x = 1$

7. $2y^2 + y = 5$ **8.** $2x^2 - 5x = -1$ **9.** $2 = 3y^2 + 4y$

10. $5 = 3x^2 + 4x$ **11.** $2z^2 = 2z - 1$ **12.** $2z = z^2 + 4$

13. $5y^2 = 2y - 3$ **14.** $4y^2 = -2y - 5$ **15.** $3 + \dfrac{7}{x} = \dfrac{6}{x^2}$

16. $\dfrac{5}{x^2} = 3 + \dfrac{8}{x}$ **17.** $\dfrac{5x + 6}{2x + 3} = 3x$ **18.** $\dfrac{25 - 5x}{3x + 3} = x$

19. $\frac{1}{2}x^2 - \frac{1}{3}x + \frac{1}{4} = 0$ **20.** $\frac{1}{2}x^2 + 2x + \frac{2}{3} = 0$

In Problems 21–32, solve each equation by the method of your choice.

21. $(x - 3)(x + 3) = 12$ **22.** $5x^2 - 3 = 0$ **23.** $(5x - 1)(2x + 3) = 3x - 3$

24. $5x^2 - 2x + 4 = 8 - 2x$ **25.** $(3x - 4)^2 = 81$ **26.** $x^2 - 2x + 4 = 0$

27. $3x^2 - 4x + 2 = 0$ **28.** $(3x - 1)^2 - 121 = 0$ **29.** $2x^2 + 5x = 3$

30. $2x^2 - 7x = 0$ **31.** $3x^2 = 11x - 10$ **32.** $\dfrac{35}{x^2} + 1 = \dfrac{12}{x}$

33. a. Solve: $x^2 - 6x + 7 = 0$.
   b. Graph: $y = x^2 - 6x + 7$.

34. a. Solve: $2x^2 + 4x - 3 = 0$.
   b. Graph: $y = 2x^2 + 4x - 3$.

35. a. Solve: $2x^2 - 6x - 9 = 0$.
   b. Graph: $y = 2x^2 - 6x - 9$.

36. a. Solve: $2x^2 - 6x + 1 = 0$.
   b. Graph: $y = 2x^2 - 6x + 1$.

37. a. Solve: $x^2 + x + 5 = 0$.
   b. Graph: $y = x^2 + x + 5$.

38. a. Solve: $x^2 - 4x + 13 = 0$.
   b. Graph: $y = x^2 - 4x + 13$.

In Problems 39–47, compute $b^2 - 4ac$ (the discriminant) and then, without actually solving, indicate if the solutions to the quadratic equation are rational, irrational, or not real numbers.

39. $y^2 - 4y - 5 = 0$

40. $4x^2 - 2x + 3 = 0$

41. $2x^2 - 11x + 3 = 0$

42. $2y^2 + 11y = 6$

43. $x^2 - 2x + 1 = 0$

44. $3y^2 = 2y - 1$

45. $x^2 - 3x - 7 = 0$

46. $3y^2 + 4y - 2 = 0$

47. $4y^2 + 2y + 5 = 0$

Application Problems

48. The function $N = 0.036x^2 - 2.8x + 58.14$ approximately models the number of deaths (N) per year per thousand people for people who are x years old, where $40 \leq x \leq 60$. Find, to the nearest whole number, the age at which 12 people per 1000 die annually.

49. The function $P = 0.78t^2 + 76.7t + 4449$ models world population (P, in millions) t years after 1980, where $0 \leq t \leq 10$. Find, to the nearest whole number, the year in which world population was 5000 million. State one reason why this function models reality for only a relatively short period of time (11 years).

50. The function $D = 9.2t^2 - 46.7t + 480$ approximately models our national debt (D, in billions of dollars) t years after 1970, where $0 \leq t \leq 22$. Find, to the nearest whole number, the year in which the model predicts a national debt of $1849.5 billion. If you project this model to one year ago, what does it indicate the national debt was at that time? Check an appropriate reference to find the actual national debt for that year. How closely does the model reflect what actually happened?

51. There is a relationship between the amount of one's income, (x, annual income in thousands of dollars) and the percent of this income (P) that one contributes to charities. This relationship is modeled by the quadratic function
$P = 0.0014x^2 - 0.1529x + 5.855$, where $5 \leq x \leq 100$.
   a. What annual income corresponds to a 2 percent charitable contribution?
   b. What annual income corresponds to the minimum percent given to charity? What is this minimum percent?

52. a. Find a quadratic function that fits the following data.

t (Number of Years After 1970)	N (Number of People (in millions) in the United States Holding More than One Job)
0	4.1
1	4.014
2	3.95

   b. Use the function to find in what year 8 million people held more than one job.
   c. What year corresponds to the minimum number of people in the United States holding more than one job? What is the minimum number?

53. a. Find a quadratic function that fits the following data.

t Number of Years After 1975	E Total School Expenditures (Elementary Through Graduate School) in the United States (in millions)
0	107,298
1	98,588
2	90,798

   b. Use the function to predict in what year total school expenditures will be $165,548 million.
   c. What year corresponds to minimum total school expenditures? What is this minimum amount?

Use the discriminant to answer Problems 54–55.

54. A ball is thrown directly upward from the top of a 200-foot building. The ball's height above the ground (*s*, in feet) *t* seconds after it is thrown is modeled by $s = -16t^2 + 40t + 200$.
   a. Will the ball ever reach a height of 240 feet?
   b. Will the ball ever reach a height of 225 feet? How often does this occur? Explain what this means in terms of the path that the ball follows.

55. The model $P = -5I^2 + 80I$ describes the power (*P*) of an 80-volt generator subject to a current (*I*) of electricity given in amperes. Can there ever be enough current to generate 340 volts of power?

True–False Critical Thinking Problems

56. Which one of the following is true?
   a. The quadratic formula is developed by applying factoring and the zero product principle to the quadratic equation $ax^2 + bx + c = 0$.
   b. In using the quadratic formula to solve the quadratic equation $5x^2 = 2x - 7$, we let $a = 5, b = 2$, and $c = -7$.
   c. The quadratic formula cannot be used to solve the equation $x^2 - 9 = 0$.
   d. Any quadratic equation that can be solved by completing the square can be solved by the quadratic formula.

57. Which one of the following is true?
   a. Any quadratic equation that can be solved by the quadratic formula can also be solved by factoring.
   b. The fastest way to solve any quadratic equation is by using the quadratic formula.
   c. If the discriminant is 0, then the quadratic equation cannot have imaginary solutions.
   d. If $ax^2 - bx + c = 0$, then
   $$x = \frac{-b \pm \sqrt{(-b)^2 - 4ac}}{2a}.$$

Technology Problems

58. Use a graphing utility to verify the real solutions that you obtained for any four problems in Problems 1–32.

59. Use a graphing utility to verify the graphs that you drew by hand in Problems 33–38.

60. Use a graphing utility to graph the given models in Problems 48–51. Then $\boxed{\text{TRACE}}$ along each curve until you reach the point whose coordinates correspond to the problem's solution.

61. Use a graphing utility to graph the models that you obtained in Problems 52–53. Then $\boxed{\text{TRACE}}$ along each curve until you reach the point that corresponds to the solution in part (b) and then part (c) of each problem.

62. Use a graphing utility to graph the given models in Problems 54–55. For Problem 54a, also graph $y_2 = 240$ and for Problem 55 also graph $y_2 = 340$ in the same viewing rectangle as the graph of the model. Explain how these graphs represent the problem's solution.

63. Most graphing utilities can fit curves to data points. Enter the data in the table below in a graphing utility. Use its quadratic fit program to find a quadratic model for the data. Then use the model to predict public school enrollment for 1999–2000.

School Year	1980–1981	1981–1982	1982–1983	1983–1984	1984–1985	1985–1986	1986–1987	1987–1988	1988–1989
x	0	1	2	3	4	5	6	7	8
Public School Enrollment (in millions) *y*	41.5	40.6	40	39.5	39.1	39	39.5	39.7	40.1

Writing in Mathematics

64. We stated that every quadratic equation can be solved using the quadratic formula
$$x = \frac{-b \pm \sqrt{b^2 - 4ac}}{2a}.$$

However, because division by 0 is undefined, then $a \neq 0$. Doesn't that mean that some quadratic equations cannot be solved using the quadratic formula, namely, those in which a is 0? Isn't this a contradiction to our original statement? Explain.

65. We now have four methods for solving quadratic equations: factoring, the square root method, complet-ing the square, and the quadratic formula. If you are given a quadratic equation to solve, how do you deter-mine which method to use? If possible, include exam-ples in your explanation.

Critical Thinking Problems

Solve Problems 66–67 using the quadratic formula.

66. $x^2 - 3\sqrt{2}x = -2$

67. $ix^2 = 5x - 2i$

Reread the Discover for Yourself on page 627. Then find all values of h that will cause each function in Problems 68–69 to have a graph with one x-intercept.

68. $y = hx^2 + 3x + 2$

69. $f(x) = 5x^2 + hx + 3$

70. Find the values of a for which $ax^2 - 2x + 3 = 0$ has no real solutions. (*Hint:* Set $b^2 - 4ac < 0$.)

71. Find the values of a for which $3x^2 - 6x + a = 0$ has two distinct real roots.

72. Use the fact that

$$\frac{-b - \sqrt{b^2 - 4ac}}{2a} \quad \text{and} \quad \frac{-b + \sqrt{b^2 - 4ac}}{2a}$$

are the solutions to the quadratic equation to answer this problem.

a. Show that the sum of the two solutions is equal to $\dfrac{-b}{a}$.

b. A student solved the quadratic equation $x^2 - 4x - 21 = 0$ and discovered the solution set to be $\{3, -7\}$.

Show that this result is incorrect by showing that the sum of the two solutions is *not* equal to $\dfrac{-b}{a}$.

(*Hint:* Because $x^2 - 4x - 21 = 0$, $a = 1$, $b = -4$, and $c = -21$.)

c. Show that the product of the two solutions is equal to $\dfrac{c}{a}$.

d. A student solved the quadratic equation $x^2 - 2x + 5 = 0$ and discovered the solution set to be $\{1 - 2i, 1 + 2i\}$. Show that this result is correct by showing that the sum of the two solutions is $\dfrac{-b}{a}$ and the product of the two solutions is $\dfrac{c}{a}$.

(*Hint:* Because $x^2 - 2x + 5 = 0$, $a = 1$, $b = -2$, and $c = 5$.)

In Problem 72, you showed that if $ax^2 + bx + c = 0$, the sum of the solutions is $\dfrac{-b}{a}$ and the product of the solutions is $\dfrac{c}{a}$.

Divide both sides of $ax^2 + bx + c = 0$ and show that the equation takes on the form

$$x^2 + (\text{Sum of solutions with the sign changed})\,x + (\text{Product of the solutions}) = 0.$$

Use this result to write each quadratic equation in Problems 73–76 with integers for a, b, and c having the given numbers as solutions.

73. 3 and 7

74. -2 and 5

75. $2 + \sqrt{3}$ and $2 - \sqrt{3}$

76. $3 + 2i$ and $3 - 2i$

77. We can use factoring to show that $(x - 3)(x - 7) = 0$ has solutions 3 and 7. Use this principle in reverse to solve Problem 74. Now try Problems 75 and 76 by this method. When is this method more difficult than the one you previously used to write an equation from its solutions?

78. Solve without using a calculator:

$$10^{-4}x^2 + 2 \cdot 10^{-3}x + 10^{-2} = 0.$$

79. If $x^2 + px + q = 0$ is to have only real solutions and if $|p| = |q| = 1$, find p and q.

80. If $4x^4 + 1$ can be expressed as $(2x^2 + 1)^2 - (2x)^2$, solve the equation $4x^4 + 1 = 0$.

81. If $c > a > 0$ and $b = a + c$, find the larger solution of $ax^2 + bx + c = 0$.

Group Activity Problem or Individual Project

82. Consult a reference source and find data that can be closely approximated using a quadratic model. Find the model using a graphing utility. Then compare val-ues generated by the model with actual data values, using a graphic and numerical approach.

Review Problems_____

83. Simplify: $\dfrac{\dfrac{1}{x} + 1}{\dfrac{1}{x^2} - 1}$.

84. Write the point-slope equation and the slope-intercept equation for the line passing through the points $(1, -2)$ and $(5, 2)$.

85. Solve for y: $5x - 2y = 10$.

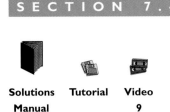

SECTION 7.4

Solutions Manual   **Tutorial**   **Video 9**

Problem Solving and Equations That Are Quadratic in Form

Objectives

1 Solve problems using quadratic models.
2 Solve equations that are quadratic in form.

In this section, our focus is on writing quadratic equations in standard form when the given equations do not appear to be quadratic. Once we have expressed an equation in the form $ax^2 + bx + c = 0$, we can solve by factoring, the quadratic formula, or the square root method. We begin by looking at problems that give rise to such equations.

1 Solve problems using quadratic models.

Solving Problems by Creating Quadratic Models

In the last section, we created quadratic models that fit given data values. Now we create models that translate given conditions. We will use our five-step problem-solving strategy. Let's begin with a situation that calls for geometric modeling.

| **EXAMPLE 1** | **Solving a Geometry Problem with a Quadratic Model** |

A rectangular swimming pool is 12 meters long and 8 meters wide. A tile border of uniform width is to be built around the pool, using 120 square meters of tile. The tile is from a discontinued stock (so no additional materials are available), and all 120 square meters are to be used. How wide should the border be? If zoning laws require at least a 2-meter-wide border around the pool, can this be done with the available tile?

Solution

As is often the case with geometric modeling, a sketch is helpful to translate the implied conditions into an equation. Such a sketch is shown in Figure 7.21. By examining the figure, and using the fact that the area of a rectangle is the product of its length and width, we can write an equation that models the area of the tile border.

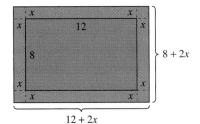

Figure 7.21

Building a tile border around a 12 meter by 8 meter pool

Steps 1 and 2. Represent unknowns in terms of x.

Let x = the width of the border.

Step 3. Write an equation that describes the problem's conditions.

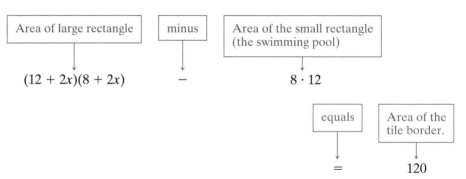

| Area of large rectangle | minus | Area of the small rectangle (the swimming pool) |

$(12 + 2x)(8 + 2x)$ $-$ $8 \cdot 12$

| equals | Area of the tile border. |

$=$ 120

This equation might not look quadratic, but by simplifying the left side and setting it equal to 0, we will obtain a quadratic equation in standard form.

Step 4. Solve the equation and answer the problem's question.

$(12 + 2x)(8 + 2x) - 8 \cdot 12 = 120$ This is the equation implied by the problem's conditions ($A = LW$).

$96 + 40x + 4x^2 - 96 = 120$ Multiply using the FOIL method.

$40x + 4x^2 = 120$ $96 - 96 = 0$

$4x^2 + 40x - 120 = 0$ Write the equation in standard form.

$x^2 + 10x - 30 = 0$ Simplify by dividing both sides by 4.

Geometric forms are a characteristic feature in the paintings of Wassily Kandinsky (1866–1944).

Wassily Kandinsky "Upward Tension" (Spannung Nach Oben) 1924, Watercolor on paper, 49 × 34 cm. Musee National d'Art Moderne, Centre George Pompidou, Paris.

Because $x^2 + 10x - 30$ is prime, we will use the quadratic formula with $a = 1$, $b = 10$, and $c = -30$.

$$x = \frac{-b \pm \sqrt{b^2 - 4ac}}{2a}$$ Use the quadratic formula.

$$= \frac{-10 \pm \sqrt{10^2 - 4(1)(-30)}}{2(1)}$$ Substitute the values for a, b, and c.

$$= \frac{-10 \pm \sqrt{220}}{2}$$ $10^2 - 4(1)(-30) = 100 - (-120) =$ $100 + 120 = 220$

$$= \frac{-10 \pm 2\sqrt{55}}{2}$$ $\sqrt{220} = \sqrt{4(55)} = 2\sqrt{55}$

$$= \frac{2(-5 \pm \sqrt{55})}{2}$$ Factor 2 from the numerator. (Do you find yourself working this step mentally?)

$$= -5 \pm \sqrt{55}$$ Divide the numerator and denominator by 2.

Using a calculator, $\sqrt{55} \approx 7.4$. Thus,

$x \approx -5 + 7.4$ or $x \approx -5 - 7.4$
≈ 2.4 ≈ -12.4

Because a negative width is geometrically impossible, the border should be approximately 2.4 meters wide. This complies with the zoning laws requiring at least a 2-meter-wide border.

Step 5. Check the solution in the original wording of the problem.

Let's check this approximate proposed answer against the condition that the area of the tile border should be 120 square meters. The area of the pool is $8(12) = 96$ square meters. The area of the rectangle containing the pool and the tile border is

$$(12 + 2 \cdot 2.4)(8 + 2 \cdot 2.4) = 215.04 \text{ square meters}$$

The area of the tile border is the difference of these areas:

215.04 − 96 = 119.04 square meters

Since our approximate solution gives an area for the tile border that is close to 120 square meters, this verifies that the border should be about 2.4 meters wide. ∎

EXAMPLE 2 Writing an Equation Based on a Verbal Model

A large boat is rented for a weekend trip at a cost of $1000. When 10 additional people join the excursion on the day of the trip, the fare per person is decreased by $5.00. How many people were originally signed up to rent the boat?

Solution

A verbal model for this problem is that

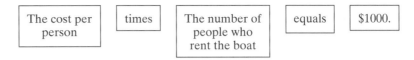

| The cost per person | times | The number of people who rent the boat | equals | $1000. |

Steps 1 and 2. Represent unknown quantities in terms of x.

The problem contains a number of unknown quantities.

Number of people originally signed up = x

Total number of people who rent the boat = $x + 10$

Original cost per person = $\dfrac{1000}{x}$

Cost per person when 10 more people join the excursion $= \dfrac{1000}{x} - 5$ We are given that the fare per person decreases by $5.00.

Step 3. Write an equation that models the problem's conditions.

Now we return to our verbal model.

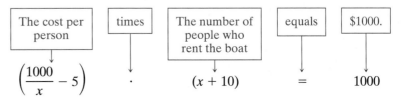

| The cost per person | times | The number of people who rent the boat | equals | $1000. |

$$\left(\frac{1000}{x} - 5\right) \qquad \cdot \qquad (x+10) \qquad = \qquad 1000$$

Step 4. Solve the equation and answer the problem's question.

This equation does not look quadratic, but by clearing fractions and setting it equal to 0, we will obtain a quadratic equation in standard form.

$$\left(\frac{1000}{x} - 5\right)(x+10) = 1000 \qquad \text{This is the equation implied by the problem's conditions.}$$

$$x\left(\frac{1000}{x} - 5\right)(x+10) = 1000x \qquad \text{Multiply both sides by } x \text{ and clear fractions.}$$

$$\left(x \cdot \frac{1000}{x} - 5x\right)(x+10) = 1000x \qquad \text{Distribute throughout the first parentheses on the left.}$$

$$(1000 - 5x)(x + 10) = 1000x \qquad \text{Simplify.}$$
$$1000x + 10{,}000 - 5x^2 - 50x = 1000x \qquad \text{Multiply using FOIL.}$$
$$-5x^2 - 50x + 10{,}000 = 0 \qquad \text{Subtract } 1000x \text{ from both sides.}$$
$$x^2 + 10x - 2000 = 0 \qquad \text{Simplify, dividing both sides by } -5.$$
$$(x - 40)(x + 50) = 0 \qquad \text{Factor.}$$
$$x - 40 = 0 \quad \text{or} \quad x + 50 = 0 \qquad \text{Set each factor equal to 0.}$$
$$x = 40 \qquad\qquad x = -50 \qquad \text{Solve.}$$

Selecting the positive solution tells us that 40 people were originally signed up to rent the boat.

Step 5. Check the proposed solution in the original wording of the problem.

Let's check our proposed solutions.

The original fare per person is $\dfrac{\$1000}{40} = \$25$.

The fare per person when 10 additional people join the excursion is $\dfrac{\$1000}{50} = \$20$.

This checks with the given condition that the fare per person is decreased by $5.00. ■

EXAMPLE 3   Designing a Web Site

Working together, two people can design a Web site in 5 hours. If each worked alone, one of them could design the site in 2 hours less time than the other. How long would it take each person to design the Web site working alone?

Solution

Steps 1 and 2. Represent unknowns in terms of x.

Let

$$x = \text{Number of hours for the slower person to complete the job alone}$$
$$x - 2 = \text{Number of hours for the faster person to complete the job alone}$$

We summarize this information in the following table.

	Fractional Part of the Job Completed in 1 Hour	Time Spent Working Together	Fractional Part of the Job Completed in 5 Hours
Slower Person	$\dfrac{1}{x}$	5	$\dfrac{5}{x}$
Faster Person	$\dfrac{1}{x - 2}$	5	$\dfrac{5}{x - 2}$

Step 3. Write an equation that models the problem's conditions.

Because the two people can complete the job in 5 hours,

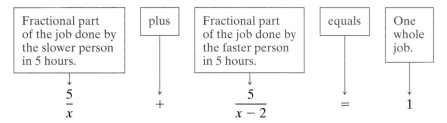

The equation implied by the problem's conditions is a rational equation that is actually quadratic. We can see this by clearing fractions.

Step 4. Solve the equation and answer the problem's question.

$$\frac{5}{x} + \frac{5}{x-2} = 1$$

This is the equation implied by the problem's conditions.

$$x(x-2)\left(\frac{5}{x} + \frac{5}{x-2}\right) = x(x-2) \cdot 1$$

Multiply both sides by the LCM.

$$\cancel{x}(x-2)\left(\frac{5}{\cancel{x}}\right) + x(\cancel{x-2})\left(\frac{5}{\cancel{x-2}}\right) = x(x-2)$$

Apply the distributive property and simplify.

$$5(x-2) + 5x = x(x-2)$$

The equation is now cleared of fractions.

$$5x - 10 + 5x = x^2 - 2x$$

Multiply.

$$10x - 10 = x^2 - 2x$$

Combine like terms on the left.

$$0 = x^2 - 12x + 10$$

Write the equation in standard form.

Because $x^2 - 12x + 10$ is prime, we will use the quadratic formula with $a = 1$, $b = -12$, and $c = 10$:

$$x = \frac{-b \pm \sqrt{b^2 - 4ac}}{2a}$$

Use the quadratic formula.

$$= \frac{-(-12) \pm \sqrt{(-12)^2 - 4(1)(10)}}{2(1)}$$

Substitute the values for a, b, and c.

$$= \frac{12 \pm \sqrt{104}}{2}$$

$(-12)^2 - 4(1)(10) = 144 - 40 = 104$

$$= \frac{12 \pm 2\sqrt{26}}{2}$$

$\sqrt{104} = \sqrt{4(26)} = 2\sqrt{26}$

$$= \frac{2(6 \pm \sqrt{26})}{2}$$

Factor 2 from the numerator.

$$= 6 \pm \sqrt{26}$$

Divide the numerator and denominator by 2.

Using a calculator, $\sqrt{26} \approx 5.1$. Thus,

$$x \approx 6 + 5.1 \qquad \text{or} \qquad x \approx 6 - 5.1$$
$$\approx 11.1 \qquad\qquad\qquad \approx 0.9$$

Because x represents the number of hours for the slower person to do the job alone and the faster person takes 2 hours less time, we reject 0.9 because $0.9 - 2$ results in negative time.

Step 5. Check the proposed solution in the original wording of the problem.

It would take approximately 11.1 hours for the slower person to design the Web site alone and approximately 9.1 hours for the faster person to design the site working alone.

The fractional part of the job done by the two people, working together, in 5 hours is

$$\frac{5}{11.1} + \frac{5}{9.1} \approx 0.45 + 0.55 = 1.$$

This checks with the given condition that working together the two people can design the Web site in 5 hours. ■

2 Solve equations that are quadratic in form.

Solving Equations That Are Quadratic in Form

Some equations that are not quadratic can be written in the form

$$au^2 + bu + c = 0$$

where u is an algebraic expression and $a \neq 0$. For example, consider the fourth-degree equation

$$x^4 - 10x^2 + 9 = 0.$$

The *structure* of the equation is quadratic, which can be seen by replacing x^2 by u.

$$
\begin{array}{ccc}
x^4 & - & 10x^2 & + 9 = 0 \\
\downarrow & & \downarrow & \\
(x^2)^2 & - & 10(x^2) & + 9 = 0 \quad \text{Since we will replace } x^2 \text{ by } u, \text{ write } x^4 \text{ as } (x^2)^2. \\
\downarrow & & \downarrow \quad \downarrow & \\
u^2 & - & 10u & + 9 = 0 \quad \text{Replace } x^2 \text{ with } u.
\end{array}
$$

The equation $u^2 - 10u + 9 = 0$ is a quadratic equation that can be solved by factoring. Then, using $u = x^2$, we can solve for x. Our next example illustrates this process.

EXAMPLE 4 Solving an Equation That Is Quadratic in Form

Solve: $x^4 - 10x^2 + 9 = 0$

Solution

$$
\begin{array}{ccc}
x^4 - 10x^2 & + 9 = 0 & \text{This is the given equation.} \\
\downarrow \quad \downarrow & & \\
(x^2)^2 - 10(x^2) & + 9 = 0 & \text{Let } u = x^2. \\
\downarrow \quad \downarrow & \downarrow \quad \downarrow & \\
u^2 - 10u & + 9 = 0 & \text{Replace } x^2 \text{ with } u.
\end{array}
$$

Now we can solve the equation $u^2 - 10u + 9 = 0$.

$$(u - 9)(u - 1) = 0 \qquad \text{Solve this quadratic equation by factoring.}$$

$$u - 9 = 0 \quad \text{or} \quad u - 1 = 0 \qquad \text{Set each factor equal to 0, and solve for } u.$$

$$u = 9 \qquad\qquad u = 1 \qquad \text{These are } not \text{ the solutions for } x.$$

We have not completed the solution since our goal is to solve for x. Since we introduced the substitution $u = x^2$, we replace u by x^2 in these last two equations.

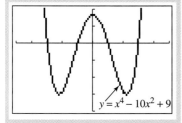
$$x^2 = 9 \quad \text{or} \quad x^2 = 1$$ Replace u with x^2 and solve. This will give solutions for the original equation.

$$x = \pm\sqrt{9} \qquad x = \pm\sqrt{1}$$ Use the square root method.

$$x = \pm 3 \qquad x = \pm 1$$

The equation $x^4 - 10x^2 + 9 = 0$, a fourth-degree equation, has four solutions, verified by substituting into the equation. The solution set is $\{-3, 3, -1, 1\}$. ∎

Before solving additional equations that are quadratic in form, let's summarize the solution method.

Equations Quadratic in Form

An equation that can be written in the form $au^2 + bu + c = 0$, for $a \neq 0$ and u an algebraic expression, is called *quadratic in form*.

Solution Method	Example: $2x^{-2} = 1 - x^{-1}$
1. If necessary, write the equation so that the power of the middle term is half that of the leading term and the final term is a constant.	$2x^{-2} + x^{-1} - 1 = 0$ $\frac{1}{2}(-2) = -1$ Constant
2. Write the variable in the leading term as the square of the variable in the middle term.	$2(x^{-1})^2 + x^{-1} - 1 = 0$
3. Let u equal the middle term without its coefficient and write the equation in the form $au^2 + bu + c = 0$.	Let $u = x^{-1}$. The equation becomes $2u^2 + u - 1 = 0$.
4. Solve the quadratic equation in step 3 for u.	$(2u - 1)(u + 1) = 0$ $u = \dfrac{1}{2} \quad \text{or} \quad u = -1$
5. Replace u with the expression containing the given equation's variable from step 3.	$x^{-1} = \dfrac{1}{2} \quad \text{or} \quad x^{-1} = -1$
6. Solve the equations in step 5 for the given equation's variable.	$\dfrac{1}{x} = \dfrac{1}{2} \quad \text{or} \quad \dfrac{1}{x} = -1$ $x = 2 \qquad\qquad x = -1$
7. Check by direct substitution or by using a graphing utility.	A check will show that 2 and -1 are solutions to the original equation.

The procedure outlined in the box is illustrated in Examples 5 through 8.

EXAMPLE 5 **Solving an Equation That Is Quadratic in Form**

Solve: $x - 5\sqrt{x} = -4$

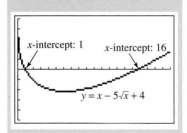
Solution

$$x - 5\sqrt{x} = -4 \qquad \text{This is the original equation.}$$

$$x - 5\sqrt{x} + 4 = 0 \qquad \text{Set the equation equal to 0.}$$

$$(\sqrt{x})^2 - 5\sqrt{x} + 4 = 0 \qquad \text{Write the variable in the highest-degree term as the square of the variable factor in the middle term.}$$

$$u^2 - 5u + 4 = 0 \qquad \text{Let } u = \sqrt{x}. \text{ Notice that } u \text{ is the middle term without its coefficient.}$$

$$(u - 4)(u - 1) = 0 \qquad \text{Solve this quadratic equation for } u. \text{ Factor.}$$

$$u - 4 = 0 \quad \text{or} \quad u - 1 = 0 \qquad \text{Set each factor equal to 0.}$$

$$u = 4 \qquad\qquad u = 1 \qquad \text{Solve the resulting equations.}$$

We're not finished! Remember that we want to solve for x, the variable in the given equation. Since we introduced the substitution $u = \sqrt{x}$, substitute $\sqrt{x}$ for u in these last two equations.

$$\sqrt{x} = 4 \qquad \sqrt{x} = 1 \qquad \text{Replace } u \text{ with } \sqrt{x}.$$

$$(\sqrt{x})^2 = 4^2 \qquad (\sqrt{x})^2 = 1^2 \qquad \text{Square both sides.}$$

$$x = 16 \qquad\quad x = 1$$

Because we squared each member of each equation to obtain these *possible* solutions, we *must* check both numbers in the original equation for extraneous solutions.

Check 16:   **Check 1:**

$$x - 5\sqrt{x} = -4 \qquad\qquad x - 5\sqrt{x} = -4$$

$$16 - 5\sqrt{16} \stackrel{?}{=} -4 \qquad 1 - 5\sqrt{1} \stackrel{?}{=} -4$$

$$16 - 5(4) \stackrel{?}{=} -4 \qquad\quad 1 - 5 \stackrel{?}{=} -4$$

$$16 - 20 \stackrel{?}{=} -4 \qquad\qquad -4 = -4 \quad \checkmark$$

$$-4 = -4 \quad \checkmark$$

Both proposed solutions check. The solution set is {16, 1}. ■

Some equations that are quadratic in form require the use of the quadratic formula in the solution process, as illustrated by Example 6.

EXAMPLE 6   **Using the Quadratic Formula on an Equation Quadratic in Form**

Solve: $(y^2 + 3y)^2 + 3(y^2 + 3y) + 2 = 0$

Solution

The first term is the square of the variable factor of the second term. This means that the equation is quadratic in form. We will let u equal the middle term without its coefficient. Thus, the substitution that will be introduced is $u = y^2 + 3y$.

$(y^2 + 3y)^2 + 3(y^2 + 3y) + 2 = 0$ This is the given equation.

$u^2 + 3u + 2 = 0$ Let $u = y^2 + 3y$. Introduce this substitution into the equation and solve.

$(u + 2)(u + 1) = 0$ Factor.

$u + 2 = 0$ or	$u + 1 = 0$	Set each factor equal to 0.
$u = -2$	$u = -1$	Solve for u. We still must find y.
$y^2 + 3y = -2$ or	$y^2 + 3y = -1$	Replace u with $y^2 + 3y$.
$y^2 + 3y + 2 = 0$	$y^2 + 3y + 1 = 0$	Write each quadratic equation in standard form.

$(y + 2)(y + 1) = 0$ $y = \dfrac{-b \pm \sqrt{b^2 - 4ac}}{2a}$ Use factoring on the first equation and the quadratic formula on the nonfactorable equation.

$y = -2$ or $y = -1$

$= \dfrac{-3 \pm \sqrt{3^2 - 4 \cdot 1 \cdot 1}}{2 \cdot 1}$ Since $y^2 + 3y + 1 = 0$, $a = 1$, $b = 3$, and $c = 1$.

$= \dfrac{-3 \pm \sqrt{5}}{2}$ No further simplification is possible.

The solution is

$$\left\{ -2, -1, \frac{-3 - \sqrt{5}}{2}, \frac{-3 + \sqrt{5}}{2} \right\}.$$

Notice that the original equation actually is a fourth-degree equation. This is why there are four solutions. ∎

Using technology

To verify the solution set to Example 6 graphically, first find decimal approximations for the proposed irrational solutions.

$\dfrac{-3 - \sqrt{5}}{2} \approx -2.6$ and $\dfrac{-3 + \sqrt{5}}{2} \approx -0.4$

The graph of

$y = (x^2 + 3x)^2 + 3(x^2 + 3x) + 2$

has exact or approximate x-intercepts at -2.6, -2, -1, and -0.4. This verifies graphically the solutions we obtained algebraically.

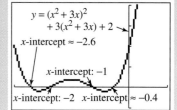

EXAMPLE 7   **Solving an Equation That Is Quadratic in Form**

Solve: $(x - 1)^{-2} + 2(x - 1)^{-1} - 3 = 0$

Solution

The first term is the square of the variable factor of the second term. We will introduce the substitution $u = (x - 1)^{-1}$.

$$(x - 1)^{-2} + 2(x - 1)^{-1} - 3 = 0 \qquad \text{This is the given equation.}$$

$$[(x - 1)^{-1}]^2 + 2(x - 1)^{-1} - 3 = 0 \qquad \text{Write the variable in the first term as the square of the variable factor in the middle term.}$$

$$u^2 + 2u - 3 = 0 \qquad \text{Let } u = (x - 1)^{-1}.$$

Solve the equation $u^2 + 2u - 3 = 0$.

$$(u + 3)(u - 1) = 0 \qquad \text{Factor.}$$

$u + 3 = 0$	or	$u - 1 = 0$

Set each factor equal to 0.

$$u = -3 \qquad\qquad u = 1 \qquad \text{Solve for } u. \text{ These are } not \text{ the solutions for } x.$$

$$(x - 1)^{-1} = -3 \quad \text{or} \quad (x - 1)^{-1} = 1 \qquad \text{Replace } u \text{ with } (x - 1)^{-1}.$$

$$\frac{1}{x - 1} = -3 \qquad\qquad \frac{1}{x - 1} = 1 \qquad b^{-1} = \frac{1}{b} \quad (b \neq 0)$$

$$(x - 1) \cdot \frac{1}{x - 1} = (x - 1)(-3) \qquad (x - 1)\frac{1}{x - 1} = (x - 1) \cdot 1 \qquad \text{Multiply by the LCD. } x \neq 1.$$

$$1 = -3x + 3 \qquad\qquad 1 = x - 1 \qquad \text{Clear equations of fractions.}$$

$$-2 = -3x \qquad\qquad 2 = x \qquad \text{Solve each equation for } x.$$

$$\tfrac{2}{3} = x$$

The given equation could be written without negative exponents as

$$\frac{1}{(x - 1)^2} + \frac{2}{x - 1} - 3 = 0.$$

Neither proposed solution causes a denominator to be 0, and so the solution set is $\{\tfrac{2}{3}, 2\}$. ∎

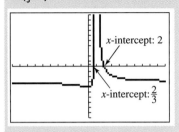
Discover for yourself

The equation in Example 7,

$$(x - 1)^{-2} + 2(x - 1)^{-1} - 3 = 0$$

can be expressed as

$$\frac{1}{(x - 1)^2} + \frac{2}{x - 1} - 3 = 0.$$

Try solving the second form of the equation by multiplying both sides by $(x - 1)^2$ and clearing fractions. Compare your solution to the one shown in Example 7. Which method do you prefer?

Our final example involves an equation with rational exponents.

EXAMPLE 8 **Solving an Equation Containing Rational Exponents That Is Quadratic in Form**

Solve: $2y^{2/3} + 3y^{1/3} - 2 = 0$

Solution

Like all equations that are quadratic in form, the variable factor of the first term is the square of the variable factor of the second term.

$$(y^{1/3})^2 = y^{2/3}$$

As always, the third term is a constant. We will introduce the substitution $u = y^{1/3}$.

<table>
<tr><td>$2y^{2/3} + 3y^{1/3} - 2 = 0$</td><td>This is the given equation.</td></tr>
<tr><td>$2(y^{1/3})^2 + 3(y^{1/3}) - 2 = 0$</td><td>Write the variable in the highest-degree term as the square of the variable factor in the middle term.</td></tr>
<tr><td>$2u^2 + 3u - 2 = 0$</td><td>Let $u = y^{1/3}$.</td></tr>
</table>

Solve the equation $2u^2 + 3u - 2 = 0$.

$(u + 2)(2u - 1) = 0$		Factor.
$u + 2 = 0$ $\quad$ or $\quad$ $2u - 1 = 0$		Set each factor equal to 0 and solve for u.
$u = -2$ $\qquad\qquad$ $u = \frac{1}{2}$		
$y^{1/3} = -2$ $\quad$ or $\quad$ $y^{1/3} = \frac{1}{2}$		Replace u by $y^{1/3}$.
$(y^{1/3})^3 = (-2)^3$ $\qquad$ $(y^{1/3})^3 = \left(\frac{1}{2}\right)^3$		Solve for y by cubing both sides of each equation.
$y = -8$ $\qquad\qquad$ $y = \frac{1}{8}$		$(y^{1/3})^3 = y^{(1/3)\cdot 3} = y^1 = y$

Checking these values, we see that the solution set is $\left\{-8, \frac{1}{8}\right\}$. ■

PROBLEM SET 7.4

Practice Problems

Problems 1–12 contain equations similar to those that we used to model verbal conditions. Simplify each equation and express it in standard form $ax^2 + bx + c = 0$. Then solve the quadratic equation. Express irrational solutions in simplified radical form.

1. $5(x^2 - 4) = 12(x - 2) + 12(x + 2)$

2. $6(x - 2) + 6x = x^2 - 2x$

3. $\dfrac{x}{x - 2} + \dfrac{2}{x} = 4$

4. $\dfrac{300}{x - 1} = \dfrac{300}{x} + 10$

5. $\dfrac{6}{x + 12} + \dfrac{6}{12 - x} = \dfrac{16}{15}$

6. $\dfrac{2}{x + 1} = \dfrac{7}{2} - \dfrac{3}{x + 2}$

7. $\dfrac{1}{x - 2} + \dfrac{1}{x + 2} = 1$

8. $\dfrac{1}{2x + 2} = \dfrac{4}{3x} - 1$

9. $\dfrac{1}{x + 2} = \dfrac{1}{3} - \dfrac{1}{x}$

10. $\dfrac{16}{x} + \dfrac{90}{x + 5} = 6$

11. $(x + 5)\left(\dfrac{50}{x} - \dfrac{1}{2}\right) = 50$

12. $\left(\dfrac{520}{x} - 5.2\right)(x + 5) = 520$

In Problems 13–44, solve the equations by making an appropriate substitution.

13. $x^4 - 5x^2 + 4 = 0$
14. $x^4 - 13x^2 + 36 = 0$
15. $9x^4 = 25x^2 - 16$
16. $4z^4 = 13z^2 - 9$

17. $y - 7\sqrt{y} + 10 = 0$
18. $z - 2\sqrt{z} - 15 = 0$
19. $x^{-2} - x^{-1} - 12 = 0$
20. $x^{-2} - 7x^{-1} + 6 = 0$

21. $x - 13\sqrt{x} + 40 = 0$
22. $2x - 7\sqrt{x} - 30 = 0$
23. $x^{-2} + 6x^{-1} = 16$
24. $2x^{-2} + 3 = -7x^{-1}$

25. $6(w - 1)^{-2} + (w - 1)^{-1} - 2 = 0$

26. $4(w - 2)^{-2} - 29(w - 2)^{-1} + 25 = 0$

27. $(x^2 + 3x)^2 - 8(x^2 + 3x) - 20 = 0$

28. $(x^2 - 2x)^2 - 11(x^2 - 2x) + 24 = 0$

29. $(x^2 + 2x - 3)^2 + 6(x^2 + 2x - 3) + 8 = 0$

30. $(x^2 - 5x + 7)^2 - 5(x^2 - 5x + 7) + 6 = 0$

31. $y^{-2} - 4y^{-1} - 3 = 0$

32. $y^4 - 5y^2 - 2 = 0$

33. $y^{2/3} - y^{1/3} - 6 = 0$

34. $2y^{2/3} + 7y^{1/3} - 15 = 0$

35. $2y - 3y^{1/2} + 1 = 0$

36. $y + 3y^{1/2} - 4 = 0$

37. $x^{3/2} + 1 = 2x^{3/4}$

38. $x^{2/5} = 6 - x^{1/5}$

39. $x^{2/5} = -8x^{1/5} - 16$

40. $x^{1/3} = 3x^{1/6} - 2$

41. $\left(\dfrac{x + 3}{x + 1}\right)^2 - 12\left(\dfrac{x + 3}{x + 1}\right) + 27 = 0$

42. $\left(\dfrac{x - 5}{x + 2}\right)^2 + 8\left(\dfrac{x - 5}{x + 2}\right) - 33 = 0$

43. $2\left(\dfrac{x - 1}{x - 3}\right)^2 - 7\left(\dfrac{x - 1}{x - 3}\right) + 5 = 0$

44. $2\left(\dfrac{x - 4}{x - 7}\right)^2 + 3\left(\dfrac{x - 4}{x - 7}\right) - 5 = 0$

Application Problems

Solve Problems 45–54 by making a sketch (if one is not given) and then creating a model that translates the geometric conditions.

45. The length of a rectangle exceeds twice its width by 3 inches. If the area of the rectangle is 10 square inches, find the measure of the width.

46. The width of a rectangle is 4 meters less than its length. If the area of the rectangle is 4 square meters, find the measure of the length.

47. A rectangular swimming pool is 5 meters wide and 9 meters long. A tile border of uniform width is to be built around the pool, using 40 square meters of tile. How wide should the border be?

48. A square swimming pool measures 8 meters on a side. A tile border of uniform width is to be built around three sides of the pool, using 70 square meters of tile. How wide should the border be?

49. One leg of a right triangle exceeds the shorter leg by 1 inch, and the hypotenuse exceeds the longer leg by 7 inches. What is the length of the shorter leg?

50. One leg of a right triangle exceeds the shorter leg by 2 centimeters. If the hypotenuse is 6 centimeters, determine the lengths of the legs.

51. A box with a square base and no top is to be made from a square piece of cardboard by cutting 5-inch squares from each corner and folding up the sides, as shown in the figure. If the box is to hold a volume 520 cubic inches, find the length of the piece of cardboard that is needed.

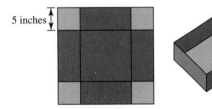

52. A strip of metal 20 inches wide is to be shaped into a trough by bending up the two sides, as shown in the figure. The cross-sectional area is to be 13 square inches. How many inches should be turned up? Show that there are two different solutions to the problem.

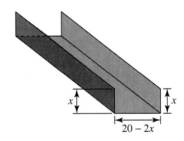

53. Many of the buildings of ancient Greece incorporated the proportions of the golden rectangle. As shown in the figure, $ABCD$ is a golden rectangle when $APQD$ is a square and

$$\frac{L}{W} = \frac{W}{L - W}.$$

If the area of square $APQD$ is 100 square feet, find L, the length of this golden rectangle.

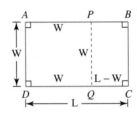

54. The sum of the areas of a rectangle and a square is 68 square feet. The length of the rectangle is twice its width, and a side of the square exceeds the width of the rectangle by 2 feet. Find the length of a side of the square.

In Problems 55–58, use a verbal model to write an equation that models the problem's conditions. Then solve the problem.

55. A boat is chartered for a fishing trip at a cost of $480. When 4 additional people join the trip, the fare per person is decreased by $4.00. How many people were originally going to charter the boat?

56. A group of people decide to purchase a boat at a cost of $6200. Each person shares equally in the boat's cost. When 6 additional people decide to be part of the group purchase, the cost per person is decreased by $930. How many people were originally going to purchase the boat?

57. Working together, two people can mow a large lawn in 4 hours. One person can do the job alone 1 hour faster than the other person. How long does it take each person working alone to mow the lawn?

58. A pool has an inlet pipe to fill it and an outlet pipe to empty it. It takes 2 hours longer to empty the pool than it does to fill it. The inlet pipe is turned on to fill the pool, but the outlet pipe is accidentally left open. Despite this, the pool fills in 8 hours. How long does it take the outlet pipe to empty the pool?

Use the verbal model

The selling price per item	minus	The dealer's cost per item	equals	The dealer's profit per item.

to solve Problems 59–60.

59. A car dealer purchased a group of identical cars from a rental agency for $64,000. After selling all but 2 of the cars at an average profit of $1600 each, the investment of $64,000 had been regained. How many cars did the dealer purchase?

60. The manager of an appliance store purchased a number of identical lamps for $160. After selling all but 4 of the lamps at an average profit of $2 each, the investment of $160 had been regained. How many lamps did the manager purchase?

61. A boat travels 7 miles upstream and then 7 miles downstream. The trip upstream and back takes 3 hours. If the speed of the current is 2 miles per hour, what is the speed of the boat in still water?

62. A boat travels 10 miles upstream and 10 miles back. The trip upstream and back takes 3 hours. If the speed of the current is 5 miles per hour, what is the speed of the boat in still water?

True–False Critical Thinking Problems

63. Which one of the following is true?
a. The polynomial equation $2x^{2/3} + 7x^{1/3} - 15 = 0$ is quadratic in form.
b. The equation $(x^2 + 3x)^4 - 8(x^2 + 3x)^3 - 20 = 0$ is quadratic in form.
c. Solving $x^6 - 9x^3 + 8 = 0$, we let $u = x^3$. We obtain $u^2 - 9u + 8 = 0$, which factors as $(u - 8)(u - 1) = 0$. Thus, $\{8, 1\}$ is the solution set of the original equation.
d. None of the above is true.

64. Which one of the following is true?
a. If an equation is quadratic in form, there is only one method that can be used to obtain its solution.
b. An equation that is quadratic in form must have a variable factor in one term that is the square of the variable factor in another term.
c. Since x^6 is the square of x^3, the equation $x^6 - 5x^3 + 6x = 0$ is quadratic in form.
d. To solve $x - 9\sqrt{x} + 14 = 0$, we let $\sqrt{u} = x$.

Technology Problems

65. Use a graphing utility to verify your solutions to any four problems in Problems 1–12.

66. Use a graphing utility to verify your solutions to any four problems in Problems 13–44.

Writing in Mathematics

67. Explain how to recognize an equation that is quadratic in form.

Critical Thinking Problems

68. Two cubes have the same surface area despite the fact that one cube is open on top and the other is closed. If each edge of the closed cube is 1 inch shorter than each edge of the open cube, find the length of an edge of the closed cube.

69. A rectangle is drawn inside a triangle whose height is represented by $2y$, as shown in the figure. If the area of the shaded region is 10 square yards, find the value of y.

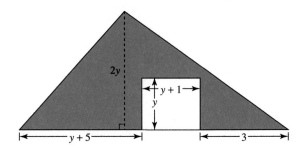

70. A racetrack is 1.25 miles long. Two horses run the track, with the slower horse averaging 1.5 miles per hour less than the faster horse. If the slower horse crosses the finish line 5 seconds after the faster horse,

does the faster horse break the track record of 1 minute, 56 seconds?

71. Solve

$$\sqrt{\frac{x+4}{x-1}} + \sqrt{\frac{x-1}{x+4}} = \frac{5}{2}$$

by introducing the substitution $u = \sqrt{\dfrac{x+4}{x-1}}$.

72. If n is a positive integer, find (in terms of n) all values of x greater than 1 that satisfy $4x^{1/(2n)} - x^{1/n} - 3 = 0$.

73. Use the substitutions $u = \sqrt{x+y}$ and $v = \sqrt{x-y}$ to find the ordered pair of real numbers (x, y) satisfying

$$x + y + \sqrt{x+y} = 12$$
$$x - y + \sqrt{x-y} = 6.$$

74. Solve

$$(1 - \sqrt{x^2 - 1} - \sqrt{x^2 + 2x + 1}) + \frac{1}{1 - \sqrt{x^2 - 1} - \sqrt{x^2 + 2x + 1}} = 2$$

by introducing the substitution

$$u = 1 - \sqrt{x^2 - 1} - \sqrt{x^2 + 2x + 1}.$$

Review Problems

75. Solve using Cramer's rule:

$$2x + 3y = -2$$
$$x - 4y = 6$$

76. Solve: $3(x + 2) + 3x = 4(2x + 3) + 2.$

77. Simplify: $\sqrt[3]{48x^7y^2}.$

S E C T I O N   7 . 5

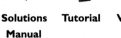

Solutions Manual   **Tutorial**   **Video 9**

Solving Quadratic and Rational Inequalities

Objectives

1 Solve quadratic inequalities.
2 Solve inequalities involving quotients.

In this section, we consider solution methods for quadratic inequalities in the form

$$ax^2 + bx + c > 0, \quad ax^2 + bx + c \geq 0, \quad ax^2 + bx + c < 0, \text{ and}$$
$$ax^2 + bx + c \leq 0.$$

After solving several quadratic inequalities, we will use one of these solution methods to solve inequalities containing quotients.

❙ Solve quadratic inequalities.

Solving Quadratic Inequalities

One method for solving a quadratic inequality such as

$$x^2 - 7x + 10 > 0$$

is to consider the graph of the related quadratic function

$$y = x^2 - 7x + 10.$$

This forms the basis of our first example.

EXAMPLE I   Using a Parabola to Solve a Quadratic Inequality

Solve: $x^2 - 7x + 10 > 0$

Solution

We begin by graphing the parabola whose equation is $y = x^2 - 7x + 10$. We find the x-intercepts by setting y equal to 0 and solving for x.

$$x^2 - 7x + 10 = 0$$

$(x - 2)(x - 5) = 0$ Factor.

$x - 2 = 0$ or $x - 5 = 0$ Set each factor equal to 0.

     $x = 2$ $x = 5$ Solve.

The parabola has x-intercepts at 2 and 5.

Now we find the y-intercept by setting x equal to 0 and solving for y.

$$y = x^2 - 7x + 10 = 0^2 - 7 \cdot 0 + 10 = 10$$

The parabola has a y-intercept at 10. Finally, we find the vertex, starting with the x-coordinate.

$$x = -\frac{b}{2a} = -\frac{(-7)}{2(1)} = \frac{7}{2}$$

Since $y = x^2 - 7x + 10$, then

$a = 1$ and $b = -7$

Now we find the y-coordinate of the vertex.

$$y = x^2 - 7x + 10 = \left(\frac{7}{2}\right)^2 - 7\left(\frac{7}{2}\right) + 10 = -\frac{9}{4}$$

The vertex is at $(\frac{7}{2}, -\frac{9}{4})$ or, equivalently, $(3\frac{1}{2}, -2\frac{1}{4})$.

The graph of $y = x^2 - 7x + 10$ is shown in Figure 7.22. Since we are interested in solving $x^2 - 7x + 10 > 0$, we are looking for x-values with positive y-values. Values of y are positive to the left and right of the x-intercepts, as shown. Thus, the solution set of

$$x^2 - 7x + 10 > 0$$

is $\{x \mid x < 2 \text{ or } x > 5\}$, or $(-\infty, 2) \cup (5, \infty)$. ∎

We actually showed more detail than what we needed to solve Example 1. Once we know the x-intercepts for $y = x^2 - 7x + 10$, since the leading coefficient is positive we know that the parabola must open upward. A rough sketch of this situation is shown in Figure 7.23. Since the graph lies above the x-axis when $x < 2$ or when $x > 5$, we can immediately reason that the solution set of $x^2 - 7x + 10 > 0$ is $\{x \mid x < 2 \text{ or } x > 5\}$.

Let's summarize these observations.

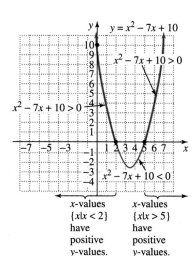

Figure 7.22

Values of x for which the graph lies above the x-axis

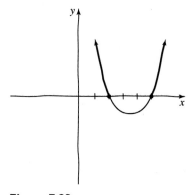

Figure 7.23

y is positive for $x < 2$ or $x > 5$. The exact location of the vertex does not matter.

Solving a quadratic inequality graphically

   1. If necessary, express the inequality in the form

$$ax^2 + bx + c > 0 \quad \text{or} \quad ax^2 + bx + c < 0.$$

2. Make a rough sketch of $y = ax^2 + bx + c$. Find the parabola's x-intercepts by solving $ax^2 + bx + c = 0$. If $a > 0$, the parabola opens upward and if $a < 0$, the parabola opens downward.

3. The solution of $ax^2 + bx + c > 0$ consists of all x-values for which the graph lies above the x-axis. The solution of $ax^2 + bx + c < 0$ consists of all x-values for which the graph lies below the x-axis.

In the next example, we need the quadratic formula to determine the x-intercepts.

EXAMPLE 2   Solving a Quadratic Inequality Graphically

Solve: $x^2 \leq 4x - 2$

Solution

Write the quadratic inequality in standard form.

$$x^2 \leq 4x - 2 \qquad \text{This is the given inequality.}$$

$$x^2 - 4x + 2 \leq 0 \qquad \text{Write the inequality in standard form by subtracting } 4x \text{ and adding 2 on both sides.}$$

We now make a rough sketch of $y = x^2 - 4x + 2$. All we really need are the x-intercepts, so we set y equal to 0 and solve $x^2 - 4x + 2 = 0$. Because $x^2 - 4x + 2$ is prime, we will solve $x^2 - 4x + 2 = 0$ using the quadratic formula. With $a = 1$, $b = -4$, and $c = 2$, the solutions are given by

$$x = \frac{-b \pm \sqrt{b^2 - 4ac}}{2a} = \frac{-(-4) \pm \sqrt{(-4)^2 - 4(1)(2)}}{2(1)}$$

$$= \frac{4 \pm \sqrt{8}}{2} = \frac{4 \pm 2\sqrt{2}}{2} = 2 \pm \sqrt{2}$$

The x-intercepts are $2 - \sqrt{2}$ and $2 + \sqrt{2}$, located approximately at 0.6 and 3.4. Since a, the leading coefficient, is positive, the parabola opens upward. A rough sketch is shown in Figure 7.24. The figure immediately reveals that the graph lies below the x-axis (that is, $x^2 - 4x + 2 < 0$) between the two x-intercepts. Since equality is included in $x^2 - 4x + 2 \leq 0$, we also include $2 - \sqrt{2}$ and $2 + \sqrt{2}$ in the solution set.

The solution set of $x^2 \leq 4x - 2$ is $\{x \mid 2 - \sqrt{2} \leq x \leq 2 + \sqrt{2}\}$, or $[2 - \sqrt{2}, 2 + \sqrt{2}]$. ∎

Once we have determined the x-intercepts for the graph of $y = ax^2 + bx + c$, we can almost visualize what the resulting parabola looks like without actually drawing it. This gives us another, perhaps faster, method for solving quadratic inequalities. The x-intercepts divide the x-axis into intervals. In each interval, values of y are all positive or all negative. We can test one real number in each of these intervals to determine whether $ax^2 + bx + c$ is positive or negative in that interval. If it is positive for one number in an interval, it will be positive for all the numbers in the interval. Similarly, if it is negative for one number in an interval, it will be negative for all the numbers in the interval.

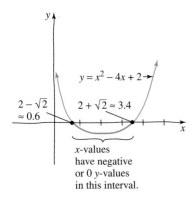

Figure 7.24

sing technology

We've verified our roughly-sketched graph in Figure 7.24 with a graphing utility. If we relied only on this graph, we could not determine the exact values of the x-intercepts, namely $2 - \sqrt{2}$ and $2 + \sqrt{2}$.

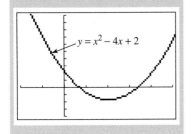

This method, using test numbers from the intervals determined by the x-intercepts, is illustrated in our next example.

EXAMPLE 3 **Using Test Numbers to Solve a Quadratic Inequality**

Solve: $2x^2 + 7x - 4 < 0$

Solution

As before, we begin by finding the x-intercepts for the parabola whose equation is $y = 2x^2 + 7x - 4$. We set $y = 0$ and solve for x.

$$2x^2 + 7x - 4 = 0$$

$$(x + 4)(2x - 1) = 0 \qquad \text{Factor.}$$

$$x + 4 = 0 \quad \text{or} \quad 2x - 1 = 0 \qquad \text{Set each factor equal to 0.}$$

$$x = -4 \qquad\qquad x = \tfrac{1}{2} \qquad \text{Solve.}$$

The x-intercepts divide the number line into three intervals: A, B, and C. Shown below is the test number that we have selected in each of these intervals.

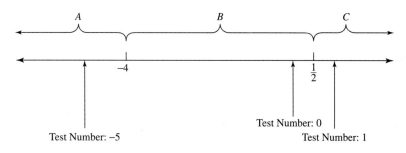

Study tip

Any number in an interval may be selected as a test number. If possible (and it usually is), choose a relatively small integer.

Using technology

A graphing utility confirms that the graph of $y = 2x^2 + 7x - 4$ lies below the x-axis between -4 and $\tfrac{1}{2}$. The solution set of $2x^2 + 7x - 4 < 0$ is $(-4, \tfrac{1}{2})$.

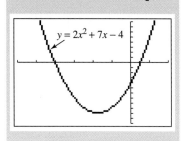

We substitute each of these test numbers into the quadratic function $y = 2x^2 + 7x - 4$ and draw a conclusion about whether the function is positive or negative in that interval.

The Function: $y = 2x^2 + 7x - 4$

Interval A	Interval B	Interval C
Testing -5: $y = 2(-5)^2 + 7(-5) - 4$ $= 11 > 0$	Testing 0: $y = 2 \cdot 0^2 + 7 \cdot 0 - 4$ $= -4 < 0$	Testing 1: $y = 2 \cdot 1^2 + 7 \cdot 1 - 4$ $= 5 > 0$
Conclusion: y is positive everywhere in interval A.	Conclusion: y is negative everywhere in interval B.	Conclusion: y is positive everywhere in interval C.

Since we are interested in solving

$$2x^2 + 7x - 4 < 0$$

we see from the above test-number analysis that any number in interval B is a solution. As verified in the Using Technology, the solution set of $2x^2 + 7x - 4 < 0$ is

$$\{x \mid -4 < x < \tfrac{1}{2}\}, \quad \text{or} \quad (-4, \tfrac{1}{2}).$$

Here is a summary of the test-number method for solving a quadratic inequality.

Solving a quadratic inequality by the test-number method

1. If necessary, express the inequality in the form
$$ax^2 + bx + c > 0 \quad \text{or} \quad ax^2 + bx + c < 0.$$

2. Solve $ax^2 + bx + c = 0$ to find the x-intercepts.
3. Use the x-intercepts to divide the number line into intervals.
4. Select a test number from each interval and substitute it into $y = ax^2 + bx + c$.
5. If y is positive for the test number, then y is positive everywhere in the interval. If y is negative for the test number, then y is negative everywhere in the interval.
6. For $ax^2 + bx + c > 0$, select the interval(s) where y is positive everywhere. For $ax^2 + bx + c < 0$, select the interval(s) where y is negative everywhere.

EXAMPLE 4 **Using Both Methods to Solve a Quadratic Inequality**

Solve: $4x^2 - 8x + 7 > 0$

Solution

Method 1. Solving Graphically
We make a rough sketch of $y = 4x^2 - 8x + 7$ or visualize what the parabola looks like without drawing it. We begin by setting y equal to 0 and finding the x-intercepts.

$$4x^2 - 8x + 7 = 0 \quad \text{We cannot factor, so we use the quadratic formula with } a = 4, b = -8, \text{ and } c = 7.$$

$$x = \frac{-b \pm \sqrt{b^2 - 4ac}}{2a} = \frac{-(-8) \pm \sqrt{(-8)^2 - 4(4)(7)}}{2(4)} = \frac{8 \pm \sqrt{-48}}{8}$$

We can stop at this point. The presence of imaginary solutions (the negative number under the square root) means that the parabola has no x-intercepts.

Since the leading coefficient is positive, the parabola opens upward, so without x-intercepts it has to look something like this:

The parabola lies above the x-axis everywhere. Thus, the solution set of $4x^2 - 8x + 7 > 0$ is the set of all real numbers, indicated by $\{x \mid x \in R\}$, or $(-\infty, \infty)$.

Using technology

A graphing utility confirms that the graph of $y = 4x^2 - 8x + 7$ lies above the x-axis everywhere. The solution set of $4x^2 - 8x + 7 > 0$ is $(-\infty, \infty)$.

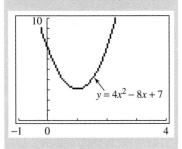

Method 2. The Test-Number Method

As above, we begin by solving $4x^2 - 8x + 7 = 0$. There are no real numbers that will divide the number line into intervals, so there is only one interval—the entire number line.

We substitute the test number 0 into the quadratic function $y = 4x^2 - 8x + 7$.

$$y = 4 \cdot 0^2 - 8 \cdot 0 + 7 = 7 > 0$$

We can conclude that y is positive everywhere on the number line. As we found using method 1, all real numbers satisfy $4x^2 - 8x + 7 > 0$. ∎

2 Solve inequalities involving quotients.

Solving Rational Inequalities

Inequalities that involve quotients can be solved in the same manner as quadratic inequalities. For example, the inequalities

$$(x + 3)(x - 7) > 0 \quad \text{and} \quad \frac{x + 3}{x - 7} > 0$$

are very similar in that both are positive under the same conditions. Each inequality must have two positive factors

$$x + 3 > 0 \quad \text{and} \quad x - 7 > 0$$

or two negative factors

$$x + 3 < 0 \quad \text{and} \quad x - 7 < 0.$$

Since rational functions are a bit more difficult to graph than quadratic functions, we will use the test-number method to solve inequalities with rational expressions. Let's see precisely what this involves.

EXAMPLE 5 Using Test Numbers to Solve a Rational Inequality

Solve: $\dfrac{x + 3}{x - 7} > 0$

Solution

To solve the quadratic inequality $(x + 3)(x - 7) > 0$, we begin by setting each factor equal to 0. We then use the resulting values of x to establish intervals on the number line. We do precisely the same thing with

$$\frac{x + 3}{x - 7} > 0$$

finding values of x that make the numerator and denominator 0.

$$x + 3 = 0 \quad x - 7 = 0 \quad \text{Set the numerator and denominator equal to 0.}$$
$$x = -3 \quad x = 7 \quad \text{Solve.}$$

These x-values divide the number line into three intervals: A, B, and C. Shown below is the test number that we have selected in each of these intervals.

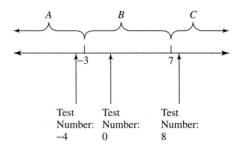

We substitute each of these test numbers into the given rational function and draw a conclusion about whether the function is positive or negative in that interval.

The Function: $y = \dfrac{x + 3}{x - 7}$

Interval A	Interval B	Interval C
Testing -4: $y = \dfrac{-4 + 3}{-4 - 7}$ $= \dfrac{-1}{-11} = \dfrac{1}{11} > 0$	Testing 0: $y = \dfrac{0 + 3}{0 - 7}$ $= -\dfrac{3}{7} < 0$	Testing 8: $y = \dfrac{8 + 3}{8 - 7}$ $= \dfrac{11}{1} = 11 > 0$
Conclusion: y is positive everywhere in interval A.	Conclusion: y is negative everywhere in interval B.	Conclusion: y is positive everywhere in interval C.

Since we are interested in solving

$$\frac{x + 3}{x - 7} > 0$$

we select the intervals in which y is positive everywhere, namely, intervals A and C. The solution set for the rational inequality is

$$\{x \,|\, \underbrace{x < -3}_{\substack{\text{Interval} \\ A}} \quad \text{or} \quad \underbrace{x > 7}_{\substack{\text{Interval} \\ B}}\}, \quad \text{or} \quad (-\infty, -3) \cup (7, \infty)$$

Here is a summary of the test-number method for solving a rational inequality.

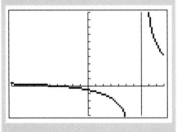
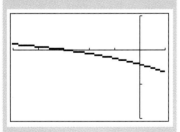

Solving a rational inequality using the test-number method

1. If necessary, express the inequality in the form

$$\frac{p(x)}{q(x)} > 0.$$

One side should be 0 and the terms on the other side should be combined into a single quotient.

2. Set the numerator and denominator equal to 0 and solve $p(x) = 0$ and $q(x) = 0$ for x.
3. Use these solutions to divide the number line into intervals.
4. Select a test number from each interval and substitute it into the rational function.
5. If y is positive for the test number, then y is positive everywhere in the interval. If y is negative for the test number, then y is negative everywhere in the interval.
6. Select the appropriate interval(s) depending on whether the given inequality involves $>$ or $<$.

After completing the solution process, the use of a graphing utility is a helpful way to verify the solution set.

EXAMPLE 6 **Solving a Rational Inequality**

Solve: $\dfrac{2x}{x + 1} \leq 1$

Solution

The right side of the inequality should be 0 and the left side should be combined into a single quotient so that we can use the method of Example 5.

$$\frac{2x}{x + 1} \leq 1 \quad \text{This is the given rational inequality.}$$

$$\frac{2x}{x + 1} - 1 \leq 0 \quad \text{Subtract 1 from both sides.}$$

$$\frac{2x}{x + 1} - 1\frac{(x + 1)}{(x + 1)} \leq 0 \quad \text{Rewrite } -1 \text{ as an equivalent fraction with } x + 1 \text{ as the denominator.}$$

$$\frac{2x - x - 1}{x + 1} \leq 0 \quad \text{Subtract the fractions.}$$

$$\frac{x - 1}{x + 1} \leq 0 \quad \text{Simplify.}$$

From this point on, the solution process will parallel that of the previous example. We set the numerator and denominator equal to 0 and solve for x.

$$x - 1 = 0 \qquad x + 1 = 0$$
$$x = 1 \qquad x = -1$$

Now we use 1 and -1 to divide the number line into intervals.

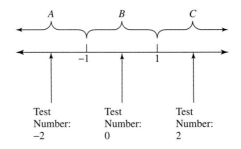

ENRICHMENT ESSAY

From Reason to Uncertainty

Early in the 16th century a mathematics professor from Venice found a general formula to solve cubic equations. Niccolo Tartaglia proved that if $x^3 + px + q = 0$, then a solution is given by

$$x = \sqrt[3]{-\frac{q}{2} + \sqrt{\frac{q^2}{4} + \frac{p^3}{27}}}$$

$$+ \sqrt[3]{-\frac{q}{2} - \sqrt{\frac{q^2}{4} + \frac{p^3}{27}}}.$$

Although this formula appears to be more complicated than the quadratic formula, it reduces the process of solving the cubic equation to a finite arithmetic process involving addition, subtraction, multiplication, division, powers, and roots.

The notion of describing any aspect of the world through equations and then developing a formula to solve them was further enhanced by the Italian mathematician Ferrari (after whom the sports car was named), who discovered a general procedure to solve the quartic equation $ax^4 + bx^3 + cx^2 + dx + e = 0$. However, a defeat for describing and controlling nature through equations came in 1824 when Niels Henrik Abel, a 22-year-old Norwegian mathematician, *proved* that it is impossible to solve fifth-degree equations using a formula that involves a finite arithmetic process. Then, in 1832, a 20-year-old Frenchman, Evariste Galois, wrote down a proof showing there is no general formula or procedure to solve equations of degree 5 or

Alfredo Castañeda, "Sir, When You Enter You Obligate Me" 1983, oil on canvas. Mary-Anne Martin/Fine Art, New York.

greater. (An ironic footnote: The day after his brilliant proof, Galois was killed in a duel.)

This does not mean that quintic, sextic, and n-degree equations cannot be solved. There are *methods* for solving polynomial equations such as

$$5x^{17,326} + 4x^{374} + 2x^3 + 9x + 13 = 0$$

an equation with 17,326 solutions! The solutions, however, cannot be obtained by substituting numbers into a formula. The absence of such formulas hung as a depressing cloud over a period of enormous intellectual electricity, one of a series of many shock waves that ushered out the age of reason, raising the curtain on a new era of uncertainty.

We substitute each test number that we have chosen from each interval into the given rational function and draw a conclusion about whether the function is positive or negative in that interval.

The Function: $y = \dfrac{x - 1}{x + 1}\left(\text{or } y = \dfrac{2x}{x + 1} - 1\right)$

Interval A	Interval B	Interval C
Testing -2: $y = \dfrac{-2 - 1}{-2 + 1} = 3 > 0$	Testing 0: $y = \dfrac{0 - 1}{0 + 1} = -1 < 0$	Testing 2: $y = \dfrac{2 - 1}{2 + 1} = \dfrac{1}{3} > 0$
Conclusion: y is positive everywhere in interval A.	Conclusion: y is negative everywhere in interval B.	Conclusion: y is positive everywhere in interval C.

Since we are interested in solving

$$\frac{2x}{x+1} \leq 1 \quad \text{or, equivalently,} \quad \frac{2x}{x+1} - 1 \leq 0$$

we select the interval in which y is negative everywhere, namely, interval B. Since equality is included in the problem, we must also include the value of x that causes the quotient

$$\frac{x-1}{x+1}$$

to be zero. The value 1 does just this, so it is included in the solution set. The solution set for the given inequality is $\{x \mid -1 < x \leq 1\}$, or $(-1, 1]$. ∎

Using technology

The graphs of

$$y_1 = \frac{2x}{x+1} \quad \text{and} \quad y_2 = 1$$

are shown below. The solution set for

$$\frac{2x}{x+1} \leq 1$$

corresponds to where the graphs intersect (at $x = 1$) or where the rational function lies below the line $y_2 = 1$ (between -1 and 1). The graphs enable us to see that the solution to

$$\frac{2x}{x+1} \leq 1$$

is $(-1, 1]$.

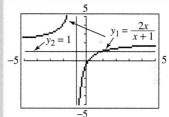

PROBLEM SET 7.5

Practice Problems

Find the solution set for the inequalities in Problems 1–26. Use both set-builder and interval notation.

1. $(x-4)(x+2) > 0$ **2.** $(x+3)(x-5) > 0$ **3.** $(x-7)(x+3) \leq 0$ **4.** $(x+1)(x-7) \leq 0$

5. $x^2 - 5x + 4 > 0$ **6.** $x^2 - 4x + 3 < 0$ **7.** $x^2 + 5x + 4 > 0$ **8.** $x^2 + x - 6 > 0$

9. $x^2 - 6x + 9 < 0$ **10.** $x^2 - 2x + 1 > 0$ **11.** $x^2 - 6x + 8 \leq 0$ **12.** $x^2 - 2x - 3 \geq 0$

13. $3x^2 + 10x - 8 \leq 0$ **14.** $9x^2 + 3x - 2 \geq 0$ **15.** $2x^2 + x < 15$ **16.** $6x^2 + x > 1$

17. $4x^2 + 7x < -3$ **18.** $3x^2 + 16x < -5$ **19.** $5x \leq 2 - 3x^2$ **20.** $4x^2 + 1 \geq 4x$

21. $x^2 - 4x \geq 0$ **22.** $x^2 + 2x < 0$ **23.** $2x^2 + 3x > 0$ **24.** $3x^2 - 5x \leq 0$

25. $-x^2 + x \geq 0$ **26.** $-x^2 + 2x \geq 0$

Find the solution set for the inequalities in Problems 27–42, using both set-builder and interval notation.

27. $\dfrac{x-4}{x+3} > 0$ **28.** $\dfrac{x+5}{x-2} > 0$ **29.** $\dfrac{x+3}{x+4} < 0$ **30.** $\dfrac{x+5}{x+2} < 0$ **31.** $\dfrac{-x+2}{x-4} \geq 0$

32. $\dfrac{-x-3}{x+2} \le 0$ **33.** $\dfrac{4-2x}{3x+4} \le 0$ **34.** $\dfrac{3x+5}{6-2x} \ge 0$ **35.** $\dfrac{x}{x-3} > 0$ **36.** $\dfrac{x+4}{x} > 0$

37. $\dfrac{x+1}{x+3} < 2$ **38.** $\dfrac{x}{x-1} > 2$ **39.** $\dfrac{x+4}{2x-1} \le 3$ **40.** $\dfrac{1}{x-3} < 1$

41. $\dfrac{x-2}{x+2} \le 2$ **42.** $\dfrac{x}{x+2} \ge 2$

Application Problems

43. An object is propelled straight up from ground level with an initial velocity of 80 feet per second, as shown in the figure. Its height at any time t is modeled by $s = -16t^2 + 80t$, where the height s is measured in feet and the time t is measured in seconds. In what time interval will the object be more than 64 feet above the ground?

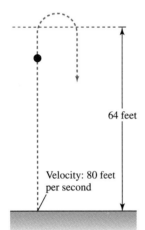

64 feet

Velocity: 80 feet per second

44. The revenue (R) from the sale of x units of a product is given by $R = 3x^2 + x$. The cost of producing x units of the product is given by $C = 2x^2 - 2x + 18$. Describe the range of units the company must produce and sell to generate a profit.

45. The function $N = 0.4x^2 - 36x + 1000$ approximates the number of accidents (N) per 50 million miles for a driver who is x years old. The formula models reality for $16 \le x \le 74$. In what age range will drivers have more than 250 accidents per 50 million miles driven?

46. The function $N = 0.036x^2 - 2.8x + 58.14$ models the number of deaths (N) per year per thousand people for people who are x years old, where $40 \le x \le 60$. Find the age range for which fewer than 20 people per 1000 die annually.

True–False Critical Thinking Problems

47. Which one of the following is true?
  a. If $(2x + 1)(x - 5) \ge 0$, then we obtain $2x + 1 \ge 0$ or $x - 5 \ge 0$.
  b. If $\dfrac{2x}{(x-4)} > 8$, then (multiplying by $x - 4$) we obtain $2x > 8(x - 4)$ and so $x < \frac{16}{3}$.
  c. If $f(x) = \sqrt{x^2 - 6x + 10}$, then the domain of f consists of all real numbers.
  d. If $f(x) = \sqrt{x^2 - 6x + 9}$, then the domain of f is $(-\infty, -3) \cup (3, \infty)$.

48. Which one of the following is true?
  a. An example of a function whose domain is the empty set is
  $$f(x) = \dfrac{1}{\sqrt{-2x^2 + 4x - 2}}.$$
  b. Every quadratic inequality has a solution set consisting of a single interval or the union of two intervals.
  c. The solution set of $x^2 + 2x + 1 \le 0$ is $(-1, 1)$.
  d. The solution set of the inequality $x^2 + 2x + 4 \ge 0$ is $(-\infty, -2] \cup [2, \infty)$.

Technology Problems

49. Use a graphing utility to verify your solutions to any four problems in Problems 1–26 and any two in Problems 27–42.

50. Graph each of the models in Problems 43–46 with a graphing utility. Then use the graph to visualize the problem's solution.

Solve each inequality in Problems 51–56 using a graphing utility.

51. $x^2 + 3x - 10 > 0$

52. $2x^2 + 5x - 3 \le 0$

53. $\dfrac{x - 4}{x - 1} \le 0$

54. $\dfrac{x + 2}{x - 3} \le 2$

55. $\dfrac{1}{x + 1} \le \dfrac{2}{x + 4}$

56. $x^3 + 2x^2 - 5x - 6 > 0$

Writing in Mathematics

57. Without actually solving, what is the solution set to the inequality $(2x - 4)^2 > -1$ if x is restricted to the set of all real numbers? Explain how you determined this.

58. Without actually solving, what is the solution set to the inequality $(2x - 4)^2 < -1$ if x is restricted to the set of all real numbers? Explain why this must be the case.

59. Why can a rational inequality have at most one endpoint of an interval included in the solution set?

Critical Thinking Problems

60. The length of a rectangle is 5 meters more than twice the width. If the area is to be at least 33 square meters, what are the possibilities for the width?

61. Find the domain for $f(x) = \sqrt{x^2 - x - 12}$. Verify your result with a graphing utility.

62. Solve: $(x - 3)(x + 5)(x + 8) \ge 0$.
(*Hint:* Solve $(x - 3)(x + 5)(x + 8) = 0$ and use test numbers in each of the four resulting intervals.)

63. Solve: $\dfrac{x^2 - x - 2}{x^2 - 4x + 3} > 0$.

64. A rectangular area is to be enclosed with 64 meters of fencing. What interval for the rectangle's length will result in an area of at least 240 square meters?

65. Solve: $x^3 + 5x^2 - 4x - 20 \ge 0$.

Group Activity Problem

66. Members of your group act as the marketing research division for a large corporation. Your division estimates that at a price of p dollars per unit, the weekly cost C and revenue R (in thousands of dollars) will be given by the equations

$$C = 14 - p \qquad \text{Cost equation}$$
$$R = 8p - p^2 \qquad \text{Revenue equation}$$

Begin by finding a price range for p that describes when a profit will result. (A profit will result if revenue is greater than cost.) Once you find this price range, decide on the kind of product manufactured and sold by the corporation that will be consistent with this price range. Once the product is determined,

present a written report to the corporation, heading the report as "Profit Analysis." Include the following in the report:

a. The price range for p, describing when a profit will result

b. A price that will result in the maximum profit

c. Graphs that the corporation might find helpful

d. Factors that might change the cost and revenue equations in the future (dependent on the product chosen)

e. A discussion of the possible limitations of the given equations

f. Anything else that you think would be helpful

Review Problems

67. Solve: $1 - 2x \ge 5 - x$.

68. Solve: $\sqrt{5x - 1} - \sqrt{x + 2} = 1$.

69. If $A = \{3, 7, 8, 9\}$ and $B = \{8, 9, 10\}$, find $A \cup B$ and $A \cap B$.

CHAPTER PROJECT

How Far?

We developed a formula in this chapter to find the distance between two points, assuming we know the coordinates of the points on the Cartesian plane. It seems that this technique for finding distance will work wherever we can lay down a coordinate grid similar to the Cartesian system. We used this approach on page 603, when we superimposed a coordinate system over a map of Thailand to give us an estimate of how far a plane would fly between two cities. However, one of the fundamental aspects of the Cartesian coordinate system was missing … we were no longer measuring distance on a flat plane.

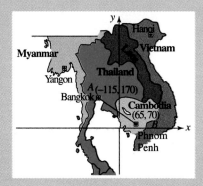

1. Select a city far from where you live as a destination for a trip. Using a globe with a scale for distances, locate the starting point and ending point for your trip. Use a thin wire or pipe cleaner to connect the two points on the globe, making sure you follow the contours of the globe. After you have marked the distance, carefully lift the wire off the globe and examine it. What kind of curve do you see? Does it look like a line would approximate this distance? Consult an atlas or an airline guide to see what is listed as the mileage between the selected cities. Does it match your mileage approximation?

2. Use a flat, scaled map of the world and repeat your measurement. How does this estimate compare to the one from the globe and other sources? If you can, find other styles of scaled maps and compare the measurements on each. Does the style of flat map used affect your measurements? How do you think cruiselines or airlines determine the mileage from one destination to another?

3. Using the globe, select numerous cities as destinations from your original starting point and measure the distances. Compare the global measures to a flat measure and see if you can determine at what distance the curvature of the earth seems to play a great part in affecting the measurement.

4. Obtain a scaled map of your city and the surrounding area and select two places far apart and over different terrain. Plan a route from one location to the other, on the map, and measure the distance using the scale. Follow the same route in your car, taking along another member of the class to record the mileage shown on the odometer. Does the map give the same result as the mileage from your car? Do you think the map considered differences in elevation, such as hills or valleys, when listing the scale you should use to calculate distances?

We know a straight line is the shortest distance between two points. If we mark two points on a sphere, the shortest distance between them, in space, would go through the sphere and join one point to another. However, if we restrict ourselves to moving on the surface of a sphere, the shortest distance between two points is found by traveling on a great circle. A great circle can be defined as a circle on a sphere that divides the sphere into two portions of equal area. The equator of the earth is a great circle, as are lines of longitude. Lines of latitude, although circles, are not great circles.

5. Convince yourself that a great circle will be the shortest path to travel by measuring the distances on a globe. Select two cities at approximately the same latitude, for example New York and Rome, and measure the distance between them following a line of latitude. Repeat the measurement, this time following a great circle. To create a great circle, wrap a wire or string around the globe at the equator and fasten the ends to create a permanent measure. Lay your string or wire over the globe so it passes through the two cities and fits snugly all around the globe. For New York and Rome, notice how the great circle route arcs up, north of the cities, and then comes back down. How does this compare to an airline route between the two cities? Repeat this procedure with several cities.

If we change our surface from a sphere to a cube and mark two points on different faces of the cube, we will find that the shortest distance between the points is once again a straight line. As with the sphere, we cannot go through the cube to join the points, so we move on the surface of the cube to trace a line.

6. Construct a cube by using six identical squares taped together. Do not tape all of the sides to each other because we will also lay the cube flat, in an unfolded form. While the cube is flat, place two marks on the surface on two different faces. Fold the cube into its three dimensional shape and use a pencil to lightly mark paths from one of the marks to the other, trying to find the shortest possible route. Unfold the cube and study the paths you have sketched. What do you observe about your paths? Place a ruler on the flattened cube and connect the two points with a straight line. How does this line compare with your paths?

To answer the question "How far?" we not only need to know precisely where we begin and end a journey, but also across what terrain we travel. We have looked at this question for traveling across the surface of the earth, but the same question may be asked as we contemplate travel to other places in space. How far to the next star? How far to the next galaxy? How far to the edge of where our telescopes can see? The larger question is: What is the large-scale geometry of the universe? Is our universe flat, like a plane, extending infinitely in all directions? Is our universe spherical, so we would travel on the curved surface of a sphere as we move through space? Does our universe have some other shape entirely? Answering our very small question may require some very great amounts of work.

Worldwide Web Resources

Go to the Prentice Hall website (http://www.prenhall.com/blitzer) to access other locations on the Internet that will allow you to further explore the concepts presented in this project.

Chapter Review

SUMMARY

1. **Solving Quadratic Equations by the Square Root Method**
   a. If $x^2 = d$, then $x = \sqrt{d}$ or $x = -\sqrt{d}$.

   b. The square root method can also be used to solve equations of the form $(ax + b)^2 = d$, resulting in the two equations $ax + b = \sqrt{d}$ or $ax + b = -\sqrt{d}$.

2. **Distance and Midpoint Formulas**
   a. The distance d between the points (x_1, y_1) and (x_2, y_2) is given by
   $$d = \sqrt{(x_2 - x_1)^2 + (y_2 - y_1)^2} \quad \text{or}$$
   $$d = \sqrt{(x_1 - x_2)^2 + (y_1 - y_2)^2}.$$
   b. The midpoint of the line segment whose endpoints are (x_1, y_1) and (x_2, y_2) is given by
   $$\left(\frac{x_1 + x_2}{2}, \frac{y_1 + y_2}{2} \right).$$

3. **Solving Quadratic Equations by Completing the Square**
   a. Rewriting $ax^2 + bx + c = 0$ as $(x + d)^2 = e$ is called completing the square.
   b. To solve $ax^2 + bx + c = 0$ by completing the square:
      1. Divide both sides by a.
      2. Put the constant term on the right side.
      3. Complete the square by adding the square of half the coefficient of the x-term to both sides of the equation.
      4. Factor the left side as the square of a binomial, and simplify the right side if possible.
      5. Apply the square root method and solve.

4. **Graphing Quadratic Functions**
   a. The graph of a quadratic function is called a parabola.
   b. The forms of the equation of a parabola opening upward or downward are
   $$y = ax^2 + bx + c \quad \text{The vertex is } \left(-\frac{b}{2a}, f\left(-\frac{b}{2a} \right) \right).$$
   $$y = a(x - h)^2 + k \quad \text{The vertex is } (h, k).$$
   If $a > 0$, the parabola opens upward; if $a < 0$, the parabola opens downward.
   c. The graph of $y = a(x - h)^2 + k$ looks like the graph of $y = ax^2$, but is shifted horizontally and vertically.
      1. The graph is shifted h units to the right if h is positive, and $|h|$ units to the left if h is negative.
      2. The graph is shifted k units up if k is positive, and $|k|$ units down if k is negative.

5. **Solving Quadratic Equations by the Quadratic Formula**
   The formula is derived by completing the square. If $ax^2 + bx + c = 0$, then
   $$x = \frac{-b \pm \sqrt{b^2 - 4ac}}{2a}.$$

6. **The Discriminant**
   The discriminant, $b^2 - 4ac$, determines the nature of the solutions to a quadratic equation.
   a. If $b^2 - 4ac < 0$, the solutions are not real numbers.
   b. If $b^2 - 4ac$ is a positive perfect square, the solutions are rational numbers.

   c. If $b^2 - 4ac$ is a positive number that is not a perfect square, the solutions are irrational numbers.

7. **Equations Quadratic in Form**
   a. Equations quadratic in form can be expressed as $au^2 + bu + c = 0$, where $a \neq 0$ and u is an algebraic expression.
   b. To solve such an equation:
      1. If necessary, write the equation so that the power of the middle term is half that of the leading term and the final term is a constant.
      2. Write the variable factor in the leading term as the square of the variable factor in the middle term.
      3. Let u equal the middle term without its coefficient and write the equation in the form $au^2 + bu + c = 0$.
      4. Solve for u.
      5. Replace u with the expression containing the given equation's variable from step 3.
      6. Solve the equations in step 5 for the given equation's variable.
      7. Check by direct substitution or by using a graphing utility.

8. **Quadratic Inequalities**
   a. Quadratic inequalities can be expressed as $ax^2 + bx + c > 0$, $ax^2 + bx + c \geq 0$, $ax^2 + bx + c < 0$, and $ax^2 + bx + c \leq 0$.
   b. Solve by one of two solution methods.
      1. Solving Graphically: Find x-intercepts for $y = ax^2 + bx + c$ by solving $ax^2 + bx + c = 0$. Graph or visualize the parabola. The solution of $ax^2 + bx + c > 0$ consists of all x-values for which the graph lies above the x-axis. The solution of $ax^2 + bx + c < 0$ consists of all x-values for which the graph lies below the x-axis.
      2. The Test-Number Method: Solve $ax^2 + bx + c = 0$, using the x-intercepts to divide the number line into intervals. Use test numbers from the intervals to select intervals where $ax^2 + bx + c$ is positive when solving $ax^2 + bx + c > 0$, or intervals where $ax^2 + bx + c$ is negative when solving $ax^2 + bx + c < 0$.

9. **Rational Inequalities**
   a. Rational inequalities involve quotients with variable denominators.
   b. Use the test-number method to solve rational inequalities.
      1. Express the inequality with 0 on one side and the other side as a single quotient.
      2. Use values of x where the numerator and denominator of the quotient equal 0 to divide the number line into intervals. Use a test number from each interval, selecting the appropriate interval(s) depending on whether the given inequality involves $>$ or $<$.

REVIEW PROBLEMS

In this problem set, express irrational solutions in simplified radical form and imaginary solutions in the form a + bi. Use both set-builder and interval notation to represent solution sets of inequalities.

In Problems 1–6, solve using the square root method.

1. $x^2 - 50 = 0$

2. $2x^2 - 3 = 0$

3. $(2x - 3)^2 = 32$

4. $(x - 4)^2 = -36$

5. Solve for a, using only the positive root: $a^2 + b^2 = c^2$.

6. Solve for r, using only the positive root: $A = P(1 + r)^2$.

7. The formula $s = 16t^2$ serves as a model for approximating the distance (s, in feet) that an object falls freely from rest in t seconds. The Chrysler Building in New York City is 1046 feet tall. How long will it take an object falling from the top to strike the ground? Express your answer in simplified radical form, and then approximate to the nearest tenth.

8. A 12-meter wire is attached to the ground 4 meters from the base of a building. How far up the building does the wire reach? Express your answer in simplified radical form, and then approximate to the nearest tenth.

9. A line segment connects the points $(3, 7)$ and $(-4, 6)$.
   a. What is the line segment's length?
   b. Find the line segment's midpoint.

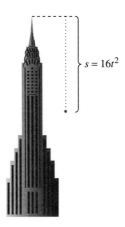

$s = 16t^2$

In Problems 10–11, complete the square and write the resulting perfect square trinomial in factored form.

10. $x^2 + 20x$

11. $x^2 - 3x$

12. Completing the square can be modeled geometrically, as shown in the figure. What is the area of the small shaded square on the right?

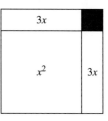

Solve each quadratic equation in Problems 13–14 by completing the square.

13. $x^2 - 7x - 1 = 0$

14. $2x^2 + 3x - 4 = 0$

Graph each parabola in Problems 15–16 using x-intercepts, the y-intercept, the vertex, and if necessary one or two additional points. Use the formula $x = -\dfrac{b}{2a}$ for the x-coordinate of the vertex.

15. $y = x^2 - 2x - 8$

16. $f(x) = -2x^2 - 4x + 1$

Graph the parabola whose equation is given in Problems 17–18.

17. $y = (x - 1)^2 + 3$

18. $f(x) = -(x + 1)^2 + 4$

19. Complete the square and write the following function's equation in the form $y = a(x - h)^2 + k$. Then use this form to graph the function $y = x^2 - 2x - 2$.

20. The parabola shown on the left in the figure has the same shape as the one on the right. What is the parabola's equation?

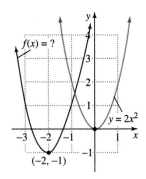

21. The number of inches that a young redwood tree grows per year can be modeled by the quadratic function $f(x) = -0.02x^2 + x + 1$, where x represents annual rainfall in inches, and $f(x)$ is the tree's annual growth, also measured in inches.
  a. Find the x-coordinate of the vertex. Describe what this means in terms of the variables modeled by the function.
  b. Find the y-coordinate of the vertex. Describe what this means in practical terms.

22. The land shown in the figure is to be enclosed with 120 yards of fencing. No fencing is to be used along the river. The area of the fenced rectangle is a function of the width that is fenced off, represented by x, and given by $A(x) = -2x^2 + 120x$. Also shown is the graph of the area function.
  a. What is the vertex of the parabola? Describe what this means in terms of the width of the plot that will result in a maximum area. What is the maximum area?
  b. The graph shows an x-intercept of 60. What does this mean in terms of the width of the plot and its resulting area? Describe exactly how the fencing is being used in this situation.

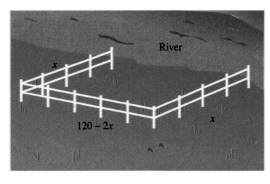
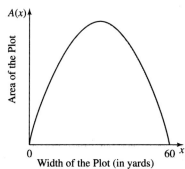

In Problems 23–25, solve each equation using the quadratic formula.

23. $x^2 = 2x + 4$

24. $x^2 - 2x + 19 = 0$

25. $2x^2 = 3 - 4x$

In Problems 26–31, solve each quadratic equation by the method of your choice.

26. $2x^2 - 3x - 1 = 0$

27. $\dfrac{3}{x} + \dfrac{10}{x + 6} = 1$

28. $(5x - 2)^2 - 9 = 0$

29. $3x^2 + 2x = 4$

30. $\dfrac{5}{x+1} + \dfrac{x-1}{4} = 2$

31. $x(x-2) = -5$

For Problems 32–34, compute the discriminant and then indicate whether the solutions are rational, irrational, or imaginary.

32. $x^2 - 3x + 7 = 0$

33. $2x^2 + 5x - 3 = 0$

34. $2x^2 + 4x - 3 = 0$

35. The personnel manager of a roller skate company knows that the company's weekly revenue is a function of the price of each pair of skates, modeled by $R = -2x^2 + 36x$, where x represents the dollar price of a pair of skates and R represents weekly revenue in tens of thousands of dollars. A job applicant promises the personnel manager an advertising campaign guaranteed to generate $190,000 in weekly revenue. Substitute 19 for R in the given model, compute the discriminant, and then explain why the applicant will or will not be hired in the advertising department.

36. The model $N = 0.337x^2 - 2.265x + 3.962$ approximates the number of mountain bike owners (N, in millions) in the United States x years after 1980, where $3 \leq x \leq 10$. In what year (to the nearest whole number) did 10.9 million Americans own mountain bikes?

37. The graph shows the number of farms in the United States from 1850 through 2010 (projected). Explain why a quadratic function is an appropriate model for the given data values.

Number of U.S. Farms, 1850–2010

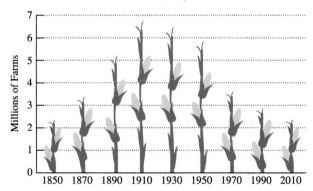

Source: U.S. Bureau of the Census

38. The height (s, in feet) of a falling object above the ground after t seconds can be modeled by a quadratic function. A ball is thrown directly upward from the top of an 80-foot building. The ball's position above the ground after 0, 1, and 2 seconds is shown in the following table.

t (Time)	s (Position above Ground)
0	80
1	128
2	144

a. Find a quadratic function that fits the data.
b. How long after the ball is thrown, to the nearest tenth of a second, will it be 75 feet above the ground?
c. How long after the ball is thrown will it hit the ground?
d. Graph the quadratic function that models this situation.

39. A rectangular pool is 12 meters wide and 20 meters long. A tile border of uniform width is to be built around the pool using 160 square meters of tile. How wide should the border be?

40. The longer leg of a right triangle exceeds the shorter leg by 2 centimeters. If the hypotenuse measures 8 centimeters, what is the length of the shorter leg?

41. A building casts a shadow that is double the length of its height. If the distance from the end of the shadow to the top of the building is 300 meters, how high is the building?

42. Two people can mow a lawn in 2 hours. If each worked alone, one of them could do the job in 1 hour less time than the other. How long would it take each person to mow the lawn working alone?

43. The length of a rectangle is 3 times its width. If the width is decreased by 1 yard and the length is increased by 3 yards, the area will be 72 square yards. Find the dimensions of the original rectangle.

44. A time-share that consists of two weeks yearly in London is purchased by a group of people at a cost of $12,000. When 2 additional people join the time-share, the cost per person is decreased by $1600. How many people originally purchased the time-share?

45. The area of the shaded region in the figure is 150 square centimeters. Find the dimensions of the large and small rectangles.

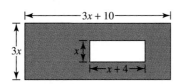

46. What is the length of line segment AG in the figure?

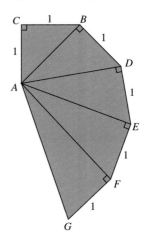

In Problems 47–51, solve the equations that are quadratic in form by making an appropriate substitution.

47. $x^4 - 5x^2 + 4 = 0$

48. $(x^2 + 2x)^2 - 14(x^2 + 2x) = 15$

49. $x^{2/3} - x^{1/3} - 12 = 0$

50. $x + 7\sqrt{x} = 8$

51. $x^{-2} + x^{-1} - 56 = 0$

Find the solution set for each inequality in Problems 52–55.

52. $2x^2 + 5x - 3 < 0$

53. $2x^2 + 9x + 4 \geq 0$

54. $\dfrac{x + 7}{x - 3} > 0$

55. $\dfrac{x - 3}{x + 4} \leq 2$

56. The length of a rectangle is 2 meters more than its width. For what values of the width is the area numerically greater than the perimeter?

Use graph of the function to solve each equation or inequality in Problems 57–60.

57. $x^2 - 2x - 3 = 0$

58. $x^2 - 3x - 28 > 0$

59. $x^4 - 13x^2 + 36 \leq 0$

60. $x^2 - 4x + 4 \leq 0$

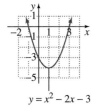

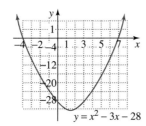

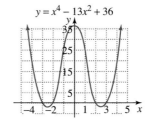

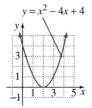

CHAPTER 7 TEST

1. The graph of $f(x) = x^2 - 6x + 7$ is shown in the figure. Using the graph, explain why $\{2, 4\}$ cannot be the solution set for $x^2 - 6x + 7 = 0$.

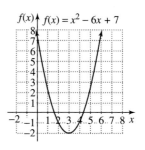

Solve Problems 2–7. Express irrational solutions in simplified radical form and imaginary solutions in the form a + bi.

2. $(3x - 2)^2 = 50$

3. $x(3x - 8) = -4$

4. $(x + 4)(x - 8) = -42$

5. $\dfrac{1}{x} + \dfrac{1}{x+2} = \dfrac{1}{3}$

6. $(x^2 + 2x)^2 - 11(x^2 + 2x) + 24 = 0$

7. $x^{1/6} - x^{1/3} + 2 = 0$

8. Solve for v, using only the positive root: $E = \frac{1}{2}mv^2$.

9. Solve by completing the square: $x^2 - 6x + 7 = 0$.

10. Find the length of the line segment connecting $(5, 3)$ and $(-5, -3)$. Express this distance in simplified radical form.

In Problems 11–12, graph each parabola whose equation is given.

11. $y = -x^2 + 2x + 3$

12. $f(x) = (x - 2)^2 + 3$

13. Compute the discriminant and then indicate if the solutions are rational, irrational, or imaginary: $2x^2 - x - 2 = 0$.

14. Find the quadratic function $f(x) = ax^2 + bx + c$ that fits the data points $(0, 0), (2, 0),$ and $(3, 3)$.

In Problems 15–16, find the solution set for each inequality.

15. $x^2 \le 2x + 35$

16. $\dfrac{4}{x-2} \ge 2$

17. Use the measurements determined by the surveyor in the figure to find the width of the pond. Express the answer in simplified radical form.

18. The figure shows a toy rocket that is propelled directly upward from ground level with an initial velocity of 64 feet per second. The model $s(t) = -16t^2 + 64t$ describes the rocket's height above the ground ($s(t)$, in feet) after t seconds. After how many seconds does the rocket reach its maximum height? How high does it go?

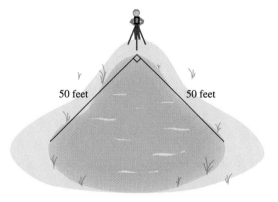

50 feet   50 feet

19. As shown in the figure, an open box is made by cutting equal squares from the four corners of a 10 inch by 14 inch piece of cardboard and folding up the sides to form an open-top box. If the area of the base is to be 32 square inches, what size square should be cut from each corner?

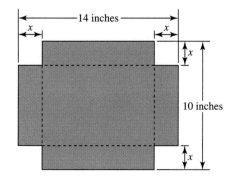

20. A rectangular pool is 20 feet wide and 30 feet long. A tile border of uniform width is to be built around the pool using 336 square feet of tile. How wide should the border be?

CUMULATIVE REVIEW PROBLEMS (CHAPTERS 1–7)

In Problems 1–9, solve for x. Express irrational solutions in simplified radical form and complex nonreal solutions in the form a + bi. Express solutions of inequalities in both set-builder and interval notation.

1. $|x - 1| > 3$

2. $x^{2/3} = 2x^{1/3} + 35$

3. $8(2x - 1) + 3(1 - x) + 2 = 5 - 2x(1 + x)$

4. $(x - 8)(x + 7) = 5$

5. $\sqrt{x + 4} - \sqrt{x - 4} = 2$

6. $-3 \le \frac{2}{5}x - 1 \le 1$

7. $|6x - 4| + 6 = 1$

8. $x - 4 \ge 0$ and $-3x \le -6$

9. $|5x + 2| = |4 - 3x|$

10. Solve the system:

$$-3x + 2y + 4z = 6$$
$$7x - y + 3z = 23$$
$$2x + 3y + z = 7$$

Perform the operations in Problems 11–16, simplifying final answers. Exponential expressions should be written with positive exponents only.

11. $\dfrac{-5x^3y^7z^4}{15x^3y^9z^{-2}}$

12. $(5x - 2)(2x^2 + 3xy - y^2)$

13. $\dfrac{x + 2}{x^2 - 6x + 8} + \dfrac{3x - 8}{x^2 - 5x + 6}$

14. $(5x^3 - 24x^2 + 9) \div (5x + 1)$

15. $\dfrac{\sqrt[3]{32xy^{10}}}{\sqrt[3]{2xy^2}}$

16. $\dfrac{3 - \sqrt{5}}{2 + \sqrt{5}}$

17. Factor completely: $x^2 - 8x + 16 - 25y^2$.

18. a. Use the square root method to solve for y: $x^2 + y^2 = 4$.

  b. Does $x^2 + y^2 = 4$ define y as a function of x? Use your answer from part (a) to decide.

  c. The graph of $x^2 + y^2 = 4$ is a circle, as shown in the figure. Does the graph define y as a function of x? Explain your answer.

19. A fastball is hit straight up over home plate. The ball's height, $(s(t)$, in feet) after t seconds is modeled by $s(t) = -16t^2 + 80t + 5$. The graph of this quadratic function is shown in the figure. Precisely when does the ball strike the ground? Does the graph furnish you with this information? Explain.

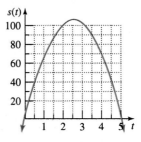

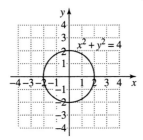

20. An open box is to be constructed from a rectangular sheet of metal measuring 10 inches by 7 inches. A square whose length and width are represented by x is cut from each corner, and the sides are then folded up. Write a function, $V(x)$, that models the volume of the resulting box in terms of the size of the piece x cut from each corner of the metal.

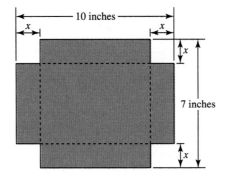

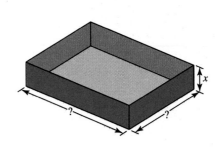

21. In 1982 the average annual salary for elementary and secondary teachers in the United States was $17,547. If this amount has been increasing steadily by approximately $1527 per year, when will the average salary reach $51,141?

22. The graph of the model for teachers' salaries in Problem 21 is a straight line. What must the slope of this line be? Explain your answer.

23. The graph shows the height of an eagle's flight over a period of time.
   a. What is the slope of the line through points A and B? Describe what this means in practical terms.
   b. What is the slope of the line through points P and Q? What does this mean in terms of the variables in the graph?
   c. What is the slope of the line through points C and D? What does this indicate about the eagle's flight?

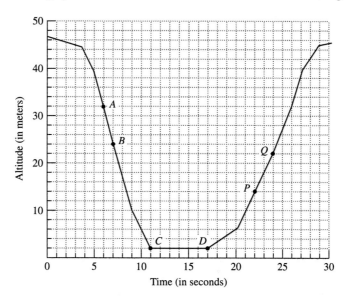

24. Three people have telephone area codes whose three digits have the same sum. One of the area codes is 252. None of the area codes contains a digit that is in one of the other area codes. No area code has a first digit of 4. One of the area codes begins with 6. Another area code ends with 1. What is the area code that ends with 1?

25. Find the minimum value of the objective function $z = 2x + 3y$ on the region shown in the figure.

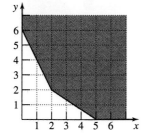

26. According to *Money Magazine,* the amount that your annual maximum Social Security benefit will be is a function of your present age, as shown in the table.

x (Your Present Age)	y (Amount Your Annual Maximum Social Security Benefit Will Be)
20	$20,151
30	$18,301

This data can be modeled by a linear function.
   a. Write the point-slope form of the line that models this data.
   b. Use the point-slope form of the line to write the slope-intercept equation of the line that fits the data.
   c. Use your equation from part (b) to answer this question: If a person is currently 55 years old, what is the amount of that person's annual maximum Social Security benefit starting from the time of eligibility?

27. The Sears Tower in Chicago is to date the world's tallest building. Its height is 582 feet less than twice that of the Library Tower in Los Angeles. The mean (average) height of the two buildings is 1236 feet. How far above street level does each building rise?

Sears Tower, Chicago   Library Tower, Los Angeles

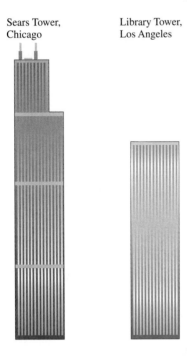

28. A rectangular room measuring 12 feet by 20 feet has wall-to-wall carpeting. A uniform strip of carpeting is removed along the edges and replaced by tile. If the area of the remaining carpet is 180 square feet, what is the width of the strip that was removed?

29. A cyclist rode 2 hours with a wind of 4 miles per hour and then turned around along the same route to ride against the same wind. The return trip took 3 hours. Find the cyclist's average speed with no wind. What distance did the cyclist travel in each direction?

30. The figure indicates that the quadratic function

$$y = \frac{1}{9000} x^2 + 5$$

can be used to model the suspension cables on the Golden Gate Bridge connecting San Francisco and Marin County, California. The road lies directly on the x-axis. How high above the road are the suspension cables connected to the towers?

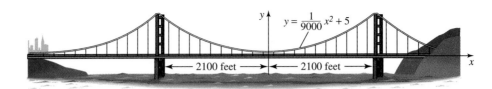

Exponential and Logarithmic Functions

Magdalena Abakanowicz "Crowd 1" 1986–87, 50 standing figures, burlap resin, $66\frac{7}{8} \times 23\frac{5}{8} \times 11\frac{3}{4}$ in. ($170 \times 60 \times 30$ cm). © 1998 Magdalena Abakanowicz/Licensed by VAGA, New York, NY/Marlborough Gallery, NY. Photography © Artur Starewicz.

How long will it be before world population reaches a level where we no longer live with any real degree of comfort or individual choice? Exponential and logarithmic functions provide a frightening answer to this question (see Example 5 on page 729), raising serious concerns about the kind of world we are leaving for our children. Modeling with these remarkable functions helps us to both predict the future and rediscover the past.

Exponential Functions

Objectives

1 Evaluate exponential functions.
2 Graph exponential functions.
3 Use compound interest formulas.
4 Solve applied problems using exponential functions.

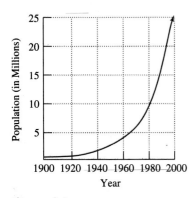

Figure 8.1

The growth of Mexico City. The model $f(t) = 67.38(1.026)^t$ describes the entire country's growth t years after 1980.

Source: U.N. Population Council

Throughout this book we have seen examples of polynomial functions such as $f(x) = x^2$ and $g(x) = x^3$. These functions involve a variable raised to a constant power. We will now interchange the variable and the constant, obtaining functions such as $F(x) = 2^x$ and $G(x) = 3^x$. These functions, with constant bases and variable exponents, are called *exponential functions.*

Exponential functions are used to model many situations. For example, an exponential function models the rate of population growth. The graph in Figure 8.1 shows the growth of the Mexico City metropolitan area from 1900 through 2000. If this growth trend continues, Mexico City could have 100 million inhabitants by the middle of the next century!

The population in millions for the entire country of Mexico t years after 1980 can be modeled by the function

$$f(t) = 67.38(1.026)^t.$$

Notice that the variable, t, is in the exponent, making this an example of an exponential function with base 1.026. All exponential functions contain a constant base and a variable exponent.

Definition of the exponential function

The exponential function f with base b is defined by

$$f(x) = b^x \quad \text{or} \quad y = b^x$$

where b is a positive constant other than 1 ($b > 0$ and $b \neq 1$) and x is any real number.

$\mathcal{S}$ **tudy tip**

Notice the difference between polynomial and exponential functions.

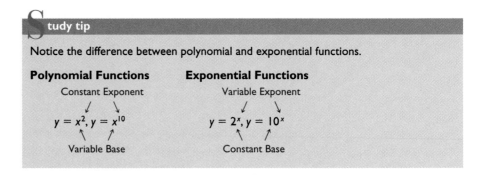

Evaluate exponential
functions.

Photri, Inc.

EXAMPLE 1   **Evaluating an Exponential Function**

The exponential function $f(t) = 67.38(1.026)^t$ describes the population ($f(t)$, in millions) of Mexico t years after 1980.

a. What was the 1980 population?
b. Predict Mexico's population in 2007, 2034, and 2061.

Solution

a. Since 1980 is zero years after 1980, we use the given exponential model and substitute 0 for t.

$f(t) = 67.38(1.026)^t$ This is the given function.

$f(0) = 67.38(1.026)^0$ Substitute 0 for t.

$f(0) = 67.38(1)$ $b^0 = 1$. A calculator is not necessary.

$f(0) = 67.38$

Mexico's 1980 population was 67.38 million.

b. We will begin with the year 2007. Since 2007 is 27 years after 1980, we substitute 27 for t.

$f(t) = 67.38(1.026)^t$ This is the given function.

$f(27) = 67.38(1.026)^{27}$ Substitute 27 for t.

$f(27) \approx 134.74$ Use a calculator: 67.38 $\boxed{\times}$ 1.026 $\boxed{\wedge}$ 27 $\boxed{\text{ENTER}}$

The model predicts Mexico's population to be approximately 134.74 million in the year 2007.

We now follow the same procedure for the years 2034 and 2061.

2034:	**2061:**
54 years after 1980:	**81 years after 1980:**
$f(54) = 67.38(1.026)^{54}$	$f(81) = 67.38(1.026)^{81}$
≈ 269.46	≈ 538.85

We can summarize these results in a table.

Year	Mexico's Predicted Population
1980	67.38 million
2007	134.74 million
2034	269.46 million
2061	538.85 million

27 years later — Population doubles.
27 years later — Population approximately doubles.
27 years later — Population approximately doubles.

The numbers in the table indicate that Mexico's population will double every 27 years. Consequently, the *doubling time* of Mexico's population is 27 years. All populations that grow exponentially have a doubling time; the doubling time of the 1990 world population is approximately 38 years. (See Table 8.1.) ■

TABLE 8.1 World Population Growth and Doubling Times

Date	Population	Doubling Time
5000 B.C.	50 million	?
800 B.C.	100 million	4200 years
200 B.C.	200 million	600 years
1200 A.D.	400 million	1400 years
1700 A.D.	800 million	500 years
1900 A.D.	1600 million	200 years
1965 A.D.	3200 million	65 years
1990 A.D.	5300 million	38 years
2020 A.D. (estimate)	8230 million	55 years

Source: Population Reference Bureau, Inc., Washington, D.C.

Using technology

The graph of Mexico's population model is shown as follows.

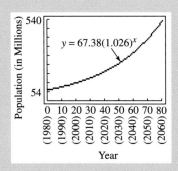

To graph the population function from 1980 to 2060, we used Xmin = 0, Xmax = 80, Xscl = 10. Our calculations in Example 1 indicated a population of 538.85 million in 2061. Since the population is growing and the function is increasing, we used the range setting Ymin = 0, Ymax = 540, and Yscl = 54.

Like all increasing exponential functions, the graph initially increases relatively slowly but ultimately climbs with increasing steepness. Populations grow extremely rapidly as they get larger because there are more people to have children. This exponential growth poses a major environmental problem to our planet.

2 Graph exponential functions.

Graphing Exponential Functions

We are familiar with expressions involving b^x, where x is a rational number. For example,

$$b^3 = b \cdot b \cdot b \qquad b^{1.7} = b^{17/10} = \sqrt[10]{b^{17}}$$

$$b^0 = 1 \qquad b^{1.73} = b^{173/100} = \sqrt[100]{b^{173}}$$

$$b^{-2} = \frac{1}{b^2} \qquad b^{-3/4} = \frac{1}{b^{3/4}} = \frac{1}{\sqrt[4]{b^3}}$$

$$b^{1/2} = \sqrt{b}$$

However, because the definition of $f(x) = b^x$ includes all real numbers for x, you may wonder what b^x means when x is an irrational number, such as $b^{\sqrt{3}}$ or b^π. A precise definition for $b^{\sqrt{3}}$ is really impossible without a background in calculus. But if we use the nonrepeating and nonterminating approximation 1.73205 for $\sqrt{3}$, we can think of $b^{\sqrt{3}}$ as the value that has the successively closer approximations

$$b^{1.7},\ b^{1.73},\ b^{1.732},\ b^{1.73205},\ \ldots$$

In this way, we will be able to graph the exponential function with no holes or points of discontinuity at the irrational domain values. Let's consider the graphs of some exponential functions.

EXAMPLE 2   **Graphing an Exponential Function**

Graph: $f(x) = 2^x$

Solution

We begin by setting up a table of coordinates.

x	$f(x) = 2^x$
-3	$f(-3) = 2^{-3} = \frac{1}{8}$
-2	$f(-2) = 2^{-2} = \frac{1}{4}$
-1	$f(-1) = 2^{-1} = \frac{1}{2}$
0	$f(0) = 2^0 = 1$
1	$f(1) = 2^1 = 2$
2	$f(2) = 2^2 = 4$
3	$f(3) = 2^3 = 8$

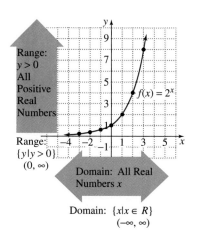

Range:
$y > 0$
All Positive Real Numbers

Range: $\{y|y > 0\}$ $(0, \infty)$

Domain: All Real Numbers x

Domain: $\{x|x \in R\}$ $(-\infty, \infty)$

Figure 8.2

The graph of $f(x) = 2^x$

We plot these points, connecting them with a continuous curve. Figure 8.2 shows the graph of $f(x) = 2^x$. Observe that the graph approaches but never touches the negative portion of the x-axis, indicating a range of all positive real numbers. There are no values of x that will result in 0 or negative numbers in the range of $f(x) = 2^x$. Furthermore, we chose integers for x in our table of coordinates. However, you can use a calculator to find additional coordinates. For example, $f(0.3) = 2^{0.3} \approx 1.231$. $f(0.95) = 2^{0.95} \approx 1.932$, and the points $(0.3, 1.231)$ and $(0.95, 1.932)$ really do fit the graph. ∎

EXAMPLE 3   **Graphing an Exponential Function**

Graph: $f(x) = \left(\dfrac{1}{2}\right)^x$

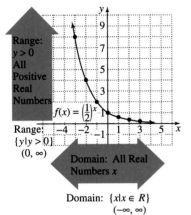

Figure 8.3

The graph of $f(x) = (\frac{1}{2})^x$

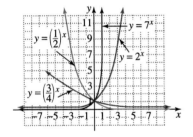

Figure 8.4

Graph of four exponential functions

Solution

We again begin with a table of coordinates.

x	$f(x) = (\frac{1}{2})^x$
-3	$f(-3) = (\frac{1}{2})^{-3} = \dfrac{1}{(\frac{1}{2})^3} = \dfrac{1}{\frac{1}{8}} = 8$
-2	$f(-2) = (\frac{1}{2})^{-2} = \dfrac{1}{(\frac{1}{2})^2} = \dfrac{1}{\frac{1}{4}} = 4$
-1	$f(-1) = (\frac{1}{2})^{-1} = 2$
0	$f(0) = (\frac{1}{2})^0 = 1$
1	$f(1) = (\frac{1}{2})^1 = \frac{1}{2}$
1.2	$f(1.2) = (\frac{1}{2})^{1.2} \approx 0.435$ $\leftarrow$ using a calculator: $\boxed{(}\ 1\ \boxed{\div}\ 2\ \boxed{)}\ \boxed{\wedge}\ 1.2\ \boxed{\text{ENTER}}$
2	$f(2) = (\frac{1}{2})^2 = \frac{1}{4}$
3	$f(3) = (\frac{1}{2})^3 = \frac{1}{8}$

We plot these points and connect them with a smooth continuous curve. Figure 8.3 shows the graph of $f(x) = (\frac{1}{2})^x$. ■

Four exponential functions have been graphed in Figure 8.4. Compare the graphs of functions where $b > 1$ to those where $0 < b < 1$. When $b > 1$, the value of y increases as the value of x increases. When $0 < b < 1$, the value of y decreases as the value of x increases.

These graphs illustrate the following general properties of exponential functions.

Characteristics of exponential functions

1. The domain of $f(x) = b^x$ consists of all real numbers. The range of $f(x) = b^x$ consists of all positive real numbers.
2. The graphs of all exponential functions pass through the point $(0, 1)$ because $f(0) = b^0 = 1\ (b \neq 0)$.
3. If $b > 1$, $f(x) = b^x$ has a graph that goes up to the right and is an increasing function.
4. If $0 < b < 1$, $f(x) = b^x$ has a graph that goes down to the right and is a decreasing function.
5. The graph of $f(x) = b^x$ approaches but does not cross the x-axis. The x-axis is an *asymptote*.

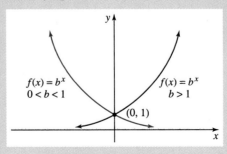

The graphs of exponential functions can be translated vertically or horizontally.

EXAMPLE 4 **Graphing an Exponential Function**

Graph: $f(x) = 3^{x+1}$

Solution

We begin with a table of coordinates.

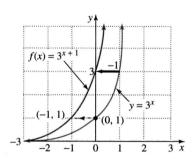

Figure 8.5

The graph of $f(x) = 3^{x+1}$ shifts the graph of $y = 3^x$ one unit to the left.

3 Use compound interest formulas.

x	$f(x) = 3^{x+1}$
-3	$f(-3) = 3^{-3+1} = 3^{-2} = \frac{1}{9}$
-2	$f(-2) = 3^{-2+1} = 3^{-1} = \frac{1}{3}$
-1	$f(-1) = 3^{-1+1} = 3^0 = 1$
0	$f(0) = 3^{0+1} = 3^1 = 3$
1	$f(1) = 3^{1+1} = 3^2 = 9$
2	$f(2) = 3^{2+1} = 3^3 = 27$

The graph of $f(x) = 3^{x+1}$ is shown in Figure 8.5. Also shown is the graph of $y = 3^x$. Notice that the graph of $f(x) = 3^{x+1}$ looks just like the graph of $y = 3^x$, but it is shifted horizontally one unit to the left. The y-intercept of $y = 3^x$ is 1, whereas the y-intercept of $y = 3^{x+1}$ is 3. The line $y = 0$ or the x-axis is a horizontal asymptote for both graphs. ∎

Compound Interest and the Irrational Number e

Compound interest provides another case in which exponential functions can be used. Suppose a sum of money P (called the principal) is invested at an annual percentage rate r (in decimal form), compounded once a year. Because the interest is added to the principal at year's end, the accumulated value (A) is

$$A = P + Pr = P(1 + r).$$

The accumulated amount of money follows this pattern of multiplying the previous principal by $(1 + r)$ for each successive year, as indicated in Table 8.2.

TABLE 8.2

Time in Years	Accumulated Value after Each Compounding
0	$A = P$
1	$A = P(1 + r)$
2	$A = P(1 + r)(1 + r) = P(1 + r)^2$
3	$A = P(1 + r)^2(1 + r) = P(1 + r)^3$
4	$A = P(1 + r)^3(1 + r) = P(1 + r)^4$
⋮	⋮
t	$A = P(1 + r)^t$

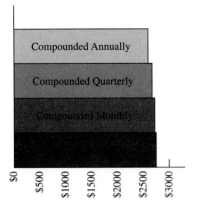

Future value of $1000 invested at 10% interest in 10 years. The increasing bar lengths show greater accumulated values as the number of compounding periods in a year increases.

If money invested at a specified rate of interest is compounded more than once a year, then the model $A = P(1 + r)^t$ can be adjusted to take into account the number of compounding periods in a year. If n represents the number of compounding periods in a year, the mathematical model becomes

$$A = P\left(1 + \frac{r}{n}\right)^{nt}$$

EXAMPLE 5 Using the Compound Interest Model

A sum of $8000 is invested at an annual interest rate of 6.5%, compounded monthly. Find the balance in the account after 10 years.

Solution

$$A = P\left(1 + \frac{r}{n}\right)^{nt} \qquad \text{This is the compound interest model.}$$

We are given the following values: P (the principal) = $8000, r (the interest rate) = 6.5% = 0.065, n (the number of compounding periods with monthly compounding) = 12, and t (time, in years) = 10. We substitute these values into the model.

$$A = 8000\left(1 + \frac{0.065}{12}\right)^{12(10)} \approx 15{,}297.47$$

After 10 years, the balance in the account is $15,297.47. ∎

The Irrational Number e

Suppose you invest $1 at an interest rate of 100% for 1 year, and the compounding periods increase infinitely (compounding interest every trillionth of a second, every quadrillionth of a second, etc.). As the compounding periods increase, the amount you would receive at the end of the year gets closer and closer to $2.718281828459045. . . . Mathematicians represent this irrational number by the symbol e, where e is approximately equal to 2.72 ($e \approx 2.72$).

Let's rewrite our compound interest model so that we can visualize what all this means.

$$A = P\left(1 + \frac{r}{n}\right)^{nt} \qquad \text{This is the compound interest model.}$$

$$A = 1\left(1 + \frac{1}{n}\right)^{n \cdot 1} \qquad \begin{array}{l} P \text{ (the principal)} = \$1, r \text{ (the interest rate)} = 100\% = 1, \\ \text{and } t \text{ (time)} = 1 \text{ year.} \end{array}$$

$$A = \left(1 + \frac{1}{n}\right)^{n} \qquad \text{Simplify.}$$

Remembering that A represents the balance in the account, we've used a graphing utility to graph $A = \left(1 + \frac{1}{n}\right)^n$ and $A = e$ (approximately 2.72) in

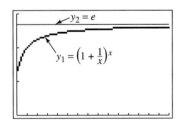

Figure 8.6

As x increases to the right, $\left(1 + \frac{1}{x}\right)^x$ gets closer to e.

the same viewing rectangle. Figure 8.6 illustrates that as *n* (the number of compounding periods) increases, the balance in the account gets closer and closer to *e*, or approximately $2.72.

The natural exponential function

The function

$$f(x) = e^x$$

is called the *natural exponential function*. The irrational number *e* is called the *natural base,* where $e \approx 2.7183$.

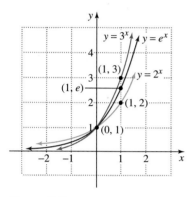

Figure 8.7

Graphs of three exponential functions

You will need to use a calculator to evaluate the natural exponential function. We used the natural exponential key  to obtain the following approximations.

$f(x) = e^x$

Function Value	Keystrokes	Approximate Display
$f(2) = e^2$	$\boxed{e^x}$ 2 $\boxed{\text{ENTER}}$	7.389056
$f(-3) = e^{-3}$	$\boxed{e^x}$ $\boxed{(-)}$ 3 $\boxed{\text{ENTER}}$	0.049787
$f(4.6) = e^{4.6}$	$\boxed{e^x}$ 4.6 $\boxed{\text{ENTER}}$	99.4843156

Because $2 < e < 3$, the graph of $y = e^x$ is between the graphs of $y = 2^x$ and $y = 3^x$, as shown in Figure 8.7.

Since the number *e* is related to compounding periods increasing infinitely, or *continuous compounding,* the number appears in the formula for the balance in an account subject to this type of compounding. The formulas for both types of compounding are given as follows.

Formulas for compound interest

After *t* years, the balance *A* in an account with principal *P* and annual interest rate *r* (in decimal form) is given by the following mathematical models.

1. For *n* compoundings per year: $A = P\left(1 + \dfrac{r}{n}\right)^{nt}$

2. For continuous compounding: $A = Pe^{rt}$

EXAMPLE 6   **Finding the Balance for Compound Interest**

In 1626 Peter Minuit convinced the Wappinger Indians to sell him Manhattan Island for $24. If the Native Americans had put the $24 into a bank account paying 5% interest, how much would the investment be worth in the year 2000 if interest were compounded   **a.** monthly?   **b.** daily?   **c.** continuously?

Solution

The time period for all three computations is $2000 - 1626$, or 374 years.

a. $A = P\left(1 + \dfrac{r}{n}\right)^{nt}$ This is the formula for n compoundings per year.

$= 24\left(1 + \dfrac{0.05}{12}\right)^{12(374)}$ $P = \$24, r = 5\% = 0.05, n = 12$ (monthly compounding), and $t = 374$ (years).

$\approx 3{,}052{,}428{,}614$ Use a calculator.

With monthly compounding, the \$24 would be worth approximately \$3,052,428,614.

b. $A = P\left(1 + \dfrac{r}{n}\right)^{nt}$ This is the formula for n compoundings per year.

$= 24\left(1 + \dfrac{0.05}{365}\right)^{365(374)}$ $n = 365$ (daily compounding, with 365 compounding periods); other values are the same as those in part (a).

$\approx 3{,}169{,}289{,}063$ Use a calculator.

With daily compounding, the \$24 would be worth approximately \$3,169,289,063.

c. $A = Pe^{rt}$ This is the model for continuous compounding.

$= 24e^{0.05(374)}$ $P = \$24, r = 5\% = 0.05,$ and $t = 374$ (years).

$\approx 3{,}173{,}350{,}574$ Use a calculator: $24\ \boxed{e^x}\ \boxed{(}\ .05\ \boxed{\times}\ 374\ \boxed{)}\ \boxed{\text{ENTER}}$

With continuous compounding, the \$24 would be worth approximately \$3,173,350,574. As the number of compounding periods increases without bound, the balance \$3,173,350,574 is a limiting amount that cannot be exceeded because we cannot increase the number of compounding periods any further.   ◼

Figure 8.8 shows the graphs of

$y_1 = 24\left(1 + \dfrac{0.05}{x}\right)^{374x}$ This is the model for the amount y_1 that \$24 will accumulate to over 374 years at 5% interest with x compoundings per year.

and

$y_2 = 24e^{0.05(374)}$ This constant function represents the accumulated amount over 374 years with continuous compounding.

Look at the graphs to see what happens as x, the number of yearly compoundings, increases. The balance in the account keeps getting closer to the amount limited by the continuous compounding model, graphed as a horizontal line.

Javne Quick-to-See Smith "Trade (gifts for trading land with white people)" 1992, oil, collage, mixed media on canvas with objects, triptych. 152.4 × 431.8 cm (60 × 170 in) Collection: Chrysler Museum, Norfolk, VA. Museum purchase. Courtesy Steinbaum Krauss Gallery, NYC.

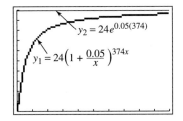

Figure 8.8

As x increases, the balance y_1 gets closer to y_2.

4 Solve applied problems using exponential functions.

Applications of Exponential Functions

Exponential functions form mathematical models describing population growth, radioactive decay, the amount of learning that takes place over time, the number of AIDS cases in the United States among intravenous drug users, and even the familiar bell-shaped curve. Many of these situations will be presented in the problem set that follows.

Our final example applies modeling with exponential functions to the Chernobyl disaster.

V. Ivelva/Magnum Photos, Inc.

Chernobyl

The spread of radioactive fallout from the Chernobyl accident is expected to result in at least 100,000 deaths from cancer in the Northern Hemisphere.

EXAMPLE 7 Modeling the Chernobyl Disaster with an Exponential Function

The 1986 disaster in the former Soviet Union at the Chernobyl nuclear power plant explosion sent about 1000 kilograms of radioactive cesium-137 into the atmosphere. The formula

$$A = 1000(0.5)^{t/30}$$

models the amount (A, in kilograms) of cesium-137 remaining after t years in the area surrounding Chernobyl. If even 100 kilograms of cesium-137 remain in the atmosphere, the area is considered unsafe for human habitation. Will people be able to live in the area 80 years after the accident?

Solution

We substitute 80 for t in the given model. If the resulting value of A is 100 or greater, the area will still be unsafe for human habitation.

$$
\begin{aligned}
A &= 1000(0.5)^{t/30} &&\text{This is the given model.}\\
&= 1000(0.5)^{80/30} &&\text{Substitute 80 for } t.\\
&\approx 157 &&\text{Use a calculator.}
\end{aligned}
$$

After 80 years, 157 kilograms of cesium-137 will be in the atmosphere. Since this exceeds 100, even by the year 2066 the Chernobyl area will remain a ghost town. ∎

Using technology

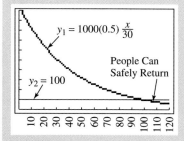

Figure 8.9

More than a century must pass before the area surrounding Chernobyl can be safely resettled.

Figure 8.9 shows the graphs of

$$y_1 = 1000(0.5)^{x/30}$$

This is the model for the number of kilograms of cesium-137 in the Chernobyl area x years after the 1986 accident.

and

$$y_2 = 100$$

The amount of cesium-137 must be lower than this for safe human habitation.

We used a graphing utility with the following range settings:

Xmin = 0, Xmax = 120, Xscl = 10,

Ymin = 0, Ymax = 1000, Yscl = 100

Current estimates are that 3 million people will eventually have to be moved from the Ukraine and Byelorussia area. They can safely return only when the graph of the exponential function lies underneath the graph of the constant function. Since this seems to occur somewhere to the right of $x = 100$, in practical terms it will be more than a century before the area can be safely resettled.

PROBLEM SET 8.1

Practice Problems

In Problems 1–4, fill in the values in the given tables of coordinates and graph each pair of functions in the same rectangular coordinate system.

1. $y = 2^x$

x	−2	−1	0	1	2
y					

$y = 2^{x+1}$

x	−3	−2	−1	0	1
y					

2. $y = 3^x$

x	−2	−1	0	1	2
y					

$y = 3^{x+2}$

x	−4	−3	−2	−1	0
y					

3. $y = 3^x$

x	−2	−1	0	1	2
y					

$y = 3^{x-2}$

x	0	1	2	3	4
y					

4. $y = 2^x$

x	−2	−1	0	1	2
y					

$y = 2^{x-1}$

x	−1	0	1	2	3
y					

5. What does the graph of $y = b^{x+c}$ do to the graph of $y = b^x$ if c is a positive integer? What does the graph of $y = b^{x+c}$ do to the graph of $y = b^x$ if c is a negative integer?

6. Graph $f(x) = 2^x$, $g(x) = 2^x + 1$, and $h(x) = 2^x - 3$ in the same rectangular coordinate system. Describe the relationship among the graphs of f, g, and h.

7. Graph $f(x) = 2^x$, $g(x) = 2^x + 3$, and $h(x) = 2^x - 1$ in the same rectangular coordinate system. Describe the relationship among the graphs of f, g, and h.

8. Graph $f(x) = 3^x$ and $g(x) = \left(\frac{1}{3}\right)^x$ in the same rectangular coordinate system. Describe the relationship between the graphs of f and g.

9. Graph $f(x) = \left(\frac{3}{2}\right)^x$ and $g(x) = \left(\frac{2}{3}\right)^x$ in the same rectangular coordinate system. Describe the relationship between the graphs of f and g.

10. Fill in the missing values in the given tables. Then graph $y = 2^x$ and $x = 2^y$ in the same rectangular coordinate system. Describe any relationships that you observe between the graphs.

$y = 2^x$

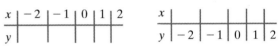

x	−2	−1	0	1	2
y					

$x = 2^y$

x					
y	−2	−1	0	1	2

11. Fill in the missing values in the given tables. Then graph $y = 3^x$ and $x = 3^y$ in the same rectangular coordinate system. Describe any relationships that you observe between the graphs.

$y = 3^x$

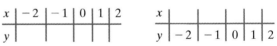

x	−2	−1	0	1	2
y					

$x = 3^y$

x					
y	−2	−1	0	1	2

In Problems 12–21, graph each exponential function.

12. $y = 2^{-x}$ **13.** $y = 3^{-x}$ **14.** $f(x) = 3^{x/2}$ **15.** $g(x) = 2^{x/2}$ **16.** $y = 2^{x+1} - 3$

17. $y = 2^{x-1} - 1$ **18.** $y = 2^{x-1} + 3$ **19.** $y = 2^{x+1} + 1$ **20.** $f(x) = 3^{-x} + 2$ **21.** $f(x) = 3^{-x/2} - 1$

Application Problems

Use the compound interest models

$$A = P\left(1 + \frac{r}{n}\right)^{nt} \quad \text{and} \quad A = Pe^{rt}$$

to answer Problems 22–25.

22. Find the accumulated value of an investment of $10,000 for 5 years at an interest rate of 5.5% if the money is a. compounded semiannually; b. compounded monthly; c. compounded continuously.

23. Find the accumulated value of an investment of $5000 for 10 years at an interest rate of 6.5% if the money is a. compounded semiannually; b. compounded monthly; c. compounded continuously.

24. Suppose that you have $12,000 to invest. What investment yields the greatest return over 3 years: 7% compounded monthly or 6.85% compounded continuously?

25. Suppose that you have $6000 to invest. What investment yields the greatest return over 4 years: 8.25% compounded quarterly or 8.3% compounded semiannually?

26. The function $f(x) = 10^6 2^x$ represents a model for the number of bacteria in a petri dish grown by a biologist at the end of x days.
    a. Graph the function, letting each unit on the y-axis represent 1 million.

b. The bacteria count at the end of 5 days is how many times as great as the count 2 days after the beginning of the experiment? $\left[Hint\text{: Find } \dfrac{f(5)}{f(2)}. \right]$

27. The function $f(x) = 5 \cdot 2^{-x}$ represents a model for the value of a truck (in thousands of dollars) at the end of x years.
    a. Graph the model for $0 \le x \le 4$.
    b. The value of the truck at the end of 1 year is how many times as great as the value of the truck at the end of 4 years?

The mathematical model for exponential growth is given by $y = Ae^{kt}$, $(k > 0)$. For this model, t is the time, A is the original amount of the quantity, and y is the amount after time t. The number k is a constant that describes the rate of growth. Use this model to answer Problems 28 and 29.

28. The exponential growth model can be used to describe the number of AIDS cases in the United States among intravenous drug users. In 1989, there were 24,000 cases of AIDS among intravenous drug users ($A = 24,000$) with a growth rate of 21% ($k = 0.21$). If this growth rate doesn't change, how many cases of AIDS will be expected for this population by the end of   a. 2000?   b. 2020?

29. Repeat Problem 28 for all AIDS cases in the United States if in 1989 there were 100,000 diagnosed cases of AIDS and a growth rate of 9% ($A = 100,000$ and $k = 0.09$).

30. The total number of AIDS cases in the United States can also be approximated by the exponential function $y = 100,000(1.4)^t$ where t represents the number of years after 1989. How many total AIDS cases are predicted by this model one year ago? How many AIDS cases are predicted for the same time by the model in Problem 29? Consult the reference section of your library or the Internet to determine the actual number. Which is the better model for that year?

31. In a report entitled *Resources and Man,* the U.S. National Academy of Sciences concluded that a world population of 10 billion "is close to (if not above) the maximum that an intensely managed world might hope to support with some degree of comfort and individual choice." For this reason, 10 billion is called the *carrying capacity* of Earth. At the time the report was issued in 1969, the world population was approximately 3.6 billion, with a relative growth rate of 2% per year. If exponential growth continues at this rate, use the model $y = 3.6e^{0.02t}$, where y is the world population (in bil-

lions) t years after 1969, to show that Earth's carrying capacity will be reached in the year 2020.

32. The model $f(x) = 14.7e^{-0.2x}$ describes atmospheric pressure (in pounds per square inch) for any locale that is situated x miles above sea level.
    a. The peak of Mount Everest is approximately $5\frac{1}{2}$ miles above sea level. What is the atmospheric pressure at the peak?
    b. The Dead Sea has an altitude that is approximately $\frac{1}{4}$ of a mile below sea level. What is the atmospheric pressure?

33. Concern about the exponential growth of the world's population began in 1798 with the publication of *An Essay on the Principle of Population as It Affects Future Improvement of Society* by British economist Thomas Malthus. With exponential population growth and linear growth in the available food supply, Malthus gloomily forecasted a planet beset by war, poverty, and disease attributable to a population that exceeded the food supply. Mathematicians have studied the world's population at different times in history, using a computer to elicit the functional model $f(x) = 6164e^{0.00667x}$ to describe world population in the year x (A.D.).
    a. Verify the three values in the following table for *estimated world population* in 1650, 1950, and 1970.

Year	Estimated Population	Actual Population
1650	371,000,000	470,000,000
1950	2,745,000,000	2,501,000,000
1970	3,137,000,000	3,610,000,000

Interestingly enough, comparing actual population with population estimated by the growth function, mathematicians discovered that population growth is increasing faster than that given by the exponential model. World population has *not* followed an exponential growth pattern, where exponential growth implies that the rate of change of the population is directly proportional to the amount of population at any given time. Exponential growth is a technical term and does not simply mean that rapid change is taking place.

b. The Malthusian model for exponential population growth does appear to apply to third-world countries and to world population over shorter intervals of time (30 to 40 years). Using the model $f(x) = 4.2e^{0.02x}$ to describe world population (in billions) x years after 1980, estimate the world's population by the year 2000 to the nearest tenth of a billion.

34. The formula $S = C(1 + r)^t$ models inflation, with C = the value today, r = the annual inflation rate, and S = the inflated value t years from now. If there is an inflation rate of 6%, what is the value of a $65,000 house in 10 years?

35. The function

$$N(t) = \frac{30,000}{1 + 20e^{-1.5t}}$$

describes the number of people ($N(t)$) who become ill with influenza t weeks after its initial outbreak in a town with 30,000 inhabitants. The horizontal asymptote in the graph at the top of the next column indicates that there is a limit to the epidemic's growth.

a. How many people became ill with the flu when the epidemic began? (When the epidemic began, $t = 0$.)

b. How many people were ill by the end of the fourth week?

c. Why can't the spread of an epidemic simply grow exponentially indefinitely? What does the horizon-

tal asymptote shown in the graph indicate about the limiting size of the population that becomes ill?

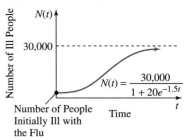

Number of People Initially Ill with the Flu

36. Another situation in which the exponential function serves as a model involves *learning curves*. Psychologists have discovered that when we learn new tasks, learning is initially quite rapid but tends to taper off as time increases. Once we have mastered a task, our performance level approaches peak efficiency, with additional practice having minimal effect on performance.

Here's a particular example of a model for learning new tasks. After t months of training, the function $N(t) = 800 - 500e^{-0.5t}$ describes the number of letters per hour ($N(t)$) that a postal clerk can sort. The horizontal asymptote in the graph shown below indicates that there is a level of peak efficiency that is being approached.

a. How many letters per hour can be sorted with no training? (With no training, $t = 0$.)

b. How many letters per hour can be sorted after 4 months of training?

c. What does the horizontal asymptote shown in the graph indicate about the level of peak efficiency? Why can't the learning curve simply grow exponentially indefinitely?

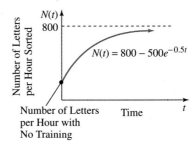

Number of Letters per Hour with No Training

True–False Critical Thinking Problems _____

37. Which one of the following is true?

a. As the number of compounding periods increases on a fixed investment, the amount of money in the account over a fixed interval of time will increase without bound.

b. The functions $f(x) = 3^{-x}$ and $g(x) = -3^x$ have the same graph.

c. $e = 2.718$

d. The functions $f(x) = (\frac{1}{3})^x$ and $g(x) = 3^{-x}$ have the same graph.

38. Which one of the following is true?
 a. The function $f(x) = 1^x$ is an exponential function.
 b. The number of exponential functions is limitless.
 c. If the population of a country is increasing steadily from year to year, the population over time can be modeled by an exponential function.
 d. The number e is not a real number.

Technology Problems

39. An important function used in statistical decision making is the *normal distribution probability density function,* a model developed by German mathematician Carl Gauss. The model

$$f(x) = \frac{1}{\sqrt{2\pi}} e^{-x^2/2}$$

has a graph that is the familiar bell-shaped curve. Measurements of data collected in nature often follow the pattern determined by this curve. The heights and weights of large populations of human beings are examples of distributions that are approximately normal. In these distributions, data items tend to cluster around the mean (the average) and become more spread out as they differ from the mean. Graph the normal distribution function using a graphing utility and the following range setting:

 Xmin = -4, Xmax = 4, Xscl = 1,
 Ymin = 0, Ymax = 0.5, Yscl = 0.05

40. You have $10,000 to invest. One bank pays 5% interest compounded quarterly and the other pays 4.5% interest compounded monthly.
 a. Use the formula for compound interest to write a model for the balance in each account at any time t.
 b. Use a graphing utility to graph both models in an appropriate viewing rectangle. Based on the graphs, which bank offers the better return on your money?

41. a. Graph $y = e^x$ and $y = 1 + x + \dfrac{x^2}{2}$ in the same viewing rectangle.

 b. Graph $y = e^x$ and $y = 1 + x + \dfrac{x^2}{2} + \dfrac{x^3}{6}$ in the same viewing rectangle.

 c. Graph $y = e^x$ and $y = 1 + x + \dfrac{x^2}{2} + \dfrac{x^3}{6} + \dfrac{x^4}{24}$ in the same viewing rectangle.

 d. Describe what you observe in parts (a)–(c). Try generalizing this observation.

42. According to the U.S. Bureau of the Census, the population of the United States in 1990 was about 250,000,000. The rate of growth of the population is about 0.7% per year. The function $y = 250,000,000(1.007)^x$ serves as a model for U.S. population (y) x years after 1990.
 a. Use your graphing utility to graph the model.
 b. $\boxed{\text{TRACE}}$ along the curve and determine the population in the year 2000.
 c. $\boxed{\text{TRACE}}$ along the curve and determine how long it will take for the population to double.

43. Use a graphing utility to verify your hand-drawn graphs in Problems 1–21.

44. Use a graphing utility to graph the model in Problem 35. $\boxed{\text{TRACE}}$ along the curve until you reach the point that illustrates the solution to part (b) of the problem.

45. Use a graphing utility to graph the model in Problem 36. $\boxed{\text{TRACE}}$ along the curve until you reach the point that illustrates the solution to part (b) of the problem.

46. Use a graphing utility to graph $y = e^x$ and $y = e^{x-2} + 2$ in the same viewing rectangle. Describe the transformations of the first function's graph that result in the graph of $y = e^{x-2} + 2$.

Writing in Mathematics

47. Explain why the graph of the exponential function $f(x) = 2^x$ gets closer and closer to the x-axis for negative values of x. Why is it that the graph doesn't intersect the x-axis?

48. In the text, an exponential function was defined as $f(x) = b^x$, where $b > 0$ and $b \neq 1$. Why do we not allow the base of an exponential function to be 1?

Critical Thinking Problems

49. Describe how you could use the graph of $y = 2^x$ to obtain a decimal approximation for $\sqrt{2}$.

50. The graphs labeled a–d in the accompanying figure represent $y = 3^x$, $y = 5^x$, $y = (\frac{1}{3})^x$, and $y = (\frac{1}{5})^x$, but not necessarily in that order. Which is which? Describe the process that enables you to make your decision.

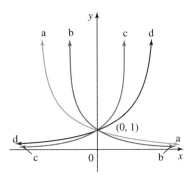

51. Each function in the table is increasing. The graphs labeled a–c in the accompanying figure represent those functions. Which is which? Describe the process that enables you to make your decision.

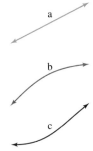

x	$f(x)$	$g(x)$	$h(x)$
1	23	15	2.7
2	25	25	3.1
3	28	34	3.5
4	32	42	3.9
5	37	49	4.3
6	43	55	4.7

Review Problems

52. When the wind is blowing at 10 mph, a marathon bike rider rides 40 miles against the wind and 40 miles back along the same route with the wind. If the total trip takes 3 hours, find the rider's speed in still air.

53. Solve for b: $D = \dfrac{ab}{a + b}$.

54. Evaluate: $\begin{vmatrix} 3 & -2 \\ 7 & -5 \end{vmatrix}$.

Composite and Inverse Functions

Objectives

1 Form composite functions.
2 Interpret the composition of functions in an applied setting.
3 Use the horizontal line test to determine if a function has an inverse.
4 Verify inverse functions.
5 Find the inverse of a one-to-one function.

Our focus in this chapter is on exponential and logarithmic functions. Logarithmic functions reverse the coordinates of exponential functions, and for this reason the functions are called inverses of one another. In this section we discuss inverse functions in general.

Form composite functions.

The Composition of Two Functions

In a free-market economy, a person's status (S) in society is a function of annual income (I), expressed as

$$S = f(I).$$

Income, in turn, is a function of the number of years of education (E), expressed as

$$I = g(E).$$

Because status is a function of income and income is a function of education, status is ultimately a function of education. In symbolic notation, because $S = f(I)$ and $I = g(E)$, then

$$S = f(g(E)).$$

We call $f(g(E))$ (read "f of g of E") the *composition* of functions f and g, or a *composite function*. This composition is written as $f \circ g$.

The dependence of status on education is illustrated in Figure 8.10. Generalizing from this situation, we can define the composition of any two functions.

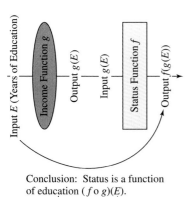

Conclusion: Status is a function of education ($f \circ g)(E)$.

Figure 8.10

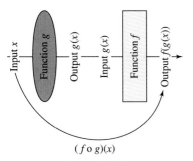

$(f \circ g)(x)$

Figure 8.11

The composition of functions

The *composition* of the functions f and g is denoted by $f \circ g$ and is defined by the equation

$$(f \circ g)(x) = f(g(x)).$$

The domain of the *composite function* $f \circ g$ is the set of all x such that

  1. x is in the domain of g and
  2. $g(x)$ is in the domain of f. (See Figure 8.11.)

EXAMPLE 1 **Forming Composite Functions**

Given $f(x) = x^2 + 6$ and $g(x) = 3x - 4$, find:

a. $(f \circ g)(3)$ **b.** $(g \circ f)(3)$ **c.** $(f \circ g)(x)$ **d.** $(g \circ f)(x)$

Solution

a. Because $(f \circ g)(3) = f(g(3))$,

  1. First find $g(3)$.   2. Put $g(3)$ into f.
  $g(3) = 3 \cdot 3 - 4 = 5$ $f(g(3)) = f(5) = 5^2 + 6 = 31$

Thus, $(f \circ g)(3) = 31$.

b. Because $(g \circ f)(3) = g(f(3))$,

  1. First find $f(3)$.   2. Put $f(3)$ into g.
  $f(3) = 3^2 + 6 = 15$ $g(f(3)) = g(15) = 3 \cdot 15 - 4 = 41$

Thus, $(g \circ f)(3) = 41$.

Study tip

Read the notation for composite functions from *right to left*.

Notation	What to Do
$f(g(3))$	Put $g(3)$ into f.
$g(f(3))$	Put $f(3)$ into g.
$f(g(x))$	Put $g(x)$ into f.
$g(f(x))$	Put $f(x)$ into g.

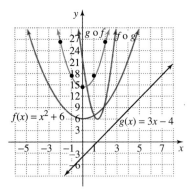

Figure 8.12
$(f \circ g)(x) \neq (g \circ f)(x)$

c. $(f \circ g)(x) = f(g(x))$ Use $f(x) = x^2 + 6$ and replace x with $g(x)$.

   Put $g(x)$ into f.

   $= [g(x)]^2 + 6$ Replace x in $x^2 + 6$ with $g(x)$.

   $= (3x - 4)^2 + 6$ Since $g(x) = 3x - 4$, replace $g(x)$ with $3x - 4$.

   $= 9x^2 - 24x + 16 + 6$ Square $3x - 4$ using the form $(A - B)^2 = A^2 - 2AB + B^2$.

   $= 9x^2 - 24x + 22$ Simplify.

Thus, $(f \circ g)(x) = 9x^2 - 24x + 22$.

d. $(g \circ f)(x) = g(f(x))$ Use $g(x) = 3x - 4$ and replace x with $f(x)$.

   Put $f(x)$ into g.

   $= 3[f(x)] - 4$ Replace x in $3x - 4$ with $f(x)$.

   $= 3(x^2 + 6) - 4$ Since $f(x) = x^2 + 6$, replace $f(x)$ with $x^2 + 6$.

   $= 3x^2 + 18 - 4$ Use the distributive property.

   $= 3x^2 + 14$ Simplify.

Thus, $(g \circ f)(x) = 3x^2 + 14$.   ■

Notice that $(f \circ g)(3) \neq (g \circ f)(3)$ and that, in general, $(f \circ g)(x) \neq (g \circ f)(x)$. The graphs of the functions, shown in Figure 8.12 illustrate this point.

Our next example further illustrates that the composition of f with g is generally not the same as the composition of g with f.

EXAMPLE 2   Comparing the Composition of Functions

Given $f(x) = \sqrt{x}$ and $g(x) = 2x + 4$, find:

a. $(f \circ g)(x)$ **b.** $(g \circ f)(x)$

Solution

a. We begin by finding the composition of f with g.

$(f \circ g)(x) = f(g(x))$ This is the definition of $f \circ g$.

   $= \sqrt{g(x)}$ Put $g(x)$ into f.

   $= \sqrt{2x + 4}$ Since $g(x) = 2x + 4$, replace $g(x)$ with $2x + 4$.

b. Now we find the composition of g with f.

$(g \circ f)(x) = g(f(x))$ This is the definition of $g \circ f$.

   $= 2[f(x)] + 4$ Put $f(x)$ into g.

   $= 2\sqrt{x} + 4$ Since $f(x) = \sqrt{x}$, replace $f(x)$ with $\sqrt{x}$.

We see that $(f \circ g)(x) = \sqrt{2x + 4}$ and $(g \circ f)(x) = 2\sqrt{x} + 4$. Once again $(f \circ g)(x) \neq (g \circ f)(x)$, reinforced by the graphs shown in the Using Technology box at the top of the next page.   ■

Discover for yourself

Use the functions in Example 2, namely,

$$f(x) = \sqrt{x} \quad \text{and} \quad g(x) = 2x + 4$$

and find $(f \circ g)(-2)$ and $(g \circ f)(-2)$. Interpret your answers graphically using the graphs shown in the Using Technology box at the top of the next page.

U sing technology

The graphs of

$$y_1 = \sqrt{2x + 4}$$
$$(\text{or } (f \circ g)(x) = \sqrt{2x + 4})$$

and

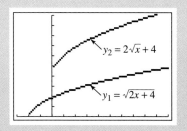

$$y_2 = 2\sqrt{x} + 4$$
$$(\text{or } (g \circ f)(x) = 2\sqrt{x} + 4)$$

were obtained with a graphing utility and are shown above. Describe how these graphs differ in terms of

a. their shapes.
b. their domains.
c. their ranges.

---

2 Interpret the composition of functions in an applied setting.

Modeling Using the Composition of Functions

In our next example, we are interested in describing the meaning of two functions and their composition in an applied setting.

EXAMPLE 3 **Interpreting the Composition of Functions**

The regular price of a computer system is x dollars.

a. If $f(x) = x - 500$, describe what this function models. Use the word *rebate* in your description.
b. If $g(x) = 0.85x$, describe what this function means in terms of the regular price of the computer system.
c. Find $(f \circ g)(x)$ and describe what this models in terms of the regular price of the computer system.
d. Repeat part (c) for $(g \circ f)(x)$.

Solution

a. Since $f(x) = x - 500$, we see that $500 is subtracted from the system's regular price. Thus, the function models the price of the system with a $500 reduction or rebate.
b. Since $g(x) = 0.85x$, this formula involves 85% of the system's regular price. Thus, the function models the price of the system at 85% of its usual cost (or the price subject to a 15% reduction).
c. Let's begin by finding $(f \circ g)(x)$, the composition of f with g, to see what it models.

$$
\begin{aligned}
(f \circ g)(x) &= f(g(x)) && \text{This is the definition of } f \circ g. \\
&= g(x) - 500 && \text{Put } g(x) \text{ into } f. \\
&= 0.85x - 500 && \text{Since } g(x) = 0.85x, \text{ replace } g(x) \text{ with } 0.85x.
\end{aligned}
$$

Since $(f \circ g)(x) = 0.85x - 500$, the function models the price of the system subject to a 15% reduction followed by a $500 rebate.

Discover for yourself

Describe the computer system's discounted price expressed in the form $0.85(x - 500)$. Use the phrases "15% reduction" and "$500 rebate" in your description.

d. By finding $(g \circ f)(x)$, the composition of g with f, we'll see what this describes in terms of the computer system's regular price.

$$
\begin{aligned}
(g \circ f)(x) &= g(f(x)) && \text{This is the definition of } g \circ f. \\
&= 0.85(f(x)) && \text{Put } f(x) \text{ into } g. \\
&= 0.85(x - 500) && \text{Since } f(x) = x - 500, \text{ replace } f(x) \text{ with } x - 500. \\
&= 0.85x - 425 && \text{Apply the distributive property.}
\end{aligned}
$$

Since $(g \circ f)(x) = 0.85x - 425$, the function models the price of the system subject to a 15% reduction followed by a $425 rebate.

With $(f \circ g)(x) = 0.85x - 500$ and $(g \circ f)(x) = 0.85x - 425$, the system's discounted price modeled by $f \circ g$ is the better deal. ∎

Inverse and One-to-One Functions

Although it is generally not the case, there are functions f and g such that the composition of f with g *is the same as* the composition of g with f.

Discover for yourself

Let $f(x) = 4x^3 - 1$ and $g(x) = \sqrt[3]{\dfrac{x + 1}{4}}$.

a. Find $(f \circ g)(x)$ and $(g \circ f)(x)$. What do you observe?

b. Find

$$f(1) \text{ and } g(3)$$

$$f(3) \text{ and } g(107)$$

$$f(-1) \text{ and } g(-5).$$

What does the function g do to the order of the coordinates of the function f?

In the Discover for Yourself, did you find that

$$f(g(x)) = x \quad \text{and} \quad g(f(x)) = x$$

so that both compositions are the same? Furthermore, the function g reverses the order of the coordinates of the function f.

$$f(1) = 3 \text{ and } g(3) = 1; f(3) = 107 \text{ and } g(107) = 3;$$
$$f(-1) = -5 \text{ and } g(-5) = -1.$$

Under these conditions, we say that f and g are *inverses* of each other.

To better understand the idea of inverses, let's concentrate on two applied situations in which we reverse the order of a function's coordinates. We begin with the graph in Figure 8.13, which shows that the rate at which one's heart beats (in beats per minute) is a function of the number of minutes that elapse after exercising. Let's use the letter f as the name for this function. The inverse of the function, denoted f^{-1}, is a function that reverses the coordinates of the function f. Since $f(4) = 120$, then $f^{-1}(120) = 4$, shown in the figure. If after exercising one's heart is beating at a rate of 120 beats per minute, approximately 4 minutes have elapsed since exercising.

In some cases, the resulting relation after reversing coordinates is not a function. Consider Figure 8.14, which shows U.S. gas consumption. The graph

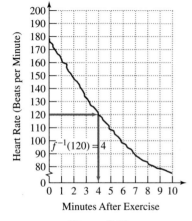

Minutes After Exercise

Figure 8.13

U. S. Gas Consumption
(billions of gallons per year)

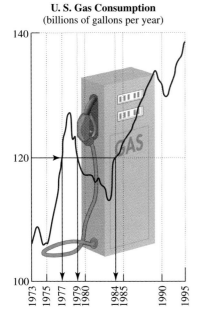

Figure 8.14
Source: U. S. Department of Energy

3 Use the horizontal line test to determine if a function has an inverse.

indicates that in 1977, 1979, and 1984 gas consumption was approximately 120 billion gallons, represented by the ordered pairs (1977, 120), (1979, 120), and (1984, 120).

Reversing these coordinates, we obtain (120, 1977), (120, 1979), and (120, 1984). A function provides exactly one output for each input, but this is not happening here. The input 120 is associated with three outputs. Thus, the U.S. gas consumption function does not have an inverse that is a function.

We can distinguish between functions whose inverses are functions and functions whose inverses are not functions with the following definition.

> **Definition of a one-to-one function**
>
> A function has an inverse that is a function if no two different numbers in the domain correspond to the same number in the range. Such a function is called *one-to-one*. For a one-to-one function, no two different ordered pairs have the same second component.

How can you tell from a function's graph if it has an inverse that is a function? Compare the horizontal lines in Figures 8.13 and 8.14. In Figure 8.13, the horizontal line intersects the graph only once. In Figure 8.14, the horizontal line intersects the graph in three points. This suggests the following test.

> **The horizontal line test for inverse functions**
>
> A function is one-to-one and has an inverse that is a function if there is no horizontal line that intersects the graph of the function at more than one point.

In Figure 8.15, all the graphs are functions since they all pass the vertical line test. However, the graphs in parts (b), (c), and (e) are not one-to-one and do not have inverses that are functions because horizontal lines can be drawn that intersect the graphs more than once. They do not pass the horizontal line test.

Now that we know which functions can have inverses, we are ready for a precise definition of the inverse of a function. The definition given on the next page shows that functions for which $(f \circ g)(x) = (g \circ f)(x)$ are inverse functions.

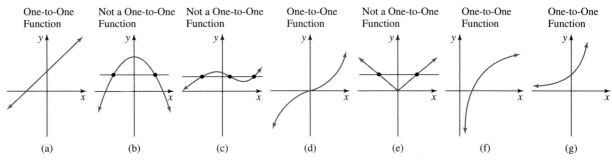

Figure 8.15
Using the horizontal line test

tudy tip

The notation f^{-1} represents the inverse function of f. The -1 is *not* an exponent. The notation f^{-1} does *not* mean $\dfrac{1}{f}$.

Definition of the inverse of a function

Let f and g be two functions such that

$$f(g(x)) = x \qquad \text{for every } x \text{ in the domain of } g$$

and

$$g(f(x)) = x \qquad \text{for every } x \text{ in the domain of } f.$$

The function g is the *inverse* of the function f, and is denoted by f^{-1} (read "f-inverse"). Thus, $f(f^{-1}(x)) = x$ and $f^{-1}(f(x)) = x$. The domain of f is equal to the range of f^{-1}, and vice versa.

4 Verify inverse functions.

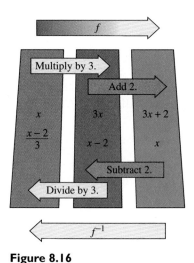

Figure 8.16

f^{-1} undoes the changes produced by f.

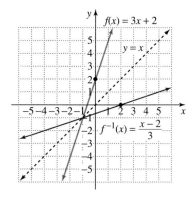

Figure 8.17

The graphs of f and f^{-1} are mirror images about $y = x$.

EXAMPLE 4   Verifying Inverse Functions

Show that each function is an inverse of the other:

$$f(x) = 3x + 2 \qquad \text{and} \qquad g(x) = \frac{x-2}{3}$$

Solution

To show that f and g are inverses of each other, we must show that $f(g(x)) = x$ and $g(f(x)) = x$.

$$
\begin{aligned}
f(g(x)) &= 3(g(x)) + 2 \\
\text{Put } g(x) \text{ into } f. \quad &= 3\left(\frac{x-2}{3}\right) + 2 \\
&= x - 2 + 2 \\
&= x
\end{aligned}
\qquad
\begin{aligned}
g(f(x)) &= \frac{f(x) - 2}{3} \\
\text{Put } f(x) \text{ into } g. \quad &= \frac{3x + 2 - 2}{3} \\
&= \frac{3x}{3} \\
&= x
\end{aligned}
$$

Because g is the inverse of f (and vice versa), we can use the notation

$$f(x) = 3x + 2 \qquad \text{and} \qquad f^{-1}(x) = \frac{x-2}{3}.$$

Notice how f^{-1} undoes the changes produced by f: f changes x by *multiplying* by 3 and *adding* 2, and f^{-1} undoes this by *subtracting* 2 and *dividing* by 3. This "undoing" process is illustrated in Figure 8.16. ■

Graphing f and f^{-1}

Figure 8.17 shows the graphs of $f(x) = 3x + 2$ and $f^{-1}(x) = \dfrac{x-2}{3}$. The graphs are mirror images of each other with respect to the line whose equation is $y = x$.

In general, if f is a one-to-one function and $f(a) = b$, then $f^{-1}(b) = a$ because the inverse function reverses the coordinates of the function. Thus, if the point (a, b) belongs to the graph of f, then the point (b, a) belongs to the graph of f^{-1}. Since the points (a, b) and (b, a) are symmetric with respect to the line $y = x$, the graph of f^{-1} is a reflection of the graph of f with respect to the line $y = x$. This is illustrated in Figure 8.18.

5 Find the inverse of a one-to-one function.

Finding Formulas for Inverses

If a function is one-to-one, its inverse function may be obtained by reversing the first and second coordinates in each ordered pair of the function. Because the inverse function interchanges the roles of x and y, we can obtain the equation for f^{-1}, the inverse of f, by interchanging the role of x and y in the equation for f.

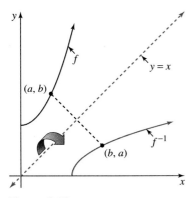

Figure 8.18

The graph of f^{-1} is a reflection of f about $y = x$.

Finding the inverse of a one-to-one function

To find the equation for the inverse of f, a one-to-one function:

    a. Replace $f(x)$ with y in the equation for $f(x)$.
    b. Interchange x and y.
    c. Solve for y.
    d. Replace y with $f^{-1}(x)$.

This process can be checked by showing that $f(f^{-1}(x)) = x$ and $f^{-1}(f(x)) = x$.

EXAMPLE 5   **Finding the Inverse of a Function**

Find the inverse of: $f(x) = 7x - 5$

Solution

Step 1. Replace $f(x)$ with y: $y = 7x - 5$

Step 2. Interchange x and y: $x = 7y - 5$ This is the inverse function.

Step 3. Solve for y: $x + 5 = 7y$ Add 5 to both sides.

$$\frac{x + 5}{7} = y$$ Divide both sides by 7.

Step 4. Replace y with $f^{-1}(x)$: $f^{-1}(x) = \dfrac{x + 5}{7}$ The equation is written with $f^{-1}(x)$ on the left.

Thus, $f(x) = 7x - 5$ and $f^{-1}(x) = \dfrac{x + 5}{7}$. Once again, f^{-1} undoes the changes produced by f. The one-to-one function f changes x by multiplying by 7 and subtracting 5. The inverse function f^{-1} undoes this by adding 5 and dividing by 7. ∎

Study tip

The procedure for finding a function's inverse uses a *switch-and-solve* strategy. Switch x and y, then solve for y.

EXAMPLE 6   **Finding the Equation of the Inverse**

a. Find the inverse of $f(x) = x^3 + 1$.
b. Verify that $f(f^{-1}(x)) = x$ and $f^{-1}(f(x)) = x$.

Solution

a. ***Step 1.*** Replace $f(x)$ with y: $y = x^3 + 1$
   Step 2. Interchange x and y: $x = y^3 + 1$
   Step 3. Solve for y: $x - 1 = y^3$
   $$\sqrt[3]{x - 1} = \sqrt[3]{y^3}$$
   $$\sqrt[3]{x - 1} = y$$

   Step 4. Replace y with $f^{-1}(x)$: $f^{-1}(x) = \sqrt[3]{x - 1}$.
   Thus, the inverse of $f(x) = x^3 + 1$ is $f^{-1}(x) = \sqrt[3]{x - 1}$.

b. To verify algebraically that $f(x) = x^3 + 1$ and $f^{-1}(x) = \sqrt[3]{x - 1}$ are inverses of each other, we must show that $f(f^{-1}(x)) = x$ and $f^{-1}(f(x)) = x$.

$$f(f^{-1}(x)) = (f^{-1}(x))^3 + 1$$
$$\underbrace{}_{\text{Put } f^{-1}(x) \text{ into } f.} = (\sqrt[3]{x - 1})^3 + 1$$
$$= x - 1 + 1$$
$$= x$$

$$f^{-1}(f(x)) = \sqrt[3]{f(x) - 1}$$
$$\underbrace{}_{\text{Put } f(x) \text{ into } f^{-1}(x).} = \sqrt[3]{x^3 + 1 - 1}$$
$$= \sqrt[3]{x^3}$$
$$= x$$ ∎

Using technology

The graphs of

$$y_1 = x^3 + 1$$
$$y_2 = \sqrt[3]{x - 1}$$

and $y_3 = x$

shown above were obtained with a graphing utility. We used the ZOOM SQUARE feature to equalize the distance between the marks on the axes. The fact that y_1 and y_2 are reflections of each other about the line $y = x$ visually confirms that they are inverses.

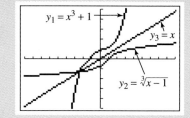

PROBLEM SET 8.2

Practice Problems _____

In Problems 1–16, let $f(x) = 2x - 5, g(x) = x^2 + 4x$, and $h(x) = \dfrac{x + 5}{2}$. Find each composition.

1. $(f \circ g)(2)$
2. $(f \circ g)(3)$
3. $(g \circ f)(2)$
4. $(g \circ f)(3)$
5. $(g \circ h)(-1)$
6. $(g \circ h)(1)$
7. $(h \circ g)(-1)$
8. $(h \circ g)(1)$
9. $(f \circ h)(11)$
10. $(f \circ h)(13)$
11. $(h \circ f)(11)$
12. $(h \circ f)(13)$
13. $(f \circ f)(6)$
14. $(f \circ f)(5)$
15. $(g \circ g)(-2)$
16. $(g \circ g)(-3)$

In Problems 17–28, find $(f \circ g)(x)$ and $(g \circ f)(x)$.

17. $f(x) = x^2 + 4, g(x) = 2x + 1$
18. $f(x) = x^2 - 2, g(x) = 3x + 1$
19. $f(x) = 5x + 2, g(x) = 3x^2 - 4$
20. $f(x) = 4x - 3, g(x) = 5x^2 - 2$
21. $f(x) = x^2 + 2, g(x) = x^2 - 2$
22. $f(x) = x^2 + 1, g(x) = x^2 - 3$

23. $f(x) = \sqrt{x}, g(x) = x - 1$

24. $f(x) = \sqrt{x}, g(x) = x + 2$

25. $f(x) = 2x - 3, g(x) = \dfrac{x + 3}{2}$

26. $f(x) = 6x - 3, g(x) = \dfrac{x + 3}{6}$

27. $f(x) = \dfrac{1}{x}, g(x) = \dfrac{1}{x}$

28. $f(x) = \dfrac{1}{x}, g(x) = \dfrac{2}{x}$

In Problems 29–34, use the graph to determine whether each function is a one-to-one function.

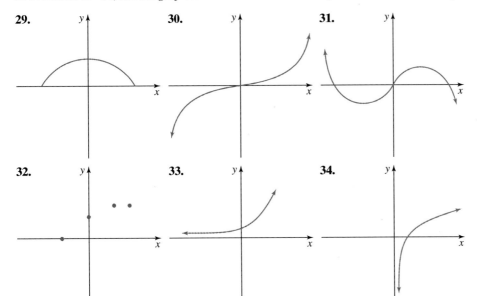

29.

30.

31.

32.

33.

34.

In Problems 35–44, determine whether each pair of functions f and g are inverses of each other.

35. $f(x) = 4x$ and $g(x) = \dfrac{1}{4}x$

36. $f(x) = 6x$ and $g(x) = \dfrac{x}{6}$

37. $f(x) = 3x + 8$ and $g(x) = \dfrac{x - 8}{3}$

38. $f(x) = 4x + 9$ and $g(x) = \dfrac{x - 9}{4}$

39. $f(x) = 5x - 9$ and $g(x) = \dfrac{x + 5}{9}$

40. $f(x) = 3x - 7$ and $g(x) = \dfrac{x + 3}{7}$

41. $f(x) = \dfrac{3}{x - 4}$ and $g(x) = \dfrac{3}{x} + 4$

42. $f(x) = \dfrac{2}{x - 5}$ and $g(x) = \dfrac{2}{x} + 5$

43. $f(x) = -x$ and $g(x) = -x$

44. $f(x) = \sqrt[3]{x - 4}$ and $g(x) = x^3 + 4$

The functions in Problems 45–60 are all one-to-one. For each problem,

> ***a.*** *Find a formula for $f^{-1}(x)$, the inverse function.*
> ***b.*** *Verify that your formula is correct by showing that $f(f^{-1}(x)) = x$ and $f^{-1}(f(x)) = x$*
> ***c.*** *Graph the function and its inverse using the same set of axes.*

45. $f(x) = x + 3$

46. $f(x) = x + 5$

47. $f(x) = 2x$

48. $f(x) = 4x$

49. $f(x) = 2x + 3$

50. $f(x) = 3x - 1$

51. $f(x) = x^3 + 2$

52. $f(x) = x^3 - 1$

53. $f(x) = (x + 2)^3$

54. $f(x) = (x - 1)^3$

55. $f(x) = \dfrac{1}{x}$

56. $f(x) = \dfrac{2}{x}$

57. $f(x) = \sqrt{x}$

58. $f(x) = \sqrt[3]{x}$

59. $f(x) = x^2 + 1$, for $x \geq 0$

60. $f(x) = x^2 - 1$, for $x \geq 0$

Use the graphs of functions f and g, shown below, to answer Problems 61–62.

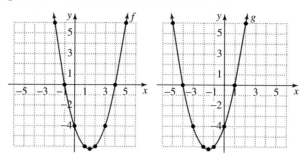

61. Find $(f \circ g)(1)$.

62. Find $(g \circ f)(-1)$.

Problems 63–66 show the graphs of one-to-one functions. Three points are clearly shown on each graph. Reverse the coordinates of these points to find points on the graph of the inverse function. Then use these three points and the fact that the graph of the inverse function is a reflection of the graph of the given function about the line y = x to graph the inverse function on the same set of axes as the function.

63. 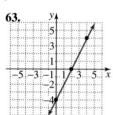   **64.** 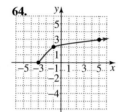   **65.** 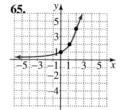   **66.**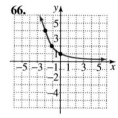

Application Problems

67. The regular price of a pair of jeans is x dollars. Let $f(x) = x - 5$ and $g(x) = x - 0.4x$.

  a. Describe what functions f and g model in terms of the price of the jeans.

  b. Find $(f \circ g)(x)$ and describe what this models in terms of the price of the jeans.

  c. Repeat part (b) for $(g \circ f)(x)$.

  d. Which composite function models the greater discount on the jeans, $f \circ g$ or $g \circ f$? Explain.

68. Skin temperature (T, in Celsius) is a function of the Celsius temperature (C) of the environment, described by $f(C) = 0.27C + 27.4$. Celsius temperature C is a function of Fahrenheit temperature F given by the function $g(F) = \frac{5}{9}(F - 32)$.

  a. If skin temperature is a function of environmental Celsius temperature and Celsius environmental temperature is a function of Fahrenheit environmental temperature, what can be said about skin temperature and Fahrenheit environmental temperature?

  b. What is the temperature of the skin when the environmental temperature is 68° Fahrenheit? (Find $f(g(68))$.)

  c. Find and interpret $f(g(41))$.

  d. Find $f(g(F))$.

69. The line graph shown is based on data from the World Health Organization.

  a. Explain why f has an inverse that is a function.

  b. Describe in practical terms the meaning of $f^{-1}(20)$.

  c. At this point, can you tell if the data in the graph is best modeled by a linear, quadratic, or exponential function? What might help you to decide?

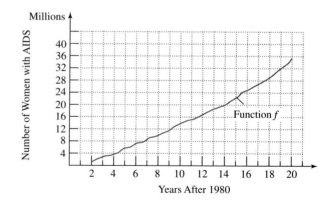

70. The line graph shows the property crime rate per 1000 people in the United States from 1975 through 1992.
   a. Does the graph have an inverse that is a function? What does this mean in terms of time and the property crime rate?
   b. What years had a property crime rate of 230 per 1000 people? Could we use the notation $f^{-1}(230)$ in this situation? Explain.

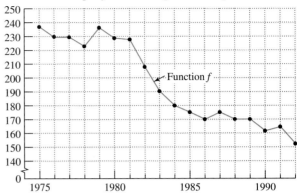

Property Crime Rate per 1000 People

Source: Bureau of Justice Statistics, U.S. Department of Justice

True–False Critical Thinking Problems

71. Which one of the following is true?
   a. The inverse of $\{(1, 4), (2, 7)\}$ is $\{(2, 7), (1, 4)\}$.
   b. The function $f(x) = 5$ is one-to-one.
   c. If $f(x) = 3x$, then $f^{-1}(x) = \dfrac{1}{3x}$.
   d. The domain of f is the same as the range of f^{-1}.

72. Which one of the following is true?
   a. If $f(x) = 5x - 4$, then $f^{-1}(x) = \dfrac{x}{5} + 4$.
   b. Any graph that passes the horizontal line test must be the graph of a function.
   c. If $f(x) = -x$, then $f^{-1}(x) = -x$.
   d. All linear functions are one-to-one.

Technology Problems

73. Use a graphing utility to verify the graphs that you drew by hand in Problems 45–60.

In Problems 74–77, use a graphing utility to graph f and g in the same viewing rectangle. In addition, graph the line y = x and visually determine if f and g are inverses.

74. $f(x) = 4x + 4, g(x) = 0.25x - 1$

75. $f(x) = \dfrac{1}{x} + 2, g(x) = \dfrac{1}{x - 2}$

76. $f(x) = \sqrt[3]{x} - 2, g(x) = (x + 2)^3$

77. $f(x) = e^x, g(x) = \ln x$ (We'll be discussing the meaning of this function in the next section. For now, use the $\boxed{\ln}$ key on your graphing utility.)

In Problems 78–85, use a graphing utility to graph each function and determine whether the function is one-to-one.

78. $f(x) = x^2 - 1$

79. $f(x) = \sqrt[3]{2 - x}$

80. $f(x) = \dfrac{x^3}{2}$

81. $f(x) = \dfrac{x^4}{4}$

82. $f(x) = \dfrac{1}{x}$

83. $f(x) = \dfrac{1}{x^2}$

84. $f(x) = \dfrac{1}{x - 1}$

85. $f(x) = \dfrac{1}{(x - 1)^2}$

Writing in Mathematics

86. Describe a procedure for finding $(f \circ g)(x)$.

87. The function $y = 15 + 0.5x$ describes the hourly wage (y) of a math-lab tutor earning a flat fee of $15 per hour plus $0.50 for each student the tutor assists during that hour. Find the inverse of this function. Write a sentence or two telling what this inverse describes, discussing what each variable in the inverse function represents.

Critical Thinking Problems

For Problems 88–91, find functions f and g so that h(x) = (f ∘ g)(x). Answers may vary.

88. $h(x) = (2x + 5)^3$

89. $h(x) = \sqrt{3x^2 + 5}$

90. $h(x) = 5(6x - 7)^2 + 11$

91. $h(x) = \dfrac{2}{x} + 8$

92. Suppose $f(x) = x^2$ and $(f \circ g)(x) = x^2 + 2x + 1$. Find $g(x)$.

93. If $f(x) = 3x - 2$ and $g(x) = x^5$, find $f(g(x) - f(x))$ and $f(g(x)) - f(f(x))$.

94. Consider the two linear functions defined by $f(x) = m_1x + b_1$ and $g(x) = m_2x + b_2$. Prove that the slope of the composite function of f with g is equal to the product of the slopes of the two linear functions.

95. If $f(x) = 2x - 5$ and $g(x) = 3x + b$, find b such that $f(g(x)) = g(f(x))$.

96. If $f(x) = 3x$ and $g(x) = x + 5$, find $(f \circ g)^{-1}(x)$ and $(g^{-1} \circ f^{-1})(x)$.

97. If $f(x) = 6 - 5x$, find $(f^{-1})^{-1}(x)$.

98. If $f(x) = 9x + 1$ and $g(x) = x^2$, find all values of x satisfying $f(g(x)) = g(f(x))$.

99. Find $f^{-1}(x)$ if

$$f(x) = 3 - \frac{2\sqrt[3]{3x + 2}}{3}.$$

100. If $f(2) = 6$, find x satisfying $7 + f^{-1}(x - 1) = 9$.

101. If $f(5) = 13$, find x satisfying $1 + f^{-1}(2x + 3) = 6$.

102. Show that

$$f(x) = \frac{3x - 2}{5x - 3}$$

is its own inverse.

103. Show that $g(f(x)) = x$ if

$$f(x) = \frac{x + 1}{x - 2} \qquad \text{and} \qquad g(x) = \frac{2x + 1}{x - 1}.$$

104. The one-to-one property for a function can be stated as follows: If $f(a) = f(b)$, then $a = b$. Use this condition to show that

$$f(x) = \frac{4}{3x + 5}$$

is one-to-one.

Review Problems

105. Use x- and y-intercepts and the vertex to graph the parabola whose equation is given by $y = x^2 - 4x + 3$.

106. Factor: $a^4 - 16b^4$.

107. Divide: $\dfrac{4.3 \times 10^5}{8.6 \times 10^{-4}}$.

Write the quotient in scientific notation.

SECTION 8.3

Solutions   Tutorial   Video
Manual   9

❘

Logarithmic Functions

Objectives

1 Graph the inverse of an exponential function.
2 Write exponential statements in logarithmic form and vice versa.
3 Evaluate logarithms.
4 Evaluate common and natural logarithms.
5 Use logarithmic functions in applications.

In this section, our focus is on the inverse of the exponential function, the *logarithmic function*. The logarithmic function serves as a model for understanding phenomena as seemingly unrelated as musical scales and the Richter scale for measuring earthquakes.

❘ Graph the inverse of an exponential function.

Graphs of Inverse Exponential Functions

The graph of $f(x) = 2^x$ is shown in blue in Figure 8.19. Because no horizontal line can be drawn that intersects this blue curve in more than one point, this

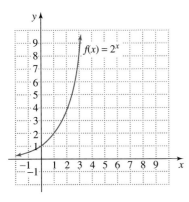

Figure 8.19

The graph of $f(x) = 2^x$

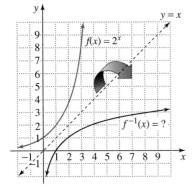

Figure 8.20

The graph of $f(x) = 2^x$ and its inverse

2 Write exponential statements in logarithmic form and vice versa.

exponential function is one-to-one. Thus, $f(x) = 2^x$ has an inverse that is also a function. We can use the procedure of Section 8.2 to find a formula for f^{-1}.

$$f(x) = 2^x$$

Step 1. Replace $f(x)$ by y: $y = 2^x$
Step 2. Interchange x and y: $x = 2^y$
Step 3. Solve for y: ?

The question mark indicates that we need some sort of notation to solve $x = 2^y$ for y. In words, we can describe y as

$y = $ the power on 2 that gives x

Even without this notation, we can graph the inverse function. First, we set up a table of coordinates for $f(x) = 2^x$. Then we reverse these coordinates to obtain the coordinates for the inverse function.

$f(x) = 2^x$ or $y = 2^x$						
x	-2	-1	0	1	2	3
y	$\frac{1}{4}$	$\frac{1}{2}$	1	2	4	8

Reverse coordinates.

Coordinates for the Inverse Function						
x	$\frac{1}{4}$	$\frac{1}{2}$	1	2	4	8
y	-2	-1	0	1	2	3

To graph $f(x) = 2^x$ and its inverse function, we can plot the ordered pairs from both tables, connecting them with smooth curves, as shown in Figure 8.20. The graph of the inverse can also be drawn by reflecting the graph of $f(x) = 2^x$ across the line $y = x$.

Logarithmic Notation

The graphs of exponential functions $f(x) = b^x$ all pass through $(0, 1)$ and have the general shape shown in Figure 8.21 on the next page. These functions have inverse functions, for which we will now introduce a notation. To find the inverse of

$$f(x) = b^x:$$

Step 1. Replace $f(x)$ by y: $y = b^x$
Step 2. Interchange x and y: $x = b^y$
Step 3. Solve for y: We do this with the following definition.

> **Definition of the logarithmic function**
>
> For all positive numbers b, where $b \neq 1$, $y = \log_b x$ is equivalent to $x = b^y$.

We use the abbreviation *log* for *logarithm*. The function $y = \log_b x$ (equivalently, $x = b^y$) is called the *logarithmic function with base b*. We read $y = \log_b x$ as "y is the logarithm of x with base b."

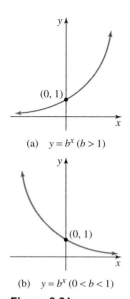

(a) $y = b^x$ $(b > 1)$

(b) $y = b^x$ $(0 < b < 1)$

Figure 8.21

Shapes for the graphs of $f(x) = b^x$

S **tudy tip**

A logarithm is an exponent.

When an expression is in the form $x = b^y$, we say that it is in *exponential form*. The equivalent form $y = \log_b x$ is called *logarithmic form*. You should memorize the location of the base and exponent in each form.

Location of base and exponent in exponential and logarithmic forms

Logarithmic form: $y = \log_b x$

Exponential form: $x = b^y$

Exponent ↓

Base ↑

Table 8.3 illustrates the two forms.

TABLE 8.3 Equivalent Exponential and Logarithmic Forms

Exponential Form	Logarithmic Form
$b^{\boxed{y}} = x$	$\log_b x = \boxed{y}$
$4^{\boxed{2}} = 16$	$\log_4 16 = \boxed{2}$
	("The logarithm of 16 with base 4 is 2." "Log, base 4, of 16 is 2.")
$5^{\boxed{-3}} = \frac{1}{125}$	$\log_5(\frac{1}{125}) = \boxed{-3}$
$(\frac{1}{3})^{\boxed{4}} = \frac{1}{81}$	$\log_{1/3}(\frac{1}{81}) = \boxed{4}$
$10^{\boxed{-4}} = 0.0001$	$\log_{10} 0.0001 = \boxed{-4}$
$e^{\boxed{2}} = 7.3891$	$\log_e 7.3891 = \boxed{2}$

Although Table 8.3 moves from exponential to logarithmic form, you may often find yourself going in the opposite direction. This direction takes you from the unfamiliar logarithmic form back to the more familiar exponential form. It also enables you to find the value of certain logarithms.

3 Evaluate logarithms.

EXAMPLE I **Evaluating Logarithms**

Evaluate: **a.** $\log_3 9$ **b.** $\log_2 32$ **c.** $\log_{25} 5$

Solution

Keep in mind that

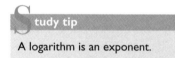

$$\log_b x = y \qquad \text{means} \qquad b^y = x.$$

In each case, we can set the given logarithm equal to y (or any other letter), convert to exponential form, and solve for y. We can also answer the question, "To what power must the base be raised to get the number that follows the log?" We'll use both methods.

a. $\log_3 9$
Method 1. Let $\log_3 9 = y$. Then

$3^y = 9$ Convert to exponential form.
$y = 2$ Since 3 squared is 9, $y = 2$.

Therefore, $\log_3 9 = 2$.
Method 2. $\log_3 9$ is the exponent to which we raise 3 to get 9. To what power must we raise base 3 to obtain 9? The answer is 2. Once again, $\log_3 9 = 2$.

b. $\log_2 32$
Method 1. Let $\log_2 32 = y$. Then

$2^y = 32$ Convert to exponential form.
$y = 5$ Use inspection to determine y.

Therefore, $\log_2 32 = 5$.
Method 2. $\log_2 32$ is the exponent to which we must raise 2 to get 32. The exponent is 5. As with method 1, $\log_2 32 = 5$.

c. $\log_{25} 5$
Method 1. Let $\log_{25} 5 = y$. Then

$25^y = 5$ Convert to exponential form.
$y = \dfrac{1}{2}$ Since the principal square root of 25 is 5 ($\sqrt{25} = 5$ or $25^{1/2} = 5$) $y = \frac{1}{2}$.

Therefore, $\log_{25} 5 = \frac{1}{2}$.
Method 2. $\log_{25} 5$ is the exponent to which we must raise 25 to get 5. The exponent is the rational exponent $\frac{1}{2}$, or the square root. As with method 1, $\log_{25} 5 = \frac{1}{2}$. ∎

Each logarithmic evaluation in Example 2 involves an important characteristic of logarithms.

EXAMPLE 2   **Evaluating Logarithms**

Evaluate, if possible: **a.** $\log_3 1$ **b.** $\log_7 7$ **c.** $\log_5(-1)$ **d.** $\log_6 0$

Solution

a. The power to which 3 must be raised to get 1 is 0.

$3^0 = 1$,   so $\log_3 1 = 0$.

b. The power to which 7 must be raised to get 7 is 1.

$7^1 = 7$,   so $\log_7 7 = 1$.

c. To evaluate $\log_5(-1)$, we must find the power to which 5 must be raised to get -1. Since 5 to any power is always positive, no such power exists. Thus, $\log_5(-1)$ is undefined.

d. To evaluate $\log_6 0$, we must find the exponent on 6 that gives 0. There is no such power. (Recall that $6^0 = 1$.) Thus, $\log_6 0$ is undefined. ∎

We can generalize from Example 2 and obtain the following logarithmic properties.

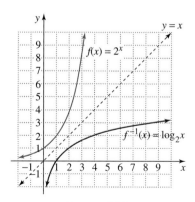

Figure 8.22

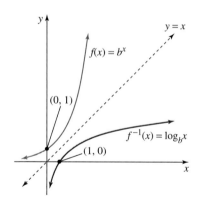

Figure 8.23

Shapes for the graphs of exponential and logarithmic functions, $b > 1$

Properties of logarithms

Let b be a positive base that is not equal to 1.

1. $\log_b 1 = 0$ because $b^0 = 1$.
2. $\log_b b = 1$ because $b^1 = b$.
3. $\log_b b^x = x$ because $b^x = b^x$.
4. $\log_b x$ is undefined if $x = 0$ or $x < 0$ because there is no power that b can be raised to that gives 0 or a negative number.

Logarithmic Functions

Earlier we graphed the inverse function for $f(x) = 2^x$. Now that we have introduced logarithmic notation, we can write the equation for this inverse function as $f^{-1}(x) = \log_2 x$. This is shown in Figure 8.22.

If we generalize the results of this situation from base 2 to base b (with $b > 1$), we see that the inverse of $f(x) = b^x$ is $f^{-1}(x) = \log_b x$, as shown in Figure 8.23. These graphs illustrate the following important points about $y = b^x$ and $y = \log_b x$.

Characteristics of exponential and logarithmic functions

1. The domain of $y = b^x$ is the set of all real numbers, and the range is the set of all real numbers > 0. Domain $= (-\infty, \infty)$ and range $= (0, \infty)$.
2. The domain of $y = \log_b x$ is the set of all real numbers > 0, and the range of $y = \log_b x$ is the set of all real numbers. Domain $= (0, \infty)$ and range $= (-\infty, \infty)$.
3. The graph of $y = b^x$ passes through $(0, 1)$, and the graph of $y = \log_b x$ passes through $(1, 0)$.
4. If $b > 1$, the graph of $y = b^x$ approaches but never touches the negative x-axis. The graph of $y = \log_b x$ approaches but never touches the negative y-axis.
5. The x-axis is an asymptote for the exponential function, and the y-axis is an asymptote for the logarithmic function.

EXAMPLE 3   **Graphing a Logarithmic Function**

Graph: $y = \log_4 x$

Solution

First we rewrite the equation in exponential form.

$$\log_b x = y \qquad \text{means} \qquad b^y = x$$
$$\log_4 x = y \qquad \text{means} \qquad 4^y = x$$

We can find ordered pairs satisfying this exponential equation by selecting values for y and computing the x-values.

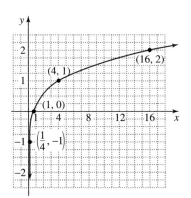

Figure 8.24

The graph of $y = \log_4 x$

4 Evaluate common and natural logarithms.

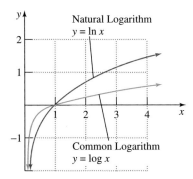

Figure 8.25

Graphs of $y = \log x$ and $y = \ln x$

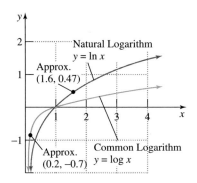

Figure 8.26

y	$x = 4^y$	(x, y)
-2	$x = 4^{-2} = \frac{1}{16}$	$\left(\frac{1}{16}, -2\right)$
-1	$x = 4^{-1} = \frac{1}{4}$	$\left(\frac{1}{4}, -1\right)$
0	$x = 4^0 = 1$	$(1, 0)$
1	$x = 4^1 = 4$	$(4, 1)$
2	$x = 4^2 = 16$	$(16, 2)$

Now we plot the ordered pairs from the third column and connect them with a smooth curve. The graph of $y = \log_4 x$ is shown in Figure 8.24. ■

Common Logarithms and Natural Logarithms

The bases that most frequently appear on logarithmic functions are e and 10. If $f(x) = 10^x$, then $f^{-1}(x) = \log_{10} x$. Similarly, if $f(x) = e^x$, then $f^{-1}(x) = \log_e x$.

> **Notations for common and natural logarithms**
>
> We write $y = \log_{10} x$ as $y = \log x$. Logarithms to the base 10 are called *common logarithms*. We write $y = \log_e x$ as $y = \ln x$. Logarithms to the base e are called *natural logarithms*.

Figure 8.25 shows the graphs of $y = \log x$ and $y = \ln x$.

You will need to use a calculator to evaluate most common and natural logarithms. Using the common logarithm key $\boxed{\log}$ and the natural logarithm key $\boxed{\ln}$, here are some examples:

Logarithm	Keystrokes	Rounded Display
$\log 1.75$	$\boxed{\log}$ 1.75 $\boxed{\text{ENTER}}$	0.24304
$\log 0.192$	$\boxed{\log}$ 0.192 $\boxed{\text{ENTER}}$	-0.7167
$\ln 1.6$	$\boxed{\ln}$ 1.6 $\boxed{\text{ENTER}}$	0.47000
$\ln(-3)$	$\boxed{\ln}$ $\boxed{(-)}$ 3 $\boxed{\text{ENTER}}$	ERROR

Although we write $\log 1.75 = 0.243\,04$, it turns out that $0.243\,04$ is an *approximate* value when the domain of the common logarithmic function is 1.75. We can reinforce this fact by writing

$$\log 1.75 \approx 0.243\,04$$

reading $\approx$ as "is approximately equal to." Similarly, although our calculator tells us that $\ln 1.75 = 0.5596$, a more accurate statement is $\ln 1.75 \approx 0.5596$.

We can locate points on the graphs of $y = \log x$ and $y = \ln x$ corresponding to the values of $\log 0.192$ and $\ln 1.6$ that we found with a calculator.

$\log 0.192 = -0.7167$ and the point $(0.2, -0.7)$ can be approximately located on the graph of $y = \log x$. (See Figure 8.26.)

$\ln 1.6 = 0.47$ and the point $(1.6, 0.47)$ can be approximately located on the graph of $y = \ln x$. (See Figure 8.26.)

Notice that $\ln(-3)$ results in an error message in a calculator's display. Remember that the domain of $y = \log_b x$ is the set of positive real numbers, which means that $\ln(-3)$ is undefined. There is no graph for $y = \ln x$ at $x = -3$. Turn back to Figure 8.26 on the previous page and verify this observation. When the domain of the natural log function is -3, there is no corresponding range value.

The basic properties of logarithms that were listed earlier in the section can be applied to common and natural logarithms.

Properties of common and natural logarithms		
General Properties	**Common Logarithms**	**Natural Logarithms**
1. $\log_b 1 = 0$	**1.** $\log 1 = 0$	**1.** $\ln 1 = 0$
2. $\log_b b = 1$	**2.** $\log 10 = 1$	**2.** $\ln e = 1$
3. $\log_b b^x = x$	**3.** $\log 10^x = x$	**3.** $\ln e^x = x$

We can use the property $\ln e^x = x$ to identify points on the graph of the natural logarithmic function.

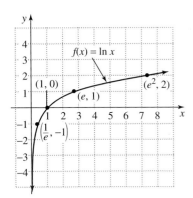

Figure 8.27

The graph of the natural logarithmic function

x	$f(x) = \ln x$	$(x, f(x))$	Conclusion (see Figure 8.27)
$\dfrac{1}{e}$	$f\left(\dfrac{1}{e}\right) = \ln \dfrac{1}{e} = \ln e^{-1} = -1$	$\left(\dfrac{1}{e}, -1\right)$	$\left(\dfrac{1}{e}, -1\right) \approx (0.36, -1)$ lies on the graph of $f(x) = \ln x$.
e	$f(e) = \ln e = \ln e^1 = 1$	$(e, 1)$	$(e, 1) \approx (2.72, 1)$ lies on the graph of $f(x) = \ln x$.
e^2	$f(e^2) = \ln e^2 = 2$	$(e^2, 2)$	$(e^2, 2) \approx (7.4, 2)$ lies on the graph of $f(x) = \ln x$.

5 Use logarithmic functions in applications.

Modeling with Logarithmic Functions

Numerous mathematical models contain natural and common logarithmic functions, as illustrated in Examples 4 and 5.

EXAMPLE 4 **Walking Speed and City Population**

The formula

$$W = 0.35 \ln P + 2.74$$

is a model for the average walking speed (W, in feet per second) for a resident of a city whose population is P thousand. Find the average walking speed for people living in the following cities.

City	1990 Population (in thousands)
Jackson, Mississippi	197
New York City, New York	7323

ENRICHMENT ESSAY

Logarithms and Music

Each note on the keyboard of a piano has a frequency, the number of vibrations per second. As shown, the frequency of each C note is written in terms of powers of 2. Consequently, the musical scale is logarithmic with base 2. For each note (D, E, F, F sharp, G, etc.), raising an octave doubles the frequency.

The figure below shows that guitar frets (the ridges to guide the fingers) are separated by greater distances for lower notes than for higher notes. What scale is used in the spacing of the frets? What kind of function is indicated by the shape of the graph?

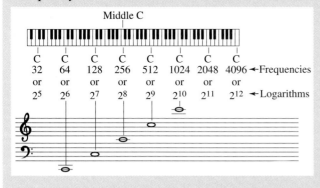

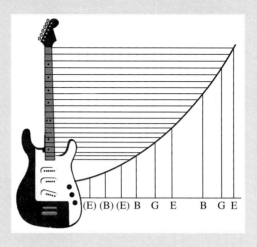

Solution

$$W = 0.35 \ln P + 2.74 \quad \text{This is the given model.}$$
$$= 0.35 \ln 197 + 2.74 \quad \text{Start with Jackson, with } P = 197 \text{ (thousand).}$$
$$\approx 4.6 \quad \text{Use a calculator:} \quad .35 \boxed{\times} \boxed{\ln} 197 \boxed{+} 2.74 \boxed{\text{ENTER}}$$

The average walking speed in Jackson is 4.6 feet per second.

$$W = 0.35 \ln 7323 + 2.74 \quad \text{For New York City, } P = 7323 \text{ (thousand).}$$
$$\approx 5.9 \quad \text{Use a calculator.}$$

The average walking speed in New York City is 5.9 feet per second. ∎

Using technology

The graph of $y = 0.35 \ln x + 2.74$ was obtained using a graphing utility. Our computations in Example 3 suggested the range setting

$$\text{Xmin} = 0, \text{Xmax} = 7500, \text{Xscl} = 375,$$
$$\text{Ymin} = 0, \text{Ymax} = 6, \text{Yscl} = 0.5$$

By tracing along the graph, you can estimate the walking speed for people living in a city of known population.

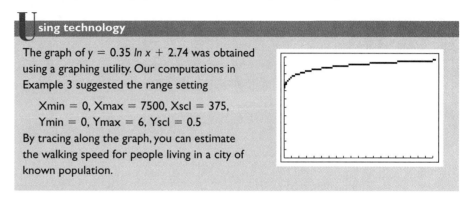

EXAMPLE 5 Earthquake Intensity

The Richter scale, used to measure the magnitudes of earthquakes, is named for U.S. seismologist Charles Richter (1900–1985). The scale is a common

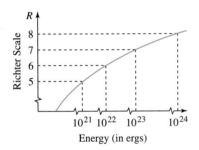

Figure 8.28

For each increase of *I* on the Richter scale, an earthquake's energy increases tenfold.

logarithmic scale: For each increase of one unit on the Richter scale, there is a tenfold increase in the energy in an earthquake, illustrated in Figure 8.28.

The magnitude *R* on the Richter scale of an earthquake of intensity *I* is given by

$$R = \log \frac{I}{I_0},$$

where I_0 is the intensity of a barely felt zero-level earthquake, that is, one that can barely be recorded on a seismograph.

a. Compare the magnitudes on the Richter scale of earthquakes that are 10 times, 100 times, 1000 times, and 10,000 times as intense as a zero-level earthquake.

b. Northern California's 1989 earthquake was $10^{7.1}$ times as intense as a zero-level earthquake. What was its magnitude on the Richter scale?

Solution

a. $R = \log \dfrac{I}{I_0}$ This is the given model.

     $= \log \dfrac{10 I_0}{I_0}$ $I = 10 I_0$ for an earthquake ten times as intense as a zero-level earthquake. I_0 is the intensity of the weakest earthquake that can be recorded on a seismograph.

     $= \log 10$ Simplify.

     $= 1$ log 10 = 1 because the power to which 10 must be raised to get 10 is 1. You can also use a calculator.

We continue to use the given model.

$$R = \log \frac{I}{I_0}$$

100 Times as Intense as a Zero-Level Quake	**1000 Times as Intense as a Zero-Level Quake**	**10,000 Times as Intense as a Zero-Level Quake**
$I = 100 I_0$, so	$I = 1000 I_0$, so	$I = 10{,}000 I_0$, so
$R = \log \dfrac{100 I_0}{I_0}$	$R = \log \dfrac{1000 I_0}{I_0}$	$R = \dfrac{10{,}000 I_0}{I_0}$
$= \log 100$	$= \log 1000$	$= \log 10{,}000$
$= 2$	$= 3$	$= 4$

Thus, for earthquakes that are 10 times, 100 times, 1000 times, and 10,000 times as intense as a zero-level earthquake, the magnitudes on the Richter scale are 1, 2, 3, and 4, respectively.

b. $R = \log \dfrac{I}{I_0}$ Use the given model.

     $= \log \dfrac{10^{7.1} I_0}{I_0}$ Northern California's earthquake was $10^{7.1}$ times the intensity of I_0, so $I = 10^{7.1} I_0$.

     $= \log 10^{7.1}$ Simplify.

     $= 7.1$ Use a calculator or the property $\log 10^x = x$.

Northern California's 1989 earthquake registered 7.1 on the Richter scale. ∎

Roger Brown "San Andreas Fault Line" 1995. Photo courtesy Phyllis Kind Gallery, New York & Chicago.

PROBLEM SET 8.3

Practice Problems

In Problems 1–16, write each logarithmic equation in exponential form.

1. $\log_2 16 = 4$

2. $\log_2 64 = 6$

3. $\log_5(\frac{1}{125}) = -3$

4. $\log_3(\frac{1}{81}) = -4$

5. $\log_{25} 5 = \frac{1}{2}$

6. $\log_{16} 64 = \frac{3}{2}$

7. $y = \log_3 8$

8. $y = \log_5 12$

9. $\log_m P = c$

10. $\log_n R = a$

11. $\ln 5 = 1.6094$

12. $\ln 7 = 1.9459$

13. $\log_b b^x = x$

14. $\log_b b = 1$

15. $\log 10 = 1$

16. $\log 10,000 = 4$

In Problems 17–36, write each exponential equation in logarithmic form.

17. $2^3 = 8$

18. $5^4 = 625$

19. $2^{-4} = \frac{1}{16}$

20. $5^{-3} = \frac{1}{125}$

21. $\sqrt[3]{8} = 2$

22. $\sqrt[3]{64} = 4$

23. $8^{1/3} = 2$

24. $32^{1/5} = 2$

25. $16^{3/4} = 8$

26. $125^{2/3} = 25$

27. $10^2 = 100$

28. $10^3 = 1000$

29. $e^4 = 54.5982$

30. $e^5 = 148.4132$

31. $e^{-2} = 0.1353$

32. $e^{-3} = 0.0498$

33. $e^y = x$

34. $10^y = x$

35. $P^a = m$

36. $Q^k = n$

In Problems 37–70, evaluate each expression. If evaluation is not possible, state the reason.

37. $\log_3 9$

38. $\log_9 81$

39. $\log_2 32$

40. $\log_3 27$

41. $\log_7 \sqrt{7}$

42. $\log_6 \sqrt{6}$

43. $\log_7(\frac{1}{7})$

44. $\log_3(\frac{1}{27})$

45. $\log_2(\frac{1}{32})$

46. $\log_{36} 6$

47. $\log_{81} 9$

48. $\log_4 4$

49. $\log_8 8$

50. $\log_b b$

51. $\log_4 1$

52. $\log_5 1$

53. $\log_5(-5)$

54. $\log_5 0$

55. $\log_{16} 8$

56. $\log_{125} 5$

57. $\log 1000$

58. $\log 100,000$

59. $\log 0.01$

60. $\log 0.001$

61. $\ln e^4$

62. $\ln e^6$

63. $\ln(-1)$

64. $\log(-1)$

65. $\log_5 5^6$

66. $\log_4 4^7$

67. $\log 10^{21}$

68. $\log 10^{32}$

69. $\log_5(\log_2 32)$

70. $\log_2(\log_3 81)$

71. a. Graph $f(x) = 3^x$.
b. Show that the inverse of $f(x) = 3^x$ is $f^{-1}(x) = \log_3 x$.
c. Graph $f^{-1}(x) = \log_3 x$ on the same set of axes as $f(x) = 3^x$.

72. a. Graph $f(x) = 5^x$.
b. Show that the inverse of $f(x) = 5^x$ is $f^{-1}(x) = \log_5 x$.
c. Graph $f^{-1}(x) = \log_5 x$ on the same set of axes as $f(x) = 5^x$.

73. a. Graph $f(x) = (\frac{1}{2})^x$.
b. Show that the inverse of $f(x) = (\frac{1}{2})^x$ is $f^{-1}(x) = \log_{1/2} x$.
c. Graph $f^{-1}(x) = \log_{1/2} x$ on the same set of axes as $f(x) = (\frac{1}{2})^x$.

74. a. Graph $f(x) = (\frac{1}{3})^x$.
b. Show that the inverse of $f(x) = (\frac{1}{3})^x$ is $f^{-1}(x) = \log_{1/3} x$.
c. Graph $f^{-1}(x) = \log_{1/3} x$ on the same set of axes as $f(x) = (\frac{1}{3})^x$.

Application Problems

75. Students in a psychology class took a final examination. To find out how much of the subject matter they remembered over a period of time, they volunteered to take equivalent forms of the exam in monthly intervals thereafter. The average score ($f(t)$) for the group after t months was given by the human memory model $f(t) = 88 - 15 \ln(t + 1)$, where $0 \le t \le 12$.

a. What was the average score on the original exam?
b. What was the average score after 2 months? 4 months? 6 months? 8 months? one year?
c. Use the results from parts (a) and (b) to sketch the graph of f. Describe what the shape of the graph indicates in terms of the material retained by the students.

76. The model

$$t = \frac{1}{c} \ln\left(\frac{A}{A - N}\right)$$

describes the time (t, in weeks) that it takes to achieve mastery of a portion of a task, where

A = Maximum learning possible

N = Portion of the learning that is to be achieved

c = Constant used to measure an individual's learning style

The formula is also used to determine how long it will take chimpanzees and apes to master a task. For example, a typical chimpanzee learning sign language can master a maximum of 65 signs. How many weeks will it take a chimpanzee to master 30 signs if c for that chimp is 0.03?

77. Archaeologists, anthropologists, and geologists use a technique called *radiocarbon dating* to estimate when a particular organism died. In living organic material, the ratio of radioactive carbon-14 to the total number of carbon atoms is approximately $\frac{1}{10^{12}}$. However, once an organism dies, carbon-14 is no longer replaced in its tissues but continues to decay. The age of a fossil is consequently a function of the ratio of carbon-14 to carbon-12 in the fossil, described by

$$f(R) = \frac{5600 \log R}{\log 0.5}$$

where R is the ratio of carbon-14 to carbon-12 in the fossil and $f(R)$ is the fossil's age.

What is the age of a fossil in which the ratio of carbon-14 to carbon-12 is 0.25?

78. The Weber–Fechner law in psychophysics relates the way we perceive the intensity of a sensation to the actual intensity of the physical stimulus causing it. Two heavy objects must differ in weight by considerably more than two light objects if we are to perceive a difference between them. The same is true for loudness of sounds, brightness of light, and pitches of musical tones.

The perceived loudness of a sound D (in decibels) of intensity I (in watts per meter^2) is given by

$$D = 10 \log \frac{I}{I_0}$$

where I_0 is the intensity of a sound barely audible to the human ear. The threshold of human hearing is 10^{-12} watt per meter^2, so we can express the model as

$$D = 10 \log \frac{I}{10^{-12}}.$$

a. What is the decibel level (I in the given model) for the threshold of human hearing?

b. A decibel level of 160 can result in a ruptured eardrum. The sound of a blue whale can be heard 500 miles away, reaching an intensity of 6.3×10^6 watts per meter^2. Determine the decibel level of this sound. At close range, can the sound of a blue whale rupture the human eardrum?

79. Recall that a prime number is a natural number greater than 1 divisible only by itself and 1. The number of prime numbers less than or equal to a given number x can be approximated by the expressions

$$\frac{x}{\ln x} \quad \text{and} \quad \frac{x}{\ln x - 1.083\,66}$$

when x is large. Fill in the last two columns of the table, rounding results to the nearest whole number. Then answer the following questions: Which of the two expressions provides a better approximation for the numbers in the second column? What happens to each expression, compared with the numbers in the second column, as x gets larger?

x	Number of Prime Numbers That are Less Than x	$\dfrac{x}{\ln x}$	$\dfrac{x}{\ln x - 1.083\,66}$
100	25		
10,000	1229		
10^6	78,498		
10^8	5,761,455		
10^9	50,847,534		
10^{10}	455,052,512		

True–False Critical Thinking Problems

80. Which one of the following is true?
  a. If (a, b) satisfies $y = 2^x$, then (a, b) satisfies $y = \log_2 x$.
  b. $\log(-100) = -2$
  c. The domain of $f(x) = \log_2 x$ is $(-\infty, \infty)$.
  d. $\log_b x$ is the exponent to which b must be raised to obtain x.

81. Which one of the following is true?
  a. $\log_{36} 6 = 2$
  b. There is no relationship between the graphs of $f(x) = 3^x$ and $g(x) = \log_3 x$.
  c. $\dfrac{\log_2 8}{\log_2 4} = \dfrac{8}{4}$
  d. $\log_{1/2} 32 = -5$

82. Which one of the following is true?
  a. $\log_b 0 = 0$, for $b > 0$ and $b \neq 1$.
  b. If $\log_b 9 = 2$, then $b = 3$ or $b = -3$.
  c. $\log_5 5^7 = 7$
  d. The inverse of $y = 4^x$ is $y = x^4$.

Technology Problems

In Problems 83–86, graph f and g in the same viewing rectangle. Then describe the relationship of the graph of g to the graph of f.

83. $f(x) = \ln x, \quad g(x) = \ln(x + 3)$

84. $f(x) = \ln x, \quad g(x) = \ln x + 3$

85. $f(x) = \log x, \quad g(x) = -\log x$

86. $f(x) = \log x, \quad g(x) = \log(x - 2) + 1$

87. Students in a mathematics class took a final examination. They took equivalent forms of the exam in monthly intervals thereafter. The average score $(f(t))$ for the group after t months was given by the human memory model $f(t) = 75 - 10 \log(t + 1)$, where $0 \leq t \leq 12$. Use a graphing utility to graph the model. Then determine how many months will elapse before the average score falls below 65.

88. Graph f and g in the same viewing rectangle.
  a. $f(x) = \ln(3x), \quad g(x) = \ln 3 + \ln x$
  b. $f(x) = \log(5x^2), \quad g(x) = \log 5 + \log x^2$
  c. $f(x) = \ln(2x^3), \quad g(x) = \ln 2 + \ln x^3$
  d. Describe what you observe in parts (a)–(c). Generalize this observation by writing an equivalent expression for $\log_b(MN)$, where $M > 0$ and $N > 0$.
  e. Complete this statement: The log of a product is equal to _____.

89. Graph f and g in the same viewing rectangle.
  a. $f(x) = \ln \dfrac{x}{2}, \quad g(x) = \ln x - \ln 2$
  b. $f(x) = \log \dfrac{x}{5}, \quad g(x) = \log x - \log 5$
  c. $f(x) = \ln \dfrac{x^2}{3}, \quad g(x) = \ln x^2 - \ln 3$
  d. Describe what you observe in parts (a)–(c). Generalize this observation by writing an equivalent expression for $\log_b\left(\dfrac{M}{N}\right)$, where $M > 0$ and $N > 0$.
  e. Complete this statement: The log of a quotient is equal to _____.

90. Graph f and g in the same viewing rectangle.
  a. $f(x) = \ln x^2, \quad g(x) = 2 \ln x$
  b. $f(x) = \log x^3, \quad g(x) = 3 \log x$
  c. $f(x) = \ln x^{1/2}, \quad g(x) = \frac{1}{2} \ln x$
  d. Describe what you observe in parts (a)–(c). Generalize this observation by writing an equivalent expression for $\log_b M^p$, where $M > 0$.
  e. Complete this statement: The log of a number with an exponent is equal to _____.

91. Graph each of the following functions in the same viewing rectangle and then place the functions in order from the one that increases most slowly to the one that increases most rapidly.

$$y = x, \ y = \sqrt{x}, \ y = e^x, \ y = \ln x, \ y = x^x, \ y = x^2$$

Writing in Mathematics

92. Describe how to graph $y = \log_4 x$ using points on the graph of $y = 4^x$.

93. Suppose that we use the model $W = 0.35 \ln P + 2.74$ relating the population (P) of a city and average walking speed (W) of the city's residents. If we know the average walking speed is 4.6 feet per second, what difficulty will we encounter by attempting to approximate the city's population? What skill is needed to overcome this difficulty?

94. Discuss similarities and differences between the graphs of $y = \log_b x$ for $0 < b < 1$ and $y = \log_b x$ for $b > 1$.

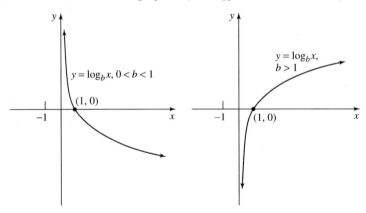

Critical Thinking Problems

In Problems 95–98, find each indicated value.

95. $\dfrac{\log_{25} 5 - \log_{1/16} 8}{\log_9(\frac{1}{27}) + \log_4 1}$

96. $\dfrac{\log_3 81 - \log_\pi 1}{\log_{2\sqrt{2}} 8 - \log_{10} 0.001}$

97. $\log_5 1 + \log_8(4 \cdot \sqrt[5]{16})$

98. $\dfrac{\log_9 81^{-3/8}}{\log_{49} 7^{2/3} - \log_{64} \sqrt[5]{\frac{1}{16}}}$

99. In the figure, what are the coordinates of the four points $A, B, C,$ and D?

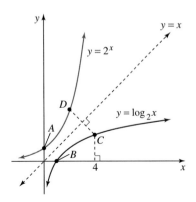

100. Graph: $y = \log_2 x + 1$.

101. Graph: $y = -\log_2 x$.

102. Solve for x: $\log_4[\log_3(\log_2 x)] = 0$.

103. Evaluate: $\log_{1/25} 25\sqrt[3]{25}$.

Review Problems

104. Find the distance between $(-2, -9)$, and $(1, -3)$, expressing the answer in simplified radical form.

105. The measurements given in the table have a linear relationship.

x (Parts of Sulfur Dioxide in the Air per Cubic Meter)	y (Number of People Who Die in a Particular Year in a Small European Town)
129	98
323	104

a. Write the linear equation (in slope-intercept form) that is the model for these variables.

b. Predict the number of people who will die in a year when the sulfur dioxide content is 800 parts per cubic meter.

106. Solve the following system:

$$2x = 11 - 5y$$
$$3x - 2y = -12$$

S E C T I O N   8 . 4

Solutions Manual   **Tutorial**   **Video 10**

Properties of Logarithmic Functions

Objectives

1 Use the product rule for logarithms.
2 Use the quotient rule for logarithms.
3 Use the power rule for logarithms.
4 Expand logarithmic expressions.
5 Condense logarithmic expressions.
6 Use the change-of-base property.
7 Use logarithmic properties in applications.

In this section, we consider four important properties of logarithms. These properties can be used to answer questions about variables that appear in logarithmic models.

Three Basic Properties of Logarithms

There are three properties of exponents that have corresponding logarithmic properties.

1. $b^M \cdot b^N = b^{M+N}$

2. $\dfrac{b^M}{b^N} = b^{M-N}$

3. $(b^M)^p = b^{Mp}$

These identities, coupled with an awareness that a logarithm is an exponent, suggest the following.

Discover for yourself

You can discover three basic logarithmic properties on your own with your graphing utility. If you wish to do so, work Problems 88–90 in Problem Set 8.3 before continuing.

Properties of logarithms

Let b, M, and N be positive real numbers with $b \neq 1$, and let p be any real number.

Property	The Property in Words	Example
The Product Rule $\log_b(MN) = \log_b M + \log_b N$	The log of a product is the sum of the logs.	$\ln(4x) = \ln 4 + \ln x$
The Quotient Rule $\log_b\left(\dfrac{M}{N}\right) = \log_b M - \log_b N$	The log of a quotient is the difference of the logs.	$\log \dfrac{x}{2} = \log x - \log 2$
The Power Rule $\log_b M^p = p \log_b M$	The log of a number with an exponent is the product of the exponent and the log of that number.	$\ln\sqrt{x} = \ln x^{1/2} = \frac{1}{2}\ln x$

U sing technology

The relationships in the third column of the properties table can be confirmed using a graphing utility. The graphs on each side of the three equalities are the same, verifying these special cases of the product, quotient, and power rules.

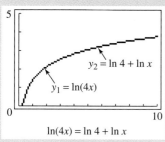

$\ln(4x) = \ln 4 + \ln x$

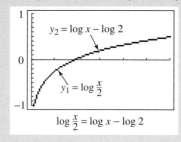

$\log \frac{x}{2} = \log x - \log 2$

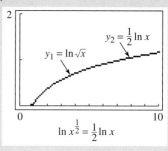

$\ln x^{\frac{1}{2}} = \frac{1}{2} \ln x$

Use the product rule for logarithms.

Product rule

$\log_b(MN)$
$\quad = \log_b M + \log_b N$

EXAMPLE I   Using the Product Rule

Use the product rule to expand: **a.** $\log_4(7 \cdot 9)$ **b.** $\log(10x)$

Solution

a. $\log_4(7 \cdot 9) = \log_4 7 + \log_4 9$ The log of a product is the sum of the logs.

b. $\log(10x) = \log 10 + \log x$ The log of a product is the sum of the logs. These are common logs with base 10 understood.

$\quad\quad\quad\quad = 1 + \log x$ Since $\log_b b = 1$, then $\log_{10} 10 = 1$.   ■

---

2 Use the quotient rule for logarithms.

Quotient rule

$\log_b\left(\dfrac{M}{N}\right)$
$\quad = \log_b M - \log_b N$

EXAMPLE 2   Using the Quotient Rule

Use the quotient rule to expand: **a.** $\log_7\left(\dfrac{14}{x}\right)$ **b.** $\ln\left(\dfrac{e^3}{7}\right)$

Solution

a. $\log_7\left(\dfrac{14}{x}\right) = \log_7 14 - \log_7 x$ The log of a quotient is the difference of the logs.

b. $\ln\left(\dfrac{e^3}{7}\right) = \ln e^3 - \ln 7$ The log of a quotient is the difference of the logs. These are natural logs with base e understood.

$\quad\quad\quad\quad = 3 - \ln 7$ $\ln e^x = x$ ■

---

3 Use the power rule for logarithms.

Power rule

$\log_b M^p = p \log_b M$

EXAMPLE 3   Using the Power Rule

Use the power rule to expand: **a.** $\log_5 \sqrt[3]{7^4}$ **b.** $\ln x^3$ **c.** $\log \sqrt[3]{x}$

Solution

a. $\log_5 \sqrt[3]{7^4} = \log_5 7^{4/3}$ Rewrite the radical using a rational exponent.

$\quad\quad\quad\quad = \dfrac{4}{3} \log_5 7$ The log of a number with an exponent is the exponent times the log of the number.

b. $\ln x^3 = 3 \ln x$ The power rule lets you bring the exponent to the front:

$\quad\quad\quad\quad \ln x^3 = 3 \ln x$

c. $\log \sqrt[3]{x} = \log x^{1/3}$ Rewrite the radical using a rational exponent.

$\quad\quad\quad\quad = \dfrac{1}{3} \log x$ Use the power rule to bring the exponent to the front.   ■

Proving the Properties of Logarithms

All three of the properties of logarithms can be proved in the same way. Let's look at the proof for the product rule,

$$\log_b(MN) = \log_b M + \log_b N.$$

We begin by letting $\log_b M = R$ and $\log_b N = S$. Now we write each logarithm in exponential form.

$$\log_b M = R \quad \text{means} \quad b^R = M$$
$$\log_b N = S \quad \text{means} \quad b^S = N$$

By substituting and using a property of exponents, we see that

$$MN = b^R b^S = b^{R+S}.$$

Now we change $MN = b^{R+S}$ to logarithmic form.

$$MN = b^{R+S} \quad \text{means} \quad \log_b(MN) = R + S$$

Finally, substituting $\log_b M$ for R and $\log_b N$ for S gives us

$$\log_b(MN) = \log_b M + \log_b N$$

the property that we wanted to prove.

The proofs of the quotient and power rules are nearly identical, and are left for the problem set.

4 Expand logarithmic expressions.

Using the Properties Together to Expand Logarithmic Expressions

When we expand a logarithmic expression, it is possible to change the domain of a function. For example, the power rule tells us that

$$\log_b x^2 = 2 \log_b x.$$

If $f(x) = \log_b x^2$, then the domain of f is the set of all nonzero numbers. However, if $g(x) = 2 \log_b x$, the domain is the set of positive real numbers. Consequently, we should write

$$\log_b x^2 = 2 \log_b x \qquad \text{for } x > 0.$$

In short, when expanding or condensing a logarithmic expression, observe whether the rewriting has changed the domain of the expression.

In summary, most of our work will be based on the three properties listed in the table on p. 713. We repeat them here for convenience.

Properties for expanding logarithmic expressions

1. $\log_b(MN) = \log_b M + \log_b N$ Product rule

2. $\log_b\left(\dfrac{M}{N}\right) = \log_b M - \log_b N$ Quotient rule

3. $\log_b M^p = p \log_b M$ Power rule

In all cases, $M > 0$ and $N > 0$.

Using technology

The graphs of $y = \ln x^2$ and $y = 2 \ln x$ have different domains.

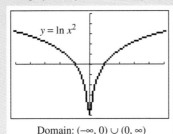

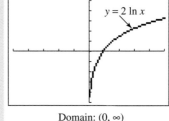

Domain: $(-\infty, 0) \cup (0, \infty)$ Domain: $(0, \infty)$

Notice that the graphs are only the same if $x > 0$. Thus, we should write

$\ln x^2 = 2 \ln x$ for $x > 0$.

EXAMPLE 4   **Expanding Logarithmic Expressions**

Use logarithmic properties to expand each expression as much as possible. Expanded forms should involve logarithms of x, y, and z.

a. $\log_b x^3 y^2$ **b.** $\log_5\left(\dfrac{\sqrt[3]{x}}{25y^2}\right)$ **c.** $\log \sqrt[4]{\dfrac{1000z}{xy^3}}$

Solution

We will have to use two or more of the properties of logarithms in each part of this example.

a. $\log_b x^3 y^2 = \log_b x^3 + \log_b y^2$ Use the product rule.
$\qquad\qquad\quad = 3 \log_b x + 2 \log_b y$ Use the power rule.

b. $\log_5\left(\dfrac{\sqrt[3]{x}}{25y^2}\right) = \log_5 x^{1/3} - \log_5 25y^2$ Use the quotient rule.

$\qquad\qquad = \log_5 x^{1/3} - (\log_5 25 + \log_5 y^2)$ Use the product rule on $\log_5 25y^2$.

$\qquad\qquad = \tfrac{1}{3} \log_5 x - (\log_5 25 + 2\log_5 y)$ Use the power rule.

$\qquad\qquad = \tfrac{1}{3} \log_5 x - \log_5 25 - 2 \log_5 y$ Apply the distributive property.

$\qquad\qquad = \tfrac{1}{3} \log_5 x - 2 - 2 \log_5 y$ $\log_5 25 = 2$ because 2 is the power to which we must raise 5 to get 25. $(5^2 = 25)$

c. $\log \sqrt[4]{\dfrac{1000z}{xy^3}} = \log\left(\dfrac{1000z}{xy^3}\right)^{1/4}$ Use exponential notation.

$\qquad\qquad = \dfrac{1}{4} \log\left(\dfrac{1000z}{xy^3}\right)$ Use the power rule.

$\qquad\qquad = \dfrac{1}{4}(\log 1000z - \log xy^3)$ Use the quotient rule.

$\qquad\qquad = \dfrac{1}{4}(\log 1000 + \log z - (\log x + \log y^3))$ Use the product rule.

$$= \frac{1}{4}(\log 1000 + \log z - \log x - \log y^3) \qquad \text{Apply the distributive property.}$$

$$= \frac{1}{4}(\log 1000 + \log z - \log x - 3 \log y) \qquad \text{Use the power rule.}$$

$$= \frac{1}{4}(3 + \log z - \log x - 3 \log y) \qquad \begin{array}{l}\log 1000 \text{ (or} \\ \log_{10}1000) = 3 \text{ because} \\ 10^3 = 1000.\end{array} \blacksquare$$

5 Condense logarithmic expressions.

Using the Properties Together to Condense Logarithmic Expressions

In Example 4, we wrote each expression as the sum and difference of simpler logarithmic expressions. We can also use the logarithmic properties to reverse the direction of this procedure, writing the sum and/or difference of logarithms as a single logarithm. Most of our work will use the following three properties.

Properties for condensing logarithmic expressions

1. $\log_b M + \log_b N = \log_b(MN)$ Product rule

2. $\log_b M - \log_b N = \log_b\left(\dfrac{M}{N}\right)$ Quotient rule

3. $p \log_b M = \log_b M^p$ Power rule

In all cases, $M > 0$ and $N > 0$.

EXAMPLE 5   **Condensing Logarithmic Expressions**

Write each as a single logarithm:

a. $\log 25 + \log 4$ **b.** $\ln 3 - 3 \ln x$ **c.** $\frac{1}{2} \log_b x - 5 \log_b y + \log_b z$

Solution

a. $\log 25 + \log 4 = \log(25 \cdot 4)$ Use the product rule.

$\qquad\qquad\qquad\quad = \log 100$ We now have a single logarithm. However, we can simplify.

$\qquad\qquad\qquad\quad = 2$ $\log 100$ or $\log_{10}100 = 2$ because $10^2 = 100$.

b. $\ln 3 - 3 \ln x = \ln 3 - \ln x^3$ Use the power rule.

$\qquad\qquad\qquad = \ln \dfrac{3}{x^3}$ Use the quotient rule.

c. $\dfrac{1}{2} \log_b x - 5 \log_b y + \log_b z$

$\quad = \log_b x^{1/2} - \log_b y^5 + \log_b z$ Use the power rule so that log coefficients are all 1.

$\quad = (\log_b x^{1/2} - \log_b y^5) + \log_b z$ This optional step emphasizes the order of operations.

$\quad = \log_b \dfrac{x^{1/2}}{y^5} + \log_b z$ Use the quotient rule.

$\quad = \log_b \dfrac{x^{1/2}z}{y^5}$ or $\log_b \dfrac{\sqrt{x}z}{y^5}$ Use the product rule. $\blacksquare$

> **S** tudy tip
>
> Coefficients of logarithms must be 1 before you can condense them using the product and quotient rules. Thus, we must rewrite $3 \ln x$ as $\ln x^3$.

Shown at the right are the graphs of $y_1 = \ln(x + 3)$ and $y_2 = \ln x + \ln 3$. Our graphing utility indicates that the graphs are *not the same*. The graph of $y_1 = \ln(x + 3)$ is actually the graph of the natural log function shifted 3 units to the left. The graph of $y_2 = \ln x + \ln 3$ is the graph of the natural log function shifted up by ln 3, which is an approximate vertical shift of 1.1 units.

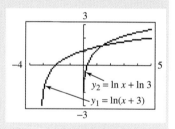

We see, then, that $\ln(x + 3) \neq \ln x + \ln 3$. In general

$$\log_b(M + N) \neq \log_b M + \log_b N.$$

Try to avoid each of the following incorrect procedures.

Incorrect

$$\overline{\log_b(M + N)} = \log_b M + \log_b N$$

$$\log_b(M - N) = \overline{\log_b M - \log_b N}$$

$$\log_b(M \cdot N) = (\log_b M)(\log_b N)$$

$$\log_b \frac{M}{N} = \frac{\log_b M}{\log_b N}$$

$$\frac{\log_b M}{\log_b N} = \log_b M - \log_b N$$

Use your graphing utility, replacing $\log_b$ by ln, M by x, and N by 3 to show that all five of these equations are not true in general.

6 Use the change-of-base property.

The Change-of-Base Property

We have seen that calculators give the values of both common logarithms (base 10) and natural logarithms (base e). To find a logarithm with any other base, we can use the following change-of-base property.

> **The change-of-base property**
>
> For any logarithmic bases a and b, and any positive number M,
>
> $$\log_b M = \frac{\log_a M}{\log_a b}.$$

Base a is a new base that we introduce. The logarithm on the left side has a base of b, allowing us to change from base b to any other base a, as long as the newly introduced base is a positive number between 0 and 1, or greater than 1.

To prove the change-of-base property, we let x equal the logarithm on the left side.

$$\log_b M = x$$

Now we rewrite this logarithm in exponential form.

$$\log_b M = x \qquad \text{means} \qquad b^x = M.$$

Since b^x and M are equal, the logarithms with base a for each of these expressions must be equal. This means that

$$\log_a b^x = \log_a M$$

$$x \log_a b = \log_a M \qquad \text{Apply the power rule for logarithms on the left side.}$$

$$x = \frac{\log_a M}{\log_a b} \qquad \text{Solve for } x, \text{ dividing both sides by } \log_a b.$$

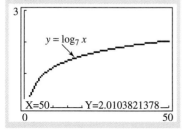
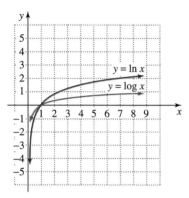

Figure 8.29

Partial graphs of natural and common logarithmic functions

In our very first step we let x equal $\log_b M$. Replacing x on the left side by $\log_b M$ gives us

$$\log_b M = \frac{\log_a M}{\log_a b}$$

which is the change-of-base property.

Because calculators contain keys for common and natural logarithms, we will frequently introduce base 10 or base e.

EXAMPLE 6 Changing Base to Common Logarithms

Use common logarithms to evaluate $\log_7 50$.

Solution

$$\log_7 50 = \frac{\log_{10} 50}{\log_{10} 7}$$

Use $\log_b M = \dfrac{\log_a M}{\log_a b}$. The newly introduced base is 10.

$$= \frac{\log 50}{\log 7}$$

We omit the 10 with common logs, writing $\log_{10}$ as log.

$$\approx 2.01$$

Use a calculator:
[log] 50 [÷] [log] 7 [ENTER]

This means that $\log_7 50 \approx 2.01$. (How can you convince yourself that this approximation is reasonable?)

EXAMPLE 7 Changing Base to Natural Logarithms

Use natural logarithms to evaluate $\log_7 50$.

Solution

$$\log_7 50 = \frac{\log_e 50}{\log_e 7}$$

Use $\log_b M = \dfrac{\log_a M}{\log_a b}$. The newly introduced base is e.

$$= \frac{\ln 50}{\ln 7}$$

We write $\log_e$ as ln to designate natural logarithms.

$$\approx 2.01$$

Use a calculator:
[ln] 50 [÷] [ln] 7 [ENTER]

We have again shown that $\log_7 50 \approx 2.01$.

We can use the change-of-base property to determine a relationship between $y = \ln x$ and $y = \log x$, whose partial graphs are shown in Figure 8.29. Begin with

$$\log_{10} x = \frac{\log_e x}{\log_e 10}$$

We've introduced base e.

ENRICHMENT ESSAY

The First Book on Logarithms

John Napier published his work on logarithms in 1614. Another person associated with the invention of logarithms is Jobst Bürgi, a Swiss mathematician who invented logarithms independently of Napier. With Napier's publication coming first, credit is given to him for creating logarithms. Exponential notation and laws of exponents were invented at the end of the 17th century, after the deaths of Napier and Bürgi.

Napier could not have known that logarithms would have a major impact on science and mathematics. Ironically, he believed that he would be remembered for another one of his books, a popular work in which he claimed that the pope was the Antichrist and that the world would end sometime between 1688 and 1700. Fortunately, this was not the book that ensured Napier's place in history.

Culver Pictures, Inc.

iscover for yourself

Read the last sentence in the text prior to objective 7. If this statement is true, why doesn't the graph of $y = \ln x$ always lie above the graph of $y = \log x$? Refer to Figure 8.29 on page 719.

Using the agreed-upon notation for common and natural logarithms, we can write

$$\log x = \frac{\ln x}{\ln 10}.$$

Multiplying both sides by $\ln 10$ results in

$$(\log x)(\ln 10) = \ln x.$$

Because $\ln 10$ is approximately 2.3026, this equation gives us a relationship between the natural and common logarithmic functions:

$$\ln x \approx (\log x)(2.3026).$$

In terms of the graphs in Figure 8.29 on page 719, the range for the natural logarithm of any (positive) number is approximately equal to the range of the common logarithm of that number multiplied by 2.3026.

7 Use logarithmic properties in applications.

Using Logarithmic Properties in Applications

Logarithmic properties can be used to answer questions about variables contained in mathematical models.

EXAMPLE 8   **Comparing the Richter Scale Measures of Earthquakes**

In Section 8.3, we saw that the magnitude R on the Richter scale of an earthquake or intensity I is given by

$$R = \log \frac{I}{I_0}$$

where I_0 is the intensity of a barely felt zero-level earthquake. If an earthquake has an intensity 1000 times the intensity of a smaller earthquake, how much larger is the Richter scale measure of the larger earthquake than that of the smaller?

Solution

Let

I = the intensity of the smaller earthquake

Since the more violent earthquake is 1000 times more intense,

$1000I$ = the intensity of the larger earthquake

Now we can use the model $R = \log \dfrac{I}{I_0}$.

Rene Magritte "La Poitrine"
(The Beast) 1961, oil on canvas,
90×110 cm. © 1997 C. Herscovici,
Brussels/Artists Rights Society
(ARS), New York. Private Collection.
Art Resource, NY.

Magnitude of Smaller Earthquake	Magnitude of the More-violent Earthquake

$$R_1 = \boxed{\log \dfrac{I}{I_0}}$$

$$
\begin{aligned}
R_2 &= \log \dfrac{1000I}{I_0} \\
&= \log \left(1000 \cdot \dfrac{I}{I_0} \right) \\
&= \log 1000 + \log \dfrac{I}{I_0} \qquad \text{Use the product rule.} \\
&= \log 1000 + R_1 \qquad \text{Substitute } R_1 \text{ for } \log \dfrac{I}{I_0}. \\
&= 3 + R_1 \qquad \log_{10} 1000 = 3 \text{ because } \\
&\qquad\qquad\qquad 10^3 = 1000.
\end{aligned}
$$

Since $R_2 = 3 + R_1$, the earthquake 1000 times as intense as the smaller earthquake will have a magnitude on the Richter scale that is 3 more than that of the smaller earthquake. ∎

PROBLEM SET 8.4

Practice Problems

In Problems 1–12, verify each equation by evaluating each side separately.

1. $\log(10 \cdot 100) = \log 10 + \log 100$

2. $\log_2(8 \cdot 4) = \log_2 8 + \log_2 4$

3. $\log_3(9 \cdot \tfrac{1}{3}) = \log_3 9 + \log_3(\tfrac{1}{3})$

4. $\log_4(64 \cdot \tfrac{1}{16}) = \log_4 64 + \log_4(\tfrac{1}{16})$

5. $\log_3(\tfrac{81}{3}) = \log_3 81 - \log_3 3$

6. $\log_2(\tfrac{64}{8}) = \log_2 64 - \log_2 8$

7. $\ln\left(\dfrac{e^{17}}{e^4}\right) = \ln e^{17} - \ln e^4$

8. $\log_6\left(\dfrac{6^{13}}{6^9}\right) = \log_6 6^{13} - \log_6 6^9$

9. $\log_5 5^3 = 3 \log_5 5$

10. $\log_2 4^3 = 3 \log_2 4$

11. $\log_5 25^{1/2} = \tfrac{1}{2} \log_5 25$

12. $\log_3 81^{1/2} = \tfrac{1}{2} \log_3 81$

In Problems 13–42, use logarithmic properties to expand each expression as much as possible. Where possible, evaluate logarithmic expressions.

13. $\log_3 3x$

14. $\log_7 7y$

15. $\log 1000x$

16. $\log 100a$

17. $\log_4 Bx$

18. $\log_8 CM$

19. $\log_7\left(\dfrac{7}{x}\right)$

20. $\log_3\left(\dfrac{x}{3}\right)$

21. $\log\left(\dfrac{x}{100}\right)$

22. $\log\left(\dfrac{x}{100,000}\right)$

23. $\log_4\left(\dfrac{64}{y}\right)$

24. $\log_5\left(\dfrac{125}{r}\right)$

25. $\log_b x^3$

26. $\log_b x^7$

27. $\log N^{-6}$

28. $\log M^{-8}$

29. $\ln \sqrt[3]{x}$

30. $\ln \sqrt[5]{x}$

31. $\log_b x^2 y$

32. $\log_b xy^3$

33. $\log_4 \left(\dfrac{\sqrt{x}}{16} \right)$

34. $\log \left(\dfrac{\sqrt[5]{y}}{1000} \right)$

35. $\log_b \dfrac{x^2 y}{z^4}$

36. $\log_b \dfrac{x^3 y}{z^2}$

37. $\log \sqrt[3]{100x}$

38. $\ln \sqrt[4]{e^2 x}$

39. $\ln \sqrt[4]{\dfrac{e^3 x}{yz^2}}$

40. $\log \sqrt[3]{\dfrac{10{,}000x}{y^2 z}}$

41. $\log_b \sqrt[5]{\dfrac{x^{10} y^{15}}{b^3 z^6}}$

42. $\log_b \sqrt[7]{\dfrac{x^7 y^{14}}{b^2 z^5}}$

In Problems 43–64, write each expression as a single logarithm, and simplify if possible.

43. $\log 5 + \log 2$

44. $\log 250 + \log 4$

45. $\log_2 96 - \log_2 3$

46. $\log_3 405 - \log_3 5$

47. $\log 5 - 5 \log x$

48. $\ln 7 - 7 \ln x$

49. $2 \log_b x + 3 \log_b y$

50. $5 \log_b x + 7 \log_b y$

51. $5 \ln x - 2 \ln y$

52. $7 \ln x - 3 \ln y$

53. $2 \ln x - \dfrac{1}{2} \ln y$

54. $3 \ln x - \dfrac{1}{3} \ln y$

55. $\dfrac{1}{2} \log x + \dfrac{1}{3} \log y$

56. $\dfrac{1}{3} \log x + \dfrac{1}{2} \log y$

57. $\dfrac{1}{2} \log x + 3 \log y - 2 \log z$

58. $\dfrac{1}{3} \log x + 4 \log y - 3 \log z$

59. $\log_3(x^2 - 9) - \log_3(x - 3)$

60. $\log_5(x^3 + 8) - \log_5(x + 2)$

61. $\ln x + 2(\ln y - \ln z)$

62. $\ln x^2 - 2 \ln \sqrt{x}$

63. $\log_b \dfrac{b}{\sqrt{x}} - \log_b \sqrt{bx}$

64. $\ln \dfrac{e}{\sqrt{x}} - \ln \sqrt{ex}$

For Problems 65–80, use common logarithms or natural logarithms and a calculator to evaluate to four decimal places.

65. $\log_5 13$

66. $\log_6 17$

67. $\log_3 1.87$

68. $\log_5 8.21$

69. $\log_9 9.63$

70. $\log_4 4.72$

71. $\log_{14} 87.5$

72. $\log_{16} 57.2$

73. $\log_6 0.724$

74. $\log_{12} 0.839$

75. $\log_{0.1} 17$

76. $\log_{0.3} 19$

77. $\log_{200} 40$

78. $\log_{100} 50$

79. $\log_\pi 63$

80. $\log_\pi 400$

Application Problems

81. The perceived loudness of a sound (D, in decibels) at the human eardrum is given by

$$D = 10 \log \dfrac{I}{I_0}$$

where I is the intensity of the sound (in watts per meter^2) and I_0 is the intensity of a sound barely audible to the human ear. If a loud sound has an intensity 100 times the intensity of a softer sound, how much larger on the decibel scale is the loudness level of the more intense sound?

82. The pH of a solution is a measure of its acidity. A low pH indicates an acidic solution, and a high pH indicates a basic solution. Neutral water has a pH of 7.

Most people have a blood pH of approximately 7.4. The precise pH of a person's blood is modeled by the Henderson–Hasselbach formula, given by pH = $6.1 + \log B - \log C$. In the model, B represents the concentration of bicarbonate, which is a base, and C represents the concentration of carbonic acid, which is acidic.

a. Rewrite the model so that the common logarithm appears only once.

b. Use the form of the model in part (a) and a calculator to answer this question: If the concentration of bicarbonate is 25 and the concentration of carbonic acid is 2, what is the pH for this person's blood?

True–False Critical Thinking Problems

83. Which one of the following is true?

a. $\dfrac{\log_7 49}{\log_7 7} = \log_7 49 - \log_7 7$

b. $\log_b(x^3 + y^3) = 3 \log_b x + 3 \log_b y$

c. $\log_b (xy)^5 = (\log_b x + \log_b y)^5$

d. $\ln \sqrt{2} = \dfrac{\ln 2}{2}$

84. Which one of the following is true?

a. $\log_7 49 \cdot \log_7 7 = \log_7 49 + \log_7 7$

b. $\dfrac{\log_b x^2}{\log_b y^2} = 2 \log_b x - 2 \log_b y$

c. $(\log 10)^2 = 2 \log 10$

d. $\ln 8 = 3 \ln 2$

85. Which one of the following is true?

a. $\dfrac{\log_2 8}{\log_2 2} = \log_2 4$

b. $\ln e^x = x$

c. $\log_b(x + y) = \log_b x + \log_b y$

d. $\sqrt{\log 10{,}000} = \frac{1}{2} \log 10{,}000$

86. Which one of the following is true?

a. $\dfrac{\log 1000}{10} = \log 100$

b. $\ln 1 = e$

c. $\log_5(\log_5 5) = 0$

d. $\log_2 \dfrac{1}{16} = \dfrac{1}{\log_2 16}$

87. Which one of the following is true?

a. $\log_5 7 = \dfrac{\log_3 7}{\log_5 3}$

b. $\log_b \sqrt{\dfrac{xy}{z}} = \dfrac{1}{2}(\log_b x + \log_b y - \log_b z);$

$\quad x > 0, y > 0, z > 0$

c. $\log_8 e = \dfrac{\log e}{\ln 8}$

d. $\frac{1}{3} \log_b x = \sqrt[3]{x}; x > 0$

Technology Problems

88. a. Use a graphing utility (and the change-of-base property) to graph $y = \log_3 x$.

b. Graph $y = (\log_3 x) + 2$, $y = \log_3(x + 2)$, and $y = -\log_3 x$ in the same viewing rectangle as $y = \log_3 x$. Then describe the change or changes that need to be made to the graph of $y = \log_3 x$ to obtain each of these three graphs.

89. Graph $y = \log x$, $y = \log(10x)$, and $y = \log(0.1x)$ in the same viewing rectangle. Describe the relationship among the three graphs. What logarithmic property accounts for this relationship?

90. Use a graphing utility and the change-of-base property to graph $y = \log_3 x$, $y = \log_{25} x$, and $y = \log_{100} x$ in the same viewing rectangle.

a. Which graph is on the top in the interval $(0, 1)$? Which is on the bottom?

b. Which graph is on the top in the interval $1, \infty)$? Which is on the bottom?

c. Generalize by writing a statement about which graph is on top, which is on the bottom, and in which intervals, using $y = \log_b x$ where $b > 1$.

Disprove each statement in Problems 91–95 by:

a. letting y equal a positive constant of your choice.

b. using a graphing utility to graph the function on each side of the equal sign. The two functions should have different graphs, showing that the equation is not true in general.

91. $\log (x + y) = \log x + \log y$

92. $\log \dfrac{x}{y} = \dfrac{\log x}{\log y}$

93. $\ln (x - y) = \ln x - \ln y$

94. $\ln (xy) = (\ln x)(\ln y)$

95. $\dfrac{\ln x}{\ln y} = \ln x - \ln y$

Writing in Mathematics

96. Why is there no logarithmic property for $\log_b(M + N)$?

97. Without using a calculator, describe how you can decide if $\log 60 > \ln 60$ or if $\ln 60 > \log 60$.

98. What is the purpose of the change-in-base rule?

99. Find $\ln 2$ using a calculator. Then find each of the following: $1 - \frac{1}{2}$; $1 - \frac{1}{2} + \frac{1}{3}$; $1 - \frac{1}{2} + \frac{1}{3} - \frac{1}{4}$; $1 - \frac{1}{2} + \frac{1}{3} - \frac{1}{4} + \frac{1}{5}$; . . . . Describe what you observe.

100. Use a calculator to complete the following ordered pairs satisfying $y = \log x$: $(3, \ )$, $(4, \ )$, and $(5, \ )$. Calculate the slope of the line connecting the first two ordered pairs. Do the same for the second two ordered pairs. Describe what information this provides about the graph of $y = \log x$.

Critical Thinking Problems

101. Prove that

$$\log_b\!\left(\frac{M}{N}\right) = \log_b M - \log_b N$$

by letting $\log_b M = R$ and $\log_b N = S$ and following the procedure used to derive the product rule on page 715.

102. Prove that $\log_b M^p = p\log_b M$.

103. If $\log 3 = A$ and $\log 7 = B$, find $\log_7 9$ in terms of A and B.

104. If $\log 80 = A$ and $\log 45 = B$, find $\log 36$ in terms of A and B.

105. If $\log(\sqrt{13} + \sqrt{3}) = A$, find $\log(\sqrt{13} - \sqrt{3})$ in terms of A.

106. If

$$A = \log\frac{1}{2} + \log\frac{2}{3} + \log\frac{3}{4} + \log\frac{4}{5} + \cdots$$

$$+ \log\frac{n}{n+1} + \cdots + \log\frac{99}{100}$$

what integer does A represent?

107. Find the positive value of x satisfying $x^2 + 4^{\log_2 x} = 8$.

108. What is the greatest integer that is less than the number $\log_4 9 + \log_9 28$? (Do not use a calculator.)

109. Without using a calculator, prove that $\log_9 16 = \log_3 4$.

Review Problems

110. If $f(x) = 3x + 17$, find $f^{-1}(x)$ and then verify that $f(f^{-1}(x)) = x$ and $f^{-1}(f(x)) = x$.

111. Solve the following system:

$$x - 5y - 2z = 6$$
$$2x - 3y + z = 13$$
$$3x - 2y + 4z = 22$$

112. Graph the inequality: $5x - 2y > 10$.

Solving Exponential and Logarithmic Equations

Objectives

1 Solve exponential equations.
2 Solve applied problems involving exponential equations.
3 Solve logarithmic equations.
4 Solve applied problems involving logarithmic equations.

In this section, we discuss procedures for solving equations with variables in the exponents and variables in logarithmic expressions. These procedures will give us further insight into exponential and logarithmic models.

Solve exponential equations.

Solving Exponential Equations

An *exponential equation* is an equation containing a variable in an exponent, such as $2^{3x} = 32$. In the case of this equation, we can express both sides with a common base.

$$2^{3x} = 2^5$$

Since the two sides are equal and the base 2 appears on each side, the exponents must represent the same number. We set the exponents on each side of the equation equal to each other and then solve for x.

$$3x = 5$$

$$x = \frac{5}{3}$$

This process is based on the fact that $f(x) = b^x$ is a one-to-one function. If the outputs b^x and b^y are equal, then the inputs x and y must also be equal. We can state this observation as follows.

> **The one-to-one property of exponential functions**
>
> For any positive number b such that $b \neq 1$,
>
> $$b^x = b^y \text{ is equivalent to } x = y$$

Here's a solution process for solving exponential equations where it is possible to write each side as a power of the same base.

> **Solving exponential equations using the one-to-one exponential function property**
>
> 1. Express each side of the equation as a power of the same base.
> 2. Set the exponents equal to each other.
> 3. Solve the resulting equation.
> 4. Check solution(s) in the original equation.

sing technology

The graphs of

$$y_1 = 2^{3x+1}$$
$$\text{and} \quad y_2 = 16$$

have an intersection point whose x-coordinate is 1. This verifies that $\{1\}$ is the solution set for $2^{3x+1} = 16$.

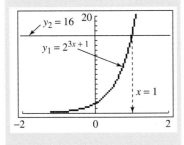

EXAMPLE 1 Solving Exponential Equations Using the One-to-One Property

Solve: **a.** $2^{3x+1} = 16$ **b.** $16^x = 64$

Solution

For each equation, we write each side as a power of the same base.

a. Because 16 is 2^4, we express each side in terms of base 2.

$2^{3x+1} = 16$	This is the given equation.
$2^{3x+1} = 2^4$	Write each side as a power of the same base.
$3x + 1 = 4$	If two powers of the same base are equal, then the exponents are equal.
$3x = 3$	Subtract 1 from both sides.
$x = 1$	Divide both sides by 3.

Check:

$2^{3x+1} = 16$	This is the original equation.
$2^{3 \cdot 1 + 1} \stackrel{?}{=} 16$	Substitute 1 for x.
$2^4 \stackrel{?}{=} 16$	
$16 = 16$ ✓	

This true statement verifies that the solution set is $\{1\}$.

b. Because $16 = 4^2$ and $64 = 4^3$, we express each side in terms of base 4.

$16^x = 64$	This is the given equation.
$(4^2)^x = 4^3$	Write each side as a power of the same base.

iscover for yourself

The equation $16^x = 64$ can also be solved by writing each side in terms of base 2. Do this. What solution method do you prefer?

Using technology

The graphs of

$$y_1 = 16^x$$
and $y_2 = 64$

have an intersection point whose x-coordinate is $\frac{3}{2}$. This verifies that $\left\{\frac{3}{2}\right\}$ is the solution set for $16^x = 64$.

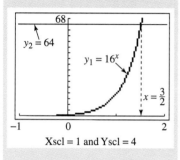

Xscl = 1 and Yscl = 4

$$4^{2x} = 4^3 \qquad (b^n)^m = b^{nm}, \text{ so } (4^2)^x = 4^{2x}.$$
$$2x = 3 \qquad \text{If two powers of the same base are equal, then the exponents are equal.}$$
$$x = \frac{3}{2} \qquad \text{Divide both sides by 2.}$$

Check

$$16^x = 64 \qquad \text{This is the original equation.}$$
$$16^{3/2} \stackrel{?}{=} 64 \qquad \text{Substitute } \tfrac{3}{2} \text{ for } x.$$
$$(\sqrt{16})^3 \stackrel{?}{=} 64 \qquad b^{m/n} = (\sqrt[n]{b})^m$$
$$4^3 \stackrel{?}{=} 64 \qquad \sqrt{16} = 4$$
$$64 = 64 \quad \checkmark$$

This true statement verifies that the solution set is $\left\{\frac{3}{2}\right\}$.

A more general method is needed for solving exponential equations such as

$$4^x = 15 \quad \text{or} \quad 2e^x = 8.$$

For these equations, it does not seem possible to express each side as a power of the same base. Logarithms are extremely useful in solving such equations. The solution begins by taking the logarithm of both sides of the equation, using the following property.

If $M = N$ $(M > 0 \text{ and } N > 0)$, then $\log_b M = \log_b N$.

Any base can be used when taking the logarithm on both sides. Since the natural logarithm function appears most frequently in mathematical models, we will always use natural logarithms. Let's see precisely how this is done.

Using technology

A graphing utility's intersection feature shows that

$$y_1 = 4^x \quad \text{and} \quad y_2 = 15$$

have an intersection point whose x-coordinate is approximately 1.95. Since

$$\frac{\ln 15}{\ln 4} \approx 1.95$$

this verifies that $\left\{\dfrac{\ln 15}{\ln 4}\right\}$ is the solution set for $4^x = 15$.

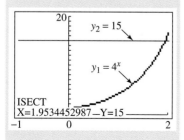

EXAMPLE 2 **Solving Exponential Equations**

Solve: **a.** $4^x = 15$ **b.** $2e^x = 8$

Solution

a. $4^x = 15$ This is the given equation.
$$\ln 4^x = \ln 15 \qquad \text{Take the natural logarithm on both sides.}$$
$$x \ln 4 = \ln 15 \qquad \text{Use the power rule for logarithms.}$$
$$x = \frac{\ln 15}{\ln 4} \qquad \text{Solve for } x \text{ by dividing both sides by } \ln 4.$$

We now have an exact value for x. We can obtain a decimal approximation using a calculator.

$x \approx 1.95$ Calculator keystroke sequence: $\boxed{\ln}\ 15\ \boxed{\div}\ \boxed{\ln}\ 4\ \boxed{\text{ENTER}}$

Since $4^2 = 16$, it seems reasonable that the solution to $4^x = 15$ is approximately 1.95. The equation's solution set is $\left\{\dfrac{\ln 15}{\ln 4}\right\}$. We use the exact value for x in expressing the solution set.

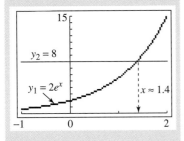

b.

$2e^x = 8$	This is the given equation.
$e^x = 4$	Isolate the exponential factor by dividing both sides by 2.
$\ln e^x = \ln 4$	Take the natural logarithm on both sides.
$x = \ln 4$	Use the property $\ln e^x = x$ on the left.
≈ 1.39	Use a calculator to obtain this decimal approximation.

As shown in the Using Technology, 1.39 or 1.4 is an approximate solution, verifying that the solution set is {ln 4}. ∎

EXAMPLE 3 Solving an Exponential Equation

Solve: $40e^{0.6x} = 240$

Solution

We will isolate $e^{0.6x}$ on one side of the equation and then take the natural logarithm on both sides.

$40e^{0.6x} = 240$	This is the given equation.
$e^{0.6x} = 6$	Isolate the exponential factor by dividing both sides by 40.
$\ln e^{0.6x} = \ln 6$	Take the natural logarithm on both sides.
$0.6x = \ln 6$	Use the property $\ln e^x = x$ on the left.
$x = \dfrac{\ln 6}{0.6} \approx 2.99$	Divide both sides by 0.6.

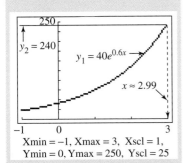
Thus, the solution of the equation is $\dfrac{\ln 6}{0.6} \approx 2.99$. Try checking this approximate solution in the original equation, verifying that $\left\{\dfrac{\ln 6}{0.6}\right\}$ is the solution set. ∎

Here are some guidelines for solving exponential equations where both sides cannot easily be written with a common base.

Using natural logarithms to solve exponential equations

1. Isolate the exponential expression.
2. Take the natural logarithm on both sides of the equation.
3. Simplify using one of the following properties:

   $\ln b^x = x \ln b$ or $\ln e^x = x$

4. Solve for the variable.

2 Solve applied problems involving exponential equations.

Applications Involving Exponential Equations

Solving exponential equations enables us to make predictions about variables that appear in exponential models. This idea is illustrated in the following example on compound interest.

| EXAMPLE 4 | **Revisiting the Formula for Compound Interest** |

The model

$$A = P\left(1 + \frac{r}{n}\right)^{nt}$$

describes the accumulated value A of a sum of money P, the principal, after t years at annual percentage rate r (in decimal form), compounded n times a year (rather than compounded continuously). How long will it take $1000 to grow to $3600 at 8% annual interest compounded quarterly?

Solution

$$A = P\left(1 + \frac{r}{n}\right)^{nt} \qquad \text{This is the given model.}$$

$$3600 = 1000\left(1 + \frac{0.08}{4}\right)^{4t} \qquad \begin{array}{l}\text{Substitute the given values: } A \text{ (the given accumulated} \\ \text{value)} = \$3600, P \text{ (the principal)} = \$1000, r \text{ (the interest} \\ \text{rate)} = 8\% = 0.08, \text{ and } n = 4 \text{ (quarterly compounding).}\end{array}$$

Our goal is to solve for t. Let's first simplify within parentheses and reverse the two sides of the equation.

$$1000(1.02)^{4t} = 3600$$

$$(1.02)^{4t} = 3.6 \qquad \begin{array}{l}\text{Isolate the exponential expression by dividing both sides by} \\ 1000.\end{array}$$

$$\ln (1.02)^{4t} = \ln 3.6 \qquad \text{Take the natural logarithm on both sides.}$$

$$4t \ln 1.02 = \ln 3.6 \qquad \text{Use the power rule for logarithms.}$$

$$t = \frac{\ln 3.6}{4 \ln 1.02} \qquad \text{Solve for } t \text{ by dividing both sides by } 4 \ln 1.02.$$

$$\approx 16.2 \qquad \text{Use a calculator.}$$

After (approximately) 16.2 years, the $1000 will grow to an accumulated value of $3600.   ∎

One aim of algebra is to predict the behavior of variables. This can be done with *exponential growth* and *decay models*.

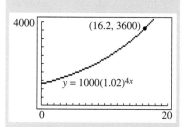
Exponential growth and decay models

The mathematical model for exponential growth or decay is given by

$$f(t) = Ae^{kt} \qquad \text{or} \qquad y = Ae^{kt}.$$

Within this model, t represents time, A is the original amount (or size) of the growing (or decaying) quantity, and y or $f(t)$ represents the amount (or size) of the quantity at time t. The number k represents a constant that depends on the rate of growth or decay. If $k > 0$, the formula represents *exponential growth;* if $k < 0$, the formula represents *exponential decay.*

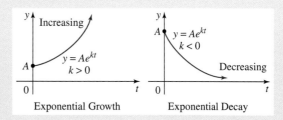

The model for continuous compounding is of the form $y = Ae^{kt}$. Consequently, we say that money grows exponentially under continuous compounding of interest. Another common application of exponential growth is in predicting population growth, illustrated in Example 5.

> **EXAMPLE 5** **Population Growth and the Carrying Capacity of Earth**

In a report entitled *Resources and Man,* the U.S. National Academy of Sciences concluded that a world population of 10 billion "is close to (if not above) the maximum that an intensely managed world might hope to support with some degree of comfort and individual choice." For this reason, 10 billion is called the carrying capacity of Earth. At the time the report was issued in 1969, the world population was approximately 3.6 billion, with a relative growth rate of 2% per year. If exponential growth continues at this rate, in what year will Earth's carrying capacity be reached?

Solution

We will use $y = Ae^{kt}$, the exponential growth model, where

A (original 1969 population) = 3.6 billion

k (given growth rate) = 2% = 0.02

y (population x years after 1969) = 10 billion

We must now solve the model for t.

$y = Ae^{kt}$	This is the exponential growth model.
$10 = 3.6e^{0.02t}$	Substitute the given values.
$3.6e^{0.02t} = 10$	Reverse the sides of the equation. This step is optional.
$e^{0.02t} = \dfrac{10}{3.6}$	Isolate the exponential expression by dividing both sides by 3.6.
$\ln e^{0.02t} = \ln \dfrac{10}{3.6}$	Take the natural logarithm on both sides.
$0.02t = \ln \dfrac{10}{3.6}$	Use the property $\ln e^x = x$ on the left.
$t = \dfrac{\ln \dfrac{10}{3.6}}{0.02}$	Solve for t by dividing both sides by 0.02.
≈ 51	Use a calculator.

Arnoldo Pomodoro "Sphere No. 4" 1963–64 (cast 1964), bronze and patina, 185 cm: circumference. Peggy Guggenheim Collection, Venice. © 1998 Arnoldo Pomodoro/Licensed by VAGA, New York, NY/ The Solomon R. Guggenheim Foundation, New York. (FN 76.2553 PG 214). Photograph by Julian Kayne.

ENRICHMENT ESSAY

Radon

Radon is a colorless, radioactive gaseous element produced by the natural decay of radium. Emitted by rocks in the soil, radon enters buildings through cracks in the foundation. Radon-induced lung cancer is one of today's most serious health issues. Surveys done by the U.S. Environmental Protection Agency in 1990 indicate that 25% of American homes exceed the maximum allowable concentration of radon, posing about the same lung-cancer threat as smoking half a pack of cigarettes each day.

The exponential decay model for radon gas is $y = Ae^{-0.18t}$, with t in days. The model indicates that it will take about $\frac{3}{5}$ of a day for radon in a sealed sample of air to fall to 90% of its original value.

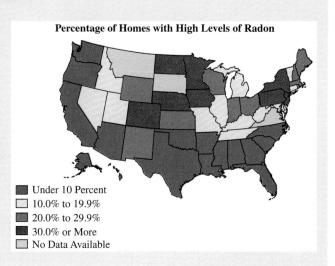

Percentage of Homes with High Levels of Radon

- ■ Under 10 Percent
- □ 10.0% to 19.9%
- ■ 20.0% to 29.9%
- ■ 30.0% or More
- □ No Data Available

Earth's carrying capacity will be reached in approximately 51 years after 1969, or in the year 2020. ■

3 Solve logarithmic equations.

Solving Logarithmic Equations

A *logarithmic equation* is an equation containing a variable in a logarithmic expression. An example of such an equation is

$$\log_4(x + 3) = 2.$$

We can solve this equation by rewriting it as an equivalent exponential equation. Our next example illustrates how this is done.

EXAMPLE 6 **Solving a Logarithmic Equation**

Solve: $\log_4(x + 3) = 2$

Solution

First we rewrite the equation as an equivalent exponential equation.

$$\log_b M = N \quad \text{means} \quad b^N = M.$$

$$\log_4(x + 3) = 2 \quad \text{means} \quad 4^2 = x + 3$$

Now we solve the equivalent exponential equation for x.

$$4^2 = x + 3$$
$$16 = x + 3$$
$$13 = x$$

Check

$$\log_4(x + 3) = 2 \qquad \text{This is the original equation.}$$

$$\log_4(13 + 3) \overset{?}{=} 2 \qquad \text{Substitute 13 for } x.$$

$$\log_4 16 \overset{?}{=} 2$$

$$2 = 2 \quad \checkmark \qquad \log_4 16 = 2 \text{ because } 4^2 = 16.$$

This true statement indicates that the solution set is {13}. ■

Using technology

The graphs of

$$y_1 = \log_4(x + 3)$$

and $\;y_2 = 2$
have an intersection point whose x-coordinate
is 13. This verifies that {13} is the solution set
for $\log_4(x + 3) = 2$. Note: Since

$$\log_b x = \frac{\ln x}{\ln b} \qquad \text{Change-of-base property}$$

we entered y_1 using

$$y_1 = \log_4(x + 3) = \frac{\ln(x + 3)}{\ln 4}.$$

The solution used in Example 6 works when one side of the equation contains a single logarithm and the other side contains a constant. Once the equation is in the form $\log_b M = N$, we rewrite it in the exponential form $b^N = M$. In the next example, we use the product rule for logarithms to obtain a single logarithmic expression on the left side.

EXAMPLE 7 **Using the Product Rule to Solve a Logarithmic Equation**

Solve: $\log_2 x + \log_2(x - 7) = 3$

Solution

$$\log_2 x + \log_2(x - 7) = 3 \qquad \text{This is the given equation.}$$

$$\log_2 x(x - 7) = 3 \qquad \text{Use the product rule to obtain a single logarithm.}$$

$$2^3 = x(x - 7) \qquad \text{If } \log_b M = N, \text{ then } b^N = M.$$

$$8 = x^2 - 7x \qquad \text{Apply the distributive property on the right.}$$

$$0 = x^2 - 7x - 8 \qquad \text{Set the equation equal to 0.}$$

$$(x - 8)(x + 1) = 0 \qquad \text{Factor.}$$

$$x - 8 = 0 \quad \text{or} \quad x + 1 = 0 \qquad \text{Set each factor equal to 0.}$$

$$x = 8 \qquad\qquad x = -1 \qquad \text{Solve for } x.$$

The graphs of

$$y_1 = \log_2 x + \log_2(x - 7),$$

entered as

$$y_1 = \frac{\ln x}{\ln 2} + \frac{\ln(x - 7)}{\ln 2}$$

and $y_2 = 3$

have only one intersection point whose x-coordinate is 8. This verifies that $\{8\}$ is the solution set for

$$\log_2 x + \log_2(x - 7) = 3.$$

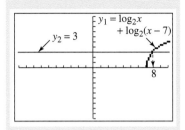

Check

Checking 8:

$$\log_2 x + \log_2(x - 7) = 3$$
$$\log_2 8 + \log_2(8 - 7) \stackrel{?}{=} 3$$
$$\log_2 8 + \log_2 1 \stackrel{?}{=} 3$$

$$3 + 0 \stackrel{?}{=} 3$$
$$3 = 3 \quad \checkmark$$

The solution set is $\{8\}$.

Checking − 1:

$$\log_2 x + \log_2(x - 7) = 3$$
$$\log_2(-1) + \log_2(-1 - 7) \stackrel{?}{=} 3$$

The number − 1 does not check. Negative numbers do not have logarithms.

Solving an equation containing logarithmic expressions and constants

1. Combine all logarithmic expressions on one side and all constants on the other side of the equation.
2. Use logarithmic properties to rewrite the logarithmic expression as a single logarithm whose coefficient is 1. The form of the equation is $\log_b M = N$.
3. Rewrite step 2 in exponential form $b^N = M$ and solve the resulting equation for the variable.
4. Check each solution in the original equation, rejecting values that produce the logarithm of a negative number or the logarithm of 0.

EXAMPLE 8 **Using the Quotient Rule to Solve a Logarithmic Equation**

Solve: $\log(9x - 2) - \log(x - 4) = 1$

Solution

$$\log(9x - 2) - \log(x - 4) = 1$$

This is the given equation.

$$\log_{10}\left(\frac{9x - 2}{x - 4}\right) = 1$$

Use the quotient rule to obtain a single logarithm. Since we have common logarithms, we wrote in the base, 10.

$$10^1 = \frac{9x - 2}{x - 4}$$

If $\log M = N$, then $10^N = M$.

$$10(x - 4) = 9x - 2$$

Clear fractions, multiplying both sides by $x - 4$.

$$10x - 40 = 9x - 2$$

Apply the distributive property.

$$x = 38$$

Solve for x, subtracting $9x$ and adding 40 on both sides.

Discover for yourself

Use a graphing utility to verify the solution set in Example 8.

Check

$$\log(9x - 2) - \log(x - 4) = 1$$

This is the given equation.

$$\log[9(38) - 2] - \log(38 - 4) \stackrel{?}{=} 1$$

Substitute 38 for x.

$$\log 340 - \log 34 \stackrel{?}{=} 1$$

Simplify within grouping symbols.

$$\log\left(\frac{340}{34}\right) \stackrel{?}{=} 1$$

Since the evaluation of the logarithms is not obvious without a calculator, use the quotient rule.

$$\log 10 \stackrel{?}{=} 1$$

$$1 = 1 \quad \checkmark \quad \log_{10} 10 = 1 \text{ because } 10^1 = 10.$$

The solution set is $\{38\}$.

4 Solve applied problems involving logarithmic equations.

Modeling Applications Involving Logarithmic Equations

The solution method for logarithmic equations provides further insight into variables contained in logarithmic models.

EXAMPLE 9 **Acid Rain**

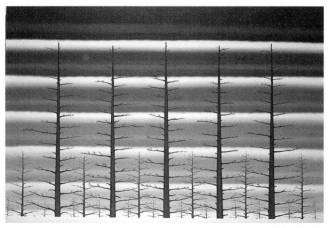

Roger Brown "Acid Rain" 1984, oil on canvas, 48 × 72 in. Photo Courtesy Phyllis Kind Gallery, New York & Chicago.

Chemists have defined the pH (hydrogen potential) of a solution by

$$pH = -\log[H^+]$$

where $[H^+]$ represents the concentration of the hydrogen ion in moles per liter. The pH scale ranges from 0 to 14, depending on a solution's acidity or alkalinity. Values below 7 indicate progressively greater acidity, while those above 7 are progressively more alkaline.

Normal, unpolluted rain has a pH of about 5.6. The acidity of precipitation over Canada and the United States is shown in Figure 8.30. Since the pH

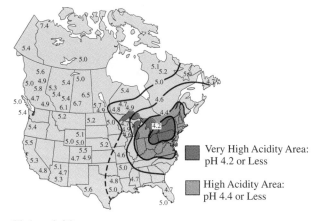

Very High Acidity Area: pH 4.2 or Less

High Acidity Area: pH 4.4 or Less

Figure 8.30

The acidity of precipitation over Canada and the United States

Sources: Data from the National Atmospheric Deposition Program and the Canadian Network for Sampling of Precipitation

scale is logarithmic, keep in mind that there is a tenfold difference in hydrogen ion concentration for each pH unit.

An environmental concern involves the destructive effects of acid rain, which is caused primarily by sulfur dioxide emissions. The most acidic rainfall ever had a pH of 2.4. What was the hydrogen ion concentration?

Solution

$$pH = -\log [H^+] \qquad \text{This is the given model.}$$
$$-\log [H^+] = 2.4 \qquad \text{Substitute 2.4 for the pH.}$$
$$\log [H^+] = -2.4 \qquad \text{We want a single logarithmic expression with coefficient 1.}$$
$$\qquad\qquad\qquad\qquad \text{Multiply both sides by } -1.$$
$$10^{-2.4} = [H^+] \qquad \text{Rewrite in exponential form.}$$

The hydrogen ion concentration of the most acidic rainfall ever is $10^{-2.4}$, or approximately 0.004 mole per liter.   ■

PROBLEM SET 8.5

Practice Problems

Solve each equation in Problems 1–16 by expressing each side as a power of the same base and then equating exponents.

1. $2^x = 64$
2. $3^x = 81$
3. $5^x = 125$
4. $5^x = 625$

5. $2^{2x-1} = 32$
6. $3^{2x+1} = 27$
7. $4^{2x-1} = 64$
8. $5^{3x-1} = 125$

9. $32^x = 8$
10. $4^x = 32$
11. $9^x = 27$
12. $125^x = 625$

13. $9^x = \frac{1}{81}$
14. $25^x = \frac{1}{125}$
15. $4^{2x-7} = \frac{1}{128}$
16. $81^{3x-4} = \frac{1}{243}$

Solve each equation in Problems 17–34 by taking the natural logarithm of both sides. Express the answer in terms of natural logarithms. Then use a calculator to obtain a decimal approximation, correct to the nearest thousandth, for the solution.

17. $2^x = 7$
18. $3^x = 11$
19. $e^x = 5$
20. $e^x = 7$

21. $(2.7)^x = 31$
22. $(3.2)^x = 90$
23. $e^{0.5x} = 9$
24. $e^{0.7x} = 13$

25. $5^{-0.03t} = 0.07$
26. $7^{-0.04t} = 0.06$
27. $e^{-0.03t} = 0.09$
28. $e^{0.08t} = 4$

29. $30 - (1.4)^x = 0$
30. $135 - (4.7)^x = 0$

31. $1250e^{0.055t} = 3750$
32. $1000\left(1 + \frac{0.045}{4}\right)^{4t} = 3600$

33. $800 - 500 \cdot 2^{-0.5t} = 733$
34. $600 - 300e^{-0.2t} = 450$

Solve each logarithmic equation in Problems 35–58. Be sure to reject any value that produces the logarithm of a negative number or the logarithm of 0.

35. $\log_3 x = 4$
36. $\log_5 x = 3$
37. $\log_2 x = -4$
38. $\log_9 x = \frac{1}{2}$

39. $\log x = 2$
40. $\log x = -2$
41. $\ln x = 3$
42. $\ln x = -2$

43. $\log_4(x + 5) = 3$
44. $\log_5(x - 4) = 2$
45. $\log_3(x - 2) = -3$
46. $\log_7(x + 1) = -2$

47. $\log_4(2x - 1) = 3$
48. $\log_2(3x + 1) = 5$
49. $\log_5 x + \log_5(4x - 1) = 1$

50. $\log_6(x + 5) + \log_6 x = 2$
51. $\log_3(x - 5) + \log_3(x + 3) = 2$
52. $\log_2(x - 1) + \log_2(x + 1) = 3$

53. $\log_4(x + 2) - \log_4(x - 1) = 1$

54. $\log_2(x + 2) - \log_2(x - 5) = 3$

55. $\log(3x - 5) - \log 5x = 2$

56. $\log(2x - 1) - \log x = 2$

57. $\log x - \log(x + 5) = -1$

58. $\log(x + 4) - \log x = -1$

Application Problems

59. Use the model

$$A = P\left(1 + \frac{r}{n}\right)^{nt}$$

describing the accumulated value of a sum of money P after x years at annual percentage rate r (in decimal form), compounded n times a year to answer this question. How long will it take $1000 to grow to $2600 at 6% annual interest compounded quarterly?

60. Use the model in Problem 59 to answer this question. What interest rate is necessary for $200 to accumulate to $350 over a 5-year period if the interest is to be compounded annually? (Round your answer to the nearest tenth of a percent.)

61. The model $A = Pe^{rt}$ describes the accumulated value of a sum of money P after t years at annual percentage rate r (in decimal form), compounded continuously. What interest rate is required for an investment of $8000 to grow to $12,000 in 5 years? (Round your answer to the nearest tenth of a percent.)

62. The model $A = Pe^{rt}$ describes the accumulated value of a sum of money P after t years at annual percentage rate r (in decimal form), compounded continuously. How long will it take $8000 to double if it is invested at 8% per year compounded continuously? (Round your answer to the nearest tenth of a year.)

63. The percentage of information that a particular person retains after x weeks is described by the memory model $f(x) = 80e^{-0.5x} + 20$. After how many weeks is half the material retained? (Round your answer to the nearest tenth of a week.)

64. After x months of training, the learning function $f(x) = 800 - 500e^{-0.5x}$ describes the number of letters per hour that a postal clerk can sort. How many months of training are required for a clerk to sort 733 letters per hour? (Round your answer to the nearest month.)

Use the exponential growth model $y = Ae^{kt}$ to solve Problems 65–66.

65. These statistics were issued in 1995 by the Population Reference Bureau in Washington, D.C.:

Region	1995 Population (billions)	Percentage of Population in 1995	Growth Rate (k)
World	5.702	100	1.5% = 0.015
More developed regions	1.169	20.5	0.2% = 0.002
Less developed regions	4.533	79.5	1.9% = 0.019

Let $t = 0$ correspond to 1995 and assume that the indicated growth rates are valid for at least 15 years. By what year (that is, how many years after 1995) will the world population be 6.37 billion? At that time, what will the percentage of the world's population be for less developed regions?

66. These statistics were issued in 1995 by the Population Reference Bureau in Washington, D.C.:

Country	1995 Population (millions)	Growth Rate (k)
United Kingdom	58.6	0.2% = 0.002
Iraq	20.6	3.7% = 0.037

The given growth rates (percentage per year) indicate that the United Kingdom has one of the lowest growth rates in the world and Iraq has one of the highest. Let $t = 0$ correspond to 1995 and assume that the indicated growth rates are valid for at least 15 years. By what year (that is, how many years after 1995) will the United Kingdom have a population of 59.2 million? What will Iraq's population be at that time?

The model $P(x) = 95 - 30 \log_2 x$ is a hypothetical formula describing the percentage ($P(x)$) of students who could recall the important features of a classroom lecture as a function of time, where x represents the number of days that have elapsed since the lecture was given. The figure shows the graph of the model. Use this information to solve Problems 67–68.

67. After how many days do only half the students recall the important features of the classroom lecture? (Let $P(x) = 50$ and solve for x.) Can you approximately locate the point on the graph that conveys this information?

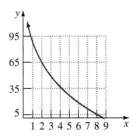

68. After how many days have all students forgotten the important features of the classroom lecture? (Let $P(x) = 0$ and solve for x.) Can you approximately locate the point on the graph that conveys this information?

69. The pH of a substance is given by pH $= -\log [H^+]$, where $[H^+]$ is the hydrogen ion concentration in moles per liter. The pH of human blood ranges between 7.37 and 7.44. What are the corresponding bounds for hydrogen ion concentration?

70. Altimeters used on most aircraft determine altitude (a, in meters above sea level) by measuring the outside barometric pressure (P, in centimeters of mercury) and temperature (T, in degrees Celsius). The model used is

$$a = (30T + 8000) \ln\left(\frac{P_0}{P}\right)$$

in which P_0 is atmospheric pressure at sea level. If the atmospheric pressure at sea level is 76 centimeters of mercury and the outside temperature at an altitude of 8830 meters is $-3°$ Celsius, what is the outside barometric pressure?

True–False Critical Thinking Problems

71. Which one of the following is true?
   a. If $2^{x+1} = 3^{5x-7}$, then $x + 1 = 5x - 7$.
   b. If $7^x = 19$, then $x \ln 7 = 19$.
   c. If $\log(x + 3) = 2$, then $e^2 = x + 3$.
   d. If $\log(x + 1) + \log(x - 1) = 5$, then $10^5 = x^2 - 1$.

72. Which one of the following is true?
   a. An earthquake measuring 8.6 on the Richter scale is twice as intense as one measuring 4.3 on the Richter scale.
   b. If $x = \dfrac{1}{k} \ln y$, then $y = e^{kx}$.
   c. If $\log(7x + 3) - \log(2x + 5) = 4$, then in exponential form $10^4 = (7x + 3) - (2x + 5)$.
   d. If $\ln(x + 3) = 2$, then $10^2 = x + 3$.

Technology Problems

73. Use a graphing utility to verify the solutions that you obtained for any ten equations in Problems 1–58.

74. Use a graphing utility to graph the given models in Problems 63 and 64. Then $\boxed{\text{TRACE}}$ along each curve until you reach the point whose coordinates correspond to the problem's solution.

75. The model $P = 145e^{-0.092t}$ describes a runner's pulse (P, in beats per minute) t minutes after a race, where $0 \le t \le 15$. Graph the function using a graphing utility. $\boxed{\text{TRACE}}$ along the graph and determine after how many minutes the runner's pulse will be 70 beats per minute. Verify your observation algebraically.

76. The model $W = 2600(1 - 0.51e^{-0.075t})^3$ describes the weight (W, in kilograms) of a female African elephant at age t years. (1 kilogram $\approx$ 2.2 pounds) Use a graphing utility to graph the weight function. Then $\boxed{\text{TRACE}}$ along the curve to estimate the age of an adult female elephant weighing 1800 kilograms.

Writing in Mathematics

77. The exponential equation $2^{4-5x} = 8$ can be solved by writing each side with a common base, resulting in $2^{4-5x} = 2^3$. The equation can also be solved by taking $\log_2$ on both sides. Solve the equation using both methods and then compare and contrast the two methods.

78. Discuss the significance of the following remark in terms of the exponential population growth model (Example 5).

The exponential growth of population and its attendant assault on the environment is so recent that it is difficult for people to appreciate how much damage is being done.
Nathan Keyfitz in "The Growing Human Population," *Scientific American* 261 (September 1989):119–126.

Critical Thinking Problems

79. Use the table in Problem 66 to estimate the year in which Iraq and the United Kingdom will have equal populations. Round your answer to the nearest 5 years.

80. The model

$$A = \frac{A_0}{k}(e^{kt} - 1)$$

describes the amount A of nonrenewable-energy resources (oil, natural gas, coal, uranium) from time $= 0$ to time $= t$, where A_0 is the amount of the resource consumed during the year $t = 0$ and k is the relative growth rate of annual consumption. Use this model to answer the question in this problem. In the Global 2000 Report to President Jimmy Carter, the 1976 worldwide consumption of oil was 21.7 billion barrels, and the predicted growth rate for oil consumption was 3% per year ($k = 0.03$). At that time,

the total remaining oil resources ultimately available were 1661 billion barrels. Using these figures, by what year will the planet be depleted of its oil resources?

81. Solve the following system of equations:

$$x + y = 25$$
$$\log x + \log y = 2$$

82. If $f(x) = \log_3 x$, find x such that $f(f(f(x))) = 3$. Generalize this result by replacing 3 by b, where $b > 0$ and $b \neq 1$.

83. Solve for x: $\log_b(\log_b Ax) = 1; A > 0$.

84. If $f(x) = x + 2$ and $g(x) = x$, find $x > -2$ satisfying $3^{g(x)\log_3 f(x)} = f(x)$.

85. Solve for x: $\log_x 25 - \log_x 4 = \log_x \sqrt{x}$.

86. Solve for x: $\log_2 x + \dfrac{1}{\log_x 2} = 4$.

Group Activity Problem

87. This problem is appropriate for discussion in a small group. India is currently one of the world's fastest-growing countries, with a growth rate of about 2.6% annually. Assuming that the relative growth rate is constant, the population will reach 1.5 billion sometime in the year 2010. Treating the year 1974 as year 0, the growth model is $y = 574,220,000e^{0.026t}$.

One problem with this model is that nothing can grow exponentially indefinitely. Although India grew at an average rate of 2.6% per year from 1974 to 1984, the model is based on the assumption that India's growth rate will remain the same at least until the year 2010.

Is this a likely assumption? Can you suggest factors that influence population growth rates for India and other countries? What factors might limit the size of a

population? Can you suggest factors that might lead to a greater population growth rate than the one predicted by an exponential growth model? How might these factors be accounted in a mathematical model predicting population growth or in the model's graph? (In answering this last question, consider the model

$$y = \frac{5000}{1 + e^{-t/6}}$$

that describes the population y of a certain species t years after it is introduced into a new habitat.)

If the assumption of an indefinite growth at a constant rate is unrealistic, why do you think models based on this assumption are widely used? What does this tell you about making predictions based on mathematical models?

Review Problems

88. Write the equation of the line passing through $(1, 5)$ and perpendicular to the line passing through $(4, 2)$ and $(-2, 3)$. Express the equation in slope-intercept form.

89. Solve for x: $\sqrt{x + 4} - \sqrt{x - 1} = 1$.

90. Solve: $\dfrac{3}{y + 1} - \dfrac{5}{y} = \dfrac{19}{y^2 + y}$.

SECTION　8.6

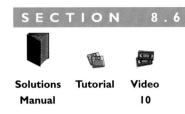

Solutions　Tutorial　Video
Manual　　　　　10

Modeling with Exponential and Logarithmic Functions

Objectives

1　Model exponential growth and decay.
2　Fit exponential functions to data.
3　Fit logarithmic functions to data.

In this section, we continue our work with applications of exponential and logarithmic functions. All examples and problems involve modeling actual real world situations.

①　Model exponential growth and decay.

Exponential Growth and Decay Models

In the previous section, we introduced the exponential growth and decay models

$$f(t) = Ae^{kt} \quad \text{or} \quad y = Ae^{kt}$$

where A is the original amount (or size) of a growing (or decaying) entity at $t = 0$, y (or $f(t)$) is the amount at time t, and k is a growth ($k > 0$) or decay ($k < 0$) constant. Some situations require us to use given data in growth and decay models to determine k. Once this value is computed, it is substituted into the model $y = Ae^{kt}$ so that the value of a variable in the model can be determined. This idea is illustrated in our first two examples.

EXAMPLE 1　Modeling the Spread of AIDS

Through the end of 1991, 200,000 cases of AIDS had been reported to the Center for Disease Control in the United States. By the end of 1995, the number had grown to 513,000.

a. Find the exponential growth function that models this data.
b. In what year will the cumulative number of AIDS cases in the United States reach one million?

Solution

a. We use the exponential growth model

$$y = Ae^{kt}$$

where t is the number of years since 1991. This means that 1991 corresponds to $t = 0$. At that time there were 200,000 cases, so A, the original amount, is 200,000. We substitute 200,000 for A in the growth model.

$$y = 200{,}000e^{kt}$$

We are given that there were 513,000 cases at the end of 1995. Since 1995 is 4 years after 1991, when $t = 4$ the value of y is 513,000. Substituting these numbers into the growth model will enable us to find k, the growth constant or the growth rate.

$$y = 200{,}000e^{kt} \qquad \text{This is the growth model with } A = 200{,}000.$$

$$513{,}000 = 200{,}000e^{k \cdot 4} \qquad \text{When } t = 4, y = 513{,}000, \text{ so substitute these numbers into the model.}$$

$$e^{4k} = 2.565 \qquad \text{Isolate the exponential expression by dividing both sides by 200,000. We also reversed the sides.}$$

$$\ln e^{4k} = \ln 2.565 \qquad \text{Take the natural logarithm on both sides.}$$
$$4k = \ln 2.565 \qquad \text{Simplify the left side using } \ln e^x = x.$$
$$k = \frac{\ln 2.565}{4} \qquad \text{Divide both sides by 4 and solve for } k.$$
$$\approx 0.235 \qquad \text{Use a calculator.}$$

The exponential growth function is

$$y = 200{,}000e^{0.235t}$$

where t is measured in years since 1991.

b. To find the year in which there will be one million AIDS cases, we substitute 1,000,000 for y in the model and solve for t.

$$y = 200{,}000e^{0.235t} \qquad \text{This is the model from part (a).}$$
$$1{,}000{,}000 = 200{,}000e^{0.235t} \qquad \text{Substitute 1,000,000 for } y.$$
$$e^{0.235t} = 5 \qquad \text{Divide both sides by 200,000.}$$
$$\ln e^{0.235t} = \ln 5 \qquad \text{Take the natural logarithm on both sides.}$$
$$0.235t = \ln 5 \qquad \text{Simplify on the left using } \ln e^x = x.$$
$$t = \frac{\ln 5}{0.235} \qquad \text{Solve for } t \text{ by dividing both sides by 0.235.}$$
$$\approx 7 \qquad \text{Use a calculator.}$$

Since 7 is the number of years after 1991, the model (perhaps unrealistically) indicates that in 1998 there will be one million AIDS cases in the United States.   ■

The model in Example 1 is based on uncontrolled exponential growth. Perhaps we need to look at more data values to see if AIDS is increasing in an exponential way. For example, the World Health Organization predicts between 30 to 40 million HIV-infected people worldwide by the year 2000, where this estimate is a compromise between a conservative linear model and an exponential growth model.

Furthermore, being HIV-infected is not the same thing as having full-blown AIDS. The huge increase in the number of AIDS cases reported in 1993 was due primarily to an expanded definition of what constitutes AIDS. We ignored this factor in our model.

Reported AIDS Cases per Annum*

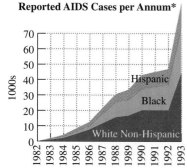

*Increase in 1993 due mostly to expanded case definition

Source: Center for Disease Control

Discover for yourself

Use a linear model on the data in Example 1 shown below. Use the point-slope formula with the given two data points and write the equation in slope-intercept form for the line modeling the data. Then use the linear model to predict the year in which the cumulative number of AIDS cases will reach one million. What year would be a good compromise between the predictions made by the exponential and the linear models?

x (Years after 1991)	y (Cumulative Reported AIDS Cases)
0	200,000
4	513,000

Slim Films

EXAMPLE 2 **Nuclear Energy and Exponential Decay**

Strontium-90 is a waste product from nuclear reactors. Its half-life is 28 years, which means that after 28 years a given amount of strontium-90 will have de-cayed to half the original amount. Unfortunately, strontium-90 in the atmosphere enters the food chain and is absorbed into our bones. As a consequence of fallout from atmospheric nuclear tests, we all have a measurable amount of strontium-90 in our bones.

a. Show that the decay model for strontium-90 is given by $y = Ae^{-0.024755t}$.
b. Suppose a nuclear accident occurs and releases 60 grams of strontium-90 into the atmosphere. How long will it take for strontium-90 to decay to a level of 10 grams?

Solution

a. We use the exponential decay model
$$y = Ae^{kt}.$$

Our goal is to find k, the decay constant or the decay rate. Strontium-90 has a half life of 28 years. In the decay model, A is the amount of strontium-90 at the beginning, so after 28 years the amount present is half of A, or $\frac{1}{2}A$.

$y = Ae^{kt}$	This is the exponential decay model.
$\dfrac{1}{2}A = Ae^{k28}$	After 28 years ($t = 28$), half the substance remains, so the amount present is $\frac{1}{2}A$.
$e^{28k} = \dfrac{1}{2}$	Isolate the exponential expression by dividing both sides by A. We also reversed the sides.
$\ln e^{28k} = \ln \dfrac{1}{2}$	Take the natural logarithm on both sides.
$28k = \ln \dfrac{1}{2}$	Simplify the left side using $\ln e^x = x$.
$k = \dfrac{\ln \frac{1}{2}}{28}$	Solve for k, dividing both sides by 28.
≈ -0.024755	Use a calculator.

The exponential decay function is
$$y = Ae^{-0.024755t}.$$

b. The nuclear accident releases 60 grams of strontium-90 into the atmosphere, so A, the amount initially present, is 60. We substitute 60 for A into the decay model.
$$y = 60e^{-0.024755t}$$

We can now use this model to determine how long it will take for strontium-90 to decay to a level of 10 grams. Substitute 10 for y in the model and solve for t.

$60e^{-0.024755t} = 10$	
$e^{-0.024755t} = \dfrac{1}{6}$	Isolate the exponential expression by dividing both sides by 60.

ENRICHMENT ESSAY

Carbon Dating

Carbon dating is a method for estimating the age of any organic material. Every living thing on Earth contains carbon. The half-life of carbon-14 is 5730 years, meaning that after 5730 years half of the carbon-14 atoms in any organic sample will have decayed into atoms of nitrogen-14. By comparing the amount of carbon-14 to nitrogen-14, archaeologists can establish the age of artifacts or fossils.

Prehistoric cave paintings were discovered in the Lascaux cave in France. Charcoal from the site was analyzed, and the paintings were estimated to be 15,505 years old.

After generations of controversy, the shroud of Turin, long believed to be the burial cloth of Jesus Christ, was shown by carbon-14 dating in 1988 to have been made later than A.D. 1200.

The shroud of Turin.
Gianni Tortori/Photo
Researchers, Inc.

Carbon dating is useful for artifacts or fossils up to 80,000 years old. Older objects do not have enough carbon-14 left to accurately age-date.

When you work Problem 5 in Problem Set 8.6, which uses carbon dating to determine the age of the Dead Sea Scrolls, model your solution using the method of Example 2.

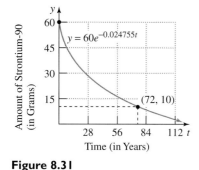

Lascaux cave painting. C. Seghers, 2/Photo Researchers, Inc.

Figure 8.31

The point on the graph indicates that it will take 72 years for 60 grams of strontium-90 to decay to 10 grams.

$$\ln e^{-0.024755t} = \ln \frac{1}{6} \qquad \text{Take the natural logarithm on both sides.}$$

$$-0.024755t = \ln \frac{1}{6} \qquad \ln e^x = x$$

$$t = \frac{\ln \frac{1}{6}}{-0.024755} \qquad \text{Solve for } t.$$

$$\approx 72 \qquad \text{Use a calculator.}$$

This means that it will take approximately 72 years for 60 grams of strontium-90 to decay to 10 grams. Figure 8.31 shows the graph of the exponential function that models this situation.

2 Fit exponential functions to data.

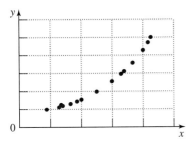

Figure 8.32

Data points suggest an exponential model.

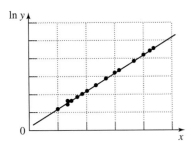

Figure 8.33

Data points suggest a linear model.

Fitting Exponential Functions to Data

Throughout this chapter, we have been working with models that were given. However, we also can create models that fit data by observing patterns in the graph of the data points.

For example, the data in Figure 8.32 suggest that we model the values with an exponential function. Since $\ln e^y = y$, if ordered pairs (x, y) lie on an exponential curve, then the ordered pairs $(x, \ln y)$ will lie on a straight line. This is shown in Figure 8.33, where we transformed the data in Figure 8.32 by taking the natural logarithm of each y value.

These observations give us a procedure for fitting exponential functions to data such as the points shown in Figure 8.32.

Modeling with exponential functions

1. Represent all data points using (x, y).
2. Take the natural logarithm of each y-coordinate.
3. Write an equation of the line that best fits the data points $(x, \ln y)$. The equation for this line, called the regression line, can be obtained with a graphing utility.
4. Since the equation in step 3 is for $\ln y$, use $e^{\ln y} = y$ to find y. Write each side of the equation as an exponent with base e.

EXAMPLE 3 **Fitting an Exponential Model to Data**

The data below indicate world population (in billions) for five years.

Year	World Population (in billions)
1950	2.6
1960	3.1
1970	3.7
1980	4.5
1989	5.3

Find a model for the data. Then use the model to predict world population in the year 2020.

Solution

The graph of the five data points in Figure 8.34 has a shape that suggests modeling with an exponential function.

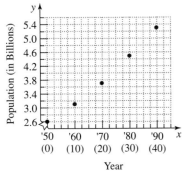

Figure 8.34

The graph of world population over time suggests exponential modeling.

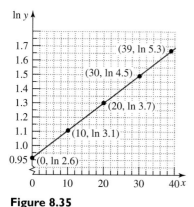

Figure 8.35

Data points suggest linear modeling.

Letting $x = 0$ represent 1950, we transform the points by taking the natural logarithm of each y-coordinate.

x (Year)	y (World Population)	$\ln y$
(1950) 0	2.6	0.9555
(1960) 10	3.1	1.1314
(1970) 20	3.7	1.3083
(1980) 30	4.5	1.5041
(1989) 39	5.3	1.6677

The five points for $(x, \ln y)$ suggest using a linear model, as shown in Figure 8.35. Using the linear regression option on a graphing utility, we find that the equation of the line that best fits the $(x, \ln y)$ transformed data is

$$\ln y = 0.0183x + 0.9504$$

Since we are interested in a model with y on the left, we use the property for inverse functions

$$e^{\ln y} = y$$

and write both sides of the equation as an exponent on base e.

$$e^{\ln y} = e^{0.0183x + 0.9504}$$

$$e^{\ln y} = e^{0.0183x} \cdot e^{0.9504} \qquad b^{M+N} = b^M \cdot b^N$$

$$y = e^{0.0183x} \cdot 2.5867 \qquad e^{\ln y} = y. \text{ Use a calculator to approximate } e^{0.9504}.$$

$$y = 2.5867e^{0.0183x} \qquad \text{Apply the commutative property on the right.}$$

The exponential model that fits the given data is

$$y = 2.5867e^{0.0183x}.$$

Since we often use t as the variable to represent time, we can equivalently write the model as

$$y = 2.5867e^{0.0183t} \quad \text{or} \quad f(t) = 2.5867e^{0.0183t}.$$

We are now ready to use the model to predict population in the year 2020.

$$f(t) = 2.5867e^{0.0183t} \qquad \text{This is one of the forms of the model we obtained.}$$

$$f(70) = 2.5867e^{0.0183(70)} \qquad \text{The year 2020 is 70 years after 1950, so substitute 70 for } t.$$

$$\approx 9.3 \qquad \text{Use a calculator.}$$

Using the model that we constructed from the data, predicted world population in the year 2020 is approximately 9.3 billion. ∎

3 Fit logarithmic functions to data.

Fitting Logarithmic Functions to Data

Suppose that the data we are attempting to model falls into a pattern like the one in Figure 8.36. The graph is increasing, and though it increases at a slower and slower rate, it does not seem to be leveling out to approach a horizontal asymptote. The shape suggests a logarithmic function, so we will attempt to model the data with such a curve. We can fit a logarithmic model to a set of points of the form (x, y) by fitting a linear model to points of the form $(\ln x, y)$. This idea is illustrated in our next example.

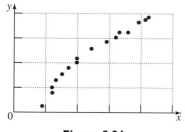

Figure 8.36

EXAMPLE 4 Fitting a Logarithmic Model to Data

Earlier in the chapter we encountered a model relating the size of a city with the average walking speed, in feet per second, of pedestrians. A study collected the following data.

Population (in thousands)	Walking Speed (in feet per second)
5.5	3.3
14	3.7
71	4.3
138	4.4
342	4.8

Source: Mark and Helen Bornstein, "The Pace of Life," *Nature*, 259 (19 Feb. 1976) 557–559.

Find a model that fits the given data.

Solution

The graph of the five data points in Figure 8.37 has a shape that suggests modeling the data with a logarithmic function. Represent each data value by (x, y) and transform the points by taking the natural logarithm of each x coordinate.

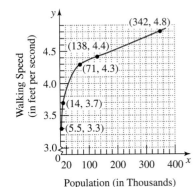

Figure 8.37

The graph of walking speed as a function of population suggests logarithmic modeling.

x	$\ln x$	y
5.5	1.7047	3.3
14	2.6391	3.7
71	4.2627	4.3
138	4.9273	4.4
342	5.8348	4.8

The five points for $(\ln x, y)$ are graphed in Figure 8.38. The points approximately fall along a line, suggesting a linear model.

Using the linear regression option on a graphing utility, we find that the equation of the regression line that models the transformed $(\ln x, y)$ data is

$$y = 0.3525 \ln x + 2.735.$$

This is the logarithmic model that fits the data. Compare this model with the one in Example 4 on page 706. ∎

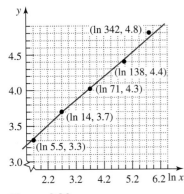

Figure 8.38

Data points suggest linear modeling.

Using the statistical menu of a graphing utility, you can immediately draw each data point as a coordinate. The resulting drawing is called a *scatter plot* of the data. The shape of the scatter plot determines what sort of function best fits the data. Possibilities include linear functions, quadratic functions, polynomial functions of degree greater than 2, exponential functions, and logarithmic functions. By telling your graphing utility the function that you want to use to model the data, the utility will find the equation of the function for you.

In this era of technology, since graphing utilities have capabilities that fit linear and nonlinear functions to data, the process of creating mathematical models involves decision-making more than computation.

tudy tip

If any data values fall approximately along a line, the equation of the regression line (or the line of best fit) can be obtained without a graphing utility using the following formulas:

Regression Line: $y = mx + b$

$$m = \frac{n(\Sigma xy) - (\Sigma x)(\Sigma y)}{n(\Sigma x^2) - (\Sigma x)^2}$$

$$b = \frac{(\Sigma y)(\Sigma x^2) - (\Sigma x)(\Sigma xy)}{n(\Sigma x^2) - (\Sigma x)^2}$$

where

n = the number of data points (x, y)

Σx = the sum of the x-values

Σy = the sum of the y-values

$\Sigma(xy)$ = the sum of the product of x and y in each pair

Σx^2 = the sum of the squares of the x-values

$(\Sigma x)^2$ = the square of the sum of the x-values

PROBLEM SET 8.6

Practice Problems and Application Problems

Use the exponential growth and decay models $y = Ae^{kt}$ or $f(t) = Ae^{kt}$, where A is the original amount (or size) of a growing (or decaying) entity at t = 0, y (or f(t)) is the amount at time t, and k is a growth k > 0) or decay(k < 0) constant to answer Problems 1–9.

1. The 1940 population of California was 6,907,387, and in 1950 the population was 10,586,223. Let $t = 0$ correspond to the year 1940.

a. Since 1940 corresponds to $t = 0$, at that time there were 6,907,387 people in California. This means that A, the original amount in the growth model, is 6,907,387. Now use the fact that there were 10,586,223 people in 1950 (when $t = 10$, $y = 10,586,223$) to find k, the growth rate for California during this decade.

b. Substitute the values for A and k into the exponential growth model and write the function that models California's population t years after 1940.

c. Use the model in part (c) to determine California's 1994 population ($t = 54$) if the growth rate from 1940 through 1950 had remained in effect from 1940 through 1994. How does this compare with the actual 1994 population of 31,431,000?

d. What are some of the factors that might account for the difference between the number predicted by the exponential growth model and the actual population?

2. The 1940 population of New York was 13,479,142 and in 1950 the population was 14,830,192. Let $t = 0$ correspond to the year 1940.

a. Since 1940 corresponds to $t = 0$, at that time there were 13,479,142 people in New York. This means that A, the original amount in the growth model, is 13,479,142. Now use the fact that there were 14,830,192 people in 1950 (when $t = 10$, $y = 14,830,192$) to find k, the growth rate for New York during this decade.

b. Substitute the values for A and k into the exponential growth model and write the function that models New York's population t years after 1940.

c. Use the model in part (c) to determine New York's 1994 population ($t = 54$) if the growth rate from 1940 through 1950 had remained in effect from 1940 through 1994. How does this compare with the actual 1994 population of 18,169,000?

d. What are some of the factors that might account for the difference between the number predicted by the exponential growth model and the actual population?

3. According to the World Health Organization Global Program on AIDS, in 1989 there were 40,637 reported AIDS cases in Africa. The number of reported cases for the year 1992 was 56,299.
   a. Find the exponential function that models the number of cases for year t, where t is measured in years since 1989.
   b. In what year will 500,000 AIDS cases be reported in Africa?
   c. The model for this data deals with *reported* AIDS cases. In what way is this a limitation of the model to describe what is really happening? What other limitations are there for this model?
   d. According to the World Health Organization, as of June 30, 1995 there were more than 11 million people infected with the HIV virus in Africa. Is this information conveyed by the model that you wrote in part (a)? Explain.

4. a. Use the exponential growth model to show that the time it takes a population to double (to grow from an initial number of A to $y = 2A$) is given by

   $$t = \frac{\ln 2}{k}.$$

   b. As of 1995, listed below are the ten countries in the world with populations exceeding 100 million. Excluding Russia, use the formula for doubling time

   $$t = \frac{\ln 2}{k}$$

   to estimate how long it will take each of these countries to double their population. (Can you see why demographers are concerned?)

Country	1995 Population (in millions)	Relative Growth Rate (% per year)
1. China	1218.8	1.1 ($k = 0.011$)
2. India	930.6	1.9 ($k = 0.019$)
3. United States	263.2	0.7 ($k = 0.007$)
4. Indonesia	198.4	1.6 ($k = 0.016$)
5. Brazil	157.8	1.7 ($k = 0.017$)
6. Russia	147.5	-0.6 ($k = -0.006$)
7. Pakistan	129.7	2.9 ($k = 0.029$)
8. Japan	125.2	0.3 ($k = 0.003$)
9. Bangladesh	119.2	2.4 ($k = 0.024$)
10. Nigeria	101.2	3.1 ($k = 0.031$)

   Source: *1995 World Population Data Sheet* (Washington, D.C.: Population Reference Bureau, Inc.)

5. a. Carbon-14 decays exponentially with a half-life of approximately 5730 years, meaning that after 5730 years a given amount of carbon-14 will have decayed to half the original amount. Use this information to show that the decay model for carbon-14 is given by $y = Ae^{-0.000121t}$.

   b. In 1947 earthenware jars containing what we now know as the Dead Sea Scrolls were found by an Arab Bedouin herdsman in search of a stray goat. Analysis indicated that the scroll wrappings contained 76% of their original carbon-14. Let $y = 0.76A$ in the model in part (a) and estimate the age of the Dead Sea Scrolls at the time they were found in 1947.

A portion of the Dead Sea Scrolls
Copyright, John C. Trever

6. The August 1978 issue of *National Geographic* described the 1964 find of dinosaur bones of a newly discovered dinosaur weighing 170 pounds, measuring 9 feet, with a 6-inch claw on one toe of each hind foot. The age of the dinosaur, called *Deinonychus* ("terrible claw"), was estimated using potassium-40 dating of rocks surrounding the bones.
   a. Potassium-40 decays exponentially with a half-life of approximately 1.31 billion years, meaning that after 1.31 billion years a given amount of potassium-40 will have decayed to half the original amount. Use this information to show that the decay model for potassium-40 is given by $y = Ae^{-0.52912t}$, where t is given in billions of years.
   b. Analysis of the surrounding rock indicated that 94.5% of the original amount of potassium-40 was still present. Let $y = 0.945A$ in the model in part (a) and estimate the age of the bones of *Deinonychus*.

7. The half-life of iodine-131 is 8 days. Suppose a country, through illegal above-ground nuclear testing, contaminates an island with debris containing 3000 grams of iodine-131. This is considered 30,000 times above a safe level for a team of scientists to visit the island and possibly identify the country violating the ban. How

long must the scientists wait before there is only 0.1 gram of iodine-131 and the level of radioactivity is safe for visiting the island?

8. Rework Problem 7 based on a radioactive substance of strontium-90 with a half-life of 28 years.

9. a. Use the exponential decay model to show that the time it takes for a substance to decay to half its original amount (to decay from an initial amount of A to $y = \dfrac{A}{2}$) is given by

$$t = \frac{\ln 2}{k}$$

Observe that this expression, called the halving time or half-life of a decaying substance, is the same expression as the doubling time for the exponential growth model. (See Problem 4(a).)

b. If the decay rate of krypton-85 is 6.3% per year ($k = 0.063$), find its half-life.

Suppose a set of data in the form (x, y) is graphed as a set of points. The resulting shape of the graph suggests modeling with an exponential function. The natural logarithm of each y-coordinate is taken and an equation of the line that best fits the data points $(x, \ln y)$ is obtained. The equation of this line is given in Problems 10–11. Write each side of each equation as an exponent with base e, simplify, and find the exponential function that fits the original data.

10. $\ln y = 0.3782 + 1.0745x$

11. $\ln y = -1.2074 + 0.873x$

12. The amount of carbon dioxide (CO_2) in the atmosphere has increased by half again its original amount since 1900, as a result of the burning of oil and coal. The buildup of gases and particles trap heat and raise the planet's temperature, a phenomenon called the greenhouse effect. Carbon dioxide accounts for about half of the warming.

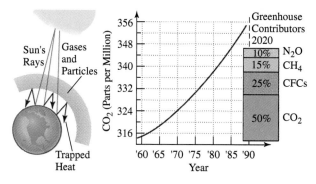

The data below shows the average CO_2 concentration in parts per million (ppm) x years after 1900.

(Year) x	y (Average CO_2 Concentration in ppm)
(1900) 0	280
(1950) 50	310
(1975) 75	331
(1980) 80	338
(1988) 88	351

a. Take the natural logarithm of each y-value.
b. Graph the five data points $(x, \ln y)$. Notice that the points lie nearly on a straight line, suggesting the use of a linear model.

c. Use the linear regression option on a graphing utility or the procedure outlined in the Study Tip on page 745 to write the equation of the line that best fits the $(x, \ln y)$ transformed data.

d. Write both sides of the equation in part (c) as exponents on base e. Simplify and obtain a model for carbon dioxide concentration in parts per million x years after 1900.

e. Use your model from part (d) to estimate when the concentration might reach 600 ppm. (This would be approximately double the level estimated to exist prior to the Industrial Revolution.)

13. In a study entitled "Reaction Time and Speed of Movement of Males and Females of Various Ages," (*Research Quarterly: American Association of Health and Physical Education and Recreation*, 1963), the following data appeared.

x (Age)	19	27	45	60
y (Reflex Time, in Seconds)	0.18	0.20	0.23	0.25

a. Graph the four data points. What function might reasonably model the data?
b. Take the natural logarithm of each x-value and graph the four data points $(\ln x, y)$. Notice that the points lie nearly on a straight line, suggesting the use of linear model.
c. Use the linear regression option on a graphing utility or the procedure outlined in the Study Tip on page 745 to find the logarithmic model for the given data.
d. Use the model from part (c) to predict the reflex time for a 75-year-old person.

True–False Critical Thinking Problem

14. Which one of the following is true?

a. The data points $(1, 2)$, $(2, 3)$, $(3, 3.5)$, $(4, 4)$, $(5, 4.1)$, $(6, 4.2)$, and $(7, 4.4)$ should be modeled using an exponential function.

b. The data points $(1, 44)$, $(1.5, 4.7)$, $(2, 5.5)$, $(4, 9.9)$ $(6, 18.1)$, and $(8, 34)$ should be modeled using a logarithmic function.

c. The points shown in the figure can be modeled by an exponential function, taking the set of points of the form (x, y) and fitting a linear model to points of the form $(\ln x, \ln y)$.

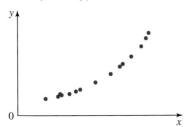

d. The points shown in the figure can be modeled by a logarithmic function, taking the set of points of the form (x, y) and fitting a linear model to points of the form $(\ln x, y)$.

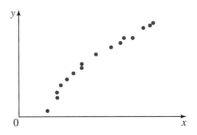

Technology Problems

15. Graphing utilities have the ability to fit nonlinear models to data. Use the exponential regression and logarithmic regression features of a graphing utility to duplicate your models in Problems 12–13.

16. The data below indicate the number of new AIDS cases among American women.

x (Years after 1985)	y (Number of New AIDS Cases Among American Women)
0	522
2	1682
4	3370
6	5375

Source: National Center for Health Statistics

a. Use a graphing utility to model the data with a linear model, a quadratic model, an exponential model, and a logarithmic model.

b. Eight years after 1985, in 1993, there were 12,789 new cases of AIDS among American women. Which of the models best predicts this number?

Writing in Mathematics

17. Describe some of the decisions that must be made in modeling data.

18. Suppose that $(60, 250)$ and $(70, 300)$ are two points for a set of data that can be modeled exponentially. Explain why $(80, 400)$ cannot be a data point in the scatter plot.

19. What does the model in Example 4 on page 744 imply about urban growth and stress?

Group Activity Problem

20. Group members should consult the current edition of *The World Almanac* or an equivalent reference. Select data that is of interest to the group that can be modeled with an exponential or logarithmic function. Model the data. Each group member should make one prediction based on the model, and then discuss a consequence of this prediction. What factors might change the accuracy of each prediction?

Review Problems

21. Multiply: $(3x^2 - 7y)^2$.

22. Solve: $\dfrac{63}{y^2 - 3y} - 11 = \dfrac{7y}{y - 3} - \dfrac{21}{y}$.

23. Graph: $3x - 2y < -6$.

HAPTER PROJECT

Reflections of our World

We have seen how the graphs of the logarithmic and exponential functions are reflections of each other across the line $y = x$. Reflection, in this sense, is not simply a mathematical term, it is the literal description. Use the graph on page 701 (Figure 8.20) and place a small mirror along the line $y = x$. The graph you create with your mirror matches the illustration (Figure 8.19) above. When we place a mirror on a line drawn through an image and recreate the whole image, the image is said to have mirror symmetry or bilateral symmetry.

1. Some of the simplest examples of bilateral symmetry may be seen by observing the letters of the alphabet. Write a capital A on a sheet of paper and place a mirror between the left and right halves. What do you observe? Repeat this procedure for the letters R and H, this time putting the mirror across the middle of the letter as well as between the left and right halves. Which of these letters exhibit mirror symmetry? Where you place the mirror to see an identical reflection is called the plane of symmetry for a particular image.
2. Write the letters of the alphabet and discover which letters have symmetries. After you have used the mirror for a time, you may find that simply imagining it to be present will be enough to tell you if the letter is symmetrical. Generally, the written word is difficult to read in a mirror. However, by selecting letters with the proper type of symmetry, we can create words that are entirely symmetrical and readable. Write the word CHOICE, using all capitals, and place a mirror a little above the word. What do you observe? Place the mirror in the middle of the word and look again. Write the word TOMATO as it appears here and also by stacking the letters in a column. Use a mirror to examine both versions of the word. Keeping these examples in mind, work in a small group to create your own symmetric words to share with the class. Can you create a sentence with mirror symmetry?
3. The human body is an example of an object with bilateral symmetry. A plane of reflection could be run down the center and recreate our left or right side. Some athletes, such as gymnasts and divers, try to keep a mirror symmetry between both halves of their bodies as they perform certain movements. Research in your library or on the Worldwide Web to obtain pictures of athletes which show symmetry and share them with your class.
4. Our faces alone also show symmetry, although usually with small, noticeable variations between the right and left side. Use a picture of your face and study the symmetry by checking both sides reflected in a mirror. Study your symmetric reflections and decide which one best represents your face. Some people consider the right side of the face as showing a more public image to the world and the left side as being representative of our private selves. Does this hold true in your case?

We can also study reflections in curved mirrors, rather than flat mirrors. Funhouse mirrors are curved to distort any reflection, but other curved mirrors may be used to correct distortions. In anamorphic art, the view on a flat canvas is distorted or twisted and may only be viewed correctly in a curved mirror, usually a mirrored cylinder or cone set on a specific place on the image. These paintings were popular in the seventeenth and eighteenth centuries throughout Europe and Asia. The highly polished, mirrored objects used for viewing, called anamorphoscopes, were sold with the paintings. Today, we can construct an inexpensive anamorphoscope using a cylinder or cone and covering the surface with Mylar paper.

5. Using resources in your library, local museum, or on the Worldwide Web, obtain reproductions of anamorphic art to share with your class. Construct the correct anamorphoscope needed to view the painting. One of the methods used to create this style of art was to draw on paper by looking at the sketch only in the mirrored anamorphoscope. This is similar to writing while looking in a mirror. Try to create your own sample of an anamorphic image using this method.

Multiple reflections of an object may be created by placing two flat mirrors at an angle to each other and setting the object between them. No matter how an object is placed, we see a symmetrical pattern. This is the basic idea behind the construction of a toy created in 1816 by David Brewster, called the kaleidoscope.

6. To create a simple kaleidoscope, you will need two mirrors to stand on edge and a few colorful objects to be scattered between the mirrors. You may wish to use colorful bits of paper, combinations of stones and glass, or even small images jumbled together. Notice how the angle of separation between the mirrors creates differing numbers of reflections.

7. We can also experiment to discover how many images we can create. Select one object to set between the mirrors. Place the mirrors, successively, at 60, 45, and 36 degrees. How many reflections do you count at each angle? Can you discover a relationship between the number of reflections and the angle of the mirrors? Try setting three mirrors on a table so they meet to form an equilateral triangle. How many reflections of an object placed towards a corner can you count by viewing from above and studying all the mirrors? Repeat this for an object placed between four mirrors forming a square. Can you discover any pattern?

Many objects do not have any type of mirror symmetry, but they do occur in pairs as mirror images of each other. One obvious pair is our hands. If we hold one hand up to a mirror, we will see the other hand. We need to place them together to form a mirror symmetry. Each hand alone has no symmetry. Many chemical and biological molecules exist in both right-handed and left-handed versions. When a drug is synthesized in a laboratory we see a fifty-fifty mix of right-handed and left-handed forms. The biological activity of the two forms may be very different.

Most of the time, our bodies cannot use one form of the molecule. Research into new low-calorie foods takes advantage of this property to create fat substitutes which taste like the "real thing" but pass harmlessly and unused through our system. Occasionally, our bodies react disastrously to different versions. This happened in the 1960s with a drug called thalidomide. The right-handed molecules were used as a safe sedative, but the left-handed molecules caused tragic birth defects. Reflections can literally be a matter of life and death.

Worldwide Web Resources

Go to the Prentice Hall website (http://www.prenhall.com/blitzer) to access other locations on the Internet that will allow you to further explore the concepts presented in this project.

Chapter Review

SUMMARY

1. Exponential Functions

a. The exponential function with base b is defined by $f(x) = b^x$, where $b > 0$ and $b \neq 1$.

b. Graphs for $f(x) = b^x$ are shown below.

Graph of $f(x) = b^x$, $b > 1$
- Domain: $(-\infty, \infty)$
- Range: $(0, \infty)$
- Intercept: $(0, 1)$
- Increasing

Graph of $f(x) = b^x$, $0 < b < 1$
- Domain: $(-\infty, \infty)$
- Range: $(0, \infty)$
- Intercept: $(0, 1)$
- Decreasing

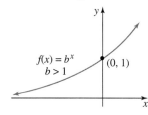

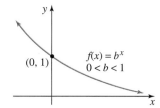

c. The natural exponential function is $f(x) = e^x$. The irrational number e is called the natural base, where $e \approx 2.7183$.

2. Formulas for Compound Interest

After t years, the balance A in an account with principal P and annual interest rate r (in decimal form) is given by the following mathematical models.

a. For n compoundings per year: $A = P\left(1 + \dfrac{r}{n}\right)^{nt}$

b. For continuous compounding: $A = Pe^{rt}$

3. Composite and Inverse Functions

a. *The Composition of Functions*
1. The composition of the functions f and g is denoted by $f \circ g$ and is defined by $(f \circ g)(x) = f(g(x))$.
2. The composition of the functions g and f is denoted by $g \circ f$ and is defined by $(g \circ f)(x) = g(f(x))$.
3. In general, $(f \circ g)(x) \neq (g \circ f)(x)$.

b. *Inverse Functions*
1. A function has an inverse that is a function if no two different numbers in the domain correspond to the same number in the range. Such a function is called one-to-one.
2. *Horizontal line test:* A function is one-to-one if there is no horizontal line that intersects its graph at more than one point.
3. The equation for the inverse of a one-to-one function f is found by replacing $f(x)$ by y, interchanging x and y, solving for y, and replacing y by $f^{-1}(x)$. The process can be checked by showing $f(f^{-1}(x)) = x$ and $f^{-1}(f(x)) = x$.

4. Logarithmic Functions

a. $y = \log_b x$ is equivalent to $x = b^y$. Thus, a logarithm is an exponent. $(b > 0, b \neq 1)$.

b. $y = b^x$ and $y = \log_b x$ are inverses. Thus, if $f(x) = b^x$, then $f^{-1}(x) = \log_b x$.

c. The graphs of $f(x) = b^x$ and $f^{-1}(x) = \log_b x$ are shown below (with $b > 1$).

d. *Common logarithms and natural logarithms:* $y = \log x$ means $y = \log_{10} x$, called the common logarithmic function. $y = \ln x$ means $y = \log_e x$, called the natural logarithmic function. If $f(x) = 10^x$, then $f^{-1}(x) = \log x$. If $f(x) = e^x$, then $f^{-1}(x) = \ln x$.

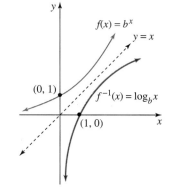

For $f(x) = b^x$:
Domain $= \{x \mid x \in R\}$ or $(-\infty, \infty)$
Range $= \{y \mid y > 0\}$ or $(0, \infty)$

For $f^{-1}(x) = \log_b x$:
Domain $= \{x \mid x > 0\}$ or $(0, \infty)$
Range $= \{y \mid y \in R\}$ or $(-\infty, \infty)$

5. Properties of Logarithmic Functions

a. *Basic Properties*

Base b ($b > 0$, $b \neq 1$)	Base 10 (Common Logarithms)	Base e (Natural Logarithms)
$\log_b 1 = 0$	$\log 1 = 0$	$\ln 1 = 0$
$\log_b b = 1$	$\log 10 = 1$	$\ln e = 1$
$\log_b b^x = x$	$\log 10^x = x$	$\ln e^x = x$

b. *The Product Rule:*
$$\log_b(MN) = \log_b M + \log_b N$$

c. *The Quotient Rule:*
$$\log_b\left(\frac{M}{N}\right) = \log_b M - \log_b N$$

d. *The Power Rule:* $\log_b M^p = p \log_b M$

e. *The Change-of-Base Property*

The General Property	Common Logarithms	Natural Logarithms
$\log_b M = \dfrac{\log_a M}{\log_a b}$	$\log_b M = \dfrac{\log M}{\log b}$	$\log_b M = \dfrac{\ln M}{\ln b}$

6. Exponential Equations
  a. Exponential equations contain a variable in an exponent.
  b. Some exponential equations can be solved using the one-to-one exponential function property.
     1. Express each side of the equation as a power of the same base.
     2. Set the exponents equal to each other.
     3. Solve the resulting equation.
  c. When both sides of an exponential equation cannot easily be written with a common base, use natural logarithms.
     1. Isolate the exponential expression.
     2. Take the natural logarithm on both sides of the equation.
     3. Simplify using one of the following properties:
        $\ln b^x = x \ln b$ or $\ln e^x = x$
     4. Solve for the variable.

7. Logarithmic Equations
  a. Logarithmic equations contain a variable in a logarithmic expression.
  b. To solve a logarithmic equation, use properties of logarithms to write the equation in the form $\log_b M = N$. Then rewrite in exponential form $b^N = M$ and solve for the variable. Reject values that produce the logarithm of a negative number or the logarithm of 0.

8. Growth and Decay Models
  a. The mathematical model for exponential growth or decay is given by

$$f(t) = Ae^{kt} \quad \text{or} \quad y = Ae^{kt}$$

Within this model, t represents time, A is the original amount (or size) of the growing (or decaying) quantity, and y or $f(t)$ represents the amount (or size) of the quantity at time t. The number k represents a constant that depends on the rate of growth or decay. If $k > 0$, the formula represents *exponential growth;* if $k < 0$, the formula represents *exponential decay.*
  b. Graphs of the Growth and Decay Models

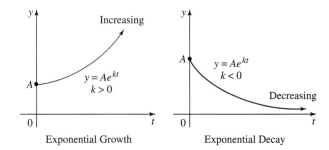

Exponential Growth   Exponential Decay

9. Fitting Functions to Data
  a. If the scatter plot appears to follow an exponential pattern, fit a linear model to $(x, \ln y)$ and then exponentiate both sides to find the model for the data points (x, y).
  b. If the scatter plot appears to follow a logarithmic pattern, fit a linear model to $(\ln x, y)$. The model for the data points (x, y) is the equation of the regression line for $(\ln x, y)$.

<hr />

REVIEW PROBLEMS

Graph the exponential functions in Problems 1–4.

1. $y = 2^x$

2. $y = 2^{x-2}$

3. $y = 3^x - 1$

4. $f(x) = \left(\dfrac{1}{2}\right)^x$

Use the compound interest models

$$A = P\left(1 + \frac{r}{n}\right)^{nt} \quad \text{and} \quad A = Pe^{rt}$$

to answer Problems 5–6.

5. Suppose that you have $5000 to invest. What investment yields the greatest return over 5 years: 5.5% compounded semiannually or 5.25% compounded monthly?

6. Suppose that you have $14,000 to invest. What investment yields the greatest return over 10 years: 7% compounded monthly or 6.85% compounded continuously?

7. Use the exponential growth model $y = Ae^{kt}$ (where $k > 0$) to answer this question. In 1990 China had a population of 1119.9 million with a growth rate of 1.3% per year ($k = 0.013$). India's 1990 population was 853.4 million with a growth rate of 2.1% per year. The United States Bureau of the Census, in their 1990 *World Population Profile,* stated that "The latest projections suggest that India's population may surpass China's in less than 60 years." Estimate the population of the two countries in the year 2050. Do these estimations reinforce the prediction by the census bureau?

8. The function

$$f(t) = 100\left(\tfrac{1}{2}\right)^{\frac{t}{5600}}$$

approximates the amount $f(t)$ of carbon-14 remaining in a fossil that is t years old and that originally started with 100 grams of carbon-14. Complete the following table and then graph the function with values of t along the x-axis and values of $f(t)$ along the y-axis. Take one of the ordered pairs on the graph and describe its significance in practical terms.

t	0	2800	5600	11,200	15,000	16,800
$f(t)$						

9. A cup of coffee is taken out of a microwave oven and placed in a room. The temperature (T, in degrees Fahrenheit) of the coffee after t minutes is modeled by the function $T = 70 + 130e^{-0.04855t}$. The graph of the function is shown in the figure. Use the graph to answer each of the following questions.

a. What was the temperature of the coffee when it was first taken out of the microwave?

b. What is a reasonable estimate of the temperature of the coffee after 20 minutes? Use your calculator to verify this estimate.

c. What is the limit of the temperature to which the coffee will cool? What does this tell you about the temperature of the room?

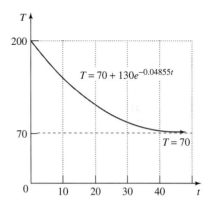

In Problems 10–11, find $(f \circ g)(x)$ and $(g \circ f)(x)$.

10. $f(x) = x^2 + 3$, $g(x) = 4x - 1$

11. $f(x) = \sqrt{x}$, $g(x) = x + 1$

12. Use the graph to determine whether each function in parts (a)–(d) is a one-to-one function.

a.

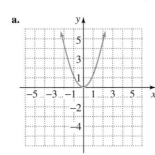

b.

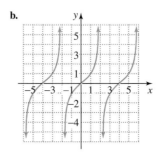

c.

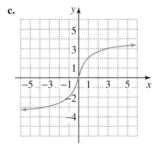

d.

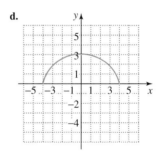

The functions in Problems 13–14 are both one-to-one. For each problem:

a. *Find a formula for $f^{-1}(x)$, the inverse function.*
b. *Verify that the formula is correct by showing that $f(f^{-1}(x)) = x$ and $f^{-1}(f(x)) = x$.*
c. *Graph the function and its inverse using the same set of axes.*

13. $f(x) = 2x - 4$ **14.** $f(x) = x^3 - 2$

15. The line graph shown is based on data from the U.S. National Center for Health Statistics.
  a. Explain why f has an inverse that is a function.
  b. Describe in practical terms the meaning of $f^{-1}(60)$.

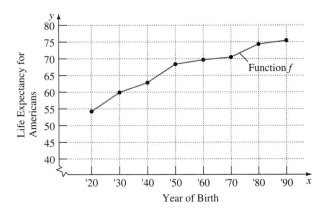

In Problems 16–17, write each logarithmic equation in exponential form.

16. $\log_2 64 = 6$ **17.** $\log_{49} 7 = \frac{1}{2}$

In Problems 18–19, write each exponential equation in logarithmic form.

18. $6^3 = 216$ **19.** $10^{-2} = \frac{1}{100}$

In Problems 20–23, evaluate each expression. If evaluation is not possible, state the reason.

20. $\log_4 64$ **21.** $\log_{125} 5$ **22.** $\log_3(-9)$ **23.** $\log_3 \frac{1}{81}$

24. Graph $f(x) = 2^x$ and $f^{-1}(x) = \log_2 x$ on the same set of axes.

25. Students in a psychology class took a final examination. They took equivalent forms of the exam in monthly intervals thereafter. The average score ($f(t)$) for the group after t months was given by the human memory model $f(t) = 76 - 18 \log(t + 1)$, where $0 \le t \le 12$.
  a. What was the average score when the exam was first given?
  b. What was the average score after 2 months? 4 months? 6 months? 8 months? one year?
  c. Use the results from parts (a) and (b) to graph f. Describe what the shape of the graph indicates in terms of the material retained by the students.

26. The model

$$t = \frac{1}{c} \ln\left(\frac{A}{A - N}\right)$$

describes the time (t, in weeks) that it takes to achieve mastery of a portion of a task. In the model, A represents maximum learning possible, N is the portion of the learning that is to be achieved, and c is a constant used to measure an individual's learning style. A 50-year-old man decides to start running as a way to maintain good health. He feels that the maximum rate he could ever hope to achieve is 12 miles per hour. How many weeks will it take before the man can run 5 miles in 1 hour if $c = 0.06$ for this person?

In Problems 27–30, use logarithmic properties to expand each expression as much as possible. Where possible, evaluate logarithmic expressions.

27. $\log_6(36x^3)$ **28.** $\log_2 \frac{xy^2}{64}$ **29.** $\log_4 \frac{\sqrt{x}}{64}$ **30.** $\log \sqrt[3]{\frac{1000}{x^2 y}}$

In Problems 31–35, write each expression as a single logarithm, and simplify if possible.

31. $\log_b 7 + \log_b 3$

32. $\log_5 250 - \log_5 2$

33. $\log 3 - 3 \log x$

34. $3 \log_b x + 4 \log_b y$

35. $\dfrac{1}{3} \ln x - \dfrac{1}{2} \ln y$

36. Express $\log_3 7$:
  a. As the quotient of two common logarithms.
  b. As the quotient of two natural logarithms.

37. Which one of the following is true?
  a. $\log_b(x + y) = \log_b x + \log_b y$, if $x > 0$ and $y > 0$.
  b. $\ln(\log_5 60 - \log_5 12) = 0$
  c. $\dfrac{\log_2 8}{\log_2 16} = \dfrac{1}{2}$
  d. $\dfrac{\log x^3}{\log y^2} = 3 \log x - 2 \log y$, if $x > 0$ and $y > 0$.

Solve each equation in Problems 38–40 by expressing each side as a power of the same base and then equating exponents.

38. $2^{4x-2} = 64$

39. $125^x = 25$

40. $9^x = \frac{1}{27}$

Solve each equation in Problems 41–43 by taking the natural logarithm on both sides. Express the answer in terms of natural logarithms. Then use a calculator to obtain a decimal approximation, correct to the nearest thousandth, for the solution.

41. $5^x = 119.4$

42. $20{,}000e^{0.08t} = 40{,}000$

43. $8^{-0.04x} = 0.06$

Solve each logarithmic equation in Problems 44–50.

44. $\log_5 x = -3$

45. $\log_x 32 = 6$

46. $\log x = -3$

47. $\log_5(2x - 1) = 3$

48. $\log x + \log(x - 21) = 2$

49. $\log_2(x + 3) = 4 - \log_2(x - 3)$

50. $\log_3(x - 1) - \log_3(x + 2) = 2$

51. The model

$$A = P\left(1 + \frac{r}{n}\right)^{nt}$$

describes the accumulated value (A) of a sum of money (P, the principal) after t years at annual percentage rate r (in decimal form), compounded n times a year. How long (to the nearest tenth of a year) will it take $12{,}500 to grow to $20{,}000 at 6.5% annual interest compounded quarterly?

52. The model $A = Pe^{rt}$ describes the accumulated value (A) of a sum of money (P) after t years at an annual percentage rate r (in decimal form), compounded continuously.
  a. How long (to the nearest tenth of a year) will it take for the investment to triple in value?
  b. What interest rate is required for the investment to triple in 8 years?

53. Use the exponential growth model $y = Ae^{kt}$, where t represents time, A is the size of the population at $t = 0$, k is the relative growth rate, and y is the population at time t, to answer this question.
  We have seen that a world population of 10 billion is the carrying capacity of the Earth. In the 1969 report *Resources and Man,* world population figures for 1975 were projected at 4.043 billion (low) and 4.134 billion (high), with a relative growth rate of 1.5% per year (low) and 2% per year (high).
  a. Use the low figures ($A = 4.043$, $k = 0.015$) to find the year in which the carrying capacity might be

reached. Substitute 10 for y in the exponential growth model and solve for t. Remember that t represents the number of years after 1975.
  b. Repeat part (a) using the high figures.

54. The formula $W = 0.35 \ln P + 2.74$ is a model for the average walking speed (W, in feet per second) for a resident of a city whose population is P thousand. What is the average walking speed (to the nearest tenth) for New York City, whose population is 7323 thousand?

55. The formula $R = 0.67 \log E - 2.9$ describes an earthquake's magnitude (R) on the Richter scale as a function of the energy (E, in joules) released by the earthquake. Colombia's 1906 earthquake measured 8.6 on the Richter scale and Northern California's 1989 earthquake measured 7.1. Determine the energy released in each earthquake.

Use the exponential growth and decay models $y = Ae^{kt}$ or $f(t) = Ae^{kt}$, where A is the original amount (or size) of a growing (or decaying) entity at $t = 0$, y (or $f(t)$) is the amount at time t, and k is a growth ($k > 0$) or decay ($k < 0$) constant, to answer Problems 56–58.

56. According to the U.S. Bureau of the Census, in 1980 there were 14,609 thousand residents of Hispanic origin living in the United States. By 1994, the number had increased to 26,077 thousand.

  a. Let $t = 0$ correspond to 1980. This means that A, the original amount in the growth model is 14,609. Now use the fact that in 1994 the number was 26,077 (when $t = 14$, $y = 26,077$) to find k, the growth constant.

  b. Substitute the values for A and k into the exponential growth model and write the function that models the Hispanic resident population of the United States t years after 1980.

  c. Use the model in part (c) to project the Hispanic resident population (in thousands) for the years 2005, 2010, and 2025.

57. a. Carbon-14 decays exponentially with a half-life of approximately 5730 years, meaning that after 5730 years a given amount of carbon-14 will have decayed to half the original amount. Use this information to show that the decay model for carbon-14 is given by $y = Ae^{-0.000121t}$.

  b. Prehistoric cave paintings were discovered in the Lascaux cave in France. The paint contained 15% of the original carbon-14. Let $y = 0.15A$ in the model in part (a) and estimate the age of the paintings.

58. Show that the time it takes for a population that is growing exponentially to triple (from an initial population of A to one of $3A$) is given by

$$t = \frac{\ln 3}{k}.$$

59. The figure shows world population projections through the year 2150. The data is from the United Nations Family Planning Program and is based on optimistic or pessimistic expectations for successful control of human population growth. Suppose that you are interested in modeling this data using exponential, linear, and quadratic functions. Which one of these functions would you use to model each of the projections? Explain your choices. For the choice corresponding to a quadratic model, would your formula involve one with a positive or negative leading coefficient? Explain. What is misleading about the scale on the horizontal axis?

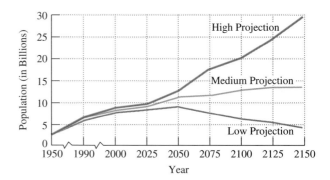

60. The data below indicate national expenditures for health care in billions of dollars for eight selected years.

Year	1960	1965	1970	1975	1980	1985	1990	1993
Health-Care Expenditures (in billions of dollars)	27.1	41.6	74.3	132.9	251.1	434.5	696.6	884.2

Source: U. S. Department of Health and Human Services

  a. Let $x = 0$ correspond to 1960 and let y correspond to each of the figures for health-care expenditure. Take the natural logarithm of each y-value.

  b. Graph the eight data points $(x, \ln y)$. Notice that the points lie nearly on a straight line, suggesting the use of a linear model.

  c. Use the linear regression option on a graphing utility or the procedure outlined in the Study Tip on page 745 to write the equation of the line that best fits the $(x, \ln y)$ transformed data.

  d. Write both sides of the equation in part (c) as exponents on base e. Simplify and obtain a model for health care expenditures x years after 1900.

  e. Use the model from part (d) to predict health care expenditures for the year 2000.

61. The data below indicate the amount of rain forest cleared for cattle ranching in the indicated year.

(Year) x	y **Rain Forest Cleared (in millions of hectares) (1 hectare = 247 acres)**
(1960) 0	0.30
(1970) 10	0.95
(1980) 20	1.36
(1988) 28	1.40

a. Graph the four data points. What function might reasonably model the data?

b. Take the natural logarithm of each x-value and graph the four data points $(\ln x, y)$. Notice that the points lie nearly on a straight line, suggesting the use of a linear model.

c. Use the linear regression option on a graphing utility or the procedure outlined in the Study Tip on page 745 to find the logarithmic model for the given data.

d. Use the model from part (c) to determine the year in which 4 million hectares of rain forest will be cleared for cattle ranching.

CHAPTER 8 TEST

1. Describe one reason why the graph shown in the figure cannot be the graph of $f(x) = 2^x$.

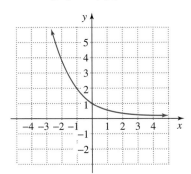

2. Graph the exponential function $f(x) = 2^{x-3}$.

3. If $f(x) = x^2 + x$ and $g(x) = 3x - 1$, find $(f \circ g)(x)$ and $(g \circ f)(x)$.

4. If $f(x) = 5x - 7$, find $f^{-1}(x)$.

5. A function f defines the amount given to charity as a function of income. The graph of f is shown in the figure on the right.

a. Can we also say that income is a function of the amount given to charity? Explain your answer using the graph of f and the horizontal line test.

b. Describe in practical terms the meaning of $f^{-1}(2000)$.

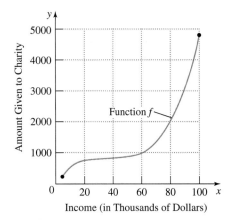

6. Write in exponential form: $\log_5 125 = 3$.

7. Write in logarithmic form: $\sqrt{36} = 6$.

8. Graph $f(x) = 3^x$ and $f^{-1}(x) = \log_3 x$ on the same set of axes.

In Problems 9–10, use logarithmic properties to expand each expression as much as possible. Where possible, evaluate logarithmic expressions.

9. $\log_4 (64x^5)$

10. $\log_3 \dfrac{\sqrt[3]{x}}{81}$

In Problems 11–12, write each expression as a single logarithm, and simplify if possible.

11. $3 \ln x + 5 \ln y$

12. $\log_3 405 - \log_3 5$

13. Express $\log_{15} 71$ as the quotient of either common or natural logarithms. Then use your calculator to obtain a decimal approximation correct to four decimal places.

In Problems 14–19, solve each equation.

14. $3^{5x-2} = 27$

15. $5^x = 1.4$

16. $400e^{0.005x} = 1600$

17. $\log_{25} x = \frac{1}{2}$

18. $\log_6(4x - 1) = 3$

19. $\log x + \log(x + 15) = 2$

20. Use the compound interest models

$$A = P\left(1 + \frac{r}{n}\right)^{nt} \quad \text{and} \quad A = Pe^{rt}$$

to answer this question. If you have $3000 to invest, what investment yields the greatest return over 10 years: 6.5% compounded semiannually or 6% compounded continuously? How much more (to the nearest dollar) is yielded by the better investment?

21. An object was removed from a furnace at a temperature of 140° Celsius and placed in a room at 20° Celsius. The object's cooling is modeled by the function $f(h) = 20[1 + 6(2^{-h})]$, where $f(h)$ is its Celsius temperature after h hours. Find and interpret $f(3)$.

22. Students in a psychology class were given an exam. As part of an experiment in memory, they were retested monthly on the same material. The function $f(t) = 75 - 16 \log(t + 1)$ models the average score ($f(t)$) for the class after t months.
   a. What was the average score when the exam was first given?
   b. What was the average score after 9 months?

23. Use the compound interest model

$$A = P\left(1 + \frac{r}{n}\right)^{nt}$$

to answer this question: How long, to the nearest year, does it take an investment of $4000 to double if it is invested at 5% compounded quarterly?

24. The model $P = 14.7e^{-0.21x}$ describes atmospheric pressure (P, in pounds per square inch) at an altitude of x miles above sea level. Find the elevation, to the nearest tenth of a mile, of a jet if the atmospheric pressure outside the jet is 5.15 pounds per square inch.

25. Medical research now indicates that the risk (R, given as a percent) of having a car accident increases exponentially as the concentration of alcohol in the blood increases. The risk is modeled by $R = 6e^{kx}$, where x is the concentration of alcohol in the blood.
   a. A concentration of alcohol in the blood of 0.06 results in a 13% risk of an accident. Substitute 13 for R and 0.06 for x in the model and find the value of k.
   b. What is the alcohol concentration that corresponds to certainty of having an accident ($R = 100\%$)?

26. The magnitude of an earthquake of intensity I is given by

$$R = \log \frac{I}{I_0}$$

where I_0 is the intensity of a barely felt zero-level earthquake. If an earthquake has an intensity of $10^{7.4}$ times the intensity of a zero-level earthquake, what is its magnitude on the Richter scale?

CUMULATIVE REVIEW PROBLEMS (CHAPTERS 1–8)

In Problems 1–12, solve for the indicated variable(s). Express irrational solutions in simplified radical form and imaginary solutions in the form a + bi. Express solutions of inequalities in both set-builder and interval notation.

1. $\sqrt{2x + 5} = \sqrt{x + 3} = 2$

2. $(x - 5)^2 = -49$

3. $(9x + 3)(x - 1) = -8$

4. $x^2 + x > 6$

5. $|2x - 3| = |x + 6|$

6. $6x - 3(5x + 2) = 4(1 - x)$

7. $\dfrac{2}{x - 3} - \dfrac{3}{x + 3} = \dfrac{12}{x^2 - 9}$

8. $(x^2 + 2x)^2 - 5(x^2 + 2x) + 6 = 0$

9. $3x + 2 < 4$ and $4 - x > 1$

10. $\dfrac{x + 1}{x - 5} > 0$

11. $4x - 3y = 7$
$3x - 2y = 6$

12. $3x - 2y + z = 7$
$2x + 3y - z = 13$
$x - y + 2z = -6$

In Problems 13–18, graph each equation or inequality in the rectangular coordinate system.

13. $f(x) = \sqrt{x - 2}$

14. $y = (x + 2)^2 - 4$

15. $y < -3x + 5$

16. $2x - 3y \leqslant 6$

17. $y = 3^{x-2}$

18. $y = \dfrac{1}{x - 2}$

In Problems 19–23, perform the indicated operations, and simplify if possible.

19. $\dfrac{2x + 1}{x - 5} - \dfrac{4}{x^2 - 3x - 10}$

20. $\dfrac{\dfrac{1}{x - 1} + 1}{\dfrac{1}{x + 1} - 1}$

21. $\dfrac{6}{\sqrt{5} - \sqrt{2}}$

22. $8\sqrt{45} + 2\sqrt{5} - 7\sqrt{20}$

23. $(2\sqrt{5} - \sqrt{7})^2$

24. The sum of the numbers along either diagonal, along any row, or along any column in the following square is the same.

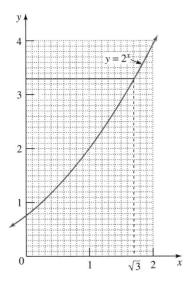

If x is the reciprocal of the solution to the equation $\log_5(x + 2) - \log_5 x = 1$, fill in the missing entries of the square.

25. Use the graph to find a reasonable estimate for $2^{\sqrt{3}}$.

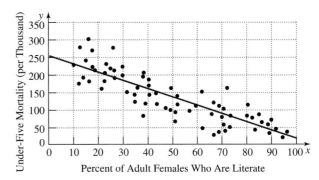

26. The graph at the top of the next column shows mortality of children under 5 in terms of the percentage of adult females who are literate in a country. Each dot represents one country. Also shown is the line that best fits the given data points.

   a. What is a reasonable estimate of the slope for this line? What does this mean in terms of the variables that appear in the graph?

   b. Estimate the coordinates of any point on the line. What does this mean in terms of the variables that appear in the graph?

   c. Use your answers from parts (a) and (b) to write the point-slope equation and the slope-intercept equation of the regression line shown in the figure.

Sources: Food and Agriculture Organization of the United Nations, United Nations Children's Fund, United Nations Development Programme, and United Nations Population Division

27. The frictional force needed to keep a car from skidding on a curve varies jointly with its mass and the square of its velocity and inversely as the radius of the curve. A 2500-pound car traveling 40 miles per hour on a curve with radius 800 feet requires 1500 pounds of friction to prevent it from skidding. Determine the frictional force necessary to keep a 4000-pound car from skidding on a curve with radius 600 feet if the car is traveling 30 miles per hour.

28. The model $f(t) = 29{,}035t^2 + 429{,}200$ describes the leading golf winnings in the United States t years after 1983. The leading golf winner for one of the years modeled by the function was Greg Norman, who won $690,515. In what year did this occur?

29. Write a polynomial function in terms of x that models the area of the shaded portion in the figure. Call the function $A(x)$ and write its equation in descending powers of x.

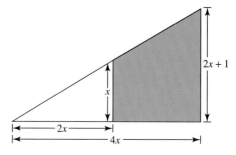

30. The diagram shows the number of telephone lines connecting survey teams for a new product in a shopping mall. Each team has a direct line to the other teams. Find a formula for the number of lines if there are n survey teams.

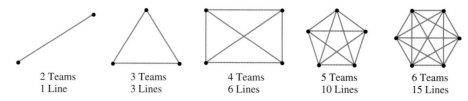

| 2 Teams | 3 Teams | 4 Teams | 5 Teams | 6 Teams |
| 1 Line | 3 Lines | 6 Lines | 10 Lines | 15 Lines |

9

Conic Sections and Nonlinear Systems of Equations

Chuck Close "Roy I" 1994, oil on canvas, 102 × 84 in. (259.1 × 213.4 cm). Photograph by: Ellen Page Wilson. Courtesy Pace Wildenstein.

Intriguing signs point out that the world is profoundly mathematical. When viewed closely, the familiar face of the ordinary world consists of shapes and forms described by algebraic equations. The mathematics of the world is present in the position of falling objects, the movement of planets, the manufacture of lenses for optical instruments, satellite communication, bridge construction, and even a procedure for disintegrating kidney stones. The mathematics behind these applications involves the conic sections, which is the focus of this chapter.

SECTION 9.1

Solutions Manual   **Tutorial**   **Video 10**

Conic Sections: Circles and Parabolas
Objectives

1 Write equations of circles.
2 Graph circles.
3 Graph parabolas opening to the left or right.

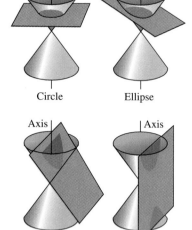

Circle   Ellipse

Parabola   Hyperbola

Figure 9.1

Conic sections were so named after their historical discovery as intersections of a plane and a cone (see Figure 9.1). Any plane perpendicular to the axis of the cone cuts a section that is a *circle*. Incline the plane, and the intersected section is an *ellipse*. Incline the plane more, and the section formed is a *parabola*. Continue tilting the plane until it is parallel to the cone's axis, and the section is a *hyperbola*, a curve with two branches.

The study of conic sections dates back over 2000 years to ancient Greece, culminating in a series of eight books by the Greek geometer Apollonius of Perga (262–190 B.C.). Until the 17th century, conic sections were studied without regard to their immediate usefulness simply because mathematicians found that they elicited ideas that were exciting, challenging, and interesting. But in a world that can be described mathematically, the geometry of ancient Greece became the practical cornerstone of modern astronomy. Conic sections, studied during the third century B.C., gave 17th-century mathematicians the tools for understanding the universe and laws of nature.

Circles

We begin our discussion with the definition of a circle.

> **Definition of a circle**
>
> A *circle* is the set of all points in a plane that are equidistant from a fixed point called the *center*. The fixed distance from the circle's center to any point on the circle is called the *radius*.

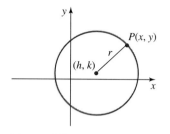

Figure 9.2

A circle with center at (h, k) and radius r

We can use this definition to derive the general equation of a circle in the rectangular coordinate system. Figure 9.2 shows a circle whose center is at (h, k) and whose radius has a length r. P represents any point on the circle. The definition of a circle tells us that all points on the circle are at a distance r from the center. Using the distance formula, we have

$$r = \sqrt{(x - h)^2 + (y - k)^2}$$

Recall that the distance between (x_1, y_1) and (x_2, y_2) is $\sqrt{(x_2 - x_1)^2 + (y_2 - y_1)^2}$.

Squaring both sides, we obtain

$$r^2 = (x - h)^2 + (y - k)^2.$$

Let's agree to write this as

$$(x - h)^2 + (y - k)^2 = r^2.$$

We can use the preceding equation to find the equation of a circle if we are given its center and radius or to find the center and radius if given the equation.

Write equations of circles.

> **Equation of a circle**
>
> The equation of the circle with the center at (h, k) and radius r is given by
>
> $$(x - h)^2 + (y - k)^2 = r^2.$$

EXAMPLE 1 **Finding the Equation of a Circle and Graphing the Circle**

Find the equation of the circle with a radius of 2 and the center at $(0, 0)$. Graph the circle.

Solution

Because $(h, k) = (0, 0)$ and $r = 2$, we have

$$(x - h)^2 + (y - k)^2 = r^2 \quad \text{This is the equation of a circle.}$$
$$(x - 0)^2 + (y - 0)^2 = 2^2 \quad \text{Substitute 0 for } h, 0 \text{ for } k, \text{ and 2 for } r.$$
$$x^2 + y^2 = 4 \quad \text{Simplify.}$$

The equation of the circle is $x^2 + y^2 = 4$. Figure 9.3 shows the graph. ∎

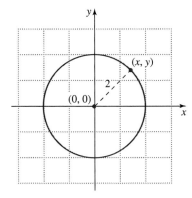

Figure 9.3

The graph of $x^2 + y^2 = 4$

> ### Using technology
>
> To graph a circle with a graphing utility, first solve the equation for y.
>
> $$x^2 + y^2 = 4$$
> $$y^2 = 4 - x^2$$
> $$y = \pm\sqrt{4 - x^2}$$
>
>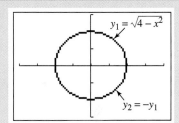
>
> Graph the two functions
>
> $$y_1 = \sqrt{4 - x^2} \quad \text{and} \quad y_2 = -\sqrt{4 - x^2}$$
>
> in the same viewing rectangle. Use a $\boxed{\text{ZOOM SQUARE}}$ setting so that the circle looks like a circle. (Many graphing utilities have problems connecting the two semicircles because the segments directly across horizontally from the center become nearly vertical.)

Observe from Example 1 that the equation of any circle with the center at the origin, $(0, 0)$, with radius r, is $x^2 + y^2 = r^2$. The equation $x^2 + y^2 = 4$, graphed in Figure 9.3, does not define y as a function of x. Vertical lines drawn between -2 and 2 intersect the graph more than once. In general, the graph of the equation of a circle fails the vertical line test and thus is not a function.

Example 2 involves finding the equation of a circle whose center is not at the origin.

EXAMPLE 2 **Finding the Equation of a Circle**

Find the equation of the circle with the center at $(-2, 3)$ and a radius of 4.

Solution

We have $(h, k) = (-2, 3)$ and $r = 4$. Thus,

$$(x - h)^2 + (y - k)^2 = r^2 \qquad \text{This is the equation of a circle.}$$
$$[x - (-2)]^2 + (y - 3)^2 = 4^2 \qquad \text{Substitute } -2 \text{ for } h, 3 \text{ for } k, \text{ and } 4 \text{ for } r.$$
$$(x + 2)^2 + (y - 3)^2 = 16 \qquad \text{Simplify.}$$

The equation of the circle is $(x + 2)^2 + (y - 3)^2 = 16$. ◼

2 Graph circles.

EXAMPLE 3 Using the Equation of a Circle to Graph the Circle

Find the center and radius of the circle whose equation is

$$(x - 2)^2 + (y + 4)^2 = 9$$

and graph the circle.

Solution

Because $(x - h)^2 + (y - k)^2 = r^2$ and we have $(x - 2)^2 + [(y - (-4)]^2 = 3^2$, the center is at $(2, -4)$, and the radius is 3. Figure 9.4 shows the graph. ◼

If we square $x - 2$ and $y + 4$ in the equation of Example 3, we obtain

$$(x - 2)^2 + (y + 4)^2 = 9 \qquad \text{This is the equation from Example 3.}$$
$$x^2 - 4x + 4 + y^2 + 8y + 16 = 9 \qquad \text{Square } x - 2 \text{ and } y + 4.$$
$$x^2 + y^2 - 4x + 8y + 20 = 9 \qquad \text{Combine numerical terms.}$$
$$x^2 + y^2 - 4x + 8y + 11 = 0 \qquad \text{Subtract 9 from both sides.}$$

This result suggests that an equation in the form $x^2 + y^2 + Dx + Ey + F = 0$ represents a circle. We can return to the familiar form $(x - h)^2 + (y - k)^2 = r^2$ by completing the square on x and y. Let's see how this is done.

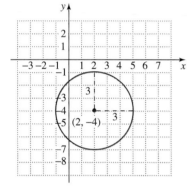

Figure 9.4

The graph of
$(x - 2)^2 + (y + 4)^2 = 9$

EXAMPLE 4 Completing the Square to Find a Circle's Center and Radius

Graph: $x^2 + y^2 + 4x - 6y - 23 = 0$

Solution

Because we plan to complete the square on both x and y, let's rearrange terms so that x terms are arranged in descending order, y terms are arranged in descending order, and the constant term appears on the right.

$$x^2 + y^2 + 4x - 6y - 23 = 0 \qquad \text{This is the given equation.}$$
$$(x^2 + 4x \quad) + (y^2 - 6y \quad) = 23 \qquad \begin{array}{l}\text{Rewrite in anticipation of completing the square.}\end{array}$$
$$(x^2 + 4x + 4) + (y^2 - 6y + 9) = 23 + 4 + 9 \qquad \begin{array}{l}\text{Complete the square on } x: \\ \frac{1}{2} \cdot 4 = 2 \text{ and } 2^2 = 4, \text{ so add 4 to} \\ \text{both sides. Complete the square} \\ \text{on } y: \frac{1}{2}(-6) = -3 \text{ and} \\ (-3)^2 = 9, \text{ so add 9 to both} \\ \text{sides.}\end{array}$$

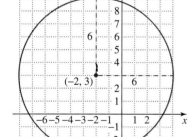

Figure 9.5

The graph of
$(x + 2)^2 + (y - 3)^2 = 36$

$$(x + 2)^2 + (y - 3)^2 = 36$$

Factor on the left and add on the right.

$$(x + 2)^2 + (y - 3)^2 = 6^2$$

This equation is in the form $(x - h)^2 + (y - k)^2 = r^2$, with the center at (h, k) and radius r.

Thus, the center is at $(-2, 3)$, and the radius is 6. Figure 9.5 on page 764 shows the graph. ■

U sing technology

To graph $x^2 + y^2 + 4x - 6y - 23 = 0$, rewrite the equation as a quadratic equation in y.

$$y^2 - 6y + (x^2 + 4x - 23) = 0$$

Now solve for y using the quadratic formula, with $a = 1$, $b = -6$, and $c = x^2 + 4x - 23$.

$$y = \frac{-b \pm \sqrt{b^2 - 4ac}}{2a} = \frac{-(-6) \pm \sqrt{(-6)^2 - 4 \cdot 1(x^2 + 4x - 23)}}{2 \cdot 1} = \frac{6 \pm \sqrt{36 - 4(x^2 + 4x - 23)}}{2}$$

Since we will enter these equations, there is no need to simplify. Enter

$$y_1 = \frac{6 + \sqrt{36 - 4(x^2 + 4x - 23)}}{2}$$

and $$y_2 = \frac{6 - \sqrt{36 - 4(x^2 + 4x - 23)}}{2}.$$

Use a ZOOM SQUARE setting. The graph is shown on the right.

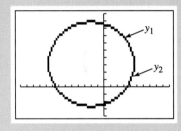

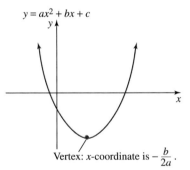

$y = ax^2 + bx + c$

Vertex: x-coordinate is $-\dfrac{b}{2a}$.

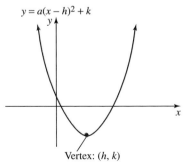

$y = a(x - h)^2 + k$

Vertex: (h, k)

Figure 9.6

Parabolas that open upward

Parabolas

In Chapter 7 we studied parabolas, viewing them as graphs of quadratic functions. Parabolas have equations that can be expressed as either

$$y = ax^2 + bx + c \quad \text{or} \quad y = a(x - h)^2 + k.$$

The parabolas corresponding to these equations are illustrated in Figure 9.6, where $a > 0$. Recall that if $a < 0$, the parabolas open downward.

The parabolas that we have studied up to this point have axes of symmetry that are parallel to the y-axis and that open upward or downward. However, the parabola can be given a geometric definition that enables us to include parabolas that open to the right and to the left.

Definition of a parabola

A *parabola* is the set of all points in a plane that are equidistant from a fixed line (the *directrix*) and a fixed point (the *focus*) that is not on the line (see Figure 9.7).

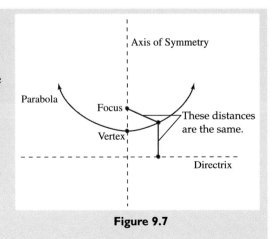

Figure 9.7

Figure 9.8

A family of four basic parabolas

Using this definition, we can expand our discussion of parabolas to include those that open left or right. Figure 9.8 shows a basic "family" of four parabolas and their equations. Notice that the parabolas that open left or right are not functions of x because they fail the vertical line test.

The equation $x = y^2$ interchanges the variables in the equation $y = x^2$. We can interchange variables in the parabola's general equation $y = ax^2 + bx + c$.

$y = ax^2 + bx + c$ This is one of the forms of a parabola's equation.

$x = ay^2 + by + c$ Interchange x and y.

The graph of $x = ay^2 + by + c$ is also a parabola. However, these parabolas open to the right when $a > 0$ and open to the left when $a < 0$.

All other facts about parabolas are modified similarly and summarized as follows.

Forms of the equation of a parabola opening to the left or right

$$x = ay^2 + by + c \quad \text{or} \quad x = a(y - k)^2 + h$$

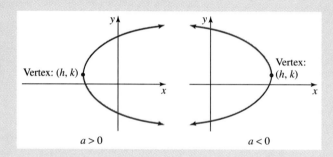

When the equation of the parabola is written in the form $x = ay^2 + by + c$, the coordinates of the vertex are found using $y = -\dfrac{b}{2a}$ and substituting y into the equation to find the x-coordinate.

We find intercepts in the usual manner. The y-intercepts are found by substituting 0 for x in the original equation and solving for y. The x-intercept is found by substituting 0 for y in the original equation and solving for x.

Let's practice graphing some of these parabolas that open to the left or right and do not define y as a function of x.

3 Graph parabolas opening to the left or right.

EXAMPLE 5   **Graphing a Parabola**

Graph: $x = y^2 + 4y - 5$

Solution

To graph the equation, we need to find the vertex and the y- and x-intercepts. The form of the equation is $x = ay^2 + by + c$, with $a = 1$, $b = 4$, and $c = -5$.

Step 1. To find the vertex, we use $y = -\dfrac{b}{2a}$, with $a = 1$ and $b = 4$.

$$y = -\frac{b}{2a} = -\frac{4}{2(1)} = -2$$

Now we find the x-coordinate.

$$x = (-2)^2 + 4(-2) - 5 \quad \text{Substitute } -2 \text{ for } y \text{ in } x = y^2 + 4y - 5.$$
$$= 4 - 8 - 5$$
$$= -9$$

The vertex is located at $(-9, -2)$.

Step 2. To find the y-intercepts (if there are any), we let $x = 0$.

$$x = y^2 + 4y - 5 \qquad \text{This is the given equation.}$$
$$0 = y^2 + 4y - 5 \qquad \text{Replace } x \text{ with 0.}$$
$$0 = (y - 1)(y + 5) \qquad \text{Factor.}$$
$$y - 1 = 0 \text{ or } y + 5 = 0 \qquad \text{Set each factor equal to 0.}$$
$$y = 1 \qquad\qquad y = -5 \qquad \text{Solve.}$$

The y-intercepts are at 1 and -5.

Step 3. To find the x-intercept, we let $y = 0$.

$$x = y^2 + 4y - 5 \qquad \text{This is the given equation.}$$
$$= 0^2 + 4 \cdot 0 - 5 \qquad \text{Replace } y \text{ with 0.}$$
$$= -5$$

The x-intercept is at -5.

Using the vertex and intercepts, we graph the parabola shown in Figure 9.9. ■

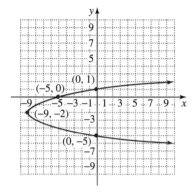

Figure 9.9

The graph of $x = y^2 + 4y - 5$

Using technology

To graph $x = y^2 + 4y - 5$ using a graphing utility, rewrite the equation as a quadratic equation in y.

$$y^2 + 4y + (-x - 5) = 0$$

Use the quadratic formula to solve for y and enter the resulting equations.

$$y_1 = \frac{-4 + \sqrt{16 - 4(-x - 5)}}{2}$$

$$y_2 = \frac{-4 - \sqrt{16 - 4(-x - 5)}}{2}$$

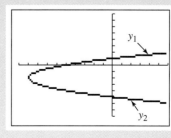

EXAMPLE 6   **Graphing a Parabola**

Graph: $x = -3(y - 1)^2 + 2$

Solution

The form of the given equation is $x = a(y - k)^2 + h$, where $a = -3$, $k = 1$, and $h = 2$. Because a is negative, the parabola will open to the left.

Step 1. The vertex (h, k) is $(2, 1)$.
Step 2. To find the y-intercepts, we let $x = 0$.

$$x = -3(y - 1)^2 + 2$$ This is the given equation.

$$0 = -3(y - 1)^2 + 2$$ Replace x with 0.

$$3(y - 1)^2 = 2$$ Add $3(y - 1)^2$ to both sides.

$$(y - 1)^2 = \frac{2}{3}$$ Divide both sides by 3.

$$y - 1 = \sqrt{\frac{2}{3}} \quad \text{or} \quad y - 1 = -\sqrt{\frac{2}{3}}$$ Apply the square root method.

$$y - 1 = \frac{\sqrt{6}}{3} \qquad y - 1 = -\frac{\sqrt{6}}{3}$$ $\sqrt{\frac{2}{3}} = \frac{\sqrt{2}}{\sqrt{3}} = \frac{\sqrt{2}}{\sqrt{3}} \cdot \frac{\sqrt{3}}{\sqrt{3}} = \frac{\sqrt{6}}{3}$

$$y = 1 + \frac{\sqrt{6}}{3} \qquad y = 1 - \frac{\sqrt{6}}{3}$$ Add 1 to both sides.

$$= \frac{3 + \sqrt{6}}{3} \qquad = \frac{3 - \sqrt{6}}{3}$$ $1 = \frac{3}{3}$

Because $\sqrt{6} \approx 2.4$, we have $y \approx 1.8$ or $y \approx 0.2$. The y-intercepts are located at approximately 1.8 and 0.2.
Step 3. To find the x-intercept, we let $y = 0$.

$$x = -3(y - 1)^2 + 2$$ This is the given equation.
$$= -3(0 - 1)^2 + 2$$ Replace y with 0.
$$= -3(-1)^2 + 2$$
$$= -1$$

The x-intercept is -1.

Figure 9.10 shows the graph.   ■

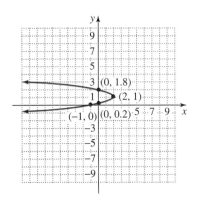

Figure 9.10

The graph of
$x = -3(y - 1)^2 + 2$

Parabolas have many applications. Cables hung between structures to form suspension bridges form parabolas. Arches whose main purpose is strength are usually parabolic in shape and constructed of steel and concrete. A projectile, such as a baseball or water from a drinking fountain, thrown directly upward, moves along a parabolic path.

If a parabola is rotated about its axis of symmetry, a parabolic surface is formed. (Refer to the figures shown at the top of the next page.) A light source originating at the focus will reflect from the surface parallel to the axis of symmetry. This property is used in the design of searchlights, automobile headlights, and parabolic microphones. The same principle is used in reverse in reflecting telescopes, radar, and T.V. satellite dishes.

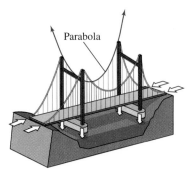

Suspension Bridge

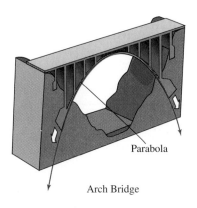

Arch Bridge

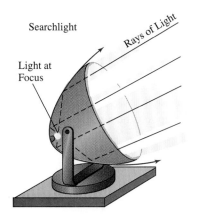

Searchlight

Light at
Focus

P R O B L E M   S E T   9 . 1

Practice Problems

In Problems 1–10, write the equation of each circle with the given center and radius.

1. Center $(3, 2)$, $r = 5$ **2.** Center $(2, -1)$, $r = 4$ **3.** Center $(-1, 4)$, $r = 2$ **4.** Center $(-3, 5)$, $r = 3$
5. Center $(-3, -1)$, $r = \sqrt{3}$ **6.** Center $(-5, -3)$, $r = \sqrt{5}$ **7.** Center $(-4, 0)$, $r = 2$ **8.** Center $(-2, 0)$, $r = 6$
9. Center $(0, 0)$, $r = 7$ **10.** Center $(0, 0)$, $r = 8$

In Problems 11–18, give the center and radius of each circle and graph its equation.

11. $x^2 + y^2 = 16$ **12.** $x^2 + y^2 = 49$ **13.** $(x - 3)^2 + (y - 1)^2 = 36$
14. $(x - 2)^2 + (y - 3)^2 = 16$ **15.** $(x + 3)^2 + (y - 2)^2 = 4$ **16.** $(x + 1)^2 + (y - 4)^2 = 25$
17. $(x + 2)^2 + (y + 2)^2 = 4$ **18.** $(x + 4)^2 + (y + 5)^2 = 36$

In Problems 19–26, complete the square and then give the center and radius of each circle. Then graph each equation.

19. $x^2 + y^2 + 6x + 2y + 6 = 0$ **20.** $x^2 + y^2 + 8x + 4y + 16 = 0$ **21.** $x^2 + y^2 - 10x - 6y - 30 = 0$
22. $x^2 + y^2 - 4x - 12y - 9 = 0$ **23.** $x^2 + y^2 + 8x - 2y - 8 = 0$ **24.** $x^2 + y^2 + 12x - 6y - 4 = 0$
25. $x^2 - 2x + y^2 - 15 = 0$ **26.** $x^2 + y^2 - 6y - 7 = 0$

In Problems 27–44, use intercepts, the vertex, and, if necessary, one or two additional points to graph each parabola.

27. $y = (x - 2)^2 - 4$ **28.** $y = (x - 3)^2 - 4$ **29.** $x = (y - 2)^2 - 4$ **30.** $x = (y - 3)^2 - 4$
31. $y = x^2 - 4$ **32.** $x = y^2 - 4$ **33.** $x = y^2 + 4$ **34.** $y = x^2 + 4$
35. $y = -2(x - 1)^2 - 1$ **36.** $y = -2(x - 1)^2 - 4$ **37.** $x = -2(y - 1)^2 - 1$ **38.** $x = -2(y - 1)^2 - 4$
39. $x = y^2 + 2y - 3$ **40.** $x = -y^2 - 4y + 5$ **41.** $x = y^2$ **42.** $x = -2y^2$
43. $x = -y^2 - 2y + 3$ **44.** $x = y^2 + 4y - 5$

Application Problems

45. The towers of a suspension bridge are 800 feet apart and rise 160 feet above the road. The cable between the towers has the shape of a parabola, and the cable just touches the sides of the road midway between the towers. What is the height of the cable 100 feet from a tower?

(400, 160)

Parabolic Arc

Road $\longleftarrow$ 800 feet $\longrightarrow$

46. An engineer is designing a flashlight so that within its cylindrical casting lies a parabolic mirror and a light source, shown in the figure below on the left. The casting has a diameter of 4 inches and a depth of 2 inches. The figure on the right positions the parabola with its vertex at the origin and opening upward.

  a. Which one of the following equations represents the parabola whose graph is shown? Explain why you chose the equation you did and not the others.

   a. $y^2 = 2x$ **b.** $y^2 = -2x$ **c.** $x^2 = 2y$ **d.** $x^2 = -2y$

  b. The light source for the flashlight should be placed at the focus. In Problem 65, you will discover that parabolas of the form $x^2 = 4py$ have a focus at $(0, p)$. Use this information to describe at what point the light source should be placed.

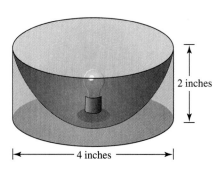

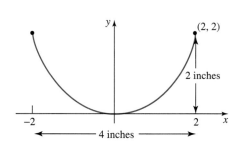

True–False Critical Thinking Problems

47. Which one of the following is true?

  a. The equation of the circle whose center is at the origin with radius 16 is $x^2 + y^2 = 16$.

  b. The graph of $(x - 3)^2 + (y + 5)^2 = 36$ is a circle of radius 6 whose center is at $(-3, 5)$.

  c. The graph of $(x - 4) + (y + 6) = 25$ is a circle of radius 5 centered at $(4, -6)$.

  d. None of the above is true.

48. Which one of the following is true?

  a. The coordinates of a circle's center satisfy the circle's equation.

  b. The radius of the circle whose equation is given by $x^2 - 6x + y^2 = 25$ is 5.

  c. If the center of a circle is at $(-1, -2)$ and the circle passes through the origin, then not enough information is provided to write the circle's equation.

  d. None of the above is true.

49. Which one of the following is true?

  a. The parabola whose equation is $x = 2y - y^2 + 5$ opens to the right.

  b. If the parabola whose equation is $x = ay^2 + by + c$ has its vertex at $(3, 2)$ and $a > 0$, then it has no y-intercepts.

  c. Some parabolas that open to the right have equations that define y as a function of x.

  d. The graph of $x = a(y - k) + h$ is a parabola with vertex at (h, k).

Technology Problems

50. Use a graphing utility to verify any two of the graphs from Problems 11–26, and two of the graphs from Problems 27–44.

51. Write the following equation as a quadratic equation in y and then use the quadratic formula to express y in terms of x: $16x^2 - 24xy + 9y^2 - 60x - 80y + 100 = 0$. Now graph the two resulting equations with your graphing utility. What effect does the xy-term have on the resulting parabola?

52. Although this is a technology problem, it is also a critical thinking problem. The figure on the right shows a line that touches the parabola at the point (x_0, y_0), called the *tangent line*. The tangent line has a y-intercept at $-y_0$. Find the equation of the tangent line to $x^2 = 4y$ at the point $(2, 1)$. Then use a graphing utility to graph the parabola and the tangent line in the same viewing rectangle. Repeat this process for a few other points on the parabola. In what way might the slope of these tangent lines be useful in describing the parabola's changing steepness?

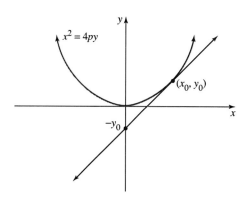

Writing in Mathematics

53. The equation $x^2 + y^2 = 25$ does not define y as a function of x. However, the semicircles whose equations are $y = \sqrt{25 - x^2}$ and $y = -\sqrt{25 - x^2}$ do represent y as a function of x. Does every semicircle formed by $x^2 + y^2 = 25$ define y as a function of x? Explain, illustrating with graphs and equations.

54. Does $(x - 3)^2 + (y - 5)^2 = 0$ represent the equation of a circle? If not, describe the graph of this equation.

55. Does $(x - 3)^2 + (y - 5)^2 = -25$ represent the equation of a circle? What sort of set is the graph of this equation?

56. Discuss similarities and differences between the graphs of the equations $y = x^2 - 6x + 5$ and $x = y^2 - 6y + 5$. Be sure to use both graphs in your discussion.

57. A parabola is defined to be the set of points in a plane that are the same distance from a fixed point as they are from a fixed line. In the figure, the fixed point is $(0, p)$ and the fixed line is $y = -p$. Explain why the 13 points in the figure must lie along a parabola.

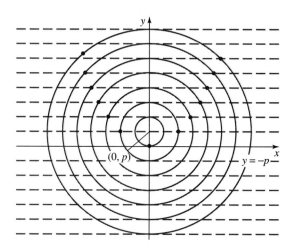

58. Match the graph in the figure with the correct equation:

$$y = (x - 1)^2 - 1; \quad x = (y - 1)^2 - 1;$$
$$x = (y - 1)^2; \quad y = (x - 1)^2$$

Now describe the process that you used to select the correct equation.

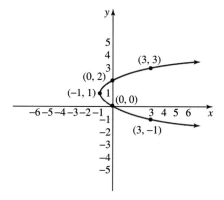

Critical Thinking Problems

59. What is the area of the region described by the inequality $(x - 3)^2 + (y + 2)^2 \leqslant 9$?

60. What is the area of the region bounded by the graphs of $x^2 + y^2 = 36$ and $x^2 + y^2 = 64$?

61. Find the equation of the circle with center at the origin that contains the point $(4, 3)$.

Use the figure shown below in solving Problems 62–63.

62. A tangent line to a circle is a line that intersects the circle at exactly one point. The tangent line is perpendicular to the radius of the circle at the point of contact. Write the point-slope equation of a line tangent to the circle whose equation is $x^2 + y^2 = 25$ at the point $(3, -4)$.

63. A rectangle centered at the origin with sides parallel to the coordinate axes is placed inside a circle whose equation is $x^2 + y^2 = 25$. If the length of the rectangle is $2x$ linear units, represent the area of the rectangle as a function of x.

64. Write the equation of a parabola whose vertex and y-intercept are the same point.

65. A parabola can be defined as the set of all points in a plane equidistant from a fixed point (the focus) and a fixed line not containing this point (the directrix).

a. As shown in the figure below, (x, y) is any point on the parabola, and the equation of the directrix is $y = -p$. Find the distance between (x, y) and the directrix. (The distance from a point to a line is the length of the perpendicular line segment from the point to the line, designated by line segment PP' in the figure.)

b. Find an algebraic expression for the distance from (x, y) to the focus $(0, p)$.

c. Equate parts (a) and (b) (using the definition of the parabola) and show that the equation of this parabola can be written in the form $x^2 = 4py$.

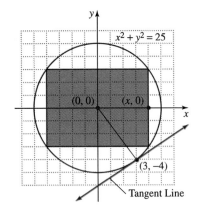

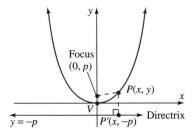

66. Use the equation derived in Problem 65 to find an equation for a parabola with focus $(0, 4)$ and directrix with the equation $y = -4$.

Review Problems

67. Solve the system:

$$3x - 2y = 1$$
$$5y + 3z = -7$$
$$2x + 5y = 45$$

68. Solve: $\sqrt{2x + 3} - \sqrt{x + 1} = 1$.

69. Solve: $(x^2 - 2x)^2 - 14(x^2 - 2x) = 15$.

Solutions   Tutorial   Video
Manual   10

S E C T I O N   9 . 2   **Conic Sections: Ellipses and Hyperbolas**

Objectives

1 Graph ellipses.
2 Solve applied problems involving ellipses.
3 Graph hyperbolas.
4 Recognize equations of conic sections.

Ellipses

The next conic section we consider is the oval-shaped curve known as the *ellipse.* Ellipses can be obtained by intersecting a plane and a cone, as shown in Figure 9.11. Figure 9.12 illustrates how to draw an ellipse. Place two thumbtacks at two fixed points, each of which is called a *focus* (plural: *foci*). If the ends of a fixed length of string are fastened to the thumbtacks and we draw the string taut with a pencil, the path traced by the pencil will be an ellipse. Notice that the sum of the distances of the pencil point from the foci remains constant, because the length of the string is fixed. This procedure for drawing an ellipse illustrates its definition.

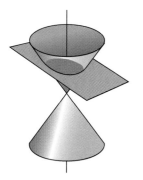

Figure 9.11

Ellipses are obtained by intersecting a plane and a cone

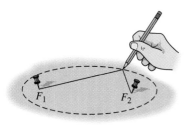

Figure 9.12

Drawing an ellipse

Definition of an ellipse

An *ellipse* is the set of all points in a plane the sum of whose distances from two fixed points F_1 and F_2 is constant (see Figure 9.13). These two fixed points are called the *foci* (plural of *focus*). The midpoint of the segment connecting the foci is the *center* of the ellipse.

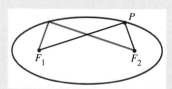

Figure 9.13

The equation of an ellipse can be derived from its definition. We will give you the opportunity to do this in the problem set. At this point, let's simply become familiar with the equation.

Discover for yourself

What happens to the equation of an ellipse

$$\frac{x^2}{a^2} + \frac{y^2}{b^2} = 1$$

if $a = b$? Substitute a for b in the equation and then clear fractions. What conic section does the resulting equation represent?

Graph ellipses.

Equation of an ellipse

The standard form for the equation of an ellipse with its center at the origin is

$$\frac{x^2}{a^2} + \frac{y^2}{b^2} = 1, \quad a, b > 0, \quad a \neq b.$$

The easiest way to graph an ellipse is to use its x- and y-intercepts. If we replace y by 0 in the equation of an ellipse, we can determine the x-intercepts.

$$\frac{x^2}{a^2} + \frac{y^2}{b^2} = 1 \qquad \text{This is the equation of an ellipse.}$$

$$\frac{x^2}{a^2} + \frac{0^2}{b^2} = 1 \qquad \text{To find } x\text{-intercepts, set } y \text{ equal to 0.}$$

$$\frac{x^2}{a^2} = 1$$

$$x^2 = a^2 \qquad \text{Multiply both sides by } a^2.$$

$$x = \pm a \qquad \text{Apply the square root method.}$$

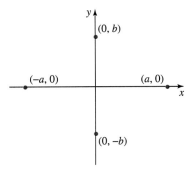

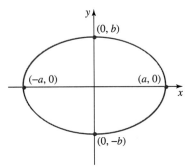

Figure 9.14

Using intercepts to graph

$$\frac{x^2}{a^2} + \frac{y^2}{b^2} = 1$$

This means that the x-intercepts are $-a$ and a. Equivalently, the ellipse passes through the points $(-a, 0)$ and $(a, 0)$.

Using a similar method, we can determine the y-intercepts for the ellipse by setting x equal to 0 in its equation. The y-intercepts are $-b$ and b, and so the ellipse passes through $(0, -b)$ and $(0, b)$.

The use of intercepts to graph the ellipse whose equation is

$$\frac{x^2}{a^2} + \frac{y^2}{b^2} = 1$$

is illustrated in Figure 9.14. Once we have graphed the four intercepts, we connect them using an oval-shaped curve.

Using intercepts to graph an ellipse with center (0, 0)

The ellipse whose equation is

$$\frac{x^2}{a^2} + \frac{y^2}{b^2} = 1$$

has x-intercepts at $-a$ and a and y-intercepts at $-b$ and b.

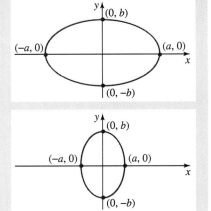

S tudy tip

If $a > b$, an ellipse is elongated horizontally.

If $a < b$, an ellipse is elongated vertically.

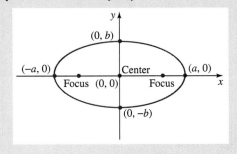

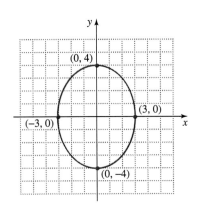

Figure 9.15

The graph of $\dfrac{x^2}{9} + \dfrac{y^2}{16} = 1$

EXAMPLE 1 **Graphing an Ellipse**

Graph: $\dfrac{x^2}{9} + \dfrac{y^2}{16} = 1$

Solution

We can express the given equation as

$$\frac{x^2}{3^2} + \frac{y^2}{4^2} = 1.$$

Using technology

We graph $\dfrac{x^2}{9} + \dfrac{y^2}{16} = 1$ with a graphing utility by solving for y and defining two functions.

$$\frac{y^2}{16} = 1 - \frac{x^2}{9}$$

$$y^2 = 16\left(1 - \frac{x^2}{9}\right)$$

$$y = \pm 4\sqrt{1 - \frac{x^2}{9}}$$

Enter

$y_1 = 4\;\boxed{\sqrt{}}\;(1\;\boxed{-}\;x\;\boxed{\wedge}\;2\;\boxed{\div}\;9)$

and

$y_2 = -y_1$

To see the true shape of the ellipse, use the $\boxed{\text{ZOOM SQUARE}}$ feature so that one unit on the x-axis is the same length as one unit on the y-axis.

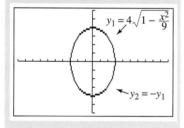

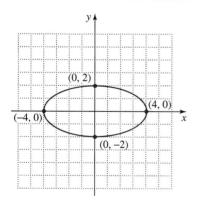

Figure 9.16

The graph of $4x^2 + 16y^2 = 64$

This means that the x-intercepts are -3 and 3, and the y-intercepts are -4 and 4. We connect the four resulting points with an oval-shaped curve, as shown in Figure 9.15 at the bottom of the previous page.

Discover for yourself

You can find other points on the ellipse in Example 1 by selecting any value for x between the x-intercepts ($-3 < x < 3$), substituting that value in the equation, and solving for y. Do this by letting $x = 1$ and solving

$$\frac{1^2}{9} + \frac{y^2}{16} = 1.$$

Multiply both sides of the equation by 144. You should obtain

$$y = \pm\sqrt{\frac{128}{9}} \approx \pm 3.8.$$

Locate the points $(1, -3.8)$ and $(1, 3.8)$ on the graph in Figure 9.15. You could also use these points to help sketch the graph.

EXAMPLE 2 **Graphing an ellipse**

Graph: $4x^2 + 16y^2 = 64$

Solution

We begin by expressing the equation in standard form. Since we want 1 on the right side, we divide both sides by 64.

$$\frac{4x^2}{64} + \frac{16y^2}{64} = \frac{64}{64}$$

$$\frac{x^2}{16} + \frac{y^2}{4} = 1$$

We can express this equation as

$$\frac{x^2}{4^2} + \frac{y^2}{2^2} = 1$$

the standard form for an ellipse. This means that the x-intercepts are -4 and 4 and the y-intercepts are -2 and 2. We plot the intercepts and connect them with an oval-shaped curve. The graph of the ellipse is shown in Figure 9.16.

Discover for yourself

Compute and plot two other points on the ellipse in Example 2.

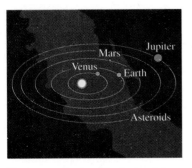

Planets move in elliptical orbits.

We graph $4x^2 + 16y^2 = 64$ with a graphing utility by solving for y and obtaining two equations.

$$y_1 = \frac{\sqrt{64 - 4x^2}}{4}$$

and $y_2 = -\frac{\sqrt{64 - 4x^2}}{4}$

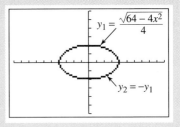

The first equation graphs as the top half of the ellipse and the second equation represents the lower half of the ellipse. Enter each equation and use a zoom square setting.

Whispering in an elliptical room

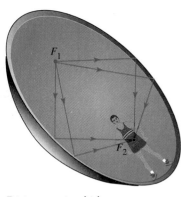

Disintegrating kidney stones

2 Solve applied problems involving ellipses.

Applications

Ellipses have many applications. German scientist Johannees Kepler (1571–1630) showed that the planets in our solar system move in elliptical orbits, with the sun as a focus. Earth satellites also travel in elliptical orbits.

One intriguing aspect of the ellipse is that a line from one focus will be reflected from the side of the ellipse exactly through the other focus. A whispering gallery is an elliptical room with an elliptical dome-shaped ceiling. People standing at the foci can whisper and hear each other quite clearly, while persons in other locations in the room cannot hear them. Statuary Hall in the U.S. Capitol building is elliptical. President John Quincy Adams, while a member of the House of Representatives, was aware of this acoustical phenomena. He situated his desk at a focal point of the elliptical ceiling, easily eavesdropping on the private conversations of other House members located near the other focus.

The elliptical reflection principle is used in a procedure for disintegrating kidney stones. The patient is placed within a device that is elliptical in shape. The patient is at one focus, while ultrasound waves from the other focus hit the walls and are reflected to the kidney stone. The convergence of the ultrasound waves at the kidney stone causes vibrations that shatter it into fragments. The small pieces can then be passed painlessly through the patient's system. The patient recovers in days, as opposed to up to six weeks if surgery is used instead.

Ellipses are often used for supporting arches of bridges and in tunnel construction. This application forms the basis of our next example.

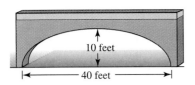

Figure 9.17
A semielliptic archway

EXAMPLE 3 | **An Application Involving an Ellipse**

A semielliptic archway over a one-way road has a height of 10 feet and a width of 40 feet (see Figure 9.17). Will a truck that is 10 feet wide and 9 feet high clear the opening of the archway?

Solution

To determine the clearance, we must find the height of the tunnel 5 feet from the center. If it's 9 feet or less, the truck will not clear the opening.

ENRICHMENT ESSAY

Aphelion and Perihelion

Aphelion and perihelion are those two points on an elliptical orbit at which a planet or comet is farthest from or closest to the sun. Earth was at aphelion on July 3, 1990, at which time it was 94,508,105 miles from the sun.

the sun and a planet sweeps equal areas in equal times. Consequently, the speed of a planet increases as it nears the sun and decreases as it recedes from the sun.

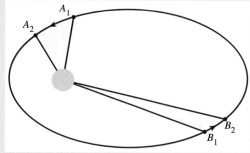

The time between A_1 and A_2 equals the time between B_1 and B_2. The speed between A_1 and A_2 is faster, as it is nearer to the sun.

German astronomer and mathematician Johannes Kepler was the first person to discern the architecture of the solar system and formulate laws governing the motions of planets. Kepler's second law of planetary motion states that a line between

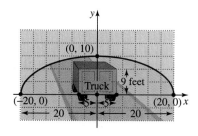

Figure 9.18

In Figure 9.18, we've constructed a rectangular coordinate system with the x-axis on the ground and the origin at the center of the archway. Also shown is the truck whose height is 9 feet.

$$\frac{x^2}{a^2} + \frac{y^2}{b^2} = 1 \qquad \text{This is the standard form for the equation of an ellipse.}$$

$$\frac{x^2}{20^2} + \frac{y^2}{10^2} = 1 \qquad \text{Substitute the } x\text{-intercept value, 20, and the } y\text{-intercept value, 10, into the equation of an ellipse.}$$

$$\frac{x^2}{400} + \frac{y^2}{100} = 1$$

As shown in Figure 9.18, the edge of the 10-foot wide truck corresponds to $x = 5$. We find the height of the archway 5 feet from the center by substituting 5 for x and solving for y.

$$\frac{5^2}{400} + \frac{y^2}{100} = 1 \qquad \text{Substitute 5 for } x.$$

$$\frac{25}{400} + \frac{y^2}{100} = 1$$

$$\frac{1}{16} + \frac{y^2}{100} = 1$$

$$1600\left(\frac{1}{16} + \frac{y^2}{100}\right) = 1600(1) \qquad \text{Clear fractions by multiplying both sides by 1600.}$$

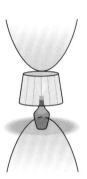

Figure 9.19

Casting hyperbolic shadows

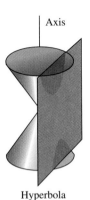

Figure 9.20

Hyperbolas are obtained by intersecting a plane and a cone.

$$100 + 16y^2 = 1600 \quad \text{Use the distributive property and simplify.}$$
$$16y^2 = 1500 \quad \text{Subtract 100 from both sides.}$$
$$y^2 = \frac{1500}{16} \quad \text{Divide both sides by 16.}$$
$$y = \sqrt{\frac{1500}{16}} \quad \text{Take only the positive square root. The archway is half of an ellipse and } y \text{ is nonnegative.}$$
$$\approx 9.68$$

Thus, the height of the opening 5 feet from the center is approximately 9.68 feet. Since the truck's height is 9 feet it will clear the tunnel. ■

Hyperbolas

Figure 9.19 shows a cylindrical lampshade casting two shadows on a wall. These shadows indicate the distinguishing feature of hyperbolas: Their graphs contain two disjoint parts called *branches*. Although each branch might look like a parabola, its shape is quite different.

Hyperbolas can be obtained by intersecting a plane and a cone, as shown in Figure 9.20. A hyperbola is the set of points in a plane the *difference* of whose distances from two fixed points (called *foci*) is a constant. The point midway between the two foci is the *center* of the hyperbola.

The definition of a hyperbola is similar to that of an ellipse. Consequently, the equations for a hyperbola look very similar to the equation of an ellipse.

Study tip

1. Like the equation for an ellipse, equations for hyperbolas have a 1 on the right.
2. The equation for an ellipse has a plus sign between terms. Equations for hyperbolas have a minus sign between terms.
3. The intercepts are found using the term without the minus sign.

Equations of a hyperbola

Hyperbolas with their centers at the origin have equations as follows.

Hyperbola with *x*-intercepts

$$\frac{x^2}{a^2} - \frac{y^2}{b^2} = 1$$

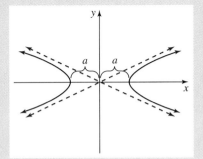

Hyperbola with *y*-intercepts

$$\frac{y^2}{b^2} - \frac{x^2}{a^2} = 1$$

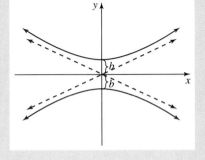

As *x* and *y* get larger and larger, the two branches of the graph of a hyperbola approach a pair of intersecting straight lines called *asymptotes*. The asymptotes pass through the center of the hyperbola and are helpful in graphing a hyperbola.

Asymptotes of a hyperbola

The hyperbolas with equations in the preceding box have asymptotes whose equations are

$$y = \frac{b}{a}x \quad \text{and} \quad y = -\frac{b}{a}x.$$

3 Graph hyperbolas.

Hyperbolas are graphed using intercepts and asymptotes.

iscover for yourself

Show that the *x*-intercepts are −5 and 5 by setting *y* equal to 0 and solving for *x*. Show that there are no *y*-intercepts by setting *x* equal to 0. Describe what happens.

EXAMPLE 4 **Graphing a Hyperbola**

Graph: $\dfrac{x^2}{25} - \dfrac{y^2}{16} = 1$

Solution

We can express the given equation as

$$\frac{x^2}{5^2} - \frac{y^2}{4^2} = 1 \quad \text{The equation is in the form } \frac{x^2}{a^2} - \frac{y^2}{b^2} = 1. \text{ Notice that } a = 5 \text{ and } b = 4.$$

This means that the *x*-intercepts are −5 and 5. The hyperbola's branches pass through $(-5, 0)$ and $(5, 0)$.

Using $a = 5$ and $b = 4$, the asymptotes are

$$y = \frac{b}{a}x \quad \text{and} \quad y = -\frac{b}{a}x.$$

or

$$y = \frac{4}{5}x \quad \text{and} \quad y = -\frac{4}{5}x.$$

We sketch the asymptotes, as shown in Figure 9.21.

Now that we have the intercepts and the asymptotes, through each intercept we draw a smooth curve that approaches the asymptotes closely. The graph of the hyperbola is shown in Figure 9.22. The asymptotes are drawn using dashed lines because they are not part of the hyperbola. ∎

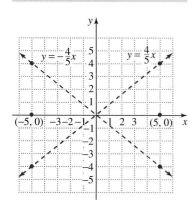

Figure 9.21

The asymptotes for

$$\frac{x^2}{25} - \frac{y^2}{16} = 1$$

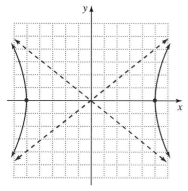

Figure 9.22

The graph of $\dfrac{x^2}{25} - \dfrac{y^2}{16} = 1$

tudy tip

Asymptotes can be drawn quickly by sketching a rectangle to use as a guide. To obtain the rectangle for

$$\frac{x^2}{5^2} - \frac{y^2}{4^2} = 1$$

use 5 and −5 on the *x*-axis (the intercepts) and 4 and −4 on the *y*-axis. The rectangle passes through these four points, shown in the figure on the right. Now draw dashed lines through the opposite corners of the rectangle to obtain the graph of the asymptotes.

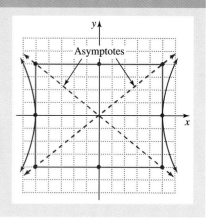

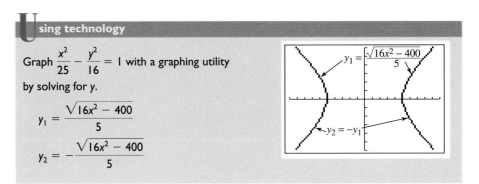

Using technology

Graph $\dfrac{x^2}{25} - \dfrac{y^2}{16} = 1$ with a graphing utility

by solving for y.

$$y_1 = \dfrac{\sqrt{16x^2 - 400}}{5}$$

$$y_2 = -\dfrac{\sqrt{16x^2 - 400}}{5}$$

EXAMPLE 5 **Graphing a Hyperbola with No x-Intercepts**

Graph: $9y^2 - 4x^2 = 36$

Solution

We begin by writing the equation in standard form. The right side should be 1, so we divide both sides by 36.

$$\dfrac{9y^2}{36} - \dfrac{4x^2}{36} = \dfrac{36}{36}$$

$$\dfrac{y^2}{4} - \dfrac{x^2}{9} = 1 \qquad \text{Simplify. The right side is now 1.}$$

We can express this equation as

$$\dfrac{y^2}{2^2} - \dfrac{x^2}{3^2} = 1 \qquad \text{The equation is in the form } \dfrac{y^2}{b^2} - \dfrac{x^2}{a^2} = 1 \text{ with } b = 2 \text{ and } a = 3.$$

The intercepts are found using the term without the minus sign. Thus, the intercepts are on the y-axis. In particular, the y-intercepts are -2 and 2.

We can construct a rectangle to find the asymptotes. Use 2 and -2 on the y-axis (the intercepts) and 3 and -3 on the x-axis. The rectangle passes through these four points. Draw dashed lines through the opposite corners to show the asymptotes. Finally, through each intercept draw a smooth curve that approaches the asymptotes. The graph of the hyperbola is shown in Figure 9.23.

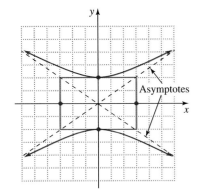

Figure 9.23
The graph of $9y^2 - 4x^2 = 36$

Hyperbolas in Nonstandard Form

Another form for a hyperbola's equation is $xy = c$, where c is a nonzero constant. Hyperbolas in this form have the x- and y-axes as asymptotes.

study tip

Hyperbolas with equations in the form $xy = c$ do not have x- and y-intercepts.

Nonstandard form for hyperbolas

Equations in the form $xy = c$ (where $c \neq 0$) are hyperbolas with the x- and y-axes as asymptotes.

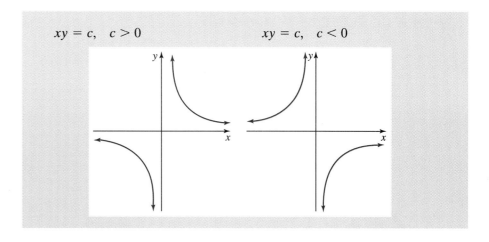

$$xy = c, \quad c > 0 \qquad\qquad xy = c, \quad c < 0$$

EXAMPLE 6 **Graphing a Hyperbola in Nonstandard Form**

Graph: $xy = -6$

Solution

We begin by solving the equation for y.

$$y = -\frac{6}{x}$$

We can graph this equation in two parts. We cannot use 0 for x because division by 0 is undefined. We can construct one table of coordinates for negative values of x and another table of coordinates for positive values of x.

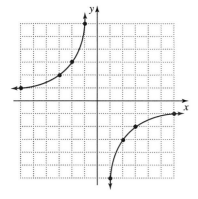

Figure 9.24

The graph of $xy = -6$

x	$y = -\dfrac{6}{x}$
-6	1
-3	2
-2	3
-1	6
$-\frac{1}{2}$	12
$-\frac{1}{6}$	36

x	$y = -\dfrac{6}{x}$
6	-1
3	-2
2	-3
1	-6
$\frac{1}{2}$	-12
$\frac{1}{6}$	-36

We plot the points and connect the points in the second quadrant with a smooth curve. In the same manner, we connect the points in the fourth quadrant with a smooth curve. Keep in mind that the x- and y-axes are asymptotes. The graph is shown in Figure 9.24. ∎

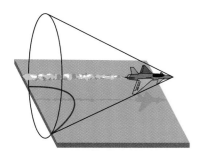

Hyperbolas have many applications. When a jet flies at a speed greater than the speed of sound, the shock wave that is created is heard as a sonic boom. The wave front has the shape of a cone. The shape formed as the cone hits the ground is one branch of a hyperbola.

ENRICHMENT ESSAY

Mathematics and Architecture

The saddle-shaped surface, called the *hyperbolic paraboloid,* has cross-sections that are perpendicular to the z-axis, which are hyperbolas. The cross-sections in planes perpendicular to the other axes are parabolas.

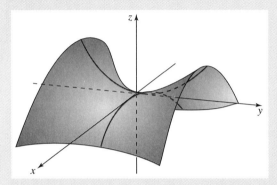

St. Mary's Cathedral. Andrea Pistolesi/The Image Bank

The hyperbolic paraboloid is used in the design of St. Mary's Cathedral in San Francisco. The top of the structure is a 2135-cubic-foot hyperbolic paraboloid cupola with walls rising 200 feet above the floor and supported by four massive concrete pylons that extend 94 feet into the ground.

At the unveiling when the designer was asked what Michelangelo would have thought of the cathedral, he responded, "Michelangelo could not have thought of it. This design comes from geometric theories not then proven."

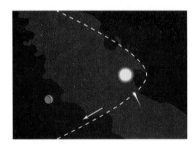

Halley's comet, a permanent part of our solar system, travels around the sun in an elliptical orbit. Other comets pass through the solar system only once, following a hyperbolic path with the sun at a focus.

Hyperbolas are of practical importance in fields ranging from architecture to navigation. Cooling towers used in the design for nuclear power plants have cross sections that are both ellipses and hyperbolas. The hyperbola is the basis for the LORAN (long-range navigation) system used to determine the location of ships.

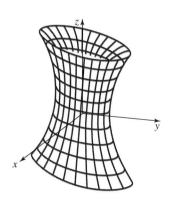

Photri, Inc.

4 Recognize equations of conic sections.

Summary of Conic Sections

Following is a summary of the equations and graphs of the conic sections that we have studied.

Conic sections

Circle

$$(x - h)^2 + (y - k)^2 = r^2$$

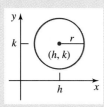

Ellipse

$$\frac{x^2}{a^2} + \frac{y^2}{b^2} = 1$$

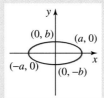

Parabola

$$y = ax^2 + bx + c$$
$$y = a(x - h)^2 + k$$

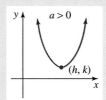

$$x = ay^2 + by + c$$
$$x = a(y - k)^2 + h$$

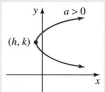

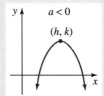

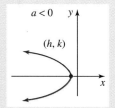

Hyperbola

$$\frac{x^2}{a^2} - \frac{y^2}{b^2} = 1$$

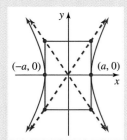

$$\frac{y^2}{b^2} - \frac{x^2}{a^2} = 1$$

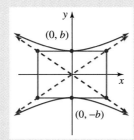

$$\text{Asymptotes: } y = \frac{b}{a}x \text{ and } y = -\frac{b}{a}x$$

Nonstandard Forms of a Hyperbola

$$xy = c, c > 0$$

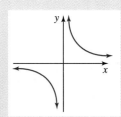

$$xy = c, c < 0$$

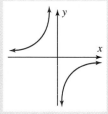

tudy tip

Here are some hints for recognizing equations of conic sections.

Conic Section	How to Identify the Equation	Example
Circle	When x^2- and y^2-terms are on the same side, they have the same coefficient.	$x^2 + y^2 = 16$
Parabola	Only one of the variables is squared.	$x = y^2 + 4y - 5$
Ellipse	When x^2- and y^2-terms are on the same side, they have different positive coefficients.	$4x^2 + 16y^2 = 64$ or (dividing by 64) $\dfrac{x^2}{16} + \dfrac{y^2}{4} = 1$
Hyperbola	When x^2- and y^2-terms are on the same side, they have coefficients with opposite signs.	$9y^2 - 4x^2 = 36$ or (dividing by 36) $\dfrac{y^2}{4} - \dfrac{x^2}{9} = 1$

EXAMPLE 7   **Recognizing Equations of Conic Sections**

Indicate whether the graph of each equation is a circle, a parabola, an ellipse, or a hyperbola.

a. $4y^2 = 16 - 4x^2$ **b.** $x^2 = y^2 + 9$
c. $x + 7 - 6y = y^2$ **d.** $x^2 = 16 - 16y^2$

Solution

If both variables are squared, the equation is not the graph of a parabola. In this case, we get the x^2- and y^2-terms on the same side of the equation.

a. The equation $4y^2 = 16 - 4x^2$ does not represent a parabola. We get the x^2- and y^2-terms on the same side by adding $4x^2$ to both sides.

$$4x^2 + 4y^2 = 16$$

Since the coefficients of x^2 and y^2 are the same, the equation's graph is a circle. This becomes more obvious if we divide both sides by 4.

$$x^2 + y^2 = 4$$
$$(x - 0)^2 + (y - 0)^2 = 2^2 \quad \text{The equation of the circle with center } (h, k) \text{ and radius } r \text{ is } (x - h)^2 + (y - k)^2 = r^2.$$

The graph is a circle with center at the origin and radius 2.

b. The equation $x^2 = y^2 + 9$ does not represent a parabola. We get the x^2- and y^2-terms on the same side by subtracting y^2 from both sides.

$$x^2 - y^2 = 9$$

Since x^2- and y^2-terms have coefficients with opposite signs, the equation's graph is a hyperbola. This becomes more obvious if we divide both sides by 9 to obtain 1 on the right.

$$\frac{x^2}{9} - \frac{y^2}{9} = 1$$

$$\frac{x^2}{3^2} - \frac{y^2}{3^2} = 1 \qquad \text{The equation of a hyperbola with } x\text{-intercepts is } \frac{x^2}{a^2} - \frac{y^2}{b^2} = 1.$$

c. The equation $x + 7 - 6y = y^2$ represents a parabola because only one of the variables is squared. We can express the equation in the form $x = ay^2 + by + c$ by isolating x on the right.

$$x = y^2 + 6y - 7$$

Since the coefficient of the y^2-term is positive, the graph is a parabola that opens to the right.

d. The equation $x^2 = 16 - 16y^2$ does not represent a parabola. We get the x^2- and y^2-terms on the same side by adding $16y^2$ to both sides.

$$x^2 + 16y^2 = 16$$

Since the x^2- and y^2-terms have different positive coefficients, the equation's graph is an ellipse. This becomes more obvious if we divide both sides by 16 to obtain 1 on the right.

$$\frac{x^2}{16} + y^2 = 1$$

$$\frac{x^2}{4^2} + \frac{y^2}{1^2} = 1 \qquad \text{An equation in the form } \frac{x^2}{a^2} + \frac{y^2}{b^2} = 1 \text{ is an ellipse.} \qquad ■$$

PROBLEM SET 9.2

Practice Problems

Graph each ellipse in Problems 1–16.

1. $\dfrac{x^2}{9} + \dfrac{y^2}{4} = 1$ **2.** $\dfrac{x^2}{25} + \dfrac{y^2}{16} = 1$ **3.** $\dfrac{x^2}{9} + \dfrac{y^2}{36} = 1$ **4.** $\dfrac{x^2}{16} + \dfrac{y^2}{49} = 1$

5. $\dfrac{x^2}{25} + \dfrac{y^2}{64} = 1$ **6.** $\dfrac{x^2}{49} + \dfrac{y^2}{36} = 1$ **7.** $\dfrac{x^2}{49} + \dfrac{y^2}{81} = 1$ **8.** $\dfrac{x^2}{64} + \dfrac{y^2}{100} = 1$

9. $25x^2 + 4y^2 = 100$ **10.** $9x^2 + 4y^2 = 36$ **11.** $4x^2 + 16y^2 = 64$ **12.** $16x^2 + 9y^2 = 144$

13. $25x^2 + 9y^2 = 225$ **14.** $4x^2 + 25y^2 = 100$ **15.** $x^2 + 2y^2 = 8$ **16.** $12x^2 + 4y^2 = 36$

Use asymptotes and intercepts to graph the hyperbola represented by each equation in Problems 17–30.

17. $\dfrac{x^2}{9} - \dfrac{y^2}{25} = 1$ **18.** $\dfrac{x^2}{16} - \dfrac{y^2}{25} = 1$ **19.** $\dfrac{x^2}{100} - \dfrac{y^2}{64} = 1$ **20.** $\dfrac{x^2}{144} - \dfrac{y^2}{81} = 1$

21. $\dfrac{y^2}{16} - \dfrac{x^2}{36} = 1$ **22.** $\dfrac{y^2}{25} - \dfrac{x^2}{64} = 1$ **23.** $\dfrac{y^2}{36} - \dfrac{x^2}{25} = 1$ **24.** $\dfrac{y^2}{100} - \dfrac{x^2}{49} = 1$

25. $9x^2 - 4y^2 = 36$ **26.** $4x^2 - 25y^2 = 100$ **27.** $9y^2 - 25x^2 = 225$ **28.** $16y^2 - 9x^2 = 144$

29. $4x^2 = 4 + y^2$ **30.** $25y^2 = 225 + 9x^2$

The equations in Problems 31–38 represent hyperbolas in nonstandard form. Graph each hyperbola.

31. $xy = 4$ **32.** $xy = 1$ **33.** $xy = 2$ **34.** $xy = 6$

35. $xy = -4$ **36.** $xy = -1$ **37.** $xy = -2$ **38.** $xy = -8$

In Problems 39–52, indicate whether the graph of each equation is a circle, a parabola, an ellipse, or a hyperbola.

39. $x = 6y^2 + 4y + 3$ **40.** $5x^2 + 5y^2 = 45$ **41.** $5x^2 - 5y^2 = 45$ **42.** $16x^2 - 9y^2 = 36$

43. $x - 7 - 8y = y^2$ **44.** $4x^2 = 36 - y^2$ **45.** $5x^2 = 12 - 10y^2$ **46.** $-x^2 + 4y^2 = 16$

47. $-3y^2 - 3x^2 = -27$ **48.** $y = \dfrac{2}{x}$ **49.** $x - \dfrac{5}{y} = 0$ **50.** $y + 4x = x^2 + 4$

51. $x + y^2 = 4y^2 - y + 7$ **52.** $5x^2 = 9y^2 + 36 + x^2$

Application Problems

53. A planet with an elliptical orbit is shown in a rectangular coordinate system in the figure. Write an equation that models the planet's orbit.

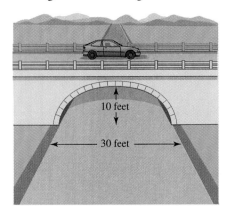

54. A truck that is 8 feet wide is carrying a load that reaches 7 feet above the ground. Will the truck clear the semielliptical arch on the one-way road that passes under the bridge shown in the figure?

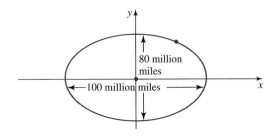

55. The elliptical chamber in the United States Capitol Building is 96 feet long and 46 feet wide, as shown in the figure at the top of the next column.
 a. Write an equation that models the shape of the room.
 b. The foci of an ellipse are located at $(-c, 0)$ and $(c, 0)$, where $b^2 = a^2 - c^2$. (See Problem 67.) John Quincy Adams discovered that he could overhear the conversations of opposing party leaders near the focus at the left side of the chamber if he situated his desk at the focus at the right side of the chamber. What were the coordinates of his desk's position?

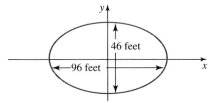

C.L. Chryslin/The Image Bank

56. An architect designs two houses that are shaped and positioned like a part of the branches of the hyperbola whose equation is $625y^2 - 400x^2 = 250,000$, where x and y are in yards. How far apart are the houses at their closest point?

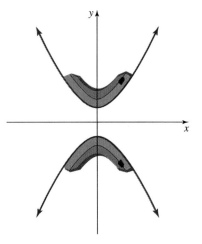

57. Scattering experiments, in which moving particles are deflected by various forces, led to the concept of the nucleus of an atom. In 1911, the physicist Ernest Rutherford (1871–1937) discovered that when alpha particles are directed toward the nuclei of gold atoms, they are eventually deflected along hyperbolic paths, illustrated in the figure. If a particle gets as close as 3 units to the nucleus, what is the equation of its path?

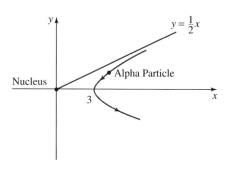

True–False Critical Thinking Problems

58. Which one of the following is true?
  a. The equation of an ellipse centered at the origin with x-intercepts of 4 and -4 and y-intercepts of 9 and -9 is

$$\frac{x^2}{4} + \frac{y^2}{9} = 1$$

  b. The circle whose equation is $x^2 + y^2 = 9$ and the ellipse described by

$$\frac{x^2}{25} + \frac{y^2}{16} = 1$$

  intersect twice.

  c. The equation of an ellipse does not define y as a function of x.

  d. A hyperbola consists of two separate parabolas.

59. Which one of the following is true?
  a. If one branch of a hyperbola is removed from a graph, then the branch that remains must define y as a function of x.

  b. Two different hyperbolas can never share the same asymptotes.

  c. The graph of

$$\frac{x^2}{25} - \frac{y^2}{9} = 1$$

  has y-intercepts at 3 and -3.

  d. The two asymptotes for the hyperbola whose equation is

$$\frac{x^2}{9} - \frac{y^2}{25} = 1$$

  have equations given by $y = \frac{5}{3}x$ and $y = -\frac{5}{3}x$.

Technology Problems

60. Use a graphing utility to verify any two of the graphs from Problems 1–16, Problems 17–30, and Problems 31–38.

61. a. Graph $x^2 + y^2 = (3950)^2$ using Xmin $= -4500$, Xmax $= 4500$, Ymin $= -4500$, and Ymax $= 4500$. This circular orbit is approximately the same as Earth's.

  b. The first artificial satellite to orbit Earth was Sputnik I, launched by the former Soviet Union in 1957. The elliptical orbit of Sputnik I had the equation

$$\frac{(x + 225.5)^2}{(4307.5)^2} + \frac{y^2}{(4301.6)^2} = 1$$

  where one of the foci was Earth's center. Graph this ellipse in the same viewing rectangle as the graph in part (a). Describe what you observe in terms of the altitude of Sputnik I's orbit compared to the size of the Earth.

62. Use a graphing utility to graph $\frac{x^2}{4} - \frac{y^2}{9} = 0$. Is the graph a hyperbola? In general, what is the graph of $\frac{x^2}{a^2} - \frac{y^2}{b^2} = 0$?

63. Graph

$$\frac{x^2}{16} - \frac{y^2}{9} = 1 \quad \text{and} \quad \frac{x|x|}{16} - \frac{y|y|}{9} = 1$$

in the same viewing rectangle. Explain why the graphs are not the same.

Writing in Mathematics _____

64. How can you distinguish an ellipse from a hyperbola by looking at their equations?

65. An elliptipool is an elliptical-shaped pool table with only one pocket. A pool shark places a ball on the table, hits it in what appears to be a random direction, and yet it constantly bounces off the cushion, falling directly into the pocket. Explain why this happens.

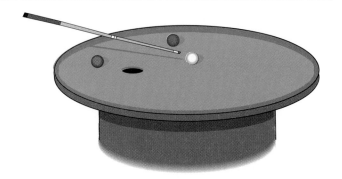

Critical Thinking Problems _____

66. An Earth satellite has an elliptical orbit described by

$$\frac{x^2}{(5000)^2} + \frac{y^2}{(4750)^2} = 1.$$

(All units are in miles.) The coordinates of the center of Earth are $(16, 0)$.

a. The perigee of the satellite's orbit is the point that is nearest Earth's center. If the radius of Earth is approximately 4000 miles, find the distance of the perigee above Earth's surface.

b. The apogee of the satellite's orbit is the point that is the greatest distance from Earth's center. Find the distance of the apogee above Earth's surface.

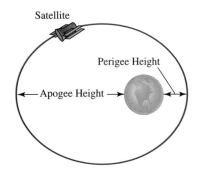

67. Consider the ellipse shown in the figure on the right above Problem 68, where the foci have coordinates designated by $(-c, 0)$, and $(c, 0)$.

a. If the sum of the distances from any point (x, y) on the ellipse to the foci is the constant $2a$, show that the equation of the ellipse is

$$\frac{x^2}{a^2} + \frac{y^2}{a^2 - c^2} = 1.$$

Hint: Begin with the distance formula, using

$$\sqrt{(x + c)^2 + (y - 0)^2} + \sqrt{(x - c)^2 + (y - 0)^2} = 2a.$$

b. Let $b^2 = a^2 - c^2$ and derive the equation

$$\frac{x^2}{a^2} + \frac{y^2}{b^2} = 1.$$

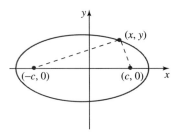

68. The equation of the ellipse in the figure below is

$$\frac{x^2}{25} + \frac{y^2}{9} = 1.$$

Write the equation for each circle shown in the figure.

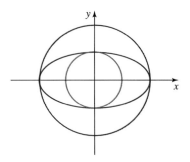

In Problems 69–75, use the graphing techniques for lines and conic sections, and point-plotting for other kinds of figures, to graph the indicated letter.

69. Graph the letter F by graphing

$y = 2$ If $0 \le x \le 1$

$y = 1$ If $0 \le x \le \frac{3}{4}$

$x = 0$ If $0 \le y \le 2$

70. Graph the letter J by graphing

$y = x^2$ If $-1 \le x \le 1$

$x = 1$ If $1 \le y \le 3$

71. Graph the letter K by graphing

$x = 0$ If $-1 \le y \le 1$

$y^2 = x^3$ If $0 \le x \le 1$

72. Graph the letter P by graphing

$x = \sqrt{1 - y^2}$

$x = 0$ If $-3 \le y \le 1$

73. Graph the letter Q by graphing

$x^2 + y^2 = 1$

$y = -x$ If $\frac{1}{2} \le x \le 1$

74. Graph the letter T by graphing

$x = 2$ If $1 \le y \le 3$

$y = 3$ If $1 \le x \le 3$

75. Graph the letter Y by graphing

$y = -x + 3$ If $1 \le x \le 2$

$y = x - 1$ If $2 \le x \le 3$

$x = 2$ If $0 \le y \le 1$

Group Activity Problem

76. Consult the research department of your library or the Internet to find an example of architecture that incorporates one or more conic sections in its design. Share this example with other group members. Explain precisely how conic sections are used. Do conic sections enhance the appeal of the architecture? In what ways?

Review Problems

In the next section, we apply the substitution and addition methods to nonlinear systems. Solve each linear system in Problems 77–79 by the indicated method. If necessary, review the substitution and addition methods for two equations in two variables (Section 3.1) before moving on to Section 9.3.

77. Solve by the substitution method:

$$x = 2y + 7$$
$$-5x + 4y = -5$$

78. Solve by the substitution method:

$$8x + 2y - 7 = 0$$
$$-2x - y + 2 = 0$$

79. Solve by the addition method:

$$12x + 5y = -2$$
$$8x - 3y = -14$$

SECTION 9.3	Nonlinear Systems of Equations

Solutions   **Tutorial**   **Video**
Manual   **11**

Objectives

1 Solve nonlinear systems by substitution.
2 Solve nonlinear systems by addition.
3 Solve problems that can be modeled by nonlinear systems.

1 Solve nonlinear systems by substitution.

An equation in which one or more terms have a variable of degree 2 or higher, such as the equation $x^2 = 2y + 10$, is called a *nonlinear equation*. All conic sections are graphs of nonlinear equations. A *nonlinear system* of equations contains at least one nonlinear equation.

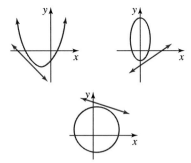

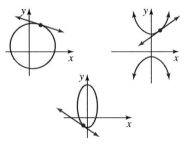

No Real Solutions

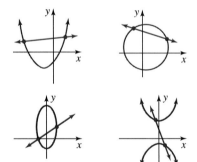

One Real Solution

Two Real Solutions

Figure 9.25

Possibilities for nonlinear systems with one linear equation

Systems Containing One Nonlinear Equation

A system consisting of one linear equation and one nonlinear equation can have no real solutions, one real solution, or two real solutions. Real solutions correspond to intersection points of the graphs of the equations in a system. Figure 9.25 illustrates possibilities for intersection points when one graph is a line.

When solving a system in which one equation is linear, it is usually easiest to use the substitution method. Before considering some examples, let's summarize the steps that we used to solve linear systems by the substitution method, for the same procedure will be applied to nonlinear systems.

The substitution method

Solving Systems of Two Equations in Two Variables (x and y)

1. Solve one equation for x in terms of y or for y in terms of x. If possible, solve for the variable whose coefficient is 1 or -1.
2. Substitute this expression for that variable into the other equation.
3. Solve the resulting equation in one variable.
4. Back-substitute the solution you get into the equation in step 1 to find the value of the other variable.
5. Check the solution in both of the given equations.

When solving a system in which one of the equations is quadratic, the substitution method often leads to a quadratic equation in one variable. Example 1 illustrates this.

EXAMPLE 1 **Solving a Nonlinear System by the Substitution Method**

Solve by the substitution method:

$$x^2 = 2y + 10 \qquad \text{The graph is a parabola.}$$
$$3x - y = 9 \qquad \text{The graph is a line.}$$

Solution

Step 1. We will solve for y in the linear equation (the second equation) to obtain

$$3x - 9 = y \qquad \text{Add } y - 9 \text{ to both sides of the second equation.}$$

Step 2. Now we substitute $3x - 9$ for y in the first equation.

$$x^2 = 2y + 10 \qquad \text{This is the first equation.}$$
$$x^2 = 2(3x - 9) + 10 \qquad \text{Replace } y \text{ by } 3x - 9.$$

Step 3. We solve this equation in one variable.

$$x^2 = 2(3x - 9) + 10 \qquad \text{Solve for } x.$$
$$x^2 = 6x - 18 + 10 \qquad \text{Use the distributive property.}$$
$$x^2 = 6x - 8 \qquad \text{Combine numerical terms on the right.}$$
$$x^2 - 6x + 8 = 0 \qquad \text{Set the quadratic equation equal to 0.}$$

$$(x - 4)(x - 2) = 0 \qquad \text{Factor.}$$
$$x = 4 \quad \text{or} \quad x = 2 \qquad \text{Set each factor equal to 0 and solve.}$$

Step 4. Now we back-substitute these values in the equation $y = 3x - 9$.

If x is 4, $y = 3(4) - 9 = 3$, so $(4, 3)$ is a solution.

If x is 2, $y = 3(2) - 9 = -3$, so $(2, -3)$ is a solution.

Step 5. Now let's check the solutions in both of the given equations. Substituting 4 for x and 3 for y, $(4, 3)$, into the first and second equations yields

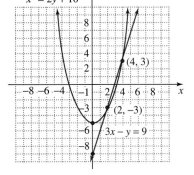

$x^2 = 2y + 10$

$x^2 = 2y + 10$	$3x - y = 9$	These are the given equations.
$4^2 \stackrel{?}{=} 2(3) + 10$	$3(4) - 3 \stackrel{?}{=} 9$	Let $x = 4$ and $y = 3$.
$16 \stackrel{?}{=} 6 + 10$	$12 - 3 \stackrel{?}{=} 9$	Simplify.
$16 = 16$ ✓	$9 = 9$ ✓	True statements result.

Thus, $(4, 3)$ is a solution to the system.

Now let's check $(2, -3)$. Substituting 2 for x and -3 for y, $(2, -3)$, into the first and second equations yields

$x^2 = 2y + 10$	$3x - y = 9$	These are the given equations.
$2^2 \stackrel{?}{=} 2(-3) + 10$	$3(2) - (-3) \stackrel{?}{=} 9$	Let $x = 2$ and $y = -3$.
$4 \stackrel{?}{=} -6 + 10$	$6 + 3 \stackrel{?}{=} 9$	Simplify.
$4 = 4$ ✓	$9 = 9$ ✓	True statements result.

Figure 9.26

Points of intersection illustrate the nonlinear system's solutions.

Thus, $(2, -3)$ is also a solution to the system. Figure 9.26 shows the graphs of the solutions. The solution set of the given system is $\{(4, 3), (2, -3)\}$.   ■

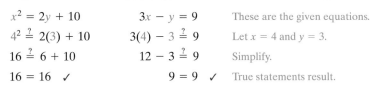

EXAMPLE 2   Solving a Nonlinear System by the Substitution Method

Solve by the substitution method:

$$x - y = 3 \qquad \text{The graph is a line.}$$
$$(x - 2)^2 + (y + 3)^2 = 4 \qquad \text{The graph is a circle.}$$

Solution

Graphically, we are finding the intersection of a line and a circle whose center is at $(2, -3)$ and whose radius measures 2.

Step 1. We will solve for x in the linear (first) equation. (We could also solve for y.)

$$x = y + 3 \qquad \text{Add } y \text{ to both sides of the first equation.}$$

Step 2. Now we substitute $y + 3$ for x in the second equation.

$$(x - 2)^2 + (y + 3)^2 = 4 \qquad \text{This is the second equation.}$$
$$(y + 3 - 2)^2 + (y + 3)^2 = 4 \qquad \text{Replace } x \text{ by } y + 3.$$

Step 3. We solve this equation in one variable for y.

$$(y + 1)^2 + (y + 3)^2 = 4 \qquad \text{Combine numerical terms in the first parentheses.}$$
$$y^2 + 2y + 1 + y^2 + 6y + 9 = 4 \qquad \text{Square } y + 1 \text{ and } y + 3.$$

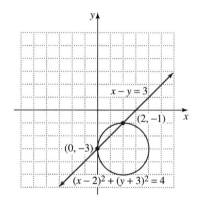

Figure 9.27

Points of intersection illustrate the nonlinear system's solutions.

$$2y^2 + 8y + 10 = 4 \qquad \text{Combine like terms on the left.}$$
$$2y^2 + 8y + 6 = 0 \qquad \text{Set the equation equal to 0.}$$
$$y^2 + 4y + 3 = 0 \qquad \text{Simplify by dividing both sides by 2.}$$
$$(y + 3)(y + 1) = 0 \qquad \text{Factor.}$$
$$y + 3 = 0 \quad \text{or} \quad y + 1 = 0 \qquad \text{Set each factor equal to 0.}$$
$$y = -3 \quad \text{or} \quad y = -1 \qquad \text{Solve for } y.$$

Step 4. Now we back-substitute these values in the equation $x = y + 3$.

If $y = -3$: $x = -3 + 3 = 0$, so $(0, -3)$ is a solution.

If $y = -1$: $x = -1 + 3 = 2$, so $(2, -1)$ is a solution.

Figure 9.27 shows the graphs of the solutions. The solution set of the given system is $\{(0, -3), (2, -1)\}$.

Step 5. Take a moment to check these solutions in both of the given equations. ∎

Discover for yourself

Use substitution to verify that the ordered pairs $(0, -3)$ and $(2, -1)$ satisfy both equations in Example 2. Then use a graphing utility to provide a check of the solutions, reproducing the graphs in Figure 9.27.

The substitution method can lead to ordered pairs that are imaginary numbers. Let's consider an example and then see how to interpret this situation graphically.

Discover for yourself

Graph the equations in this system either by hand or using a graphing utility. What do you observe? What do you think this means in terms of the number of real solutions for the system?

EXAMPLE 3   **A Nonlinear System with Ordered-Pair Solutions That Are Imaginary Numbers**

Solve by the substitution method:

$$x^2 - y = 0 \qquad \text{The graph is a parabola.}$$
$$2x - y = 2 \qquad \text{The graph is a line.}$$

Solution

Step 1. Using the linear (second) equation, we see that

$$2x - 2 = y \qquad \text{Add } y - 2 \text{ to both sides of the second equation.}$$

Step 2. Substituting $2x - 2$ for y in the first equation results in

$$x^2 - (2x - 2) = 0 \qquad \text{Replace } y \text{ with } 2x - 2 \text{ in the first equation.}$$

Step 3. We solve this quadratic equation for x.

$$x^2 - 2x + 2 = 0$$
$$a = 1 \quad b = -2 \quad c = 2$$

ENRICHMENT ESSAY

Nonlinear Systems in Art

The geometry of intersecting conics and lines appears in the paintings of Wassily Kandinsky (1866–1944). The colors are independent of the forms, creating a transparent space, which is characteristic of Kandinsky's works.

Wassily Kandinsky "Auf Spitzen" (On Points) 1928, oil on canvas, 140 × 140 cm. Paris, Musee National d'Art Moderne, Centre Georges Pompidou.

$$x = \frac{-b \pm \sqrt{b^2 - 4ac}}{2a} = \frac{2 \pm \sqrt{4 - 4(1)(2)}}{2(1)} \qquad \begin{array}{l}x^2 - 2x + 2 \text{ is} \\ \text{prime.}\end{array}$$

$$= \frac{2 \pm \sqrt{-4}}{2} = \frac{2 \pm 2i}{2} = 1 \pm i \qquad \begin{array}{l}\sqrt{-4} = \sqrt{4(-1)} \\ \quad = 2i\end{array}$$

Step 4. Now we back-substitute these values of x into $y = 2x - 2$.

If $x = 1 + i$, $y = 2(1 + i) - 2 = 2 + 2i - 2 = 2i$. $(1 + i, 2i)$ is a solution.

If $x = 1 - i$, $y = 2(1 - i) - 2 = 2 - 2i - 2 = -2i$. $(1 - i, -2i)$ is a solution.

Step 5. Let's now check to show that $(1 + i, 2i)$ satisfies both equations.

$$
\begin{array}{ll}
x^2 - y = 0 & 2x - y = 2 \\
(1 + i)^2 - 2i \stackrel{?}{=} 0 & 2(1 + i) - 2i \stackrel{?}{=} 2 \\
1 + 2i + i^2 - 2i \stackrel{?}{=} 0 & 2 + 2i - 2i \stackrel{?}{=} 2 \\
1 + (-1) \stackrel{?}{=} 0 \quad \text{\small Recall that } i^2 = -1. & 2 = 2 \\
0 = 0 \quad \checkmark &
\end{array}
$$

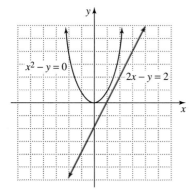

Figure 9.28

A system with no real solutions

Similarly, we can check $(1 - i, -2i)$, resulting in the solution set

$$\{(1 + i, 2i), (1 - i, -2i)\}.$$

Figure 9.28 shows the graphs of the equations in this system. We can see that the parabola and the line do not intersect in the real plane. *When an imaginary number appears as either the x-coordinate or y-coordinate of a solution, this solution is not a point of intersection for the graphs of the equations.* Thus, there is no real solution to this system of equations. ■

Systems Containing Two Nonlinear Equations

We now consider systems consisting of two equations whose graphs are both conic sections. The following figure shows that such systems can have no real solutions, one real solution, two real solutions, three real solutions, or four real solutions.

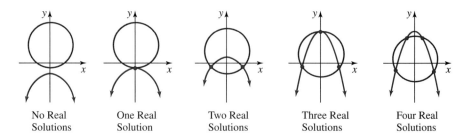

| No Real Solutions | One Real Solution | Two Real Solutions | Three Real Solutions | Four Real Solutions |

2 Solve nonlinear systems by addition.

To solve systems of two nonlinear equations, we can use either the substitution method or the addition method. The addition method works particularly well when each equation is in the form $Ax^2 + By^2 = C$. If necessary, we will multiply either equation or both equations by appropriate numbers so that the coefficients of x^2 or y^2 will have a sum of 0. We then add the equations. The sum will be an equation in one variable.

EXAMPLE 4   **Solving a Nonlinear System by the Addition Method**

Solve the system

$$4x^2 + y^2 = 13 \quad \text{Equation 1 (The graph is an ellipse.)}$$
$$x^2 + y^2 = 10 \quad \text{Equation 2 (The graph is a circle.)}$$

Solution

We can eliminate y^2 by multiplying Equation 2 by -1 and adding equations.

$$
\begin{array}{lll}
4x^2 + y^2 = 13 & \xrightarrow{\text{No change}} & 4x^2 + y^2 = 13 \\
x^2 + y^2 = 10 & \xrightarrow{\text{Multiply by } -1.} & \underline{-x^2 - y^2 = -10} \\
& \text{Add.} & 3x^2 = 3 \\
& & x^2 = 1 \\
& & x = \pm 1
\end{array}
$$

Now we back-substitute each value of x into either one of the original equations. Let's use $x^2 + y^2 = 10$, Equation 2. If $x = 1$,

$$1^2 + y^2 = 10 \quad \text{Replace } x \text{ with 1 in Equation 2.}$$
$$y^2 = 9 \quad \text{Subtract 1 from both sides.}$$
$$y = \pm 3 \quad \text{Apply the square root method.}$$

$(1, 3)$ and $(1, -3)$ are solutions. If $x = -1$,

$$(-1)^2 + y^2 = 10 \quad \text{Replace } x \text{ with } -1 \text{ in Equation 2.}$$
$$y^2 = 9 \quad \text{The steps are the same as above.}$$
$$y = \pm 3$$

$(-1, 3)$ and $(-1, -3)$ are solutions.

Each ordered pair can be checked and shown to satisfy both equations, indicating the solution set $\{(1, 3), (1, -3), (-1, 3), (-1, -3)\}$.

The geometric interpretation of this situation appears in Figure 9.29.

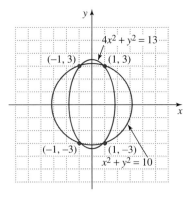

Figure 9.29

A system with four solutions

EXAMPLE 5 | **Solving a Nonlinear System by the Addition Method**

Solve the system:

$$y = x^2 + 3 \quad \text{Equation 1 (The graph is a parabola.)}$$
$$x^2 + y^2 = 9 \quad \text{Equation 2 (The graph is a circle.)}$$

Solution

We could use substitution because Equation 1 has y expressed in terms of x, but this will result in a fourth-degree equation. However, we can rewrite Equation 1 by subtracting x^2 from both sides and add the equations to eliminate the x^2-terms.

$$
\begin{array}{ll}
-x^2 + y = 3 & \text{Subtract } x^2 \text{ from both sides of Equation 1.} \\
\underline{x^2 + y^2 = 9} & \text{This is Equation 2.} \\
y + y^2 = 12 & \text{Add the equations.}
\end{array}
$$

We now solve this quadratic equation.

$$
\begin{array}{ll}
y + y^2 = 12 & \\
y^2 + y - 12 = 0 & \text{Write the quadratic equation in standard form.} \\
(y + 4)(y - 3) = 0 & \text{Factor.} \\
y = -4 \quad \text{or} \quad y = 3 & \text{Set each factor equal to 0 and solve for } y.
\end{array}
$$

We back-substitute these values into Equation 1, the equation containing the lower degree of y.

$$y = x^2 + 3$$

If $y = -4$,

$$
\begin{array}{ll}
-4 = x^2 + 3 & \text{Back-substitute } -4 \text{ for } y \text{ in Equation 1.} \\
-7 = x^2 & \text{Subtract 3 from both sides.} \\
x = \pm\sqrt{-7} & \text{If } x^2 = a, \text{ then } x = \pm\sqrt{a}. \\
x = \pm i\sqrt{7} & \sqrt{-7} = \sqrt{7(-1)} = i\sqrt{7}
\end{array}
$$

Two solutions are $(i\sqrt{7}, -4)$, and $(-i\sqrt{7}, -4)$. If $y = 3$,

$$
\begin{array}{ll}
3 = x^2 + 3 & \text{Back-substitute 3 for } y \text{ in Equation 1.} \\
0 = x^2 & \text{Subtract 3 from both sides.} \\
0 = x & \text{Solve for } x.
\end{array}
$$

The third solution is $(0, 3)$. Our solution set is $\{(i\sqrt{7}, -4), (-i\sqrt{7}, -4), (0, 3)\}$. Figure 9.30 shows that the real ordered pair is the intersection for the circle and the parabola in our system. ■

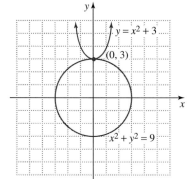

Figure 9.30

A system with one real solution

In Example 6, one equation is in the form $Ax^2 + By^2 = C$ and the other equation is in the form $xy = c$. We will solve for one of the variables in the equation with the product and then use substitution.

EXAMPLE 6   **Solving a Nonlinear System by the Substitution Method**

Solve the system:

$$x^2 + y^2 = 5 \quad \text{Equation 1 (The graph is a circle.)}$$

$$xy = 2 \quad \text{Equation 2 (The graph is a hyperbola.)}$$

Solution

We will solve for y in Equation 2 (even though we could have solved for x).

$$y = \frac{2}{x} \quad \text{Divide both sides of Equation 2 by } x.$$

We substitute $\frac{2}{x}$ for y in Equation 1 and solve for x.

$$x^2 + y^2 = 5 \quad \text{This is Equation 1.}$$

$$x^2 + \left(\frac{2}{x}\right)^2 = 5 \quad \text{Substitute } \frac{2}{x} \text{ for } y.$$

$$x^2 + \frac{4}{x^2} = 5 \quad \text{Square the numerator and denominator of } \frac{2}{x}.$$

$$x^2\left(x^2 + \frac{4}{x^2}\right) = 5x^2 \quad \text{Multiply both sides by } x^2 \text{ to clear fractions.}$$

$$x^4 + 4 = 5x^2 \quad \text{Apply the distributive property and simplify.}$$

$$x^4 - 5x^2 + 4 = 0 \quad \text{Set the equation equal to 0.}$$

This equation is quadratic in form. We introduce the substitution $t = x^2$.

$$t^2 - 5t + 4 = 0 \quad \text{Substitute } x^2 \text{ for } t \text{ in } x^4 - 5x^2 + 4 = 0.$$

$$(t - 4)(t - 1) = 0 \quad \text{Factor}$$

$$t = 4 \quad \text{or} \quad t = 1 \quad \text{Set each factor equal to 0 and solve for } t.$$

Now we substitute x^2 for t and solve these equations.

$$x^2 = 4 \quad \text{or} \quad x^2 = 1$$

$$x = \pm 2 \quad \text{or} \quad x = \pm 1$$

Finally, we back-substitute these values of x into $y = \frac{2}{x}$.

If $x = 2$: $\quad y = \frac{2}{2} = 1$, so $(2, 1)$ is a solution.

If $x = -2$: $\quad y = \frac{2}{-2} = -1$, so $(-2, -1)$ is a solution.

If $x = 1$: $\quad y = \frac{2}{1} = 2$, so $(1, 2)$ is a solution.

If $x = -1$: $\quad y = \frac{2}{-1} = -2$, so $(-1, -2)$ is a solution.

The solution set is $\{(2, 1), (-2, -1), (1, 2), (-1, -2)\}$. Check that these ordered pairs satisfy the equations in the system.

Using technology

Using your graphing utility, enter

$$y_1 = \frac{2}{x}$$

$$y_2 = \sqrt{5 - x^2}$$

$$y_3 = -y_2$$

The graph of the system is shown below. Intersection points occur at the system's four real solutions.

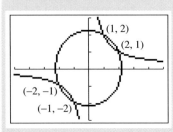

3 Solve problems that can be modeled by nonlinear systems.

Problem Solving and Modeling Using Nonlinear Systems

Many geometric problems can be modeled and solved by the use of nonlinear systems of equations.

| EXAMPLE 7 | **Solving a Problem Involving Area and Perimeter** |

A gallery director would like to exhibit Francesco Clemente's painting at an art show entitled "The Self and the Imagination." The artist informs the director that the painting has a perimeter of 342 inches and an area of 7254 square inches. Find the dimensions of the wall space that the gallery director must set aside to exhibit Clemente's painting.

Francesco Clemente "Untitled" 1983, oil on canvas. Private Collection. Photograph courtesy Thomas Ammann Fine Art, Zurich.

Solution

Step 1. Use variables to represent unknown quantities.

Let

L = the painting's length

W = the painting's width

Step 2. Write a system of equations modeling the problem's conditions.

The problem's conditions can be translated into two equations.

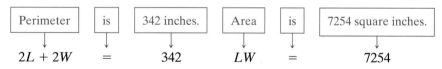

| Perimeter | is | 342 inches. | Area | is | 7254 square inches. |

$$2L + 2W = 342 \qquad LW = 7254$$

Step 3. Solve the system and answer the problem's question.

We must solve the system

$$2L + 2W = 342$$
$$LW = 7254$$

We will solve the second equation for L.

$$L = \frac{7254}{W}$$

Now we substitute $\dfrac{7254}{W}$ for L in the first equation and solve for W.

$$2\left(\frac{7254}{W}\right) + 2W = 342$$

$$14{,}508 + 2W^2 = 342W \qquad \text{Multiply both sides by } W \text{ to clear fractions.}$$

$$2W^2 - 342W + 14{,}508 = 0 \qquad \text{Set the equation equal to 0.}$$

$$W^2 - 171W + 7254 = 0 \qquad \text{Simplify by dividing both sides by 2.}$$

With these relatively large numbers, we will use the quadratic formula, where $a = 1, b = -171$, and $c = 7254$.

$$W = \frac{-b \pm \sqrt{b^2 - 4ac}}{2a} = \frac{-(-171) \pm \sqrt{(-171)^2 - 4 \cdot 1 \cdot 7254}}{2 \cdot 1}$$

$$= \frac{171 \pm \sqrt{225}}{2} = \frac{171 \pm 15}{2}$$

$$W = 93 \quad \text{or} \quad W = 78$$

Now we back-substitute these values into $L = \dfrac{7254}{W}$.

If $W = 93$, then $L = \dfrac{7254}{93} = 78$.

If $W = 78$, then $L = \dfrac{7254}{78} = 93$.

Since length is the longer dimension, we obtain $L = 93$ and $W = 78$. The gallery director must set aside wall dimensions of 93 inches by 78 inches to exhibit the painting.

Step 4. Check the proposed solutions in the original wording of the problem.

If $L = 93$ and $W = 78$, the perimeter is $2(93) + 2(78)$, or 342 inches. The area is $93(78)$, or 7254 square inches. The numbers check with those given in the problem. ◼

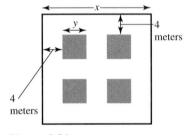

Figure 9.31

The floor plan of a library. Books are shelved in each of the four shaded square areas.

Step 1. Use variables to represent unknown quantities.

EXAMPLE 8 **Using a Geometric Figure to Set Up a Nonlinear System**

An architect is designing the floor plan for one room of a library. Figure 9.31 shows that the room is to have a square floor plan and that books are to be shelved in each of the four smaller shaded square areas. The border around each of the smaller squares is to uniformly measure 4 meters so that people have ample space to browse. The area of the larger square, the room's floor, excluding the four squares for shelving books in its interior, is to be 720 square meters. Find the dimensions of the larger square (the library's floor) and the smaller squares where the books are to be displayed.

Solution

Let

$x =$ the length of each side of the larger square

$y =$ the length of each side of the smaller squares

These variables appear in Figure 9.31.

Step 2. Write a system of equations describing the problem's conditions.

We must refer to the picture to set up a system of equations. Because the border around the smaller squares uniformly measures 4 meters, we consider the portion of the diagram shown in Figure 9.32.

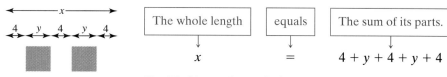

Figure 9.32

Simplify this equation to obtain $x = 2y + 12$.

We obtain our second equation by translating the following verbal model.

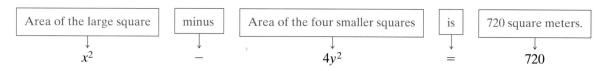

Step 3. Solve the system and answer the problem's question.

The system

$$x = 2y + 12$$
$$x^2 - 4y^2 = 720$$

can be solved by the substitution method. We will substitute $2y + 12$ for x in the second equation.

$$(2y + 12)^2 - 4y^2 = 720$$
$$4y^2 + 48y + 144 - 4y^2 = 720 \qquad \text{Square the binomial using } (A + B)^2 = A^2 + 2AB + B^2.$$
$$48y + 144 = 720 \qquad \text{Simplify.}$$
$$48y = 576 \qquad \text{Subtract 144 from both sides.}$$
$$y = 12 \qquad \text{Divide both sides by 48.}$$

Now we back-substitute 12 for y into $x = 2y + 12$, the first equation.

$$x = 2(12) + 12 = 36$$

We see that $x = 36$ and $y = 12$. This means that the larger square, the library's floor, measures 36 meters by 36 meters, and the smaller squares, the book display areas, each measure 12 meters by 12 meters.

Step 4. Check the proposed solutions in the original wording of the problem.

Take a moment to check the solutions in each of the two sentences that are boxed and that we used to obtain our equations. ■

PROBLEM SET 9.3

Practice Problems

Solve the systems in Problems 1–18 by the substitution method. Confirm solutions by direct substitution or by using graphs.

1. $x^2 + y^2 = 25$
$x - 3y = -5$

2. $x^2 + y^2 = 4$
$x - 2y = 4$

3. $\qquad y = x^2 - 1$
$4x + y + 5 = 0$

4. $\qquad y = x^2$
$2x - y + 3 = 0$

5. $\dfrac{x^2}{4} + \dfrac{y^2}{9} = 1$
  $3x + 2y - 6 = 0$

6. $\dfrac{x^2}{9} + \dfrac{y^2}{4} = 1$
  $2x + 3y - 6 = 0$

7. $xy = 6$
  $2x - y = 1$

8. $xy = -12$
  $x - 2y = -14$

9. $x^2 + y^2 = 52$
  $3x - 2y = 0$

10. $x^2 + 2y = 19$
  $2x - y = 1$

11. $x - y^2 = 1$
  $x = 2y^2$

12. $x - y = 2$
  $x^2 - 3y^2 = 8$

13. $y - 4x = 0$
  $y = x^2 + 5$

14. $y^2 = x^2 - 9$
  $2y = x - 3$

15. $y = 4 - x^2$
  $2x + y = 1$

16. $y = (x + 3)^2$
  $x + 2y = -2$

17. $x^2 + y^2 = 25$
  $x = y$

18. $x^2 - 2y^2 + 3x - 4y = -20$
  $2x - y = 1$

Solve the systems in Problems 19–30 by the addition method. Confirm solutions by direct substitution or by using graphs.

19. $x^2 + y^2 = 13$
  $x^2 - y^2 = 5$

20. $4x^2 - y^2 = 4$
  $4x^2 + y^2 = 4$

21. $x^2 + y^2 = 13$
  $2x^2 + 3y^2 = 30$

22. $x^2 - y^2 = 13$
  $2x^2 + y^2 = -1$

23. $3x^2 - 7y^2 = -15$
  $7x^2 + 9y^2 = 22$

24. $x^2 + y^2 = 1$
  $\dfrac{x^2}{4} + \dfrac{y^2}{16} = 1$

25. $x^2 + y^2 = 4$
  $\dfrac{x^2}{9} + \dfrac{y^2}{1} = 1$

26. $x^2 + 4y = 25$
  $x^2 + y^2 = 25$

27. $x^2 - y^2 = 12$
  $x^2 - 7y = 2$

28. $y = x^2 + 4$
  $x^2 + y^2 = 16$

29. $y = x^2 + 5$
  $x^2 + y^2 = 25$

30. $y = x^2 - 4$
  $x^2 + y^2 = 10$

Solve the systems in Problems 31–40 by the method of your choice.

31. $3x + y = 2$
  $2x^2 - y^2 = 1$

32. $3x^2 + 4y^2 = 16$
  $2x^2 - 3y^2 = 5$

33. $2x^2 = 10 + 3y^2$
  $x^2 + 4y^2 = -17$

34. $x + y^2 = 4$
  $x^2 + y^2 = 16$

35. $x^2 + 4y^2 = 20$
  $xy = 4$

36. $x^2 - 2y^2 = 2$
  $xy = 2$

37. $xy - 2y^2 = -6$
  $xy - y^2 = -4$

38. $6x - y = 5$
  $xy = 4$

39. $a^2 + b^2 = 25$
  $ab = 12$

40. $a^2 + b^2 + 3b = 22$
  $2a + b = -1$

Application Problems

41. An orbiting body follows the elliptical path

$$\dfrac{x^2}{4} + \dfrac{y^2}{16} = 1.$$

A comet follows the parabolic path $y = x^2 - 4$. Assume that all units are in 100,000 kilometers. Where will the comet intersect the orbiting body? Illustrate the situation graphically.

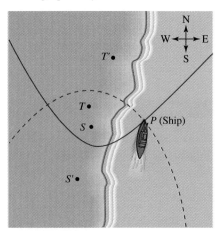

42. The hyperbola is the basis for the navigational system LORAN (for long-range navigation). Signals sent out by ratio transmitters at T and T' reach a radio receiver in a ship located at point P shown on the left. LORAN is based on the time difference between the reception of signals sent simultaneously from stations T and T'. The difference in times of arrival of the signals is used to determine that the ship lies on one branch of a hyperbola. The process is then repeated for radio transmitters at S and S'. The point of intersection of the two hyperbolas over the water is the location of the ship.

A LORAN system indicates that a ship is on the hyperbola given by $16y^2 - x^2 = 16$. The process is repeated and the ship is found to lie on the hyperbola given by $9x^2 - 4y^2 = 36$. If it is known that the ship is located in the first quadrant of the coordinate system, determine its exact location.

43. Find the length and width of a rectangle whose perimeter is 20 feet and whose area is 21 square feet.

44. Find the dimensions of a rectangle whose area is 32 square feet and whose perimeter is 24 feet.

45. The area of a rug is 108 square feet, and the length of a diagonal is 15 feet. Use the formula for the area of a rectangle and the Pythagorean Theorem to find the dimensions of the rug.

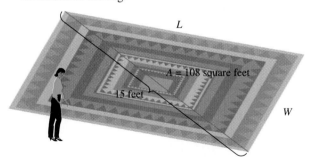

46. The perimeter of a rug is 70 feet and its diagonal length is 25 feet. Find the dimensions of the rug.

47. The figure shows a square floor plan with a smaller square area that will accommodate a combination fountain and pool. The floor with the fountain-pool area removed has an area of 21 square meters and a perimeter of 24 meters. Find the dimensions of the floor and the dimensions of the square that will accommodate the pool.

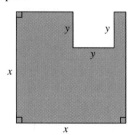

48. Use the Pythagorean Theorem and the fact that the perimeter of the trapezoid in this figure is 84 feet to find x and y.

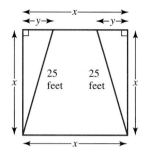

49. If $CD = 8$ inches, find the length of the sides of the right triangle shown in the figure where variables now appear.

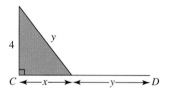

50. Find the lengths of the sides in which variables appear in the figure.

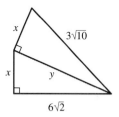

51. The area of a rectangular garden is 480 square feet. If the length of the garden is increased by 10 feet and the width of the garden is decreased by 2 feet, the area will be increased by 20 square feet. What are the dimensions of the garden?

52. A certain amount of money is saved for 1 year. The interest at year's end is $72. If the principal had been $120 more and the interest rate 2% less, the interest would have been the same. Find the principal and the interest rate.

53. The area of the rectangular piece of cardboard shown on the left is 216 square inches. The cardboard is used to make an open box by cutting a 2-inch square from each corner and turning up the sides. If the box is to have a volume of 224 cubic inches, find the length and width of the cardboard that must be used.

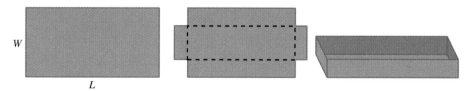

True–False Critical Thinking Problems

54. Which one of the following is true?

a. The system

$$y = x^2 - 1$$
$$x + y = 1$$

contains one linear equation and so cannot be solved by addition.

b. By visualizing the graphs of the following system, it is immediately obvious that the system has no real solutions.

$$x^2 + y^2 = 100$$
$$\frac{x^2}{9} + \frac{y^2}{4} = 1$$

c. When solving a system with a linear equation and a nonlinear equation by substitution, it is easiest to solve the nonlinear equation and substitute the resulting expression back into the same equation.

d. By visualizing the graphs of the following system, it is immediately obvious that the system has no real solutions.

$$y = x - 1$$
$$(x - 3)^2 + (y + 1)^2 = 9$$

55. Which one of the following is true?

a. A system of two equations in two variables whose graphs represent a circle and a line can have four real solutions.

b. If a nonlinear system has nonintersecting graphs, this means that there are no real number solutions.

c. The graphs of a nonlinear system cannot be used to determine the number of real solutions.

d. A nonlinear system whose graphs represent a circle and a hyperbola cannot have three real solutions.

Technology Problems

56. Use a graphing utility and its $\boxed{\text{TRACE}}$ feature to verify the solutions to any five problems that you solved algebraically in Problems 1–30.

57. The orbit of Earth around the sun is described by $\frac{x^2}{1.03} + \frac{y^2}{0.97} = 1$, where distances are measured in

astronomical units (1 AU $\approx$ 93,000,000 miles). The orbit of a comet is described by

$$x^2 - 2xy + y^2 - 2x - 2y + 1 = 0.$$

Graph the two equations and determine how many times the comet will cross Earth's orbit. At what point(s) will this occur?

Writing in Mathematics

58. Explain why the system

$$x^2 + y^2 = 9$$
$$\frac{x^2}{25} + \frac{y^2}{16} = 1$$

has no ordered pairs that are real numbers in its solution set.

59. Write a system of equations, one equation whose graph is a line and the other whose graph is an ellipse, that has no ordered pairs that are real numbers in its solution set. Describe how you determined the two equations.

60. Describe the process of solving the system

$$x^2 - y^2 = 1$$
$$9x^2 + y^2 = 9$$

graphically. As part of your description, actually solve the system.

61. Describe how to use the substitution method to solve the system

$$x^2 + y^2 = 9$$
$$2x - y = 3$$

62. Explain how to use the addition method to solve the system

$$3x^2 - 6y^2 = 9$$
$$x^2 - 2y^2 = 6$$

Critical Thinking Problems

63. The points of intersection of the graphs of $xy = 20$ and $x^2 + y^2 = 41$ are joined to form a rectangle. Find the area of the rectangle.

64. Find all ordered pairs of real numbers (a, b) for which

$$3\sqrt{x - 2y} + \frac{3}{\sqrt{x - 2y}} = 10 \text{ and } x = ay + b$$

65. If $(a + bi)^2 = b + ai$, find the ordered pair of positive numbers (a, b).

66. Solve the system

$$\frac{2}{x^2} - \frac{3}{y^2} = -6$$

$$\frac{3}{x^2} + \frac{4}{y^2} = 59$$

by introducing the substitutions $u = \dfrac{1}{x^2}$ and $v = \dfrac{1}{y^2}$.

67. Solve the system

$$x^2 - y^2 = 9$$

$$\sqrt{x + y} + \sqrt{x - y} = 4$$

(*Hint:* Square the second equation.)

68. Find a and b in this figure.

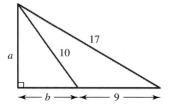

69. The sum of the perimeters of two circles is 12π meters, and the sum of the areas is 20π square meters. What is the radius of each circle?

70. Find the equation of the circle with center on the x-axis and passing through the points $(0, -2)$ and $(6, 0)$. (*Hint:* These ordered points must satisfy $(x - h)^2 + y^2 = r^2$.)

71. Before leaving for summer camp, Woody promised his worried mother that he would not take any hike that lasted more than 8 hours. His first letter home read, "Dear Mom: I went on a 36-mile hike. My time on the hike was 3 hours less than 5 times my average rate of travel." Did Woody keep his promise?

Group Activity Problem

72. To do this problem, you need the following information.
   a. An ellipse can be very long and thin, or it can be almost circular. The eccentricity of an ellipse is a measure of its roundness. The eccentricity is the ratio of the distance from the center to the focus and the length of half the major axis. In terms of the figure shown below,

$$\text{eccentricity} = \frac{c}{a}$$

   b. The relationship between the intercepts (a and b) and the location of the foci, shown in the figure, is given by $b^2 = a^2 - c^2$.

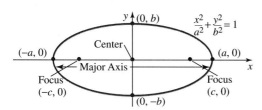

As we know, the planets in our solar system revolve around the sun in elliptical orbits with the sun at one focus. The eccentricities and the lengths of the major axes for these elliptical orbits are given in the following table.

Planet	Eccentricity	Length of Major Axis (in millions of kilometers)
Mercury	0.206	116
Venus	0.007	216
Earth	0.017	299
Mars	0.093	456
Jupiter	0.048	1557
Saturn	0.056	2854
Uranus	0.047	5738
Neptune	0.008	8996
Pluto	0.249	11,800

Find an equation for the orbits of each of the nine planets around the sun. Then consider the equations in pairs and solve the resulting systems to determine whether any two of the planets in our solar system could ever collide.

Review Problems

73. Convert to an exponential equation and solve for x: $\log_x 36 = 2$.

74. Simplify: $\dfrac{\dfrac{1}{x^2 - 9}}{\dfrac{1}{x + 3} + \dfrac{1}{x - 3}}$.

75. Solve, expressing the solution in terms of natural logarithms: $7^x = 125$.

CHAPTER PROJECT

The Geometry of Curves

As we worked with word problems throughout the book we met the challenge of translating from English to algebra. We will now look at another challenge, translating from English to geometry. Notice that each conic section in this chapter has a definition which is not algebraic. These geometric descriptions of the conic graphs are far older than the algebraic equations we derived to represent the graphs. Each of the definitions requires no knowledge of algebra whatsoever, yet gives a precise way to draw a particular conic section. In fact, by using a geometric description to draw a curve, we are able to draw complicated curves that would be described algebraically using equations well beyond the level of algebra found in this text. Although the tools used to draw the curves may lack sophistication, the curves themselves do not.

We have already seen one conic section graphed by using its definition rather than an equation. In the figure below, we see how to graph an ellipse using string, thumbtacks, and a pencil. This method utilized the geometric interpretation of an ellipse. The two thumbtacks are the "fixed points" and the length of string stretched between them is our constant (distance). When we put the point of a pencil under the string and stretch it, as in the figure, we clearly see how the two lengths of string found on either side of our pencil add to the constant length of our string. Thus, we have the sum of the distances from two fixed points (the ends of our string) is a constant, as the definition of an ellipse requires.

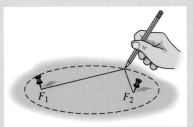

Drawing an ellipse

1. Write geometric instructions, in your own words, of how to draw an ellipse. Do not mention that the resulting figure will be an ellipse and do not use illustrations, such as the one in the figure above, to help with your description. Give your instructions to someone who is not in your class and ask them to follow the directions while you watch. Do not answer any questions as the person draws the figure, but do make a note of the questions on a piece of paper. Try the same procedure again with a second person. As a contrast, you might want to try an adult and a child, or someone who has had algebra and someone who has not. Study the drawings produced and look at any questions asked during the process. Is there a need to change any part of your directions? Would you make your directions longer or shorter? Do you think a picture would have been helpful? Would it have been easier to give an algebraic equation and a piece of graph paper and ask the person to graph the unknown curve?

2. Work with others in a small group to develop a description of how to draw an ellipse with specific measurements of length and width, using only string, thumbtacks, and a pencil. Each member of the group should try a different set of measurements. Experiment with different lengths of string until you can draw an ellipse with the measurements you have selected. Compare your results with others in the group and use the findings to develop a set of instructions on drawing an ellipse. Test the effectiveness of your group's instructions by asking someone from outside the group to draw an ellipse following the directions. After you are satisfied with the method, present the instructions to the class and compare your methods with the procedures of other groups. Compare any of the geometric descriptions to an algebraic equation for an ellipse. Which method, algebraic or geometric, would you prefer and for what reasons?

A different geometric way of creating the graph of a conic section is to use a series of straight lines, tangent to the curve, to outline the form of the conic. This collection of lines is called an *envelope of the curve*. We will create our envelope of lines by folding paper.

3. To create an ellipse, begin by drawing a circle on a sheet of paper you can easily see through, such as waxed paper or tracing paper. At any point on the inside of the circle, place a dot. Fold the paper so that a point on the circumference of the circle touches the dot inside the circle. Make a sharp crease in the paper on the fold. Continue this process round and round the edge of the circle, each time bringing the edge of the circle up to touch the point and making a sharp crease. The number of folds required before you see the outline of the ellipse will depend on the size of the original circle. When you can clearly see the ellipse, study the positions of the dot you placed in the circle and the center of the circle. How are these two points related to the ellipse?

4. The graph of a hyperbola may be created in a similar fashion to the graph of an ellipse, but this time place your dot anywhere outside the circle before you begin folding. You will still fold so that any point on the edge of the circle touches the dot. How is the dot outside the circle and the center of the circle related to the graph of the hyperbola?

5. For the graph of a parabola, we will draw a straight line on the paper and place a point somewhere off the line. We make the folds so that the line touches the point. When you see the parabola appear from the folded lines, study the line and the dot you originally marked on the paper. How would you describe the line and the dot in relation to the parabola?

6. We may also obtain the graph of a curve called a cardioid by using an envelope of straight lines. This time, we will not fold paper to obtain the lines, we will use a straightedge and a pencil to draw them. Begin by tracing a large circle on the paper. Use a protractor to mark off intervals of ten degrees around the edge of the circle, giving you 36 divisions. Choose any of the marks on the edge of the circle and label it point A. Number the marks around the outside edge of the circle from 0 to 36, beginning with point A as 0, and ending with point A also as 36. Label point 18, across the diameter of the circle from A, as point B on the inside edge of the circle. You may wish to use a different color for the labels on the inside. Number the intervals from 0 to 36 on the inside edge of the circle, moving in increments of twenty degrees. This means the inside label on point 18 is 0 and B, the label on the inside above point 20 is 1, the label on the inside above point 22 is 2, continuing back twice around the inside of the circle. Point B will be labeled 18 on the outside edge and 0, 18, and 36 on the inside edge, point 20 on the outside edge will be labeled 1 and 19 on the inside edge, and so on. Use a straightedge to connect the matching numbers on the inside and outside of the circle with a line. Can you see why this curve is called a cardioid?

7. In some cases, curves may be found as envelopes of circles, rather than lines. To draw the cardioid as an envelope of circles, you will need a compass and a pencil. Draw a circle and mark a point A on its circumference. Place the point of the compass anywhere on the edge of the circle and open it so that the pencil rests on point A. Draw a circle with this diameter by swinging the compass in a full circle. Continue around your original circle, choosing different points on the circumference and repeating this procedure. Try to maintain a uniform distance as you move on points around the edge to obtain the best graph. If you repeat this procedure with your point A taken outside the circle, or with A inside the circle, you will obtain a slightly different style of curve called a *limacon*.

Worldwide Web Resources

Go to the Prentice Hall website (http://www.prenhall.com/blitzer) to access other locations on the Internet that will allow you to further explore the concepts presented in this project.

Chapter Review

SUMMARY

1. The Circle

a. A circle is the set of all points in a plane that are equidistant from a fixed point called the center. The fixed distance from the circle's center to any point on the circle is called the radius.

b. The equation of the circle with center at (h, k) and radius r is $(x - h)^2 + (y - k)^2 = r^2$.

2. The Parabola

a. A parabola is the set of all points in the plane whose distances from a fixed line and a fixed point are equal.

b. The forms of the equation of a parabola opening upward or downward are

$y = ax^2 + bx + c$ The vertex is $\left(-\dfrac{b}{2a}, f\left(-\dfrac{b}{2a}\right)\right)$.

$y = a(x - h)^2 + k$ The vertex is (h, k).

If $a > 0$, the parabola opens upward; if $a < 0$, the parabola opens downward.

c. The forms of the equation of a parabola opening to the left or right are

$x = ay^2 + by + c$ The y-coordinate of the vertex is $-\dfrac{b}{2a}$.

$x = a(y - k)^2 + h$ The vertex is (h, k).

If $a > 0$, the parabola opens to the right; if $a < 0$, the parabola opens to the left.

3. The Ellipse

a. An ellipse is the set of all points in the plane, the sum of whose distances from two distinct fixed points is a constant. Each fixed point is called a focus.

b. The standard form for the equation of an ellipse centered at the origin is

$$\frac{x^2}{a^2} + \frac{y^2}{b^2} = 1.$$

The x-intercepts are $\pm a$, and y-intercepts are $\pm b$.

4. The Hyperbola

a. A hyperbola is the set of all points in the plane, the difference of whose distances from two distinct points is a constant. Each fixed point is called a focus.

b. The standard forms for the equation of a hyperbola centered at the origin are

$$\frac{x^2}{a^2} - \frac{y^2}{b^2} = 1 \qquad \text{The } x\text{-intercepts are } -a \text{ and } a.$$

$$\frac{y^2}{b^2} - \frac{x^2}{a^2} = 1 \qquad \text{The } y\text{-intercepts are } -b \text{ and } b.$$

Asymptotes are found by drawing lines through opposite corners of the rectangle whose sides pass through $-a$, a, $-b$, and b on the axes. The equations of the asymptotes are $y = \dfrac{b}{a}x$ and $y = -\dfrac{b}{a}x$.

c. The nonstandard form for the equation of a hyperbola is $xy = c, c \neq 0$. The x- and y-axes are asymptotes.

5. Recognizing Equations of Conic Sections Centered at the Origin

a. If only one of the variables is squared, the equation represents a parabola.

b. If both variables are squared, get the x^2- and y^2-terms on the same side of the equation.

   1. If these terms have the same coefficient, the equation represents a circle.

2. If these terms have different positive coefficients, the equation represents an ellipse.
3. If these terms have coefficients with opposite signs, the equation represents a hyperbola.

6. Solving Nonlinear Systems of Equations
Nonlinear systems can be solved algebraically by the substitution and addition methods.

REVIEW PROBLEMS

For Problems 1–2, write the equation of each circle.

1. Center at $(-2, 4)$, radius of 6

2. Center at the origin, radius of 3

In Problems 3–5, give the center and radius of each circle and graph its equation.

3. $x^2 + y^2 = 16$

4. $(x + 2)^2 + (y - 3)^2 = 9$

5. $x^2 + y^2 - 4x + 2y - 4 = 0$

In Problems 6–9, use intercepts, the vertex, and, if necessary, one or two additional points to graph each parabola.

6. $y = x^2 - 2x + 3$

7. $y = -(x + 1)^2 + 4$

8. $x = y^2 - 8y + 12$

9. $x = (y - 3)^2 - 4$

In Problems 10–11, graph the ellipse with the given equation.

10. $\dfrac{x^2}{9} + \dfrac{y^2}{25} = 1$

11. $4x^2 + 25y^2 = 100$

Use asymptotes and intercepts to graph the hyperbola represented by each equation in Problems 12–14.

12. $\dfrac{x^2}{16} - \dfrac{y^2}{9} = 1$

13. $\dfrac{y^2}{9} - \dfrac{x^2}{4} = 1$

14. $x^2 - y^2 = 4$

The equations in Problems 15–16 represent hyperbolas in nonstandard form. Graph each hyperbola.

15. $xy = 8$

16. $xy = -4$

17. A planet with an elliptical orbit is shown in a rectangular coordinate system. Write an equation that models the planet's orbit.

18. A semielliptical archway has a height of 15 feet at the center and a width of 50 feet, as shown in the figure.

The 50-foot width consists of a two-lane road. Can a truck 12 feet high drive under the archway without going over the road's centerline, if the truck is 14 feet wide?

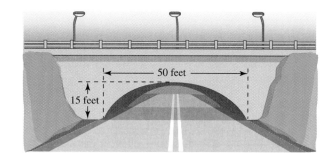

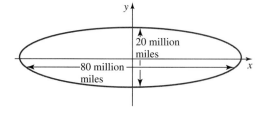

In Problems 19–24, indicate whether the graph of each equation is a circle, a parabola, an ellipse, or a hyperbola. Then graph each equation.

19. $5x^2 + 5y^2 = 180$

20. $4x^2 + 9y^2 = 36$

21. $x + 3 = -y^2 + 2y$

22. $4x^2 - 9y^2 = 36$

23. $x^2 + y^2 + 6x - 2y = -6$

24. $xy = 3$

Solve the systems in Problems 25–38 by either the substitution or the addition method. Confirm your solutions by direct substitution or by using graphs.

25. $5y = x^2 - 1$
$x - y = 1$

26. $y = x^2 + 2x + 1$
$x + y - 1 = 0$

27. $x^2 + y^2 = 2$
$x + y = 0$

28. $x^2 + y^2 = 4$
$2x - y = 0$

29. $x^2 + 2y^2 = 4$
$x - y - 1 = 0$

30. $2x^2 + y^2 = 24$
$x^2 + y^2 = 15$

31. $xy - 4 = 0$
$y - x = 0$

32. $x^2 + y^2 = 2$
$y = x^2$

33. $y^2 = 4x$
$x - 2y + 3 = 0$

34. $\dfrac{x^2}{4} + \dfrac{y^2}{9} = 1$
$2x - y = 0$

35. $4x^2 + y^2 = 1$
$x^2 + 4y^2 = 1$

36. $x^2 + y^2 = 9$
$(x - 2)^2 + y^2 = 21$

37. $x^2 + 2y^2 = 12$
$xy = 4$

38. $y = x^2 - 2$
$x^2 + y^2 = 4$

39. The perimeter of a rectangle is 26 meters, and its area is 40 square meters. Find its dimensions.

40. Two adjoining square fields with an area of 2900 square feet are to be enclosed with 240 feet of fencing. The situation is represented in the figure. Find the length of each side where a variable appears.

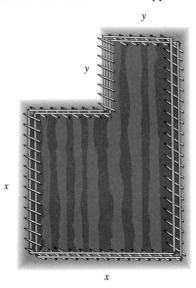

41. The owner of a sailboat wants to make a sail in the shape of a right triangle with an area of 54 square yards and a hypotenuse of 15 yards. What are the lengths of the legs?

42. A LORAN (long range navigation) system indicates that a ship is on the hyperbola given by $4y^2 - x^2 = 1$. The process of measuring the difference in time that it takes for radio signals to reach the ship is repeated, revealing that the ship lies on the hyperbola given by $x^2 - 3y^2 = 1$. If it is known that the ship is located in the first quadrant of the coordinate system, determine its exact location.

43. On a 2000-kilometer trip, the average speed (in kilometers per hour) was numerically 10 less than the time spent traveling (in hours). Find the speed and the time for the trip.

44. Find the coordinates of all points (x, y) that lie on the line whose equation is $2x + y = 8$, so that the area of the rectangle shown in the figure is 6 square units.

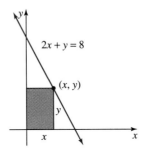

CHAPTER 9 TEST

In Problems 1–2, find the center and radius of each circle.

1. $(x - 5)^2 + (y + 3)^2 = 49$

2. $x^2 + y^2 + 4x - 6y - 3 = 0$

In Problems 3–10, classify each equation as a circle, a parabola, an ellipse, or a hyperbola. Then graph each equation.

3. $y = x^2 - 2x - 3$

4. $\dfrac{x^2}{4} - \dfrac{y^2}{9} = 1$

5. $4x^2 + 9y^2 = 36$

6. $xy = 4$

7. $x = (y - 1)^2 + 4$

8. $16x^2 + y^2 = 16$

9. $\dfrac{x^2}{4} + \dfrac{y^2}{4} = 1$

10. $25y^2 - 9x^2 = 225$

11. Explain why the graphs in the figure cannot be the graphs of a system whose solution is $\{(0, -1), (2, 1)\}$.

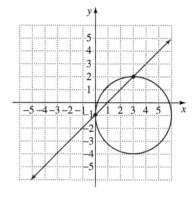

In Problems 12–13, solve each system.

12. $x^2 + y^2 = 25$
   $x + y = 1$

13. $2x^2 - 5y^2 = -2$
   $3x^2 + 2y^2 = 35$

14. A rectangle has a diagonal of 15 feet and a perimeter of 42 feet. Find the rectangle's dimensions.

15. The rectangular plot of land shown in the figure is to be fenced along three sides using 39 feet of fencing. No fencing is to be placed along the river's edge. The area of the plot is 180 square feet. What are its dimensions?

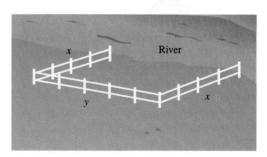

CUMULATIVE REVIEW PROBLEMS (CHAPTERS 1–9)

In Problems 1–10, solve for the indicated variable(s). Express irrational solutions in simplified radical form and imaginary solutions in the form a + bi. Express solutions of inequalities in both set-builder and interval notation.

1. $2(2x + 3) - 10 \geqslant 6(x - 2)$

2. $5x^{2/3} + 2x^{1/3} - 7 = 0$

3. $(5x - 4)^2 + 6 = 8$

4. $x^2 - 6x + 13 = 0$

5. $2x^2 - 5x - 3 > 0$

6. $x^2 + y^2 = 25$
   $x - 2y = -5$

7. $\left| 2x - 1 \right| < 3$

8. $x - 2y + 3z = 7$
    $2x + y + z = 4$
    $-3x + 2y + 3z = 19$

9. $\log (x + 3) + \log x = 1$

10. $\sqrt{2x + 3} - \sqrt{x - 2} = 2$

In Problems 11–15, graph each equation or inequality in the rectangular coordinate system.

11. $2x - 3y \geq 6$

12. $y = (\frac{1}{2})^x + 1$

13. $9x^2 + 4y^2 = 36$

14. $y = \log_3 x$

15. $f(x) = \sqrt{x + 1}$

16. Find the solution set of the following system, using graphing and points of intersection. Check all solutions.

$$x = y + 1$$
$$y = -(x + 1)^2 + 4$$

17. Use the compound interest model $A = P\left(1 + \dfrac{r}{n}\right)^{nt}$ to answer this question. Suppose that you have $10,000 to invest. What investment yields the greatest return over 10 years: 6.5% compounded semiannually or 6.25% compounded monthly?

18. If $f(x) = x^2 + 1$ and $g(x) = 2x - 3$, find $(f \circ g)(x)$ and $(g \circ f)(x)$.

19. If $f(x) = 3x + 5$, find $f^{-1}(x)$.

20. The model $P = 14.7e^{-0.21x}$ describes atmospheric pressure (P, in pounds per square inch) x miles above sea level.
   a. What is the atmospheric pressure at sea level?
   b. How many miles above sea level, to the nearest tenth of a mile, is atmospheric pressure half the pressure at sea level?

In Problems 21–23, perform the indicated operations, and simplify if possible.

21. $\dfrac{3 + 2i}{2 - i}$

22. $(x^3 + 2x^2 - 3x + 4) \div (x - 2)$

23. $2\sqrt{12} - \sqrt{\dfrac{1}{3}}$

24. A diver jumps from a diving board that is 32 feet above the water. The position of the diver above the water after t seconds is given by $s(t) = -16t^2 + 16t + 32$.
   a. When will the diver reach a maximum height above the water? What is this height?
   b. How many seconds will it take for the diver to reach the water?
   c. Construct an appropriate table of coordinates and graph the position function.

25. A rectangular pool is 9 meters wide and 12 meters long. A tile border of uniform width is to be built around the pool with 162 square meters of tile. Determine the width of the border.

26. The temperature in degrees Celsius at a depth of x meters beneath the surface of the earth is modeled by

$$T = S + 2.5\left(\dfrac{x - 3000}{100}\right)$$

where S is the surface temperature and $x \geq 3000$. If the surface temperature is 20° Celsius, at what depth is the temperature 45° Celsius?

27. There are four ways to express 96 as the difference of two squares. Find one way of doing this.

28. Write an expression that models the area of the figure shown. Then write the expression in factored form.

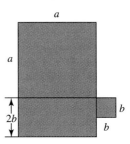

29. Two models, one linear and the other quadratic, for the number of AIDS cases (N, in thousands) reported worldwide t years after 1986 are given by

$$N = 330.8t + 218.8$$
$$\text{and} \quad N = 62t^2 + 144.8t + 280.8$$

Use each model to make projections about the number of AIDS cases in 1990 and in 1993. The actual numbers were 1872 thousand and 4820 thousand. How accurate are the given models for these years? Which formula comes closer to describing reality?

30. Solve for x and y using Cramer's rule and, if possible, simplify:

$$cx + dy = c^2$$
$$dx + cy = d^2.$$

Sequences, Series, and the Binomial Theorem

Rene Magritte "The False Mirror" (Le Faux Miroir) 1928, oil on canvas, $21\frac{1}{4} \times 31\frac{7}{8}$ in. (54 × 80.9 cm). The Museum of Modern Art, New York. Purchase. Photograph © 1997 The Museum of Modern Art, New York. © 1997 C. Herscovici, Brussels/Artists Rights Society (ARS), New York.

Infinity has been the object of speculation for millenia. Because it is so difficult to use human intuition to grasp infinite processes, one could ask, is infinity something that reason can investigate?

In this chapter, we will discover that infinite processes can be examined rationally. We will see how infinite sums can be used to model numerous situations, including tax rebates. And although we proceed as if the study of infinite processes was a smooth, peaceful parade of progress toward greater knowledge, the inquiry into infinity was actually waged with controversial fervor. One of the last victims of the Inquisition, Giordano Bruno, was burned at the stake in 1600 for his inquiry into the nature of infinity.

S E C T I O N   1 0 . 1

Solutions
Manual

Tutorial

Video
11

Sequences and Series

Objectives

1 Find the terms of a sequence given the general term.
2 Find the general term of a sequence.
3 Find partial sums.
4 Expand and evaluate sums in summation notation.
5 Write sums in summation notation.

Sequences

The creations of nature involve intricate mathematical designs, including a variety of spirals. The arrangement of the individual florets in the head of a sunflower form spirals. In some species, there are 21 spirals in the clockwise direction and 34 in the counterclockwise direction. In other species, there are 34 spirals in the clockwise direction and 55 in the counterclockwise direction. The precise numbers depend on the species of sunflower, but you always find the combinations 21 and 34, or 34 and 55, or 55 and 89, or even 89 and 144.

Dick Morton

This observation becomes even more interesting when we consider a sequence of numbers investigated by Leonardo of Pisa, also known as Fibonacci, an Italian mathematician of the 13th century. The Fibonacci sequence of numbers is an infinite sequence that begins as follows:

$$1, 1, 2, 3, 5, 8, 13, 21, 34, 55, 89, 144, 233, \ldots$$

The first two terms are 1. Every term thereafter is the sum of the two preceding terms. The number of spirals in a sunflower (21 and 34) corresponds to two consecutive Fibonacci numbers, as do the spirals in a pine cone (8 and 13) and a pineapple (8 and 13). Furthermore, numbers in the Fibonacci sequence can be found in an octave on the piano keyboard. The octave contains 2 black keys in one cluster, 3 black keys in another cluster, 5 black keys, 8 white keys, and a total of 13 keys altogether (Figure 10.1). Observe that 2, 3, 5, 8, and 13 are the third through seventh terms of the Fibonacci sequence.

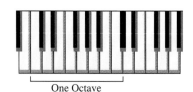

One Octave

Figure 10.1

Mario Merz "Alligator with Fibonacci Numbers to 377" 1979, Stuffed aligator and neon alligator: $27 \times 10 \times 3$ in. ($68.6 \times 25.4 \times 7.6$ cm) dimensions of numbers vary. Private Collection. Courtesy Sperone Westwater, New York.

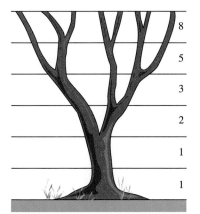

As this tree branches, the number of branches forms the Fibonacci sequence.

▌ Find the terms of a sequence given the general term.

We can think of the Fibonacci sequence as a function. The terms of the sequence

$$1, 1, 2, 3, 5, 8, 13, 21, 34, 55, 89, 144, 233, \ldots$$

are the range values for a function whose domain is the set of positive integers.

Domain: 1, 2, 3, 4, 5, 6, 7, . . .
           ↓ ↓ ↓ ↓ ↓ ↓ ↓
  Range: 1, 1, 2, 3, 5, 8, 13, . . .

Thus, $f(1) = 1$, $f(2) = 1$, $f(3) = 2$, $f(4) = 3$, $f(5) = 5$, $f(6) = 8$, $f(7) = 13$, and so on. This leads us to the following definition.

Definition of an infinite sequence

An *infinite sequence* is a function whose domain is the set of positive integers.

The letter a with a subscript is used to represent function values of a sequence, rather than the usual function notation.

Term of a Sequence	Term of the Fibonacci Sequence $1, 1, 2, 3, 5, 8, 13, \ldots$
a_1 represents the first term of a sequence.	$a_1 = 1$
a_2 represents the second term of a sequence.	$a_2 = 1$
a_3 represents the third term of a sequence.	$a_3 = 2$
$\vdots$	$\vdots$
a_n represents the nth term of a sequence.	$a_n =$ sum of previous two terms

The nth term of a sequence, a_n, is called the *general term* of the sequence. The general term is used to define the other terms of the sequence. If we are provided the formula for the general term, a_n, we can find any other term in the sequence.

EXAMPLE 1 **Writing the Terms of a Sequence from the General Term**

Find the first four terms of the infinite sequence defined by: $a_n = \dfrac{1}{2n + 1}$

Solution

$a_n = \dfrac{1}{2n + 1}$ General term

Replacing n with $1, 2, 3,$ and 4, we generate the first four terms.

$$a_1 = \frac{1}{2(1) + 1} = \frac{1}{3} \quad \text{First term}$$

$$a_2 = \frac{1}{2(2) + 1} = \frac{1}{5} \quad \text{Second term}$$

$$a_3 = \frac{1}{2(3) + 1} = \frac{1}{7} \quad \text{Third term}$$

$$a_4 = \frac{1}{2(4) + 1} = \frac{1}{9} \quad \text{Fourth term}$$

The sequence defined by $a_n = \dfrac{1}{2n + 1}$ can be written as

$$\frac{1}{3}, \frac{1}{5}, \frac{1}{7}, \dots, \frac{1}{2n + 1}, \dots .$$

Because each term in the sequence is smaller than the preceding term, this is an example of a *decreasing sequence*. ■

EXAMPLE 2 **Writing the Terms of a Sequence from the General Term**

Find the first 4 terms and the 47th term of the sequence whose general term is:

$$a_n = (-1)^n(n + 1)$$

Solution

Replacing n with $1, 2, 3,$ and 4, we generate the first four terms.

$$a_1 = (-1)^1(1 + 1) = -2 \quad \text{First term}$$
$$a_2 = (-1)^2(2 + 1) = 3 \quad \text{Second term}$$
$$a_3 = (-1)^3(3 + 1) = -4 \quad \text{Third term}$$
$$a_4 = (-1)^4(4 + 1) = 5 \quad \text{Fourth term}$$

The 47th term is found by replacing n by 47.

$$a_{47} = (-1)^{47}(47 + 1) = -48 \quad \text{47th term}$$

The sequence defined by $a_n = (-1)^n(n + 1)$ can be written as

$$-2, 3, -4, \dots, (-1)^n(n + 1), \dots .$$

The factor $(-1)^n$ causes the signs of the terms to alternate between positive and negative, depending on whether n is even or odd. ■

Many applied situations frequently involve finite sequences.

Definition of a finite sequence

A *finite sequence* is a function whose domain includes only the first n positive integers.

For example.

$$1, 8, 27, 64, 125$$

is a finite sequence whose domain is $\{1, 2, 3, 4, 5\}$. We see that $a_1 = 1$, $a_2 = 8$, $a_3 = 27$, $a_4 = 64$, and $a_5 = 125$.

 2 Find the general term of a sequence.

Finding the General Term of a Sequence

So far we have listed terms in a sequence with the general term given. In contrast, listing the first few terms of a sequence is not enough to say for certain what the general term is. However, we can make a prediction by looking for a pattern. For example, the sequence

$$1, 4, 9, 16, \ldots$$

has terms that are squares of consecutive integers. Thus, a possible formula for the general term is

$$a_n = n^2.$$

Discover for yourself

Write the first five terms for the following sequences whose general term is given.

$$a_n = n^2$$
$$a_n = n^2 + (n - 1)(n - 2)(n - 3)(n - 4)\left(\frac{\pi - 25}{1 \cdot 2 \cdot 3 \cdot 4}\right)$$

What do you observe? Can you see why listing the first few terms of a sequence is not enough to define the general term in only one way? What does this tell you about a traditional intelligence test that asks for the missing number in the sequence 1, 4, 9, 16, _____?

If we are given the first few terms of a sequence, the best we can do is to find one possible general term, where more than one formula for a_n may be correct. Let's do this in Example 3.

EXAMPLE 3   Finding a Possible General Term of a Sequence

Find a possible formula for the general term of the sequence:

a. $1, 8, 27, 64, 125, \ldots$
b. $\frac{3}{1}, \frac{4}{2}, \frac{5}{3}, \frac{6}{4}, \frac{7}{5}, \ldots$
c. $-2, 4, -8, 16, -32, \ldots$

Solution

a. n: $1, \ 2, \ 3, \ 4, \ 5, \ldots$
$$\downarrow \ \downarrow \ \downarrow \ \downarrow \ \downarrow$$
Terms: $1, \ 8, \ 27, 64, 125, \ldots$
$$\downarrow \ \downarrow \ \downarrow \ \downarrow \ \downarrow$$
$$1^3, 2^3, 3^3, 4^3, 5^3, \ldots$$

The terms of the sequence are the cubes of consecutive positive integers. Using this pattern, we can say that a possible formula for the general term is

$$a_n = n^3.$$

b. n: $1, 2, 3, 4, 5, \ldots$
$$\downarrow \downarrow \downarrow \downarrow \downarrow$$
Terms: $\frac{3}{1}, \frac{4}{2}, \frac{5}{3}, \frac{6}{4}, \frac{7}{5}, \ldots$

The terms of the sequence have denominators that are consecutive positive integers. Furthermore, the numerator is 2 more than the denominator. Using this pattern, we can say that a possible formula for the general term is

$$a_n = \frac{n + 2}{n}.$$

c. n: $1, \ 2, \ 3, \ 4, \ \ 5, \ldots$
$$\downarrow \ \downarrow \ \downarrow \ \downarrow \ \ \downarrow$$
Terms: $-2, 4, -8, 16, -32, \ldots$

The terms of the sequence are powers of 2 with alternating signs. A possible formula for the general term is

$$a_n = (-1)^n 2^n.$$ ■

3 Find partial sums.

Series and Summation Notation

Now let's focus on the *sum of the terms* of a sequence. Consider the sequence

$$\frac{1}{2}, \frac{1}{4}, \frac{1}{8}, \frac{1}{16}, \frac{1}{32}, \ldots, \left(\frac{1}{2}\right)^n, \ldots.$$

The figure below shows how we can model this sequence using parts of a square.

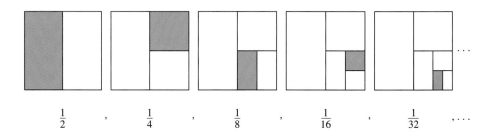

$$\frac{1}{2} \quad , \quad \frac{1}{4} \quad , \quad \frac{1}{8} \quad , \quad \frac{1}{16} \quad , \quad \frac{1}{32} \quad , \ldots$$

The sum of a number of terms in a sequence is called a *finite series*. The sum of the first n terms is denoted by S_n and is called the *nth partial sum*. The figure at the top of page 817 shows the first five partial sums for the sequence

$$\frac{1}{2}, \frac{1}{4}, \frac{1}{8}, \frac{1}{16}, \frac{1}{32}, \ldots, \left(\frac{1}{2}\right)^n, \ldots.$$

with geometric models.

S_1: First Partial Sum

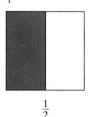

$$\frac{1}{2}$$

S_2: Second Partial Sum

$$\frac{1}{2} + \frac{1}{4} = \frac{3}{4}$$

S_3: Third Partial Sum

$$\frac{1}{2} + \frac{1}{4} + \frac{1}{8} = \frac{7}{8}$$

S_4: Fourth Partial Sum

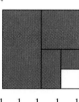

$$\frac{1}{2} + \frac{1}{4} + \frac{1}{8} + \frac{1}{16} = \frac{15}{16}$$

S_5: Fifth Partial Sum

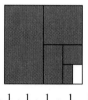

$$\frac{1}{2} + \frac{1}{4} + \frac{1}{8} + \frac{1}{16} + \frac{1}{32} = \frac{31}{32}$$

As we continue to add terms, the sum is getting closer to modeling the area of one whole square. Thus, these partial sums are getting closer and closer to 1. Although the number 1 will never be reached, we can say that the *infinite series*

$$\frac{1}{2} + \frac{1}{4} + \frac{1}{8} + \frac{1}{16} + \frac{1}{32} + \cdots + \left(\frac{1}{2}\right)^n + \cdots$$

has a sum that is equal to 1.

This discussion can be generalized with the following definitions.

Partial sums and infinite series

Consider the infinite sequence

$$a_1, a_2, a_3, \ldots, a_n, \ldots.$$

The finite series S_n is the sum of the first n terms of the sequence and is called the *nth partial sum.*

$$S_n = a_1 + a_2 + a_3 + \cdots + a_n$$

The sum of the terms

$$a_1 + a_2 + a_3 + \cdots + a_n + \cdots$$

is called an *infinite series.*

EXAMPLE 4   **Finding Partial Sums**

Consider the sequence whose general term is given by $a_n = 3n - 1$. Find:

a. S_3, the third partial sum
b. S_5, the fifth partial sum

Solution

We replace n with $1, 2, 3, 4$, and 5 to find the first five terms of the sequence.

$$a_1 = 3 \cdot 1 - 1 = 2, a_2 = 3 \cdot 2 - 1 = 5, a_3 = 3 \cdot 3 - 1 = 8,$$
$$a_4 = 3 \cdot 4 - 1 = 11, a_5 = 3 \cdot 5 - 1 = 14$$

The sequence can be expressed as

$$2, 5, 8, 11, 14, \ldots, 3n - 1, \ldots.$$

a. We find S_3, the third partial sum, by adding the first three terms of the sequence.

$$S_3 = 2 + 5 + 8 = 15$$

b. We find S_5, the fifth partial sum, by adding the first five terms of the sequence.

$$S_5 = 2 + 5 + 8 + 11 + 14 = 40$$ ∎

4 Expand and evaluate sums in summation notation.

Summation Notation

Mathematicians have a compact notation for dealing with series, called *summation notation* or *sigma notation*. The name comes from the use of a stylized version of the capital Greek letter sigma, denoted by $\sum$. An example of this notation is the fourth partial sum

$$\sum_{i=1}^{4} 16(2i - 1).$$

To find this partial sum, we replace the letter i in $16(2i - 1)$ with $1, 2, 3,$ and 4, as follows:

$$\sum_{i=1}^{4} 16(2i - 1) = 16(2 \cdot 1 - 1) + 16(2 \cdot 2 - 1)$$
$$+ 16(2 \cdot 3 - 1) + 16(2 \cdot 4 - 1)$$
$$= 16(1) + 16(3) + 16(5) + 16(7)$$
$$= 16 + 48 + 80 + 112$$
$$= 256$$

We read

$$\sum_{i=1}^{4} 16(2i - 1)$$

as "the sum as i goes from 1 to 4 of $16(2i - 1)$." The letter i is called the *index of summation* and is not related to the use of i to represent $\sqrt{-1}$. Any letter can be used for the index of summation. Furthermore, the index of summation can start at a number other than 1.

Using technology

Graphing utilities can calculate the sum of a sequence. For example, to find the sum of the sequence in Example 5a, enter

SUM SEQ (I ∧ 2 + 1, I, 1, 6, 1)

Then press ENTER, and 97 should be displayed. Use this capability to verify Examples 5b and 5c.

EXAMPLE 5 **Using Summation Notation**

Expand and evaluate the sum:

a. $\displaystyle\sum_{i=1}^{6} (i^2 + 1)$ **b.** $\displaystyle\sum_{i=4}^{7} (-2)^i$ **c.** $\displaystyle\sum_{k=0}^{3} (2^k + 4)$

Solution

a. We must replace i in the expression $i^2 + 1$ with all consecutive integers from 1 to 6 inclusively, and then add.

$$\sum_{i=1}^{6} (i^2 + 1) = (1^2 + 1) + (2^2 + 1)$$
$$+ (3^2 + 1) + (4^2 + 1) + (5^2 + 1) + (6^2 + 1)$$

$$= 2 + 5 + 10 + 17 + 26 + 37$$
$$= 97$$

b. We must replace i in the expression $(-2)^i$ with all consecutive integers from 4 to 7 inclusively, and then add.

$$\sum_{i=4}^{7} (-2)^i = (-2)^4 + (-2)^5 + (-2)^6 + (-2)^7$$
$$= 16 + (-32) + 64 + (-128)$$
$$= -80$$

c. This time the index of summation is k, but the process is the same. We evaluate $2^k + 4$ for all integers from 0 through 3, and then add.

$$\sum_{k=0}^{3} (2^k + 4) = (2^0 + 4) + (2^1 + 4) + (2^2 + 4) + (2^3 + 4)$$
$$= 5 + 6 + 8 + 12$$
$$= 31 \qquad \blacksquare$$

Because mathematicians express their ideas in compact, symbolic notation, a sum is frequently written in compact form using summation notation.

> **iscover for yourself**
>
> Expand and evaluate each of the following.
>
> $$\sum_{i=1}^{5} i \quad \text{and} \quad \sum_{i=0}^{4} (i+1)$$
>
> Describe what this means about writing a sum in summation notation.

5 Write sums in summation notation.

EXAMPLE 6 **Writing Sums in Summation Notation**

Write the sum using summation notation:

a. $1 + 4 + 9 + 16 + 25 + 36 + 49$
b. $1 + 3 + 5 + 7 + 9$
c. $-1 + 3 - 5 + 7 - 9$
d. $2 + 4 + 8 + 16 + \cdots$

Solution

a. $1 + 4 + 9 + 16 + 25 + 36 + 49$
This is the sum of squares, $1^2 + 2^2 + 3^2 + 4^2 + 5^2 + 6^2 + 7^2$. Using i for the index of summation, we can say that a possible general term is i^2. Thus,

$$1 + 4 + 9 + 16 + 25 + 36 + 49 = \sum_{i=1}^{7} i^2.$$

b. $1 + 3 + 5 + 7 + 9$
This is the sum of odd numbers. Note that

$$1 + 3 + 5 + 7 + 9 = (2 \cdot 1 - 1) + (2 \cdot 2 - 1)$$
$$+ (2 \cdot 3 - 1) + (2 \cdot 4 - 1) + (2 \cdot 5 - 1).$$

Thus, $2i - 1$ is a possible formula for the ith odd integer. Using summation notation, we have

$$1 + 3 + 5 + 7 + 9 = \sum_{i=1}^{5} (2i - 1).$$

c. $-1 + 3 - 5 + 7 - 9$
This is almost like the series in part (b), except that the signs alternate. Since $(-1)^i = 1$ when i is even and $(-1)^i = -1$ when i is odd, we write the

ENRICHMENT ESSAY

Golden Ratios and the Fibonacci Sequence

The golden ratio appears in the Fibonacci sequence

$$1, 1, 2, 3, 5, 8, 13, 21, 34, 55, 89, 144, 233, \ldots.$$

As the terms progress, the ratio of each number to its predecessor keeps getting closer to the golden ratio,

$$\frac{1 + \sqrt{5}}{2} \text{ to } 1$$

or approximately 1.618 to 1.

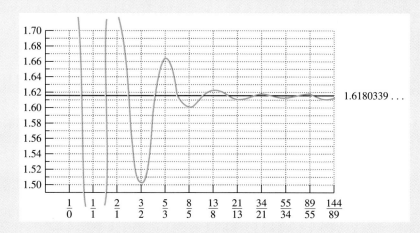

$$\frac{a_6}{a_5} = \frac{8}{5} = 1.6 \text{ to } 1$$

$$\frac{a_7}{a_6} = \frac{13}{8} = 1.625 \text{ to } 1$$

$$\frac{a_8}{a_7} = \frac{21}{13} = 1.\overline{6153846} \text{ to } 1$$

$$\frac{a_9}{a_8} = \frac{34}{21} \approx 1.619048 \text{ to } 1$$

$$\vdots$$

$$\frac{a_{99}}{a_{98}} = \frac{218922995834555169026}{135301852344706746049} \approx 1.6180339887498948482045868343656381177203$$

$$\frac{a_{100}}{a_{99}} = \frac{354224848179261915075}{218922995834555169026} \approx 1.6180339887498948482045868343656381177202$$

If a golden rectangle $\left(\text{length to width in the ratio } \dfrac{1 + \sqrt{5}}{2} \text{ to } 1\right)$ is di-

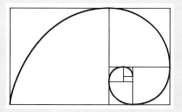

vided into a square and a rectangle, the smaller rectangle is a golden rectangle. If the smaller golden rectangle is divided again, the same is true of the yet smaller rectangle, and so on. If the corresponding points on the golden rectangles are joined as shown in the figure, the result is what is known as a logarithmic spiral—the same spiral as a snail's shell.

The Fibonacci numbers and the golden rectangle are widespread in nature.

Bruce Iverson

series in summation notation as follows:

$$-1 + 3 - 5 + 7 - 9 = \sum_{i=1}^{5} (-1)^i (2i - 1).$$

d. $2 + 4 + 8 + 16 + \cdots$

This infinite series consists of powers of 2. We use the symbol ∞ to represent infinity, writing the series in summation notation as follows:

$$2 + 4 + 8 + 16 + \cdots = \sum_{i=1}^{\infty} 2^i.$$

■

PROBLEM SET 10.1

Practice Problems

For Problems 1–16, the general term of a sequence is given. For each problem, write the first four terms, the 10th term, (a_{10}) and the 15th term (a_{15}).

1. $a_n = 2n - 1$ **2.** $a_n = 3n + 2$ **3.** $a_n = n^2 + 2$ **4.** $a_n = n^2 - 3$ **5.** $a_n = \dfrac{n}{n + 1}$ **6.** $a_n = \dfrac{n}{n + 4}$

7. $a_n = 1 + \dfrac{1}{n}$ **8.** $a_n = 1 - \dfrac{1}{n}$ **9.** $a_n = \dfrac{n - 1}{n + 2}$ **10.** $a_n = \dfrac{n + 2}{n + 3}$

11. $a_n = (-1)^n n$ **12.** $a_n = (-1)^n (n - 1)$ **13.** $a_n = (-1)^{n+1} n^2$ **14.** $a_n = (-1)^{n+1}(n^3 - 1)$

15. $a_n = (-\frac{1}{2})^n$ **16.** $a_n = (-\frac{1}{2})^{n-3}$

For Problems 17–22, find the indicated term of each sequence.

17. $a_n = 3n - 4; a_{12}$ **18.** $a_n = 2n + 5; a_{14}$ **19.** $a_n = (-2)^{n-2}(2n + 1); a_6$

20. $a_n = (1 - \frac{1}{n})^2; a_{10}$ **21.** $a_n = \ln e^n; a_{45}$ **22.** $a_n = \log 10^n; a_{376}$

For Problems 23–40, write an expression for the general, or nth term, of each sequence. More than one expression may be possible.

23. $1, 3, 5, 7, 9, \ldots$

25. $3, 6, 9, 12, 15, \ldots$

27. $\frac{2}{1}, \frac{3}{2}, \frac{4}{3}, \frac{5}{4}, \frac{6}{5}, \ldots$

29. $1, 4, 7, 10, 13, \ldots$

31. $-2, 4, -6, 8, -10, \ldots$

33. $2, -4, 6, -8, 10, \ldots$

35. $-1, 1, -1, 1, -1, \ldots$

37. $-1, -4, -7, -10, -13, \ldots$

39. $\sqrt{5}, 5, 5\sqrt{5}, 25, 25\sqrt{5}, \ldots$

24. $2, 4, 6, 8, 10, \ldots$

26. $6, 12, 18, 24, 30, \ldots$

28. $\frac{1}{2}, \frac{1}{4}, \frac{1}{8}, \frac{1}{16}, \frac{1}{32}, \ldots$

30. $\frac{1}{2}, \frac{2}{3}, \frac{3}{4}, \frac{4}{5}, \frac{5}{6}, \ldots$

32. $-3, 6, -9, 12, -15, \ldots$

34. $3, -6, 9, -12, 15, \ldots$

36. $1, -1, 1, -1, 1, \ldots$

38. $-1, -5, -9, -13, -17, \ldots$

40. $1 \cdot 2, 2 \cdot 3, 3 \cdot 4, 4 \cdot 5, 5 \cdot 6, \ldots$

In Problems 41–48, find the indicated partial sum for the given sequence.

41. $3, 6, 9, 12, \ldots; S_6$

43. $\frac{1}{3}, \frac{2}{3}, \frac{3}{3}, \frac{4}{3}, \ldots; S_8$

45. $a_n = \dfrac{3n}{n + 2}; S_3$

47. $a_n = \dfrac{(-1)^n}{2n - 1}; S_4$

42. $2, 4, 8, 16, \ldots; S_6$

44. $1, \frac{1}{2}, \frac{1}{4}, \frac{1}{8}, \ldots; S_5$

46. $a_n = \dfrac{n + 3}{2n}; S_3$

48. $a_n = \dfrac{(-1)^{n+1}}{2n}; S_4$

Expand and evaluate Problems 49–62.

49. $\sum_{i=1}^{4} 3i$ 　　　　**50.** $\sum_{i=1}^{5} (3i - 2)$ 　　　　**51.** $\sum_{i=2}^{6} (i^2 + 3)$ 　　　　**52.** $\sum_{i=4}^{7} (i^3 - 2)$ 　　　　**53.** $\sum_{i=1}^{5} i(i + 4)$

54. $\sum_{i=1}^{4} (i - 3)(i + 2)$ 　　**55.** $\sum_{j=1}^{4} (-1)^j$ 　　　　**56.** $\sum_{j=2}^{5} (-2)^j$ 　　　　**57.** $\sum_{k=1}^{5} \left(-\frac{1}{2}\right)^k$ 　　　　**58.** $\sum_{k=2}^{4} \left(-\frac{1}{3}\right)^k$

59. $\sum_{i=2}^{4} (-i)^i$ 　　　　**60.** $\sum_{i=0}^{3} \frac{(-1)^i}{i + 1}$ 　　　　**61.** $\sum_{i=3}^{5} \frac{2i - 1}{i - 1}$ 　　　　**62.** $\sum_{i=4}^{6} \frac{i^2 + 1}{2}$

In Problems 63–78, write each sum using summation notation.

63. $1 + 8 + 27 + 64 + 125$

64. $1 + 4 + 9 + 16$

65. $2 + 4 + 6 + 8 + 10$

66. $3 + 6 + 9 + 12 + 15$

67. $-2 + 4 - 6 + 8 - 10$

68. $-3 + 6 - 9 + 12 - 15$

69. $\frac{1}{1} + \frac{1}{4} + \frac{1}{9} + \frac{1}{16} + \frac{1}{25}$

70. $\frac{1}{1} + \frac{1}{8} + \frac{1}{27} + \frac{1}{64} + \frac{1}{125}$

71. $\frac{1}{1} - \frac{1}{4} + \frac{1}{9} - \frac{1}{16} + \frac{1}{25}$

72. $\frac{1}{1} - \frac{1}{8} + \frac{1}{27} - \frac{1}{64} + \frac{1}{125}$

73. $\frac{2}{3} + \frac{3}{4} + \frac{4}{5} + \frac{5}{6} + \frac{6}{7} + \frac{7}{8}$

74. $\frac{1}{3} + \frac{2}{5} + \frac{3}{7} + \frac{4}{9} + \frac{5}{11}$

75. $6 + 12 + 18 + 24 + 30 + \cdots$

76. $8 + 16 + 24 + 32 + 40 + \cdots$

77. $\frac{1}{1 \cdot 2} + \frac{1}{2 \cdot 3} + \frac{1}{3 \cdot 4} + \frac{1}{4 \cdot 5} + \cdots$

78. $\frac{1}{1 \cdot 2^3} + \frac{1}{2 \cdot 3^3} + \frac{1}{3 \cdot 4^3} + \frac{1}{4 \cdot 5^3} + \cdots$

Application Problems

79. The finite sequence whose general term is $a_n = 0.1\sqrt{9n^2 + 82}$, where $n = 1, 2, \ldots, 15$, models the U.S. debt (in trillions of dollars) from 1981 through 1995. Find the terms of this finite sequence. Then construct a bar graph with the years from 1981 through 1995 on the horizontal axis and the federal debt (in trillions) on the vertical axis.

80. A laboratory study determines that a certain culture of bacteria doubles in number every hour. If there were originally 50 bacteria in the culture, how many are present after 6 hours? n hours?

81. The number of AIDS cases reported in the United States for 1984 through 1990 is approximated by the model $a_n = -143n^3 + 1810n^2 - 187n + 2331$, where $n = 1$ corresponds to 1984, $n = 2$ to 1985, and so on. Find and interpret S_3, the third partial sum.

True–False Critical Thinking Problems

82. Which one of the following is true?
　　a. Every sequence is a relation, but some sequences are not functions.
　　b. A general term for the sequence $-1, \frac{1}{3}, -\frac{1}{5}, \frac{1}{7}, \ldots$ is

$$a_n = \frac{(-1)^n}{2n - 1}.$$

　　　There could be others.
　　c. A general term for the sequence $-1, 4, -9, 16, \ldots$ is $a_n = (-1)^{n+1} n^2$. There could be others.
　　d. The sixth term of the sequence whose general term is $a_n = (-1)^{n+1} 2^n$ is 64.

83. Which one of the following is true?
　　a. The range of an infinite sequence is the set of integers.
　　b. It is possible to find the 51st term of a sequence without knowing all 50 of the preceding terms.
　　c. There is no formula for the general term of a sequence whose first five terms are 5, 8, 11, 14, and 17.

　　d. The general term of the sequence $-1, -4, -9, -16, -25, \ldots$ is $a_n = (-1)^n n^2$.

84. Which one of the following is true?
　　a. The symbol $\sum$ represents sums and products.
　　b. There are ten terms in the series

$$\sum_{i=2}^{10} i^2$$

　　c. The sum of a series can never be negative.
　　d. The series

$$\sum_{i=1}^{6} \frac{(-1)^i}{i^2} \quad \text{and} \quad \sum_{j=0}^{5} \frac{(-1)^{j+1}}{(j + 1)^2}$$

　　have the same sum.

85. Which one of the following is true?
　　a. $\sum_{i=1}^{3} (2i + 5) = \left(\sum_{i=1}^{3} 2i \right) + 5$

b. $\displaystyle\sum_{i=1}^{5} 4 = 4$

c. The series

$$\sum_{i=0}^{6} (-1)^i (i+1)^2 \quad \text{and} \quad \sum_{j=1}^{7} (-1)^j j^2$$

have the same sum.

d. $\displaystyle\sum_{i=1}^{5} 6i = 6 \sum_{i=1}^{5} i$

86. Which one of the following is true?
  a. A sequence is the indicated sum of the terms of a series.
  b. There is only one way to write a given sum in summation notation.
  c. $\displaystyle\sum_{i=1}^{4} 3i + \sum_{i=1}^{4} 4i = \sum_{i=1}^{4} 7i$
  d. $\displaystyle\sum_{i=1}^{2} (-1)^i 2^i = 0$

Technology Problems

87. As n increases, the terms of the sequence

$$a_n = \left(1 + \frac{1}{n}\right)^n$$

get closer and closer to the number e (where $e \approx 2.7183$). Use a calculator to find a_{10}, a_{100}, a_{1000}, $a_{10,000}$, and $a_{100,000}$, comparing these terms to the decimal approximation for e.

Most graphing utilities have a sequence-graphing mode that plots the terms of a sequence as points on a rectangular coordinate system. Use this capability to graph each of the sequences in Problems 88–91. Try to duplicate the graph of the sequence shown in the indicated figure.

88. $a_n = \dfrac{n}{n+1}$

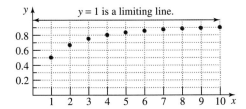

90. $a_n = \dfrac{2n^2 + 5n - 7}{n^3}$

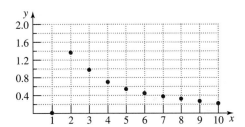

89. $a_n = \dfrac{100}{n}$

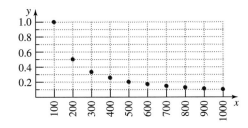

91. $a_n = \dfrac{3n^4 + n - 1}{5n^4 + 2n^2 + 1}$

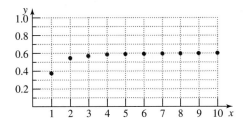

92. Use the ⌐SUM SEQ¬ capability of a graphing utility to calculate the sum of a sequence to verify your answers to Problems 49–62.

Writing in Mathematics

93. At the end of n years, a car that was purchased for $24,000 has its value given by $a_n = 24,000(\frac{3}{4})^n$. Find a_5 and write a sentence explaining what this value represents. Describe the nth term of the sequence in terms of the value of the car at the end of each year.

94. The figure shows the graph of the sequence whose general term is $a_n = \dfrac{1}{2^n}$. Explain how the four points were determined. Why is the graph not shown as a continuous exponential function? What is the difference between the functions $a_n = \dfrac{1}{2^n}$ and $f(x) = (\frac{1}{2})^x$?

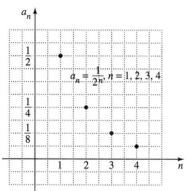

95. Describe the difference between a sequence and a series.

Critical Thinking Problems

96. a. The figure shows the triangular array of numbers known as Pascal's triangle. Each row begins and ends with the number 1. Any other number in a row is the sum of the two closest numbers above it. For example, in the seventh row, $15 = 5 + 10$. Write the first nine terms of the sequence whose terms consist of the sum of the numbers on the diagonal lines as shown. Observe that you have written the first nine terms of the Fibonacci sequence.

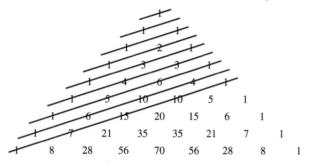

b. The Fibonacci sequence can be defined by $a_{n+1} = a_n + a_{n-1}$ with $a_1 = a_2 = 1$. Thus, we can find a_3 by substituting 2 for n in the formula above. If $n = 2$, $a_{n+1} = a_n + a_{n-1}$ becomes $a_3 = a_2 + a_1$. Because $a_1 = a_2 = 1$, we obtain $a_3 = 1 + 1 = 2$.

If $n = 3$, $a_{n+1} = a_n + a_{n-1}$ becomes $a_4 = a_3 + a_2$. Because $a_3 = 2$ and $a_2 = 1$, we obtain $a_4 = 2 + 1 = 3$.

Use the formula $a_{n+1} = a_n + a_{n-1}$ to verify the first ten terms of the Fibonacci sequence in the table in Problem 97.

97. The first ten Fibonacci numbers are as follows:

F_1	F_2	F_3	F_4	F_5	F_6	F_7	F_8	F_9	F_{10}
1	1	2	3	5	8	13	21	34	55

Complete the following table. Describe the patterns you observe.

n	$F_1 + F_2 + F_3 + \cdots + F_n$	$F_1^2 + F_2^2 + F_3^2 + \cdots + F_n^2$	$F_{n+2} - 1$	$F_n F_{n+1}$
1				
2				
3				
4				
5				

98. Use the first ten Fibonacci numbers in Problem 97 to complete the following table. Does this prove that $F_{n+1}^2 = F_n F_{n+2} + (-1)^n$? Explain.

n	F_{n+1}^2	$F_n F_{n+2} + (-1)^n$
1		
2		
3		
4		
5		

99. Write the first four terms of a sequence whose general term is given by

$$a_n = \frac{3 \cdot 5 \cdots (2n-1)(2n+1)}{2 \cdot 4 \cdots (2n-2)(2n)}.$$

100. The first seven terms of a sequence are $2, 3, 5, 7, 11, 13$, and 17. Describe at least one pattern of this sequence and use this pattern to write the next ten terms.

101. Show by writing out all the terms that the following symbols all represent the same sum.

$$\sum_{i=-3}^{2} \frac{i}{i+4} \qquad \sum_{j=-20}^{-15} \frac{j+17}{j+21} \qquad \sum_{k=14}^{19} \frac{k-17}{k-13}$$

Expand and write the answers for Problems 102–105 as a single logarithm with a coefficient of 1.

102. $\displaystyle\sum_{i=2}^{5} \log i$

103. $\displaystyle\sum_{i=1}^{4} \log 2i$

104. $\displaystyle\sum_{i=1}^{4} i \log x$

105. $\displaystyle\sum_{i=2}^{4} 2i \log x$

106. Find a and b if $\displaystyle\sum_{i=0}^{4} (ai+b) = 10$ and

$$\sum_{i=0}^{4} (ai+b) = 14$$

107. Is the following statement true?

$$\sum_{i=1}^{n} a_i b_i = \sum_{i=1}^{n} a_i \sum_{i=1}^{n} b_i$$

108. Is the following statement true?

$$\sum_{i=1}^{n} \frac{1}{a_i} = \frac{1}{\displaystyle\sum_{i=1}^{n} a_i}$$

109. Find $\displaystyle\sum_{i=1}^{6} a_i$ if $a_1 = 1$, $a_2 = 2$, and $a_n = a_{n-1}^2 + a_{n-2}^2$, for $n \geqslant 3$.

Group Activity Problem

110. Enough curiosities involving the Fibonacci sequence exist to warrant a flourishing Fibonacci Association that publishes a quarterly journal. Do some research on the Fibonacci sequence by consulting the research department of your library or the Internet, and find one property that interests you. After doing this research, get together with your group to share these intriguing properties. Depending on the size of the group, it will be interesting to see how much duplication occurs or if each member researches a unique property for this fascinating sequence.

Review Problems

111. Factor: $2y^6 + 16$.

112. Graph: $\dfrac{x^2}{16} + \dfrac{y^2}{9} = 1$.

113. Solve for x: $\log_4 x + \log_4 (x-6) = 2$.

SECTION 10.2

Solutions Manual   Tutorial   Video 11

Arithmetic Sequences and Series

Objectives

1 Find the common difference for an arithmetic sequence.
2 Find a term for an arithmetic sequence.
3 Use arithmetic sequences to model applied situations.
4 Find partial sums of arithmetic sequences.
5 Use sums of arithmetic sequences to solve applied problems.

Arithmetic Sequences

A person is considering two job offers. Company A will initially pay \$15,000 a year with a guaranteed raise of \$500 each year. Company B offers a starting salary of \$16,000 a year but will only guarantee a raise of \$200 each year.

The sequences below show the yearly salaries for each company.

Year	1	2	3	4	5	6	7	8	9	10
Co. A	15,000	15,500	16,000	16,500	17,000	17,500	18,000	18,500	19,000	19,500
Co. B	16,000	16,200	16,400	16,600	16,800	17,000	17,200	17,400	17,600	17,800

The sequences of numbers for company A and company B are called *arithmetic sequences* or *arithmetic progressions* because each term after the first differs from the preceding term by a constant amount.

> **Definition of an arithmetic sequence**
>
> An *arithmetic sequence* or an *arithmetic progression* is a sequence in which each term after the first differs from the preceding term by a constant amount. The difference between consecutive terms is called the *common difference* of the sequence.

1 Find the common difference for an arithmetic sequence.

EXAMPLE 1   **Finding Common Differences of Arithmetic Sequences**

Identify the common difference:

a. $5, 10, 15, 20, \ldots$
b. $5, 2, -1, -4, \ldots$
c. $1, \frac{3}{2}, 2, \frac{5}{2}, \ldots$
d. $4, 3\frac{3}{4}, 3\frac{1}{2}, 3\frac{1}{4}, \ldots$
e. $a_1, a_1 + 2d, a_1 + 4d, a_1 + 6d, \ldots$

Solution

We obtain the common difference by subtracting from any term its predecessor. We will subtract the second term from the first term.

Sequence	Second Term − First Term	Common Difference
a. $5, 10, 15, 20, \ldots$	$10 - 5 = 5$	5
b. $5, 2, -1, -4, \ldots$	$2 - 5 = -3$	-3
c. $1, \frac{3}{2}, 2, \frac{5}{2}, \ldots$	$\frac{3}{2} - 1 = \frac{3}{2} - \frac{2}{2} = \frac{1}{2}$	$\frac{1}{2}$
d. $4, 3\frac{3}{4}, 3\frac{1}{2}, 3\frac{1}{4}, \ldots$	$3\frac{3}{4} - 4 = -\frac{1}{4}$	$-\frac{1}{4}$
e. $a_1, a_1 + 2d, a_1 + 4d,$ $a_1 + 6d, \ldots$	$(a_1 + 2d) - a_1 = 2d$	$2d$

■

EXAMPLE 2 **Writing the Terms of an Arithmetic Sequence Using the First Term and the Common Difference**

Write the first six terms of the arithmetic sequence with first term 6 and common difference -2.

Solution

To find the second term, we add -2 to the first term 6, giving 4. For the next term, we add -2 to 4, and so on.

First term: 6
Second term: $6 + (-2) = 4$
Third term: $4 + (-2) = 2$
Fourth term: $2 + (-2) = 0$
Fifth term: $0 + (-2) = -2$
Sixth term: $-2 + (-2) = -4$

The first six terms are

$$6, 4, 2, 0, -2, -4.$$ ■

2 Find a term for an arithmetic sequence.

Let's now consider generalizing from Example 2 for an arithmetic sequence whose first term is a_1 and whose common difference is d. We start with the first term and add d to each successive term.

a_1:   First term $= a_1$
a_2:   Second term $= a_1 + d$
a_3:   Third term $= a_2 + d = (a_1 + d) + d = a_1 + 2d$
a_4:   Fourth term $= a_3 + d = (a_1 + 2d) + d = a_1 + 3d$
a_5:   Fifth term $= a_4 + d = (a_1 + 3d) + d = a_1 + 4d$
a_6:   Sixth term $= a_5 + d = (a_1 + 4d) + d = a_1 + 5d$
$\ \vdots$ $\vdots$
a_n: nth term $= ?$

iscover for yourself

Turn back the page and look at the pattern for the first six terms of the arithmetic sequence whose first term is a_1 and whose common difference is d. What relationship do you observe between the coefficient of d and the subscript of a denoting the term number? Use this relationship to write the missing formula for the nth term.

Did you notice that the coefficient of d is 1 less than the subscript of a denoting the term number? Thus, the formula for the nth term is

$$a_n: \quad n\text{th term} = a_1 + (n-1)d.$$

1 less than the
subscript of a

General term of an arithmetic sequence

The nth term (the general term) of an arithmetic sequence with first term a_1 and common difference d is

$$a_n = a_1 + (n-1)d.$$

Because the formula contains four variables ($a_n, a_1, n,$ and d), if any three of these are known, we can find the value of the fourth variable. This idea is illustrated in Examples 3 and 4.

EXAMPLE 3   Finding a Specified Term of an Arithmetic Sequence

Find a_{30}, the 30th term, in the arithmetic sequence: $-4, -1, 2, 5, \ldots$

Solution

The first term of this sequence is -4, so $a_1 = -4$. The common difference can be found by subtracting any two adjacent terms or by observation. Using the first two terms, we see that the common difference is $-1 - (-4)$ or 3, so $d = 3$. Since we are looking for the 30th term, we use $n = 30$. Using the formula for the general term of an arithmetic sequence, we obtain

$$a_n = a_1 + (n-1)d$$
$$a_{30} = -4 + (30-1) \cdot 3 = -4 + 29 \cdot 3 = -4 + 87 = 83$$

The 30th term is 83.   ■

EXAMPLE 4   Using the Formula for the General Term of an Arithmetic Sequence

In the sequence in Example 3, which term is 593?

Solution

We must find n so that $a_n = 593$. We substitute -4 for a_1 (the first term), 3 for d (the common difference) and 593 for a_n into the formula for the general term of an arithmetic sequence. Our goal is to solve for n.

sing technology

The formula for the nth term of $-4, -1, 2, 5, \ldots$ is

$$a_n = -4 + (n-1) \cdot 3$$
or $a_n = 3n - 7$

Consult the manual to see if your graphing utility is capable of generating tables. If so, use the $\boxed{\text{TABLE}}$ function and the formula for the nth term of the sequence to display the first 40 terms.

$$a_n = a_1 + (n - 1)d \qquad \text{This is the formula for the } n\text{th term of an arithmetic sequence.}$$
$$593 = -4 + (n - 1)3 \qquad \text{Substitute the given values.}$$
$$593 = -4 + 3n - 3 \qquad \text{Apply the distributive property.}$$
$$593 = -7 + 3n \qquad \text{Combine numerical terms.}$$
$$600 = 3n \qquad \text{Add 7 to both sides.}$$
$$200 = n \qquad \text{Divide both sides by 3.}$$

The term 593 is the 200th term of the sequence. ∎

3 Use arithmetic sequences to model applied situations.

Modeling Using Arithmetic Sequences

Any variable that increases or decreases at a steady rate can be modeled using the formula for the nth term of an arithmetic sequence.

EXAMPLE 5 **Modeling Nursing Home Costs**

According to *Fortune, 1992 Investor's Guide,* the average annual cost to reside in a nursing home in 1990 was \$29,930. Average increases have amounted to \$1800 yearly. Write the general term for the arithmetic sequence modeling nursing home costs where $n = 1$ corresponds to 1990.

Solution

We can model nursing home costs by the arithmetic sequence

\$29,930, \$31,730, \$33,530, ...

where a_1, the first term, represents costs in 1990. We obtain each subsequent term by adding \$1800, the yearly cost increase, so $d = 1800$.

$$a_n = a_1 + (n - 1)d \qquad \text{This is the formula for the } n\text{th term of an arithmetic sequence.}$$
$$a_n = 29,930 + (n - 1)1800 \qquad \text{Substitute the given values.}$$
$$a_n = 29,930 + 1800n - 1800 \qquad \text{Apply the distributive property.}$$
$$a_n = 28,130 + 1800n \qquad \text{Simplify.}$$

The general term for the arithmetic sequence modeling nursing home costs is

$$a_n = 28,130 + 1800n$$

where $n = 1$ corresponds to 1990. ∎

Discover for yourself

Use the model

$$a_n = 28,130 + 1800n$$

to confirm that nursing home costs in 1990, 1991, and 1992 were \$29,930, \$31,730, and \$33,530. Predict annual nursing home costs for the year 2000 by substituting 11 for n in the model.

Francesco Clemente "The Four Corners" 1985, gouache on handmade paper, 94 × 94 in. (238.8 × 238.8 cm). Private collection. Courtesy Sperone Westwater, New York.

EXAMPLE 6 **Modeling Money Spent by U.S. Travelers**

According to the U.S. Bureau of Economic Analysis, U.S. travelers spent \$12,808 million in other countries in 1984. This amount has increased by approximately \$2350 million yearly. How much will U.S. travelers spend in other countries by the year 2000?

Solution

We can model money spent abroad by the arithmetic sequence

12,808, 15,158, 17,508, ...

where a_1, the first term, represents the amount spent in millions of dollars in 1984. Each subsequent year this amount increases by 2350 million dollars, so

$d = 2350$. Since $n = 1$ corresponds to 1984, $n = 17$ corresponds to the year 2000. We are looking for the 17th term of the arithmetic sequence. Using the formula for the general term of an arithmetic sequence, we obtain

$$a_n = a_1 + (n - 1)d$$
$$a_{17} = 12{,}808 + (17 - 1)2350 = 12{,}808 + 16 \cdot 2350 = 50{,}408$$

The 17th term is 50,408. The model predicts that U.S. travelers will spend $50,408 million in other countries by the year 2000. ■

4 Find partial sums of arithmetic sequences.

Arithmetic Series

In Example 5, we modeled yearly nursing home costs using

$$a_n = 28{,}130 + 1800n.$$

Using this formula for the nth term of the sequence, we can compute partial sums such as S_3. This third partial sum describes total nursing home costs over a three-year period, beginning with 1990. Instead of adding the three values, let's derive a quicker method for calculating the partial sums of arithmetic sequences.

Let's consider the sum of the first n terms of an arithmetic sequence. This sum, denoted by S_n, is called an *arithmetic series*. Recall that the first n terms of the arithmetic sequence are

$$a_1, a_1 + d, a_1 + 2d, \dots, a_1 + (n - 1)d.$$

We proceed as follows:

$S_n = a_1 + (a_1 + d) + (a_1 + 2d) + \cdots + [a_1 + (n - 1)d]$ S_n is the sum of the first n terms of the sequence.

$S_n = a_n + (a_n - d) + (a_n - 2d) + \cdots + [a_n - (n - 1)d]$ The same series can be obtained by beginning with the last term, a_n, and subtracting d.

$2S_n = (a_1 + a_n) + (a_1 + a_n) + (a_1 + a_n) + \cdots + (a_1 + a_n)$ Add the two equations. The terms with d cancel.

$2S_n = n(a_1 + a_n)$ Simplify the right side using the fact that there are n sums of $(a_1 + a_n)$.

$S_n = \dfrac{n}{2}(a_1 + a_n)$ Solve for S_n.

We have the following result.

> **The nth partial sum of an arithmetic sequence**
>
> The sum, S_n, of the first n terms of an arithmetic sequence is given by
>
> $$S_n = \frac{n}{2}(a_1 + a_n)$$
>
> where a_1 is the first term and a_n is the nth term.

The following examples illustrate how to use this formula.

EXAMPLE 7   **Finding the 15th Partial Sum of an Arithmetic Sequence**

Find the sum of the first 15 terms of the arithmetic sequence: $3, 6, 9, 12, \dots$

Solution

Before finding the sum of the first 15 terms, we must find the 15th term.

$$d = a_2 - a_1 = 6 - 3 = 3 \quad \text{Find the common difference.}$$

$$a_n = a_1 + (n - 1)d \quad \begin{array}{l} \text{Use the formula for the } n\text{th term of an arithmetic} \\ \text{sequence to find the 15th term.} \end{array}$$

$$a_{15} = 3 + (15 - 1)3 \quad \text{Substitute 15 for } n.$$

$$= 3 + 14(3)$$

$$= 45 \quad \text{The 15th term is 45.}$$

Now we are ready to find the sum of the first 15 terms.

$$S_n = \frac{n}{2}(a_1 + a_n) \quad \begin{array}{l} \text{This is the formula for the sum of the first } n \text{ terms of an arithmetic} \\ \text{sequence. Let } n = 15, a_1 = 3, \text{ and } a_{15} = 45. \end{array}$$

$$S_{15} = \frac{15}{2}(3 + 45) \quad \text{Substitute the given values and 45 for } a_{15}.$$

$$= \frac{15}{2}(48)$$

$$= 360$$

The sum of the first 15 terms is 360. ■

EXAMPLE 8 **Using S_n to Evaluate a Summation**

Find: $\displaystyle\sum_{i=1}^{14} (2i + 3)$

Solution

$$\sum_{i=1}^{14} (2i + 3) = (2 \cdot 1 + 3) + (2 \cdot 2 + 3) + (2 \cdot 3 + 3) + \cdots + (2 \cdot 14 + 3)$$

$$= \quad 5 \quad + \quad 7 \quad + \quad 9 \quad + \cdots + \quad 31$$

Observe that we want to find the sum of the first 14 terms of the arithmetic sequence $5, 7, 9, \ldots$.

$$S_n = \frac{n}{2}(a_1 + a_n) \quad \begin{array}{l} \text{This is the formula for the sum of the first } n \text{ terms of an arithmetic} \\ \text{sequence. Let } n = 14, a_1 = 5, \text{ and } a_{14} = 31. \end{array}$$

$$S_{14} = \frac{14}{2}(5 + 31)$$

$$= 7(36)$$

$$= 252$$

Thus, $\displaystyle\sum_{i=1}^{14} (2i + 3) = 252$. ■

5 Use sums of arithmetic sequences to solve applied problems.

Modeling Using Partial Sums of Arithmetic Sequences

Let's return to the situation with which we opened this section.

EXAMPLE 9 **Modeling Total Salary Over a Ten-Year Period**

A person is considering two job offers. Company *A* will initially pay $15,000 a year with a guaranteed raise of $500 each year. Company *B* offers a starting salary of $16,000 a year but will only guarantee a raise of $200 each year.

The sequences below show the yearly salaries for each company.

Year	1	2	3	4	5	6	7	8	9	10
Co. A	15,000	15,500	16,000	16,500	17,000	17,500	18,000	18,500	19,000	19,500
Co. B	16,000	16,200	16,400	16,600	16,800	17,000	17,200	17,400	17,600	17,800

Salaries for Company A

Over a 10-year period, which company will pay the greater total amount?

Solution

Since we are interested in total salary over ten years, we need to find the tenth partial sum for each sequence. We'll use the formula for S_n with $n = 10$.

$$S_n = \frac{n}{2}(a_1 + a_n)$$

Company A:

$a_1 = 15,000$ (Starting salary)

$a_{10} = 19,500$ (Salary in year 10)

$S_{10} = \frac{10}{2}(15,000 + 19,500)$

$\qquad = 5(34,500) = 172,500$

Company B:

$a_1 = 16,000$ (Starting salary)

$a_{10} = 17,800$ (Salary in year 10)

$S_{10} = \frac{10}{2}(16,000 + 17,800)$

$\qquad = 5(33,800) = 169,000$

Over a 10-year period, company A will pay $172,500 and company B will pay $169,000. Company A will pay the greater total amount.   ■

EXAMPLE 10   **Modeling Total Nursing Home Costs Over a Six-Year Period**

Your grandmother and her financial counselor are looking at options in case nursing home care is needed in the future. One possibility involves immediate nursing home care for a six-year period beginning in 1998. Your grandmother has saved $250,000.

Using the model

$$a_n = 28,130 + 1800n$$

which describes yearly nursing home costs with $n = 1$ corresponding to 1990, does your grandmother have enough in savings to pay for the facility?

Solution

We must find the sum of an arithmetic sequence whose first term corresponds to nursing home costs in the year 1998 and whose last term corresponds to nursing home costs in the year 2003. Since $n = 1$ describes the year 1990, $n = 9$ describes the year 1998 and $n = 14$ describes the year 2003.

$a_n = 28,130 + 1800n$

$a_9 = 28,130 + 1800 \cdot 9 = 44,330$

$a_{14} = 28,130 + 1800 \cdot 14 = 53,330$

The first year the facility will cost $44,330 and by year six the facility will cost $53,330. Now we want the sum of these costs for all six years. Our focus is now on the sum of the first six terms of the arithmetic series

44,330, 46,130, ... , 53,330.

We can now call a_1, the first term, 44,330 and a_6, the last term, 53,330.

$$S_n = \frac{n}{2}(a_1 + a_n)$$

$$S_6 = \frac{6}{2}(44{,}330 + 53{,}330) = 3(97{,}660) = 292{,}980$$

Total nursing home costs for your grandmother are predicted to be $292,980 for the six-year period. Since she has saved only $250,000, her savings are not enough to pay for the facility. ■

PROBLEM SET 10.2

Practice Problems

Find the common difference for each arithmetic sequence in Problems 1–8.

1. $2, 6, 10, 14, \ldots$

2. $3, 8, 13, 18, \ldots$

3. $-7, -2, 3, 8, \ldots$

4. $-10, -4, 2, 8, \ldots$

5. $5.12, 5.25, 5.38, 5.51, \ldots$

6. $714, 711, 708, 705, \ldots$

7. $\dfrac{e}{6}, \dfrac{e}{3}, \dfrac{e}{2}, \dfrac{2e}{3}, \ldots$

8. $\dfrac{\pi}{12}, \dfrac{\pi}{6}, \dfrac{\pi}{4}, \dfrac{\pi}{3}, \ldots$

In Problems 9–14, write the first six terms of each arithmetic sequence with the given first term and common difference.

9. First term = 8, common difference = 2

10. First term = -14, common difference = 3

11. First term = -16, common difference = -4

12. First term = 26, common difference = -5

13. First term = $\frac{3}{2}$, common difference = $\frac{3}{4}$

14. First term = $\frac{3}{5}$, common difference = $-\frac{1}{2}$

For each arithmetic sequence in Problems 15–28, use the formula for the general term to find the indicated term.

15. $4, 7, 10, 13, \ldots; a_{26}$

16. $2, 7, 12, 17, \ldots; a_{20}$

17. $7, 3, -1, -5, \ldots; a_{14}$

18. $6, 1, -4, -9, \ldots; a_{14}$

19. $3.15, 3.10, 3.05, 3.00, \ldots; a_{22}$

20. $3, 3.75, 4.5, 5.25, \ldots; a_{15}$

21. $\frac{3}{2}, \frac{9}{4}, 3, \frac{15}{4}, \ldots; a_{10}$

22. $\frac{1}{12}, \frac{1}{6}, \frac{1}{4}, \frac{1}{3}, \ldots; a_{11}$

23. $a_1 = 9, d = 2; a_{16}$

24. $a_1 = 11, d = 3; a_{14}$

25. $a_1 = -\frac{1}{3}, d = \frac{1}{3}; a_{30}$

26. $a_1 = 6, d = -\frac{1}{4}; a_{10}$

27. $a_1 = 4, d = -0.3; a_{12}$

28. $a_1 = 5, d = -0.2; a_{12}$

29. In the arithmetic sequence $5, 8, 11, 14, \ldots$, which term is 32?

30. In the arithmetic sequence $-8, -13, -18, -23, \ldots$, which term is -53?

31. In the sequence $4, 1, -2, -5, -8, \ldots$, which term is -281?

32. In the sequence $\frac{5}{4}, \frac{1}{2}, -\frac{1}{4}, -1, \ldots$, which term is $-\frac{55}{4}$?

33. In the sequence $-\frac{5}{3}, -2, -\frac{7}{3}, -\frac{8}{3}, \ldots$, which term is -9?

34. In the sequence $10, 14.5, 19, 23.5, \ldots$, which term is 203.5?

35. Find the sum of the first 20 terms of the arithmetic sequence $4, 10, 16, 22, \ldots$.

36. Find the sum of the first 25 terms of the arithmetic sequence $7, 19, 31, 43, \ldots$.

37. Find the sum of the first 10 terms of the arithmetic sequence $-15, -7, 1, 9, \ldots$.

38. Find the sum of the first 15 terms of the arithmetic sequence $-12, -3, 6, 15, \ldots$.

For Problems 39–44, write out the first three terms and the last term. Then use the formula for the sum of the first n terms of an arithmetic sequence to find the indicated sum.

39. $\sum_{i=1}^{17} (5i + 3)$ **40.** $\sum_{i=1}^{20} (6i - 4)$ **41.** $\sum_{i=1}^{30} (-3i + 5)$ **42.** $\sum_{i=1}^{40} (-2i + 6)$ **43.** $\sum_{i=1}^{100} 4i$ **44.** $\sum_{i=1}^{100} -2i$

45. Find $1 + 2 + 3 + 4 + \ldots + 100$, the sum of the first 100 natural numbers.

46. Find $2 + 4 + 6 + 8 + \ldots + 200$, the sum of the first 100 positive even numbers.

47. Find the sum of the even integers between 21 and 45.

48. Find the sum of the odd integers between 30 and 54.

49. Find the sum of the first 25 terms of an arithmetic sequence whose first term is -9 and whose common difference is 5.

50. Find the sum of the first 32 terms of an arithmetic sequence whose first term is -15 and whose common difference is 8.

51. Find the sum of the first 40 terms of an arithmetic sequence whose first term is 50 and whose common difference is -3.

52. Find the sum of the first 25 terms of an arithmetic sequence whose first term is 45 and whose common difference is -5.

53. Find the sum of the first 12 terms of an arithmetic sequence whose first term is $\frac{1}{2}$ and whose common difference is $-\frac{1}{2}$.

54. Find the sum of the first 15 terms of an arithmetic sequence whose first term is $\frac{1}{3}$ and whose common difference is $-\frac{2}{3}$.

Application Problems

55. According to the National Center for Education Statistics, the total enrollment in U.S. public elementary and secondary schools in 1985 was 39.05 million. Enrollment has increased by approximately 0.45 million each year.

   a. Write the general term for the arithmetic sequence modeling public school enrollment, where $n = 1$ corresponds to 1985.

   b. Use the model to predict total enrollment for the year 2000.

   c. In what year will enrollment reach 50.75 million students?

56. In 1911, the world record time for the men's mile run was 1043.04 seconds. The world record has decreased by approximately 0.4118 seconds each year since then.

   a. Write the general term for the arithmetic sequence modeling record times for the men's mile run, where $n = 1$ corresponds to 1911.

   b. Use the model to predict the record time for the men's mile run for the year 2000.

57. Company A pays $12,000 yearly with raises of $800 per year. Company B pays $14,000 yearly with raises of $500 per year. Which company will pay more in year 10? How much more?

58. The first tread of a flight of stairs is 22 centimeters above the ground. If each tread after the first is 14.5 centimeters above the level of the preceding step, how high above ground level is the 30th step?

59. Suppose you are investigating employment opportunities and you discover that company A will start you at $20,000 a year and guarantee you a raise of $1000 each year and company B will start you at $21,000 a year but will only guarantee you a raise of $800 each year. Over a 12-year period, which company will pay you the greater total amount?

60. A theater in the round has 70 seats in the first row, 78 seats in the second row, 86 seats in the third row, and so on in an arithmetic sequence. If the theater has 24 rows of seats, find the total number of seats it contains.

61. According to the Environmental Protection Agency, in 1960 the United States generated 87.1 million tons of solid waste. This amount has increased by approximately 3.14 million tons each year.

   a. Write the general term for the arithmetic sequence modeling the amount of solid waste generated in the United States, where $n = 1$ corresponds to 1960.

   b. What is the total amount of solid waste generated from 1960 through 1998?

62. According to the Environmental Protection Agency, in 1960 the United States recovered 3.78 million tons of solid waste. Due primarily to recycling programs, this amount has increased by approximately 0.576 million tons each year.

   a. Write the general term for the arithmetic sequence modeling the amount of solid waste recovered in the United States, where $n = 1$ corresponds to 1960.

b. What is the total amount of solid waste recovered from 1960 through 1998?

63. A *degree-day* is a unit used to measure the fuel requirements of buildings. By definition, each degree that the average daily temperature is below 65°F is 1 degree-day. For example, a temperature of 42°F constitutes 23 degree-days If the average temperature on January 1 was 42°F and fell 2°F for each subsequent day up to and including January 10, how many degree-days are included from January 1 to January 10?

64. A display of 108 cans contains 24 cans in the bottom row, 21 cans in the next row, and so on in an arithmetic sequence. Find the number of rows in the display.

65. What is the sum of the perimeters of the first 14 triangles in the sequence of equilateral triangles shown in the figure?

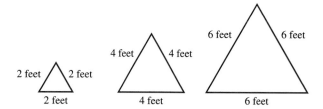

True–False Critical Thinking Problems

66. Which one of the following is true?
 a. The common difference for the arithmetic sequence given by $1, -1, -3, -5, \ldots$ is 2.
 b. The sequence $1, 4, 8, 13, 19, 26, \ldots$ is an arithmetic sequence.
 c. The nth term of an arithmetic sequence whose first term is a_1 and whose common difference is d is $a_n = a_1 + nd$.
 d. If the first term of an arithmetic sequence is 5 and the third term is -3, then the fourth term is -7.

67. Which one of the following is true?
 a. The sequence $3, 6, 3, 0, 3, 6, 3, 0, \ldots$ is an arithmetic sequence.
 b. A sequence that follows a pattern involving addition is an arithmetic sequence.
 c. If the first term of an arithmetic sequence is -3 and the third term is -2, then the fourth term is -1.

 d. In any arithmetic sequence, we can write as many terms as we want by adding the same number repeatedly.

68. Which one of the following is true?
 a. An arithmetic sequence is the indicated sum of an arithmetic series.
 b. Any series that can be expressed in sigma notation, such as $\displaystyle\sum_{i=1}^{5} (2i + 3)$, must be an arithmetic series.
 c. The sum of the even integers from 10 through 30 inclusively is $\frac{20}{2}(10 + 30)$.
 d. The formula $\dfrac{n(n + 1)}{2}$ describes the sum of the first n natural numbers.

Technology Problems

69. One way to graph a sequence using a graphing utility is to set the mode to the DOT mode rather than the connected mode and use an integer viewing rectangle. For example, the arithmetic sequence $10, 14.5, 19, 23.5, \ldots$, whose general term is $a_n = 4.5n + 5.5$ can be graphed by entering $y_1 = 4.5x + 5.5$ in the DOT mode. You could then use the TRACE feature to find a particular term. For example, the TRACE feature indicates that if $x = 26$, then $y = 122.5$, so the 26th term of the sequence is 122.5. Use this technique to graph the arithmetic sequences whose general term you found in Problems 15–28. Then TRACE along the graph and verify the indicated term that you found.

70. Use the SUM SEQ capability of a graphing utility to calculate the sum of a sequence to verify your answers to Problems 39–44.

Writing in Mathematics

71. What is meant by an arithmetic sequence?

72. Describe what information is needed to find the general term of an arithmetic sequence.

73. Describe the difference between a sequence and a series.

Critical Thinking Problems

74. Give examples of two different arithmetic sequences with $a_4 = 10$. Describe how you obtained each sequence.

75. If $\dfrac{1}{x}, \dfrac{1}{y}, \dfrac{1}{z}$ are three consecutive terms of an arithmetic sequence, show that

a. $\dfrac{x}{z} = \dfrac{x - y}{y - z}$ **b.** $y = \dfrac{2xz}{x + z}$

76. Find the value of x if $3 - x, x, \sqrt{9 - 2x}$ form an arithmetic sequence.

77. If the digits of a positive three-digit integer are reversed, the new number is 396 more than the original number. The sum of the digits is 21. Furthermore, the three digits of the number form an arithmetic sequence. What is the number?

78. A theater contains 73 seats in the first row, 79 seats in the second row, 85 seats in the third row, and so on in an arithmetic sequence. If the theater contains 4077 seats, how many rows does it contain?

79. A person wants to save $6580 over an 8-month period. If $700 is saved the first month and each month the amount saved is to be increased by a fixed sum, what should be this fixed sum?

80. How many of the first n natural numbers must be added to produce a sum of 4950?

81. Find the sum of the series $\displaystyle\sum_{i=1}^{n} (ai + b)$.

82. If the sum of the first n terms of an arithmetic sequence is described by $S_n = 3n^2 - n$, show that the formula for the nth term is $a_n = 6n - 4$.

Group Activity Problem

83. Members of your group have been hired by the Environmental Protection Agency to write a report on whether we are making significant progress in recovering solid waste. Use the models from Problems 61–62 as the basis for your report. A graph of each model from 1960 through 2000 would be helpful. What percent of solid waste generated is actually recovered on a year-to-year basis? Be as creative as you want in your report and then draw conclusions. The group should actually write up the report and perhaps even include suggestions as to how we might improve recycling progress.

Arman "Large Bourgeois Refuse" 1960. Trash in a glass box with wooden base, $25\frac{3}{4} \times 15\frac{3}{4} \times 3\frac{1}{4}$ in. (65.4 × 40 × 8.3 cm). Collection, Jeanne-Claude Christo and Christo, New York. Photograph by eeva-inkeri, courtesy the artist. © 1997 Artists Rights Society (ARS), New York/ADAGP, Paris.

Review Problems

84. Solve: $(y + 1)(2y + 3) - 3(y + 2)(y + 1) = -3(y + 5)$.

85. Solve the following system:

$$x^2 + 4y^2 = 13$$
$$x^2 - y^2 = 8$$

86. The resistance required in an electrical circuit to produce a given amount of power varies inversely with the square of the current. If a current of 0.8 ampere requires a resistance of 50 ohms, what resistance will be required by a current of 0.5 ampere?

SECTION 10.3

Solutions Manual   **Tutorial**   **Video II**

Geometric Sequences and Series

Objectives

1 Find the common ratio for a geometric sequence.
2 Find a term for a geometric sequence.
3 Use geometric sequences to model applied situations.
4 Find partial sums of geometric sequences.
5 Use sums of geometric sequences to solve applied problems.
6 Find sums of certain infinite geometric series.

Geometric Sequences

Table 10.1 contains two sequences. The sequence for total food production capacity is an arithmetic sequence; each term is 50,000 more than the previous term. However, the sequence for total population does not appear to be arithmetic. Each year the population is increasing by 7% of what it had been during the previous year. The sequence in the column representing total population is an example of a geometric sequence. In such a sequence, each term after the first is obtained by *multiplying* by a constant, in this case 1.07.

TABLE 10.1 Population and Food Production Sequences

Time	Total Population	Total Food Production Capacity
Beginning year	100,000	100,000
1	107,000	150,000
2	114,490	200,000
3	122,504	250,000
13	240,984	750,000
25	542,743	1,350,000
40	1,497,446	2,100,000
46	2,247,262	2,400,000
47	2,404,571	2,450,000
48	2,572,891	2,500,000

Economist Thomas Malthus (1766–1834) predicted that population growth increases in a geometric sequence and food production increases only in an arithmetic sequence. He concluded that eventually a point would be reached where the world's population would exceed the world's food production. If two sequences, one geometric and one arithmetic, are increasing, the geometric sequence will eventually overtake the arithmetic sequence, regardless of the head start that the arithmetic sequence might initially have. In short, according to Malthus, unchecked population growth would eventually lead to mass starvation. The numbers in the table indicate this would occur during the 48th year.

The sequence for total food production capacity in Table 10.1 is arithmetic because each term after the first is obtained by adding a constant amount, 50,000, to the preceding term. Let's contrast this with the definition of a geometric sequence.

> ### Definition of a geometric sequence
>
> A *geometric sequence* or a *geometric progression* is a sequence in which each term after the first is obtained by *multiplying* the preceding term by a nonzero constant. The amount by which we multiply each time is called the *common ratio* of the sequence.

Find the common ratio for a geometric sequence.

EXAMPLE 1 **Finding Common Ratios of Geometric Sequences**

Identify the common ratio:

a. $2, 8, 32, 128, \ldots$ **b.** $2, -2, 2, -2, \ldots$ **c.** $9, 3, 1, \frac{1}{3}, \ldots$

Solution

We obtain the common ratio by dividing any term by its predecessor. We will divide the second term by the first term.

Discover for yourself

Confirm each common ratio shown on the right by computing

$$\frac{\text{Third term}}{\text{Second term}}$$

and

$$\frac{\text{Fourth term}}{\text{Third term}}$$

for each of the three geometric sequences.

Sequence	$\dfrac{\text{Second term}}{\text{First term}}$	Common Ratio
a. $2, 8, 32, 128, \ldots$	$\frac{8}{2} = 4$	4
b. $2, -2, 2, -2, \ldots$	$\frac{-2}{2} = -1$	-1
c. $9, 3, 1, \frac{1}{3}, \ldots$	$\frac{3}{9} = \frac{1}{3}$	$\frac{1}{3}$

EXAMPLE 2 **Writing the Terms of a Geometric Sequence Using the First Term and the Common Ratio**

Write the first six terms of the geometric sequence with first term 6 and common ratio $-\frac{1}{3}$.

Solution

To find the second term, we multiply the first term by $-\frac{1}{3}$, the common ratio: $(6)(-\frac{1}{3}) = -2$. The second term is -2. For the next term, we multiply -2 by $-\frac{1}{3}$, and so on.

First term: 6
Second term: $6(-\frac{1}{3}) = -2$
Third term: $-2(-\frac{1}{3}) = \frac{2}{3}$

Study tip

When the common ratio of a geometric sequence is negative ($r < 0$), the signs of the terms alternate.

Fourth term: $\frac{2}{3}\left(-\frac{1}{3}\right) = -\frac{2}{9}$
Fifth term: $-\frac{2}{9}\left(-\frac{1}{3}\right) = \frac{2}{27}$
Sixth term: $\frac{2}{27}\left(-\frac{1}{3}\right) = -\frac{2}{81}$

The first six terms are

$$6, -2, \frac{2}{3}, -\frac{2}{9}, \frac{2}{27}, -\frac{2}{81}.$$ ■

2 Find a term for a geometric sequence.

Just as we can express the general term (or the nth term), a_n, of an arithmetic sequence in terms of the first term a_1 and the common difference d, we can write the general term of a geometric sequence in terms of the first term a_1 and the common ratio, designated by r $(r \neq 0)$. We start with the first term and multiply each successive term by r, as follows.

a_1: First term $= a_1$
a_2: Second term $= a_1 r$
a_3: Third term $= a_2 r = (a_1 r)\,r = a_1 r^2$
a_4: Fourth term $= a_3 r = (a_1 r^2)\,r = a_1 r^3$
a_5: Fifth term $= a_4 r = (a_1 r^3)\,r = a_1 r^4$
a_6: Sixth term $= a_5 r = (a_1 r^4)\,r = a_1 r^5$
⋮   ⋮
a_n: nth term $=$?

iscover for yourself

Look at the pattern for the first six terms of the geometric sequence whose first term is a_1 and whose common ratio is r. What relationship do you observe between the exponent on r and the subscript of a denoting the term number? Use this relationship to write the missing formula for the nth term.

Did you notice that the exponent on r is 1 less than the subscript of a denoting the term number? Thus, the formula for the nth term is

$$a_n = a_1 r^{n-1}.$$

1 less than the
subscript of a

General term of a geometric sequence

The nth term (the general term) of a geometric sequence with first term a_1 and common ratio r is

$$a_n = a_1 r^{n-1}.$$

Examples 3 and 4 illustrate the use of this formula.

EXAMPLE 3 **Finding a Specified Term of a Geometric Sequence**

Find the general term and the eighth term, a_8, of the geometric sequence:

$$64, -32, 16, -8, \ldots$$

Solution

To find the general term of a geometric sequence, we need the first term and the common ratio. The first term, a_1, is 64. We find the common ratio, r, by dividing any term by its predecessor. Thus,

$$r = \frac{-32}{64} = -\frac{1}{2}.$$

Now we use the formula for the general term of a geometric sequence.

$$a_n = a_1 r^{n-1}$$

Substituting 64 for a_1 and $-\frac{1}{2}$ for r, we obtain the general term of the sequence.

$$a_n = 64\left(-\frac{1}{2}\right)^{n-1}$$

Now we find the eighth term of the sequence by substituting 8 for n.

$$a_8 = 64\left(-\frac{1}{2}\right)^{8-1} = 64\left(-\frac{1}{2}\right)^7 = 64\left(-\frac{1}{128}\right) = -\frac{1}{2}$$

The eighth term of the geometric sequence is $-\frac{1}{2}$. ∎

> **study tip**
>
> Be careful with the order of operations when evaluating $a_1 r^{n-1}$. First find r^{n-1}. Then multiply the result by a_1.

3 Use geometric sequences to model applied situations.

EXAMPLE 4 **Modeling Population Growth**

In 1798 English economist Thomas Malthus claimed that populations increase in a geometric sequence. The population of Florida from 1980 through 1987 is shown in Table 10.2.

a. Show that the population is increasing geometrically.
b. Write the general term for the geometric sequence modeling population growth for Florida, where $n = 1$ corresponds to 1980.

Solution

a. Divide the population each year by the population in the preceding year.

$$\frac{10.03}{9.75} \approx 1.029, \quad \frac{10.32}{10.03} \approx 1.029, \quad \frac{10.62}{10.32} \approx 1.029$$

Continuing in this manner, we keep getting approximately 1.029. This means that the population is increasing geometrically, with $r \approx 1.029$. In this situation, the common ratio is the growth rate, indicating that the population of Florida in any year shown in the table is approximately 1.029 times the population the year before.

b. Since the population is increasing geometrically, we can use the formula for the general term of a geometric sequence to model the data. Using $a_1 = 9.75$ (the 1980 population) and $r = 1.029$, we have

$$a_n = a_1 r^{n-1} \qquad \text{This is the formula for the } n\text{th term of a geometric sequence.}$$
$$a_n = 9.75(1.029)^{n-1} \qquad \text{Substitute the given values.}$$

TABLE 10.2 Florida's Population from 1980–1987

Year	Population (in millions)
1980	9.75
1981	10.03
1982	10.32
1983	10.62
1984	10.93
1985	11.25
1986	11.57
1987	11.91

The general term for the geometric sequence modeling Florida's population growth is

$$a_n = 9.75(1.029)^{n-1}$$

where $n = 1$ corresponds to 1980. We can use this model to predict Florida's population (in millions) for the year 2000 by substituting 21 for n.

$$a_{21} = 9.75(1.029)^{21-1} \approx 17.27$$

The model predicts a population of 17.27 million in the year 2000. ■

4 Find partial sums of geometric sequences.

Finite Geometric Series

Let's now consider the sum of the first n terms of a geometric sequence. This sum, denoted by S_n, is called a *geometric series*. Recall that the first n terms of a geometric sequence are

$$a_1, a_1 r, a_1 r^2, \dots, a_1 r^{n-2}, a_1 r^{n-1}.$$

We proceed as follows:

$$S_n = a_1 + a_1 r + a_1 r^2 + \cdots + a_1 r^{n-2} + a_1 r^{n-1} \qquad \text{S_n is the sum of the first n terms of the sequence.}$$

$$rS_n = a_1 r + a_1 r^2 + a_1 r^3 + \cdots + a_1 r^{n-1} + a_1 r^n \qquad \text{Multiply both sides of the equation by r.}$$

$$S_n - rS_n = a_1 - a_1 r^n \qquad \text{Subtract the second equation from the first equation.}$$

$$S_n(1 - r) = a_1 - a_1 r^n \qquad \text{Solve for S_n (assuming that $r \neq 1$).}$$

$$S_n = \frac{a_1 - a_1 r^n}{1 - r} = \frac{a_1(1 - r^n)}{1 - r}$$

We have proved the following result.

The nth partial sum of a geometric sequence

The sum, S_n, of the first n terms of a geometric sequence is given by

$$S_n = \frac{a_1(1 - r^n)}{1 - r},$$

where a_1 is the first term and r is the common ratio ($r \neq 1$).

The following examples illustrate the application of this formula.

EXAMPLE 5   **Finding the Sixth Partial Sum of a Geometric Sequence**

Find the sum of the first six terms: $2, -8, 32, -128, \dots$

Solution

Before finding the sum of the first 6 terms, we must find the common ratio.

$$r = \frac{a_2}{a_1} = \frac{-8}{2} = -4 \qquad \text{Find the common ratio.}$$

Now we are ready to find the sum of the first 6 terms.

$$S_n = \frac{a_1(1 - r^n)}{1 - r} \qquad \text{This is the formula for the sum of the first } n \text{ terms of a finite geometric sequence. Let } n = 6, a_1 = 2, \text{ and } r = -4.$$

$$S_6 = \frac{2(1 - (-4)^6)}{1 - (-4)}$$

$$= \frac{-8190}{5}$$

$$= -1638$$

The sum of the first six terms is -1638. ∎

Geometric series can be written using summation notation. The sum indicated can then be found using the formula for S_n. Let's see exactly how this is done in Example 6.

Discover for yourself

Suppose you save $1 on the first day of the month, $2 on the second day, $4 on the third day, and so on in a geometric sequence. The total amount saved in a 30-day month is given by

$$\sum_{i=0}^{29} 2^i.$$

Use the formula for S_n to evaluate this sum. Are you surprised at the total amount saved? What does this tell you about the growth of a geometric series?

EXAMPLE 6 **Using S_n to Evaluate a Summation**

Evaluate: $\displaystyle\sum_{i=0}^{8} 2^i$

Solution

By writing out a few of the terms, we obtain

$$\sum_{i=0}^{8} 2^i = 2^0 + 2^1 + 2^2 + 2^3 + \cdots + 2^8$$

$$= 1 + 2 + 4 + 8 + \cdots + 256$$

We have a geometric series with $a_1 = 1, r = 2$, and $n = 9$. (There are *nine* terms.) We substitute these numbers into our formula for S_n.

$$S_n = \frac{a_1(1 - r^n)}{1 - r}$$

$$S_9 = \frac{1(1 - 2^9)}{1 - 2} = \frac{-511}{-1} = 511$$

Thus, $\displaystyle\sum_{i=0}^{8} 2^i = 511$. ∎

5 Use sums of geometric sequences to solve applied problems.

Modeling Using Partial Sums of Geometric Sequences

We have considered situations in which salaries increase by a fixed amount each year. A more realistic situation is one in which salaries increase by a certain percent each year. Example 7 shows how such a situation can be modeled using a geometric series.

EXAMPLE 7   Computing a Lifetime Salary

A job pays a salary of $30,000 the first year. During the next 29 years, the salary increases by 6% each year. What is the total lifetime salary over the 30-year period?

Solution

The salary for the first year is $30,000. With a 6% raise, the second-year salary is computed as follows:

Salary for year 2 $= 30,000 + 30,000(0.06) = 30,000(1.06)$

Each year, the salary is 1.06 times what it was in the previous year. Thus, the salary for year 3 is 1.06 times $30,000(1.06)$, or $30,000(1.06)^2$. The salaries for the first five years are given as follows.

Yearly Salaries

Year 1,	Year 2 ,	Year 3 ,	Year 4 ,	Year 5
↓	↓	↓	↓	↓
$30,000$,	$30,000(1.06)$,	$30,000(1.06)^2$,	$30,000(1.06)^3$,	$30,000(1.06)^4$

The numbers in the last row form a geometric sequence with $a_1 = 30,000$ and $r = 1.06$. To find the total salary over 30 years, we use the formula for the nth partial sum of a geometric sequence, with $n = 30$.

$$S_n = \frac{a_1(1 - r^n)}{1 - r}$$

$$S_{30} = \frac{30,000(1 - (1.06)^{30})}{1 - 1.06} \quad \text{Compute the total salary over 30 years.}$$

$$= \frac{30,000(1 - (1.06)^{30})}{-0.06}$$

$$\approx 2,371,746 \qquad \text{Use a calculator.}$$

The total salary over the 30-year period is $2,371,746. ■

An important application of the formula for the nth partial sum of a geometric sequence is in projecting the value of an annuity. An *annuity* is a sequence of equal payments made at equal time periods. The payments can be investments or loan payments.

EXAMPLE 8   Value of an Annuity

A deposit of $1000 is made the first day of each month in an account that pays 6% annual interest compounded monthly. What is the value of this annuity at the end of 3 years?

Solution

To solve this problem, we must use the formula for compound interest from Chapter 8.

$$A = P\left(1 + \frac{r}{n}\right)^{nt}$$

$A =$ The balance in the account
$P =$ The initial deposit ($1000)
$r =$ The annual interest rate (0.06)
$n =$ The number of compounding periods per year
     (12 with monthly compounding)
$t =$ The time in years

U.S. Bureau of Engraving and Printing

Since $t = 3$ years $= 36$ months, we can consider each of the 36 deposits separately. The first deposit of \$1000 will earn interest for 36 months, so its balance will be

$$A_{36} = 1000\left(1 + \frac{0.06}{12}\right)^{12 \cdot 3}$$

$$= 1000(1.005)^{36}.$$

We call this A_{36} because the money is invested for 36 months.

The second deposit of \$1000 will earn interest for 35 months, and amounts to

$$A_{35} = 1000\left(1 + \frac{0.06}{12}\right)^{35}$$

$$= 1000(1.005)^{35}.$$

Continuing in this manner, the last (36th) deposit will earn interest for only one month, and its balance will be

$$A_1 = 1000\left(1 + \frac{0.06}{12}\right)$$

$$= 1000(1.005).$$

The total balance in the account at the end of 3 years is the sum of the balances of the 36 deposits.

$$S_{36} = A_1 + A_2 + A_3 + \cdots + A_{35} + A_{36}$$

$$= 1000(1.005) + 1000(1.005)^2 + 1000(1.005)^3 + \cdots + 1000(1.005)^{35} + 1000(1.005)^{36}$$

$$= \frac{a_1(1 - r^n)}{1 - r}$$

The series is geometric with $r = 1.005$, so use the formula for the sum of the first n terms of a geometric series.

$$= \frac{1000(1.005)(1 - 1.005^{36})}{1 - 1.005}$$

$a_1 = 1000(1.005)$, $r = 1.005$, and $n = 36$ because we are adding 36 terms.

$$\approx 39,533$$

The value of the annuity at the end of 3 years is approximately \$39,533. ■

6 **Find sums of certain infinite geometric series.**

Infinite Geometric Series

In some cases, it is possible to find the sum of an infinite number of terms of a geometric sequence. Let's first consider a dance at which men are on one side of the room and women are on the other side. The men are not permitted to move, and the women are allowed to walk one-half the distance between themselves and the men at the conclusion of each song.

At the conclusion of the first song, the women walk half the room's length. At the conclusion of the second song, the women walk one-half of the remaining distance between themselves and the men, covering $\frac{1}{4}$ of the room's length. At the conclusion of the third song, the women walk one-half of the remaining quarter of the distance between themselves and the men, covering $\frac{1}{8}$ of the room's length. Each time a song ends, the women must cover half the remaining distance, then half the remaining distance at the conclusion of the next song, and so on. The situation is illustrated in Figure 10.2 at the top of the next page. Clearly, the women must cover a distance given by the *infinite geometric series*

$$\frac{1}{2} + \frac{1}{4} + \frac{1}{8} + \frac{1}{16} + \frac{1}{32} + \frac{1}{64} + \cdots.$$

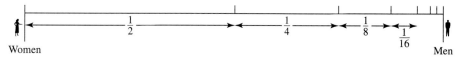

Women   Men

Figure 10.2

One possible way to investigate this infinite sum is to write a formula for the sum of the first n terms and then consider what will happen to this expression as n gets larger and larger.

$$S_n = \frac{a_1(1 - r^n)}{1 - r}$$ This is the formula for the nth partial sum.

$$= \frac{\frac{1}{2}(1 - (\frac{1}{2})^n)}{1 - \frac{1}{2}}$$ For the sum $\frac{1}{2} + \frac{1}{4} + \frac{1}{8} + \frac{1}{16} + \cdots$, $a_1 = \frac{1}{2}$ and r, the common ratio, is also $\frac{1}{2}$.

$$= \frac{\frac{1}{2}(1 - (\frac{1}{2})^n)}{\frac{1}{2}}$$

$$= 1 - (\tfrac{1}{2})^n$$

M. C. Escher
(1898–1972)
"Whirlpools" © 1997
Cordon Art-Baarn-
Holland. All rights
reserved.

Discover for yourself

What happens to $(\frac{1}{2})^n$ as n gets larger and larger? Use your calculator and find $(\frac{1}{2})^5$, $(\frac{1}{2})^{100}$, $(\frac{1}{2})^{1000}$, and so on. What does this mean in terms of the expression for S_n as n gets larger and larger?

In the Discover for Yourself box, did you observe that as n gets larger, the term $(\frac{1}{2})^n$ gets closer to 0? The expression for S_n is getting closer and closer to 1.

$$S_n = 1 - \left(\frac{1}{2}\right)^n \quad \text{Approximately 0}$$ For a very large n, $(\frac{1}{2})^n$ is approximately 0.

$$\approx 1$$ As we continue to add more terms, our sum approaches 1.

Let's see how these sums are approaching 1.

$$\frac{1}{2} + \frac{1}{4} + \frac{1}{8} + \frac{1}{16} = \frac{15}{16} = 0.9375$$

$$\frac{1}{2} + \frac{1}{4} + \frac{1}{8} + \frac{1}{16} + \frac{1}{32} = \frac{31}{32} = 0.96875$$

$$\frac{1}{2} + \frac{1}{4} + \frac{1}{8} + \frac{1}{16} + \frac{1}{32} + \frac{1}{64} = \frac{63}{64} = 0.984375$$

Consequently, we define 1 to be the sum of the infinite geometric series. Thus,

$$\frac{1}{2} + \frac{1}{4} + \frac{1}{8} + \frac{1}{16} + \frac{1}{32} + \frac{1}{64} + \cdots = 1$$

indicating that the women do cover one complete length of the room, and eventually reach the men. In short, the infinite process of addition for the given series results in a finite sum.

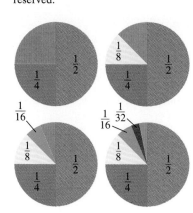

The sum $\frac{1}{2} + \frac{1}{4} + \frac{1}{8} + \frac{1}{16} + \frac{1}{32} \cdots$ is approaching 1.

iscover for yourself

If r is any number between -1 and 1, what happens to r^n as n gets larger and larger? Use your calculator to answer this question by selecting values between -1 and 1 and find larger and larger powers.

In the Discover for Yourself box, did you observe that if $-1 < r < 1$ (equivalently, $|r| < 1$) and n is large, r^n is close to 0? Thus, when considering the sum of an infinite geometric series, the expression r^n approaches 0 when $|r| < 1$. Replacing r^n with 0 in the formula $S_n = \dfrac{a_1(1 - r^n)}{1 - r}$ gives us a formula for the sum of an infinite geometric series.

Formula for the sum of an infinite geometric series

If $|r| < 1$, then the sum of the infinite geometric series

$$a_1 + a_1 r + a_1 r^2 + a_1 r^3 + \cdots$$

where a_1 is the first term and r is the common ratio, is given by

$$S_\infty = \dfrac{a_1}{1 - r}.$$

If $|r| \geq 1$, the infinite geometric series does not have a sum. For example, the sum of the infinite geometric series $3 + 9 + 27 + 81 + \cdots$ (in which $r = 3$) increases without limit.

The following examples illustrate the application of this formula.

EXAMPLE 9   Finding the Sum of the Terms of an Infinite Geometric Series

Find the sum of the infinite geometric series: $\frac{3}{8} - \frac{3}{16} + \frac{3}{32} - \frac{3}{64} + \cdots$

Solution

Before finding the sum, we must find the common ratio.

$$r = \frac{a_2}{a_1} = \frac{-\frac{3}{16}}{\frac{3}{8}} = -\frac{3}{16} \cdot \frac{8}{3} = -\frac{1}{2}$$

With $r = -\dfrac{1}{2}$, the condition that $|r| < 1$ is met, so the infinite geometric series has a sum.

$$S_\infty = \frac{a_1}{1 - r}$$

This is the formula for the sum of an infinite geometric series. Let $a_1 = \frac{3}{8}$ and $r = -\frac{1}{2}$.

$$= \frac{\frac{3}{8}}{1 - \left(-\frac{1}{2}\right)} = \frac{\frac{3}{8}}{\frac{3}{2}} = \frac{1}{4}$$

Thus, the sum of this infinite geometric series is $\frac{1}{4}$. Put in an informal way, as we continue adding more and more terms, the sum is approximately $\frac{1}{4}$. ■

We can use the formula for the sum of an infinite series to express a repeating decimal as a fraction in lowest terms.

EXAMPLE 10 **Writing a Repeating Decimal as a Fraction**

Express $0.\overline{27}$ as a fraction in lowest terms:

Solution

Observe that $0.\overline{27} = 0.272727\ldots = 0.27 + 0.0027 + 0.000027 + \cdots$ is a geometric series in which $a_1 = 0.27$ and $r = 0.01$. Therefore,

$$
\begin{aligned}
S_\infty &= \frac{a_1}{1 - r} \\
&= \frac{0.27}{1 - 0.01} \\
&= \frac{0.27}{0.99} \\
&= \frac{27}{99} \\
&= \frac{3}{11}
\end{aligned}
$$

An equivalent fraction for $0.\overline{27}$ is $\frac{3}{11}$. ∎

Modeling With Infinite Geometric Series

Infinite sums can model numerous situations, as illustrated in Example 11.

EXAMPLE 11 **Tax Rebates and the Multiplier Effect**

A tax rebate that returns a certain amount of money to taxpayers can have a total effect on the economy that is many times this amount. In economics this phenomenon is called the *multiplier effect*. Suppose, for example, the government reduces taxes so that each consumer has $1000 more income. The government feels that each person will spend 80% of this (= $800). The individuals and businesses receiving this $800 in turn spend 80% of it (= $640), creating extra income for yet other people to spend, and so on. Determine the total amount spent on consumer goods from the initial $1000 tax rebate.

Solution

The total amount spent is given by the infinite geometric series

$$800 + 640 + 512 + \cdots$$

in which $a_1 = 800$ and $r = 0.8$. Using our formula for the sum of an infinite geometric series, we obtain

$$
\begin{aligned}
S_\infty &= \frac{a_1}{1 - r} \\
&= \frac{800}{1 - 0.8} \\
&= \frac{800}{0.2} \\
&= 4000
\end{aligned}
$$

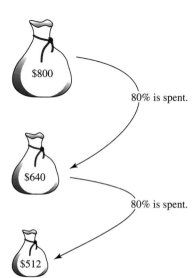

80% is spent.

80% is spent.

$800

$640

$512

This means that the total amount spent on consumer goods from the initial $1000 rebate is $4000. ■

PROBLEM SET 10.3

Practice Problems

In Problems 1–12, find the common ratio for each geometric sequence.

1. $5, 15, 45, 135, \ldots$

2. $-15, 30, -60, 120, \ldots$

3. $-8, 8, -8, 8, \ldots$

4. $6, -6, 6, -6, \ldots$

5. $5, -\frac{5}{2}, \frac{5}{4}, -\frac{5}{8}, \ldots$

6. $-\frac{4}{3}, \frac{8}{3}, -\frac{16}{3}, \frac{32}{3}, \ldots$

7. $90, 30, 10, \frac{10}{3}, \ldots$

8. $3, \frac{9}{2}, \frac{27}{4}, \frac{81}{8}, \ldots$

9. $\sqrt{5}, 5, 5\sqrt{5}, 25, \ldots$

10. $\sqrt{3}, 3, 3\sqrt{3}, 9, \ldots$

11. $\frac{a}{b}, \frac{a}{b^2}, \frac{a}{b^3}, \frac{a}{b^4}, \ldots$

12. $7, \frac{7x}{2}, \frac{7x^2}{4}, \frac{7x^3}{8}, \ldots$

In Problems 13–20, find the first five terms of each geometric sequence with the given first term and common ratio.

13. $a_1 = 10, r = \frac{1}{2}$

14. $a_1 = 12, r = \frac{1}{3}$

15. $a_1 = -\frac{1}{4}, r = -2$

16. $a_1 = -\frac{1}{16}, r = -4$

17. $a_1 = 3, r = -3$

18. $a_1 = 5, r = -5$

19. $a_1 = \frac{a^2}{b}, r = \frac{2b}{a}$

20. $a_1 = c, r = \frac{b}{c}$

In Problems 21–34, find the general term and the indicated term for each geometric sequence.

21. $-3, -15, -75, -375, \ldots; a_8$

22. $18, 54, 162, 486, \ldots; a_9$

23. $-12, -6, -3, -\frac{3}{2}, \ldots; a_{10}$

24. $18, -6, 2, -\frac{2}{3}, \ldots; a_7$

25. $3, -1.5, 0.75, -0.375, \ldots; a_6$

26. $0.0008, -0.008, 0.08, -0.8, \ldots; a_{14}$

27. $\frac{2}{3}, \frac{2}{15}, \frac{2}{75}, \frac{2}{375}, \ldots; a_8$

28. $\frac{1}{3}, \frac{1}{2}, \frac{3}{4}, \frac{9}{8}, \ldots; a_{12}$

29. $-4, -2\sqrt{2}, -2, -\sqrt{2}, \ldots; a_{11}$

30. $-2, -2\sqrt{3}, -6, -6\sqrt{3}, \ldots; a_9$

31. $222\frac{2}{9}, 22\frac{2}{9}, 2\frac{2}{9}, \ldots; a_6$

32. $333\frac{1}{3}, 33\frac{1}{3}, 3\frac{1}{3}, \ldots; a_8$

33. $c^7 d^6, c^6 d^4, c^5 d^2, \ldots; a_{14}$

34. $c^6, c^5 d, c^4 d^2, \ldots; a_{13}$

Find the general term for each graphed geometric sequence in Problems 35–36.

35.
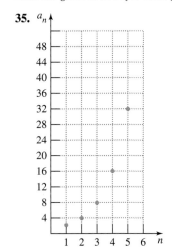

36.
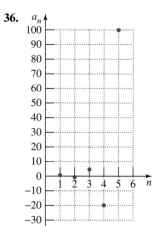

Use the formula for the nth partial sum of a geometric sequence to answer Problems 37–42.

37. Find the sum of the first six terms of the geometric sequence $2, 6, 18, \ldots$.

38. Find the sum of the first seven terms of the geometric sequence $3, 6, 12, \ldots$.

39. Find the sum of the first five terms of the geometric sequence $3, -6, 12, \dots$.

40. Find the sum of the first six terms of the geometric sequence $4, -12, 36, \dots$.

41. Find the sum of the first seven terms of the geometric sequence $-\frac{3}{2}, 3, -6, \dots$.

42. Find the sum of the first eight terms of the geometric sequence $-\frac{1}{24}, \frac{1}{12}, -\frac{1}{6}, \dots$.

Use the formula for the nth partial sum of a geometric sequence to evaluate each sum in Problems 43–50.

43. $\sum_{i=0}^{6} 3^i$

44. $\sum_{i=0}^{4} 6^i$

45. $\sum_{i=0}^{6} (-3)^i$

46. $\sum_{i=0}^{8} (-2)^i$

47. $\sum_{i=1}^{5} 2^{i-1}$

48. $\sum_{i=1}^{6} \left(\frac{1}{2}\right)^{i+1}$

49. $\sum_{i=1}^{4} \left(-\frac{2}{3}\right)^i$

50. $\sum_{i=1}^{5} \left(\frac{3}{4}\right)^i$

In Problems 51–62, find the sum of each geometric series if there is a sum.

51. $1 + \frac{1}{4} + \frac{1}{16} + \cdots$

52. $1 + \frac{1}{3} + \frac{1}{9} + \cdots$

53. $12 + 6 + 3 + \cdots$

54. $32 + 16 + 8 + \cdots$

55. $27 - 18 + 12 - \cdots$

56. $8 - 4 + 2 - \cdots$

57. $5 + 10 + 20 + \cdots$

58. $4 + 10 + 25 + \cdots$

59. $\frac{4}{3} + \frac{2}{9} + \frac{1}{27} + \cdots$

60. $\frac{4}{5} + \frac{8}{15} + \frac{16}{45} + \cdots$

61. $1 - \frac{1}{2} + \frac{1}{4} - \frac{1}{8} + \cdots$

62. $4 - 2 + 1 - \frac{1}{2} + \cdots$

Express each repeating decimal in Problems 63–74 as a fraction in lowest terms.

63. $0.\overline{5}$

64. $0.\overline{7}$

65. $0.\overline{49}$

66. $0.\overline{62}$

67. $0.\overline{241}$

68. $0.\overline{562}$

69. $5.\overline{47}$

70. $6.\overline{82}$

71. $3.\overline{285}$

72. $4.\overline{673}$

73. $0.1\overline{2}$

74. $0.1\overline{6}$

Application Problems

Use the formula for the general term of a geometric sequence to answer Problems 75–80.

75. In 1798 English economist Thomas Malthus argued that populations grow geometrically. At the time, the population of Britain was 7 million, denoted in the table by year 0. Britain's population based upon predictions of the Malthusian model are shown in the table.

a. Show that Britain's population is increasing geometrically every 5 years.

b. Write the general term for the geometric sequence modeling Britain's population growth, where $n = 1$ corresponds to 1798, $n = 2$ corresponds to 1803, $n = 3$ corresponds to 1808, and so on. This is precisely the model that Malthus used.

c. What does the Malthusian model predict for Britain's population in 1997? Consult an appropriate reference to find the actual population. How well does the Malthusian model work when extended this far from 1798?

Number of Years After 1798	Britain's Population (in millions)
0	7.00
5	8.04
10	9.23
15	10.59
20	12.16
25	13.96
30	16.03
35	18.40
40	21.13
45	24.25
50	27.85
100	110.77

76. *E. coli* is a rod-shaped bacterium approximately 10^{-6} meter long, found in the intestinal tracts of humans. The cells of *E. coli* reproduce quite rapidly, as shown in the table at the top of the next page.

a. Write the general term for the geometric sequence modeling the growth of the bacteria, where $n = 1$ corresponds to the first 20-minute time period, $n = 2$ corresponds to the second 20-minute time period, and so on.

b. Use the model to predict the number of bacteria after 8 hours, which is the 24th time period.

Initial Population = 100 Bacteria

20-Minute Time Interval	Number of E. coli Bacteria
1	200
2	400
3	800
4	1600
5	3200
6	6400

77. The body weight of a dieting person is 99% of what it was on the preceding week, indicating a loss of 1% of one's body weight.
  a. Write out the first four terms for the weight of a dieting 200-pound person.
  b. Find the general term of this sequence.

78. On the average, the value of a home increases 5% each year. Suppose a home is initially worth $200,000.
  a. Write out the first four terms for the value of the home during the first four years.
  b. Find the general term of this sequence.

79. A person is investigating two employment opportunities. They both have a beginning salary of $20,000 per year. Company A offers an increase of $1000 per year. Company B offers 5% more than during the preceding year. Which company will pay more in the sixth year?

80. In the figure, L represents the length from the bridge to the neck of a guitar. The frets on the neck form a geometric sequence, at lengths from the bridge given by Lr, Lr^2, Lr^3, Lr^4, and so on.
  a. Find a formula for the length of the 12th fret from the bridge, a_{12}.
  b. Find the common ratio r if it is known that the length of the 12th fret from the bridge is half the original length L.

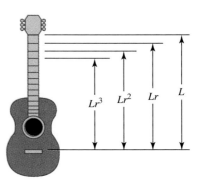

Use the formula for the nth partial sum of a geometric sequence to answer Problems 81–86.

81. A deposit of $200 is made at the beginning of each month for 6 years in an account that pays 9% annual interest compounded monthly. What is the balance in the account at the end of six years?

$$A = 200\left(1 + \frac{0.09}{12}\right) + 200\left(1 + \frac{0.09}{12}\right)^2 + \cdots$$

$$+ 200\left(1 + \frac{0.09}{12}\right)^{72}$$

82. A deposit of $2000 is made the first day of each year in an account that pays 5% interest compounded annually. What is the value of this annuity at the end of six years?

83. A person is investigating two employment opportunities. They both have a beginning salary of $20,000 per year. Company A offers an increase of $1000 per year. Company B offers 5% more than during the preceding year. Which opportunity will produce the better total income for a six-year period?

84. The length of the path of the first swing of the bob of a pendulum is 20 inches. Each succeeding swing is only 0.9 as long as the preceding one. How far will the bob travel on the first five swings?

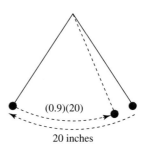

85. A bungee jumper is using a cord that stretches to 200 feet. Each rebound is 65% of the distance of the previous fall.

First fall: 200 feet
First rebound: 200(0.65) feet ⎫
Second fall: 200(0.65) feet ⎭ 2(200)(0.65) feet
Second rebound: 200(0.65)² feet ⎫
Third fall: 200(0.65)² feet ⎭ 2(200)(0.65)² feet
Third rebound: 200(0.65)³ feet ⎫
Fourth fall: 200(0.65)³ feet ⎭ 2(200)(0.65)³ feet

Find the total distance traveled by the bungee jumper up through and including the eighth fall.

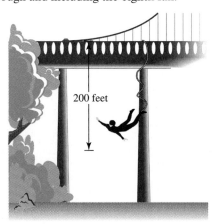

200 feet

86. How many ancestors from parents to great-great-great-great grandparents does a person have?

Use the formula for the sum of an infinite geometric series to answer Problems 87–92.

87. How much additional spending will be generated by a $40 billion tax rebate if 90% of all income is spent?

88. A square piece of paper whose sides measure 40 inches is cut into four smaller squares, each with 20 inches on a side. One of these squares is cut into four smaller squares, each with 10 inches on a side. One of these squares is cut into four smaller squares, each with 5 inches on a side. Assuming that this process is continued indefinitely, find the sum of the perimeters of all the squares, including the original square.

89. An equilateral triangle has a perimeter of 12 inches. By joining the midpoints of its sides with line segments, a second triangle is formed. This operation is continued indefinitely for each new triangle that is formed. Determine the sum of the perimeters of all triangles, including the original triangle.

90. A person insists on a salary of $10,000 after taxes in a country with a 10% income tax. A firm pays the person $10,000, but because the government takes $1000 for taxes, more money is requested from the firm. The company pays an extra $1000 for the $1000 the government took, but again the government takes $100 of this for taxes. Once again, to compensate for this, an extra $100 is given the employee to cover what the government took. But the government takes $10 for taxes, and so on. If this process is continued indefinitely, find the sum of the geometric series representing the employee's salary.

91. The basic attitude of the ancient Greeks was to avoid the infinite. One extremely strong reason why they avoided it was the irritating paradox, attributed to the Greek philosopher Zeno (495–435 B.C.), called "Achilles and the Tortoise." Zeno argued that Achilles, who can run 10 meters per second, can never pass a tortoise, who can run 1 meter per second, if the tortoise is given a 10-meter head start. Zeno's proof was as follows.

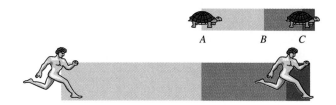

A   B   C

To pass the tortoise, Achilles must get to where the tortoise was at the start of the race. With a 10-meter head start, this will take Achilles 1 second. However, during this 1 second, the tortoise has moved ahead 1 meter. To pass the tortoise, Achilles must move this 1 meter, taking him $\frac{1}{10}$ of a second. But during this $\frac{1}{10}$ of a second, the tortoise has again moved ahead $\frac{1}{10}$ of a meter. To pass the tortoise, Achilles must move $\frac{1}{10}$ of a meter, taking him $\frac{1}{100}$ of a second.

The argument continues indefinitely. To pass the tortoise, Achilles is going to have to run an infinite number of distances $(10 + 1 + \frac{1}{10} + \frac{1}{100} + \cdots)$ in a finite amount of time. Unravel the paradox by finding the sum of the infinite geometric series $10 + 1 + \frac{1}{10} + \frac{1}{100} + \cdots$. How far must Achilles run to overtake the tortoise? How long will this take?

92. What is the total vertical distance traveled by a surfer who follows the path indicated by the ocean wave shown in the figure?

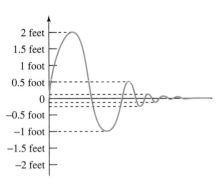

True–False Critical Thinking Problems

93. Which one of the following is true?
  a. The sequence $2, 6, 24, 120, \ldots$ is an example of a geometric sequence.
  b. Adjacent terms in a geometric sequence have a common difference.
  c. A sequence that is not arithmetic must be geometric.
  d. If a sequence is geometric, we can write as many terms as we want by repeatedly multiplying by the common ratio.

94. Which one of the following is true?
  a. If the nth term of a geometric sequence is $a_n = 6(\frac{1}{3})^{n-1}$, then the common ratio is 3.
  b. The first five terms of the geometric sequence whose nth term is $a_n = 3(-2)^{n-1}$ are $-3, -6, -12, -24,$ and -48.
  c. Not every geometric sequence is a function.
  d. If the nth term of a geometric sequence is $a_n = 3(0.5)^{1-n}$, then the common ratio is 2.

95. Which one of the following is true?
  a. The only way to find $\sum\limits_{i=1}^{6}(0.1)^i$ is to write out all terms and then find their sum.

b. The sum of the first four terms of a geometric sequence whose nth term is given by $a_n = (0.78)^n$ can be done only with a calculator.
  c. The sum of the geometric series $\frac{1}{2} + \frac{1}{4} + \frac{1}{8} + \cdots + \frac{1}{512}$ can only be estimated without knowing precisely what terms occur between $\frac{1}{8}$ and $\frac{1}{512}$.
  d. The sum of the first seven terms of $\sqrt{3}, \sqrt{6}, 2\sqrt{3}, 2\sqrt{6}, \ldots$ contains two radical terms, one of which is $15\sqrt{3}$.

96. Which one of the following is true?
  a. $3 + 6 + 12 + 24 + \cdots = \dfrac{3}{1-2}$
  b. $\sum\limits_{i=1}^{\infty} 8(\frac{3}{4})^{i-1} = 32$
  c. $10 - 5 + \frac{5}{2} - \frac{5}{4} + \cdots = \dfrac{10}{1 - \frac{1}{2}}$
  d. Any given infinite geometric sequence has a finite sum.

Technology Problems

97. a. Find a complete graph of the function
$$f(x) = \frac{1 - r^x}{1 - r}$$
  for the following values of r: $r = 0.5$, $r = 0.9$, $r = 1.1$, $r = 3$.

  b. For what values of r does the complete graph of f have a horizontal asymptote?
  c. What does part (b) indicate about when the infinite series $1 + r + r^2 + r^3 + r^4 + \cdots$ has a sum?

In Problems 98–99, use a graphing utility to graph the function. Determine the horizontal asymptote for the graph of f and discuss its relationship to the sum of the given series.

Function	Series
98. $f(x) = \dfrac{2\left[1 - \left(\frac{1}{3}\right)^x\right]}{1 - \frac{1}{3}}$	$\sum\limits_{n=0}^{\infty} 2(\frac{1}{3})^n$

Function	Series
99. $f(x) = \dfrac{4[1 - (0.6)^x]}{1 - 0.6}$	$\sum\limits_{n=0}^{\infty} 4(0.6)^n$

100. Graph $f(x) = \dfrac{4(1 - 3^x)}{1 - 3}$ in the viewing rectangle [0, 10] by [0, 1000]. Does the graph of f have a horizontal asymptote? What does the complete graph of f indicate about the sum of the infinite series $\displaystyle\sum_{n=1}^{\infty} 4 \cdot 3^{n-1}$?

Writing in Mathematics

101. What is the difference between a geometric sequence and a geometric series?

102. Would you rather have $10,000,000 and a brand new BMW or 1¢ today, 2¢ tomorrow, 4¢ on day 3, 8¢ on day 4, 16¢ on day 5, and so on, for 30 days? Explain.

103. How do you determine if an infinite geometric series has a sum or if the sum is not defined?

104. Describe a situation that can be modeled by an infinite geometric series.

105. What happens to the series $\displaystyle\sum_{i=1}^{n} 2^n$ as n increases? Why does the formula $S_\infty = \dfrac{a_1}{1 - r}$ not apply to $\displaystyle\sum_{i=1}^{\infty} 2^i$?

Critical Thinking Problems

If the sequence $a_1, a_2, a_3, \ldots, a_n$ is geometric with the common ratio r, each sequence described in Problems 106–111 is also geometric. Find the common ratio for each sequence in terms of r.

106. $\dfrac{1}{a_1}, \dfrac{1}{a_2}, \dfrac{1}{a_3}, \ldots, \dfrac{1}{a_n}$

107. $-a_1, -a_2, -a_3, \ldots, -a_n$

108. $3a_1, 3a_2, 3a_3, \ldots, 3a_n$

109. $a_1, a_3, a_5, \ldots, a_{2n+1}$

110. $a_n, a_{n-1}, a_{n-2}, \ldots, a_1$

111. $a_1, 4a_2, 16a_3, \ldots, 4^{n-1}a_n$

112. If $x, y,$ and z are three consecutive terms of a geometric sequence, show that

$$\frac{1}{x + y}, \quad \frac{1}{2y}, \quad \frac{1}{z + y}$$

are three consecutive terms of an arithmetic sequence.

113. In the sequence 6, a, b, 16, the first three terms form an arithmetic sequence and the last three terms form a geometric sequence. Find all possible ordered pairs (a, b).

114. A prize of $7380 is awarded in such a way that the winner receives $1280 the first year and a 25% increase in each subsequent year. How long will it take to collect the prize?

115. Find n if $3 + 3^2 + 3^3 + \cdots + 3^n = 120$.

116. Can two different geometric series have the same sum, the same first term, and the same number of terms? (*Hint:* Consider $1 + r + r^2 = 7$.)

117. The sum of the first six terms of the geometric sequence 1, 2, 4, 8, . . . is equal to the sum of the first six terms of an arithmetic sequence whose first term is 1. What is the sixth term of the arithmetic sequence?

118. In a pest-eradication program, sterilized male flies are released into the general population each day. Ninety percent of those flies will survive a given day. How many flies should be released each day if the long-range goal of the program is to keep 20,000 sterilized flies in the population?

119. Find the product of

$$\left(1 + \tfrac{1}{2} + \tfrac{1}{4} + \tfrac{1}{8} + \cdots\right), \left(1 + \tfrac{1}{3} + \tfrac{1}{9} + \tfrac{1}{27} + \cdots\right),$$
$$\text{and } \left(1 + \tfrac{1}{5} + \tfrac{1}{25} + \tfrac{1}{125} + \cdots\right)$$

120. Find the sum: $\displaystyle\sum_{i=1}^{\infty} \frac{7^{i-1}}{10^i}$.

121. Find the sum of the infinite series:

$$\tfrac{2}{1} + \tfrac{1}{3} + \tfrac{2}{9} + \tfrac{1}{27} + \tfrac{2}{81} + \tfrac{1}{243} + \cdots.$$

Review Problems

122. Simplify: $\sqrt[3]{54x^6y^7}$.

123. Solve algebraically and check by graphing:

$$4x^2 + y^2 = 16$$
$$2x + y = 4.$$

124. Rationalize the denominator and simplify: $\dfrac{\sqrt{5} + \sqrt{3}}{\sqrt{5} - \sqrt{3}}$.

SECTION 10.4

Solutions Manual   Tutorial   Video II

The Binomial Theorem

Objectives

1 Expand a binomial raised to a power.
2 Evaluate factorial expressions and binomial coefficients.
3 Expand binomials using the binomial theorem.
4 Use the binomial theorem to find probabilities.

In Chapter 4, we used the FOIL method to square a binomial such as $(x + 3)^2$. In this section, we study higher powers of binomials.

1 Expand a binomial raised to a power.

Patterns in Binomial Expansions

By writing out the *binomial expression* $(a + b)^n$, where n is a positive integer, a number of patterns begin to appear.

$$(a + b)^1 = a + b$$
$$(a + b)^2 = a^2 + 2ab + b^2$$
$$(a + b)^3 = a^3 + 3a^2b + 3ab^2 + b^3$$
$$(a + b)^4 = a^4 + 4a^3b + 6a^2b^2 + 4ab^3 + b^4$$
$$(a + b)^5 = a^5 + 5a^4b + 10a^3b^2 + 10a^2b^3 + 5ab^4 + b^5$$

Discover for yourself

Each expanded form of the binomial expression is a polynomial.

1. For each polynomial, describe the pattern for the exponents on a. What is the largest exponent of a? What happens to the exponent on a from term to term?
2. Describe the pattern for the exponents on b. What is the exponent on b in the first term? What is the exponent on b in the second term? What happens to these exponents from term to term?
3. Find the sum of the exponents on the variables in each term for the polynomials in the five rows. Describe the pattern.
4. How many terms are there in the polynomials on the right in relation to the power of the binomial?

In the Discover for Yourself box, how many of the following patterns were you able to discover?

1. The first term is a^n. The exponent on a decreases by 1 in each successive term.
2. The exponents on b increase by 1 in each successive term. In the first term, the exponent on b is 0. (Because $b^0 = 1$, b is not shown in the first term.) The last term is b^n.
3. The sum of the exponents on the variables in any term is equal to the exponent on $(a + b)^n$.
4. There is one more term in the polynomial expansion than there is in the power of the binomial, n. There are $n + 1$ terms in the expanded form of $(a + b)^n$.

Using these observations, we find that the variable parts of the expansion of $(a + b)^6$ are

$$a^6, a^5b, a^4b^2, a^3b^3, a^2b^4, ab^5, b^6.$$

The first term is a^6, with the exponent on a decreasing by 1 in each successive term. The exponents on b increase from 0 to 6, with the last term being b^6. The sum of the exponents in each term is equal to 6.

Generalizing from these observations, the variable parts of the expansion of $(a + b)^n$ are

$$a^n, a^{n-1}b, a^{n-2}b^2, a^{n-3}b^3, \ldots, ab^{n-1}, b^n.$$

Let's now establish a pattern for the coefficients of the terms in the binomial expansion. Notice that each row begins and ends with 1. Any other number in the row can be obtained by adding the two numbers immediately above it.

Coefficients for $(a + b)^1$:   1   1

Coefficients for $(a + b)^2$:   1   2   1

Coefficients for $(a + b)^3$:   1   3   3   1

Coefficients for $(a + b)^4$:   1   4   6   4   1

Coefficients for $(a + b)^5$:   1   5   10   10   5   1

The triangular array of coefficients is called *Pascal's triangle*. If we continue with the sixth row, the first and last numbers are 1. Each of the other numbers is obtained by finding the sum of the two closest numbers above it in the fifth row.

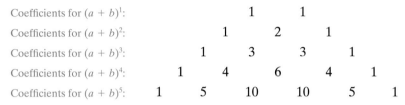

Using the numbers in the sixth row and the variable parts we found, we write the expansion for $(a + b)^6$ as

$$(a + b)^6 = a^6 + 6a^5b + 15a^4b^2 + 20a^3b^3 + 15a^2b^4 + 6ab^5 + b^6$$

Factorial Notation and Binomial Coefficients

2 Evaluate factorial expressions and binomial coefficients.

Pascal's triangle becomes cumbersome when a binomial contains a relatively large power. Another method for determining the coefficients uses the idea of a factorial.

---

n Factorial

For any positive integer n, $n!$ (n factorial) is the product of all the consecutive integers from n down to 1.

$$n! = n(n - 1)(n - 2) \cdots 3 \cdot 2 \cdot 1$$

$0!$, by definition, is 1.

$$0! = 1$$

ENRICHMENT ESSAY

A Triangular Table of Probabilities

Pascal's triangle is a ready reference for finding probabilities and odds. For example, to determine the probability of any boy–girl combination in a family of six children, add the numbers in row six, for a total of 64. The numbers at the ends of the row stand for the probability of all boys or all girls, namely, $\frac{1}{64}$. The second numbers from the ends apply to the next most likely combination (five boys, one girl, or vice versa), $\frac{6}{64}$. The center number applies to three boys and three girls, a probability of $\frac{20}{64}$.

The run of 13 boys shown in this picture has a probability of 1 chance in 8192, found by consulting row 13 of Pascal's triangle. UPI/Corbis-Bettmann

| EXAMPLE 1 | **Evaluating Factorial Expressions** |

Evaluate: **a.** 4!   **b.** 6!

Solution

$$4! = 4 \cdot 3 \cdot 2 \cdot 1 = 24 \qquad \text{We read this as "4 factorial equals 24."}$$
$$6! = 6 \cdot 5 \cdot 4 \cdot 3 \cdot 2 \cdot 1 = 720$$

The coefficients in a binomial expansion are given in terms of factorials. The coefficients are written in a special notation, which we define next.

Definition of a binomial coefficient

The expression $\binom{n}{r}$ is called a binomial coefficient and is defined by

$$\binom{n}{r} = \frac{n!}{r!(n-r)!}$$

EXAMPLE 2 **Evaluating Binomial Coefficients**

Evaluate: **a.** $\binom{7}{3}$ **b.** $\binom{5}{0}$ **c.** $\binom{8}{5}$ **d.** $\binom{7}{7}$

Using technology

Graphing utilities have the capability of computing binomial coefficients. For example, to find $\binom{7}{3}$, many utilities require the sequence

7 [nCr] 3 [ENTER]

and 35 is displayed. Consult your manual and verify the other three evaluations in Example 2.

Solution

In each case, we apply the definition for the binomial coefficient.

a. $\binom{7}{3} = \dfrac{7!}{3!(7-3)!}$

$= \dfrac{7!}{3!4!}$

$= \dfrac{7 \cdot 6 \cdot 5 \cdot 4 \cdot 3 \cdot 2 \cdot 1}{(3 \cdot 2 \cdot 1)(4 \cdot 3 \cdot 2 \cdot 1)}$

$= 35$

b. $\binom{5}{0} = \dfrac{5!}{0!(5-0)!}$

$= \dfrac{5!}{0!5!}$

$= \dfrac{5 \cdot 4 \cdot 3 \cdot 2 \cdot 1}{(1)(5 \cdot 4 \cdot 3 \cdot 2 \cdot 1)}$

$= 1$

c. $\binom{8}{5} = \dfrac{8!}{5!(8-5)!}$

$= \dfrac{8!}{5!3!}$

$= \dfrac{8 \cdot 7 \cdot 6 \cdot 5 \cdot 4 \cdot 3 \cdot 2 \cdot 1}{(5 \cdot 4 \cdot 3 \cdot 2 \cdot 1)(3 \cdot 2 \cdot 1)}$

$= 56$

d. $\binom{7}{7} = \dfrac{7!}{7!(7-7)!}$

$= \dfrac{7!}{7!0!}$

$= \dfrac{7 \cdot 6 \cdot 5 \cdot 4 \cdot 3 \cdot 2 \cdot 1}{(7 \cdot 6 \cdot 5 \cdot 4 \cdot 3 \cdot 2 \cdot 1)(1)}$

$= 1$ ∎

3 Expand binomials using the binomial theorem.

The Binomial Theorem

If we use factorials and the pattern for the variable part of each term, a formula called the *binomial theorem,* or the *binomial expansion* can now be written for any natural number power of a binomial.

A formula for expanding binomials

The Binomial Theorem

For any positive integer n,

$$(a + b)^n = \binom{n}{0}a^n + \binom{n}{1}a^{n-1}b + \binom{n}{2}a^{n-2}b^2$$

$$+ \binom{n}{3}a^{n-3}b^3 + \cdots + \binom{n}{n}b^n$$

EXAMPLE 3 **Using the Binomial Theorem**

Expand: $(a + b)^7$

Solution

$$(a + b)^7 = \binom{7}{0}a^7 + \binom{7}{1}a^6b + \binom{7}{2}a^5b^2 + \binom{7}{3}a^4b^3 + \binom{7}{4}a^3b^4 + \binom{7}{5}a^2b^5 + \binom{7}{6}ab^6 + \binom{7}{7}b^7$$

$$= \frac{7!}{0!7!}a^7 + \frac{7!}{1!6!}a^6b + \frac{7!}{2!5!}a^5b^2 + \frac{7!}{3!4!}a^4b^3 + \frac{7!}{4!3!}a^3b^4 + \frac{7!}{5!2!}a^2b^5 + \frac{7!}{6!1!}ab^6 + \frac{7!}{7!0!}b^7$$

$$= a^7 + 7a^6b + 21a^5b^2 + 35a^4b^3 + 35a^3b^4 + 21a^2b^5 + 7ab^6 + b^7$$

EXAMPLE 4 **Using the Binomial Theorem**

Expand: $(5x + 3y)^3$

Solution

$$(5x + 3y)^3 = \binom{3}{0}(5x)^3 + \binom{3}{1}(5x)^2(3y) + \binom{3}{2}(5x)(3y)^2 + \binom{3}{3}(3y)^3$$

$$= \frac{3!}{0!3!}(5x)^3 + \frac{3!}{1!2!}(5x)^2(3y) + \frac{3!}{2!1!}(5x)(3y)^2 + \frac{3!}{3!0!}(3y)^3$$

$$= 1(125x^3) + 3(25x^2)(3y) + 3(5x)(9y^2) + 1(27y^3)$$

$$= 125x^3 + 225x^2y + 135xy^2 + 27y^3$$

EXAMPLE 5 **Using the Binomial Theorem**

Expand: $\left(\dfrac{1}{2} - 2b\right)^6$

Solution

$$\left(\frac{1}{2} - 2b\right)^6 = \left[\frac{1}{2} + (-2b)\right]^6$$

$$= \binom{6}{0}\left(\frac{1}{2}\right)^6 + \binom{6}{1}\left(\frac{1}{2}\right)^5(-2b) + \binom{6}{2}\left(\frac{1}{2}\right)^4(-2b)^2$$

$$+ \binom{6}{3}\left(\frac{1}{2}\right)^3(-2b)^3 + \binom{6}{4}\left(\frac{1}{2}\right)^2(-2b)^4$$

$$+ \binom{6}{5}\left(\frac{1}{2}\right)(-2b)^5 + \binom{6}{6}(-2b)^6$$

$$= \frac{6!}{0!6!}\left(\frac{1}{2}\right)^6 + \frac{6!}{1!5!}\left(\frac{1}{2}\right)^5(-2b) + \frac{6!}{2!4!}\left(\frac{1}{2}\right)^4(-2b)^2$$

$$+ \frac{6!}{3!3!}\left(\frac{1}{2}\right)^3(-2b)^3 + \frac{6!}{4!2!}\left(\frac{1}{2}\right)^2(-2b)^4$$

$$+ \frac{6!}{5!1!}\left(\frac{1}{2}\right)(-2b)^5 + \frac{6!}{6!0!}(-2b)^6$$

ENRICHMENT ESSAY

The Universality of Mathematics

Pascal's triangle, credited to French mathematician Blaise Pascal (1623–1662), appeared in a Chinese document printed in 1303. The binomial theorem was known in Eastern cultures prior to its discovery in Europe. The same mathematics is often discovered by independent researchers separated by time, place, and culture.

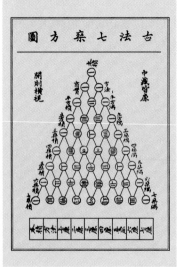

$$= 1\left(\frac{1}{64}\right) + 6\left(\frac{1}{32}\right)(-2b) + 15\left(\frac{1}{16}\right)(4b^2) + 20\left(\frac{1}{8}\right)(-8b^3)$$

$$+ 15\left(\frac{1}{4}\right)(16b^4) + 6\left(\frac{1}{2}\right)(-32b^5) + 1(64b^6)$$

$$= \frac{1}{64} - \frac{3}{8}b + \frac{15}{4}b^2 - 20b^3 + 60b^4 - 96b^5 + 64b^6 \qquad ∎$$

4 Use the binomial theorem to find probabilities.

Modeling Using the Binomial Theorem

Suppose that a student did not study for a quiz with five multiple-choice questions. Each question has five possible choices, with exactly one correct choice. The student knows that three questions must be answered correctly to pass the quiz, and wonders about the probability of guessing and answering exactly three questions correctly.

The binomial theorem provides a model for finding probabilities in situations with n independent trials, where p is the probability of success on any one trial, and $q = 1 - p$ is the probability of failure. For our unprepared student, there are five independent trials ($n = 5$) because there are five questions on the quiz. The probability of success on any one question is $\frac{1}{5}$ because there are five choices and only one is correct. Thus, $p = \frac{1}{5}$. The probability of failure on a question is $1 - \frac{1}{5}$ or $\frac{4}{5}$, meaning that four of the five choices are incorrect. Thus $q = \frac{4}{5}$.

Our student wants to answer three questions successfully, hoping for three successes in five trials. The probability of k successes in n trials is given by a term in the expansions of $(p + q)^n$.

Probability and the binomial theorem

Suppose that p is the probability of success on any one trial, and $q = 1 - p$ is the probability of failure. The probability of k successes in n trials is given by

$$\binom{n}{k} p^k q^{n-k}.$$

EXAMPLE 6 Using the Binomial Theorem to Compute Probability

A student randomly guesses at five multiple-choice questions. Each question has five possible choices. What is the probability of getting three questions correct?

Solution

The probability of k successes in n trials $= \binom{n}{k} p^k q^{n-k}$

$$\text{The probability of 3} \atop \text{successes in 5 trials} = \binom{5}{3}\left(\frac{1}{5}\right)^3\left(\frac{4}{5}\right)^{5-3}$$

p, the probability of guessing correctl. is $\frac{1}{5}$.
q, the probability of failure (guessing incorrectly) is $1 - \frac{1}{5} = \frac{4}{5}$.

$$= 10 \cdot \frac{1}{125} \cdot \frac{16}{25}$$

$$= 0.0512$$

The probability of randomly guessing and getting three questions correct is 0.0512.   ■

PROBLEM SET 10.4

Practice Problems

In Problems 1–14, evaluate each expression.

1. $3!$ **2.** $5!$ **3.** $2!$ **4.** $1!$ **5.** $\dfrac{10!}{8!2!}$ **6.** $\dfrac{9!}{6!3!}$ **7.** $\dfrac{7!}{6!1!}$

8. $\dfrac{8!}{6!2!}$ **9.** $\dbinom{6}{3}$ **10.** $\dbinom{7}{2}$ **11.** $\dbinom{12}{1}$ **12.** $\dbinom{11}{1}$ **13.** $\dbinom{6}{6}$ **14.** $\dbinom{15}{2}$

In Problems 15–36, expand each binomial and express the result in simplified form.

15. $(c + 2)^5$ **16.** $(a + 3)^4$ **17.** $(x + y)^6$ **18.** $(x + y)^7$ **19.** $(a - 2)^4$

20. $(3 - b)^5$ **21.** $(x + 2y)^6$ **22.** $(x + 3y)^5$ **23.** $\left(\dfrac{a}{2} + 1\right)^4$ **24.** $(y^4 - 1)^4$

25. $(2x + 3y)^4$ **26.** $(x^2 + 2y)^4$ **27.** $(2x^2 - y^2)^3$ **28.** $(3x^2 + 2y^4)^3$ **29.** $\left(\dfrac{a}{3} + 2\right)^3$

30. $\left(\dfrac{a}{3} + \dfrac{b}{2}\right)^4$ **31.** $(\sqrt{2} - x)^6$ **32.** $(\sqrt{3} - y)^6$ **33.** $(a^{1/2} + 2)^4$ **34.** $(a^{1/3} + 3)^3$

35. $(a^{-1} + b^{-1})^3$ **36.** $(a^{-1} - b^{-1})^3$

In Problems 37–48, write the first three terms in each binomial expansion, and express the result in simplified form.

37. $(x^2 + x)^8$ **38.** $(a^2 + a)^{14}$ **39.** $(2a + b)^8$ **40.** $(a + 3b)^9$ **41.** $(a - 2b)^8$ **42.** $(2a - 5)^7$

43. $(a^2 + b^2)^{10}$ **44.** $(a^2 - b^2)^{14}$ **45.** $(a + b)^{42}$ **46.** $(a - b)^{93}$ **47.** $\left(y + \dfrac{1}{y}\right)^7$ **48.** $\left(y - \dfrac{1}{y}\right)^8$

Application Problems

The expression

$$\binom{n}{k} p^k q^{n-k}$$

describes the probability of k successes in n trials, where p is the probability of success on any one trial, and $q = 1 - p$ is the probability of failure. Use this expression to solve Problems 49–52.

49. If you have to guess at all true–false questions on a ten-question quiz, what is the probability that exactly half of your guesses are correct?

50. A football player has an average success rate of two completions in three passes. What is the probability of exactly five completions in six attempts?

51. A survey from Teenage Research Unlimited, Northbrook, Ill., found that 30% of teenagers received their spending money from part-time jobs. If five teenagers are selected at random, find the probability that exactly three of them receive spending money from part-time jobs.

52. If 40% of people over the age of 62 wear hearing aids, find the probability that for a sample of seven people over 62 exactly four wear hearing aids.

True–False Critical Thinking Problems

53. Which one of the following is true?
   a. There are 14 terms in the expansion of $(a + b)^{14}$.
   b. For all values of x, $(x + 2)^6 = x^6 + 64$.
   c. The sum of the coefficients in the expansion of $(a + b)^n$ is 2^n.
   d. The factorial symbol, as in $n!$, is a way of writing special addition problems.

54. Which one of the following is true?
   a. The rth term of the expansion of $(a + b)^n$ is

$$\binom{n}{r - 1}a^{n-r+1}b^{r-1}.$$

   b. $\binom{n}{n} \neq \binom{n + 1}{n + 1}$

   c. $\binom{n}{n - 1}$ can never be equal to n.

   d. There are no values of a and b such that $(a + b)^4 = a^4 + b^4$.

Technology Problems

55. Use a graphing utility to confirm the binomial coefficients that you evaluated by hand in Problems 9–14.

In Problems 56–57, graph each function in the same viewing rectangle. Describe how the graphs illustrate the binomial theorem.

56. $f_1(x) = (x + 2)^3$
   $f_2(x) = x^3$
   $f_3(x) = x^3 + 6x^2$
   $f_4(x) = x^3 + 6x^2 + 12x$
   $f_5(x) = x^3 + 6x^2 + 12x + 8$

57. $f_1(x) = (x + 1)^4$
   $f_2(x) = x^4$
   $f_3(x) = x^4 + 4x^3$
   $f_4(x) = x^4 + 4x^3 + 6x^2$
   $f_5(x) = x^4 + 4x^3 + 6x^2 + 4x$
   $f_6(x) = x^4 + 4x^3 + 6x^2 + 4x + 1$

In Problems 58–60, use the binomial theorem to find a polynomial expansion for each function. Then use a graphing utility and an approach similar to the one in Problems 56–57 to verify the expansion.

58. $f_1(x) = (x - 1)^3$ **59.** $f_1(x) = (x - 2)^4$ **60.** $f_1(x) = (x + 2)^6$

Writing in Mathematics

61. Are there situations under which it is easier to use Pascal's triangle than the binomial coefficients? Describe these situations.

62. Write 11^6 as $(10 + 1)^6$, and describe how the binomial theorem can be used to find 11^6 without the use of a calculator. Give some examples that are similar. Now give an example where this technique would not be particularly useful. Describe how this example differs from your previous two examples.

63. Add the entries in the first four rows of Pascal's triangle. Describe the pattern that you observe.

Critical Thinking Problems

64. For a certain rare disease, the probability of recovery without treatment is $\frac{3}{10}$. In a group of ten people afflicted with the disease, what is the probability that exactly eight will recover without treatment? What is the probability that eight or more will recover without treatment? If a drug company claims that an experimental drug cured eight out of ten people afflicted with the disease, does the drug warrant further experimentation? Explain.

65. Evaluate: $\dfrac{(3!)^2 + (3!)!}{\binom{3!}{2!}}$.

66. Use the binomial theorem to expand and then simplify the result: $(x^2 + x + 1)^3$. (*Hint:* Write $x^2 + x + 1$ as $x^2 + (x + 1)$.)

67. Simplify: $\dfrac{(n + 1)!}{n!}$.

68. If $f(x) = x^5$, find $\dfrac{f(a + h) - f(a)}{h}$ and simplify.

69. Prove that

$$\binom{n}{r} = \binom{n}{n - r}$$

70. In the binomial theorem, $\binom{n}{0}a^n b^0$ gives us the first term, $\binom{n}{1}a^{n-1}b^1$ gives us the second term, $\binom{n}{2}a^{n-2}b^2$ gives us the third term, and so on.

a. Use these patterns to write an expression for the $(r + 1)$st term of $(a + b)^n$.

b. Use your formula from part (a) to find the third term in the expansion of $(2x + y)^6$.

c. Use your formula from part (a) to find the fifth term in the expansion of $(3x - 2y)^7$.

d. Use your formula from part (a) to find the term having y^5 as a factor in the expansion of $(x^2 + y)^9$.

Review Problems

71. Solve the system: $4x^2 + 4y^2 = 65$
$6x - 2y = 5$

72. If $f(x) = 3x + 5$, find $f^{-1}(x)$. Then verify that $f(f^{-1}(x)) = x$ and $f^{-1}(f(x)) = x$.

73. Express as a single logarithm and simplify:
$\log 5x + 2 \log x$.

HAPTER PROJECT

Infinite Diversions

Working with sums containing an infinite number of terms may sometimes lead us to intriguing results. This is especially true when we combine geometric representations with infinite series. The geometry we study need not be complicated. In fact, we can see an interesting problem by investigating one of the simplest of all geometric forms: a line.

1. We will describe a process to create a Cantor set.
  a. Begin with a line one unit long
  b. Remove the middle third of the line
  c. Remove the middle third of each of these
  d. Continue the process infinitely many times, always removing the middle third and making sure the endpoints of the lines stay behind.

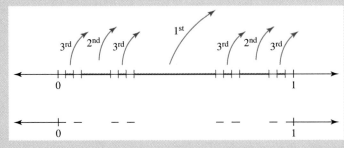

Try this process using graph paper to measure a line 81 grids wide. You will be able to repeat the process four times. If we did wish to continue on paper, notice that after a certain number of repetitions, you will see essentially the same graph, or picture, over and over. The graph appears as a scattering of dots toward the ends. Although we repeat this process an infinite number of times, there will always be something left behind to divide because we have an infinite number of points on a line. The points left behind, when we do our divisions, or erasures, are the Cantor set.

2. We can find a measure for the length of the Cantor set using our knowledge of infinite geometric series. We may not be able to draw the Cantor set completely, but we can add up the lengths of the pieces we are removing.

The first time: We removed a length of $\frac{1}{3}$, leaving behind two lengths measuring $\frac{1}{3}$ each.

The second time: We removed $\frac{1}{3}$ of $\frac{1}{3}$ or $\frac{1}{9}$ from each of our pieces. Thus we removed $\frac{2}{9}$ and leave behind 4 pieces, each measuring $\frac{1}{9}$.

The third time: We removed $\frac{1}{3}$ of $\frac{1}{9}$ or $\frac{1}{27}$ from each piece. We had 4 pieces, so we removed a total of $\frac{4}{27}$.

Continuing the process, we have the sequence of lengths we are removing: $\frac{1}{3}, \frac{2}{9}, \frac{4}{27}, \frac{8}{81}, \ldots$. You may wish to verify this by studying your graph from Problem 1. The sum of these numbers is the length we are removing from our original line of length 1. Thus, we need to find the sum of: $\frac{1}{3} + \frac{2}{9} + \frac{4}{27} + \frac{8}{81} + \ldots$.

Use your knowledge of geometric series to find this sum. If you subtract your solution from 1, you should have the length of the Cantor Set. Remember, in our line above, at each step, we leave behind two-thirds of the line. Be prepared for a surprise.

3. We began with a line containing an infinite number of points and subtracted an infinite number of points from the line. How many points remain behind?

The answer to Problem 3 may be found in a series of works concerning the concept of infinity published between 1874 and 1884 by Georg Cantor. We tend to take for granted the idea that the *whole* is always greater than any of its *parts*. However, this is not true for an infinite collection of things. Cantor realized that proper subsets of infinite sets will certainly be as large as the original infinite set.

4. Consider the set of natural numbers {1, 2, 3, 4, … } and the set of perfect squares of natural numbers {1, 4, 9, 16, … }. Which set contains more numbers? If we compare the set of natural numbers to a set consisting of perfect cubes of the natural numbers does your conclusion change?

Infinite lists which can be *counted* by pairing each thing on the list with a natural number, Cantor called *countable* or *denumerably infinite*. We may visualize this pairing by using an empty auditorium with a particular number of seats, and letting a crowd of people sit down. We know when each person has a seat, even if we do not know how many seats were in the original auditorium. We will also know when the crowd is too large for the auditorium, because some people will remain standing, unmatched with any seat.

5. Cantor used an elegant *trick* to show the number of rational numbers may be counted using the set of natural numbers, thus proving there are as many rational numbers as there are natural numbers. Research and discover how Cantor accomplished his proof and compare your findings with others in the class.

Cantor did much more than describe how to think of infinity. He also showed there was more than one kind of infinity. The next statements, all true, are left for exploration in small groups. Discover how these statements can be true, and then work together, as a class, to come to your own conclusions on the nature of infinity.

6. There are more irrational numbers than rational numbers.
7. There are more points on the real number line between 0 and 1 than there are natural numbers.
8. There are exactly as many points on the real number line from 0 to 1 as there are on the entire real number line.

Worldwide Web Resources

Go to the Prentice Hall website (http://www.prenhall.com/blitzer) to access other locations on the Internet that will allow you to further explore the concepts presented in this project.

Chapter Review

SUMMARY

1. **Sequences**
   a. An infinite sequence is a function whose domain is the set of positive integers.
   b. A finite sequence is a function whose domain includes only the first n positive integers.

2. **Series**
   a. The nth partial sum of the infinite sequence
   $$a_1, a_2, a_3, \ldots, a_n, \ldots$$
   is the sum of the first n terms of the sequence.
   $$S_n = a_1 + a_2 + a_3 + \cdots + a_n$$
   b. An infinite series is the sum of the terms
   $$a_1 + a_2 + a_3 + \cdots + a_n + \cdots.$$
   c. Series are frequently expressed in summation, or sigma, notation using Σ.
   $$\sum_{i=1}^{n} a_i = a_1 + a_2 + a_3 + \cdots + a_n$$
   $$\sum_{i=1}^{\infty} a_i = a_1 + a_2 + a_3 + \cdots + a_n + \cdots$$

3. **Arithmetic Sequences and Series**
   a. An arithmetic sequence, or arithmetic progression, is one in which each term after the first differs from the preceding term by a constant amount. The difference between consecutive terms is called the common difference of the sequence.
   b. *Formula for the nth term of an arithmetic sequence:* The nth term (the general term) of an arithmetic sequence with first term a_1 and common difference d is given by $a_n = a_1 + (n - 1)d$.

   c. *Formula for the sum of the first n terms of an arithmetic sequence:* The sum, S_n, of the first n terms of a finite arithmetic sequence is given by
   $$S_n = \frac{n}{2}(a_1 + a_n).$$
   where a_1 is the first term and a_n is the nth term.

4. **Geometric Sequences and Series**
   a. A geometric sequence, or geometric progression, is one in which each term after the first is obtained by multiplying the preceding term by a nonzero constant. This common multiple is called the common ratio of the sequence, obtained by dividing any term by its predecessor.
   b. *Formula for the nth term of a geometric sequence:* The nth term (the general term) of a geometric sequence with first term a_1 and common ratio r is given by $a_n = a_1 r^{n-1}$.
   c. *Formula for the sum of the first n terms of a finite geometric sequence:* The sum, S_n, of the first n terms of a geometric sequence with first term a_1 and the common ratio r is given by
   $$S_n = \frac{a_1(1 - r^n)}{1 - r}. \quad (r \neq 1)$$
   d. *Formula for the sum of an infinite geometric series:* The sum of an infinite geometric series in which a_1 is the first term and r is the common ratio, where $|r| < 1$, is given by
   $$S_\infty = \frac{a_1}{1 - r}.$$

5. The Binomial Theorem

  a. $n! = n(n - 1)(n - 2) \cdots 3 \cdot 2 \cdot 1; 0! = 1$

  b. $\binom{n}{r} = \dfrac{n!}{r!(n - r)!}$

  c. *Binomial theorem:*

$$(a + b)^n = \binom{n}{0}a^n + \binom{n}{1}a^{n-1}b$$

$$+ \binom{n}{2}a^{n-2}b^2 + \cdots + \binom{n}{n}b^n$$

  d. *Probability*

  The probability of k successes in n trials is $\binom{n}{k}p^k q^{n-k}$, where p is the probability of success on any one trial and $q = 1 - p$ is the probability of failure.

REVIEW PROBLEMS

For Problems 1–3, the general term of a sequence is given. For each problem, write the first four terms, the 10th term (a_{10}), and the 15th term (a_{15}).

1. $a_n = 5n - 2$

2. $a_n = \dfrac{n - 2}{n^2 + 1}$

3. $a_n = (-1)^{n+1}(n^2 - 1)$

For Problems 4–5, write an expression for the general, or nth term, of each sequence. More than one expression may be possible.

4. $-3, -6, -9, -12, -15, \ldots$

5. $\frac{1}{3}, -\frac{1}{4}, \frac{1}{5}, -\frac{1}{6}, \frac{1}{7}, -\frac{1}{8}, \ldots$

In Problems 6–7, find the indicated partial sum for each sequence.

6. $1, 4, 9, 16, \ldots; S_7$

7. $a_n = \dfrac{n - 1}{3n}; S_5$

8. Expand and evaluate: $\displaystyle\sum_{i=2}^{6}(i^3 - 4)$.

In Problems 9–10, write each sum using summation notation.

9. $1 + 4 + 9 + 16 + 25 + 36 + 49$

10. $-\frac{1}{2} + \frac{1}{4} - \frac{1}{8} + \frac{1}{16} - \frac{1}{32} + \frac{1}{64}$

11. The finite sequence whose general term is $a_n = 0.0012n^2 - 0.027n + 1.09$ models the ratio of men to women in the United States, where $n = 1$ corresponds to 1910, $n = 2$ to 1920, $n = 3$ to 1930, and so on.

  a. Find and interpret $a_3, a_4, a_9,$ and a_{10}.

  b. Projections for the United States indicate that the population in the year 2000 will be 275 million. How many men and how many women will there be at that time?

For each arithmetic sequence in Problems 12–15, use the formula for the general term to find the indicated term.

12. $-7, -3, 1, 5, \ldots; a_{15}$

13. $9, 4, -1, -6, \ldots; a_{22}$

14. $\frac{3}{5}, \frac{1}{10}, -\frac{2}{5}, \ldots; a_{22}$

15. $a_1 = -12, d = -\frac{1}{2}; a_{19}$

16. In the arithmetic sequence $\frac{5}{6}, \frac{9}{6}, \frac{13}{6}, \frac{17}{6}, \ldots$, which term is $\frac{41}{6}$?

17. After 20, the human heart becomes less adept at accelerating in response to exertion. The bar graph shows maximum rate in beats per minute. Write the general term for the arithmetic sequence modeling this data, where $n = 1$ corresponds to age 20, $n = 2$ to age 30, $n = 3$ to age 40, and so on.

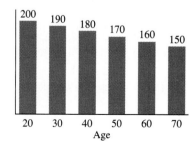

18. According to the National Center for Education Statistics, in 1985 the average salary for public school teachers in the United States was $27,966. Average increases have amounted to $553 yearly.

a. Write the general term for the arithmetic sequence modeling average salary for public school teachers, where $n = 1$ corresponds to 1985.

b. Use the model to predict average salary for the year 2010.

Use the formula for the nth partial sum of an arithmetic sequence to answer Problems 19–26.

19. Find the sum of the first 22 terms of the arithmetic sequence $5, 12, 19, 26, \ldots$

20. Find the sum of the first 16 terms of an arithmetic sequence whose first term is 3 and whose common difference is 5.

21. Find the sum of the arithmetic series $\sum_{i=1}^{16}(3i + 2)$.

22. Find the sum of the arithmetic series $\sum_{i=1}^{25}(-2i + 6)$.

23. Find $3 + 6 + 9 + 12 + \cdots + 300$, the sum of the first 100 positive multiples of 3.

24. *The Toronto Star* contained a photograph of students at a Canadian high school laying on the ground in the form of a human pyramid. There were 16 students in the first row, 15 in the second row, 14 in the third row, with one fewer student in each successive row. Only one student was in the last row. Find the total number of students in the pyramid.

25. A company offers a starting salary of $23,000 with raises of $1200 per year. Find the total salary over a 10-year period.

26. According to the Campbell Soup Company, in 1984 they generated approximately 2421 million dollars in annual sales. This amount has increased steadily by 327 million dollars per year. This situation is modeled by the sequence $a_n = 327n + 2094$, where $n = 1$ corresponds to 1984, $n = 2$ corresponds to 1985, and so on. Find the total sales from 1984 through 1998.

For each geometric sequence in Problems 27–29, use the formula for the general term to find the indicated term.

27. $12, 4, \frac{4}{3}, \frac{4}{9}, \ldots; a_8$

28. $\frac{1}{3}, \frac{1}{2}, \frac{3}{4}, \frac{9}{8}, \ldots; a_{11}$

29. $3, 3\sqrt{3}, 9, 9\sqrt{3}, \ldots; a_{10}$

30. Find the general term for the geometric sequence $3, -12, 48, -192, \ldots$.

31. Find the general term for the geometric sequence whose first five terms are graphed in the figure.

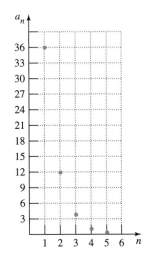

32. Strontium-90 is a by-product of nuclear fission. Its half-life is 28 years, meaning that it takes 28 years for half of it to decay, but half still remains. If the initial amount of Strontium-90 is 100 milligrams, the amount present over time is shown in the table below. Write the general term for the geometric sequence modeling the decay of the Strontium-90, where $n = 1$ corresponds to the amount initially present, $n = 2$ corresponds to the amount present after 28 years, $n = 3$ corresponds to the amount present after 56 years, and so on. Use the model to find the number of milligrams remaining after 196 years.

t, Time in Years	Strontium-90 (in milligrams)
0	100.000
28	50.000
56	25.000
84	12.500

33. A car is worth $10,000 the first year, 80% of its first-year value the second year, 80% of its second-year value the third year, and so on. What is the car's value during the seventh year?

Use the formula for the nth partial sum of a geometric sequence to answer Problems 34–42.

34. Find the sum of the first six terms of the geometric sequence $7, -14, 28, -56, \ldots$.

35. Find the sum of the first seven terms of the geometric sequence $\frac{3}{5}, 1, \frac{5}{3}, \frac{25}{9}, \ldots$.

36. Find S_6 for a geometric series in which $a_1 = 84$ and $r = -\frac{1}{4}$.

37. Find the sum of the geometric series $\sum_{i=1}^{6} 3 \cdot 4^{i-1}$.

38. A job pays a salary of \$30,000 the first year, with a guaranteed raise of 8% each year. The salaries for the first five years are given by the sequence

$$30{,}000, \; 30{,}000(1.08), \; 30{,}000(1.08)^2,$$
$$30{,}000(1.08)^3, \; 30{,}000(1.08)^4$$

What is the salary at the end of year 20? What is the total salary over a 20-year period?

39. A deposit of \$100 is made at the beginning of each month in an account that pays 6% annual interest compounded monthly. What is the balance in this annuity at the end of four years?

$$A = 100\left(1 + \frac{0.06}{12}\right) + 100\left(1 + \frac{0.06}{12}\right)^2 + \cdots$$
$$+ \; 100\left(1 + \frac{0.06}{12}\right)^{48}$$

40. A pendulum moves 12 meters on its first swing. Each swing thereafter is $\frac{3}{4}$ the length of the previous swing. What is the total distance covered after 8 swings?

41. As shown in the figure, we all have 2 biological parents, 4 grandparents, 8 great-grandparents, 16 great-great-grandparents, and so on. If the word *great* is put in front of the word *grandparents* 10 times, how many of this type of relative do you have?

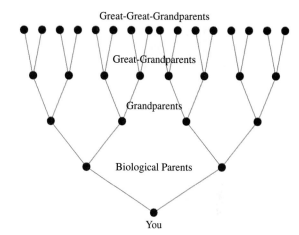

Great-Great-Grandparents

Great-Grandparents

Grandparents

Biological Parents

You

42. After winning the lottery, a person goes to Nevada with \$100,000, deciding to triple the bet with each loss. With how much money will that person return home after eight straight losses and an initial bet of \$10?

In Problems 43–47, use the formula for S_∞, the sum of an infinite geometric series, to find the sum of each series, if there is one.

43. $7 + \frac{7}{2} + \frac{7}{4} + \frac{7}{8} + \cdots$

44. $5 - 1 + \frac{1}{5} - \frac{1}{25} + \cdots$

45. $5 + \frac{10}{3} + \frac{20}{9} + \frac{40}{27} + \cdots$

46. $0.75 + 1.5 + 3 + 6 + \cdots$

47. $6 + 3 + 1.5 + 0.75 + \cdots$

In Problems 48–49, use the formula for S_∞, the sum of an infinite geometric series, to express each repeating decimal as a fraction in lowest terms.

48. $0.\overline{8}$

49. $0.\overline{23}$

Use the formula for the sum of an infinite geometric series to answer Problems 50–54.

50. A sequence of equilateral triangles contains triangles whose sides each measure 6 meters, 2 meters, $\frac{2}{3}$ meter, $\frac{2}{9}$ meter, and so on. If the sequence of triangles continues indefinitely, find the sum of the perimeters of all the triangles in the sequence.

51. The first swing of a pendulum measures 25 centimeters. The lengths of successive swings form the geometric sequence $25, 20, 16, 12.8, \ldots$. If the pendulum continues to swing back and forth indefinitely, find the total distance that it travels.

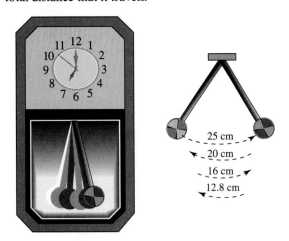

52. Each mirror image in the figure shown is half the area of the previous one. If the areas of all the rectangles excluding the first are combined, how does this total area compare to the area of the first and largest rectangle?

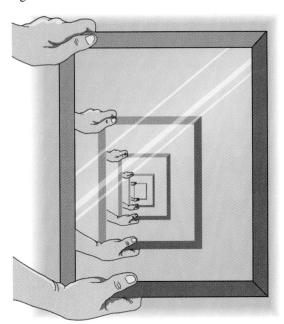

53. A factory in an isolated town has an annual payroll of $4 million. It is estimated that 70% of this money is spent within the town, that people in the town receiving this money will again spend 70% of what they receive in the town, and so on. What is the total annual economic impact of the factory?

54. As shown in the figure, the radius of the largest circle (A_1B) is 1 meter. Each circle then has a radius that is half the radius of the larger circle, so that $A_2B = \frac{1}{2}$ meter, $A_3B = \frac{1}{4}$ meter, and so on. If the sequence of circles continues indefinitely, find the sum of the areas of all the circles in the sequence.

In Problems 55–56, evaluate each expression.

55. $6!$

56. $\binom{9}{2}$

In Problems 57–60, use the binomial theorem to expand each binomial and express the result in simplified form.

57. $(3x + y)^4$

58. $(x - 2y)^5$

59. $(x^2 + 2y^3)^6$

60. $(2x - y^4)^7$

The expression

$$\binom{n}{k}p^k q^{n-k}$$

describes the probability of k successes in n trials, where p is the probability of success on any one trial, and q = 1 − p is the probability of failure. Use this expression to solve Problems 61–62.

61. The probability that a child has had German measles by age 12 is 0.6. If six 12-year-olds are selected at random, find the probability that exactly half of them have had German measles.

62. The batting average of a baseball player is the number of hits divided by the number of times the player was at bat. Babe Ruth had 2873 hits and was at bat 8399 times, giving him a batting average of $\frac{2873}{8399}$, or approximately 0.342. Suppose Ruth came to bat five times in a game. Find the probability of his getting exactly three hits.

CHAPTER 10 TEST

1. Find the first 5 terms and the 10th term of a sequence with general term

$$a_n = \frac{n+1}{n^2}.$$

2. Write an expression for the general, or nth term, of the sequence

$$\frac{1}{4}, \frac{4}{9}, \frac{9}{16}, \frac{16}{25}, \dots .$$

3. Expand and evaluate:

$$\sum_{i=1}^{5}(2^i - 2).$$

4. Rewrite using summation notation:

$$\frac{5}{4(1)+1} + \frac{5}{4(2)+1} + \frac{5}{4(3)+1} + \dots + \frac{5}{4(13)+1}.$$

5. Find a formula for the nth term of the arithmetic sequence

$$11, 7, 3, -1, \dots .$$

Then use the formula to find a_{19}, the 19th term.

Use the formula for the sum of the first n terms of an arithmetic sequence to answer Problems 6–8.

6. Find the sum of the first 18 terms of the sequence

$$-16, -11, -6, -1, \dots .$$

7. Find the sum of all multiples of 15 from 15 to 150 inclusively.

8. Find the sum:

$$\sum_{i=1}^{50}(3i - 26).$$

9. Explain why the graph in the figure cannot be the graph of an arithmetic sequence.

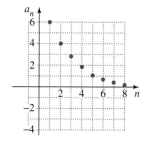

10. Find a formula for the nth term of the geometric sequence

$$2, \frac{2}{3}, \frac{2}{9}, \frac{2}{27}, \dots .$$

Then use the formula to find a_9, the 9th term.

Use the formula for the sum of the first n terms of a geometric sequence to answer Problems 11–12.

11. Find the sum of the first 15 terms of the sequence

$$5, -15, 45, -135, \ldots.$$

12. Find the sum:

$$\sum_{i=1}^{10} 12\left(\frac{3}{2}\right)^i.$$

Use the formula for the sum of an infinite geometric series to answer Problems 13–14.

13. Find the sum:

$$\frac{1}{4} + \frac{1}{12} + \frac{1}{36} + \frac{1}{108} + \cdots.$$

14. Express $0.\overline{73}$ as a fraction in lowest terms.

15. Evaluate:

$$\binom{17}{14}.$$

16. Expand:

$$(x^3 - 2y)^5.$$

17. The model $a_n = 7.7n + 55$ describes the annual advertising expenditures in billions of dollars by U.S. companies beginning with the year 1980, where $n = 0$ corresponds to 1980. Find $\sum_{n=0}^{20} a_n$ and describe what this number represents in practical terms.

18. A theater has 30 seats in the first row, 38 seats in the second row, 46 seats in the third row, and so on in an arithmetic sequence. Find the total number of seats if there are 25 rows.

19. A job pays a salary of $22,000 the first year, with a guaranteed raise of 6% each year. The salaries for the first five years are given by the sequence

$$22{,}000, \ 22{,}000(1.06), \ 22{,}000(1.06)^2, \ 22{,}000(1.06)^3,$$
$$22{,}000(1.06)^4.$$

What is the total salary over a 10-year period?

20. A factory in an isolated town has an annual payroll of $10 million. It is estimated that half of this money will be spent within the town, that people in the town receiving this money will again spend half of what they receive in the town, and so on. What is the total annual economic impact of the factory?

(A collection of review problems covering the entire book can be found in the appendix.)

Appendix

Review Problems Covering the Entire Book

If your course included the use of graphing utilities, use your grapher to verify as many of your answers as possible.

Solving Equations and Inequalities

Systematic procedures for solving certain equations and inequalities are an important component of algebra. Problems 1–32 give you the opportunity to review these procedures. Solve each problem, expressing irrational solutions in simplified form and imaginary solutions in the form a + bi.

1. $2[3 - 2(x + 4)] = 3(4 - x)$

2. $\dfrac{4y - 2}{3} - \dfrac{y + 2}{4} = \dfrac{7y - 2}{12}$

3. $-4(x - 2) \leqslant 6x + 4$

4. $3x + 7 > 4$ or $6 - x < 1$

5. $4x - 3 < 13$ and $-3x - 4 \geqslant 8$

6. $-7 < \dfrac{4 - 2x}{3} \leqslant \dfrac{1}{3}$

7. $\left| \dfrac{2x - 3}{5} \right| = 1$

8. $|6x| = |3x - 9|$

9. $|3 - 2x| < 7$

10. $|2x - 1| \geqslant 7$

11. $\begin{aligned} 3x + 4y &= 2 \\ 2x + 5y &= -1 \end{aligned}$

12. $\begin{aligned} 5x - 2y &= -7 \\ y &= 3x + 5 \end{aligned}$

13. $\begin{aligned} x - 2y + z &= -4 \\ 2x + 4y - 3z &= -1 \\ -3x - 6y + 7z &= 4 \end{aligned}$

14. Write the augmented matrix for the following system and solve using row-equivalent matrices:

$$\begin{aligned} x - 2y + z &= 16 \\ 2x - y - z &= 14 \\ 3x + 5y - 4z &= -10 \end{aligned}$$

15. Solve for x only, using determinants (Cramer's rule):

$$\begin{aligned} 2x - z &= 1 \\ 3y + 2z &= 0 \\ x - y &= -3 \end{aligned}$$

16. $x(2x - 7) = 4$

17. $x^3 + 3x^2 - 9x - 27 = 0$

18. $\dfrac{2}{3x + 1} + \dfrac{4}{3x - 1} = \dfrac{6x + 8}{9x^2 - 1}$

19. $\dfrac{5}{x - 3} = 1 + \dfrac{30}{x^2 - 9}$

20. $\sqrt{2x + 4} - \sqrt{x + 3} - 1 = 0$

21. $2x^2 = 5 - 4x$

22. $\dfrac{1}{y + 2} - \dfrac{1}{3} = \dfrac{1}{y}$

23. $x^{2/3} - x^{1/3} - 6 = 0$

24. $(x^2 + x)^2 - 5(x^2 + x) = -6$

25. $3x^2 + 8x + 5 < 0$

26. $\dfrac{x - 1}{x + 3} \leqslant 0$

27. $\log_5(x - 2) = 3$

28. $\log_2 x + \log_2(2x - 3) = 1$

29. $3^{2x-1} = 81$

30. Solve by taking the natural logarithm on both sides. Express the answer in terms of a natural logarithm. Then use a calculator to obtain a decimal approximation, correct to the nearest thousandth, for the solution.

$$30e^{0.7x} = 240$$

31. $3x^2 + 4y^2 = 39$
$5x^2 - 2y^2 = -13$

32. $2x^2 - y^2 = -8$
$x - y = 6$

Graphs and Graphing

Throughout the book, we have used graphs to help visualize a problem's solution. We also studied graphing equations and inequalities in the rectangular (Cartesian) coordinate system. Problems 33–56 focus on problem solving with graphs.

33. The annual homicide rates per 100,000 workers is shown for five selected occupations. The average for all workers is 0.7 per 100,000. If x represents the homicide rate per 100,000 workers, write the occupation or occupations described by each of the following.
  a. $x \in [3.6, 5.1)$
  b. $x > 9.3$
  c. $x \leqslant 4.5$

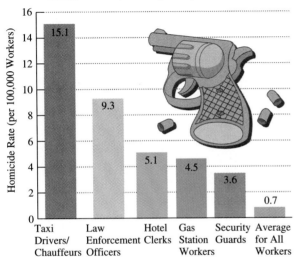

Annual Homicide Rates per 100,000 Workers

Source: National Institute for Occupational Safety and Health

34. The graph at the top of the next column shows the sales, in billions of dollars, of exercise equipment in the United States from 1984 through 1993.
  a. For the period shown in the graph, when did the sale of exercise equipment reach a maximum? What is a reasonable estimate of the sales for that year?

b. Describe the trend in the sale of exercise equipment in the United States from 1984 through 1993.

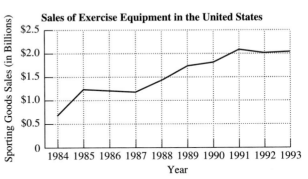

Source: Based on National Sporting Goods statistics

35. The box to the left of the graph shown below is constructed to adhere to the postal regulation that the perimeter of the square base plus the height cannot exceed 108 inches.
  a. Write a polynomial in descending powers of x that models the volume of the box.
  b. The graph shows the volume of the box as a function of the length of the side of its square base. Use the graph to determine the length of the side of the square that will maximize the volume of the box. What is a reasonable estimate of this maximum volume?
  c. Substitute the length of the side of the square that results in maximum volume into your formula from part (a). What is the exact maximum volume of the box?

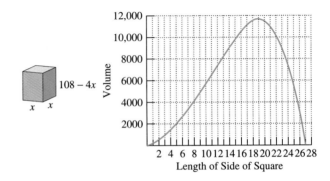

36. The graphs in the left column on the next page show verbal SAT scores for males and females from 1967 through 1993.

a. Estimate the coordinates for the intersection point of the graphs and interpret your answer in practical terms.
b. Describe the trends shown by the graphs.

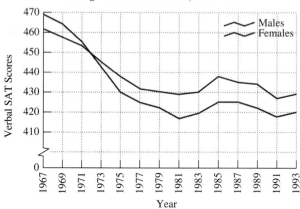

Average Verbal SAT Scores, 1967–1993

Source: SAT

37. Graph $f(x) = |x|$, $g(x) = |x| - 1$, and $h(x) = |x| + 2$ in the same rectangular coordinate system by first completing the accompanying table. Describe the relationship among the graphs using the phrase "vertical shift."

x	$f(x) = |x|$	$g(x) = |x| - 1$	$h(x) = |x| + 2$
-3			
-2			
-1			
0			
1			
2			
3			

38. The graph at the top of the next column shows the average sentence imposed, in months, by U.S. district courts for both violent offenses and drug offenses.

Estimate the slope of the line segments for the violent offenses graph between 1980 and 1985, 1985 and 1990, and 1990 and 1995. Compare the three slopes and explain what this means in terms of average rate of change.

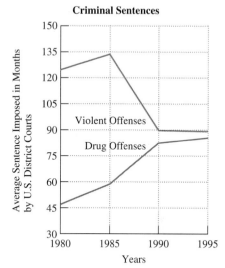

Criminal Sentences

Source: U.S. Department of Justice

39. The graphs below show the percent of U.S. men and women in the labor force. If y represents the percentage of women in the labor force, write an interval describing the years for which:
a. $-2y + 5 \leq -75$
b. $85 \leq 3y - 5 \leq 115$

Percentage of U.S. Men and Women in the Labor Force

Year (*= Projected)
Source: U.S. Bureau of the Census, *Statistical Abstract*, 1995.

40. Which one of the following represents the graph of a system of three equations in three variables that has a solution? Explain your answer.

a. **b.** **c.** **d.**

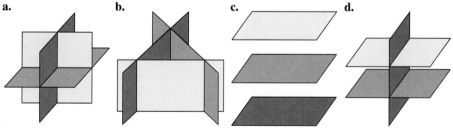

41. The elliptical orbit of a comet that orbits the sun is described by the equation $\frac{x^2}{9.61} + \frac{y^2}{4.84} = 1$. If NASA were to launch a satellite whose orbit would be $\frac{x^2}{7.84} + \frac{y^2}{5.16} = 1$, would the satellite be in danger of colliding with the comet? Use the graphs in the figure shown to explain your answer.

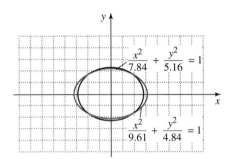

Graph Problems 42–46.

42. $y = -\frac{2}{3}x + 4$

43. $3x - y > 6$

44. $y \geq \frac{3}{4}x - 5$

(In Problems 45–46, find the coordinates of any vertices that are formed.)

45. $2x - y \geq 4$
  $x \leq 2$

46. $x \leq 5$
  $y \leq 6$
  $14x + 8y \geq 56$
  $2x + 3y \geq 12$

47. Solve by graphing:

  $y = -2x - 1$

  $x + 2y = 4$

48. Use x-intercept, y-intercept, the vertex, and, if necessary, one or two additional points to graph the quadratic function $f(x) = x^2 - 4x - 5$.

49. Graph $f(x) = \sqrt{x}$ and $g(x) = \sqrt{x - 1}$ in the same rectangular coordinate system. State each function's domain and range. Then describe the relationship of the graph of g to the graph of f.

50. Graph the parabola whose equation is $y = (x - 1)^2 - 4$.

51. Graph: $y = 2^{x/2} + 1$.

52. The graph on the right shows the costs of Social Security from 1970 through 2000 (projected).
  a. Explain why f has an inverse that is a function.
  b. Describe in practical terms the meaning of $f^{-1}(300)$.

c. Is this data best modeled by a linear, quadratic, or exponential function? Explain.

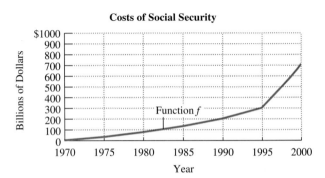

Source: U.S. Bureau of the Census, *Statistical Abstracts*.

53. Graph: $y = \log_2 x$.

54. Find the center and radius of the circle whose equation is $x^2 + y^2 + 4x - 6y + 9 = 0$. Then graph the circle.

55. Graph the ellipse: $\frac{x^2}{9} + \frac{y^2}{4} = 1$.

56. Use asymptotes and intercepts to graph the hyperbola whose equation is

$$\frac{x^2}{4} - \frac{y^2}{9} = 1.$$

Mathematical Models

Describing the world compactly and symbolically using formulas is one of the most important aspects of algebra. Problems 57–80 concentrate on mathematical models.

57. The formula

$$t = \sqrt{\frac{2h}{g}}$$

describes the time (t, in seconds) it takes for an object dropped from a height of h feet to reach the ground, where g represents the acceleration due to gravity. All objects in free fall near the earth's surface have an acceleration due to gravity of 32 feet per second every second. If a ball is dropped from a window 64 feet high, how long will it take for it to reach the ground?

58. The underground temperature of rocks varies with their depth below the surface. The deeper the rocks are, the hotter they are. The temperature (t, in degrees Celsius) is approximated by the model $t = 35d + 20$, where d is the depth in kilometers.
   a. At what depth is the temperature of the rocks 125° Celsius?
   b. Identify your solution to part (a) on the accompanying graph.

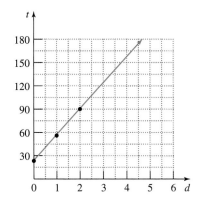

59. If a car hits a tree, the passengers in the car keep moving forward until the impact occurs. After impact, passengers are repelled. Seatbelts and airbags limit how far passengers are jolted in this manner. The formula for the velocity that a passenger is thrown back is

$$V_f = \frac{m_1 - m_2}{m_1 + m_2} \cdot v_i$$

where m_1 and m_2 are the masses of the two objects meeting and v_i is the initial velocity. Solve the model for m_1.

60. The number of motor vehicle registrations (in millions) in the United States x years after 1960 is modeled by the function $f(x) = 0.002x^2 + 4.05x + 71.78$. Find and interpret $f(10)$.

61. Suppose that the following pairs of data are modeled by a linear function. In each case, is the slope of the linear function positive or negative? Explain your answer.
   a. Interest rates (x) and the number of new home construction loans (y)
   b. Calories consumed (x) and weight (y)

62. Mailings in the United States increased by more than 40% from 1983 to 1993.

x (Number of Years after 1983)	y (Number of Pieces of Mail, in Billions)
0	119.4
10	171.1

   a. Write the point-slope form of the line on which these measurements fall.
   b. Use the point-slope form of the equation to write the slope-intercept form of the equation.
   c. Use the slope-intercept model from part (b) to extrapolate from the data and predict the pieces of mail, in billions, for the year 2000.

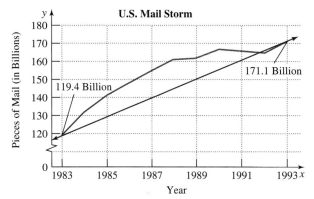

Source: U.S. Postal Service

63. Rent-a-Heap advertises their daily car rental rate at $30 plus $0.25 a mile.
   a. Write a formula that models the daily rental rate (r) if d miles are driven.
   b. The graph of this model is shown in the figure at the top of the next page. Use the graph to find the cost of driving 100 miles, and then verify this amount using your model from part (a).
   c. A competing company, Lease-a-Lemon, charges a daily rate of $70 for driving 100 miles. How does this compare with the offer by Rent-a-Heap?

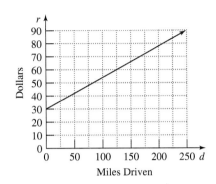

Miles Driven

64. A company manufactures tape decks and amplifiers. The profit per tape deck is $8 and the profit per amplifier is $12.

a. Let x = the number of tape decks manufactured in a day and let y = the number of amplifiers manufactured in a day. Write the objective function that models the total daily profit.

b. The manufacturer is bound by two constraints:

1. They cannot produce more than 200 units of tape decks and amplifiers combined in a day.

2. Each day they must produce at least 10 tape decks and 80 amplifiers.

Write a series of inequalities that models these constraints.

c. Graph the system of inequalities in part (b). Use only the first quadrant, since x and y must both be positive.

d. Evaluate the objective function at each of the three vertices of the graphed region. Use these evaluations to complete the following statements:

The manufacturer will make the greatest profit by manufacturing ____ tape decks and ____ amplifiers each day. The maximum daily profit is $____.

65. In the figure shown, *ACED* is a rectangle and *ABFG* is a square. The dimensions of the rectangle are given in terms of x. Point B is the midpoint of line segment AC. Write a polynomial function, $A(x)$ in descending powers of x, that models the area of the region outside the square and inside the rectangle.

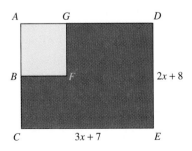

66. A 12 inch by 18 inch rectangular picture is to have a frame of uniform width surrounding it. If the width of the frame is represented as x inches, write a polynomial function, $A(x)$ in descending powers of x, that models the area of the frame.

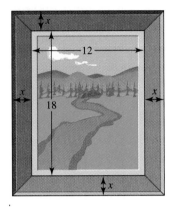

67. A ball is thrown vertically upward from the top of a 96-foot tall building with an initial velocity of 80 feet per second. The height of the ball above ground is modeled by the position function

$$s(t) = -16t^2 + 80t + 96.$$

a. After how many seconds will the ball strike the ground?

b. When does the ball reach its maximum height? What is the maximum height?

c. Use your answers from parts (a) and (b) to graph the position function, with meaningful values of t along the horizontal axis and values of $s(t)$ along the vertical axis.

68. In situations involving environmental pollution, cost-benefit models express the cost of cleanup as a function of the percentage of pollutants removed. For example, the rational function

$$C(x) = \frac{250x}{100 - x}$$

models the cost ($C(x)$, in millions of dollars) to remove x percent of the pollutants that are discharged into a river.

a. Find $C(90) - C(40)$ and describe what this means in this situation.

b. What value of x must be excluded from the domain of C? What does this mean in terms of removing all of the river's pollutants?

69. The function

$$\overline{C}(x) = \frac{40x + 50,000}{x}$$

models the average cost per pair of running shoes ($\overline{C}(x)$) for a business to produce x pairs of shoes. How many pairs must be produced to bring the average manufacturing cost for each pair of running shoes down to $45? If the business manufactures ten times this amount, what is the average manufacturing cost per pair? What pattern do you observe?

Solve each formula in Problems 70–72 for the given variable.

70. $S = \dfrac{p}{1 - nd}$, for d

71. $\dfrac{1}{p} + \dfrac{1}{q} = \dfrac{1}{f}$, for q

72. $r = \sqrt[3]{\dfrac{3w}{4\pi d}}$, for d

73. The model $N = 1220\sqrt[3]{t - 42} + 4900$ describes the number of congressional aides in the House of Representatives t years after 1930. In what year were approximately 7340 aides assigned to the House of Representatives?

74. The function $f(x) = 2x^2 + 22x + 320$ models the number of inmates ($f(x)$, in thousands) in federal and state prisons x years after 1980. In what year will the number of inmates reach 2120 thousand? Describe one change that might take place between now and then that would make this prediction inaccurate.

75. The regular price of a graphing utility is x dollars. Let
$$f(x) = x - 25$$
$$\text{and} \quad g(x) = x - 0.3x.$$

a. Describe what functions f and g model in terms of a discount on the utility's regular price.
b. Find $(f \circ g)(x)$ and describe what this models in terms of a discount on the utility's regular price.
c. Repeat part (b) for $(g \circ f)(x)$.
d. Which composite function models the greater discount on the utility, $f \circ g$ or $g \circ f$? Explain.

The mathematical model for exponential growth is given by $y = Ae^{kt}$, where A is the amount or size at t = 0, y (or f(t)) is the amount or size at time t, and k is a growth constant. Use this model to answer Problems 76–77.

76. In 1995, the world population was approximately 5.702 billion, with a relative growth rate of 1.6% per year ($k = 0.016$), down from the high of 2% per year in 1969.
a. What will be the population of the world in the year 2000?
b. In what year will Earth's carrying capacity of 10 billion be reached?

Population Growth

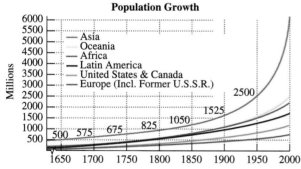

Source: U.S. Bureau of the Census

77. In 1990, the population of the United States was 250 million. By the end of 1995, the population had grown to 263 million.
a. Find the exponential growth function that models the data from 1990 to 1995.

b. In what year will the U.S. population reach 317 million?

National Censuses
(Resident U.S. Population Reported by Census)

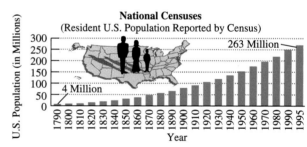

Source: U.S. Bureau of the Census

78. In an experiment on memory, subjects were shown a picture of a store shelf with six items. The picture was then removed and the subjects were asked to recall each of the items on the shelf every t minutes thereafter. The function $f(t) = 80 - 27 \ln t$ models the percent ($f(t)$) of subjects who remembered all of the items after t minutes, for $t \geq 1$. After how many minutes (to the nearest minute) do approximately 15% of the subjects remember all the items?

79. A person accepts a position with a company and will receive a salary of $28,600 for the first year and a guaranteed raise of $2250 per year.
a. Write the general term for the arithmetic sequence modeling yearly salary, where $n = 1$ corresponds to the first year.

b. What will be the salary during the eighth year of employment?

c. Find the total salary that the company pays this person by the end of the eighth year.

80. A job pays a salary of $20,000 the first year. The salary increases by 6% each year, with yearly salaries shown by the following geometric sequence:

$$20{,}000, 20{,}000(1.06), 20{,}000(1.06)^2, 20{,}000(1.06)^3, \ldots$$

$\quad\uparrow\qquad\quad\uparrow\qquad\qquad\uparrow\qquad\qquad\uparrow$

Year 1   Year 2   Year 3   Year 4

a. Write the general term for the geometric sequence modeling yearly salary, where $n = 1$ corresponds to the first year.

b. What will be the salary during the eighth year of employment?

c. Find the total salary that the company pays this person by the end of the eighth year.

Factoring

Factoring is a skill needed when working with rational expressions and solving certain polynomial equations. Factor the polynomials in Problems 81–84.

81. $12x^3 - 36x^2 + 27x$

82. $27x^3 - 125$

83. $x^3 - 2x^2 - 9x + 18$

84. $15x^4 - 35x^2 - 100$

85. Factor out the GCF: $10x^{1/2} - 30x^{-1/2}$

Algebra's Simplifications

The word "simplify" in algebra has a variety of meanings ranging from performing indicated operations, removing grouping symbols and combining like terms, rewriting exponential expressions with positive exponents, reducing rational expressions to lowest terms, and rationalizing denominators. Simplify in Problems 86–110.

86. $-2(3^2 - 12)^3 - 45 \div 9 - 3$

87. $3x - [6 - 7(3 - 4x) - 5x]$

88. $(-4x^2y^{-5})(-3x^{-1}y^{-4})$

89. $\dfrac{30x^2y^4}{-5x^{-3}y^6}$

90. $(7.2 \times 10^{-3})(5.0 \times 10^{-5})$(Express the product in scientific notation.)

91. $(8x^2 - 9xy - 11y^2) - (7x^2 - 4xy + 5y^2)$

92. $(2x - 7)(3x + 11)$

93. $(5x^2 - 7y)^2$

94. $\dfrac{x + 2}{3x + 9} \cdot \dfrac{x^2 - 9}{x^2 - x - 12}$

95. $\dfrac{3x^2 + 17xy + 10y^2}{6x^2 + 13xy - 5y^2} \div \dfrac{6x^2 + xy - 2y^2}{6x^2 - 5xy + y^2}$

96. $\dfrac{2y - 6}{3y^2 - 14y - 5} - \dfrac{y - 3}{y^2 - 5y}$

97. $\dfrac{1 - \dfrac{14y - 45}{y^2}}{\dfrac{y}{9} - \dfrac{9}{y}}$

98. $(3x^3 - 19x^2 + 17x + 4) \div (3x - 4)$

99. Use synthetic division: $(3x^3 - 5x^2 + 2x - 1) \div (x - 2)$

100. $\sqrt[8]{16x^4y^2}$

101. $\sqrt{3} \cdot \sqrt[3]{3}$

102. $(16x^{1/2}y^{4/3})^{3/2}$

103. $\sqrt[3]{16x^7y^{11}}$

104. $\sqrt{5xy} \cdot \sqrt{10x^2y}$

105. $\sqrt[3]{4x^2y^5} \cdot \sqrt[3]{4xy^2z^2}$

106. $\dfrac{\sqrt[3]{54}}{\sqrt[3]{2y^2}}$

107. $7\sqrt{18x^5} - 3x\sqrt{2x^3}$

108. $\dfrac{1 - \sqrt{y}}{1 + \sqrt{y}}$

109. $(7 + 3i)(9 - 4i)$

110. $\dfrac{6}{3 + 5i}$

A Potpourri of Skills

Problems 111–131 give you the opportunity to review a number of course objectives presented throughout the book.

111. Evaluate $\dfrac{xy - (y - z)^2}{xz}$ when $x = 3$, $y = -2$, and $z = -4$.

In Problems 112–113, write the point-slope equation and then the slope-intercept equation of each line.

112. Passing through $(1, -4)$ and $(-5, 8)$

113. Passing through $(3, -2)$ and perpendicular to the line whose equation is $-\frac{1}{4}x + y = 5$

In Problems 114–119, if $f(x) = x^2 - 4$ and $g(x) = x + 2$, find each function.

114. $(f + g)(x)$ **115.** $(f - g)(x)$ **116.** $(fg)(x)$ **117.** $\left(\dfrac{f}{g}\right)(x)$

118. $(f \circ g)(x)$ **119.** $(g \circ f)(x)$

120. If $f(x) = 3x^2 - 7x - 2$, find $\dfrac{f(a + h) - f(a)}{h}$.

121. Find the distance (in simplified radical form) from $(4, 3)$ to $(2, -1)$.

122. Suppose that $f(x) = 4x - 3$.
  a. Find a formula for $f^{-1}(x)$, the inverse function.
  b. Verify that your formula is correct by showing that $f(f^{-1}(x)) = x$ and $f^{-1}(f(x)) = x$.

123. Write in exponential form: $\log_b x = 4$.

124. Expand using logarithmic properties. Where possible, evaluate logarithmic expressions.

$$\log_5 \frac{x^3 \sqrt{y}}{125}$$

125. Write as a single logarithm whose coefficient is 1:

$$2 \log_b x + 3 \log_b y - \tfrac{1}{2} \log_b z.$$

126. Expand and evaluate: $\displaystyle\sum_{i=2}^{5} (i^3 - 4)$.

127. Find the sum of the first 30 terms of the arithmetic sequence $2, 6, 10, 14, \ldots$.

128. Find the sum of the first eight terms of the geometric sequence $\tfrac{1}{2}, 2, 8, 32, \ldots$.

129. Find the sum of the infinite geometric series $1 + \tfrac{1}{2} + \tfrac{1}{4} + \tfrac{1}{8} + \cdots$.

130. Express $0.\overline{34}$ as a fraction in lowest terms.

131. Expand and simplify: $(2x - y^3)^5$.

Problem Solving by Writing an Equation

Problem solving is the central theme of algebra. Some problems can be solved by translating given conditions or modeling implied conditions into linear equations, linear inequalities, quadratic equations, or systems of equations. Use this technique and the five-step strategy for solving problems discussed throughout the book to solve Problems 132–153.

132. The price of a computer is reduced by 30% to $784. What was the original price?

133. Rent-a-Truck charges a daily rental rate for a truck of $39 plus $0.16 a mile. A competing agency, Ace Truck Rentals, charges $25 a day plus $0.24 a mile for the same truck. How many miles must be driven in a day to make the daily cost of both agencies the same? What will be the cost?

134. A long-distance telephone service charges $19.95 a month with 50 minutes of free service. A charge of $0.25 is added for each long-distance minute over 50. If a customer does not want a bill that exceeds $82.45 a month, how many minutes can be spent talking long distance by phone?

135. The area of the shaded region shown in the figure is 17 square meters. Find the dimensions of the larger rectangle.

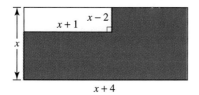

136. The perimeter of the rectangular floor plan shown in the figure on the right is 144 feet. The room on the lower left has a length that is 6 feet longer than its width. What are the plan's dimensions and what is the area of the floor for all three rooms combined?

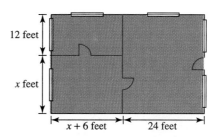

137. Two cars leave from the same place at the same time, traveling in opposite directions. The rate of the faster car exceeds that of the slower car by 10 miles per hour. After 3 hours, the cars are 342 miles apart. Find the speed of each car.

138. The longest-lived U.S. presidents are John Adams (age 90), Herbert Hoover (also 90), and Harry Truman (88). Behind them are James Madison, Thomas Jefferson, and Richard Nixon. The latter three men lived a total of 249 years, and their ages at the time of death form consecutive odd integers. For how long did Nixon, Jefferson, and Madison live?

139. Three apples and two bananas provide 354 calories, and two apples and three bananas provide 381 calories. Find the number of calories in one apple and one banana.

140. A rectangular lot whose perimeter is 320 feet is fenced along three sides. An expensive fencing along the lot's length costs $9 per foot, and an inexpensive fencing along the two side widths cost only $5 per foot. The total cost of the fencing along all three sides comes to $1510. What are the lot's dimensions?

141. Find the measures of the angles marked x and y in the triangle shown.

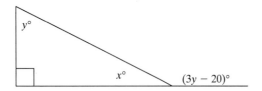

142. One solution is 15% acid and a second solution is 40% acid. How many liters of each solution should be used to create a 6-liter mixture that is 25% acid?

143. A motor boat traveling with the current can cover 84 miles in 2 hours. When the boat travels in the opposite direction against the current, it takes 3 hours to travel the 84 miles. Find the speed of the boat in calm water and the speed of the current.

144. The bar graph indicates the number of recruits in each branch of the armed forces. There are 35,600 more recruits in the Army than in the Marines. The number of recruits in the Army is also 55,600 less than twice that of the Navy. Find the number of recruits in the Army, Navy, and Marines.

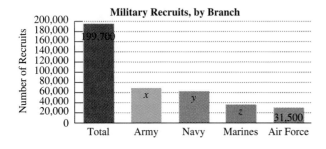

145. In a triangle, the largest angle is 29° greater than the smallest angle. The largest angle is also 58° less than twice the remaining angle. Find the measure of each angle.

146. A rectangular garden measuring 8 meters by 10 meters is to be made smaller so that a path of uniform width can be placed along one length and one width, as shown in the figure at the top of the next column.

If 35 square meters is still desired for gardening area, what should the path's width be?

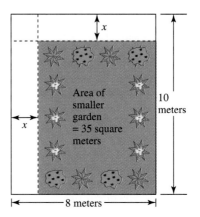

147. The base of a triangular roof truss is 1 foot more than twice the height. The area of the triangle is 39 square feet. Find the height and the base.

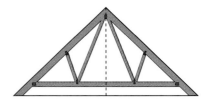

148. A piece of wire measuring 17 feet is attached to a telephone pole as a guy wire. The distance along the ground from the bottom of the pole to the end of the wire is 7 feet less than the height where the wire is attached to the pole. How far up the pole does the wire reach?

149. When a car's brakes are suddenly applied and it goes into a skid, the length of the skid marks varies directly as the square of its speed. A car traveling along a road at 30 miles per hour leaves skid marks that are 40 feet long. Is it possible that a car that left skid marks 250 feet long on the same road under the same conditions could have been traveling at the legal limit of 55 miles per hour before the brakes were suddenly applied?

150. The box shown in the figure in the left column at the top of the next page has a square base, a height of 5 inches, and no top. The box is to be constructed with 341 square inches of material, meaning that the area of its square base and four rectangular sides combined must be 341. What should be the dimensions of the box?

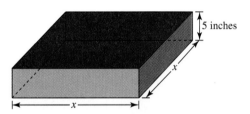

151. A box with a rectangular base 1 inch longer than its width is to be made from a piece of cardboard by cutting 2-inch squares from each corner and folding up the sides, as shown in the figure. If the box is to hold 40 cubic inches, find the dimensions of the piece of cardboard that is needed.

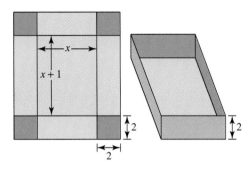

152. The two corrals shown in the figure have the same size and shape. Their combined area is 8400 square feet and they can be enclosed as shown with 450 feet of fencing. What are the dimensions of each corral?

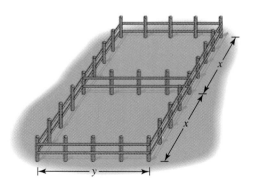

153. The perimeter of the cover of a rectangular book is 14 inches. If the cover's diagonal measures 5 inches, what are its dimensions?

Critical Thinking

Problems 154–160 cannot be solved by translating conditions into equations. These problems require strategies such as looking for a pattern, eliminating possibilities, making a systematic list, working backward, using a drawing, or guessing at an answer and then checking the guess against the problem's conditions. Use one or more of these strategies to solve each problem.

154. What number is less than 100, odd, a multiple of 5, divisible by 3, and has a sum of digits that is odd?

155. Consider the first four triangular numbers (1, 3, 6, and 10), shown in the figure.

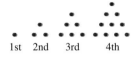

1st   2nd   3rd   4th

Use patterns to complete the following table.

1st Triangular Number	2nd	3rd	4th	5th	6th	12th	nth
1	3	6	10				

156. Suppose w, x, y, and z each represent a different, nonzero, one-digit positive number satisfying

$$\frac{10x + y}{w} = z.$$

a. Find z and x if $w = 3$ and $y = 1$.
b. Find two possible values for z if $x = y - 3$.
c. Find x, y, z, and w if $w = 2z$ and $x > y$.
d. Find x and y if $w + z = x + y$.

157. Insert parentheses to make the equation true: $2 \cdot 1 + 2 \cdot 3 - 2 \div 2 - 1 = 12$.

158. If three darts are thrown in one round of play and each dart hits a target, in how many ways can a player score 90 points? (Assume that no dart lands directly on a circle that divides the regions.)

160. The figure shows the number of cubes necessary to construct the first three solids in a sequence. How many cubes are needed to build the tenth solid in the sequence?

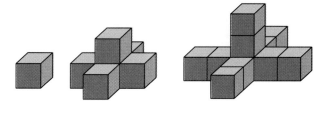

159. The figure shows the areas of three figures of the same type. What is the area of the same type of figure with a radius of 9 and a height of 10?

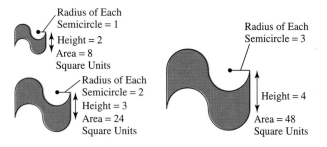

Answers to Selected Exercises

Chapter 1

PROBLEM SET 1.1

1. $\{14, 16, 18, 20\}$ **3.** $\{5, 10, 15, 20, ...\}$ **5.** $\{0\}$ **7.** $\{-1, 1\}$ **9.** $\{\sqrt{5}\}$ **11.** $\{..., -4, -3, -2, -1, 0\}$

13. $\{1, 4\}$ **15.** $\left\{\frac{2}{3}\right\}$ **17.** $\left\{\sqrt{4}, 7, \frac{18}{2}, 100\right\}$ **19.** $\left\{-10, 0, \sqrt{4}, 7, \frac{18}{2}, 100\right\}$ **21.** $\left\{-\sqrt{2}, \sqrt{3}, \pi\right\}$ **23.** –13 is less than or equal to –2; true

25. –6 is greater than 2; false   **27.** 4 is greater than or equal to –7; true   **29.** –13 is less than –5; true   **31.** $-\pi$ is greater than or equal to $-\pi$; true
33. $-\sqrt{2}$ is less than $-\sqrt{2}$; false   **35.** $(0, 3)$

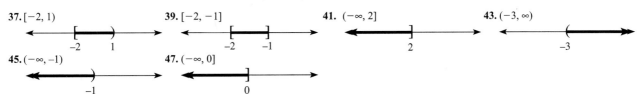

37. $[-2, 1)$ **39.** $[-2, -1]$ **41.** $(-\infty, 2]$ **43.** $(-3, \infty)$

45. $(-\infty, -1)$ **47.** $(-\infty, 0]$

49. $\{x\,|\,5 < x < 12\} = (5, 12)$ **51.** $\{x\,|\,2 < x \leq 13\} = (2, 13]$ **53.** $\{x\,|\,x \leq 6\} = (-\infty, 6]$ **55.** $\{x\,|\,2 \leq x \leq 5\} = [2, 5]$
57. $\{x\,|\,x \leq 60\} = (-\infty, 60]$ **59.** $\{x\,|\,-2 \leq x < 0\} = [-2, 0)$ **61.** Commutative property of addition   **63.** Distributive property
65. Commutative property of addition   **67.** Associative property of multiplication   **69.** Identity property of addition
71. *Home Alone 2: Lost in New York, The Fugitive, Dances with Wolves,* and *Terminator 2: Judgement Day*   **73.** *Aladdin, Ghost, Home Alone, Lion King,* and *Jurassic Park*   **75.** *Terminator 2: Judgement Day, Aladdin,* and *Ghost*   **77.** *Robin Hood, Prince of Thieves*   **79.** None   **81.** d
83–85. Answers may vary.   **87.** Explanations may vary; .14159265358979323846264383383279   **89.** $[250, 300]$ **91.** $[0, 150]$ or $[300, 350]$
93. Not commutative

PROBLEM SET 1.2

1. 25   **3.** -2 **5.** 13   **7.** 33   **9.** 32   **11.** -24 **13.** 0   **15.** -32 **17.** 45   **19.** 1   **21.** 36   **23.** $\frac{9}{10}$ **25.** -6 **27.** -2

29. $-\frac{79}{4}$ **31.** 5.8   **33.** $-\frac{2}{5}$ **35.** 14   **37.** $-\frac{8}{3}$ **39.** -26 **41.** 42   **43.** -24 **45.** $-2x$ **47.** $-4x^2y$ **49.** $7x + 7y$

51. $7x^2y - 3xy^2$ **53.** $-7x^2 - y^2$ **55.** $13x^3y - 11$ **57.** $5x + 12$ **59.** $-6b + 10$ **61.** $-2x^2 + 43$ **63.** $-4y^2 + 10$ **65.** $-14x^2 + 22$
67. $-10x + 32y$ **69.** $-3x + 30y$ **71.** $-10x + 19y - 3z$ **73.** $48x - 75$ **75.** $-4x + 24y$ **77.** $-8x + 9y^2$ **79.** 640 W   **81.** 89.2 feet
83. $R = 194.57$ **85.** Approximately 102   **87.** 112.5%   **89.** Yes   **91.** d   **93.** d   **95–97.** Answers may vary   **99.** $8 - 2 \cdot (3 - 4) = 10$

101. $\left(2 \cdot 5 - \frac{1}{2} \cdot 10\right) \cdot 9 = 45$ **103.** $[(2x + 9) + x] \div 3 + 4 - x = 7$ **105.** Group activity

Review Problems

107. $\{x \mid x \leq 2\}, (-\infty, 2]$ **108.** $\{..., -6, -4, -2\}$ **109.** Inverse property of addition

PROBLEM SET 1.3

Art for **1, 3, 5,** and **7.**   **1.** I   **3.** III   **5.** On x-axis   **7.** III

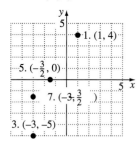

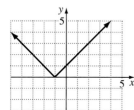

9. $A((0, 6), B(2, 4), C(-2, 1), D(2, -4), E(0, -5)$ **11.** $A(-2, 0), B(-1, 3), C(0, 4), D(1, 3), E(2, 0)$

13. $y = x^2 - 2$ **15.** $y = 2x + 1$ **17.** $y = -\frac{1}{2}x$ **19.** $y = |x| + 1$

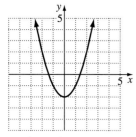

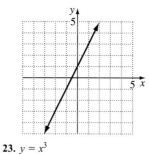

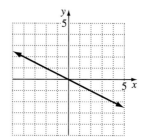

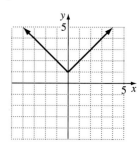

21. $y = |x + 1|$ **23.** $y = x^3$

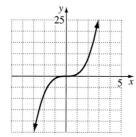

25. a. 1965; 60,000   **b.** 1985; 22,000   **c.** The number of TB cases decreased dramatically from a high in the mid 1960's to a low in the mid 1980's and has remained fairly constant at about 25,000 cass during the 1990's.   **27. a.** 64 ft after 2 seconds   **b.** After 4 seconds
29. a. (2070, 19); approximately 19% of each age group in population in year 2070   **b.** The population will have a greater percentage of people aged 65 or older than people under age 15.
31. a. A:(50, 22.5), B:(50, 17.5)   **b.** 5 mpg
c. Compact cars: 35 mph; medium sized cars: 40 mph

33. 2331, 3811, 8053, 14,199, 21,391, 28,771, 35,481, 40,663, 43,459   **35. a.** $x = 40$ **b.** $y \approx 55$ **c.** $y = 54$

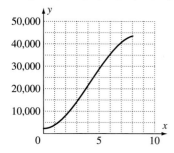

37. a. $y = -3.1x^2 + 51.4x + 4024.5$ **b.** No

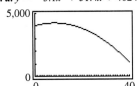 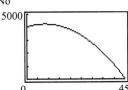

39. a. $y = \dfrac{80{,}000x}{100 - x}$

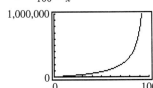 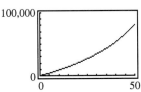 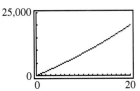

b.

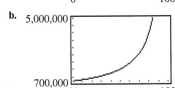

41. Correct

43. Incorrect; $4x^2 - x^2 - 3x - 3x^2 + 4 = 4 - 3x$

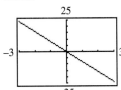 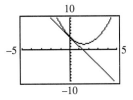

45. Answers may vary.

47. $y = \dfrac{1}{x + 2}$

49. $y = \sqrt{x - 2}$

51. $(-220, 0)$

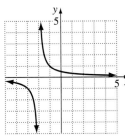 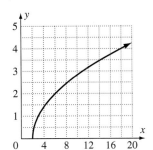

Review Problems

53. $\left\{ -\dfrac{1}{2}, -\dfrac{2}{4}, -\dfrac{3}{6} \right\}$ **54.** 3   **55.** $\{x \mid 0 < x \le 7\}$; $(0, 7]$

PROBLEM SET 1.4

1. 2^6 **3.** 5^4 **5.** x^7 **7.** $6x^{13}$ **9.** $-10x^4 y^{19}$ **11.** 2^{12} **13.** x^{32} **15.** $64x^3$ **17.** $81x^4 y^4$ **19.** $8x^3 y^6$ **21.** $9x^4 y^{10}$ **23.** $32x^3 y$

25. $9x^3 y^2$ **27.** $x^4 y$ **29.** $\dfrac{x^6}{y^6}$ **31.** $\dfrac{81x^4}{y^4}$ **33.** $\dfrac{x^{12}}{y^6}$ **35.** $-\dfrac{125x^9}{8y^3}$ **37.** 5^3 **39.** x^8 **41.** $4x^3$ **43.** $-4x$ **45.** $-10xy^3$

47. $-8a^{11}b^8 c^4$ **49.** 1   **51.** 1   **53.** 1   **55.** $\dfrac{1}{25}$ **57.** $-\dfrac{1}{64}$ **59.** $\dfrac{1}{16}$ **61.** $-\dfrac{1}{16}$ **63.** $\dfrac{16}{9}$ **65.** 125   **67.** 81   **69.** -81

71. $4x^3$ **73.** $\dfrac{1}{x^6}$ **75.** $-\dfrac{4}{x}$ **77.** $\dfrac{2}{x^7}$ **79.** $\dfrac{10a^2}{b^5}$ **81.** b^5 **83.** $\dfrac{1}{x^9}$ **85.** $-\dfrac{6}{a^2}$ **87.** $-6a^4$ **89.** $\dfrac{1}{x^{18}}$ **91.** x^{18} **93.** x^{10} **95.** $3y^{10}$

97. $-\dfrac{x^{17}}{5}$ **99.** $\dfrac{1}{x^{10}}$ **101.** $\dfrac{1}{x^4}$ **103.** $-\dfrac{5y^8}{x^6}$ **105.** $-\dfrac{1}{3a^5 b^2}$ **107.** x^{16} **109.** $-\dfrac{27b^{15}}{a^{18}}$ **111.** $14y$ **113.** 184 pounds   **115.** d   **117.** d

119. a. $y_m = 67.0166\,(1.00308)^x$, $y_f = 74.9742(1.00201)^x$ **b.** 75.8 years; 81.2 years   **c.** Both curves increase.   **121.** Answers may vary.

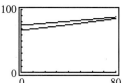

123. x^{9n} **125.** $x^{3n}y^{6n+3}$ **127. a.** True   **b.** Gigaplex

Review Problems

128. $4xy - 2$ **129.** $-\dfrac{4}{3}$ **130.** $-\dfrac{3}{4}$

P R O B L E M   S E T   1 . 5

1. 0.006   **3.** 5,930,000   **5.** 0.0000006284   **7.** 40,030,000,000   **9.** 9.8×10^{10} **11.** 7.46×10^{17} **13.** 2.3×10^{-8} **15.** 7.924×10^{-4}
17. 8.96×10^{7} **19.** 2.795×10^{5} **21.** 1.21203×10^{-2} **23.** 2.42004×10^{-23} **25.** 9×10^{3} **27.** 2×10^{11} **29.** $2.7\overline{27} \times 10^{-12}$
31. 5×10^{2} **33.** 2.46×10^{17} **35.** 3.071×10^{3} **37.** 4.5×10^{5} **39.** 1.9×10^{4} **41.** 4×10^{15} **43.** 6×10^{-12} **45.** 1×10^{4}
47. 4×10^{-1} **49.** 1.5×10^{11} dollars   **51.** $6.\overline{6} \times 10^{9}$ times as large   **53.** $1.\overline{1} \times 10^{24}$ times greater   **55.** 2.5×10^{4} years
57. 6 years   **59.** d   **61.** 1.985×10^{20} newtons   **63.** 7.63×10^{-4} **65–67.** Answers may vary.   **69.** $cd \times 10^{n+m}; \dfrac{c}{d} \times 10^{n-m}$

Review Problems

71. $y = x^2 + 3$

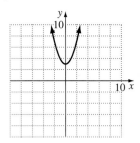

72. $\{x \mid -3 \le x < 5\}$

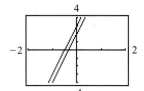

73. -24

P R O B L E M   S E T   1 . 6

1. $\{3\}$ **3.** $\{11\}$ **5.** $\{-17\}$ **7.** $\{11\}$ **9.** $\left\{\dfrac{7}{5}\right\}$ **11.** $\left\{\dfrac{25}{3}\right\}$ **13.** $\{8\}$ **15.** $\{-3\}$ **17.** $\{2\}$ **19.** $\{-4\}$ **21.** $\{26.7\}$ **23.** $\{-1\}$

25. $\{1\}$ **27.** $\{6\}$ **29.** $\{-1\}$ **31.** $\left\{-\dfrac{7}{2}\right\}$ **33.** $\{19\}$ **35.** $\{-12\}$ **37.** $\left\{-\dfrac{13}{5}\right\}$ **39.** $\{10\}$ **41.** $\left\{\dfrac{3}{2}\right\}$ **43.** $\{-1600\}$ **45.** $\{2100\}$

47. $\{200\}$ **49.** $\{x \mid x \in R\}$ **51.** $\varnothing$ **53.** $\{x \mid x \in R\}$ **55.** $\varnothing$ **57.** $\varnothing$ **59.** $\{0\}$ **61.** d

63. $y_1 = 6x + 3; y_2 = 6x + 2$; inconsistent   **65.** $y_1 = \dfrac{x+1}{2}; y_2 = \dfrac{x-3}{4}$; conditional; $\{-5\}$ **67–73.** Answers may vary.

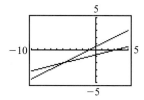

Review Problems

75. $y = 2x - 4$ **76.** $-\dfrac{3x^6}{y^3}$ **77.** 2.5×10^{-4}

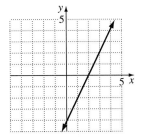

PROBLEM SET 1.7

1. 100 grams   **3.** 205 mg/dl; 125,000 lives   **5.** Region 1, Region 3, Region 2   **7.** 261°F   **9. a.** 45   **b.** $n = \dfrac{L - a + d}{d}$

11. $M = \dfrac{P - C}{C}$ or $M = \dfrac{P}{C} - 1$ **13. a.** $y = \dfrac{7}{3}x + 238\dfrac{1}{3}$ **b.** 254,666,667 people   **15.** 10,562.5 feet   **17.** $y = 2x + 2$ **19.** $y = \dfrac{1}{6}x - \dfrac{5}{2}$

21. $y = \dfrac{3x - 2}{4 - x}; x \neq 4$ **23.** $y = \dfrac{5}{7 - a}$ **25.** $r = \dfrac{26}{x - y}$ **27.** $b = \dfrac{3A - cd}{c}$ **29.** $x = \dfrac{3y + 8z}{5}$ **31.** $x = \dfrac{3R - ac}{a}$ **33.** $a = \dfrac{7c}{3b}$

35. $x = \dfrac{a}{b - 1}$ **37.** $n = \dfrac{2S}{a + 1}$ **39.** $x = \dfrac{21}{2c}$ **41.** d   **43.** 1989   **45.** Yes   **47.** $359.98

49. a. $y = -0.00949x + 0.479$ **b.** Percentage of smokers decreases linearly   **c.** 2009   **d.** 2025

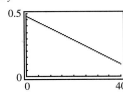

51. a. Bank: $y = 8 + 0.05x$; credit union: $y = 2 + 0.08x$ **b.** 200 checks
53. 86 pounds   **55.** Answers may vary.

Review Problems

56. $\{x \mid x \geq 7\}$ **57.** $x = 78$ **58.** $\{3, 6, 9, 12, 15\}$

PROBLEM SET 1.8

1. 6   **3. a.** $x + 0.52$ **b.** 1993: $1.9 billion, 1994: $2.42 billion   **5.** Cabinet members: $148,400; Vice President: $171,500; Senators: $133,600
7. Pink Floyd: $56 million; Spielberg: $165 million; Cosby: $34 million   **9.** $182.5 billion   **11.** $886.67   **13.** $373.33; 25%   **15.** $15,000
17. a. Men: $6300 + 1600x$; women: $2100 + 1200x$ **b.** 18 yr   **c.** 2100; 8100; 14,100; 20,100; 26,100; No   **d.** Answer may vary.

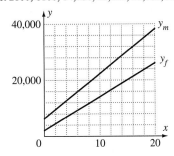

19. 1992   **21.** 9.62 yrs; $27,360   **23.** 11 minutes   **25.** 70 yds by 250 yds   **27.** 4 ft wide $\times$ 7 ft tall   **29.** 11 in., 6 in., 13 in.
31. 224 cm^2 **33.** 3 yd, 8 yd, 8 yd, 17 yd   **35.** 5 hours   **37.** Slower truck: 57.5 mph; faster truck: 62.5 mph
39. Outgoing: 6 hr; return: 4 hr; 120 miles   **41.** $16,000 at 8%; $4000 at 5%   **43.** $20,000 at 9%; $15,000 at 6%; $7200

45. $x = 3$

15	4	8	3
2	9	5	14
1	10	6	13
12	7	11	0

47. 10,652   **49.** 11   **51.** All 4 yd   **53.** One woman is a daughter and a mother
55. 3 spelling errors, false claim

57. a. $y = 3.82 + 0.30x$ **b.** 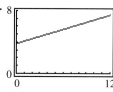   **c.** Students should use TRACE feature.   **d.** 1988

59–61. Answers may vary.   **63.** 66 yr   **65.** Mrs. Ricardo, $4000; son, $8000; daughters, $2000   **67.** $29   **69.** $54   **71.** 36 plants
73. Running, 4 mph; biking, 8 mph; walking, 1.6 mph   **75.** $40,000 at 5%; $10,000 at 16%   **77.** Answers may vary.

Review Problems

78. $-\dfrac{5}{3}$ **79.** $n = \dfrac{D}{A} + 1$ or $n = \dfrac{D + A}{A}$ **80.** 1000.9

CHAPTER 1 REVIEW PROBLEMS

1. {4, 8, 12, 16, ...}   **2.** {0}   **3.** $(-\infty, 1]$ **4.** $[-2, \infty)$ **5.** $(-1, 2]$

6. $\{x \mid -3 \le x < 6\}$; $[-3, 6)$ **7.** $\{x \mid x \le 12\}$; $(-\infty, 12]$ **8.** (0, 23,350]; (23,350, 56,550]; (56,550, 117,950]; (117,950, 256,500]; (256,500, ∞)
9. ServiceMaster, Jazzercise, Dairy Queen, Century 21   **10.** Subway, McDonald's, 7-Eleven   **11.** Baskin-Robbins   **12.** McDonald's, 7-Eleven
13. Baskin-Robbins, Jani-King, ServiceMaster   **14.** Commutative property of addition   **15.** Associative property of multiplication
16. Distributive property   **17.** -39 **18.** 30   **19.** 9   **20.** -3 **21.** 0.1   **22.** -24 **23.** -48 **24.** -2 **25.** $\dfrac{2}{3}$ **26.** -18
27. 55   **28.** 1   **29.** $\dfrac{23}{17}$ **30.** 16   **31.** $\dfrac{5}{34}$ **32.** $\dfrac{5}{8}$ **33.** 1   **34.** -2.1 **35.** 39   **36.** -4 **37.** 43   **38.** -3 **39.** $-3x - 8$
40. $114b - 18a$ **41.** $5x - 4y$ **42.** $x^2 - 4y^2$ **43.** $-2x + 4x^2$ **44.** $-2xy - 14$ **45.** $\dfrac{1}{4}x - 4$ **46.** 107.5 people per square mile
47. 68°F   **48.** $864   **49. a.** 19 heartbeats per minute; vertical difference between the two lines   **b.** 37.5 bpm
50. $y = x^2 - 1$ **51.** $y = |2x|$ **52.** $y = -2x + 3$ **53–55.** Answers may vary.

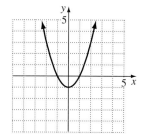   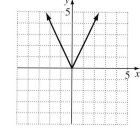   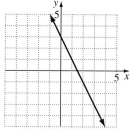

56. a. 5 PM, -4°F   **b.** 8 PM, 16°F   **c.** $x = 4, 6$; answers may vary.   **d.** 12°F; answers may vary.   **e.** 300%

57. $y = \dfrac{10,000x}{100 - x}$;

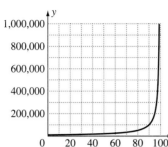

The cost of removing the pollutants from the lake increases rapidly as the percent of removed pollutants approaches 100%.

58. a. $y = 553x + 27,996$ **b.** Salaries increase linearly.   **c.** $33,496   **d.** After 1994   **59.** $24y^{13}$ **60.** $49x^6y^2$ **61.** $-24x^7y$

62. $\dfrac{9}{4}$ **63.** $-7x^3y^2$ **64.** $-\dfrac{8}{y^7}$ **65.** $\dfrac{12}{x^7}$ **66.** $-3x^{10}$ **67.** $-\dfrac{a^8}{2b^5}$ **68.** $\dfrac{5}{8}$ **69.** $-\dfrac{32}{a^5b^5}$

70. $-\dfrac{6x^9}{y^5}$ **71.** 9.37×10^{13} **72.** 4.09×10^{-7} **73.** 1.176×10^8 **74.** 3.75×10^1

75. 5.91×10^{12} m   **76. a.** Defense budget: 3.02×10^{11}; total budget: 1.41×10^{12}, 26%

b. Defense budget: 2.6×10^{11}; total budget: 1.41×10^{12}, 18%   **c.** No   **77.** $\{2\}$ **78.** $\{4\}$

79. $\{-6\}$ **80.** $\left\{\dfrac{3}{5}\right\}$ **81.** $\varnothing$ **82.** $\left\{\dfrac{72}{11}\right\}$ **83.** $\left\{\dfrac{77}{15}\right\}$ **84.** $\{600\}$ **85.** $\{x \mid x \in R\}$

86. $y = \dfrac{2}{3}x - 3$ **87.** $x = \dfrac{1}{m}(y - b)$; $m \neq 0$ **88.** $F = \dfrac{9}{5}C + 32$ **89.** $\dfrac{2A}{b} = h$; $b \neq 0$

90. $H = \dfrac{A - 2LW}{2(W + L)}$ when $L + W \neq 0$ **91.** $y = \dfrac{2x + 5}{x - 3}$; $x \neq 3$

92. $x = \dfrac{17}{2a - 3b}$ for $2a - b \neq 0$ **93.** 1997   **94.** 8 million gallons/day   **95. a.** $C = 4F - 160$ **b.** 90°F   **96. a.** 200 hours

b. (200, 800)   **97. a.** $y = 3.4x + 155$ **b.** 206 thousand   **98.** 5   **99.** Black males: 113; black females: 94;

white males: 90; white females: 52   **100.** $2600   **101. a.** 17.5 years; $24,250

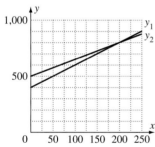

b. $y_e = 5000 + 1100x$, $y_g = 12,000 + 700x$ **102.** 85.7 months   **103.** 1993   **104.** $10,000 in sales   **105.** more than 20 trips

106. 4 ft × 12 ft   **107. a.** 5.875 in. × 7.125 in.   **b.** $1,800,000   **108.** 5 yards   **109.** 6 hours

110. Slower train: 40 mph; faster train: 60 mph   **111.** $3500 at 6%; $1500 at 7%

112. $x = 8$;

10	3	8
5	7	9
6	11	4

113. 12 ways   **114.** 2, 3, and 16   **115.** $\dfrac{x}{2x + 1}$

C H A P T E R I T E S T

1. $\{x \mid x \geq 6\}$, $[6, \infty)$ **2.** 1980, 1990, and 1993   **3.** -3 **4.** -8 **5.** $-10x + 17$ **6.** $32y^2 - 10y$ **7.** 72,000

8. $y = 4 - x^2$

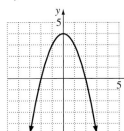

9. $A(6, 90)$; the number of nuclear power plants in the U.S. in 1984 was about 90.   **10.** About 120, 1990
11. About age 60   **12.** As age increases, death rates per 1000 increase   **13.** 1989, 5 billion dollars
14. $\dfrac{14}{x^6}$ **15.** $\dfrac{4x^8}{y^4}$ **16.** 3.33×10^{18} **17.** 5×10^5 **18.** $\{2\}$ **19.** $\{-6\}$ **20.** $r = \dfrac{A - P}{Pt}$

21. $n = \dfrac{2s}{f + \ell}$ **22.** 2003   **23.** 1994, \$220 billion; 1995, \$206 billion; 1996, \$197 billion
24. \$32,000   **25.** 1986   **26.** 60 hours   **27.** Width is 50 yards, length is 101 yards.
28. \$2000 at 6%, \$5000 at 8%   **29.** 18

PROBLEM SET 2.1

1. Not a function; domain $= \{1\}$; range $= \{3, 7, 10\}$ **3.** Not a function; domain $= \{2, 3\}$; range $= \{3, 4\}$
5. Function; domain $= \{-1, 0, 1, 2\}$; range $= \{-1, 0, 1, 2\}$
7. Function; domain $= \{1, 2, 3, 4\}$; range $= \{-4, -3, -2, -1\}$
9. a. -2 **b.** -8 **c.** 19   **d.** 0   **e.** $6a - 2$ **11. a.** 5   **b.** 8   **c.** 53   **d.** 32   **e.** $48b^2 + 5$
13. a. -1 **b.** 26   **c.** 19   **d.** 1   **e.** $50r^2 + 15r - 1$
15. a. $\dfrac{3}{4}$ **b.** -3 **c.** $\dfrac{11}{8}$ **d.** $\dfrac{13}{9}$ **e.** $\dfrac{2a + 2h - 3}{a + h - 4}$ **f.** Denominator $= 0$ **17.** Yes   **19.** No
21. $f(x) = x^2$; $g(x) = x^2 - 20$;   **23.** $f(x) = \sqrt{x}$; $g(x) = \sqrt{x - 2}$;   **25.** $f(x) = x^3$; $g(x) = x^3 + 1$;   **27.** $f(x) = \sqrt{x}$; $g(x) = -\sqrt{x}$;
$h(x) = x^2 + 1$ $h(x) = \sqrt{x + 1}$ $h(x) = (x + 1)^3$ $h(x) = \sqrt{-x}$

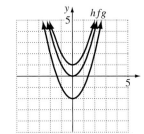
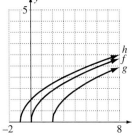
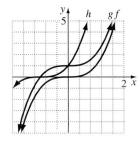
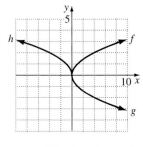

$f(x) \pm c$ is the graph of $f(x)$ $f(x \pm c)$ is the graph of f shifted   For $c > 0$, $f(x) + c$ is $f(x)$ shifted $-f(x)$ is $f(x)$ reflected about the
shifted vertically up ($+c$) or   horizontally c units, to the left   up c units and $f(x + c)$ is $f(x)$ x-axis and $f(-x)$ is $f(x)$ reflected
down ($-c$) c units.   if "$+c$" and of the right if "$-c$".   shifted to the left c units.   about the y-axis.
29. Not a function   **31.** Function   **33.** Function   **35.** Not a function   **37.** Function   **39.** Not a function
41. 11.5; In 1950 there were approximately 11.5 million union members in the U.S.   **43. a.** $f(2) = 46$ **b.** No   **c.** $f(3.2) - f(1.83)$
45. 14.69 yr; difference in life expectancy at birth for those born in 1920 and those born in 2000
47. $f(t) = -t^4 + 12t^3 - 58t^2 + 132t$ **49. a.** Each year maps onto exactly one consumption figure   **b.** about 4.3 in 1965   **c.** about 3.0 in 1990
d. ≈ 0.15 thousand per capita   **e.** ≈ 1995 **51.** d   **53.** Students should verify.

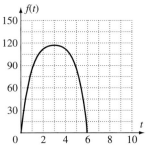

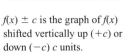
Concentration increases, then decreases. 6 hours; point (6, 0)

55. a. $l = x$, $w = 40 - x$ **b.** 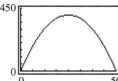   **c.** 20 meters   **57** Answers may vary.   **59. a.** $-\dfrac{x}{2}$ **b.** $\dfrac{x+1}{2}$ **c.** 3

61. Answers may vary.

Review Problems

63. 2010   **64.** 40 years   **65.** 2.15×10^{13}

PROBLEM SET 2.2

1. $\dfrac{2}{5}$ **3.** -1 **5.** -3 **7.** $-\dfrac{3}{2}$ **9.** Parallelogram

11.

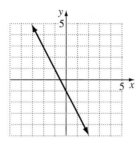

a. Opposite sides AB and CD have slope -1; opposite sides BC and AD have slope 4.

b. Product of slopes AB and BC is -4, not -1.   **13.** Slopes of diagonals AC and BD are negative reciprocals.

15. $m_1 > m_2 > m_3 > m_4$ **17.** $y = -2x - 1$

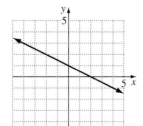

19. $y = \dfrac{1}{2}x + 1$ **21.** $y = \dfrac{1}{2}x - 1$ **23.** $y = -\dfrac{1}{2}x + 1$ **25.** $y = -\dfrac{1}{2}x - 1$

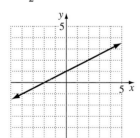   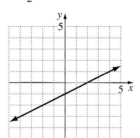   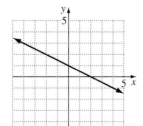   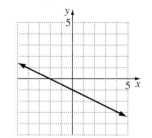

27. $x + y = 3$

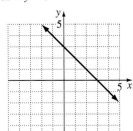

29. $2x + y = -1$

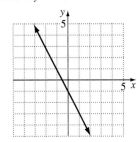

31. $2x + 3y = 6$

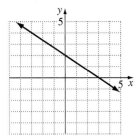

33. $-2x + 3y = 6$

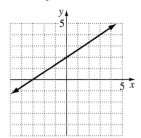

35. $y = -3x + 5, m = -3, b = 5$ **37.** $y = -\frac{2}{3}x + 6, m = -\frac{2}{3}, b = 6$ **39.** $y = 2x - 3, m = 2, b = -3$ **41.** $y = \frac{7}{3}x - \frac{13}{3}, m = \frac{7}{3}, b = -\frac{13}{3}$

43. Parallel   **45.** Perpendicular   **47.** Neither

49. $y = 3$

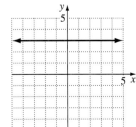

51. $y = -2$

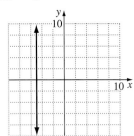

53. $3y = 18$

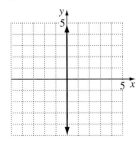

55. $f(x) = 2$

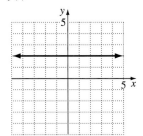

57. $x = -5$

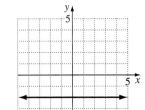

59. $3x - 12 = 0$

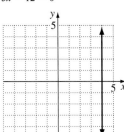

61. $x = 0$

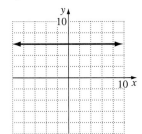

63. $m = 34, b = 1549$; There were 1,549,000 nurses in 1985 and the number grew by 34,000 each year thereafter.   **65.** $m = -0.5, b = 100$; The percentage of hikers found is 100% when the searchers are 0 meters apart and decreases by 0.5% for every meter increase the searchers are apart.

67. a. $m = 380.3125$; From 1980 to 1996, the cost of education has increased approximately $380 per year.   **b.** $14,600

69. $m_{AB} = -\frac{7}{10}, m_{BC} = -\frac{3}{2}, m_{CD} = -\frac{4}{5}$, and $m_{DE} = -\frac{4}{5}$. While the percentage of men with high sperm counts continues to decrease, the rate of decreasing is slowing.   **71.** d   **73.** a

75. $y = 2x + 4; m = 2$

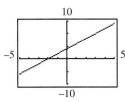

77. $y = -\frac{1}{2}x - 5; m = -\frac{1}{2}$

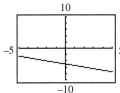

79. a. $y_1 = \frac{1}{3}x + 1, y_2 = -3x - 2$, no   **b.** Yes; rescaled the window.

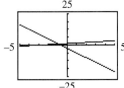

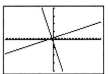

81–83. Answers may vary.   **85.** Let $y = 0$; $0 = mx + b$; $mx = -b$; $x = -\frac{b}{m}$; x-intercept is $-\frac{b}{m}$.   **87.** $\frac{5}{2}$

Review Problems

89. $y = \dfrac{3r}{a+b}$ **90.** $y = |x| + 2$ **91.** $\{3\}$

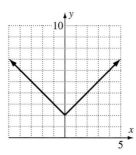

PROBLEM SET 2.3

1. $y - 1 = 3(x - 2)$ **3.** $y - 2 = -3(x - 6)$ **5.** $y + 5 = 4(x + 3)$ **7.** $y + 3 = 0(x - 0)$

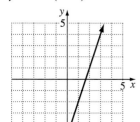   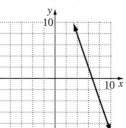   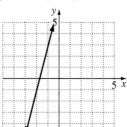   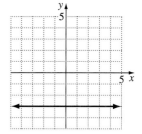

9. $y + 2 = \dfrac{4}{5}(x - 4)$

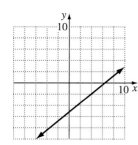

11. $y - 2 = 2(x + 3)$; $y = 2x + 8$ **13.** $y - 0 = \dfrac{1}{2}(x - 1)$; $y = \dfrac{1}{2}x - \dfrac{1}{2}$

15. $y - 4 = \dfrac{1}{2}(x + 3)$; or $y - 6 = \dfrac{1}{2}(x - 1)$; $y = \dfrac{1}{2}x + \dfrac{11}{2}$

17. $y + 2 = 1(x + 4)$ or $y - 11 = 1(x - 9)$; $y = x + 2$ **19.** $y - 3 = 2(x - 1)$; $y = 2x + 1$

21. $y + 3 = -\dfrac{1}{2}(x - 1)$; $y = -\dfrac{1}{2}x - \dfrac{5}{2}$

23. $y = -7x + 5$ **25.** $y = -8$ **27.** $y = -\dfrac{1}{5}x + \dfrac{27}{5}$

29. $y - 2 = 2(x + 3)$; $2x - y = -8$ **31.** $y - 0 = \dfrac{1}{2}(x - 1)$; $x - 2y = 1$

33. $y - 6 = -1(x - 5)$; $x + y = 11$ **35.** $y + 5 = -2(x + 2)$; $2x + y = -9$

37. $y + 3 = -2(x - 1)$; $2x + y = -1$

39. $y - 546 = 60(x - 3)$ or $y - 666 = 60(x - 5)$; $y = 60x + 366$; 1.866 trillion dollars

41. $y = -\dfrac{1}{500}x + 59$; 4500 shirts   **43.** $y - 58.9 = \dfrac{1}{45}(x - 0)$ or $y - 59.3 = \dfrac{1}{45}(x - 18)$; $y = \dfrac{1}{45}x + 58.9$; $\approx 60°F$ **45–47.** Answers may vary.

49. $y + 2 = -\dfrac{2}{3}(x - 0)$; $y = -\dfrac{2}{3}x - 2$; $2x + 3y = -6$ **51.** 74, 97   **53.** $\dfrac{4}{3}$ **55.** $(n, 8n + 9)$ **57.** Answers may vary.

Review Problems

58. $\{-2\}$ **59.** $\dfrac{16x^{36}}{81y^8}$ **60.** 124

PROBLEM SET 2.4

1. $\{x \mid x > 3\}$; $(3, \infty)$ **3.** $\{x \mid x \le -2\}$; $(-\infty, -2]$ **5.** $\{x \mid x < -5\}$; $(-\infty, -5)$ **7.** $\{x \mid x > 8\}$; $(8, \infty)$

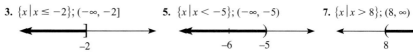   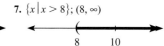

9. $\{x \mid x \geq -6\}; [-6, \infty)$

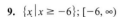

11. $\{x \mid x > 3\}; (3, \infty)$

13. $\{x \mid x \geq 1\}; [1, \infty)$

15. $\left\{x \mid x \leq -\dfrac{53}{6}\right\}; \left(-\infty, -\dfrac{53}{6}\right]$

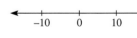

17. $\{x \mid x \geq 0\}; [0, \infty)$

19. $\left\{x \mid x < \dfrac{18}{5}\right\}; \left(-\infty, \dfrac{18}{5}\right)$

21. $\{x \mid x \leq -4\}; (-\infty, -4]$

23.

$\{x \mid x \geq 2\}; [2, \infty)$

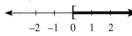

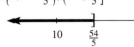

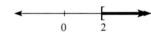

25. $\left\{x \mid x \leq \dfrac{54}{5}\right\}; \left(-\infty, \dfrac{54}{5}\right]$

27. $\{x \mid x > 10\}; (10, \infty)$

29. $\{y \mid y < -2\}; (-\infty, -2)$

31. $\varnothing$

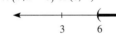

33. 1997   **35.** Men 18–25, men 26–34   **37.** All groups except men or women age 35 and older   **39.** More than 720 miles per week
41. More than \$32,000   **43.** 79   **45.** b   **47.** $x < 4$ **49.** No x-values   **51. a.** $y_A = 4 + 0.10x; y_B = 2 + 0.15x$
b.

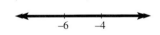

c. Students should find intersection.   **d.** When the number of checks written is greater than 40.
53. Answers may vary.   **55.** \$8

Review Problems

57. $\{-14\}$ **58.** $h = \dfrac{2}{11}(w + 220)$; 68 inches   **59.** $x^2 - 8y^2$

PROBLEM SET 2.5

1. $\{2, 4\}$ **3.** $\varnothing$ **5.** $\{7, 9, 10\}$ **7.** $\{x \mid x > 6\}$ or $(6, \infty)$ **9.** $\{x \mid 3 \leq x \leq 6\}$ or $[3, 6]$ **11.** $\varnothing$

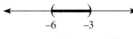

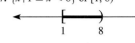

13. $\{x \mid -6 < x < -4\}$ or $(-6, -4)$ **15.** $\{x \mid -3 < x \leq 6\}$ or $(-3, 6]$ **17.** $\{x \mid 2 < x < 5\}$ or $(2, 5)$ **19.** $\varnothing$, no solution

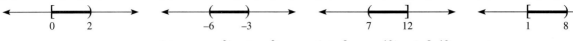

21. $\{x \mid 0 \leq x < 2\}$ or $[0, 2)$ **23.** $\{x \mid -6 < x < -3\}$ or $(-6, -3)$ **25.** $\{x \mid 7 < x \leq 12\}$ or $(7, 12]$ **27.** $\{x \mid 1 \leq x < 8\}$ or $[1, 8)$

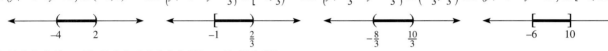

29. $\{y \mid -4 < y < 2\}$ or $(-4, 2)$ **31.** $\left\{y \mid -1 \leq y < \dfrac{2}{3}\right\}$ or $\left[-1, \dfrac{2}{3}\right)$ **33.** $\left\{y \mid -\dfrac{8}{3} < y < \dfrac{10}{3}\right\}$ or $\left(-\dfrac{8}{3}, \dfrac{10}{3}\right)$ **35.** $\{y \mid -6 \leq y \leq 10\}$ or $[-6, 10]$

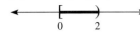

37. $\{1, 2, 3, 4, 5\}$ **39.** $\{1, 2, 3, 4, 5, 6, 7, 8, 10\}$ **41.** $\{7, 9, 10\}$
43. $\{x \mid x > 3\}$ or $(3, \infty)$ **45.** $\{x \mid -\infty < x < \infty\}$ or $(-\infty, \infty)$ **47.** $\{x \mid x < -3 \text{ or } x > 1\}$ or $(-\infty, -3) \cup (1, \infty)$

49. $\{x \mid -\infty < x < \infty\}$ or $(-\infty, \infty)$ **51.** $\{x \mid -\infty < x < \infty\}$ or $(-\infty, \infty)$ **53.** $\left\{x \mid x \leq -\dfrac{3}{2} \text{ or } x > 4\right\}$ or $\left(-\infty, -\dfrac{3}{2}\right] \cup (4, \infty)$

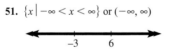

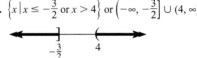

55. $\left\{x\,\middle|\,x<\frac{5}{2}\right\}$ or $\left(-\infty,\frac{5}{2}\right)$

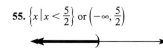

$\frac{5}{2}$

57. $\{x\,|\,x\in R\}$ or $(-\infty,\infty)$

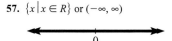

0

59. $\{y\,|\,y\in R\}$ or $(-\infty,\infty)$

0

61. $\{x\,|\,x<9\}$ or $(-\infty,9)$

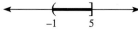

9

63. $25°\le C\le 40°$

65. a. Between 1964 and 1984   **b.** The model is a good fit.   **67.** Between 1988 and 1992   **69.** U.S. total, U.S. Latino, U.S. African American
71. None   **73.** U.S. Whites, Canada, Australia, United Kingdom, France, Netherlands, Germany, Sweden

75. Greater than 18 meters but not more than 20 meters   **77.** Between 1 hr and $1\frac{1}{3}$ hr (or 1 hr 20 min)   **79.** a   **81.** Answers may vary.

83. $\{y\,|\,-1<y\le 5\}$ or $(-1,5]$ **85.** $\{x\,|\,-1\le x<2\}=[-1,2)$

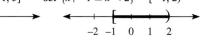

−1   5   −2 −1 0 1 2

Review Problems

87. 15   **88.** 99 feet   **89.** 9, 2, 1, 0, −7; $y=-x^3+1$

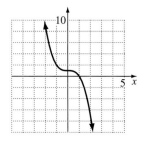

PROBLEM SET 2.6

1. $\{-8,8\}$ **3.** $\varnothing$ **5.** $\{-5,9\}$ **7.** $\{-3\}$ **9.** $\{-2,3\}$ **11.** $\{-1,2\}$ **13.** $\{-10,4\}$ **15.** $\varnothing$ **17.** $\left\{\frac{3}{2},\frac{7}{2}\right\}$ **19.** $\left\{-\frac{4}{3},0\right\}$

21. $\left\{-\frac{1}{4},\frac{7}{2}\right\}$ **23.** $\left\{\frac{1}{4}\right\}$ **25.** $\{-1,15\}$ **27.** $\{x\,|\,x\in R\}$

29. $\{x\,|\,-4<x<4\}$
$(-4,4)$

31. $\{x\,|\,x\le -4\text{ or }x\ge 4\}$
or $(-\infty,-4]\cup[4,\infty)$

33. $\{x\,|\,1<x<3\}$ or $(1,3)$

35. $\{x\,|\,-3\le x\le -1\}$ or $[-3,-1]$

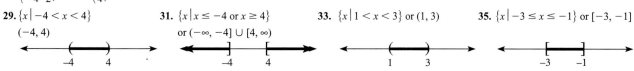

−4   4   −4   4   1   3   −3   −1

37. $\{x\,|\,x<-4\text{ or }x>-2\}$
or $(-\infty,-4)\cup(-2,\infty)$

39. $\{x\,|\,x\le 2\text{ or }x\ge 6\}$
or $(-\infty,2]\cup[6,\infty)$

41. $\{y\,|\,-1<y<7\}$ or $(-1,7)$

43. $\{x\,|\,-5\le x\le 0\}$ or $[-5,0]$

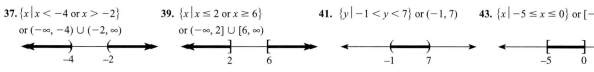

−4   −2   2   6   −1   7   −5   0

45. $\left\{x\,\middle|\,\frac{1}{4}<x<\frac{5}{12}\right\}$ or $\left(\frac{1}{4},\frac{5}{12}\right)$

47. $\left\{x\,\middle|\,x<-4\text{ or }x>\frac{14}{3}\right\}$
or $\left(-\infty,-4\right)\cup\left(\frac{14}{3},\infty\right)$

49. $\left\{x\,\middle|\,x\le\frac{4}{5}\text{ or }x\ge 6\right\}$
or $\left(-\infty,\frac{4}{5}\right]\cup[6,\infty)$

51. $\{y\,|\,y<-5\text{ or }y>3\}$
or $(-\infty,-5)\cup(3,\infty)$

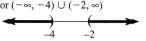 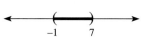 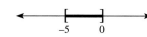

$\frac{1}{4}$ $\frac{5}{12}$ −4 $\frac{14}{3}$ $\frac{4}{5}$ 6   −5   3

53. $\left\{x \mid x \le -\frac{3}{7} \text{ or } x \ge 1\right\}$ **55.** $\{x \mid -9 \le x \le 5\}$ or $[-9, 5]$ **57.** $\{x \mid x < 1 \text{ or } x > 2\}$ **59.** $\varnothing$

or $\left(-\infty, \frac{3}{7}\right] \cup [1, \infty)$

or $(-\infty, 1) \cup (2, \infty)$

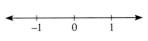

61. $\{x \mid -\infty < x < \infty\}$ or $(-\infty, \infty)$ **63.** High: 2,695,000 barrels; low: 2,425,000 barrels

65. Adults with 12 years of education in 1990, adults with 13–15 years of education in 1987 or 1990, all adults with 16 or more years of education
67. Adults with less than 12 years of education in 1974 or adults with 13–15 years of education in 1974   **69.** c   **71.** Answers may vary.

73. $\{x \mid x \ge 0\}$ **75.** $\{x \mid x < -1\}$ **77.** $\varnothing$ **79.** $\left\{x \mid x \le \frac{2}{3}\right\}$

Review Problems

81. 5 ft wide $\times$ 8 ft tall   **82.** $\left\{\frac{7}{9}\right\}$ **83.** $4a + 54b - 2$

PROBLEM SET 2.7

1. Yes   **3.** No   **5.** $x + y \ge 2$ **7.** $3x - y \ge 6$ **9.** $2x + 3y > 12$

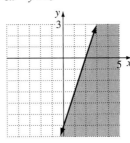 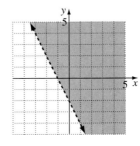 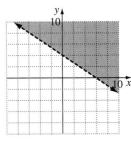

11. $5x + 3y \le -15$ **13.** $2y - 3x > 6$ **15.** $5x + y > 3x - 2$ **17.** $\frac{x}{2} + \frac{y}{3} < 1$

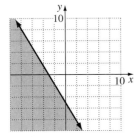 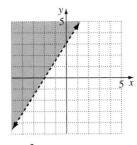 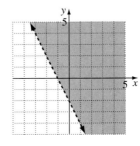 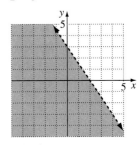

19. $y > 2x - 1$ **21.** $y \ge \frac{2}{3}x - 1$ **23.** $y < x$ **25.** $y > -\frac{1}{2}x$

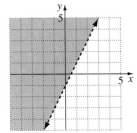 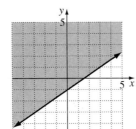 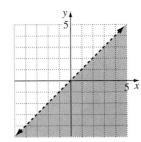 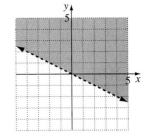

27. $x \le 1$

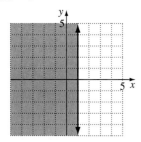

29. $y > 1$

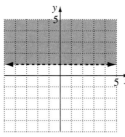

31. $x \ge 0$

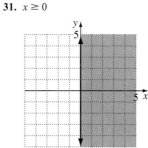

33. $x > -5$

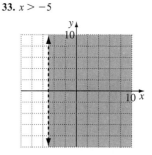

35. (x, y) number of eggs x and the number of ounces of meat y that satisfy the dietary requirements

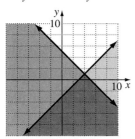

37. $50x + 150y > 2000$ results in an overload.

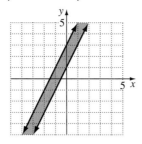

39. b

41. $y \le 6 - 1.5x$

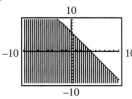

43. $3x + 6y \le 6$

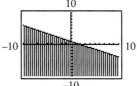

45. Students should verify.   **47.** Answers may vary.

49. $x - y \ge 3$ and $x + y \le 5$

51. $y \ge 2x + 1$ and $y \le 2x + 3$

53. $-2 < x \le 5$

Review Problems

54. $\{x \mid x < -7\}$ **55.** $f = \dfrac{dw}{l - d}$ **56.** $8x + 8y - 6$

CHAPTER 2 REVIEW PROBLEMS

1. Function; domain $= \{2, 3, 5\}$; range $= \{7\}$ **2.** Function; domain $= \{1, 2, 13\}$; range $= \{10, 500, \pi\}$

3. Function; domain $= \{1, 3, 5\}$; range $= \{2, 4, 6\}$ **4.** Not a function; domain $= \{12, 14\}$; range $= \{13, 15, 19\}$

5. a. 5   **b.** 19   **c.** -30 **d.** 2   **e.** $-7b - 16$ **6. a.** 2   **b.** 24   **c.** 30   **d.** $\dfrac{1}{4}$ **e.** $27a^2 - 15a + 2$

7. $f(x) = x^2$, $g(x) = x^2 + 2$,

$h(x) = x^2 - 1$

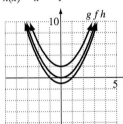

8. $f(x) = |x|$, $g(x) = |x| + 2$,

$h(x) = |x| - 1$

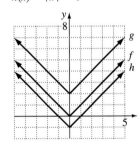

9. $f(x) = \sqrt{x}$, $g(x) = \sqrt{x + 2}$,

$h(x) = \sqrt{x - 1}$

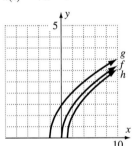

10. $f(x) = x^2 - 3x$;

10, 4, 0, −2, −2, 0, 4, 10

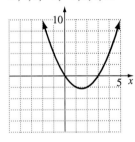

$f(x) \pm c$ is $f(x)$ shifted up (+)
or down (−) c units ($c > 0$).

For $c > 0$, $f(x + c)$ is $f(x)$ shifted
left c units and $f(x - c)$ is $f(x)$
shifted right c units.

11. Yes   **12.** Not a function   **13.** Function   **14.** Not a function   **15.** Function   **16.** Not a function
17. 9.811 is the number of infant deaths (per thousand live births) in 1990.   **18. a.** $912.22 is the weekly salary in 2000.
b. ≈$230 is the weekly salary between 1980 and 1990.   **19. a.** Because each year maps into one and only one violent crime rate.
b. 35 (per thousand people) in 1981   **c.** Approximately 28 per thousand in 1986   **d.** Cannot expect to accurately predict the value in 2009,
17 years beyond the range of the data and 34 years after 1975 from the information given.
20. a.

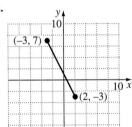

b. $m = -2$ **c.** $m = -2$; same slope   **d.** Choice of points will vary; $m = -2$.   **21. a.** $f(x) = 60$ (percent)
b. Answers may vary.   **c.** $g(x) = 45$ (percent)   **22.** $m_{1950 - 1995} \approx 2$; the percentage of households with TV's
increased by an average of 2% per year over the 45 years. $m_{1955-1960} = 2.8$; from 1955 to 1960,
the percentage of homes with TV's increased an average of 2.8% per year. $m_{1990-1995} = 0$; there was
virtually no change in the percentage of households with TV's during the period 1990–1995.
23. a. Parallelogram   **b.** Rectangle

24. $y = \dfrac{1}{2}x - 4$

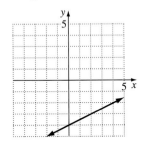

25. $f(x) = -\dfrac{3}{4}x + 5$

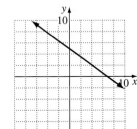

26. $2x - 3y = 6$

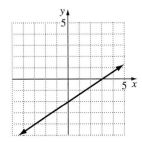

27. $x - 3 = 2$

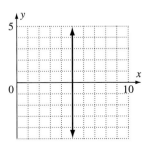

28. $y + 3 = -2(x - 1)$; $y = -2x - 1$; $2x + y = -1$ **29.** $y - 3 = -3(x - 1)$ or $y - 18 = -3(x + 4)$; $y = -3x + 6$; $3x + y = 6$
30. $y + 4 = \dfrac{2}{3}(x + 1)$; $y = \dfrac{2}{3}x - \dfrac{10}{3}$; $2x - 3y = 10$ **31.** $y + 3 = -2(x - 2)$; $y = -2x + 1$; $2x + y = 1$ **32. a.** $y - 16 = -0.13(x - 1900)$ or
$y - 12.1 = -0.13(x - 1930)$ **b.** $y = -0.13x + 263$ **c.** 6.9 feet in 1970; 5.6 feet in 1980
d. No; 4.3 feet in 1990 is probably unreasonably short; $1900 \leq x \leq 1980$

33. $\left\{x \mid x > \dfrac{3}{2}\right\}$, $\left(\dfrac{3}{2}, \infty\right)$

$\dfrac{3}{2}$

34. $\{x \mid x \geq -2\}$, $[-2, \infty]$

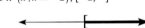

−2

35. $\left\{x \mid x \geq \dfrac{3}{5}\right\}$ or $\left[\dfrac{3}{5}, \infty\right)$

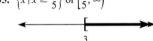

$\dfrac{3}{5}$

36. $\{x \mid x > -3\}$, $(-3, \infty)$

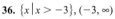

−3

37. $\{x \mid x \geq 3\}$, $[3, \infty]$

3

38. $\{x \mid x > -3\}$, $(-3, \infty)$

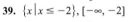

−3

39. $\{x \mid x \leq -2\}$, $[-\infty, -2]$

−2

40. More than 20 years

41. More than 50 checks   **42.** $59° \leq F \leq 95°$ **43.** $\{a, c\}$ **44.** $\{a\}$ **45.** $\{a, b, c, d, e\}$ **46.** $\{a, b, c, d, f, g\}$

47. $\{x \mid x \le 3\}, (-\infty, 3]$

48. $\{x \mid x < 6\}, (-\infty, 6)$

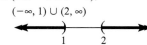

49. $\{x \mid 6 < x < 8\}, (6, 8)$

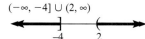

50. $\{x \mid x \le 1\}, (-\infty, 1]$

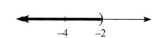

51. $\varnothing$

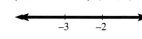

52. $\{x \mid x < 1 \text{ or } x > 2\},$
$(-\infty, 1) \cup (2, \infty)$

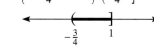

53. $\{x \mid x \le -4 \text{ or } x > 2\},$
$(-\infty, -4] \cup (2, \infty)$

54. $\{x \mid x < -2\}, (-\infty, -2)$

55. $\{x \mid -\infty < x < \infty\}, (-\infty, \infty)$

56. $\left\{x \mid -\dfrac{3}{4} < x \le 1\right\}, \left(-\dfrac{3}{4}, 1\right]$

57. $\left\{x \mid -\dfrac{5}{2} < x \le 4\right\}, \left(-\dfrac{5}{2}, 4\right]$

58. $3 \le t \le 7$ years old

59. $5 < t \le 9$ years old   **60.** $t \le 10$ years old   **61.** The line does not adequately represent the data.   **62.** $w(30) = 191$; a male 30 years old would weigh 191 pounds.   **63.** Answers may vary.   **64.** $\{-4, 3\}$ **65.** $\varnothing$ **66.** $\{-5.5, 11.5\}$ **67.** $\left\{-4, -\dfrac{6}{11}\right\}$

68. $\{x \mid -9 \le x \le 6\}, [-9, 6]$ **69.** $\{x \mid x < -6 \text{ or } x > 0\}, (-\infty, -6) \cup (0, \infty)$ **70.** $\{x \mid -2 < x < 5\}, (-2, 5)$

71. $\{x \mid x \le -3 \text{ or } x \ge -2\}, (-\infty, -3] \cup [-2, \infty)$ **72.** 90% of the population sleep between $5\frac{1}{2}$ and $7\frac{1}{2}$ hours per day (inclusive).

73. $x - 3y \le 6$

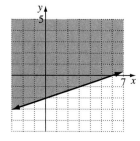

74. $y > -\dfrac{1}{2}x + 3$

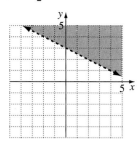

75. $x \le 2$

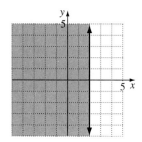

76. $y > -3$

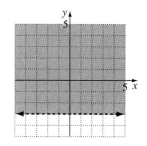

C H A P T E R   2   T E S T

1. $12a^2 - 14a + 3$ **2.** g is f shifted one unit to the right.   **3.** $y = x^2 + 2x + 1$ **4.** b, c

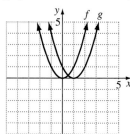

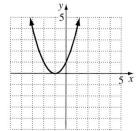

5. 55; There is 55 board feet in a 16-foot log whose diameter is 10 inches.   **6.** d   **7.** The average change per year in world population from 1976 to 2016 is $\dfrac{1}{10}$ billion increase.

8. $y = -\frac{1}{3}x + 2$ **9.** $4x - 3y = 12$ **10.** $y + 3 = 1(x + 1)$ or $y - 2 = 1(x - 4)$, $y = x - 2$, $x - y = 2$

11. $y - 3 = 2(x + 2)$, $y = 2x + 7$, $2x - y = -7$ **12.** Yes

13. $y - 1.98 = 2(x - 0)$ or $y - 7.98 = 2(x - 3)$, 35.98 million

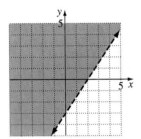   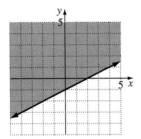

14. $\{x \mid x \le -2\}$ **15.** $\{x \mid x \ge 9\}$ **16.** less than 100 hours   **17.** $\{2, 4, 10\}$ **18.** $\{2, 4, 5, 6, 7, 9, 10\}$

19. $\{x \mid 3 < x \le 12\}$ or $(3, 12]$ **20.** $\{x \mid -12 < x < 5\}$ or $(-12, 5)$ **21.** $\{x \mid x \le -4$ or $x > 2\}$ or $(-\infty, -4] \cup (2, \infty)$

22. $\{x \mid x \ge -2\}$ or $[-2, \infty)$ **23.** $\{x \mid x < 4\}$ or $(-\infty, 4)$ **24.** $\{x \mid -2 < x < -1\}$ or $(-2, -1)$ **25.** $\{6, 12\}$ **26.** $\{-2, 12\}$

27. $\{x \mid -3 < x < 4\}$ or $(-3, 4)$ **28.** $\{x \mid x \le -1$ or $x \ge 4\}$ or $(-\infty, -1] \cup [4, \infty)$

29. $3x - 2y < 6$ **30.** $y \ge \frac{1}{2}x - 1$

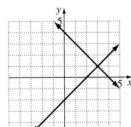   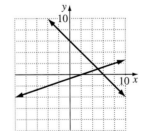

Chapter 3

PROBLEM SET 3.1

1. $\{(3, 1)\}$ **3.** $\{(5, 1)\}$ **5.** $\{(3, 4)\}$ **7.** $\{(3, -1)\}$

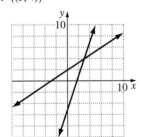   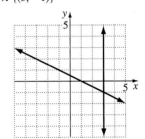

9. $\varnothing$

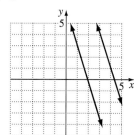

11. $\varnothing$

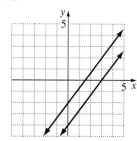

13. Dependent

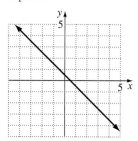

15. Dependent

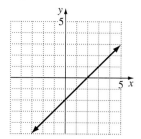

17. $\{(-31, -13)\}$ **19.** $\{(2, -3)\}$ **21.** $\{(5, -5)\}$ **23.** $\{(5, 4)\}$ **25.** $\{(5, 2)\}$ **27.** $\{(2, -3)\}$ **29.** $\{(-1, -3)\}$ **31.** $\left\{\left(4, \frac{1}{3}\right)\right\}$

33. $\{(-2, -1)\}$ **35.** $\{(1, -2)\}$ **37.** $\{(5, -10)\}$ **39.** $\{(1, -3)\}$ **41.** $\{(2, 8)\}$ **43.** Dependent; $\{(x, y)|x + 2y - 3 = 0\}$

45. Dependent; $\{(x, y)|2x - y - 5 = 0\}$ **47.** $\varnothing$ **49.** $\left\{\left(-5, -\frac{10}{3}\right)\right\}$ **51.** $\left\{\left(-\frac{1}{2}, 4\right)\right\}$ **53.** $\{(-5, 7)\}$ **55.** $\varnothing$ **57.** 40,000 units

59. a.

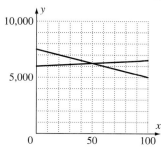

b. Achilles overtakes the tortoise in 2.2 minutes. This occurs approximately 110 yards from the starting point.   **c.** $\{(2.\overline{2}, 1111.\overline{1})\}$ **61.** A   **63.** Students should verify problems.

65–69. Answers may vary.   **71.** D   **73.** $k = -4$ **75.** $\varnothing$ **77.** $\left\{\left(\frac{c_1 b_2 - c_2 b_1}{a_1 b_2 - a_2 b_1}, \frac{c_1 a_2 - c_2 a_1}{b_1 a_2 - b_2 a_1}\right)\right\}$

Review Problems

79.

17	10	15
12	14	16
13	18	11

80. $2.75   **81.** $\left\{\frac{1}{3}, 4\right\}$

PROBLEM SET 3.2

1. 3 and -3 **3.** Sponge cake has 162 mg; pound cake has 68 mg   **5.** 27   **7.** Tutors: $6.50; graders: $3.35
9. $600 weekly salary; 15% commission   **11.** 12 day-care centers; 200 miles of road repair   **13.** 80 ft by 100 ft
15. length AE, 7 m; length EB, 9 m; length of DC, 16 m   **17.** $\angle A = 100°$; $\angle B = 40°$; $\angle C = 40°$ **19.** 79° and 40°   **21.** 30°, 150°, and 30°
23. 60° and 120°   **25.** Plane is 180 mph; wind is 20 mph   **27.** 20 gallons solution A; 40 gallons solution B   **29.** 500 students
31. a. $m = 2833, b = 67,933$ **b.** Each year the value of a house increases by $2833.   **c.** $124,593   **33. a.** $m = -25, b = 7500$
b. $m = 5, b = 6000$ **c.** ($50, 6250); When fans are priced at $50, the supply and demand both equal 6250 fans.
d. $y = -25x + 7500, y = 5x + 6000$;

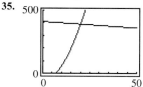

35. $20, 390 units

37. Mangos: $45; Avocados: $22   **39.** State: $800; Federal: $1800   **41.** Each person is traveling at 40 miles per hour.

Review Problems

43. $x \geq -1$ or $x \leq -\frac{7}{3}$;

$-\frac{7}{3}$

-4 -3 -2 -1 0

44. $x = 5$ **45.** $f(x) = -x^2 - x + 2$;

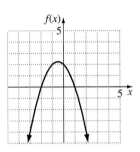

PROBLEM SET 3.3

1. Solution   **3.** Solution   **5.** $\{(2, 3, 3)\}$ **7.** $\{(2, -1, 1)\}$ **9.** $\left\{\left(\frac{1}{3}, -\frac{2}{5}, \frac{1}{2}\right)\right\}$ **11.** $\{(3, 1, 5)\}$ **13.** $\{(1, 0, -3)\}$ **15.** $\{(1, -5, -6)\}$

17. $\{(2, 2, 2)\}$ **19.** Inconsistent; $\varnothing$ **21.** $\{(8, 12, -9)\}$ **23.** Inconsistent; $\varnothing$ **25.** $\left\{\left(\frac{1}{2}, \frac{1}{3}, -1\right)\right\}$ **27.** Dependent; $\{(x, y, z)|x = y, z = 0\}$

29. $\{(4, 6, 8)\}$ **31.** 7, 4, and 5   **33.** Missouri is 2540 miles, Mississippi is 2340 miles, Yukon is 1980 miles   **35.** 100°, 60°, 20°
37. 13 triangles, 21 rectangles, 6 pentagons   **39.** Length, 5 cm; width, 3 cm; height, 4 cm   **41.** 3 servings of A, 2 servings of B, 4 servings of C
43. $1200 at 8%; $2000 at 10%; $3500 at 12%   **45.** 5 $2 packages, 3 $3 packages, 4 $4 packages
47. Truck I: 5 cu yd; Truck II: 6 cu yd; Truck III: 8 cu yd   **49.** C   **51.** Students should verify results.   **53.** Answers may vary.
55. Two planes could be parallel with one equation in one plane and two equations in the other plane

57. $\{(-2, 1, 4, 3)\}$ **59.** 5 adults tickets, 7 adult with one child, 3 adult with 2 children   **61.** $\left\{\left(\frac{a-b+c}{2}, \frac{a+b-c}{2}, \frac{-a+b+c}{2}\right)\right\}$
63. Group activity

Review Problems

65. $y - 5 = -3(x - 3)$ or $y - 2 = -3(x - 4)$; $y = -3x + 14$; $3x + y - 14 = 0$
66.

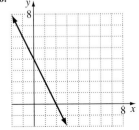

4 square units

67. a. Not a function   **b.** Function   **c.** Not a function   **d.** Function

PROBLEM SET 3.4

1. $x - 3y = 11$, $y = -3$; $\{(2, -3)\}$ **3.** $x - 3y = 1$, $y = -1$; $\{(-2, -1)\}$ **5.** $x + 2y + z = 0$, $y = -2$, $z = 3$; $\{(1, -2, 3)\}$
7. $x + y = 3$, $y + \frac{3}{2}z = -2$, $z = 0$; $\{(5, -2, 0)\}$

9. $\begin{bmatrix} 1 & -\frac{3}{2} & 5 \\ 2 & \frac{2}{5} & 5 \end{bmatrix}$ **11.** $\begin{bmatrix} 1 & -\frac{4}{3} & 2 \\ 3 & \frac{5}{5} & -2 \end{bmatrix}$ **13.** $\begin{bmatrix} 1 & -3 & 5 \\ 0 & 12 & -6 \end{bmatrix}$ **15.** $\begin{bmatrix} 1 & -\frac{3}{2} & \frac{7}{2} \\ 0 & \frac{17}{2} & -\frac{17}{2} \end{bmatrix}$ **17.** $\begin{bmatrix} 1 & -1 & 5 & -6 \\ 0 & 6 & -16 & 28 \\ 0 & 4 & -3 & 11 \end{bmatrix}$ **19.** $\begin{bmatrix} 1 & 1 & -1 & 6 \\ 0 & -3 & 3 & -15 \\ 0 & -4 & 2 & -14 \end{bmatrix}$
21. $\{(4, 2)\}$ **23.** $\{(3, -3)\}$ **25.** $\{(2, -5)\}$ **27.**
Inconsistent; $\varnothing$ **29.** Dependent; $\{(x, y)|x - 2y = 1\}$ **31.** $\{(1, -1, 2)\}$
33. $\{(3, -1, -1)\}$ **35.** $\{(1, 2, -1)\}$ **37.** $\{(1, 2, 3)\}$ **39.** $\{(2, 7, 3)\}$ **41.** Dependent; $\{(x, y, z)|x - 2y + z = 4\}$ **43.** $\{(-1, 2, -2)\}$
45. a. $x + y = 90$, $-2x + y = 15$ **b.** 25° and 65°   **47.** 85°, 65°, and 30°   **49.** D   **51.** Students should verify results.
53. Answers may vary.   **55.** $\{(1, 2, 3, -2)\}$ **57. a.** $y = -\frac{36}{5}x^3 + 286x^2 - 3339x + 25{,}130$

b. 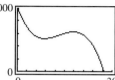   It starts to decrease rapidly.

Review Problems

59. a. $\{1, 2, 7, 8, 9, 11\}$ **b.** $\{8, 9\}$ **60.** $\{x \mid x \le -2\}$ or $(-\infty, -2]$

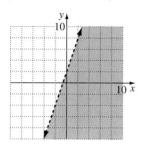

61. 140 miles;

P R O B L E M   S E T   3 . 5

1. 1   **3.** -29 **5.** 0   **7.** 10   **9.** -30 **11.** $\{(5, 2)\}$ **13.** $\{(2, -3)\}$ **15.** $\{(3, -1)\}$ **17.** Dependent; $\{(x, y) \mid x + 2y = 3\}$
19. $\{(3, 1, 2)\}$ **21.** $\{(2, 0, 1)\}$ **23.** $\{(6, -2, 3)\}$ **25.** $\{(-2, 1, 3)\}$ **27. a.** 28 square units
b.

29. Yes   **31.** $-11x - 5y = -8$ **33.** D   **35.** Students should verify results   **37–41.** Answers may vary.
43. $x = \dfrac{3}{14}$ **45.** Use Cramer's Rule for proof.
47. Show that the determinant is equivalent to $x_2y - x_2y_1 - x_1y = y_2x - y_2x_1 - xy_1$
49. a. a^2 **b.** a^3 **c.** a^4 **d.** Diagonal is made up of a's and below the a's are zeros.
e. Number of a's in diagonal is exponent on a in evaluation of determinant.   **51.** Explanations may vary.

Review Problems

52. 21 miles   **53.** $y < 3x + 2$ and $2x + 3y > 12$ **54.** $x = -\dfrac{1}{2}$

P R O B L E M   S E T   3 . 6

1. $3x + 6y \le 6$
$2x + y \le 8$

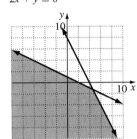

$\left(\dfrac{14}{3}, -\dfrac{4}{3}\right)$

3. $y < -2x + 3$
$x - y > 2$

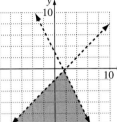

$\left(\dfrac{5}{3}, -\dfrac{1}{3}\right)$

5. $y < -2x + 4$
$y < x - 4$

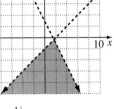

$\left(\dfrac{8}{3}, -\dfrac{4}{3}\right)$

7. $-2 < x \le 4$

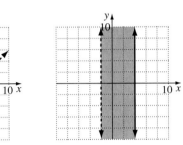

$(3, 3)$

9. $-4 \le y < 2$

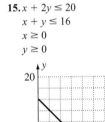

11. $4x - 5y > -20$
$x > -3$

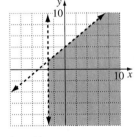

$\left(-3, \dfrac{8}{5}\right)$

13. $x + y \ge 0$
$y \le 3$
$3x - y \le 6$

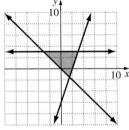

$(-3, 3), \left(\dfrac{3}{2}, -\dfrac{3}{2}\right), (3, 3)$

15. $x + 2y \le 20$
$x + y \le 16$
$x \ge 0$
$y \ge 0$

17. $x + 2y \le 8$
$y \ge 2x$
$y \le 3x$

19. max: 70, min: 17   **21.** max: 610, min: 0
23. max: 19 at (6, 13); min: 1 at (0, 1)
25. max: 30 at (6, 12); min: −24 at (6, −6)

$(12, 4), (0, 10), (16, 0), (0, 0)$

$(0, 0), \left(\dfrac{8}{5}, \dfrac{16}{5}\right), \left(\dfrac{8}{7}, \dfrac{24}{7}\right)$

27. a. $z = 125x + 200y$ **b.** $2x + 3y \le 1200, x \le 450, y \le 200, x \ge 0, y \ge 0$
c.   **d.** 0, 40,000, 77,500, 76,250, 56,250   **e.** 300; 200; $77,500

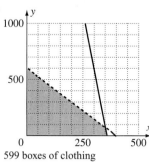

29.

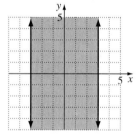

599 boxes of clothing

31.

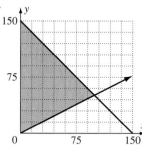

100 parents and 50 students

33. b   **35.** Answers may vary.

37. $|x| \le 3$

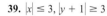

39. $|x| \le 3, |y + 1| \ge 3$ **41.** Proofs may vary.

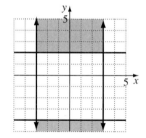

Review Problems

43. -2 **44.** $x = 10$ **45.** $x \ge 7$

CHAPTER 3 REVIEW PROBLEMS

1. $\{(2, 3)\}$ **2.** Dependent   **3.** $\{(-5, -6)\}$ **4.** Inconsistent   **5.** $\{(3, 4)\}$ **6.** Dependent; $\{(x, y) | y = 4 - x\}$ **7.** $\{(0, -1)\}$

8. Inconsistent; $\varnothing$ **9.** $\left\{ \left(3, \frac{8}{3} \right) \right\}$ **10.** 9.4 million in 1992 and 12.4 million in 1993   **11.** 42 mg in shrimp and 15 mg in scallops

12. Cost of one pen: $1.45; cost of one pad: 75¢   **13.** $A = 40$ feet, $B = 25$ feet   **14.** 68°, 56°, 56°   **15.** 4.5 mph, 1.5 mph

16. 4 liters of 90%, 16 liters of 75%   **17.** 12 rabbits, 23 pheasants   **18. a.** $y = -60x + 1000$ **b.** $y = 4x + 200$

c. (12.5, 250); At a video price of $12.50, the number sold equals the number supplied, 250.   **d.** $y = -60x + 1000$, $y = 4x + 200$;

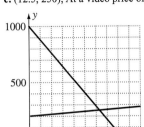

19. a. 5400 gallons   **b.** $x < 5400$ gallons   **c.** Sales cannot be negative   **d.** $520 per week
20. $\{(2, -2, -1)\}$ **21.** $\{(0, 2, 3)\}$ **22.** $\{(1, 2, 3)\}$
23. New York City: 30,531; Chicago: 12,598; Los Angeles: 7631   **24.** 37°, 40°, and 103°

25. 3 oz of A, 2 oz of B, 4 oz of C   **26.** $\{(-5, 3)\}$ **27.** $\left\{\left(1, \frac{5}{3}\right)\right\}$ **28.** $\{(1, 3, -2)\}$ **29.** $\{(3, 5, -1)\}$

30. $\{(x, y, z) | 2x - y - 3z = 1\}$ **31.** Inconsistent   **32.** 17   **33.** 4   **34.** -97 **35.** -236
36. $\{(3, 4)\}$ **37.** $\{(3, 0)\}$ **38.** $\{(-1, 2, 3)\}$ **39.** $\{(-4, 4, 4)\}$ **40.** Inconsistent; $\varnothing$
41. Dependent; $\{(x, y, z) | x - y - 2z = 1\}$

42. $3x - y \leq 6$
$x + y \geq 2$

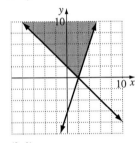

$(2, 0)$

43. $y < -x + 4$
$y > x - 4$

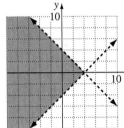

$(4, 0)$

44. $-3 \leq x < 5$

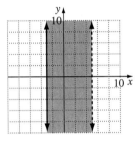

45. $-2 < y \leq 6$

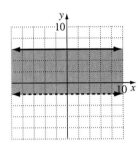

46. $x \geq 3$
$y \leq 0$

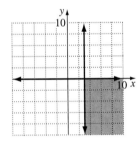

$(3, 0)$

47. $2x - y > -4$
$x \geq 0$

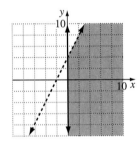

$(0, 4)$

48. $x + y \leq 6$
$y \geq 2x - 3$

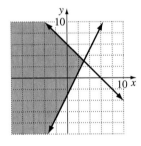

$(3, 3)$

49. $3x + y \leq 6$
$4x + 6y \geq 24$
$x \geq 0, y \geq 0$

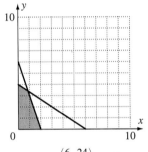

$(0, 0), (0, 4), \left(\frac{6}{7}, \frac{24}{7}\right), (2, 0)$

50. max: 10, min: 2   **51.** max: 1130; min: 1050   **52. a.** $z = 500x + 350y$ **b.** $x + y \leq 200, x \geq 10, y \geq 80$
c.

d. (10, 80): $z = 33,000$; (10, 190): $z = 71,500$; (120, 80): $z = 88,000$ **e.** 120; 80; $88,000

CHAPTER 3 TEST

1. $\{(2, 4)\}$ **2.** $\{(6, -5)\}$ **3.** $\{(1, -3)\}$ **4.** Dependent, infinite solutions   **5.** Teachers is $34,027, attorneys is $66,271.
6. 43°, 47°, and 90°   **7.** Orchestra, $23; mezzanine, $14   **8.** $30; 400   **9.** $\{(-9, 5, 5)\}$ **10.** $\{(2, 1, 1)\}$ **11.** A, $60; B, $75, C, $90

12. $\{(4, -2)\}$ **13.** $\{(-1, 2, 2)\}$ **14.** 17   **15.** 38   **16.** $\{(-1, -6)\}$ **17.** 4

18. $\left(\frac{7}{2}, -\frac{3}{2}\right)$ **19.** $(0, 0), (0, 3), (4, 0), \left(\frac{8}{3}, \frac{5}{3}\right)$ **20.** $(0, 1), (1, 3), (6, 3), (10, 1)$

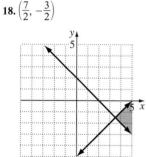

CUMULATIVE REVIEW PROBLEMS (CHAPTERS 1–3)

1. $x = -15$ **2.** $x = \frac{1}{5}$ **3.** $-\frac{2y^7}{3x^5}$ **4.** 4.743×10^6 **5.** $\{x \mid x < -2 \text{ or } x \geq 3\}; (-\infty, -2) \cup [3, \infty)$ **6.** 2006

7. $f(10) = 3.58$; The average number of people per family in 1950 was 3.58.   **8.** 89 dots; $\frac{n^2 + 7n + 8}{2}$ dots

9. $f(x) = x^2 - 1$; for each value of x there is exactly one value of y.   **10.** $\left\{-\frac{1}{4}, \frac{7}{4}\right\}$

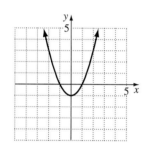

11. a. For each value of x there is exactly one value for y.   **b.** $\left(\frac{1}{2}, 119\right)$; Approximately $\frac{1}{2}$ hr after eating a candy bar, the glucose level is at a maximum of about 119 mg per deciliter.   **c.** $\left(2\frac{1}{2}, 104\right)$; Approximately $2\frac{1}{2}$ hours after eating an apple, the glucose level is at a maximum of about 104 mg per deciliter.   **d.** $\left(1\frac{1}{3}, 96\right)$; Approximately $1\frac{1}{3}$ hours after eating either food the glucose is equal at about 96 mg per deciliter.   **e.** Apple; The graph of the candy bar rises quickly, but drops off quickly.

12. $y - 4 = -3(x - 2)$ or $y + 2 = -3(x - 4)$; $y = -3x + 10$; $3x + y = 10$

13. $y - 3 = \frac{2}{3}(x - 2)$; $y = \frac{2}{3}x + \frac{5}{3}$; $2x - 3y = -5$

14. $2x - 4y = -8$ **15.** $f(x) = \frac{3}{4}x - 6$ **16.** $y = 4$ **17.** $6x - 3y < 12$ and $x < 2$

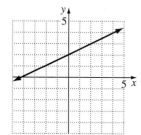   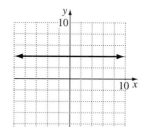   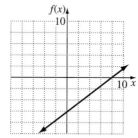   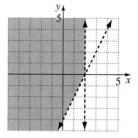

18. 141   **19.** $\{x \mid x < -5 \text{ or } x > 6\}; (-\infty, -5) \cup (6, \infty)$ **20.** $a = \frac{R + 2b}{2}$ **21. a.** 48,144.9; The average rate of change per year from 1900 to 1910 is an increase of 48,144.9 people   **b.** -4743.9; The average rate of change from 1910 to 1920 is a decrease of 4,743.9 people   **22.** 15

23. $\{(3, -4)\}$ **24.** $\{(6, -5)\}$ **25.** $\{(3, 0, -2)\}$ **26.** $\{(2, -3, 2)\}$ **27.** $\left\{\left(-\frac{27}{4}, \frac{9}{2}\right)\right\}$ **28.** $\{(5, -4, 3)\}$ **29.** 50 ft by 94 ft

30. $y - 124 = \frac{6}{5}(x - 30)$ or $y - 136 = \frac{6}{5}(x - 40)$; $y = \frac{6}{5}x + 88$; 160

Chapter 4

PROBLEM SET 4.1

1. $-12x^4 - 4x^3 + x^2 + 9x - 6$; Deg. of term 4, 3, 2, 1, 0; Degree of polynomial: 4

3. $x^4 + 6x^3y^2 + 5x^2 + 7xy - 3$; Deg. of term: 4, 5, 2, 2, 0; Degree of polynomial: 5

5. $-3x^6y + 8x^5 + 4x^3yz^2 + 5x^2y^4z^3 - 12y$; Deg. of term: 7, 5, 6, 9, 1; Degree of polynomial: 9

7. Binomial, degree 2   **9.** Five terms, degree 3   **11.** Trinomial, degree 3

13. Vertex: $(1, 4)$ **15.** Vertex: $\left(\frac{1}{2}, -2\frac{1}{4}\right)$ **17.** Vertex: $\left(-\frac{1}{2}, 2\frac{1}{4}\right)$ **19.** Vertex: $(-1, 0)$

$y = -x^2 + 2x + 3$ $y = x^2 - x - 2$ $y = -x^2 - x + 2$ $y = 4x^2 + 8x + 4$

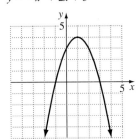 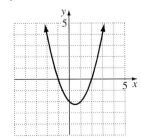 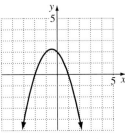 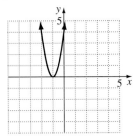

Range: $\{(f(x)|f(x) \le 4\}$ Range: $\left\{f(x)|f(x) \ge -2\frac{1}{4}\right\}$ Range: $\left\{f(x)|f(x) \le 2\frac{1}{4}\right\}$ Range: $\{f(x)|f(x) \ge 0\}$

21. $f(x) = -2x^3 + 6x^2 + 2x - 6$ **23.** $f(x) = x^3 + 2x - 1$

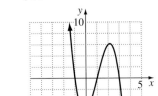 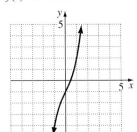

25. $11x^2 - 10x - 8$ **27.** $-7r^3 + 5r^2 + 10r + 2$ **29.** $-5x^3y + x^2y + 9xy$ **31.** $12x^3 + 4x^2 + 12x - 14$

33. $22r^5 + 9r^4 + 7r^3 - 13r^2 + 3r - 5$ **35.** $-13x^3y^2 - 3x^2y + 2xy$ **37.** $6x^2 - 6x + 2$ **39.** $7y^4 - 9y^3 - 5y^2 + 9y - 2$

41. $(f + g)(x) = x^2 + 4x + 3$; $(f - g)(x) = x^2 - 3$; $f(1) = 3$; $g(1) = 5$; $(f + g)(1) = 8$

43. $(f + g)(x) = -2x^2 + 3$; $(f - g)(x) = -4x^2 + 10x + 11$; $f(1) = 9$; $g(1) = -8$; $(f + g)(1) = 1$

45. $(f + g)(x) = -11x^3 + 11x^2 - 14x$; $(f - g)(x) = 3x^3 + 5x^2 + 4x$; $f(1) = -1$; $g(1) = -13$; $(f + g)(1) = -14$

47. $(f + g)(x) = -4x^3 - 3x^2 - 3x - 5$; $(f - g)(x) = 10x^3 - 11x^2 + 13x + 1$; $f(1) = -1$; $g(1) = -14$; $(f + g)(1) = -15$

49. a.   **b.** At each value of x, the graph of $h(x)$ is the sum of the values of $f(x)$ and $g(x)$.

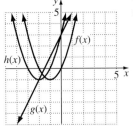

51. $13.3t^2$ **53.** 1957, 12.4 miles per gallon   **55.** d   **57.** Students should verify problems.

59. $y = -0.25x^2 + 40x$
Vertex: $(80, 1600)$

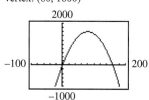

61. $y = 5x^2 + 40x + 600$
Vertex: $(-4, 520)$

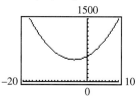

63. c

65. $y = -x^3 + 8x^2 + 10x - 15$

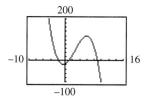

67. $y = x^2 - 8x - 2$

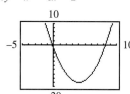

69. $y = 7x^3 + x^2 + 5x$

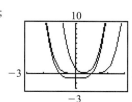

71. Answers may vary.   **73.** $3x^{2n} - x^n + 4$ **75.** $9x^2 - 3x + 5$
77. $g(-6)$

79. a. $f(x) = x^4$, $g(x) = (x - 1)^4$, $h(x) = x^4 - 1$;

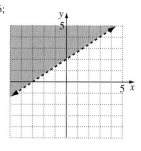

b. They have similar shapes, but different vertex points.
c. Rises, rises

81. From left to right: $f(x) = x$, $g(x) = x^2$, $h(x) = x^3$, $F(x) = x^4$, $G(x) = x^5$, $H(x) = x^6$ **83.** 125, 216, 343; x^3 **85.** 123, 214, 314; $x^3 - 2$
87. 130, 222, 350; $x^3 + x$

Review Problems

88. $x = \dfrac{2}{3}$ **89.** $2x - 3y < -6$;

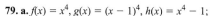

90. $y - 5 = -\dfrac{1}{3}(x + 2)$; $y = -\dfrac{1}{3}x + \dfrac{13}{3}$; $x + 3y = 13$

PROBLEM SET 4.2

1. y^{12} **3.** $15x^{14}$ **5.** $15x^3y^{11}$ **7.** $-6x^2y^9z^9$ **9.** $12x^3 - 8x^2$ **11.** $-10x^5 + 16x^4 - 14x^3 + 6x^2$ **13.** $8x^3y^2 + 14x^2y - 4xy^2 - 6xy$
15. $\dfrac{1}{6}x^4y^{13} + \dfrac{2}{15}x^7y^9 + 2x^3y^7$ **17.** $-2x^5y^5z^4 + x^8y^6z^5 - \dfrac{1}{2}x^{10}y^7z^4$ **19.** $-30u^6v^6w^2 + 40u^6v^4w - 5u^5v^3w^5$ **21.** $x^3 - x^2 - x - 15$
23. $x^3 + 125$ **25.** $a^3 - b^3$ **27.** $x^4 + 5x^3 + x^2 - 11x + 4$ **29.** $x^3y^3 + 8$ **31.** $5a^3b^2 + 5a^3b - 18a^2b^2 - 3a^2b + 11ab^2 + 2ab^3 - 6b^3$
33. $x^2 + 11x + 28$ **35.** $y^2 - y - 30$ **37.** $35x^2 + 26x + 3$ **39.** $6y^2 - 31y + 33$ **41.** $21x^2 + 20x - 96$ **43.** $27x^4 + 33x^2 - 20$
45. $40x^6 - 31x^4 + 6x^2$ **47.** $15z^3 - 20z^2 + 3z - 4$ **49.** $36x^2 + 43x - 35$ **51.** $81x^5 + 36x^3 - 36x^2 - 16$ **53.** $6x^2 - xy - 35y^2$
55. $15x^4y^2z^2 + 11x^2y^2z - 14y^2$ **57.** $x^2 + 6x + 9$ **59.** $y^2 - 10y + 25$ **61.** $4x^2 + 4xy + y^2$ **63.** $25x^2 - 30xy + 9y^2$

65. $4x^4 - 12x^2y + 9y^2$ **67.** $16x^6 + 16x^3y^5 + 4y^{10}$ **69.** $16a^2b^4 + 8a^2b^3 + a^2b^2$ **71.** $y^2 - 49$ **73.** $25x^2 - 9$ **75.** $9x^2 - 49y^2$
77. $x^4 - y^2z^2$ **79.** $49x^4y^2 - 9z^2$ **81.** $9x^6y^4z^2 - 1$ **83.** $y^4 - x^2y^2$ **85.** $9x^2 + 42x + 49 - 25y^2$ **87.** $25y^2 - 4x^2 - 12x - 9$
89. $4x^2 + 4xy + y^2 + 4x + 2y + 1$ **91.** $9x^2 - 6x + 1 + 6xy - 2y + y^2$ **93.** $9x^2 - 6x + 1 - y^2$ **95.** $x^4 - 16$ **97.** $81x^4 - y^4$
99. $(f + g)(x) = 3x^2 - 1; (f - g)(x) = x^2 + 7; (fg)(x) = 2x^4 - 5x^2 - 12$ **101.** $(f + g)(x) = x^3 + 2x^2 - x + 4; (f - g)(x) = -x^3 - 3x - 2;$
$(fg)(x) = x^5 - x^4 + 2x^2 - 5x + 3$ **103.** $f(a + 3) = a^2 + 4a + 8; f(2a - 1) = 4a^2 - 8a + 8; f(a + h) - f(a) = 2ah + h^2 - 2h$
105. $f(a + 3) = 3a^2 + 23a + 40; f(2a - 1) = 12a^2 - 2a - 4; f(a + h) - f(a) = 6ah + 3h^2 + 5h$ **107.** $6x^2 + 3x$ **109.** $20x^2 + 16x$
111. $20x + 150$ cm^3 **113.** $\dfrac{x^2y - 8xy + 16y}{16}$ or $\dfrac{x^2y}{16} - \dfrac{xy}{2} + y$ **115.** b
117. $y_1 = y_2$ **119.** $y_1 = y_2$

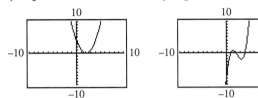

121. Answers may vary.   **123.** $6y^n - 13$ **125.** $12x^2 - 11x - 5$ **127.** $x^2 + 2x$ **129.** $3x^2 + x - 4$ **131.** $x^3 + 7x^2 - 3x$
133. a. $y^2 - 1$ **b.** $y^3 - 1$ **c.** $y^4 - 1$ **d.** $y^5 - 1$ **135.** $121 - 110 + 25$ **137.** Group activity

Review Problems

138. $\left\{x \mid x \le -2 \text{ or } x \ge \dfrac{14}{3}\right\}; (-\infty, -2] \cup \left[\dfrac{14}{3}, \infty\right)$ **139.** $4y^2 - 5y - 1$ **140.** $\{(1, -1, 2)\}$

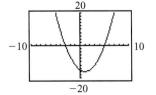

P R O B L E M   S E T   4 . 3

1. $4(x - 5)$ **3.** $9(2a + 3)$ **5.** $4x(3x + 1)$ **7.** $9(x - 2y)$ **9.** $4xy(3x - 2y)$ **11.** $xy(4 - 7y)$ **13.** $9x^2(2x^2 + x - 3)$
15. $2y^3(5y^4 - 8y + 4)$ **17.** $5x^3y^3(3x^2 - 5y)$ **19.** $ab(2a - 5b + 7ab)$ **21.** $12xy^2(2y - 3x^2 + xy^2)$ **23.** $13x^5y^2(2y + 4x^2 - 3x^3y^3)$
25. $10x^2(3x^2y + 5y^2 + 2xz^3)$ **27.** $11x^2y^2z(5z^3 - 7x)$ **29.** $7xy(x^2y + 2x - 6x^4y^2 + 3y^3)$ **31.** $14xy(5x^2 + 3x - 2xy - 6y^2)$
33. $(x - 3)(7 + y)$ **35.** $-2(x + 7y)(a - 2c)$ or $2(x + 7y)(2c - a)$ **37.** $2(a + b - c)(2x - y)$ **39.** $(x - 4y)(z^2 + z + 1)$
41. $(x - 4y)(z^2 + z - 1)$ **43.** $5x^2(2a - 7b)(x + 3)$ **45.** $11x^2y(7a - 5b)(7xy + 1)$ **47.** $4(-x + 3); -4(x - 3)$ **49.** $3x(-x + 9); -3x(x - 9)$
51. $3x(-x^2 + 5x - 7); -3x^2(x^2 - 5x + 7)$ **53.** $1(-x^2 + 7x - 5); -1(x^2 - 7x + 5)$ **55.** $b(-b^3 + 3b^2 - 15); -b(b^3 - 3b^2 + 15)$
57. $(x - 3)(x^2 + 4)$ **59.** $(y - 1)(y^2 + 2)$ **61.** $(y - 1)(y^2 - 5)$ **63.** $(a + 5)(ac + 2)$ **65.** $(3Y + 4Z)(Y + 8)$ **67.** $(a - 3b)(2 + c)$
69. $(b + c)(3a - 1)$ **71.** $(x^2 + 3)(5 - y)$ **73.** $(2b + c)(a - 3)$ **75.** $(a - b)(a - 1)$ **77.** $P(1 + r)^2$ **79.** $2\pi r(h + r)$
81. a. $s(t) = -16t(t - 4)$ **b.** $(2, 64);$ **83.** c

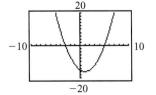

85. $y = (x + 3)(x - 5)$ **87.** $y = (x + 1)(x^2 - 1)$

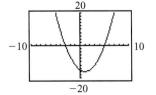

89. Answers may vary.   **91.** $x^{2m}y^m(3x^m + 7y^m)$ **93. a.** $0.95x - 0.05(0.95x)$ **b.** $0.9025x$

Review Problems

94. Length: 8 feet; Width: 3 feet   **95.** $L = \dfrac{12A + B}{2}$ **96.** $\{(5, -1, -2)\}$

P R O B L E M   S E T   4 . 4

1. $(x + 3)(x + 2)$ **3.** $(a + 5)(a + 3)$ **5.** $(x + 5)(x + 4)$ **7.** $(d + 8)(d + 2)$ **9.** $(t + 3)^2$ **11.** $(Y - 7)^2$ **13.** $(x - 5)(x - 3)$
15. $(y - 10)(y - 2)$ **17.** $(y + 7)(y - 2)$ **19.** $(x + 6)(x - 5)$ **21.** $(d - 7)(d + 4)$ **23.** $(R - 9)(R + 4)$ **25.** $(x - 7)(x - 5)$
27. Prime   **29.** $(X + 10Y)(X + 2Y)$ **31.** $(W - 11Z)(W + Z)$ **33.** Prime   **35.** $2x(x + 2)(x + 1)$ **37.** $(2M + 7)(M + 1)$
39. $(4x - 1)(x + 2)$ **41.** $(5T + 2)(T + 3)$ **43.** $(3b + 5)(2b + 3)$ **45.** $(3x + 4)(2x + 3)$ **47.** $(3y - 5)^2$ **49.** $(4a - 3)(a - 6)$
51. $(2S - 3)(8S + 9)$ **53.** $(4y - 9)(y + 2)$ **55.** $(3M + N)(3M - 2N)$ **57.** $(5x - 3y)(2x + 7y)$ **59.** Prime   **61.** $5w(3w^2 - 5w + 2)$
63. $(-x + 9)(x + 3)$ **65.** $(-12x - 1)(x - 3)$ **67.** $4a^2b(a + 2)(a - 8)$ **69.** $2xy(6x - 5)(3x + 2)$ **71.** $(2x - 3y)^2$
73. $(7a + 3b)(5a - 8b)$ **75.** $(4x - 13y)(2x + 3y)$ **77.** $(2xy + 3)(3xy + 2)$ **79.** $13x^3y(y + 4)(y - 1)$ **81.** $(y^2 + 3)(y^2 + 2)$
83. $(5m^2 + 6)(m + 1)(m - 1)$ **85.** $(2n^2 - 3m)(n^2 + 2m)$ **87.** $(y^3 - 12)(y^3 + 3)$ **89.** $(y^4 + 13)(y^4 - 3)$ **91.** $(a - 3b + 4)(a - 3b - 9)$
93. $(5a + 5b + 7)(a + b + 1)$ **95.** $(6x + 6y + 7b)(3x + 3y - 4b)$ **97.** $(3a^2 - 3ab - 2b)(2a^2 - 2ab - 3b)$

99. a. $s(t) = -16(t - 2)(t + 1)$ **b.** Vertex: $\left(\dfrac{1}{2}, 36\right)$ **c.** The diver rises to 36 feet at $\dfrac{1}{2}$ second then drops to the water at 2 seconds.   **101.** b

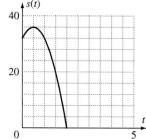

103. $y = (x + 4)(x + 3)$ **105.** $y = -(3x + 2)(x - 6)$ **107.** $y = x(3x + 4)(2x - 1)$

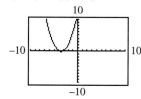   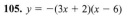   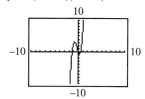

109. Answers may vary.   **111.** $(x^n - 3)(x^n - 5)$ **113.** $(4y^m - 5)(y^m - 1)$ **115.** $6, -6, 9,$ and -9 **117.** $8, -8, 16,$ and -16

Review Problems

119. $P = \dfrac{I}{1 + rt}$ **120.** $\{x | x > 5\}; (5, \infty)$ **121.** $y + 2 = 2(x - 3); y = 2x - 8; 2x - y = 8$

P R O B L E M   S E T   4 . 5

1. $(B + 1)(B - 1)$ **3.** $(5 + a)(5 - a)$ **5.** $(6x + 7)(6x - 7)$ **7.** $(6x + 7y)(6x - 7y)$ **9.** $(xy + ab)(xy - ab)$ **11.** $(xy^3 + a^2b)(xy^3 - a^2b)$
13. $(2xy^3 + 5a^2b)(2xy^3 - 5a^2b)$ **15.** $(9ab^2c^3 + 7x^4y)(9ab^2c^3 - 7x^4y)$ **17.** $(x + 3 + y)(x + 3 - y)$ **19.** $(x + y + 6)(x + y - 6)$
21. Prime   **23.** $(4y + 3x - 1)(4y - 3x + 1)$ **25.** $-4(x + 2)$ **27.** $-(5x + 1)(x + 3)$ **29.** $(a^7 + 3)(a^7 - 3)$ **31.** Prime
33. $(x + 4)(x - 4)$ **35.** $(xy + 5A)(xy - 5A)$ **37.** $(y^2 + 1)(y + 1)(y - 1)$ **39.** $(1 + 9b)(1 - 9b)$ **41.** $y[(xy + 4)(xy - 4)]$
43. $3x(x + 1)(x - 1)$ **45.** $3xy(x + 2)(x - 2)$ **47.** $(p + q)(p^2 - pq + q^2)$ **49.** $(3x - 4y)(9x^2 + 12xy + 16y^2)$
51. $(3R - 1)(9R^2 + 3R + 1)$ **53.** $(5b + 4)(25b^2 - 20b + 16)$ **55.** $(5 + 4d)(25 - 20d + 16d^2)$ **57.** $(2ab - 3)(4a^2b^2 + 6ab + 9)$
59. $(4 - 3YZ)(16 + 12YZ + 9Y^2Z^2)$ **61.** $(2x + 3y^4)(4x^2 - 6xy^4 + 9y^8)$ **63.** $(ab^2 - c^2d^4)(a^2b^4 + ab^2c^2d^4 + c^4d^8)$
65. Prime   **67.** $2x(x^2 + 3y^2)$ **69.** $-2y(12x^2 + y^2)$ **71.** $(y + 2)^2$ **73.** $(z - 5)^2$ **75.** $(3x - 2y)^2$ **77.** $(x + y + 1)^2$ **79.** $(v - w + 2)^2$
81. $(x - 3 + y)(x - 3 - y)$ **83.** $(3x - 5 + 6y)(3x - 5 - 6y)$ **85.** $(r + 4s - 3)(r - 4s + 3)$ **87.** $(y + x + 2)(y - x - 2)$
89. $(z + x - 2y)(z - x + 2y)$ **91.** $(3a + b)(3a - b)$ **93. a.** $A^2 - B^2$ **b.** $(A + B)(A - B)$ **c.** $A^2 - B^2 = (A + B)(A - B)$ **95.** True
97. 1521   **99.** d

101. $y = (x + 2)^2$

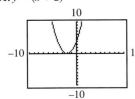

103. $y = -(x + 7)(x - 3)$

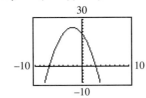

105. $y = (x + 1)^2$

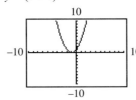

107. $y = (x + 2)(x^2 + x + 1)$

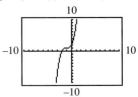

109. Error is in step that says, "Divide both sides by $(a - b)$." Since $a = b$, $a - b = 0$ and division by 0 is undefined.
111. $(2x^n + 5y^m)^2$ **113.** $k = 9$ **115.** $k = 49$ **117.** $(x + y)(x^2 - xy + y^2 + 1)$ **119.** Group activity

Review Problems

121. $y > x - 2$ and $x + y > -2$;   **122.** 4 hours   **123.** $x \leq -14$

PROBLEM SET 4.6

1. $c(c + 4)(c - 4)$ **3.** $3(x + 3)^2$ **5.** $3(3x - 1)(9x^2 + 3x + 1)$ **7.** $(B + 4)(B - 4)(C - 2)$ **9.** $-(x - 9)(x - 3)$ **11.** $2b(2a + 5)(a - 3)$
13. $(y + 2)(y - 2)(a - 4)$ **15.** $11x(x^2 + y)(x^2 - y)$ **17.** $3(x + y)(x - y + 1)$ **19.** Prime   **21.** $a(x + 2)(x^2 - 2x + 4)$
23. $(s - 6 + 7t)(s - 6 - 7t)$ **25.** $(2m^5 + 3n^3)^2$ **27.** $9t^2(s + 2)(s - 2)$ **29.** $(x + y)(a + b)$ **31.** $5yz(x + y)(x - y)$
33. $5ab(2a + 7b)(2a - 7b)$ **35.** $3(7y - 6)(3y + 4)$ **37.** $(r^3 + 2s)^2$ **39.** $4a(x - 2)(x^2 + 2x + 4)$ **41.** $4x^2(5x + 3y)^2$ **43.** $(7x + 9y)^2$
45. $71b(x^2 + 1)(x + 1)(x - 1)$ **47.** Prime   **49.** $(r - s)(r^2 + rs + s^2 + 1)$ **51.** $(x + 3)(x - 3)(x + 1)(x - 1)$
53. $(a + 2 + 4b)(a + 2 - 4b)$ **55.** $3rs(3r + 4s)^2$ **57.** $-6by(y + 2)(y - 2)$ **59.** Prime   **61.** $(3x - y + 10a)(3x - y - 10a)$
63. $(5x + 3y - 3)^2$ **65.** $(2y^2 - 5)(3y^2 + 2)$ **67.** $3(4y^2 + 9)(2y + 3)(2y - 3)$ **69.** $5b(2x^2 + 11y)^2$ **71.** $9x(2x - 1)(x + 4)$
73. $4x(x^2 + 2y)(x^4 - 2x^2y + 4y^2)$ **75.** $(x - 3)(x + 2)(x + 3)(x - 2)$ **77.** Prime   **79.** $(r^2 + 1)(r + 1)(r - 1)(s - 3)$ **81.** $(y - 2)^2(y + 2)$
83. $(y - 1)(y^2 + y + 1)(a - b)$ **85.** Prime   **87.** $2x(2 - x)(4 + 2x + x^2)$ **89.** $(y - 2)(y + 1)(y - 1)$ **91.** $a^3b^3(ab - 1)(a^2b^2 + ab + 1)$
93. $x(10x^2 - 6x - 21)$ **95.** $x^2 - y^2$ **97.** $\pi h(R - r)(R + r)$ **99.** d
101. Substitute various values of a into each side of the equation. Graph each to see if they are identical.   **103.** $2(x - 3a)^2$
105. $xa(x + 4a)(x - 4a)$ **107.** Answers may vary.   **109.** Answers may vary.   **111.** $5rs^2(r - s)(r + s - 1)$ **113.** $(5x - y + 6z)(5x - y - 4z)$
115. $(x + y)(x - y)^3$ **117.** $(x^2 + 1)(2x^2 + x + 2)$

Review Problems

119. 6   **120.** $x = -1$ **121.** $\dfrac{8x^9}{y^3}$

PROBLEM SET 4.7

1. $\{-3, 7\}$ **3.** $\left\{-\dfrac{3}{2}, \dfrac{1}{5}\right\}$ **5.** $\{-4, 3\}$ **7.** $\{-7, 1\}$ **9.** $\left\{-4, \dfrac{2}{3}\right\}$ **11.** $\left\{\dfrac{3}{5}, 1\right\}$ **13.** $\left\{-\dfrac{7}{3}, \dfrac{5}{2}\right\}$ **15.** $\left\{-\dfrac{1}{5}, -5\right\}$ **17.** $\left\{-\dfrac{2}{5}, 1\right\}$

19. $\left\{-2, \dfrac{1}{3}\right\}$ **21.** $\left\{-\dfrac{7}{5}, 3\right\}$ **23.** $\{-1, 2\}$ **25.** $\left\{\dfrac{2}{3}, 5\right\}$ **27.** $\{-6, 9\}$ **29.** $\left\{-\dfrac{3}{2}, 1\right\}$ **31.** $\left\{-\dfrac{1}{2}, \dfrac{4}{3}\right\}$ **33.** $\{-2, 8\}$ **35.** $\{-3\}$

37. $\{-14, 9\}$ **39.** $\left\{-\dfrac{1}{3}, 5\right\}$ **41.** $\left\{-\dfrac{1}{3}\right\}$ **43.** $\left\{\dfrac{5}{3}\right\}$ **45.** $\{-3, 6\}$ **47.** $\{1, 2\}$ **49:** $\{-1, 1\}$ **51.** $\{-5, -4, 5\}$ **53.** $\{-5, 1, 5\}$

55. $\{-2, -1, 0, 2\}$

57. $y = x^2 + 6x + 5$

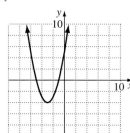

59. $y = x^2 + 4x + 3$

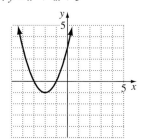

61. $y = -x^2 - 4x - 5$

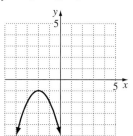

63. $y = -x^2 - 4x - 3$

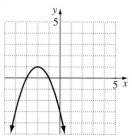

65. a. 5 seconds   **b.** 144 ft; 2 seconds **c.**   **d.** 4 seconds; (4, 80); same height as (0, 80)

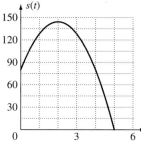

67. a. Arousal level should be 50.   **b.** 50   **69.** After 10 days   **71.** $20   **73.** 5 meters   **75.** 12 inches $\times$ 12 inches

77. Distance up the wall is 24 m; length of the ladder is 26 m   **79.** 8 inches, 15 inches and 17 inches   **81. a.** $y = -3x^2 + 51x + 4025$

b. 1968; 4242 per capita   **c.**   **d.** Consumption decreases   **83.** c

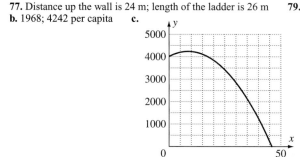

85.

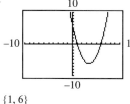

$\{1, 6\}$

87.

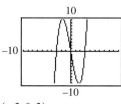

$\{-3, 0, 3\}$

89.

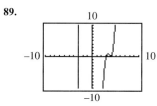

$\{-3, 3, 4\}$

91. Answers may vary.   **93.** $\{-8, -6, 4, 6\}$ **95.** $\left[-1, \frac{1}{2}\right]$ **97.** 30 hectometers $\times$ 40 hectometers or 20 hectometers $\times$ 60 hectometers

99. 10 meters   **101.** $\{-1\}$

Review Problems

102. $(2x - 5y)(2x + 5y - 2)$ **103.** 68   **104.** $\{(2, 0, -4)\}$

CHAPTER 4 REVIEW PROBLEMS

1. a. Trinomial; degree 3   **b.** Monomial; degree 8   **c.** Binomial; degree 7   **d.** 5 terms; degree 5   **2.** $x^3 - 6x^2 - x - 9$

3. $12x^3y - 2x^2y - 14y - 17$ **4.** $11x^2 - 11x + 8$ **5.** $15y^3 - 10y^2 + 11y + 6$ **6.** $12x^5y^2 - 4x^3y - 8$ **7.** $7a^4b^3 + 3ab^3 - 9$

8. $(4, -1)$ **9.** In 1990, the number of mountain bike owners in the U.S. was 15.012 million.   **10.** 1968, 4238

11. b., c., and d., are all slightly different from the basic graph of a.; As x goes to $-\infty$, y goes to $-\infty$ or ∞. As x goes to ∞, y goes to $-\infty$ or ∞;

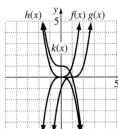

12. Answers may vary. **13.** $-12x^6y^2z^7$ **14.** $2x^8 - 24x^5 - 12x^3$ **15.** $21x^{10}y^5 - 35x^7y^7 - 42x^3y^4$
16. $6x^3 + 29x^2 + 27x - 20$ **17.** $15x^4 + x^3 - 42x^2 + 46x - 16$ **18.** $2x^3y^2 + 2x^2y - 6xy^2 + 8xy - 6y + 8$
19. $12x^2 - 26x + 10$ **20.** $12x^2y + 11xyz - 5z^2$ **21.** $56x^6 + 51x^4 - 27x^2$ **22.** $9x^2 + 42xy + 49y^2$
23. $4x^4y^2 - 12x^2yz + 9z^2$ **24.** $9x^2 - 30x + 25 + 48xy - 80y + 64y^2$ **25.** $9x^2 + 6xy - 12x - 4y + y^2 + 4$
26. $4x^2 - 49y^2$ **27.** $49a^4b^2 - 9b^2$ **28.** $25y^2 - 4x^2 - 28x - 49$ **29.** $(f + g)(x) = 8x^2 - 2x + 6$;
$(f - g)(x) = -8x^2 + 12x - 12$; $(fg)(x) = 40x^3 - 59x^2 + 66x - 27$ **30.** $V(x) = 4x^3 - 18x^2 + 20x$
31. $4x^2 - 88x + 468$ **32.** 3 m $\times$ 7 m **33. a.** $4a^2 + 43a + 116$ **b.** $4h^2 + 8ah - 5h$

34. 72.75 feet **35.** $3x(5x + 1)$ **36.** $5x^3y^2(x - 4y + 3x^3)$ **37.** $(x + 5)(x^2 - 2)$ **38.** $(y^2 + 9)(x - 2)(x + 2)$ **39.** $(c - d)(b + 1)$
40. $(x + 36)(x + 1)$ **41.** $x(x - 13)(x - 2)$ **42.** $-2x(x - 2)(x - 16)$ **43.** $(4y^2 + 3)(2y^2 - 5)$ **44.** $(3x + 5)(2x - 7)$
45. $-2a^2(a - 9)(a - 3)$ **46.** $(3y^3 - 1)(2y^3 + 5)$ **47.** $(3x + 8)(x + 6)$ **48.** $4(x - 2)(x + 2)$ **49.** $(9x^2 + 10y^2)(9x^2 - 10y^2)$
50. $(-2x + 5)(4x + 3)$ **51.** $(b + 3)(b - 3)(x + 5)(x - 5)$ **52.** $(x - 2)(x + 2)(x^2 + 4)$ **53.** $4(a + 2)(a^2 - 2a + 4)$
54. $(1 - 4y)(1 + 4y + 16y^2)$ **55.** $(2x + 3)^2$ **56.** Prime **57.** $y(y + 1)(y - 1)(y^2 + 1)$ **58.** $(3x - 5)^2$ **59.** $(x + 3 - 2a)(x + 3 + 2a)$
60. $(3x - 5y)(3x - 2y)$ **61.** $(a + b - 2)(a - b + 2)$ **62.** $(3x + 5y)^2$ **63.** $2a(a + 3)^2$ **64.** $-(x - 7)(x + 3)$
65. $5x(x - 2)(x^2 + 2x + 4)$ **66.** $(x - 3)(x + 3)(2x - 1)$ **67.** $(x + 3)(x - 3)(x + 2)(x - 2)$ **68.** $(x^2 - 3)^2$
69. $(x + y)(x^2 - xy + y^2 + 1)$ **70.** $(3b - 5c)(9b^2 + 15bc + 25c^2)$ **71.** $2xy(5x - 4)(x + 3)$ **72.** $(x - 1)^2(x^2 + x + 1)$

73. $-x(7x^2 + 9xy + 3y^2)$ **74.** $\{-5, -1\}$ **75.** $\{2, 10\}$ **76.** $\left\{\frac{1}{2}, 1\right\}$ **77.** $\{0, 4\}$ **78.** $\{3\}$ **79.** $\{-5, -3, 3\}$

80. $y = x^2 + 5x + 4$ **81.** $y = -2x^2 + 12x - 10$ **82. a.** 5 seconds **b.** 640 feet; 0 seconds
c.

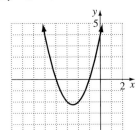
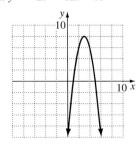
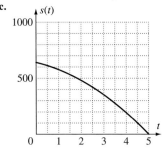

83. 4 milligrams **84.** 12 meters **85.** 30 feet by 40 feet **86.** 2 cm **87.** 1 cm **88.** 14 in. $\times$ 28 in.
89. $y = 104.5x^2 - 1501.5x + 6016$; 5 hours: 1121; 6 hours: 769; 11 hours: 2144

C H A P T E R 4 T E S T

1. $7x^3y - 18x^2y - y - 9$ **2.** $11x^2 - 13x - 6$ **3.** $84x^7y^3$ **4.** $21x^2 - 20xy - 9y^2$ **5.** $15a^2b^2 + 7abc - 4c^2$ **6.** $x^3 - 4x^2y + 2xy^2 + y^3$
7. $25x^2 + 90xy + 81y^2$ **8.** $4x^2 - 28x + 49 - 12xy + 42y + 9y^2$ **9.** $49x^2 - 9y^2$ **10.** $x^3 + 4x^2 + 3x$ **11.** $8x^2 + 8x + 2$
12. Deficit is increasing. **13.** $-\$269.09$ billion, not very good **14.** $x^2(14x - 15)$ **15.** $(9y - 5)(9y + 5)$ **16.** $(x - 5)(x + 5)(x + 3)$
17. $(5x - 3)^2$ **18.** $(x + 5 - 3y)(x + 5 + 3y)$ **19.** Prime **20.** $(y - 18)(y + 2)$ **21.** $(7x + 3)(2x + 5)$ **22.** $5(p - 1)(p^2 + p + 1)$
23. $3(2x - y)(2x + y)$ **24.** $2(3x - 1)(2x - 5)$ **25.** $3(x - 1)(x + 1)(x^2 + 1)$ **26.** $b^6(3a - 2)(9a^2 + 6a + 4)$ **27.** $\left\{-\frac{1}{3}, 2\right\}$
28. $\{-5, -1\}$ **29.** $\{-1, 0, 1\}$ **30.** $y = x^2 - 6x + 8$ **31.** 2 seconds **32.** 5 seconds, 400 feet **33.** Width is 4 m, length is 14 m

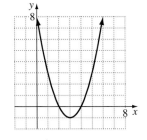

CUMULATIVE REVIEW PROBLEMS (CHAPTERS 1–4)

1. Associative property of addition   **2.** 4×10^6 or 4,000,000   **3.** 2   **4.** $\left\{-\frac{1}{4}, \frac{15}{8}\right\}$ **5.** {4}   **6.** $\{x \mid 2 < x < 3\}$, (2, 3)

7. $x = \dfrac{b}{c - a}, c \neq a$ **8.** 2   **9.** 2015; Answers may vary.   **10.** $y = 2x + 1$ **11.** 1   **12.** {(−12, −1)}   **13.** {(1, 2, −3)}

14. $\left\{\left(-\frac{1}{38}, -\frac{8}{19}\right)\right\}$ **15.** $\left\{\left(-\frac{1}{3}, -2, -\frac{2}{3}\right)\right\}$ **16.** 20,000 books;

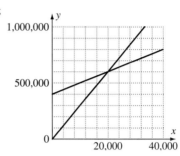

17. Can of paint: $16; each paint brush: $7   **18.** $\dfrac{1}{4}$ **19.** Loser: 1320 votes; winner: 1480 votes   **20.** 2133   **21.** $12x^3 + 14x^2 - 40x - 7$

22. $5x^2 + 20x$ **23.** $\left\{x \mid x \leq -2 \text{ or } x \geq 7\right\}$, $(-\infty, -2] \cup [7, \infty)$ **24.** $-\dfrac{y^{10}}{2x^6}$ **25.** $2x - y < -4$ and $x < -2$;

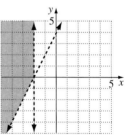

26. $(x + 2)(x - 2)(x^2 + 3)$ **27.** $(x + 3)(x - 3)^2$ **28.** $(a + b + c)^2 = a^2 + 2ab + 2ac + b^2 + 2bc + c^2$ **29.** $3200

30. a. $\{6, \sqrt{9}, 1\}$ **b.** $\{6, 0, \sqrt{9}, 1\}$ **c.** $\{-5, 6, 0, \sqrt{9}, 1\}$ **d.** $\left\{\frac{1}{4}, -5, 6, 0, \sqrt{9}, -3.72, 1\right\}$ **e.** $\{\sqrt{10}\}$

f. $\left\{\frac{1}{4}, -5, 6, 0, \sqrt{9}, \sqrt{10}, -3.72, 1\right\}$

Chapter 5

PROBLEM SET 5.1

1. $\{x \mid x \neq 4\}$ or $(-\infty, 4) \cup (4, \infty)$ **3.** $\{x \mid x \neq 0, 3\}$ or $(-\infty, 0) \cup (0, 3) \cup (3, \infty)$ **5.** $\{x \mid x \neq -1, 1\}$ or $(-\infty, -1) \cup (-1, 1) \cup (1, \infty)$

7. $\{x \mid x \neq -3, 4\}$ or $(-\infty, -3) \cup (-3, 4) \cup (4, \infty)$ **9. a.** $h(x) = \dfrac{3}{x^2 - 1}$ **b.** $g(x) = \dfrac{3}{x^2 + 1}$ **c.** $f(x) = \dfrac{2}{(x - 1)^3}$

11. a. Domain of $f = \{x \mid x \neq 2\}$ or $(-\infty, 2) \cup (2, \infty)$ **b.** $-2, -4, -20, -200, -2000; 2000, 200, 20, 4, 2$ **c.**

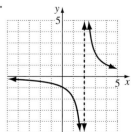

13. $\dfrac{1}{y+1}$ **15.** $\dfrac{3}{a+3}$ **17.** $\dfrac{c-6}{4}$ **19.** $\dfrac{a-1}{a+1}$ **21.** $\dfrac{x^2-y^2}{x^2+y^2}$ **23.** $\dfrac{a+9}{a-1}$ **25.** $\dfrac{x-2}{x-3}$ **27.** $\dfrac{c-7}{c+7}$ **29.** $\dfrac{x+3}{x^2+3x+9}$

31. $\dfrac{y-6}{y^2+3y+9}$ **33.** $\dfrac{x+5}{x-3}$ **35.** $\dfrac{x-3}{x+2}$ **37.** Cannot be reduced.   **39.** Cannot be reduced   **41.** $\dfrac{x+2y}{x-5}$ **43.** $-a-2$

45. $x-2\ (x\neq -1)$ **47.** $x-7\ (x\neq -2)$

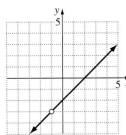   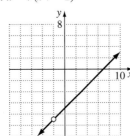

49. $\dfrac{y+2}{y+3}$ **51.** $\dfrac{3(x+y)}{2}$ **53.** $\dfrac{b(b+3)}{(b-2)(b-1)}$ **55.** $\dfrac{2}{m+3}$ **57.** $\dfrac{b+2}{b+6}$ **59.** $\dfrac{x^2+2x+4}{3x}$ **61.** $\dfrac{a-2}{a-3}$ **63.** 1   **65.** $\dfrac{2a}{3}$ **67.** $\dfrac{1}{3x}$

69. 4   **71.** $\dfrac{x^2+3x+9}{a^2-2a+4}$ **73.** $\dfrac{4b(2b+5)}{5(2b-5)}$ **75.** $\dfrac{a-7}{a-5}$ **77.** $\dfrac{(3x-2)(x+1)}{(x-1)(x-2)}$ **79.** $\dfrac{(b+3)^2}{x-y}$ **81. a.** $x-7$ **b.** -4 **c.** -4

83. a. $\dfrac{x(x+4)}{x+2}$ **b.** 9.6   **c.** 9.6   **85. a.** \$16,000   **b.** 76; The cost to remove 95% of pollutants is \$76,000.

c. $\{x|x\neq 100\}$; Impossible to remove 100%   **d.** Goes to infinity   **87.** $\dfrac{0.54t^2+12.64t+107.1}{-0.14t^2+0.51t+31.6}$ billion/million; 1994: \$5294/person;

1995: \$5997/person; 1996: \$6833/person; 1997: \$7843/person; Average: \$6492/person   **89. a.** $T(x)=\dfrac{600}{x}+\dfrac{600}{x-10}$

b. 27; The total time for a trip traveling 600 miles at 50 mph and 600 miles at 40 mph is 27 hours.   **91. a.** 35 words, 11 words, 7 words

b. 35; 11; 7   **c.** Decreases   **d.** No matter how much time goes by, the number of words remembered will always be 5.   **93.** $\dfrac{3}{x}$ **95.** c   **97.** d

99. 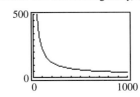   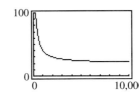

The graph shows how the price approaches \$20 as the number of canoes produced increases.

101. a. Exactly the same   **b.** $f(x)=x+1; x\neq 2$; No   **c.** $g(x)=3$ at $x=2$, $f(x)$ is undefined at $x=2$

103–107. Answers may vary.   **109.** $\dfrac{6x^{2n}}{x^n-y^n}$ **111.** $y(y-3)$ **113.** $\dfrac{-c^2}{4(a-b)}$ **115.** $c^2-4cd+8d^2$ **117.** $\dfrac{7x}{6}$ **119.** Group activity

Review Problems

120. $6x^3-25x^2+33x-20$ **121.** 8×10^5 or 800,000   **122.** $y\geq 3x-2$

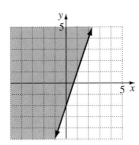

PROBLEM SET 5.2

1. $\dfrac{5}{x}$ **3.** $\dfrac{1}{xy^2}$ **5.** 2   **7.** $\dfrac{2(x+4)}{x-4}$ **9.** 1   **11.** $\dfrac{a-2}{a+2}$ **13.** $\dfrac{x}{2}$ **15.** 1   **17.** $\dfrac{a^2+ab+b^2}{a+b}$ **19.** $\dfrac{17}{6x}$ **21.** $\dfrac{35-4x}{14x^2}$ **23.** $\dfrac{9-4x}{18x^3}$

25. $\dfrac{8ad-9bc}{12c^2d}$ **27.** $\dfrac{9ad^2-4acd+ac^2}{6c^2d^2}$ **29.** $\dfrac{b^2+bc-2c^2}{b^2c^2}$ **31.** $\dfrac{5by-6y^2+4b^2}{12b^2y^2}$ **33.** $\dfrac{22z-21y}{24yz}$ **35.** $\dfrac{8x-68}{(x+4)(x-6)}$

37. $\dfrac{b^2+c^2}{(b-c)(b+c)}$ **39.** $\dfrac{-3}{a(a+1)}$ **41.** $\dfrac{4x^2+13x-2}{(x-2)(x+2)}$ **43.** $\dfrac{2x^2+5x+12}{(x-3)(x+2)}$ **45.** $\dfrac{-4ab}{(a+b)(a-b)}$ **47.** $\dfrac{3x^2+4x+4}{x(x+1)(x+2)}$

49. $\dfrac{13b - 17}{2(b - 4)(2b - 1)}$ **51.** $\dfrac{4x + 16}{(x + 3)^2}$ **53.** $\dfrac{5x - 17}{x - 5}$ **55.** $\dfrac{3x^2 + x - 16}{x^2 - 4}$ **57.** $\dfrac{-c^2 + 11c - 20}{2(c - 5)^2}$ **59.** $\dfrac{5a^2 + 2ab + 5b^2}{6(a + b)(a - b)}$

61. $\dfrac{b + 3}{(b + 1)(b - 1)}$ **63.** $\dfrac{7y + 1}{(y + 1)^2(y - 2)}$ **65.** $\dfrac{1}{(x - 1)(x^2 + x + 1)}$ **67.** 0   **69.** $y + 1$ **71.** $\dfrac{x^2 - 2x - 3}{x - 2}$

73. $\dfrac{12xy}{(x - 4y)(x - 3y)(x + 3y)}$ **75.** $\dfrac{3}{y^2 - 1}$ **77.** $\dfrac{3b - 2}{5}$ **79.** $\dfrac{1}{y}$ **81.** $\dfrac{x^2 + y^2}{x + y}$ **83.** $b(b + c)$ **85.** $\dfrac{1}{x}$ **87.** $\dfrac{y + 2}{y - 3}$ **89.** $\dfrac{a - 1}{a + 1}$

91. $\dfrac{1}{xy}$ **93.** $bc(b - c)$ **95.** $\dfrac{1}{c}$ **97.** $\dfrac{b - 2}{b + 1}$ **99.** $\dfrac{y - 4}{y - 7}$ **101.** $\dfrac{-y + 14}{7}$ **103.** $\dfrac{(b + 2)^2}{b^2 - 1}$ **105.** $\dfrac{-x + 1}{(x - 3)(x + 6)}$ **107.** $\dfrac{3x - 2y}{5x + 13y}$

109. a. $t = \dfrac{24}{10 - x} + \dfrac{24}{10 + x}$ **b.** $t = \dfrac{480}{100 - x^2}$ **c.** 5; If the current is 2 mph, then the total trip will take 5 hours.   **111.** $\dfrac{2r_1 r_2}{r_2 + r_1}$; 24 mph

113. d   **115. a.** $\dfrac{2x^2 + 5000}{x}$ **b.**   **c.** (50, 200)   **d.** 50 feet; square   **117–123.** Answers may vary.

125. a. $\dfrac{1}{42}$; $\dfrac{1}{42}$; Equal   **b.** Show sum = product   **127.** $\dfrac{2x}{x^2 - y^2}$ **129.** $\dfrac{4x^2 + 8x - 2\pi}{x^2}$ **131.**

1260	840	630
504	420	360
315	280	252

Review Problems

133. $\left\{x \mid -\dfrac{13}{3} \le x \le 5\right\}$ **134.** $x^2(2x^2 - 9)$ **135.** $\{3, 9\}$

P R O B L E M   S E T   5 . 3

1. y^2 **3.** $3x^2$ **5.** $-5x^2y^2$ **7.** $-18x^4y^3z^3$ **9.** $8x^6 - 5x^3 + 6x^2$ **11.** $4x^2 - 2x - 5$ **13.** $-3x^5 + 2x^3 + 6x$ **15.** $2x^4 - x + \dfrac{1}{3x} - \dfrac{5}{3x^2} + \dfrac{2}{3x^3}$

17. $1 - xy^2 - x^3y^4$ **19.** $4x - 7y - 5y^3$ **21.** $6x - \dfrac{3}{y} - \dfrac{2}{xy^2}$ **23.** $-2x^2 + \dfrac{3y}{2x} + \dfrac{7x}{2y} + \dfrac{3}{xy}$ **25.** $a + 3$ **27.** $b + 2$ **29.** $c + 12$

31. $b^2 + b - 2$ **33.** $2b^2 + 3b + 5$ **35.** $a^2 + 2ab$ **37.** $4x + 3 + \dfrac{2}{3x - 2}$ **39.** $2y^2 + y + 6 - \dfrac{38}{y + 3}$

41. $4x^3 + 16x^2 + 60x + 246 + \dfrac{984}{x - 4}$ **43.** $x^2 + x + 1$ **45.** $2a + 5$ **47.** $y^2 - 2y + 1$ **49.** $6y^2 + 3y - 1 + \dfrac{-3y + 1}{3y^2 + 1}$ **51.** $2x + 5$

53. $3x - 8 + \dfrac{20}{x + 5}$ **55.** $4x^2 + x + 4 + \dfrac{3}{x - 1}$ **57.** $6y^4 + 12y^3 + 22y^2 + 48y + 93 + \dfrac{187}{y - 2}$ **59.** $x^3 - 10x^2 + 51x - 260 + \dfrac{1300}{x + 5}$

61. $z^4 + z^3 + 2z^2 + 2z + 2$ **63.** $y^3 + 4y^2 + 16y + 64$ **65.** $2y^4 - 7y^3 + 15y^2 - 31y + 64 - \dfrac{129}{y + 2}$ **67.** $8t^2 + 4t$; 144 gallons

69. $3x - 5$ **71.** $2y - 5$ **73.** c

75. $y = x^2 + 4x + 1$ **77.** $y = 3x^3 - 8x^2 - 5$

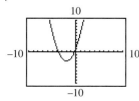

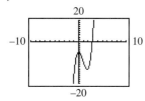

79–81. Answers may vary.   **83.** $x - 2$ **85.** $9y^{2n} + 3y^n + 1$ **87.** $K = 10$ **89.** $K = -1$

Review Problems

91. $\{(-5, 3)\}$ **92.** $2(2x + 5)(4x^2 - 10x + 25)$ **93.** $\{(-3, -1, 4)\}$

PROBLEM SET 5.4

1. {2}   **3.** {3}   **5.** {5}   **7.** {6}   **9.** $\left\{-\dfrac{1}{4}\right\}$ **11.** {3}   **13.** {7}   **15.** $\left\{-\dfrac{9}{2}\right\}$ **17.** No solution   **19.** No solution   **21.** {−6, 3}

23. {−1, 8}   **25.** $\left\{\dfrac{8}{5}, 5\right\}$ **27.** {2}   **29.** $\left\{-\dfrac{43}{2}, 2\right\}$ **31.** {−2, 2}   **33.** {−6, 2}   **35.** {2}   **37.** {2, 5}   **39.** {−1}   **41.** {2}

43. {4}   **45.** {1, 5}   **47.** 98%; The height of the bar representing 98% is at 196 million dollars.   **49.** 10 mph   **51.** 1 minutes; 2 minutes

53. 1993   **55.** 25 species per thousand individuals   **57.** $p = \dfrac{100C}{C + 4}$; 98   **59. a.** $n = \dfrac{32 - 2t}{t - 6}$ **b.** 18 trials   **61.** $\dfrac{-fp}{f - p}$ or $\dfrac{fp}{p - f}$

63. $\dfrac{Fd^2}{m_1 m_2}$ **65.** $\dfrac{-dw}{d - l}$ or $\dfrac{dw}{l - d}$ **67.** $\dfrac{d_2 - d_1 + vt_1}{v}$ **69.** $\dfrac{F_s x}{x - r - p}$ **71.** $\dfrac{3.78p}{t^2 N}$ **73.** $R_1 = 180$ ohms, $R_2 = 90$ ohms   **75.** 40,000 calculators

77. 15   **79.** 4 mph   **81.** 10 mph   **83.** 10 days   **85.** 10 minutes   **87.** $x = 5$ **89.** d   **91.** Students should verify results.

93–95. Students should verify results.   **97.** Answers may vary.   **99.** $N = \dfrac{TC}{C - V}$ **101.** 5   **103.** $\left\{-\dfrac{19}{5}, 13\right\}$ **105.** 0

107. $b = 20$ **109.** 8% and 9%   **111.** 54 mph

Review Problems

113. $f(x) = x^2 - 6x + 8$

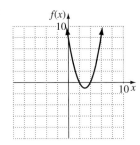

114. $14x^2 - 18$ **115.** $16x^4 - 8x^2 y + y^2$

PROBLEM SET 5.5

1. a. 7   **b.** $y = 7x$ **c.** 84   **3. a.** $5.\overline{3}$ **b.** $y = 5.\overline{3}x^2$ **c.** $261.\overline{3}$ **5. a.** 50   **b.** $y = \dfrac{50}{x}$ **c.** 25   **7. a.** 75   **b.** $y = \dfrac{75}{x^2}$ **c.** $\dfrac{3}{4}$

9. a. $\dfrac{1}{15}$ **b.** $z = \dfrac{1}{15}\left(\dfrac{x}{y}\right)$ **c.** $\dfrac{2}{5}$ **11. a.** $\dfrac{5}{2}$ **b.** $y = \dfrac{5}{2}xz$ **c.** 240   **13. a.** $\dfrac{1}{9}$ **b.** $y = \dfrac{1}{9}\left(\dfrac{x^2 z}{R^3}\right)$ **c.** 2   **15.** 16π ft   **17.** $555\dfrac{5}{9}$ feet

19. 6.4 lb   **21.** $11\dfrac{1}{9}$ foot candles   **23.** 15 m³   **25.** approximately $746.67   **27.** 67.5 units   **29.** 32; No   **31.** 0.5 ohm   **33.** b   **35.** a

37–41. Answers may vary.   **43.** $\dfrac{K_1}{K_2}$ **45.** 16 times original force   **47.** Group activities

Review Problems

49. $y(y^2 + 2)(y - 3)$ **50.** $3x - 6y < 12$

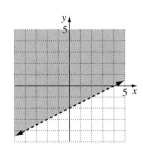

51. {(1, 2, −1)}

CHAPTER 5 REVIEW PROBLEMS

1. Domain of $f = \{x | x \neq 2\}$ or $(-\infty, 2) \cup (2, \infty)$ **2.** Domain of $f = \{x | x \neq -5, 1\}$ or $(-\infty, -5) \cup (-5, 1) \cup (1, \infty)$

3. Domain of $f = \left\{x \mid x \neq -3, \frac{1}{2}\right\}$ or $(-\infty, -3) \cup \left(-3, \frac{1}{2}\right) \cup \left(\frac{1}{2}, \infty\right)$ **4.** Domain of $f = \{x \mid x \in R\}$ or $(-\infty, \infty)$

5. a. Domain of $f = \{x \mid x \neq -2\}$ or $(-\infty, -2) \cup (-2, \infty)$ **b.** $-1, -2, -10, -100, -1000, 1000, 100, 10, 2, 1$

c.

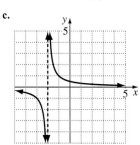

6. c   **7. a.** $C(x) = 250{,}000 + 3x$ **b.** $\overline{C}(x) = \dfrac{250{,}000 + 3x}{x}$ **c.** 253; 28; 5.5; Cost per clock drops from
$253 to $28 to $5.50 as the number of clocks produced increases from 1000, to 10,000, to 100,000, respectively.
Cost drops as number produced increases.   **8.** $\dfrac{x}{3} - \dfrac{7}{3x}$ **9.** $\dfrac{x-1}{x-7}$ **10.** $\dfrac{3m+2}{m-5}$ **11.** $\dfrac{y^2+2y+4}{y+2}$

12. $\dfrac{x+y}{x(x-y)}$ **13.** $x - 3 \ (x \neq 4)$ **14.** $\dfrac{5(x+1)}{3}$ **15.** $\dfrac{x-7}{x^2}$ **16.** $4(y+3)$ **17.** $\dfrac{15}{x}$ **18.** $\dfrac{5x-1}{x-5}$

19. $\dfrac{1}{6}$ **20.** $\dfrac{1}{b+c}$ **21.** $\dfrac{1}{y+2}$ **22. a.** $2x - 5$ **b.** 7   **c.** 7   **23. a.** $t = \dfrac{30}{x} + \dfrac{30}{x+10}$

b. $\dfrac{60x+300}{x^2+10x}$ **24.** $\dfrac{x^3+5}{x^3+125}$ **25.** $\dfrac{1}{x+3}$ **26.** $\dfrac{3x-5}{x(x-5)}$ **27.** $\dfrac{4x}{9x^2-16}$ **28.** $\dfrac{7-5x+4y^2}{x^2y^3}$

29. $\dfrac{y-3}{(y+3)(y+1)}$ **30.** $\dfrac{2x^2-9}{(x+3)(x-3)}$ **31.** $\dfrac{5a^2-7a+6}{(a-2)^2(a+2)(a-1)}$ **32.** $\dfrac{4x+1}{x^3}$ **33.** $\dfrac{2y-1}{2y+1}$ **34.** $\dfrac{3x+8}{3x+10}$ **35.** $\dfrac{(y+6)(y-3)}{1-y}$

36. $7xy - 5x - 3x^3y$ **37.** $3x - 7 + \dfrac{26}{2x+3}$ **38.** $2x^2 - 4x + 1 - \dfrac{10}{5x-3}$ **39.** $2x^2 + 3x - 1$ **40.** $4x^2 - 7x + 5 - \dfrac{4}{x+1}$

41. $3y^3 + 6y^2 + 10y + 10 + \dfrac{20}{y-2}$ **42.** No solution   **43.** $\{7\}$ **44.** $\left\{-\dfrac{9}{2}\right\}$ **45.** $\left\{-\dfrac{1}{2}, 3\right\}$ **46.** $\{-23, 2\}$ **47.** $\{-3, 2\}$ **48. a.** 80%

b. $\dfrac{100C}{C+4}$ **c.** 100; Impossible to remove 100%   **d.** Increases rapidly as x approaches 100; cost increases to infinity as percent approaches 100%.

49. 1993   **50. a.** $\dfrac{1}{2}$ hr and 2 hr   **b.** 1.5 mg/L   **c.** Rises rapidly, then decreases slowly   **51.** $R - pn$ **52.** $\dfrac{A}{rT+1}$ **53.** $\dfrac{fq}{q-f}$

54. $\dfrac{Ir}{E-IR}$ **55.** 9 cm   **56.** 5000 desks   **57.** 12 mph   **58.** 9 mph   **59.** Mother: 45 days; Norman: 36 days   **60.** 12 hours   **61.** 180 kg
62. 40 pounds   **63.** 784 feet   **64.** 360 pounds; Yes

CHAPTER 5 TEST

1. $\{x \mid x \neq 2, 5\}, \dfrac{x}{x-5}, x \neq 2, x \neq 5$ **2.** Domain of $f(x)$ does not include $x = -2$ and $f(-2) = \dfrac{1}{4}$.   **3.** $\dfrac{x^2+4x}{x^2-6}$ **4.** $\dfrac{x-3y}{3}$ **5.** $\dfrac{x+1}{x-1}$

6. $\dfrac{x^2+2x+15}{x^2-9}$ **7.** $\dfrac{3x-4}{(x-3)(x+2)}$ **8.** $\dfrac{5x^2-7x+4}{(x-2)(x+2)(x-1)}$ **9.** $\dfrac{y^2}{x^2-y^2}$ **10.** $\dfrac{a-1}{3a}$

11.

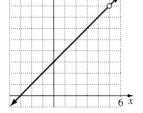

12. $\dfrac{x-2}{8}$ **13.** $\dfrac{(x+2)(x-3)}{(x+3)(5x-12)}$ **14** $4y^2 - \dfrac{5y}{2} + 3x^2y^2$ **15.** $3x^2 - 3x + 1 + \dfrac{2}{3x+2}$

16. $3x^3 - 4x^2 + 7$ **17.** $\{-3, -2\}$ **18.** $\{5\}$ **19.** 750 manuals   **20.** 2000   **21.** $\dfrac{Rs}{s-R}$ **22.** -4

23. 12 hours   **24.** 4 mph   **25.** 45 foot candles

CUMULATIVE REVIEW PROBLEMS (CHAPTERS 1–5)

1. $\left\{-\dfrac{15}{4}\right\}$ **2.** 3   **3.** $\{x \mid x < -6\}$ or $(-\infty, -6)$ **4. a.** 4 seconds   **b.** 1.5 seconds, 100 feet

c. $s(t) = -16t^2 + 48t + 64$

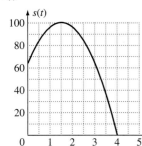

5. 2.5; The average number of people in HMO has risen by 2.5 million each year from 1976 to 1995.
6. Mike weighs the most, which is necessarily the case, since Ana's weight < Juan's weight < Mike's weight
7. a. No; Answers may vary.   **b.** (1935, 200); In the mid- to late-1930s, there were about 200 executions.
c. About 2040   **8.** d   **9.** {(3, 2)}   **10.** {(−2, 0, 4)}   **11.** Speed of plane in still air: 812.5 kph;
speed of wind: 162.5 kph   **12.** −9   **13.** {(2, 1, 2)}   **14.** $14\frac{22}{27}$ footcandles, or approximately
14.815 footcandles   **15.** $6x^3 - 23x^2 + 29x - 12$ **16.** $\left\{-\frac{1}{2}, 4\right\}$ **17.** $3x^2 - 3x + 12$ **18.** 500 copies
19. $(3x - 2)(2x + 5)$ **20.** $(3x + 16)(x + 1)$ **21.** Width: 5 yards; Length: 13 yards
22. 130,000 more people hold more than one job in 1980 than in 1970.   **23.** $(a + b)^2 = a^2 + 2ab + b^2$

24. $\dfrac{3x^2 - 8x + 28}{(x + 2)(x - 2)}$ **25.** $\dfrac{3 - x}{3x^2 + 9x}$ **26.** $-8x^3y^2 - 7xy^3 + 11x^2y^2$ **27.** $\{-4, -3\}$ **28.**

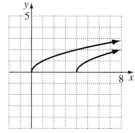

29. $3x + 4 + \dfrac{2}{x + 2}$ **30.** No; x-coordinate 3 has two different y-coordinates, 4 and 11

Chapter 6

PROBLEM SET 6.1

1. 7   **3.** −7   **5.** Not a real number   **7.** 2   **9.** −1   **11.** $\dfrac{1}{3}$ **13.** $-\dfrac{1}{4}$ **15.** $\dfrac{1}{2}$ **17.** Not a real number   **19.** 2   **21.** −3

23. $f(x) = \sqrt{x}, g(x) = \sqrt{x} + 3$ **25.** $f(x) = \sqrt{x}, g(x) = \sqrt{x + 3}$ **27.** $f(x) = \sqrt{x}, g(x) = \sqrt{x} - 4$ **29.** $f(x) = \sqrt{x}, g(x) = \sqrt{x - 4}$

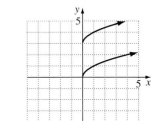

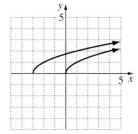

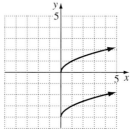

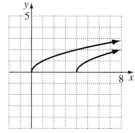

Domain of $f = [0, \infty)$;
Range of $f = [0, \infty)$;
Domain of $g = [0, \infty)$;
Range of $g = [3, \infty)$;
g is f shifted 3 units up.

Domain of $f = [0, \infty)$,
Range of $f = [0, \infty)$;
Domain of $g = [-3, \infty)$,
Range of $g = [0, \infty)$;
g is f shifted 3 units left.

Domain of $f = [0, \infty)$,
Range of $f = [0, \infty)$,
Domain of $g = [0, \infty)$,
Range of $g = [-4, \infty)$;
g is f shifted 4 units down.

Domain of $f = [0, \infty)$,
Range of $f = [0, \infty)$,
Domain of $g = [4, \infty)$,
Range of $g = [0, \infty)$;
g is f shifted 4 units right.

31. $f(x) = \sqrt[3]{x}$, $g(x) = \sqrt[3]{x} - 1$ **33.** $f(x) = \sqrt[3]{x}$, $g(x) = \sqrt[3]{x - 1}$ **35.** $f(x) = \sqrt[3]{x}$, $g(x) = \sqrt[3]{x + 3}$ **37.** $f(x) = \sqrt[4]{x}$, $g(x) = \sqrt[4]{x - 1}$

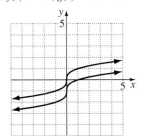
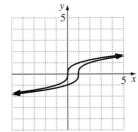
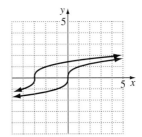
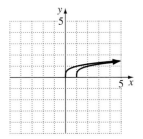

Domain of $f = (-\infty, \infty)$,
Range of $f = (-\infty, \infty)$,
Domain of $g = (-\infty, \infty)$,
Range of $g = (-\infty, \infty)$;
g is f shifted 1 unit down.

Domain of $f = (-\infty, \infty)$,
Range of $f = (-\infty, \infty)$,
Domain of $g = (-\infty, \infty)$,
Range of $g = (-\infty, \infty)$;
g is f shifted 1 unit right.

Domain of $f = (-\infty, \infty)$,
Range of $f = (-\infty, \infty)$,
Domain of $g = (-\infty, \infty)$,
Range of $g = (-\infty, \infty)$;
g is f shifted 3 units left.

Domain of $f = [0, \infty)$,
Range of $f = [0, \infty)$,
Domain of $g = [1, \infty)$,
Range of $g = [0, \infty)$;
g is f shifted 1 unit right.

39. $[-5, \infty)$ **41.** $(-\infty, 4]$ **43.** $[3, \infty)$ **45.** $(-\infty, 2]$ **47.** $(-\infty, \infty)$ **49.** $[-5, \infty)$ **51.** 18   **53.** -1 **55.** $-\dfrac{8}{5}$

57. 346.4; The length of a 400-meter tall starship moving at 148,800 miles/second from the perspective of an observer at rest is about 346.4 meters.
59. 13 syllables   **61.** $21,396; The yearly income for a person with 16 years of education is $21,396.
63. Since $872.45 < 2000$, exposed flesh will not freeze under these conditions.   **65.** d   **67.** Students should verify results.
69. Answers may vary.   **71.** 2   **73.** The numbers approach 1. The decimal part is halved.
75. $f(x) = \sqrt{x}$, $g(x) = \sqrt{-x}$ **77.** $h(x) = \sqrt{x - 3}$

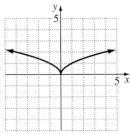

Domain of $f = [0, \infty)$,
Range of $f = [0, \infty)$,
Domain of $g = (-\infty, 0]$,
Range of $g = [0, \infty)$;
g is f reflected about the y-axis.

Review Problems

79. $\dfrac{x^2 - 4x + 16}{x - 4}$ **80.** $(5 - 2x)(25 + 10x + 4x^2)$ **81.** $\{(3, 2, 4)\}$

PROBLEM SET 6.2

1. 4   **3.** -2 **5.** 2   **7.** 2   **9.** -2 **11.** -9 **13.** 9   **15.** -8 **17.** 8   **19.** $\sqrt[4]{x}$ **21.** $\sqrt{36} = 6$ **23.** $\sqrt[3]{8} = 2$ **25.** $\sqrt[6]{xy}$
27. $(\sqrt{16})^5 = 1024$ **29.** $(\sqrt[3]{8x})^2 = (2\sqrt[3]{x})^2 = 4\sqrt[3]{x^2}$ **31.** $-(\sqrt[3]{27})^4 = -81$ **33.** $\dfrac{1}{\sqrt{64}} = \dfrac{1}{8}$ **35.** $\sqrt[3]{64} = 4$ **37.** $-\dfrac{1}{4}$ **39.** $(\sqrt[5]{32})^3 = 8$

41. $\dfrac{1}{(\sqrt{16})^5} = \dfrac{1}{1024}$ **43.** $7^{1/3}$ **45.** $x^{3/2}$ **47.** $x^{3/5}$ **49.** $(ab)^{1/4}$ **51.** $(x^2yz^4)^{1/5}$ **53.** $(19xy)^{3/2}$ **55.** $(11xy^2)^{5/6}$ **57.** 3   **59.** 4

61. $x^{13/12}$ **63.** $x^{3/5}$ **65.** x^2 **67.** $\dfrac{1}{x^{5/12}}$ **69.** $14y^{7/12}$ **71.** $-15x^{1/4}$ **73.** $4x^{1/4}$ **75.** $\dfrac{8}{y^{1/12}}$ **77.** $32xy^{10}z^2$ **79.** $5x^2y^3$

81. $2x^{1/4}y^{1/16}z^{1/6}$ **83.** $\dfrac{8x^{3/4}}{125y}$ **85.** $\dfrac{y^{5/2}}{x^{3/2}}$ **87.** $\sqrt[3]{a}$ **89.** $2x^2$ **91.** x^2y^3 **93.** $\sqrt[3]{2xy^2}$ **95.** $\sqrt[3]{3xy^2}$ **97.** $\sqrt[6]{3^5}$ or $\sqrt[6]{243}$ **99.** $\sqrt[4]{8}$

101. $\sqrt[6]{3}$ **103.** $x^{1/2}(x + 1)$ **105.** $3x^{1/3}(2 + x)$ **107.** $\dfrac{x + 1}{x^{3/5}}$ **109.** $\dfrac{5(3x^2 - 4)}{x^{5/2}}$ **111.** $300,000   **113.** $\dfrac{5}{32}$ units of pollution

115. 10.7%   **117.** The duration of a storm whose diameter is 9 miles is 1.89 hours.   **119.** b   **121. a.** $x \geq 0$ **b.** $x < 0$

123. a. Graphs for $x \geq 0$ **b.** Graphs for all real numbers x.

125. a. 4   **b.**   No; As the temperature drops, the number of O-rings that fail increases.

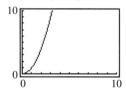

127–129. Answers may vary.   **131.** The birthday boy ate $\frac{1}{2}$ of the cake.

Professor ate $\frac{1}{4}$ of the cake   **133.** $(3x^{1/3} - 4)(2x^{1/3} - 3)$ **135.** 61

137. From line 5 to 6; if $\left(\frac{1}{2}\right)^2 = \left(-\frac{1}{2}\right)^2$ it is incorrect to say $\frac{1}{2} = -\frac{1}{2}$ **139.** $b^{1/m}$ **141.** 4

Review Problems

143. $\{(3, -2, -1)\}$ **144.** $x^2(x - 1)(x + 1)(x^2 + 1)$ **145.** No solution

P R O B L E M   S E T   6 . 3

1. $\sqrt{15}$ **3.** $\sqrt[3]{18}$ **5.** $\sqrt[4]{33}$ **7.** $\sqrt{x^2 - 9}$ **9.** $\sqrt{14xy}$ **11.** $\sqrt[7]{77x^5y^3}$ **13.** $2\sqrt{5}$ **15.** $4\sqrt{5}$ **17.** $5\sqrt{10}$ **19.** $14\sqrt{7}$ **21.** $14\sqrt{2}$
23. $3\sqrt[3]{2}$ **25.** $2\sqrt[5]{2}$ **27.** $12\sqrt[3]{2}$ **29.** $x^3\sqrt{x}$ **31.** $y^2\sqrt[3]{y^2}$ **33.** $z^3\sqrt[5]{z}$ **35.** $x^4y^4\sqrt{y}$ **37.** $x^4y^3\sqrt{x^2z}$ **39.** $2\sqrt{3y}$ **41.** $4x\sqrt{3x}$
43. $2x^4\sqrt[3]{4x}$ **45.** $3x^2y^2\sqrt[3]{3x^2}$ **47.** $2x^2\sqrt[4]{5x^2}$ **49.** $2xy^3\sqrt[5]{2xy^2}$ **51.** $x^5y^2\sqrt[3]{18yz^2}$ **53.** $12x^4y^2z^2\sqrt[4]{2x^3z}$ **55.** $(x + y)\sqrt[3]{x + y}$ **57.** $3\sqrt{2}$
59. 60   **61.** $3\sqrt[3]{2}$ **63.** $2x^2\sqrt{10x}$ **65.** $2x^2y^5\sqrt{3y}$ **67.** $5xy^4\sqrt[3]{x^2y^2}$ **69.** $2xyz\sqrt[4]{x^2z^3}$ **71.** $2xy^2\sqrt[5]{2y^3}$ **73.** $(x + 2)^2\sqrt[3]{x + 2}$
75. $4x^6y^2\sqrt[3]{x}$ **77.** $20\sqrt{2}$ mph   **79.** d
81. $\sqrt{x^4} = x^2$ $(x \geq 0)$ **83.** $\sqrt{18x^2} = 3x\sqrt{2}$ $(x \geq 0)$

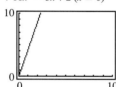

85. Answers may vary.   **87.** Answers may vary.   **89.** Multiplies the square root by $\sqrt{3}$.   **91.** Multiply the number by 27.

Review Problems

93. $\dfrac{-y^7}{3x^4}$ **94.** $2y^2 - 6y + 7$ **95.** $s = \dfrac{33{,}000H}{62.4N}$; 165 feet

P R O B L E M   S E T   6 . 4

1. $2\sqrt{2}$ **3.** $2\sqrt[3]{2}$ **5.** $3x$ **7.** $5\sqrt[3]{3}$ **9.** $-2\sqrt[4]{4}$ **11.** $2x^2\sqrt{5}$ **13.** $3x\sqrt[3]{2y^2}$ **15.** $7x^2\sqrt{x}$ **17.** $\sqrt{x + y}$ **19.** $\dfrac{\sqrt{11}}{2}$ **21.** $\dfrac{\sqrt[3]{19}}{3}$

23. $\dfrac{x}{6y^4}$ **25.** $\dfrac{2x\sqrt{2x}}{5y^3}$ **27.** $\dfrac{x\sqrt[3]{x}}{2y}$ **29.** $\dfrac{x^2\sqrt[3]{75x^2}}{3y^4}$ **31.** $\dfrac{y\sqrt[4]{9y^2}}{x^2}$ **33.** $\dfrac{2x^2\sqrt[5]{2x^3}}{y^4}$ **35.** $\dfrac{\sqrt{2}}{2}$ **37.** $\dfrac{\sqrt{70}}{10}$ **39.** $\dfrac{\sqrt{10}}{4}$ **41.** $\dfrac{\sqrt[3]{4}}{2}$

43. $3\sqrt[3]{2}$ **45.** $\dfrac{\sqrt[3]{18}}{3}$ **47.** $\dfrac{\sqrt[3]{6}}{3}$ **49.** $\dfrac{\sqrt[4]{27}}{3}$ **51.** $\dfrac{\sqrt[4]{40}}{2}$ **53.** $\dfrac{\sqrt{6}}{2}$ **55.** $\dfrac{3\sqrt{x}}{x}$ **57.** $\dfrac{5\sqrt{3x}}{3x}$ **59.** $\dfrac{\sqrt{15x}}{5x}$ **61.** $\dfrac{\sqrt{7xy}}{7y}$ **63.** $\dfrac{\sqrt{21xy}}{7y}$

65. $\dfrac{4\sqrt[3]{x^2}}{x}$ **67.** $\dfrac{\sqrt[3]{2x}}{x}$ **69.** $\dfrac{5\sqrt{3y}}{6y}$ **71.** $\dfrac{7\sqrt[4]{4x}}{2x}$ **73.** $\dfrac{3\sqrt[4]{x^3}}{x}$ **75.** $\dfrac{7\sqrt[5]{y^3}}{y}$ **77.** $\dfrac{\sqrt[3]{28x^2}}{2x}$ **79.** $\dfrac{\sqrt[3]{4xy^2}}{2xy}$ **81.** $\dfrac{\sqrt[5]{12x^2}}{2x}$

83. $\dfrac{\sqrt[4]{135x^3}}{3x}$ **85.** $\dfrac{\sqrt{6}}{8y}$ **87.** $\dfrac{\sqrt{55y}}{10y^2}$ **89.** $\dfrac{\sqrt{6xy}}{8y^2}$ **91.** $\dfrac{3x\sqrt{2x}}{4x^2}$ **93.** $\dfrac{7\sqrt[3]{4x}}{2}$ **95.** $\dfrac{7\sqrt[5]{x^3y}}{2x^2y}$ **97.** $\dfrac{2\sqrt{x}}{xy}$ **99.** $\dfrac{\sqrt[3]{30x^2y^2}}{6x^2y}$

101. b   **103.** $\sqrt{\dfrac{2}{x}} = \dfrac{\sqrt{2x}}{x}, x > 0$ **105.** $\dfrac{x}{\sqrt[4]{2}} = \dfrac{x\sqrt[4]{8}}{2}$

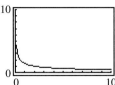

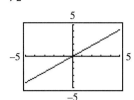

107. Answers may vary.   **109.** $\dfrac{5\sqrt{2x-3y}}{2x-3y}$

Review Problems

111. $\dfrac{10+y}{10y}$ **112.** $8x^3 - 22x^2 + 13x + 3$ **113.** $\{-5\}$

PROBLEM SET 6.5

1. $15\sqrt{3}$ **3.** 0 **5.** $\sqrt{13} + 2\sqrt{5}$ **7.** $7\sqrt{2} + 3\sqrt{11}$ **9.** $\sqrt{15}$ **11.** $8\sqrt{2}$ **13.** $-16\sqrt{2}$ **15.** $20\sqrt{2} - 5\sqrt{3}$ **17.** $3\sqrt{2}$

19. $\sqrt{7} + 7\sqrt{3}$ **21.** $9\sqrt{x}$ **23.** $17\sqrt[4]{2}$ **25.** $6\sqrt[3]{5}$ **27.** $\dfrac{19\sqrt{2}}{12}$ **29.** $-\dfrac{31\sqrt{5}}{12}$ **31.** $17\sqrt[3]{2}$ **33.** $-3\sqrt[3]{2}$ **35.** $8\sqrt{3x}$ **37.** 0

39. $7\sqrt{10}$ **41.** $8\sqrt{6}$ **43.** $\sqrt{6} + \sqrt{14}$ **45.** $8\sqrt{15} + 12\sqrt{21}$ **47.** $140\sqrt{3} - 60\sqrt{2}$ **49.** $28\sqrt{2x} - 12\sqrt{xy}$ **51.** $2x\sqrt{3} - 3x\sqrt{2}$

53. $\sqrt[3]{12} + 4\sqrt[3]{10}$ **55.** $\sqrt{6} + \sqrt{10} + \sqrt{21} + \sqrt{35}$ **57.** $\sqrt{6} - \sqrt{10} - \sqrt{21} + \sqrt{35}$ **59.** $8\sqrt{6} + 12\sqrt{10} + 10\sqrt{21} + 15\sqrt{35}$

61. $4\sqrt{10} - 2\sqrt{6}$ **63.** -44 **65.** -71 **67.** $8 + 2\sqrt{15}$ **69.** $124 - 16\sqrt{21}$ **71.** $x + \sqrt{2x} + \sqrt{3x} + \sqrt{6}$ **73.** $x^2 + 2x\sqrt{y} + y$

75. $x + 2\sqrt{xy} + y$ **77.** $2x^2 + x\sqrt[3]{y^2} - y\sqrt[3]{y}$ **79.** $8\sqrt{5} + 16$ **81.** $\dfrac{13\sqrt{11} - 39}{2}$ **83.** $3\sqrt{5} - 3\sqrt{3}$ **85.** $\dfrac{11\sqrt{7} + 11\sqrt{3}}{4}$

87. $\dfrac{\sqrt{35} - \sqrt{15}}{4}$ **89.** $\dfrac{\sqrt{14} + 2}{5}$ **91.** $24 - 16\sqrt{2}$ **93.** $25\sqrt{2} + 15\sqrt{5}$ **95.** $\dfrac{-2 + \sqrt{15}}{11}$ **97.** $\dfrac{7\sqrt{x} + 35}{x - 25}$ **99.** $\dfrac{y - 3\sqrt{y}}{y - 9}$

101. $\dfrac{13 - 4\sqrt{3}}{11}$ **103.** $4 + \sqrt{15}$ **105.** $\dfrac{x - 2\sqrt{x} - 3}{x - 9}$ **107.** $\dfrac{\sqrt{xy} - y}{x - y}$ **109.** $\dfrac{56 + 5\sqrt{10}}{78}$ **111.** $\dfrac{8x - 2\sqrt{xy} - 3y}{16x - 9y}$

113. $(x^2 + xy + y^2)(\sqrt{x} - \sqrt{y})$ **115.** $P = 12\sqrt{5}; A = 25$

117. $\dfrac{15\sqrt{238}}{4}$ cubic units   **119.** c

121. $x\sqrt{8} + x\sqrt{2} = 3x\sqrt{2}$ **123.** $8\sqrt{x} + 2\sqrt{x} = 10\sqrt{x}, x \geq 0$ **125.** $\dfrac{\sqrt{x^7} + \sqrt{x^3}}{\sqrt{x}} = x^3 + x, x > 0$

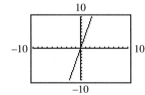

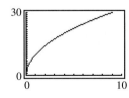

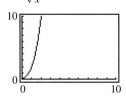

127–133. Answers may vary.   **135.** $\left\{2\sqrt{2}, \dfrac{-14}{3}\sqrt{2}\right\}$ **137.** $(4 + \sqrt{6})(4 - \sqrt{6}) = 16 - 6 = 10$

139. Squaring both sides shows $2\sqrt{15} + 8 = 2\sqrt{15} + 8$ **141.** $\dfrac{3\sqrt{3} + 5\sqrt{2} - 4\sqrt{6} + 2}{23}$ **143.** 1 m^2 **145.** $\dfrac{1}{\sqrt{a+h} + \sqrt{a}}$

Review Problems

147. $\sqrt{5(2) + 6} = \sqrt{3(2) - 2} = 6; \sqrt{16} + \sqrt{4} = 6; 4 + 2 = 6$ **148.** $\left\{-\dfrac{5}{4}, 2\right\}$ **149.** 60 minutes (or 1 hour)

PROBLEM SET 6.6

1. $\left\{\dfrac{17}{3}\right\}$ **3.** $\{16\}$ **5.** No solution   **7.** No solution   **9.** $\{-1, 3\}$ **11.** $\{5\}$ **13.** $\{8\}$ **15.** $\{5\}$ **17.** $\{2\}$ **19.** $\left\{1, -\dfrac{1}{4}\right\}$

21. $\left\{-\dfrac{1}{2}, -1\right\}$ **23.** $\{9\}$ **25.** $\{0\}$ **27.** $\{4\}$ **29.** No solution   **31.** No solution   **33.** $\{2\}$ **35.** $\{0\}$ **37.** $\{1\}$ **39.** $\{1, 3\}$

41. $\{2\}$ **43.** $\{-1\}$ **45.** No solution   **47.** $\{3\}$ **49.** To age 36   **51.** 18 feet   **53.** 128 ft^3 **55.** 1952   **57.** $\dfrac{S^2}{20}$ **59.** $\dfrac{4\pi^2 r^3}{Gt^2}$

61. c   **63.** Students should verify results.   **65.** $\{-5\}$ **67.** $\{-6\}$ **69–71.** Answers may vary.   **73.** $93,000\sqrt{3}$ miles/second   **75.** 81
77. 4   **79.** Group activity

Review Problems

80. $\sqrt[12]{5}$ **81.** $\dfrac{6}{23}$ amperes   **82.** $(x + y - 3)(x - y - 3)$

PROBLEM SET 6.7

1. $2i$ **3.** $i\sqrt{17}$ **5.** $2i\sqrt{7}$ **7.** $3i\sqrt{5}$ **9.** $\dfrac{2}{3}i$ **11.** $10i\sqrt{3}$ **13.** $8 + 3i$ **15.** $8 - 2i$ **17.** $-2 + i$ **19.** $6 + 6i$ **21.** $9 + 5i\sqrt{3}$

23. $16 - 2i\sqrt{2}$ **25.** $-6 + 18i\sqrt{2}$ **27.** $17i$ **29.** $3i$ **31.** $71i$ **33.** $11i\sqrt{2}$ **35.** $i\sqrt{2}$ **37.** $5i\sqrt{2}$ **39.** $29 + 29i$ **41.** $40 - 5i$
43. $13 - 47i$ **45.** $-29 - 11i$ **47.** 34 **49.** 34 **51.** $-5 + 12i$ **53.** $24 + 28i$ **55.** $-\sqrt{14}$ **57.** -6 **59.** $-5\sqrt{7}$ **61.** $-2\sqrt{6}$
63. $-24\sqrt{3}$ **65.** $24\sqrt{15}$ **67.** $4 + 3i\sqrt{2}$ **69.** $-6 + 6i\sqrt{2}$ **71.** $16 - 5i\sqrt{2}$ **73.** $16 - 3\sqrt{2} - (8\sqrt{3} + 2\sqrt{6})i$ **75.** Yes   **77.** Yes

79. No   **81. a.** -1 **b.** $-i$ **c.** -1 **d.** i **83.** $-\dfrac{1}{5} + \dfrac{3}{5}i$ **85.** $-\dfrac{1}{2} - \dfrac{3}{2}i$ **87.** i **89.** $\dfrac{3}{10} + \dfrac{11}{10}i$ **91.** $\dfrac{8}{5} + \dfrac{1}{5}i$ **93.** $-\dfrac{27}{29} - \dfrac{34}{29}i$

95. $-\dfrac{5}{2} - 4i$ **97.** $-\dfrac{7}{3} + \dfrac{4}{3}i$ **99.** $-\dfrac{7}{3}i$ **101.** $5i$ **103.** 2 **105.** $2i\sqrt{10}$ **107.** $(47 + 13i)$ volts   **109.** Sum: 10; Product: 40

111. $31i$ ohms   **113.** b   **115–121.** Answers may vary.   **123.** $23 + 10i$ **125.** $\dfrac{6}{5}$ **127.** $-5, -4,$ and -3

129. The opposite and reciprocal of i are the same number   **131.** $a + bi$ when $b = 0$ **133.** Line 4, since $\sqrt{\dfrac{a}{b}} = \dfrac{\sqrt{a}}{\sqrt{b}}$ only if $a \geq 0$ and $b > 0$.

Review Problems

134. $\dfrac{x + 1}{2x + 3}$ **135.** $\dfrac{500}{3}\pi$ m^3 **136.** $\{(1, -1, 2)\}$

CHAPTER 6 REVIEW PROBLEMS

1. 9   **2.** $\dfrac{2}{7}$ **3.** -3 **4.** $\dfrac{4}{5}$ **5.** -2 **6.** 9.25; The height of a bamboo plant 25 weeks after it comes through the soil is 9.25 inches.

7. $[2, \infty)$ **8.** $(-\infty, 25]$ **9.** $(-\infty, \infty)$

10. $f(x) = \sqrt{x}$, $g(x) = \sqrt{x} + 2$ **11.** $f(x) = \sqrt{x}$, $g(x) = \sqrt{x + 2}$ **12.** $f(x) = \sqrt[3]{x}$, $g(x) = \sqrt[3]{x - 1}$

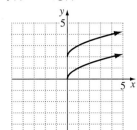 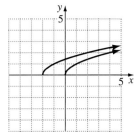 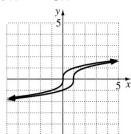

Domain of $f = [0, \infty)$;   Domain of $f = [0, \infty)$;   Domain of $f = (-\infty, \infty)$,
Range of $f = [0, \infty)$;   Range of $f = [0, \infty)$;   Range of $f = (-\infty, \infty)$;
Domain of $g = [0, \infty)$;   Domain of $g = [-2, \infty)$;   Domain of $g = (-\infty, \infty)$;
Range of $g = [2, \infty)$;   Range of $g = [0, \infty)$;   Range of $g = (-\infty, \infty)$;
g is f shifted 2 units up. g is f shifted 2 units left. g is f shifted one unit right.
13. 217.9; The length of a 500-meter starship moving at 167,400 miles/second from the perspective of an observer at rest is about 217.9 meters.

14. 5   **15.** -5 **16.** 5   **17.** -5 **18.** -2 **19.** -2 **20.** 64   **21.** $\frac{1}{4}$ **22.** $\frac{1}{16}$ **23.** $28x^{7/12}$ **24.** $-4y^{11/20}$ **25.** $\frac{27y^{3/4}}{x^3}$

26. $\frac{2x}{y^{2/15}}$ **27.** $a^{2/3}$ **28.** $x^{1/4}y^{1/2}$ **29.** $2^{2/3}x^{2/3}y^{1/3}$ **30.** $3^{5/6}$ **31.** $2^{1/12}$ **32.** $x^{1/2}(1 + x^{1/4})$ **33.** $\frac{4(1 - 2x)}{x^{1/2}}$ **34.** 40 miles

35. 4.48; The duration of a storm whose diameter is 16 miles is 4.48 hours.   **36.** $2x\sqrt{5x}$ **37.** $3x^2y^2\sqrt[3]{2x^2}$ **38.** $2y^2\sqrt[4]{2x^3y^2}$
39. $2y^2z\sqrt[3]{2x^3y^2z}$ **40.** $2x^2\sqrt{6x}$ **41.** $2xy^2\sqrt[3]{2}$ **42.** $2xy\sqrt[5]{2y^4}$ **43.** $\sqrt{x^2 - 1}$ **44.** $15\sqrt{3}$ feet/second   **45.** $2\sqrt{6}$ **46.** $-10\sqrt[3]{2}$
47. $2x\sqrt[4]{2x}$ **48.** $10x^2\sqrt{xy}$ **49.** $\frac{2\sqrt{2}}{5}$ **50.** $\frac{x\sqrt{x}}{10y^2}$ **51.** $\frac{y\sqrt[4]{3y}}{2x^5}$ **52.** $\frac{2\sqrt{6}}{3}$ **53.** $\frac{\sqrt{14}}{7}$ **54.** $4\sqrt[3]{3}$ **55.** $\frac{\sqrt{10xy}}{5y}$ **56.** $\frac{7\sqrt[4]{4x}}{x}$
57. $\frac{\sqrt[4]{189x^3}}{3x}$ **58.** $\frac{5\sqrt[5]{xy^4}}{2xy}$ **59.** $\frac{\sqrt{10x}}{4x^2}$ **60.** $\frac{x^2\sqrt{6xy}}{y^2}$ **61.** $\frac{2\sqrt[3]{4x}}{x^2}$ **62.** $\frac{9\sqrt[5]{x^3y}}{x^2y}$ **63.** $20\sqrt{2}$ **64.** $27x\sqrt{2x} - 7x$ **65.** $12\sqrt[3]{6}$
66. $\sqrt[3]{2x}(14x - 3)$ **67.** $-5xy\sqrt[4]{2y} + xy\sqrt[4]{2}$ or $xy\sqrt[4]{2}(1 - 5\sqrt[4]{y})$ **68.** $2\sqrt{3} + 2\sqrt{2}$ **69.** $30\sqrt{2} + 60\sqrt{5}$ **70.** $2x\sqrt{3} + 3x\sqrt{2}$
71. $30\sqrt[3]{2} - 8\sqrt[3]{10}$ **72.** $34 - 13\sqrt{6}$ **73.** $\sqrt{xy} + \sqrt{11x} + \sqrt{11y} + 11$ **74.** $12 + 2\sqrt{35}$ **75.** $22 - 4\sqrt{30}$ **76.** -6 **77.** 4
78. $3\sqrt{3} + 3$ **79.** $\frac{\sqrt{35} - \sqrt{21}}{2}$ **80.** $\frac{-14\sqrt{5} - 21\sqrt{7}}{43}$ **81.** $\frac{y + 8\sqrt{y} + 15}{y - 9}$ **82.** $\frac{5 + \sqrt{21}}{2}$ **83.** $\frac{2x - 2\sqrt{xy}}{x - y}$
84. $\frac{a\sqrt{15} + \sqrt{5ab} - \sqrt{6ab} - b\sqrt{2}}{5a - 2b}$ or $\frac{a\sqrt{15} - b\sqrt{2} + (\sqrt{5} - \sqrt{6})\sqrt{ab}}{5a - 2b}$ **85.** $\frac{-3\sqrt{21} - 12\sqrt{35} - \sqrt{3} - 4\sqrt{5}}{77}$ **86.** $\frac{3\sqrt{2}}{20}$ inch
87. $P = 10\sqrt{6}$; $A = 36$ **88.** $P = 12\sqrt{10}$; $A = 50$ **89.** $\{16\}$ **90.** No solution   **91.** $\{2\}$ **92.** $\{-4, -2\}$ **93.** $\{8\}$ **94.** $\{1, 5\}$
95. a. 24π ft^3 **b.** Will not   **96.** $s = \frac{gt^2}{2}$ **97.** $c = \frac{2mM}{r^3}$ **98.** 150 feet   **99.** $9i$ **100.** $3i\sqrt{7}$ **101.** $-2i\sqrt{2}$ **102.** $12 + 2i$

103. $10 - 5i$ **104.** $29 + 11i$ **105.** $-21 + 20i$ **106.** $\frac{3}{26} + \frac{15i}{26}$ **107.** $\frac{1}{5} - \frac{11}{10}i$ **108.** $\frac{1}{3} - \frac{5i}{3}$ **109.** $-i$ **110.** $38i$ **111.** $-3\sqrt{5}$
112. 2   **113.** $2 + 8i\sqrt{2}$ **114.** $(1 - 2i)^2 - 2(1 - 2i) + 5 = 0$; $1 - 4i - 4 - 2 + 4i + 5 = 0$; $0 = 0$ **115.** $(10 + 2i)$ ohms

C H A P T E R   6   T E S T

1. $-\frac{2}{5}$ **2.** $(-\infty, 4]$ **3.** $f(x) = \sqrt{x}$, $g(x) = \sqrt{x - 2}$ g is f shifted 2 units right.   **4.** $\frac{1}{16}$ **5.** $\frac{125y^{3/4}}{x^6}$

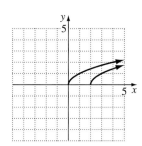

6. $x^{1/2}$ **7.** $2^{5/6}$ **8.** $\dfrac{6(x+2)}{x^{1/2}}$ **9.** 1.89; The duration of a storm whose diameter is 9 miles is 1.89 hours.   **10.** $2x^3y^3\sqrt[3]{2xy^2}$ **11.** $2x^3\sqrt[3]{x^2}$

12. $5\sqrt{30}$ mph   **13.** $2x\sqrt[3]{2x}$ **14.** $\dfrac{\sqrt{15}}{5}$ **15.** $\dfrac{\sqrt[3]{25x}}{x}$ **16.** $\dfrac{\sqrt{2x}}{2x^3}$ **17.** $15\sqrt{2}$ **18.** $62 - 14\sqrt{6}$ **19.** $52 - 14\sqrt{3}$ **20.** $2(\sqrt{5} + \sqrt{2})$

21. $\{7\}$ **22.** $\{5\}$ **23.** $10\sqrt{2}$ in.   **24.** $\dfrac{1}{2}$ ft   **25.** $5i\sqrt{3}$ **26.** $11 + 12i$ **27.** $26 + 7i$ **28.** $45 - 28i$ **29.** $\dfrac{1}{5} + \dfrac{7}{5}i$ **30.** -10

CUMULATIVE REVIEW PROBLEMS (CHAPTERS 1–6)

1. 9×10^{-3} **2.** $\dfrac{15}{4}$ **3.** $\left\{x\,\middle|\,x < \dfrac{2}{3} \text{ or } x > 2\right\}, \left(-\infty, \dfrac{2}{3}\right) \cup (2, \infty)$ **4.** $\{-2\}$ **5. a.** 1.15; The average annual rate of increase from 1960 to 1980

for lung cancer deaths is 1.15 per 100,000 people.   **b.** $y = 10$ per 100,000 people   **6. a.** $4x^2 - 64x + 256$ cubic inches   **b.** 16 in. by 16 in.

7. $y + 1 = \dfrac{5}{7}(x + 4)$ or $y - 4 = \dfrac{5}{7}(x - 3)$; $y = \dfrac{5}{7}x + \dfrac{13}{7}$ **8.** $\{(-2, -1, -2)\}$ **9.** (80, 700); At age 80, there are 700 deaths per 1,000 population

for each, men and women. The rate for men is larger than women as they both get older.

10. $x + 2y < 2$ and $2y - x > 4$ **11.** $W = \dfrac{S - 2LH}{2L + 2H}$ **12.** 11   **13.** $3y^4(x - 4y)(x + 4y)$ **14.** 11.64; 11.64% of the American population

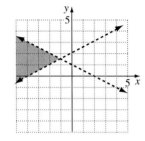

graduated from college in 1970.   **15.** 2004   **16.** $R = \dfrac{2V - 2Ir}{I}$

17. 40 units of corn and 10 units of soybeans.   **18.** $-\dfrac{x^2}{15(x + 2)}$ **19.** $\dfrac{x}{y}$ **20.** No solution

21. Width = 18 in., Length = 37 in.   **22.** $8x^3 - 22x^2 + 11x + 6$

23. $y = x^2 + 2x - 3$

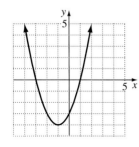

24. a. $C(x) = 20x + 50,000$ **b.** $\overline{C}(x) = \dfrac{20x + 50,000}{x}$ **c.** 70, 25, 20.5; The cost per chair to produce 1000,

10,000, and 100,000 chairs is $70, $25, $20.50, respectively. As the number of chairs produced increases, the cost per chair approaches $20.   **25.** 112 decibels   **26.** (1934, 17); In 1934, the suicide rate was at a maximum (from 1900 to 1990) at 17 per 100,000 people; Answers may vary.   **27.** $r^2(\pi - 1)$

28. $4x^2 + 1 - \dfrac{15}{x + 3}$ **29.** $-3y\sqrt{2y}$ **30.** $3 - 2i$

Chapter 7

PROBLEM SET 7.1

1. $y = x^2 + 2x - 3$ **3.** $y = x^2 - 9x + 14$ **5.** $y = 3x^2 - 8x + 4$ **7.** $y = 2x^2 + 5x - 7$

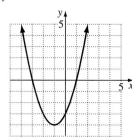 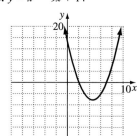 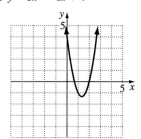 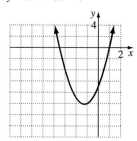

9. $\left\{-\frac{5}{2}, \frac{3}{4}\right\}$ **11.** $\left\{-\frac{5}{6}, 2\right\}$ **13.** $\left\{-1, \frac{2}{3}\right\}$ **15.** $\left\{-\frac{3}{2}, 4\right\}$ **17.** $\{0, 1\}$ **19.** $\{-10, 10\}$ **21.** $\{-\sqrt{7}, \sqrt{7}\}$ **23.** $\{-5\sqrt{3}, 5\sqrt{3}\}$

25. $\{-2i, 2i\}$ **27.** $\{-5, 5\}$ **29.** $\left\{-\frac{5\sqrt{3}}{3}, \frac{5\sqrt{3}}{3}\right\}$ **31.** $\left\{-\frac{\sqrt{77}}{7}, \frac{\sqrt{77}}{7}\right\}$ **33.** $\{-2i, 2i\}$ **35.** $\{-i\sqrt{5}, i\sqrt{5}\}$ **37.** $\{-5i\sqrt{3}, 5i\sqrt{3}\}$

39. $\{-10, -4\}$ **41.** $\left\{-1, \frac{5}{3}\right\}$ **43.** $\left\{\frac{-7 - \sqrt{5}}{2}, \frac{-7 + \sqrt{5}}{2}\right\}$ **45.** $\left\{\frac{4 - 2\sqrt{6}}{5}, \frac{4 + 2\sqrt{6}}{5}\right\}$ **47.** $\{3 - 2i, 3 + 2i\}$

49. $\left\{\frac{-5 - i\sqrt{5}}{2}, \frac{-5 + i\sqrt{5}}{2}\right\}$ **51.** $\left\{\frac{2 - 5i\sqrt{2}}{3}, \frac{2 + 5i\sqrt{2}}{3}\right\}$ **53.** $\left\{\frac{4 - 3\sqrt{3}}{5}, \frac{4 + 3\sqrt{3}}{5}\right\}$ **55.** $\left\{\frac{-5 - 5i}{2}, \frac{-5 + 5i}{2}\right\}$

57. $\left\{\frac{1}{2}, \frac{7}{10}\right\}$ **59.** $\left\{\frac{1 - \sqrt{3}}{3}, \frac{1 + \sqrt{3}}{3}\right\}$ **61.** $5\sqrt{5}$ **63.** $\sqrt{34}$ **65.** 13   **67.** $3\sqrt{5}$ **69.** $4\sqrt{5}$ **71.** $22 + 2\sqrt{85}$ **73.** Proof

75. Show points are collinear.   **77.** $(-5, -4)$ **79.** $\left(1, \frac{5}{2}\right)$ **81.** $(-0.2, 1.9)$ **83.** $(0, 0)$ **85.** $\left(\frac{1}{24}, \frac{1}{12}\right)$ **87.** $(5\sqrt{3}, -4)$ **89.** $(0, \sqrt{3})$

91. After 4 hours   **93.** 10 ft   **95.** Width: 5 yd, Length: 15 yd   **97.** $c = \sqrt{\dfrac{E}{m}}$ **99.** $r = \sqrt{\dfrac{A}{\pi}}$ **101.** $s = 2\sqrt{15(C - 12)}$

103. 3 ft by 3 ft   **105.** $90\sqrt{2} \approx 127.3$ ft   **107.** $6\sqrt{5}$ ft   **109.** 25 ft   **111. a.** AD: $\frac{5}{9}$; BC: $\frac{5}{9}$; AB: $-\frac{9}{5}$; DC: $-\frac{9}{5}$ **b.** AD and AB, $d = \sqrt{106}$

c. AB: $-\frac{9}{5}$; BC: $\frac{5}{9}$ **113. c**   **115.** 1976   **117–119.** Students should verify results.   **121–125.** Answers may vary.   **127.** $\frac{5}{2}$

129. Proof   **131. a.** Both distances are equal to $\frac{1}{2}\sqrt{(x_2 - x_1)^2 + (y_2 - y_1)^2}$ **b.** Proof

Review Problems

132. $\dfrac{y + 1}{(y + 4)(y - 3)}$ **133.** $\sqrt{3} - 1$ **134.** Width: 8 yd; Length: 12 yd

PROBLEM SET 7.2

1. $x^2 + 12x + 36 = (x + 6)^2$ **3.** $x^2 - 16x + 64 = (x - 8)^2$ **5.** $x^2 + 7x + \frac{49}{4} = \left(x + \frac{7}{2}\right)^2$ **7.** $x^2 - 3x + \frac{9}{4} = \left(x - \frac{3}{2}\right)^2$

9. $x^2 + \frac{1}{3}x + \frac{1}{36} = \left(x + \frac{1}{6}\right)^2$ **11.** $x^2 - \frac{2}{3}x + \frac{1}{9} = \left(x - \frac{1}{3}\right)^2$ **13.** $\{-3, 7\}$ **15.** $\{-2, 8\}$ **17.** $\{3 + \sqrt{7}, 3 - \sqrt{7}\}$

19. $\left\{\frac{-1 + \sqrt{5}}{2}, \frac{-1 - \sqrt{5}}{2}\right\}$ **21.** $\left\{-\frac{1}{2}, 3\right\}$ **23.** $\left\{\frac{5}{3}\right\}$ **25.** $\left\{\frac{-1 + \sqrt{41}}{4}, \frac{-1 - \sqrt{41}}{4}\right\}$ **27.** $\left\{\frac{1 + \sqrt{7}}{2}, \frac{1 - \sqrt{7}}{2}\right\}$ **29.** $\{-1 + i, -1 - i\}$

31. $\left\{\frac{1 + i\sqrt{3}}{2}, \frac{1 - i\sqrt{3}}{2}\right\}$ **33.** $\left\{\frac{1 + i}{4}, \frac{1 - i}{4}\right\}$ **35.** $\left\{\frac{-1 + i\sqrt{11}}{3}, \frac{-1 - i\sqrt{11}}{3}\right\}$

37. a. $(2, 1)$
  b. 2 units right, 1 unit up
  c. $y = (x - 2)^2 + 1$

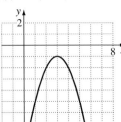

39. a. $(-1, -2)$
  b. 1 unit left, 2 units down
  c. $y = (x + 1)^2 - 2$

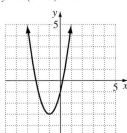

41. $f(x) = -(x - 3)^2 - 1$

43. $y = 2(x + 1)^2 - 3$

45. $g(x) = -2(x - 4)^2 + 3$

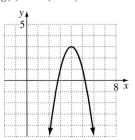

47. $y = -3(x + 1)^2 + 4$

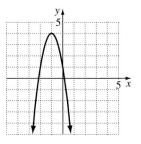

49. a. $y = (x + 3)^2 - 4$
b.

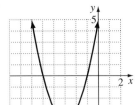

51. a. $y = -(x + 2)^2 + 1$
b.

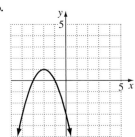

53. a. $y = 2\left(x + \dfrac{3}{2}\right)^2 + \dfrac{7}{2}$
b.

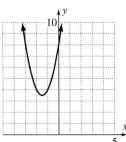

55. a. $y = 3(x - 2)^2 + 1$
b.

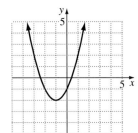

57. a. $y = (x + 2)^2$ **b.** $y = x^2 + 4x + 4$ **59. a.** $y = (x - 2)^2 + 2$ **b.** $y = x^2 - 4x + 6$
61. a. 55; In 1985, the number of union members was at a maximum.   **b.** The maximum number of members was about 25 million.   **63.** c
65–67. Students should verify results.   **69.** For $a > 1$, the larger a is the steeper the graph rises. For $0 < a < 1$, the closer a is to 0,
the flatter the graph is.   **71–73.** Answers may vary.   **75.** $y = ax^2 + c$

Review Problems

76. $\{(5, 7, -14)\}$ **77.** $\{-1, 3\}$ **78.** $\dfrac{x - y}{x}$

PROBLEM SET 7.3

1. $\{4, 5\}$ **3.** $\{-6, 10\}$ **5.** $\left\{1, \dfrac{5}{2}\right\}$ **7.** $\left\{\dfrac{-1 + \sqrt{41}}{4}, \dfrac{-1 - \sqrt{41}}{4}\right\}$ **9.** $\left\{\dfrac{-2 + \sqrt{10}}{3}, \dfrac{-2 - \sqrt{10}}{3}\right\}$ **11.** $\left\{\dfrac{1 + i}{2}, \dfrac{1 - i}{2}\right\}$

13. $\left\{\dfrac{1 + i\sqrt{14}}{5}, \dfrac{1 - i\sqrt{14}}{5}\right\}$ **15.** $\left\{-3, \dfrac{2}{3}\right\}$ **17.** $\left\{\dfrac{-1 - \sqrt{10}}{3}, \dfrac{-1 + \sqrt{10}}{3}\right\}$ **19.** $\left\{\dfrac{2 - i\sqrt{14}}{6}, \dfrac{2 + i\sqrt{14}}{6}\right\}$ **21.** $\{-\sqrt{21}, \sqrt{21}\}$

23. $\left\{-1, 0\right\}$ **25.** $\left\{-\dfrac{5}{3}, \dfrac{13}{3}\right\}$ **27.** $\left\{\dfrac{2 - i\sqrt{2}}{3}, \dfrac{2 + i\sqrt{2}}{3}\right\}$ **29.** $\left\{-3, \dfrac{1}{2}\right\}$ **31.** $\left\{\dfrac{5}{3}, 2\right\}$

33. a. $\{3 - \sqrt{2}, 3 + \sqrt{2}\}$

b. $y = x^2 - 6x + 7$

35. a. $\left\{\dfrac{3 - 3\sqrt{3}}{2}, \dfrac{3 + 3\sqrt{3}}{2}\right\}$

b. $y = 2x^2 - 6x - 9$

37. a. $\left\{\dfrac{-1 - i\sqrt{19}}{2}, \dfrac{-1 + i\sqrt{19}}{2}\right\}$

b. $y = x^2 + x + 5$

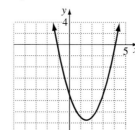

39. 36; Rational solutions   **41.** 97; Irrational solutions   **43.** 0; Rational solutions   **45.** 37; Irrational solutions   **47.** -76; Not real numbers

49. 1987; Population starts to increase much more rapidly.   **51. a.** Approximately \$39,000 and \$70,000   **b.** $\approx$ \$54,600; $\approx$ 1.68%

53. a. $y = 460x^2 - 9170x + 107,298$ **b.** 2000   **c.** About 1985, $\approx$ \$61,600 million   **55.** No   **57.** c   **59–61.** Students should verify results.

63. $y = -0.163x + 40.5$; \$37.4 million   **65.** Answers may vary.   **67.** $\left\{\left(\dfrac{-5 - \sqrt{33}}{2}\right)i, \left(\dfrac{-5 + \sqrt{33}}{2}\right)i\right\}$ **69.** $h = \pm 2\sqrt{15}$ **71.** $a < 3$

73. $x^2 - 10x + 21 = 0$ **75.** $x^2 - 4x + 1 = 0$ **77.** Students should show equalities.   **79.** $p = \pm 1$ and $q = -1$ **81.** -1

Review Problems

83. $\dfrac{x}{1 - x}$ **84.** $y + 2 = x - 1$ or $y - 2 = x - 5$; $y = x - 3$ **85.** $y = \dfrac{5}{2}x - 5$

PROBLEM SET 7.4

1. $5x^2 - 24x - 20 = 0$; $\left\{\dfrac{12 - 2\sqrt{61}}{5}, \dfrac{12 + 2\sqrt{61}}{5}\right\}$ **3.** $3x^2 - 10x + 4 = 0$; $\left\{\dfrac{5 - \sqrt{13}}{3}, \dfrac{5 + \sqrt{13}}{3}\right\}$ **5.** $x^2 - 9 = 0$; $\{-3, 3\}$

7. $x^2 - 2x - 4 = 0$; $\{1 - \sqrt{5}, 1 + \sqrt{5}\}$ **9.** $x^2 - 4x - 6 = 0$; $\{2 - \sqrt{10}, 2 + \sqrt{10}\}$ **11.** $x^2 + 5x - 500 = 0$; $\{-25, 20\}$

13. $\{-2, -1, 1, 2\}$ **15.** $\left\{-\dfrac{4}{3}, -1, 1, \dfrac{4}{3}\right\}$ **17.** $\{4, 25\}$ **19.** $\left\{-\dfrac{1}{3}, \dfrac{1}{4}\right\}$ **21.** $\{25, 64\}$ **23.** $\left\{-\dfrac{1}{8}, \dfrac{1}{2}\right\}$ **25.** $\left\{-\dfrac{1}{2}, 3\right\}$ **27.** $\{-5, -2, -1, 2\}$

29. $\{-1 - \sqrt{2}, -1, -1 + \sqrt{2}\}$ **31.** $\left\{\dfrac{1}{2 - \sqrt{7}}, \dfrac{1}{2 + \sqrt{7}}\right\}$ **33.** $\{-8, 27\}$ **35.** $\left\{\dfrac{1}{4}, 1\right\}$ **37.** $\{1\}$ **39.** $\{-1024\}$ **41.** $\left\{-\dfrac{3}{4}, 0\right\}$

43. $\left\{\dfrac{13}{3}\right\}$ **45.** $\dfrac{-3 + \sqrt{89}}{4}$ in. $\approx$ 1.6 in.   **47.** $\dfrac{-7 + \sqrt{89}}{2}$ m $\approx$ 1.2 m   **49.** $7 + 4\sqrt{7}$ in. $\approx$ 17.6 in.   **51.** $10 + 2\sqrt{26}$ in. $\approx$ 20.2 in.

53. $5 + 5\sqrt{5}$ ft $\approx$ 16.2 ft   **55.** 20   **57.** Slower: $\dfrac{9 + \sqrt{65}}{2}$ hr $\approx$ 8.5 hr; Faster: $\dfrac{7 + \sqrt{65}}{2}$ hr $\approx$ 7.5 hr   **59.** 10   **61.** $\dfrac{7 + \sqrt{85}}{3}$ mph $\approx$ 5.4 mph

63. d   **65.** Students should verify results.   **67.** Answers may vary.   **69.** $y = -4 + \sqrt{26}$ yd $\approx$ 1.1 yd   **71.** $\left\{-\dfrac{17}{3}, \dfrac{8}{3}\right\}$ **73.** $\left\{\left(\dfrac{13}{2}, \dfrac{5}{2}\right)\right\}$

Review Problems

75. $\left\{\left(\dfrac{10}{11}, -\dfrac{14}{11}\right)\right\}$ **76.** $\{-4\}$ **77.** $2x^2\sqrt[3]{6xy^2}$

PROBLEM SET 7.5

1. $\{x | x < -2 \text{ or } x > 4\}$, $(-\infty, -2) \cup (4, \infty)$ **3.** $\{x | -3 \le x \le 7\}$, $[-3, 7]$ **5.** $\{x | x < 1 \text{ or } x > 4\}$, $(-\infty, 1) \cup (4, \infty)$

7. $\{x | x < -4 \text{ or } x > -1\}$, $(-\infty, -4) \cup (-1, \infty)$ **9.** No solution   **11.** $\{x | 2 \le x \le 4\}$, $[2, 4]$ **13.** $\left\{x | -4 \le x \le \dfrac{2}{3}\right\}$, $\left[-4, \dfrac{2}{3}\right]$

15. $\left\{x | -3 < x < \dfrac{5}{2}\right\}$, $\left(-3, \dfrac{5}{2}\right)$ **17.** $\left\{x | -1 < x < -\dfrac{3}{4}\right\}$, $\left(-1, -\dfrac{3}{4}\right)$ **19.** $\left\{x | -2 \le x \le \dfrac{1}{3}\right\}$, $\left[-2, \dfrac{1}{3}\right]$ **21.** $\{x | x \le 0 \text{ or } x \ge 4\}$, $(-\infty, 0] \cup [4, \infty)$

23. $\left\{x | x < -\dfrac{3}{2} \text{ or } x > 0\right\}$, $\left(-\infty, -\dfrac{3}{2}\right) \cup (0, \infty)$ **25.** $\{x | 0 \le x \le 1\}$, $[0, 1]$ **27.** $\{x | x < -3 \text{ or } x > 4\}$, $(-\infty, -3) \cup (4, \infty)$

29. $\{x | -4 < x < -3\}$, $(-4, -3)$ **31.** $\{x | 2 \le x < 4\}$, $[2, 4)$ **33.** $\left\{x | x < -\dfrac{4}{3} \text{ or } x \ge 2\right\}$, $\left(-\infty, -\dfrac{4}{3}\right) \cup [2, \infty)$

35. $\{x \mid x < 0 \text{ or } x > 3\}$, $(-\infty, 0) \cup (3, \infty)$ **37.** $\{x \mid x < -5 \text{ or } x > -3\}$, $(-\infty, -5) \cup (-3, \infty)$ **39.** $\left\{x \mid x < \frac{1}{2} \text{ or } x \geq \frac{7}{5}\right\}$, $\left(-\infty, \frac{1}{2}\right) \cup \left[\frac{7}{5}, \infty\right)$

41. $\{x \mid x \leq -6 \text{ or } x > -2\}$, $(-\infty, -6] \cup (-2, \infty)$ **43.** Between 1 and 4 seconds   **45.** 16 to about $32\frac{1}{2}$ years and about $57\frac{1}{2}$ to 74 years   **47.** c

49. Students should verify results.   **51.** $\{x \mid x < -5 \text{ or } x > 2\}$ **53.** $\{x \mid 1 < x \leq 4\}$ **55.** $\{x \mid -4 < x < -1 \text{ or } x \geq 2\}$

57–59. Answers may vary.   **61.** Domain of $f(x) = \{x \mid x \leq -3 \text{ or } x \geq 4\} = (-\infty, -3] \cup [4, \infty)$

63. $\{x \mid x < -1 \text{ or } 1 < x < 2 \text{ or } x > 3\}$, $(-\infty, -1) \cup (1, 2) \cup (3, \infty)$ **65.** $\{x \mid -5 \leq x \leq -2 \text{ or } x \geq 2\}$, $[-5, -2] \cup [2, \infty)$

Review Problems

67. $\{x \mid x \leq -4\}$ **68.** $\{2\}$ **69.** $\{3, 7, 8, 9, 10\}$; $\{8, 9\}$

CHAPTER 7 REVIEW PROBLEMS

1. $\{-5i\sqrt{2}, 5i\sqrt{2}\}$ **2.** $\left\{-\sqrt{\frac{3}{2}}, \sqrt{\frac{3}{2}}\right\}$ **3.** $\left\{\frac{3 - 4\sqrt{2}}{2}, \frac{3 + 4\sqrt{2}}{2}\right\}$ **4.** $\{4 - 6i, 4 + 6i\}$ **5.** $a = \sqrt{c^2 - b^2}$ **6.** $r = \sqrt{\frac{A}{P}} - 1$

7. $\frac{\sqrt{1046}}{4} \approx 8.1$ seconds   **8.** $8\sqrt{2} \approx 11.3$ m   **9. a.** $5\sqrt{2}$ **b.** $\left(-\frac{1}{2}, \frac{13}{2}\right)$ **10.** $x^2 + 20x + 100 = (x + 10)^2$

11. $x^2 - 3x + \frac{9}{4} = \left(x - \frac{3}{2}\right)^2$ **12.** 9   **13.** $\left\{\frac{7 - \sqrt{53}}{2}, \frac{7 + \sqrt{53}}{2}\right\}$ **14.** $\left\{\frac{-3 - \sqrt{41}}{4}, \frac{-3 + \sqrt{41}}{4}\right\}$

15. $y = x^2 - 2x - 8$ **16.** $y = -2x^2 - 4x + 1$ **17.** $y = (x - 1)^2 + 3$ **18.** $f(x) = -(x + 1)^2 + 4$

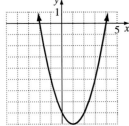
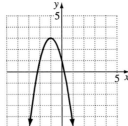
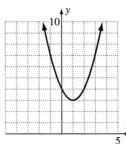
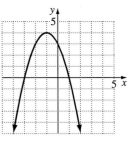

19. $y = (x - 1)^2 - 3$

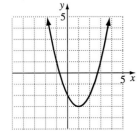

20. $y = 2(x + 2)^2 - 1$ **21. a.** 25; 25 inches of rain is optimal for tree growth.   **b.** 13.5; In a year with 25 inches of rain, the tree grows a maximum of 13.5 inches.   **22. a.** (30, 1800); A width of 30 yards yields the greatest area of 1800 square yards.   **b.** If the width is 60 yards, the area is 0 square yards. The fencing runs in two parallel lines right next to each other in a 60 yard line perpendicular to the river.

23. $\{1 - \sqrt{5}, 1 + \sqrt{5}\}$ **24.** $\{1 - 3i\sqrt{2}, 1 + 3i\sqrt{2}\}$ **25.** $\left\{\frac{-2 - \sqrt{10}}{2}, \frac{-2 + \sqrt{10}}{2}\right\}$ **26.** $\left\{\frac{3 - \sqrt{17}}{4}, \frac{3 + \sqrt{17}}{4}\right\}$ **27.** $\left\{-2, 9\right\}$

28. $\left\{-\frac{1}{5}, 1\right\}$ **29.** $\left\{\frac{-1 - \sqrt{13}}{3}, \frac{-1 + \sqrt{13}}{3}\right\}$ **30.** $\{4 - \sqrt{5}, 4 + \sqrt{5}\}$ **31.** $\{1 - 2i, 1 + 2i\}$ **32.** -19; Not real, imaginary

33. 49; Rational   **34.** 40; Irrational   **35.** 1144; Will be hired   **36.** 1989   **37.** Bell-shaped with one peak   **38. a.** $s = -16t^2 + 64t + 80$

b. about 4.1 seconds   **c.** 5 seconds   **d.** $s = -16t^2 + 64t + 80$;

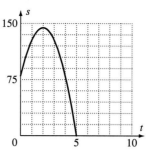

39. $-8 + 2\sqrt{26}$ m ≈ 2.2 m   **40.** $-1 + \sqrt{31}$ cm ≈ 4.6 cm   **41.** $60\sqrt{5}$ m ≈ 134.2 m   **42.** Slower person: $\dfrac{5 + \sqrt{17}}{2} \approx 4.6$ hr;

Faster person: $\dfrac{3 + \sqrt{17}}{2} \approx 3.6$ hr   **43.** Width: 5 yd; Length: 15 yd   **44.** 3   **45.** 9 cm by 19 cm and 3 cm by 7 cm   **46.** $\sqrt{6}$

47. $\{-2, -1, 1, 2\}$ **48.** $\{-5, -1, 3\}$ **49.** $\{-27, 64\}$ **50.** $\{1\}$ **51.** $\left\{-\dfrac{1}{8}, \dfrac{1}{7}\right\}$ **52.** $\left\{x\middle|-3 < x < \dfrac{1}{2}\right\}, \left(-3, \dfrac{1}{2}\right)$

53. $\left\{x\middle| x \le -4 \text{ or } x \ge -\dfrac{1}{2}\right\}, (-\infty, -4] \cup \left[-\dfrac{1}{2}, \infty\right)$ **54.** $\{x|x < -7 \text{ or } x > 3\}, (-\infty, -7) \cup (3, \infty)$

55. $\{x|x \subseteq -11 \text{ or } x > -4\}, (-\infty, -11] \cup (-4, \infty)$ **56.** Width $> 1 + \sqrt{5}$ m ≈ 3.24 m   **57.** $\{-1, 3\}$ **58.** $\{x|x < -4 \text{ or } x > 7\}$
59. $\{x|-3 \le x \le -2 \text{ or } 2 \le x \le 3\}$ **60.** $\{2\}$

CHAPTER 7 TEST

1. At $x = 2$, the graph shows $y = -1$, not $y = 4$.   **2.** $\left\{\dfrac{2 - 5\sqrt{2}}{3}, \dfrac{2 + 5\sqrt{2}}{3}\right\}$ **3.** $\left\{\dfrac{2}{3}, 2\right\}$ **4.** $\{2 - i\sqrt{6}, 2 + i\sqrt{6}\}$ **5.** $\{2 - \sqrt{10}, 2 + \sqrt{10}\}$

6. $\{-4, -3, 1, 2\}$ **7.** $\left\{\dfrac{(1 - i\sqrt{7})^3}{8}, \dfrac{(1 + i\sqrt{7})^3}{8}\right\}$ **8.** $v = \sqrt{\dfrac{2E}{m}}$ **9.** $\{3 - \sqrt{2}, 3 + \sqrt{2}\}$ **10.** $2\sqrt{34}$

11. $y = -x^2 + 2x + 3$ **12.** $f(x) = (x - 2)^2 + 3$

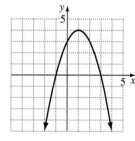   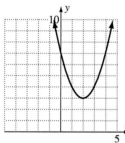

13. 17; irrational   **14.** $f(x) = x^2 - 2x$ **15.** $\{x|-5 \le x \le 7\}, [-5, 7]$ **16.** $\{x|2 < x \le 4\}, (2, 4]$ **17.** $50\sqrt{2}$ ft   **18.** 2 seconds; 64 ft
19. 3 in. by 3 in.   **20.** 3 ft

CUMULATIVE REVIEW PROBLEMS (CHAPTERS 1-7)

1. $\{x|x < -2 \text{ or } x > 4\}, (-\infty, -2) \cup (4, \infty)$ **2.** $\{-125, 343\}$ **3.** $\left\{-8, \dfrac{1}{2}\right\}$ **4.** $\left\{\dfrac{1 - 7\sqrt{5}}{2}, \dfrac{1 + 7\sqrt{5}}{2}\right\}$ **5.** $\{5\}$ **6.** $\{x \mid -5 \le x \le 5\}, [5, -5]$

7. No solution   **8.** $\{x|x \ge 4\}, [4, \infty)$ **9.** $\left\{-3, \dfrac{1}{4}\right\}$ **10.** $\{(2, 0, 3)\}$ **11.** $-\dfrac{z^6}{3y^2}$ **12.** $10x^3 + 15x^2y - 5xy^2 - 4x^2 - 6xy + 2y^2$

13. $\dfrac{4x - 13}{(x - 4)(x - 3)}$ **14.** $x^2 - 5x + 1 + \dfrac{8}{5x + 1}$ **15.** $2y^2\sqrt[3]{2y^2}$ **16.** $\dfrac{11 - 5\sqrt{5}}{-3}$ **17.** $(x - 4 - 5y)(x - 4 + 5y)$

18. a. $\pm\sqrt{(2 - x)(2 + x)}$ **b.** No   **c.** No; It does not pass the vertical line test.   **19.** $\dfrac{10 + \sqrt{105}}{4}$ seconds; No; the graph indicates that it

strikes the ground at slightly more than 5 seconds.   **20.** $V(x) = 4x^3 - 34x^2 + 70x$ **21.** 2004   **22.** 1527; Each year, from 1982 on, the average
salary increases by $1527.   **23. a.** -8; The altitude decreased by an average of 8 m per second from 6 to 7 seconds.   **b.** 4; The altitude increased
by an average of 4 m per second from 22 to 24 seconds.   **c.** 0; The altitude did not change from 11 to 17 seconds.   **24.** 711 and 171   **25.** 10

26. a. $y - 20{,}151 = -185(x - 20)$ or $y - 18{,}301 = -185(x - 30)$ **b.** $y = -185x + 23{,}851$ **c.** $13{,}676
27. Library Tower: 1018 ft; Sears Tower: 1454 ft   **28.** 1 ft   **29.** 20 mph; 48 miles   **30.** 495 ft

Chapter 8

PROBLEM SET 8.1

1.

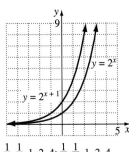

$\dfrac{1}{4}, \dfrac{1}{2}, 1, 2, 4; \dfrac{1}{4}, \dfrac{1}{2}, 1, 2, 4$

3.

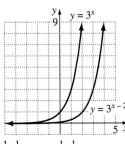

$\dfrac{1}{9}, \dfrac{1}{3}, 1, 3, 9; \dfrac{1}{9}, \dfrac{1}{3}, 1, 3, 9$

5. Shifted to the left if c is positive and shifted to the right if c is negative

7. $f(x) = 2^x$, $g(x) = 2^x + 3$, $h(x) = 2^x - 1$

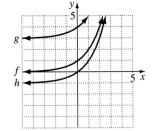

g is f shifted up 3 units and h is f shifted down 1 unit.

9. $f(x) = \left(\dfrac{3}{2}\right)^x$, $g(x) = \left(\dfrac{2}{3}\right)^x$

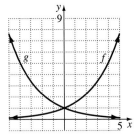

g is f reflected across the y-axis.

11. $\dfrac{1}{9}, \dfrac{1}{3}, 1, 3, 9; \dfrac{1}{9}, \dfrac{1}{3}, 1, 3, 9$

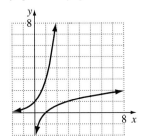

$x = 3^y$ is $y = 3^x$ reflected about the line $x = y$.

13. $y = 3^{-x}$

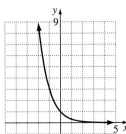

15. $g(x) = 2^{x/2}$

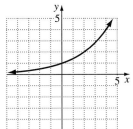

17. $y = 2^{x-1} - 1$

19. $y = 2^{x+1} + 1$

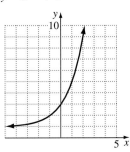

21. $f(x) = 3^{-x/2} - 1$ **23. a.** \$9479.19   **b.** \$9560.92   **c.** \$9577.70   **25.** 8.25% compounded quarterly

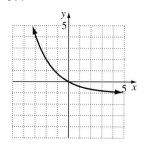

27. a. $f(x) = 5 \cdot 2^{-x}$

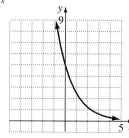

b. 8   **29. a.** 269,123   **b.** 1,628,102   **31.** $y = 3.6e^{0.02(51)} \approx 10$
33. a. Students should verify results.   **b.** 6.3 billion   **35. a.** 1429   **b.** 28,583
c. Answers may vary.   **37.** d

39. $f(x) = \dfrac{1}{\sqrt{2\pi}} e^{-x^2/2}$

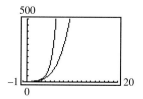

41. a. $y = e^x,\ y = 1 + x + \dfrac{x^2}{2}$ **b.** $y = e^x,\ y = 1 + x + \dfrac{x^2}{2} + \dfrac{x^3}{6}$ **c.** $y = e^x,\ y = 1 + x + \dfrac{x^2}{2} + \dfrac{x^3}{6} + \dfrac{x^4}{24}$

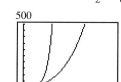
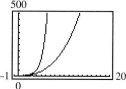

d. The second graph is approaching $y = e^x$ in (a), then (b), then (c).   **43–45.** Students should verify results.   **47.** Answers may vary.
49. Determine the y-value at $x = \dfrac{1}{2}$.   **51.** a is $h(x)$; b is $g(x)$; c is $f(x)$; Answers may vary.

Review Problems

52. 30 mph   **53.** $b = -\dfrac{Da}{D - a}$ or $b = \dfrac{Da}{a - D}$ **54.** -1

PROBLEM SET 8.2

1. 19   **3.** -3 **5.** 12   **7.** 1   **9.** 11   **11.** 11   **13.** 9   **15.** 0   **17.** $4x^2 + 4x + 5;\ 2x^2 + 9$ **19.** $15x^2 - 18;\ 75x^2 + 60x + 8$
21. $x^4 - 4x^2 + 6;\ x^4 + 4x^2 + 2$ **23.** $\sqrt{x - 1};\ \sqrt{x} - 1$ **25.** $x;\ x$ **27.** $x;\ x$ **29.** Not one-to-one   **31.** Not one-to-one   **33.** One-to-one
35. Yes   **37.** Yes   **39.** No   **41.** Yes   **43.** Yes

45. a. $f^{-1}(x) = x - 3$ **47. a.** $f^{-1}(x) = \dfrac{x}{2}$ **49. a.** $f^{-1}(x) = \dfrac{x-3}{2}$ **51. a.** $f^{-1}(x) = \sqrt[3]{x-2}$

b. Students should verify results.   **b.** Students should verify results.   **b.** Students should verify results.   **b.** Students should verify results.

c. 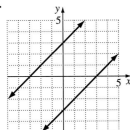   **c.** 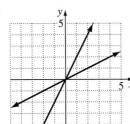   **c.** 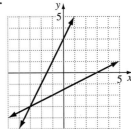   **c.**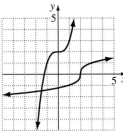

53. a. $f^{-1}(x) = \sqrt[3]{x} - 2$ **55. a.** $f^{-1}(x) = \dfrac{1}{x}$ **57. a.** $f^{-1}(x) = x^2$ **59. a.** $f^{-1}(x) = \sqrt{x-1}$

b. Students should verify results.   **b.** Students should verify results.   **b.** Students should verify results.   **b.** Students should verify results.

c. 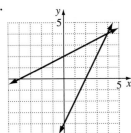   **c.** 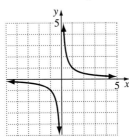   **c.** 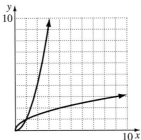   **c.**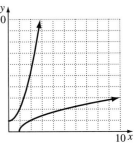

61. -4 **63.** 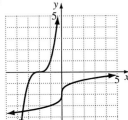   **65.**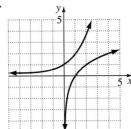

67. a. f is $5 off the regular price; g is 40% off the regular price   **b.** $0.6x - 5$; $5 off the sale price of 40% off
c. $0.6x - 3$; $3 off the sale price of 40% off   **d.** $(f \circ g)(x)$; $0.6x - 5$ is less than $0.6x - 3$ **69. a.** passes horizontal line test
b. $f^{-1}(20) = 14$; There were 20 million women with AIDS in 1994.   **c.** Answers may vary.   **71.** d   **73.** Students should verify results.
75. Yes   **77.** Yes

79. $f(x) = \sqrt[3]{2-x}$ **81.** $f(x) = \dfrac{x^4}{4}$ **83.** $f(x) = \dfrac{1}{x^2}$ **85.** $f(x) = \dfrac{1}{(x-1)^2}$

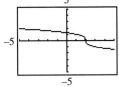   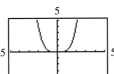   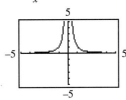   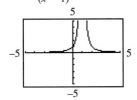

One-to-one   Not one-to-one   Not one-to-one   Not one-to-one

87–91. Answers may vary.   **93.** $3x^5 - 9x + 4$; $3x^5 - 9x + 6$ **95.** $b = -10$ **97.** $6 - 5x$ **99.** $f^{-1}(x) = \dfrac{1}{3}\left[\left(\dfrac{9-3x}{2}\right)^3 - 2\right]$ **101.** $x = 5$

103. $g[f(x)] = x$

Review Problems

105. $y = x^2 - 4x + 3$ **106.** $(a - 2b)(a + 2b)(a^2 + 4b^2)$ **107.** 5×10^8

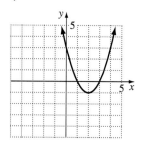

P R O B L E M   S E T   8 . 3

1. $2^4 = 16$ **3.** $5^{-3} = \dfrac{1}{125}$ **5.** $25^{1/2} = 5$ **7.** $3^y = 8$ **9.** $m^c = P$ **11.** $e^{1.6094} = 5$ **13.** $b^x = b^x$ **15.** $10^1 = 10$ **17.** $\log_2 8 = 3$

19. $\log_2 \dfrac{1}{16} = -4$ **21.** $\log_8 2 = \dfrac{1}{3}$ **23.** $\log_8 2 = \dfrac{1}{3}$ **25.** $\log_{16} 8 = \dfrac{3}{4}$ **27.** $\log_{10} 100 = 2$ **29.** $\ln 54.5982 = 4$ **31.** $\ln 0.1353 = -2$

33. $\ln x = y$ **35.** $\log_P m = a$ **37.** 2 **39.** 5 **41.** $\dfrac{1}{2}$ **43.** -1 **45.** -5 **47.** $\dfrac{1}{2}$ **49.** 1 **51.** 0

53. Not possible: No value of x will make $5^x = -5$.   **55.** $\dfrac{3}{4}$ **57.** 3 **59.** -2 **61.** 4 **63.** Not possible; No value of x makes $e^x = -1$.
65. 6 **67.** 21 **69.** 1

71. $f(x) = 3^x, f^{-1}(x) = \log_3 x$ **73.** $f(x) = \left(\dfrac{1}{2}\right)^x, f^{-1}(x) = \log_{1/2} x$

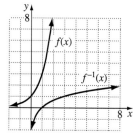   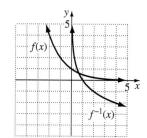

75. a. 88 points   **b.** 71.5 points; 63.9 points; 58.8 points; 55.0 points; 49.5 points
   c. $f(t) = 88 - 15 \ln(t + 1)$; The students retained less material as time passed.

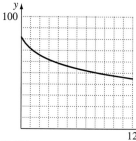

77. 11,200 years   **79.** The fourth column provides the better estimate. They get further away.   **81.** d

83. $f(x) = \ln x, g(x) = \ln(x + 3)$ **85.** $f(x) = \log x, g(x) = -\log x$ **87.** $f(t) = 75 - 10 \log(t + 1)$

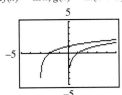

g(x) is f(x) shifted 3 units left.

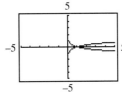

g(x) is f(x) reflected across the x-axis.

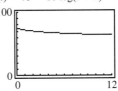

9 months

89. a. $f(x) = \ln \frac{x}{2}, g(x) = \ln x - \ln 2$ **b.** $f(x) = \log \frac{x}{5}, g(x) = \log x - \log 5$ **c.** $f(x) = \ln \frac{x^2}{3}, g(x) = \ln x^2 - \ln 3$

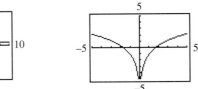

d. $f(x) = g(x); \log_b\left(\frac{M}{N}\right) = \log_b M - \log_b N$ **e.** The difference of the logs

91. $y = \ln x, y = \sqrt{x}, y = x, y = x^2, y = e^x, y = x^x$ **93.** Answers may vary.   **95.** $-\frac{5}{6}$ **97.** $\frac{14}{15}$ **99.** $A(0, 1); B(1, 0); C(4, 2); D(2, 4)$

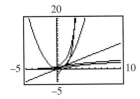

101. $y = -\log_2 x$ **103.** $-\frac{4}{3}$

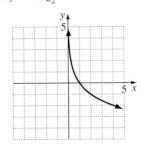

Review Problems

104. $3\sqrt{5}$ **105. a.** $y = \frac{3}{97}x + \frac{9119}{97}$ **b.** Approximately 119   **106.** $\{(-2, 3)\}$

PROBLEM SET 8.4

1. $3 = 1 + 2$ **3.** $1 = 2 + (-1)$ **5.** $4 - 1 = 3$ **7.** $13 = 17 - 4$ **9.** $3 = 3(1)$ **11.** $\frac{1}{2}(2) = 1$ **13.** $1 + \log_3 x$ **15.** $3 + \log x$

17. $\log_4 B + \log_4 x$ **19.** $1 - \log_7 x$ **21.** $\log x - 2$ **23.** $3 - \log_4 y$ **25.** $3\log_b x$ **27.** $-6\log N$ **29.** $\frac{1}{3}\ln x$ **31.** $2\log_b x + \log_b y$

33. $\frac{1}{2}\log_4 x - 2$ **35.** $2\log_b x + \log_b y - 4\log_b z$ **37.** $\frac{1}{3}(2 + \log x)$ **39.** $\frac{1}{4}(3 + \ln x - \ln y - 2 \ln z)$ **41.** $\frac{1}{5}(10 \log_b x + 15 \log_b y - 3 - 6 \log_b z)$

43. 1   **45.** 5   **47.** $\log\left(\frac{5}{x^5}\right)$ **49.** $\log_b x^2 y^3$ **51.** $\ln\left(\frac{x^5}{y^2}\right)$ **53.** $\ln\left(\frac{x^2}{\sqrt[3]{y}}\right)$ **55.** $\log(\sqrt{x})(\sqrt[3]{y})$ **57.** $\log\left(\frac{\sqrt{x} \cdot y^3}{z^2}\right)$ **59.** $\log_3(x + 3)$

61. $\ln x\left(\frac{y}{z}\right)^2$ **63.** $\log_b\left(\frac{b}{x\sqrt{b}}\right)$ **65.** 1.5937   **67.** 0.5698   **69.** 1.0308   **71.** 1.6944   **73.** -0.1802 **75.** -1.2304 **77.** 0.6962

79. 3.6193   **81.** 20 decibels   **83.** d   **85.** b   **87.** b

89. $y = \log x$, $y = \log(10x)$,   **91–99.** Answers may vary.   **101.** Proof   **103.** $\dfrac{2A}{B}$ **105.** $1 - A$ **107.** $x = 2$ **109.** Proof

$y = \log(0.1x)$

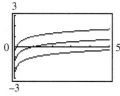

$y = \log(10x)$ is $y = \log x$
shifted 1 unit up.

$y = \log(0.1x)$ is $y = \log x$ shifted
1 unit down; The Product Rule

Review Problems

110. $f^{-1}(x) = \dfrac{x - 17}{3}$ **111.** $\{(2, -2, 3)\}$ **112.** $5x - 2y > 10$

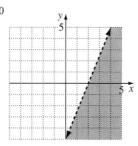

P R O B L E M   S E T   8 . 5

1. $\{6\}$ **3.** $\{3\}$ **5.** $\{3\}$ **7.** $\{2\}$ **9.** $\left\{\dfrac{3}{5}\right\}$ **11.** $\left\{\dfrac{3}{2}\right\}$ **13.** $\{-2\}$ **15.** $\left\{\dfrac{7}{4}\right\}$ **17.** $x = \dfrac{\ln 7}{\ln 2} \approx 2.807$ **19.** $x = \ln 5 \approx 1.609$

21. $x = \dfrac{\ln 31}{\ln(2.7)} \approx 3.457$ **23.** $x = \dfrac{\ln 9}{0.5} \approx 4.394$ **25.** $t = \dfrac{\ln 0.07}{-0.03 \ln 5} \approx 55.076$ **27.** $t = \dfrac{\ln 0.09}{-0.03} \approx 80.265$ **29.** $x = \dfrac{\ln 30}{\ln(1.4)} \approx 10.108$

31. $t = \dfrac{\ln 3}{0.055} \approx 19.975$ **33.** $t = \dfrac{\ln 0.134}{-0.5 \ln 2} \approx 5.799$ **35.** $\{81\}$ **37.** $\left\{\dfrac{1}{16}\right\}$ **39.** $\{100\}$ **41.** $\{e^3\}$ **43.** $\{59\}$ **45.** $\left\{\dfrac{55}{27}\right\}$ **47.** $\left\{\dfrac{65}{2}\right\}$

49. $\left\{\dfrac{5}{4}\right\}$ **51.** $\{6\}$ **53.** $\{2\}$ **55.** No solution   **57.** $\left\{\dfrac{5}{9}\right\}$ **59.** 16 years   **61.** 8.1%   **63.** After 2 weeks   **65.** 2002; 81.3%

67. ≈ 2.83 days   **69.** 3.631×10^{-8} to 4.266×10^{-8} **71.** d   **73.** Students should verify results.

75. $P = 145e^{-0.092t}$ **77.** Answers may vary.   **79.** 2025   **81.** $\{5, 20\}$ **83.** $x = \dfrac{b^b}{A}$ **85.** $\left\{\dfrac{625}{16}\right\}$ **87.** Group activity

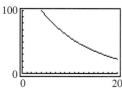

About 7.9 minutes

Review Problems

88. $y = 6x - 1$ **89.** $\{5\}$ **90.** $\{-12\}$

PROBLEM SET 8.6

1. a. $k \approx 0.0427$ **b.** $y = 6,907,387e^{0.0427t}$ **c.** 69,296,295; Not very close   **d.** Answers may vary.   **3. a.** $y = 40,637e^{0.1087t}$ **b.** 2012

c–d. Answers may vary.   **5. a.** $\frac{A}{2} = Ae^{k(5730)}$, $\ln\left(\dfrac{\frac{A}{2}}{A}\right) = \ln e^{5730k}$, $k = \dfrac{\ln\left(\frac{1}{2}\right)}{5730} = -0.000121$, $y = Ae^{-0.000121t}$ **b.** 2268 years

7. 119 days   **9. a.** $\frac{A}{2} = Ae^{kt}$, for $k < 0$, $\ln\left(\frac{1}{2}\right) = \ln e^{kt}$, for $k < 0$, $t = \dfrac{\ln\left(\frac{1}{2}\right)}{-k}$, for $k > 0$, $t = \dfrac{-\ln 2}{-k} = \dfrac{\ln 2}{k}$ **b.** 11 years   **11.** $y = 0.29897e^{0.873x}$

13. a. logarithmic model   **15.** Students should verify results.   **17–19.** Answers may vary.

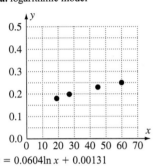

b.

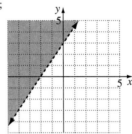

c. $y = 0.0604\ln x + 0.00131$ **d.** 0.26 second

Review Problems

21. $9x^4 - 42x^2y + 49y^2$ **22.** No solution   **23.** $3x - 2y < -6$;

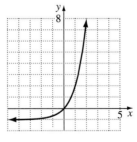

CHAPTER 8 REVIEW PROBLEMS

1. $y = 2^x$ **2.** $y = 2^{x-2}$ **3.** $y = 3^x - 1$ **4.** $f(x) = \left(\dfrac{1}{2}\right)^x$

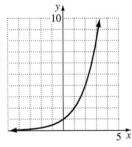   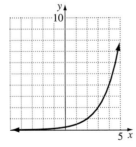   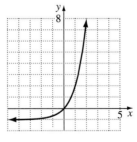   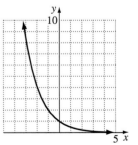

5. 5.5% compounded semiannually   **6.** 7% compounded monthly   **7.** China: ≈ 2443; India: ≈ 3009; Yes

8. $f(t) = 100\left(\frac{1}{2}\right)^{t/5600}$; 100, 70.7, 50, 25, 15.6, 12.5

9. a. 200°F   **b.** About 120°F   **c.** 70°F; room is 70°F

10. $(f \circ g)(x) = 16x^2 - 8x + 4$; $(g \circ f)(x) = 4x^2 + 11$

11. $(f \circ g)(x) = \sqrt{x+1}$; $(g \circ f)(x) = \sqrt{x}+1$ **12.** (c) is one-to-one

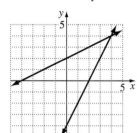

Answers may vary.

13. a. $f^{-1}(x) = \dfrac{x+4}{2}$

b. Students should verify results.

c.

14. a. $f^{-1}(x) = \sqrt[3]{x+2}$

b. Students should verify results.

c.

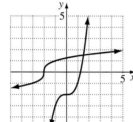

15. a. passes horizontal line test.   **b.** $f^{-1}(60) = 1930$; The life expectancy was 60 years for someone born in 1930   **16.** $2^6 = 64$

17. $49^{1/2} = 7$ **18.** $\log_6 216 = 3$ **19.** $\log \dfrac{1}{100} = -2$ **20.** {3}   **21.** $\left\{\dfrac{1}{3}\right\}$

22. No solution   **23.** {−4}

24. $f(x) = 2^x, f^{-1}(x) = \log_2 x$;

25. a. 76   **b.** 67; 63; 61; 59; 56

c. $f(t) = 76 - 18\log(t+1)$

The material retained decreases over time.

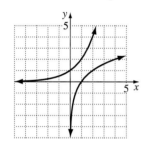

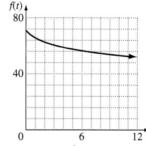

26. Approximately 9 weeks   **27.** $2 + 3\log_6 x$ **28.** $\log_2 x + 2\log_2 y - 6$ **29.** $\dfrac{1}{2}\log_4 x - 3$ **30.** $\dfrac{1}{3}(3 - 2\log x - \log y)$ **31.** $\log_b 21$ **32.** 3

33. $\log \dfrac{3}{x^3}$ **34.** $\log_b x^3 y^4$ **35.** $\ln\left(\dfrac{\sqrt[3]{x}}{\sqrt{y}}\right)$ **36. a.** $\log_3 7 = \dfrac{\log 7}{\log 3}$ **b.** $\log_3 7 = \dfrac{\ln 7}{\ln 3}$ **37.** b   **38.** {2}   **39.** $\left\{\dfrac{2}{3}\right\}$ **40.** $\left\{-\dfrac{3}{2}\right\}$

41. $x = \dfrac{\ln 119.4}{\ln 5} \approx 2.972$ **42.** $t = \dfrac{\ln 2}{0.08} \approx 8.664$ **43.** $x = \dfrac{\ln 0.06}{-0.04 \ln 8} \approx 33.824$ **44.** $\left\{\dfrac{1}{125}\right\}$ **45.** $\{\sqrt[6]{32} \approx 1.782\}$ **46.** $\left\{\dfrac{1}{1000}\right\}$

47. {63}   **48.** {25}   **49.** {5}   **50.** No solution   **51.** 7.3 years   **52. a.** $\dfrac{\ln 3}{r}$ yr   **b.** About 13.7%   **53. a.** Year 2035   **b.** Year 2019

54. 5.9 feet/second   **55.** Columbia: 1.459×10^{17} joules; California: 8.421×10^{14} joules   **56. a.** $k \approx 0.04139$ **b.** $y = 14{,}609 e^{0.04139t}$

c. 41,116 thousand; 50,569 thousand; 94,084 thousand   **57. a.** $\dfrac{A}{2} = Ae^{k(5730)}$, $\ln\dfrac{1}{2} = \ln e^{5730k}$, $k = \dfrac{\ln\dfrac{1}{2}}{5730} = -0.000121$, $y = Ae^{-0.000121t}$

b. 15,680 years   **58.** $3A = Ae^{kt}$, $\ln\left(\dfrac{3A}{A}\right) = \ln e^{kt}$, $kt = \ln 3$, $t = \dfrac{\ln 3}{k}$ **59.** High: exponential; Medium: linear; Low: quadratic, negative; Explanations may vary; There is not equal spacing representing the same time period.

60. 3.3, 3.7, 4.3, 4.9, 5.5, 6.1, 6.5, 6.8   **b.**

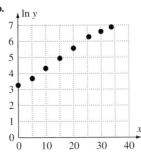

c. $\ln y = 0.110x + 3.25$
d. $y = 25.79e^{0.110x}$ **e.** \$2,101 billion

61. a.

b.

c. $y = 0.460\ln x - 0.087$ **d.** 9181

Logarithmic function

CHAPTER 8 TEST

1. $f(1) = \dfrac{1}{2}$ on the graph, not $f(1) = 2^1 = 2$ **2.** $f(x) = 2^{x-3}$;

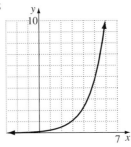

3. $(f \circ g)(x) = 9x^2 - 3x$; $(g \circ f)(x) = 3x^2 + 3x - 1$ **4.** $f^{-1}(x) = \dfrac{x + 7}{5}$ **5. a.** Yes; It passes the horizontal line test.   **b.** $f^{-1}(2000) = 80$;

A person who gives \$2000 earns \$80,000.   **6.** $5^3 = 125$ **7.** $\log_{36} 6 = \dfrac{1}{2}$

8. $f(x) = 3^x, f^{-1}(x) = \log_3 x$ **9.** $3 + 5\log_4 x$ **10.** $\dfrac{1}{3}\log_3 x - 4$ **11.** $\ln x^3 y^5$ **12.** 4   **13.** $\dfrac{\log 71}{\log 15} \approx 1.5741$ **14.** {1}

15. {0.209062}   **16.** {277}   **17.** {5}   **18.** $\left\{\dfrac{217}{4}\right\}$ **19.** {5}

20. 6.5% compounded semiannually; \$221   **21.** After 3 hours, the object's temperature is 35°C.
22. a. 75   **b.** 59   **23.** 14 years   **24.** 5.0 miles   **25.** $k \approx 12.8865$ **b.** 0.218   **26.** 7.4

CUMULATIVE REVIEW PROBLEMS (CHAPTERS 1-8)

1. $\{22\}$ **2.** $\{5 - 7i, 5 + 7i\}$ **3.** $\left\{\dfrac{1}{3} - \dfrac{2}{3}i, \dfrac{1}{3} + \dfrac{2}{3}i\right\}$ **4.** $\{x \mid x < -3 \text{ or } x > 2\}, (-\infty, -3)\cup(2, \infty)$ **5.** $\{-1, 9\}$ **6.** $\{-2\}$ **7.** No solution

8. $\{-3, 1, -1 - \sqrt{3}, -1 + \sqrt{3}\}$ **9.** $\left\{x \mid x < \dfrac{2}{3}\right\}, \left(-\infty, \dfrac{2}{3}\right)$ **10.** $\{x \mid x < -1 \text{ or } x > 5\}, (-\infty, -1) \cup (5, \infty)$ **11.** $\{(4, 3)\}$ **12.** $\{(4, 0, -5)\}$

13. $f(x) = \sqrt{x - 2}$ 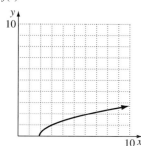   **14.** $y = (x + 2)^2 - 4$ 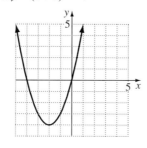   **15.** $y < -3x + 5$ 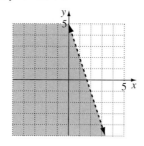   **16.** $2x - 3y \leq 6$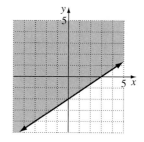

17. $y = 3^{x-2}$ 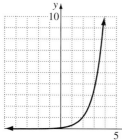   **18.** $y = \dfrac{1}{x - 2}$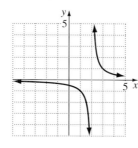

19. $\dfrac{2x^2 + 5x - 2}{(x - 5)(x + 2)}$ **20.** $\dfrac{x + 1}{1 - x}$ **21.** $2(\sqrt{5} + \sqrt{2})$

22. $12\sqrt{5}$ **23.** $27 - 4\sqrt{35}$

24.

4	9	2
3	5	7
8	1	6

25. 3.3

26. a. About -2.39; The mortality of children under 5 decreases by about $2\dfrac{1}{3}$ for each percent increase in the number of adult females who are literate. **b.** Answers may vary.
  c. $y - 80 = -2.39(x - 73); y = -2.39x + 254.47$

27. 1800 lb **28.** 1986 **29.** $A(x) = 3x^2 + 2x$ **30.** $\dfrac{n^2 - n}{2}$

Chapter 9

PROBLEM SET 9.1

1. $(x - 3)^2 + (y - 2)^2 = 25$ **3.** $(x + 1)^2 + (y - 4)^2 = 4$ **5.** $(x + 3)^2 + (y + 1)^2 = 3$ **7.** $(x + 4)^2 + y^2 = 4$ **9.** $x^2 + y^2 = 49$

11. Center: $(0, 0)$; $r = 4$

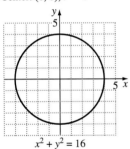

$x^2 + y^2 = 16$

13. Center: $(3, 1)$; $r = 6$

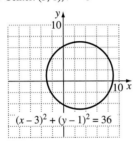

$(x - 3)^2 + (y - 1)^2 = 36$

15. Center: $(-3, 2)$; $r = 2$

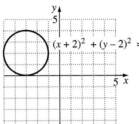

$(x + 2)^2 + (y - 2)^2 = 4$

17. Center: $(-2, -2)$; $r = 2$

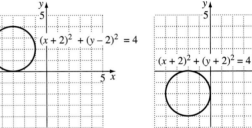

$(x + 2)^2 + (y + 2)^2 = 4$

19. Center: $(-3, -1)$; $r = 2$

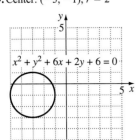

$x^2 + y^2 + 6x + 2y + 6 = 0$

21. Center: $(5, 3)$; $r = 8$

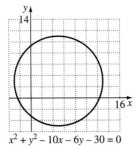

$x^2 + y^2 - 10x - 6y - 30 = 0$

23. Center: $(-4, 1)$; $r = 5$

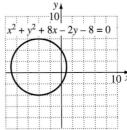

$x^2 + y^2 + 8x - 2y - 8 = 0$

25. Center: $(1, 0)$; $r = 4$

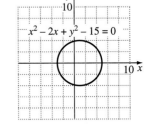

$x^2 - 2x + y^2 - 15 = 0$

27. $y = (x - 2)^2 - 4$

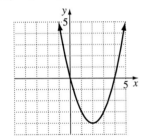

29. $x = (y - 2)^2 - 4$

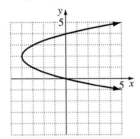

31. $y = x^2 - 4$

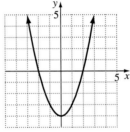

33. $x = y^2 + 4$

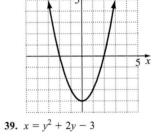

35. $y = -2(x - 1)^2 - 1$

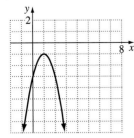

37. $x = -2(y - 1)^2 - 1$

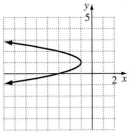

39. $x = y^2 + 2y - 3$

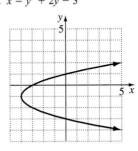

41. $x = y^2$

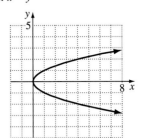

43. $x = -y^2 - 2y + 3$

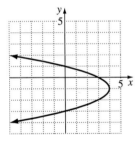

45. 90 ft   **47.** d   **49.** b

51. $y_1 = \dfrac{12x + 40 + 10\sqrt{15x + 7}}{9}$

$y_2 = \dfrac{12x + 40 - 10\sqrt{15x + 7}}{9}$

53–57. Answers may vary.   **59.** 9π **61.** $x^2 + y^2 = 25$ **63.** $4x\sqrt{25 - x^2}$

65. a. $|y + p|$ **b.** $\sqrt{x^2 + (y - p)^2}$ **c.** $|y + p| = \sqrt{x^2 + (y - p)^2}$; $x^2 = 4py$

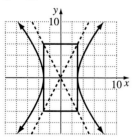

Answers may vary.

Review Problems

67. $\{(5, 7, -14)\}$ **68.** $\{-1, 3\}$ **69.** $\{-3, 1, 5\}$

PROBLEM SET 9.2

1. $\dfrac{x^2}{9} + \dfrac{y^2}{4} = 1$

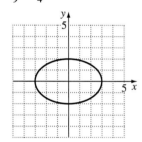

3. $\dfrac{x^2}{9} + \dfrac{y^2}{36} = 1$

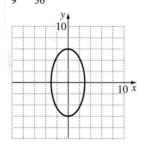

5. $\dfrac{x^2}{25} + \dfrac{y^2}{64} = 1$

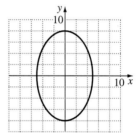

7. $\dfrac{x^2}{49} + \dfrac{y^2}{81} = 1$

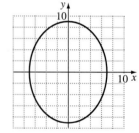

9. $25x^2 + 4y^2 = 100$

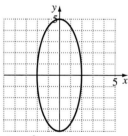

11. $4x^2 + 16y^2 = 64$

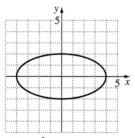

13. $25x^2 + 9y^2 = 225$

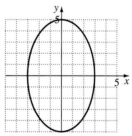

15. $x^2 + 2y^2 = 8$

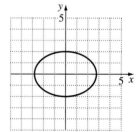

17. $\dfrac{x^2}{9} - \dfrac{y^2}{25} = 1$

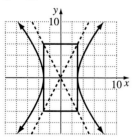

19. $\dfrac{x^2}{100} - \dfrac{y^2}{64} = 1$

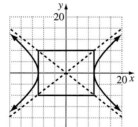

21. $\dfrac{y^2}{16} - \dfrac{x^2}{36} = 1$

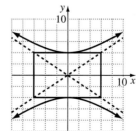

23. $\dfrac{y^2}{36} - \dfrac{x^2}{25} = 1$

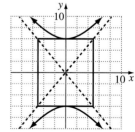

25. $9x^2 - 4y^2 = 36$

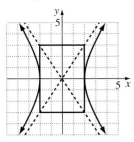

27. $9y^2 - 25x^2 = 225$

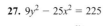
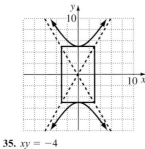

29. $4x^2 = 4 + y^2$

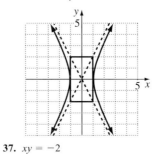

31. $xy = 4$

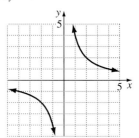

33. $xy = 2$

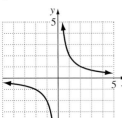

35. $xy = -4$

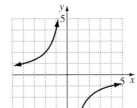

37. $xy = -2$

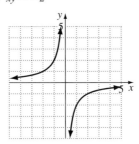

39. parabola   **41.** hyperbola   **43.** parabola   **45.** ellipse   **47.** circle   **49.** hyperbola   **51.** parabola   **53.** $\dfrac{x^2}{2500} + \dfrac{y^2}{1600} = 1$

55. a. $\dfrac{x^2}{2304} + \dfrac{y^2}{529} = 1$ **b.** $(5\sqrt{71}$ ft, 0 ft)   **57.** $\dfrac{x^2}{9} - \dfrac{4y^2}{9} = 1$ **59.** d

61. a. $x^2 + y^2 = (3950)^2$ **b.** $\dfrac{(x + 225.5)^2}{(4307.5)^2} + \dfrac{y^2}{(4301.6)^2} = 1$ **63.** $\dfrac{x^2}{16} - \dfrac{y^2}{9} = 1$ and $\dfrac{x|x|}{16} - \dfrac{y|y|}{9} = 1$

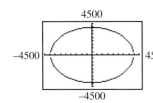

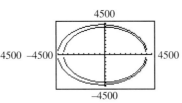

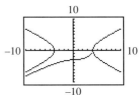

Altitude is higher

65. Answers may vary.   **67. a.** $\dfrac{x^2}{a^2} + \dfrac{y^2}{a^2 - c^2} = 1$ **b.** $\dfrac{x^2}{a^2} + \dfrac{y^2}{b^2} = 1$

69. $y = 2$ if $0 \le x \le 1$
$y = 1$ if $0 \le x \le 0.75$
$x = 0$ if $0 \le y \le 2$

71. $x = 0$ if $-1 \le y \le 1$
$y = \pm x\sqrt{x}$ if $0 \le x \le 1$

73. $x^2 + y^2 = 1$ or $y = \pm\sqrt{1 - x^2}$
$y = -x$ if $\dfrac{1}{2} \le x \le 1$

75. $y = -x + 3$ if $1 \le x \le 2$
$y = x - 1$ if $2 \le x \le 3$
$x = 2$ if $0 \le y \le 1$

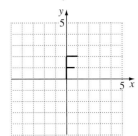

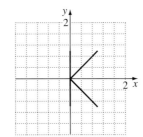

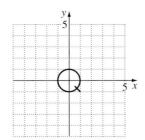

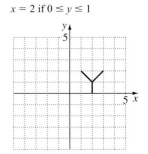

Review Problems

77. $\{(-3, -5)$ **78.** $\left\{\left(\frac{3}{4}, \frac{1}{2}\right)\right\}$ **79.** $\{(-1, 2)\}$

PROBLEM SET 9.3

1. $\{(-5, 0), (4, 3)\}$ **3.** $\{(-2, 3)\}$ **5.** $\{(2, 0), (0, 3)\}$ **7.** $\left\{\left(-\frac{3}{2}, -4\right), (2, 3)\right\}$ **9.** $\{(-4, -6), (4, 6)\}$ **11.** $\{(2, 1), (2, -1)\}$

13. $\{(2 + i, 8 + 4i), (2 - i, 8 - 4i)\}$ **15.** $\{(3, -5), (-1, 3)\}$ **17.** $\left\{\left(-\frac{5}{\sqrt{2}}, -\frac{5}{\sqrt{2}}\right), \left(\frac{5}{\sqrt{2}}, \frac{5}{\sqrt{2}}\right)\right\}$ **19.** $\{(-3, -2), (-3, 2), (3, -2), (3, 2)\}$

21. $\{(-3, -2), (3, -2), (-3, 2), (3, 2)\}$ **23.** $\left\{\left(-\frac{1}{2}, -\frac{3}{2}\right), \left(-\frac{1}{2}, \frac{3}{2}\right), \left(\frac{1}{2}, -\frac{3}{2}\right), \left(\frac{1}{2}, \frac{3}{2}\right)\right\}$

25. $\left\{\left(-\frac{3}{2}\sqrt{\frac{3}{2}}, -\sqrt{\frac{5}{8}}\right), \left(\frac{3}{2}\sqrt{\frac{3}{2}}, -\sqrt{\frac{5}{8}}\right), \left(-\frac{3}{2}\sqrt{\frac{3}{2}}, \sqrt{\frac{5}{8}}\right), \left(\frac{3}{2}\sqrt{\frac{3}{2}}, \sqrt{\frac{5}{8}}\right)\right\}$ **27.** $\{(\sqrt{37}, 5), (-\sqrt{37}, 5), (4, 2), (-4, 2)\}$

29. $\{(i\sqrt{11}, -6), (-i\sqrt{11}, -6), (0, 5)\}$ **31.** $\left\{\left(\frac{5}{7}, -\frac{1}{7}\right), (1, -1)\right\}$ **33.** $\{(-i, -2i), (i, -2i), (-i, 2i), (i, 2i)\}$

35. $\{(-2, -2), (2, 2), (-4, -1), (4, 1)\}$ **37.** $\{(-\sqrt{2}, \sqrt{2}), (\sqrt{2}, -\sqrt{2})\}$ **39.** $\{(-4, -3), (4, 3), (-3, -4), (3, 4)\}$

41. $\{(-2, 0), (2, 0), (0, -4)\}$ **43.** 7 ft by 3 ft   **45.** Width is 9 ft; Length is 12 ft   **47.** $x = 5$ ft, $y = 2$ ft   **49.** $x = 3$, $y = 5$

51. 40 ft by 12 ft   **53.** Width: 12 in.; Length: 18 in.   **55.** b

57. $\frac{x^2}{1.03} + \frac{y^2}{0.97} = 1$; $x^2 - 2xy + y^2 - 2x - 2y + 1 = 0$; 2 times; $(0, \sqrt{0.97})$ and $(\sqrt{1.03}, 0)$

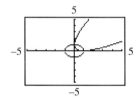

59–61. Answers may vary.   **63.** $\sqrt{324}$ or 18 square units   **65.** $\left(\frac{\sqrt{3}}{2}, \frac{1}{2}\right)$ **67.** $\{(5, -4), (5, 4)\}$ **69.** 2 m and 4 m   **71.** No

Review Problems

73. $\{6\}$ **74.** $\frac{1}{2x}$ **75.** $\frac{\ln 125}{\ln 7}$

CHAPTER 9 REVIEW PROBLEMS

1. $(x + 2)^2 + (y - 4)^2 = 36$ **2.** $x^2 + y^2 = 9$

3. Center: $(0, 0)$; $r = 4$ **4.** Center: $(-2, 3)$; $r = 3$ **5.** Center: $(2, -1)$; $r = 3$ **6.** $y = x^2 - 2x + 3$

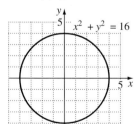
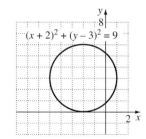
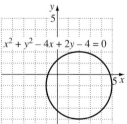
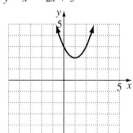

7. $y = -(x + 1)^2 + 4$

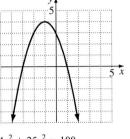

8. $x = y^2 - 8y + 12$

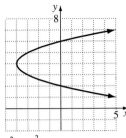

9. $x = (y - 3)^2 - 4$

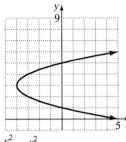

10. $\dfrac{x^2}{9} + \dfrac{y^2}{25} = 1$

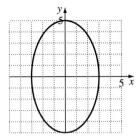

11. $4x^2 + 25y^2 = 100$

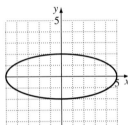

12. $\dfrac{x^2}{16} - \dfrac{y^2}{9} = 1$

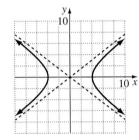

13. $\dfrac{y^2}{9} - \dfrac{x^2}{4} = 1$

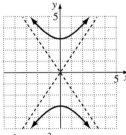

14. $x^2 - y^2 = 4$

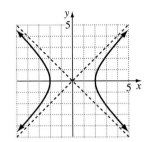

15. $xy = 8$

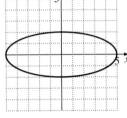

16. $xy = -4$

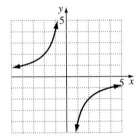

17. $\dfrac{x^2}{1600} + \dfrac{y^2}{100} = 1$ **18.** Yes

19. $5x^2 + 5y^2 = 180$; Circle

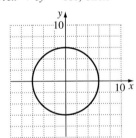

20. $4x^2 + 9y^2 = 36$; Ellipse

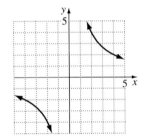

21. $x + 3 = -y^2 + 2y$; Parabola

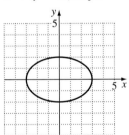

22. $4x^2 - 9y^2 = 36$; Hyperbola

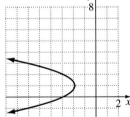

23. $x^2 + y^2 + 6x - 2y = -6$; Circle

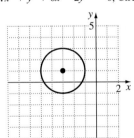

24. $xy = 3$; Hyperbola

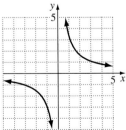

25. $\{(1, 0), (4, 3)\}$ **26.** $\{(0, 1), (-3, 4)\}$ **27.** $\{(-1, 1), (1, -1)\}$

28. $\left\{\left(-\dfrac{2\sqrt{5}}{5}, -\dfrac{4\sqrt{5}}{5}\right), \left(\dfrac{2\sqrt{5}}{5}, \dfrac{4\sqrt{5}}{5}\right)\right\}$ **29.** $\left\{\left(\dfrac{2 - \sqrt{10}}{3}, \dfrac{-1 - \sqrt{10}}{3}\right), \left(\dfrac{2 + \sqrt{10}}{3}, \dfrac{-1 + \sqrt{10}}{3}\right)\right\}$

30. $\{(-3, -\sqrt{6}), (-3, \sqrt{6}), (3, -\sqrt{6}), (3, \sqrt{6})\}$ **31.** $\{(-2, -2), (2, 2)\}$

32. $\{(1, 1), (-1, 1), (-i\sqrt{2}, -2), (i\sqrt{2}, -2)\}$ **33.** $\{(1, 2), (9, 6)\}$

34. $\left\{\left(-\dfrac{6}{5}, -\dfrac{12}{5}\right), \left(\dfrac{6}{5}, \dfrac{12}{5}\right)\right\}$

35. $\left\{\left(-\dfrac{\sqrt{5}}{5}, -\dfrac{\sqrt{5}}{5}\right), \left(-\dfrac{\sqrt{5}}{5}, \dfrac{\sqrt{5}}{5}\right), \left(\dfrac{\sqrt{5}}{5}, -\dfrac{\sqrt{5}}{5}\right), \left(\dfrac{\sqrt{5}}{5}, \dfrac{\sqrt{5}}{5}\right)\right\}$

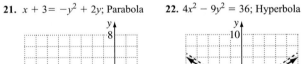
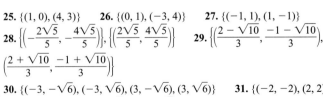
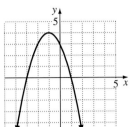
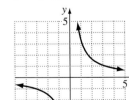

36. $\{(-2, -\sqrt{5}), (-2, \sqrt{5})\}$ **37.** $\{(-2, -2), (2, 2), (-2\sqrt{2}, -\sqrt{2}), (2\sqrt{2}, \sqrt{2})\}$ **38.** $\{(0, -2), (\sqrt{3}, 1), (-\sqrt{3}, 1)\}$
39. 5 meters $\times$ 8 meters   **40.** $\{(46 \text{ ft}, 28 \text{ ft}), (50 \text{ ft}, 20 \text{ ft})\}$ **41.** 9 yards and 12 yards   **42.** $(\sqrt{7}, \sqrt{2})$ **43.** rate: 40 kph; time: 50 hr
44. $(1, 6)$ and $(3, 2)$

CHAPTER 9 TEST

1. Center: $(5, -3)$; $r = 7$ **2.** Center: $(2, 3)$; $r = 4$

3. $y = x^2 - 2x - 3$; Parabola   **4.** $\dfrac{x^2}{4} - \dfrac{y^2}{9} = 1$; Hyperbola   **5.** $4x^2 + 9y^2 = 36$; Ellipse   **6.** $xy = 4$; Hyperbola

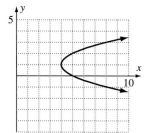   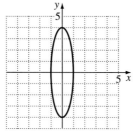   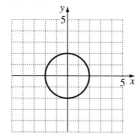   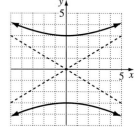

7. $x = (y - 1)^2 + 4$; Parabola   **8.** $16x^2 + y^2 = 16$; Ellipse   **9.** $\dfrac{x^2}{4} + \dfrac{y^2}{4} = 1$ **10.** $25y^2 - 9x^2 = 225$

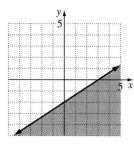   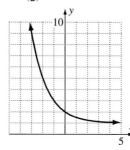   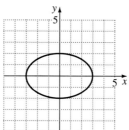   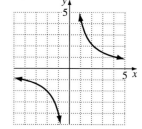

11. The graphs cross at $(0, -1)$ and $(3, 2)$.   **12.** $\{(4, -3), (-3, 4)\}$ **13.** $\{(-3, -2), (-3, 2), (3, -2), (3, 2)\}$ **14.** Width: 9 ft; Length: 12 ft
15. $y = 15$ ft, $x = 12$ ft or $y = 24$ ft, $x = 7\frac{1}{2}$ ft

CUMULATIVE REVIEW PROBLEMS (CHAPTERS 1–9)

1. $\{x | x \leq 4\}, (-\infty, 4]$ **2.** $\left\{-\dfrac{343}{125}, 1\right\}$ **3.** $\left\{\dfrac{4 - \sqrt{2}}{5}, \dfrac{4 + \sqrt{2}}{5}\right\}$ **4.** $\{3 - 2i, 3 + 2i\}$ **5.** $\left\{x | x < -\dfrac{1}{2} \text{ or } x > 3\right\}, \left(-\infty, -\dfrac{1}{2}\right) \cup (3, \infty)$
6. $\{(-5, 0), (3, 4)\}$ **7.** $\{x | -1 < x < 2\}, (-1, 2)$ **8.** $\{(-1, 2, 4)\}$ **9.** $\{2\}$ **10.** $\{3, 11\}$

11. $2x - 3y \geq 6$ **12.** $y = \left(\dfrac{1}{2}\right)^x + 1$ **13.** $9x^2 + 4y^2 = 36$ **14.** $y = \log_3 x$

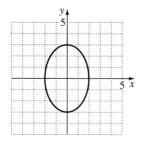

15. $f(x) = \sqrt{x + 1}$

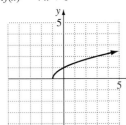

16. $\{(-4, -5), (1, 0)\}$

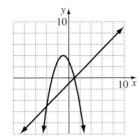

17. 6.5% compounded semiannually   **18.** $(f \circ g)(x) = 4x^2 - 12x + 10$; $(g \circ f)(x) = 2x^2 - 1$ **19.** $f^{-1}(x) = \dfrac{x - 5}{3}$

20. a. 14.7 pounds per square inch   **b.** 3.3 miles   **21.** $\dfrac{4 + 7i}{5}$ **22.** $x^2 + 4x + 5 + \dfrac{14}{x - 2}$ **23.** $\dfrac{11\sqrt{3}}{3}$

24. a. 36 feet at 0.5 sec   **b.** 2 seconds   **c.** $s(t) = -16t^2 + 16t + 32$

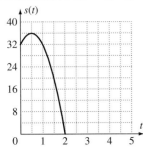

25. 3 meters   **26.** 4000 meters   **27.** $10^2 - 2^2 = 96$ **28.** $a^2 + 2ab + b^2 = (a + b)^2$

29. First model: 1542 thousand, 2534.4 thousands; Second model: 1852 thousand, 4332.4 thousand; Second model comes closer

30. $x = \dfrac{c^2 + cd + d^2}{c + d}$; $y = \dfrac{-cd}{c + d}$

Chapter 10

PROBLEM SET 10.1

1. 1, 3, 5, 7, 19, 29   **3.** 3, 6, 11, 18, 102, 227   **5.** $\dfrac{1}{2}, \dfrac{2}{3}, \dfrac{3}{4}, \dfrac{4}{5}, \dfrac{10}{11}, \dfrac{15}{16}$ **7.** $2, \dfrac{3}{2}, \dfrac{4}{3}, \dfrac{5}{4}, \dfrac{11}{10}, \dfrac{16}{15}$ **9.** $0, \dfrac{1}{4}, \dfrac{2}{5}, \dfrac{1}{2}, \dfrac{3}{4}, \dfrac{14}{17}$

11. $-1, 2, -3, 4, 10, -15$ **13.** $1, -4, 9, -16, -100, 225$ **15.** $-\dfrac{1}{2}, \dfrac{1}{4}, -\dfrac{1}{8}, \dfrac{1}{16}, \dfrac{1}{1024}, -\dfrac{1}{32,768}$ **17.** 32   **19.** 208   **21.** 45

23. $a_n = 2n - 1$ **25.** $a_n = 3n$ **27.** $a_n = \dfrac{n + 1}{n}$ **29.** $a_n = 3n - 2$ **31.** $a_n = (-1)^n 2n$ **33.** $a_n = (-1)^{n+1} 2n$ **35.** $a_n = (-1)^n$

37. $a_n = 2 - 3n$ **39.** $a_n = (\sqrt{5})^n$ **41.** 63   **43.** 12   **45.** $\dfrac{43}{10}$ **47.** $-\dfrac{76}{105}$ **49.** 30   **51.** 105   **53.** 115   **55.** 0   **57.** $-\dfrac{5}{16}$

59. 233   **61.** $\dfrac{85}{12}$ **63.** $\displaystyle\sum_{i=1}^{5} i^3$ **65.** $\displaystyle\sum_{i=1}^{5} 2i$ **67.** $\displaystyle\sum_{i=1}^{5} (-1)^i 2i$ **69.** $\displaystyle\sum_{i=1}^{5} \dfrac{1}{i^2}$ **71.** $\displaystyle\sum_{i=1}^{5} \dfrac{(-1)^{i+1}}{i^2}$ **73.** $\displaystyle\sum_{i=1}^{6} \dfrac{i+1}{i+2}$ **75.** $\displaystyle\sum_{i=1}^{\infty} 6i$ **77.** $\displaystyle\sum_{i=1}^{\infty} \dfrac{1}{i(i+1)}$

79. 1.0, 1.1, 1.3, 1.5, 1.8, 2.0, 2.3, 2.6, 2.8, 3.1, 3.4, 3.7, 4.0, 4.3, 4.6

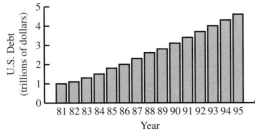

81. 26,063; The total number of AIDS cases reported for 1984, 1985, and 1986 approximated by this model is 26,063.   **83.** b   **85.** d

87. $a_{10} \approx 2.59374$; $a_{100} \approx 2.70481$; $a_{1000} \approx 2.71692$; $a_{10,000} \approx 2.71815$; $a_{100,000} \approx 2.71827$ **89–91.** Students should verify results.

93–95. Answers may vary.   **97.**

1	1	1	1	1
2	2	2	2	2
3	4	6	4	6
4	7	15	7	15
5	12	40	12	40

99. $a_2 = \dfrac{15}{8}$; $a_3 = \dfrac{35}{16}$; $a_4 = \dfrac{315}{128}$; $a_5 = \dfrac{693}{256}$

Column 2 = Column 4; Column 3 = Column 5

101. The sums are the same: $\dfrac{-3}{1} + \dfrac{-2}{2} + \dfrac{-1}{3} + 0 + \dfrac{1}{5} + \dfrac{2}{6}$ **103.** $\log 384$ **105.** $\log x^{18}$ **107.** Not true   **109.** 751,700

Review Problems

111. $2(y^2 + 2)(y^4 - 2y^2 + 4)$ **112.** $\dfrac{x^2}{16} + \dfrac{y^2}{9} = 1$;   **113.** $\{8\}$

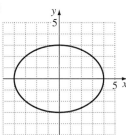

P R O B L E M   S E T   1 0 . 2

1. 4   **3.** 5   **5.** 0.13   **7.** $\dfrac{e}{6}$ **9.** 8, 10, 12, 14, 16, 18   **11.** $-16, -20, -24, -28, -32, -36$ **13.** $\dfrac{3}{2}, \dfrac{9}{4}, 3, \dfrac{15}{4}, \dfrac{9}{2}, \dfrac{21}{4}$ **15.** 79   **17.** -45

19. 2.10   **21.** $\dfrac{33}{4}$ **23.** 39   **25.** $\dfrac{28}{3}$ **27.** 0.7   **29.** 10th   **31.** 96th   **33.** 23rd   **35.** 1,220   **37.** 210   **39.** 816   **41.** $-1,245$

43. 20,200   **45.** 5050   **47.** 396   **49.** 1275   **51.** -340 **53.** -27 **55. a.** $a_n = 0.45n + 38.6$ **b.** 45.8 million   **c.** 2011

57. Company A pays $700 more than Company B is year 10.   **59.** Company A will pay the greater amount.   **61. a.** $a_n = 83.96 + 3.14n$

b. 5723.64 million tons   **63.** 320 degree days   **65.** 630 feet   **67.** d   **69.** Students should verify results.   **71–73.** Answers may vary.

75. Students should show the equality.   **77.** 579   **79.** $35   **81.** $\dfrac{n(n + 1)a}{2} + nb$ or $\dfrac{n}{2}(a + na + 2b)$ **83.** Group activity

Review Problems

84. $\{-4, 3\}$ **85.** $\{(3, 1), (3, -1), (-3, 1), (-3, -1)\}$ **86.** 128 ohms

P R O B L E M   S E T   1 0 . 3

1. 3   **3.** -1 **5.** $-\dfrac{1}{2}$ **7.** $\dfrac{1}{3}$ **9.** $\sqrt{5}$ **11.** $\dfrac{1}{b}$ **13.** $10, 5, \dfrac{5}{2}, \dfrac{5}{4}, \dfrac{5}{8}$ **15.** $-\dfrac{1}{4}, \dfrac{1}{2}, -1, 2, -4$ **17.** $3, -9, 27, -81, 243$

19. $\dfrac{a^2}{b}, 2a, 4b, \dfrac{8b^2}{a}, \dfrac{16b^3}{a^2}$ **21.** $a_n = -3(5)^{n-1}$; $a_8 = -234,375$ **23.** $a_n = -12\left(\dfrac{1}{2}\right)^{n-1}$; $a_{10} = -\dfrac{3}{128}$ **25.** $a_n = 3\left(-\dfrac{1}{2}\right)^{n-1}$; $a_6 = -0.09375$

27. $a_n = \dfrac{2}{3}\left(\dfrac{1}{5}\right)^{n-1}$; $a_8 = \dfrac{2}{234,375}$ **29.** $a_n = -4\left(\dfrac{\sqrt{2}}{2}\right)^{n-1}$; $a_{11} = -\dfrac{1}{8}$ **31.** $a_n = 222\dfrac{2}{9}\left(\dfrac{1}{10}\right)^{n-1}$; $a_6 = 0.00\overline{2}$

33. $a_n = c^7 d^6\left(\dfrac{1}{cd^2}\right)^{n-1}$; $a_{14} = \dfrac{1}{c^6 d^{20}}$ **35.** $a_n = 2(2)^{n-1} = 2^n$ **37.** 728   **39.** 33   **41.** $-\dfrac{129}{2}$ **43.** 1093   **45.** 547   **47.** 31   **49.** $-\dfrac{26}{81}$

51. $\dfrac{4}{3}$ **53.** 24   **55.** $\dfrac{81}{5}$ **57.** No finite sum   **59.** $\dfrac{8}{5}$ **61.** $\dfrac{2}{3}$ **63.** $\dfrac{5}{9}$ **65.** $\dfrac{49}{99}$ **67.** $\dfrac{241}{999}$ **69.** $\dfrac{542}{99}$ **71.** $\dfrac{1094}{333}$ **73.** $\dfrac{11}{90}$

75. a. $r \approx 1.148$ **b.** $a_n = 7(1.148)^{n-1}$ **c.** 1523 million   **77. a.** 200, 198, 196.02, 194.0598   **b.** $w_n = 200(0.99)^{n-1}$

79. Company B will pay more in the sixth year (approximately $526 more)   **81.** $19,143.92   **83.** Company B produces the better total income.

85. About 906.44 ft   **87.** $360 billion   **89.** 24 inches   **91.** $11\frac{1}{9}$ meters; $1\frac{1}{9}$ seconds   **93.** d   **95.** d

97. a. $f(x) = \dfrac{1 - 0.5^x}{0.5}$ $f(x) = \dfrac{1 - 0.9^x}{0.1}$ $f(x) = \dfrac{1 - (1.1)^x}{-0.1}$ $f(x) = \dfrac{1 - 3^x}{-2}$

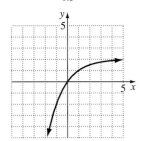   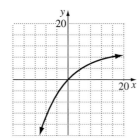   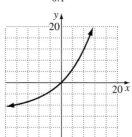   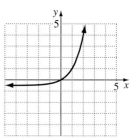

b. For $r = 0.5$, as $x \to \infty$, asymptote $= 2$; for $r = 0.9$, as $x \to \infty$, asymptote $= 10$; for $r = 1.1$, as $x \to -\infty$, asymptote $= -10$;

for $r = 3$, as $x \to -\infty$, asymptote $= -\dfrac{1}{2}$ **c.** when $0 < r < 1$ **99.** Horizontal asymptote $= 10$; $\displaystyle\sum_{n=0}^{\infty} 4(0.6)^n = 10$ **101–105.** Answers may vary.

107. r **109.** r^2 **111.** $4r$ **113.** $(9, 12)$ and $(1, -4)$ **115.** $n = 4$ **117.** 20   **119.** $\dfrac{15}{4}$ **121.** $\dfrac{21}{8}$

Review Problems

122. $3x^2y^2\sqrt[3]{2y}$ **123.** $\{(0, 4), (2, 0)\}$ **124.** $4 + \sqrt{15}$

PROBLEM SET 10.4

1. 6   **3.** 2   **5.** 45   **7.** 7   **9.** 20   **11.** 12   **13.** 1   **15.** $c^5 + 10c^4 + 40c^3 + 80c^2 + 80c + 32$
17. $x^6 + 6x^5y + 15x^4y^2 + 20x^3y^3 + 15x^2y^4 + 6xy^5 + y^6$ **19.** $a^4 - 8a^3 + 24a^2 - 32a + 16$
21. $x^6 + 12x^5y + 60x^4y^2 + 160x^3y^3 + 240x^2y^4 + 192xy^5 + 64y^6$ **23.** $\dfrac{a^4}{16} + \dfrac{a^3}{2} + \dfrac{3a^2}{2} + 2a + 1$

25. $16x^4 + 96x^3y + 216x^2y^2 + 162xy^3 + 81y^4$ **27.** $8x^6 - 12x^4y^2 + 6x^2y^4 - y^6$ **29.** $\dfrac{a^3}{27} + \dfrac{2a^2}{3} + 4a + 8$

31. $8 - 24\sqrt{2}x + 60x^2 - 40\sqrt{2}x^3 + 30x^4 - 6\sqrt{2}x^5 + x^6$ **33.** $a^2 + 8a^{3/2} + 24a + 32a^{1/2} + 16$ **35.** $\dfrac{1}{a^3} + \dfrac{3}{a^2b} + \dfrac{3}{ab^2} + \dfrac{1}{b^3}$
37. $x^{16} + 8x^{15} + 28x^{14} + \ldots$ **39.** $256a^8 + 1024a^7b + 1729a^6b^2 + \ldots$ **41.** $a^8 - 16a^7b + 112a^6b^2 + \ldots$ **43.** $a^{20} + 10a^{18}b^2 + 45a^{16}b^4 + \ldots$
45. $a^{42} + 42a^{41}b + 861a^{40}b^2 + \ldots$ **47.** $y^7 + 7y^5 + 21y^3 + \ldots$ **49.** About 0.246   **51.** 0.1323   **53.** c   **55.** Students should verify results.
57.

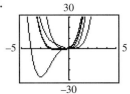

$f_2, f_3, f_4,$ and f_5 are approaching $f_1 = f_6$.
59. $x^4 - 8x^3 + 24x^2 - 32x + 16$ **61–63.** Answers may vary.   **65.** $\dfrac{252}{5}$ **67.** $n + 1$ **69.** Proof

Review Problems

71. $\left\{\left(2, \dfrac{7}{2}\right), \left(-\dfrac{1}{2}, -4\right)\right\}$ **72.** $f^{-1}(x) = \dfrac{x - 5}{3}$ **73.** $\log 5x^3$

CHAPTER 10 REVIEW PROBLEMS

1. 3, 8, 13, 18, 48, 73   **2.** $-\dfrac{1}{2}, 0, \dfrac{1}{10}, \dfrac{2}{17}, \dfrac{8}{101}, \dfrac{13}{226}$ **3.** 0, -3, 8, -15, -99, 224   **4.** $a_n = -3n$ **5.** $a_n = (-1)^{n+1}\left(\dfrac{1}{n + 2}\right)$ **6.** 140

7. $\frac{163}{180}$ **8.** 420   **9.** $\sum_{i=1}^{7} i^2$ **10.** $\sum_{i=1}^{6} \left(-\frac{1}{2}\right)^n$ **11. a.** $a_3 = 1.0198$; $a_4 = 1.0012$; $a_9 = 0.9442$; $a_{10} = 0.94$; The ratio of men to women in the U.S. is 1.0198, 1.0012, 0.9442, and 0.94 in years 1930, 1940, 1990, and 2000, respectively.   **b.** Approximately 141.75 million women and 133.25 million men   **12.** 49   **13.** -96 **14.** $-\frac{99}{10}$ **15.** -21 **16.** 10th   **17.** $a_n = 210 - 10n$ **18. a.** $a_n = 27{,}413 + 553n$ **b.** \$41,791

19. 1,727   **20.** 648   **21.** 440   **22.** -500 **23.** 15,150   **24.** 136   **25.** \$284,000   **26.** \$70,650 million   **27.** $\frac{4}{729}$ **28.** $\frac{19{,}683}{1024}$

29. $243\sqrt{3}$ **30.** $a_n = 3(-4)^{n-1}$ **31.** $a_n = 36\left(\frac{1}{3}\right)^{n-1}$ **32.** $a_n = 100\left(\frac{1}{2}\right)^{n-1}$; $\frac{25}{32}$ mg   **33.** \$2621.44   **34.** -147 **35.** ≈ 31.25

36. ≈ 67.18 **37.** 4095   **38.** \$129,471.03; \$1,372,858.90   **39.** \$5436.83   **40.** 43.19 m   **41.** 4096   **42.** \$67,200   **43.** 14   **44.** $\frac{25}{6}$

45. 15   **46.** Has no sum   **47.** 12   **48.** $\frac{8}{9}$ **49.** $\frac{23}{99}$ **50.** 27 m   **51.** 125 cm   **52.** The same size   **53.** $\$13\frac{1}{3}$ million

54. $\frac{4\pi}{3}$ square meters   **55.** 720   **56.** 36   **57.** $81x^4 + 108x^3y + 54x^2y^2 + 12xy^3 + y^4$ **58.** $x^5 - 10x^4y + 40x^3y^2 - 80x^2y^3 + 80xy^4 - 32y^5$
59. $x^{12} + 12x^{10}y^3 + 60x^8y^6 + 160x^6y^9 + 240x^4y^{12} + 192x^2y^{15} + 64y^{18}$
60. $128x^7 - 448x^6y^4 + 672x^5y^8 - 560x^4y^{12} + 280x^3y^{16} - 84x^2y^{20} + 14xy^{24} - y^{28}$ **61.** 0.27648   **62.** 0.1731929

CHAPTER 10 TEST

1. $2, \frac{3}{4}, \frac{4}{9}, \frac{5}{16}, \frac{6}{25}, \frac{11}{100}$ **2.** $a_n = \frac{n^2}{(n+1)^2}$ **3.** 52   **4.** $\sum_{i=1}^{13} \frac{5}{4i+1}$ **5.** $a_n = 15 - 4n$; $a_{19} = -61$ **6.** 477   **7.** 825   **8.** 2525

9. Difference is not constant   **10.** $a_n = 2\left(\frac{1}{3}\right)^{n-1}$; $a_9 = \frac{2}{6561}$ **11.** 17,936,135   **12.** ≈ 2039.9 **13.** $\frac{3}{8}$ **14.** $\frac{73}{99}$ **15.** 680

16. $x^{15} - 10x^{12}y + 40x^9y^2 - 80x^6y^3 + 80x^3y^4 - 32y^5$
17. 2772; The total annual advertising expenditures by U.S. companies from 1980 to 2000 is \$2772 billion.   **18.** 3150   **19.** \$289,977.49
20. \$20 million

Appendix

REVIEW PROBLEMS COVERING THE ENTIRE BOOK

1. $\{-22\}$ **2.** $\{2\}$ **3.** $\left\{x \middle| x \geq \frac{2}{5}\right\}$, $\left[\frac{2}{5}, \infty\right)$ **4.** $\{x|x > -1\}$, $(-1, \infty)$ **5.** $\{x|x \leq -4\}$, $(-\infty, -4]$ **6.** $\left\{x \middle| \frac{3}{2} \leq x < \frac{25}{2}\right\}$, $\left[\frac{3}{2}, \frac{25}{2}\right)$ **7.** $\{-1, 4\}$

8. $\{-3, 1\}$ **9.** $\{x|-2 < x < 5\}$, $(-2, 5)$ **10.** $(-\infty, -3] \cup [4, \infty)$ **11.** $\{(2, -1)\}$ **12.** $\{(-3, -4)\}$ **13.** $\left\{\left(-2, \frac{3}{2}, 1\right)\right\}$ **14.** $\{(6, -4, 2)\}$

15. $x = -1$ **16.** $\left\{-\frac{1}{2}, 4\right\}$ **17.** $\{-3, 3\}$ **18.** $\left\{\frac{1}{2}\right\}$ **19.** $\{-1, 6\}$ **20.** $\{6\}$ **21.** $\left\{\frac{-2 - \sqrt{14}}{2}, \frac{-2 + \sqrt{14}}{2}\right\}$

22. $\{-1 - i\sqrt{5}, -1 + i\sqrt{5}\}$ **23.** $\{-8, 27\}$ **24.** $\left\{\frac{-1 - \sqrt{13}}{2}, \frac{-1 + \sqrt{13}}{2}, -2, 1\right\}$ **25.** $\left\{x \middle| -\frac{5}{3} < x < -1\right\}$, $\left(-\frac{5}{3}, -1\right)$

26. $\{x|-3 < x \leq 1\}$, $(-3, 1]$ **27.** $\{127\}$ **28.** $\{2\}$ **29.** $\left\{\frac{5}{2}\right\}$ **30.** $\left\{\frac{\ln 8}{0.7} \approx 2.971\right\}$ **31.** $\{(-1, -3), (-1, 3), (1, -3), (1, 3)\}$

32. $\{(2, -4), (-14, -20)\}$ **33. a.** Security guards, Gas station workers   **b.** Taxi drivers/chauffers   **c.** Gas station workers, Security guards
34. a. 1991, \$2.1 B   **b.** Rising   **35. a.** $-4x^3 + 108x^2$ **b.** $x = 18$ in., Volume $\approx 11{,}700$ in.^{3} **c.** 11,664 in.^{3}
36. a. (1972, 450); In 1972, males and females had equal verbal SAT scores of 450.   **b.** Scores are decreasing over time.

37.

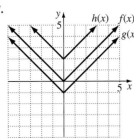

g is f with a vertical shift of −1.
h is f with a vertical shift of 2.

38. 2; −9; $-\frac{3}{5}$; The average rate of change per year in the average number of months imposed for violent offenses from 1980 to 1985 was 2, from 1985 to 1990 was −9, and from 1990 to 1995 was $-\frac{3}{5}$.

39. a. Years 1970–2000; Years 1943, 1948–1970   **40.** a; All three planes intersect at one point.
41. Yes; they intersect at 4 points.

42. $y = -\frac{2}{3}x + 4$

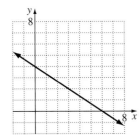

43. $3x - y > 6$

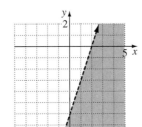

44. $y \geq \frac{3}{4}x - 5$

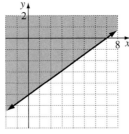

45. $2x - y \geq 4$
$x \leq 2$

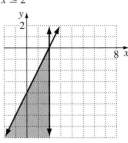

46. $x \leq 5, y \leq 6$
$14x + 8y \geq 56$
$2x + 3y \geq 12$

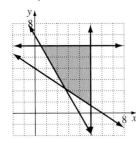

$(5, 6), \left(5, \frac{2}{3}\right), \left(\frac{36}{13}, \frac{28}{13}\right), \left(\frac{4}{7}, 6\right)$

47. $\{(-2, 3)\}$

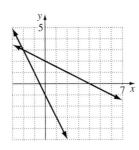

48. $y = x^2 - 4x - 5$

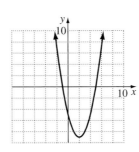

49. $f(x) = \sqrt{x}, g(x) = \sqrt{x - 1}$

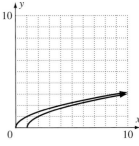

Domain of f = [0, ∞)
Range of f = [0, ∞)
Domain of g = [1, ∞)
Range of g = [0, ∞)
g is f shifted one unit to the right.

50. $y = (x - 1)^2 - 4$

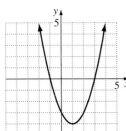

51. $y = 2^{x/2} + 1$

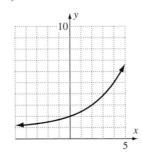

52. a. There is only one value of y for every x.
b. $f^{-1}(300) = 1995$; $300 billion was spent in 1995.
c. quadratic

53. $y = \log_2 x$

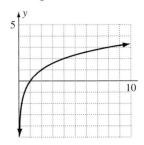

54. Center: $(-2, 3)$; $r = 2$ **55.** $\dfrac{x^2}{9} + \dfrac{y^2}{4} = 1$ **56.** $\dfrac{x^2}{4} - \dfrac{y^2}{9} = 1$

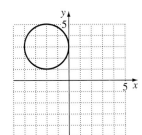   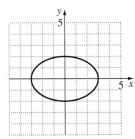   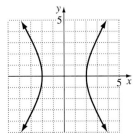

57. 2 seconds   **58. a.** 3 km   **b.** point $(3, 125)$ **59.** $m_1 = \dfrac{m_2(V_f + v_i)}{v_i - V_f}$ **60.** 112.48; In 1970, there were 112.48 million motor vehicle registrations in the U.S.   **61. a.** Negative; As x increases, y decreases   **b.** Positive; As x increases, y increases

62. a. $y - 119.4 = 5.17x$ or $y - 171.1 = 5.17(x - 10)$ **b.** $y = 5.17x + 119.4$ **c.** 207 billion   **63. a.** $r = 0.25d + 30$ **b.** $55
c. Rent-a-Heap has a better deal.   **64. a.** $P = 8x + 12y$ **b.** $x + y \le 200, x \ge 10, y \ge 80$

c.

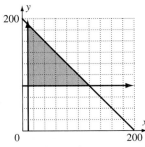

d. 10, 190, $2360   **65.** $A(x) = 5x^2 + 30x + 40$ **66.** $A(x) = 4x^2 + 60x$ **67. a.** 6 seconds

b. 2.5 seconds; 196 ft   **c.** $s(t) = -16t^2 + 80t + 96$

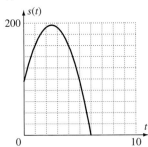

68. a. $2083\frac{1}{3}$; The difference in cost to remove 90% of the pollutants versus 40% of the pollutants is $2083\frac{1}{3}$ million.

b. 100; It is impossible to remove 100% of the pollutants.   **69.** 10,000; $40.50; Cost decreases as the number produced increases.

70. $\dfrac{S - p}{Sn}$ **71.** $\dfrac{fp}{p - f}$ **72.** $\dfrac{3w}{4\pi r^3}$ **73.** 1980   **74.** 2005   **75. a.** f is $25 off the regular price while g is 30% off the regular price

b. $0.7x - 25$; The discount price is $25 off the sale price of 30% off.   **c.** $0.7x - 17.5$; The discount price is 30% off the sale price of $25 off.

d. $(f \circ g)(x)$; $0.7x - 25$ is less than $0.7x - 17.5$ **76. a.** 6.176903 billion   **b.** 2030   **77. a.** $y = 250e^{0.0101386t}$ **b.** 2013   **78.** 11 minutes

79. a. $a_n = 26{,}350 + 2250n$ **b.** $44,350   **c.** $291,800   **80. a.** $a_n = 20{,}000(1.06)^{n-1}$ **b.** $30,072.61   **c.** $197,949.37

81. $3x(2x - 3)^2$ **82.** $(3x - 5)(9x^2 + 15x + 25)$ **83.** $(x - 3)(x + 3)(x - 2)$ **84.** $5(3x^2 + 5)(x + 2)(x - 2)$ **85.** $10x^{-1/2}(x - 3)$ **86.** 46

87. $15 - 20x$ **88.** $\dfrac{12x}{y^9}$ **89.** $-\dfrac{6x^5}{y^2}$ **90.** 3.6×10^{-7} **91.** $x^2 - 5xy - 16y^2$ **92.** $6x^2 + x - 77$ **93.** $25x^4 - 70x^2y + 49y^2$

94. $\dfrac{(x + 2)(x - 3)}{3(x + 3)(x - 4)}$ **95.** $\dfrac{x + 5y}{2x + 5y}$ **96.** $\dfrac{-y^2 + 2y + 3}{y(3y + 1)(y - 5)}$ **97.** $\dfrac{9(y - 5)}{y(y + 9)}$ **98.** $x^2 - 5x - 1$ **99.** $3x^2 + x + 4 + \dfrac{7}{x - 2}$

100. $\sqrt[8]{16}\sqrt{x}\sqrt[4]{y}$ **101.** $(\sqrt[6]{3})^5$ **102.** $64y^2(\sqrt[4]{x})^3$ **103.** $2x^2y^3\sqrt[3]{2xy^2}$ **104.** $5xy\sqrt{2x}$ **105.** $2xy^2\sqrt[3]{2yz^2}$ **106.** $\dfrac{3\sqrt[3]{y}}{y}$

107. $18x^2\sqrt{2x}$ **108.** $\dfrac{1 - 2\sqrt{y} + y}{1 - y}$ **109.** $75 - i$ **110.** $\dfrac{9 - 15i}{17}$ **111.** $\dfrac{5}{6}$ **112.** Point-slope: $y + 4 = -2(x - 1)$ or $y - 8 = -2(x + 5)$;

Slope-intercept: $y = -2x - 2$ **113.** Point-slope: $y + 2 = -4(x - 3)$; Slope-intercept: $y = -4x + 10$ **114.** $x^2 + x - 2$ **115.** $x^2 - x - 6$

116. $x^3 + 2x^2 - 4x - 8$ **117.** $x - 2$ **118.** $x^2 + 4x$ **119.** $x^2 - 2$ **120.** $6a + 3h - 7$ **121.** $2\sqrt{5}$ **122. a.** $f^{-1}(x) = \dfrac{x+3}{4}$

b. $f(f^{-1}(x)) = 4\left(\dfrac{x+3}{4}\right) - 3 = x; f^{-1}(f(x)) = \dfrac{4x - 3 + 3}{4} = x$ **123.** $x = b^4$ **124.** $3 \log_5 x + \dfrac{1}{2} \log_5 y - 3$ **125.** $\log_b\left(\dfrac{x^2 y^3}{\sqrt{z}}\right)$ **126.** 208

127. 1800 **128.** 10,922.5 **129.** 2 **130.** $\dfrac{34}{99}$ **131.** $32x^5 - 80x^4 y^3 + 80x^3 y^6 - 40x^2 y^9 + 10xy^{12} - y^{15}$ **132.** \$1120 **133.** 175; \$67

134. 300 minutes **135.** 3 m by 7 m **136.** 27 ft by 45 ft; 1215 ft^2 **137.** 52 mph and 62 mph **138.** 81 yr, 83 yr and 85 yr

139. Apple has 60 calories, banana has 87 calories **140.** 90 ft $\times$ 70 ft **141.** 35° and 55°
142. 3.6 liters of 15% acid and 2.4 liters of 40% acid **143.** Boat's speed is 35 mph; Current's speed is 7 mph
144. Army: 70,400; Navy: 63,000; Marines: 34,800 **145.** 43°, 72° and 65° **146.** 3 m **147.** Height: 6 ft; Base: 13 ft **148.** 15 ft
149. No **150.** 11 in. by 11 in. by 5 in. **151.** 8 in. by 9 in. **152.** 52.5 ft by 80 ft or 60 ft by 70 ft

153. 3 in $\times$ 4 in **154.** 45 **155.** 15, 21, 78, $\dfrac{n(n+1)}{2}$ **156. a.** $x = 2$ and $z = 7$

b. $y = 4, x = 1, w = 7$, so $z = 2$; $y = 4, x = 1, w = 2$, so $z = 7$ **c.** $x = 3, y = 2, z = 4, w = 8$ **d.** $x = 1, y = 8$ (with $w = 3$ and $z = 6$)
157. 12 **158.** 3 ways **159.** 360 square units **160.** 46 cubes

Index

Definitions, Rules, and Formulas

The Real Numbers

Natural Numbers: $\{1, 2, 3, \ldots\}$
Whole Numbers: $\{0, 1, 2, 3, \ldots\}$
Integers: $\{\ldots, -3, -2, -1, 0, 1, 2, 3, \ldots\}$
Rational Numbers: $\{\frac{a}{b} \mid a$ and b are integers, $b \neq 0\}$

Irrational Numbers: $\{x \mid x$ is real and not rational$\}$

Basic Rules of Algebra

Commutative: $a + b = b + a$; $ab = ba$
Associative: $(a + b) + c = a + (b + c)$;
$(ab)c = a(bc)$
Distributive: $a(b + c) = ab + ac$;
$a(b - c) = ab - ac$
Identity: $a + 0 = a$; $a \cdot 1 = a$
Inverse: $a + (-a) = 0$; $a \cdot \frac{1}{a} = 1$ $(a \neq 0)$
Multiplication Properties: $(-1)a = -a$;
$(-1)(-a) = a$; $a \cdot 0 = 0$; $(-a)(b) = (a)(-b) = -ab$;
$(-a)(-b) = ab$

Order of Operations

1. Perform operations above and below any fraction bar, following steps (2) through (5).

2. Perform operations inside grouping symbols, innermost grouping symbols first, following steps (3) through (5).

3. Simplify exponential expressions.

4. Do multiplication or division as they occur, working from left to right.

5. Do addition and subtraction as they occur, working from left to right.

Set-Builder Notation, Interval Notation, and Graphs

$(a, b) = \{x \mid a < x < b\}$

$[a, b) = \{x \mid a \leq x < b\}$

$(a, b] = \{x \mid a < x \leq b\}$

$[a, b] = \{x \mid a \leq x \leq b\}$

$(-\infty, b) = \{x \mid x < b\}$

$(-\infty, b] = \{x \mid x \leq b\}$

$(a, \infty) = \{x \mid x > a\}$

$[a, \infty) = \{x \mid x \geq a\}$

$(-\infty, \infty) = \{x \mid x$ is a real number$\} = \{x \mid x \in R\}$

Slope Formula

$$\text{slope } (m) = \frac{\text{change in } y}{\text{change in } x} = \frac{y_2 - y_1}{x_2 - x_1} \quad (x_1 \neq x_2)$$

Equations of Lines

1. *Slope-intercept form:* $y = mx + b$
   m is the line's slope and b is its y-intercept.

2. *Standard form:* $Ax + By = C$

3. *Point-slope form:* $y - y_1 = m(x - x_1)$
   m is the line's slope and (x_1, y_1) is a fixed point on the line.

4. *Horizontal line parallel to the x-axis:* $y = b$

5. *Vertical line parallel to the y-axis:* $x = a$

Absolute Value

1. $|x| = \begin{cases} x \text{ if } x \geq 0 \\ -x \text{ if } x < 0 \end{cases}$

2. If $|x| = c$, then $x = c$ or $x = -c$. $(c > 0)$

3. If $|x| < c$, then $-c < x < c.$ $(c > 0)$
4. If $|x| > c$, then $x < -c$ or $x > c.$ $(c > 0)$

Systems of Equations

Consistent system

One solution

Inconsistent system

No Solution

Dependent system

An Infinite Number of Solutions

A system of linear equations may be solved: (a) graphically: (b) by the substitution method, (c) by the addition or elimination method, (d) by matrices, or (e) by determinants.

$$\begin{vmatrix} a_1 b_1 \\ a_2 b_2 \end{vmatrix} = a_1 b_2 - a_2 b_1$$

Cramer's Rule: Given a system of equations of the form

$$\begin{matrix} a_1 x + b_1 y = c_1 \\ a_2 x + b_2 y = c_2 \end{matrix} \text{ then } x = \frac{\begin{vmatrix} c_1 b_1 \\ c_2 b_2 \end{vmatrix}}{\begin{vmatrix} a_1 b_1 \\ a_2 b_2 \end{vmatrix}} \text{ and } y = \frac{\begin{vmatrix} a_1 c_1 \\ a_2 c_2 \end{vmatrix}}{\begin{vmatrix} a_1 b_1 \\ a_2 b_2 \end{vmatrix}}.$$

Special Factorizations

1. *Difference of two squares:*

$$A^2 - B^2 = (A + B)(A - B)$$

2. *Perfect square trinomials:*

$$A^2 + 2AB + B^2 = (A + B)^2$$
$$A^2 - 2AB + B^2 = (A - B)^2$$

3. *Sum of two cubes:*

$$A^3 + B^3 = (A + B)(A^2 - AB + B^2)$$

4. *Difference of two cubes:*

$$A^3 - B^3 = (A - B)(A^2 + AB + B^2)$$

Variation

English Statement	**Equation**
y varies directly as x.	$y = kx$
y is proportional to x.	
y varies directly as x^n.	$y = kx^n$
y is proportional to x^n.	
y varies inversely as x.	$y = \dfrac{k}{x}$
y is inversely proportional to x.	
y varies inversely as x^n.	$y = \dfrac{k}{x^n}$
y is inversely proportional to x^n.	
y varies jointly as x and z.	$y = kxz$

Exponents

1. $b^{1/n} = \sqrt[n]{b}$

2. $b^{m/n} = \sqrt[n]{b^m} = (\sqrt[n]{b})^m$

3. $b^{-m/n} = \dfrac{1}{b^{m/n}}$

4. $b^r \cdot b^s = b^{r+s}$

5. $(b^r)^s = b^{rs}$

6. $(ab)^r = a^r b^r$

7. $\left(\dfrac{a}{b}\right)^r = \dfrac{a^r}{b^r}$

8. $\dfrac{b^r}{b^s} = b^{r-s}$

Radicals

1. $\sqrt[n]{x^n} = |x|$ if n is even.
$\sqrt[n]{x^n} = x$ if n is odd.

2. $\sqrt[n]{M} \cdot \sqrt[n]{N} = \sqrt[n]{MN}, M > 0, N > 0$

3. $\sqrt[n]{\dfrac{M}{N}} = \dfrac{\sqrt[n]{M}}{\sqrt[n]{N}}, \quad M > 0, N > 0$

Imaginary Numbers

1. $i = \sqrt{-1} \quad i^2 = -1 \quad i^3 = -i \quad i^4 = 1$

2. If $a > 0$, then $\sqrt{-a} = \sqrt{-1} \sqrt{a} = i\sqrt{a}$

Distance and Midpoint Formulas

1. The distance from (x_1, y_1) to (x_2, y_2), is

$$\sqrt{(x_2 - x_1)^2 + (y_2 - y_1)^2}.$$

2. The midpoint of the line segment with endpoints (x_1, y_1) and (x_2, y_2) is

$$\left(\frac{x_1 + x_2}{2}, \frac{y_1 + y_2}{2}\right).$$

Quadratic Formula

The solutions to $ax^2 + bx + c = 0$ with $a \neq 0$ are

$$x = \frac{-b \pm \sqrt{b^2 - 4ac}}{2a}.$$

Functions

1. Linear Function: $f(x) = mx + b$
Graph is a line with slope m.

2. Quadratic Function: $f(x) = ax^2 + bx + c, a \neq 0$
Graph is a parabola with vertex at $x = -\dfrac{b}{2a}$.
Quadratic Function: $f(x) = a(x - h)^2 + k$
In this form, the parabola's vertex is (h, k).

3. Cubic Function: $f(x) = ax^3 + bx^2 + cx + d$
If $a > 0$, the graph falls to the left and rises to the right.
If $a < 0$, the graph rises to the left and falls to the right.

4. nth-Degree Polynomial Function: $f(x) = a_n x^n + a_{n-1}x^{n-1} + a_{n-2}x^{n-2} + \cdots + a_1 x + a_0, a_n \neq 0$
The graph is a smooth, continuous curve.

5. Rational Function: $f(x) = \dfrac{p(x)}{q(x)}$, $p(x)$ and $q(x)$ are polynomials, $q(x) \neq 0$